Proceedings of the 6th International Asia Conference on Industrial Engineering and Management Innovation

Ershi Qi
Editor

Proceedings of the 6th International Asia Conference on Industrial Engineering and Management Innovation

Innovation and Practice of Industrial Engineering and Management (volume 2)

ATLANTIS
PRESS

Editor
Ershi Qi
Tianjin
China

ISBN 978-94-6239-144-4 ISBN 978-94-6239-145-1 (eBook)
DOI 10.2991/978-94-6239-145-1

Library of Congress Control Number: 2015946069

Printed on acid-free paper

Preface

It is my great pleasure to welcome all the delegates coming all the way for the 6th International Asia Conference on Industrial Engineering and Management Innovation 2015 (IEMI 2015). It is your great efforts that bring out the proceedings of IEMI 2015 which records the new research findings and development in the domain of IEMI. What is more exciting, you are the experts or scholars with significant achievement in the field. I believe that the proceedings will serve as the guidebook for potential development in IEMI and play great role in promoting the IEMI development.

With the ongoing dramatic paradigm shifts of industrial engineering theories and applications, more and more enterprises have realized it is the key to innovate their products by utilizing advanced technology to enhance their core competitiveness. It is quite imperative to bring professionals from both academia and business together to share their new findings and experience.

IEMI 2015 caters to the purpose by providing a platform to exchange the state-of-the-art research, achievement exhibition, case study and development in the field of IEMI, as well as promoting its application. The papers selected center on the main themes of the conference: Industrial Engineering Theory, Industrial Engineering Technology Practice, Information Technology Application and Development, Automation System Theory and Application, and Management Innovation Method and Practice. All the papers included in the proceedings have undergone rigid peer review. We have also invited some prominent experts as our keynote speakers.

The conference is sponsored by Chinese Industrial Engineering Institution, CMES, and organized by Tianjin University, China. We would like to extend our sincerest thanks to Atlantis Press for their generous support in the compilation of the proceedings. We also would like to extend sincerest thanks to

Tianjin University for holding such an excellent event, and to all the delegates, keynote speakers, and staff of the organizing committee for their contribution to the success of the conference in various ways.

Thank you very much!

Chinese Industrial Engineering Institution, CMES
Ershi Qi
Zhen He
Jiang Shen
Runliang Dou

Contents

Part I
Information Technology Application and Development

The Research of Reducing Energy Consumption Method on WSN Based on the Data Correlation

Su-jing Sun, Cui-jian Zhao, Peng Guo and Xiao-qing Li

Abstract Application of wireless sensor network is very broad. It usually works under mal-environment, so reducing energy consumption is very important in the aspect of network management because of limited energy supply. The wireless sensor network is usually divided into DWSN and CWSN. CWSN is evolved from DWSN. On the basis of the distributed sensor network structure, the nodes' clustering is mainly according to the regional relationship, without considering the data relationship. But so many nodes in WSN make the time and spatial relationship existence between them. That is data correlation. According to DWSN structure, the nodes clustering are based on the data correlation in this paper. Some nodes hibernate by turns in the same cluster. This method can bring the redundant data uploading and the energy consumption down. The experiment proves that the energy consumption reduction is very obvious.

Keywords Clustering · Data correlation · Reducing energy consumption · WSN

1 Introduction

Wireless sensor network is applied in a variety of different important fields, such as military, environmental and safety monitoring [1]. The study of it has attracted great attention of many scholars. The network is composed of mass sensor nodes which

S. Sun
Department of Computer Science and Technology, Shijiazhuang Tiedao University, Shijiazhuang, China

C. Zhao (✉) · P. Guo
Department of Physical and Electrical Information Engineering, Shijiazhuang University, Shijiazhuang, China
e-mail: zhaocuijian@126.com

X. Li
Unit 69223 of PLA, Xijiang, China

E. Qi (ed.), *Proceedings of the 6th International Asia Conference on Industrial Engineering and Management Innovation*,
DOI 10.2991/978-94-6239-145-1_1

power supply is by batteries [2]. The wireless sensor network has many advantages, such as, all of sensor nodes that there are no connecting line and no interference between them has a high level integration. The cost is lower because the nodes are random putted in the every region when they are installed [3]. But because of the spatial construction, there are many difficulties in design the network. Such as [4]:

(1) The sensor node's power is limited on account of it is providing by batteries. And the sensor nodes often worked in the hostile environment, energy equipments are generally unable to replace, so they are often invalided because of draining the energy. Meanwhile there are many new nodes join in the network, therefore the network structure often changes dynamically.
(2) The core function of WSN is detecting data rather than its hardware structure. This characteristic requires that perceiving data and processing data is the core of designing the network.

In conclusion, the energy conservation is the primary target of the WSN design, and the designer use eclectic mechanism to choose good program between the extending the survival time and improving the network throughput and reducing the communication delay. Secondly they also consider the scalability and robustness of the network [5]. With the increase of sensor nodes, the survival time and handling capacity is enhanced correspondingly, the new nodes can be effective fusion, and join in the global application [6]. At the same time the invalided node can be timely discovered and eliminated.

2 DWSN and CWSN Structure

WSN can be divided into DWSN (distributed wireless sensor network) and CWSN (clustering wireless sensor network). There is no fixed infrastructure in DWSN [7]. All nodes randomly distribute in a certain monitoring region and some or all of nodes can shift. These nodes which form the network in self-organized manner gather and regroup automatically based on the changing the task and network topology requirements to create a temporary network. This approach can adapt to the node failure and performance deterioration. Each node is able to collecting data and manages the data using the built-in processor, and then they transmit related data to routing control node by mean of multi hop relay [8].

With the development of WSN, hierarchical routing technology brings CWSN out. According to the different measuring signal or the function and distinctive energy of the each node, all the normal nodes can be divided into several clusters. Every cluster has a cluster header which can preprocess and fuse the data to achieve the purpose of saving energy [9]. The data are transmitted to the base station by cluster head and handled ultimately by the base station. The normal nodes don't exchange the data between them [10]. The nodes cluster is based on regionalism, without considering the data correlation between nodes in CWSN.

The main purpose of WSN is using the data from sensors to monitor the region covered by network reliably [11]. There is so much sensor nodes in the network, so the data correlation are inevitably exist from the time and space [12]. The typical WSN are high density arrangement in the monitoring region which can ensure the reliability of the monitoring data. The data spatial correlation exists between nods on account of the high density arrangement. The periodical data transmission of nodes can assure the efficiency in the network [13]. Nodes in the network are to transmit data periodically, to ensure the timeliness of data, nodes adjacent time data have some similarity. So there is time correlation data between nodes [14]. Data correlation is actually redundant data exist between nodes [15].

This paper purposes a method of nodes clustering according to the data correlation based on the DWSN. The nodes in the same cluster can hibernate by turn. This method can save energy, and meanwhile it can contribute to the efficient data management and processing.

3 The Data Correlation

The data correlation among the WSN nodes is divided three classes: the data spatial correlation between the nodes, the data time correlation between the nodes and the data time correlation within the same node.

3.1 Data Spatial Correlation Between the Nodes

The sensor nodes are randomly arranged in the certain monitoring region in the network, so the detected data by the nodes have similarities and some data are the same.

So for a pair of nodes (v_x, v_y), the detected data at some point t is x_t and y_t. Supposing that there is a function $F(x_t, y_t) < M_t$, in the formula M_t is preset limit value, then we say that it is the data spatial correlation between the nodes v_x and v_y. If $F(x_t, y_t) < M_t/2$, we say that it is the data strong spatial correlation.

3.2 Data Time Correlation Between the Nodes

Within a certain period of time, the detected data by the nodes are the similarity. So for any pair of nodes (v_x, v_y), supposing that τ is the time period of nodes uploading data,$\{x_1, x_2, \ldots x_n\}$, is the detected data series of vx which is in the time period $(\tau, n\tau)$(n is a positive integer), $\{y_1, y_2, \ldots, y_n\}$ is the detected data series of v_y which is in the time period $(\tau, n\tau)$. If there is a function $G(\{x_1, x_2, \ldots x_n\},\{y_1, y_2, \ldots, y_n\}) < N_t$, in the formula N_t is preset limit value, then it is to say that data time correlation between the nodes v_x and v_y is existing. If G({x_1, x_2, … x_n},{y_1, y_2, …, y_n} < N_t/2, then the data strong time correlation is existing.

3.3 Time Correlation of the Same Node

The detected data by a certain node in the WSN has similarity, so for a node v_x,{x_1, x_2, … x_n} is the detected data series which is in the time period $(\theta, m\theta)$(m is positive integer). For a pair of data (x_i, x_j)∈ $(1, m)$ there are a function T(x_i, x_j) < P_t, then it is to say that node x has data correlation within the time period $(\theta, m\theta)$. If T(x_i, x_j) < P_t/2, then it is to say that data time strong correlation is existing.

4 Network Nodes Cluster Based on the Data Correlation

The nodes cluster based on the data correlation from the initial formation to the network update is divided into several stages:

4.1 WSN Formation Stage

In this stage,all nodes of the network is $A = \{v_1, v_2, \ldots v_n\}$(n is the total numbers), they form a network by means of self-organization. For every node v_i($1 \le i \le n$), it upload the data to the routing control node and the data information include the ID, location and adjacent nodes.

4.2 Cluster Based on the Data Correlation Stage

After formation of the initial topological structure, the detected data of any two nodes (x, y) or a single node x is not correlative. So each node form a cluster. After several rounds of data uploading by nodes, the routing control node analyzes the correlation of every two nodes (x, y) and a single node according to detected data.

In accordance with the data correlation definition in 3.1, for every pair of two nodes (v_i, v_j) ($1 \le i, j \le n$) in the node set A, the routing node analyzes the data correlation relation between them. According to the spatial correlation of all nodes of the cluster formation spatial correlation matrix S; according to the strong correlation node clusters formed strong spatial correlation matrix SS:

S=

	v_1	v_2	v_3	····	v_n
v_1	0				
v_2	0	0			
v_3	1	1	0		
⋮	⋮	⋮	⋮		
v_n	0	1	0	····	0

SS=

	v_1	v_2	v_3	····	v_n
v_1	0				
v_2	0	0			
v_3	1	0	0		
⋮	⋮	⋮	⋮		
v_n	0	1	0	····	0

In the matrix S, if two nodes exit spatial correlation, then the corresponding bit is 1. Otherwise the bit is 0. In the matrix SS, if two nodes exit spatial strong correlation, then the corresponding bit is 1, if not the bit is 0. In the two matrices, the corresponding bit of node itself is 0, this indicates that the node itself does not exit the spatial and spatial strong correlation.

According to the time correlation between all nodes of the cluster formation time correlation matrix T; according to the strong correlation between the time node cluster formation time strong correlation matrix TS:

T=

	v_1	v_2	v_3	····	v_n
v_1	1				
v_2	0	0			
v_3	1	1	1		
⋮	⋮	⋮	⋮		
v_n	0	1	0	····	1

TS=

	v_1	v_2	v_3	····	v_n
v_1	0				
v_2	0	1			
v_3	1	1	0		
⋮	⋮	⋮	⋮		
v_n	0	1	0	····	1

In the matrix T, if two nodes exit time correlation, then the corresponding bit is 1. Otherwise the bit is 0. In the matrix TS, if two nodes exit time strong correlation, then the corresponding bit is 1, if not the bit is 0. In the two matrices, the corresponding bit of node itself is 1 or 0, which is decided by node if there is data time correlation or data time strong correlation. If there is data time correlation, then the bit is 1, otherwise the bit is 0. In the process of the network clustering, and firstly, the clusters of all nodes are generated according to the matrices S and T. Those nodes which can be divided in cluster A and in cluster B, according to SS and TS to judge its ownership. Meanwhile, each node must store the information of nodes which is spatial strong correlation or time strong correlation or both of them with it. In order to save the storage space, the stored information is ID of correlative node. Every routing control node uploads the data to total control node, which processes the data finally and stores the final data into database.

4.3 *Management of the Network Data After Clustering*

Through the above operations, the whole network has been divided into a number of clusters. There is data correlation of every node in one cluster. In fact, one node exist the data correlation, means it will upload a large number of redundant information. By this time, the method of node alternate hibernation can be adopted to manage the data uploading.

For the nodes with in the same cluster, supposing that the nodes series $\{v_1, v_2, \ldots, v_k\}$(k is less than total cluster number) are detecting and uploading data periodically, while the nodes which are correlative with $\{v_1, v_2, \ldots, v_k\}$ is hibernating during the period of time (t, mt)(m is a positive integer). When the data correlative nodes is waking up, the node which exists data strong correlation with $\{v_1, v_2, \ldots, v_k\}$ is chosen to start working. There may be multiple data strong correlation nodes with each node in $\{v_1, v_2, \ldots, v_k\}$ set, at this moment the random or by-turn method can be adopted to decide which node starts working. If there is not node which is data strong correlative with one or some of $\{v_1, v_2, \ldots, v_k\}$, then the node in $\{v_1, v_2, \ldots, v_k\}$ set will continue to work. The waking up nodes which is not chosen will be continued to hibernate.

Such alternate dormant method can save the node itself energy consumption, but also reduces the redundant data uploading. And the method can optimize the data management of routing control node.

For an individual data correlation node, because the detected data is the same or similar within a period of time, the node can hibernate. The hibernating time is (t, nt) (n is a positive integer and it is less than 3). This node can work sequentially. This approach can extend the uploading cycle of nodes, thereby the node energy can be saved and the redundant data uploading is reduced.

4.4 *Update Clustering*

The new nodes often join in WSN and the old ones quit because of the malfunction or energy exhausting, so cluster must be updated periodically, that is the data correlation between nodes and it must be renewably determined periodically to update cluster.

5 Test

The WSN of 60 nodes is chose in this paper. Supposing that data accuracy requirement is not very accurate, then the preset limit value can be a bigger data. When the time is 10th upload cycle, all nodes have started to work. When it is 20th, all the data are foundation of initial clustering. The renewably clustering is executed every 50 cycles. The function of different data correlation is as follows:

(1) The spatial correlation: the function

$$F(x_t,y_t) = |x_t - y_t|;M_t = 1/2_\circ$$

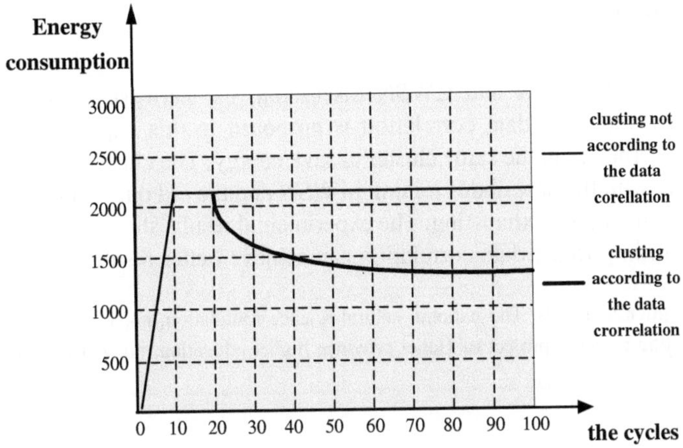

Fig. 1 Energy consumption comparison

(2) The time correlation: supposing that τ is 5 min, initially clustering n is 20, then data series of the node v_x and v_y detecting within $(\tau, 20\tau)$ is $\{x_1, x_2, \ldots, x_{20}\}$ and $\{y_1, y_2, \ldots, y_{20}\}$. The function $G(\{x_1, x_2, \ldots, x_{20}\}, \{y_1, y_2, \ldots, y_{20}\} = |x_i - y_j|$, $i,j \in (1,20)$, $N_t = 1/2$. When the cluster is renewed, n is 50. The data series is $\{x_1, x_2, \ldots, x_{50}\}$ and $\{y_1, y_2, \ldots, y_{50}\}$. $G(\{x_1, x_2, \ldots, x_{50}\}, \{y_1, y_2, \ldots, y_{50}\} = |x_i - y_j|$, $i,j \in (1,50)$, $N_t = 1/2$.

(3) The node itself time correlation: supposing that θ is 5 min, initially clustering m is 20, the detected data series of the node is $\{x_1, x_2, \ldots, x_{20}\}$, the function $T(x_i, x_j) = |x_i - x_j|\ i,j \in (1,20)$, $P_t = 1/2$. When the cluster renews, m is 50, the data series is $\{x_1, x_2, \ldots, x_{50}\}$, the function $T(x_i, x_j) = |x_i - x_j|\ i,j \in (1,50)$, $P_t = ½$.

According to the above clustering algorithms, in the network operation process, energy consumption of whole network presents a trend of significantly lower. As shown in Fig. 1.

From the figure, when the network runs initially, energy consumption shows a rising trend. After all nodes have already started to work, the maximum energy consumption of network is achieved. Hereafter, the energy consumption is little changed in the network which cluster is not according to the data correlation, but the energy consumption is a significant decline in the network which cluster is according to the data correlation, and later the consumption is flat. However, with the aging of the equipment, the energy consumption still rises some time later of network working.

6 Conclusion

On the basis of DWSN and CWSN discussion, the network clustering method according to the node data correlation is proposed in this paper. The nodes can hibernate alternately in the same cluster to save energy. The cluster must be renewed periodically with the new nodes joining in WSN and the old ones quit because of the malfunction or energy exhausting. The experimental results show that: the method of cluster according to the data correlation can achieve better results.

Acknowledgments Funds: The national natural science foundation, no. 11372197.
The scientific research project of Hebei province higher education, no. Z2012186.

References

1. Intanagonwiwat C, Govindan R, Estrin D (2003) Directed diffusion for wireless sensor networking. IEEE/ACM Trans Netw 11(1):2–16
2. Li JZ, Li JB (2003) Sensor networks and it concept, question and advance of data supervise. J Softw 14(10):1717–1726
3. Hartog AH (1983) A distributed temperature sensor based on liquid-core optical fibers. IEEE J Lightwave Technol 3(1):498–5096
4. Hartog AH, Leach AP, Gold MP (1985) Distributed temperature sensing in solid-core fibers. Electron Lett 23(21):1061–1062
5. Perring A, Szewczyk R, wen V (2001) Secure protocols for sensor networks. In: Proceeding of the 7th annual ACM/IEEE international conference on mobile computing and networking (Mobi-Com), Rome, Italy, pp 189–199
6. Pi Y,Yang J (2007) Principle of synthetic aperture radar imaging. University of Electronic Science and Technology of China press, Chengdu, pp 110–120
7. Le C, Chan S (2004) Onboard FPGA-based SAR processing for future spaceborne systems. In: Proceedings of the IEEE radar conference, pp 15–20
8. Jain S, Shan RC (2006) Exploiting mobility for energy efficient data collection in wireless sensor networks. Mobile Netw Appl 11(3):27–39
9. Marta M, Cardei M (2008) Using sink mobility to increase wireless sensor networks lifetime. In: 2008 international symposium on a world of wireless, mobile and multimedia networks, pp 10–18
10. Buratti C,Verdone R (2008) A hybrid hierarchical architecture: from a wireless sensor network to the fixed infrastructure. In: 2008 European wireless conference (EW), pp 1–7
11. Chakrabarti A, Sabharwal A (2006) Communication power optimization in a sensor network with a path-constrained mobile observer. ACM Trans Sens Netw 2(3):297–324
12. Xu J, Bi W, Zhu J, Zhao H (2011) Design & simulation of WSN equal-cluster-based multi-hop routing algorithm. J Syst Simul 23(5)
13. Ghidini G, Das SK (2011) An energy-efficient Markov chain-based randomized duty cycling scheme for wireless sensor networks. In: IEEE 31st international conference on distributed computing systems, pp 67–76
14. Akyildiz IF, Su W, Sankarasubramaniam Y, Cayirci E (2002) Wireless sensor networks: a survey. Comput Netw 38(4):393–422
15. Zhu Y, Zhang J, Lifen, Peng W (2010) Multiple ant colony routing optimization based on cloud model for WSN with long-chain structure. In: International conference on wireless communications networking and mobile computing, pp 1–4

Research on Situation and Key Issues of Smart Mobile Terminal Security

Hao-hao Song, Jun-bing Zhang, Lei Lu and Jian Gu

Abstract As information technology continues to develop, smart mobile terminal has become the electronic equipment most closely with people life and work. Meanwhile, the security threats are also increasing to smart mobile terminal when it conducts the business application. Based on the analysis security situation on smart mobile terminals, the key security issues of smart mobile terminals are analyzed detailed: program security analysis, data security protection and defect research. Finally, we point out the development direction on research of smart mobile terminals.

Keywords Dynamic analysis · Data security protection · Defect research · Smart mobile terminal · Security situation

1 Introduction

Currently, the ordinary mobile phones on the market can be classified as the ordinary mobile terminals, and the smart mobile phones are classified as the smart mobile terminals. The smart mobile terminals refers to the mobile terminals having the operating system platform with open (the flexible development, installation and

Research presented in this paper is sponsored by National Science and Technology Significant Project "Security Evaluation Technology Research on Smart Mobile Terminals" (No. 2012ZX03002011) and 2013 Annual Technical Standards Special Project of Science and Technology Commission of Shanghai Municipality "Research on key technical standards of testing on Information security product" (No. 13DZ0500501).

H. Song (✉) · J. Gu
The MPS Quality Supervision and Testing Center of Security Products for Computer Information System, The Third Research Institute of Ministry of Public Security, Shanghai, China
e-mail: haohaosong@gmail.com

J. Zhang · L. Lu
Network Security Protection Bureau of Ministry of Public Security, Beijing, China

E. Qi (ed.), *Proceedings of the 6th International Asia Conference on Industrial Engineering and Management Innovation*,
DOI 10.2991/978-94-6239-145-1_2

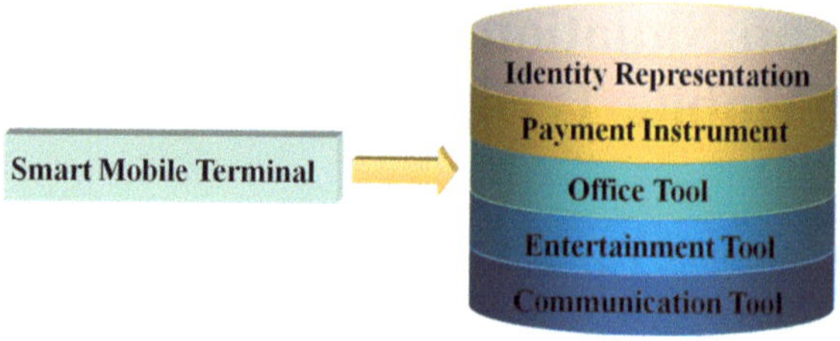

Fig. 1 Development trend of smart mobile terminals

operation in application programs), the PC-class processing power, the high-speed access capabilities and the rich interactive interface, including the smart mobile phones and the tablet computers. The mainstream operating system platforms include two categories—Android and iOS on the market.

With the development of encryption chip, secure communication protocols, authentication technologies, the future development trend of smart mobile terminals is the smart mobile terminal not only is as a communication tool, entertainment tool and office tool, but also is as a representative of payment instruments and identity. Using it, you can carry out the small amounts of payment or credit card shopping. The smart mobile terminal can also be the access card and the membership card. Figure 1 shows the development trend of smart mobile terminals.

2 Security Situation on Smart Mobile Terminals

With the development of 3G network and other wireless, smart mobile terminal is not just communications equipment that used to call and send SMS, the users that access the Internet through smart mobile terminals increase significantly [1]. Smart mobile terminal has become to the key strength to promote mobile internet business quickly develop. According to the 30th "China Internet Network Development Statistics Report" released by China Internet Network Information Center on July 2012 recently shows that at the end of June 2012, China's mobile phone users reached 388 million, increased by about 32.7 million compared with the end of 2011. The percent of mobile internet user increases from 69.3 to 72.2 % [2].

Meanwhile, with the development of technologies and applications, the terminal that stores private information and economic benefits become the primary attack target of the black chain. Both R&D capabilities such as Mobile malware, network attacks and misuse of resources and implementation of environmental already exist. Security issues that the mobile intelligent terminal and operating system facing, such as Android, IOS, Symbian, Windows Mobile and Rim, increasingly prominent.

In the end of December 2013, more than 6000 Mobile malwares were developed in Chinese mainland. As shown in the report of CNCERT, 6249 Mobile Internet malicious programs were captured in 2011, increasing by more than twice as in 2010. The most is deductions malicious programs, with the amount of 1317, which is 21.08 %, following by spread malicious programs, information theft programs, Hooliganism programs and Remote control programs. From the view of Mobile platform, about 60.7 % of malicious programs aim at Symbian platform, which is lower as in 2010. About 7.12 million Mobile intelligent terminals were infected by malicious programs in Chinese mainland, which brings serious threat and damage to the Mobile intelligent terminals [3]

With the development of technology and application, smart mobile terminals will be confronted with many kinds of security threat in the industrial applications, such as virus, disclosure of confidential information, illegal tampering of code, malicious replace of key components, and so on.

These security events show that attacks may occur in any part of the smart mobile terminal, while the ultimate goal is to get the value of the smart mobile terminal. If any security risk of the mobile intelligent terminal is used by an attacker, it will cause the loss of profits of developers, users and operators, and finally influences the development of smart mobile terminal industry.

We conclude that the threats to the smart mobile terminals can be categorized into the following five categories: 1. The hardware lacks protection measures, resulting from the physical user information stolen and destroyed; 2. The operating system has vulnerabilities, leading to malicious code infection and hacking; 3. The application software is easily hijacked, causing the user accounts and passwords stolen; 4. The communication processing lacks safeguards, call information is easily eavesdropped, leading to user privacy leaks; 5. The user data lacks protection, resulting in the leakage of user privacy. The threats to the smart mobile terminals are shown as Fig. 2.

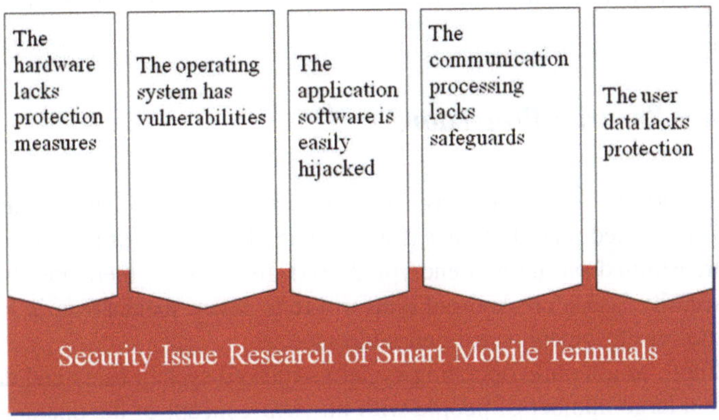

Fig. 2 The threats to the smart mobile terminals

3 Key Security Issue Research of Smart Mobile Terminals

At this stage, we consider that the security research on smart mobile terminal should be focused on three aspects: program security analysis, data security protection and defect research.

3.1 Program Security Analysis

In the application security analysis tools, the majority are the static analysis tools. The static analysis tools can quickly analyze the program features. However, the static analysis tools cannot make good use of information in runtime, so it cannot guarantee the accuracy of the analysis. Dynamic program analysis techniques can be very good to make up for the inadequacies of static analysis.

The current dynamic analysis tools on Android platform including its own Dalvik Debug Monitor Server (DDMS), as well as some of the dynamic monitoring system, such as DroidBox, TaintDroid and so on. However, the existing dynamic analysis tools on Android platform lack the fine-grained control, and lack the flexibility analysis strategy at the same time. In addition, for some heavyweight dynamic analysis (such as the instruction-level tracking), it not enough for the existing tools to complete efficiently.

Currently, the analysis for iOS system and its application is mainly the static analysis and testing. For some added confusion and hidden deep act, it is difficult to find only by a static method, they can be found only in a dynamic perform process. Therefore, the dynamic analysis techniques for iOS system can greatly compensate for the lack of existing detection methods. But it is difficult for the dynamic analysis techniques for iOS system to achieve, and requires the long-term studies.

3.2 Data Security Protection

For the current data security on Android system, there are many non-platform-related security issues (for example, the passwords and other sensitive data are transmitted on the non-encrypted text). In addition, there are also a lot of specific security issues on Android platform (e.g. Intent mechanism is imperfect). These issues are to be enhanced and addressed.

In addition, as a mobile operating system, Android system users and developers maybe are a large number of people with no clear concept of the operating system security, software security. That the users ignore security issues is likely to cause leakage of private data. For example, a considerable majority of Android users in the process of installation program does not care whether the authority applied for the program is necessary; they also do not care about whether their privacy data stored in the phone has been protected. As developers, some developers have the unskilled development or insufficient attention to security issues, in the process of the programming, they may make some mistakes to give the opportunities for the attackers (for example, passwords do not be encrypted store, and important data stored on the SD card). Such applications into the market would cause great threats to user's privacy and security.

In addition, the traditional Android platforms are mostly based password authentication mechanism, gestures password. This traditional authentication mechanism is simple, vulnerable to violence to crack. For the defect, the future research on authentication could focus on the recognition based on facial, fingerprint, hardware and other non-traditional way.

Current Android applications are vulnerable to attack by the reverse, tampering, and repackaging. Current mainstream Android application obfuscation tool Proguard is unable to meet the growing security needs in confusion intensity. Therefore, more advanced Android code obfuscation program need to be considered to research and develop.

3.3 Defect Research

Android system is based on the Linux Kernel; the common security threats for Linux (primarily security vulnerabilities) are capable to threaten Android system. At the same time, Android system supports the program execution based on the native codes (C language and C++ language) to, which left a hidden danger for the underlying security [4].

The top security system of Android is built based on the security sandbox of Linux Kernel and the checking mechanism of installation permission. Because of the vulnerability of Android kernel, in the event of root right of Android system is taken, the upper security will be completely lost.

The libraries layer in Android system is consist of a number of frequently used system functions, and due to the performance requirements, they is based on Native Code. This layer handles many important functions in Android system, such as database operations, SSL network transmission and so on. It is found based on research that the attack, caused by improper using of HTTPS/SSL in a large number of Android applications, may make the instant messaging content stored in online banking, social networking sites, email to be exposed to the danger. In addition, the system library is responsible for maintaining a variety of data storage and management on Android system (phone, SMS, email, GPS information, etc.). However, because data protection mechanisms as Android are not perfect, private information on the mobile phone cannot be well protected. Figure 3 shows the password is stored unencrypted in Android system.

There is the risk of data loss in Android system. For example, because Google account synchronization function is built on the Android, it will synchronize a lot of information on mobile phone, such as contact lists, system installation application,

```
config.xml — Edited
<string name="account_number">1860[redacted]2</string>
<string name="account_number_service">1860[redacted]2</string>
<boolean name="account_create_flag" value="true" />
<boolean name="account_rememberpw" value="true" />
<int name="account_net_type" value="0" />
<boolean name="account_auto_register" value="false" />
<string name="account_password">NT[redacted]g==</string>
<boolean name="account_enterprise" value="false" />
<boolean name="account_accept_protocol" value="true" />
<boolean name="account_login_status" value="true" />
<boolean name="account_first_flag" value="false" />
<string name="account_email_service">xiao[redacted]@163.com</
string>
<string name="account_email">xiao[redacted]@163.com</string>
```

Fig. 3 The password is stored unencrypted in Android system

schedule management, directly to Google's cloud platform. At the same time, in Android system customized by a third party, it is possible that there are other data will be synchronized to an external system.

4 Development Direction on Security Research of Smart Mobile Terminals

In the future, we believe that the security research on smart mobile terminals can be divided into three directions: research and development, testing and standards compilation. In the aspect of security research and development, the advanced code

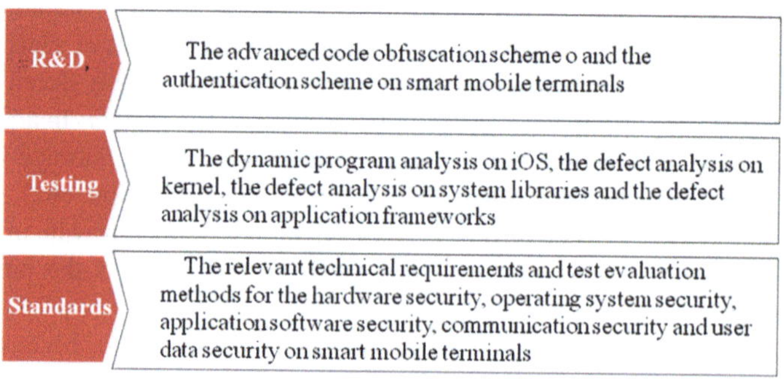

Fig. 4 Development direction on security research of smart mobile terminals

obfuscation scheme on smart mobile terminals, the authentication scheme on smart mobile terminals should be research as soon as possible. In the aspect of security testing, the dynamic program analysis on iOS, the defect analysis on kernel, the defect analysis on system libraries and the defect analysis on application frameworks should be strengthened. In the aspect of standards compilation, the relevant technical requirements and test evaluation methods for the hardware security, operating system security, application software security, communication security and user data security on smart mobile terminals should be accelerated to development. Development direction on security research of smart mobile terminals is presented as Fig. 4.

References

1. La Polla M, Martinelli F, Sgandurra D (2013) A survey on security for mobile devices. IEEE Commun Surv Tutor 15(1):446–471
2. China Internet Network Information Center (2012) The 30th China internet network development statistics report, July 2012
3. CNCERT (2013) 2012 China internet network security report
4. Ongtang M, McLaughlin S, Enck W, McDaniel P (2012) Semantically rich application-centric security in Android. Secur Commun Netw 6:658–673

A Hybrid Slot Allocation Algorithm for TDMA Network

Wei-zhen Tian, Ming-rui Guo and Xiong-chang Pang

Abstract To improve the performance of TDMA Network, we propose a new slot allocation algorithm called Traffic Characteristics-Hybrid Time Division Multiple Access (TC-HTDMA). In this method, we add dynamic slot reservation (DSR) mechanism to traditional static slot assignment and take node priority into consideration. In the DSR mechanism, each node employs ARMA prediction method to compute the slot number that it will need in the next frame stage, then, sends the value to the Virtual Network Controller (VNC) which will execute the slots assignment algorithm by summarizing all nodes' requirements and priority levels and give the final slot allocation scheme. Furthermore, we verify the effectiveness of the proposed algorithm through computer simulation using OPNET and compare the results with another hybrid slot allocation method MOHSAP. The results show that TC-HTDMA performs better than MOHSAP in terms of slot utilization, throughput and end to end delay, guarantees the quality of service (QoS) of different traffic, and realizes rational allocation of network bandwidth resources.

Keywords Dynamic slot reservation · Qos · TC-HTDMA · Traffic prediction

1 Introduction

TDMA (Time Division Multiple Access) is a medium access control protocol widely used in digital cellular telephone systems, satellite communication systems and TDL (Tactical Data Link) systems. In TDMA networks, by using time slot as the basic transmission unit, only one transmitter can occupy the slot for transmitting

W. Tian (✉) · X. Pang
Department of Communication Command, Xi'an Communications Institute, Xi'an, China
e-mail: twz-flyaway@163.com

M. Guo
Aviation Electronics Co. Ltd, Aviation Industry Corporation of China, Beijing, China

E. Qi (ed.), *Proceedings of the 6th International Asia Conference on Industrial Engineering and Management Innovation*,
DOI 10.2991/978-94-6239-145-1_3

without conflict, while others are listening. The system slot utilization, quality of service and network capacity can be affected by the slot allocation algorithm directly; therefore, how to find an efficient slot allocation scheme to improve the system performance has been an important research issue in the decades.

Take TDL system as an example, static slot allocation method is mostly adopted, which has the advantage of excellent stability and anti-interference ability, in the while, lack of high system utilization and flexibility in contrast [1, 2]. In recent years, dynamic slot allocation methods have gradually been proposed. [3, 4] improve the dynamic TDMA scheme by changing the frame length dynamically. [5] puts forward the algorithm of bintree average allocation in the block which allocates slots for users according to the degree of emergency or user's priority. However, most of the papers donnot combine the static slot allocation and dynamic slot allocation together. Wang et al. [6] adopts different slot allocation method based on the characteristics of message stream, but it is not suitable to apply the scheme to a network in changeable environment owing to the lack of traffic prediction.

In this paper, we propose a hybrid TDMA algorithm TC-HTDMA that assigns part of the slots in one frame to network members fixedly, allocating the remaining slots through dynamic reservation according to nodes' requirement. In the scheme, VNC acts as the center controller which is responsible for receiving reservation request and applying the slot allocation algorithm based on the summarized information it has collected. Further more, we add ARMA model to our algorithm to accomplish the traffic prediction for all nodes, so as to achieve the Demand Assigned Multiple Access (DAMA) and QoS guarantee for different types of traffic.

The remaining parts of the article are organized as follows. Section 2 summarizes related works. Section 3 illustrates the proposed algorithm including slot allocation scheme and traffic prediction in detail. Section 4 shows the performance evaluation and simulation results. Finally, Sect. 5 gives conclusions.

2 Related Works

In 2003, Kanzaki proposed a dynamic TDMA algorithm named ASAP (Adaptive Dynamic Slot Assignment Protocol) that is applicable to ad hoc self-organizing network [3]. In ASAP, the frame length is increased exponentially as the growth of the number of competing nodes, with an upper limit value of the frame length, but there is no rule for shrinking the length as nodes number decreases that leads to a waste of slots resource when competing nodes number declines. Wang [4] improved the mechanism above with consideration to shrinking side and the traffic conditions of nodes. That is called Adaptive-frame-based Dynamic Slot Assignment Protocol (ADSAP). ADSAP improves slot utilization to a certain extent, but does not supply priority categories for various traffic types in detail.

Dynamic slot assignment is flexible indeed, but cannot be wholly used in some TDMA network, such as TDL, because the nodes have basic demand for fixed slots to guarantee the basic information exchange in the network. In order to make full use of the remaining slots, [6] proposed Message-streams Oriented Hybrid Slot Allocation Protocol (MOHSAP) that combines static allocation and dynamic allocation methods. For different characteristics of packets, the slot allocation process is divided into two parts: static time slot allocation process based on the fixed packet flow and dynamic slot allocation process based on random packet flow. Although ADSAP retains both the advantages of two processes and is also based on frame length expansion and shrinking mechanisms, it lacks of the traffic prediction mechanism for changeable traffic burst amounts that may bring about delay jitter.

Compared with the improvement methods mentioned above, our proposed protocol can outcome in the following aspects:

(1) A flexible frame structure has been designed that divide three stages; the frame length can be adjusted periodically;
(2) Traffic prediction model ARMA is set up to pre-determine the slot requirement for each node;
(3) A hybrid slot allocation algrithm is proposed which combine static slot allocation and dynamic reservation scheme together, which maintain both of their advantages;
(4) Nodes in the network are divided into different teams according to their priority so as to guarantee the quality of service.

3 Proposed Algorithm TC-HTDMA

The proposed algorithm contains dynamic allocation and static assignment. In the static part, each node in the network can obtain basic slots, while the dynamic part needs reserving remaining slots for nodes according to the priority and required slot number which will be acquired by ARMA prediction method for their traffic amount in the next frame. Compared with completely static slot assignment, TC-HTDMA can not only make full use of slots resources, but also ensure the QoS requirement of different traffic type.

As in Fig. 1, a TDMA frame is constituted of static slots (SS), pre-allocation (PA) slots and dynamic executing (DE) slots. The detailed composition of a slot can refer to [7].

3.1 Frame Structure

The PA stage contains request slots (RS), actual allocating slots (AS) and broadcasting slots (BS). The PA stage takes only little resource; two slots are enough for VNC. Section 3.2 shows how every nodes request slots from VNC and how the allocation algorithm implements specifically.

Fig. 1 Frame structure

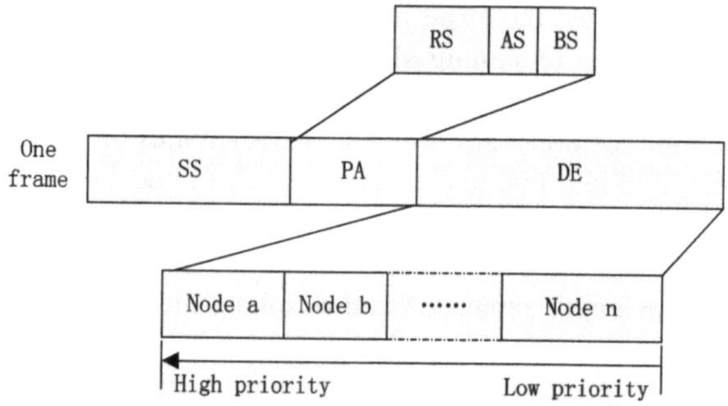

The dynamic executing (DE) stage is the last part of one frame following the slots pre-allocation and broadcasting. In this stage, each node occupies the slots according to the tag when the slots allocated to it arrive.

3.2 ARMA Prediction Model

Traffic forecasting methods that are used commonly include AR, ARMA, Kalman, wavelet, neural network and so on. Among these methods, wavelet and neural network methods are effective for most traffic models, but they all need complex estimation procedure, only AR and ARMA are suitable for time-stationary series forecasting [8]. Obviously, the actual data in some tactical communication system cannot be collected easily, so we use time-stationary series to model the traffic and employ ARMA model to accomplish short-term traffic forecasts with less parameters than Kalman model and smaller prediction error variance than AR model.

ARMA (Autoregressive moving average) model contains p entries of autoregressive term and q entries of moving average term. ARMA (p, q) model can be expressed as in (1):

$$X_t = \mu_t + \sum_{i=1}^{p} \varphi_i X_{t-i} + \sum_{j=1}^{q} \theta_j \varepsilon_{t-j} \tag{1}$$

In the second section of [9], AIC (Akaike Information Criterion) was used to determine the order of ARMA model [10]. Combing the traffic characteristics, we also use ARMA (2, 1) with a lower complexity to model the traffic; the prediction model is as follows:

$$\varphi(B)X_i = \theta(B)a_i \tag{2}$$

$$\varphi(B) = 1 - \varphi_1 B - \varphi_2 B^2 \tag{3}$$

$$\theta(B) = 1 - \theta_1 B \tag{4}$$

In Eqs. (2–4), B is the backward shift operator, α_i is WGN. We can use LMS method to estimate the parameters $\varphi_1, \varphi_2, \theta_1, \sigma_a^2$ (the variance of WGN). The one-step prediction model can be expressed as in (5):

$$\widehat{X}_t(1) = \sum_{j=1}^{m} I_j X_{t+1-j} \tag{5}$$

m is the number of observed values before that can be determined by prediction accuracy, α_i is the inverse function of ARMA, the multi-step predictive model is (6):

$$\widehat{X}_t(l) = \widehat{\varphi_1}\widehat{X}_t(l-1) + \widehat{\varphi_2}\widehat{X}_t(l-2) \tag{6}$$

3.3 *Slot Allocation Process*

3.3.1 Static Slot Assignment (SA)

Allocating static slots can guarantee the nodes have basic chance for joining in the network, synchronizing, and message exchange and so on.There are two basic methods that can be used in SA: uniform or continuously assignment. In view of that dynamic slots reservation occupies a leading position in TC-HTDMA, we choose continuous slots assignment in the static slots stage that cannot lead to dispersed static slots in each frame for producing higher slot utilization.

3.3.2 Slots Request and Broadcast

VNC is responsible for providing the time benchmark for the synchronization of the network without a center node. In PA stage, VNC receives request information, executing dynamic slots reservation and broadcast the result of dynamic slots assignment.

In the slot request stage, each node converts the traffic forecasting result in its queue to number of slots in demand according to the ARMA prediction method, then, sends the request packet to VNC. In this way, VNC will get a demand table that lists all the requests of the active nodes in the network. We use $S_t(i, j)$, $S_s(i, j)$, $S_r(i, j)$ to represent the total slots demand, static slots that have assigned to, the reserved dynamic slots for node *j* belong to priority level *i*. The node tag *i* and *j* are useful in resource allocation when the priority is considered. The relationship is: $S_t(i, j), = S_s(i, j) + S_r(i, j)$.

3.3.3 Pre-allocation Algorithm

The total slots in the next frame must be forecasted by the packets number and burst state. If the assignment static slots cannot meet the demand of the node, the reservation mechanism will work. After collecting the summary demand of all the nodes, VNC will reserve dynamic slots for higher priority, then, the total slots belong to each priority level will be its own members according to the requests. The allocation result is effective just in current frame and updated in next frame periodically.

S, S_0 and S_1 respectively denote the total slot number, the static slots and the rest for dynamic reservation in a frame, we can conclude that $S_1 = S\text{-}S_0$. In the PA stage, VNC starts reservation as soon as it collects all the requests information. The allocation method is considered in two conditions:

$$(a) \qquad \sum_i \sum_j S_r(i,j) \le S_1$$

In this condition, the remaining slots can satisfy the demand of all the nodes, so, the reservation will begin from the higher priority. The low demand also indicates that the current frame length is too long that we can finish the dynamic stage right after the last occupied slot to improve slot utilization.

$$(b) \qquad \sum_i \sum_j S_r(i,j) > S_1$$

In this case, the surplus slots are not enough to let the traffic of all the nodes send out in time. To solve the problem, a suitable scheduling method has to be brought out. If we adopt the entirely fair queue scheduling, the emergency traffic with higher priority may be delayed a lot; if we use the traditional priority queuing totally, the traffic with the lowest priority may "starve" due to long-term lack of chance for transmitting [11]. How to trade off? A reasonable allocation method is proposed in the following, which can ensure the priority level while fairness is guaranteed.

First, we set corresponding weights $\delta_0, \delta_1, \ldots \delta_{m-1}$, the constraints for dynamic slots reservation are listed in (7).

$$\begin{cases} \sum_j S_r(0,j) \le \dfrac{\delta_0}{\sum_{i=0}^{m-1} \delta_i} S_1 \\ \sum_j S_r(m-1,j) \ge \dfrac{\delta_{m-1}}{\sum_{i=0}^{m-1} \delta_i} S_1 \end{cases} \tag{7}$$

m stands for the number of priority levels. δ_0 is the upper limit of the proportion that occupied by the highest priority traffic, while δ_m accounts for the lower limit for the lowest priority traffic.

In link-16 system, application functions contain many kinds of formatted messages (J-series message) that can complete tactical missions by combining together. These messages can also be divided into various priority levels to satisfy their quality of service [12, 13].

The above equation shows that lower priority traffic must gives way to higher one. The value of $\delta_0, \delta_1, \cdots \delta_{m-1}$ are set according to the specific needs and network features. The weights for lower priority need to be set smaller, to avoid competing abundant slots with delay-sensitive traffic. If the slots that can be reserved are far less than requirement, the leftover traffic will be arranged for transmission in next frame as much as possible.

After allocating slots in priority level, nodes with the same priority will acquire their ratio on demand.

4 Performance Evaluation

4.1 Simulation Enviornment

To simulate a changeable network environment by OPNET simulation software, we use 10 stations, change the traffic volume to model different network load. The frame length is 128 ms, so the number of frames transmitted in one second is 1/0.128. One epoch contains 256 frames each containing 64 slots. The length of one time slot is 2 ms. The traffic model is ON/OFF with ON distribution Weibull (5, 3) and OFF distribution Exponential (5). We use 30 slots for static assignment as the initial value, 2 slots for operating reservation algorithm [14].

The system throughput is the product of slot number in one frame, bits number transmitted in one slot and frames number in one second. Under the configuration in this paper, the theoretical value of saturation throughput should be $64 \times 120 \times 8\,\text{bits} \times 1/0.128\,\text{s} = 600000\,\text{bps}$. The delay is directly related to the setting of traffic volume and priority. In practical communications, the packet header and control packets overhead may lead to a drop in throughput.

4.2 Performance Analysis and Criteria

From the content above, we grasp the idea that the dynamic slots adjust in each frame and slots allocated to each member are continuous one by one. Next frame starts as soon as the last member finishes its occupancy. Concrete analysis is as follows:

Suppose that traffic with different priorities is delivered to different queues. The number of nodes is N, and the traffic style is Poisson distribution with rates $\lambda_1, \lambda_2, \cdots, \lambda_m$, m is number of priority levels. The procedure of multiple slots occupied can be seen as a simple service process in queuing theory. Assumes that the processing rate for every packet is the same, so the delay of the packet with priority $\sigma_i (1 \leq i \leq m)$ contains two parts: the remaining processing time required for the higher priority traffic that are receiving services, the processing time for this queue.

S_{left} denotes the remaining slots higher priority traffic has not occupied in current time, traffic (message) amount for all members of priority level i is M_i, the time of processing for each message is T_i, other time consumed is negligible. The average queuing time for nodes with priority σ_i is as in (8):

$$\mathrm{d_i} = \delta \mathrm{E}(\mathrm{S_{left}}) + \mathrm{E}\left(\sum_{\mathrm{i=1}}^{\mathrm{M_i}} \mathrm{T_i}\right) \tag{8}$$

The dynamic slots have been planned rationally and fully used that illustrate the probability δ of a busy system is equal to 1 under the condition that there are always packets waiting to send. N_i is the nodes number of priority level i. In order to simplify the analysis, we use Poisson distribution as the traffic model. From Little Theorem used in [15], the average packet number of priority level i can be expressed in (9):

$$E\{M_i\} = N_i \lambda_i d_i \tag{9}$$

The processing time for each message has the same value $E\{T_i\} = 1/\mu$. If the system is processing high priority messages currently, S is the service time, the remaining processing time obeys to time stationary distribution, and the mean value of S_{left} can be expressed as (10):

$$E\{S_{left}\} = \frac{E\{S^2\}}{2E\{S\}} = \frac{\frac{1}{\mu} + \frac{1}{\mu^2}}{2\frac{1}{\mu}} = \frac{\mu + 1}{2\mu} \tag{10}$$

From the formulas above, we can deduce (11):

$$d_i = \frac{\mu + 1}{2(\mu - N_i \lambda_i)} \tag{11}$$

The delay for a message with priority σ_i is the sum of queuing delay and processing delay in (12):

$$D_i = d_i + \frac{1}{\mu} = \frac{\mu^2 + 3\mu - 2N_i \lambda_i}{2\mu(\mu - N_i \lambda_i)} \tag{12}$$

4.3 Simulation Results

Figures 2 and 3 shows the contrasting results of the network delay and slot utilization for MOHSAP and the proposed hybrid algorithm when no priority is set and other conditions above are the same. In order to simulate the network performance under traffic saturated conditions, we use heavy load configuration to model the traffic.

In static assignment, all the slots are pre-allocated that there are no spare time slots for stations in the need of more resource. The performance is worse than both of the two curves in the figures above that are not displayed. In MOHSAP, although the slots are used dynamically, the protocol can not grasp the trend of the traffic amount that lead to lower slot utilization. In proposed TC-HTDMA, the slots are made full use as the frame length and requirements are changeable frame by frame.

Fig. 2 End to end delay

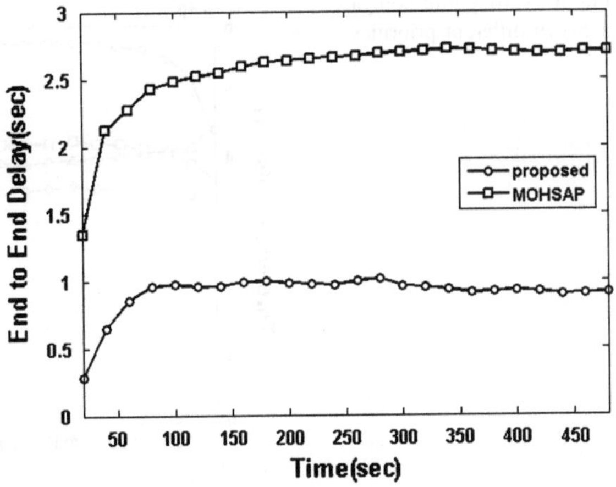

Fig. 3 Slot utilization

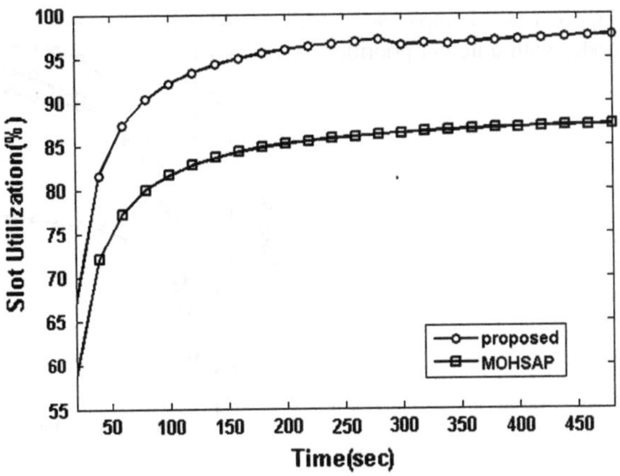

So, the corresponding slot utilization is improved by 10 % totally under this condition comparing with MOHSAP, while the average delay of packet is reduced obviously.

Figures 4 and 5 show the results of throughput and delay for nodes with different priorities when the other conditions are the same except the priority level.

In order to verify the influence of priority setting to a single node's performance, we use heavy volume traffic in the simulation, because only under this condition will the slots be allocated by priority levels. In the contrary, when the traffic is light, the priority has no effect on the allocation results in terms of the slot number.

In MOHSAP, there are no priority settings, each node, either with emergent traffic, or with background traffic has the same chance to access to time slots. In proposed algorithm with dynamic reservation scheme as in FigS. 4 and 5, the node with higher priority receives increased throughput and dropped delay obviously,

Fig. 4 Throughput compare between different priorities

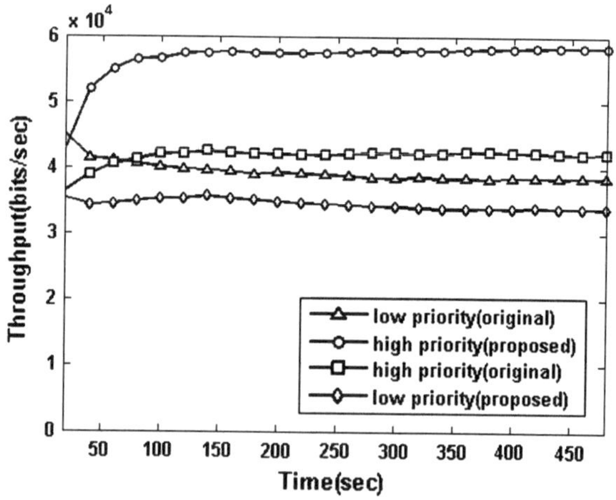

Fig. 5 Delay compare for nodes with different priorities

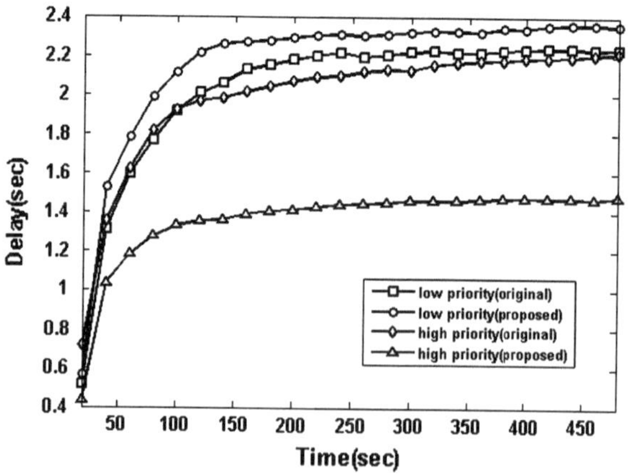

while the degradation amplitude of nodes with lower priority is small and acceptable. The traffic has different priority setting, and their QoS can be guaranteed, although these improvements are at the cost of performance of lower priority nodes.

5 Conclusions

A new hybrid TDMA algorithm that is suitable in TDMA network has been proposed, and the effectiveness of the method has been verified by OPNET simulation. ARMA prediction method is introduced for forecasting traffic amount of each node in next frame stage. VNC here plays a role of center controller that collects requirement information and allocating dynamic slots according to priority rules. The slot utilization of the frame has been greatly improved, the QoS requirement guaranteed in the same time.

The new algorithm can be used in small operational networks for allocating slots resources rationally. But, with more nodes joining in the network, even relay nodes, the application will be restricted, in that the static slots assigned will grow unlimitedly and lead to more delay for all nodes. The study on traffic classification and is only at the preliminary stage; we will study the detailed traffic requirement and attempt to apply the algorithm to actual network environment in our future work.

References

1. Mei W, Cai S (2007) JTIDS/Link16 Data Link. National defense industry Press, pp 165–189
2. Zhaoxiong H, Xing L, Xuemin L, Jinglun Z (2010) Link-16 model Architecture for multiple nets simulation in NS-2. Ind Eng Eng Manage (IEEM) 1645–1649
3. Lewis AM, Pizzi SV (2005) Quality of service for tactical data links: TDMA with dynamic scheduling. In: Military communications conference (MILCOM), MITRE Corp., Bedford, MA, USA, vol 4, pp 2350–2359
4. Wenzheng W, Pengcheng L, Jinglun Z, Pei R (2009) Adaptive-frame-based dynamic slot assignment protocol for tactical data link system. In: International conference on networks security, wireless communications and trusted computing, NSWCTC '09, vol 1, pp 709–714
5. Li J, Fan X, Zhang Z, Wan M (2011) TDMA dynamic slot allocation algorithm based on priority. Comput Eng 37(14):288–290
6. Wang W, Zhou J, Zheng L, Luo P (2009) Message-streams oriented hybrid slot allocation protocol for tactical data link system. In: Communication networks and services research conference, 2009. CNSR '09. Seventh Annual, pp 201–208
7. Shin JB, Roh B (2010) TDMA frame format analysis for applying Policy based management to TDL system. In: 2010 Proceedings of the 5th international ubiquitous information technologies and applications (CUTE), vol 1, pp 1–7
8. Jian W, Yifeng W, Juanjun X, Hangzao D (2009) Short-time prediction method based on fractal theory for traffic flow. In: Control and decision conference, CCDC '09, pp 458–461
9. Li J, Liu X, Han Z (2007) Research on the ARMA-based traffic prediction algorithm for wireless sensor network. J Electron Inf Technol 29(5):1224–1227

10. Zhang S, Qi L (2003) Concise guide to time series analysis. Tsinghua University Press, Beijing, pp 98–122
11. Zhang Y, Zhang Y, Tang J (2009) A new slot allocation model in STDMA data link SYstems. Commun Technol 42:59–61
12. Long-Ya W, Wei X, Wu-Zhong L (2011) Model analysis of link-16 based on global grid reference model. Int Conf Comput Sci Netw Technol 2011:910–913
13. Wilson WJ (2011) Applying layering principles to legacy systems: link 16 as a case study. IEEE
14. Long H (2006) OPNET modeler and computer network simulation. Xi'an University of Electronic Science and Technology Press, Xi'an, pp 40–48 (in Chinese)
15. Li X, Lun P (2010) Research and improvement of real-time queue scheduling algorithm. Int Forum Inf Technol Appl (IFITA) 1:102–104

The Research of Optimum Route Algorithm in Time Measure Space

Kai-guang Zhang, Hong-ling Meng, Ming-ting Ba and Yan-min Sun

Abstract The measuring space of city streets traffic on consuming time is a NonEuclidean distance space. From the condition of city traffic attributes, an optimum route algorithm based on NonEuclidean time distance space was designed. For the turn consuming time on the streets intersection being the major section of the route using time, the virtual nodes were defined according to the azimuths of leaving the Streets intersection. Considered the consuming time passing through street sections and streets intersections, the algorithm was completed by calculating the accumulated time from the start-point to the virtual node. The algorithm was a decision-making course with some phases and realized by recursive methods. Finally, the feasibility was proved through an example.

Keywords Decision-making graph distance space · NonEuclidean · Optimum route · Time measure

1 Introduction

Urban streets, which are the Carrier of urban logistics and the flow of people, is an important part of the urban infrastructures, the blood vessels and the lifeline of a city, responses for the main functions of the urban transport [1]. In recent years, with the rapid development of city's economic, and the rapid increase in urban car ownership, the urban traffic congestion is becoming a serious problem. In order to solve the trouble and meet the needs of economic globalization, urbanization and urban expansion, many cities rapidly pull the cities framework, this also greatly increases the complexity of urban traffic [2].

K. Zhang (✉) · H. Meng · M. Ba · Y. Sun
Institute of 3S Technology, Zhengzhou Normal University, Zhengzhou, China
e-mail: zzgis@sina.com

H. Meng
Institute of Curriculum and Instruction, East China Normal University, Shanghai, China

E. Qi (ed.), *Proceedings of the 6th International Asia Conference on Industrial Engineering and Management Innovation*,
DOI 10.2991/978-94-6239-145-1_4

The choice of optimum route in urban traffic could largely ensure the orderly running of the city streets network, lighten the pressure on urban streets, meet the travel needs of the people [3]. Here the Optimum has different meanings, including the least consuming time, the shortest route distance, and the least cost consumption [4, 5].

The measure space, based on consuming time and cost consumption, is not a Euclidean distance space, the optimum route algorithm, based on the Euclidean distance (or Manhattan distance), could meet the computing requirements. For the actual characteristics of the city communication, a new algorithm should be developed.

2 Time Measure Space

Urban traffic time measure space, based on the traveling time of the vehicles on the road section and the time consumption across through the road intersections, is a NonEuclidean distance space. In order to complete the calculation of optimum route, need to establish the time-partition database of urban street traffic flow, urban street segment time consumptions and streets intersection time consumptions, which build up the measure space of the two points distance. The accumulated time, traveling from one point to another, is called time distance of the two points [6–8].

In urban traffic, for the reason of people's travel habits and traffic regulation, the time distance between two points does not satisfy symmetry, triangle inequality, and other constraints in Euclidean distance space [5–7].

The Fig. 1 is the traffic regulation schematic diagram for the district of Zhou King City Square in Luoyang City, east Renmin road is an one-way street with reverse driving, and west Renmin road is an one-way street with forward traveling [6, 7]. Subject to traffic regulatory restrictions, the optimum route AD from A to D is constituted by sections AB and BCD, the Euclidean distance is 853.21 m, the time distance in the Seventh Period [6] is 192.00 s. Including the traveling time in section AB 74 s, the time consumption across the intersection B (middle Zhongzhou road and east Renmin Road) 20 s, the traveling time in sections BCD is 98 s.

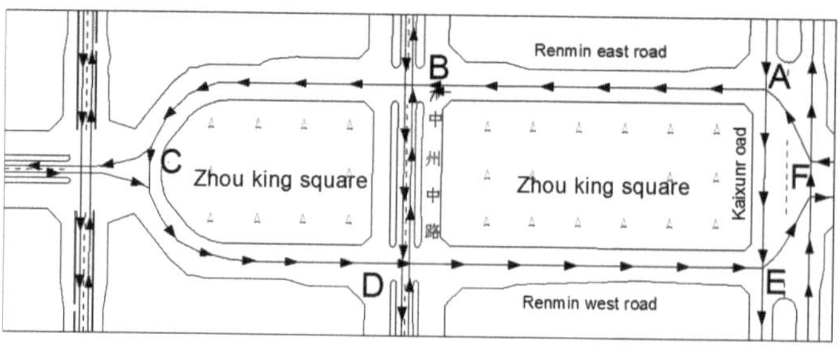

Fig. 1 Zhou King City Square district in Luoyang City traffic regulations

$$D(A,D) = D(A,B) + D(\widehat{BCD}) = 851.21 \tag{1}$$

$$T(A,D) = T(A,B) + T(B) + T(\widehat{BCD}) = 192.00 \tag{2}$$

The optimum route DA, from D to A, is constituted by sections DE and EFA. The Euclidean distance is 483.33 m, the time distance in the Seventh Period is 106 s. Including the time across the intersection D (middle Zhongzhou road and west Renmin road) 20 s, the traveling time in sections DE 44 s, the time consumption for turning left on the intersection E (Kaixuan road and west Renmin road) 10 s, the time consumption for turning left on the intersection A (east Renmin road and Kaixuan road) 10 s.

$$D(D,A) = D(D,E) + D(\widehat{EFA}) = 483.33 \tag{3}$$

$$T(D,A) = T(D) + T(D,E) + T(E) + T(\widehat{BCD}) + T(A) = 106 \tag{4}$$

$$D(D,A) \ll D(A,D)\ T(D,A) \ll T(A,D) \tag{5}$$

3 Optimum Route Algorithm

The principle of the optimum route in the urban streets network, based on NonEuclidean distance, is a simple path, a road segment to be appeared in the path only once. If the optimum route T_0 from E to H passing any order two points F and G, then the optimum route T_1, for F to G, should be included in the T_0 [9–12].

The calculation of an optimum route is a multi-stages decision process, and need to select a road segment in every stage, once a road segment is chosen, it should directly affect the select of the next segments, and sequentially affect the entire process of path selection [13]. The serial road segments, to be selected in every stage, constitute a linked road segment sequence, is called a path. Since each stage has many choices for selected, there are also many paths, with different time distance and Euclidean distance, from one point to another. The optimum route is the one with the shortest time distance under some predetermined criterions in the selected path set.

Denotes Point A as start-point, H as destination H, when A and H are determined, using street network and intersections construct decision diagram. Suppose the decision-making process is divided into n stages, k and x_k respectively denote the stage variable and a certain state of k-th stage, if the k-th stage has r kinds of status, then the set of all status is

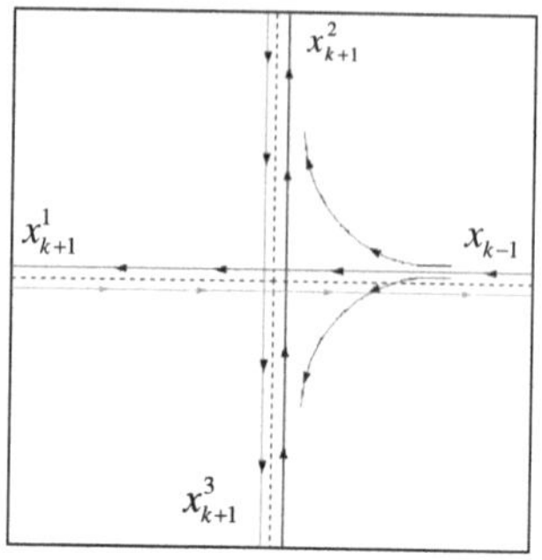

Fig. 2 The passage way in the road junctions

$$X_k = \{x_k^1, x_k^2, x_k^3 \ldots x_k^r\} \tag{6}$$

After a state of the stage is given, to select a section from the set, the evolution of a state to the next stage is a making-decision process.

In Fig. 2, there are three possible ways could reach the state x_{k+1}, from the state x_{k-1}, across the state x_k, such as going straight to x_{k+1}^1, turning right to x_{k+1}^2 and turning left to x_{k+1}^3, therefore there are different time consumption corresponding to the their ways. $x_{k+1}^i, i = 1, 2, 3$ called virtual status they do not exist in the real world, their definition is only for calculate requires.

For an intermediate state x_k^j, define its sub-state sequence as $x_k^j(\alpha)$, where α is leaving azimuth, j is the sequence number of the sub-state (In Fig. 3, $\alpha = 90, 180, 270$, $j = 1, 2, 3$). Using $T_k^\alpha(A, x_2, x_3, \ldots, x_k)$ denotes the accumulated time consumption from the starting point x_k^j to, $t_{k-1,k}(x_{k-1}, x_k)$ denotes the time consumption from x_{k-1} to x_k [2]. $r_{k-1,k}^\alpha(x_{k-1}, x_k^j(\alpha))$ denotes the time consumption shift from x_{k-1} to $x_k^j(\alpha)$ [2]. $u_k(x_k)$ denotes the decision variable of k-th state, $S_k(x_k)$ represents the set of all $u_k(x_k)$, therefore

$$u_k(x_k) \in S_k(x_k) \tag{7}$$

$$\begin{cases} T_1^{\alpha(1)}(A) = 0 \\ T_2^{\alpha(2)}(A, x_2) = T_1(A) + t_{1,2}(A, x_2) + r_{1,2}^{\alpha(2)}(A, x_3^j(\alpha(2))) \\ T_3^{\alpha(3)}(A, x_2, x_3) = T_2^{\alpha(2)}(x_2) + t_{2,3}(x_2, x_3) + \\ \qquad r_{2,3}^{\alpha(3)}(x_2, x_3^j(\alpha(3))) \\ \cdots\cdots\cdots \\ T_k^{\alpha(k)}(A, x_2, x_3, \ldots, x_k) = T_{k-1}^{\alpha(k-1)}(x_{k-1}) + \\ \qquad t_{k-1,k}(x_{k-1}, x_k) + r_{k-1,k}^{\alpha(k)}(x_{k-1}, x_k^j(\alpha(k))) \end{cases} \tag{8}$$

where $\alpha(k-1)$ is the azimuth of x_{k-1} traffic x_k.

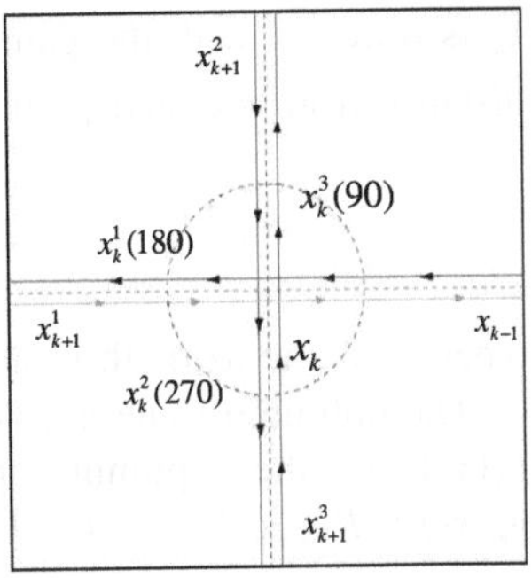

Fig. 3 The state of road junctions

The path, from the start-point A to the k-th stage, is a decision-making process, all consecutive decisions $u_i(x_i)$, $i = 1, 2, 3 \ldots k$, make up the decision-making sequence

$$R_{1,k} = \{A, u_1(x_1), u_2(x_2) \ldots u_{k-1}(x_{k-1})\} \tag{9}$$

that is a path from the start-point A to state $u_{k-1}(x_{k-1})$. Every decision-making sequence has a time distance, recorded as T_k, that is,

$$\begin{cases} T_1(A) = 0 \\ T_2(A, x_2) = T_1(A) + t_{1,2}(A, x_2) \\ T_3(A, x_2, x_3) = T_2^{\alpha}(x_2) + t_{2,3}(x_2, x_3) \\ \cdots\cdots\cdots \\ T_k(A, x_2, x_3, \ldots, x_k) = T_{k-1}^{\beta}(x_{k-1}) + t_{k-1,k}(x_{k-1}, x_k) \end{cases} \tag{10}$$

When $k = n$, $R_{1,n} = \{A, u_1(x_1), u_2(x_2) \ldots u_{n-1}(x_{n-1})\} = \{A, x_2, \; x_3 \ldots H\}$ and $T_n(R_{1,n})$ respectively are the time distance and the path from the start-point to the destination.

In Formulas (8) and (10) x_k has different meaning, in formula (8) x_k denotes the intermediate state in the path, is calculating intermediate process, in formula (10) x_k denotes the end point of the path.

Denotes $\Gamma(n) = \{R_{1,n}\}$ as the set of $R_{1,n}$, which is all possible path from start-point to destination.

The purpose of the optimum route calculation is looking for the minimum element R in the set of $\Gamma(n)$, that is,

$$R \in \Gamma(n), T(R) = \min\{T_n(R_{1,n}) | R_{1,n} \in \Gamma(n)\} \tag{11}$$

$T(R)$ is shortest time distance from A to H, corresponding the sequence of decision-making $R_{1,n} = \{A, x_2, x_3 \ldots H\}$ is the optimum route.

For the reason of traffic situation on the road intersection, different passing methods have different time consumption, therefore, the optimum route from A to

x_k is different with the path from A to $x_k^j(\alpha)$. Assume $f_k^\alpha(x_k)$ is the shortest time distance from the start point A to $x_k^j(\alpha)$, then

$$f_k^\alpha(x_k) = \min\left\{f_{k-1}^\beta(A, x_2, x_3 \ldots x_{k-1}) + t_{k-1,k}(x_{k-1}, x_k) + r_k^\alpha(x_{k-1}, x_k^j(\alpha))\right\} \quad (12)$$

where β is the azimuth running into x_k.

The optimum route is calculated using the following recursive method, assume $g_k(x_k)$ is the optimum path from the A to the state x_k, and meets $g_k(x_k) \in R_{1,k}, g_k(x_k) = \min\{T(R_{1,n})\}$, that is,

$$\begin{cases} g_1(A) = 0 \\ g_2(x_2) = \min\limits_{u_1}\{t_{1,2}(A, x_2) + f_1(A)\} \\ g_3(x_3) = \min\limits_{u_2}\{t_{2,3}(x_2, x_3) + f_2^\alpha(x_2) \\ \qquad |\alpha \text{ is the azimuth running into } x_3\} \\ \ldots \\ g_k(x_k) = \min\limits_{u_{k-1}}\{t_{k-1,k}(x_{k-1}, u_k(x_k)) + f_{k-1}^\alpha(x_{k-1}) \\ \qquad |\alpha \text{ is the azimuth running into } x_k\} \end{cases} \quad (13)$$

4 Simulation

Figures 4 and 5 are the decision-making diagram based on Xigong district (part) street network and street intersections in Luoyang city. Start-point A is the intersection of north Jiefang road with Hantun Road, destination H is the intersection of Kaixuan road with King City avenue, the word in the circle (street node, see Fig. 4) represents state, line does not have meaning, the arrow on the line represents the direction of road segment, the word above (or on the right) the line represents road name, the data under line (or on the left) represents the azimuth of road segment. In Fig. 5, the data above the line represents the Euclidean distance of this segment (unit: m), the data under the line represents the time distance passing through this segment in the Seventh Time period on Monday (unit: s) [14].

For conveniently calculating, use the averages of the turning time consumption in the district as the time-consumption for every road-intersection. Suppose r_{st} (go straight), r_{tr} (turn right), r_{tl} (turn left) respectively is 12,9, 25 s [15].

The process of the optimum route calculation form A to H is as followings:

Then $X_1 = \{A\}$, $X_2 = \{B_1, B_2\}$, $X_3 = \{C_1, C_2, C_3\}$, $X_4 = \{D_1, D_2, D_3, D_4\}$, $X_5 = \{E_1, E_2, E_3\}$, $X_6 = \{F_1, F_2\}$, $X_7 = \{H\}$.

(1) when $k = 1$, $g(A) = T_1(A) = 0$

(2) when $k = 2$, $X_2 = \{B_1, B_2\}$

$T_2(A, B_1)$ Means the shortest time distance from A to B_1, have

Fig. 4 Decision diagram of Luoyang city road network (part)

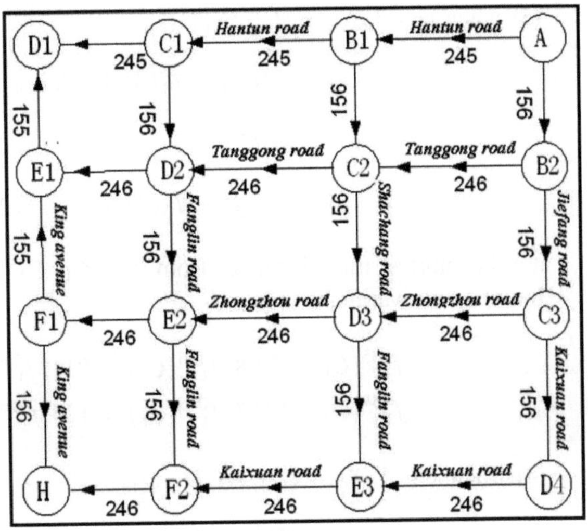

Fig. 5 Euclidean distance and time distance

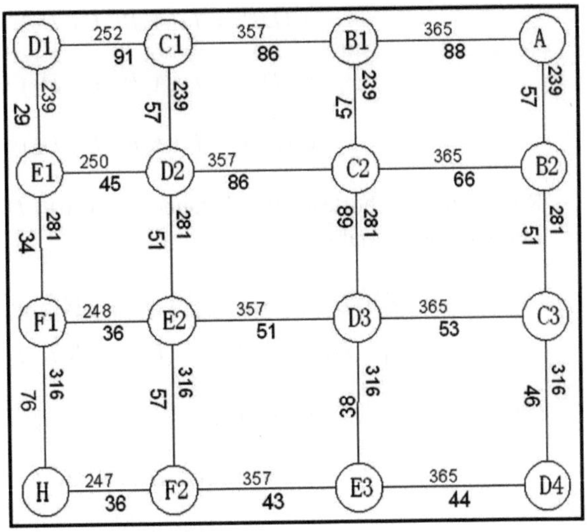

$$g_1(B_1) = T_2(A, B_1) = 88, \text{ Similarly,}$$

$$g_1(B_2) = T_2(A, B_2) = 57,$$

their Intermediate variables respectively are

$$f_2^{245}(B_1) = 100, f_2^{156}(B_1) = 113, f_2^{246}(A, B_2) = 66,$$

$$f_2^{156}(A, B_2) = 69.$$

(3) when $k = 3$, $X_3 = \{C_1, C_2, C_3\}$

$$g_3(C_1) = T_3(A, B_1, C_1) = 186,$$

$$g_3(C_2) = T_3(A, B_2, C_2) = 132,$$

$$g_3(C_3) = T_3(A, B_2, C_3) = 120,$$

are the shortest time distances from A to C_1, C_2, C_3.
Intermediate variables are:

$$f_3^{245}(C_1) = 198, f_3^{156}(C_1) = 211, f_3^{246}(C_2) = 144,$$
$$f_3^{156}(C_2) = 157, f_3^{246}(C_3) = 129, f_3^{156}(C_3) = 132$$

(4) when $k = 4$, $X_4 = \{D_1, D_2, D_3, D_4\}$

$$g_4(D_1) = T_4(A, B_1, C_1, D_1) = 289,$$

$$g_4(D_2) = T_4(A, B_2, C_2, D_2) = 230,$$

$$g_4(D_3) = T_4(A, B_2, C_3, D_3) = 182,$$

$$g_4(D_4) = T_4(A, B_2, C_3, D_4) = 178,$$

Represent the shortest time distance from A to D_1, D_2, D_3, D_4.
Intermediate variables are

$$g_4^{155}(D_1) = 314, g_4^{246}(D_2) = 242, g_4^{156}(D_2) = 255,$$
$$g_4^{246}(D_3) = 194, g_4^{156}(D_3) = 207, g_4^{246}(D_4) = 187.$$

(5) when $k = 5$, $X_5 = \{E_1, E_2, E_3\}$

$$g_5(E_1) = T_4(A, B_2, C_2, D_2, E_1) = 287,$$

$$g_5(E_2) = T_5(A, B_2, C_3, D_3, E_2) = 245,$$

$g_5(E_3) = T_5(A, B_2, C_3, D_4, E_3) = 231$, represent the shortest time distance from A to E_1,E_2,E_3. Intermediate variables are

$$f_5^{155}(E_1) = 312, f_5^{246}(E_2) = 257, f_5^{156}(E_2) = 270,$$
$$f_5^{256}(E_3) = 243.$$

(6) when $k = 6$, $X_6 = \{F_1, F_2\}$

$g_6(F_1) = T_6(A, B_2, C_3, D_3, E_2, F_1) = 293$, $g_6(F_2) = T_5(A, B_2, C_3, D_4, E_3, F_2) = 286$ represent the shortest time distance from A to F_1,F_2. Intermediate variables are

$$f_6^{156}(F_1) = 315, f_5^{246}(E_2) = 298.$$

(7) when $k = 7$, $X_7 = \{H\}$
$f_7(F_1) = T_7(A, B_2, C_3, D_4, E_3, F_2, H) = 334$, represent the shortest time distance from A to H is 334 s, the optimum decision sequence is $\{A, B_2, C_3, D_4, E_3, F_2, H\}$, the optimum route is $A \rightarrow B_2 \rightarrow C_3 \rightarrow D_4 \rightarrow E_3 \rightarrow F_2 \rightarrow H$.

5 Conclusions

In NonEuclidean distance space, the algorithm of the optimum route, based on urban traffic time measure, is a multi-stage decision-making process, every decision is the foundation of next decision, the order of the decision-making chain does not have returnable, the optimum Route from A to H and H to A contain different road segments. The algorithm also applies to the distance, profits, costs, goods distribution, resources consumption and other spatial analysis based on road network system.

Acknowledgments This paper is support by the Natural Science Foundation of Henan (No.122102310379).

References

1. Zhang KG, Meng HL, Sun YM, Ba MT, Wang JY (2012) The design and development of Luoyang smart traffic system. Henan Sci 30(2):217–221 (in Chinese)
2. Wang JY, Liu R, Cheng Y, Sun LN (2011) Making the city smarter. J Geomatics Sci Technol 28(2):79–83 (in Chinese)
3. Wang JY (2012) Smartness make city better. Chin J Nat 34(3):139–142 (in Chinese)
4. Sha ZY, Yu Z, Huang M, Xiao ML (2011) Non-planar network data model for urban road guide sign system. J Geomatics Sci Technol 28(6):442–445 (in Chinese)
5. Wang GF, Song PF, Zhang YL (2012) Review on development status and future of intelligent transportation system. Highway 5:217–222 (in Chinese)
6. Zhang KG, Meng HL, Ba MT, Sun YM, Ba MT, Wang JY (2012) The development and application of database in LYSTS. Henan Sci 30(4):461–464 (in Chinese)
7. Huang M, Zhao JH, Wu HB, Sha ZR (2012) Transportation network database construction based on traffic markings deployment. Sci Surv Mapp 37(1):189–191 (in Chinese)
8. Tang JJ, Liu F (2010) A driver route prediction based map-matching algorithm integrating uncertain reasoning. Acta Geodaetica Cartogr Sin 39(5):546–550 (in Chinese)
9. Chang DH, Lin B, Zhang M (2011) The Chinese description of GIS path based on continuous polar coordinates. J Geomatics Sci Technol 28(4):307–312 (in Chinese)

10. Zhu W, Xu KL, Sun Y, Gao L (2011) Logistics distribution route planning with fusion algorithm of petri net and ant colony. J Zhejiang Univ (Eng Sci) 45(12):2229–2234 (in Chinese)
11. Gong EC, Li LQ (2011) The optimal path algorithm design based on Bellman-Ford algorithm. Bull Surv Mapp 8:26–41 (in Chinese)
12. Zheng NP, Lu F, Li QQ, Duan YY (2010) The adaption of A* algorithm for least-time paths in time-dependent transportation networks with turn delays. Acta Geodaetica Cartogr Sin 39 (5):534–539 (in Chinese)
13. Ledoux C (1997) An urban traffic flow model integrating network. Transp Res 5(5):287–300
14. Zhang KG, Meng HL, Ba MT, Song GX, Wang JY (2012) The research of address geocoding technology in LYSTS. Henan Sci 30(7):921–925 (in Chinese)
15. Zhang KG, Ba MT, Meng HL, Sun YM, Wang JY (2012) The research of vehicles guidance system in LYSTS. Henan Sci 30(8):1121–1125 (in Chinese)

Cell Selection and Reselection Schemes in Multi-RAT Networks

Fauzi Fanny, Dengkun Xiao and Dongkai Yang

Abstract In order to perform cell selection and reselection in Multi-RAT (Multi-Radio Access Technology) 2G, 3G and 4G networks, we use combination between priority and ranking based. Specifically for reselection, Treselection variance is utilized. We borrow cell selection and reselection criterion algorithm as measurement procedure. We have used the combination above to observe the number of selection and reselection occurred, failure rate and energy saving based on the number of reselection to target rat-cell and have optimized the parameters performance.

Keywords Cell selection · Cell reselection · Multi-RAT

1 Introduction

Next-generation mobile devices will be equipped with multiple network interfaces, they can support multiple network technology depend on the application service or user on demand and in network side, the deployment of 4G is envisioned to be in the same cell sites of 2G and 3G [1].

In life cycle cellular connection, it has two state: idle and connected. This paper focus on idle mode. Idle mode consists of two procedures which are cell selection and reselection. Cell selection is a process when mobile is switched on.

F. Fanny (✉) · D. Yang
School of Electronic and Information Engineering, Beihang University, Beijing, China
e-mail: fauzi.fanny@yahoo.com

D. Yang
e-mail: edkyang@buaa.edu.cn

D. Xiao
Research Dept of Beijing R&D Subdivision, WN, Huawei Technologies Co., Ltd, Beijing, China
e-mail: xpeter@huawei.com

E. Qi (ed.), *Proceedings of the 6th International Asia Conference on Industrial Engineering and Management Innovation*,
DOI 10.2991/978-94-6239-145-1_5

An UE (User Equipment) measure nearby rat-cell and select a suitable rat-cell before entering cell reselection. UE performs cell reselection to aim better quality cell by re-measuring nearby rat-cell. The new target rat-cell could be non-serving cell or keep stay on serving cell in case UE does not perform long movement and does not have experience signal degradation below the threshold during reselection time.

The significant factors which affect to the number of selection and reselection which are received cell signal, UE velocity, timer reselection and thresholds value as in [2]. It has significant contribution to the success and failure connection during idle mode procedure.

In Multi-RAT networks, there are much quite challenges in idle mode state, especially parameterization in different UE velocity. UE shall measure three technologies that is 2G, 3G and 4G. UE should able to detect higher or lower RAT priority and then perform measurement during interval *Treselection*.

There are related issue should be taken into consideration especially about energy saving as in [3] mention that more energy saving can be achieved by implementing dynamic threshold on neighbour cell, however we only set threshold as static value and also for kind of service, only utilize full buffer which is not distinguish the different service as in [4].

This paper is organized as follows. Section 2 presents cell selection and reselection algorithm. Section 3 gives the simulation assumptions and result. Finally, we concluded the paper in Sect. 4.

2 Cell Selection and Reselection

2.1 Cell Selection

According to [5–7], GSM, UTRAN and E-UTRAN, respectively, UE select serving cell rely on signal strength by Srxlev formula or quality signal by Squal. In particular, a brief introduction as follows:

(1) *Cell Selection criterion in GSM*: The pathloss criterion parameter C1 which is used for cell selection as given in [1] is:

$$C1 = (A - \max(B, 0)) \quad (1)$$

where:

$$A = RLA\_C - RXLEV\_ACCESS\_MIN \quad (2)$$

$$B = MS\_TXPWR\_MAX\_CCH - P \quad (3)$$

RXLEV_ACCESS_MIN is the minimum received signal level which is used by the MS to acccess the system, while *MS_TXPWR_MAX_CCH* is the maximum TX power level an MS to access the system. The pathloss criterion is satisfied if $C1 > 0$

(2) *Cell Selection criterion in UTRA FDD and E-UTRAN*: The cell selection criterion S as explained in [6] and [7] is:

$$Srxlev > 0\, AND\, Squal > 0 \tag{4}$$

where:

$$Srxlev = Qrxlevmeas - (Qrxlevmin + Qrxlevminoffset) - Pcompensation \tag{5}$$

$$Squal = Qqualmeas - (Qqualmin + Qqualminoffset) \tag{6}$$

Srxlev is cell selection RX level value, *Squal*is cell selection quality value. *Qrxlevmeas* is measured cell RX level value (RSCP on UTRA FDD or RSRP on LTE), $P_{compensation}$ can be obtained from $\max(P_{EMAX} - P_{PowerClass}, 0)$.

2.2 Cell Reselection

As specified in [7], The cell-ranking criterion R_s for serving cell and R_n for neighboring cells are defined by Eqs. (7) and (8) is:

$$R_s = Q_{meas,serving} + Q_{hyst}, \tag{7}$$

$$R_n = Q_{meas,neighbour} - Q_{offset}, \tag{8}$$

where:

Q_{meas} is signal strength or quality measurement used in cell reselection. Q_{hyst} is hysterisis value and Q_{offset} for offset value.

Based on the formulation (7), UE perform selection to certain rat-cell. We assume as follows:

Open RAT_i, perform cell selection. If the selection isn't successfull, then goes to select another RAT. Assuming finally selection on $cell_i$.

The selection model can be presented as:

$$SuitCell = argmax_{i=1}^{N,Ratname}(Srxlev(Qrxlevmeas[i], Ratn \\ ||\boldsymbol{Srxlev}(\boldsymbol{Qrxlevmeas}[\boldsymbol{i}], \boldsymbol{Ratname}) > \boldsymbol{Srxlev}(\boldsymbol{Qrxlevmeas}[\boldsymbol{i}-1], \boldsymbol{RatName})|| \; \{Srxlev > 0\} \tag{9}$$

where:

SuitCell is suitable cell to be selected. *Srxlev* is srxlev value target cell *i*. Qrxlevmeas[i] is receiving signal strength target cell *i*, could be RXLEV for 2G or RSCP for UTRA FDD or RSRP for LTE. *Ratname* is current RAT to be measured, *i* is target cell i and *N* is maximum rat-cell which is it signal is received by an UE. The simple code is as follows:

```
Algorithm to obtain suitable rat-cell
Meas_used_val = 0
for every cell in RAT, do
    /*get RSSI or RSCP or RSRP */
    meas_val = get_meas_value(cell[i], current_rat)
    /* calculate srxlev or C1 */
    current_used_val = calculate_quality_value( meas_val, current_rat)
    if (current_used_val > meas_used_val), then
        meas_used_val = current_used_val
        best_cell_id = cell[i]
    end if
end for
return best_cell_id
```

Figure 1 shows UE performance measurement cell reselection during Treselection. RSS (Received Signal Strength) serving cell have degradation and RSS target cell rising. UE prepare to select the target cell. UEs always need to be associated to the base stations (enhanced Node Bs, eNBs) which provide the best radio link [8] and good cell selection schemes not only guarantees the QoS but also able to reduce the interference level [9] for intra-RAT case.

UE receive SI (System Information) [10, 11] from the $cell_i$, SI contain information suggested which is better/suitable RAT to be selected, could be priorities. The suggested RAT priority in our system level simulation from first to last are 4G, 3G and 2G, respectively. The workflow cell reselection procedure for reselection to higher priority RAT depict as follow:

In Fig. 2, assume RAT serving cell is RAT_i and RAT target cell is RAT_j

1. UE receive broadcast SI from $\boldsymbol{cell_i}$ which contain priority RAT. Reselection to a cell on higher priority inter-RAT frequency than the serving frequency shall be performed if Step 2 is fullfilled.
2. If ***threshServingLowQ*** is provided:
   Yes: Goto Step 3, otherwise Goto Step 11

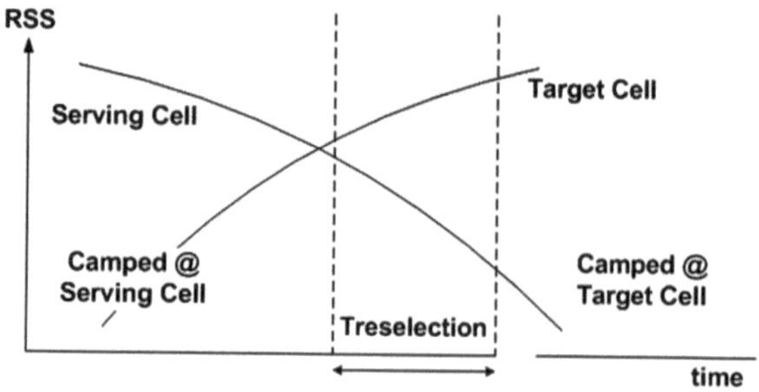

Fig. 1 Cell reselection

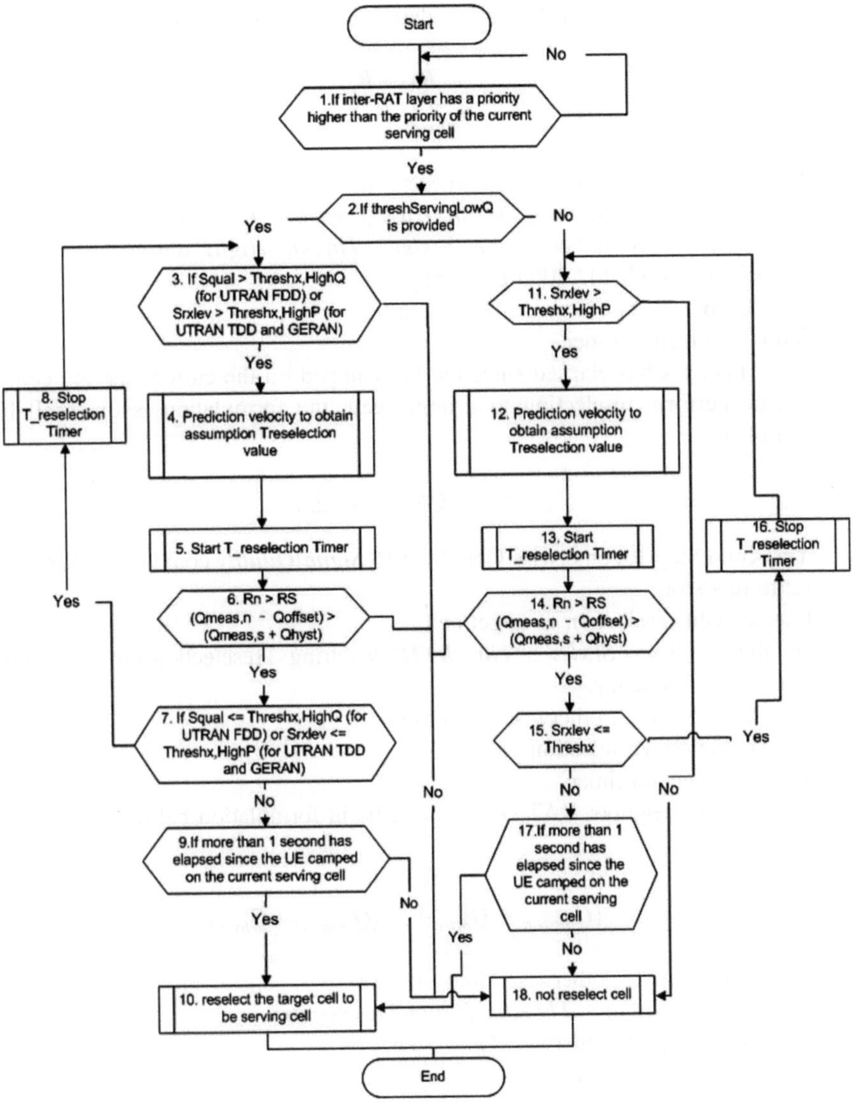

Fig. 2 Cell reselection to higher priority

3. Neighbor cell has ***Squal*** > ***ThreshxHighQ*** during Treselection on E-UTRAN or UTRAN or, Neighbor cell has ***Srxlev*** > ***ThreshXHighP*** during Treselection on GERAN or CDMA2000 frequency.
   Yes: Goto Step 4, otherwise Goto Step 18
4. Get prediction Treselection value.
5. Start Treselection timer.

6. UE measure neighbor RAT cell as describe in formulation below:

$$R_n > R_s$$
$$(Q_{meas,n} - Q_{offset}) > (Q_{meas,s} - Q_{hyst})$$

   Yes: Goto Step 7, otherwise Goto Step 18
7. Neighbor cell has ***Squal*** $\leq$ ***ThreshxHighQ*** during Treselection on E-UTRAN or UTRAN or, Neighbor cell has ***Srxlev*** $\leq$ ***ThreshXHighP*** during Treselection on GERAN or CDMA2000 frequency.
   Yes: Goto Step 8, otherwise Goto Step 9
8. Stop Treselection timer.
9. More than 1 s has elapsed since the UE camped on the current serving cell. Before perform reselection to a target cell, the formulation below shall be fullfilled.

$$\mathit{SignalQuality} > Q_{out}$$

   Yes: Goto Step 10, otherwise Goto Step 18 ***SignalQuality*** could be 2G SINR or Ec/Io or EsIot.
10. Execute cell reselection to target cell.
11. Neighbor cell has ***Srxlev*** > ***ThreshXHigh*** during Treselection on higher priority RAT frequency.
    Yes: Goto Step 12, otherwise Goto Step 18
12. Get prediction Treselection value.
13. Start Treselection timer.
14. UE measure neighbor RAT cell as describe in formulation below:

$$R_n > R_s$$
$$(Q_{meas,n} - Q_{offset}) > (Q_{meas,s} - Q_{hyst})$$

    Yes: Goto Step 15, otherwise Goto Step 18
15. Neighbor has ***Srxlev*** $\leq$ ***Threshx*** during Treselection.
    Yes: Goto Step 16, otherwise Goto Step 17
16. Stop Treselection timer.
17. More than 1 s has elapsed since the UE camped on the current serving cell. Before perform reselection to a target cell, the formulation below shall be fullfilled.

$$\mathit{Signal\,Quality} > Q_{out}$$

    Yes: Goto Step 10, otherwise Goto Step 18 ***SignalQuality*** could be 2G SINR or Ec/Io or EsIot.

Based on the suggested RAT, close RAT_i and open RAT_j then reselection is executed.

Figure 3 shows cell reselection to lower priority RAT. The step is similar to cell reselection to higher priority RAT however parameters threshold is changed to *ThreshXLowQ* and *ThreshXLowP*.

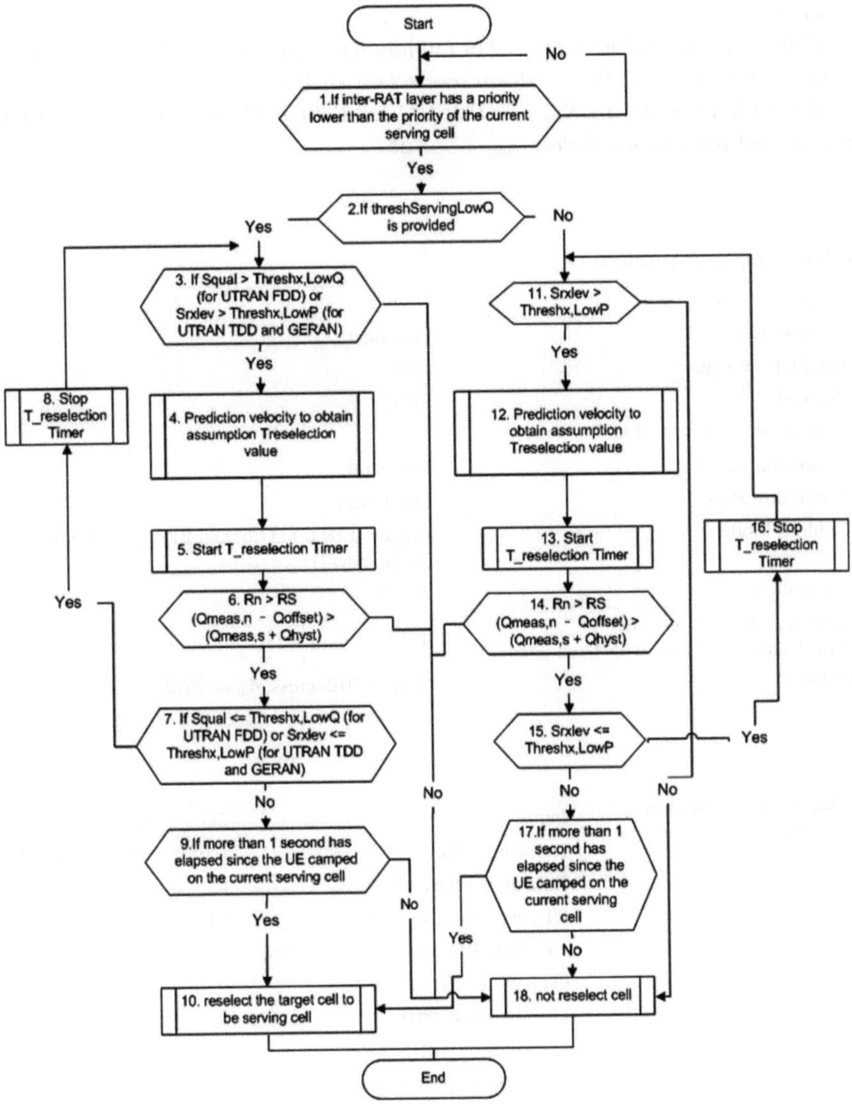

Fig. 3 Cell reselection to lower priority

3 Simulation results

In this section, we construct topology for radius Multi-RAT networks, 570 UEs are dropping randomly in overlap RAT network.

Table 1 describes the parameters for network topology in our system level simulation.

Table 2 summaries the parameters of the UE for selection and reselection procedure.

Cell reselection failure threshold or minimum requirement signal quality for used RAT (serving RAT cell) to perform reselection shall be fulfilled as follows:

If used RAT is 2G, $SNIR_{serving} < 9\,\text{db}$, if used RAT is 3G, $Ec/Io_{serving} < -15\,\text{db}$ and if used RAT is 4G, $EsIot_{serving} < -8\,\text{db}$.

Table 1 Topology parameters

Items	Values
Scenario case	Homonet
Total UE number	570
Channel	SCM
Number of carrier per RAT	1
Simulation mode	Downlink
Traffic downlink	Full buffer
Pathloss Model	L(dB)= 128.1 + 37.6*log(d(km)) + 20*log (frequency/2)
Penetration	20 dB
Antenna pattern, horizontal (For 3-sector cell sites with fixed antenna patterns)	$A(\theta) = -\min\left[12\left(\frac{\theta}{\theta_{3\,\text{dB}}}\right)^2, A_m\right]$ $\theta_{3\,\text{dB}} = 70\,\text{degress}, A_m = 20\,\text{dB}$

Table 2 2G/3G/4G UE parameters

Items	2G	3G	4G
Rxlevmin (dBm)	−100	−114	−131.7
Rxlevmin_offset (dBm)	0	0	0
UE_max_tx_power (dBm)	24	21	23
UE_max_RF_output_power (dBm)	24	21	23
Qqualmin (dB)	–	–	–
Threshx,High (dB)	–	32	40
Threshx,Low (dB)	12	34	–
ThreshServing,Low (dB)	–	28	36
Qhyst	0	0	0
Qoffset s,n	2	2	2

Figure 4 depicts the number of selection occured. Since 2G network radius coverage greater than other RAT, UEs having greater chance to select 2G cells, followed by 3G and then 4G.

Figure 5 depicts failure rate during cell selection in percentage. It shows small failure rate occurence, for 2G less than 1.5 %, for 3G around 0.5 % and for 4G about 3.5 %. The requirement for achieving success rate for 2G [12], a PLMN (Public LAN Mobile Network) should be understood to be received with high quality signal level which is above −85 dBm. For 3G FDD cell [6], the measured primary CPICH RSCP value shall be greater than or equal to −95 dBm. For 4G cell [7], the measured RSRP value shall be greater than or equal to −110 dBm.

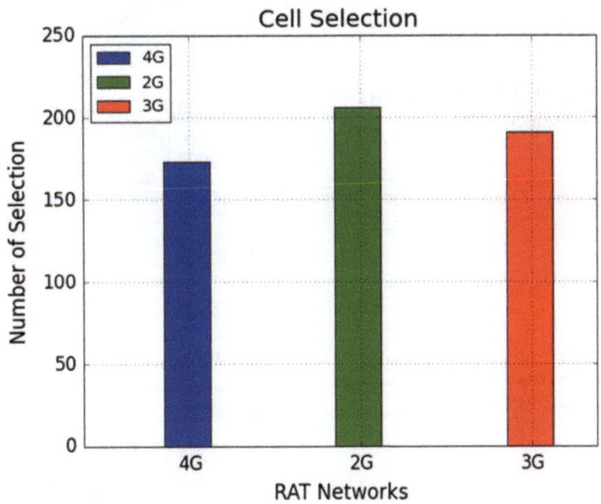

Fig. 4 UEs cell selection

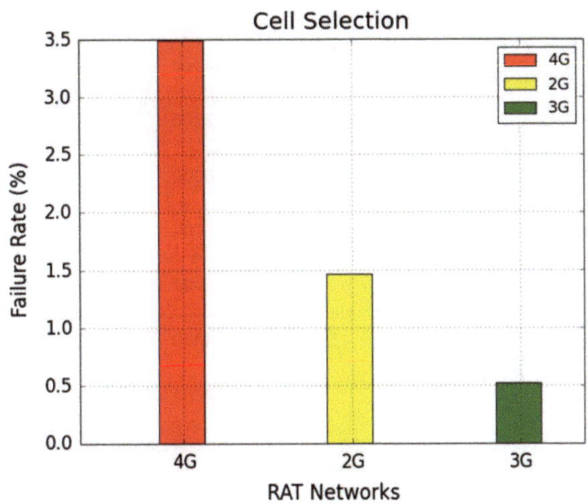

Fig. 5 UEs cell selection failure rate (%)

In cell reselection, total simulation time for UEs movement is 200 s. For velocity, we select 3 different velocity as in [13, 14]. We perform 4 cases rely on Treselection variance which are 1s, 2s, 3s, and 4s. The details for each case is shown in Table 3.

Figures 6, 7, 8 and 9 depict number of reselection occured during certain Treselection.

Table 3 Parameters case

Case	L1 measurement period	Treselection	Total simulation time
#1		1s	
#2	1280 ms	2s	200 s
#3		3s	
#4		4s	

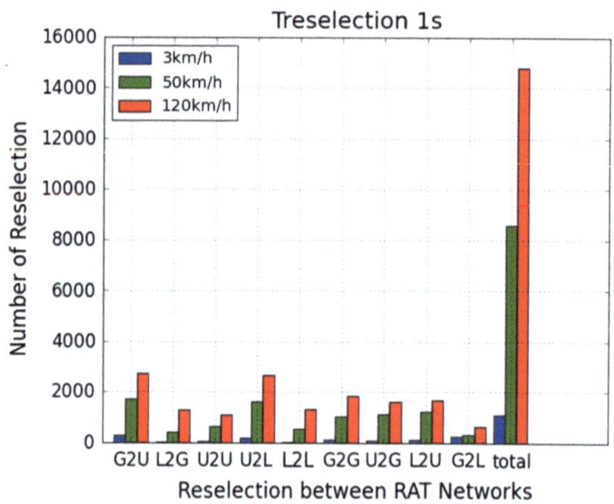

Fig. 6 UEs reselection for Treselection 1s

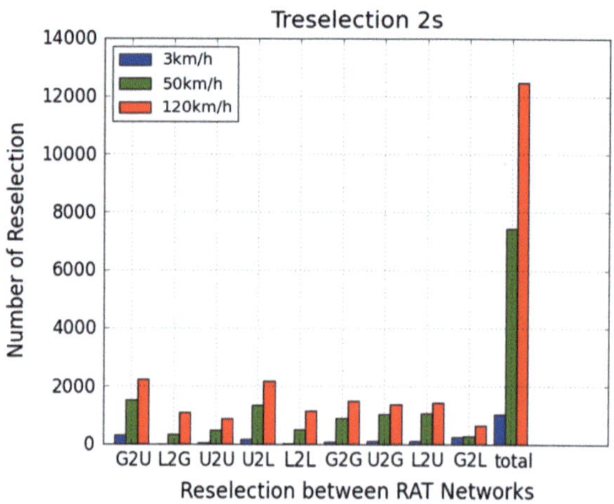

Fig. 7 UEs reselection for Treselection 2s

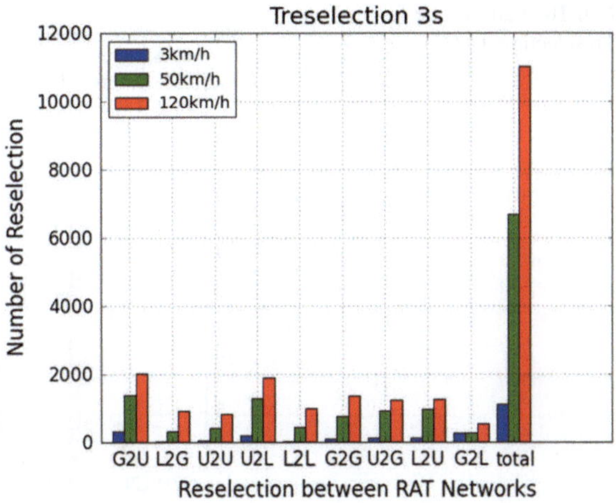

Fig. 8 UEs reselection for Treselection 3s

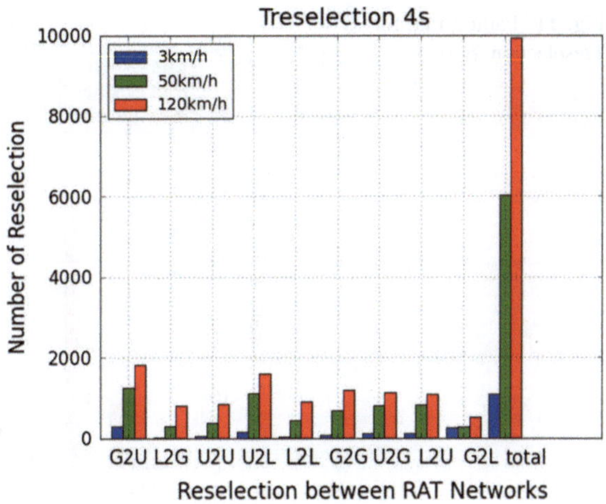

Fig. 9 UEs reselection for Treselection 4s

During Treselection 1 second, total reselection occured for UEs velocity at 3 km/h is 1087, for 50 km/h is 8566 and for 120 km/h is 14795.

During Treselection 2 second, total reselection occured for UEs velocity at 3 km/h is 1037, for 50 km/h is 7434 and for 120 km/h is 12464.

During Treselection 3 second, total reselection occured for UEs velocity at 3 km/h is 1100, for 50 km/h is 6666 and for 120 km/h is 11024.

During Treselection 4 second, total reselection occured for UEs velocity at 3 km/h is 1081, for 50 km/h is 6024 and for 120 km/h is 9920.

Figures 10, 11, 12 and 13 depict failure rate during reselection action on certain Treselection.

During Treselection 1 second, total failure rate for UEs velocity at 3 km/h is 0.00 %, for 50 km/h is 1.33 % and for 120 km/h is 6.83 %.

Fig. 10 Failure rate for Treselection 1s (%)

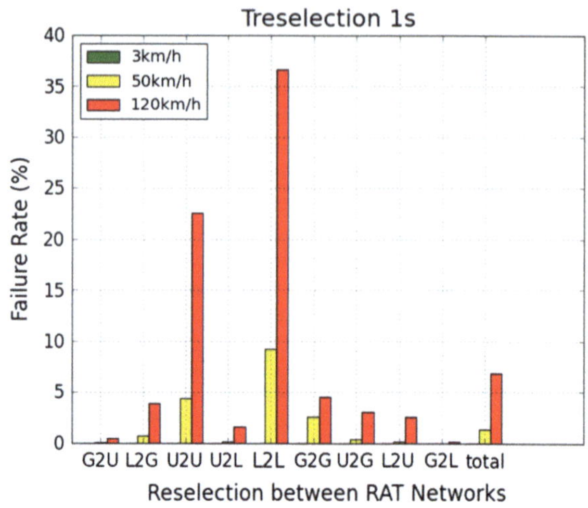

Fig. 11 Failure rate for Treselection 2s (%)

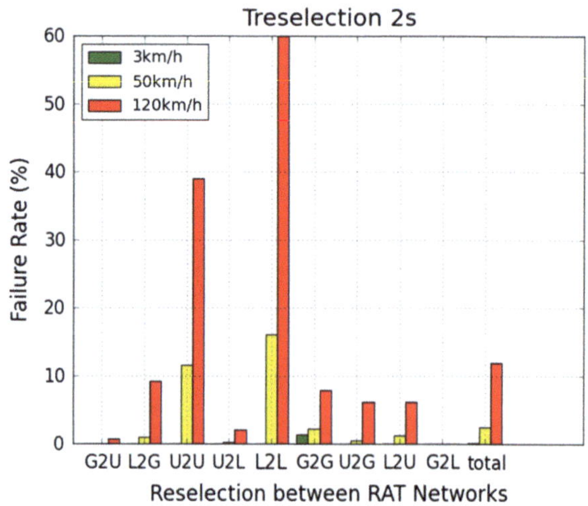

During Treselection 2 second, total failure rate for UEs velocity at 3 km/h is 0.10 %, for 50 km/h is 2.37 % and for 120 km/h is 11.83 %.

During Treselection 3 second, total failure rate for UEs velocity at 3 km/h is 0.00 %, for 50 km/h is 3.95 % and for 120 km/h is 16.12 %.

During Treselection 4 second, total failure rate for UEs velocity at 3 km/h is 0.09 %, for 50 km/h is 5.86 % and for 120 km/h is 21.93 %.

Based on the Figs. 6–13, we can conclude it into Table 4.

NR is the total Number of Reselection and FR is the total Failure Rate during reselection. Case #1 to #4 are representing Treselection 1s–4s, respectively.

Refer to the NR column, we assume that power consumption have decreased by increasing Treselection time from 1s to 4s step by step due to NR decreased. Its therefore NR value is inversely to power consumption.

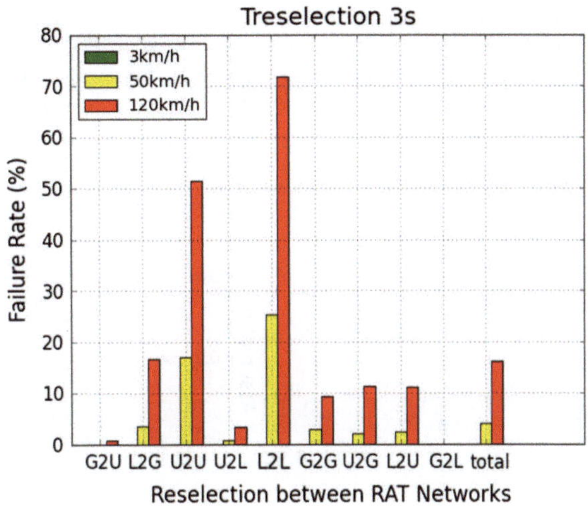

Fig. 12 Failure rate for Treselection 3s (%)

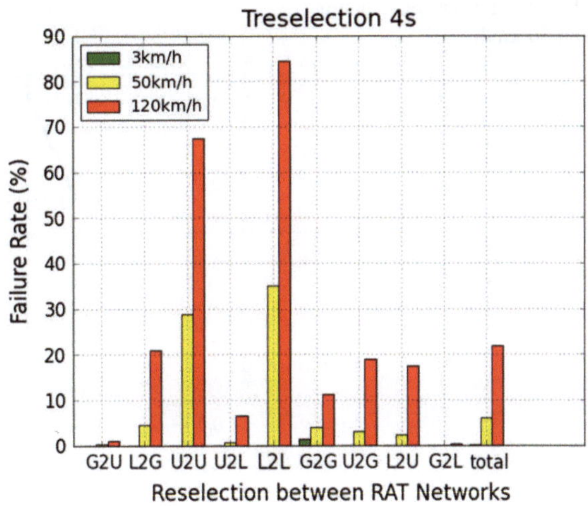

Fig. 13 Failure rate for Treselection 4s (%)

Table 4 Simulation result of cell reselection

Case		3 km/h		50 km/h		120 km/h	
		NR	FR	NR	FR	NR	FR
#1	Total	1087	0.00	8566	1.33	14795	6.83
#2	Total	1037	0.10	7434	2.37	12464	11.83
#3	Total	1100	0.00	6666	3.95	11024	16.12
#4	Total	1081	0.09	6024	5.86	9920	21.93

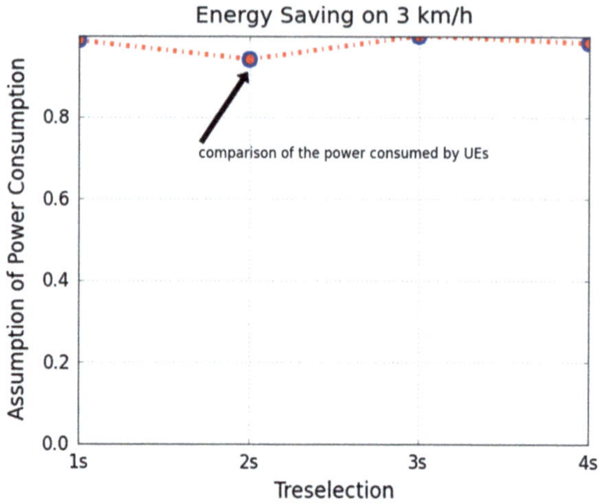

Fig. 14 Power consumption on 3 km/h

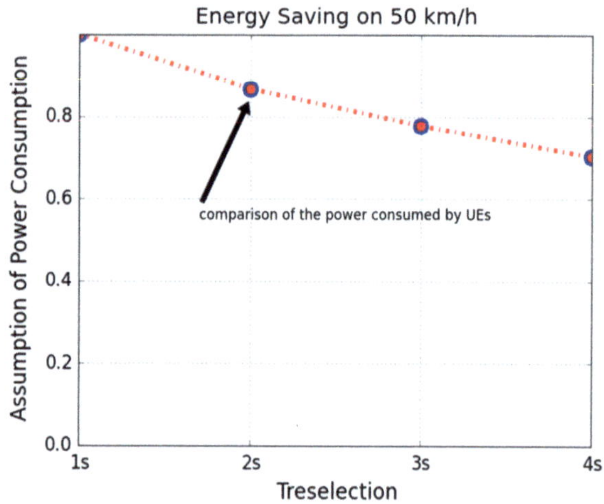

Fig. 15 Power consumption on 50 km/h

According to [15], $T_{detect,EUTRAN\_intra}[s]$ is 11.52 ms. Its representing the number of DRX cycle length time. Based on that case, we consider the probability of 1/2 wake up state to obtain effective DRX cycle is 5.76 ms.

Figures 14, 15, 16 represent assumption energy saving which is affected by number of reselection.

Figure 14 is not much represent the different of the power consumed by UEs since the NR occurred in UEs velocity 3 km/h are not much different.

Figure 15 shows the different of the power consumed by UEs in velocity 50 km/h. Assume that during Treselection 1s, the number of reselection occurred is 8566 and the effective DRX cycle is 5.76 ms, then the comparison is shown in Table 5.

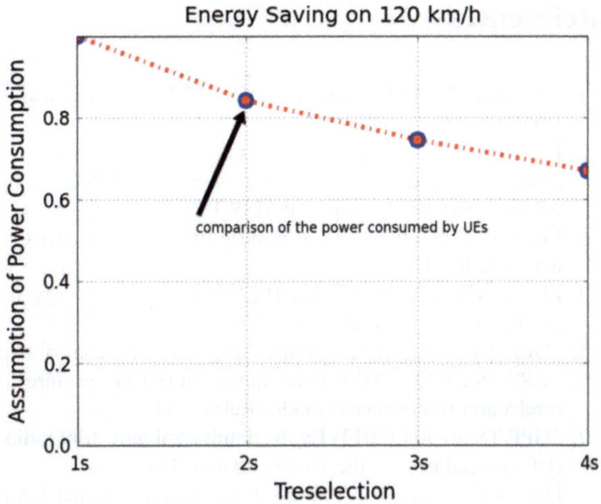

Fig. 16 Power consumption on 120 km/h

Table 5 Assumption power consumption on UEs 3 km/h

Parameters	t1	t2	t3	t4
Number of reselection	8566	7434	6666	6024
Effective DRX cycle detection period	5.76 ms	4.99 ms	4.48 ms	4.05 ms
Power consumption	1	0.87	0.78	0.70

Table 5, refers to Treselection 1s (t1), the power consumed by UEs is 1, then energy saved by Treselection 2s (t2) is 87 %, and then Treselection 3s (t3) could save 78 % and last Treselection 4s (t4) is 70 %.

Figure 16 is similar to Fig. 15 except the UEs velocity is 120 km/h. It shows more significant energy saving than the lower speed.

4 Conclusion

In this paper, we evaluate the number of selection and reselection which occur during idle mode in Multi-RAT networks. The result is very interesting with variance of parameter values. In reselection case, we utilize variance Treselection on variance velocity. By increasing Treselection and velocity value, the number of reselection decreasing but failure rate rising. If the speed is low, Treselection value does not much give effect to the number of failure rate. Increasing number of reselection causes increasing UE power consumption.

Acknowledgments This work is supported by cooperation project between Huawei Technologies Co.,Ltd and Beihang University.

References

1. Abuhaija B, Al-Begain K (2010) LTE capacity and service continuity in multi radio environment. In: IEEE
2. Kolehmainen N, Puttonen J, Henttonen T, Kaikkonen J (2010) Performance of idle mode mobility state detection schemes in evolved UTRAN. In: 5th international symposium on wireless pervasive computing (ISWPC)
3. Mach T, Tafazolli R (2010) Battery life idle parameter optimization of UE in self organizing network. In: IEEE
4. Zhu Y, Chen Y, Li W, Yu P (2011) A novel energy-saving cell selection mechanism for cellular access networks. In: IEEE
5. 3GPP TS 45.008 (2013) Radio subsystem link control (Release 11)
6. 3GPP TS 25.304 (2013) User equipment (UE) procedures in idle mode and procedures for cell reselection in connected mode (Release 11)
7. 3GPP TS 36.304 (2013) Evolved universal terrestrial radio access (E-UTRA); user equipment (UE) procedures in idle mode (Release 11)
8. Dimou K © Ericsson AB 2011 | Ericsson External | KTH | Wireless Seminar In Signals, Sensors, Systems | Mobility Management Within 3GPP Rel. 9 | 2011- 03- 03, http://www.kth.se/polopoly_fs/1.225520!/Menu/general/ column-content/attachment/Mobility_In_3GPP_LTE_ver5_ PartI. pdf, 2013
9. Ismail M, Aripin N (2005) Downlink soft handover performance for different cell selection schemes in WCDMA system. In: IEEE
10. 3GPP TS 36.331 (2012) Evolved universal terrestrial radio access (E-UTRA), radio resource control (RRC)
11. Mach T, Tafazolli R (2009) Mass mobility signaling congestion avoidance mechanism using randomized time distribution of cell reselections. In: IEEE
12. 3GPP TS 43.022 (2011) Technical specification group GSM/EDGE radio access network; functions related to mobile station (MS) in idle mode and group receive mode (Release 10). p 14
13. Puttonen J, Kolehmainen N, Henttonen T, Kaikkonen J (2009) On idle mode mobility state detection in evolved UTRAN. In: IEEE
14. Kastell K (2011) Challenges and improvements in communication with vehicles and devices moving with high-speed. In: IEEE
15. 3GPP TSG-RAN, R4-081081 (2008) Idle mode performance requirements for 36.133

A New Classification Algorithm Based on Emerging Patterns

Jun Pei and Min Zhang

Abstract As one of the most important part in classification branches, emerging patterns has become one hot research area. Classification methods based on emerging patterns are put forward for large datasets in recent years. One of them is DeEPs and it has been proved a good classification method. This paper proposes a novel classification method-MDeEPs. On the base of DeEPs algorithm to mining EPs, allocating misclassified test instances to the correspond class for adjusting deviation. Sequentially processing each test instance and at the same time counting error number. While meeting the condition that test instances has reached a certain number and within the scope of error rate, the algorithm ends. Otherwise, perform the entire process cyclically. Each class of EPs derived from different test instance is recorded. All of them will be aggregated to build classifier. Experiments show that the improved algorithm has a good performance.

Keywords Classification · Class label · Dataset · Emerging patterns

1 Introduction

In the past few decades, as one of basic disciplines of data mining, classification has been widely studied in learning statistics, machine, neural networks and expert systems. The input data is training data instance; each data record has several

---

This paper is supported by Science and Technology of Liaoning, China (Grant No. 201205534).

J. Pei · M. Zhang (✉)
Department of Information &Engineering, University of Dalian, Dalian, China
e-mail: zhangmin@dlu.edu.cn

J. Pei
e-mail: pj353588026@163.com

E. Qi (ed.), *Proceedings of the 6th International Asia Conference on Industrial Engineering and Management Innovation*,
DOI 10.2991/978-94-6239-145-1_6

attributes. There is a remarkable property called class label. Generally speaking, given a database record, each record has class label. Classifier containing all attributes of each class generates an accurate and meaningful model [1]. This model is used to predict unknown object class label. Classification, one method of supervised learning, each training instance is labeled to its class. In contrast, the other is non-supervised learning, which is based on the value of their property to identify class of the same class objects and there is no class label pointing to the object. Classification has been widely applied in various applications, such as scientific experiments, genetic diagnosis, weather forecasting, credit assessment, customer segmentation, market research and fraud investigation [2–5].

Classification based on model is a relatively new research method. Jumping emerging patterns, which is defined as a support from one class to another class, as a support of one itemset from zero to non-zero, and it can quickly identify differences between different classes.

For example, to mushroom data sets in UCI. Suppose the pattern A = {(odor = none), (gill-size = broad), (ring-number = one)}, B = {(bruises = no), (gill-spacing = close), (veil-color = color)}. Pattern A is a jumping emerging pattern, which appears in edible mushrooms in large numbers, but does not appear in the poisonous mushrooms [6–8]. Pattern B is not a jumping emerging pattern, though it has a high growth rate in poisonous mushrooms. To unknown class mushrooms, if this test case contains such a model, then we can believe that this mushroom is edible. Bring the most expressive jumping mode together, excellent accuracy of classification beat out the other categories, such as C4.5 and CBA.

Although the most effective mining algorithms in the literature [9] are mentioned, the accuracy and efficiency is not impressive. To overcome these difficulties, we propose a new algorithm for mining emerging patterns. After adjusting algorithm, join error rate and aggregate EPs, its efficiency and performance is quite good.

The rest part of the paper is organized as follows: Sect. 2 outlines the basic ideas of emerging patterns. Section 3 demonstrates essential features of the DeEPs approach. Section 4 presents detailed steps for instance based pattern discovery, including an explanation on how to discover EPs by the improved method, and why this method is highly efficient and then showing its pseudo code. Sections 5 describe a case to illustrate the entire process of MDeEPs. Section 6 describes experiments results. Section 7 is the summary of the paper.

2 Emerging Patterns

Emerging patterns (EPs) are those itemsets or characteristics whose supports change significantly from one data class to another [9, 10]. Its classification is useful, because it can capture differences between different data sets. EP which cited as a new KDD pattern has the following characteristics: validity, novelty, potential usefulness and understandability. These characteristics make emerging patterns success in classification.

The following is specific concepts.

Assuming training data set is a standard relational table, which contains N instances (samples), each instance has m different properties. All continuous attributes has been discredited, and mapped to a contiguous set of positive integers. Assume N instances are divided into K known classes $C_1, C_2, \ldots, C_k$, and its class label. Item is the dual, which contains attribute names and values. Let $I = \{I_1, I_2, \ldots, I_n\}$ are complete items appearing in the sample. I_n is called a subset of I. Using the concept of item, each sample is a set of n elements containing items, while the training data set is a collection of itemsets [11, 12].

Definition 1 Define the support of itemsets X in the data set D, denoted $\sup_D(X)$, $\sup_D(X) = count_D(X)/|D|$, where $count_D(X)$ is the number of samples contained X in the data set, $|D|$ is the total number of samples in D.

Definition 2 Given two different data sets D' and D, the growth rate of itemset X from D' to D is $GR(X, D', D)$ defined as follows:

$$\text{GrowthRate(X)} = \begin{cases} 0, \text{when support1(X)} = \text{support2(X)} = 0 \\ \infty, \text{when support1(X)} = 0\,\text{support2(X)} \neq 0 \\ \text{support2(X)}\,/\text{support1(X)}, \text{other} \end{cases} \quad (1)$$

If the data set D' and D are sample collection of Class C' and Class C respectively. The growth rate is a measure for the itemset X whose support changes significantly from class C' to Class C.

Definition 3 Given growth rate threshold $\rho > 1$, if X is an item whose growth rate from D' to D is $GR(X, D', D) \geq \rho$, then X is called ρ-EP or (EP) from D' to D, or simply referred to X is a EP of D. If the growth rate of X is ∞, then X is called JEP (Jumping EP) of D.

Definition 4 Error rate, the total number of errors instances divided by the total number of instances of the statistical test instances.

EPs are those itemsets whose support change dramatically from one data set to another. Such itemsets can capture on different target classes and non-target classes in multiple properties and therefore EP has a good discrimination [13, 14].

3 Deeps Discription

DeEPs take a lazy way of learning as a measure, using a case-based approach to deal with new training data without requirement for training the classifier. DeEPs abandoned those attributes whose attributes and properties are not appearing in test instance S [15, 16].

Assuming classification has a positive training set $D_p = \{P_1, P_2, \ldots, P_m\}$, negative training set $D_n = \{N_1, N_2, \ldots, N_m\}$, and a test instance S. DEePs perform the following steps:

Select a test instance T and then intersect with the positive and negative training set respectively.

Select itemsets which is the intersection of T and positive samples then take use of the method of frequent itemsets to obtain its maximum frequent itemsets as a new positive sample, denoted max_D_p. Similarly, after the election the intersection of a negative sample itemsets and of T, its maximum frequent item sets as a new set of negative samples, denoted max_D_n.

Use boundary algorithm to mining all EPs from max_D_p and max_D_n.

Select the left boundary of the JEP to calculate score for each class.

Comparing to CAEP, the function score of DeEPs only consider a data set that contains at least one instance of the total number of EP; do not consider the growth rate. The function score is as follows:

$$\text{score}\,(c_j) = \frac{count_{D_i}(E(c_j))}{|D_i|} \tag{2}$$

Note: $E(c_j)$ is the set of all the JEP from c_j, $count_{D_i}(E(c_j))$ contains at least one or more EPs in $E(c_j)$, this method avoids the reuse of training data.

DeEPs classification requires three steps to determine a test instance of the class label:

(1) Found the boundary EP.
(2) Select the better ability to distinguish JEP.
(3) Determined a comprehensive score for classification.

4 Mdeeps Discription

Based on the ideas of DeEPs algorithm, we put forward a new algorithm-MDeEPs (Modified DeEPs emerging patterns). Here, we follow the instance-based approach of DeEPs, but the method of classification is different. After deleting unrelated attributes to S, DeEPs try to classificate without considering misclassified instances. For each derived EPs, it did not do any treatment, but only proposed a classification method. The same above process is done by MDeEPs, but not finishing after it. The algorithm will judge if each classification is consistent with its class label, then recorded separately for each class of EPs. If the class label of test instance is inconsistent with their classification, demonstrating that test instance is misclassification, so according to its class label to determine the real class, add it to the correct category in the training instances, and then update the data set. The inconsistent situations do not need to record EPs, because it is not true. Each misclassified instance will be statistic until the error rate is in the control of a threshold value range. By integrating EPs of each correctly classified instance, create a classifier. We can use the classifier to classify directly using general score function (e.g. CAEP) without re-training classifiers. The algorithm has higher accuracy and efficiency. The core pseudo code is as follows:

A and *B* are given data sets, $\{A_1, \ldots, A_{k_1}\}$ and $\{B_1, \ldots, B_{k_2}\}$ are maximum frequent itemsets.

JEPPRODUCER $(<\{\Phi\}, \{A_1, \ldots, A_{k_1}\}, \{\Phi\}, \{B_1, \ldots, Bk_2\}>)$
//return $< L, R >$

(1) initialize $L = \{\}; R = \{\}$;
(2) for j from 1 to k_1 do
(3) if some B_{k_i} is a superset of A_j then continue;
(4) border = BORDER-DIFF $((<\{\Phi\}, \{A_j\}>, <\{\Phi\}, \{B_1, \ldots, Bk_2\}>))$;
(5) R = R $\cup$ the right bound of border;
(6) L = L $\cup$ the left bound of border;
(7) return $< L, R >$;

BORDER-DIFF $(<\{\Phi\}, \{U\}>, <\{\Phi\}, \{S_1, S_2, \ldots, S_k\}>)$
//return border of $[\{\Phi\}, \{U\}], [\{\Phi\}, \{S_1, S_2, \ldots, S_k\}]$

(1) initialize L to $\{\{x\}| = U - S_j\}$
(2) for j = 2 to k do
(3) $L = \{X \cup x | X \subseteq L, x \in U\text{-}S_j\}$;
(4) remove all Y in L that are not minimal;
(5) return $<L,R>$;

COMPACTSCORE (*L*)
//obtain each score in different classes and divide into correspond class;

(1) JEPPRODUCE $(<\{\Phi\}, \{A_1, \ldots, A_{k_1}\}, \{\Phi\}, \{B_1, \ldots, Bk_2\}>)$
(2) initialize $X = \{\{x\}|x \in L\}$, count = 0;
(3) for every record of each class do
(4) if the record $\supseteq X$ then count++;
(5) count/the total record number of each class;
(6) return class label;

UPDATE ()
//return new records in two class, *T* is a test instance which belongs to *I* class

(1) initialize H = COMPACTSCORE(L), count1 = 0;
(2) if *H* is not equals to *I* then add *T* to *I*, count1 ++;
(3) else reserve *X* for each class;

AGGREGATE (count1,σ)
//obtain final EPs for each class, σ is the threshold of error rate.

(1) UPDATA (T, I)
(2) if count1/sum(T) $\leq \sigma$ then stop training;

(3) return EPs;
(4) else repeat the whole process;

5 Instance Analysis

In order to make the description process algorithm clearer, the following is an analysis of a specific example. Two data sets about weather condition on Saturday morning are shown in Table 1. O, T, H, W means outlook temperature, humidity, windy respectively.

Given a test instance *S*, *S* = {sunny, mild, high, true}, the label is not suitable class. Now filter out the related attribute value. The data set is shown in Table 2.

After filtering, JEP which is suitable for activity {sunny, mild, true}, not suitable are {sunny, high}, {sunny, mild, high}, {sunny, high, true}. By compact score

Table 1 Weather condition for activity

Suitable for activity				Not suitable			
O	T	H	W	O	T	H	W
overcast	hot	high	false	sunny	hot	hot	false
rain	mild	high	false	sunny	hot	hot	true
rain	cool	normal	false	rain	cool	normal	true
overcast	cool	normal	true	sunny	mild	high	false
sunny	cool	normal	false	rain	mild	high	true
rain	mild	normal	false				
sunny	mild	normal	true				
overcast	mild	high	true				
overcast	hot	normal	false				

Table 2 Weather condition after filtering

Suitable for activity				Not suitable			
O	T	H	W	O	T	H	W
–	–	high	–	sunny	–	high	–
–	mild	high	–	sunny	–	high	true
–	–	–	–	–	–	–	true
–	–	–	true	sunny	mild	high	–
sunny	–	–	–	–	mild	high	true
–	mild	–	–	–	–	–	–
sunny	mild	–	true	–	–	–	–
–	mild	high	true	–	–	–	–
–	–	–	–	–	–	–	–

Table 3 Update the data sets

Suitable for activity				Not suitable			
O	T	H	W	O	T	H	W
overcast	hot	high	false	sunny	hot	hot	false
rain	mild	high	false	sunny	hot	hot	true
rain	cool	normal	false	rain	cool	normal	true
overcast	cool	normal	true	sunny	mild	high	false
sunny	cool	normal	false	rain	mild	high	true
rain	mild	normal	false	*rain*	*cool*	*high*	*true*
sunny	mild	normal	true				
overcast	mild	high	true				
overcast	hot	normal	false				

function, the test instance belongs to suitable class is 1/9, the other is 3/5. So S is judged to the other class. And it's true.

Take another case for example, S_1 = {rain, cool, high, true}. Its label is suitable class. The score of JEP for suitable class if 2/9; the other class is 1/5. According to the principle, it should be classified to suitable class. But, its real class label is not suitable class. So, we need to put the instance the right place to adjusting deviation, see the italic part. Table 3 is the update data sets.

At the same time, the misclassified number increase by one. Then go on to the next test instance for classification. We will finish the algorithm until the percent of misclassified number in total test instances number scarifies the threshold of error rate σ. Reserve those EPs which is classified correctly.

In the future classification, we can use the existing EPs to classify class directly. Using general score function as is mentioned in CAEP.

6 Experiment

The data sets derive from UCI machine learning, Table 4 displays the result of experiment. Here, we choose $\sigma = 0.001$.

Table 4 point out the correct rate of above algorithms. MDeEPs is our experiment result, the others is obtained by weka. As can be see, MDeEPs is winner at ten data sets. Compare to DeEPs MDeEPs has thirteen successes. Experimental results show that the algorithm is more rational.

Table 4 Result of experiment in UCI machine learning

Data Sets	NB	C4.5	DeEPs	MDeEPs
Balancescale	90.20	77.56	81.66	81.61
Breat-w	95.01	93.84	95.74	94.19
Credit-a	77.76	85.77	85.26	85.67
Crx	84.01	84.44	85.32	86.82
Diabetes	75.95	74.24	75.87	76.42
Glass7A	56.02	63.01	86.91	85.74
Heart-c	83.34	78.65	83.87	86.29
Hungarian	82.97	80.34	82.56	84.93
Iris	95.42	94.21	89.33	87.98
Labor	93.27	77.16	78.89	77.80
Led	74.31	73.01	65.52	66.08
Lymph	82.30	75.92	74.76	73.44
Mfeat-mor	68.75	68.69	69.47	69.52
Mushhroom	95.76	100.00	98.01	99.33
Pid	70.43	64.72	69.26	68.90
Segment	80.18	95.74	87.57	89.75
Sign	46.71	60.67	68.28	69.04
Solarflare	81.74	83.03	83.51	84.73
Sonar	68.96	72.07	77.36	79.18
Ttt	69.64	85.05	78.13	79.41
vehicle	47.02	72.51	54.45	53.46
vote	89.79	79.98	85.14	80.67
average	77.71	79.12	79.86	80.04

7 Conclusion

This paper proposes an improved algorithm—MdeEPs. By recitifying the misclassified instance to the right class to improve the accuracy of the algorithm, obtained an ideal pre-classification results. However, DeEPs did not do anything after classifying. Introduce the concept of error rate to control the end time of the algorithm and the same time integrating EPs for the future classification. Though experiment has shown the algorithm has a better accuracy and efficiency than DeEPs. The result mostly depends on specific test instances. Whether those patterns we obtained can coverage most of EPs is significant. This is question we need to study the next stage.

Acknowledgments First and foremost, J.P. Author would like to show deepest gratitude to everyone who has provided with valuable guidance in every stage of the writing of this thesis and thanks the support of Science and Technology of Liaoning, China (Grant No. 201205534).

References

1. Takizawa A (2011) Classification and feature extraction of criminal occurrence points using CAEP with transductive clustering. Selection 07:36–42 (Elsevier Ltd.)
2. Huang C, Li J (2010) Analysis on fault tree of theft cases. J Jiang Su Plice Officer Coll 135–141(2010) (In Chinese)
3. Du W, Li M, Peng H (2010) Research on data mining algorithm and application based on decision tree. Guangdong Gongan Keji 40–43 (In Chinese)
4. Quantang D, Zhang Y, He Y, Xiao Z (2011) Research and application on crime rule based on graph data mining alogorithm. Comput Technol Devel 89–92 (In Chinese)
5. Lu XG, Li D,Wang HJ (2010) Cancer classification algorithm with improved emerging pattern. Comput Eng Appl 233–237 (In Chinese)
6. Wang L, Wang Y, Zhao D (2010) Building emerging pattern (EP) random forest for recognition 26–29
7. Milton G-B (2011) José Fco Martínez-Trinidad. Jesús Ariel Carrasco-Ochoa.Fuzzy emerging patterns for classifying hard domains.Knowl, Inf Syst, pp 473–489
8. LI J, Dong G, Wong L (2004) DeEPs: a new instance-based lazy discovery and classification system. Mach Learn 54–99
9. Hiroya I, Takahisa A, Akira S, Seishi O (2002) Discovery of emerging patterns from neareat neighbors. Mach Teaming Cybern, Beijing, pp 4–5
10. Liu Q, Dong G (2009) A contrast pattern based clustering quality index for categorical data. In: Proceedings of IEEE ICDM, pp 101–106
11. Novak PK, Lavrač N (2009) Supervised descriptive rule discovery: a unifying survey of contrast set, emerging pattern and subgroup mining. J Mach Learn Res 377–403
12. García-Borroto M, Martínez-Trinidad JF, Carrasco-Ochoa JA (2010) A new emerging pattern mining algorithm and its application in supervised classification. Springer, Berlin, pp 150–157
13. Ngo TS, Feng M, Liu G, Wong L (2010) Efficiently finding the best parameter for the emerging pattern-based classifier PCL*. Springer, Berlin, pp 121–133
14. Bailey J, Loekito E (2010) Efficient incremental mining of contrast patterns in changing data 88–92
15. Nohuddin PNE, Christley R (2011) Social network trend analysis using frequentp pattern mining and self organizing maps. Springer, London, p 311
16. Li JY, Dong G, Wong L (2004) DeEPs: a new instance-based lazy discovery and classification system. Mach Learn 54:99–124

Study on Data Fusion Techniques in Wireless Sensor Networks

Man-tao Wang, Jiang-shu Wei, Yong-hao Pan and Zhe Wei

Abstract In wireless sensor networks or WSNs, to improve the accuracy and correctness of the target sensing and monitoring, individual nodes often have overlapping sensing ranges. This usually generates data redundancy and results in both time and power consumption especially when large amount of similar data are to be transmitted. Data fusion scheme takes the advantage of this overlapping feature and can reduce, delete and refine the redundant data in the process of data transmission. A survey is presented in this article studying on the data fusion techniques in WSNs from the aspects of cluster based schemes, chain based schemes to statistics based schemes. Their advantages and disadvantages are also discussed in this research.

Keywords Data fusion · Data redundancy · Target sensing · Wireless sensor networks

1 Introduction

Wireless sensor networks (WSNs) consist of large number of small sensors and several base stations or data sinks. With limited computing abilities and storage capacities [1], these battery-powered small sensors are usually equipped with sense unit and signal transmission unit. Sensors are grouped together to undertake a certain task such as temperature, humidity, gas concentration, pulse, oxygen, and so on [2].

M. Wang · J. Wei (✉) · Y. Pan
College of Information and Engineering Technology,
Sichuan Agricultural University, Ya'an, China
e-mail: sicauwjs@163.com

Z. Wei
College of Information Technology, Sichuan Normal University, Guanghan, China

E. Qi (ed.), *Proceedings of the 6th International Asia Conference on Industrial Engineering and Management Innovation*,
DOI 10.2991/978-94-6239-145-1_7

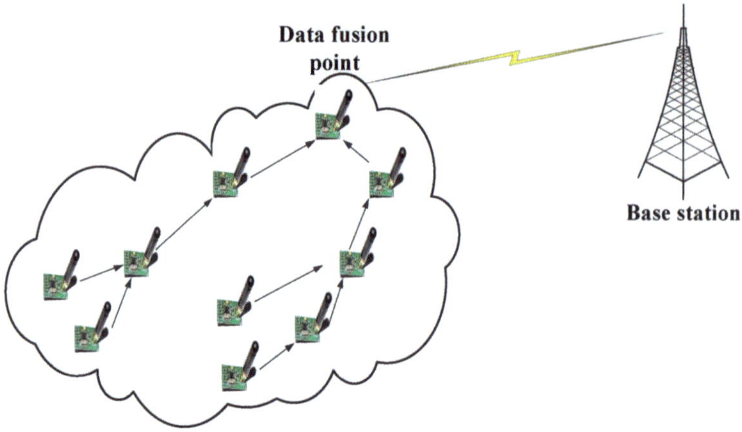

Fig. 1 Data fusion in wireless sensor networks

As is shown in Fig. 1, these collected data will be transmitted through single-hop or multiple-hop relays among sensors to the nearest data sink for further procession. Compared with the sensor node, the data sink is more powerful in computing and has longer communication range. It usually directly linked to the outside of the sensor network such as the Internet, through which users can issue task commands and check the corresponding results. According to the different tasks or sense units carried by individual sensors, WSNs can monitor the physical metrics such as temperature, pressure, vibration, radiation, and humidity, to name a few.

One of the advantages of a WSN lies in its flexible organization and extensible scale, namely topology of WSNs changes depending on specific regions for the target sensing and numbers of WSN nodes can be up to hundreds or even thousands. In addition, the self-independence and wireless communication ability enable individual nodes to be deployed in an ad hoc manner without any infrastructures in unattended or harsh areas that people cannot reach. Thus, WSNs are ideal for applications such as battlefield surveillance and wildlife monitoring.

Due to the large density of the nodes that are deployed in the monitoring region, how to reduce the data traffic is an effective method to minimize the power consumption of individual nodes as well as to improve the longevity of the whole network. Additionally, because of the limited communication range, individual nodes often have overlapping sensing ranges in order to improve the accuracy and correctness of the object monitoring. Data fusion technology takes advantage of this feature and can reduce and delete the redundant data when sensor nodes send the collected data to the other sensors or base station. WSN nodes are often deployed in unattended or even hostile regions, therefore security issue must be taken into consideration when the networks are being designed. Further, lack of physical security makes sensor nodes easy to be compromised by intruders who will later attack the whole network. Thus, compromised sensor nodes pose a challenging constraint for the designing of data fusion, and the data fusion algorithm must be energy-efficient and can operate safely under the existence of compromised nodes.

The basic idea of data fusion schemes [3–9] in WSNs is to combine the data from different sensor nodes at the data fusion point which then processes these data according to the data fusion algorithms and eliminates the redundancy before sending them to the data sink or the outside of the network. The aim of data fusion algorithms is how to aggregate the data in a safe, reliable, and efficient manner. So far, data fusion schemes in WSNs have been studied extensively. These schemes can be categorized into three parts: cluster-based schemes [10–12], chain based schemes [13, 14] and tree based schemes [15, 16].

In the following sections, we first present some representative data fusion schemes and then give an overview of statistics based schemes in WSNs.

2 Cluster Based Schemes

In cluster based data fusion schemes, the network is organized hierarchically and senor nodes are grouped in clusters. Special nodes named cluster heads are elected to perform the data fusion locally and transmit the corresponding results directly to the data sink via multiple hops. This results in great energy savings for the whole network. LEACH [10] is the first dynamic cluster head protocol for WSNs where sensor nodes are randomly deployed and cluster heads act as data fusion point. The protocol works in rounds and each round contains two sub-rounds, namely a setup phase to organize the clusters and a steady-state phase to transfer the data to the fusion point. LEACH works in a completely distributed manner and it outperforms traditional clustering algorithms in that cluster heads rotate in turn to balance the energy expenditure, but LEACH requires that the fusion data should be sent to the data sink through a single link, which may be quite energy-consuming especially when the data sink is far away from the cluster head. CAG [11] uses spatial and temporal properties of data to compute the approximate answers to queries. When forming clusters of nodes, CAG ignores the redundant data by using the spatial and temporal correlations, thus it can provide significant energy savings. Compared with LEACH, CAG provides a mechanism to compute the fusion data by using cluster head values, while LEACH does not. EECDA [12] is used for heterogeneous WSNs and can enhance the life span of the network by a factor of 51 %. Data fusion in EECDA consists of three phases, i.e. cluster head election phase, routing selection phase, and data communication phase. EECDA is more stable and energy efficient when compared with LEACH.

3 Chain Based Schemes

Instead of transmitting the fusion data from cluster members to the cluster head in cluster based data fusion schemes, chain based data fusion schemes only send the data to the closest neighbor. The chain is created either in a centralized or distributed manner. PEGASIS [13] organizes the sensor nodes in a chain manner and nodes act as the chain leader in turn. In a given time, only the chain leader has the authority to transmit the fusion data to the data sink, thus energy expenditure can be evenly distributed among the nodes in the network. In PEGASIS, every chain member aggregates its own data with those forwarded by its neighbor in a single packet and transmits the packet to the next neighbor until the packet reaches the current chain leader who will add its own data into the packet and sends it to the data sink. EPEGASIS [14] is an enhance variation of PEGASIS. One of the main differences between EPEGASIS and PEGASIS is that in EPEGASIS, the sensing area that is centered at the data sink is circularized into several concentric cluster levels and for each cluster level, a node chain is created based on the greedy algorithm of PEGASIS.

4 Tree Based Schemes

Tree-based data fusion schemes are organized in a tree topology and use a data sink to denote the tree root. Data fusion is performed by intermediate nodes and transferred to the data sink. Tree-based scheme is suitable for in-network data fusion, and how to construct an energy efficient tree is the focus of this scheme. In TREEPSI [15], root re-election mechanism is implemented. When the root node finishes the data fusion, it will directly send the result to the data sink. Then, election of a new root node is held by the network. Note that the tree path will not change until the current root node is replaced. Although the procedure of transmitting data from leaf node to the chain leader or root node is same in both TREEPSI and PEGASIS, TREEPSI uses a shorter path and outperforms PEGASIS by about 30 %. Due to the channel errors or battery dead, link failure and node malfunction may occur. PERLA [16] considers the link failure and node malfunction and by means of specific procedures it can identifies and recovers from link failures. Robustness of the network can be enhanced at the expense of energy consumption in error detection and recovery procedure in PERLA.

5 Statistics Based Schemes

Traditional security solutions such as cryptography and intrusion detection have shown some promises in the computer networks, but when it comes to the issue of internal compromised nodes, these methods may not that effective. The reason is that compromised nodes still have the access to the cryptographic keys that are used to secure the communication links within the network, and additionally, a compromised node pretending to be an authorized one cannot be detected by using cryptographic primitives. If the compromised nodes cannot be identified in time, secret information may be revealed and the whole network could be under the control of the adversaries [17]. In order to detect and delete the compromised nodes from the networks, reputation mechanisms [18–28] have been gradually studied by researchers. Through the evaluation and storing the reputation value of other nodes in WSNs, it is feasible to compute how much those nodes can be trusted when they are performing a certain task such as data relay and routing information response. The basis of reputation mechanism is that its calculation is either directly based on the historical behaviors of participating nodes or indirectly based on the recommendation from other nodes. In reputation based mechanisms of WSNs, behaviors of nodes are usually measured by the neighboring nodes using Watchdog [29, 30], and thus nodes will be punished for not being cooperative regarding a certain task. Many reputation models have been proposed by researchers in the past several decades. Among these models, classical ones include Bayesian theory model [18–28], subjective logic model [31, 32], entropy model [33, 34], fuzzy model [20], and game theory model [35, 36]. The first two models, on which our method is based, are the focus of this research.

Generally, the mechanism of a reputation system depends on the historical records of a participating node and Bayesian theory fundamentally complies with the procedure of reputation evaluation. Bayesian theory based reputation system attempts to discover the behavior patterns through historical actions [18]. It first calculates the prior probability of an event, then applies the prior probability into the binomial distribution, and finally modifies or updates such probability by using a posterior inference according to the relevant evidences.

As a representative application of Bayesian theory in WSNs, RFSN [19] uses a similar Bayesian formulation. In RFSN, each sensor holds reputation metrics representing past behavior of other nodes in order to predict these nodes' future behavior. According to reputation metrics built for other nodes by the behavior monitoring, a sensor node can rate them as *cooperative* or *non-cooperative* and evaluate the trustworthiness of these nodes. There are two building blocks in RFSN, namely Watchdog and Reputation. The Watchdog part is responsible for monitoring the actions of other nodes and classifying the actions into cooperation or non-cooperation. The Reputation part is responsible for maintaining the reputation of a node as well as other tasks such as managing reputation representation, updating reputation, integrating the reputation information, aging the reputation, and creating an output of reputation value. RFSN uses a completely decentralized manner and can run on each sensor node, the latter of which in RFSN only interacts with nodes within its wireless communication range and thus only maintains the reputation of nodes in its vicinity. In RFSN, a transaction is defined as two nodes making an exchange of information or participating in a collaborative process. After each transaction, one partner will rate the other as *cooperative* or not.

6 Conclusion

With the development of sensor technologies, wireless sensor networks have been used in a variety of of real applications. Due to the intrinsic sensing features of wireless sensor networks, data fusion has received wide attention by researchers. In this article, we give a survey of current data fusion techniques used in WSNs, especially the statistics based schemes. Pros and cons of these schemes are also discussed in this article.

References

1. Ali S, Madani S (2011) Distributed grid based robust clustering protocol for mobile sensor networks. Int Arab J Inf Technol 8(4):414–422
2. Xiao F, Wu M, Huang H et al (2012) Novel node localization algorithm based on nonlinear weighting least square for wireless sensor networks. Int J Distrib Sens Netw 2012:1–6
3. Fei J (2011) A data fusion strategy of wireless sensor networks based on specific applications. *In*: The 2nd International Conference on Theoretical and Mathematical Foundations of Computer Science, Singapore, pp 161–166
4. Yao X, Wang X, Li S (2011) A new data fusion routing algorithm for wireless sensor networks. In: IEEE international conference on computer science and automation engineering, Shanghai, China, pp 676–680
5. Yue J et al (2011) A novel cluster-based data fusion algorithm for wireless sensor networks. In: The 7th international conference on wireless communications, networking and mobile computing, Wuhan, China, pp 1–5
6. Kalpana B, Sangeetha R (2011) A novel framework for target tracking and data fusion in wireless sensor networks using kernel based learning algorithm. In: The 2nd international conference on advances in communication, network, and computing, Bangalore, India, pp 170–174
7. Lu Z et al (2011) A power-aware framework for distributed data fusion application in wireless senor networks. In: The 36th annual IEEE conference on local computer networks, Bonn, Germany, pp 127–130, 2011
8. Singh B, Sinha A, Narwal P (2011) An integrated fusion protocol for congregating sensor data in wireless sensor network. In: 2011 IEEE symposium on computers and informatics, Kuala Lumpur, Malaysia, pp 588–592
9. Damianos G et al (2010) An Approach for near-optimal distributed data fusion in wireless sensor networks. Wireless Networks, vol 16, no 5, pp 1407–1425
10. Heinzelman WB et al (2002) An application-specific protocol architecture for wireless sensor microsensor networks. IEEE Trans Wireless Commun 1(4):660–670

11. Sun HY, Shahabi C (2007) The clustered aggregation technique leveraging spatial and temporal correlations in wireless sensor networks. ACM Trans Sens Netw 3(1):1–39
12. Kumar D et al (2011) EECDA: energy efficient clustering and data aggregation protocol for heterogeneous wireless sensor networks. In: International journal of computers, communications and control, vol 6, no 1, pp 113–124
13. Lindsey S et al (2002) Data gathering algorithms in sensor networks using energy metrics. IEEE Trans Parallel Distrib Sys 13(9):924–935
14. Jung SM et al. The concentric clustering scheme for efficient energy consumption in the PEGASIS. In: Proceedings of the 9th international conference on advanced communication technology, Gangwon-Do, Korea, pp 260–265
15. Satapathy SS, Sarma N (2006) TREEPSI: tree based energy efficient protocol for sensor information. In: 2006 IFIP international conference on wireless and optical communications networks, Bangalore, India, pp 1–4
16. Messina D et al (2007) A network protocol to enhance robustness in tree-based WSNs using data aggregation. In: IEEE international conference on mobile adhoc and sensor systems, Pisa, Italy, pp 1–4
17. Feng R et al (2011) A trust evaluation algorithm for wireless sensor networks based on node behaviors and D-S theory. Sensors 11:1345–1360
18. Wang F, Wei Z (2012) Energy-saving reputation method for wireless sensor networks. J Shanghai Jiaotong Univ (Science) 17(2):223–227
19. Ganeriwal S et al (2008) Reputation based framework for high integrity sensor networks. ACM Trans Sens Netw 4(3):15–37
20. Boukerche A, Ren Y (2008) A trust-based security system for ubiquitous and pervasive computing environments. Comput Commun 31(18):4343–4351
21. Zouridaki C et al (2009) E-Hermes: A robust cooperative trust establishment scheme for mobile Ad Hoc networks. Ad Hoc Netw 7(6):1156–1168
22. Omar M et al (2009) Reliable and fully distributed trust model for mobile Ad Hoc networks. Comput Secur 28(3):199–214
23. Marmol F, Perez G (2010) Towards pre-standardization of trust and reputation model for distributed and heterogeneous systems. Comput Stan Interfaces 32:185–196
24. Wang F et al (2012) SONR: a reliable reputation system of self-organized network. J Netw Comput Appl 35(3):914–926
25. Mukherjee P, Sen S (2011) Comparing reputation schemes for detecting malicious nodes in sensor networks. Comput J 3(54):482–498
26. Momani M et al (2007) RBATMWSN: recursive bayesian approach to trust management in wireless sensor networks. In: The third international conference on intelligent sensors, sensor networks and information, Melbourne, Australia, pp 347–352
27. Alzaid H et al. RSDA: reputation-based secure data aggregation in wireless sensor networks. In: The ninth international conference on parallel and distributed computing, applications and technologies, Dunedin, New Zealand, pp 419–424
28. Pissinou N, Crosby GV (2007) Cluster-based reputation and trust for wireless sensor networks. In: The fourth IEEE consumer communications and networking conference, Las Vegas, United states, pp 604-608
29. Yan G et al (2009) Mobi-Watchdog: you can steal, but you can't run!. In: Proceedings of the 2nd ACM conference on wireless science conference, Zurich, Switzerland, pp 139–150
30. Lee JW, Lee Y, Syrotiuk VR (2007) The performance of a watchdog protocol for wireless sensor network security. Int J Wireless Mobile Comput 2(1):28–36
31. Liu Y et al (2011) A novel reputation computation model based on subjective logic for mobile and Ad Hoc networks. Future Gener Comput Syst 27(5):547–554
32. Tian J, Li C, He X (2011) Trust model based on the multinomial subjective logic and risk mechanism for P2P network of file sharing. J Electron 28(1):108–117
33. Caticha A, Giffin A (2006) Updating probabilities. In: The 26th international workshop on bayesian inference and maximum entropy methods, Paris, France, pp 31-42

34. Sun YL et al. (2006) A trust evaluation framework in distributed networks: vulnerability analysis and defense against attacks. In: IEEE INFOCOM, Barcelona, Spain, pp 1–13
35. Kornathyk K, Narayanasamy P (2008) Trust-based evolutionary game model Assisting Aodv routing against selfishness. J Netw Comput Appl 31(4):446–471
36. Papaioannou T, Stamoulis G (2008) Achieving honest ratings with reputation-based fines in electronic markets. In: IEEE INFOCOM, Phoenix, United states, pp 1714–1722

An Algorithm for Mining Association Rules Based on the Database Characteristic Matrix

YU Tong and XU Meide

Abstract This paper proposes a new algorithm for mining association rules. In order to calculate itemsets support, this paper puts forward the concept of database characteristic matrix and characteristic vector, and emerges an algorithm for mining association rules based on the characteristic matrix. This algorithm needs to traverse the database one time only, and the database operation has been reduced greatly. Based on the characteristic vector inner product, an itemset support can be obtained and the efficiency of the algorithm has been improved.

Keywords Association rules · Data mining · Database characteristic matrix · Database traversal

1 Introduction

In the research field of data mining, the research on association rules is deeper and more extensive [1–3]. The focus of the research is to find frequent item sets (frequent-sets). There are numbers of typical algorithm, such as Apriori algorithm [4] and DHP algorithm [5] proposed by R. Agrawal et al. They have improved the production process of candidate set C_k by using Hash technology. These algorithms are database traversal algorithm. A method of database segmentation algorithm was proposed by Savasere in 1995, the algorithm has reduced the number of times of I/O in the mining process and has lightened the burden of CPU. H. Toivonen has found out the association rules from large databases by sampling method, this method costs less [6]. These algorithms have improved mining process of association rules more or less. But they usually need to traverse the database to obtain the support, the execution of the algorithm needs to operation of the database continuously by using SQL statements. Frequent communications between many I/O

Y. Tong (✉) · X. Meide
Automation Engineering Institute, Beijing Polytechnic, Beijing 100176, China
e-mail: yut707@163.com

E. Qi (ed.), *Proceedings of the 6th International Asia Conference on Industrial Engineering and Management Innovation*,
DOI 10.2991/978-94-6239-145-1_8

operations and the ODBC interfaces are limits of the efficiency of the system. This paper presents a method of association rules mining algorithm based on the characteristic matrix of database, it calculates the itemset support by the calculation of feature vectors. It simplified the process of calculation of the support so that the efficiency of association rules mining algorithms has been improved.

2 The Mathematical Model of Association Rules

Association rules can be described with mathematical model [7–9]: A set of letters $I = \{i_1, i_2, \cdots, i_m\}$ is called the set of data items, the data items collection is called item-set, D is a collection of all the affairs, business T is a subset of I, i.e. $T \subseteq I$, each transaction consists of an only TID mark. Association rules have the following form:$X \Rightarrow Y$, and $X \subseteq I$, $Y \subseteq I$, $X \cap Y = \Phi$, X is the conditions of rules, Y is the results of rules. If a record contains X, Y, the record will met rules $X \Rightarrow Y$. For $X \subset I$, if the record number of X which is included in D is s, then the support of X: $\text{support}(X) = s$. Confidence can be described as strength of rules, defined as:

$$confidence(X \Rightarrow Y) = \frac{\text{support}(X \cup Y)}{\text{support}(X)}$$

Association rules mining algorithm is usually divided into the following two steps:

(1) Find frequent itemsets (frequent item-sets) which is a set of data items that its support is greater than a given value.
(2) Generate candidate association rules by using the frequent itemsets, and verify the credibility of the rule.

Step (1) is the key to improve the efficiency of the algorithm [10, 11]. First of all, there are definitions as follows:

Definition 1 the attribute domain r of relational database is bool variables, matrix M is the vector matrix of a database:

$$M_{I\times J} = (m_{ij})_{I\times J} = \left[\vec{m}_1, \vec{m}_2, \cdots, \vec{m}_J\right] \begin{cases} m_{ij} = 1 & \text{attribute J in records I is true} \\ m_{ij} = 0 & \text{attribute J in records I is false} \end{cases}$$

Which I is the number of records in database, J is the number of bool variables, each column vector $\vec{m}_j (j = 1, 2, \cdots, J)$ which corresponds to the attribute is the feature vector of attribute.

Definition 2 the inner product is described as k n-dimensional characteristic vectors in K elements function:

$$\left\langle \overrightarrow{R_1}, \overrightarrow{R_2}, \cdots, \overrightarrow{R_3} \right\rangle = \sum_{i=1}^{k} r_{1i} \times r_{2i} \times \cdots \times r_{ki}$$

Definition 3 If 1-frequent item sets L_1 of the database D is not empty, then the matrix composed by order of the feature vector of element in L_1 is defined as the characteristic matrix of database.

Theorem 1 support(X) of element X in k frequency set of database D is equal to k elements inner product of feature vector which k attributes form X, that is:

$$\text{support}(X) = \langle \overrightarrow{x_1}, \overrightarrow{x_2}, \cdots, \overrightarrow{x_k} \rangle$$

The proof of theorem1 If there is k frequent item sets in record U of X, then $U(x_1), \cdots, U(x_k)$ is real at the same time, so $U(x_1) \times U(x_2) \times \cdots \times U(x_k) = 1$; Otherwise it is zero. $n = \langle \overrightarrow{x_1}, \overrightarrow{x_2}, \cdots, \overrightarrow{x_k} \rangle$ is the number of k property set in D. Theorem 1 has been proved.

Theorem 2 property set A belongs to the k- frequent item sets L_k, if $X \subset A$, then support(X) will not less than the stated threshold of k project set.

The proof of theorem2 According to $X \subset A \subset I$ (I is the attribute collection in database D), D satisfies with the functional dependency $A \to X$, then record U,V are in D.If $U(A) = V(A)$, then $U(X) = V(X)$ [8]. So if A meets the conditions,X will also meet them, otherwise it won't bring into existence. So support$(X) \geq$ support(A), support$(X) \geq$ support$_{\text{min}}$ because of support$(A) \geq$ support$_{\text{min}}$.

3 Design of the Association Rules Algorithm Based on the Characteristic Matrix

According to the above definitions and theorems, a new algorithm for mining association rules has been built, the algorithm is given below:

Step 1. Write vector matrix based on the database:

$$M_{I \times J} = (m_{ij})_{I \times J} = \left[\overrightarrow{m_1}, \ \overrightarrow{m_2}, \ \cdots \ , \overrightarrow{m_j} \right]$$

Construct the candidate set C_1 which its elements are the field name, and calculate support set $S(C_1)$ of C_1:

$$S(C_1) = \left\{\text{support}(x_i)|\text{support}(x_i) = \left\langle \overrightarrow{m_j}, \vec{I} \right\rangle\right\}$$

$x_i \in C_1, \overrightarrow{m_j}$ is the feature vector which corresponding properties of x_i, $\vec{I}$ is the unit vector.

Delete properties of C_1 when $\text{support}(x_i) < \text{support}_{\min}$ and obtain L_1.

Step 2. According to the theorem 2 and L_1, construct candidate C_2, and calculate support set $S(C_2)$ of C_2:

$$S(C_2) = \left\{\text{support}(x_i)|\text{support}(x_i) = \left\langle \overrightarrow{m_j}, \overrightarrow{m_k} \right\rangle\right\}$$

$x_i \in C_2, \overrightarrow{m_j}, \overrightarrow{m_k}$ is the feature vector which corresponding properties of x_i.

Delete properties of C_2 when $\text{support}(x_i) < \text{support}_{\min}$ and obtain L_2.

Step 3. According to the theorem 2 and L_{k-1}, construct candidate C_k, and calculate support set $S(C_k)$ of C_k:

$$S(C_k) = \left\{\text{support}(x_i)|\text{support}(x_i) = \left\langle \overrightarrow{m_1}, \overrightarrow{m_2}, \cdots, \overrightarrow{m_k} \right\rangle\right\}$$

$x_i \in C_k, \overrightarrow{m_j}(j = 1, 2, \cdots k)$ is the feature vector which corresponding properties of x_i.

Delete properties of C_k when $\text{support}(x_i) < \text{support}_{\min}$ and obtain L_k.

Step 4. According to the theorem 2 and L_k, construct candidate L_{k+1}, if $C_{k+1} \neq \Phi$ then to step 3.

Step 5. Generate association rules.

4 Algorithm Evaluation

This algorithm is based on the direct data instead of a database management system (DBMS). The first is to traverse the database, make the database vector. Identify records in the database by computing the vector product of vectors. By Theorem 1, we can obtain support degree of candidate elements. The calculation of support degree steps is simplified, the speed of discovery the frequent item sets can be improved; it leads to improved of algorithm of mining association rules. Because data mining in large scale database (Very large database), database vectorization can only make the data more compact, at the same time, it requires large memory, so this algorithm to compute K frequent item set L, is based on the reduced dimension feature matrix, The algorithm occupied a smaller memory. If the database is too large, we can also use the database segmentation vectorization.

5 Example Analysis

The following is the illustration of the new algorithm with an example from literature [6], the threshold is set to 2, find all frequent item sets. Database D is shown in Table 1:

Step 1: the vector matrix in Table 1:

$$M_{4\times5} = \begin{pmatrix} 1 & 0 & 1 & 1 & 0 \\ 0 & 1 & 1 & 0 & 1 \\ 1 & 1 & 1 & 0 & 1 \\ 0 & 1 & 0 & 0 & 1 \end{pmatrix}$$

Let $C_1 = \{A, B, C, D, E\}$,

$$S(c_1) = (1 \quad 1 \quad 1 \quad 1) \times \begin{pmatrix} 1 & 0 & 1 & 1 & 0 \\ 0 & 1 & 1 & 0 & 1 \\ 1 & 1 & 1 & 0 & 1 \\ 0 & 1 & 0 & 0 & 1 \end{pmatrix} = [2 \quad 3 \quad 3 \quad 1 \quad 3]$$

The threshold is 2, then $L_1 = \{A, B, C, E\}$, then the feature matrix in Table 1 is $\begin{bmatrix} 1 & 0 & 1 & 0 \\ 0 & 1 & 1 & 1 \\ 1 & 1 & 1 & 1 \\ 0 & 1 & 0 & 1 \end{bmatrix}$.

Step 2: According to the theorem 2 and L_1, construct candidate $C_2 = \{AB, AC, BC, BE, CE\}$. In order to express the operations of seeking support in matrix form, we define operation $\otimes$: A, B are both N dimensional column vector

$$A^{\tau} \otimes B = a_1 \times b_1 \wedge a_2 \times b_2 \wedge \cdots \wedge a_n \times b_n$$

$$S(C_2) = (1 \quad 1 \quad 1 \quad 1) \times \left(\begin{pmatrix} 1 & 0 & 1 & 0 \\ 0 & 1 & 1 & 1 \\ 1 & 1 & 1 & 1 \\ 0 & 1 & 0 & 1 \end{pmatrix} \otimes \begin{pmatrix} 1 & 1 & 1 & 0 & 0 & 0 \\ 1 & 0 & 0 & 1 & 1 & 0 \\ 0 & 1 & 0 & 1 & 0 & 1 \\ 0 & 0 & 1 & 0 & 1 & 1 \end{pmatrix} \right) = [1 \quad 2 \quad 1 \quad 2 \quad 3 \quad 2]$$

The threshold is 2, then $L_2 = \{AC, BC, BE, CE\}$

Table 1 Database D

TID	Items
100	ACD
200	BCE
300	ABCE
400	BE

Setp 3: According to the theorem 2 and L_1, construct candidate $C_3 = \{BCE\}$.

$$S(C_3) = (1 \quad 1 \quad 1 \quad 1) \times \left(\begin{pmatrix} 1 & 0 & 1 & 0 \\ 0 & 1 & 1 & 1 \\ 1 & 1 & 1 & 1 \\ 0 & 1 & 0 & 1 \end{pmatrix} \otimes \begin{bmatrix} 0 \\ 1 \\ 1 \\ 1 \end{bmatrix} \right) = [2]$$

$$L_3 = \{BCE\}$$

Step 4: C_4 is null, then the algorithm ends.

6 Conclusion

This paper presents a new algorithm for mining association rules. In order to calculate the itemset support, the concept of feature matrix in database and the feature vector is put forward. As this algorithm only requires one traversal of the database, it greatly reduces the operation of the database. Based on the characteristic vector inner product to get an itemset support, the efficiency of the algorithm is improved. Algorithm analysis has proved the rapidity of the algorithm. Algorithm example has proved the efficiency of the algorithm.

References

1. Mannila H, Toivonen H, Verkamo A (1994) Efficient algorithm for discovering association rules. In: AAAI Workshop on knowledge discovery in databases, pp 181–192
2. Srikant R, Agrawal R (1995) Mining generalized association rulers. In: Proceedings of the 21th international conference on very large database, pp 407–419
3. Toivonen H, Klemettinen M, Ronkaine P et al (1995) Pruning and grouping discovery association rules. In: Mlnet workshop on statistics, machine learning and discovery in database, Herakloon, Crete, Greece
4. Agrawal R, Srikant R (1994) Fast algorithms for mining. Association rules in large databases. In: Proceedings 20th international conference very large databases, pp 478–499
5. Spark J (ed) (1995) An effective hash based algorithm for mining association rules. In: Proceedings ACM SIGMOD, pp 175–186
6. Toivonen H (1996) Sampling large databases for association rules. In: Proceedings of the 22th international conference on Very Large Databases (VLDB), Bombay, india. Morgan Kaufmann, pp 134–145
7. Agrawal R, Imielinski T, Swami A (1993) Mining association rules between sets of items in large databases. In: Proceedings of the ACM SIGMOD conference on management of data, pp 207–216
8. Wang NB (2000) Database management system. Publishing House of electronics industry, Beijing, pp 435–439
9. Jiawei H (2001) Micheline, translated by Meng xiao feng. Data mining-concepts and techniques. Machinery Industry Press, p 8
10. Zhong Zhi S (2002) Knowledge discovery. Tsinghua University press, p 1
11. Ye Xin T, Qi C, Rui Zhao Y (2000) Survey of association rule mining. Appl Res Comput, p 1

Offline Password Guessing Attacks on Smart-Card-Based Remote User Authentication Schemes

Xue-lei Li, Qiao-yan Wen, Hua Zhang, Zheng-ping Jin and Wen-min Li

Abstract Password as an easy-to-remember credential plays an important role in remote user authentication schemes, while drawing from a space so small that an adversary may exhaustively search all possible candidate passwords to guess the correct one. In order to enhance the security of the password authentication scheme, smart card is introduced as the second factor to construct two-factor authentication scheme. However, we find out that two latest smart-card-based password authentication schemes are vulnerable to offline password guessing attacks under the definition of secure two-factor authentication. Furthermore, in order to show the serious consequence of offline password guessing attacks, we illustrate that the password compromise impersonation attacks as further threats are effective to break down the authentication schemes. Finally, we conclude the reasons why these weaknesses exist and present our improved ideas to avoid these problems in the future.

Keywords Authentication · Password · Security · Smart card

1 Introduction

This paper focuses on the problems of offline password guessing attacks in smart-card-based password authentication schemes. We consider the scenario in which there are two entities, a client U_i and a server S, where U_i holds a password PW_i, a smart card SC, and S holds a secret key x. After confirming the identity of each other, the participants would like to engage a fresh session key sk, which is known to nobody but the participants. Meanwhile there is present an active adversary who has various targets, including impersonating any of the participants,

X. Li (✉) · Q. Wen · H. Zhang · Z. Jin · W. Li
State Key Laboratory of Networking and Switching Technology,
Beijing University of Posts and Telecommunications, Beijing 100876, China
e-mail: will8898@163.com

E. Qi (ed.), *Proceedings of the 6th International Asia Conference on Industrial Engineering and Management Innovation*,
DOI 10.2991/978-94-6239-145-1_9

obtaining the fresh session key, guessing the low entropy password and cryptanalyzing the high entropy long term secret key.

Password authentication schemes with insecure communication have been developed since Lamport's [1] primitive contribution in 1981. However, in 1999, Halevi and Krawczyk [2] pointed out that constructing secure password protocols using only symmetric key technologies (e.g., block ciphers and hash functions) was extremely unlikely to resist offline password guessing attacks. Therefore, they presented a protocol for strong authentication and key exchange in asymmetric scenario, where public key infrastructure was required as additional technology. On the other side, Hwang and Li [3] proposed a novel remote user authentication scheme using smart cards based on ElGamal's public key cryptosystem in 2000. Their scheme did not require the password verifier tables any more to prevent the malicious server from compromising the password verifier tables. Later on, Sun [4] further put forward a new efficient and practical scheme using smart cards without ElGamal's public key cryptosystem. After that, the researches [5–12] have been presented to enhance the security, efficiency and flexibility of smart-card-based password authentication scheme using symmetric key cryptosystem. Recently, Chen et al. [13] analyzed Sood et al.'s [11] and Song's [12] improvements, and found that their improvements were insecure to resist several attacks, including offline password guessing attacks. Chen et al. also introduced their improved scheme to remedy the security flaws. Quickly, Ma et al. [14] pointed out that Chen et al.'s [13] improvement was vulnerable to offline password guessing attack and failed to provide forward secrecy. In 2013, Li et al. [15] also analyzed Chen et al.'s proposal and further presented their modified scheme, which was shown to be more user friendly and secure than related schemes. Almost at the same time, Li [22] raised a new password authentication and user anonymity scheme based on smart card and elliptic curve cryptography. Li tried to solve the hard security threats, mentioned in [16–22], especially for offline password guessing attacks, stolen-verifier attacks and insider attacks.

In this paper, nevertheless, we illustrate that both Li et al.'s [15] and Li's [22] schemes are vulnerable to offline password guessing attacks under the non-tamper resistance assumption of smart cards. Once the password guessing attacks are effective to compromise user's correct password, it means that all the credentials of the user are compromised to the adversary. Consequently, as a trivial attack, the adversary can obtain the highest authority of the user. Against all expectations, the adversary can also masquerade as the remote server to cheat the user, contrary. We denote it as enhanced password compromise impersonation (E-PCI) attack and show it as further threats to break down the authentication schemes. Finally, we propose our improved ideas to avoid offline password guessing attacks according to the related contributions [2, 23, 24], while concluding the reasons why offline password guessing attacks are effective in these schemes.

2 Preliminaries

In this section, we present the definition of secure smart-card-based password authentication scheme and related assumptions for security.

Definition 1 A smart-card-based password authentication scheme consists of two basic phases after system initialization [23].

User Registration: U_i with initial ID_i and PW_i registers in S, S generates the unique smart card and issues it to U_i. After this phase, U_i and S share the same credentials used for authenticating each other. Note that the registration phase is executed over secure channels, and the adversary cannot get any private information (transmitted messages) in this phase.

Login and Authentication: U_i and S interact with the partner over public channels to authenticate each other by proofing that they have the valid credentials. After this phase, the participants confirm the identity of their partners. Note that the login and authentication phase is executed over insecure channels, all the adversaries can get the messages transmitted between U_i and S.

Definition 2 The adversary *Adv* is classified into the following two types [10].

Adv with Smart Card: Adv_1 denotes such an adversary, who not only has the full control over public channels, but also obtains U_i's smart card and extracts the information stored in the memory of the smart card [25, 26].

Adv with Password: Adv_2 denotes such an adversary, who not only has the full control over public channels, but also obtains U_i's password.

Definition 3 A smart-card-based password authentication scheme is secure to resist various attacks if and only if all the following requirements are satisfied.

Authentication: *Adv* cannot masquerade as U_i (or S) to cheat S (or U_i). In other words, *Adv* cannot impersonate the legal entities to participate in the scheme without being detected.

Secrecy: All the credentials in the scheme cannot be deduced by *Adv*. In details, Adv_1 cannot deduce user's password PW_i and Adv_2 cannot deduce the information stored in user's smart card, while *Adv* cannot deduce any private keys of S.

Assumption 1 All the underlying components (hash function, random number generator and symmetric/asymmetric encryption/decryption algorithms) involved in the smart-card-based password authentication schemes are invulnerable with regard to message secrecy and integrity, i.e., the security of cryptographic primitives is out of consideration in this paper.

Assumption 2 All the passwords are selected in the password dictionary D with the space size $|D|$.

Definition 4 Offline password guessing attacks are effective if the adversary can exhaustively search the password dictionary and obtain the correct password with enough redundant information by a probabilistic polynomial-time algorithm in an offline manner.

3 Review and Analysis

3.1 *Li et al.'s Scheme in [15]*

3.1.1 Review

There are four phases in Li et al.'s scheme [15], we briefly review only three of them for relativity and simplicity: registration, login and authentication phase.

Phase 1. Registration

(1) U_i chooses ID_i with PW_i, and registers in S.
(2) S computes $A_i = h(ID_i||PW_i)^{PW_i}, B_i = h(ID_i)^{x+PW_i}$, where $h(\cdot)$ is one-way hash function and x is the secret key of S.
(3) S initializes the smart card with $\{A_i, B_i, h(\cdot), p, q\}$, where p, q are large prime numbers satisfied $p = 2q + 1$. After that, S sends the smart card to U_i.

Phase 2. Login

(1) U_i inserts the smart card into card reader and inputs ID_i with PW_i.
(2) SC computes $A_i^* = h(ID_i||PW_i)^{PW_i}$ and compares it with A_i to detect whether the inputs are correct. If $A_i^* = A_i$, SC confirms the card holder. Otherwise, terminates the current procedures.
(3) SC generates $\alpha \in_R Z_q^*$ and computes $C_i = B_i/h(ID_i)^{PW_i}, D_i = h(ID_i)^{\alpha}, M_i = h(ID_i||C_i||D_i||T_i)$, where T_i is the timestamp.
(4) SC sends the login request $\{ID_i, D_i, M_i, T_i\}$ to S.

Phase 3. Authentication

(1) S confirms the validity of ID_i and T_i in the login request.
(2) S computes $C_i^{'} = h(ID_i)^x, M_i^{'} = h(ID_i||C_i^{'}||D_i||T_i)$.
(3) S compares $M_i^{'}$ with M_i to authenticate the remote user. If $M_i^{'} = M_i$, S accepts the login request and continues the next step. Otherwise, rejects.
(4) S generates $\beta \in_R Z_q^*$ and computes $V_i = h(ID_i)^{\beta}, sk = D_i^{\beta}$.
5) S computes $M_S = h(ID_i||C_i^{'}||V_i||sk||T_S)$ and sends the mutual authentication message $\{ID_i, V_i, M_S, T_S\}$ to U_i.
(6) U_i checks the validity of ID_i and T_S.
(7) U_i computes $sk' = V_i^{\alpha}, M_S^{'} = h(ID_i||C_i||V_i||sk'||T_S)$ and compares $M_S^{'}$ with M_S. If $M_S^{'} = M_S$, mutual authentication is finished successfully and the participants share $sk = sk'$ as the fresh session key.

3.1.2 Analysis

Theorem 1 Adv_1 can launch offline password guessing attacks on Li et al.'s scheme [15].

Proof According to Definition 2, Adv_1 can record U_i's login request messages $\{ID_i, D_i, M_i, T_i\}$ and obtain the information $\{A_i, B_i, h(\cdot), p, q\}$ stored in U_i's smart card. We demonstrate that Adv_1 can launch two types of offline password guessing attacks. □

Type I. Adv_1 with $\{ID_i \in OnLine, A_i, h(\cdot), p \in SC\}$ can perform the following operations to guess U_i's password offline.

(1) Selects the candidate password $PW_i^* \in D$.
(2) Computes $A_i^* = h(ID_i||PW_i^*)^{PW_i^*}$.
(3) Compares A_i^* with A_i.
(4) If $A_i^* = A_i$, password guesses correctly. Otherwise, Adv_1 repeats the operations (1–3) with next candidate password.

Type II. Adv_1 with $\{ID_i, D_i, M_i, T_i \in OnLine, B_i, h(\cdot), p \in SC\}$ can perform the following operations to guess U_i's password offline.

(1) Selects the candidate password $PW_i^* \in D$.
(2) Computes $C_i^* = B_i/h(ID_i)^{PW_i^*}, M_i^* = h(ID_i||C_i^*||D_i||T_i)$.
(3) Compares M_i^* with M_i.
(4) If $M_i^* = M_i$, password guesses correct. Otherwise, Adv_1 repeats the operations (1-3) with next candidate password.

Therefore, Adv_1 can launch offline password guessing attacks to break down the security in Definition 3 of Li et al.'s scheme [15] in terms of Definition 4.

Theorem 2 Adv_1 can launch E-PCI attack if Theorem 1 holds.

Proof If Theorem 1 holds, Adv_1 intercepts $\{ID_i, D_i, M_i, T_i \in OnLine\}$, obtains $\{A_i, B_i, h(\cdot), p, q \in SC\}$ and guesses the correct password PW_i. Adv_1 can deduce $C_i^* = h(ID_i)^x = B_i/h(ID_i)^{PW_i}$, thus all the credentials of U_i are compromised to Adv_1. Therefore, Adv_1 can forge the mutual authentication message $\{ID_i, V_i^*, M_S^*, T_S^*\}$ to masquerade as S, where $V_i^* = h(ID_i)^{\beta^*}, \beta^* \in_R Z_q^*$, $M_S^* = h(ID_i||C_i^*||V_i^*||sk^*||T_S^*), sk^* = D_i^{\beta^*}$. Finally, Adv_1 can masquerade as S to cheat U_i without being detected under the assumption of Theorem 1. □

3.2 B. Li's Scheme in [22]

We briefly review two phases in Li's et al.'s scheme [22] for simplicity.

3.2.1 Review

Phase 1. Registration:

(1) U_i chooses ID_i with PW_i, and registers in S with his/her ID_i and password-verifier $U_C = PW_i \cdot r_i \cdot G$, where G is the generator of the elliptic curve group and r_i is a random number which is entered to U_i's smart card in step (3).
(2) S stores ID_i, U_C and the *status-bit* in its database, and issues a smart card containing $\{G, U_S, h(\cdot), E_K(.)/D_K(.)\}$ to U_i, where $U_S = d_S \cdot G$ is the public key and d_S is the private key of S, $E_K(.)/D_K(.)$ is the encryption/decryption algorithm.
(3) U_i enters r_i to his/her smart card to initialize it. After that the smart card contains $\{G, U_S, h(\cdot), E_K(.)/D_K(.), r_i\}$.

Phase 2. Authentication:

(1) U_i inserts the smart card into card reader and inputs ID_i with PW_i.
(2) SC retrieves r_i, generates $r_i^{'}$ and computes $R_i = r_i \cdot U_S, W_i = r_i \cdot r_i \cdot PW_i \cdot G$, $U_C^{'} = PW_i \cdot r_i^{'} \cdot G$, $E_{K_x}(ID_i, R_i, W_i, U_C^{'})$, where $K = PW_i \cdot r_i \cdot U_S = (K_x, K_y)$. After that, SC sends the login request $\{ID_i, E_{K_x}(ID_i, R_i, W_i, U_C^{'})\}$ to S.
(3) S computes $K = d_S \cdot U_C = PW_i \cdot r_i \cdot d_S \cdot G = (K_x, K_y)$ and decrypts $E_{K_x}(ID_i, R_i, W_i, U_C^{'})$ to reveal $ID_i, R_i, W_i, U_C^{'}$. S authenticates U_i if the equation $\hat{e}(R_i, U_C) = \hat{e}(W_i, U_S)$ holds, where *hate* is the bilinear paring computation.
(4) S sends $W_i + W_S, h(W_S, U_C^{'})$ to U_i, where $W_S = r_S \cdot U_S = r_S \cdot d_S \cdot G$ and r_S is a random number generated by S.
(5) U_i retrieves W_S by subtracting W_i from $W_i + W_S$ and checks whether the computed value $h(W_S, U_C^{'})$ is equal to received value $h(W_S, U_C^{'})$. If so, U_i computes $h(W_i, W_S, U_C^{'})$ and sends it to S.
(6) S checks the validity of $h(W_i, W_S, U_C^{'})$. If it is valid, S grants U_i's login request and replaces U_C with $U_C^{'}$. Otherwise, S denies U_i's login request.

Finally, U_i's smart card replaces r_i with $r_i^{'}$ if all the conditions are satisfied.

3.2.2 Analysis

Theorem 3 Adv_1 can launch offline password guessing attack on Li's scheme [22].

Proof: According to Definition 2, Adv_1 can record U_i's login request messages $\{ID_i, E_{K_x}(ID_i, R_i, W_i, U_C)\}$ and obtain the information $\{G, U_S, h(\cdot), E_K(.)/D_K(.), r_i\}$ stored in U_i's smart card. We demonstrate that Adv_1 can launch offline password guessing attack by performing the following operations.

(1) Selects the candidate password $PW_i^* \in D$.
(2) Computes $K^* = PW_i^* \cdot r_i \cdot U_S = (K_x^*, K_y^*)$ and decrypts $E_{K_x}(ID_i, R_i, W_i, U_C)$ with K_x^* to retrieve ID_i^*
(3) Compares ID_i^* with ID_i.
(4) If $ID_i^* = ID_i$, password guesses correct. Otherwise, Adv_1 repeats the operations (1–3) with next candidate password.

Therefore, Adv_1 can launch offline password guessing attack to break down the security in Definition 3 of Li's scheme [22] in terms of Definition 4. □

Theorem 4 Adv_1 can launch E-PCI attack if Theorem 3 holds in Li's scheme [22].

Proof If Theorem 3 holds, Adv_1 intercepts $\{ID_i, E_{K_x}(ID_i, R_i, W_i, U_C) \in OnLine\}$, obtains $\{G, U_S, h(\cdot), E_K(.)/D_K(.), r_i \in SC\}$ and guesses the correct password PW_i. Adv_1 can deduce $K = PW_i \cdot r_i \cdot U_S = (K_x, K_y)$, thus all the credentials of U_i in this session are compromised to Adv_1. Therefore, Adv_1 can masquerade as S to cheat U_i without being detected under the assumption of Theorem 3, because K is the key value shared by S and U_i as the credential for authenticating each other. □

4 Discussion

The reason why offline password guessing attacks are always effective is that the secret information related to (protected by) password is traceable by the adversaries, who can find enough redundant information to confirm whether their guessed values are correct. Offline password guessing attacks are serious security flaws in remote authentication schemes. In details, once the offline password guessing attacks are successful to compromise users' secret passwords, the adversary not only can obtain the highest authority of the user to intrude on the remote server, but also can launch enhanced password compromise impersonation attack as further threats by masquerading as the remote server to cheat the user.

Although, the additional technologies, e.g., public key infrastructures (long term secret keys) [2], biometrics verification (the third authentication factor) [23] and nonces (ephemeral private random numbers) [24], have been introduced to prevent offline password guessing attacks, there are still several problems, which block the development of authentication technologies, e.g., high performance cost, lacking privacy preserving, low user friendly and flexibility.

5 Conclusion

In this paper, we demonstrate two smart-card-based password authentication schemes are vulnerable to offline password guessing attacks under the definition of secure two-factor authentication. In addition, the enhanced password compromise impersonation attacks are provided to show the consequences of offline password guessing attacks are very serious. The discussions about the reasons and improved ideas recommend us that avoiding the redundant information appearing in the scheme is a key issue to resist offline password guessing attacks. Therefore, we will address ourselves to find a better approach to solve these problems in the future.

Acknowledgments This work is supported by NSFC (Grant Nos. 61300181, 61272057, 61202434, 61170270, 61100203, 61121061), the Fundamental Research Funds for the Central Universities (Grant No. 2012RC0612, 2011YB01).

References

1. Lamport L (1981) Password authentication with insecure communication. Commun ACM 24(11):770–772
2. Halevi S, Krawczyk H (1999) Public-key cryptography and password protocols. ACM Trans Inf Syst Secur (TISSEC) 2(3):230–268
3. Hwang MS, Li LH (2000) A new remote user authentication scheme using smart cards. IEEE Trans Consum Electron 46(1):28–30
4. Sun HM (2000) An efficient remote use authentication scheme using smart cards. IEEE Trans Consum Electron 46(4):958–961
5. Chien HY, Jan JK, Tseng YM (2000) An efficient and practical solution to remote authentication: smart card. Comput Secur 21(4):372–375
6. Wu ST, Chieu BC (2003) A user friendly remote authentication scheme with smart cards. Comput Secur 22(6):547–550
7. Hsu CL (2004) Security of Chien et al.'s remote user authentication scheme using smart cards. Comput Stan Interfaces 26(3):167–169
8. Lee NY, Chiu YC (2005) Improved remote authentication scheme with smart card. Comput Stan Interfaces 27(2):177–180
9. Lee SW, Kim HS, Yoo KY (2005) Improvement of Chien et al.'s remote user authentication scheme using smart cards. Comput Stan Interfaces 27(2):181–183
10. Xu J, Zhu WT, Feng DG (2009) An improved smart card based password authentication scheme with provable security. Comput Stan Interfaces 31(4):723–728
11. Sood SK, Sarje AK, Singh K (2010) An improvement of Xu et al.'s authentication scheme using smart cards. In: Proceedings of the third annual ACM Bangalore conference
12. Song R (2010) Advanced smart card based password authentication protocol. Comput Stan Interfaces 32(5):321–325
13. Chen BL, Kuo WC, Wuu LC (2012) Robust smart-card-based remote user password authentication scheme. Int J Commun Syst. doi:10.1002/dac.2368
14. Ma CG, Wang D, Zhao SD (2012) Security flaws in two improved remote user authentication schemes using smart cards. Int J Commun Syst. doi:10.1002/dac.2468
15. Li X, Niu J, Khan MK, Liao J (2013) An enhanced smart card based remote user password authentication scheme. J Netw Comput Appl. doi:10.1016/j.jnca.2013.02.034

16. Peyravian M, Zunic N (2000) Methods for protecting password transmission. Comput Secur 19(5):466–469
17. Hwang JJ, Yeh TC (2002) Improvement on Peyravian-Zunic's password authentication schemes. IEICE Trans Commun 85(4):823–825
18. Lin CL, Hwang T (2003) A password authentication scheme with secure password updating. Comput Secur 22(1):68–72
19. Islam SKH, Biswas GP (2011) Design of improved password authentication and update scheme based on elliptic curve cryptography. Math Comput Model 57(11–12):2703–2717. doi:10.1016/j.mcm.2011.07.001
20. He D (2011) Comments on a password authentication and update scheme based on elliptic curve cryptography. Cryptology ePrint Archive. https://eprint.iacr.org/2011/411.pdf
21. Wang D, Ma CG, Shi L, Wang YH (2012) On the security of an improved password authentication scheme based on ECC. Inf Comput Appl 7473:181–188
22. Li CT (2013) A new password authentication and user anonymity scheme based on elliptic curve cryptography and smart card. IET Inf Secur 7(1):3–10
23. Huang X, Xiang Y, Chonka A, Zhou J, Deng RH (2011) A generic framework for three-factor authentication: preserving security and privacy in distributed systems. IEEE Trans Parallel Distrib Syst 22(8):1390–1397
24. Li X, Wen Q, Zhang H, Jin Z (2013) An improved authentication with key agreement scheme on elliptic curve cryptosystem for global mobility networks. Int J Network Manage 23(5):311–324
25. Kocher P, Jaffe J, Jun B (1999) Differential power analysis, advances in cryptology-CRYPTO'99. LNCS 1666:388–397
26. Messerges TS, Dabbish EA, Sloan RH (2002) Examining smart-card security under the threat of power analysis attacks. IEEE Trans Comput 51(5):541–552

Laboratory Measurement of Suspended Sediment Concentration by Using ADV

Sheng-qi Lv, Hong-wu Tang, Yuan Xue, Sai-yu Yuan and Xue-bing Bai

Abstract When an Acoustic Doppler Velocimetry (ADV) is conducted to measure the flow velocities, it can record some information of particle concentration in signal amplitude. This phenomenon suggests that a new method to measure the suspended sediment concentration (SSC) by using ADV might come true. In this paper, the working principle of the ADV was analyzed, and then a series of experiments were carried out to clarify the relationship between the SSC and the signal amplitude. Experimental results show that when SSC is lower than 0.014 (ratio by volume), ADV can work effectively; meanwhile, SSC is a linear function of the signal amplitude. Based on experimental data, several linear formulas between SSC and the signal amplitude were established. In the formulas, the constant coefficients differ from each other for the various flow conditions, which may cause calculation errors. Accordingly, in order to improve the calculation precision, it is necessary to conduct additional experimental calibration. In this study, ADV is proposed to determine the SSC. In some experimental conditions of continuous and long-lasting SSC measuring, the method has great advantages and will be of great importance to study on the interaction mechanism between turbulence and sediment particles.

Keywords ADV · Plastic sand · Signal amplitude · Suspended sediment concentration (SSC)

S. Lv (✉) · H. Tang · Y. Xue · S. Yuan
State Key Laboratory of Hydrology-Water Resources and Hydraulic Engineering, Hohai University, Nanjing, China
e-mail: cnjlvsq@hhu.edu.cn

X. Bai
College of Harbor, Coastal, and Offshore Engineering, Hohai University, Nanjing, China

E. Qi (ed.), *Proceedings of the 6th International Asia Conference on Industrial Engineering and Management Innovation*,
DOI 10.2991/978-94-6239-145-1_10

1 Introduction

In the studies on river dynamics, problems about sediment-laden flow are especially important. Related studies include the suspension and transportation of sediment, the vertical distribution rule of the suspended sediment concentration (SSC), the resistance and the turbulent structure of the sediment-laden flow, and the impacts of particles upon turbulent structure, etc. Among all these studies, the accurate measurements of the flow velocity and the SSC are needed. But this problem has not been thoroughly solved nowadays due to the limitation of the measuring instruments. Wren et al. [1] investigated the techniques of measuring the SSC, in terms of different measurement techniques with their operating principles. Also, the advantages and disadvantages of each technique were discussed. The techniques can be divided into two types: the direct measurement and the indirect measurement. The direct measurement method includes the bottle sampling and the pump sampling, which were used by many researchers and the measurement accuracies can be ensured. The indirect method includes acoustic, focused beam reflectance, laser diffraction, nuclear, optical backscatter, optical transmission, and spectral reflectance. These methods have high spatial and temporal resolution; furthermore, the probes will not disturb the water column at the measured location. However, the calibration against the bottle samplers is needed, and sometimes, the measurement accuracy may not be satisfied. In the measurement of velocity and SSC in sediment-laden flow, the acoustic method has been widely applied. It has the potential to measure non-intrusively, with high temporal and spatial resolution. Information such as profiles of suspended sediment concentration, profiles of flow, and the bedform morphology [2–4] can also be get from the measurement. Acoustic Doppler current profiler (ADCP) can be used to perform SSC combined with high-resolution velocity and SSC measurements in the highly turbid near-bed zone [5–8]. Acoustic Doppler Velocimetry (ADV) is a sophisticated acoustical instrument designed for flow velocity measurement, which has been widely used in field and laboratory experiments [9–13]. When ADV is used to conduct velocity measurement in flows, information of particle concentration could be reflected from acoustic backscattered signal strength, which could be used to determine SSC. However, due to the complexity of impact factors related to acoustic backscatters, previous results are not desirable. In this paper, laboratory experiments have been conducted to study the influence of SSC on ADV and explore the possibility of SSC measurement with ADV.

2 Measurement Principle and Calculation Method

2.1 Measurement Principle of ADV

ADV determines the flow velocity according to the principle of the Doppler shift. The transmitting probe sends out a beam of acoustic waves at a fixed frequency (Fig. 1). These waves bounce off moving particles (suspended solids, zooplankton,

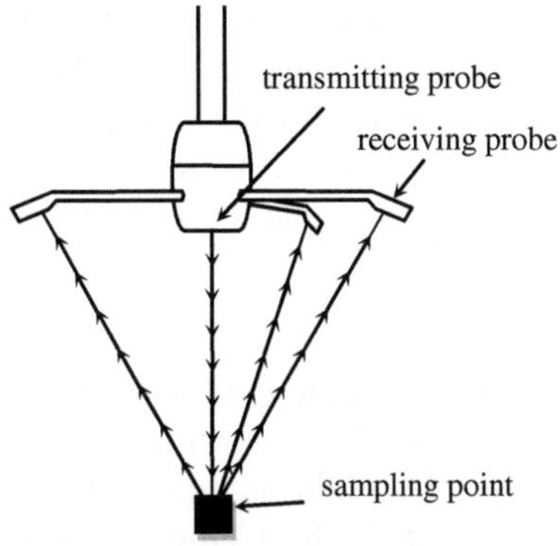

Fig. 1 Sketch of the probes and the acoustic waves

bubbles, etc.) in the flow and three receiving probes capture the frequency change of the returning waves (Fig. 1). ADV then calculates velocities of the flow in the x, y and z directions. Since the distance from the measured point to the transmitting probe is either 5 cm or 10 cm approximately, ADV is regarded as a type of non-contact measuring, high-accuracy equipment, which doesn't need calibration. The newly-developed Nortek Vectrino Velocimetry represents as a leap forward in performance compared with the previous model of ADV, allowing the data collection rates up to 200 Hz. It also contains a separate echo sounder mode, by which the distance to the bottom can be measured with the sub-millimeter accuracy at 10 Hz. The latest version of ADV is an ideal instrument adopted in both the velocity measurement and the distance detection of the sand bed, clay bed or a pulverized coal bed. ADV makes use of acoustic waves reflected from suspended particles in flows to gather signals. In this way, it is required that there are enough particles in flows, however, if the amount of particles is out of range, the measuring results will be negatively influenced, which will also make the ADV off the truth. For example, when the particle concentration goes up beyond a normal scope, the acoustic backscatters will be reflected before the acoustic wave reaches the bottom, which makes the result of the detected distance lower than the truth.

2.2 Correlation Between Signal Amplitude and Particle Concentration

In flows with various particle concentrations, different signal strengths of acoustic backscatters will be induced correspondingly. ADV's ability of possessing the particle concentration in signal amplitude makes it possible to estimate SSC value. Some researchers have measured the flow velocities and the fluctuation of sediment concentration with ADV [14–16]. Lin et al. [15] analyzed the signal of the sediment-laden flow received by ADV and revealed the correlation between the sediment concentration and the flow velocities in the microcosmic structure. Based on the measurements of SSC and theoretical analysis, Cheng et al. [8] found the relationship between the SSC and back-scattering strength and also, the back-scattered ADCP signal could be used to extract the information of SSC. In light of the former studies, the following empirical equation between the strength of acoustic wave and the sediment concentration has been assumed:

$$E = K_1 + K_2 \lg S \tag{1}$$

where E is strength of acoustic backscatters; K_1 is a constant related to the gravitational acceleration, sediment density and particle size; K_2 is a constant and S is the suspended sediment concentration. In Eq. (1), it can be seen that E is obviously related to S, which shows an absolute logarithmic relationship. Actually, the relationship between E and S should not be simplified as Eq. (1). When S is very low, E will also be very weak for particles at the measuring point. Therefore, when S increases gradually, E will be strengthened because more particles at the measuring point will reflect more signals. In such case, E and S have a positive relationship. However, when S increases to a certain extent, large quantities of particles outside the measuring sample volume will gradually disturb the signals, resulting in obvious reduction of E, which means that E and S have a non-positive relationship. Accordingly, the defined relationship between E and S needs to be confirmed by further experimental studies.

3 Experimental Arrangement and Measurement Method

SSC is an important impact factor for the operation of ADV, in other words, an out-of-range sediment concentration will cause the failure of ADV. To verify this fact, experiments were designed and conducted in a glass tank, which were 158 mm in diameter and 245 mm in height. In order to perform a stable turbulent flow, a stirrer was placed at the bottom of the tank. Under the action of the stirrer, the turbulent flow will be induced and the sediment particles will be suspended. In the experiments, the probe of the ADV will be fixed at a certain position under the water. When the ADV works normally, the distance from the probe to the bottom will be detected and should be maintained as a "stable" value. By increasing the SSC gradually, five series of records of the detected distance were obtained with different particle sizes (Fig. 2). In Fig. 2, h is the distance from the probe to the bottom, S_v is the sediment concentration (ratio by volume) and d is the diameter of the particles. The experimental results show that ADV can work normally when the SSC is lower than 0.014; otherwise, the detected value of the distance becomes lower significantly. This phenomenon indicates that a high SSC will affect the ADV's performance. Therefore, all SSC in the following experiments should be no more than 0.014.

Experiments were conducted in a rectangular flume, which is 0.42 m wide and 12 m long. Slope of the flume could be adjusted to achieve a uniform flow and

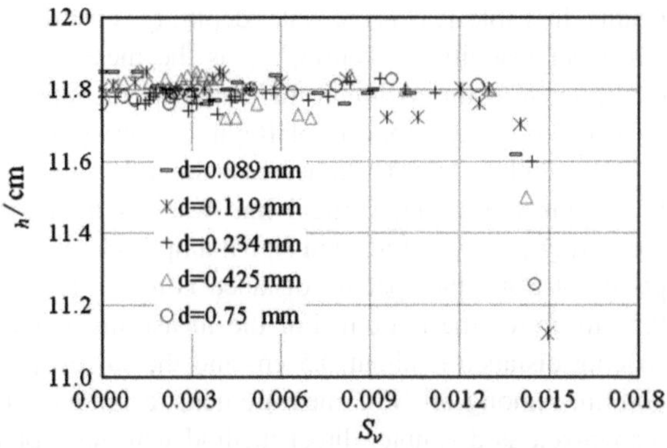

Fig. 2 Influence of SSC on the operation of ADV

different magnitudes of SSC could be achieved with the help of a sand feeder. ADV was placed at the middle part of the flume, with which the signal amplitude could be measured, and then the relationship between SSC and the signal amplitude could be analyzed. In order to enlarge the range of SSC, a kind of uniform plastic sands with the diameter of 0.217 mm and the density of 1.082 g/cm^3 were used. Each time after changing the SSC, the information of the amplitude from the ADV could be automatically recorded and the SSC could be obtained via the sampling method. In this way, a series of data related to the SSC and the signal amplitude could be collected.

4 Analysis of Experimental Results

4.1 Hydraulic Conditions of the Experiments

In this paper, five groups of experiments of the steady and uniform sediment-laden flow with different water depths were conducted. The hydraulic conditions are shown in Table 1. The first three groups of experiments (A1–A3) are used to calibrate the quantitative relationship between SSC and the signal amplitude, and the following two groups of experiments (A4 and A5) are used to test the previous

Table 1 Flow conditions of the experiments

Code	H	Q	J	U_m	S_{vL}	u_*	R_e	F_r
	cm	L/s	‰	cm/s	%	cm/s	$\times 10^4$	
A1	12	16.17	0.20	32.08	0.130	1.23	2.9	0.30
A2	12	26.17	0.50	51.92	0.433	1.85	4.7	0.48
A3	18	30.17	0.25	39.90	0.431	1.54	4.6	0.30
A4	15	26.83	0.33	42.59	0.374	1.65	4.4	0.35
A5	15	33.65	0.46	53.41	4.97	1.97	5.4	0.43

results. In Table 1, H is the water depth; Q is the discharge; J is the bed slope; U_m is the mean velocity in section; S_{vL} is the mean sediment concentration in vertical direction; R is the hydraulic radius; u_* is the shear velocity, which could be calculated via the distribution of Reynolds stress; Re is Reynolds number and Fr is Froude number. In order to decrease the influence on the flow by the sidewall of the flume, the measuring vertical line was set in the middle of the section and all measuring points of SSC and signal amplitude were on this line. For the measuring points of SSC, the spacing distance is about 1 cm and the lowest point is about 0.5 cm above the bottom. For the measuring points of the signal amplitude, the spacing distance is about 0.5 cm and the lowest point is about 0.1 cm above the bottom. Among all SSC measurement techniques today, the sampling method is recognized as a simple, direct method with high precision. This paper figures out SSC at different measuring points based on the sampling method and acquires the signal amplitude of ADV correspondingly at the same time.

4.2 Analysis of Experimental Results

In open channel flows, SSC changes with water depths and generally presents as a distribution with "small top" and "big bottom" layout in terms of the magnitude of SSC. This can also be tested by the experimental results. In Fig. 3, y/H is the relative water depth of measuring points; S_v is the sediment concentration (ratio by volume); AMP (of which the unit is count and a count represents 0.43 dB) is the signal amplitude of ADV.

It can be seen in Fig. 3 that under the given three flow conditions, the signal amplitude shows a distribution with "big top" and "small bottom", which suggests an opposite tendency with the SSC. The correlation relation between the SSC and the signal amptitude are analyzed in (1), (2) and (3) of Fig. 4, from which it can be found that in a specific flow condition, there is a fine linear correlation relation between the SSC and the signal amplitude. Three empirical equations were established as Eq. (2), Eq. (3) and Eq. (4). By comparing the three fomulas, it can be seen that the structures of the three fomulas are similar, the constant coefficients of Eq. (3) and Eq. (4) show little difference but a large deviation with that of Eq. (2). In the three flow conditions, the correlation coefficients between SSC and amplitude are 0.849, 0.987 and 0.997, respectively. The least coefficient lies in flow condition A1, because the SSC gradient in this flow condition is much lower, which may increase the measuring error. In order to further investigate the analysis on the correlation relation between the SSC and the signal amplitude, the overall experimental data were ploted together in (4) of Fig. 4. Using the trend analysis method, it is easy to reach the conclusion that there is also a fine linear correlation relation between the SSC value and the signal amplitude. The quantitive correlation formula was established as Eq. (5) which has the similar structure with Eq. (2)–Eq. (4).

Fig. 3 Vertical distributions of the SSC and the signal amplitude

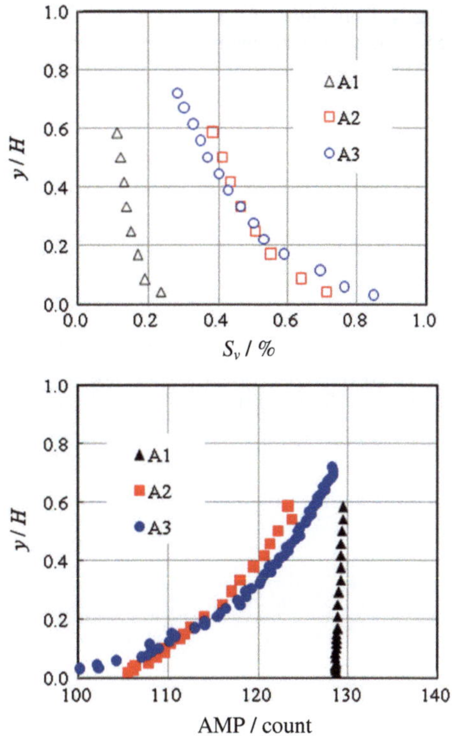

$$S_{v=} - 0.116 \times \mathrm{AMP} + 15.13 \tag{2}$$

$$S_v = -0.017 \times \mathrm{AMP} + 2.585 \tag{3}$$

$$S_v = -0.019 \times \mathrm{AMP} + 2.810 \tag{4}$$

$$S_v = -0.022 \times \mathrm{AMP} + 3.168 \tag{5}$$

4.3 Test and Discussion

In order to test the application of Eq. (5), two experiments (flow condition A4 and A5 in TABLE I) were conducted. The testing methods are as follows. Firstly, the SSC and the signal amplitudes of the two flow conditions will be measured by sampling method and by ADV. Secondly, the measured signal amplitude will be put into the Eq. (5) and the SSC value will be calculated. Finally, the SSC of the measured and the calculated value will be compared. Figure 4 shows the comparison between the measured SSC value and the calculated value. It can be seen that the calculated value and the measured results share the same trends with each

Fig. 4 Correlation of the SSC and the signal amplitude

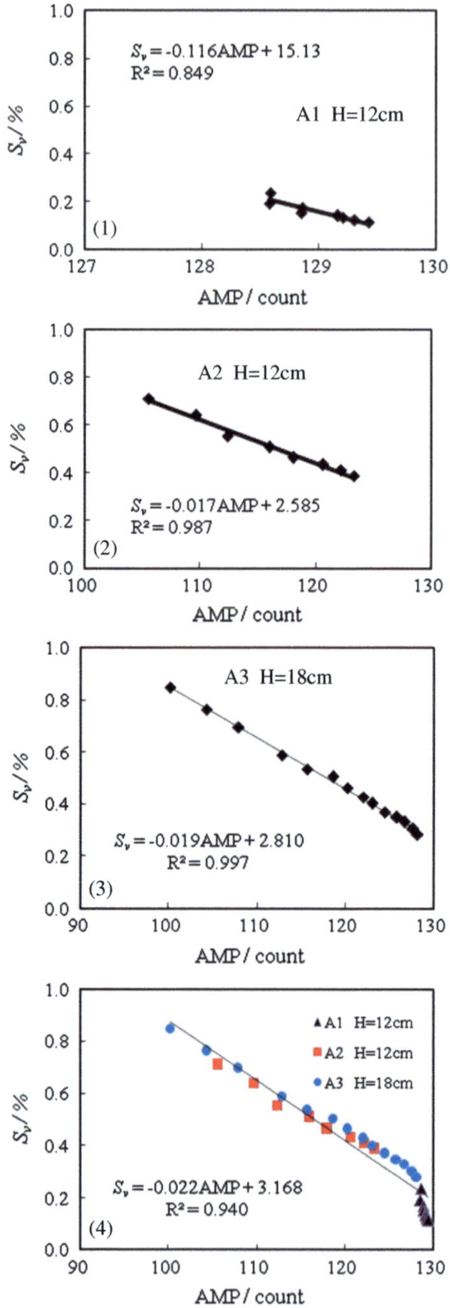

other. The relative error in the flow condition A4 is −18.2–27.2%, and that in the flow condition A5 is -8.9–18.2%. Experiments and analysis indicate that many factors may cause the error, such as the grain gradation and the composition of the sediment, the environmental noise and the turbulent characteristics of the flow. Therefore, SSC measurement using acoustic method, the calibration test is necessary and must be further considered.

5 Conclusions

This paper analyzed the working principle of ADV. It can record some information of particle concentration in signal amplitude. The finding of this phenomenon can be applied to measure the SSC. In order to achieve the goal, a series of experiments were carried out to clarify the influence of the SSC on the performance of ADV. The quantitative relationships between SSC and the signal amplitude were established.

Experimental results show that ADV can work efficiently when the SSC is lower than 0.014. Otherwise, ADV will work abnormally, and the distance detected from the transducer to the bed will decrease obviously. On the flow conditions of this study, the calibration between SSC and the signal amplitude indicates that the signal amplitude is a linear function of SSC, which means that it is possible and feasible to measure SSC using ADV, especially for those experiments of continuous and long-lasting SSC measuring. Under various flow conditions, the constant coefficients in the linear function may differ from each other, which will cause errors in estimating SSC. Therefore, in order to improve the calculation precision, it is necessary to conduct additional experimental calibration.

Using ADV, the widely used velocity measurement instrument, to determine the SSC in flows still needs further studies. If this method can be applied appropriately, the information of both the turbulent fluctuation and SSC fluctuation in flows can be acquired at the same time. This will be of great importance to study on the interaction mechanism between turbulence and the sediment particles.

Acknowledgments This work was financially supported by the National Natural Science Foundation of China Youth Science Fund Project (51109065, 51309083), the Fundamental Research Funds for the Central Universities (2009B08614), China National Funds for Distinguished Young Scientists (51125034).

References

1. Wren DG, Barkdoll BD, Kuhnle RA, Derrow RW (2000) Field Techniques for Suspended-Sediment Measurement. J Hydraul Eng 126(2):97–104
2. Thorne PD, Hanes DM (2002) A review of acoustic measurement of small-scale sediment processes. Cont Shelf Res 22(4):603–632

3. Strom KB, Papanicolaou AN (2007) ADV measurements around a cluster microform in a shallow mountain stream. J Hydraul Eng 133(12):1379–1389
4. Islam Md R, Zhu DZ (2013) Kernel density–based algorithm for despiking ADV data. J Hydraul Eng 139(7):785–793
5. Sottolichio A, Hurther D, Gratiot N, Bretel P (2011) Acoustic turbulence measurements of near-bed suspended sediment dynamics in highly turbid waters of a macrotidal estuary. Cont Shelf Res 31(10):36–49
6. Kuang L, Yu H, Ding Y, Li X, Kuang H (2011) A review on quantitative analysis of suspended sediment concentrations. Mar Sci Bull 13(2);51–61
7. Lemmin U, Rolland T (1997) Acoustic Velocity Profiler for Laboratory and Field Studies. J Hydraul Eng 23(12):1089–1098
8. Cheng P, Gao S (2001) Suspended sediment concentration measurements with ADCP: feasibility analysis and in Situ calibration (in Chinese). Oceanologia Et Limnologia Sinica 32(2):168–176
9. Parsheh M, Sotiropoulos F, Porté-Agel F (2010) Estimation of power spectra of acoustic-doppler velocimetry data contaminated with intermittent spikes. J Hydraul Eng 136(6):368–378
10. Nikora VI, Goring DG (2002) Fluctuations of suspended sediment concentration and turbulent sediment fluxes in an open-channel flow. J Hydraul Eng 128(2):214–224
11. Yang XIAO, Hong-wu TANG, Ye MAO, Zhi-liang WANG (2002) New acoustic-doppler velocimeter and its application (in Chinese). J Hohai Univ 33(3):15–18
12. Maa JP-Y, Kwon J-I (2007) Using ADV for cohesive sediment settling velocity measurements. Estuar Coast Shelf Sci 73:351–354
13. Cea L, Puertas J, Pena L (2007) Velocity measurements on highly turbulent free surface flow using ADV. Exp Fluids 42(3):333–348
14. Goring DG, Nikora VI (2002) Despiking Acoustic Doppler Velocimeter Data. J Hydraul Eng 128(1):117–126
15. Peng LIN, Li CHEN, Xiao-yun YE (2002) Study on turbulent fluctuation characteristics of temporal signal of sediment concentration in sediment-laden flow(in Chinese). J Wuhan Univ Hydraul Electr Eng 35(2):36–39
16. Ye X, Chen L, Wang Y, Zou J (2003) Experimental study of vertical distribution of sediment concentration in “sediment-laden” flow(in Chinese). J Wuhan Univ Hydraul Electr Eng 36(2):10–14

A High Precision Positioning Algorithm with the Assistance of the Topology Information

Zhan-qi Dong

Abstract In view of the problem existing in range-free node positioning algorithms, the paper proposes a topology-assisted high precision positioning algorithm. To achieve the node positioning, the algorithm first conducts preliminary positioning estimation by ascertaining the overlap area of the transmitting power rings of the relative anchor nodes, then do corresponding correction by computing the logical correlation between the logical coordinate sequence (corresponding to the local topology) and the estimating sequence. The proposed algorithm makes full use of the active information which can be provided both by the anchor node and the neighbor (unknown) node. Computer simulations show that the proposed positioning method is precise and robust. It can adapt to the actual sensing application, in which the anchor nodes may be scarce and the distribution of the anchor nodes may be asymmetrical.

Keywords Centroid algorithm · Logical coordinates · Position adjusting · Topology-assisted

1 Introduction

The sensing node positioning is very significant for the development and application of the IOT technology, and the relative study is paid great attention by the field experts. For now, lots of methods are proposed to achieve the positioning of the sensing nodes, but, in general, all of them can be divided into two main categories: range-free and range-based, and the range-free approach is paid much attention with the consideration of the hardware cost and energy consumption. However the positioning precision of the range-free approach is unsatisfactory, so how to

Z. Dong (✉)
School of Electrics & Information Engineering, Xuchang University, Xuchang, Henan, China
e-mail: dongzqxxgc@126.com

E. Qi (ed.), *Proceedings of the 6th International Asia Conference on Industrial Engineering and Management Innovation*,
DOI 10.2991/978-94-6239-145-1_11

improve the precision is becoming the emphasis. The study of the paper aims to improve the precision of centroid algorithm, which is a kind of range-free approach.

Centroid positioning algorithm takes the geometric centroid of the overlap region, which is formed by those anchor nodes whose broadcast information can be received by the unknown node, as the unknown node's position estimating. And all of the processing is based on the network connectivity, the corresponding algorithm is simple and the realization is easy. However, because the unknown node may be anywhere of the overlap region, the centroid positioning algorithm can only achieve large scale positioning. To improve the positioning precision, some paper does some kind of modification on centroid algorithm. Qiu proposes to gain the overlap region by using the ring ROCRSSI-based Instead of triangle, and takes the geometric center of the overlap region as the position estimating [1]. Liu et al. points out that the high precision positioning may be achieved by broadcasting the anchor node position information with multi-energy-level [2]. Deng and Huang select the anchor node, which is closed to the unknown node, as the distribution center of the anchor node's homogenization evaluation, and look the geometric centroid of the corresponding triangles or polygon as the position estimating [3]. Zhu and Chen propose to improve the positioning precision by doing some certain of weighting based on the diverse RSSI corresponding to every anchor node [4]. Sun et al. does some certain of reverse modification based on the range factor [5, 6]. After studying the research of other paper, we try to utilize the local topology information of the sensing network for the node positioning, and also do some analyses on the performance.

2 Positioning Principle and the Algorithm Description

2.1 Positioning Principle

When the transmitting power remains unchanged, the radio signal is gradually weakened with the increase of the propagation distance, and the corresponding signal cannot be detected while the propagation distance exceeds a certain scope. The proposed topology-assisted positioning algorithm obtains the effective positioning overlap region, which is determined by the received broadcast information (Containing the position information and the transmitting power information) of the anchor nodes; And the corresponding processing is as follows: For an unknown node, for anyone anchor node whose broadcast information can be received, it determines the maximum distance from the anchor node according to the corresponding minimum detectable power, and takes the anchor node as the center of a circle, the corresponding maximum distance as the outer circle radius, the effective propagation distance corresponding to the transmitting power that can not only be used, but also is lower and just close to the minimum detectable power as the inner circle radius, then the unknown node can get a circular ring. Thus, with multiple anchor nodes whose broadcasting information can be received by an unknown node, the unknown node can obtain multiple circular rings, and the common covering part of the circular rings does be the effective positioning overlap region of the unknown node. As example, Fig. 1 shows two kinds of the effective positioning overlap region.

Obviously, the overlap region must be the area that the unknown node located, and taking the geometric centroid of the area as the positioning estimation of the unknown node is appropriate and it is easily realized; yet there are some prominent problems for the positioning of the unknown node. For instance, when the anchor node is scarce or the distribution of the anchor nodes is uneven, there may be some unknown nodes, which can only receive the broadcast information from one or two anchor nodes, and even can not receive any broadcast information, so the positioning of some unknown nodes may not be achieved for the overlap region is larger or there exist more than one overlap region. And Fig. 2 gives several typical cases.

In order to avoid the problem stated above, for the case shown in Fig. 2a, here scheme to achieve the high precision positioning of the unknown node with the help of the neighbor nodes (that have achieved self-localization), by finding the overlap region of the power circular ring of the anchor node and the effective communication coverage rings of the neighbor nodes (which are decided just as the anchor node's power-ring); For the case shown in Fig. 2b, the unknown node may make sure which one of the two independent overlapping area is the area that the unknown node does locate according to the local topology information and the positioning results of its neighbor nodes; For the case shown in Fig. 2c, the unknown node can realize the positioning by finding the common overlap region of the effective communication coverage rings of its neighbors who have achieved self-localization and taking the geometric centroid of the overlap region as its position estimation. In essence, the principle of the processing stated here is quite simple, it just takes the local topology information of the sensing network into account in the positioning, but the topology-assisted positioning algorithm proposed here gives great stress on the utilizing of the local topology information to

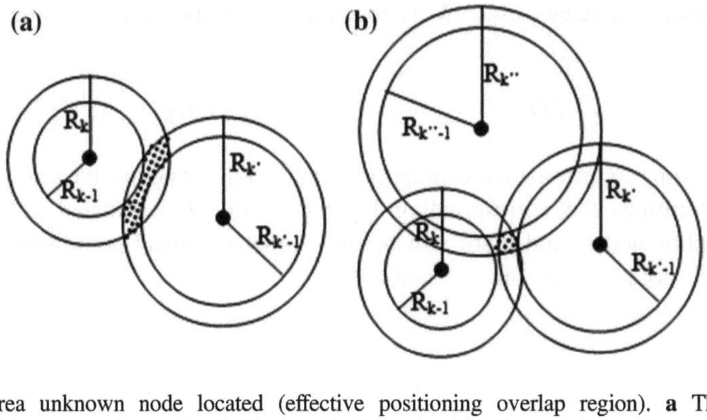

Fig. 1 Area unknown node located (effective positioning overlap region). **a** The area is determined by two anchor nodes. **b** The area is determined by three anchor nodes

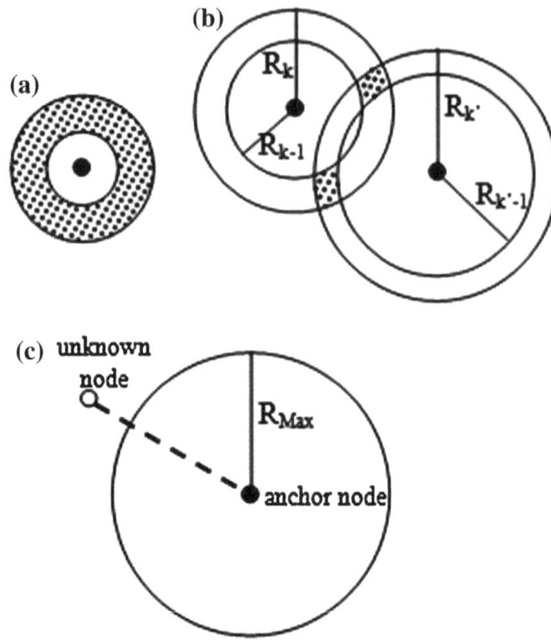

Fig. 2 The typical problem existing in the centroid algorithm based on multi-power-level division. **a** The effective area determined by only one anchor node. **b** The effective area determined by two anchor nodes. **c** Unknown node can not receive the broadcasting information of any anchor node

find the inaccurate positioning and do corresponding correction. The corresponding processing is shown below.

(1) Taking itself as the origin of coordinates, the unknown node decides the logical coordinates of the neighbor nodes in a way that the ordinate is assigned 1 when the neighbor node is north of it, −1 when the neighbor node is south of it, and the horizontal ordinate is assigned 1 when the neighbor node is east of it, −1 when the neighbor node is west of it, the ordinate or/and the horizontal ordinate is assigned 0 when the deviation is less than the corresponding minimum distinguishable range. And then for node i, there will be a logical coordinate sequence SQ_i (K is the number of the corresponding neighbor nodes).

$$SQ_i = ((x_{i,1}, y_{i,1}), (x_{i,2}, y_{i,2}), \ldots, (x_{i,K}, y_{i,K})) \tag{1}$$

(2) According to the position estimation of the neighbor nodes, the unknown node i determines the corresponding logical coordinate estimated value of the neighbor nodes, and there will be an estimated logical coordinate sequence SQ'_i (whose element are ± 1 and 0).

$$SQ'_i = ((x'_{i,1}, y'_{i,1}), (x'_{i,2}, y'_{i,2}), \ldots, (x'_{i,K}, y'_{i,K})) \tag{2}$$

(3) Do bitwise correlation operation between SQ_i and SQ_i', if the elements of the result sequence is "all 1" or "1 and 0", it will indicate that the node positioning estimation of the local area, especially the positioning estimation of node i, is accurate; If there exists only a few of "−1", it will denote the positioning of node i is accurate in some degree. When there are quite a few of "−1", it will suggest that the positioning of node i is inaccurate, and it is needed to do re-positioning estimation with the help of the positioning information of the neighbor node who has achieved the positioning precisely. The corresponding processing is that the unknown node determines the common overlap region of the power circular ring of the anchor nodes (if there are anchor nodes whose broadcast information can be received) and the effective communication coverage rings of the neighbor nodes, searches the geometric centroid of the overlap region, and the position of the centroid located can be looked as the position estimation of the unknown node.

$$SQ_i \cdot *SQ_i' = ((x_{i,1} \cdot x_{i,1}', y_{i,1} \cdot y_{i,1}'), (x_{i,2} \cdot x_{i,2}', y_{i,2} \cdot y_{i,2}'), \ldots, (x_{i,K} \cdot x_{i,K}', y_{i,K} \cdot y_{i,K}')) \quad (3)$$

After having processed or corrected as above, the positioning will merge the topology information of the sensing network, and the positioning precision will be improved significantly, thus the sensing information of the sensing node provided will be more effective.

2.2 *The Implementation of the Positioning Algorithm*

The positioning algorithm merging the topology information of the sensing network may be implemented as follows:

Step 1 Scatter the nodes randomly and does necessary initialization, every unknown node grasps the local topology information of the sensing network by transmitting and receiving broadcasting information.

Step 2 Anchor node transmits broadcast information, which contains the anchor node ID, the position and the transmitting power information, with diverse power periodically.

Step 3 Unknown node senses the transmitting of the anchor nodes, and saves the corresponding broadcast information.

Step 4 For every anchor node whose broadcast information may be received, unknown node determines the corresponding effective ring area, whose circle center is located at the position of the corresponding anchor node, and the outer circle radius is equal to the maximum distance from the anchor node according to the corresponding minimum detectable power, the inner circle radius is equal to the effective propagation distance corresponding to the transmitting power that is lower and just close to the minimum detectable power.

Step 5 For any unknown node, compute the overlap area of the ring area corresponding to every anchor node whose broadcast information can be received by it, and the centroid of the overlap area will be the preliminary positioning estimation of the unknown node.

Step 6 Any unknown node, which has achieved the preliminary positioning, transmits broadcast information that contains the node ID, the positioning and transmitting power information, and at the same time senses the transmitting of other unknown nodes. Then the unknown nodes combine the positioning estimation of the neighbor node with the local topology information of the sensing network, find the inaccurate positioning and do corresponding correction by computing the corresponding logical correlation. The corresponding processing is as follows:

(a) Decides the logical coordinates of the neighbor nodes according to the local topology information;
(b) Decides the logical coordinate estimation according to the position estimation of the neighbor nodes;
(c) Do bitwise correlation operation between the logical coordinates and the corresponding logical coordinate estimations;
(d) Do judgment according to the rule stated at (3) in part of 2.1, and carry out necessary correction.

Step 7 The unknown node, which doesn't achieve self-localization, computes the common overlap region of the power circular ring of the anchor node, if existed (the broadcast information of the anchor node can be received), and the effective communication coverage rings of the neighbor nodes (which have achieved self-localization), and takes the geometric centroid of the common overlap region as its positioning estimation.

3 Computer Simulation and Performance Analysis

To test the performance of the positioning algorithm the paper stated, a lot of experiments are carried out through Matlab simulation software in windows environment. And the basic conditions of the experiment is that a number of unknown nodes and anchor nodes are distributed randomly in an area of 100 m × 100 m; And the unknown nodes can transmit information in three different emission power, the corresponding effective communication range is 10, 15 and 20 m; the anchor nodes can transmit broadcast information in five different emission power, whose effective communication range is 8, 16, 24, 32 and 40 m respectively. In the positioning, the

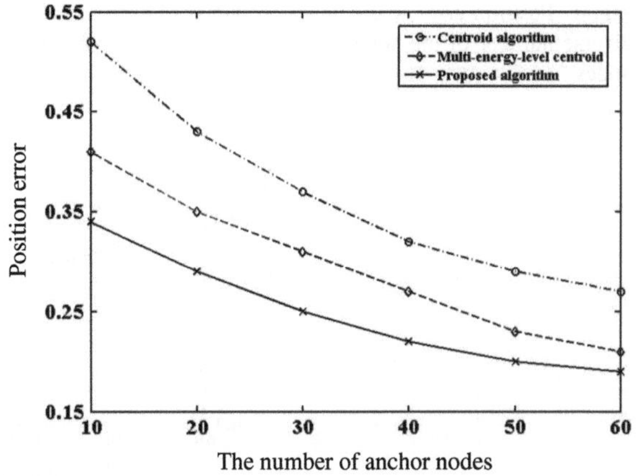

Fig. 3 Position error with diverse number of anchor nodes

anchor nodes transmit broadcast information with diverse power periodically, the unknown nodes only take the one-hop neighbor nodes into account while utilizing the topology information of the sensing network. Figures 3 and 4 show some of the typical simulation results.

The results shown in Fig. 3 are the position errors that are the relative values of the actual errors and the effective communication distance difference of two adjacent transmitting powers of the anchor nodes; And the results are gotten under the condition of 100 unknown nodes existing and the number of the anchor nodes being variable. With Fig. 3, it is clear that, whether for general centroid algorithm, multi-energy-level centroid algorithm or the positioning algorithm proposed, the position precision is improved with the increasing of the anchor nodes; Comparing with centroid algorithm and multi-energy-level centroid algorithm, the proposed algorithm has obvious superiority, it can achieve the positioning with higher precision, especially when the anchor nodes is scarce. And when there are only 10 anchor nodes, the proposed algorithm can get the same positioning precision as the multi-energy-level centroid algorithm whose anchor nodes needed is no less than 25 or so.

Figure 4 exhibits the position errors that are gotten while 30 anchor nodes exist and the number of the unknown nodes is variable. The concrete error curve suggests that the position precision of the proposed algorithm is related with the number of the unknown nodes, and the precision is improved with the increasing of the unknown nodes when the number of the anchor nodes is constant. In contrast, by and large the position precision of the general centroid algorithm and the multi-energy-level centroid algorithm are not affected by the varying of the number of the unknown nodes. And especially, the proposed algorithm performs well, when there are 120 unknown nodes, the position error of the proposed algorithm is about half of the general centroid algorithm and the multi-energy-level centroid algorithm.

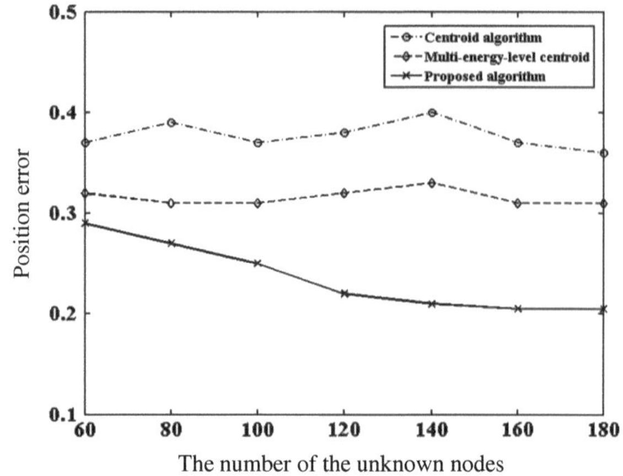

Fig. 4 Position error with diverse number of unknown nodes

4 Conclusion

Based on comprehensive study of the centroid algorithm and the improved algorithm, here proposes a new positioning algorithm. The algorithm emphasizes on the utilization of the local topology information of the sensing network in the positioning processing; it can find and correct the inaccurate positioning with the help of the local topology information. And especially, for the unknown nodes, which can not achieve self-localization only with the anchor node information, the positioning can be achieved by determining the common overlap region of the effective communication coverage rings of its neighbors who have achieved self-localization and the power circular ring of the anchor node, if existed (the broadcast information of the anchor node can be received). All in all, the proposed positioning algorithm is more comfortable with the uneven distribution and the scarce of anchor nodes, and it is unlimited with the scale of the sensing network, the development space is gratifying.

References

1. Qiu M, Xu H (2007) A distributed range-free localization algorithm based on clustering for wireless sensor networks. In: Proceedings of the international conference on wireless communications, networking and mobile computing, Shanghai China, pp 2633–2636, 21–25 Sept 2007
2. Liu K, Yu Q, Wang S, Hu F (2009) Multi-energy-level centroid localization scheme for wireless sensor networks. J Huazhong Univ Sci Tech 37(8):61–64 (Natural Science Edition)
3. Deng B, Huang G (2010) Study on centroid localisation algorithm of WSN. Comput Appl Softw 27(1):213–214
4. Zhu B, Chen S (2010) An improved centroid localization algorithm for wireless sensor network. Chin J Sens Actuators 23(6):868–872

5. Sun W, Chi W, Wang Q (2012) Error back correction for weighted centroid localization. Chin J Sens Actuators 25(1):114–118
6. Ai Y (2013) A fine-grained localization scheme based on effect of multi-energy level for sensor networks. J Wuhan Univ Technol 35(2):157–160 (Information and Management Engineering)

Intrusion Detection Network Based on Fuzzy C-Means and Particle Swarm Optimization

Zhongxing Zhang and Baoping Gu

Abstract Based on the Fuzzy C-Means algorithm, we proposed the PSO-FCM algorithm combining the Fuzzy C-Means with PSO. Then KDD cup 99 dataset was applied to algorithm, the experiments indicate that the PSO-FCM algorithm can avoid the inherent shortcomings of the FCM algorithm, and has higher detection performance with detection rate rising and false alarm rate falling. In addition, we compare the performance of PSO-FCM with other clustering; it can be more satisfactory results.

Keywords Fuzzy C-Means · Intrusion detection · PSO · PSO-FCM

1 Introduction

Intrusion detection was introduced by James Anderson in 1980 [1], after that, intrusion detection plays major role next to firewall. An Intrusion is method of comprising confidentiality, integrity, scalability and availability of network resources. It monitors and analyzes the user and network traffic, verifies system configurations and vulnerabilities and alerts the administrator through alarms. Intrusion detection network (IDS) have been developed due to the dramatic growth of attacks by the hackers. The hacking software also developed in parallel to attack the IDS. So there are some difficulties to provide safeguard mechanism to data. But still researchers are trying so hard to develop efficient IDS. Firewall is different from IDS because through firewall we cannot able to prevent the attack but in case of IDS the administrator can identify the attack and also we take preventive

Z. Zhang · B. Gu (✉)
Department of Information Engineering, Henan Radio & Television University, Zhengzhou, China
e-mail: gubp@open.ha.cn

Z. Zhang
e-mail: hnddzzx@163.com

E. Qi (ed.), *Proceedings of the 6th International Asia Conference on Industrial Engineering and Management Innovation*,
DOI 10.2991/978-94-6239-145-1_12

measures to avoid those attacks to be happen further. IDS include the profiles for user behavior over network. According to those patterns it identifies the intruders and passes response. In IDS it includes two types of detection approaches namely misuse and anomaly detection approaches.

Data mining is to extract relevant information from large database or data warehouse by the user. Information can be represented as concept, rules law and model. An idea of behind using data mining is to help the decision maker to differentiate between data as useful or irrelevant. Mining techniques are classification, Regression, and Deviation detection are used to predict unknown or future values from another variables. Although many kinds of clustering methods, such as Fuzzy C-Means (FCM) [2], K-means [3], are widely used in intrusion detection, few clustering algorithms guarantee a global optimal solution. Therefore, to find global optimal clusters instead of local optimal results, this article presents a new approach for network anomaly intrusion detection, the normal behaviors of a user or a program by using clustering the training data set and the optimal clustering results are obtained by means of Particle swarm optimization (PSO) algorithm [4]. PSO has many advantages over other evolutionary computation techniques such as simple implementation, faster convergence rate and fewer parameters to adjust. PSO is presented so as to combine it with clustering techniques for finding the minimum of the fitness function, producing a good result and enhancing the detection rate of intrusion. PSO-FCM can combine the advantages of both PSO with clustering. Results are also compared with FCM and K-means.

The remainder of this paper is organized as follows. Section 2 discusses the related work of FCM and Particle swarm optimization. Section 3 describes the proposed method of PSO-FCM in detail. Section 4 shows the experiment results on KDD cup. Section 5 gives some conclusions.

2 Related Work

2.1 Fuzzy C-Means Clustering

The process of organizing similar objects into groups is called clustering. In a cluster there may be many groups according to the dataset it differs. But objects in one group are not similar to the objects in other groups and vice versa. Hence the goal of clustering is to include the essential group in a set of unlabeled data. Clustering algorithm has different types of clustering namely partitioning-based clustering, hierarchical algorithms, density based clustering and grid based clustering.

Dunn developed FCM clustering algorithm and it is improved by Bezdekin 1981 which has been extensively used in pattern recognition. Different initialization involves many iteration steps. Fuzzy C-Means algorithm is based on the minimization of an objective function called c-means functional. Fuzzy C-Means algorithm is one of the well known relational clustering algorithms. It partitions the sample data of each input variable into a number of clusters which have fuzzy boundaries, so that each data value belongs to each cluster to some degree or other. Membership is not constrained. Based upon the number of cluster involved and mid-points of those clusters we can determine the associated membership functions and the degree of membership. Fuzzy clustering methods permit improbability in the cluster assignments. FCM is to find cluster centre's that minimize a dissimilarity function.

FCM Algorithm [5]

Input: n data objects and number of clusters
Output: Membership value of each object in each cluster
Step 1: Select the initial location for the cluster centre.
Step 2: Generate a new partition of the data by assigning each data point to its nearby centre.
Step 3: Estimate the membership value of each object in each cluster.
Step 4: Evaluate new cluster centers.
Step 5: If the cluster partition is stable then terminate, or go to step 2.

Iteratively update the cluster centre and membership function for each data point moves the cluster centre to the right location within a dataset. FCM does not meet an optimal solution, because the cluster centers are being randomly initialized. Since the performance depends upon the initial centroids which as follows:

- By Using an algorithm determines all centroids.
- Run Fuzzy C Means clustering algorithm for some times using different initial centroids.

2.2 *Particle Swarm Optimization*

In PSO, each single solution is a "bird" in the search space. We call it "particle". All of particles have fitness values which are evaluated by the fitness function to be optimized, and have velocities which direct the flying of the particles. The particles fly through the problem space by following the current optimum particles.

PSO is initialized with a group of random particles (solutions) and then searches for optima by updating generations. In every iteration, each particle is updated by following two "best" values. The first one is the best solution (fitness) it has achieved so far (The fitness value is also stored.). This value is called pBest. Another "best" value that is tracked by the particle swarm optimizer is the best value, obtained so far by any particle in the population. This best value is a global best and called gBest. When a particle takes part of the population as its topological neighbors, the best value is a local best and is called lbest.

After finding the two best values, the particle updates its velocity and positions with following Eqs. (1) and (2).

$$v[] = v[] + c1 * rand() * (pbest[] - present[]) + c2 * rand() * (gbest[] - present[]) \tag{1}$$

$$present[] = present[] + v[] \tag{2}$$

v[] is the particle velocity, persent[] is the current particle (solution). pbest[] and gbest[] are defined as stated before. Rand () is a random number between (0,1). c1, c2 are learning factors, usually c1 = c2 = 2.

The pseudo code of the procedure is as follows:

```
The pseudo code of the procedure is as follows:
For each particle {
Do {
    Initialize particle
     }
Do {
For each particle {
Calculate fitness value
If the fitness value is better than the best fitness value
(pBest) in history
Set current value as the new pBest
}
Choose the particle with the best fitness value of all the
particles as the gBest
For each particle {
Calculate particle velocity according equation (1)
  Update particle position according equation (2)
     }
 } While maximum iterations or minimum error criteria is
not attained
```

Particles' velocities on each dimension are clamped to a maximum velocity Vmax. If the sum of accelerations would cause the velocity on that dimension to exceed Vmax, which is a parameter specified by the user. Then the velocity on that dimension is limited to Vmax. It can be seen from above PSO algorithm that there are two important steps to solve optimization problems using PSO: fitness function and coding of particle.

3 PSO-FCM Algorithm

3.1 Coding and Fitness

The core of FCM algorithm is to determine the clustering center, so PSO codes the clustering center. Each particle represents a collection of clustering center, the position of each particle is composed of k clustering center, in addition to position of each particle, each particle has velocity and fitness, The position of any particle x_i is defined by m clustering centers, so the position is m × d dimensional vector, the velocity is also m × d dimensional vector. So the particle can be coded as follows:

$P_{11}P_{12}\ldots P_{1d}\ldots P_{m1}P_{m2}\ldots P_{md}$	$V_1V_2\ldots V_{md}$	f(x)

If the scale of the particle swarm is m, there are d types of clustering. We define the fitness function of each particle:

$$J_w(U,E) = \sum_{k=1}^{n}\sum_{i=1}^{c} \mu_{ik}\|x_k - e_i\|^2$$

The better the clustering result is, the smaller $J_w(U,E)$ is, and then the larger individual fitness is $f(x) = 1/J_w(U,E)$

3.2 PSO-FCM Algorithm

Step 1: Initialize the particle swarm, Initialize a particle on randomly selected m criminal records from the n criminal records, and iterate N times.
Step 2: Clustering this particle according to FCM algorithm
Step 3: Compute the fitness of each particle according to f(x)
Step 4: Compare each particle's fitness with its pbest. If the current value is better than pbest, then set pbest equal to the current value
Step 5: Compare pbest of particles with each other then update gbest.
Step 6: Change the velocity and position of the particle according to (1) and (2) respectively
Step 7: Repeat Step 2 to Step 6 until to reach the end condition
Step 8: Determine each cluster according to the method of maximum subjection principle

PSO-FCM algorithm has great randomness in producing the next generation, not easy to fall into local minima, because of sharing mechanism and individual fitness optimization, each generation of particle has dual advantages of self learning and learning from others, so it has fast convergence speed, convergence is relatively stable, little fluctuation.

4 Experimental Results

In this section, we describe the experiments conducted to evaluate the proposed system. The proposed system is tested on an AMD Processor 2.60 GHz with 2G RAM running Windows 7 and coded by Matlab 7.01.

4.1 Dataset Source

KDD CUP 99 Dataset [6] is developed based on DARPA 98 dataset in a MIT Lincoln Laboratory. Protocol such as TCP, UDP and ICMP has been used in this dataset to evaluate the anomaly detection methods. The dataset contains 24 different training attacks and 14 types in the test data. The attacks fall in one of the following four categories, as is shown in Table 1.

- DoS Attack: It is kind of attack in which the attacker makes traffic busy over the network in order to restrict the authorized user access to the systems.
- User to Root Attack: Here, the attacker access to a normal user account on the system by compromising it through sniffing password and gain access to the remote system.
- Remote to Local Attack: In R2L, an attacker sends packets to a machine over a network since who does not have any rights to access a system and make some violations.
- Probe Attack: This type of attack is carried out to get the systems and network information to make some attacks in future.

In order to reduce the size of the dataset, we select some records including training dataset and testing dataset. Table 2 shows detailed information about the number of all records. It is important to note that the test data includes specific attack types not present in the training data. This makes the intrusion detection task more realistic.

For evaluating the IDS outputs (in the test phase), Detection Rate (DR) and False Alarm Rate (FAR) [7], were used to measure the accuracy of method. The Detection Rate shows the percentage of true intrusions that have been successfully detected. The False Alarm Rate is defined as the number of normal instances incorrectly labeled as intrusion divided by the total number of normal instances. A good method should provide a high Detection Rate together with a low False Alarm Rate.

Table 1 Specifies different attack types involved in KDD cup 99 dataset

Denial of service attacks	Back, land, Neptune, pod, smurf, teardrip
Probes	Satan, ipsweep, nmap, portsweep
Remote to local attacks	Ftp_write, guess_pwd, imap, multihop, phf, spy,wareaclient, warezmaster
User to root attacks	Buffer_overflow, load module, Peri, root kit

Table 2 Number and distribution of training and testing dataset

Different types	Training dataset	Testing dataset
DoS	99	294
Probes	78	218
R2L	53	157
U2R	42	126

Table 3 Performance analysis of three algorithms

DR%	DoS	Probe	R2L	U2R
K-means	88.03	56.72	83.71	41.47
FCM	82.31	92.13	91.52	78.29
PSO-FCM	90.25	95.46	94.27	82.92

Table 4 Performance analysis of three algorithms

FAR%	DoS	Probe	R2L	U2R
K-means	7.42	8.21	12.79	22.40
FCM	6.48	7.8	10.91	19.43
PSO-FCM	5.32	5.89	9.75	12.11

4.2 Performance Evaluation

To evaluate and compare the performance of the proposed algorithm, this paper presents the performance analysis of three algorithms, they are PSO-FCM, K-means clustering and Fuzzy C-Means, among them, K-means and Fuzzy C-Means are selected as standard algorithm, PSO-FCM is compared with the others. Table 3 gives the accuracy percentage for the all above 3 algorithms; Table 4 gives error percentage for the three algorithms. From Tables 3 and 4 we can see that regardless of FAR or DR PSO-FCM is superior to the other two algorithms (K-means, FCM), this proves that our proposed algorithm is feasible.

Figures 1 and 2 show the chart for accuracy of three algorithms in terms of DoS, Probe, R2L, and U2R attacks. From the result obtained we can conclude that, PSO-FCM algorithm outperforms than other two algorithms both in DR and FAR.

In this paper, in order to estimate PSO-FCM algorithm's comprehensive performance, we do another experiment with the four groups of data (DoS, Probe, R2L and U2R) together for the intrusion detection. KDD CUP 99 data is selected from "kddcup.data10.percent", there are 494,021 records in this dataset, the normal

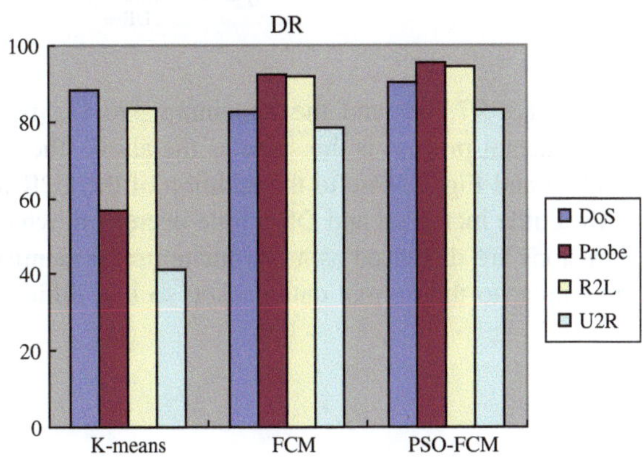

Fig. 1 DR of three algorithms

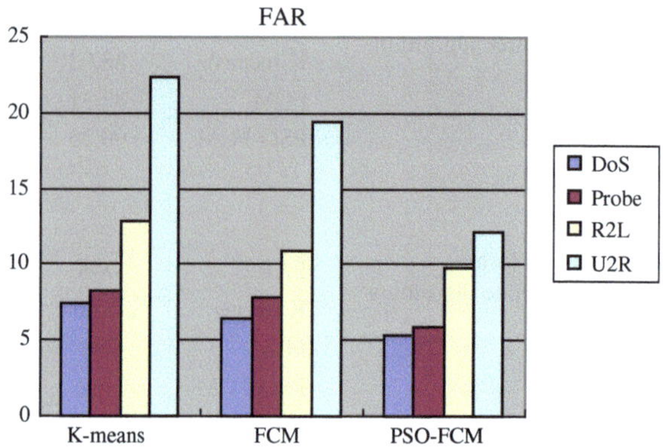

Fig. 2 FAR of three algorithms

Table 5 Comparison of for the attacks

DR and FAR	Four attack types	
	DR%	FAR%
K-means	62.12	18.86
FCM	72.33	16.79
PSO-FCM	89.46	13.55

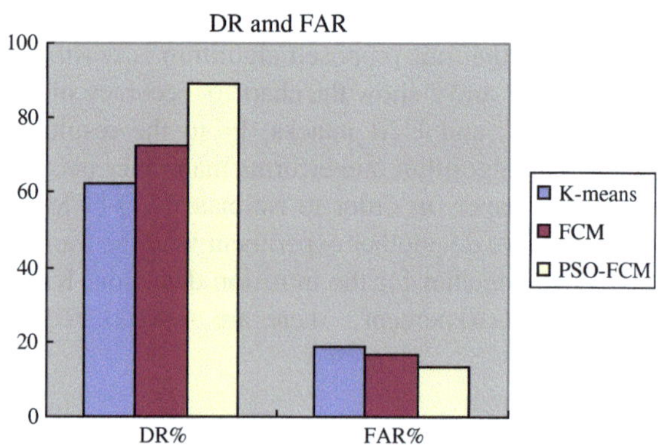

Fig. 3 DR and FAR of four attack types

records are 97,278, and the remaining 396,473 records are abnormal data. The experimental process is the same as the above, the experiment results are given in Table 5 and Fig. 3. Due to the addition of the U2R and R2L two types of attacks, FAR a little increased and DR a little decreased, which is why many R2L and U2R intrusions are disguised as a legitimate user's identity to attack, the characteristics are similar to the normal data packet, so it is difficult to detect attack.

5 Conclusions

In this paper, we have proposed a Fuzzy C-Means Clustering based on Particle swarm Optimization for intrusion detection system. The main advantage of this algorithm is that it uses clustering to identify anomaly intrusion and classification to find intruder. The simulation experiments compared with K-means and FCM in intrusion detection indicate the following aspects: we can obtain the global optimal value by using the PSO-FCM algorithm; DR is higher and FAR are lower than those of FCM and K-means.

For future work, we have planned to improve the detection rate on predicting attacks, And due to the parameters' impact on the algorithm, we will consider introducing the other intelligent algorithms for the optimal parameters' selection.

Acknowledgments The authors would like to thank the editors and the anonymous reviewers whose invaluable comments and suggestions led to greatly improved manuscript.

This work was supported in part by the Education Department of Henan Province Science & Technology Research Project (Grant No. 14B520052).

References

1. Anderson JP (1980) Computer security threat monitoring & surveillance. Technical report, James P Anderson co., Fort Washington, Pennsylvania
2. Nascimento S, Mirkin B, Moura-Pires F (2000) A fuzzy clustering model of data and fuzzy c-means. In: FUZZ IEEE 2000. The ninth IEEE international conference on Fuzzy systems, vol 1, pp 302–307
3. Zhang D, Chen S (2004) A novel kernelized fuzzy c-means algorithm with application in medical image segmentation. Artif Intell Med 32(1):37–50
4. Kennedy J, Eberhart RC (1995) Particle swarm algorithm. In: Proceedings of the 1995 IEEE international conference on neural networks, IEEE Press, New York, pp 1942–1948
5. Izakian H, Abraham A, Snášel V (2009) Fuzzy clustering using hybrid fuzzy c-means and fuzzy particle swarm optimization. In: 2009 World congress on nature & biologically inspired computing (NaBIC 2009), pp 1690–1694
6. KDD Cup 1999 dataset. http://kdd.ics.uci.du/databases/kddcup99/kdd-cup99.html
7. Elbasiony RM, Sallam EA (2013) A hybrid network intrusion detection framework based on random forests and weighted k-means. Ain Shams Eng J 4(4):753–762

OSTP: Optimization Selection of Path Transmission Routing Algorithm in Opportunistic Networks

Jia Wu and Zhi-gang Chen

Abstract Opportunistic network is a random network and do not communicate with each other among the respective communication areas. This situation leads to the message transfer process become very difficult. In this paper a reducing energy consumption optimization selection of path transmission routing algorithm (OSPT) in opportunistic networks is proposed. Algorithm is applied to the design of dynamic random network topology, created a dynamic link, optimized and selected the path. It solves the problem of undeliverable messages for a long time in opportunistic networks. According to the simulation experiment and compare with epidemic algorithm and spray and wait algorithm. Experimental results show that OSPT algorithm improves deliver ratio, reduce energy consumption, cache time and transmission delay.

Keywords Cache time · Deliver ratio · Energy consumption · Opportunistic networks · Routing algorithm · Transmission delay

1 Introduction

Opportunistic network [1] is not required link to exist between the source node and the destination node. It is used by the mobile node the opportunity to encounter brought to communicate. Opportunistic networks belong to one of branches in DTN [2] (delay tolerant networks). Opportunistic networks emphasize the delivery of messages whether the opportunity network nodes carry messages are passed to the next node. In opportunistic networks, it is very important to analysis of message transfer.

The traditional routing algorithm [3] must request the network topology is fully connected before looking routing. Hence there routing algorithms are no longer

J. Wu (✉) · Z. Chen
School of Software, Central South University, Changsha, China
e-mail: jiawu5110@163.com

E. Qi (ed.), *Proceedings of the 6th International Asia Conference on Industrial Engineering and Management Innovation*,
DOI 10.2991/978-94-6239-145-1_13

suitable opportunistic networks and it must be found the establishment of a new routing algorithm method. The current study focused on opportunistic networks routing algorithm: in the one hand, the process of using the mobile node, so that mobile nodes can provide communications for different communication area. In the other hand, how to select the appropriate node is a problem when it is carried message to communication. Related research, in opportunistic networks, is mainly based on single-copy and multi-copy format. Using a multi-copy mode flood broadcast packet through intermediate nodes to copy the data which packet forwarding efficiency can be improved. To reduce the overhead of packet forwarding delay, but the excessive consumption of resources and energy nodes of the network itself. The single-copy approach mainly from energy savings throughout the network node departure, targeted to select the appropriate next hop to forward data packets, and to ensure the successful transmission rate. Then it may improve network performance accordingly. This article is also a single copy of the study focuses on the way.

This paper is divided into blocks by regional connectivity to the opportunistic network and formation of sub-networks. Combined with the idea of graph theory, sub-network topology composed over a period time to analyze, to find a suitable path. The message can be more likely to pass out from the sub-network, reach different communication areas, and ultimately the message a wide range of delivery. According to this viewpoint, we propose a new routing algorithm OSPT, passing by selecting the appropriate node message, the formation of the optimal path, at the same time saving energy consumption to ensure the success rate of message transmission. Finally, the simulation results show the effectiveness by ONE [4]. Then it is compared with the Epidemic algorithm and Spray and Wait algorithm.

2 Related Work

Related research focuses on the use of node mobility model to select routes.

Zhu [5] and Wang [6] proposed ferry route data packets MF (message ferrying) method. MF controlled using special mobile nodes called ferry nodes. They deployed in a particular area and were responsible for transmission of data packets between ordinary nodes.

Segui [7] proposed a method for determining when an empty case diagram (space-time graph) routing framework of node mobility patterns. Space-time diagram routing table records the time dimension structure and increases the target address space and time. Message which reaches the next hop from the routing table can be selected by the next hop and future neighbors.

Nguyen [8] and Tian [9] proposed MV (meetings and visits) routing algorithm. MV learns by moving the nodes. The probability that the contact between the node and the node to access the probability of a particular place, in order to assist the routing and buffer allocation of resources. MV routing node sends a message to a distribution having a higher probability.

Li [10] proposed route of infection (epidemic routing). Routing can infect almost all the data correctly distribute almost all of the data, thus overcoming the traditional routing protocols due to the limit connectivity and it does not distribute any shortcomings message. Due to infection route exhaustive of all possible transmission paths, is guaranteed to find the shortest path possible in the case of bandwidth, buffer space is not competition, but severe consumption of resources, and in the real scene, the energy, bandwidth and buffer space resources may be caused by a lack of competition for resources, performance will be severely downgraded the route of infection.

PROPHET (probabilistic routing protocol using history of encounters and transitivity) [11, 12] is based on the probability estimation, with the route of infection blindly forwarded to all neighbor nodes with different messages. Then the probability of each node has in the route to the destination node estimates will probability.

According to previous research, we will solve how to build and design models and to reduce energy consumption in the process of passing messages.

3 Model Design

Node mobility is a difficult research in opportunistic networks. Setting the appropriate scene analysis is a very important issue for opportunistic networks communication. In practice, the objects of studying are nodes which involved in the communication area. Because the active nodes involved in a high degree of communication, high active illustrate the information may be transmitted for a long time. On the contrary, non-participation nodes would be isolated and could not participate in corresponding. They carry messages may lose when transmitting. Consequently, in the design of the algorithm, to select the communication node involvement is very important.

3.1 Research Model

A typical opportunistic network structure is formation of different regional connectivity networks with a certain period of time. Then these areas are random and variable. As shown in Fig. 1 in the time period T, network is divided into multiple connected regions. In those regions, nodes can communicate, part of the region can only be one way transferred of information, but it does not deliver the message to the network in all regions.

According to Fig. 1, in the time period T, regional A information can only be passed to B, C D E areas do not receive the message from A. Only nodes from region A move, message can be delivered out of. How to choose the node in region

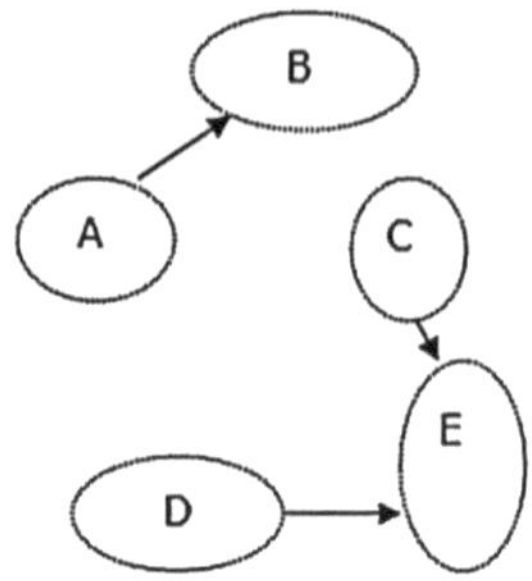

Fig. 1 Opportunistic network communications structure

A, protecting the message may be carried and delivered other regions. These are the next step is to study the work.

3.2 Establish Topology

According to opportunistic networks research model, arbitrarily select a sub-network research. Figure 2 is a sub-network of the opportunistic network in the time period T.

According to Fig. 2, there are a lot of nodes in the sub-network. Assuming T periods, node A carry message, node moving speed is less than the speed of message transfer, topology remains unchanged in the period T. Then node A sends the message can be delivered to all nodes in sub-network. It can constitute a topology about sub-network shown in Fig. 3.

According to the topology of Fig. 3, the process of information transmission can be determined to establish the topology of a directed graph.

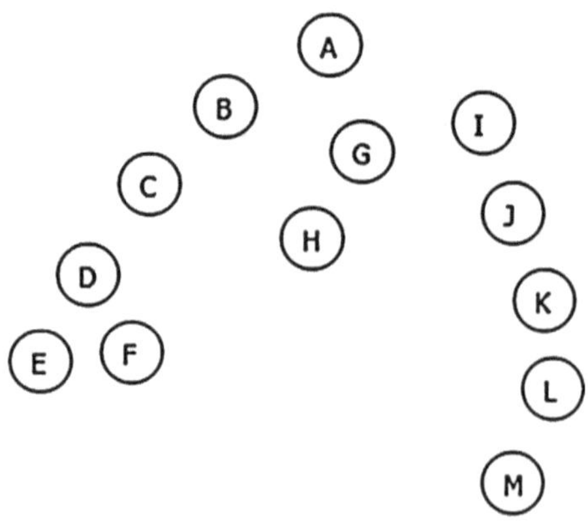

Fig. 2 A sub-network of the opportunistic network

Fig. 3 Message deliver topology in opportunistic networks

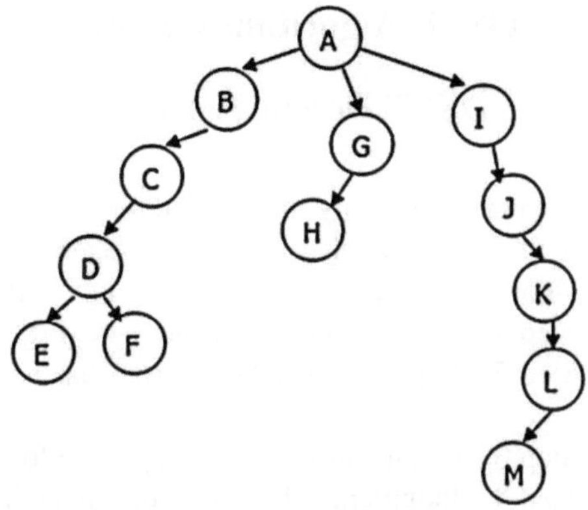

3.3 *Establish Digraph*

The digraph during the time T could be defined according to topology structure.

Definition: the digraph of opportunistic network

$$G = (V, E)$$

In which, V = {A, B, C, D, E, F, G, H, I, J, K, L, M}.

$$E = \{<A, B>, <B, C>, <C, D>\ <C, E>, <D, F>, <A, G>, <G, H>, <A, I>, <I, J>, <J, K>, <K, L>, <L, M>\}$$

According to digraph, it can be seen the process of information transfer, determining the transmission path, and selecting the path of travel. Message via multi-hop transmission, can be transmitted to the distant nodes, it guaranteed communication between different regions.

3.4 *Transfer Model Analysis*

According to analyze digraph, nodes can find their neighbors by themselves and choose next hop in opportunistic networks. Supposed going through the multi-hop in the network, nodes can pass to the information carried by a sub-network. The more hops, the higher the level of activity on the path junction. Then, select a maximum number of hops the path, we can achieve the transfer of message. In this case, the maximum number of hops path message becomes optimal path.

4 OSTP Algorithm Design

4.1 OSPT Design Process

According to analysis and modeling, it got a message in the network transmission mode. The model can pass messages to each node. Therefore, we designed the algorithm through the model. Design steps are as follows:

Step 1 Message in i layer, node A (Fig. 4a)
Step 2 Judgment whether node A has neighbors in layer i+1. It defines neighbors BCD
Step 3 Pointer traversal p jump once to layer i+1 (Fig. 4b)
Step 4 Judgment whether node BCD have neighbors in layer i+2. It defines neighbors EFG
Step 5 Pointer traversal p jump twice to layer i+2 (Fig. 4c)
Step 6 After jumping j times, pointer traversal p arrived at layer i+j and node M. Then there is no neighbor (Fig. 4d)
Step 7 Selecting the M node path. Obviously, the nodes on the path have higher neighbor, and level of activity, for all messages on the path most likely to be portable, so the path is set to the optimal path (Fig. 4e).

On the path, nodes deliver the message to the next communication area. Ensure that the information can have the opportunity to pass to the entire opportunistic networks. According to algorithm design process, it gets a message delivery process. At the same time, it is determining the critical path to reduce the energy consumption during transmission.

4.2 OSPT Routing Algorithm Design

According to the process of design, we can plan program flow diagram.

OSPT routing algorithm design is as Table 1:

4.3 Delivery Rate Calculation

In opportunistic networks, r is reached at target node parameter. Concerning reached delivery ratio function is defined $d(r_i)$. These are m messages delivered by nodes. Then during the transmission, the reached nodes of message $D(r)$ are:

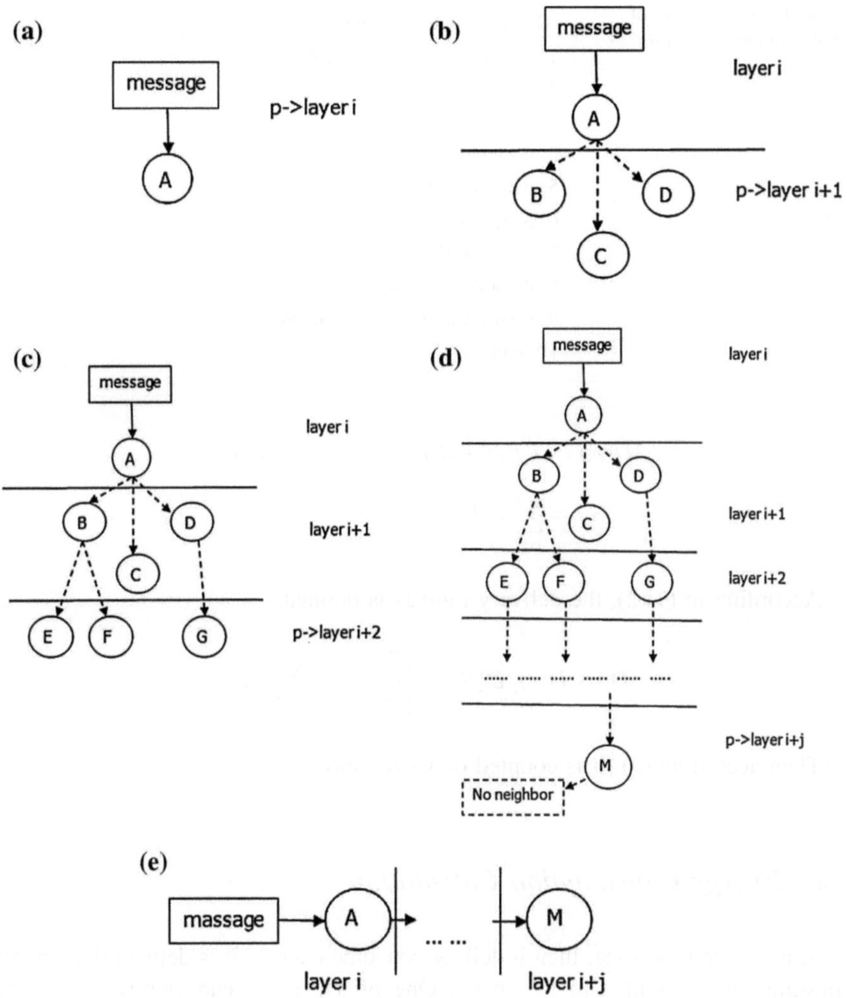

Fig. 4 **a** Message in layer i, **b** message in layer i+1, **c** message in layer i+2, **d** message in layer i+j, **e** select optimal path

$$D(r) = d(r_1) + d(r_2) + \ldots\ldots + d(r_n) = \sum_{i=1}^{m} d(r_i) \tag{1}$$

There are n nodes sent in opportunistic networks. R is parameter about delivery. Each message is delivered as $d(R_i)$. Then the whole messages are delivered"

Table 1 Pseudo-code of OSPT routing algorithm

Algorithm 1 OSPT routing algorithm
1: OSPT (message,n,layer[m])
2: node<-message
3: index p
4: p->layer[i]
5: while (neighbor!=NULL)
6: for j 1 to m
7. p->lay[i+j]
8: if (node->neighbor==NULL)
9: send message->node->neighbor
10: end

$$D(R) = d(R_1) + d(R_2) + \ldots\ldots + d(R_n) = \sum_{i=1}^{n} d(R_i) \tag{2}$$

According to (1)(2), the delivery ratio D is defined:

$$D = D(r)/D(R) = \sum_{i=1}^{m} d(r_i) / \sum_{i=1}^{n} d(r_i) \tag{3}$$

Then according (3), it is counted delivery ratio.

4.4 Energy Consumption Calculation

Each node send message, then it deliver consume energy. It is defined that energy consumption is divided to two parts. One of a part is send energy, others are delivery energy.

There is a road has k nodes to send message, each node consume the same energy e_c to send. Then the path consumes energy E_c is:

$$E_c = k \times e_c \tag{4}$$

If there are m roads taken part in sending messages, each road may have p nodes, Sending messages consume energy E_s is:

$$E_s = m \times p \times e_c \tag{5}$$

A node needs the same energy e_d in delivery by itself. There are q nodes joined in delivery. Then energy consumption E_d in delivery is:

$$E_d = e_d + e_d + \ldots + e_d = q \times e_d \tag{6}$$

From (4)(5), a single node needs energy consumption E_a is:

$$E_a = E_s + E_d = m \times p \times e_c + q \times e_d \tag{7}$$

Then according to (7), it is counted energy consumption.

4.5 Cache Time Calculation

In opportunistic networks, cache time is taken to be the total time being used to cache nodes during the life circle of message. Supposed the cache time of a single node is t_i, these are n nodes taken part in delivery. Then the cache time T is:

$$T = t_1 + t_2 + \ldots + t_n = \sum_{i=1}^{n} t_i \tag{8}$$

According to (8), it is counted cache time.

5 Simulation and Analysis

5.1 Simulation Environment

In order to tested in a real environment, evaluating the performance of OSTP routing algorithm simulation and implementation of the algorithm. Therefore, this paper used Helsinki University of Technology ONE1.4 (Opportunistic Network Environment simulator) to develop simulation tools for simulation tests. The software uses different mobile models, depicting the trajectory of the node. It can be expanded to various opportunistic networks routing algorithms, and provides interactive interface to view real-time status of the nodes of the motion, the advantages of data packet forwarding. This article compared OSTP, spray and wait and epidemic routing protocols. It chose Helsinki city topology as the background, in different streets, parks, shops are nearby and in a different position on the map the model to simulate the distribution, which is more than the realistic.

5.2 Analysis of Simulation Results

Through the simulation environment and performance parameters of the analysis, the simulation can be done to operate.

Figure 5a shows deliver rate of several algorithms. The abscissa represents the number of nodes, and the vertical axis indicates the transmission success rate. A number of nodes increase gradually over time when the success rate of transmission of different algorithms increases. You can clearly see that with the higher number of nodes increases and Spray and Wait OSTP algorithm algorithm algorithm transmission success rate, while lower Epidemic algorithm transmission success rate. Through experiments can be seen, the traditional epidemic algorithm will not increase the success rate of transmission nodes increased. Instead, spay and wait algorithm and OSTP algorithm after increasing transmission nodes, message passing rate increased significantly, the rate of increase is greater than the spay and wait OSTP algorithm.

Figure 5b shows the variation of several algorithms cache time. Abscissa indicates the node, the vertical axis represent the cache time. It can be seen from the figure, the node is gradually increased, and cache time is gradually increased. OSTP algorithm cache time reduce less than the epidemic and spay and wait algorithm. Figure 5b also shows by the space OSTP algorithm consumes less than the network and spay and wait epidemic algorithm.

Figure 5c shows the variation of several algorithms energy consumption. The abscissa represents the number of nodes, and the vertical axis indicates the energy consumption. It can be seen from Figure, OSTP algorithm node show low energy consumption, energy consumption algorithm obvious trends. The result is better than spray and wait algorithm, the increase in the junction node in the process, epidemic algorithms maximum energy consumption.

Figure 5d shows the variation of several algorithms energy consumption. The abscissa represents the number of nodes, and the vertical axis indicates the energy consumption. It can be seen from Figure, OSTP algorithm node show low energy consumption, energy consumption algorithm obvious trends. The result is better than spray and wait algorithm, the increase in the junction node in the process, epidemic algorithms maximum energy consumption.

Figure 5d shows a transmission delay of several algorithms. The abscissa represents the number of nodes, and the vertical axis indicates the propagation delay. OSTP algorithm is, in the beginning, the more nodes increases, the higher transmission delay, to reach 400 after the node reaches the maximum propagation delay, and then as the number of nodes increases, the transmission delay starts to decrease. Epidemic algorithm and the algorithm delay spay and wait nodes increases and the delay increases.

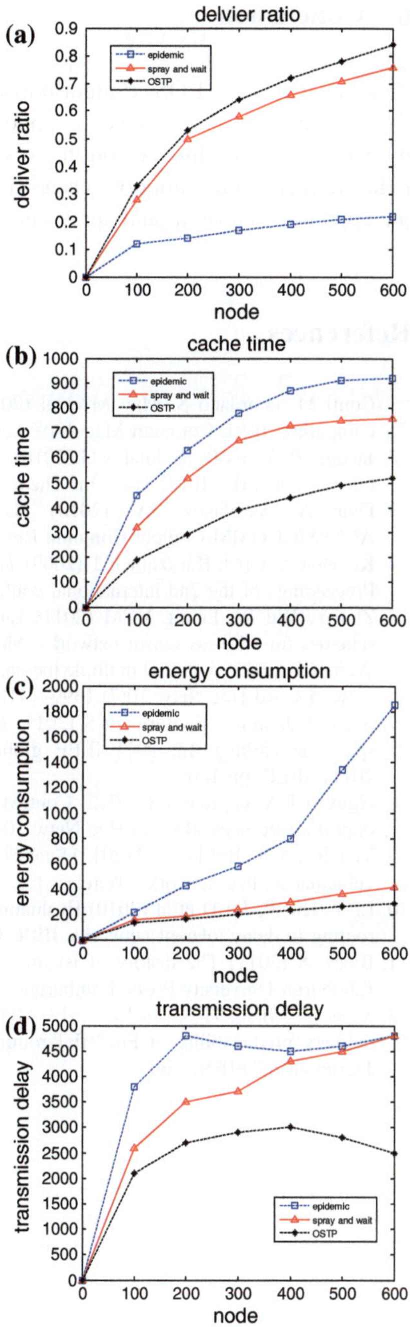

Fig. 5 **a** Deliver ratio, **b** cache time, **c** energy consumption, **d** transmission delay

6 Conclusion

This paper designed OSTP algorithm model, generated a model which facilitates the transmission of message in opportunistic networks. According to designed model, it got algorithm. According to simulation, algorithm improved deliver ratio, reduced energy consumption, cache time and transmission delay. Compared with the epidemic algorithm and spray and wait algorithm, better results are produced.

References

1. Conti M, Giordano S, May M et al (2010) From opportunistic networks to opportunistic computing. IEEE Commun Mag 48(9):126–139
2. Jacquet P, Mans B, Rodolakis G (2010) Information propagation speed in mobile and delay tolerant networks. IEEE Trans Inf Theory 56(10):5001–5015
3. Dvir A, Vasilakos AV (2010) Backpressure-based routing protocol for DTNs. ACM SIGCOMM Comput Commun Rev 40(4):1–9
4. Keränen A, Ott J, Kärkkäinen T (2009) The ONE simulator for DTN protocol evaluation. In: Proceedings of the 2nd international conference on simulation tools and techniques
5. Zhu Y, Wu W, Leung VCM (2011) Energy-efficient tree-based message ferrying routing schemes for wireless sensor networks. Mob Netw Appl 16(1):58–70
6. Wang CF (2012) A virtual multiple message ferry backbone routing scheme for mobile ad-hoc networks. Ad Hoc Netw 10(7):1399–1418
7. Seguí J, Jennings E, Burleigh S (2011) Enhancing contact graph routing for delay tolerant space networking. In: 2011 IEEE global telecommunications conference (GLOBECOM 2011), IEEE, pp 1–6
8. Nguyen HA, Giordano S (2012) Context information prediction for social-based routing in opportunistic networks. Ad Hoc Netw 10(8):1557–1569
9. Tian R, Zhang B, Li C et al (2013) Sparsely-deployed relay node assisted routing algorithm for vehicular ad hoc networks. Wireless Commun Mob Comput 5:1309–1319
10. Li Y, Hui P, Jin D et al (2010) Evaluating the impact of social selfishness on the epidemic routing in delay tolerant networks. IEEE Commun Lett 14(11):1026–1028
11. Black A (2011) The history of islamic political thought: from the prophet to the present. Edinburgh University Press, Edinburgh
12. Yang SY, Jiang JT, Chen PZ (2013) OOPProPHET: a new routing method to integrate the delivery predictability of ProPHET-routing with OOP-Routing in delay tolerant networks. J Comput 8(7):1656–1663

A Rapid Payment Confirm Scheme Base on QRCode

Ming-xin Li, Wei-ping Zhu, Xiao-hui Jiang and Wen-xia Geng

Abstract This paper presents a rapid payment confirm method based on QRCode barcode, the program combines QRCode barcode and public key cryptography. The server get reliable and secure keypair, and encrypt product information with the public key, and then using cipher text to generate QRCode barcode, and then send the QRCode image to customer handset device through the wireless network, then the user use the QRCode barcode on handset device to the specified shop to consume, the specified shop use QRCode reader to decode QRCode image, to obtain the decrypted payment confirmation, confirm payment. This scheme combines QRCode barcode and the advantages of public-key cryptosystem, and give the key codes, can help shop to rapid confirm payment information with security, and this scheme has good industrial efficiency.

Keywords Crypto · ECC · Payment · QRCode

1 Introduction

With the development of Internet of things, 2D code is widely used. It can achieve the data collection fast, accurately and reliably, to help manufacturers monitor and control their products in the logistics system, not only improving the logistics management level, reducing and avoiding logistics risk, also realizing the rapid confirmation of the goods payment.

The traditional consumer code is generally to send a string of specially coded character to the user, then send the consumer code to the user through the wireless network, and the user use the specified consumption codes to consume. Because the traditional consumption code is the plaintext, it exists the high hidden danger. If the

M. Li (✉) · W. Zhu · X. Jiang · W. Geng
Computer and Engineering Department, Chengdu Aeronautic Polytechnic College, Chengdu, China
e-mail: starlee2008@163.com

E. Qi (ed.), *Proceedings of the 6th International Asia Conference on Industrial Engineering and Management Innovation*,
DOI 10.2991/978-94-6239-145-1_14

consumer code is copied, it can easily cause a loss to the users. So this article is introducing the achieving rapid confirmation of payment by using the 2D bar code.

Because the standard of 2D bar code is open, it can fast realize the 2D bar code according to certain norms. Due to 2D bar code code itself does not have the anti-counterfeiting function, in order to achieve security goals, the public key encryption system is introduced into the 2D bar code in this paper. The key of the system is the generation of the public key pair, generating 2D bar code according to the commodity information encrypted ciphertext,then generate a reliable secret key pair relying on the advantages of the java language. The 2D bar code can be quickly verified by the ordinary bar code reader in the client. The encrypted data get by the QRCode scanner, then getting the private key to decrypte the ciphertext in the server, getting the relevant product transaction information.

In this paper, we first introduce some basic theories, advantages of QRCode barcode and the basic theory of the elliptic curve encryption in Sect. 2. In Sect. 3, we display the framework of this system, then detailed describe the system implementation, including client-side and server-side implementation. In Sect. 4, we give a concrete example. Finally, in Sect. 5, we point out some future work to do.

2 QRCode

The QRCode is a matrix 2D code with the large information capacity, high reliability, support Chinese characters and image information and strong security characters etc., designed by Japan Denso Company in 1994 September [1, 2].

Features of Qrcode:

(1) Ultra high speed reading. When reading QRCode bar code,the QRCode code symbol reading information is detected by QRCode graphics symbol position using hardware to realize. Therefore, information identification is required for a short time. When using CCD 2D bar code reader device, it can read 30 QRCode code symbols including 100 characters every second. Because of the ultra high speed reading properties of 2D QRCode code, it can be widely used in the field of industrial automation production line management etc.

(2) A full range (360°) reading. QRCode is square, only black and white, with the help of decoding software localization "back"-shaped affirmative pattern in its 3 corners,so users do not need to align the scanner and regardless of the scan to any angle, who can also correctly read the code information.

(3) Efficient representation of Chinese characters. Because the QRCode code represents the Chinese characters with specific data compression pattern, it can express a Chinese characters with 13 bit other than other 2D bar code which the efficiency is increased by 20 %. A QRCode can represent up to 1817 Chinese characters.

3 Elliptic Curve

Public key cryptosystem, that is, asymmetric key cryptography system, its greatest feature is encryption and decryption use different keys, using public key that open to user to encrypt and verify, while the private key is confidential, using private key to decrypt and sign.

Elliptic curve cryptosystem is based on elliptic curve discrete logarithm problem (ECDLP), which is based on the assumption: E (Fp) there are two points G and Q on elliptic curves over E, G is the starting point, Q = kG, by the G and Q to calculate K is a difficult problem in Mathematics.

Definition 1 Elliptic curves over prime fields [3, 4]

Let p be an odd prime numbers greater than 3, the parameters of a, b meet $4a^3 + 27b^2 \neq 0(\bmod p)$, then the elliptic curve finite prime field Fp can be defined as E:

$$y^2 \equiv x^3 + ax + b(\bmod p) \tag{1}$$

Prime field elliptic curve parameter of the Fp is a six tuple. They are T = {p, a, b, G, n, h}.p, a, b is used to determine an elliptic curve point; $G = (x_G, y_G)$ is the base point of elliptic curve; prime n is the order of G, that's n is #E (Ep), then it can be calculation of elliptic curve by using the Hasse (1) theorem of order:

$$p + 1 - 2\sqrt{p} \leq \#\text{E(Fp)} \leq \text{p} + 1 + 2\sqrt{\text{p}} \tag{2}$$

h is a cofactor about n in #E (Ep), which is namely for h = #E (Ep)/n. Elliptic curve parameters prime field Fp specify exactly the elliptic curve and the base point, which is EC cryptosystem must meet.

The elliptic curve encryption and decryption process is described as follows: Suppose that Alice is to use ElGamal elliptic curve encryption system and the elliptic curve E (Fp) is given, then he will perform the following steps:

select a random integer d,and meet $2 \leq d \leq (\#E - 1)$;
Calculation of Q = dG = (x_Q, y_Q);
Using (G, Q, p, #E) as public key and d is private key that generated by the algorithm of 1.

Next, suppose that Bob wants to send a message (x_1, x_2) to Alice, then he will perform the following steps:

1. Select a random number k, meet $2 \leq k \leq (\#E - 1)$;
2. Calculate the following calculation:

$$\text{R} = \text{KG}$$

$$(c_1, c_2) = KQ$$
$$(y_1, y_2) = (c_2 x_1, c_2 x_2)(\mathrm{mod}\, p);$$

3. Send (R, y_1, y_2) to Alice.When receiving ciphertext, Alice will follow the steps below to decrypt the ciphertext (x_1, x_2):

$$(c_1, c_2) = \mathrm{dR};$$
$$(x_1, x_2) = (c_1^{-1} y_1, c_2^{-1} y_2)(\mathrm{mod} p)$$

4 Implementation of Elliptic Curve Crypto System

According to the theory of the elliptic curve, the elliptic curve encryption system is designed with the object orient programming method. The definition of related variables given the elliptic curve, and then the key algorithm implementation is given. This system can be divided into four main classes.

1. Class EllipticCurve: the elliptic curve finite element based on domain
2. Class ECPoint: elliptic curve point finite element domain based class
3. Class ECKey: elliptic curve key to the class
4. Class ECCryptoSystem: elliptic curve cryptography class

Among them, EllipticCurve provides the parameters and structure of elliptic curve operations. The ECPoint provides a point operation on the elliptic curve. The ECKey provides key pair generated. The ECCryptoSystem class provides encryption, and decryption functions.

4.1 System Architecture

The traditional B2C transaction model is generally as follows: users browse the website and select the satisfied goods, send the purchase request to the merchant's datacenter. The businessmen check the order information, and submit it to the payment system. Then the user pays a specified account which is the amount of consumption to the third party agencies; Then the merchant send out (preparing) goods. When users receiving the goods (or consumer products) confirms the payment. At last the third party agencies will consume amount payment to the business.

The main propose of system is to achieve payment confirmation after the user paid. When user payment is successful, the system will send a message to user's hand held devices in timely.

The information embedded in QRCode was encrypted by elliptic curve encryption system (ECC). The barcode contains transaction information, including Transaction id and the user's mobile phone number. When the user go to the

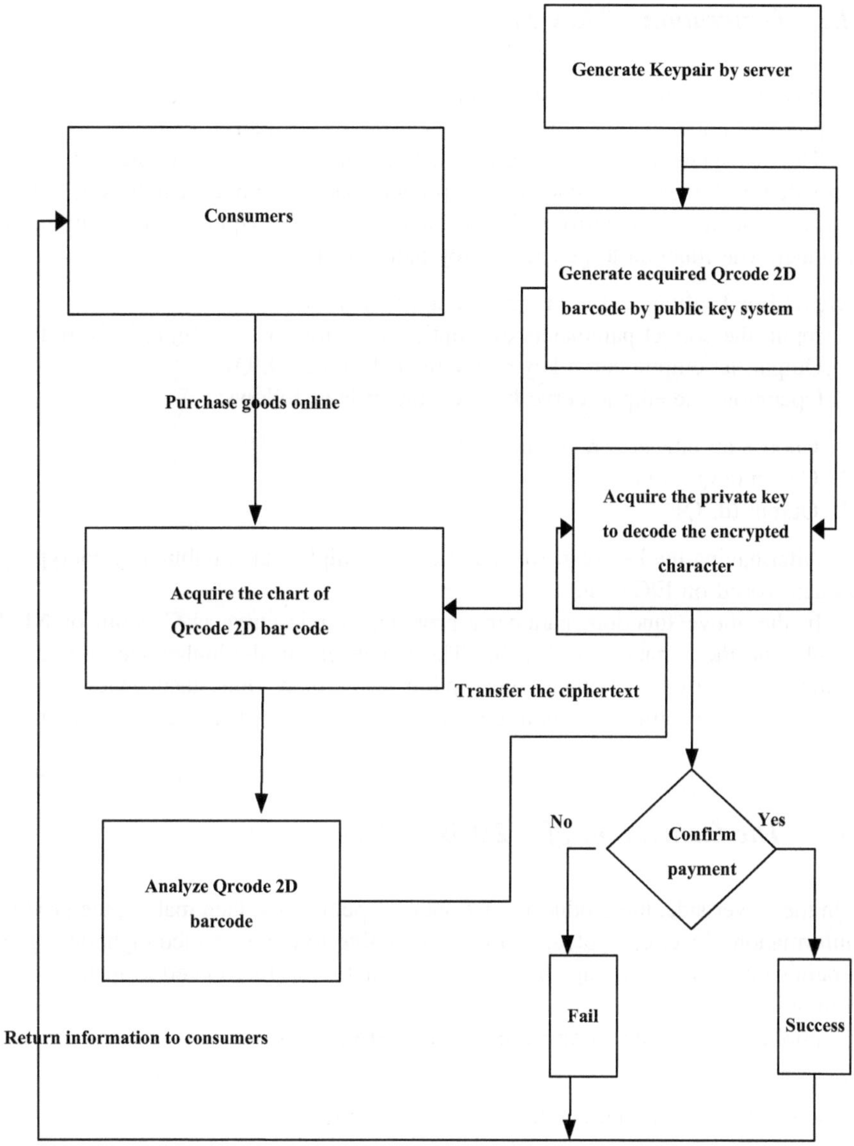

Fig. 1 System flow chart

specified shop or receive the goods through express delivery, Servant will use device to read the barcode information that decrypt by the private key, then checking whether the barcode information and consumer information are matched. Finally finish the payment confirmation. Figure 1 shows the flow chart of the system.

4.2 Generation of Key Pair

In this system, the key pair is generated by the main server, and the server is responsible for the generation and maintenance of key pairs.

The management of the key pair is making full use of the large integer function provided by Java, so the generation of pseudo-random numbers can be safe, which is the foundation of security and generated to meet the requirements of the prime number. The function to generate a key pair is as follows:

Algorithm 1 elliptic curve key generation is as follows:

Input: the correct parameters of elliptic curve domain, T = (p, a, b, G, n, h).

Output: an elliptic curve key and T related to the (D, Q).

Operation: the elliptic curve key generation is as follows:

1. Choose an integer d randomly in the interval [1, n−1];
2. Compute Q = dG;
3. Output (d, Q);

After having the key pair, we can introduce elliptic curve public key encryption system based on ElGamal.

In the above function, parameter provider elliptic curve is Ceticom or NIST [4–9], and the parameter is 192 bit. The larger length, the higher security, at the same time, the encryption and decryption needs more time, then generating the keypairs which satified the requirement of the system, and saving the keypairs.

4.3 The Generation of a 2D Bar Code

On the server side, the producers first obtain a public key, then making the product information object encrypted, and then according to the encrypted ciphertext generating the barcode. Using the encrypted text to generate Qrcode function is as follows:

public void encode (String encode_data, String filePath)

{
Initialize QRcode which is object of Class Qrcode;
Set the parameters of qrcode, such as fault tolerant rate, character mode, and version number;
Call ECC cryptosystem to encrypt the product information;
Call the calQrcode method to generate Boolean array;
Create a Graphics2D g object, and set the background and foreground, and draw the Qrcode 2D bar code according to the Boolean data generation;
According to the parameter filePath to save the barcode graphics.
}

In the above function, first generate Qrcode class object qrcode, and then setting its fault tolerance rate, character mode, version number, and then calling the ECC encryption algorithm to encrypt and product information, and generate Qrcode code according to the ciphertext, and save, and finally send to the users through the wireless network.

4.4 Payment Confirmation

Customers go to the shop with the barcode, and the Shop identify the encrypted information barcode by reader, and get the server's private key through the network, based on the private key to decode the ciphertext. In order to confirm payment, the servant will ask consumers confirmation information. If the phone number etc. are successful, send a message to the server. Make the confirmation of payment, and the money will be transferred to the account designated merchants.

To verify the function as follows:

public Productverify(String filePath)

{
Initialize decoder which is object of class QRCodeDecoder;
Call the decode method to decode the information of the goods;
Call the elliptic curve ECC decryption decryption algorithm to obtain product information;
Verify;
Finally return information goods to the users.
}

5 Experimental Result

Generate the reliable keypairs, call ECC system, to generate the reliable Qrcode, The existing commodity information as follows:

"TransId: 1000123Tel: 13550302123"

Transaction code used to uniquely identify the transaction information, using transaction code can get the related transaction information, such as product name, commodity prices, commodity trading time. The phone number is used for consumption confirmation by user or service waiter input.

QRcode barcode is as follows, the former Qrcode is not encrypted, show in Fig. 2. The latter Qrcode is encrypted, show in Fig. 3.

Based on the test of the server platform, the time of generation and decoding is about 1 s, it can meet the needs most practical application, and has good industrial benefits.

Fig. 2 QRCode code is not encrypted

Fig. 3 QRCode code is encrypted

6 Conclusion

In this paper, in order to improve the security of barcode, we combine qrcode and public key crypto system, which is used for e-commerce payment confirmation. After the test, it can satisfy the need of practical application, and has good industrial benefits, but also has some problems such as the man in the middle attack. The next step will be to solve the problem of man in the middle attack and to improve the efficiency of system.

References

1. QRCode. http://en.wikipedia.org/wiki/QR_Code
2. Open Source QR Code Library. http://qrcode.sourceforge.jp/
3. Koblitz N (1987) Elliptic curve cryptosystems. Mathematics of computation
4. SEC1-V2, SECG Released Standards. http://www.secg.org
5. St Denis T, Johson S (2007) The Programmer cryptology. Machinery Industry Press, Beijing (Chinese)
6. Knuth DE (2002) The art of computer programming, vol 2. National Defense Industry Press (Chinese)

7. Wang Y (2001) Current status and prospect: studies on elliptic curve cryptography. J PLA Univ Sci Technol 3(6) (Nat Sci Ver) (Chinese)
8. Liu Y, Liu M (2005) Research on QR code data encoding. J Beijing Inst Technol 25(4) (Chinese)
9. Xu QL, Li DX (1999) Elliptic curve cryptosystems. J Comput Res Dev 36(11) (Chinese)

Arcball Algorithm in a 3-D Laser Scanning Imaging System

Chao-xuan Fu, Tong-gang Zhao and Zhi-hong Bai

Abstract 3-D rotation interaction is an indispensable function of 3-D imaging system of laser scanning, therefore it is also a hot topic of 3-D interaction research. The drag of mouse can be reflected by the on-screen movement of the cursor however, when the dragged object is a 3-D object, the contradiction between it and the 2-D screen exposes. Aiming to resolve this problem, we establish a new rotation algorithm basing on quaternion, called the "arcball", which we will elaborate in more details later in this article. In order to get a comprehensive understanding of the "arcball", we will first illustrate the mathematical bases of using quaternion to describe rotations, and then move onto the mathematical foundation of the "arcball" algorithm, with the provision of pseudo code of it. At last, we try to apply this algorithm into research projects and provide of an analysis of the results obtained in our researches.

Keywords 3-D imaging · "arcball" algorithm · Laser scanning · Rotation interaction

1 Introduction

Laser scanning imaging technology is a technology that aims to obtain distance, angle and other observational data as laser scanning is able to produce three-dimensional coordinates of the object surface points, so as to achieve target surface contour extraction and three-dimensional imaging. Currently, 4 techniques including Euler angles, rotation matrices and quaternion are widely used in

C. Fu (✉) · T. Zhao
School of Electronic Engineering, Beijing University of Posts and Telecommunication, Beijing 100876, China
e-mail: lsfcx1992@gmail.com

Z. Bai
Technical Service Division, China Petroleum Pipeline Bureau, Hebei, China

E. Qi (ed.), *Proceedings of the 6th International Asia Conference on Industrial Engineering and Management Innovation*,
DOI 10.2991/978-94-6239-145-1_15

3-D graphics to achieve rotation [1–3]. Because of the order in accordance with an independent rotation about an axis in Euler angle rotation, there may be gimball lock, and this might be a limit in practical use. Both matrix Rotation and quaternion rotation can be achieved, but there will be relatively large gap concerning computational efficiency, especially for the case of more frequent rotation operations. On the other hand, quaternion can be easily calculated given a particular angle of rotation, while the result of rotation matrix is converted by a complex calculation. Therefore, the main purpose of this article focuses on Quaternion rotation operation, discusses how a rotation quaternion based interactive algorithm, curvature algorithm, the algorithm can be easily achieved through mouse drag-dimensional image rotation operation [4, 5].

2 Methodology

A rotation problem is more complicated in a three-dimensional space, as to obtain the original definition about the rotation, we need to define an equation (vector representation) and the rotation angle of the size (scalars). One solution is to use the combination of vector and scalar representation, which is the plural promotion, usually called as quaternion. We can represent and this process provides another rotational transformation method, having many advantages in rotation animation hardware implementation [6, 7].

Quaternion form as, for example:

$$q = w + xi + yj + zk$$

where w is a scalar quantity, x, y, z represent three-dimensional coordinates of a point in space, and i, j, k satisfy the following relation:

$$i * i = -1,\ j * j = -1,\ k * k = -1,\ ijk = -1$$

Can also be expressed as: q = [w, v], where v = (x, y, z) vector, w is a scalar. As mentioned earlier, you can use a quaternion rotation, then discuss the rotation quaternion form definition.

Figure 1: Suppose there is an arbitrary axis of rotation vector,

$$\vec{OA} = [X_A, Y_A, Z_A] = (ax, ay, az)$$

With either a rotating angle of rotation of the entire rotation matrix is defined as follows. Each of the above variables can be expressed as:

Fig. 1 3-D Vector

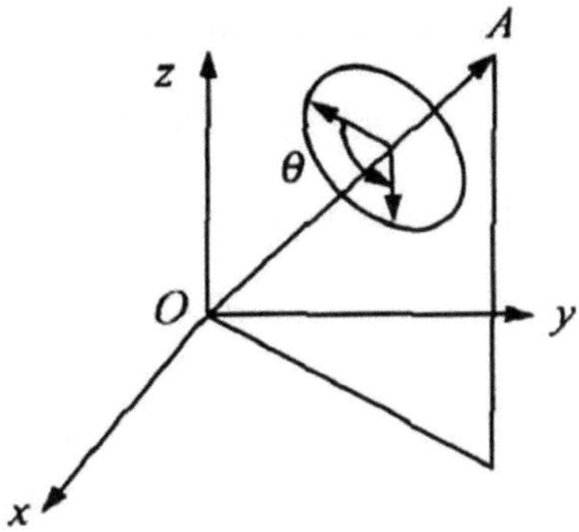

$$w = \cos(\theta/2)$$
$$x = ax * \sin(\theta/2)$$
$$y = ay * \sin(\theta/2)$$
$$z = az * \sin(\theta/2)$$

Make $c = \cos(\theta/2)$, $s = \sin(\theta/2)$, so that

$$x = cX_A; y = cY_A; z = cZ_A; w = s$$

Order q = [w, x, y, z], a quaternion. Through this form we can know that any rotation in space can be easily showed by a rotating shaft and a corner.

Quaternion model, defined as:

$$|q| = N(q) = sqrt(w^2 + x^2 + y^2 + z^2)$$

Quaternion normalization, defined as:

$$Normlize(q) = q/N(q) = q/sqrt(w^2 + x^2 + y^2 + z^2)$$

To meet quaternion, called unit quaternion, Quaternion conjugate is defined as:

$$q^* = w - xi - yj - zk$$

Quaternion with its conjugate reciprocal satisfies the following relationship:

$$1/q = q^*/N(q)$$

Suppose two quaternion:

$$q_1 = w_1 + x_1 i + y_1 j + z_1 k$$
$$q_2 = w_2 + x_2 i + y_2 j + z_2 k$$

The addition of two quaternion is defined as follows:

$$q_1 + q_2 = (w_1 + w_2) + (x_1 + x_2)i + (y_1 + y_2)j + (z_1 + z_2)k$$

Two quaternion multiplication definition:

$$\begin{aligned} q_1 * q_2 = &(w_1w_2 - x_1x_2 - y_1y_2 - z_1z_2) \\ &+ (w_1x_2 + x_1w_2 + y_1z_2 - z_1y_2)i \\ &+ (w_1y_2 - x_1z_2 + y_1w_2 + z_1x_2)j \\ &+ (w_1z_2 + x_1y_2 - y_1x_2 + z_1w_2)k \end{aligned}$$

According quaternion arithmetic nature of the equivalent, unit quaternion matrix, defined as:

$$\begin{aligned} p' &= \begin{pmatrix} \cos\theta \\ \begin{pmatrix} n_x \sin\theta \\ n_y \sin\theta \\ n_z \sin\theta \end{pmatrix} \end{pmatrix} \begin{pmatrix} 0 \\ \begin{pmatrix} x \\ y \\ z \end{pmatrix} \end{pmatrix} \begin{pmatrix} \cos\theta \\ \begin{pmatrix} -n_x \sin\theta \\ -n_y \sin\theta \\ -n_z \sin\theta \end{pmatrix} \end{pmatrix} \\ &= \begin{pmatrix} -xn_x \sin\theta - yn_y \sin\theta - zn_z \sin\theta \\ \begin{pmatrix} x\cos\theta + yn_z \sin\theta - zn_y \sin\theta \\ y\cos\theta + zn_z \sin\theta - xn_z \sin\theta \\ z\cos\theta + xn_y \sin\theta - yn_x \sin\theta \end{pmatrix} \end{pmatrix} \begin{pmatrix} \cos\theta \\ \begin{pmatrix} -n_x \sin\theta \\ -n_y \sin\theta \\ -n_z \sin\theta \end{pmatrix} \end{pmatrix} \\ &= \begin{pmatrix} 0 \\ \begin{pmatrix} n_x^2 \sin^2\theta + \cos^2\theta - n_y^2 \sin^2\theta - n_z^2 \sin^2\theta & 2n_xn_y \sin^2\theta + 2n_z \sin\theta\cos\theta & 2n_xn_z \sin^2\theta - 2n_y \sin\theta\cos\theta \\ 2n_xn_y \sin^2\theta - 2n_z \sin\theta\cos\theta & n_y^2 \sin^2\theta + \cos^2\theta - n_x^2 \sin^2\theta - n_z^2 \sin^2\theta & 2n_yn_z \sin^2\theta + 2n_x \sin\theta\cos\theta \\ 2n_xn_z \sin^2\theta + 2n_y \sin\theta\cos\theta & 2n_yn_z \sin^2\theta - 2n_x \sin\theta\cos\theta & n_z^2 \sin^2\theta + \cos^2\theta - n_x^2 \sin^2\theta - n_y^2 \sin^2\theta \end{pmatrix} \begin{pmatrix} x \\ y \\ z \end{pmatrix} \end{pmatrix} \end{aligned}$$

Seen, $p' = qpq^* = qpq^{-1}$ can be described as the result of a vector P rotating 2θ about a rotation axis q.

According to quaternion arithmetic definition of quaternion multiplication, we can merge two rotating:

$$Q = Q_1 * Q_2$$

Q is the result expressed on a rotating Q_1 first, then rotating Q_2, so quaternion multiplication can represent multiple rotation, do it again on the basis of the X, P can be expressed as: $s = rp'r^{-1}$

where $r = [\cos\theta', \sin\theta'(n_x, n_y, n_z)]$, P can be achieved by the rotation of S:

$$\mathrm{s} = \mathrm{r}(\mathrm{qpq}^{-1})\mathrm{r}^{-1} = (\mathrm{rq})\mathrm{p}(\mathrm{rq})^{-1}$$

So that we achieve a point P with a rotation quaternion Q, and then with a quaternion rotation R again.

$$\begin{bmatrix} 1-2(y^2+z^2) & 2(xy-wz) & 2(xz+wy) & 0 \\ 2(xy+wz) & 1-2(x^2+z^2) & 2(yz-wx) & 0 \\ 2(xz-wy) & 2(yz+wx) & 1-2(x^2+y^2) & 0 \\ 0 & 0 & 0 & 1 \end{bmatrix} = \begin{bmatrix} & R_Q & & 0 \\ & & & 0 \\ & & & 0 \\ 0 & 0 & 0 & 1 \end{bmatrix}$$

X being orthogonal matrix can be proved, and in X, only three parameters are independent, for the unit quaternion meeting:

$$N(q) = w^2 + x^2 + y^2 + z^2 = 1$$

Therefore, the calculation only needs three parameters.

2.1 *The Proof of Quaternion Rotation*

After discussing the basics of quaternion, we move on to deduce how to calculate rotation matrix through quaternion. Suppose a vector $p = [0, (x, y, z)]$ facing a pair of unit quaternion $q = [\cos\theta, \sin\theta(n_x, n_y, n_z)]$, having rotating change to be $p' = [0, (x', y', z')]$, satisfy the following relation:

$$p' = qpq^* = qpq^{-1}$$

According unit quaternion conjugate reciprocal relationship, so that:

$$q^{-1} = q^* = [\cos\theta, -\sin\theta(n_x, n_y, n_z)]$$

Using quaternion multiplication to expand $p' = qpq^{-1}$:

Note that this sequence of multiple rotary connections is "right to left", which is different from our habits of using multiple rotations from left to right [8, 9].

Quaternion rotation can be easily achieved around a fixed axis and rotate around any axis operation. Taking these derivation processes can be summed up with the quaternion rotation to achieve the general process, as shown in Fig. 2:

2.2 *Arc of the Ball Algorithm and Implementation*

The base of algorithm

Any three-dimensional object can be considered as the direction around a rotation axis determined in accordance with a reference direction. In addition, one rotation is determined by the composition of any number of times of rotation. According to different rotation parameters, the combination rule is not always the same. Although the combination formula varied, the essence is the application of a simple spherical

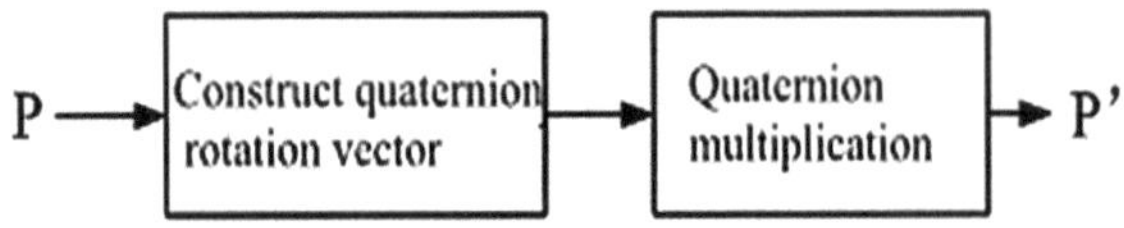

Fig. 2 Procedure of rotation based on quaternion

geometry. Suppose that R represents about a rotation axis a, as shown in Figure, according to knowledge of spherical geometry, a rotation axis a perpendicular to the plane with a circle centered on the length of the directed arcs, the so called predetermined angle is a positive indication is made about the counterclockwise rotation of a shaft [10, 11].

During the rotation of Fig. 3 can be seen as a combination of two arcs process, that by strings. When the first R1 and the second R2 endpoint of the arc coincides with the beginning of the arc, the two arcs of the spherical triangle on both sides, arc R1; R2 is the third side of the spherical triangle. As shown in Fig. 4:

Fig. 3 Arc interpretation

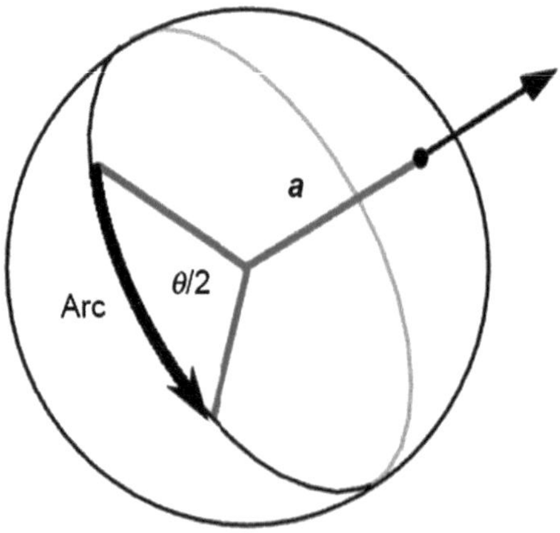

Fig. 4 Arc combination

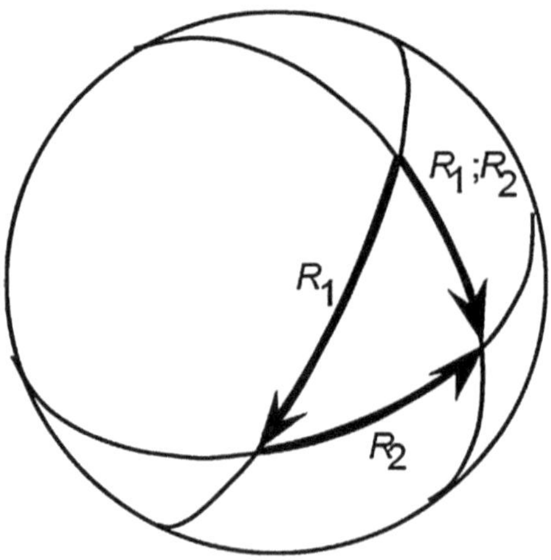

Arc R1; R2 represents the starting point of the arc from the first to the second endpoint of the arc, this combination represents a time of rotation from R1 to R2. Because the rotation is not commutative, and when combined sequence in turn, it will form a different arc. Now prove the above rotation using quaternion. Assuming a three-dimensional rotation vector P to do rotation R1 first, and then do the R2 rotation; wherein R1, R2, respectively, you can use two unit quaternion q1 and q2, said the effect of rotation in order to achieve the above, you need to do multiplication q1 and q2, [12, 13]

$$
\begin{aligned}
& q_2 * (q_1 * P * q_1^{-1}) * q_2^{-1} \\
& = (q_2 * q_1) * P * (q_1^{-1} * q_2^{-1}) \\
& = (q_2 * q_1) * P * (q_2 * q_1)^{-1}
\end{aligned}
$$

The formula is constructed by a quaternion transformation matrix, to achieve rotation.

The above proof of that arc of the ball Algorithm is mathematical basis. The algorithm is implemented using the quaternion rotation operation, the use of two degrees of freedom to express three degrees of freedom mouse spatial rotation. Dragging the mouse will imagine a sphere in a projection drawing arcs and arc of this movement is mapped to empty the actual scene, which changes the viewpoint and perspective in order to achieve an interactive roaming action. This innovative idea into a combination of mathematics and with interactive learning two fields, you can easily rotate freely express or three-dimensional object bound rotation, its flexibility and stability is fully guaranteed.

2.3 Algorithm Pseudo-code Description

Ball rotation is a way through a graphical display of the rotating side input side, is a technology that allows users to use the mouse to adjust the orientation of the three-dimensional interaction techniques. In this section will mainly discuss the algorithm implementation process and gives the pseudo-code description [14].

Assuming a hemispherical unit window center is located, the center of the hemisphere of the unit as the origin, (X, Y) of a point in space, the two-dimensional coordinates into three-dimensional coordinates of the basic pseudo-code is implemented as:

```
IF x2 + y2 < 1
THEN z     Sqrt[1 – x2 – y2];
ELSE z  0;
```

Fig. 5 Mouse input

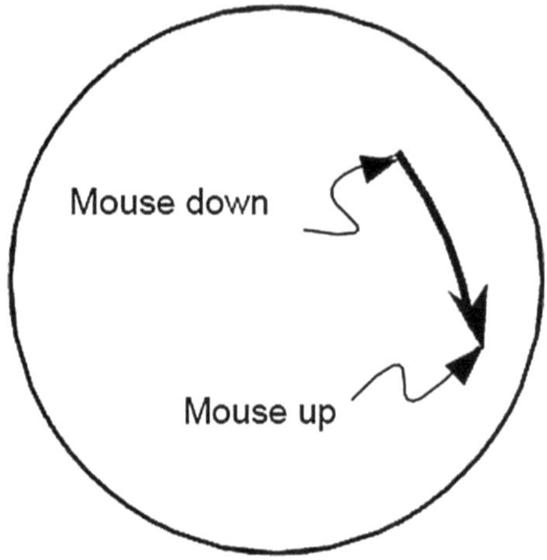

Generated when the mouse is pressed starting point, the end point is generated when released, these two algorithms can be converted through the relevant two points on the hemisphere, as shown in Fig. 5:

When you drag the mouse, used to calculate the coordinates of the current arc, then, this is called fast-moving objects or avatars start with the rotation. The direction of rotation and speed of the ball is the model previously described with the half-length arcs. The best half is the long arc spherical shortest path between two points, depending on the length of two of the plane and containing the center of the sphere. Discuss a special case, if the end of the starting point falls exactly, but in opposite direction, then the plane of the arc will not be defined. At this point, the ball arc algorithm can also be a good deal. Two opposite points mean that, the end of the starting point of a circle around a rotation made points obtained, the direction will not change, that is, the starting point of the "packaging point", the arc can be in, and the points corresponding to the "packaging point" generated. As shown in Fig. 6:

Fig. 6 Wrapping

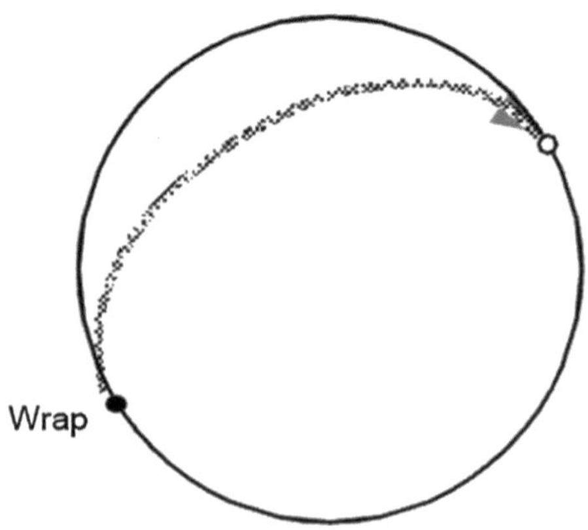

Fig. 7 Conversion of space

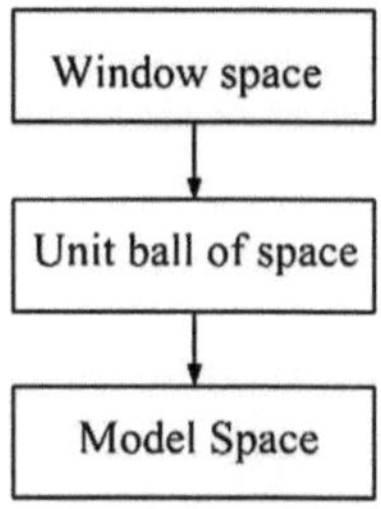

Figure 6 shows that over a half circular arc, the direction is changed smoothly naturally. Therefore, the algorithm has no restrictions on the number of rotations. Generated when the mouse button into the starting point of the end point is generated when release the mouse, these two can be converted through the conversion algorithm to two points on the hemisphere, assuming that the center points and the two points is θ in radians. According quaternion rotation calculation process can be introduced when these two fall hemisphere (Z > 0), then the rotation axis and these two points and the center perpendicular to the plane constituted, the rotation angle of 2θ radians. When these two points fall outside the dome (Z = 0), then the rotation axis is the Z axis, for the same rotation angle 2θ in radians. Thus under the arc of the ball algorithm theory, we can know the basic realization of the rotation process shown in Fig. 7:

The first is from the initial window space is converted to the unit ball space, and then by the unit sphere space into model space (transformation matrix). Then we discuss this process in accordance with each of the specific implementation process [15, 16].

2.3.1 Window Space to the Unit Sphere Space Conversion

Point of the window unit of the computer can be viewed as a sphere point of the hemispherical surface of the coordinate conversion. Assuming pt for three-dimensional coordinates of the sphere, screen for coordinates of a point on the screen, center for coordinates of said center of the screen, radius as the radius, r is the distance from the origin to the screen, the window space to the space conversion unit sphere pseudo-code implementation:

```
pt.x  (screen.x − center.x)/radius;
pt.y  (screen.y − center.y)/radius;
r  pt.x * pt.x + pt.y * pt.y;
IF r > 1.0
THEN s  1.0/Sqrt[r];
pt.x  s * pt.x;
pt.y  s * pt.y;
pt.z  0.0;
ELSE pt.z  Sqrt[1.0 − r];
```

When rotating around a fixed shaft, prior to use, the point of the sphere will be projected to a plane perpendicular thereto, if necessary, it will flip to the front hemisphere and then normalized. If the point on the axis, to a point on the plane. Packaging as previously described using flip effect, is described as pseudo-code:

```
dot  V3_Dot[pt,  axis];
proj  V3_Sub[pt, V3_Scale[axis,  dot]];
norm  V3_Mag[proj];
IF  norm > 0
THEN s  1.0/norm;
IF proj.z<0 THEN s   − s;
pt  V3_Scale[proj, s];
ELSE  IF  axis.z = 1.0;
THEN  pt  [1.0, 0.0, 0.0];
ELSE  pt  V3_Unit[[−axis.y, axis.x, 0]];
```

2.3.2 Unit Sphere Space to Model Space Conversion

Calculating the rotation axis and rotation angle, thereby obtain the transformation matrix. According to the geometrical meaning of quaternions, we can know that this conversion can be achieved through a quaternion. In the quaternions, the rotation is done by the end of the point and the initial point of the conjugate product, that is

$$q = p_1 p_0^*$$

Defined by a quaternion known, q = [v, w] = [x, y, z, w] consists of two parts, one is a scalar w, which value is $\cos(\theta/2)$, another is the vector v, which is equal to a vector along the rotation axis $\sin(\theta/2)$ times, where θ is the angle of rotation.

Suppose q is quaternion, p0 and p1 represents the start and end points of rotation (can be obtained by converting a); V3_Cross represents the cross product of vectors, which records the direction of the axis of rotation; V3_Dot represents the dot product of vectors, which records the angle of rotation; TM indicates rotation matrix, QuatToMatrix convert the quaternion matrix. So how to get from a three-dimensional coordinates of the points in space corresponding quaternion, calculated pseudo-code achieved as follows:

```
[q.x, q.y, q.z]    V3_Cross[p0, p1]
q.w  V3_Dot[p0, p1];
```

When the mouse drag is a continuous action, object rotation direction is determined by the current quaternion when dragging begins with quaternary vector product obtained. And over quaternion multiplication to get the angle of rotation of the object, qnow QuatMul[q, qstart]; To do such an operation requires only 16 multiplications and 12 additions, while at the same overhead as the result is converted into a matrix, that is, TM = QuatToMatrix(q);

In summary, the user can drag the mouse to draw a like a rubber band-like from the starting point of the arc to the current display is updated to show the object or camera canvas rotation, easy to see graphical changing fast feedback, ease of use and simplicity. Consistency is another highlight of the algorithm's performance.

3 Results

In participating research and development of laser scanning three-dimensional imaging system control software, we use mouse drag to rotate 3-D real-time to display the scanned object and the degree of rotation needs, arc ball algorithm can

Fig. 8 Initial status of system

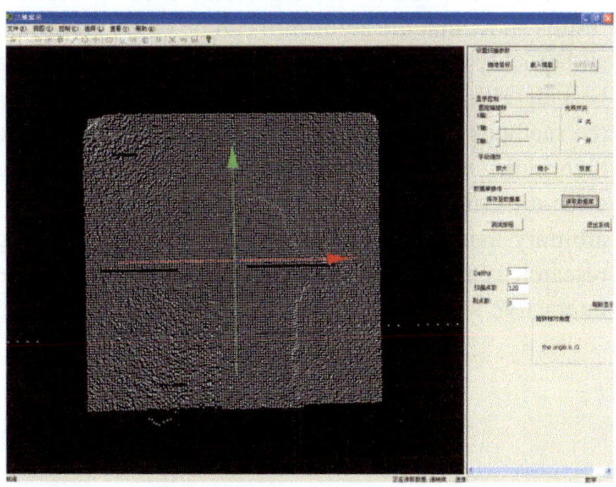

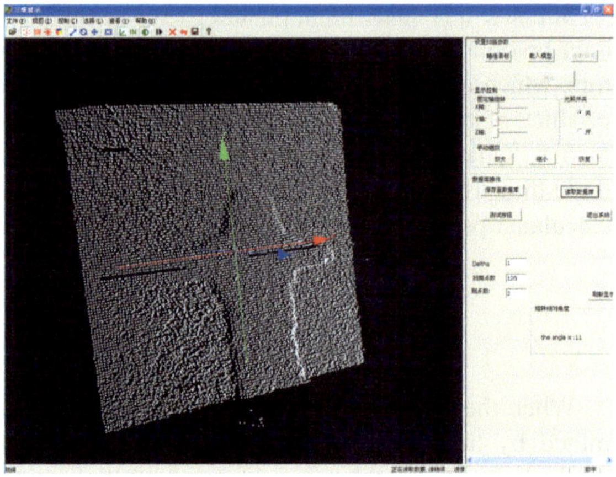

Fig. 9 Rotation of mouse drag

facilitate the achievement of this requirement. The arc of the ball algorithm is applied to the software encoded results, shown in Figs. 8 and 9.

Press the left mouse button when you start dragging, Fig. 9 shows that this process can be fast and smooth implementation, the angle of rotation can be observed simultaneously as we moves our mouse.

In summary, the arc of the ball algorithm is fully capable of rotating three-dimensional objects for mouse drag function, with good universality.

4 Conclusion

In this paper, we discussed the three-dimensional imaging laser scanning rotating interactive control from technical aspect, the specific requirements for the research project is proposed based on quaternion rotation interaction algorithm, arc seeking algorithm, and gives the algorithm used in research experiments. Experimental results show that the algorithm can be easily arc of the ball, drag the mouse to rotate and quickly deal with real-time display rotation angle. Arc of the ball might be the best algorithm, and it is likely to be used in the broadest sense of the mouse to rotate the methods, but also more researches in mathematics are needed. Arc of the ball algorithm is now used only for controlling rotation, the traditional three-dimensional interaction techniques is used in addition to rotation about an arbitrary axis, but there might be a need of zooming at the same time, so further research is needed to make it applicable for a wider range of three-dimensional interactive scenes.

References

1. Schilling BW, Barr DN, Templeton GC et al (2002) Multiple-return laser radar for three-dimensional imaging through obscurations [J]. Appl Opt 41(15):2791–2799
2. Albota MA, Heinrichs RM, Kocher DG et al (2002) Three-dimensional imaging laser radar with a photon-counting avalanche photodiode array and microchip laser[J]. Appl Opt 41(36):7671–7678
3. Liu X, Li L (2009) Design of the optical system of flash lidar based on an APD array[J]. Infrared Laser Eng 38(5):893–896
4. Wang H, Liu Z (2012) 3-D laser imagine technology and applications [J]. Electron Des Eng 20(12):1674–6236
5. Mao M, Bu H, Jin H, Shu R (2010) Design and test method for scan system of 3D lidar [J]. Chin J Lasers 37(S1):110–114
6. Xiao W, Liang J, Chen W (2012) 3D object Rotation and Motion Interpolation Based on Quaternion [J]. J Syst Simul 24(3):1004–731X
7. Yoshikado S, Aruga T (1998) Feasibility study of synthetic aperture infrared laser radar techniques for imaging of static and moving objects[J]. Appl Opt 37(24):5631–5639
8. Raut JC, Chazette P, Fortain A (2008) New approach using lidar measurements to characterize spatiotemporal aerosol mass distribution in an underground railway station in Paris[J]. Atmos Environ 43(3):575–583
9. Qiu JH, Lu DR (1991) On lidar application for remote sensing of the atmosphere[J]. Advances in atmospheric sciences, 8(3):369–378
10. Kamerman GW (1996) Laser radar signals the civilian sector. Laser Focus World (4): 81–87
11. Marc LS, Roger KR, Don PH et al (1997) Application of coherent 10 micron imaging lidar[C]. In: 9th conference on coherent laser radar. Linkoping, Sweden, 1997
12. Anthes JP, Garcia P, Piercs JT et al (1993) Non-scanned ladar imaging and applications[C]. In: Proceedings of SPIE 1936, pp 11–22
13. Gao J, Sun J, Wei J, Wang Q (2011) Research of underwater target detection using a slit streak tube imaging lidar[C]. AISOMT. Harbin, pp 240–243
14. Grasso RJ. Ackleson JE, Stimson CG et al (2000) Imaging laser radar performance assessment against various naturally occurring and manmand terrestrial objects[C]. CLEO. Nice, France, pp 21–23
15. do Carmo JP (2011) Imaging LIDAR technology developments at the European Space Agency. In: International conference on applications of optics and photonics, 2011
16. Barry S, Ahmed AA, Wiiliam R et al (2000) Line imaging ladar using a laser diode transmitter and FM/CW radar principle for submunition applications[C]. In: Proceedings of SPIE 4035, pp 277–286

Effect of Linewidth Enhancement Factor in Semiconductor Laser on Fiber Dispersion Transmission System

Zhe-yu Liu, Tong-gang Zhao and Zhi-hong Bai

Abstract The distance of information transmission in fiber system can be affected by Linewidth Enhancement Factor. In this paper, the influence of chirp parameter in optical fiber communication system on fiber dispersion is analyzed as emphases. Taking Gaussian pulse as signal source, the impact of frequency chirp together with the Linewidth Enhancement Factor on the performance of optical fiber communication system is studied by solving nonlinear Schrödinger equation. Finally, it is concluded that the efficiency of transmission is improved dramatically by reducing the Linewidth of laser.

Keywords Chromatic dispersion · Frequency chirp · Gaussian pulse · Linewidth · Lasers

1 Introduction

In recent years, an IP-based Internet business showed explosive growth, for promoting the development of the main forces of global information. Optical fiber communication system as the carrier of information transfer is forward to high speed and network-oriented development. Large-capacity communication systems, intelligent network management technology constantly introduce new source performance requirements. Semiconductor lasers have an advantage of small size, long life

This work is supported by Research Innovation Fund for College Students of Beijing University of Posts and Telecommunications.

Z. Liu (✉) · T. Zhao
School of Electronic Engineering, Beijing University of Posts and Telecommunication, Beijing 100876, China
e-mail: liuzheyubupt@gmail.com

Z. Bai
Technical Service Division, Petroleum Pipeline Bureau, Hebei, China

E. Qi (ed.), *Proceedings of the 6th International Asia Conference on Industrial Engineering and Management Innovation*,
DOI 10.2991/978-94-6239-145-1_16

and can be directly modulated, low cost. Semiconductor lasers have been widely used in the field of optical fiber communication [1, 2]. Because laser performance will directly affect the transmission quality of optical fiber communication systems, this makes the semiconductor laser of some of the parameters is particularly important, such as Linewidth Enhancement Factor. To analysis clearly produce physical origin of these parameters and context, not only in the research field of semiconductor lasers, and the optical communication system network optimization, optical devices and transmission medium of choice are of great theoretical significance [3].

2 Semiconductor Laser Frequency Chirp Causes of Physical Mechanism

Semiconductor lasers can be used as a way of directly modulated. Direct modulation causes changes in the number of carriers. Changes in the number of carriers lead to changes in gain. Thus the refractive index of laser cavity alters. As a result, in a modulation period, a periodic movement occurred in the mode frequency of laser, which is the frequency chirp. Frequency chirp is intensity-modulated optical communication system capacity constraints. However, if the frequency chirp is used directly to frequency modulation, coherent optical communication system can be applied. Its in-depth analysis has important reference value. Linewidth Enhancement Factor β_c established the intensity modulation and frequency chirp contact [4, 5].

Definition of Linewidth Enhancement Factor β_c:

$$\beta_c = \frac{d[\mathrm{Re}(\chi_p)]/dN}{d[\mathrm{Im}(\chi_p)]/dN} = 2\beta\frac{dn/dN}{dg/dN} \approx 2\beta\frac{\Delta n}{\Delta g} \tag{1}$$

Analytical methods with classical theory [3] can be obtained:

$$\beta_c = -\frac{\Delta n'}{\Delta n''} \tag{2}$$

Annotate: χ_p is electric susceptibility, N is the number of carriers, β is propagation constant, $\Delta n'$ is the change in the real part of the refractive index, $\Delta n''$ is the imaginary part of the refractive index variation.

The change of $\Delta n'$ causes phase offset. The change of $\Delta n''$ causes the change of optical gain. The argument of semiconductor laser Linewidth Enhancement Factor β_c is usually negative. It is not only related to the number of carriers, respectively, also the temperature, the semiconductor laser structure, such as β_c of the quantum well lasers is often—1.5–2, however, β_c of transverse junction stripe semiconductor laser structure is—6–8 [6, 7]. The coupling between the phase fluctuations and intensity fluctuations will cause linewidth broadening what affecting the performance of the communication system.

In a semiconductor laser resonator, wave field transmission satisfies the wave equation without considering nonlinear effects:

$$\nabla^2 E - \frac{1}{c^2} \cdot \frac{\partial^2}{\partial t^2}(\varepsilon E) = 0 \tag{3}$$

Annotate: ε is permittivity.

Make: $E = A\exp[i(wt - \beta z)] + c.c.$ Plug into (3). According to the theory of Henry:

$$\dot{A} = (\frac{G - \gamma}{2})(1 + \beta_c i)A \tag{4}$$

$$\dot{A}^* = (\frac{G - \gamma}{2})(1 - \beta_c i)A^* \tag{5}$$

Annotate: v_g is group velocity, G is gain factor, γ is loss coefficient, $\dot{A}$ is the derivative of the time of A, A^* is the complex conjugate of A.

To solving (4) and (5) two differential equations can obtain the following set of solutions:

$$I = AA^* = \exp[(G - \gamma)t] \tag{6}$$

$$\dot{I} = (G - \gamma)I \tag{7}$$

$$\phi = \frac{1}{2i}\ln(A/A^*) = \frac{1}{2}\beta_c(G - \gamma)t \tag{8}$$

$$\dot{\phi} = \frac{1}{2}\beta_c(G - \gamma) \tag{9}$$

Annotate: I is light intensity, ϕ is phase, $\dot{I}$ is the time rate of change of the light intensity, $\dot{\phi}$ is the time rate of change of the phase.

Take (6) to (9) into the expression of E, we get:

$$E = \exp[(\frac{G - \gamma}{2})(1 + \beta_c i)t]\exp[i(wt - \beta z)] + c.c. \tag{10}$$

To (7) and (9) using the small-signal analysis methods, we get [8]:

$$\Delta f = (1 + \beta_c^2) \cdot \Delta f_0 = \frac{v_g^2(1 + \beta_c^2)h\nu g_{th} n_{sp} \alpha_m}{8\pi P_0} \tag{11}$$

Annotate: v_g is group velocity, h is Planck's constant, g_{th} is threshold gain, α_m is terminal mirror losses, v is frequency, n_{sp} is the statistical factor carrier distribution which is not completely reversed.

Δf_0 comes from the spontaneous emission noise. When the CW operation, there is always a certain laser linewidth. This is an expression of phase fluctuation occurred within the laser phase. $\beta_c^2 \Delta f_0$ is created δN. Intensity modulation brings the number of injected carriers' changes.

The active region refractive index and the gain coefficient of the semiconductor laser all will be changed. It results phase change. The role of phase fluctuation cause Spectral width broadening 3 times. The reducing of β_c makes Light source spectral width reduces. This increase in system capacity is important. Simultaneously the reducing of β_c is also conducive to the application of coherent communication systems. In practice, to adopt effective measures can not only make the semiconductor laser pressure narrow linewidth, also achieve single-mode oscillation. For example: now widely used DFB (Distributed Feedback) semiconductor laser, FBG-ECL (Fiber Bragg Grating External Cavity Semiconductor Laser) are a good example [9, 10].

3 The Influence of Frequency Chirp in the Optical Transmission System

The wave equation describing the optical fiber transmission can find from Maxwell's equations. Leading into the group velocity v_g of reference we got Nonlinear Schrödinger equation [11] which described the light pulses in monomode optical fiber transmission.

$$i\frac{\partial A}{\partial z} = -\frac{i}{2}\alpha A + \frac{1}{2}\beta_2\frac{\partial^2 A}{\partial T^2} - \gamma|A|^2 A \tag{12}$$

Annotate: A is the slowly varying pulse envelope amplitude, T is timescale in the reference system changing with group rate with pulses, α is loss coefficient of optical fiber, β_2 is the group velocity dispersion parameter, γ is fiber nonlinearity parameter.

3.1 The Consistency of Linewidth Enhancement Factor and Gaussian Pulse Chirp Parameter

To the case of investigate optical pulses in optical fiber transmission case we as long as put the (10) into (12). In the actual optical communication system, actual light output pulse waveform approximates Gauss shape. If we let

$$A = \sqrt{I_0}U(z,T) \tag{13}$$

Introducing normalized amplitude: $U(0,T) = \exp(-\frac{1+iC}{2}\frac{T^2}{T_0^2})$. T_0 is pulse half width. In the study single-mode fiber transmission problems, C is usually called Chirp parameter.

Expressions obtained from the $U(0,T)$:

$$I = \exp(-\frac{T^2}{T_0^2}) \tag{14}$$

$$\phi = -\frac{iC}{2}\frac{T^2}{T_0^2} \tag{15}$$

$$\dot{\phi} = -\frac{\beta_c}{T_0^2}T = \frac{C}{2I}\dot{I} \tag{16}$$

In the semiconductor laser resonant cavity, wave equation obtained by the expression E (9) and (10) excluding the non-linear effects to give:

$$\dot{\phi} = \frac{1}{2}\beta_c(G - \gamma) = \frac{\beta_c}{2I}\dot{I} \tag{17}$$

To comparison (16) and (17) two formula, in the Laser field equations Linewidth Enhancement Factor is consistent with the Chirp parameter of using Gaussian pulses when discussing Fiber Transport. This is $C = \beta_c$. To study on the chirp parameter is not only meaningful for laser performance analysis, but also very important in the dispersion limited optical fiber transmission systems for pulse broadening discussion.

3.2 *The Regular Pattern of Gaussian Pulse in the Single-Mode Fiber Transmission Systems*

Substituting (13) into the formula nonlinear Schrodinger equation of optical fiber transmission without considering the impact of loss we can get:

$$i\frac{\partial U}{\partial z} = \frac{\beta_2}{2}\frac{\partial^2 U}{\partial T^2} \tag{18}$$

Fourier method (18) is solved. We can examine the frequency-domain and time-domain optical pulse propagation.

To Fourier transform of the $U(0,T) = \exp(-\frac{1+iC}{2}\frac{T^2}{T_0^2})$, we can get:

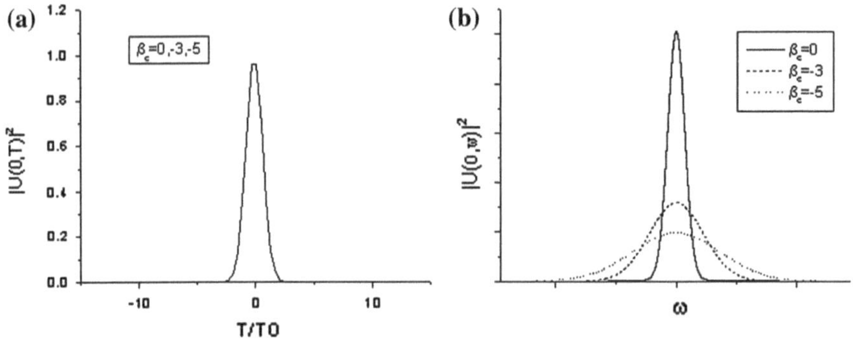

Fig. 1 **a** z = 0, the waveform of Gaussian pulse in time field. **b** z = 0, the waveform of Gaussian pulse in frequency field

$$\tilde{U}(0,w) = (\frac{2\pi T_0^2}{1+i\beta_c})^{\frac{1}{2}} \exp[-\frac{w^2 T_0^2}{2(1+i\beta_c)}] \tag{19}$$

According to (19), frequency domain waveform can be drawn at z = 0, $\beta_c = 0$, −3, −5 in the Gaussian pulse, shown in Fig. 1a.

Analysis: seen from Fig. 1b, the initial conditions, time domain pulse width and line width is independent of growth factors; Frequency domain, the pulse width in the same condition, Linewidth Enhancement Factor is greater, spectral width is wider, Initial chirp makes spectral width increased $(1+\beta_c^2)^{1/2}$.

Get through the transmission distance z, Gaussian pulse analytical solution is:

$$U(z,T) = \frac{T_0}{[T_0^2 - i\beta_2 z(1+i\beta_c)]^{1/2}} \exp\{-\frac{(1+i\beta_c)T^2}{2[T_0^2 - i\beta_2 z(1+i\beta_c)]}\} \tag{20}$$

So that $L_D = T_0^2/|\beta_2|$. Using (20), we can draw different β_2 respectively, the case of β_c, evolution of pulse waveform graph we can obtain relationship between the amount of pulse broadening and β_c, β_2. Figures 2a, b and 3a, b as follows:

Analysis:

(1) To see Fig. 2, when $\beta_c = 0$, no initial chirped Gaussian pulse along the optical fiber transmission, pulse width increases with z. Pulse broadening is dependent on the propagation distance z and the dispersion ratio of the length, And the group velocity dispersion parameter is unrelated with symbols. When $z > L_D$, broadening amount increases as the transmission distance increases rapidly. In the optical fiber communication system, pulse width broadening during the transmission gap to the bit, the interfering with each other can produce errors. This will greatly limit the transmission capacity. From this point, with dispersion compensation to suppress dispersion on pulse broadening is most importance [12, 13].

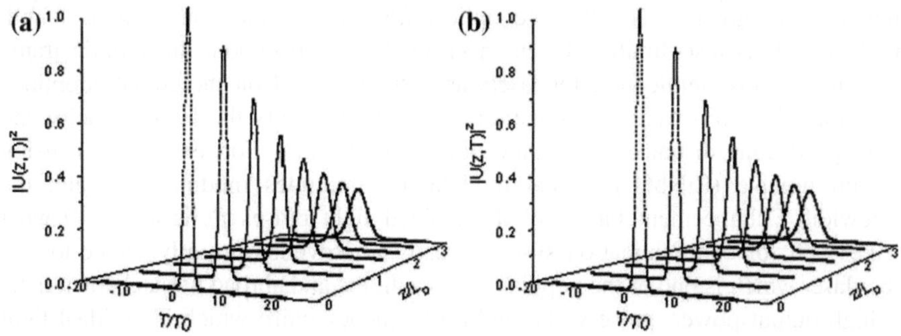

Fig. 2 **a** $\beta_c = 0$, $\beta_2 > 0$, variation of Gaussian pulse with propagated distance. **b** $\beta_c = 0$, $\beta_2 < 0$, variation of Gaussian pulse with propagated distance

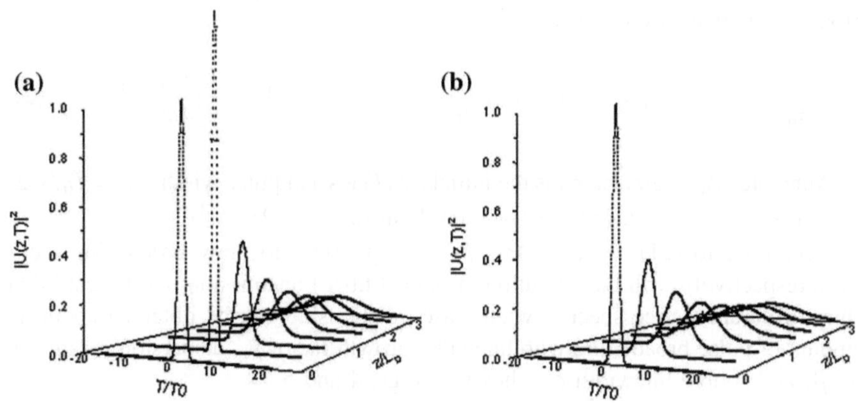

Fig. 3 **a** $\beta_c = -3$, $\beta_2 > 0$, variation of Gaussian pulse with propagated distance. **b** $\beta_c = -3$, $\beta_2 < 0$, variation of Gaussian pulse with propagated distance

(2) From Fig. 3, $\beta_c \neq 0$, pulse broadening depends on the group velocity dispersion parameter β_2 and initial chirp parameter together. The amount of pulse broadening

Is not only based the amount of β_2, β_c, but also related with their relative symbols. The Linewidth Enhancement Factor of directly modulated semiconductor laser β_c is usually negative. Seen from the figure, Gaussian pulse is different in normal dispersion fiber and anomalous dispersion fiber. Shown in Fig. 2b, $\beta_2\beta_c < 0$, and with increasing distance chirp of net pulse is decreased what firstly cause the pulse narrowing, however, as the transmission distance increases chirp reduced by dispersion is exceed initial chirp what play a major role and pulse begin broadening. Figure 3a, $\beta_2\beta_c > 0$, light pulse is monotonically increasing with the increase of transmission distance. Comparing Figs. 2a and 3a, 2b and 3b, it has a larger propagation distance z broadening amount that chirped pulse with initial than no initial chirp. Obviously, since the initial chirp directly modulated bring dispersion is limited more greatly. The bit rate optical fiber communication systems length product B*L is also limited. In engineering, for the backbone fiber optic transmission systems, semiconductor lasers are usually based on the actual adoptions consumer chirp or external modulation technique to satisfy the large capacity and high-speed long-distance communication requirements. For example, if using lithium niobate ($LiNbO_3$) optical modulator amplitudes modulated signal, the Linewidth Enhancement Factor of the emitted light pulses β_c is almost 0 what makes B*L of communication system greatly improved. Currently, mode-locked fiber laser which is one of the popular issues at home and abroad has a characteristic of high output power, pulse width and no frequency chirp which is an ideal light source of optical fiber communication system [14].

In the optical communication system, general compliance guidelines: $4\sigma < T_B$ [15]. σ is rms pulse width. When taking into account the group velocity dispersion parameter β_2, linewidth growth factor β_c and light source spectral width of the pulse broadening effect, we can get:

$$\frac{\sigma}{\sigma_0} = [(1+\frac{\beta_c\beta_2 z}{2\sigma_0^2})^2 + (1+V_w^2)(\frac{\beta_2 z}{2\sigma_0^2})^2 + (1+\beta_c^2+V_w^2)^2\frac{1}{2}(\frac{\beta_3 z}{4\sigma_0^3})^2]^{1/2} \tag{21}$$

Annotate: $V_w = 2\sigma_w\sigma_0$. σ_0 is the initial rms Gaussian pulse width ($\sigma_0 = T_0/\sqrt{2}$). σ_w is rms spectral width of the Gaussian light ($\sigma_w = \sigma_\lambda 2\pi c/\lambda^2$).

According to (21), if $\Delta\lambda = 4\sigma_\lambda$, $\beta_3 = 0.1\,\mathrm{ps}^3/\mathrm{km}$, to draw broadening factor σ/σ_0 respectively with the evolution of optical fiber transmission in different initial chirp, different source spectral width case, relationship can be obtained about the amount of pulse broadening with linewidth growth factor β_c, dispersion parameters β_2, β_3 and source linewidth σ_λ, shown in Figs. 4 and 5.

Analysis:
Seen from Fig. 4, (1) if $V_w \approx 0$, $\beta_2\beta_c < 0$, pulse being transmitted, first compressed and then broadening, and if $z > L_D$, optical pulse with initial chirp broadening in the transmission case is much larger than no chirped. (2) The amount of light pulse

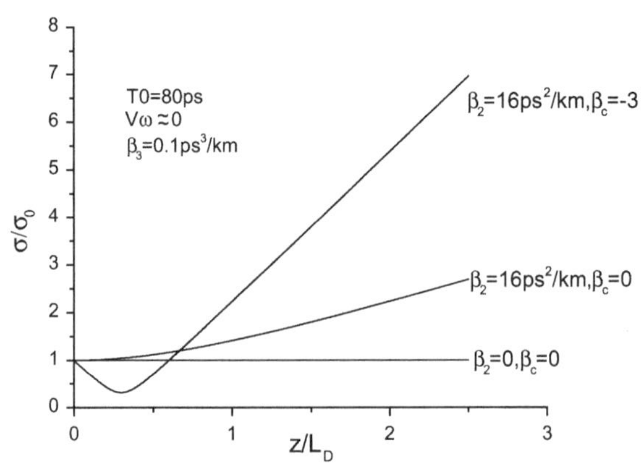

Fig. 4 Variation of σ/σ_0 with propagated distance in $V_w \approx 0$ and different values of β_2, β_c

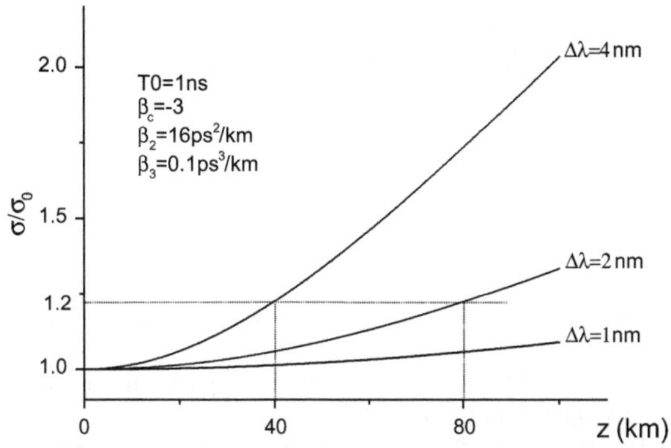

Fig. 5 Variation of σ/σ_0 with propagated distance in $V_w \neq 0$ and different values of $\Delta\lambda$

broadening of zero dispersion shifted fiber ($\beta_2 = 0$) is far less than the work in the general case of single-mode fiber.

Be seen from Fig. 5, it can make much larger pulse broadening to select wide linewidth semiconductor laser than to adopt narrow linewidth light source. Because in addition to the initial frequency chirp, group velocity dispersion common cause pulse broadening, the mode partition noise of the wide spectral width of semiconductor lasers will play a leading role at pulse broadening. If system requires that broadening the amount of the initial width of optical pulses in the transmission process is less than 20 %, as shown, for $T_0 = 1$ ns, light pulse selects light source when $\Delta\lambda = 2$ nm and satisfies system requirements that the following can be transmitted 80 km. If $\Delta\lambda = 4$ nm, you can only transfer 40 km. Wide spectral width of the semiconductor laser mode noise greatly limits the bit rate—length product B * L.

In short, to achieve ultra-high-speed, high-capacity, long-distance communication purposes, be sure to use linewidth growth factor β_c which is small, line width, single longitudinal mode semiconductor laser as a light source. In addition, choose low dispersion coefficient optical fiber, can also reduce the pulse broadening effect due to the dispersion effect in the optical fiber transmission.

4 Summary

Semiconductor laser has a good small-signal frequency response. It is small, flexible, and simple and now widely used in optical fiber communication systems. However, direct modulation of a semiconductor laser has a large frequency chirps that increases the line width of the light source. And in the dispersion-limited optical transmission system the role of the initial chirp makes the pulse broadening after long distance transmission larger than no initial chirp pulse that thereby reduces the information transmission capacity of the system. To select narrow linewidth and small linewidth growth factor semiconductor laser and improving external modulation technique has far-reaching significance for the construction of long-distance, high-capacity, intelligent, low-cost optical fiber communication system.

References

1. Lee TP (1991) Recent advances in long-wavelength semiconductor lasers for optical fiber communication. Proc IEEE 79(3):253–276
2. Chan S-C, Diaz R, Liu J-M (2008) Novel photonic applications of nonlinear semiconductor laser dynamics. Opt Quantum Electron 40:2–4
3. Clarke Bruce R (1991) The effect of reflections on the system performance of intensity modulated laser Diodes. IEEE J Lightwave Technol 9(6):741–749
4. Valhalla K, Yariv A (1983) Semi classical theory of noise in semiconductor lasers-part 1. IEEE J Quantum Electron 19(6):1096–1101
5. Marcuse D (1981) Pulse distortion in single-mode fibers.3: chirped pulses. Appl Opt 20 (20):3573–3579
6. Eriksson S (2002) Dependence of the experimental stability diagram of an optically injected semiconductor laser. Opt Commun 210(3):343–353
7. Osinski M, Buus J (1987) Linewidth Enhancement Factor in semiconductor lasers-an overview. IEEE J Quantum Electron 23(1): 9–29
8. Henry CH (1982) Theory of the linewidth of semiconductor lasers. IEEE J Quantum Electron 18(2):259–264
9. Hong J, Huang WP, Makino T (1995) Static and dynamic simulation for ridge-waveguide MQW DFB lasers. IEEE J Quantum Electron 31(1):49–59
10. Schunk N, Petermann K (1988) Numerical analysis of the feedback regimes for a single-mode semiconductor laser with external feedback. IEEE J Quantum Electron 24(7):1242–1247
11. Agrawal GP (1995) Nonlinear fiber optics, 2nd edn. Academic, Boston, pp 45–46
12. Qin Y, Zhao Y, Jin-song Zhang et al (1999) Experimental investigation on uniform fiber Bragg grating chirplization and its application in dispersion compensation. Chin J Lasers 26(10):935–939 (in Chinese)
13. Jian-xin L, Yong-qing H, Xiao-min R (2000) A tunable dispersion compensation scheme based on the transmission dispersion of the fiber gratings. J Optoelectron Laser 11(2):140–142 (in Chinese)
14. Zhi-jian H, Jun-qiang S, De-xiu H (1998) Theoretical analysis of fast and slow saturable absorber mode locking erbium-doped fiber lases. Acta Phys Sin 47(1):9–18 (in Chinese)
15. Agrawal GP (1997) Fiber-Optic communication systems, 3rd edn. Wiley, New York, pp 49–51

Initial Phase Estimation of Single-Tone Signal Based on Fast Filter Bank

Jin-guang Hao, Wen-jiang Pei and Kai Wang

Abstract In this paper, an initial phase estimation of single-tone signal is proposed based on the Fast Filter Bank (FFB). With the help of characteristics of FFB such as narrower transition width and higher stopband attenuation and low complexity, the proposed initial phase estimation based on FFB is more accurate than that of based on FFT, and the phase estimation can be also estimated without correction regardless of whether there is deviation of frequency. Simulation explains the better performance of estimating initial phase based on FFB and the method can be extended to estimate other parameters such as magnitude of the single-tone signal in a similar way.

Keywords Cascaded structure · FFB · Low complexity · Phase estimation · Stopband attenuation

1 Introduction

Initial phase estimation is an important issue in many applications such as communication, mechanical diagnosis and radar. In these fields, many algorithms are applicable based on FFT, because of low complexity and easily implementation. In [1], a parameter estimation method is proposed with interpolated FFT. Qi and Jia [2] introduces a phase estimation method based on phase of DFT. Huai et al. [3] gives a method to estimation phase using all phase FFT spectrum analysis. Matsuoka and Ulrych [4] provides the parameters estimation based on bispectrum. All the mentioned methods based on FFT improve the performance of estimation.

However, these methods based on FFT have the inherent drawbacks of FFT such as poor passband response and high side-lobes in its stopband with approximately −13 dB irrespective of the FFT length. In order to improve the performance of FFT,

J. Hao (✉) · W. Pei · K. Wang
School of Information Science and Engineering, Southeast University, Nanjing, China
e-mail: Jinguang.hao@gmail.com

E. Qi (ed.), *Proceedings of the 6th International Asia Conference on Industrial Engineering and Management Innovation*,
DOI 10.2991/978-94-6239-145-1_17

and new class of filter bank is first introduced in [5], which is based on the Frequency-Response Masking approach in [6, 7], and regarded as a generalized form of sliding FFT filter bank with higher order prototype filter. As a result, fast filter bank has a better frequency response than sliding FFT.

Furthermore, FFB has the similar complexity with FFT, many literatures [8–10] researching on the complexity of FFB shows that compared to the complexity of FFT, complexity of FFB increases slightly per sample per channel and compared to other single-rate filter banks, it has very lower complexity because of his sparse coefficients. Hence, FFB can be used in many applications. For example, in [11], FFB is used for adaptive Fourier analysis. In [12, 13], musical notes identification is using FFB, and other applications such as audio signal processing and spectrum sensing etc. can be found in literatures [14, 15].

Because of the advantages of FFB in comparison to FFT, in this paper, an initial phase estimation method based on the fast filter bank is proposed and the structure of the paper is arranged as follows. Section 2 reviews fast filter bank briefly. Section 3 introduces initial phase estimation algorithm based on FFB. Section 4 exemplifies the proposed method and gives the performance comparison with that of FFT. Section 5 concludes the paper.

2 Review of FFB

Fast filter bank is first introduced in [5] and it can be regarded as a generalized form sliding FFT filter bank with cascaded structure with higher order prototype subfilter, which results in better performance than FFT filter bank with similar low complexity. In this section, a brief overview is given to illustrate the FFB characteristics. A N-channel FFB is shown in Fig. 1 with the number of stage L. The number of channel N must be the power of 2, and relationship between N and number of stage L is or $N = 2^L$. i.e. in Fig. 1 $L = 3, N = 8$.

Fig. 1 Structure of FFB

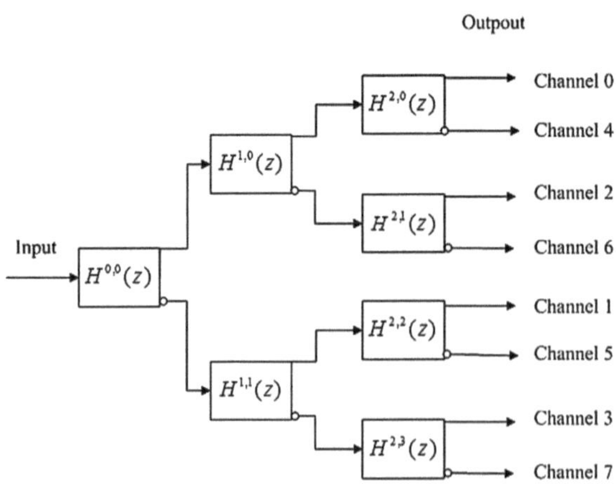

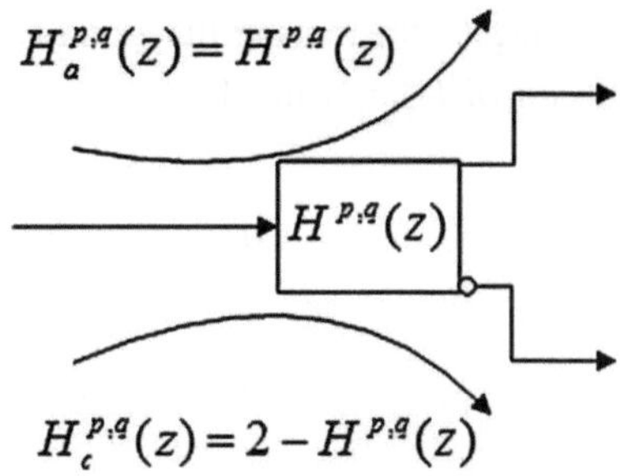

Fig. 2 Structure of subfilter

The structure of FFB is cascaded and the transfer function of each sub-filter is denoted by $H^{p,q}(z)$ in z domain, where $p(p = 0, 1, \ldots, L-1)$ stands for the stage index, and q for the index of sub-filter at each stage, where q ranges from 0 to $2^p - 1$ as non-negative integer at p^{th} stage.

All the sub-filters have the similar structure with one input and two outputs shown as in Fig. 2.

The upper path output is termed as primary output or low frequency output component denoted by $H_a^{p,q}(z)$ with the transfer function $H_a^{p,q}(z) = H^{p,q}(z)$, and the lower path output as complementary output or high frequency output component denoted by $H_c^{p,q}(z)$ with $H_c^{p,q}(z) = 2 - H^{p,q}(z)$, respectively.

The primary output and the complementary output satisfy the following condition

$$H_a^{p,q}(z) + H_c^{p,q}(z) = 2 \tag{1}$$

The transfer function $H^{p,q}(z)$ of each subfilter at p^{th} stage can be obtained by replacing delay factor z of the prototype low-pass filter for stage p. In specific, the prototype lowpass filter is given for stage p. In order to decrease the complexity of the implementation; we'd like to design the filter with odd length symmetric halfband design method so as to decrease the number of the nonzero coefficients, which are about one quarter of the total filter length. And the prototype filter can be expressed by

$$H_a^p(z) = \sum_{m=-N_p}^{N_p} h_a^p(m) z^{-m} \tag{2}$$

where $H_a^p(z)$ denotes the transfer function of prototype for stage p,integer $N_p > 0$, $h_a^p(n)$ is the real-valued symmetric coefficient of the impulse response in time domain and satisfies

$$\begin{aligned} h_a^p(m) &= h_a^p(-m), \, for \, m \in [-N_p, -1] \cup [1, N_p] \\ h_a^p(0) &= 1 \end{aligned} \tag{3}$$

At stage p, the transfer function of qth sub-filter can be formulated by replacing the z factor of $H_a^p(z)$ with $W_N^{\tilde{q}} z^{2^{L-1-p}}$.

$$\begin{aligned} H^{p,q}(z) &= \sum_{m=-N_p}^{N_p} h_a^p(m) W_N^{-m\cdot\tilde{q}} z^{-m\cdot 2^{L-1-p}} \\ &= 1 + \sum_{\substack{m=-N_p \\ m\neq 0}}^{N_p} h_a^p(m) W_N^{-m\cdot\tilde{q}} z^{-m\cdot 2^{L-1-p}} \end{aligned} \tag{4}$$

where $\tilde{q}$ is the bit-versed version of q in $L-1$ bits and W_N represents $e^{-\frac{2\pi}{N}}$.

With the expression in (4), the primary output and complementary output can be rewritten in unified expression with

$$\begin{aligned} H_{ac}^{p,q}(z) &= \sum_{m=-N_p}^{N_p} h_a^p(m) W_N^{-m\cdot\tilde{q}} z^{-m\cdot 2^{L-1-p}} \\ &= 1 + (-1)^{B_q} \sum_{\substack{m=-N_p \\ m\neq 0}}^{N_p} h_a^p(m) W_N^{-m\cdot\tilde{q}} z^{-m\cdot 2^{L-1-p}} \end{aligned} \tag{5}$$

where B_q is the sign of the output, i.e. when output is primary output B_q equals to 0, otherwise B_q equals to 1.

The unified expression in (5) shows that transfer function of a subfilter is also symmetrical with odd length and its center occurs at time zero.

3 The Proposed Algorithm

In this section, we propose a method of initial phase estimation based on the FFB. Considering a single-tone signal passing through FFB discussed in Sect. 2, the single-tone signal is expressible in complex-valued form

$$s(n) = e^{j(2\pi f_0 n + \phi_0)} \tag{6}$$

where f_0 is the normalized frequency and ϕ_0 is the initial phase of the signal. The proposed estimation method aims to estimate the initial phase ϕ_0 more accurate.

For simplicity, we define the symbol $H^{p,k_p}(z)$ for sub-filters, which constitute a specific channel k, here p denotes the stage index, for $0 \leq p \leq L-1$, and k_p is the sub-filter index of channel-k at stage p. For example the transfer functions of the sub-filters of channel-5 shown in Fig. 1 are $H^{0,0}(z)$ complementary output, $H^{1,1}(z)$ primary output and $H^{2,2}(z)$ complementary output and the indices $k_p (p = 0, 1, 2)$ of channel-5 at each stage is 0, 1 and 2. With the unified expression of primary output and complementary output of sub-filter in (5), the channel-k transfer function can be determined by the product of transfer function of each sub-filter

$$H_k(z) = \prod_{p=0}^{L-1} H_{ac}^{p,k_p}(z) \tag{7}$$

Because of the symmetrical with odd coefficients of the each component of $H_k(z)$ in (7), $H_k(z)$ is also symmetrical with odd length coefficients and expanded with the form

$$H_k(z) = 1 + \sum_{\substack{p=-N_g \\ p\neq 0}}^{N_g} g(p,k)z^{-p} \tag{8}$$

The complex-valued coefficients in (8), which can be obtained by the convolution by the coefficients of the sub-filter of channel-k, exhibit symmetric property, i.e.

$$g(p,k) = g^*(-p,k) \tag{9}$$

where "*" is used to denote the complex conjugate function.

Then the output $y_k(n)$ of channel-k with z transformation $Y_k(z)$ can be expressed by the multiplication of transfer function of channel-k $H_k(z)$ and the input signal $E(z)$ in z domain

$$Y_k(z) = H_k(z) \cdot E(z) \tag{10}$$

The $Y_k(z)$ can be converted from time domain to z domain using inverse z-transformation, and (10) can be expressible with equivalent expression in time domain

$$y_k(n) = e(n) + \sum_{\substack{p=-N_g \\ p\neq 0}}^{N_g} g(p,k)e(n-p) \tag{11}$$

The single-tone input signal

$$s(n) = e^{j(2\pi f_0 n+\phi_0)}(n \in [-N_g, N_g])$$

are chosen as the observed sequence and substituted in (11), one obtains

$$\begin{aligned} y_k(n) =& e^{j\phi_0} + \sum_{\substack{p=-N_g \\ p\neq 0}}^{N_g} g(p,k)e^{j(-2\pi f_0 p+\phi_0)} \\ =& e^{j\phi_0}[1 + \sum_{p=1}^{N_g} 2 \cdot real[g(p,k)]\cos(2\pi f_0 p)] \end{aligned} \tag{12}$$

where *real*[] represents the real component of the complex-valued $g(p,k)$. Equation (12) infers that the phase of the output $y_k(n)$ of channel-k is dependent on the initial phase ϕ_0 of single-tone input signal irrespective to the frequency f_0 and the deviation of the frequency, i.e. when the component

$$[1+\sum_{p=1}^{N_g} 2\cdot real[g(p,k)]\cos(2\pi f_0 p)]$$

is positive real-valued, the phase of output $y_k(n)$ is ϕ_0 and $\phi_0 \pm \pi$ otherwise.

Hence, the initial phase ϕ_0 of input signal can be obtained with the amplitude of channel-k_0, whose amplitude is the maximum of that of all channels and determined by

$$k_0 = \arg_k\{\max(|y_k(n)|\}, (0 \le k \le L-1) \quad (13)$$

4 Simulation and Analysis

In this section, we choose 128-channel FFB to illustrate initial phase estimation described in Sect. 3. There are seven stages for 128-channel FFB and the coefficients $h_a^p(n)(p \in [0,6])$ of the prototype sub-filters are listed in Table 1.

The magnitude of frequency response of channel 20 of channel-128 FFB is shown in Fig. 3 with narrower transition width and higher stopband attenuation.

The observed single-tone single is given with normalized frequency 10.2/128, which deviates 0.2/128 from integer time's frequency resolution, and with the initial phase ϕ_0 equals to 0.4.

The magnitude frequency response and phase frequency response of signal passing through FFB are shown in Fig. 4 and that of 128-point FFT are shown in Fig. 5, respectively.

Table 1 Coefficients of prototype subfilters

n	$h_a^0(n)$	$h_a^1(n)$	$h_a^2(n)$	$h_a^3(n)$	$h_a^4(n)$	$h_a^5(n)$	$h_a^6(n)$
0	1	1	1	1	1	1	1
−1,1	0.6283	0.6216	0.5983	0.5691	0.5703	0.5005	0.5007
−3,3	−0.188	−0.171	−0.118	−0.070	−0.071		
−5,5	0.09	0.0677	0.0204				
−7,7	−0.046	−0.024					
−9,9	0.0219	0.0062					
−11,11	−0.009						
−13,13	0.0028						

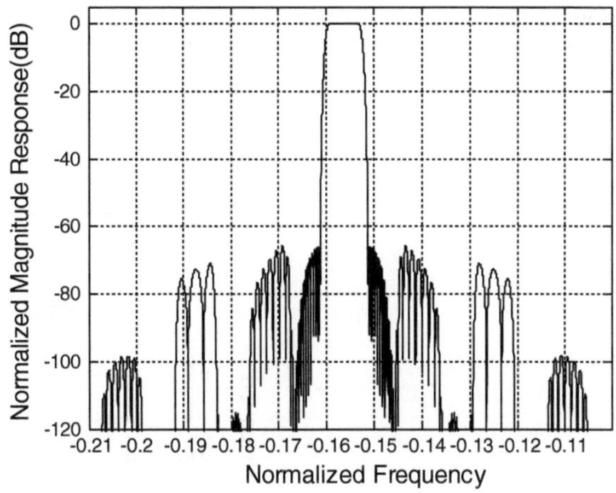

Fig. 3 The magnitude frequency response of channel 20 of 128-channel FFB

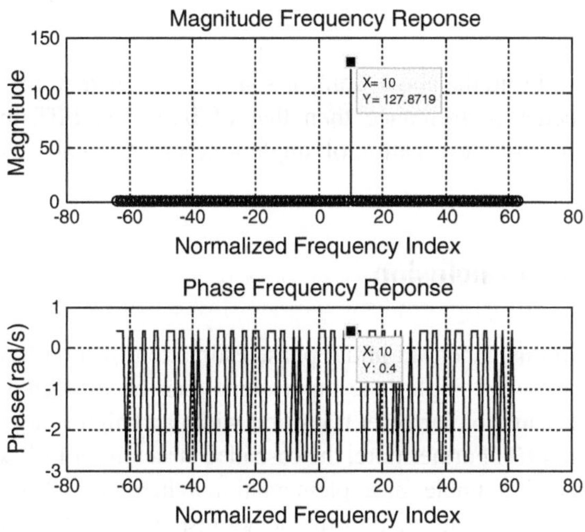

Fig. 4 The frequency response passing through FFB

From Fig. 4, it is can be observed that although there is deviation of frequency of integer times frequency resolution, there is no adjacent channel leakage, i.e. for single-tone signal there is only one spectrum line in frequency domain in theory, and from the magnitude frequency response of Figs. 4 and 5, the magnitude frequency response keeps one spectrum line through FFB while the spectrum line is split into many spectrum lines in adjacent channels passing through FFT.

With FFB-based proposed initial phase estimation shown in Fig. 4, the phase corresponding to the maximum magnitude of frequency response is 0.4, which is exactly same as the initial phase of the single-tone input signal, while with FFT shown in Fig. 5, the initial phase estimation at the point of the maximum magnitude of frequency response is 0.3951, which is caused by deviation of frequency of the input single.

Fig. 5 The frequency response passing through FFT

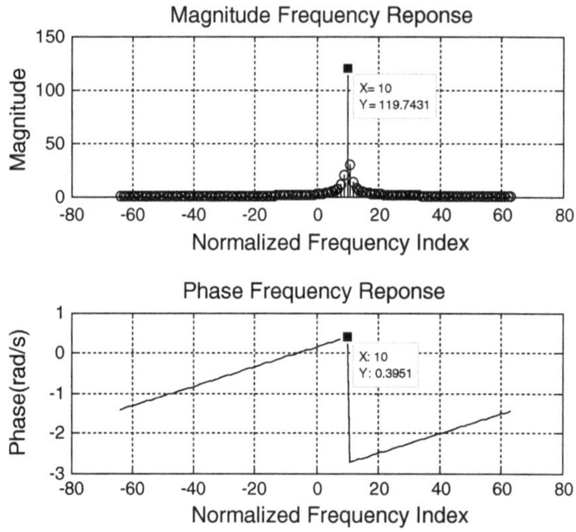

From the above comparison, we can see that the method based on FFB achieves better performance than that of based on FFT in initial phase estimation and spectrum estimation of single-tone input signal.

5 Conclusion

In this paper, an initial phase estimation method based on FFB is proposed. Initial phase of single-tone input signal can be estimated without correction with FFB, because of narrower transition width and higher stopband attenuation in comparison to FFT. Furthermore, the cascaded structure of FFB keeps low complexity like FFT, and is suitable for implementation in hardware. With this structure, other parameters such as magnitude of frequency response can be also estimated in a similar way.

Acknowledgments Support from numerous colleagues, friends and experts from Signal Processing Laboratory of Southeast University is gratefully acknowledged. Furthermore, we also thank to anonymous reviewers for their useful comments and language editing which have greatly improved the manuscript.

References

1. Jain VK, Collins WL, Davis DC (1979) High-accuracy analog measurements via interpolated FFT. IEEE Trans Instrum Meas 28(2):113–122
2. Qi G, Jia X (2001) High-accuracy frequency and phase estimation of single-tone based on phase of DFT. Acta Electronica Sinica 29(9):1164–1167

3. Huai L, Sobelman GE, Zhou X (2011) Wideband spectrum sensing using the all-phase FFT. In: 2011 IEEE 9th international conference on ASIC (ASICON), IEEE
4. Matsuoka T, Ulrych TJ (1984) Phase estimation using the bispectrum. Proc IEEE 72 (10):1403–1411
5. Lim YC, Farhang-Boroujeny B (1992) Fast filter bank (FFB). IEEE Trans Circ Syst II Analog Digit Sig Process 39(5):316–318
6. Lim Y (1986) Frequency-response masking approach for the synthesis of sharp linear phase digital filters. IEEE Trans Circ Syst 33(4):357–364
7. Yang R, Liu B, Lim YC (1988) A new structure of sharp transition FIR filters using frequency-response masking. IEEE Trans Circ Syst 35(8):955–966
8. Lim YC, Farhang-Boroujeny B (1994) Analysis and optimum design of the FFB. In: 1994 IEEE international symposium on circuits and systems, 1994, ISCAS'94, vol 2, IEEE
9. Wei LJ, Ching LY (2003) Designing the fast filter bank with a minimum complexity criterion. In: 2003 Proceedings of the seventh international symposium on signal processing and its applications, vol 2, IEEE
10. Lee JW, Lim YC (2008) Efficient fast filter bank with a reduced delay. In: IEEE Asia Pacific conference on circuits and systems, 2008, APCCAS 2008, IEEE
11. Várkonyi-Kóczy AR et al (1998) A fast filter-bank for adaptive Fourier analysis. IEEE Trans Instrum Meas 47(5):1124–1128
12. Foo SW, Lee EWT (2002) Transcription of polyphonic signals using fast filter bank. In: IEEE international symposium on circuits and systems, 2002. ISCAS 2002, vol 3, IEEE
13. Foo SW, Lee EWT (2003) Application of fast filter bank for transcription of polyphonic signals. J Circ Syst Comput 12(05): 655–674
14. dos Santos CN et al (2004) A modified constant-Q transform for audio signals. In: IEEE international conference on acoustics, speech, and signal processing, 2004, Proceedings (ICASSP'04), vol 2, IEEE
15. Smitha KG, Prasad Vinod A (2012) A multi-resolution fast filter bank for spectrum sensing in military radio receivers. IEEE Trans Very Large Scale Integr VLSI Syst 20(7): 1323–1327

A Sparsity Adaptive Compressive Sampling Matching Pursuit Algorithm

Xiang-pu Liu, Feng Yang, Xiang Yi and Li-li Guo

Abstract In the area of compressed sensing (CS), the compressive sampling matching pursuit (CoSaMP) algorithm offers a theoretical reconstruction guarantee in noise environment by exploiting a backtracking framework. But it relies on the sparsity of signal. With a stage by stage reconstruction structure, the sparsity adaptive matching pursuit (SAMP) algorithm can reconstruct signals when the sparsity is unknown. By taking both advantages of CoSaMP and SAMP, we propose a greedy algorithm for reconstruction of sparse signals, called the *sparsity adaptive compressive sampling matching pursuit* (SACoSaMP). The proposed algorithm can reconstruct signals without prior information of sparsity, and it is robust to noise. In this paper, we give a residual upper bound of the proposed SACoSaMP algorithm, and demonstrate the performance of it through simulation experiments.

Keywords Compressed sensing · Compressive sampling matching pursuit · Greedy algorithm · Sparsity adaptive matching pursuit · Sparsity adaptive

---

X. Liu (✉) · L. Guo
College of Information and Communication Engineering,
Harbin Engineering University, Harbin, China
e-mail: liuxiangpu@hrbeu.edu.cn

L. Guo
e-mail: guolili@hrbeu.edu.cn

X. Liu · F. Yang · X. Yi
Science and Technology on Electronic Information Control Laboratory,
Chengdu, China
e-mail: eyangf@sina.com

X. Yi
e-mail: easyyx@163.com

E. Qi (ed.), *Proceedings of the 6th International Asia Conference on Industrial Engineering and Management Innovation*,
DOI 10.2991/978-94-6239-145-1_18

1 Introduction

Compressed sensing (CS) is an effective method for acquisition of sparse signals at rate significantly lower than Nyquist rate [1–3]. If a length-N signal $\mathbf{x}$ can be represented by $K << N$ significant coefficients over an N-dimensional basis, $\mathbf{x}$ is called a K-sparse signal. The information of $\mathbf{x}$ can be contained in measurement vector $\mathbf{y}$ through global linear measurement

$$\mathbf{y} = \mathbf{\Phi x} + \mathbf{e} \tag{1}$$

where $\mathbf{\Phi}$ is a $M \times N$ sampling matrix with $M << N$, and $\mathbf{e}$ is an additive bounded noise vector with $\|e\|_2^2 \leq \sigma^2$ [4]. $\mathbf{x}$ can be reconstructed from $\mathbf{y}$ by complex reconstruction algorithms if $\mathbf{\Phi}$ satisfies restricted isometry property (RIP) with parameter (K, δ) [5, 6]

$$(1 - \delta)\|\mathbf{x}\|_2^2 \leq \|\mathbf{\Phi}_I \mathbf{x}\|_2^2 \leq (1 + \delta)\|\mathbf{x}\|_2^2 \tag{2}$$

where $\|\cdot\|_2$ denotes the ℓ_2 vector norm, $0 < \delta < 1$, and I is an index set with $|I| < K < M$. Thus, to a great extent, the reconstruction algorithms determine the performance of compressed sensing.

Recently, greedy algorithms have received significant attention to reconstruct sparse signals due to their low complexity and simple geometric interpretation [7]. In the field of greedy algorithm, matching pursuit (MP) [8], orthogonal matching pursuit (OMP) [9], stage-wise orthogonal matching pursuit (StOMP) [10] and regularized orthogonal matching pursuit (ROMP) [11] have been proposed. But these algorithms cannot effectively reconstruct signals in noise environment. Compressive sampling matching pursuit (CoSaMP) [12] is a robust reconstruction algorithm for signals with noise, but it assumes that signal sparsity level K is known [13]. In many practical applications, this parameter is often unavailable. Subspace pursuit (SP) [7] is similar to CoSaMP, also depends on sparsity. Sparsity adaptive matching pursuit (SAMP) [14] does not rely on sparsity any more, but it is more complex than other greedy algorithms when the sparity level K is large.

The framework of proposed algorithm in this paper is similar to that of CoSaMP, but the real sparsity of signal is replaced by a variable to eliminate the dependence on sparsity. A sparsity estimation method is introduced to SACoSaMP to obtain a suitable initial value of the variable. This method can avoid unnecessary iterations when the variable is much smaller than real sparsity. In order to reconstruct signal exactly, we exploit the variation characteristics of residual to adjust the value of the variable. Simulations on the 0-1 random sparse signal show that the proposed algorithm sacrifices anti-noise performance and computational complexity for sparsity adaptation. But compared with the SAMP algorithm, which is widely used in the case that the sparsity is unknown, the proposed algorithm achieves better reconstruction performance.

The remainder of this paper is organized as follows. Section 2 describes details of the proposed SACoSaMP algorithm and analyzes it. Simulation results and discussion are shown in Sect. 3, and in Sect. 4 we conclude this paper.

2 The Proposed Algorithm

2.1 Sparsity Estimation

When the variable that used to reconstruct signal is much smaller than real sparsity, algorithm iteration makes little sense to obtain supports of signal. A sparsity estimation method is introduced to solve this problem [15].

Defining $\mathbf{g}_i$ to be the i-th element in vector $\mathbf{g}$, and

$$\mathbf{g} = \mathbf{\Phi}^{\mathrm{T}}\mathbf{y} \tag{3}$$

then vector $\mathbf{\Gamma}_0$ contains indices of the $K_0 (1 \le K_0 \le N)$ largest elements in $|\mathbf{g}_i|$. If sampling matrix $\mathbf{\Phi}$ satisfies RIP with parameter (K, δ_K), then the following inequality will be established in the condition of $K_0 < K$.

$$\left\| \mathbf{\Phi}_{\mathbf{\Gamma}_0}^{\mathrm{T}} \mathbf{y} \right\|_2 < \frac{1 - \delta_K}{\sqrt{1 + \delta_K}} \|\mathbf{y}\|_2 \tag{4}$$

The sparsity estimation method can be obtained from this proposition. Parameter K_0 increases by step one until inequality (4) is not satisfied, and then we can get an estimated value K_s' of sparsity level K to start iteration.

Simulation in Sect. 4 shows that the value of K_s' is inversely proportional to that of δ_K, and K_s' is often smaller than K.

2.2 Algorithm Overview

In CoSaMP, each iteration increases the reconstruction SNR by about 3 dB when the sampling matrix $\mathbf{\Phi}$ has a restricted isometry constant $\delta_{4K} \le 0.1$ [12]. In the following paragraph, we have proven that residual of each iteration can also be decreased even using a smaller K' replace K in CoSaMP. When K' is close to K, the residual will converge to a minimum value. Based on the property mentioned above, we propose the SACoSaMP algorithm, and the pseudo code of it is presented in Table 1.

Table 1 Pseudo code of SACoSaMP

Input: measurement vector $\mathbf{y}$, sampling matrix $\mathbf{\Phi}$.
Output: signal approximation $\mathbf{a}$.
Initialization: $\mathbf{a}^0 = \mathbf{0}$, $\mathbf{r} = \mathbf{y}$, $\mathbf{g} = \mathbf{\Phi}^{\mathrm{T}}\mathbf{y}$, $K_0 = 0$, $k = 0$.
Sparsity estimation
While $\left\Vert \mathbf{\Phi}_{\mathbf{\Gamma}_0}^{\mathrm{T}}\mathbf{y} \right\Vert_2 < \frac{1-\delta_K}{\sqrt{1+\delta_K}} \left\Vert \mathbf{y} \right\Vert_2$ **do**
$K_0 = K_0 + 1$
$\mathbf{\Gamma}_0$ ={ indices of the largest K_0 value of $\left\vert \mathbf{g}_i \right\vert$ }
End
$K'_s = K_0$ (K'_s is the initial value of K')
Signal approximation
While $\left\Vert \mathbf{r} \right\Vert_2 > \sigma$ **do**
$\mathbf{v} = \mathbf{\Phi}^{\mathrm{T}}\mathbf{r}$ {Signal proxy}
$\mathbf{\Omega} = \mathrm{supp}(\mathbf{v}, 2K')$ {Primary supports selection}
$\mathbf{T} = \mathbf{\Omega} \cup \mathrm{supp}(\mathbf{a}^{k-1}, K')$ {Merge supports}
$\mathbf{b}_{\mathbf{T}} = \mathbf{\Phi}_{\mathbf{T}}^{\dagger}\mathbf{y}$, $\mathbf{b}_{\mathbf{T}^{\mathrm{C}}} = 0$ {Signal estimation}
$\mathbf{a}^k = \mathbf{b}_{K'}$ {Prune}
$\mathbf{r}_{pre} = \mathbf{r}$, $\mathbf{r} = \mathbf{y} - \mathbf{\Phi}\mathbf{a}^k$ {Update residual}
if $\left\Vert \mathbf{r} \right\Vert_2 \geq \left\Vert \mathbf{r}_{pre} \right\Vert_2$, $K' = K' + K'_{step}$ {Adjust sparsity}
else $K' = K'$
End

where supp($\mathbf{v}$, $2K'$) obtains the supports of the largest $2K'$ elements in vector $\mathbf{v}$. $\mathbf{b}_{\mathbf{T}}$ denotes a subvector of $\mathbf{b}$ with elements indexed by the set $\mathbf{T}$, and it is similar to matrix $\mathbf{\Phi}_{\mathbf{T}}$. $\mathbf{b}_{K'}$ keeps the largest K' elements in vector $\mathbf{b}$. $\mathbf{\Phi}^{\dagger} = (\mathbf{\Phi}^{\mathrm{T}}\mathbf{\Phi})^{-1}\mathbf{\Phi}^{\mathrm{T}}$ is the pseudo-inverse matrix of $\mathbf{\Phi}$, and $\mathbf{T}^{\mathrm{C}}$ is the complementary set of $\mathbf{T}$.

We use a variable K' to replace the sparsity of signal in SACoSaMP. The initial value of this variable can be obtained by the sparsity estimation method. For each value of K', the SACoSaMP retains the backtracking framework of CoSaMP. In signal proxy building phase, residual of the previous iteration is projected to sampling matrix $\mathbf{\Phi}$. Then the new $2K'$ matching supports are added into set $\mathbf{T}$. All the supports in set $\mathbf{T}$ will be checked again in signal estimation and prune phase to find out the most K' matching supports of signal. The backtracking process mentioned above can improve the anti-noise performance of SACoSaMP. When residual increases between consecutive iterations, it implies that the most K' matching supports have been found in the previous iteration, and wrong supports are chosen in this iteration. In this case, K' increases by step K'_{step} to update the number of supports need to be chosen in the next stage, which is similar to SAMP. Finally, the original signal can be reconstructed when the variable is closed to the real sparsity.

2.3 Algorithm Performance

In order to prove the algorithm convergence of SACoSaMP, refer to the proof of CoSaMP in [12], it is necessary to introduce the following two conclusions.

(1) In CoSaMP algorithm, if $\mathbf{x}$ is a K-sparse signal, then the approximation $\mathbf{a}^k$ is also K-sparse, and residual meets the following inequality in the $(k+1)$-th iteration.

$$\left\|\mathbf{x}-\mathbf{a}^{k+1}\right\|_2 \leq 0.5\left\|\mathbf{x}-\mathbf{a}^k\right\|_2 + 7.5\|\mathbf{e}\|_2 \tag{5}$$

(2) If $||\mathbf{x}||_0 \leq r$, and sampling matrix $\mathbf{\Phi}$ satisfies the RIP, then $\mathbf{x}$ satisfies the following inequality.

$$\|\mathbf{\Phi x}\|_2 \leq \sqrt{1+\delta_r}\left[\|\mathbf{x}\|_2 + \frac{1}{\sqrt{r}}\|\mathbf{x}\|_1\right] \tag{6}$$

Without loss of generality, we assume SACoSaMP algorithm has completed k iterations, and the $(k+1)$-th iteration will be analyzed in the following.

K-sparse signal $\mathbf{x}$ can be divided into two sparse signals $\mathbf{x}_{K'}$ and $\mathbf{x}-\mathbf{x}_{K'}$, then the measurement vector $\mathbf{y}$ can be represented as

$$\mathbf{y} = \mathbf{\Phi x}_{K'} + \tilde{\mathbf{e}} \tag{7}$$

where

$$\tilde{\mathbf{e}} = \mathbf{\Phi}(\mathbf{x}-\mathbf{x}_{K'}) + \mathbf{e} \tag{8}$$

and $\mathbf{x}_{K'}$ is a K'-sparse signal, $\tilde{\mathbf{e}}$ is a new noise measurement vector.

SACoSaMP is the same as CoSaMP in each iteration, thus approximation $\mathbf{a}^k$ of $\mathbf{x}_{K'}$ is K'-sparsity from the conclusion (1), and

$$\left\|\mathbf{x}_{K'}-\mathbf{a}^{k+1}\right\|_2 \leq 0.5\left\|\mathbf{x}_{K'}-\mathbf{a}^k\right\|_2 + 7.5\|\tilde{\mathbf{e}}\|_2 \tag{9}$$

Since $||\mathbf{x}-\mathbf{x}_{K'}||_0 < 4K$, and sampling matrix $\mathbf{\Phi}$ satisfies RIP with parameter $(4K, 0.1)$. From the conclusion (2), we can get the following inequality through taking $\delta_{4K} = 0.1$.

$$\|\mathbf{\Phi}(\mathbf{x}-\mathbf{x}_{K'})\|_2 \leq 1.05\left[\|\mathbf{x}-\mathbf{x}_{K'}\|_2 + \frac{1}{\sqrt{4K}}\|\mathbf{x}-\mathbf{x}_{K'}\|_1\right] \tag{10}$$

By the definition of $\tilde{\mathbf{e}}$ and the triangle nature of inequality, we can obtain

$$\|\tilde{\mathbf{e}}\|_2 = \|\mathbf{\Phi}(\mathbf{x} - \mathbf{x}_{K'}) + \mathbf{e}\|_2 \le \|\mathbf{\Phi}(\mathbf{x} - \mathbf{x}_{K'})\|_2 + \|\mathbf{e}\|_2 \tag{11}$$

In the $(k+1)$-th iteration, residual satisfies

$$\begin{aligned} \left\|\mathbf{x} - \mathbf{a}^{k+1}\right\|_2 &= \left\|\mathbf{x} - \mathbf{x}_{K'} + \mathbf{x}_{K'} - \mathbf{a}^{k+1}\right\|_2 \\ &\le \|\mathbf{x} - \mathbf{x}_{K'}\|_2 + \left\|\mathbf{x}_{K'} - \mathbf{a}^{k+1}\right\|_2 \end{aligned} \tag{12}$$

Using the Eqs. (9)–(12), we can get the residual upper bound of SACoSaMP as follows

$$\begin{aligned} \left\|\mathbf{x} - \mathbf{a}^{k+1}\right\|_2 \le\ & 0.5\left\|\mathbf{x} - \mathbf{a}^{k}\right\|_2 + 9.375\|\mathbf{x} - \mathbf{x}_{K'}\|_2 \\ & + \frac{7.875}{\sqrt{4K}}\|\mathbf{x} - \mathbf{x}_{K'}\|_1 + 7.5\|\mathbf{e}\|_2 \end{aligned} \tag{13}$$

where $\mathbf{x}_{K'}$ is the best K'-sparse approximation of $\mathbf{x}$, and residual will converge to a minimum value when K' approximates to real sparsity K.

Comparing Eqs. (5) with (13), we can find that residual of CoSaMP in each iteration is smaller than that of SACoSaMP, which means that the reconstruction and anti-noise performance of CoSaMP will be better than that of SACoSaMP. The reason is that CoSaMP chooses exact K supports to approximate signal in each iteration, while SACoSaMP approximates real sparsity by increasing variable K'.

In the SACoSaMP, the speed of reconstruction will improve when step K'_{step} is longer or δ_K is smaller. But the probability of wrong supports selection will increase in that case because the number of iterations has been reduced, this will influence the reconstruction result. So we can achieve a balance between reconstruction accuracy and complexity of algorithm through changing the value of K'_{step} or δ_K.

3 Simulation Results

The signals to be reconstructed are 0-1 random sparse signals with $N = 256$, and $M \times N$ sampling matrix $\mathbf{\Phi}$ is a Gaussian random matrix with $M = 128$. In the experiments of comparing algorithm performance, the steps of sparsity increase in SACoSaMP and SAMP are both set to one, and $\delta_K = 0.1$. Successful signal reconstruction must satisfy two conditions at the same time. One is that index set of reconstructed signal supports must be the same as that of original signal supports, and the other is that residual is smaller than 10^{-4} in noise free environment or σ in noise environment.

Fig. 1 Sparsity estimation

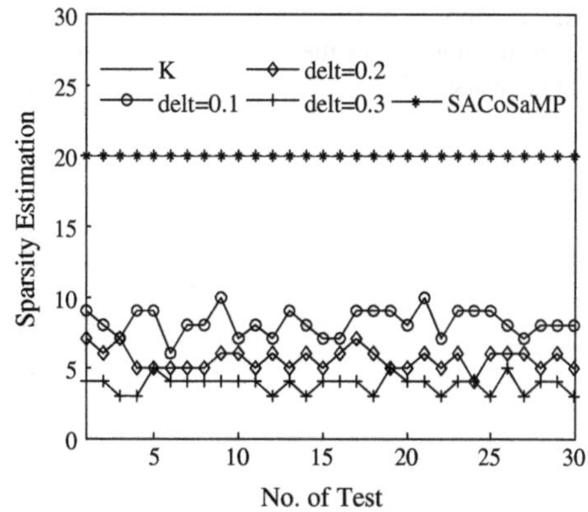

3.1 *Performance of Sparsity Estimation*

In order to test the performance of sparsity estimation, a sparse signal **x** with $K = 20$ is estimated by 30 times. Figure 1 presents the estimated value of sparsity. It can be seen that estimated value is generally smaller than real sparsity, and changes in a small range. This property makes $K_s' \leq K$ at the beginning of reconstruction, which promises the number of iterations and avoids unnecessary iterations.

When SACoSaMP has reconstructed signal successfully, the values of K' are also presented in Fig. 1. As can be seen, SACoSaMP can obtain real sparsity K exactly in this case.

3.2 *Comparison of Signal Reconstruction Without Noise*

Because the proposed algorithm is based on CoSaMP, and commonly used SAMP also does not depend on sparsity, Figs. 2 and 3 demonstrate the reconstructed results of CoSaMP, SAMP and SACoSaMP with different sparsity K and different measurement number M.

In Fig. 2, K increases from 10 to 70 by step 5, and 500 simulations are made to calculate the probabilities of exact reconstruction for different algorithms. It can be seen, probabilities of exact reconstruction of SACoSaMP and SAMP are almost the same, which verifies the feasibility of SACoSaMP. In each iteration of SACoSaMP, K' is smaller than real sparsity K, so the performance of CoSaMP exceeds that of SACoSaMP, but SACoSaMP does not need real sparsity K as input any more.

Figure 3 presents the reconstructed results of CoSaMP, SAMP and SACoSaMP with M increased from 40 to 120 by step 5. Signal sparsity K is fixed to 20, and reconstruction is repeated 500 times for each value of M. We can find that SACoSaMP needs fewer numbers of measurements than CoSaMP and SAMP under the same probability of exact reconstruction.

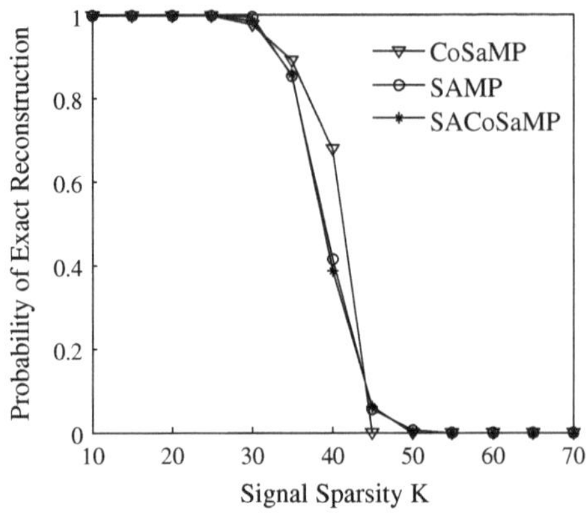

Fig. 2 Prob. of exact reconstruction versus the signal sparsity K

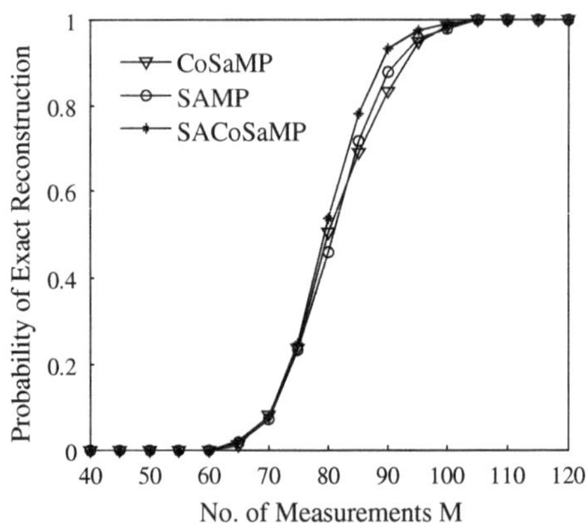

Fig. 3 Prob. of exact reconstruction versus the number of measurements M

3.3 Comparison of Signal Reconstruction with Noise

In greedy algorithms, CoSaMP is robust to noise. The framework of proposed algorithm in this paper is similar to that of CoSaMP, so it is necessary to analyze the performance of SACoSaMP in noise environment. These experiments add white Gaussian noise into sparse signal, and then compare the reconstructed probabilities and relative errors of different greedy algorithms. Figures 4 and 5 present the compared results.

In Fig. 4, sparsity K increases from 10 to 70 by step 5, and SNR is set to 18 dB. Probabilities of exact reconstruction are calculated through 500 reconstructions for different algorithms. Reconstructed residual upper bound of CoSaMP is smaller than that of SACoSaMP in noise environment. Thus, the noise robustness of CoSaMP is better than that of SACoSaMP. On the other hand, because the SACoSaMP retains the framework of CoSaMP, its ability to resist noise is better than that of SAMP.

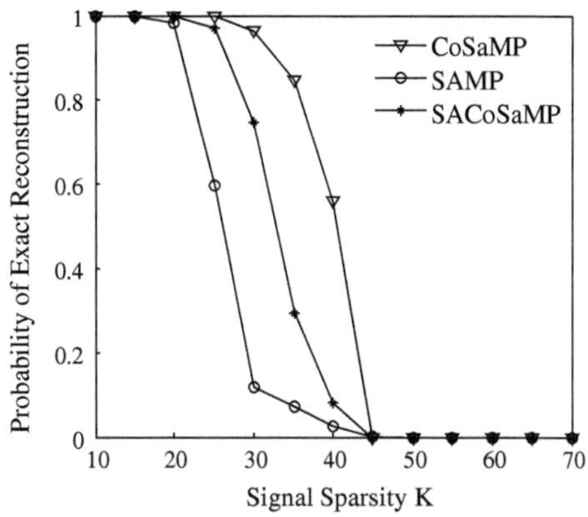

Fig. 4 Prob. of exact reconstruction versus the signal sparsity K (noise)

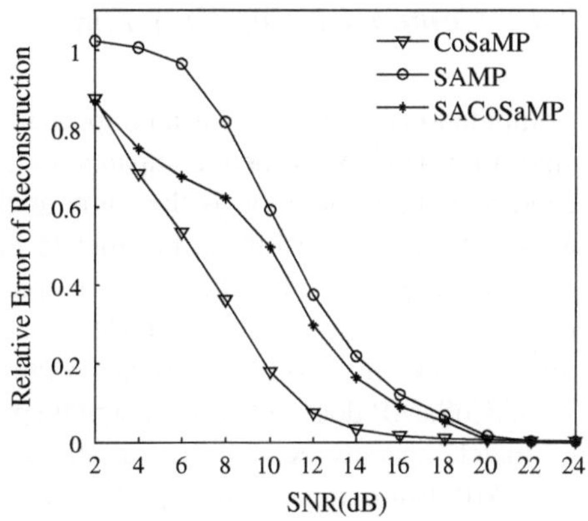

Fig. 5 Relative error of reconstruction versus the SNR

Figure 5 presents the relative error of different algorithms versus different SNR. Signal sparsity is fixed to 20. Reconstructed relative error is presented as

$$e_{rror} = \frac{||\mathbf{x} - \hat{\mathbf{x}}||_2^2}{||\mathbf{x}||_2^2} \tag{14}$$

where $\hat{\mathbf{x}}$ is the reconstructed signal. Each reconstructed relative error is the average value of 500 reconstructions for different algorithms. It can be seen that the CoSaMP algorithm achieves the highest reconstruction accuracy, and the SACoSaMP is better than the SAMP. When the SNR is low, the reconstruction accuracy of SACoSaMP is more close to that of CoSaMP, which implies the anti-noise property of SACoSaMP. This is because SACoSaMP retains the framework of CoSaMP. When the SNR is high, the reconstruction performance of SACoSaMP is more close to that of SAMP, which implies the sparsity adaptive property of SACoSaMP. This is because the sparsity adjustment of SACoSaMP is similar to that of SAMP.

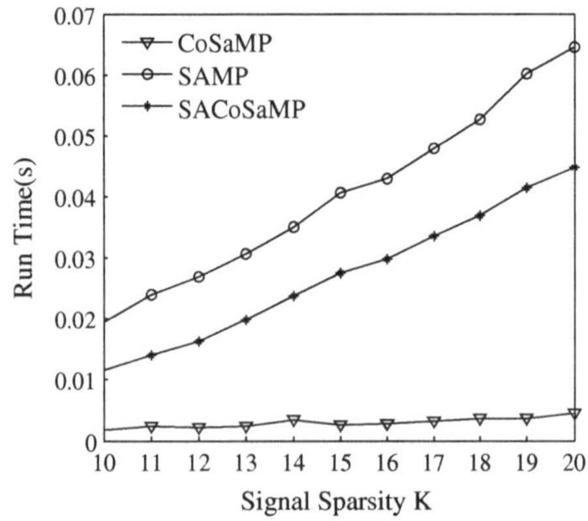

Fig. 6 Run time versus the signal sparsity K

3.4 *Comparison of Run Time*

Computational complexity is an important indicator to represent the performance of algorithm. This experiment compares run time of different algorithm under the same condition. Test condition is the same as the former two types of experiments (no noise). According to Fig. 2, in order to ensure that all algorithms can reconstruct original signal exactly, the rang of sparsity K is set from 10 to 20, and 500 reconstructed experiments are made in each sparsity K to calculate the average run time of different algorithms. Figure 6 presents the experiment results.

SACoSaMP does not rely on signal sparsity K, and needs to iterate step by step to match sparsity K, so the run time of SACoSaMP is longer than that of CoSaMP. But compared with SAMP, which also does not rely on sparsity, the run time of SACoSaMP is shorter than that of SAMP in the same condition.

4 Conclusion

CoSaMP needs signal sparsity K as input, but this parameter is hard to obtain in practical application. In this paper, the proposed sparsity adaptive compressive sampling matching pursuit (SACoSaMP) algorithm does not rely on prior information of signal sparsity any more. This improves the feasibility of greedy algorithms in practical application when the significant coefficients of a signal are not available. Because of the lack of signal sparsity information, reconstruction experiments on 0-1 random sparse signal show that SACoSaMP has worse anti-noise performance and higher computational complexity than CoSaMP. But compared with SAMP, which is also sparsity adaptive, the proposed algorithm achieves higher reconstruction accuracy, better anti-noise performance and shorter run time.

Acknowledgments Project supported by the National Natural Science Foundation of China (No. 61271263).

References

1. Candès EJ, Romberg J, Tao T (2006) Robust uncertainty principles: exact signal reconstruction from highly incomplete frequency information. IEEE Trans Inf Theory 52 (2):489–509
2. Donoho DL (2006) Compressed sensing. IEEE Trans Inf Theory 52(4):1289–1306
3. Baraniuk RG (2007) Compressive sensing. IEEE Signal Process Mag 24(4):118–121
4. Giryes R, Elad M (2012) Cosamp and SP for the cosparse analysis model. In: Proceedings of the 20th European signal processing conference, EUSIPCO′12, Bucharest, Rumania, pp 964–968
5. Candès EJ, Tao T (2005) Decoding by linear programming. IEEE Trans Inf Theory 51(12):4203–4215
6. Candès EJ, Tao T (2006) Near-optimal signal recovery from random projections: universal encoding strategies? IEEE Trans Inf Theory 52(12):5406–5425
7. Dai W, Milenkovic O (2009) Subspace pursuit for compressive sensing signal reconstruction. IEEE Trans Inf Theory 55(5):2230–2249
8. Mallat SG, Zhang Z (1993) Matching pursuits with time-frequency dictionaries. IEEE Trans Sign Process 41(12):3397–3415
9. Tropp JA, Gilbert AC (2007) Signal recovery from random measurements via orthogonal matching pursuit. IEEE Trans Inf Theory 53(12):4655–4666
10. Donoho DL, Tsaig Y, Drori I, Starck J-L (2012) Sparse solution of underdetermined linear equations by stagewise orthogonal matching pursuit. IEEE Trans Inf Theory 58(2):1094–1121
11. Needell D, Vershynin R (2010) Signal recovery from incomplete and inaccurate measurements via regularized orthogonal matching pursuit. IEEE J Sel Top Sign Proces 4(2):310–316
12. Needell D, Tropp JA (2008) CoSaMP:iterative signal recovery from incomplete and inaccurate samples. Appl Comput Harmonic Anal 26(3):301–321
13. Giryes R, Elad M (2011) Denoising with greedy-like pursuit algorithms. In: Proceedings of the 19th European signal processing conference, EUSIPCO′11, Barcelona, Spain, pp 1475–1479
14. Do TT, Gan L, Nguyen N, Tran TD (2008) Sparsity adaptive matching pursuit algorithm for practical compressed sensing. In: Proceedings of 2008 42nd Asilomar conference on signals, system and computers, Pacific Grove, CA, pp 581–587
15. Cheng Y, Wei F, Hui F, Tao Y, Bo H (2010) A sparsity subspace pursuit algorithm for compressive sampling. Acta Electronica Sinica 38(8):1914–1917

Design and Implementation of Distributed Stage DB: A High Performance Distributed Key-Value Database

Hui-jun Wu, Kai Lu and Gen Li

Abstract With the development of high performance computing and Web 2.0 applications, unstructured data storage becomes more and more important. Traditional RDBMS isn't efficient for big data management. Meanwhile, RDBMS's scalability is weak. Databases' expansion often leads to a large scale of data migration. This paper designs and implements a high performance distributed key-value database, that is Distributed StageDB. Database servers are organized by a consistent hashing ring and distributed with the support of Zookeeper, a distributed service framework. It has a high single-node read/write performance. The route information is calculated by clients, which reduces the expense of expansion.

Keywords Distributed system · Database · Key-value · Zookeeper

1 Introduction

In recent years, high-performance computing, Web 2.0 applications is developing rapidly, big data analysis and processing, social networking applications proposed new requirements for database. These new scenarios require a database system which has higher read/write performance, more efficiency on big data management. Meanwhile, the new system should be expanded at low prices when it is needed.

For some applications represented by a large number of data processing applications, PB and even ZB-level data analysis and mining are often necessary. For instance, social network's data is able to be used for mining user behavior. Data to be analyzed is usually simple unstructured data like text, video, pictures which are not associated tightly. In addition, websites such as Taobao.com suffers a mass of concurrent access at every moment, so the system should have a high-speed

H. Wu (✉) · K. Lu · G. Li
Science and Technology on Parallel and Distributed Processing Laboratory,
National University of Defense Technology, Changsha, People's Republic of China
e-mail: whj_nudt@foxmail.com

E. Qi (ed.), *Proceedings of the 6th International Asia Conference on Industrial Engineering and Management Innovation*,
DOI 10.2991/978-94-6239-145-1_19

read/write performance and good concurrency support. With the increase of user data, the system must be easy to expand in order to keep service. One of the most important standards is whether the system performance has basically linear growth with the increase of node numbers.

However, sometimes the traditional relational database system (RDBMS) is not very well-performed in massive data management. Moreover, when the system-scale increases, a large amount of data movement is inevitable, and sometimes it even leads to a termination of service. For scenarios like social networks, big data mining, the natural relationship between data isn't clear; RDBMS' superiority for transaction becomes a performance bottleneck.

Nowadays, Key-Value database and other types of NoSQL (Not only SQL) [1] storage appear more frequently. Their concurrency support, scalability is far better than traditional relational database. These superiority are mainly based on the characteristics of Key-Value database. Firstly, by means of memory caching and persistence technology, Key-Value database system's CRUD operation speed is higher. Secondly, Key-Value database system has streamlined data structure design; it can be more efficient on massive data management. Thirdly, Key-Value database system's good data expandability makes it easy to construct large-scale systems.

Existing Key-Value database includes Amazon Dynamo systems [2], Redis [3], Google levelDB [4], Kyoto Cabinet, Tair [5], Nuclea [6] etc. Meanwhile, there are still some deficiencies in these systems. First of all, the single-node operation speed isn't high enough, for example, the official test result of Nuclear at Intel Xeon E5405 is 25,000 operations/sec [6]. In some scenarios, data generates at a very quick speed, systems without high operation rate may be difficult to meet the demand. What's more, most of these systems' Key-Value request routing is calculated by database server, which will cause excessive pressure for servers. On the other hand, they did not take full advantage of the increasing performance of multicore client. In addition, the existing systems' concurrent load processing isn't so sufficient. For instance, Kyoto Cabinet's concurrent processing mechanism uses Linux's original epoll asynchronous event handling mechanism. In some real-time applications, clients' Key-Value requests can't be handled in time.

This paper designs and implements a distributed database system Distributed StageDB. Distributed StageDB's server uses a Key-Value database framework called StageDB, which responds user's Key-Value request by a pipeline. In each processing stage, multithreading is used to deal with requests. This paper solves the request dependency problems in multi-thread environment by CAAC algorithm [7]. Server nodes are organized by consistent hashing ring. Clients calculate the Key-Value request routing, reducing the pressure of limited servers. Based on ZooKeeper [8], we realize a distributed server management mechanism.

2 Design and Implementation of Distributed StageDB

This section introduces the design of Distributed StageDB's framework, followed by the introduction of various parts of the system's implementation.

2.1 Framework Design

Shown in Fig. 1, Distributed StageDB system framework is contains database services cluster, clients, metadata management cluster and distributed node management system.

Single-node Key-Value database server organizes process phases into a pipeline to improve system throughput, using levelDB as the underlying database engine. System maintains a consistent hashing ring of servers. Network and single-node database operation interfaces are provided, supporting C++ and Python. Metadata management system works with the support of distributed service framework ZooKeeper, which maintains consistent hashing ring etc. metadata.

2.2 Consistent Hashing

Distributed Key-Value database needs to meet the high performance and high scalability goals, so workload must be allocated to servers averagely. When new nodes are added or existing nodes are deleted, performance jitter should be controlled at a local scale. Therefore Distributed StageDB uses consistent hashing [9] algorithm to organize the servers. Unlike ordinary hash algorithm, consistent hashing algorithm has many characteristics on storage and workload balance.

Fig. 1 Distributed StageDB system framework

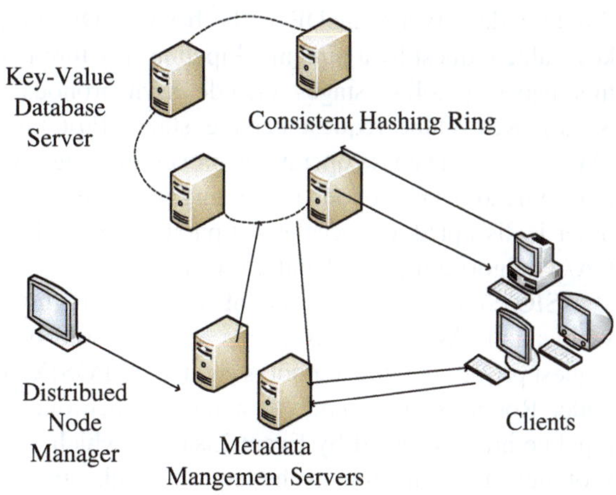

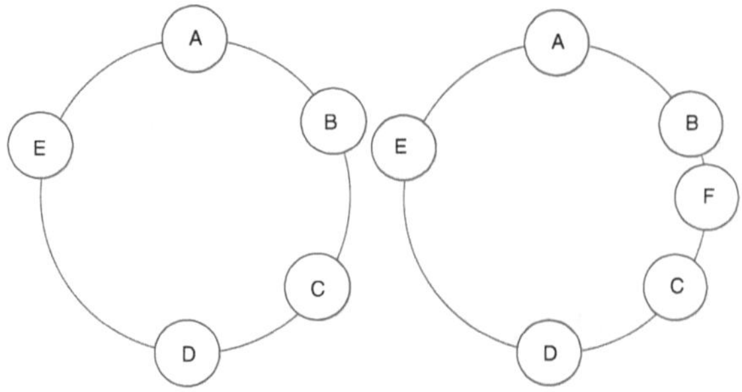

Fig. 2 The condition of hash ring before and after F is added

In load distribution, firstly, Distributed StageDB calculates the hash value of a certain server node, which decides its position in the hash ring with a size of 2^{128}. As shown in Fig. 2, there are A, B, C, D, E five machines totally. After calculating their hash value, they are put on different position (Fig. 2). Then use the same algorithm to calculate the hash value of user request's key field, and mapped onto the ring. The process to handle user request start from hash (Key), go clockwise to find the first server. The advantage of doing so is that when the size of the system needed to be increased.

When an additional server F add in (Shown in Fig. 2) after node increases, a new user request arrives, F will take over the original requests between B and C which are handled by C before. Data migration is only needed to be done from C to F. It is the same when a server is damaged. The system failure is controlled at a local scale.

After testing some of common consistent hashing algorithms like FNV, SuperFastHash and Murmurhash3, we find that MurmurHash3 [10] has high performance. So this paper uses MurmurHash3 algorithms.

2.3 Design of Distributed StageDB's Single-Node Server

Single-node server StageDB is the basis of Distributed StageDB, its handles user key-value request by a pipeline. Pipeline structure manager maintains the stages. In this paper, pipeline stages include asynchronous event handling stage (ASIO Stage), Key-Value request parsing stage (Protobuf Stage), access control stage (ACL Stage), database operations processing stage (CRUD Stage). In order to solve multi-thread asynchronous processing problem in which the user requests' original order is disrupted and accelerates parallel processing of the request, StageDB uses CAAC algorithm [2] to handle user requests.

ASIO Stage is a front end of server, receiving a user request and going on verification. Asynchronous events are processed by Boost C++ ASIO library; the request processing thread pool is created by POSIX [11] thread library which works under Round-Robbin mode to handle asynchronous requests. Different stages of pipeline are assembled by Event-Listener, which are completed by stage manager. Protobuf Stage implements the request serialization. User-defined messages will be compiled into C++ classes, which provide some class method. This stage prepares an asynchronous queue for user requests' processing. ACL Stage takes charge of permission checks. To be specific, the request from unauthenticated users will be prevented to access database. CRUD Stage ultimately achieves the user request's database operations.

CAAC algorithm for parallel environment aims at four kinds of dependency issues (write–write, write–read, read–write, read–read). The key process of CAAC algorithm is mixing up user key-value requests at first, and then serializing the requests with inevitable dependency. Those requests without dependency and requests with evitable dependency can be processed parallelly.

Distributed StageDB's underlying database engine is Google levelDB, in which key-value data is organized by Skiplist (jump table). That data structure provides a high read and write performance.

2.4 ZooKeeper Based Distributed Node Manager

Distributed StageDB put the metadata like information of very server and consistent hashing ring into ZooKeeper cluster. These metadata changes with the condition of servers (online, offline and work status), Distributed StageDB must ensure that the clients and servers have a common view of metadata. Therefore, this paper uses ZooKeeper to support atomic operations and data consistency maintenance mechanism. So it can provide a global consistent metadata view. ZooKeeper's atomic operations are encapsulated in ZAB [12] (ZooKeeper Atomic Broadcast) Protocol, the key point is to ensure that there is only one leader in a cluster; all transactions' update can only be initiated by the leader.

In this paper, distributed node manager are developed based on zkpython package. Node initialization step includes connecting zookeeper server, creating and setting some of management information znode (a type of data structure in zookeeper) and initializing server's information.

Manager provides commands to manage server nodes. System administrator's management is based on monitoring the changes of system status. For instance, the manger watches on critical znode (e.g. /info, /group) by Watcher of ZooKeeper to subscribe the message when server node status changes. Its work process is like followings, when a new server node is connected to ZooKeeper, the manager will get the message, thereby put the node into a certain group and set some configuration. Similarly, when a node is off-line, /group will return an event message. Then the manger will know that and handle with it accordingly.

In order to obtain reliable ZooKeeper service, ZooKeeper take cluster deployment, as long as most of the service node startup, the service will not be interrupted. In addition, preferably an odd machine, machines such as 5 two nodes can tolerate downtime.

2.5 Implementation of Distributed Server

Distributed Key-Value database server is based on single node servers. Each server node handles user key-value requests by a pipelined structure. Servers are logically organized by consistent hashing ring. With the support of ZooKeeper distributed service framework, it realizes nodes' distribution. Distributed server's operation sequence is shown in Fig. 3.

After Distributed StageDB server starts, it registers on ZooKeeper server and establishes a TCP connection. Subsequently, the server completes the creation of relevant znode. When all of these are done, the server will be configured and run with the support of ZooKeeper's node event subscription and message mechanism.

When Distributed StageDB server starts, it connects to ZooKeeper actively, ZooKeeper' configuration is organized by a XML file, which can be decoded through the xml_parser method in boost C++ property_tree library. Each server node creates a znode at the /servers/db znode, the distributed node manager set Watcher in the corresponding node. Therefore, the manager could get known right after the server node was created. The administrator assigns IP addresses and ports, organizing a server group, by which a consistent hashing ring can be created. As far as each server subscripts the message of /control/db/command znode, it can complete these operations after the administrator sends command.

The user's requests are sent to the levelDB database engine for final database operation. For high-performance server-side program, as frequent allocating and releasing memory space will occupy a lot of system resources, frequent object instantiation will lead to low performance. In this paper, all objects' instantiation are done when Distributed StageDB server is initialized.

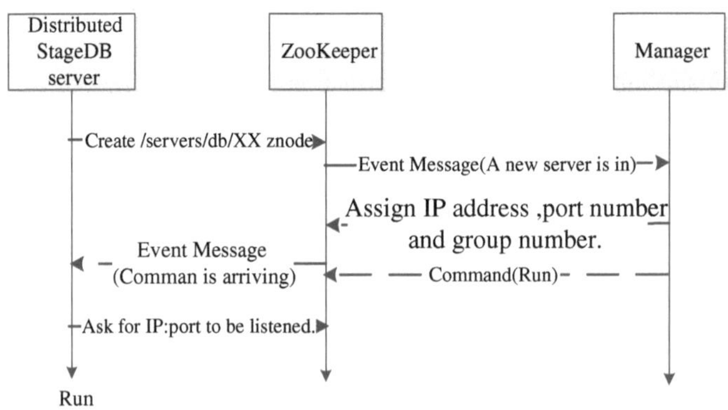

Fig. 3 Distributed server's operation sequence

2.6 *Implementaion of Distributed StageDB client*

Distributed StageDB clients obtain the information of global consistent hashing ring from ZooKeeper cluster. User's key-value requests's routing and distribution are calculated at clients directly. This design scheme can make a full use of client's performance. Meanwhile, it reduces the computing pressure of servers, spare more server resource for requests' processing capability.

To make client libraries more compatible for existing program, the client is developed by Python, Python's C++ support generates C++ interfaces easily. Clients are initialized with servers' information, so they need to register to ZooKeeper at first, and provide users with the interface of showing groups. Client can read the server and group information to achieve initialization after connection is set. Each client is connected to only one instance of the server node. The instance pointers are managed by boost shared_ptr. In order to ensure the system balance, the server's index is distributed value averagely on consistent hashing ring with a size of 2^{128}. Client library provides both synchronous and asynchronous operations, including the get, set, append, delete, etc.

3 System Test and Evaluation

In order to gain realistic result, we use the user behavior statistical model of key-value storage concluded by Facebook. On the basis of the statistical model, we realize a key-value request generator. Then we test the key-value performance of Distributed StageDB with different request size. Finally, we analyze the result.

Facebook's Berk Atikoglu et al. studied key-value storage system Memcached's user behavior and analyzed the workload [13]. They pointed that key-value requests in social network application could be divided into five types (shown in Table 1) including USR request, APP request, different requests varies a lot in size and key-value's distribution [13]. For APP requests, 90 % of the key's size is 31B. While in SET request, more than 30 % are 270B. USR requests' distribution is more extreme, key has only two kind of size, 16B and 21B, value size is 2B. In ETC requests, the request less than 11B are more than 40 %. But occasionally there are very large (over 1 MB) requests which distorts the distribution. Values with a size under 70B make up

Table 1 The classification of user request in social network applications

Request	Description	Component
USR	User accesses other user's information	99.8 % are GET
APP	User accesses program data	GET(main) and DELETE
ETC	Application program accesses configuration data	GET(main) and DELETE
VAR	Short-term key-value storage (e.g. config)	UPDATE
SYS	Servers and related request	GET

more than 97 % of all. According to the characteristics of the user request, user key-value's statistical models are concluded. Key size obeys generalized extreme values distribution [14], the parameters are $\mu = 30.7984$, $\sigma = 8.20499$, $k = 0.78688$, the probability density function is shown in Fig. 4.

$$F(x) = e^{-[1+k(\frac{x-\mu}{\sigma})]^{(-1/k)}} \tag{1}$$

Value Size obey generalized Pareto distribution [15], parameters are $\theta = 0$, $\sigma = 214.476$, $k = 0.348238$, the probability density function shown in Fig. 5.

$$F(x) = 1 - \frac{1}{\Gamma(\theta)}(\frac{k}{k+x})^{\theta}\left(\frac{\Gamma(\theta)}{\Gamma(1)} + \frac{\Gamma(\theta+1)}{\Gamma(2)}\frac{x}{x+k} + \ldots + \frac{\Gamma(\theta+k-1)}{\Gamma(\sigma)}\left(\frac{x}{x+k}\right)^{\sigma-1}\right) \tag{2}$$

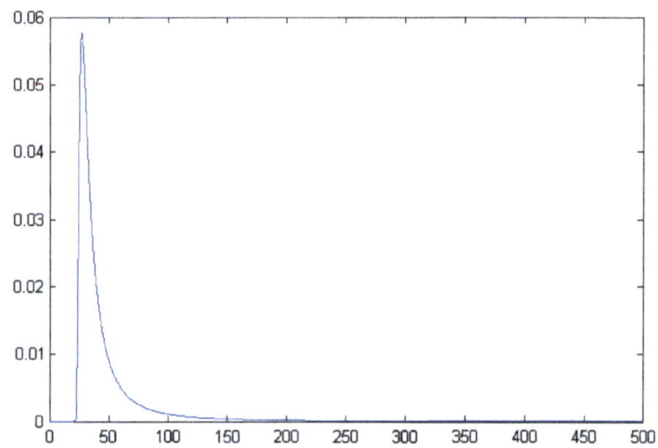

Fig. 4 Key size distribution's probability density X-axis is key size (byte), y-axis is probability density

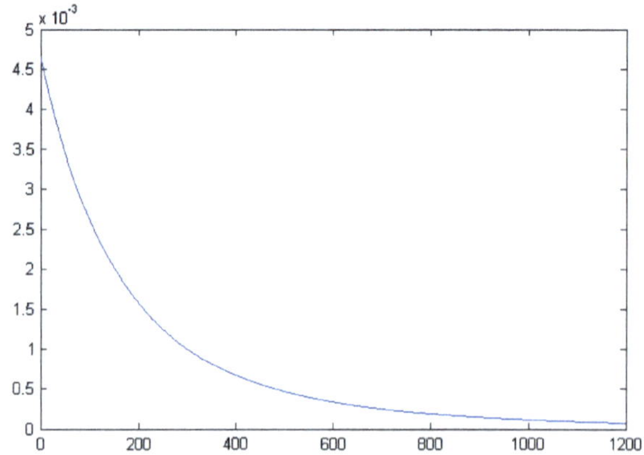

Fig. 5 Value size distribution's probability density, X-axis is value size (byte), and y-axis is probability density

Table 2 System performance (1,000,000 requests generated by request generator)

Concurrency	Server1 (ops/sec)	Server2 (ops/sec)
5	5081	5202
10	11434	11009
20	22092	23989
40	47018	44821
60	57112	56691

Table 3 System performance (10,000,000 requests generated by request generator)

Concurrency	Server1 (ops/sec)	Server2 (ops/sec)
5	5177	4983
10	10994	11235
20	23987	21853
40	55101	56329
60	54980	56062

Moreover, in most applications, distributed database systems' main purpose is for querying. In Memcached system, the ratio of GET and SET is about 30:1. In key-value database, DELETE requests often follow a GET request, and thus it should be included in the GET request.

We tested the system performance on two machines with 1,000,000 and 10,000,000 respectively. Configuration of the test machines is Intel Core i5 3230, 4 GB RAM.

We test the single-node performance at first. The peak speed of single-node performance is about 57000 req/sec.

Then we add one more server. Here are the results (Tables 2 and 3).

As we see, because of the limited sending speed of each client, with the increment of concurrency, the servers' performance is improving. When concurrency arrives 40, the peak performance appears. Because the routing is calculated by clients, the server performance barely reduces when the server number changes. Thus, the system' total performance can get an approximately linear improvement with the expansion of server numbers.

References

1. Stonebraker M (2010) SQL databases v. NoSQL data-bases. Commun ACM 53(4):10–11
2. DeCandia G, Hastorun D, Jampani M, Kaklapati G, Lakshman A, Pilchin A, Vogels W (2007, October) Dynamo: amazon's highly available key-value store. In: SOSP, vol 7, pp 205–220
3. Sanfilippo S, Noordhuis P (2010) Redis. http://redis.io. Accessed 19 Mar 2011
4. Google levelDB (May 2011). http://code.google.com/p/leveldb/
5. Ruohai (2010, June) Tair: Taobao's key-value storage solution. http://rdc.taobao.com
6. Leng H (2010, Jan) Nuclear:mass storage system for Ren-ren.com. http://ugc.renren.com/2010/01/21/ugc-nuclear-guide-use

7. Hu H, Lu K, Li G, Wang X, Xu T (2012) CAAC: a key-value database performance boosting algorithm. In: Computational and information sciences (ICCIS), 2012 fourth international conference on IEEE, pp 437–440
8. Hunt P, Konar M, Junqueira FP, Reed B (2010, June). ZooKeeper: wait-free coordination for internet scale systems. In: Proceedings of the 2010 USENIX conference on USENIX annual technical conference, vol 8, pp 11–11
9. Karger D, Sherman A, Berkheimer A, Bogstad B, Dhanidina R, Iwamoto K, Yerushalmi Y (1999) Web caching with consistent hashing. Comput Netw 31(11):1203–1213
10. Appleby A (2012) SMHasher & MurmurHash
11. Butenhof DR (1997) Programming with POSIX threads. Addison-Wesley Professional
12. Junqueira FP, Reed BC, Serafini M (2011, June). Zab: high-performance broadcast for primary-backup systems. In: Dependable systems & networks (DSN), 2011 IEEE/IFIP 41st international conference on IEEE, pp. 245–256
13. Atikoglu B, Xu Y, Frachtenberg E, Jiang S, Paleczny M (2012, June) Workload analysis of a large-scale key-value store. In: ACM SIGMETRICS performance evaluation review, 40(1):53–64 (ACM)
14. Beirlant J, Matthys G (2006) Generalized etreme value distribution. Encyclopedia of envirometrics
15. del Castillo J, Daoudi J (2009) Estimation of the generalized Pareto distribution. Stat Probab Lett 79(5):684–688

The Application of Adaptive Kalman Filter to Dynamic Tracking System

Bei Huo, Min-le Wang and Min Liang

Abstract An idea and method about improving Kalman filter is proposed to solve the difficulties of data fuse and filtration for tracking objects. The model of tracking objects comprises the models of linear dynamic system, detecting, Kalman filter and equations about measurement conversion. The improved algorithm initializes value based on ascertaining the area of turning. The performance evaluation of this algorithm is presented based on the algorithm example. The experiment results show that the improved algorithm is accurate and effective.

Keywords Kalman filter · Track fuse · Tracking system

1 Introduction

Kalman filter is an effective filter calculation that is suitable for the application in the flight track fusion, but the dynamic system of linear kalman is described as discrete and linear. When the flying item has uniform rectilinear flight, the filter effect is quite fantastic. Meanwhile, when the flying item adjusts its direction, obvious deviation occurs. Using the Kalman filter algorithm can improve the filter effect, but it greatly increases the workload and complexity of computation, and the steering of flying is unpredictable in advance. Filter effect can't be guaranteed even though the filter algorithm is adopted. This article aims at the improvement of Kalman filter algorithm by determining the direction and controlling the initial value of filter with the premise that slight increasing of system computation and complexity.

B. Huo (✉) · M. Liang
9th Department, Xi'an Research Institute of Hi-Tech, Xi'an, People's Republic of China
e-mail: lion1109@yeah.net

M. Wang
Department of Science, Xi'an Research Institute of Hi-Tech, Xi'an,
People's Republic of China

E. Qi (ed.), *Proceedings of the 6th International Asia Conference on Industrial Engineering and Management Innovation*,
DOI 10.2991/978-94-6239-145-1_20

2 The Target Track Fusion Model

Flying target track fusion is to process the flight tracking information detected by flying detector by using discrete, incomplete, imprecise information data to gain the complete, precise flying target track. The precise flying target, flying status and threat prediction can be gained. The whole flying target fusion model is made up of three parts. They are state model, detector observation model and filtering algorithm model.

2.1 *Flight State Model [1–3]*

The flight State of flying target is:

$$X(t_k) = \Phi(t_k/t_{k-1})X(t_{k-1}) + W(t_{k-1}),\ k = 1, 2, \ldots, n$$

$W(t_k)$ is state noise at period of t_k, setting as Gauss White noise.

$X(t_k)$ is flying status at period of t_k. This article is about the research in two dimension right-angle coordinate system. $X(t_k)$ is :

$$X(t_k) = \left(x_x(t_k), x_x'(t_k), x_y(t_k), x_y'(t_k)\right)^T$$

$\Phi(t_k/t_{k-1})$ is state transfer matrix. When the target is flying in a straight line and uniform, the transfer matrix is:

$$\Phi(t_k/t_{k-1}) = \begin{bmatrix} 1 & \Delta t & 0 & 0 \\ 0 & 1 & 0 & 0 \\ 0 & 0 & 1 & \Delta t \\ 0 & 0 & 0 & 1 \end{bmatrix}$$

2.2 *The Observational Model of Detector*

(1) Observational model

The observational model of detector is:

$$Y(t_k) = HX(t_k) + V(t_k),\ k = 1, 2, \ldots, n$$

H is observation matrix. In two-dimensional right-angle coordinate, H is:

$$H = \begin{bmatrix} 1 & 0 & 0 & 0 \\ 0 & 0 & 1 & 0 \end{bmatrix}$$

$V(t_k)$ is observatory noise at the period of t_k, setting as Gauss White Noise. $Y(t_k)$ is observatory at the period of t_k.

(2) Coordinate conversion calculation

In general, the data from detector is usually obtained in the condition of polar or spherical coordinate, but the state model and filtering model in right-angle coordinate description are convenient. So it is necessary to convert.

The observatory value of flying target from detector at period of t_k is:

$$Y(t_k) = (y_D(t_k), y_\alpha(t_k))^T$$

$y_D(t_k)$ is distance observatory value. $y_\alpha(t_k)$ is azimuth observatory value.

The observatory values change reference frame from polar coordinate to right-angle coordinate.

$$Y(t_k) = \left(y_x(t_k), y_y(t_k)\right)^T$$

$$\begin{Bmatrix} y_x(t_k) = x_{rad} + y_D(t_k) \cdot \cos y_\alpha(t_k) \\ y_y(t_k) = y_{rad} + y_D(t_k) \cdot \sin y_\alpha(t_k) \end{Bmatrix}$$

(x_{rad}, y_{rad}) is detector coordinate

(3) The calculation error after conversion

The real value of flying $X(t_k)$ and observatory value $Y(t_k)$ can be calculated by using Taylor formula. The approximate right-angle coordinate observatory error is:

$$V(t_k) = \left(v_x(t_k), v_y(t_k)\right)^T$$

$$\begin{cases} v_x(t_k) = \cos y_\alpha(t_k) \cdot v_D(t_k) - y_D(t_k) \cdot \sin y_\alpha(t_k) \cdot v_\alpha(t_k) \\ v_y(t_k) = \sin y_\alpha(t_k) \cdot v_D(t_k) + y_D(t_k) \cdot \cos y_\alpha(t_k) \cdot v_\alpha(t_k) \end{cases}$$

$(v_D(t_k), v_\alpha(t_k))$ is detector's polar coordinate observatory error. All is mean zero Gauss noise.

(4)Variance after conversion

Polar coordinate system observatory value $y_D(t_k)$ and $y_\alpha(t_k)$ are mutual independent. Detector angle mean variance is σ_α^2, measurement distance variance is σ_D^2. After conversion, right-angle coordinate system observatory value is not independent, so the variance matrix is:

$$R(t_k) = \begin{bmatrix} \sigma_x^2(t_k) & \sigma_{xy}(t_k) \\ \sigma_{yx}(t_k) & \sigma_y^2(t_k) \end{bmatrix}$$

$$\begin{aligned} \sigma_x^2(t_k) &= E[v_x(t_k) \cdot v_x(t_k)] \\ &= \sigma_D^2 \cdot \cos^2 y_\alpha(t_k) + y_D^2(t_k) \cdot \sigma_\alpha^2 \cdot \sin^2 y_\alpha(t_k) \end{aligned}$$

$$\sigma_y^2(t_k) = E\left[y_y(t_k)\cdot y_y(t_k)\right]$$
$$= \sigma_D^2\cdot\sin^2 y_\alpha(t_k) + y_D^2(t_k)\cdot\sigma_\alpha^2\cdot\cos^2 y_\alpha(t_k)$$

$$\sigma_{xy}(t_k) = \sigma_{yx}(t_k) = E\left[v_x(t_k)\cdot v_y(t_k)\right]$$
$$= \left[\sigma_D^2 - y_D^2(t_k)\cdot\sigma_\alpha^2\right]\cdot\sin y_\alpha(t_k)\cdot\cos y_\alpha(t_k)$$

$R(t_k)$ feature value is :

$$\Lambda(t_k) = \left(\lambda_x(t_k), \lambda_y(t_k)\right)^T$$

$$\begin{cases} \lambda_x(t_k) = \sigma_x^2(t_k) + \sigma_{xy}(t_k)\tan\alpha(t_k) \\ \lambda_y(t_k) = \sigma_y^2(t_k) - \sigma_{yx}(t_k)\tan\alpha(t_k) \end{cases}$$

The standard deviation is:

$$\sigma(t_k) = \sqrt{\sqrt{\lambda_x(t_i)}\cdot\sqrt{\lambda_y(t_i)}}$$

2.3 *Kalman Filter Calculation Model*

Kalman filter Calculation Model is based on flying status model and detector observatory model. The basic process [4–7] is shown in Fig. 1.

In the model: $K(t_k)$ is filter gaining. $P(t_k/t_k)$ is filter variance matrix. $P(t_k/t_{k-1})$ is covariance matrix for extrapolation.

3 Improvement of filter algorithm

3.1 *The Idea of Improvement*

Through the analysis of flight tracking model, it indicates that removing the measurement noise filtering calculation is established on the basis of flight state model. The quality of filter is closely linked to the accuracy of predicting the future target status by the flight state model. When the flight target status changes, it must adjust the flight state model in order to ensure the quality of filter.

It is not difficult to identify that the flying target adopts uniform motion in a straight line at the most part of time when it in the state of cruise. Only steering maneuver in rare region. In other words, the entire flight target track can be regarded as a number of uniform linear tracks spliced. Therefore, the flying turning identification can be added to track fusion calculation for the improvement of filter

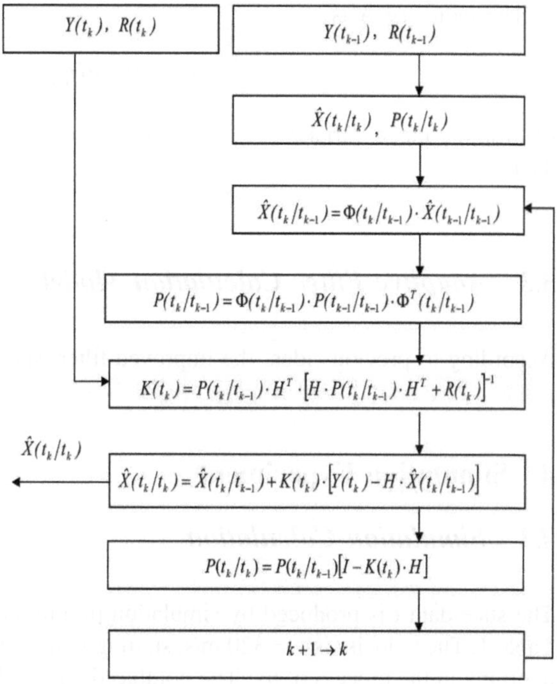

Fig. 1 Kalman filter calculation model

quality. It means the flying turning identification should be conducted before the filter calculation. Slight adjustment to filter initial value can ensure the accuracy of filter as the target track steering identified.

3.2 *Turning Identification*

The method of turning identification is: Setting an identification area A, when observatory result falls into the area A, it means there is no turning of flying target. When observatory result doesn't fall into area A, it means turning. It is necessary to set a new flying condition model. Identification area A is a round area regarding to the tracking fusion prediction point $Q(t_k)$ as center.

Center of Circle of A :

$$Q(t_k) = (\hat{x}_x(t_k/t_{k-1}), \hat{x}_y(t_k/t_{k-1}))^T$$

$$\begin{cases} \hat{x}_x(t_k/t_{k-1}) = \hat{x}_x(t_{k-1}/t_{k-1}) + \hat{x}'_x(t_k/t_{k-1}) \cdot \Delta t \\ \hat{x}_y(t_k/t_{k-1}) = \hat{x}_y(t_{k-1}/t_{k-1}) + \hat{x}'_y(t_k/t_{k-1}) \cdot \Delta t \end{cases},$$

$$\Delta t = t_k - t_{k-1}$$

The radius of A is

$$r_A = W \cdot \sigma(t_k)$$

$\sigma(t_k)$ is the mean variance; W is coefficient adjustment. It can be adjusted by test data.

3.3 Adaptive Filter Calculation Model

According to previous idea, the improved filter calculation is shown in Fig. 2.

4 Simulation Experiment

4.1 Simulation Calculation

The state data t is produced by simulation program of flying target. It is shown in Table 1. The velocity is v = 320 m/s, starting is $x_0 = 5000$ m, $y_0 = 1200$ m. There is a turning in the process of fly. The boldfaced data is the point of turning in Table 1.

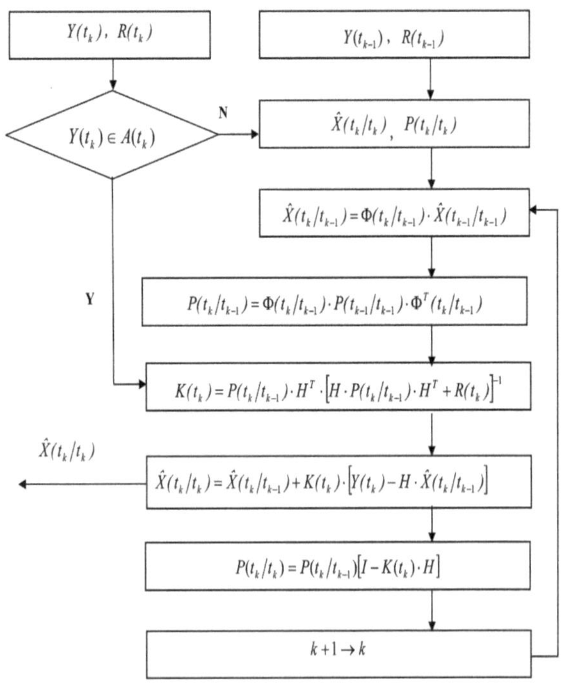

Fig. 2 Adaptive filter calculation model

Table 1 Tracking fusion real data

No.	1	2	3	4	5
T	0	10	20	30	40
X	5000	7000	9000	11000	13000
Y	1200	3700	6200	8700	11200
No.	6	7	8	9	10
T	50	60	70	80	90
X	15000	17000	19000	21000	**23500**
Y	13700	16200	18700	21200	**23200**
No.	11	12	13	14	15
T	100	110	120	130	140
X	26000	28500	31000	33500	36000
Y	25200	27200	29200	31200	33200
No.	16	17	18	19	20
T	150	160	170	180	190
X	38500	41000	43500	46000	48500
Y	35200	37200	39200	41200	43200

Observatory data detector simulation program is produced by track data and capacity of the detector simulation. It is shown in Table 2. The distance error of detector is $\alpha_D = 650\,\mathrm{m}$, angle error is $\sigma_\alpha = 1^\circ$, measurement cycle is T = 20S, Coordinate is $x_R = -50000\,\mathrm{m}$, $y_R = -50000\,\mathrm{m}$.

The filter value is produced by simulation program adaptive filter calculation model. It is shown in Table 3.

Table 2 Tracking fusion observatory data

No.	1	2	3	4	5
T	0	10	20	30	40
X	4818	6842	8968	11058	13124
Y	1112	3616	6178	8748	11308
No.	6	7	8	9	10
T	50	60	70	80	90
X	15160	17166	19140	21088	**23354**
Y	13848	16358	18838	21288	**23054**
No.	11	12	13	14	15
T	100	110	120	130	140
X	25904	28474	31054	33628	36178
Y	25104	27174	29254	31326	33374
No.	16	17	18	19	20
T	150	160	170	180	190
X	38692	41168	43608	46022	48428
Y	35392	37368	39308	41222	43126

Table 3 Track fusion filter value

No.	1	2	3	4	5
T	0	10	20	30	40
X	4818	6842	8966	11009	13045
Y	1112	3616	6166	8690	11218
No.	6	7	8	9	10
T	50	60	70	80	90
X	15073	17091	19097	21092	**23354**
Y	13742	16261	18769	21268	**23054**
No.	11	12	13	14	15
T	100	110	120	130	140
X	25904	28485	31024	33562	36093
Y	25104	27186	29224	31261	33291
No.	16	17	18	19	20
T	150	160	170	180	190
X	38613	41118	43610	46088	48558
Y	35311	37318	39309	41288	43258

4.2 Analysis of Results

Tracking fusion data observatory error and filter error is shown in the Figs. 3 and 4 as below.

The analysis of data indicates the three track fusion points of initial track or turning point are filter model adjustment period, so the filter error is big. Later, the filter error tends to be sable, and observatory errors are greatly restrained.

Detector error variance is :σ_1 =155 m.

Filter value error variance is: σ_2 =72 m.

The variance value is:

$$B = \frac{\sigma_2}{\sigma_1} = 0.47$$

Fig. 3 Observatory error

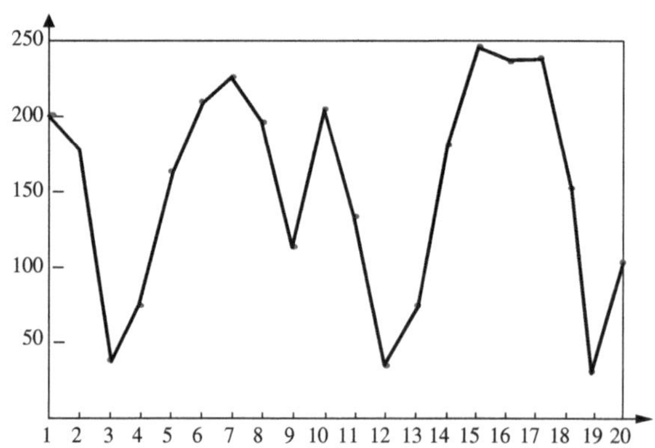

Fig. 4 Filter error

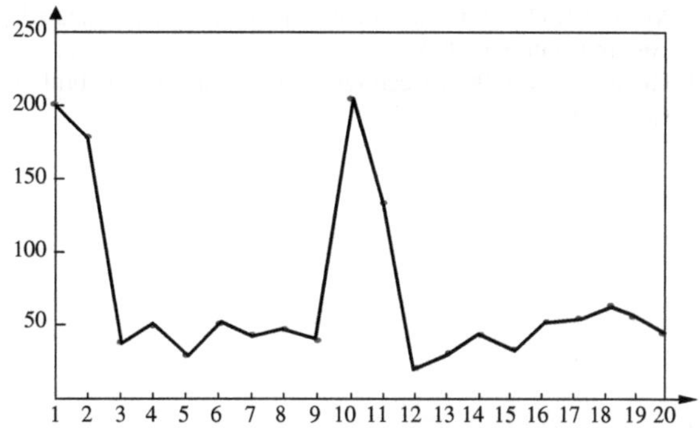

Results based on plenty of calculations show that this filter calculation model has a perfect filter effect that can eliminate observatory noise of 60–70 %. For flying target fusion turning, the responding duration of this calculation is short. Timely and precise adjustment can be carried out to ensure the accuracy of filter calculation.

5 Conclusion

Because the filter effect is easily affected by cruise track steering, the solution of adjusting filter initial value based on the turning identification is put forward. The improvement Kalman filter calculation can be highly satisfied the auto-adjustment of cruise target turning. It not only has a quick response to track turning, but also keep the traditional calculation method that is mostly ensure the filter effect in the state of linear dynamic system. Besides, this improvement only adds slight workload of calculation. The project operation is available.

Acknowledgments The authors would like to thank the anonymous reviewers and editor for their careful reading. The comments and suggestions have helped us to improve the clarity of the paper.

References

1. Gao W, Zhang X (2011) An algorithm based on Kalman filtering for estimating kinematic model error. Sci Surv Mapp 36:56–58
2. Sun L, Jiang Y, Wang X (2011) Research into motive target tracking method based on Kalman filtering. Shipboard Electron Countermeasure 34:32–35
3. Nie H, Guan J, Bian H (2011) Fuzzy adaptive Kalman filtering in INS/GPS integrated navigation system. Shipboard Electron Countermeasure 31:69–70
4. Mehra RK (1970) On the identification of variances and adaptive Kalman filtering. IEEE Trans Autom Control 15:175–184
5. Singer RA (1973) Estimating optimal tracking filter performance for manned maneuvering targets. IEEE Trans Aerosp Electron Syst 9

6. Mehra RK (1970) On the identification of variances and adaptive Kalman filtering. IEEE Trans Autom Control 15:175
7. Grossman W (1994) Bearings—only tracking: a hybrid coordinate system approach. J Guid Control Dyn 17:451–457

An Improved Geometrical Attack Robust Digital Watermarking Algorithm Based on SIFT

Xing-jun Wang and Wei Tan

Abstract The paper presents an effective algorithm, which is robust against geometric attack. The proposal achieves watermark synchronization using SIFT (Scale Invariant Feature Transform). We built some feature regions based on suitable SIFT feature points. The main orientation and scale of the feature point determines the size of the corresponded region. In this paper we argue that the feature regions contain semantic information of the image. Thus we can lay the foundation of our watermark embedding on these regions. Most of these regions can be detected using the improved algorithm. We present an improvement work by using some improved ways. We experimentally show an increase in detection rate and robustness of the watermarks after geometric attacks.

Keywords Circle region · Feature region · Geometrical attack · SIFT · Watermark

1 Introduction

Digital watermarking is a supplement of traditional encryption system. At present, most digital watermarking scheme is resistance to conventional signal processing, such as compression, low-pass filtering, enhance, but most of them are weak to geometric transformation, such as rotate, scaling, translating, cropping. It is not easy for us to detect watermark without original image. Synchronization is not only the first step, but also the most important step. We can't detect anything if synchronization fails.

X. Wang
Department of Electronic Engineering/Graduate School at Shenzhen, Tsinghua University, Beijing, China
e-mail: Robert.wang@unitend.com

W. Tan (✉)
Shenzhen Key Laboratory of Information Science and Technology/Shenzhen Engineering Laboratory of IS & DRM, Shenzhen, China
e-mail: Tanwei0404@163.com

E. Qi (ed.), *Proceedings of the 6th International Asia Conference on Industrial Engineering and Management Innovation*,
DOI 10.2991/978-94-6239-145-1_21

At present, there are three main ideas to solve the problem: (a) Watermarking is embedded in geometric invariant domain. O'Ruanaidh [1] proposes Fourier-Mellin transform domain. Watermarking is embedded in Fourier-Mellin transform domain. It is effective for RST (rotation, scaling, transforming). Sang [2] uses 2D CDMA watermark to hide the multi-bit message. Lin [3] uses DFT to generate a geometrical invariant domain. (b) Some extra information is embedded in the image, extra information are used to help us synchronize. Pereira [4] uses a kind of template to help us estimate geometric transformation. Fleet [5] embeds invisible information in color images. (c) The position of the watermark embedded in is based on image content. Bas uses Harris detector extract the feature points in [6], but Harris feature point is not stable enough in some cases, such as scaling. It makes watermarking algorithm weak if the image is scaling one. Tang [7] uses Mexican Hat wavelet to find image feature points and then embed watermark in a circle region around the feature point. However, the radius is regular, so it is difficult to resist the attacks of imagine scaling. Li et al. [8–14] use SIFT extract the feature point in Article. Obviously, it can be improved by some ways.

Lowe [15] proposes the idea of SIFT. In this paper, watermarking is embedded in the feature region. The feature region is formed by SIFT point. Each SIFT point has two parameters: main orientation and scale. The feature region is square instead of circle. Some paper [8–14] use circle region here. But it is not very proper here. Most of watermarking algorithm use square region to embed watermark, so we should convert circle area to square area first in some ways, such as add zeros. It will bring us extra crop. However, we use square region in this paper. It makes a better result. Watermarks are embedded in several regions. We can detect watermark in each region. We use a better way to improve result in our paper. Feature regions are selected carefully. The robustness of the algorithm is influenced by the selection of feature regions. In this paper, more feature regions are selected in an image. Then we use exhaustive search method to search feature regions which are embedded watermark. The algorithm improve the resistance to geometric attacks watermark robustness.

2 SIFT Feature Region

SIFT is a theory that based on the scale space to extract local feature. The algorithm finds extreme points which are invariant to location, scale and rotation in scale space. Steps are shown below:

(a) Generate DOG (difference-of-Gaussian) space

$$D(x, y, k\sigma) = (G(x, y, k\sigma) - (G(x, y, \sigma)) \cdot I(x, y) \tag{1}$$

In formula (1) $G(x, y, \sigma) = \frac{1}{2\pi\sigma^2} e^{-(x^2+y^2)/2\sigma^2}$. σ is scale coordinate, $I(x, y)$ is space coordinate.

$$\text{Let } G(x, y, \sigma) \cdot I(x, y) = L(x, y, \sigma) \tag{2}$$

then we get:

$$D(x, y, k\sigma) = L(x, y, k\sigma) - L(x, y, \sigma) \tag{3}$$

(b) Find the extreme points
Find the extreme points which are the maximum ones or minimum ones in scale space and image space. So each points should be compared with 26 points including 9 points in the upper scale, 9 points in the lower scale and 8 points in the same scale.

(c) Locate the feature points accurately
Some points which have low contrast should be rejected, so do the points which are poorly localized along the edge.

(d) Distribute the main orientation
We use the gradient to specify the main orientation of the feature points in order to assure the rotation invariance.

$$\begin{aligned} m(x, y) &= \sqrt{(L(x+1, y) - L(x-1, y))^2 + (L(x, y+1) - L(x \cdot y - 1))^2} \\ \theta(x, y) &= \arctan\begin{pmatrix} (L(x, y+1) - L(x, y-1)) \\ /(L(x+1, y) - L(x-1, y)) \end{pmatrix} \end{aligned} \tag{4}$$

Formula (4) is the value of modulo and orientation. the scale of L is the scale which the feature point located. Then we should get the main orientation. In Lowe's article, firstly, choose a 16 × 16 window around the point and divide it into 16 small pieces. In each small piece, we count the gradient histogram of 8 directions and draw the following steps accumulated value for each gradient direction, then a seed point is formatted. A feature points is described with 16 seed points which have information vectors with 8 directions.

3 Embed Watermark in Feature Regions

3.1 Select Feature Regions to Embed Watermark

There are so many SIFT feature points in an image. For example, in a 500 × 500 image 'Lena', we detect about 200 feature points. We can not embed watermark around every feature point. Because feature regions can't overlap. We can only select some feature regions to embed watermark. A simple method is described below:

Firstly, we should remove some unstable feature points. According to the result of [4], the small-scale and large-scale feature points are not very stable yet. We also take into account that the regions which are based on the small-scale feature points are not large enough to embed watermark because of the limited capacity. While the regions which are based on the large-scale feature points are too large to embed watermark. So we remove small-scale feature points and large-scale feature points first.

Even excluding this part of the feature points, the rest of the feature points is still a lot. Obviously these feature points can not be determined as feature regions to embed watermark, we must choose the proper points to embed watermark. In order to the make the algorithm more robust, watermark should be embedded in an area which is large enough. The steps are shown below:

(a) Firstly, we select the point whose scale is the maximum one
(b) Then remove all the overlapping region with the selected point.
(c) If there are some feature points left, back to step (a). If there are not any feature points left yet. The algorithm comes to an end.

The left feature points can build several feature regions. In [9], feature region is circle. It is improper in a way. So far, most of the watermarking algorithm is carried out based on rectangle image. Circular area must first fill in the corresponding rectangular area at first. Then we must remove the area which is outside the circle. This step makes a cropping here. It will bring us some data loss.

So we can provide a better way to embed watermark. We can use square region to embed watermark. After we get the SIFT feature point, consider that each of the SIFT feature point has several parameters: location (h, k), scale r, a main direction. We can build a feature region in use of these parameters. The center of the square region is located in (h, k). $m \times r$ is the length of the square. And the most important thing is the orientation of the square. Take advantage of the main orientation of SIFT feature point to rotate the region to normalize the orientation of the region in order to assure that the square is invariant.

Let us discuss the following fact which is according to Kutter's [16] article:

(1) The feature region is not sensitive to noise. Additive noise is common in fact. SIFT is insensitive for additive noise, so we can conclude that our feature region is insensitive for additive noise.
(2) The feature region is covariant in geometric transformation.
(3) The feature region is based on local information.

In theory, our algorithm works after some geometric attacks. Results are shown in the following pictures.

3.2 Embed the Watermark in Feature Regions

After getting feature regions in the previous step, we can embed watermark in these regions one by one. In a simple way, we embed the same watermark in every feature region. We use discrete cosine transform (DCT) to embed watermark. Watermarking algorithm based on DCT is different from the air-plane watermarking

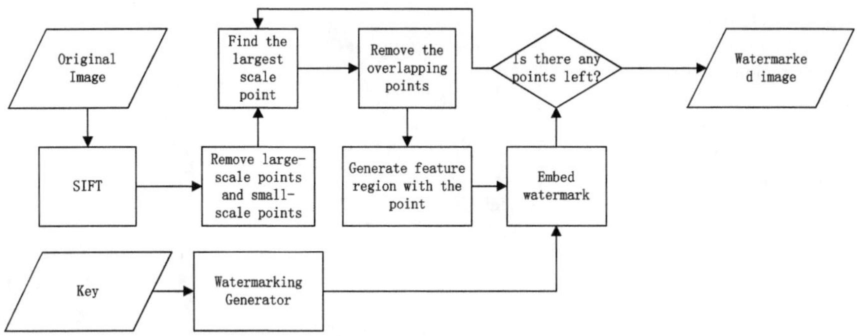

Fig. 1 Proposed watermark embedding scheme

algorithm. The image which is embedded watermark is no longer a small fraction of the area, but spread in the whole image area. In this way, digital watermark won't be damage even if part of the image pixels are changed.

Most of the energy is concentrated in the low frequency part. The human eye is more sensitive to the low frequency part, and is not sensitive to the high frequency part. Therefore, we should embed the watermark into the high frequency part as far as possible. But it is easy to damage the watermark if we embed the image into the high frequency, such as low-pass filtering [17]. We consider a compromise approach: embedding the watermark into intermediate frequency. In this way, we can make sure the transparency of the watermark and the robustness of watermark.

Before embedding watermark, we get a series of square areas. But we should rotate the square area according to the main orientation first. However, it will bring us a problem. Obviously, interpolation is needed when we rotate a digital image. In Matlab, function "imrotate" is implemented in 3 interpolation ways: The nearest neighbor interpolation, the bilinear interpolation and bicubic interpolation. According to the result of the experiment, we can get a better watermarking result if we use bicubic interpolation.

The steps are shown in Fig. 1.

4 Detect Watermark

Consider that the SIFT algorithm is based on the local gradient information, we conclude that SIFT feature points can be detected accurately. In order to make sure that the area which is embed with watermark we can use the exhaustive search method.

In order to check whether there is watermark in a given region, we can set a flag in the watermark image. After some signal processing, the watermark we detect will be changed in some pixels. We can design a special watermark to help us check whether the watermark is existed in the given region. Watermark is shown in Fig. 2a and the tag information as shown in Fig. 2b. Of course, we can other watermark to help us check if watermark is existed.

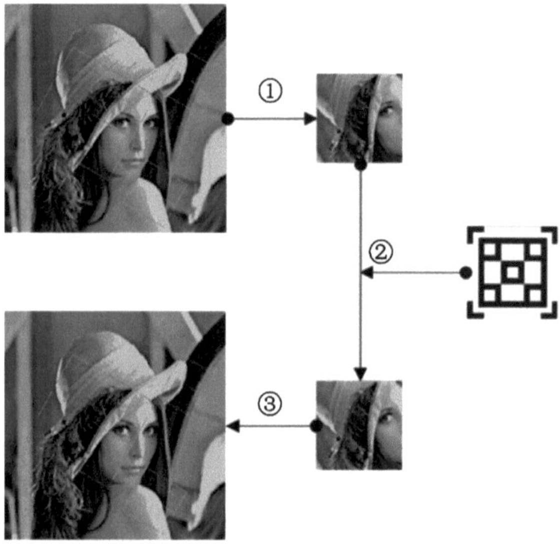

Fig. 2 Watermark embedding steps. *1* Get the feature regions and rotate it according to the main orientation. *2* Embed the watermarking. *3* Copy the regions into the original image

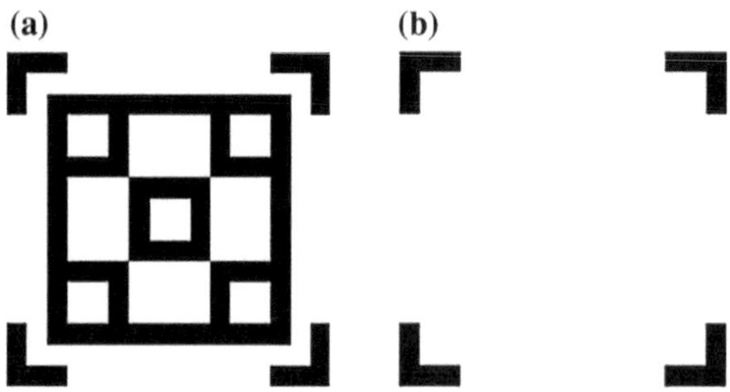

Fig. 3 **a** Watermarking with tag, **b** tag

It is similar to the watermark embedding, the first step is detecting the SIFT feature points of the image. Then, using the exhaustive search method to find our which religions are embedded watermark. The searching process can take advantage of the characteristics of embedding algorithm. Process is designed as follows (Figs. 3 and 4):

(1) First remove large scale and small scale feature points, because according to embedding algorithm we have to get rid of these feature points. So the watermark will not be embedded in these regions.
(2) Find the point whose scale is the largest. If there is no point to (6)
(3) The watermark detection is divided into two steps: first of all, according to the main direction for the direction of the characteristics of regional normalization. And then to extract the watermark.
(4) Check whether the region is embedded with watermark. If there is not watermark embedded in the feature region we should remove the feature point and then jump to (2). If there is watermark embedded in the feature region, save the watermark we detect in this step.

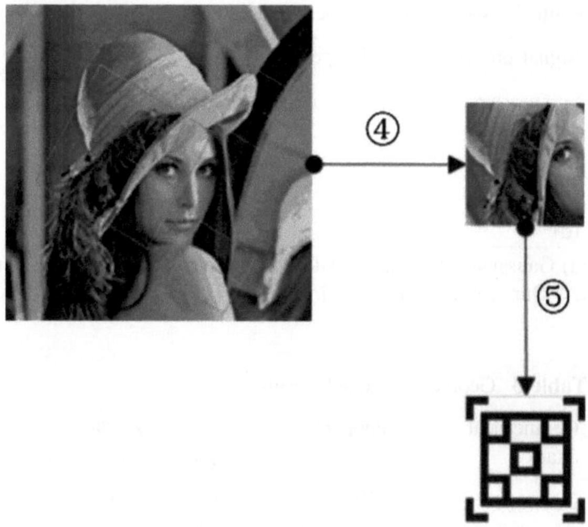

Fig. 4 Watermark embedding steps. *1* Get the feature region. *2* Detect watermark

Table 1 Transparency of watermark (embed coefficient is 20)

Image	PSNR
Lena	49.1973
Zelda	46.3066
Boats	44.6126
Barbara	46.4889
Airplane	46.8552
Sailboat	42.1720

(5) According to the embed algorithm before, we can remove some feature points after we find a feature region which is embedded watermark. Remove all the feature points whose regions are overlap with the one which is embedded watermark. Then back to (2)

(6) If there is not any regions embedded watermark, it means that the image is not embed watermark. But there is many feature points in an image, we should set a threshold to reduce computation. We needn't check all the points in some cases. If there are some regions embedded watermarking, We can analysis the watermark to improve results. As a general rule, we can get more than one region embedded watermark.

5 Results

We use 'Lena' to show the images in every step. The size of 'Lena' is 512 × 512.

From the result we get, we conclude that it is suitable that the embed coefficient is set to 20. It makes sure that watermark is transparent and robustness.

Table 2 Normal signal processing result

Signal processing	Peppers			Zelda			House		
	(A)	(B)	(C)	(A)	(B)	(C)	(A)	(B)	(C)
(a)	5/8	5/8	1/8	6/8	3/8	5/8	6/8	6/8	5/8
(b)	6/8	2/8	4/8	7/8	2/8	3/8	7/8	3/8	3/8
(c)	3/8	2/8	2/8	3/8	5/8	1/8	3/8	2/8	4/8
(d)	5/8	1/8	2/8	5/8	1/8	2/8	5/8	1/8	3/8

(a) Gaussian filtering, (b) Addition noise, (c) Jpeg compression, (d) Gaussian filtering
(A) Our work, (B) Ref. [12], (C) Ref. [14]

Table 3 Geometrical attack results

Geometrical attack	Peppers			Zelda			House		
	(A)	(B)	(C)	(A)	(B)	(C)	(A)	(B)	(C)
(a)	6/16	2/16	3/16	8/16	2/16	3/16	7/16	2/16	1/16
(b)	4/16	0/16	1/16	5/16	3/16	2/16	3/16	1/16	2/16
(c)	5/16	1/16	2/16	6/16	4/16	4/16	7/16	1/16	1/16
(d)	6/16	2/16	3/16	7/16	4/16	5/16	6/16	2/16	1/16
(e)	7/16	4/16	3/16	5/16	3/16	4/16	6/16	1/16	2/16

(a) Cropping 2.5 %, (b) Cropping 1 %, (c) Scaling 0.7, (d) Scaling 1.4, (e) Rotating 20°
(A) Our work, (B) Ref. [12], (C) Ref. [14]

Some results are shown in the Tables 1, 2 and 3:

6 Conclusion

The watermarking algorithm based on SIFT feature region is a useful one according to the result. As a second generation watermarking algorithm, it has all the advantages which general second generation watermarking algorithm has. The algorithm is well-worked in many cases.

Firstly, SIFT has many advantages to improve watermarking algorithm. It is stable while image is processed by normal processing. Feature regions can be found after some normal processing. Secondly, SIFT feature points make the algorithm robustness on rotation, scaling and transforming. Then, SIFT feature points are local information. While image cropping will not damage local information in the rest image, we can find out feature regions which are embedded watermark.

Although our algorithm is well worked, the algorithm can be improved in some places. It spends lots of time on detecting regions which are embedded watermark. Exhaustive search method is not the best choice obviously. If we embed watermark in a better way, we can reduce the time. In our algorithm here, we do not have redundancies. We can use channel coding to improve the result.

References

1. O'Ruanaidh JJK, Pun T (1998) Rotation, scale and translation invariant spread spectrum digital image watermarking. Sig Process 66(3):303–317
2. Sang MD, Zhao Y (2007) Moment based multibit digital image watermarking resisting to RST attacks. J Electron Inf Technol 29(4):76–80
3. Lin CY, Chang SF (1999) Distortion modeling and invariant extraction for digital image print and scan process. In: International symposium on multimedia information processing (ISMIP 99), Taipei
4. Pereira S, Pun T (1999) Fast robust template matching for affine resistant image watermarks. In Proceedings 3rd international information hiding workshop, Dresden, Germany, pp 207–218
5. Fleet DJ, Heger DJ (1997) Embedding invisible information in color images. In Proceedings IEEE international conference on image processing (ICIP-97), vol 1. Santa Barbara, pp 532–535
6. Bas P, Chassery JM, Macq B (2000) Robust watermarking based on the warping of pre-defined triangular patterns. In: Proceedings of SPIE: security and watermarking of multimedia contents II, vol 3971. San Jose, California, pp 99–109
7. Tang CW, Hang H (2003) M.A feature-based robust digital image watermarking scheme. IEEE Trans Sig Process 51(4):950–959
8. Li. LD, Guo BL, Shao K (2007) Geometrically robust image watermarking using scale-invariant feature transform and Zernike moments. Chin Opt Lett 5(6):332–335
9. Lee H-Y, Kim H, Lee H-K (2006) Robust image watermarking using local invariant features. Opt Eng 45(3):037002
10. Deng C, Gao X (2009) Geometrically robust image watermarking based on SIFT feature regions. Acta Photonica Sin 4:056
11. Verstrepen L, Meesters T, Dams T, Dooms A, Bardyn D (2009) Circular spatial improved watermark embedding using a new global SIFT synchronization scheme. In: 2009 16th international conference on digital signal processing, pp 1–8
12. Sun J, Lan S (2010) Geometrical attack robust spatial digital watermarking based on improved SIFT. In: 2010 international conference on innovative computing and communication, and 2010 Asia-Pacific conference on information technology and ocean engineering (CICC-ITOE)
13. Sharma G, Coumou DJ (2006) Watermark synchronization: perspectives and a new pardigm. In Proceedings 40th annual conference on Information Sciences and Systems (CISS). Princeton, NJ, 22–24 Mar 2006, pp 1182–1187
14. Deng C, Gao X (2009) Geometrically robust image watermarking based on SIFT feature regions. Acta Photonica Sin 4:056
15. Lowe D (2004) Distinctive image features from scale-invariant keypoints. Int J Comput Vis 60 (2):63–86
16. Kutter M, Bhattacharjee S, Ebrahimi T (1999) Towards second generation watermarking schemes. In: Proceedings of the 6th international conference on image processing (ICIP'99), vol 1 of controlling chaos and bifurcations in engineering systems, IEEE, pp 320–323
17. Briassouli A, Tsakalides P, Stouraitis A (2005) Hidden messages in heavy-tails: DCT-domain watermark detection using alpha-stable models. IEEE Trans Multimedia 7(4):700–715

Design of Wireless Monitoring System for Environment Monitoring in Greenhouse Cultivation

Sheng Jiang, Wei-xing Wang, Yue-ming Hu and Yun Ye

Abstract A wireless monitoring system was set up in greenhouse at Xining city, Qinghai province, for the application of WSN in greenhouse environment monitoring. Within the system, one type of sensing node and one type of gateway were designed and developed. The system's hardware platform was composed of a MSP430F149 micro-processor for data processing and controlling, an AM2301 sensor module for data measurements, an nRF905 transceiver module for data transmission and reception, and a GPRS module for remote data transmission. The wireless routing protocol, time synchronization algorithm and application program were compiled based on C programming language. A WEB mode remote data management platform was developed based on GPRS network, ASP.NET, HTML and C#. The system solved the problems in greenhouse cultivation such as the informatization of greenhouse environment, provided a new method to guarantee safety and high-productivity.

Keywords Greenhouse environment · Network performance · Wireless monitoring system · Wireless sensor network

---

S. Jiang · W. Wang (✉) · Y. Hu
College of Engineering South China Agricultural University, Guangzhou, China
e-mail: weixing@scau.edu.cn

S. Jiang
e-mail: jiangsheng@scau.edu.cn

Y. Hu
e-mail: ymhu@scau.edu.cn

Y. Ye
Guangdong Youyuan Land Information Technology CO. LTD, Guangzhou, China
e-mail: inthefly@163.com

E. Qi (ed.), *Proceedings of the 6th International Asia Conference on Industrial Engineering and Management Innovation*,
DOI 10.2991/978-94-6239-145-1_22

1 Introduction

Wireless sensor network (WSN) technology has been integrated into people's lives and production in all areas. It is listed as one of the most influential technologies in the 21st century. Recently, WSN has been widely used at various fields including agriculture [1–5]. Intel set up a wireless sensor networks in Oregon vineyards [6], and analyzed the performance of the network. Aline Baggio established a WSN system based on T-MAC protocol to monitor temperature, humidity and leaf surface temperature of potatoes [7]. In-field WSN systems were designed for collecting agricultural information such as soil water content, air temperature and humidity, soil salinity etc. [8–11]. Nodes were deployed in farmland to gather temperature, humidity, soil water content, sunshine intensity and the concentration of CO_2 [12–15]. Large planted area and high requirement for greenhouse environment are considered to be the important features of greenhouse cultivation. Although keeping suitable temperature and humidity is quite critical for crop growth, but the existing measurement of greenhouse environment still rely upon human. This paper aimed at design of wireless sensor network hardware platform and set up a wireless monitoring system of greenhouse cultivation.

2 Design of the Wireless Monitoring System

The wireless monitoring system is composed of sensor nodes, gateway node and remote manage platform. The sensor nodes are used to collect the environment data of greenhouse, while the gateway node is used to transport network data to the remote manage platform through the GPRS network, and the wireless sensor network is set up by the form of Ad hoc network. The system structure is shown in Fig. 1.

Fig. 1 System structure

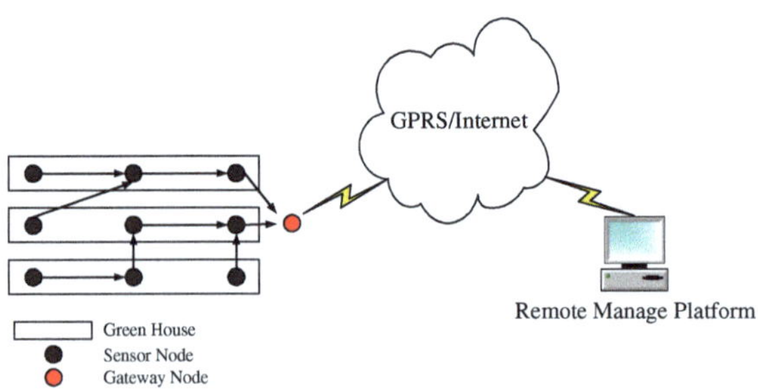

Fig. 2 Sensor nodes

2.1 The Hardware Platform of the System

The senor node hardware platform was composed of a micro controller module, a transceiver module, a sensor module, a serial port communication module and a power module, the node platform is shown in Fig. 2.

- The micro-controller module is the core component. Equipment control, task schedule, data integration and communication protocol could be achieved under the support of micro controller module. AVR series MCU MSP430F149 was selected to be the main control chip.
- WSN is working based upon the requirements of low power and the transceiver module is the key part. As a result, the transceiver module must be designed to consume low power and could change its transmitting power with different requirements. The transceiver nRF905 works at 433/868/915 MHz ISM frequency band with controllable transmitting power was selected to be the wireless communication chip.
- The sensor modules are responsible of finishing the collection of interested physical parameters. The ambient temperature and humidity were collected by AM2301 transducers.
- The JTAG download module is the interface for program download between the embedded system and computer.

The gateway node hardware platform was composed of a micro controller module, a transceiver module, a GPRS module, a data storage module, a serial port communication module and a power module, the gateway node platform is shown in Fig. 3.

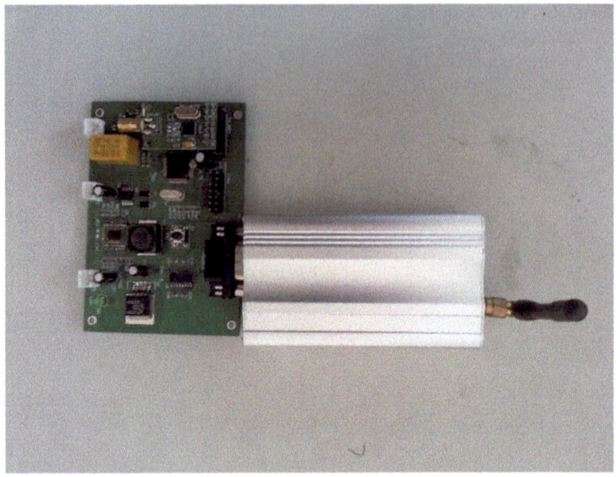

Fig. 3 Gateway node

- As the same of sensor node, the gateway node employed MSP430F149 and nRF905 as the main control chip and RF chip respectively.
- The GPRS module was mainly responsible for data transmission between the gateway node and the external network, and the sending of warning message. The gateway node used MC55 as the main control chip of GPRS module, and the module communicated with controller module through the serial port.
- For the data integrity, a data storage module was used to store network data. The data were stored in to a SD card before each transport to the remote manage platform, and the production manager could get the whole network data to analyze the change law of greenhouse environment.
- The power module employed two Li-batteries to supplied power for the each module of gateway, and charged the batteries by using solar energy. The energy self-sufficient was realized by the battery charging, and prolonged the lifetime of gateway node effectively.

2.2 *The Software Platform of the Nodes*

The design of nodes software platform may directly affect node performance and stability of network. In order to enable the system nodes complete the network task effectively, reduce the package loss rate and guarantee the synchronization of network, the design of system software includes application programs of sensor node and of gateway node.

- The application programs of sensor node realized the calling of bottom drivers, data collection and transmission. The design ideas are as follows, the sensor node collect environment data of greenhouse and broadcast to network when the set time arrives. Then, the sensor nodes enter receive mode to receive other's

Fig. 4 Flowchart of sensor node

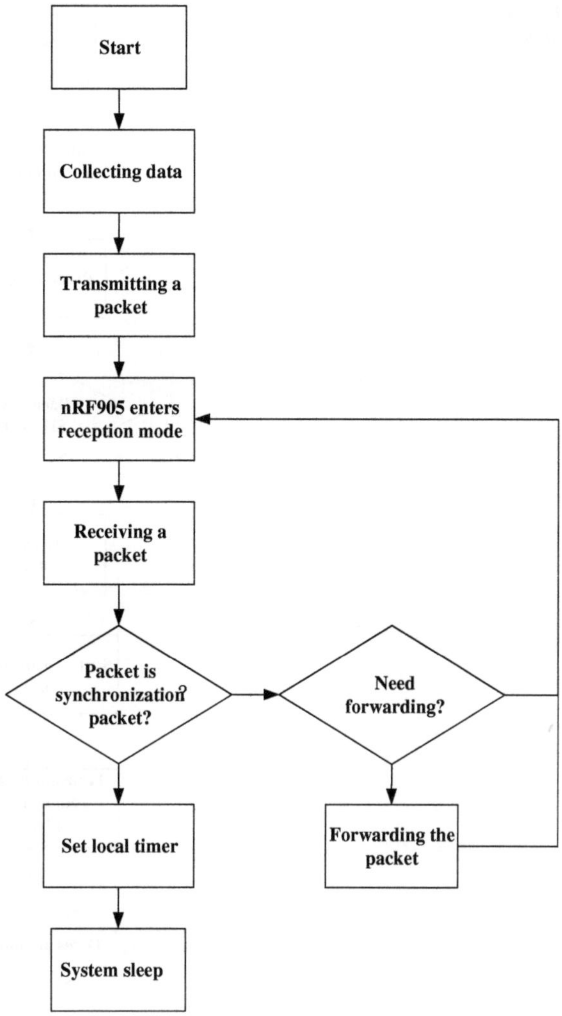

data package. The data package from other nodes will be judged whether need forwarding or discarding. After receiving synchronization packet from gateway node, the sensor nodes set local time and forward the synchronization packet to network, then enter sleep mode waiting for the next cycle period. The flowchart is shown in Fig. 4.

The application programs of gateway node realized the data remote transmission and backup, and send warning message while data abnormal. The design ideas are as follows, after system starting, the gateway node sent synchronization packet to network firstly, to sleep the whole network. When the set time arrives, the gateway node enter receive mode to receive data package from network and backup the data

Fig. 5 Flowchart of gateway node

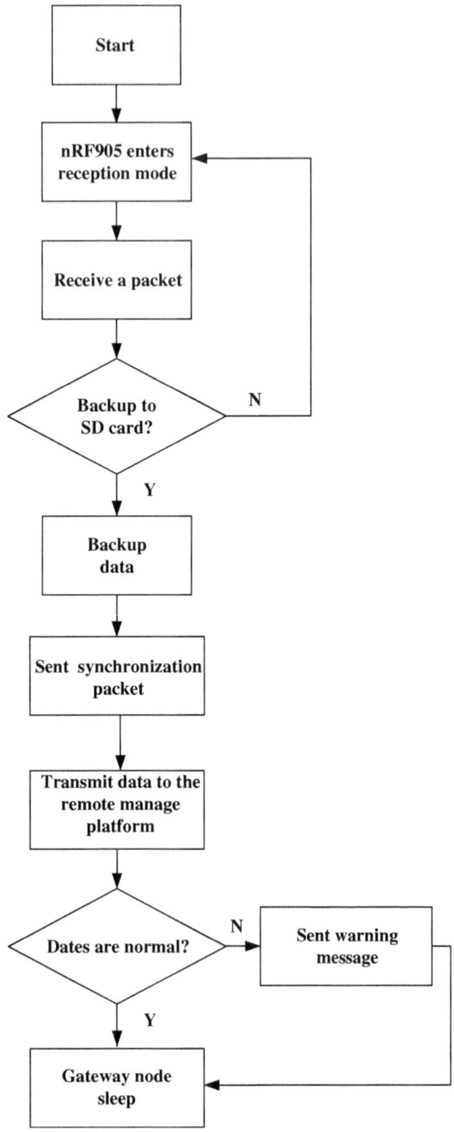

into SD card. Then, sent the synchronization packet to sleep the network, and transmit data to the remote manage platform by GPRS network and internet. The flowchart is shown in Fig. 5.

2.3 Design of Remote Manage Platform

For the convenience of data management and consulting, a remote manage platform was designed, which based on .NET development platform with SQL Server 2000 and C#. Users could manage and consult data remotely by computer and mobile phone that connected to the internet or Wap network.

The functions of platform mainly include data real-time display, data statistic analysis, real-time warning system and historical data inquiry etc.

3 Networking Experiments and Analysis of Node Performance

The system was set up in HuiFeng vegetables plant base in Xining city, Qinghai province. 5 greenhouses were selected to online monitor environment dates include ambient temperature and humidity. Each greenhouse placed 1 sensor node and 3 sensors to measure data of 3 points of greenhouse. The on-site pictures of sensor nodes deployment is shown in Fig. 6.

Fig. 6 Picture of node installation

Table 1 PLR values of nodes

ID	Local pakts	Forwarding pakts	Rec pakts	Lost
Node 1	336	15	345	6
Node 2	336	12	339	9
Node 3	336	7	337	6
Node 4	336	5	329	12
Node 5	336	1	328	9
Total	1680	40	1678	42

3.1 Networking Test

The packet loss rate (PLR) was tested to evaluate network reliability. The data acquisition cycle was set to 30 min, and the PLR test period was set to 168 h. The result of PLR test are shown in Table 1, in the table Local pakts means the number of local packets generated by each node, Forwarding pakts means the number of packets which the node forwarded, Rec pakts means the number of packet that the PC received, and Lost means the number of packets that each node lost. The PLR of system network computed as follow, and the value is 2.4 %.

$$\mathrm{PLR} = \frac{\sum \mathrm{Droped}}{\sum \mathrm{Local\_Pakts} + \sum \mathrm{Forwarding\_Pakts}} (\%)$$

Ambient temperature and humidity was measured and delivered during the networking with AM2301 transducers. The average data of ambient temperature and humidity from nodes 1, 2, 3, 4 and 5 on August 19th were drawn in Figs. 7 and 8 respectively.

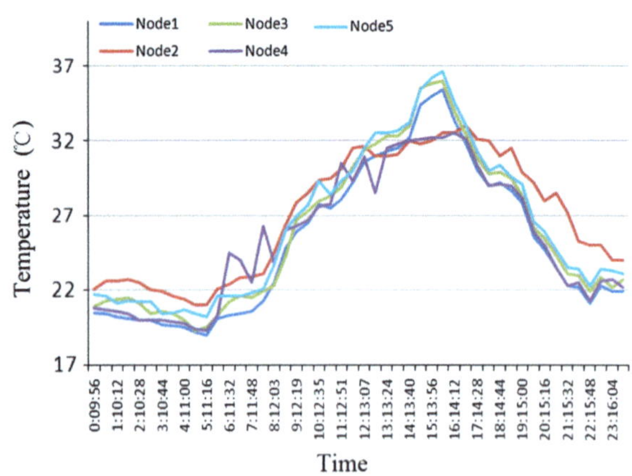

Fig. 7 Temperature in greenhouses on August 19th

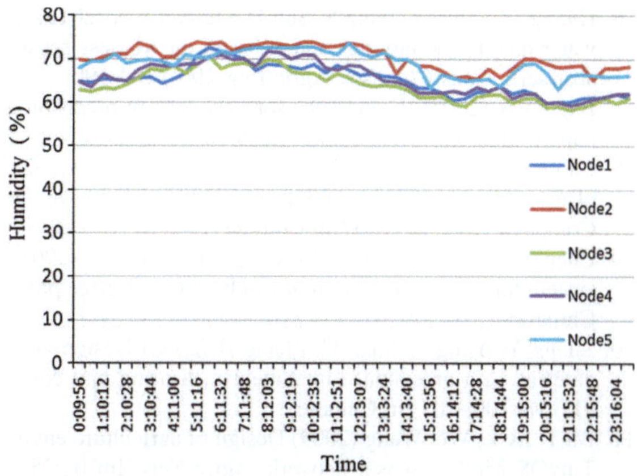

Fig. 8 Humidity in greenhouses on August 19th

4 Conclusion

WSN is an effective tool for measuring the environmental parameters. The nodes designed in this paper could work with low power and measured ambient temperature and humidity precisely. Analysis of test results indicated that the delivery of the measured in-field data to the remote manage platform could be achieved through the established WSN, and the PLR of network was examined.

Acknowledgments This research was supported by the Projects in the National Science & Technology Pillar Program during the Twelfth Five-year Plan Period (Grant No. 2012BAK17B15). We appreciate the continued and enthusiastic support.

References

1. Beckwith R, Teibel C, Bowen P (2004) Report from the field: results from an agricultural wireless sensor network. In Proceedings 29th annual IEEE international conference. IEEE, Piscataway, New Jersey, pp 241–249
2. Li Z, Wang N, Franzen A, Venkateshwaran A, Godsey C, Li X (2008) Development of a wireless sensor network for field soil moisture monitoring. ASABE Paper No. 083835. St. Joseph. Mich.: ASABE, pp 135–141
3. Andrade-Sanchez P (2007) Performance assessment of wireless sensor networks in agricultural setting. In: 2007 ASABE annual international meeting. Minneapolis, USA, pp 311–319
4. Jiang S, Wang W, Sun D, Zhen L (2012) Design of energy self-sufficient wireless sensor network node for orchard information acquisition (in Chinese with English abstract). Trans Chin Soc Agric Eng 28(9):153–158
5. Wang W, Luo X, Sun D, Jiang S, Xiao G (2011) Design of wireless sensor network node for data transmission in tea plantations (in Chinese with English abstract). Trans Chin Soc Agric Eng 27(5):169–173

6. Huang J, Wang W, Jiang S, Sun D, Ou G, Lu K (2013) Development and test of aquacultural water quality monitoring system based on wireless sensor network (in Chinese with English abstract). Trans Chin Soc Agric Eng 29(4):183-190
7. Baggio A (2007) Wireless sensor networks in precision agriculture. www.sics.se/realwsn05/papers/baggio05wireless.pdf. Accessed 9 July 2007
8. Bao-hua Zhang, Shi-ning Li, Wen-xing Teng, Yao-wei Zhang (2008) Development and design of greenhouse testing and control system based on wireless sensor networks. Microeletronics Computer 25(5):154–157 in Chinese
9. Chen X, Xue M, Wang J, Wu G, Qin L, Cheng R (2007) Wireless measurement and control system for greenhouse environment based on ZigBee protocol. Autom Instrum 3:39–41, 50 (in Chinese)
10. Zhang Y, Yang Z, Shen C, Zhang B (2008) Design and implementation of the WSN-based gateway in a greenhouse intelligent measuring and controlling system. Comput Eng Sci 30 (06):98–100, 105 (in Chinese)
11. Bu T, Lv L, Wei Wang (2009) Design of agriculture environment monitoring system based on TinyOS wireless sensor network. Agric Netw Inf 02:25–27 in Chinese
12. Li L-W, Yeo T-S, Kooi P-S, Leong M-S (1998) Radio wave propagation along mixed paths through a four-layered model of rain forest: an analytic approach. IEEE Trans Antennas Propag 46(7):1098–1111
13. Ayday C, Safak S (2009) Application of wireless sensor networks with GIS on the soil moisture distribution mapping. In: Proceedings of 16th international symposium GIS Ostrava 2009. Seamless Geoinformation Technologies, Ostrava, Czech Republic, pp 123–132
14. Akyildiz IF, Su W, Sankarasubramaniam Y, Cayirci E (2002) Wireless sensor networks: a survey. Comput Netw 38(4):393–422
15. Camilli A, Cugnasca CE, Saraiva AM, Hirakawa AR, Corrêa LP (2007) From wireless sensor to field mapping: anatomy of an application for precision agriculture. Comput Electron Agric 58:25–36

An Energy-Balanced Clustering Algorithm for Wireless Sensor Networks Based on Distance and Distribution

Li-zhen Zheng, Lu Gao and Tong-gang Yu

Abstract Organizing wireless sensor networks into clustered architectures is an effective approach for energy balancing so as to prolong the network lifetime. This paper proposes an Energy-Balanced Clustering Algorithm based on Distance to the base station and neighbor Distribution (EBCADD) to generate clusters in wireless sensor networks with random distribution. In order to minimize the energy consumption of entire network, a cluster has a larger cluster size as increasing distance from the base station. The optimal node became cluster head on the basis of the residual energy and neighbor distribution in cluster head election process. Besides the algorithm controls the size of cluster based on threshold size in terms of the number of the nodes per cluster to form an appropriate architecture of the network topology. Simulation results demonstrate that the new algorithm can establish more balanceable clustering structure effectively in uniform and nonuniform distribution of nodes, and prolong the network life.

Keywords Clustering · Energy balance · Neighbor distribution · Wireless sensor networks

1 Introduction

Wireless Sensor Networks (WSNs) consist of a number of small, self-powered sensing devices (nodes) that communicate with each other using a wireless medium [1]. The routing energy efficiency of a wireless sensor network is a crucial issue for the network lifetime.

Clustering provides an effective way to achieve energy efficiency. The network is divided into clusters in cluster-based WSNs. Each sensor node exchanges its

L. Zheng (✉) · L. Gao · T. Yu
Department of Equipment Command and Management,
Ordnance Engineering College, Shijiazhuang, China
e-mail: zhlzh1981@126.com

E. Qi (ed.), *Proceedings of the 6th International Asia Conference on Industrial Engineering and Management Innovation*,
DOI 10.2991/978-94-6239-145-1_23

information only with its cluster head (CH), which transmits the aggregated information to the Base Station (BS) via onehop or multihop. The energy consumed on intracluster communication increases in proportion to the number of nodes within a cluster, while the energy spent on intercluster communication is a function of the expected load from the clusters further away. If we assume that the sensors are uniformly deployed in a sensing field, and an equal cluster scheme is adopted, cluster heads further away from the BS have to consume more energy than those closer to the BS for one-hop communication, while in multihop communication, the cluster heads closer to the BS are burdened with heavy relay traffic and tend to die earlier (i.e., the 'hotspot' problem). So the "equal clustering problem" occurs whether in onehop or multihop networks.

Many clustering based protocols have been proposed for WSNs in recent years. We mention some of the most relevant papers. The LEACH [2] protocol is the first typical hierarchy routing protocol, which is a distributed, onehop clustering algorithm for homogeneous WSNs. In LEACH, cluster head role is periodically rotated among the sensor nodes to balance energy dissipation. However, owing to selecting the cluster heads stochastically in disregard of the size of clusters and the residual energy of nodes and adopting the onehop forwarding model in inter-cluster communication, it cannot guarantee the even distribution of cluster heads and the cluster heads further away die much faster. Some improvements are made, but the cluster heads are elected by global information, bearing large computation, and need to know the location information [3, 4].

HEED [5] extends LEACH by incorporating communication range limits and intra-cluster communication cost information. The initial probability for each node to become a tentative cluster head depends on its residual energy, and final heads are selected according to the cost. Cluster headers generate in a distributed and iterative manner, which avoid the uneven distribution of cluster heads. But when the nodes are in random distribution, the cluster heads whose loads are heavier die much earlier.

An unequal clustering model is first investigated to balance the energy consumption of cluster heads in multihop wireless sensor networks [6]. The work focuses on a heterogeneous network where cluster heads (super nodes) are deterministically deployed at some precomputed locations, thus it's easy to control the actual sizes of clusters. But that assumption is not suitable for the randomly deployed network.

EEUC [7] is a distributed cluster heads competitive algorithm, where the competitive range is determined on the basis of relative value of maximum distance between the sensors and the BS. Simulation results in literature show that EEUC clearly balances the energy consumption and improves the network lifetime over LEACH an HEED. However, the four parameters selection strategy is not given in that paper. So the algorithm is difficult to achieve optimal performance in practice, and its main disadvantage is that it can generate no balanced clusters since there is no relationship between the diameter of cluster and its size.

ESAC [8, 9] has considered combined weight of residual energy and nodes density, but have not considered distances to the base station, which lead to the hot problem.

In this paper, we propose an Energy-Balanced Clustering Algorithm based on Distance and Distribution (EBCADD) in wireless sensor networks, which divide the network into different size of clusters according to the distance of cluster heads to the base station. Considering the clusters nearby the BS take more intercluster relay task, the proposed protocols design them to have smaller range, in order to decrease the innercluster consumptions. Furthermore, the algorithm adjusts the ownership of the node to form an appropriate size of clusters if the generation of clusters with a great number of nodes and the generation of clusters with a small number of nodes occur. So the proposed clustering approach are suitable both in uniform and random distribution networks. The simulation results show that the protocols efficiently balance the energy consumption of the whole network and prolong the network lifetime.

The remainder of this paper is organized as follows. Section 2 describes the network and radio models used in our protocol. Section 3 gives a detailed description of the EBCADD Algorithm. We present the simulation study of the proposed protocol in Sect. 4. Finally, Sect. 5 draws conclusions.

2 Preliminaries

2.1 Network Model

We consider a sensor network consisting of N sensor nodes randomly deployed over a square sensing field. We make some assumptions [1], with the following properties:

(1) All nodes are randomly deployed over the sensing field, and the BS is located far away from the square sensing field. All nodes are stationary and energy constrained.
(2) All nodes are homogeneous and have the same capabilities. Each node is assigned a unique identifier (ID).
(3) The nodes have power control capabilities to vary their amount of transmitted power which depends on the distance to the receiver.
(4) Nodes needn't to be equipped with GPS-capable unit to get precise location information.
(5) Data fusion is used to reduce the total data message sent.

2.2 Radio Energy Model

Our energy model for the sensors is based on the first order radio model as used in Ref. [2]. In this model, the transmitter has power control abilities to dissipate minimal energy to send data to the receiver. If the distance is less than a threshold

d_0, the free space (FS) model is used; otherwise, the multipath (MP) model is used. This concept can be expressed mathematically as following:

$$E_{Tx}(k,d) = \begin{cases} kE_{elec} + k\varepsilon_{fs}d^2; & d < d_0 \\ kE_{elec} + k\varepsilon_{mp}d^4 & d \geq d_0 \end{cases} \quad (1)$$

where E_{elec} is the energy spent in transmitting and receiving data for a sensor; ε_{fs} and ε_{mp} are the energy spent in amplifying, and the energy consumption of the receiver is defined in (2), which is irrelative with the distance.

$$E_{Rx}(k) = kE_{elec} \quad (2)$$

The data fusion model used in our simulations assumes that the overall information collected by a cluster of n nodes, where each node collects k bits of data, can be compressed to k bits regardless of the number of nodes. Cluster heads consumes E_{DA} amount of energy for data aggregation.

$$E_{fusion}(k) = kE_{DA} \quad (3)$$

3 The EBCADD Algorithm

In this section, the EBCADD algorithm and its pseudo code are illustrated. The cluster formation phase, which is divided into three phases: initialization phase, main iterative clustering phase and finalization phase. The steady data transmission phase is broken up into TDMA frames. During each frame, every regular node, at the time of its respective time slot, sends sensing data to its CH. At the end of each TDMA frame, every CH forwards the aggregated data to the BS via multi hop routing.

3.1 Compute the Initial Competition Radius of Clusters

In cluster preparation process, the algorithm determined the radius of clusters according to relative distance of the base station to balance energy consumption of clusters.

First, if N sensor nodes uniformly deployed in the square sensing field, when the cluster is equal to the optimal cluster radius, the total energy spent in the network is minimal [10]. The optimal radius of cluster is given as follows:

$$R_{opt} = 2\sqrt[4]{\frac{2\pi||A||E_{DA}}{27N\varepsilon_{fs}}} \quad (4)$$

where ||A|| denote the size of sensing field. The proof is omitted here. As an example, if the size of sensing field $||A|| = 100 \times 100$, and the number of sensor nodes $N = 100$, $\varepsilon_{fs} = 10$ pJ/bit/m^2, $E_{DA} = 5$ nJ/bit/signal, the optimal radius of cluster $r_{opt} = 20.7719$ m.

Then compute the competition range according to relative distance to the base station to balance energy consumption of clusters. The cluster heads closer to the BS should have smaller cluster sizes because of higher energy consumption during the intercluster multihop forwarding communication. Thus the node's competition radius should be increase proportionally to its distance to the BS. We set R_{comp} of node s_i as a function of its distance to the BS,

$$R_{comp} = \frac{2d(s_i, BS)}{d_{\max} + d_{\min}} R_{opt} \tag{5}$$

where d_{max} and d_{min} denote the maximum and minimum distance between sensor nodes and $d(s_i, BS)$ is the distance between s_i and the BS.

3.2 Weigh Based on Distance and Distribution

In the EBCADD algorithm, each node computes its weight according to (6). The nodes with greater weight have a higher priority of being elected as CH. On the other hand, every regular node joins the CH which has the highest weight in its vicinity.

$$\begin{gathered} Weight(i) = \alpha \cdot E(i) + \beta \cdot D(i) \\ \text{With}(0 \leq \alpha, \beta \leq;\ \alpha + \beta = 1) \end{gathered} \tag{6}$$

The weight of each sensor node is a combination of the residual energy, and neighbor distribution. Moreover the coefficient of each parameter can be chosen depending on the application.

The first item considers the energy distribution, which is the radio of the node residual energy to the average neighbor residual energy. The computing equation is

$$E(i) = V_i \cdot E_{redidual} / \frac{\sum_{j=1}^{m} V_j \cdot E_{redidual}}{m} \tag{7}$$

where m denotes the number of node neighbor. This function considers not only the node energy, but also the neighbors' energy distribution, which balance the energy distribution.

The second item consider the combination of two main features, node centrality (the centrality of a node among its neighbors) and node degree (the number of neighbors), which computed as follows:

$$D(i) = \sum_{j=1}^{m} \frac{1}{dist^2(s_i, s_j)} \quad (8)$$

The lower the value of the neighbors' distances from node s_i, the higher the value of the weight. This means that the location of node s_i is more central among its neighbors and is thus more suitable for election as a CH. The node centrality feature in CH election has some advantages for consideration.

The greater the number of node s_i's neighbors, the higher the value of weight. In other words, node s_i is more suitable for election as a CH and so, dense clusters are constructed in the area. The aim is to exploit the correlation and eliminate the redundancies that often exist in the sensor readings.

3.3 Cluster Heads Election Phase

The clustering process of EBCADD is divided into three phases:

(1) *Initialization Phase*: In the beginning of this phase, neighbor information must be updated. Afterwards, each node may compute its weight independently. This weight will not be broadcasted to neighbors as it is exchangeable through *CH_msg* messages. As previously mentioned, note that updating the neighbor information and computing weights are not required every time clustering is triggered. At first, the protocol sets an initial percentage of CHs among all sensor nodes, C_{prob}. Here, C_{prob} is set to 0.05. Each sensor node sets its own probability of becoming a CH, CH_{prob}, (the same as [3]):

$$CH_{prob} = \max(C_{prob} \times E_{res}/E_{\max}, P_{\min}) \quad (9)$$

In the above, E_{res} is the current energy of the sensor node and E_{max} is the maximum energy corresponding to a fully charged battery. CH_{prob} is not allowed to fall below a certain threshold p_{min}, which has been selected to be inversely proportional to E_{max}. The threshold p_{min} limits the number of iterations of the Main processing phase to O(1). See the Initialization phase in Fig. 1.

(2) *Main Processing Phase*: S_{CH} shall be defined as follows: $S_{CH} = S_{final\_CH} \cup S_{tentative\_CH}$, where S_{final_CH} = {All *final_CHs* from iteration 1 to iteration i} and $S_{tentative\_CH}$ = {All *tentative_CHs* from iteration 1 to iteration i}. At the beginning of this repetitive phase, a node with higher amount of residual energy has a greater chance, CH_{prob}, of becoming a *tentative_CH*. If a node becomes a *tentative_CH*, it broadcasts a message announcing its new status to all other sensor nodes within its cluster range. In the next iterations, if this particular node has the highest weight

Fig. 1 Pseudo code of EBCADD protocol

Phase Ⅰ. Initialization
For every node in the network
compute R_{comp} according to Equation (5);
obtain the neighbor node set (S_{nbr}) throught broadcasting,
S_{nbr}={node *i*: *i* lies in the competition range };

$$Weight(i) = \alpha \cdot E(i) + \beta \cdot D(i);$$

$$CH_{prob} = \max(C_{prob} \times E_{res} / E_{\max}, P_{\min});$$

is_final_CH = FALSE;

Phase Ⅱ. Main Processing

```
REPEAT
  IF((S_CH←(v: v is a tentative_CH or final_CH)≠ Φ )
  IF(S_fialCH←(v: v is a final_CH)≠ Φ )
   IF(NodeID = MOST_WEIGHT (S_CH ) AND
CH_prob=1)
      CH_msg (NodeID, final_CH, weight)
      is_final_CH =TRUE;
 Else IF(CH_prob= =1)
   CH_msg (NodeID, final_CH, weight)
   is_final_CH =TRUE;
 Else IF(Random(0,1)< CH_prob)
   CH_msg (NodeID, tentativel_CH, weight)
   is_final_CH =TRUE;
 END IF
 CH_previus= CH_prob
 CH_prob=MIN(CH_prob×2, 1);
UNTIL CH_prob=1
```

PhaseⅢ Finalization

```
IF(is_final_CH = FALSE)
  IF((S_fialCH←(v: v is a final_CH)≠ Φ )
   my_CH←MOST_WEIGHT (S_final_CH )
   JOIN_CLUSTER(my_CH, NodeID)
  ELSE
    CH_msg (NodeID, final_CH, weight);
```

among the *tentative_CHs* in its proximity and its CH_{prob} reaches one, it will become a *final_CH* and shall broadcast a *final_CH* message within its cluster range. On the other hand, if a node receives a *final_CH* message, it can no longer be elected as a CH. Therefore, in the following phase, it must choose to connect to one of the *final_CHs* in its cluster radius, based on the weight of that *final_CH*. Finally, each sensor node doubles its CH_{prob} value and continues to the next iteration of the repetitive phase. When its $CH_{previous}$ reaches one, the sensor node stops executing this phase. As a result, the nodes with a higher amount of remaining energy will finish the execution of EBCADD earlier than nodes with a lower amount of remaining energy. This permits the nodes with lower amount of energy to join their clusters. Observe that, in this phase, each *tentative_CH* (or *final_CH*) node can send a *CH_msg* only once. See the Main processing phase in Fig. 1.

(3) *Finalization Phase*: During this phase, each sensor node makes a final decision about its status. If the node is not a *final_CH* and has received at least one *final_CH* message, it will elect the *final_CH* with the highest weight to join it. If a node completes the clustering process and has not yet received any *final_CH* message, it will find itself uncovered and so shall introduce itself as the *final_CH*. See the Finalization phase in Fig. 1.

3.4 Adjust the Size of Clusters

The initial radius are computed on the assumption that the nodes are uniformly deployed, but in fact the nodes distribution is uneven, the neighbor number may vary greatly, so it may generate an important number of clusters with a reduce number of nodes or the clusters with an important size in the case of the dense network. Since there is no relationship between the diameter of the cluster and its size, so we need to control the size of clusters in terms of the number of the nodes per cluster to form an appropriate architecture of the network topology. The expected number of clusters is

$$K_{opt} = \frac{4\,||A||}{3\sqrt{3}r^2} \tag{10}$$

and the expected size of each cluster is

$$N_{opt} = \frac{N}{K_{opt}} \tag{11}$$

In order to generated K_{opt} number of even clusters, and guarantee every cluster has N_{opt} member, the neighbor number of every node should satisfy

$$N_{nei} = \frac{\alpha N}{K_{opt}} \tag{12}$$

where the proper value of α will be determined in simulations.

The execution of adjusting the size of clusters proceeds in the following way. Cluster-heads that belong to clusters whose size if strictly lower than N_{nei}, broadcast a new message called *RE-AFF_CH* to re-affiliate nodes belonging to the small clusters to them. Then, each node that receives this message and belongs to a small cluster should re-affiliate to the nearest cluster-heads whose weight is greater than its own. This additional measure can eliminate small clusters so long as to create stable number and balanced clusters.

4 Simulation Results and Analysis

In this section, a comparison between the simulation results in EBCADD, LEACH and HEED protocols is performed via Matlab software. The parameters are listed in Table 1. In the following, the proper value of α is discussed, and then network lifetime, which is defined as the number of execution rounds until the first node dies (FND), are evaluated in uniform and uneven distribution of nodes.

4.1 Setting α Variable

In order to obtain the proper α, EBCADD was run for different scenarios. In Fig. 2, α differs from 1.3 to 1.5 and each plot demonstrates the average of three executions for 400, 500 and 600 number of nodes until the first node dies.

Table 1 Parameter settings

Parameter	Value
Network coverage	(0,0)–(400, 400)
BS location	(200, 450)
ε_{fs}	10 pJ/bit/m^2
ε_{mp}	0.0013 pJ/bit/m^4
E_{elec}	50 nJ/bit
E_{DA}	5 nJ/bit/signal
Threshold distance(d_0)	75 m
Initial energy per node	2 J
Data packet size	100 byte
Control packet size	25 byte
P_{min}	0.005
C_{prob}	0.05

Fig. 2 Influence of the value of parameter α on the performance of EBCADD

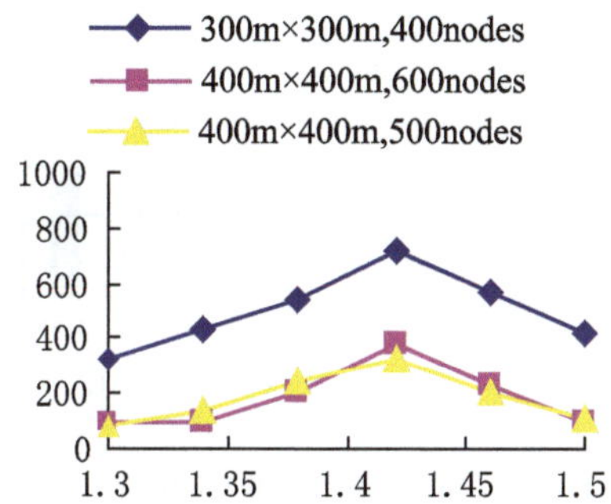

Figure 2 shows that as the number of nodes increase, the network lifetime increase. From the simulation results under different node density, $\alpha \in [1.39, 1.43]$ perform better in network lifetime.

4.2 *Simulations in Uniform Distribution*

Figure 3 shows the simulation results when the nodes are in uniform distribution. In the simulation experiment, node sizes of 400, 500 and 4600 are chosen. It is demonstrated that the proposed protocol is able to outperform the HEED and the LEACH protocols with respect to network lifetime by 30 and 80 % at most, respectively. It is also shows that as the node density increase, the network lifetime of HEED and EBCADD increase, but LEACH decrease. That because HEED and EBCADD adjust the cluster heads distribution to balance the energy consumption. When the node density increase, they make more nodes to periodically rotate to became cluster heads, and shorten the time of every node to be a cluster head, so as to prolong the network lifetime. The LEACH generates CHs in a random way, and there is no guarantee of a fair distribution of CHs in the field. It is possible that some CHs are located far away from their respective members. So EBCADD is a scalable clustering algorithm in terms of the number of nodes density.

4.3 *Simulations in Non-uniform Distribution*

In this section, the simulation results when the nodes are in non-uniform distribution are discussed. From the comparison between Figs. 3 and 4, the performances of three algorithms are all declined. But the EBCADD are declined only 10 %, while HEED and LEACH are declined 25 and 50 %, respectively. Because the EBCADD adjust the size of the clusters, avoid the generation of clusters with a great number of nodes and the generation of clusters with a small number of nodes.

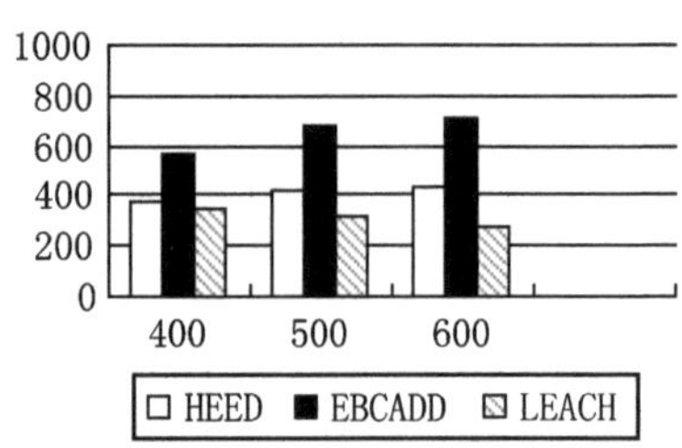

Fig. 3 Comparison of the simulation results when the nodes uniformly distributed

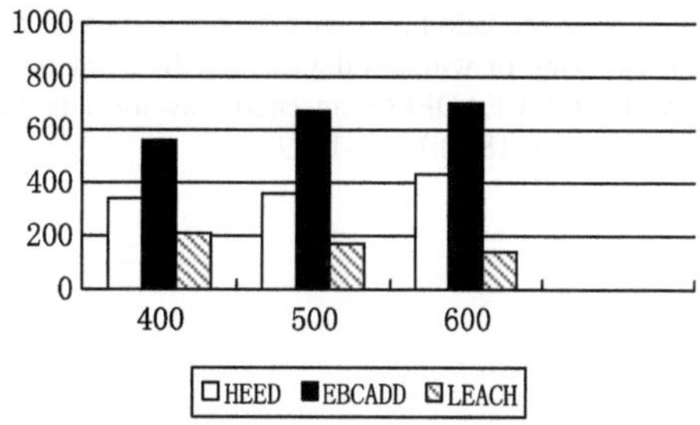

Fig. 4 Comparison of the simulation results when the nodes non-uniformly distributed

5 Conclusion

In this paper, we tackle the equal clustering problem and use the optimal cluster size based on the distance between CH and the BS. We propose a clustering algorithm which has unequal cluster size, where the cluster has the larger size as increasing the distance from the BS. Additionally, the algorithm controls the size of cluster based on threshold size in terms of the number of the nodes per cluster to form an appropriate architecture of the network topology. Simulation results demonstrate that the new algorithm can establish more balanceable clustering structure effectively in uniform and non-uniform distribution of nodes, and prolong the network life.

References

1. Akyildiz I, Su W, Sankarasubramaniam Y, Cayirci E (2002) A survey on sensor networks. Commun Mag 40(8):102–114
2. Heinzelman WB, Chandrakasan AP, Balakrishnan H (2002) An application-specific protocol architecture for wireless micro sensor networks. IEEE Trans Wirel Commun 1(4):660–670
3. Chang RS, Kuo CJ (2006) An energy efficient routing mechanism for wireless sensor networks. In: Proceedings of the 20th international conference on advanced information networking and applications (AINA'06), pp 308– 312
4. Heinzelman W, Chandrakasan A, Balakrishnan H (2002) An application-specific protocol architecture for wireless microsensor networks. IEEE Trans Wirel Commun 4:660–670
5. Younis O, Fahmy S (2004) HEED: a hybrid, energy-efficient, distributed clustering approach for ad hoc sensor networks. IEEE Trans Mob Comput 3(4):366–379
6. Soro S, Heinzelman WB (2005) Prolonging the lifetime of wireless sensor networks via unequal clustering. In: Proceeding of the 19th IEEE international parallel and distributed processing symposium (IPDPS), pp 236–244
7. Li CF, Ye M, Chen GH, Wu J (2005) An energy-efficient unequal clustering mechanism for wireless sensor networks. In: Proceeding of the IEEE international conference on mobile ad hoc and sensor systems, Washington, pp 597–604
8. Lehsaini M, Guyennet H, Feham M (2008) A novel cluster-based self-organization algorithm for wireless sensor networks. In: Proceedings of international symposium on collaborative technologies and systems, Irvine: IEEE Press, pp 19–26

9. Olascuaga-Cabrera JG, Lopez-Mellado E, Mendez-Vazquez A et al (2011) A self-organization algorithm for robust networking of wireless devices [J]. IEEE Sens J 11(3):771–780
10. Ming L, Cao J, Chen G (2007) EADEEG: an energy-aware data gathering protocol for w ireless sensor networks. J Softw 18(5):1092–1109

Coded Excitation System for Stationary Target Detection Using Multi Segment Coding

Jie Li, Kai Zhan, Pan-pan Liu, Xiao-nian He and Zheng-di Qin

Abstract In ultrasound systems, the analysis of inherent signal-to-noise ratio (SNR) is one of the eternal topics. Desired signal improvements of approximately 10 times are theoretically possible for ultrasonic testing (UT) using coded excitation. When applying the coded excitation into the ultrasound pulse-echo system, code selection is important as the limited time duration of the transmission pulse. Any selected code sequences can hardly get a good auto-correlation property. In earlier reports, a technique called Truncated Long Code (TLC) was proposed for dynamic signal detection [1–3]. Good auto-correlation property has been acquired as the parameter estimation time window can be much longer than the single pulse duration. In this report, we use the same approach to form a multi-segment coding scheme to detect the stationary objects in ultrasound pulse-echo system. The importance of this study are to confirm the efficacy of this new scheme and try to find possible uses in biomedical and industrial applications as well.

Keywords Coded excitation · Multi-segment coding · Pulse-echo · Ultrasound system

J. Li · K. Zhan · P. Liu · X. He · Z. Qin
National-Regional Key Technology Engineering Laboratory of China for Medical Ultrasound, Shenzhen, China

J. Li · K. Zhan · P. Liu · X. He · Z. Qin
Guangdong Province Key Laboratory for Biomedical Information Detection and Ultrasound Imaging, Shenzhen, China

J. Li · K. Zhan · P. Liu · X. He · Z. Qin (✉)
Department of Biomedical Engineering, Health Science Center, Shenzhen University, Shenzhen, China
e-mail: finyan@szu.edu.cn

E. Qi (ed.), *Proceedings of the 6th International Asia Conference on Industrial Engineering and Management Innovation*,
DOI 10.2991/978-94-6239-145-1_24

1 Introduction

Ultrasound pulse-echo system for stationary target detection have a lot of applications in many fields, such as medical ultrasound imaging, non-destructing testing (NDT) and ultrasonic ranging and positioning. Compared with pulse-echo system for dynamic target detection, the former might be relatively simple. But some typical problems also exist such as the tradeoff between resolution and penetration and SNR. There is a close relationship between the bandwidth and resolution of ultrasound (As shown in Fig. 1). Wide bandwidth, short wavelength can achieve a good axial resolution, but reduce the penetration power. Narrow bandwidth, long wavelength can increase penetration power, but lose the resolution performance. So it is unreasonable to simultaneously improve the axial resolution and penetration only by changing the pulse duration. For improving SNR, increasing transmit power is a direct method which result in much energy consumption. The system simplification and miniaturization will be more difficult [4]. Also the peak power must follow some standard (e.g. ECI standard enacted by Chinese government [5]). Consequently, coded excitation, one of the methods greatly increase the transmit energy without affecting the peak power. The SNR of ultrasound system can be improved [6].

Coded excitation can increase the penetration and improve the SNR in the ultrasound pulse-echo system. This technique has been widely used in radar and radio communication for many years. These systems operate at large time-bandwidth products resulting in small range side lobe levels. It was successfully introduced for ultrasound system, especially in clinical ultrasound systems only within the last 10 years. The application of coded excitation method to ultrasound system is limited because of the restricted time-bandwidth products available brings large range side lobe levels, as well as the complexity of the electronics needed to generate coded signals.

As is well known, such kind of short codes cannot give good appearances in ultrasound testing systems. It could suffer from the artifacts by the side lobes. So code selection is a critical important topic all the time. Extensive efforts have been made to search for codes with good auto-correlation properties. Even though carefully selected codes given by computer have many good performances, while they don't have any advantage over simple a code sequence such as m-sequence when they penetrate through inhomogeneous and nonlinear medium such as metal

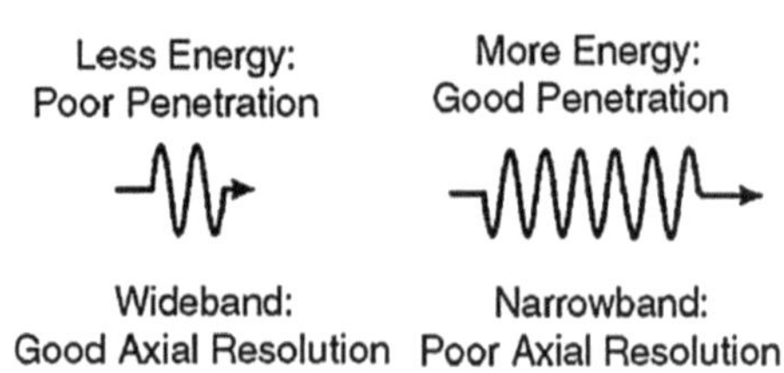

Fig. 1 Tradeoff between bandwidth and axial resolution

materials and human body. In fact, not like show on the computer, it is true that a carefully selected code sequence might have even poorer performance compared with a random picked-up code sequences.

A coding system is highly coherent. The phase change results from the target movement will destroy the system coherence. So, when a sequence of longer codes is used, the target should be kept as stationary as possible. While for a moving target, a phase compensation filter should be added into the system. The key point of this multi-segment coding concept is to try to make the optimized use of the system resources (time and bandwidth). In many applications, the time period for getting a result or what is called the estimation window could be much longer than the single pulse duration. A relatively longer code could be considered to be filled into the estimation window. To do so, a long code is spilled into many short segments and they are modulated into every individual pulse. In this report, simulations and measurements are used to demonstrate the efficacy of the method in the application of stationary target detection.

2 The Principle of Multi-segment Coding

The principle of the multi-segment coding technique is illustrated in Fig. 2. The estimation window contains N bursts. The long code is truncated into N segments. Each segment has a short code sequence of m chips, and they are modulated one transmission burst. Totally, m × N chips yielded in the estimation window.

In the conventional coded excitation for ultrasound pulse-echo system, each pulse is modulated by a same code sequences such as Barker sequences [7–9]. Consequently, the approach results in serious interference among each other. For all bursts, the same side lobes overlay in radio frequency (RF) signals after decoding. Due to the poor correlation property, artifacts will greatly decrease the system qualities.

The long code is truncated into many different segments. Each segment is modulated one burst. These side lobes can offset each other as they are different

Fig. 2 The principle of multi-segment coding

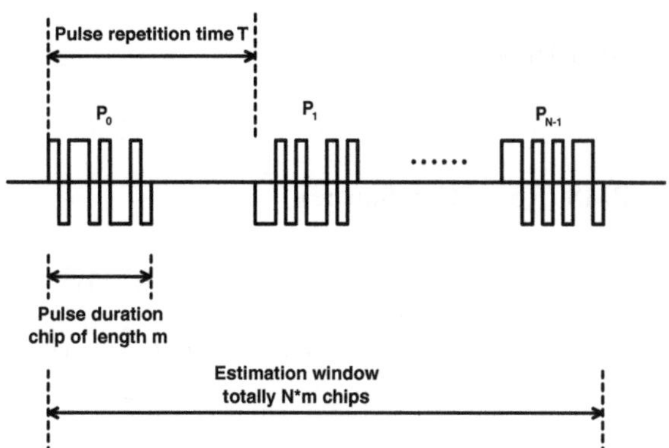

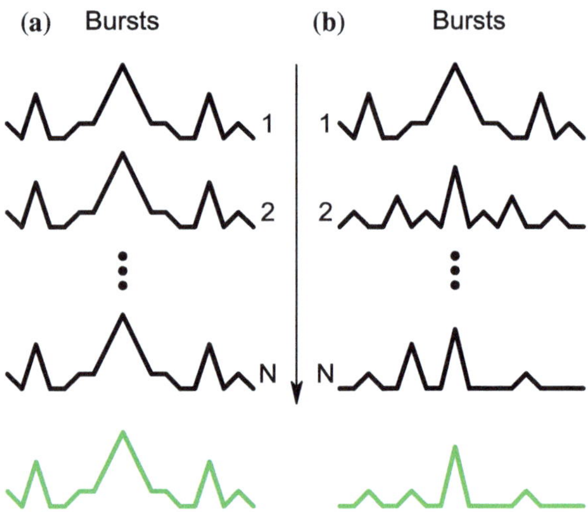

Fig. 3 Auto-correlation property in coding system. **a** Coding with the same code. **b** Coding with different codes

from burst to burst. Therefore, the side lobes effects can be dramatically reduced during the recombination process (see Fig. 3b). This phenomenon reminds us of so-called complementary codes. But in the complementary coding, we collect pulse-echo only from a stationary target. Any kind of movement is not allowed. While in the multi-segment coding scheme, we can not only get echoes from stationary objects, but also can be well applied in the moving case. What's more, the number of the segments can be as many as possible in this approach. Based on the earlier simulation, good performance has been obtained in the research of ultrasound Doppler pulse-echo detection.

After the transmission of each coded pulse, pulse-echoes collection is followed [10]. The analog RF signal is acquired in the memory device by Adlink DAQ card and store the data in the computer. These data are processed using MATLAB after acquisition.

We assume that the received RF signal uniform without any additive noise:

$$s(t) = x(t)cos[2\pi f_0 t + \varphi(t)]. \tag{1}$$

After Hilbert Transform:

$$\hat{s}(t) = x(t)sin[2\pi f_0 t + \varphi(t)]. \tag{2}$$

RMS (Root Mean Square):

$$\begin{aligned} &\sqrt{s(t)^2 + \hat{s}(t)^2} \\ &= \sqrt{x(t)^2(sin[2\pi f_0 t + \varphi(t)]^2) + \cos[2\pi f_0 t + \varphi(t)]^2)} \\ &= x(t). \end{aligned} \tag{3}$$

where $s(t)$ is the data which is acquired from analog RF signal; $x(t)$ is the signal without modulation; f_0 is center frequency of the transducer. These processes are some of the approaches of demodulation. It can gain approximately 3 dB higher than absolute square approach.

The impulse response of Hilbert Transform is shown as:

$$h(n) = \frac{2}{n\pi} \sin^2(\frac{n\pi}{2}). \tag{4}$$

After demodulation and low-pass filter, $x_n(i)$ is the data with code sequences which has been orthogonally demodulated, and correlated by the same codes in the transmission to get the depth information by shifting the codes for different depths:

$$y_n(d) = \sum_{i=1}^{m} x_n(i) c_n^*(i-d), n = 0, 1, 2, \cdots, N-1. \tag{5}$$

where n is the number of the pulse sequence; *d* is the depth in the axial direction; N is the total number of the pulse within the estimation window; *m* is the number of the code; $x_n(i)$ is the base band signal of the echo from the nth transmitted pulse. $c_n(i)$ is the short code sequence for the *n*th coded pulse; $*$ refers a conjugate operation.

The following is combination after all echoes are collected and processed. Without the phase shift because of the echo from stationary target, the combination operation process is relatively simple. It can be expressed as:

$$Z(d,f) = \sum_{n=0}^{N-1} y_n(d) e^{-j2\pi fnT}. \tag{6}$$

where *T* is the pulse repetition time. As:

$$fT = k/N. \tag{7}$$

We have:

$$Z(d,k) = \sum_{n=0}^{N-1} y_n(d) e^{-j2\pi kn/N}, k = 0, 1, 2, \cdots, N-1 \tag{8}$$

It turns out to be a typical Fourier Transform with a constant coefficient for the fixed depth. This process fulfills the spectrum estimation.

3 Experiment Set-up

The experiment conducts in a phantom which out of work. The tube without liquid flow is irradiated by the transducer. The stationary object is the tube wall. The pulse-echoes returned from the tube wall are collected and processed. The functional block diagram of the ultrasound system is illustrated in Fig. 4.

A random selected long code of length 1024 is truncated into 256 segments and placed in a flash memory which represents 256 different short codes. The long random codes are generated by computer and select it based on its auto-correlation (better auto-correlation is needed). The process is shown on Fig. 5, and the Fig. 6 shows the result of the procedure of codes selection we selected. All segments modulate the transmitted pulses from burst to burst in turn. Each segment contains 4 chips for modulating each transmitted pulse. The carrier frequency is 2 MHz, and a transducer with the center frequency of 2 MHz.

4 Experiment Results

Using the oscilloscope, we can see the RF signals from conventional ultrasound system and coding system as shown in Fig. 7. The elements circled by red ellipse are the pulse-echoes returned from the tube wall. The left wall signal was detected by conventional ultrasound system without coded excitation and the right was detected by the coding system using multi-segment coding method.

After processing of demodulation and auto-correlation, we get Fig. 8. As illustrated in the figure, the two main peaks show the 'Wall Signal' in Fig. 7. Figure 8a is the result from the conventional ultrasound system and Fig. 7b is the coding system using multi-segment coding. The differences appear. At first, the peak amplitude value of received waveform was normalized to 1 in the

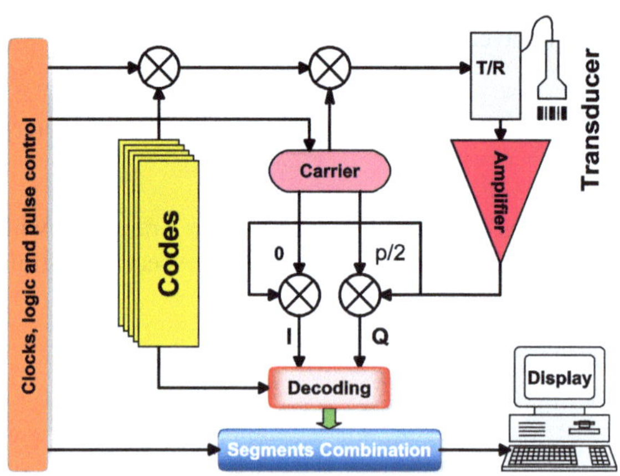

Fig. 4 Functional block diagram of ultrasound imaging system

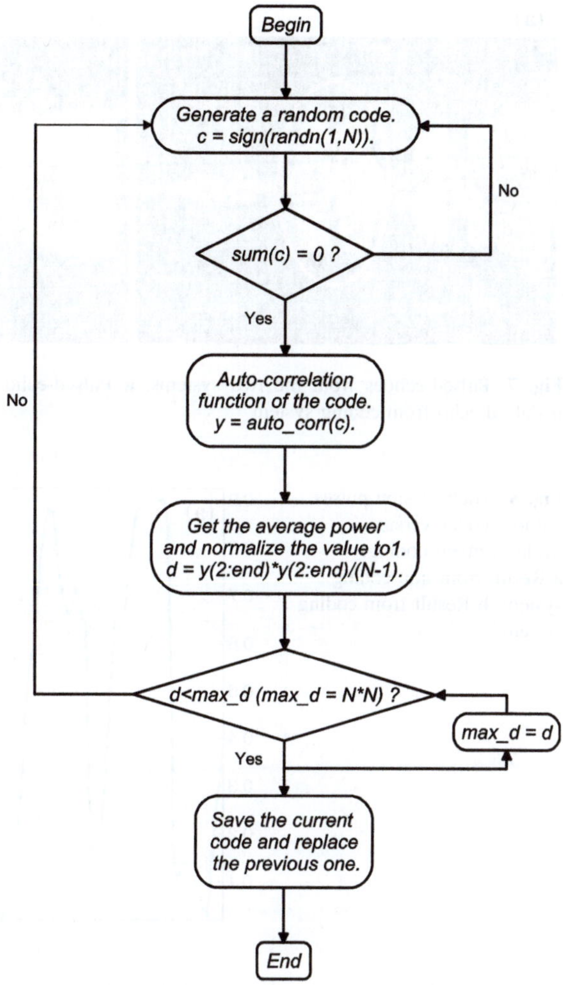

Fig. 5 Get the long random codes with the best auto-correlation property

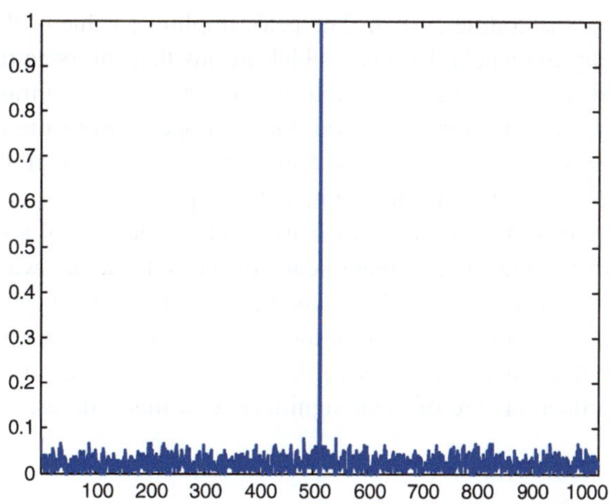

Fig. 6 A set of random codes generated by computer

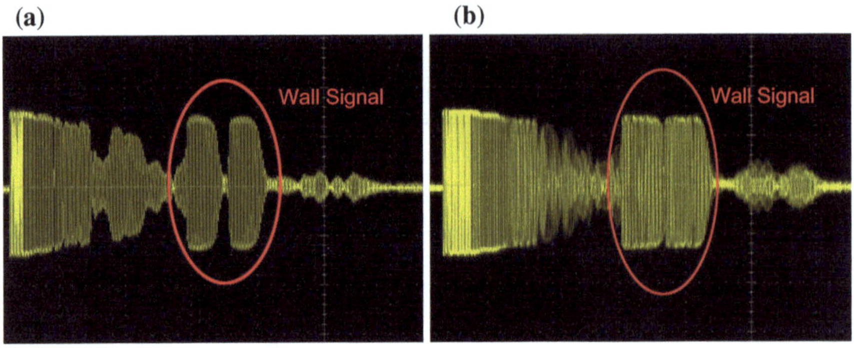

Fig. 7 Pulsed-echoes from the two systems. **a** Pulsed-echo from the system without codes. **b** Pulsed-echo from coding system

Fig. 8 Transmission pulses without codes versus Multi-segment coding. **a** Result from non-coding system. **b** Result from coding system

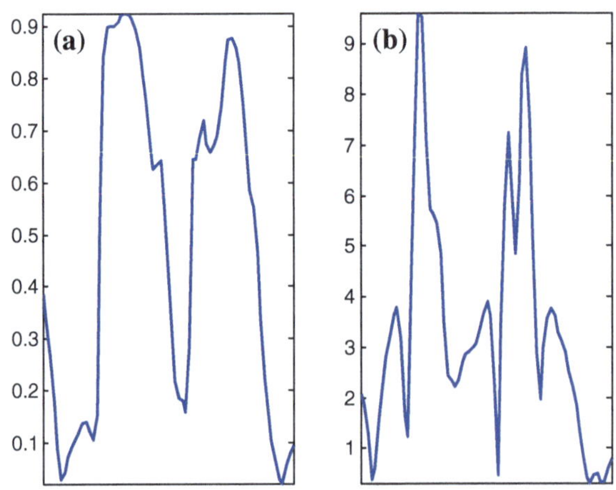

conventional ultrasound system and then the same normalization factor was used for the coding system. The peak amplitude value of the coding system is as high as approximately 10 times which means that the penetration ability is approximately 10 times stronger than that in the conventional ultrasound system when the same transmit power were used. Other property from the conclusion illustrates that the peaks width is narrower than in the coding system. Consequently, the penetration power and SNR are significantly improved.

In simulation result, we also find the element of a waveform like the wall signal. In the Fig. 9, The main peaks of the red line are wall signal. On the left of them, there is another peak (As the blue line illustrates) can be seen. Based on analysis it is a multiple reflection of ultrasound. This is a new discovery. However, the second reflection is too weak to detect in conventional ultrasound system. The multiple reflections are of great significance in many fields.

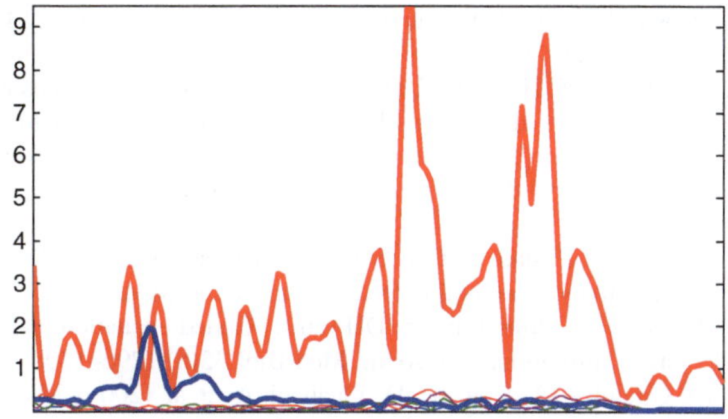

Fig. 9 Discovery of multiple reflections

5 Conclusion

Coded excitation is a powerful technique to improve system quality such as penetration, resolution and SNR. It is a well established technique permitting greatly improved SNR at modest peak power levels compared to that in conventional radar systems. This method of using the multi-segment coding technique proposed in this report can achieve a good performance. Actually, it also can be used in many other fields as well, such as ultrasound thickness measurement, NDT and ultrasonic positioning. Results to date, we have applied the technique in ultrasound Doppler detection earlier and stationary signal detection in this study. This report shows that the application for stationary target detection using multi-segment coding technique also can achieve a much better performance than conventional ultrasound systems. The method presented here appears to be an attractive approach to increasing the penetration power and SNR in the coded excitation system for stationary target detection using Multi-segment Coding.

Acknowledgments This study was supported by National Nature Science Foundation of China (No. 61031003), Natural Basic Research Priorities Program of Shenzhen (No. JC201005-280685A), Shenzhen Overseas Innovation Fund Project (No. KQC201105310020A), Improving Project of Shenzhen Key Laboratory (No. CXB201104210003A).

References

1. Qin ZD, Chen SP, Jin C (2010) A method of pulse ultrasound detection, system, device, transmitter and receiver. Chinese Patent 101819185A
2. Qin ZD, Chen SP, Chen X (2010) Coded transmission for ultrasound Doppler detection using truncated long code. In: 2010 3rd international conference on biomedical engineering and informatics (BMEI), vol 1, pp 159–161, 16–18 Oct 2010
3. Qin ZD, Liu DQ, Chen SP (2012) Experimental study on the truncated long code sequence for ultrasound Doppler detection. In: 2012 5th international conference on biomedical engineering and informatics (BMEI)
4. O'Donnell M (1992) Coded excitation system for improving the penetration of real-time phased-array imaging systems. IEEE Trans Ultrason Ferroelectr Freq Control 39(3):341–351

5. Ruo F (2002) Ultrasound handbook. Nanjing University Press, Nanjing, China, pp 873–875 (Chinese)
6. Chiao RY, Hao XH (2005) Coded excitation for diagnostic ultrasound: a system developer's perspective. IEEE Trans Ultrason Ferroelectr Freq Control 52(2):160–170
7. Zhao H, Mo LYL, Gao SK (2007) Barker-coded ultrasound color flow imaging: theoretical and practical design considerations. IEEE Trans Ultrason Ferroelectr Freq Control 54(2):319–331
8. Xiang L, Zhao H, Gao S (2009) Barker code in TCD ultrasound systems to improve the sensitivity of emboli detection. Ultrasound Med Biol 35(1):94–101
9. Cowe J, Gittins J, Evans DH (2007) Coded excitation in TCD ultrasound systems to improve axial resolution. Ultrasound Med Biol 33(8):1296–1308
10. Girault J-M, Kouame D, Ouahabi A et al (2000) Micro-emboli detection: an ultrasound Doppler signal processing viewpoint. IEEE Trans Biomed Eng 47(11):1431–1439

Word 2003 Document Password Cracking Based on the China Supercomputer

Lijuan Chen, Ying Yang, Jizhi Wang, Zhenya Chen, Liqiang Wen and Guang Yang

Abstract Microsoft Office is undoubtedly the best of the office application suites available on the planet, hence very high market share for Word 2003 processing and a certain market demand for Word 2003 document password cracking. Our past research has focused on the cracking programme of encrypted Word document. In this paper, however, we propose a scheme that can crack Word 2003 document password based on the supercomputer made in China (Sunway Bluelight MPP supercomputer) by analyzing Microsoft Office document file format, Word 2003 document encryption mechanism and vulnerability that exists in the encryption mechanism.

Keywords Brute force · MPI · RC4 · Supercomputer · SW1600

L. Chen (✉) · Y. Yang · J. Wang · Z. Chen · L. Wen · G. Yang
Shandong Computer Science Center, Jinan, China
e-mail: chenlj@sdas.org

Y. Yang
e-mail: yangy@sdas.org

J. Wang
e-mail: wangjzh@sdas.org

Z. Chen
e-mail: chenzhy@sdas.org

L. Wen
e-mail: wenlq@sdas.org

G. Yang
e-mail: yangg@sdas.org

L. Chen · Y. Yang · J. Wang · Z. Chen · L. Wen · G. Yang
National Supercomputer Center, Jinan, China

L. Chen · Y. Yang · J. Wang · Z. Chen · L. Wen · G. Yang
Key Laboratory for Computer Network of Shandong Province, Jinan, China

E. Qi (ed.), *Proceedings of the 6th International Asia Conference on Industrial Engineering and Management Innovation*,
DOI 10.2991/978-94-6239-145-1_25

1 Introduction

Microsoft Office is an office suite of desktop applications, servers and services for the Microsoft Windows and OS X operating systems. Initially a marketing term for a bundled set of applications, the first version of Office contained Microsoft Word, Microsoft Excel, and Microsoft PowerPoint [1]. Office software suite provides protection for user document through its security options, such as Word 2003.

A supercomputer is a computer that is at the frontline of current processing capacity, particularly speed of calculation. Supercomputers play an important role in the field of computational science, and are used for a wide range of computationally intensive tasks in various fields [2]. Supercomputers have also been used to break encryption using "brute force" algorithm. This paper presents a brute force attack of Word 2003 document based on the supercomputer made in China.

The outline of this paper is organized as follow. Section 1 is the INTRODUCTION of this paper; Sect. 2 introduces Office 2003 document file format; Sect. 3 reviews encryption scheme of Word 2003 documents; Sects. 4 and 5 present Brute Force Attack method to crack the encryption key with the help of Sunway Bluelight MPP supercomputer; Sect. 6 describes decrypting the encrypted contents of the file; and Sect. 7 presents possible future work.

2 Office 2003 Document File Format

Microsoft Office 2003 together with its previous version organizes documents in CFB (Compound File Binary) structure. Compound document files are similar with real file systems, which contain a number of independent data streams organized in a hierarchy of storages.

Storages and streams are named and the names of all storages and streams that are direct members of storage must be different. Names of streams or storages that are members of different storages may be the same. Each compound document file contains a root storage that is the direct or indirect parent of all other storages and streams. An example of a hierarchic storage/stream is shown in Fig. 1. The names of all direct members of a storage must be different, and it is possible that two different storages may contain a stream named "Stream1" [3].

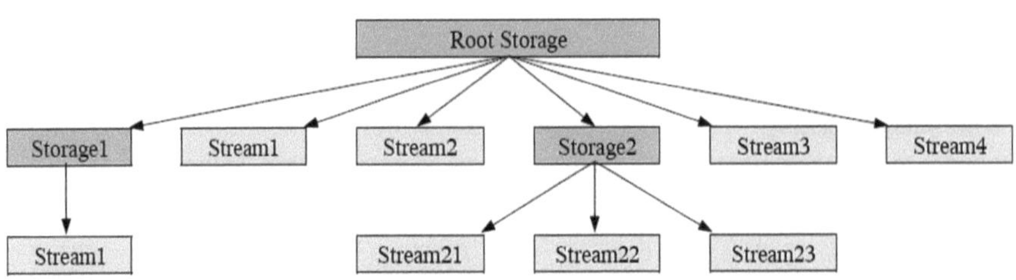

Fig. 1 Example of a hierarchic storage/stream

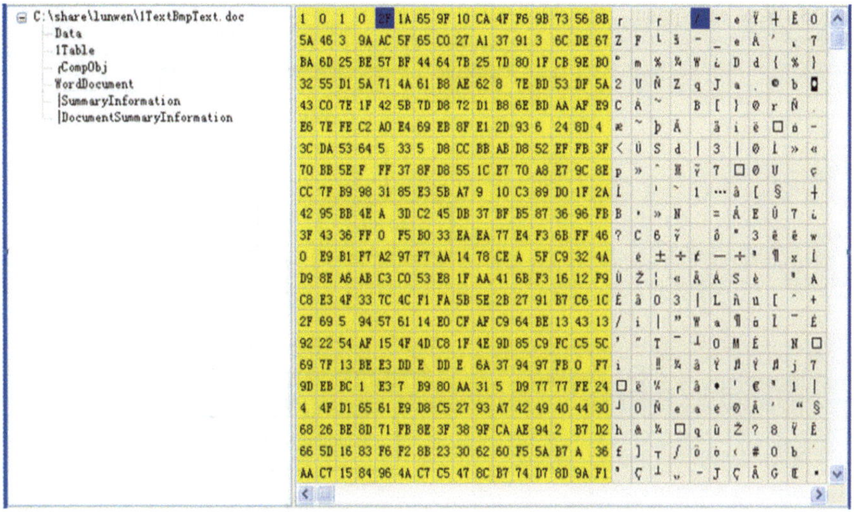

Fig. 2 Word 2003 document file format

Microsoft Office Suite includes Word, Excel, PowerPoint and other components. We consider, however, only the Word 2003 in this paper.

Microsoft Word 2003 is compound document, which means it has different sections stored in one file and each section carries different types of information. When a document is opened by Structured Storage File Viewer, each section shall be displayed (Fig. 2).

Data stream holds Word picture structures that are sequentially concatenated if the document includes pictures. Another stream, called 1Table, contains information about the fragmentation of the text part. A stream called Word Document consists of a header part and a text part. Most of the property information describing an Office document is contained in the Summary Information and Document Summary Information streams.

3 Encryption Scheme of Word 2003 Documents

Microsoft Word 2003 use RC4 stream cipher to protect the documents [4]. By default, Microsoft Word 2003 supports encryption with key size of 40-bit. Key size in algorithm RC4 which is used in MS Word 2003 could be up to 128 bit, but were artificially reduced to 40 bits, as shown in Fig. 3. Among 16 byte of MD5 digest, 11 bytes were set to zero. Thus 5 meaningful bytes and 11 zeros form the RC4 key. The default 40-bit encryption method used by Word 2003 is very weak and insecure. The 40-bit encryption scheme is vulnerable to the brute force attack.

Some streams might not be encrypted when Word 2003 document is encrypted [5]. The stream that only contains text and picture data are encrypted, such as 1Table,

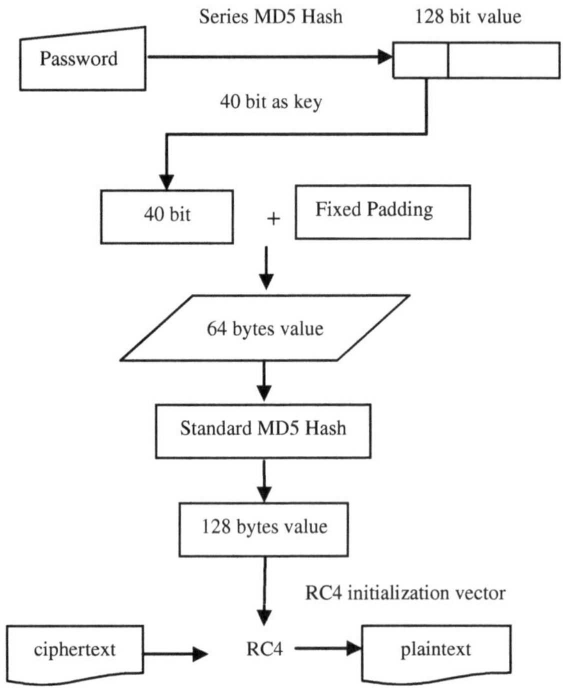

Fig. 3 Word encryption process

Data, Word Document stream. Depending on the application, the reusing of key stream with known plaintext could occur, which means certain portions of encrypted data could be extracted directly or easily retrieved. Because of the cryptographic weaknesses of the RC4 encryption algorithm in Office document, it is considered to be easily reversible [6].

Each section of the Word 2003 documents has a unique name which can be used to access that particular section. The name of RC4 header section is “1table”. We can access to this section by using this unique name. Once we read the content available in the “1table” section, we can take the first 52 bytes of ‘1table’ stream which has all required details. The first 4 bytes is Version information. The next 16 bytes is the salt, a randomly generated array of bytes, which is used during the key generation. The next 16 byte is EncryptedVerifier, an additional 16-byte verifier encrypted using a 40-bit RC4 cipher. Last 16-byte is EncryptedVerifierHash, a 40-bit RC4 encrypted MD5 hash of the verifier used to generate the EncryptedVerifier field [7].

Consider the example in Fig. 2:

Salt = 2F 1A 65 9F 10 CA 4F F6 9B 73 56 8B 5A 46 03 9A

EncryptedVerifier = AC 5F 65 C0 27 A1 37 91 03 6C DE 67 BA 6D 25 BE

EncryptedVerifierHash = 57 BF 44 64 7B 25 7D 80 1F CB 9E B0 32 55 D1 5A

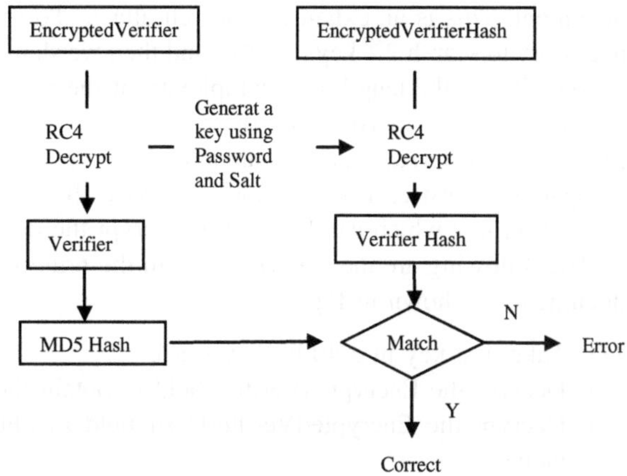

Fig. 4 Password verification

The password verification process is as follows [7], as shown in Fig. 4:

(1) Generate an encryption key using Password and Salt.
(2) Decrypt the EncryptedVerifier field of the RC4 Encryption Header structure to obtain the Verifier value. The resulting Verifier value MUST be an array of 16 bytes.
(3) Decrypt the EncryptedVerifierHash field of the RC4 Encryption Header structure to obtain the hash of the Verifier value. The number of bytes used by the encrypted Verifier hash MUST be 16.
(4) Calculate the MD5 hash value of the results of step 2.
(5) Compare the results of step 3 and step 4. If the two hash values do not match, the password is incorrect. Otherwise, the password is correct.

RC4 Algorithm

RC4 [8] was designed by Ron Rivest of RSA Security in 1987. RC4 generates a pseudorandom stream of bits (a keystream). As with any stream cipher, these can be used for encryption by combining it with the plaintext using bit-wise exclusive-or. Decryption is performed in the same way.

MD5 Algorithm

MD5 [9] is one of a series of message digest algorithms designed by Ronald Rivest of MIT (Rivest, 1992). MD5 processes a variable-length message into a fixed-length output of 128 bits. The input message is broken up into chunks of 512-bit blocks (sixteen 32-bit words); the message is padded so that its length is divisible by 512.

4 The Attack on the Microsoft Word 2003

As explained in [5], we will use brute force to attack the encryption key instead of the password. We only need to validate each possible key available in the key space against the 'verifier hash' which is stored in the RC4 encryption header in the document. This is an exhaustive search attack. To recover a Word 2003 file, it is necessary to search 2^{40} keys at most and then we shall certainly find the correct key, irrespective of the length and complexity of the password being used.

Simply, we need all these 3 fields—Salt, EncryptedVerifier and EncryptedVerifierHash to generate a final decrypted hash value which shall then be compared against each key in the key space. If a match is found, it is the actual encryption key which can be used to decrypt the document content.

The following are the steps to perform the brute force attack on the Word 2003 document, as shown in Fig. 5:

(1) Take first key in a 40-bit key space.
(2) Decrypt the EncryptedVerifier field to obtain the Verifier value.
(3) Decrypt the EncryptedVerifierHash field to obtain the hash of the Verifier value.
(4) Calculate the MD5 hash value of the results of step2.
(5) Compare the results of step 3 and step 4. If the two hash values do match, the key is correct. Exit.
(6) Take next key in a 40-bit key space. Go to step 2.

Fig. 5 Key attack

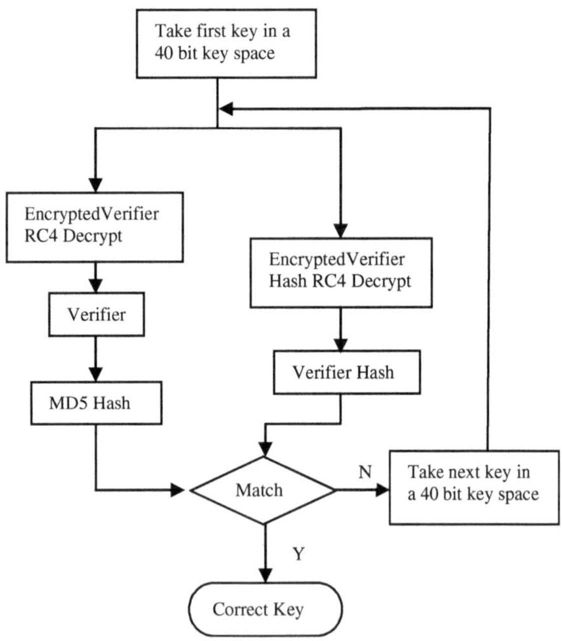

5 Brute Force Attack Based on Supercomputer

Brute force decryption has fallen out of vogue in the last few years [10]. In order to prevent someone to find the actual key which can be used to decrypt a message from using brute force attack, the key space is usually designed to be large enough to make it infeasible [11]. We can use a supercomputer to try different keys at the same time and able to search the key we need in a few minutes or hours. If it takes 400 days to try one million keys on single CPU, it shall take only 1 h to try the same number of keys on a supercomputer with 10,000 CPUs.

5.1 An overview of Sunway BlueLight MPP Supercomputer

As shown in Fig. 6, Sunway BlueLight MPP Supercomputer [12], is the first publicly announced PFLOPS supercomputer using ShenWei processors developed in China. The water-cooled 9-rack system has 8704 ShenWei SW1600 processors organized as 34 super nodes (each consisting of 256 compute nodes), 150 TB main memory, 2 PB external storage, peak performance of 1.07016 PFLOPS, sustained performance of 795.9 TFLOPS, LINPACK efficiency 74.37 %, and the total power consumption 1074 kW. The software architecture of Sunway BlueLight MPP Supercomputer is shown in Fig. 7.

5.2 MPI

MPI (Message Passing Interface) [13] is a standard library specification for message-passing, proposed as a standard by a broadly based committee of vendors, implementers, and users. MPI was designed for high performance computing on both massively parallel machines and on workstation clusters. MPI specifications

Fig. 6 Sunway BlueLight

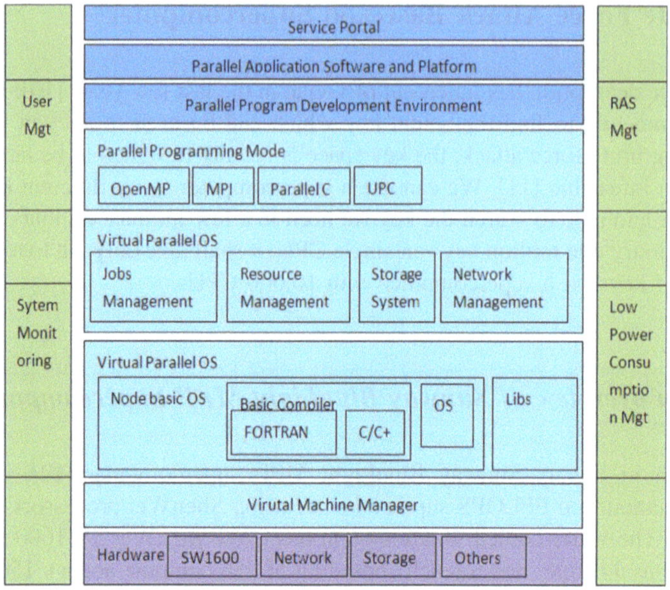

Fig. 7 Software architecture of Sunway BlueLight

have been defined for C, C++ and Fortran. General MPI Program Structure is shown in Fig. 8.

Running parallel applications can obtain the required results within a reasonable time. It requires considerable processing resources. Before parallelizing serial applications, we need to analyze whether it shall benefit from it or not.

Fig. 8 General MPI program structure [14]

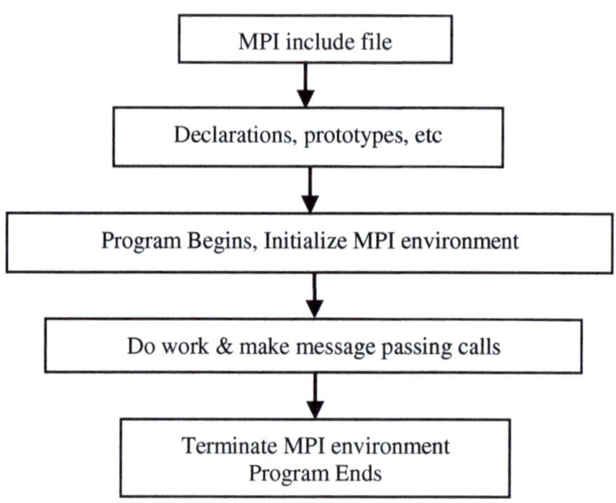

5.3 MPI-Implementation of Searching the Entire Key-Space to Find the Correct Key

The challenge in writing MPI programs is how to divide the concerned problem into smaller sub problems [15]. Our problem focuses on distributing 2^{40} keys across different parallel computing nodes. Each processor performs the same searching key task. 2^{40} keys can be divided into pieces of approximately the same size and then mapped to different processors depending on the process ID. Each processor then operates only on the portion of 2^{40} keys that is assigned to it.

Following is the C code with MPI to perform the processing:

```
//Initialise the MPI execution environment
MPI_Init(&argc,&argv);
//Determines the size of the group associated with a
communicator
MPI_Comm_size (MPI_COMM_WORLD, &nprocs);
//Determines the rank of the calling process in the
communicator
MPI_Comm_rank (MPI_COMM_WORLD,&myid);

//the first 4 bytes of '1Table' stream
for (i = 0; i < 4; i++)
        x = read_8ubit (ps->tablefd);
  // the next 16 byte of '1Table' stream
  for (i = 0; i < 16; i++)
        docid[i] = read_8ubit (ps->tablefd);
  // the next 16 byte of '1Table' stream
  for (i = 0; i < 16; i++)
        salt [i] = read_8ubit (ps->tablefd);
  // last 16 byte of '1Table' stream
  for (i = 0; i < 16; i++)
        hashedsalt [i] = read_8ubit (ps->tablefd);
// try all possible keys in a 40 bit key space
For (( i = myid; i< 16777216;i=i+nprocs){
Makekey();
// if a match is found, then that's our actual key which can be
used to decrypt the document content.
If  ( verifypwd()==0)  fprintf  (myid, key)  ;
}
// Terminates MPI execution environment
MPI_Finalize();
```

Test Results

Tests of performance and speedup were taken with 1000, 3000, 5000, 7000, 10000, 12000, 15000, 18000 and 20000 cores respectively, with results shown in Table 1. The first row of table is the number cores. The second row is the average calculating time to complete the search of entire keyspace with different number of cores. The third row refers to the parallel computing speedup based upon 1000 cores. It can be seen that for the parallel computing of MPI's massively parallel program from thousands to ten thousands of cores, the speedup is linear with the increase in the

Table 1 Real-time speedup measurement

Cores	1000	3000	5000	7000	10,000	12,000	15,000	18,000	20,000
Minutes	52,415.12	17084.1533	10066.848	7039.24285	5236.392	4212.756	3362.314	2777.9	2462.78
Speedup	1	3.068	5.206	7.446	10.009	12.442	15.589	18.868	21.282

number of computing cores. It indicates that the program possesses the high parallel accelerating performance and parallel efficiency. Results show that this system could run with high speedup ratio and parallel efficiency, as shown is Fig. 9.

We ran the script with 20,000 cores to search the entire keyspace using 40 h. It is therefore possible with the use of 100,000 cores; we can exhaust the Word 2003 keyspace in about 8 h. For the comparison, the same versions of these programs working on Pentium D CPU 2.80 GHz would spend about 80 days for a 40-bit key.

As shown in Fig. 10, the script read the value of available (Salt, EncryptedVerifier, EncryptedVerifierHash) in the “1table” section. Then, it got a 40-bit key (1c 3f 1f e0 91) which can be used to decrypt the document content on process 1413. The whole process took 8 min.

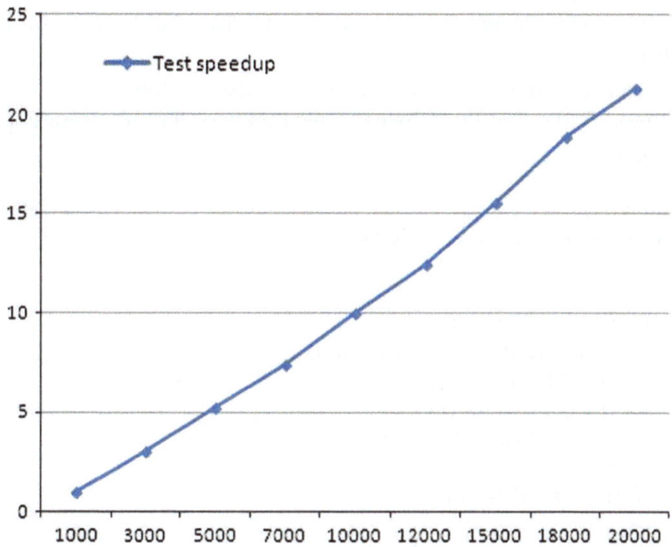

Fig. 9 Speedup

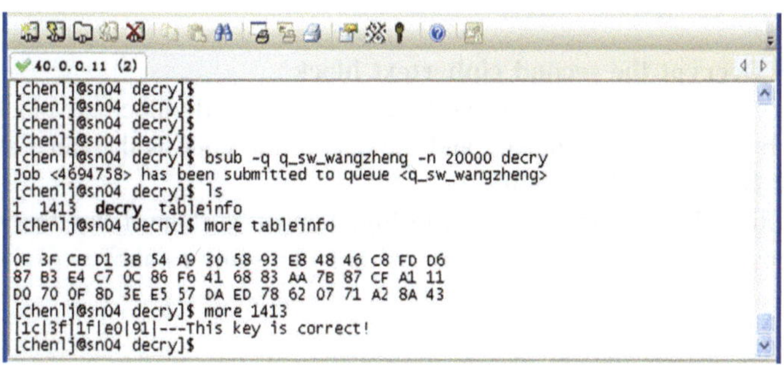

Fig. 10 Screenshot

Salt = 0F 3F CB D1 3B 54 A9 30 58 93 E8 48 46 C8 FD D6
EncryptedVerifier = 87 B3 E4 C7 0C 86 F6 41 68 83 AA 7B 87 CF A1 11
EncryptedVerifierHash = D0 70 0F 8D 3E E5 57 DA ED 78 62 07 71 A2 8A 43

6 Decrypting the Encrypted Contents of the Word 2003 Document

This is almost instantaneous when finding the actual key to decrypt the document content. We can use an RC4 decryption algorithm to decrypt the content using the found key as follows:

Step 1 Set flag bit to zero

Set the least significant bit of the value of offset 0x20b to 0. (1 means the encrypted file, 0 otherwise).

Step 2 Decrypt the first ciphertext block (offset ox200)

Do not modify the first 0x44 bytes, except encrypt flag bit. Here is part of the encrypting code:

```
blockcnt = 0;
makekey (blockcnt, &key, &valContext);
while (j < end)
 {
        for (i = 0; i < 0x10; i++)
          test[i] = read_8ubit (encStream);
        rc4 (test, 0x10, &key);
        for (i = 0; i < 0x10; i++)
          fputc (test[i], decStream);
        j += 0x10;
        if ((j % 0x200) == 0)
          {
                //rekey the rc4 algorithm
                blockcnt ++;
                makekey (blockcnt, &key, &valContext);
          }
 }
```

Step 3 Decrypt the second ciphertext block

Search “00 01 00 01” to find the header. The second ciphertext block is offset from the beginning of the header and the decryption code is the same as above.

Step 4 Decrypt data stream

Decrypting data stream as shown in Fig. 1 must be done if the decrypted document consists of pictures. Decryption code is the same as above.

7 Future Work

We plan to create a huge database of the pre-calculated values and to apply the Rainbow-attack so as to find the key in seconds. In addition, we plan to implement the parallelization of MD5 based on Sunway BlueLight MPP Supercomputer using MPI to reduce the time of brute force attack the Word 2003 document.

References

1. Microsoft Office, From Wikipedia, the free encyclopaedia. http://en.wikipedia.org/wiki/Microsoft_Office
2. Supercomputer, From Wikipedia, the free encyclopaedia. http://en.wikipedia.org/wiki/Supercomputer
3. OpenOffice.org's Documentation of the Microsoft Compound Document File Format. http://sc.openoffice.org/compdocfileformat.pdf
4. Wu H (2005) The misuse of RC4 in Microsoft Word and Excel. Institute for Infocomm Research, Singapore, p 2
5. Wang P, You L, Tian J, Gao Z (2013) Word document decryption based on GPU. Inf Technol 2013(4):52–153 (in Chinese)
6. Office Binary Document RC4 Encryption. http://msdn.microsoft.com/en-s/library/dd945044(v=office.12).aspx
7. Microsoft Office File Formats Documentation. http://download.microsoft.com/download/2/4/8/24862317-78F0-4C4B-B355-C7B2C1D997DB/[MS-OFFCRYPTO].pdf
8. RC4, From Wikipedia, the free encyclopaedia. http://en.wikipedia.org/wiki/RC4
9. MD5, From Wikipedia, the free encyclopaedia. http://en.wikipedia.org/wiki/MD5
10. What can you do with a supercomputer? http://tehparadox.com/forum/f28/what-can-you-do-supercomputer-3975364/
11. Key space, From Wikipedia, the free encyclopaedia. http://en.wikipedia.org/wiki/Key_space_(cryptography)
12. Shen W, From Wikipedia, the free encyclopaedia. http://en.wikipedia.org/wiki/ShenWei
13. Message Passing Interface (MPI). https://computing.llnl.gov/tutorials/mpi/
14. The Message Passing Interface (MPI) standard. http://www.mcs.anl.gov/research/projects/mpi/
15. El-Nashar AI (2011) To parallelize or not to parallelize, speed up issue. Int J Distrib Parallel Syst 2(2):15

An Effective Rapid Algorithm for Portable X-Ray Fluorescence Spectral Matching

Xiao Long and Yuanxiang Li

Abstract Portable X-ray fluorescence (XRF) instrumentation usually employs two approaches to alloy identification. The first is based on performing an accurate quantification of alloying elements and then comparing the results with tables of specifications listing chemical compositions along with alloy designations (grade or trade names). The second approach is based on the concept of “alloy spectral signature” or “pattern recognition” in which spectral features of the unknown alloy are compared to obtain a match with previously stored spectral features of known, reference alloys. This paper develops a new XRF spectral matching algorithm called spectral conformity feature matching (SCFM). It belongs to the second approach mentioned above. In order to evaluate its performance, two known methods, spectral angle mapper (SAM) and spectral derivative feature coding (SDFC) are used to conduct comparative analysis. Experiments results demonstrate that the proposed SCFM performs more effectively in the spectral classification and maintains the rapidity.

Keywords Alloy identification · Spectral conformity feature matching (SCFM) · Spectral matching · X-ray fluorescence

1 Introduction

Portable X-ray fluorescence (PXRF) analysis offers some unique advantages that cannot be rivalled by any other analytic technique, which arise from the availability to the operator of information on the chemical composition of a sample in the field, or the capability of the technique to undertake in situ analytic measurements [1].

X. Long (✉) · Y. Li
Department of Computer Science, Wuhan University, Wuhan, China
e-mail: long4u@126.com

Y. Li
e-mail: yxli@whu.edu.cn

E. Qi (ed.), *Proceedings of the 6th International Asia Conference on Industrial Engineering and Management Innovation*,
DOI 10.2991/978-94-6239-145-1_26

The attribute of "hand held" brings on a compromise between the capability and the speed of data processing. Least-squares fitting method using reference spectra is one common way in which quantitative XRF analysis is possible without obtaining net peak areas of the characteristic lines [2]. Spectral matching method, which is often used for hyper-spectral signature discrimination and data classification in the remote sensing field, is a traditional hyper-spectral analysis approach and concentrates on recognizing absorption band shapes in each individual spectrum [3]. The latter method is supposed to be introduced to XRF analysis since the difference between the former method using reference spectra and the latter method lies in that the reference spectra used in the former method are a combination of necessary pure element spectra and the reference spectrum used in the latter method is a single sample or standard spectrum for pairwise matching. Similarly, spectral matching method is also possible for quantitative XRF analysis in the case that no spectrum evaluation in the strict sense is performed.

Generally, Spectrum evaluation essentially comprises the mathematical procedures to extract relevant information, the most important aspect of which is the analytically important net peak areas of the element characteristic fluorescence lines, from acquired X-ray spectra, and remains a crucial step in X-ray spectrometry. Although energy-dispersive X-ray spectrometry (ED-XRF) lacks the precision of wavelength-dispersive X-ray spectrometry (WD-XRF), a correct spectrum evaluation improves the accuracy of ED-XRF to the same or even higher level than WD-XRF. This is also important from an economical perspective, i.e. in many industrial applications the more expensive WD-XRF systems might be replaced or backed-up by cheaper and more versatile ED-XRF instruments especially such as PXRF instruments [4].

Portable XRF is one of the instrumental techniques that have developed rapidly in recent years. Like other XRF instrumentation, PXRF consists of an excitation source, sample positioning facility, detector and pulse processing and analysis facility. However, portability means that in the selection and integration of these components, there is an emphasis on minimizing both mass and power consumption in an ergonomically designed instrument [1]. Consequently, this gives rise to a tradeoff between the capability of data processor and the real time demand for in situ analysis. Considering that the often-quoted inferior accuracy of ED-XRF can, to a large part, be attributed to errors associated with the evaluation of these spectra including net peak area determination and continuum estimation, and that, for least-squares fitting method using reference spectra, the background must be removed prior to the least-squares fitting, spectral matching techniques have their advantage in that they are considerably dependent on the information of entire reference spectrum or standard spectrum during pairwise matching.

2 Existing Spectral Matching Techniques

A majority of spectral matching techniques have been proposed in the hyper-spectral imaging area [5–7] or the computer vision area [8, 9], such as spectral angle mapper (SAM) [10], cross correlogram spectral matching (CCSM) [11], spectral derivative feature coding (SDFC) [12], artificial DNA-based spectral matching (ADSM) [3], etc.

Particularly, SAM, formularized as

$$\alpha = \cos^{-1}\frac{\sum XY}{\sqrt{\sum (X)^2 \sum (Y)^2}}, \tag{1}$$

α Angle formed between reference spectrum and target spectrum
X Target spectrum
Y Reference spectrum

uses the angle formed between the reference (known) spectrum and the target spectrum (unknown) as the metric for spectral matching, treating both spectra as vectors in a space with dimensionality equal to the number of bands. CCSM uses cross correlograms calculated between a reference and target spectrum shifted subsequently over m channels to shorter and longer wavelength. In other words, it uses the correlation coefficient

$$r = \frac{\sum (X-\bar{X})(Y-\bar{Y})}{\sqrt{\sum (X-\bar{X})^2}\sqrt{\sum (Y-\bar{Y})^2}} \tag{2}$$

as the metric for spectral matching. The function $\cos(\alpha)$ from (1) is similar to the correlation coefficient from (2). The big difference is that correlation coefficient standardizes the data, centralizing itself in the mean of x and y. A big limitation of SAM is the impossibility of distinguishing between negative and positive correlations because only the absolute value is considered [13]. In this regard, the measurement from CCSM is more beneficial and gathers even better estimates. Although the cross correlogram is insensitive to gain factor and random noise does not affect the shape of the correlogram, some detailed spectral information such as subtle spectral differences or misinterpreting the real position of maximum depth of the feature which results from the continuum of every spectrum will still be overlooked by CCSM [14]. So the continuum estimation and removal, which will burden the data processor especially in Portable XRF instruments, must be taken into account when CCSM is applied.

SDFC encodes the reference spectrum and target spectrum according the degrees of successive gradient changes in spectral values among three consecutive adjacent spectral bands and then uses hamming distance (HD) or absolute value distance (AVD) as the measurement for the similarity of both spectra. It groups the derivative features into four types which can be represented by graphic in the Fig. 1. Or

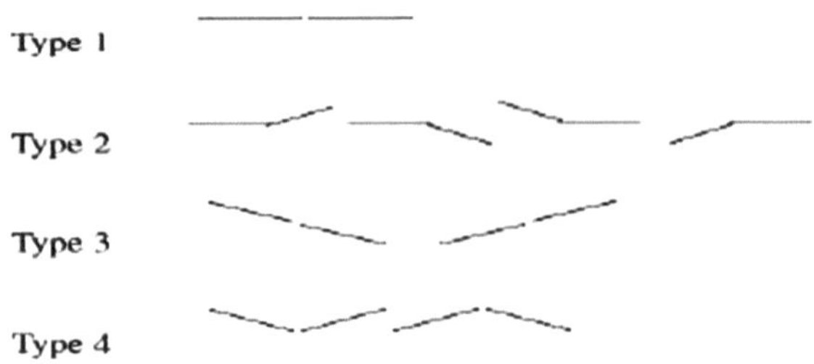

Fig. 1 Graphical representation of Types 1–4

these four types of successive gradient changes in spectral values that can be described as follows [12]:

$$\begin{aligned} &\text{Type 1}: \text{ if } |s_1 - s_{l-1}| \le \Delta \text{ and } |s_{l+1} - s_1| \le \Delta \\ &\text{Type 2}: \text{ if either } |s_1 - s_{l-1}| \le \Delta \text{ and } |s_{l+1} - s_1| > \Delta \\ &\quad \text{or } |s_1 - s_{l-1}| > \Delta \text{ and } |s_{l+1} - s_1| \le \Delta \\ &\text{Type 3}: \text{ if either } s_1 - s_{l-1} < \Delta \text{ and } s_{l+1} - s_1 < \Delta \\ &\quad \text{or } s_1 - s_{l-1} > \Delta \text{ and } s_{l+1} - s_1 > \Delta \\ &\text{Type 4}: \text{ if either } s_1 - s_{l-1} < \Delta \text{ and } s_{l+1} - s_1 > \Delta \\ &\quad \text{or } s_1 - s_{l-1} > \Delta \text{ and } s_{l+1} - s_1 < \Delta \end{aligned} \tag{3}$$

where, s is the l-th spectral band and Δ is a desired spectral value tolerance. ASDM uses the encoding scheme that combines DNA encoding and the SDFC and introduces artificial DNA computing to the spectral matching. Three distance discrimination criteria [HD, AVD and protein value distance (PVD)] are used to measure the discrepancy between the DNA strands of training samples, or pixels to be classified with the strands in the spectral DNA database [3]. Although it is demonstrated that SDFC and ASDM have their superiority over other methods in performance effectiveness and tolerance of spectral diversity, it seems that the cost of their performance time for the encoding of both and the evolutionary iterative process of ADSM is not easy to meet the real time performance speed demand for portable XRF instruments.

3 Spectral Conformity Feature Matching

Considering the feature of XRF spectrum such as complex continuum or matrix effect and the specialty of portable XRF capability especially in Alloy Identification area, this paper proposes a new spectral matching method named spectral conformity feature matching (SCFM) that has O(n) time complexity.

As known in Sect. 2, SAM and CCSM only establish an entire comparative relation in the contents of channels between the reference spectrum and the target spectrum, while SDFC and ASDM only use the relation among the channels of each spectrum to encode the information. By combining the neighborhood relations in one single spectrum and the comparative pairwise relations between two spectra, a new set of compound relations is built up based on derivative.

Assume that $X = (x_1, x_2, \ldots, x_L)^T$ is a XRF reference spectrum where L is the total number of channels and x_l is the content of channel l, and that $Y = (y_1, y_2, \ldots, y_L)^T$ is a XRF target spectrum where L is the total number of channels and y_l is the content of channel l, and that $szF = (szf_1, szf_2, \ldots, szf_L)^T$, $diF = (dif_1, dif_2, \ldots, dif_L)^T$ and $sdF = (sdf_1, sdf_2, \ldots, sdf_L)^T$ are three flag vectors as follows. There are three types of conformity feature, each of which is evaluated as a flag value (−1 or 0 or 1) by three cases and the 3rd type of which is evaluated by four cases (−1 or 0 or 1 or a positive integer), presented by pairwise comparison between the reference spectrum and target spectrum in size, direction and scale that can be described in Tables 1, 2 and 3 respectively. The measurement of discrepancy between the pair of spectra can be derived in the following paragraph.

The new method only counts the absolute difference between the pair of channels where the corresponding flag represents scale conformity case 4 or the case, where the number of channels is lesser, of case 1 and case 3. Since both cases may be regarded as the nonconformity cases, the distance between the pair of spectra can be expressed as Nonconformity Distance (NCD):

$$NCD = \begin{cases} \sum\limits_{scf_l > 1\, OR\, scf_l = -1} |y_l - x_l| \; if \; \sum\limits_{scf_l = -1} 1 < \sum\limits_{scf_l = 1} 1 \\ \sum\limits_{scf_l \geq 1} |y_l - x_l| \; if \; \sum\limits_{scf_l = -1} 1 > \sum\limits_{scf_l = 1} 1 \end{cases} \quad (4)$$

Table 1 Size conformity

Case flag	Case conditions	
	Expressional conditions	Graphical conditions[a]
$szf_l = -1$	If $y_l < x_l$	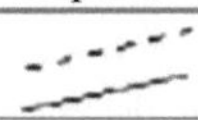
$szf_l = 0$	If $y_l = x_l$	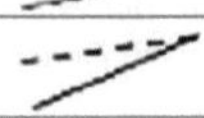
$szf_l = 1$	If $y_l > x_l$	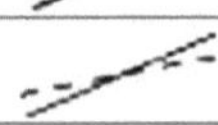

[a] - - - - - x, ——— y

Table 2 Direction conformity

Case flag	Case conditions	
	Expressional conditions[a]	Graphical conditions[b]
$dif_l = -1$	If $y_l' = 0$ and $x_l' > 0$ or $y_l' < 0$ and $x_l' > 0$ or $y_l' < 0$ and $x_l' = 0$	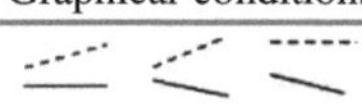
$dif_l = 0$	If $y_l' > 0$ and $x_l' > 0$ or $y_l' < 0$ and $x_l' < 0$ or $y_l' = 0$ and $x_l' = 0$	
$dif_l = 1$	If $y_l' > 0$ and $x_l' = 0$ or $y_l' > 0$ and $x_l' < 0$ or $y_l' = 0$ and $x_l' < 0$	

[a] $y_l' = y_l - y_{l-1}$, $x_l' = x_l - x_{l-1}$
[b] - - - - - x, ——— y

Table 3 Scale conformity

Case flag	Case conditions	
	Expressional conditions[a]	Graphical conditions[b]
$scf_l = -1$	If ($y_l' < x_l'$ and $y_l' > 0$ or $y_l' > x_l'$ and $y_l' < 0$) and $dif_l = 0$ and $szf_l \leq 0$	
$scf_l = 0$	If $y_l' = x_l'$	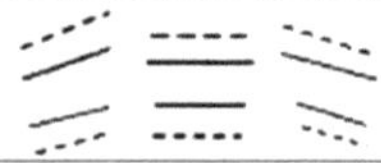
$scf_l = 1$	If ($y_l' > x_l'$ and $y_l' > 0$ or $y_l' < x_l'$ and $y_l' < 0$) and $dif_l = 0$ and $szf_l \geq 0$	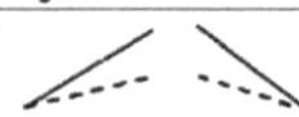
$scf_l = \mid y_l - x_l \mid$	Other conditions	Other conditions

[a] $y_l' = y_l - y_{l-1}$, $x_l' = x_l - x_{l-1}$
[b] - - - - - x, ——— y

Thus NCD considers the spectral diversity tolerance in the aspect of scale conformity. When the target spectrum is compared with a reference spectrum, the least NCD is equivalent to the most similarity. The pseudocode of SCFM is given in the following:

SCFM Algorithm:

Input: *reference spectra* {X^1, X^2, …, X^n} *(n represents index other than power) and target spectrum* Y

Output: X_{op} *which is most similar to* Y

Step 1. *Initial each spectrum*

For each X^k (k = 1, 2, …, n) *and* Y

For each channel l (l = 1, 2, …, L)

Calculate the difference between two adjacent channels l and l-1;

Step 2. *Compare each pair of spectra by getting their conformity feature*

For each pair of X^k (k = 1, 2, …, n) *and* Y

For each channel l (l = 1, 2, …, L)

Calculate the three types of conformity feature by referring to Table I and Table II and Table III;

Step 3. *Locate the optimal reference spectrum*

For each pair of X^k (k = 1, 2, …, n) *and* Y

Calculate the nonconformity distance NCD^k *by referring to the Equation (4);*

Return X_{op} *where* $NCD_{op} = \min(NCD^k)$

(k = 1, 2, …, n)

4 Experiments

In order to demonstrate the SCFM Algorithm's effectiveness in portable XRF area, two experiments concerned with XRF spectra classification were performed on a set of known alloy spectra which were classified into fourteen sorts including Albata, Alpaka, Aluminum brass, Cobalt alloy, Hastelloy, High-speed tool steel, Lead brass, Low-alloy steel, Manganese brass, Packfong, Stainless steel, Tin bronze, Tin-silver alloy and Titanium alloy. The alloy spectra were obtained under the single condition and through the Cu filter. The channels' range was from the 50th to 2000th of 2048. Figure 2 shows a graph of samples from these alloy sorts. Considering that SAM and CCSM use the similar correlations as the metrics while SDFC and ASDM use the similar coding techniques, and that CCSM and ASDM consume more time, the tests only compared SCFM with SAM and SDFC in accuracy and comparison time. The running environment is as follows: CPU is Intel Core 2 Duo 2.1 GHz, and RAM is 3.5 GB, and Operating System is Windows 7 Ultimate.

The first experiment randomly divided the 168 different known alloys from 14 categories into two parts, 60 of them as the target set and 108 of them as the reference set, or 60 of them as the reference set and 108 of them as the target set. The former way was called dividing way A and the latter was called dividing way B. The comparisons of three classifiers about overall accuracy and total time were performed in the both dividing way. In this experiment, the HD was taken as the distance measuring criteria for the SDFC algorithm. The spectral sorts and associated the numbers of the target signatures and the reference signatures in each sort in the dividing way A are shown in Table 4. Since the dividing way B only exchanges the number of the targets with that of the references, the table for it is omitted. Tables 5 and 6 show the results of comparisons between the target sample data and the reference data obtained by three spectral matching classifiers: SAM,

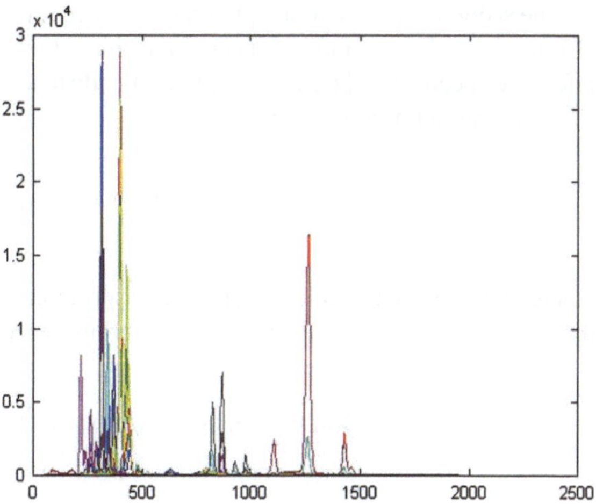

Fig. 2 Fourteen sample alloy spectra from fourteen alloy categories in one graph with different colors

Table 4 The spectral sorts and associated the numbers of the target signatures and the reference signatures in each sort in the dividing way A

Alloy sorts	Samples		
	Total samples	Reference samples	Target samples
Stainless steel	35	12	23
High-speed tool steel	8	3	5
Cobalt alloy	19	7	12
Aluminum brass	5	2	3
Manganese brass	3	1	2
Alpaka	5	2	3
Albata	5	2	3
Hastelloy	26	9	17
Lead brass	7	3	4
Titanium alloy	26	9	17
Tin bronze	4	1	3
Tin-silver alloy	6	2	4
Packfong	5	2	3
Low-alloy steel	14	5	9

Table 5 Comparison of three classifier performances in dividing way a in experiment 1

Assessment	Methods		
	SAM	SDFC	SCFM
Overall accuracy (%)	68.52	91.67	98.15
Total comparison time (s)	3.276	3.697	3.339

SDFC and ADSM algorithms in two dividing ways. From the two tables, it is apparent that the SCFM classifier produces better classification results than the other classifiers, while the SDFC exhibits better stability.

The second experiment adopted relative spectral discriminatory entropy (RSDE) [15] to assess the accuracy of spectral matching. Given the target spectra t, the set of reference spectra S = $\{s_1, s_2, \ldots, s_n\}$ and the distance $d(t, s_i)$ between t and s_i, $i = 1, 2, \ldots$, n, the definition is [15]:

$$RSDE_s^d(t) = -\sum_{i=1}^{n} RSDPB_s^d(t, s_i)\log(RSDPB_s^d(t, s_i)) \tag{5}$$

where, RSDPB is the relative spectral discriminatory probability of the measure distance d for target t with respect to S formularized as [15]:

$$RSDPB_s^d(t, s_i) = d(t, s_i) \Big/ \sum_{i=1}^{n} d(t, s_i) \tag{6}$$

Table 6 Comparison of three classifier performances in dividing way b in experiment 1

Assessment	Methods		
	SAM	SDFC	SCFM
Overall accuracy (%)	66.67	91.67	100
Total comparison time (s)	1.389	2.777	1.607

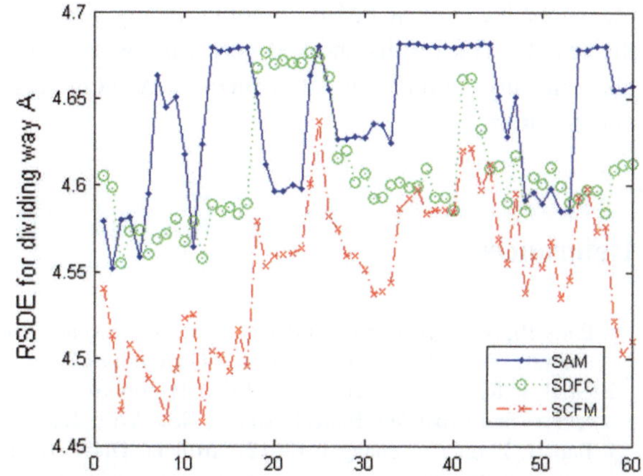

Fig. 3 Plot of the RSDE versus the targets in dividing way A

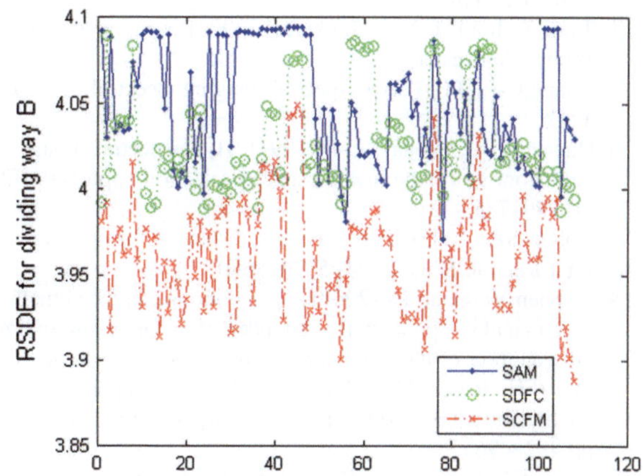

Fig. 4 Plot of the RSDE versus the targets in dividing way B

This measure RSDE is seen as a way to analyze the uncertainty with respect to the match between t and reference spectra. A lager entropy value indicates a higher degree of uncertainty with respect to t. The lower the entropy value, the higher the chance the targets will be correctly matched [15]. Figures 3 and 4 display the RSDE values for the three classifiers in two dividing ways. Note that SCFM outperform SAM and SDFC in both ways by producing RSDE values lower. The results benefit from that SCFM only takes account of nonconforming channels which decrease the degree of uncertainty for matching.

5 Conclusion

This paper presents a new spectral matching algorithm that can be used for identification when using portable X-ray fluorescence analyzer. The nonconformity distance (NCD) is the measure of the similarity between the target spectrum and the reference spectrum. Compared with SAM and HD of SDFC, another two typical measures, NCD of SCFM is computationally simpler and maintains the accuracy of the results. The results showed here can be used in real time target detection and also in any application that needs frequent computations such as spectral classification.

References

1. Potts PJ, West M (2008) Portable X-ray fluorescence spectrometry: capabilities for in situ analysis. Royal Society of Chemistry, Cambridge
2. Grieken REV, Markowicz AA (2002) Handbook of X-ray spectrometry—second edition, revised and expanded. Marcel Dekker, New York–Basel
3. Jiao H, Zhong Y, Zhang L (2012) Artificial DNA computing-based spectral encoding and matching algorithm for hyperspectral remote sensing data. IEEE Trans Geosci Remote Sens 50 (10):4085–4104
4. Tsuji K, Injuk J, Van Grieken R (2004) X-ray spectrometry: recent technological advances. Wiley, Chichester
5. Chang C-I (2003) Hyperspectral imaging: spectral detection and classification. Plenum, New York
6. Harsanyi JC, Chang C-I (1994) Hyperspectral image classification and dimensionality reduction: an orthogonal subspace projection approach. IEEE Trans Geosci Remote Sens 32 (4):779–785
7. Robila SA (2005) Using spectral distances for speedup in hyperspectral image processing. Int J Remote Sens 26(24):5629–5650
8. Stephen Viggiano JA (2001) A perception-referenced method for comparison of radiance ratio spectra and its application as an index of metamerism. In: Proceedings of AIC colour 2001: the 9th congress of the international colour association, pp 701–704
9. Imai FH, Rosen MR, Berns RS (2002) Comparative study of metrics for spectral match quality. In: CGIV 2002: first european conference on colour graphics, imaging, and vision, pp 492–496
10. Kruse FA, Lefkoff AB, Boardman JW, Heidebrecht KB, Shapiro AT, Barloon PJ et al (1993) The spectral image processing system (SIPS)—interactive visualization and analysis of imaging spectrometer data. Remote Sens Environ 44(2/3):145–163
11. Meer FVD, Bakker W (1997) CCSM: cross correlogram spectral matching. Int J Remote Sens 18(5):1197–1201
12. Chang C-I, Chakravarty S, Chen H-M, Ouyang Y-C (2009) Spectral derivative feature coding for hyperspectral signature analysis. Pattern Recogn, 42:95–408
13. De Carvalho OA, Meneses PR (2000) Spectral correlation mapper (SCM): an improvement on the spectral angle mapper (SAM). In: NASA JPL AVIRIS workshop
14. Van Der Meer F (2000) Spectral curve shape matching with a continuum removed CCSM algorithm. Int J Remote Sens 21(16):3179–3185
15. Du Y, Chang C-I, Ren H, Chang C-C, Jensen JO, D'Amico FM (2004) New hyperspectral discrimination measure for spectral characterization. Opt Eng 43(8):1777–1786

Hybrid Artificial Fish Algorithm to Solve TSP Problem

Chun-ying Cheng, Hai-Feng Li and Chun-Hua Bao

Abstract Based on the research of the Artificial Fish Swarm Algorithm, this paper put forward an improved hybrid artificial fish algorithm which involves improved preying behavior and improved swarming behavior. Then the performance of the algorithm is improved through introducing genetic crossover operator and a better scope of vision function so as to solve TSP problem. Through the simulation experiment and comparison of improved artificial fish algorithm of literature, the results show that the improved hybrid algorithm in convergence performance and the solution accuracy and convergence rate of the improved artificial fish algorithm is superior to literature.

Keywords Crossover operator · Hybrid algorithm · Improved behavior · TSP problem

1 Introduction

The TSP Problem (Traveling Salesman Problem), namely the Traveling Salesman Problem, is one of the prominent problems in mathematics, and is also one of the classic combinatorial optimization problems. Many engineering application, such as logistics distribution, network cabling and Circuit board drilling, etc., can be summed up as TSP problems and be solved. Therefore, looking for a kind of effective algorithm to solve the problem of the TSP has important theoretical value and application background. So far, for the TSP problem, people put forward many

C. Cheng (✉) · C.-H. Bao
College of Computer Science and Technology, Inner Mongolia University for Nationalities, Tongliao 028043, China
e-mail: chengchunying_80@163.com

H.-F. Li
Computer Education Department, Inner Mongolia Coal Industry Technical School, Tongliao 028021, China

E. Qi (ed.), *Proceedings of the 6th International Asia Conference on Industrial Engineering and Management Innovation*,
DOI 10.2991/978-94-6239-145-1_27

valuable method such as Simulated Annealing, SA [1], Genetic Algorithm, GA [2, 3], ant colony optimization, ACO [4, 5], Particle Swarm Optimization, PSO [6, 7], Immune Algorithm, IA [8], etc. And Artificial Fish Algorithm Swarm (AFSA) is Li [9, 10] through the observation of the Fish life habits in 2002 put forward a new random search Algorithm, this Algorithm is a kind of new ideas that based on animal behavior to seek the new way of global optimal, behaviorism is a typical application of Artificial intelligence. The Algorithm starts from the simple low-level animal behavior structure, based on the local optimization of individual animals' behavior, finally emerged in groups to make the global optimal value.

Artificial Fish Swarm algorithm has the capability to overcome the local extreme value and obtain the global extreme value, and the realization of the algorithm without gradient value of the objective function features, so the search space has certain self-adaptive ability. However, the artificial fish algorithm also focused on the main application to solve continuous optimization problems. The outcome of combinatorial optimization problems prone to "premature" or trapped in locally optimal solution, or other shortcomings, such as slow convergence speed. In basic artificial fish algorithm with the crossover operator of genetic algorithm, the artificial fish can effectively get rid of the local extreme value and accelerate the search speed, and this also can make artificial fish jump out of local extreme value, make the group eventually converge to the global extreme value quickly, and get high precision of solution.

2 Traveling Salesman Problem

Traveling salesman problem is a NP hard problem, its solution is multidimensional, and has many local extreme values. The basic description of Traveling salesman problem: suppose there is a traveling merchant to visit n cities, and he had to choose one route. Starting from a city, he will visit every city only once, and finally return to the original departure city.

The target is to find the shortest route. In short, traveling salesman problem is to find such a tourist routes: from a city, through each city once and only once, finally returned to the departure city, and find out the shortest route. The mathematical model of the TSP problem can be described as:

$$\min f(X),\ s.t. g(X) \geq 0, \quad X \in D \tag{1}$$

In formula, $f(X)$ act as the objective function, $g(X)$ act as the constraint function, X act as decision variables, D represent assemble that set composed of a finite number of points. Usually, a combinatorial optimization problem can be used to represent the three parameters, D which says the decision variable domain, and $F = \{X|X \in D, g(X) \geq 0\}$ f represents the target function, meet the feasible solution is called the optimal solution of the problem.

To facilitate the use of artificial fish algorithm to solve TSP problem, we introduce the following relevant definitions [11]:

Definition 1 to combinatorial optimization problem, the distance between the decision variables can be represented as:

$$distance(X_1, X_2) = |X_1 - X_2| + |X_2 - X_1| \quad (2)$$

It refers to the number of elements which do not belong to X_1 and X_2 at the same time.

Definition 2 for combinatorial optimization problems (D, F, f)

$$N(X, k) = \{X' | distance(X, X') < k, X' \in D\} \quad (3)$$

Called—distance field, $X' \in N(X, k)$ known as a neighbor of X.

Definition 3 for combinatorial optimization problems (D, F, f)

$$center(X_1, X_2, \ldots, X_m) = \bigcup_{i=1}^{m} \bigcup_{\substack{j-1 \\ j \neq i}}^{m} (X_i \cap X_j) \quad (4)$$

Known as the center of the decision variables of $X_1, X_2, \ldots, X_m$.

3 The Improved Hybrid Artificial Fish Algorithm

Artificial fish individual state vector which can be expressed as $X = (x_1, x_2, \ldots, x_n)$, in which $x_i (i = 1, 2, \ldots, n)$ represents optimization variables by desire, the location of the current food concentration of artificial fish is $Y = f(X)$, in which X acts as the objective function value. The distance between individuals of artificial fish is expressed as d_{ij}. *Visual* Refers to distance perception of artificial fish, X_j refers to artificial fish moving step, δ and refers to congestion level factor.

3.1 Improvement of Visual Field

Improvement of Vision range parameters of artificial fish algorithm are made as follows: visual field parameters in the initial stages of artificial fish algorithm, each of the artificial fish with a great vision to find solutions, to expand the scope of optimization. As the algorithm run, later the field of vision will be appropriately reduced to speed up the convergence speed.

3.2 Improved Preying Behavior

Preying behavior [12]: set the current position of the artificial fish is X_i, select the position X_j in its visual field randomly. If $Y_i > Y_j$ (forward condition), then move a step to this direction. Otherwise randomly select a position X_j in its visual field randomly again, and then judge whether it satisfies the forward conditions. Try number times, if the forward condition is not satisfied and the current state is not within the current group of optimal state then perform random moving behavior. If the current group in the optimal state, it is not moved.

3.3 Improved the Swarming Behavior

The following improvement actions of the swarming behavior [12]: set the current position of the artificial fish is X_i, exploring the current number of partners within the neighborhood n_f and the center position X_c, if $Y_c < Y_i$ and $n/N < \delta$, shows that partner center has less food, and are not too crowded, the partner go forward to the center of the direction, in the process of advance, if encounter food concentration less than food concentration in partner center position, was moved to the location; If not found lower than the partner center food concentration location, has moved to the central position; Otherwise performs preying behavior.

3.4 Following Behavior

Set the current position of the artificial fish is X_i, minimum of X in its neighborhood. If $Y_i > Y_j$, The number of partners in the neighborhood X_j not satisfied $n_f/N > \delta$,it shows that the food of low concentration and not crowded. Then the artificial fish move a step; otherwise performs preying behavior.

3.5 Random Behavior

The realization of the random behavior is relatively simple; it selects a random position in its visual field and then moves a step to this direction. It is a default behavior of preying behavior.

3.6 Bulletin Board

Bulletin board is used to record the optimal artificial fish. In the optimization process, each artificial fish compares its position with the bulletin board after a moving. If its position is better than the bulletin board, the bulletin board is rewritten by this artificial fish.

3.7 Behavior Choice

The choice of a kind of behavior is determined by evaluating the current environment of the artificial fish according to the nature of the problem to be solved. Artificial fish first simulate swarming behavior, following behavior, etc., and then evaluation value after the operation, so as to choose one of the largest actual executions, the default behavior for preying behavior.

3.8 To Cross Local Optimization Strategy

Determine whether two paths exist cross principle [13] is as follows: Plane two line segments likely have cross, overlapping, parallel, a cross path respectively, the coordinates of the four vertices of the ABCD (x_1, y_1), (x_2, y_2), (x_3, y_3) and (x_4, y_4), and the linear equation of two line segments are as follows:

$$\text{The line AB}: \begin{cases} x = x_1 + (x_2 - x_1) \cdot m \\ y = y_1 + (y_2 - y_1) \cdot m \end{cases} \quad 0 \le m \le 1 \tag{5}$$

$$\text{The line CD}: \begin{cases} x = x_3 + (x_4 - x_3) \cdot n \\ y = y_3 + (y_4 - y_3) \cdot n \end{cases} \quad 0 \le n \le 1 \tag{6}$$

$$\mathrm{D} = \begin{vmatrix} x_2 - x_1 & x_3 - x_4 \\ y_2 - y_1 & y_3 - y_4 \end{vmatrix}$$

When $D = 0$, the two line AB and CD parallel or overlapping, as no intersection; And when there $D \neq 0$,

$$m = \frac{1}{D} \begin{matrix} x_3 - x_1 & x_3 - x_4 \\ y_3 - y_1 & y_3 - y_4 \end{matrix}$$

$$n = \frac{1}{D} \begin{vmatrix} x_2 - x_1 & x_3 - x_4 \\ y_2 - y_1 & y_3 - y_4 \end{vmatrix}$$

If $0 \leq m \leq 1$ and $0 \leq n \leq 1$, the line AB, CD have a point. On the other hand, argues line AB, CD, no intersection point.

4 Hybrid Artificial Fish Algorithm in Solving TSP

Artificial fish algorithm has to grasp the search direction, and to a certain extent avoid falling into local optimum, but as part of the artificial fish in aimless random motion or artificial fish in the global extreme value point becomes seriously crowded, the convergence speed will slow down greatly, and will cause the search accuracy reduced. In order to overcome this drawback, we introduced crossover operator in genetic algorithm and the improved swarming and improved preying behavior. To judge whether have improved with the increase of the number of iterations search results, the bulletin board is set up in the algorithm to record the optimal artificial fish in the individual state. Each artificial fish compares its position with the bulletin board after a moving. If its position is better than the bulletin board, the bulletin board is rewritten by this artificial fish. When the best individual in the process of continuous multiple iterations did not change or change is very small, with a crossover operation, and retains the best individual state history, will other artificial fish according to certain probability to cross a small fish.

By introducing genetic algorithm crossover mechanism, realized of the jump of artificial fish in fish individuals, thereby, adjusting and optimizing group improved visual field and improvement actions to improve the algorithm convergence speed at the same time to ensure that the global search ability.

4.1 Hybrid Artificial Fish Algorithm Steps

Based on the analysis of the literature [14] on the basis of the proposed algorithm, combined with improved preying behavior, swarming behavior, combined them effectively, can get good effect. Improved hybrid artificial fish algorithm implementation steps are as follows:

Step 1 create Initialization groups: set the initial bulletin board of the optimal artificial fish state does not change or changes very small continuously iteration *Bstep* is 0, the initial number of iterations for *Gnum* to 0, don't change the maximum threshold *Maxbeststep* in a row, the largest number of iterations for *Maxnumber*, randomly generated N artificial fish in the feasible region, form the initial fish, and set the parameters of artificial fish.

Step 2 bulletin board initial value: calculating the value of each initial fish in the current status, compare the size, the minimum value into the bulletin board, and the relative state of artificial fish is assigned to the bulletin board.

Step 3 behavior choice: according to the code of conduct in artificial fish algorithm, to carry out following behavior and improved swarming behavior, the default behavior of preying behavior. If carry out following behavior and swarming behavior, the best value or there is no change to carry on a random behavior.

Step 4 after the action selection of artificial fish, check their own value and the value of the bulletin board, if is superior to the bulletin board, to replace itself, at the same time update the bulletin board on the state of artificial fish.

Step 5 besides the best individual in bulletin board, other artificial fish performs the following operations: The artificial fish for genetic crossover operation. Calculating the value of the artificial fish which newly formed, and comparing with optimal value in bulletin board, if is superior to the bulletin board, is to replace itself.

Step 6 *Determine whether the termination condition is satisfied*, not satisfied, go to Step 3, into the next step of the fish and optimization process, otherwise go to step 7.

Step 7 Algorithm is end, to output the optimal solution of artificial fish status value on the bulletin board.

4.2 Numerical Example and Simulation Results

Data from the TSP problems including the following: http://www.iwr.uni-heidelberg.de/iwr/comopt/soft-ware/TSPLIB95, the parameters of the algorithm for Try number = 20, N = 15 $\delta = 10$, the largest number of iterations *Maxnumber* = 1000, 1000, 2000, crossover probability is 0.5.

Carry out the simulation experiment of Ulesses22, Oliver30, att48 problems in literature [15], this paper use the improved hybrid artificial fish algorithm comparing with algorithm of literature [15]. Running 30 times for each algorithm, the simulation results are shown in Table 1.

Table 1 The statistical results of tsp solution

TSP	The known optimal solution at resent	Mixed AFSA			Literature [15]		
		The best value	Average value	The worst value	The best value	Average value	The worst value
Ulesses 22	75.6651	75.3097	75.9636	76.2354	75.4010	76.2874	77.0784
Oliver 30	42.7406	423.7406	424.2651	425.2460	423.7406	424.4681	426.5439
Att 48	33523.709	34728.2156	34956.9635	35753.3264	34015.2740	34937.9221	35622.4780

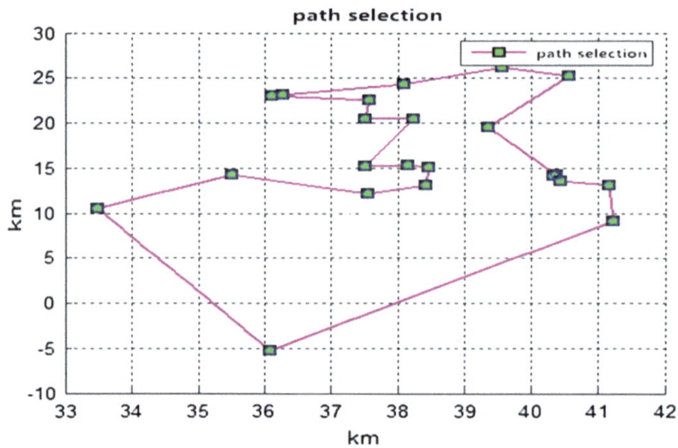

Fig. 1 The shortest Ulesses22 of the improved algorithm in this paper

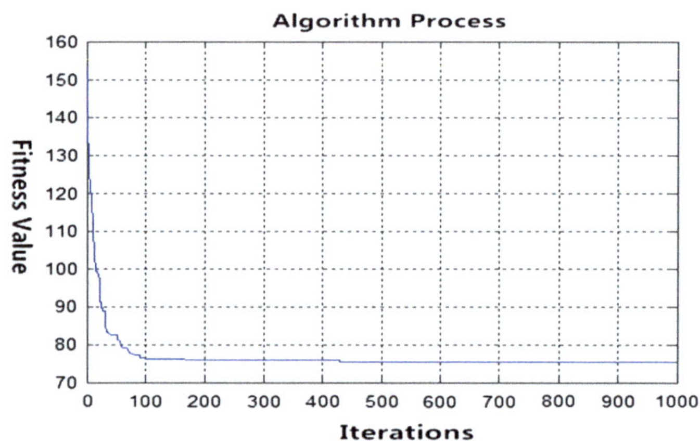

Fig. 2 Ulesses22 optimal curve of the improved algorithm in this paper

The algorithm in this paper to find the Ulesses22, Oliver30 att48 and the shortest path as shown in Figs. 1, 3 and 5, their optimal curve respectively as shown in Figs. 2, 4 and 6.

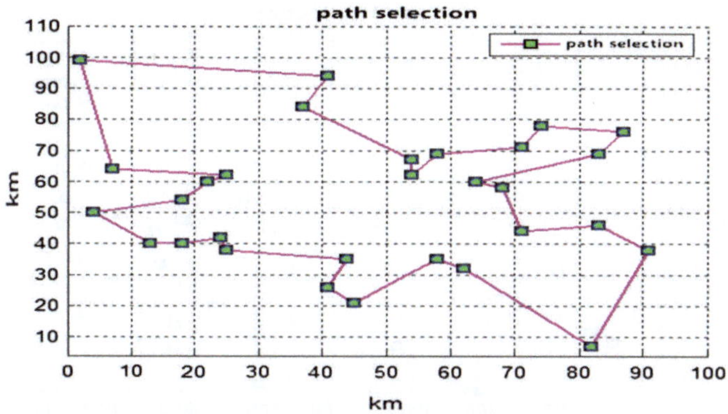

Fig. 3 the shortest Oliver30 of the improved algorithm in this paper

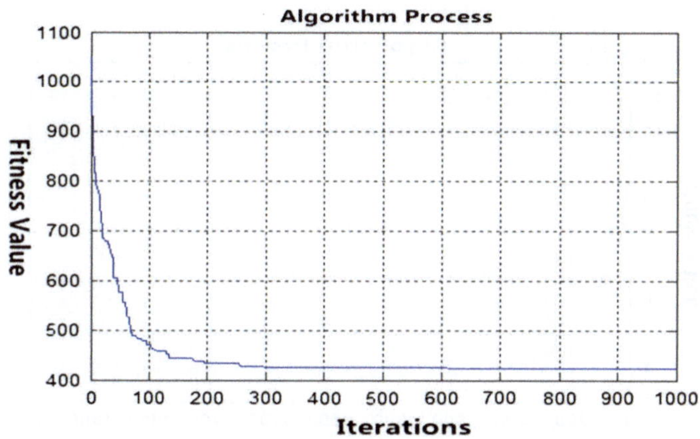

Fig. 4 Oliver30 optimal curve of the improved algorithm in this paper

From the experimental results in this paper, we can see the improved artificial fish algorithm find the optimal solution of the number of iterations is less, and get the optimal solution is better than the average of the optimal solution and obtain the best solution is to close to the one of the best known solution of TSP problem.

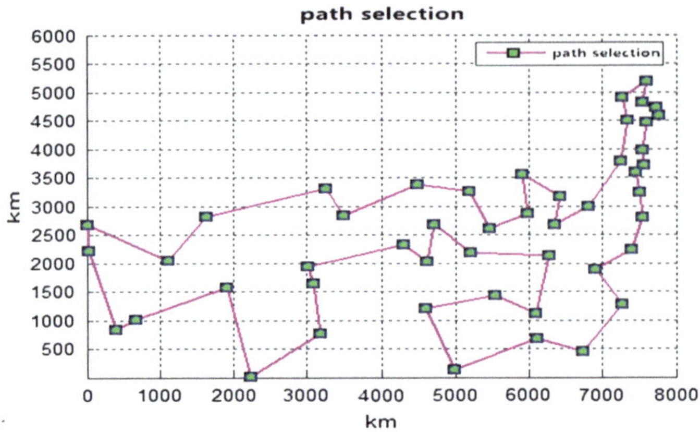

Fig. 5 The shortest att48 of the Improved algorithm in this paper

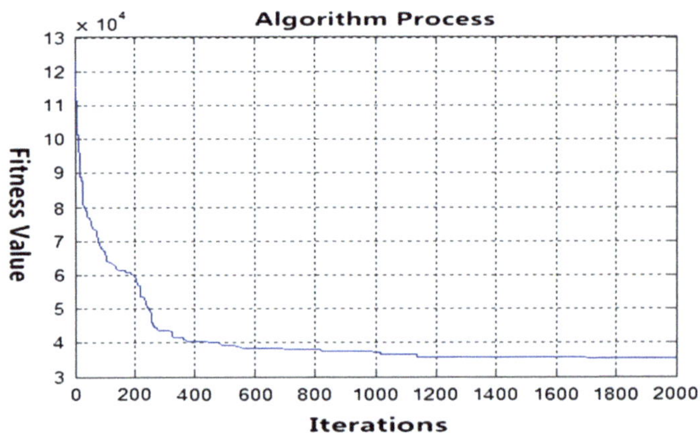

Fig. 6 att48 optimal curve of the improved algorithm in this paper

4.3 Conclusion

The simulation research shows that using the genetic algorithm and artificial fish algorithm of hybrid optimization algorithm for solving TSP is feasible, and compared with the improved artificial fish algorithm of literature, experimental results show that the algorithm convergence performance is better, the optimal solution precision and convergence speed are improved.

References

1. Kirkpatr Ick S, Gelatt CD, Vecchi MP (1983) Optimization by simulated annealing. Science 220(4598):671–680
2. Holland JH (1962) Outline for a logical theory of adaptive system. J Assoc Comput Mach 3:297–314
3. Holland JH (1969) A new kind of turnpike theorem. Bull Am Math Soc 75:1311–1317
4. Colorni A, Dorigo M, Maniezzo V (1991) Distributed optimization by ant colonies. In: Proceedings of 1st european conference on artificial life, pp 134–142
5. Dorigo M (1992) Optimization, learning and natural algorithms. Ph. D. thesis, Department of Electronics, Politecnico di Milano
6. Kennedy J, Eberhart RC (1995) Particle swarm optimization. In: Proceedings of IEEE international conference on neural networks. IEEE Service Center, Piscataway NJ, pp 1942–1948
7. Shi Y, Eberhart RC (1999) Empirical study of Particle swarm optimization. In: Proceedings of congress on evolutionary computation, Washing DC, pp 1945–1950
8. De Castro LN, Von Zuben FJ (1999) Artificial immune systems: Part I-basic theory and applications. School of Computing and Electrical Engineering Brazil: State University of Campinas .No. DCA-RT 01/99
9. Li X, Shao Z, Qian J (2002) An optimizing method based on autonomous animats: fish-swarm algorithm. Syst Eng Theory Pract 22(11):2–38
10. Li X, Lu F (2004) Applications of artificial fish school algorithm in combinatorial optimization problems. J Shan Dong Univ (Engineering Science) 34(5):65–67
11. Xing W, Xie J (1999) Modern optimization method. Tsinghua University Press, Beijing
12. Fan Y, Wang D, Sun M (2007) Improved artificial fish-school algorithm. J Chongqing Normal Univ (Natural Science Edition) 24(3):23–26
13. Haibin D (2005) Principle and application of ant colony algorithm. Science Press, Beijing
14. Zhou Y, Xie Z (2009) Improved artificial fish-school swarm algorithm for solving TSP. Syst Eng Electron 31(6):1458–1461
15. Zhu M, Ku X (2010) Improved artificial fish school algorithm to solve traveling salesman problem. Appl Res Comput 27(10):3734–3736

Identifying Abbreviations in Biological Literature Based on Reverse Text Alignment

Jing Peng, Cheng L. Jing, Peng Wang and Hong M. Sun

Abstract The number of biomedical literatures is growing rapidly, and bioinformatics literature mining is becoming essential. The identification of abbreviations in literatures is important for all text mining tools. In the paper, an abbreviation recognition algorithm based on reverse text alignment is proposed and it is easy to be implemented and has fast running time. The algorithm achieves 91 % precision and 93 % recall on the gold standard corpus "Medstract" and 96 % precision and 84 % recall on the larger test data that includes 128 full text literatures.

Keywords Abbreviation recognition · Reverse text alignment · Text mining

1 Introduction

With the development of genome sequence technology, the number of biomedical literatures is growing rapidly [1, 2]. It is difficult to read all of them and obtain some knowledge by hand. Text mining technology has been widely used as an effective method to automatically extract valuable information from the literatures in recent years [3]. The words of biomedical terminologies are usually long and researchers like to use abbreviations, acronyms or symbols to communicate with others (in the following sections, abbreviation is used to represent all the short forms including acronyms and symbols). Therefore, there are lots of abbreviations of organisms which are named by entities in biomedical literatures, for example: "soybean cyst nematode (SCN)", "quantitative trait loci (QTL)", "Recombinant inbred lines (RIL)".

Abbreviations can simplify the expressions the biomedical definitions and make the literatures easy to be understood. However, the use of abbreviations has brought

J. Peng (✉) · C.L. Jing · P. Wang · H.M. Sun
Bioinformatics Laboratory, Electrical and Information Engineering College, Northeast Agricultural University, Heilongjang, People's Republic of China
e-mail: applepj@msn.com

E. Qi (ed.), *Proceedings of the 6th International Asia Conference on Industrial Engineering and Management Innovation*,
DOI 10.2991/978-94-6239-145-1_28

difficulties to biological text mining. This is first cause that abbreviations have no semantics, but have ambiguity at a high degree. For example: "RIL" is an abbreviation of "recombinant inbred lines", and it represents "radical inguinal lymphadenectomy" at the same time. Additionally, the writing of abbreviations has no uniform standards, and researchers write them using their favorite style. For example, in the pair "Targeting induced local lesions in genomes (TILLING)", the abbreviation includes a stop words "in" and in the "efficiency of somatic embryogenesis (ESE)", the abbreviation omit the stop word "of". Writing having no standards reduces the accuracy of text mining, and is one of the reasons of the slow development in biomedical text mining. Secondly, with the rapid development of biology science, the volume of biomedical terminologies and their abbreviations is growing at a fast rate. These new unknown terms that can not be recorded in biological dictionaries timely, bring difficulties to biological name entity recognition and knowledge discovery. Therefore, how to identify and define the abbreviations in literatures has became an important key task in bioinformatics literature mining [4].

In this paper, we first investigate a large number of biomedical literatures, and propose an abbreviation recognition algorithm based on reverse text alignment. The algorithm is evaluated by the gold standard test corpus, and we discuss the future work for improving the algorithm finally.

2 Abbreviations Recognition

In the biological literatures, abbreviations and their definitions usually appear in pairs at the first time, for example, "....... six additive quantitative trait loci (QTL) and 11 pairs of epistatic QTL using an F_7 recombinant inbred line (RIL) population ......" Where "QTL" and "RIL" are the abbreviations for the definitions of "quantitative trait loci "and" recombinant inbred line". Abbreviation recognition is to extract an abbreviation and its definition in a certain literature, such as <QTL, quantitative trait loci>.

The existing approaches for identifying abbreviations are mainly divided into three categories, rule-based approach, machine learning and text alignment. Rule-based approach [5–8] use many hand-made rules after reading a lot of literatures. It is simple, and has a high accurate, but the effectiveness depends on the integrity of the rule set. And there is difficult to collect a complete rule set covering all abbreviations. Machine learning approaches [8, 9] are to build an identifier by training lots of literature texts. The identifier then decides the correct pair of abbreviation and definition from candidates. Machine learning approaches need not study the grammatical structure of texts, and they are objective. However, machine learning needs a lot of training data to make the identifier accurate, and the training data is mainly collected and annotated by hand. The work is very time-consuming and labor-consuming. Text alignment approaches [10] first determine the scope of an abbreviation and its definition, and then compare each character until a matching character is found. This kind of approach is quick, simple, and easy to be integrated into the framework of other natural language processing or text mining platform. The disadvantage is that it is not suitable for the abbreviations with non-standard writing, for example, latent transforming growth factor beta 1 (LTGF-beta 1).

In the survey of a large volume of biomedical literatures, we found that the following three types are the most common in which the occurrence rate of the first type is as high as 98 %. This paper discusses the first case, and the others are out of the scope of our research.

1. Definition is front of abbreviation, and abbreviation is in brackets or square brackets, e.g., soybean looper (SBL).
2. Abbreviation is in front of definition, and definition is in brackets or square brackets, e.g., (soybean looper).
3. Both definition and abbreviation are in brackets or square brackets, and separated by commas or semicolon, e.g., (SBL, soybean looper).

3 Method and Implementation

The approach proposed in the paper is based on text alignment using reverse search algorithm (RTA). RTA is implemented by four steps that are text preprocessing, finding abbreviations, finding definitions and judging repeat, respectively. The flow chart of the algorithm is shown in Fig. 1.

The detail of our approach is explained as follow.

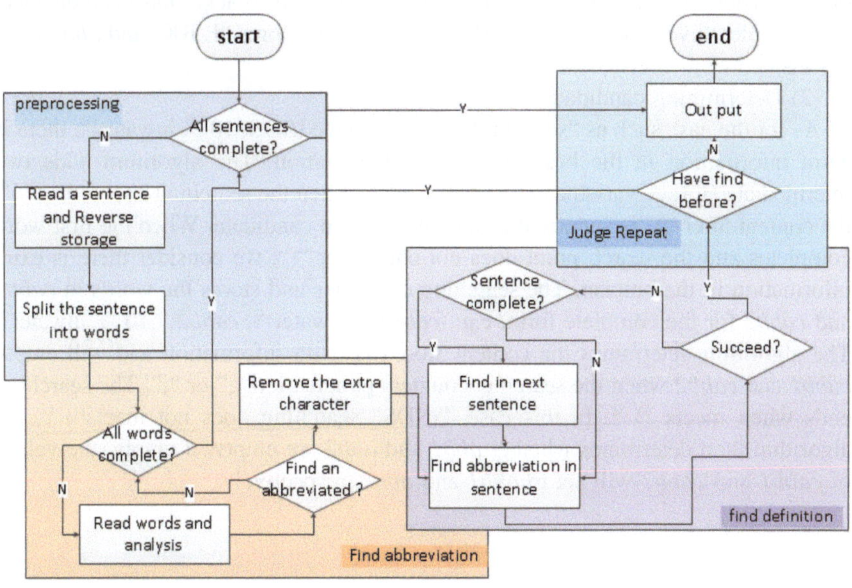

Fig. 1 The flow chart of RTA (Reverse Text Alignment)

3.1 Text Preprocessing

The sequence of text characters is stored in reverse order. For example, "pod-shell 10 days after flowering (DAF) and pod-shell 12-DAF" is stored as "FAD-21 llehs-dop dna) FAD (gnirewolf retfa syad 01 llehs-dop".

3.2 Finding Abbreviations

In general, abbreviations occur with a certain form in literatures. That is abbreviations usually appear in brackets. However, we found that besides abbreviations, biological species examples, the relevant reference information, and comment information are also written in brackets. Therefore, Finding Abbreviations process can be divided into two steps, searching brackets and determining whether the content of the bracket is an abbreviation or not.

As abbreviations are composed of uppercase and lowercase letters, and no special symbols such as punctuations and spaces are added in an abbreviation, we consider that the content that is all the letters and has no other characters, is a abbreviation candidate.

(1) Searching brackets

Search a character sequence reversely until meet the character ")". The content after ")" is supposed as an abbreviation candidate and the symbols if any are deleted. The candidate is stored in a string *abb1* and will be used in the next step. The copy *abb2* may contain non-letter symbols and will be used as a complete output. When the searching encounters "(", the scope of bracket has been decided, for example, "liver / bone / kidney (L / B / K)", *abb1* stores "L BK" and *abb2* is "L / B / K".

(2) Determining candidates

As for the case such as "sodium dodecyl sulphate (SDS, 10 % inwater)", there is extra information in the bracket besides abbreviation. The algorithm adds two intermediate storage variable *cabb1* and *cabb2*. When the searching encounters ")", the content after ")" is supposed as an abbreviation candidate. When the first word completes and the search point does not encounter "(", we consider there is extra information in the bracket. The searching continues and stores the words in *cabb1* and *cabb2* for the complete form, e.g., *cabb1*:" inwater ", *cabb2*:" 10 % inwater". The algorithm determines the content stored is extra information and will empty *cabb1* and *cabb2* when the search encounters punctuation "," or ";". The searching ends when meets ") ". In this case, "(SDS" searching does not meet ") ", the algorithm then determines whether *abb1* and *abb2* are empty, if empty, the values of *cabb1* and *cabb2* will set to *abb1* and *abb2*, respectively.

3.3 Finding Definitions

Single-character abbreviation (e.g., day(D)) is considered as a special forms and ignored in many studies. In this paper, we consider this case, and process the searching according to the length of the candidate abbreviations. Definitions of abbreviations fall into two categories, definitions for single-character and definitions for multi-character. Since the definition of single-character search is relatively simple and easy, the searching just needs to a simple determination, while the multi-character case is relatively complex, it requires some rules to limit the content.

(1) Definitions for single-character abbreviations

The previous word of a single-character abbreviation is supposed the definitions in our algorithm. If the first character of the word is the same as the single-character abbreviation, the algorithm determines the word is the target definition.

(2) Definitions for multi-character abbreviations

We use reverse search method for multi-character case and the searching starts at the last character of the word.

abb: the found abbreviation
sen: the current sentences
fn: the found definition

(1) If the first character of the current word is different from the current character, the indictor shifts downwards.
(2) If the current character of a word is matched to a character of an abbreviation, the current word is supposed as one word of the definition. The word is added into *fn*, and the indictors of definition and abbreviation both shift to the next.
(3) If the first character of the current word is matched a character of an abbreviation, but not the first one, the current word is supposed as one word of the definition. The word is added into *fn*, and the indictors of definition and abbreviation both shift to the next.
(4) If the current word fails to match the character in an abbreviation, stop word is considered to appear in the definition. In this case, the current word is added into *fn*, the indictor shift to the next word.
(5) If the first character of the current word is the same as the first character of abbreviation, then the word assed into *fn*. And *fn* is determined as the target definition.
(6) The following cases are determined as match failure.

- The current word is the first word of a sentence, and the indictor does not point the first character of abbreviation.
- “)” appears in the current word.

4 Experiments and Results

4.1 Experiment Data Preparation

(1) Medstract Gold Standard test corpus

We first apply our TCR algorithm on the Medstract Gold Standard test corpus [5], which is composed of 400 abstracts extracted from MEDLINE, and is one of the most authoritative test corpus currently used to evaluate abbreviation recognition approaches. Medstract Gold Standard corpus provides 413 pairs of abbreviation and its definition on a web page (http://www.medstract.com/index.php?f=gold-result). However, we found that there were some omissions and errors in the pair list. We then supplemented the 40 abbreviations (e,g,. human menopausal gonadotrophin (hMG); transformed rabbit corneal epithelial cells (tRCEC)) and corrected the errors. We collected a total of 453 abbreviations finally.

(2) Soybean QTL test the full text corpus (SBQTL)

All the proposed approaches in the related research about abbreviation recognition are applied for the abstract of biomedical literatures. In this paper, we prepared a larger text set of full texts to evaluate our approach. We extracted 128 full text literatures by the keywords "soybean + QTL" from MEDLINE, and then annotated all the abbreviations and their definitions in the set. The work is done and corrected by 3 junior undergraduates, and spent about 10 h each student. We obtained 1329 pairs from the 128 texts.

4.2 Evaluation Method

We used Recall, Precision and F-measure to evaluate our abbreviation recognition results. The three values reflect the completeness, the accuracy and the effectiveness of the approach, respectively. The definitions of the three criterions are shown as follow.

$Recall = TA / AN$

$Precision = TA / (TA + FA)$

$F\text{-}measure = (2 \times Precision \times Recall) / (Precision + Recall)$

in which, *TA* represents the number of abbreviations indentified correctly. *AN* represents the number of actual abbreviation in test set. *FA* represents the number of abbreviations indentified wrongly.

4.3 Baseline System

We implemented the simple abbreviation recognition algorithm proposed by Schwartz & Hearst (S&H) [10] as our baseline system. The algorithm sets up two

Table 1 Experiment results on two test sets

Algorithm	Test Corpus	Precision	Recall	F-measure
S & H (Baseline)	Medstract	88 %	87 %	0.87
	SBQTL	90 %	76 %	0.82
MBA	Medstract	91 %	88 %	0.89
	SBQTL	–	–	–
TCR*	Medstract	91 %	93 %	0.92
	SBQTL	96 %	84 %	0.88

indicators that point to the end of abbreviation and its definitions, and search the characters one by one until the match is completed. We also attempt compare our approach to literature mining system MBA [11], which is based on machine learning to improve the recognition recall.

4.4 *Experiment Results*

The experiment results are summarized in Table 1. The baseline system achieves 88 % precision and 87 % recall on the gold standard corpus "Medstract" and 90 % precision and 76 % recall on the SBQTL test data. Our approach TCR achieves 91 % precision and 93 % recall on the gold standard corpus "Medstract" and 96 % precision and 84 % recall on the SBQTL test data. We cannot implement MBA system because there has no enough details shown in its literature, therefore, we just compare the results reported in the technological paper.

From Table 1, we found our approach is better than the baseline both in the two test data and also better than MBA on the medstract corpus.

5 Discussion

Although our approach performs better results, there are some misses need to be improved. Table 2 shows the error list on Medstract corpus. We found our approach cannot perform well in the abbreviations having numbers or other symbols, e.g., <IAA, 3-indoleacetic acid>, <3-D, 3-dimensional>, <A&E, accident and emergency>. There are 8 of this kind of abbreviations, 40 % of all. For some abbreviations, such as <VIN, vulval intraepithelial neoplasia>, <CR, complete remission>, our approach did not work because the context of the pair makes the scope of definition uncertain.

Analysis of the errors mentioned above shows the further improvement of the approach.

Table 2 Experiment results on two test sets

NO	Abbreviation	Definition
1	IAA	3-indoleacetic acid
2	DMDP	4R-dihydroxypyrrolidine
3	3-D	3-dimensional
4	LTGF-beta 1	latent transforming growth factor beta 1
5	b FGF	basic fibroblast growth factor
6	A&E	accident and emergency
7	NTx	N-telopeptides of type I collagen
8	ALP	alkaline phosphatase levels
9	K9	keratin 9
10	VIN	vulval intraepithelial neoplasia
11	ALA	5-aminolaevulinic acid
12	CN	Chloracetophenone
13	BHLF	Bam HI-H, l-fragment
14	AT	Ataxia teleangiectasia
15	SH	Society for Hematopathology
16	CR	complete remission
17	EC	electrochemical detection
18	GlOx	glutamate oxidase
19	PEGDGE	Poly(ethylene glycol) diglycidyl ether
20	LC/ESI/MS/MS	HPLC/electrospray ionization tandem mass spectrometric

Firstly, besides searching based on text alignment, the algorithm requires the use of syntactic information, such as syntactic rules to filter some noise data. This idea was suggested in Pustejovsky et al. [12]. The rules can be collected based on natural language processing (NLP) such as POS, NER, and shallow parsing. Text pre-processing based NLP might help correct some errors inherent in our approach, and identify the context near the pair. Additionally, combining evidence from more than one MEDLINE abstract at a time, as was done in [13], might also prove to be beneficial for increasing both precision and recall. Finally, the algorithm currently only considers candidate definitions when the abbreviation is enclosed in brackets; finding all possible pairs is a more difficult problem and requires additional study.

6 Conclusion

The identification of abbreviations in literatures is important for all text mining tools. In the paper, an abbreviation recognition algorithm based on reverse text alignment is proposed and it is easy to be implemented and has fast running time. The algorithm achieves 91 % precision and 93 % recall on the gold standard corpus "Medstract" and 96 % precision and 84 % recall on the larger test data (SBQTL) that includes 128 full text literatures, while the baseline system achieves 88 % precision and 87 % recall on "Medstract" and 90 % precision and 76 % recall on SBQTL.

After analyzed the errors produced by our approach, the further improvement work of the approach is discussed, they focus on combine syntactic information as rules into text alignment algorithm and extend the scope our research to all possible abbreviation.

Acknowledgments The authors would like to thank the Oversea Point Foundation of Heilongjiang Province Department of Education (1253HQ001) and Doctor Start-up fund of Northeast Agricultural University (2012RCB54) for their support in this research.

References

1. US National Library of Medicine National Institutes of Health. http://www.ncbi.nlm.nih.gov/pubmed
2. Lu Z (2011) PubMed and beyond: a survey of web tools for searching biomedical literature. Database (Oxford) 2011:36
3. Yandell MD, Majoros WH (2002) Genomics and natural language processing. Nat Rev Genet 3(8):601–610
4. Torii M et al (2007) A comparison study on algorithms of detecting long forms for short forms in biomedical text. BMC Bioinf 8(Suppl 9):S5. doi:10.1186/1471-2105-8-S9-S5
5. Pustejovsky J et al (2002) Medstract: natural language tools for mining the biobibliome. http://www.medstract.org/. Accessed 20 Dec 2013
6. Ao H, Takagi T (2005) Alice: an algorithm to extract abbreviations from MEDLINE. J Am Med Inform Assoc 12:576–586. doi:10.1197/jamia.M1757
7. Larkey LS, Ogilvie P, Price MA, Tamilio B (2000) Acrophile: an automated acronym extractor and server. In: Nurnberg PJ, Hicks DL, Furuta R (eds) Proceedings of the 5th ACM international conference on digital libraries. ACM Press, San Antonio, pp 205–214
8. Chang JT, Schutze H, Altman RB (2002) Creating an online dictionary of abbreviations from MEDLINE. J Am Med Inform Assoc 9:612–620
9. Chang JT, Schutze H, Altman RB (2002) Creating an online dictionary of abbreviations from MEDLINE. J Am Med Inform Assoc 9:612–620
10. Schwartz AS, Hearst MA (2003) A simple algorithm for identifying abbreviation definitions in biomedical text. Proc Pac Sympos Biocomput 8:451–462
11. Xu Y et al (2009) MBA: a literature mining system for extracting biomedical abbreviations. BMC Bioinf 10:14. doi:10.1186/1471-2105-10-14
12. Pustejovsky J, Castano J, Cochran B (2001) Automatic extraction of acronym-meaning pairs from medline databases. Stud Health Technol Inform 10:371–375
13. Adar E (2002) S-RAD: a simple and robust abbreviation dictionary. HP Laboratories Technical Report, Sept 2002

Under the Strong Noise in Ultraviolet Missile Approach Warning to Extract the Target

Qi-na Gao, Ying Zhu, Jin-guang Wang and Jin-hui Yang

Abstract In the ultraviolet missile approach warning signal processing, strong noise and burst interference will affect the test missile target, but the classic two-dimensional filtering algorithm cannot effectively filter out the interference, because the strong noise and burst interference and spatial location spatial frequency position is uncertain. This paper presents a three-dimensional image sequences filtering algorithm, the two-dimensional and one-dimensional spatial filter IIR filter composed of the time, the two filters are stable. This article also gives the stability detection algorithm. An advantage of three-dimensional recursive filter is a small amount of computation, and can achieve rapid real-time image sequences filtering. Ultraviolet missile approach warning signal processing simulation results show that three-dimensional filtering and related algorithms presented in this paper are correct, efficient and practical.

Keywords 3-dimensional recursive filtering · Burst interference · Missile approach warning · Missile target extraction · Signal processing

Q. Gao · Y. Zhu (✉) · J. Wang · J. Yang
Beijing Research Center of Aerial Engineering Technology,
Beijing 100076, China
e-mail: zhuying713272@126.com

J. Wang
School of Electronics and Information Engineering,
Nanjing University of Aeronautics and Astronautics, Nanjing 210016, China

E. Qi (ed.), *Proceedings of the 6th International Asia Conference on Industrial Engineering and Management Innovation*,
DOI 10.2991/978-94-6239-145-1_29

1 Introduction

UV sensor can be used to detect and track missile plume of light bands in blind, missile approach warning issued [1–7]. If the amplitude of the missile target signals amplitude close to interference and noise, then the noise from the random noise and extract the missile target is not an easy task. The main difficulty lies in the strong noise and burst interference spatial location and spatial frequency position is uncertain. In general, the use of existing two-dimensional spatial filtering algorithms, including filtering Lee, Kuan filtering, Sigma filtering, Frost filtering, eliminate the random noise or interference is very difficult, the main problem is not considered filtering [8] along the time direction. Three-dimensional filtering algorithm [9] is a calculation problem faced: a three-dimensional FIR filter is not practical, because it generally requires more buffer 100 to get a better quality filtering; meanwhile, the IIR filter may be a three-dimensional stability problems [10]. To solve this problem, we will leapfrog type of two-dimensional digital filters [11] expanded into a three-dimensional IIR digital filter, the joint use of space and time directions filter to eliminate noise and interference in the ultraviolet image, and then extract the missile target signal. Three-dimensional IIR digital filter we propose is a stable three-dimensional recursive digital filter, this paper presents stability test theorems and algorithms used in the filter design, since the recursive filtering requires only two ultraviolet image frame buffer, so you can fast filtering ultraviolet image sequences.

2 Signal Model

With three-dimensional filtering algorithm Blind band in the light, ultraviolet CCD detected ultraviolet image has the following features: (1) burst interference occupy the high-frequency portion of airspace frequency burst interference in the time domain is short, do not connect. (2) Motion missiles in space and time domain are continuous. (3) In the airspace or the frequency domain, the noise in the image sequence is no obvious structural or spatial frequency structure. In the spatial image sequence, the high frequency portion of the noise occupies. the characteristics of (1) and features (2) reminds us that the proposed three-dimensional IIR should be a three-dimensional low- pass filter.

Now, we define the 3-D discrete imaging sequence to be $y(n_1, n_2, n_3)$, $x(n_1, n_2, n_3)$ to be the desired object, $w(n_1, n_2, n_3)$ to be the random noises or interferences n_1, n_2 and n_3 are horizontal variable, vertical variable, and time variable of the image sequence, respectively, where $0 \le n_1 \le N_1$, $0 \le n_2 \le N_2$, and $0 \le n_3 \le N_3$. N_1, N_2 and N_3 are the lengths of the 3-D sequence along three directions. In this paper, to solve the problem of fast filtering, we will limit $N_1 = N_2 = N_3 = 2$ for the proposed 3-D IIR filter, which means we only need buffer 2 frames for 3-D filtering.

We establish the signal model for the imaging sequence $y(n_1, n_2, n_3)$ with noise or burst interferences as following,

$$y(n_1, n_2, n_3) = x(n_1, n_2, n_3) + w(n_1, n_2, n_3) \tag{1}$$

For the CCD output $y(n_1, n_2, n_3)$, the intensity of $w(n_1, n_2, n_3)$ may be large or equal to the strength of $x(n_1, n_2, n_3)$, which will lead to a wrong warning for missile approaching.

This paper proposes following 3-D IIR digital filter to remove the noise term $w(n_1, n_2, n_3)$,

$$H(z_1, z_2, z_3) = H_s(z_1, z_2)H_t(z_3) \tag{2}$$

$H_s(z_1, z_2)$ is a 2-D IIR digital filter, which completes the spatial filtering along x and y directions,

$$H_s(z_1, z_2) = \frac{\sum\limits_{k_1=0}^{K_1} \sum\limits_{k_2=0}^{K_2} a_s(k_1, k_2) z_1^{-k_1} z_2^{-k_2}}{\sum\limits_{k_1=0}^{K_1} \sum\limits_{k_2=0}^{K_2} b_s(k_1, k_2) z_1^{-k_1} z_2^{-k_2}} \tag{3}$$

And $H_t(z_3)$ is a 1-D IIR digital filter, which completes the time filtering for the imaging sequence $y(n_1, n_2, n_3)$ along time directions,

$$H_t(z_3) = \frac{\sum\limits_{k_3=0}^{K_3} a_t(k_3) z_3^{-k_3}}{\sum\limits_{k_3=0}^{K_3} b_t(k_3) z_3^{-k_3}} = \frac{A_t(z_3)}{B_t(z_3)} \tag{4}$$

From (3) and (4), we can get the following 2-D IIR filtering algorithm and 1-D IIR filtering algorithm, Eqs. (5) and (6) are the two filtering equations of the proposed 3-D IIR digital filter. Equations (5) and (6) are recursive difference equations. The stability of them will be tested following Theorem 1.

$$\begin{aligned} & z(n_1, n_2, n_3) \\ & = \sum_{k_1=0}^{K_1} \sum_{k_2=0}^{K_2} a_s(k_1, k_3) y(n_1 - k_1, n_2 - k_2, n_3) \\ & - \sum_{k_1=0}^{K_1} \sum_{k_2=0, (k_1, k_2) \neq (0,0)}^{K_2} b_s(k_1, k_2) y(n_1 - k_1, n_2 - k_2, n_3) \end{aligned} \tag{5}$$

And

$$\begin{aligned} u(n_1, n_2, n_3) = & \sum_{k_3=0}^{K_3} a_t(k_3) z(n_1, n_2, n_3 - k_3) \\ & - \sum_{k_3=1}^{K_3} b_t(k_3) u(n_1, n_2, n_3 - k_3) \end{aligned} \tag{6}$$

Here, the problem to be solved is to design a stable 3-D IIR digital filter. We can obtain the characteristic polynomials of the 2-D IIR filter (3) and the 1-D IIR filter (4) as following,

$$B_s(z_1, z_2) = \sum_{k_1=0}^{K_1} \sum_{k_2=0}^{K_2} b_s(k_1, k_2) z_1^{-k_1} z_2^{-k_2} \tag{7}$$

$$B_t(z_3) = \sum_{k_3=0}^{K_3} b_t(k_3) z_3^{-k_3} \tag{8}$$

We have following stability theorem to the proposed 3-D IIR filter.

Theorem 1 *If the 3-D IIR digital filter in* (2) *has no non-singularities, then the 3-D IIR filter is stable if and only if*

$$B_s(z_1, z_2) \neq 0, |z_1| \geq 1, |z_2| \geq 1 \tag{9}$$

And

$$B_t(z_3) \neq 0, |z_3| \geq 1 \tag{10}$$

The proof can be obtained by referring [12].

The stability condition of $B_t(z_3)$ in (6) is easy, many well known tool can be found. While, to test the stability condition of $B_s(z_1, z_2)$ in (5) is a little difficult. To solve the problem, we provide following theorem [13].

Theorem 2 *If the 2-D polynomial* $B_s(z_1, z_2)$ *is stable if and only if*

$$B_s(z_1, 1) \neq 0, |z_1| \geq 1 \tag{11}$$

And

$$B_s(e^{j\omega_1}, z_2) \neq 0, \omega_1 \in R, |z_2| \geq 1 \tag{12}$$

There are two problems to apply the 3-D IIR filter in (2): one is stability problem, we had to design the 3-D IIR filter satisfy the conditions of Theorem 1; the other is that the designed 3-D IIR filter need to remove the noise term $w(n_1, n_2, n_3)$.

3 Design of Stable 3-D IIR Filter

From (2), we need to design stable $H_s(z_1, z_2)$ and $H_t(z_3)$. The design of $H_t(z_3)$ is easy, we can refer [14] and obtain a low-pass filter,

$$H_t(z_3) = \frac{a_{t,0} + a_{t,1}z_3^{-1} + a_{t,2}z_3^{-2}}{1 + b_{t,1}z_1^{-1} + b_{t,2}z_3^{-2}} = \frac{A_t(z_3)}{B_t(z_3)} \tag{13}$$

where

$$\begin{aligned} a_{t,0} &= 3\omega_{t,c}^2/(1 + 3\omega_{t,c} + 3\omega_{t,c}^2) \\ a_{t,1} &= 2a_{t,0} \\ a_{t,2} &= a_{t,0} \end{aligned} \tag{14}$$

$$\begin{aligned} b_{t,1} &= (6\omega_{t,c,}^2 - 2)/(1 + 3\omega_{t,c} + 3\omega_{t,c}^2) \\ b_{t,2} &= (1 - 3\omega_{t,c} + 3\omega_{t,c}^2)/(1 + 3\omega_{t,c} + 3\omega_{t,c}^2) \end{aligned} \tag{15}$$

The cut-off angular frequency of the 1-D IIR

$$\omega_{t,c} = 2\pi f_{t,c}/f_{t,s}$$

We watched $f_{t,s}$ the sampling frequency along time-direction.

Since established the relationship of filter parameters and $\omega_{t,c}$, we can adjust the filter parameters according to $\omega_{t,c}$. We select $\omega_{t,c}$ by considering the noise characteristics of UV image sequence along time axis. We design $H_t(z_3)$ as low-pass filter since the noises or the interferences of the imaging sequence $y(n_1, n_2, n_3)$ are high frequency signals.

To guarantee the designed $H_t(z_3)$ to be stable, we have following theorem from [15].

Theorem 3 *For $b_{t,1}$ and $b_{t,2}$ given in* (14), the $B_t(z_3)$ *in* (12) *is stable if and only if*

$$|1 + b_{t,2}| > b_{t,1} \tag{16}$$

The design of stable $H_s(z_1, z_2)$ is some complicated, we derive the parameter matrices $a_s(k_1, k_2)$ and $b_s(k_1, k_2)$ by following algorithms.

Let

$$\mathbf{a}_i = [\,a_{i,0} \quad a_{i,1} \quad a_{i,2}\,], i = 1, 2 \tag{17}$$

where

$$\begin{aligned} a_{i,0} &= 3\omega_{i,c}^2/(1 + 3\omega_{i,c} + 3\omega_{i,c}^2),\ i\ =\ 1, 2 \\ a_{i,1} &= 2a_{i,0} \\ a_{i,2} &= a_{i,0} \end{aligned} \tag{18}$$

and

$$\mathbf{b}_i = [\,b_{i,0} \quad b_{i,1} \quad b_{i,2}\,], i = 1, 2 \tag{19}$$

$$\begin{aligned} b_{i,1} &= (6\omega_{i,c}^2 - 2)/(1 + 3\omega_{i,c} + 3\omega_{i,c}^2), i = 1, 2 \\ b_{i,2} &= (1 - 3\omega_{i,c} + 3\omega_{i,c}^2)/(1 + 3\omega_{i,c} + 3\omega_{i,c}^2) \end{aligned} \tag{20}$$

where $\omega_{i,c} = 2\pi f_{i,c}/f_{i,s}, i = 1, 2$ the cut-off angular frequencies of 2-D IIR are, $f_{i,s}, i = 1, 2$ are the sampling frequencies along horizontal and vertical directions, which can be different in the two directions.

Then, from (17) and (19) we obtain

$$\mathbf{A}_s = \mathbf{a}_1^T\mathbf{a}_2 = \begin{bmatrix} a_s(0,0) & a_s(0,1) & a_s(0,2) \\ a_s(1,1) & a_s(1,1) & a_s(1,2) \\ a_s(2,0) & a_s(2,1) & a_s(2,2) \end{bmatrix} \tag{21}$$

$$\mathbf{B}_s = \mathbf{b}_1^T\mathbf{b}_2 = \begin{bmatrix} b_s(0,0) & b_s(0,1) & b_s(0,2) \\ b_s(1,1) & b_s(1,1) & b_s(1,2) \\ b_s(2,0) & b_s(2,1) & b_s(2,2) \end{bmatrix} \tag{22}$$

We can get the 2-D polynomials in $H_s(z_1, z_2)$,

$$\begin{aligned} A_s(z_1, z_2) & \\ &= [\,1 \quad z_1^{-1} \quad z_1^{-2}\,] \begin{bmatrix} a_s(0,0) & a_s(0,1) & a_s(0,2) \\ a_s(1,0) & a_s(1,1) & a_s(1,2) \\ a_s(2,0) & a_s(2,1) & a_s(2,2) \end{bmatrix} \begin{bmatrix} 1 \\ z_2^{-1} \\ z_2^{-2} \end{bmatrix} \\ &= \sum_{k_1=0}^{2}\sum_{k_2=0}^{2} a_s(k_1, k_2) z_1^{-k_1} z_2^{-k_2} \end{aligned} \tag{23}$$

And

$$\begin{aligned} B_s(z_1, z_2) &= \begin{bmatrix} 1 & z_1^{-1} & z_1^{-2} \end{bmatrix} \begin{bmatrix} b_s(0,0) & b_s(0,1) & b_s(0,2) \\ b_s(1,0) & b_s(1,1) & b_s(1,2) \\ b_s(2,0) & b_s(2,1) & b_s(2,2) \end{bmatrix} \begin{bmatrix} 1 \\ z_2^{-1} \\ z_2^{-2} \end{bmatrix} \\ &= \sum_{k_1=0}^{2} \sum_{k_2=0}^{2} b_s(k_1, k_2) z_1^{-k_1} z_2^{-k_2} \end{aligned} \tag{24}$$

According to Theorem 2, we need keep $B_s(z_1, 1)$ and $B_s(e^{j\omega_1}, z_2)$ to be stable,

$$\begin{aligned} B_s(z_1, 1) &= \begin{bmatrix} 1 & z_1 & z_1^2 \end{bmatrix} \begin{bmatrix} b_s(0,0) & b_s(0,1) & b_s(0,2) \\ b_s(1,0) & b_s(1,1) & b_s(1,2) \\ b_s(2,0) & b_s(2,1) & b_s(2,2) \end{bmatrix} \begin{bmatrix} 1 \\ 1 \\ 1 \end{bmatrix} \\ &= \sum_{k_1=0}^{2} \sum_{k_2=0}^{2} b_s(k_1, k_2) z_1^{-k_1} \end{aligned} \tag{25}$$

And

$$\begin{aligned} B_s(e^{j\omega_1}, z_2) &= \begin{bmatrix} 1 & e^{-j\omega_1} & e^{-j2\omega_1} \end{bmatrix} \\ &\cdot \begin{bmatrix} b_s(0,0) & b_s(0,1) & b_s(0,2) \\ b_s(1,0) & b_s(1,1) & b_s(1,2) \\ b_s(2,0) & b_s(2,1) & b_s(2,2) \end{bmatrix} \cdot \begin{bmatrix} 1 \\ z_2^{-1} \\ z_2^{-2} \end{bmatrix} \\ &= \sum_{k_1=0}^{2} \sum_{k_2=0}^{2} b_s(k_1, k_2) e^{-j\omega_1 k_1} z_2^{-k_2} \end{aligned} \tag{26}$$

Since $B_s(z_1, 1)$ in (25) is a 1-D polynomial with rank 2, the stability test is easy, we have following 1-D stability test theorem.

Theorem 4 *Let*

$$B_s(z_1, 1) = \sum_{k_1=0}^{2} b_s(k_1) z_1^{-k_1} \tag{27}$$

where

$$b_s(k_1) = \sum_{k_2=0}^{2} b_s(k_1, k_2) \tag{28}$$

Then the $B_s(z_1, 1)$ in (24) is stable if and only if

$$|1 + b_s(2)| > |b_s(1)| \tag{29}$$

Similarly, $B_s(e^{j\omega_1}, z_2)$ in (26) can be regarded as a 1-D polynomial with complex coefficients, we have following stability test theorem.

Theorem 5 *Let*

$$B_s(e^{j\omega_1}, z_2) = \sum_{k_2=0}^{2} b_s(e^{-j\omega_1}, k_2) z_2^{-k_2} \tag{30}$$

where

$$b_s(e^{j\omega_1}, k_2) = \sum_{k_1=0}^{2} b_s(k_1, k_2) e^{-jk_1\omega_1} \tag{31}$$

Then $B_s(e^{j\omega_1}, z_2)$ in (25) is stable if and only if

$$|b_s(e^{j\omega_1}, 0) + b_s(e^{j\omega_1}, 2)| - |b_s(e^{j\omega_1}, 1)| > 0 \tag{32}$$

where

$$b_s(e^{j\omega_1}, 0) = \sum_{k_1=0}^{2} b_s(k_1, 0) e^{-jk_1\omega_1} \tag{33}$$

$$b_s(e^{j\omega_1}, 1) = \sum_{k_1=0}^{2} b_s(k_1, 1) e^{-jk_1\omega_1} \tag{34}$$

$$b_s(e^{j\omega_1}, 2) = \sum_{k_1=0}^{2} b_s(k_1, 2) e^{-jk_1\omega_1} \tag{35}$$

Limited by the paper size, we could not provided the proof and the proof can be obtained by referring [13].

4 Simulations

First, we design the proposed 3-D IIR filter that composes of $H_t(z_3)$ and $H_s(z_1, z_2)$.

We design $H_t(z_3)$ by using (12)–(14), and selecting $f_{t,c} = 0.2$ in $\omega_{t,c} = 2\pi f_{t,c}/f_{t,s}$. Substitute $\omega_{t,c}$ into (13) and (14), we can obtain the parameters $a_{t,0} = 0.1323, a_{t,1} = 0.2645, a_{t,2} = 0.1323$ and $b_{t,1} = -0.6289, b_{t,2} = 0.1580$.

Applying Theorem 3, we can find

$$|1 + b_{t,2}| > |b_{t,1}|$$

Thus $H_t(z_3)$ is stable, we can use (6) to complete the time-filtering of 3-D IIR. We watched $H_t(z_3)$ amplitude frequency response in Fig. 1 and the phase frequency response in Fig. 2.

We design $H_s(z_1, z_2)$ by using (17)–(26), and selecting $f_{i,c} = 0.1, f_{i,s} = 4, i = 1, 2$ in $\omega_{i,c} = 2\pi f_{i,c}/f_{i,s}$. Substitute $\omega_{t,c}$ into (16–19), we can obtain the parameters.

$$\mathbf{a}_i = [\, a_{i,0} \quad a_{i,1} \quad a_{i,2} \,], i = 1, 2$$

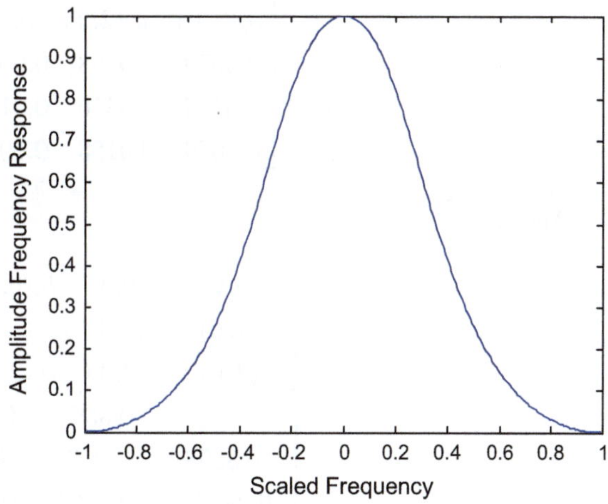

Fig. 1 The amplitude frequency response of $H_t(z_3)$

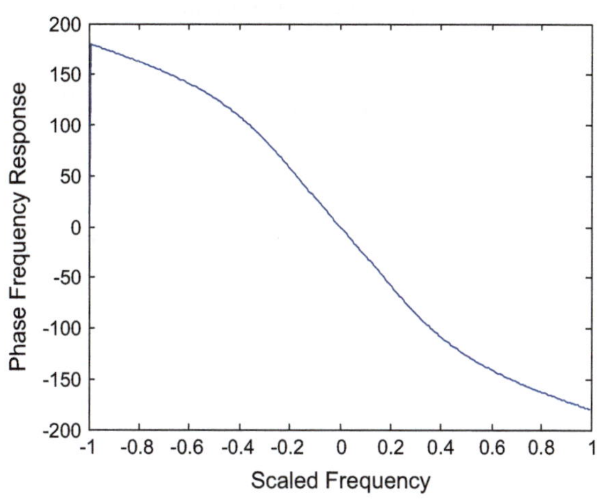

Fig. 2 The phase frequency response of $H_t(z_3)$

where

$$a_{i,0} = 0.0889, a_{i,1} = 0.1778, a_{i,2} = 0.0889, i = 1,2$$

And

$$\mathbf{b}_i = \begin{bmatrix} b_{i,0} & b_{i,1} & b_{i,2} \end{bmatrix}, i = 1,2$$

where
$b_{i,1} = -0.8898, b_{i,2} = 0.2454, i = 1,2.$
Substitute $\mathbf{a}_i$ and $\mathbf{b}_i$ into (21) and (22), we can obtain

$$\begin{aligned}\mathbf{A}_s &= \begin{bmatrix} a_s(0,0) & a_s(0,1) & a_s(0,2) \\ a_s(1,0) & a_s(1,1) & a_s(1,2) \\ a_s(2,0) & a_s(2,1) & a_s(2,2) \end{bmatrix} \\ &= \begin{bmatrix} 0.0079 & 0.0158 & 0.0079 \\ 0.0158 & 0.0316 & 0.0158 \\ 0.0079 & 0.0158 & 0.0079 \end{bmatrix}\end{aligned} \tag{36}$$

And

$$\begin{aligned}\mathbf{B}_s &= \begin{bmatrix} b_s(0,0) & b_s(0,1) & b_s(0,2) \\ b_s(1,0) & b_s(1,1) & b_s(1,2) \\ b_s(2,0) & b_s(2,1) & b_s(2,2) \end{bmatrix} \\ &= \begin{bmatrix} 1 & -0.8898 & 0.2454 \\ -0.8898 & 0.7917 & -0.2183 \\ 0.2454 & -0.2183 & 0.0602 \end{bmatrix}\end{aligned} \tag{37}$$

From (36) and (26), we can obtain

$$b_s(\mathrm{e}^{\mathrm{j}\omega_1}, 0) = \begin{bmatrix} 1 & \mathrm{e}^{-\mathrm{j}\omega_1} & \mathrm{e}^{-\mathrm{j}2\omega_1} \end{bmatrix} \cdot \begin{bmatrix} b_s(0,0) \\ b_s(1,0) \\ b_s(2,0) \end{bmatrix} \tag{38}$$

$$b_s(\mathrm{e}^{\mathrm{j}\omega_1}, 1) = \begin{bmatrix} 1 & \mathrm{e}^{-\mathrm{j}\omega_1} & \mathrm{e}^{-\mathrm{j}2\omega_1} \end{bmatrix} \cdot \begin{bmatrix} b_s(0,1) \\ b_s(1,1) \\ b_s(2,1) \end{bmatrix} \tag{39}$$

and

$$b_s(e^{j\omega_1}, 2) = \begin{bmatrix} 1 & e^{-j\omega_1} & e^{-j2\omega_1} \end{bmatrix} \cdot \begin{bmatrix} b_s(0,2) \\ b_s(1,2) \\ b_s(2,2) \end{bmatrix} \tag{40}$$

Applying Theorem 5, we can find $|b_s(e^{j\omega_1}, 0) + b_s(e^{j\omega_1}, 2)| - |b_s(e^{j\omega_1}, 1)| > 0$, This is shown in Fig. 3. Thus $H_s(z_1, z_2)$ is stable, The amplitude frequency response of $H_s(z_1, z_2)$ in Fig. 4 shows that the designed 2-D IIR can complete the low-pass filtering in the given spatial range. The phase frequency response of $H_s(z_1, z_2)$ in Fig. 5 shows that it can keep the required linear property at the given pass band.

Fig. 3 The stability test based on Theorem 5 for $H_s(z_1, z_2)$

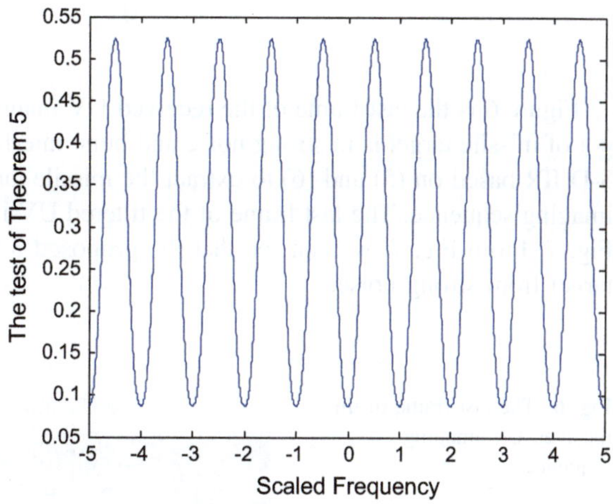

Fig. 4 The amplitude frequency response of $H_s(z_1, z_2)$

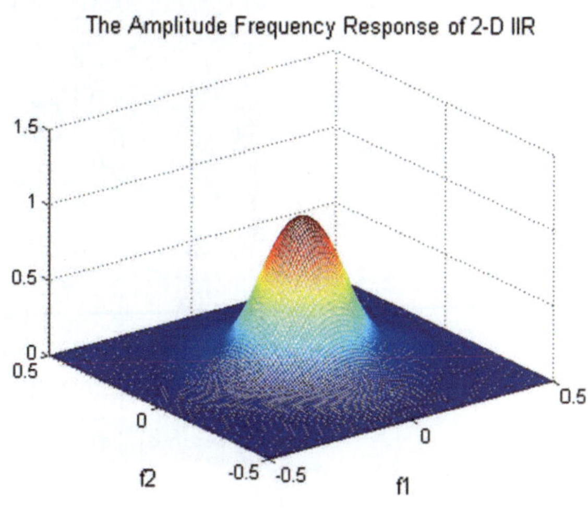

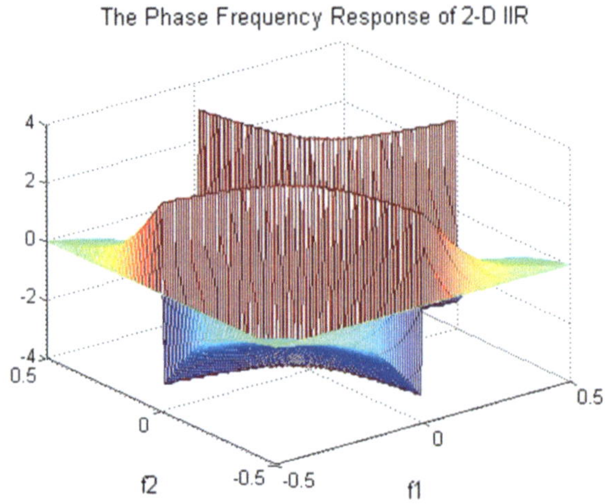

Fig. 5 The phase frequency response of $H_s(z_1, z_2)$

Figure 6 is the last frame of the received UV imaging sequence, the small white pot of missile target is in strong noise and burst interferences. We use the proposed 3-D IIR based on (5) and (6) to extract the missile target, and obtain a filtered UV imaging sequence. The last frame of the filtered UV imaging sequence is shown in Fig. 7. From Fig. 7, we can see that the proposed 3-D IIR can extract the missile target from strong noise.

Fig. 6 The last frame of the received UV imaging sequence

Fig. 7 The last frame of the filtered UV imaging sequence

5 Conclusion

This paper proposed a 3-D IIR for extracting missile targets in UV images from strong noise and burst interferences. The proposed 3-D recursive filter adopts a 2-D spatial-filtering IIR and 1-D time-filtering IIR, which can separate the missile targets from strong noise and burst interferences. Both proposed 1-D and 2-D filters are stable, the stability test algorithms for them are provided. The stability of 1-D time-filtering IIR and 2-D spatial-filtering IIR can be tested by Theorem 3 and Theorem 5, respectively. Application of 3-D filtering for UV images is simple and valid for extracting missile targets.

References

1. Clampin M (2000) UV-Optical CCDs. In: Proceedings of the space astrophysics detectors and detector technologies conference held at the STScI. Baltimore, 26–29 June 2000
2. Siegmund OHW, Welsh B (2000) Narrow-band tunable filters for use in the FUV region. In: Proceedings of the space astrophysics detectors and detector technologies conference held at the STScI. Baltimore, 26–29 June 2000
3. Joseph C (2000) UV technology overview. In: Proceedings of the space astrophysics detectors and detector technologies conference held at the STSc I. Baltimore, 26–29 June 2000
4. Ulmer MP, Wessels WB, Han B, Gregie J, Tremsin A, Siegmund OHW (2009) Advances in wide-bandgap semiconductor base photocathode devices for low light level applications. Proc SPIE 5164:144–154

5. Siegmund OHW (2000) MCP imaging detector technologies for UV instruments, in proceedings of the space astrophysics detectors and detector technologies conference held at the STScI. Baltimore, 26–29 June 2000
6. Wilkinson E (1999) Maturing and developing technologies for the next generation of UV gratings ultraviolet-optical space astronomy beyond HST: conference held at the STScI. ASP Conf Ser 164:420
7. Berk A, Anderson GP, Acharya PK, Chetwynd JH, Bernstein LS, Shettle EP, Matthew MW, Adler-Golden SM (1999) MODTRAN4 user's manual, Air Force Research Laboratory, 1 June 1999
8. Ouchi K, Wang H (2005) Interlock cross-correlation function of speckle in SAR images of sea surface processed with partially overlapped subapertures. IEEE Trans Geosci Remote Sens 43(4):695–701
9. Kuenzle B, Bruton LT (2006) 3-D IIR filtering using decimated DFT-polyphase filter bank structures. IEEE Trans Circ Syst I 53(2):394–408
10. Sekiguchi T, Takahashi S, Matsuda K (1989) Design of three-dimensional spherically symmetric IIR digital filters with real and complex coefficients using spectral transformations. In: European conference on circuit theory and design, pp 467–471
11. Boulter GA, Lampropoulos JF (1995) Filtering of moving targets using SBIR sequential frames. IEEE Trans Aerosp Electron Syst 31(4):1255–1267
12. Xiao Y, Unbehauen R (1998) New stability test algorithm for two-dimensional digital filters. IEEE Trans Circ Syst I: Fundam Theory Appl 45(7):739–741
13. Xiao Y, Unbehauen R, Du X (1999) A finite test algorithm for 2D Schur polynomials based on complex Lyapunov equation. In: Proceedings of the 1999 IEEE international symposium on circuits and systems (ISCAS '99), vol 3, pp 339–342, 30 May–2 June, 1999
14. Xiao Y, Zhang YK (2009) An approach of extracting object of SAR based 2-D hybrid transform, China's State Intellectual Property Office, Patent Application No. 20091023-7860.6 (in Chinese)
15. Xiao Y (2003) Stability analysis of multidimensional systems. Shanghai Science and Technology Press, Shanghai (in Chinese)

Runway Extraction Method Based on Rotating Projection for UAV

Zhen-yu Guan, Jie Li and Huan Yang

Abstract We bring out a runway extraction method based on rotating projection in this paper, which consists of three steps, locating the Region of interest (ROI), edge extraction and line detection. Firstly we employ template matching to locate the ROI which contains the runway area. Then we use Sobel operator to extract edges. The rotating projection algorithm is proposed to seek the potential straights in ROI, which will be integrated into the real straights by means of improved K-means clustering method. Simulations are carried out in the end, and results show that the algorithm proposed in this paper can extract the four boundaries of the runway effectively, while it can reduce 50 % of computing time compared with Hough transform.

Keywords Improved K-means clustering algorithm runway extraction · Line detection · Rotating projection method

1 Introduction

Vision based auto-landing is one of the key technologies for the development of intelligent unmanned aerial vehicle (UAV), through which, UAVs can be independently controlled at landing stage. Since such technology relies on the extraction of ground marks strongly, the extraction method of runway, which may be the most obvious markers during landing stage, has been a hot topic in this field.

Z. Guan (✉) · J. Li · H. Yang
Beijing Electro-mechanical Engineering Institute, Beijing 100074, China
e-mail: guanzhenyu@bit.edu.cn

J. Li
e-mail: Lijie@bit.edu.cn

H. Yang
e-mail: yanghuan@bit.edu.cn

E. Qi (ed.), *Proceedings of the 6th International Asia Conference on Industrial Engineering and Management Innovation*,
DOI 10.2991/978-94-6239-145-1_30

Basically, there are 2 ways to extract runway from images. The first one is the methods based on multisource images and GIS information [1–3], the second one is based on single frame information, which mainly use the parallel test and Hough transform (HT) algorithm [4–6]. These algorithms above are often designed for the satellite images and aerial images, which have great difference with the methods used on landing UAV. In the field of vision (FOV) in the camera on board of the landing aircrafts, the runway area is usually a trapezoidal area, which cannot satisfy the parallel test. Literature [7] brought a method for landing UAV, which also employs the HT algorithm. The main shortcoming of those methods based on HT is that the calculation is too enormerous to meet the demands of the real-time system. Literature [8] gave out a method based on the value of affinity, the main limitation of this algorithm is that it relies on the prior information such as the contrast information of shades between the internal and external runway, which cannot be obtained before landing probably. Literature [9] brought out an extraction method with heuristic searching by employing the parallel test. But this method only can be used in the overhead view which will not exist in the landing stage.

To avoid the above-mentioned drawbacks, we propose an extraction method based on rotating projection in this paper. This method extracts the region of runway as the ROI firstly, and then calculates the gray projection of the image with runway from different direction. Finally the K-means clustering is employed to get the location of runway. Simulation results show that this method can effectively extract the boundary lines of the runway while calculating time is reduced by more than 50 % than methods based on Hough transform.

2 Image Preprocessing

When the airplane approaches, the runway area in the image can be seen as a trapezoidal area, which is made up of 4 lines: the touchdown zone line, the ending line and 2 side lines. So the problem of runway extraction can be transformed into a line extraction problem, which usually follow these steps: first locate the region of interest (ROI), then extract the edges, and finally get the equations of lines in the image coordinate system.

The image preprocessing includes two parts, the ROI segmentation and edge extraction.

The ROI segmentation is used to extract the runway area from the original image in order to avoid tremendous global calculation. We employ template matching to deal with this problem. Template matching is a technique in digital image processing for finding small parts of an image which match a template image [10]. We call the search image S (x, y), where (x, y) represents the coordinate of each pixel in the search image. We call the template $T(x_t, y_t)$, where (x_t, y_t) represents the coordinate of each pixel in the template. Then we simply move the center of the template $T(x_t, y_t)$ over each (x, y) point in the search image and calculate the sum of products between the coefficients in S(x, y) and $T(x_t, y_t)$ over the whole area spanned by the template. As all possible positions of the template with respect to the search image are considered, the position with the highest score is the best. We use the Normalized Cross Correlation (NCC) as the coefficient; the R_{NCC} is defined as (1).

$$R_{NCC}(x,y) = \frac{\sum_{x_t,y_t} (T(x_t,y_t) \times S(x+x_t,y+y_t))}{\sqrt{\sum_{x_t,y_t} T(x_t,y_t)^2 \sum_{x_t,y_t} S(x+x_t,y+y_t)^2}} \quad (1)$$

where, R_{NCC} is the matching degree of the template and the original image.

One thing we should pay attention to is that the algorithm should be adaptive to the scale when the template matching algorithm is designed, which means it should extract the runway accurately when the runway area expands during the approaching of the UAV. To achieve the purpose, we made the template expand at a certain specified proportions when the algorithm goes on.

The edge extraction will be carried on the ROI extracted from the original image. We employed the Sobel operator, which is shown as (2)–(3).

$$G_x = \begin{bmatrix} 1 & 0 & -1 \\ 2 & 0 & -2 \\ 1 & 0 & -1 \end{bmatrix} * S,\ G_y = \begin{bmatrix} 1 & 2 & -1 \\ 0 & 0 & 0 \\ -1 & -2 & -1 \end{bmatrix} * S \quad (2)$$

$$G = \sqrt{G_x^2 + G_y^2}, \theta = \arctan(\frac{G_x}{G_y}) \quad (3)$$

where, S is the source image, Gx and Gy are the horizontal and the vertical derivative approximations at each pixel, G is the gradient magnitude, θ is the gradient's direction. We gave out a threshold T, making pixels whose gray value is larger than T as the edge pixels.

3 Rotating Projection Method for Line Detection

3.1 Rotating Projection Algorithm

Gray projection method is one of the most common used image processing methods. By calculating the sum of the gray value of the row elements and column elements, the distribution of image's gray value can be obtained, which means we can transform a two-dimensional matrix to two feature vectors.

For the source image S, whose size is $i \times j$, we have

$$S = \begin{bmatrix} a_{11} & a_{12} & \cdots & a_{1j} \\ a_{21} & \cdots & \cdots & a_{2j} \\ \cdots & \cdots & \cdots & \cdots \\ a_{i1} & a_{i2} & \cdots & a_{ij} \end{bmatrix} \tag{4}$$

Define the row gray projection of S as,

$$GP_{row} = [\sum_j a_{1j}, \sum_j a_{2j}, \ldots, \sum_j a_{ij}] \tag{5}$$

Define the column gray projection of S as,

$$GP_{column} = [\sum_i a_{i1}, \sum_i a_{i2}, \ldots, \sum_i a_{ij}] \tag{6}$$

The so-called rotating projection algorithm can be described as follow, firstly, rotate the image θ degrees counterclockwise, then calculate the GP_{column} as the projection spectrum at θ degrees. If the straight edge in the image is vertical to the horizontal after the rotation, there will be a peak in the projection spectrum as shown in Fig. 1. Figure 1 a is the source image and its projection spectrum is Fig. 1b, c is the result of rotating source image by 90°, while Fig. 1d is its projection spectrum. We can see that peaks in Fig. 1d corresponds to lines in Fig. 1c, so we can achieve the line detection.

Define $\Re^{\theta}$ as the rotating operator, which means rotate a matrix by θ counter-clockwise. For the images described as (4), its projection spectrum function is

$$GP_c(\theta, r) = \sum_i \Re^{\theta}(a_{in}) \tag{7}$$

where, θ is the rotation angle counterclockwise, $GP_c(\theta,r)$ is the sum of the *r*th column's gray value.

GP_c can be treated as the weight of the possibility of the straight line's existence. The larger GP_c presents that more dense edge points are along the vertical direction, which means it is more likely to be a straight line along the vertical direction. Set the rotating step is k to rotate the source image from 0° to π counterclockwise. We can calculate a series of projection spectrum at different rotation angel. Set the line detection threshold T, for those points (θ, r) in projection spectrum, they should satisfy

$$GP_c(\theta, r) > T \tag{8}$$

the lines corresponded to these points can be named as potential straights.

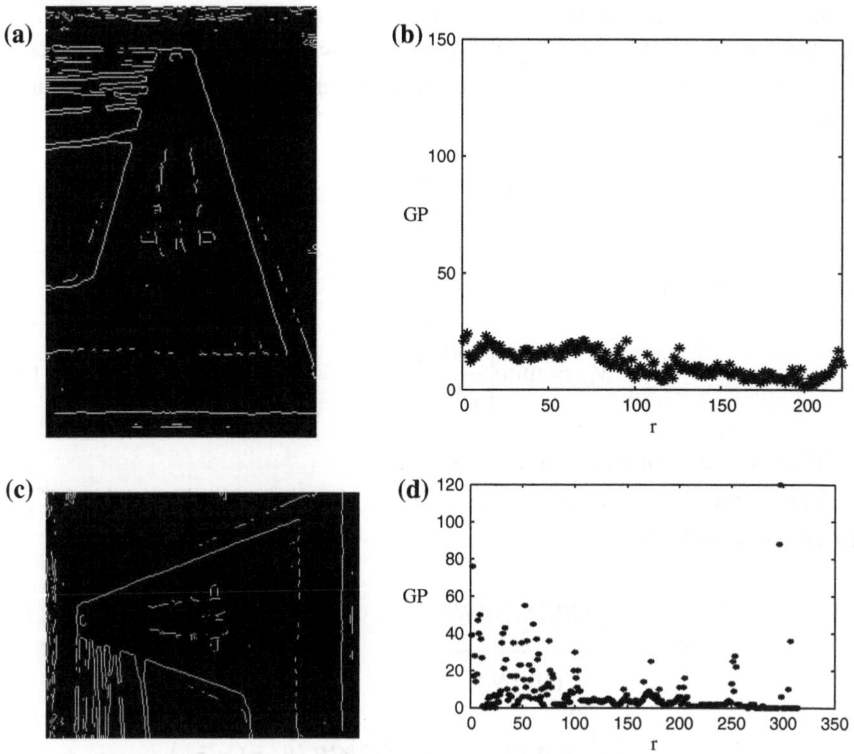

Fig. 1 Rotating image and projection spectrum. **a** Source image. **b** Projection spectrum of source image. **c** Image after rotation. **d** Projection spectrum after rotation

3.2 Improved K-Means Clustering Algorithm

Most of the time, the width of straight is larger than 1 pixel. It means that there will be some neighbor peaks existed in the projection spectrum, which corresponded to the same straight, especially when T is small. To avoid this situation, we bring out an improved K-means clustering method.

K-means clustering method is a method of vector quantization originally from signal processing, which is popular for cluster analysis in data mining. K-means clustering aims to partition n observations into k clusters in which each observation belongs to the cluster with the nearest mean, serving as a prototype of the cluster.

Traditional K-means clustering use the distances between observations to cluster centers as the similarity functions, and locate the cluster center by calculating the average of observations' Euclidean distance in the same cluster. In this paper, the observations have 3 parameters $(\theta, r, GP_c(\theta, r))$, which are not pairwise independent. To solve this problem, we propose an improved K-means clustering algorithm as follows.

In the (θ, r) space, let the observations be $N_i = (\theta_i, r_i)$, where in $i = 1, 2, \ldots, n$; let the cluster be S_j,where $j = 1, 2, \ldots, k$; define the initial cluster centers as $C_j = (\tilde{\theta}_j, \tilde{r}_j)$, where $j = 1, 2, \ldots, k$. The similarity function between N_i and S_j can be written as,

$$V_{ij} = \frac{GP_c(\theta_i, r_i)}{\left\|N_i - C_j\right\|^2} \tag{9}$$

$\forall i \in [1, n]$, $\exists \mu \in [1, k]$, if

$$V_{i\mu} = \min\left\{\frac{GP_c(\theta_i, r_i)}{\left\|N_i - C_j\right\|^2} | j \in [1, k]\right\} \tag{10}$$

Then $N_i \in S_\mu$, where $\mu = 1, 2, \ldots, k$.

Update the cluster center, let $S_\mu = \{\tilde{N}_1, \tilde{N}_2, \ldots, \tilde{N}_m\}$, the cluster center $C_\mu = (\tilde{\theta}_\mu, \tilde{r}_\mu)$ is as follows,

$$\begin{aligned} C_\mu &= \frac{\sum_{i=1}^{m} [GP_c(\tilde{N}_i) \times \tilde{N}_i]}{\sum_{i=1}^{m} GP_c(\tilde{N}_i)} \\ &= \left(\frac{\sum_{i=1}^{m} [GP_c(\tilde{\theta}_i, \tilde{r}_i) \times \tilde{\theta}_i]}{\sum_{i=1}^{m} GP_c(\tilde{\theta}_i, \tilde{r}_i)}, \frac{\sum_{i=1}^{m} [GP_c(\tilde{\theta}_i, \tilde{r}_i) \times \tilde{r}_i]}{\sum_{i=1}^{m} GP_c(\tilde{\theta}_i, \tilde{r}_i)}\right) \end{aligned} \tag{11}$$

The cluster centers obtained are the location of the straights, where θ_μ is the rotation angle and n_μ is the abscissa of line.

3.3 Seeking the Line Equation

After getting the cluster centers, we established rotation coordinate system to solve the line equation, as shown in Fig. 2

Let the image coordinate system be XOY; let the rotation image coordinate system be $X_WO_WY_W$. Establish rotation coordinate system xoy at the rotation center. The point p is a random point in the source image, and p_1 is the new position of p after rotation. Define the width and height of the source image as W and H, while these values are changed to W_W and H_W after the rotation.

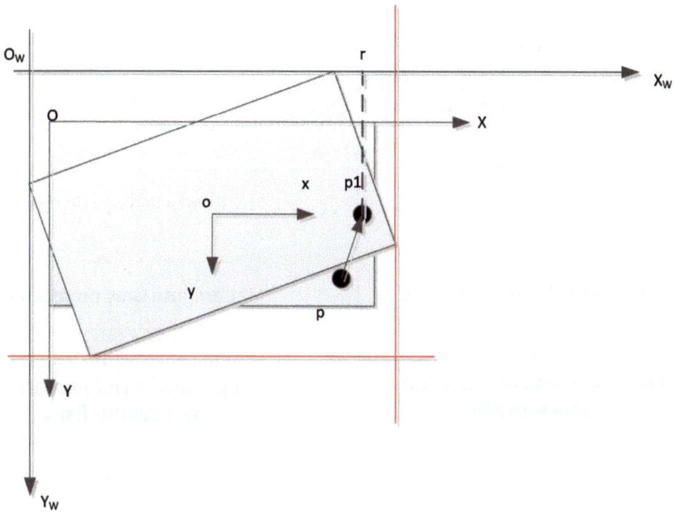

Fig. 2 Rotation coordinate system

The coordinates of the point p are (X_0,Y_0) in XOY, while in xoy, its coordinates are (x_0,y_0), which satisfy,

$$\begin{bmatrix} x_0 \\ y_0 \end{bmatrix} = \begin{bmatrix} X_0 \\ Y_0 \end{bmatrix} - \begin{bmatrix} \frac{w}{2} \\ \frac{h}{2} \end{bmatrix} \tag{12}$$

After the rotation of θ, the coordinates of p_1 are (x_1,y_1), which can be written as,

$$\begin{bmatrix} x_1 \\ y_1 \end{bmatrix} = \begin{bmatrix} \cos\theta & -\sin\theta \\ \sin\theta & \cos\theta \end{bmatrix} \begin{bmatrix} x_0 \\ y_0 \end{bmatrix} \tag{13}$$

(x_W, y_W) are the coordinates of p_1 in $X_W O_W Y_W$,

$$\begin{bmatrix} x_W \\ y_W \end{bmatrix} = \begin{bmatrix} x_1 \\ y_1 \end{bmatrix} + \begin{bmatrix} \frac{W}{2} \\ \frac{H}{2} \end{bmatrix} \tag{14}$$

Combine (12), (13) and (14), we can get

$$\begin{bmatrix} x_W - \frac{W}{2} \\ y_W - \frac{H}{2} \end{bmatrix} = \begin{bmatrix} \cos\theta & -\sin\theta \\ \sin\theta & \cos\theta \end{bmatrix} \begin{bmatrix} X_0 - \frac{w}{2} \\ Y_0 - \frac{h}{2} \end{bmatrix} \tag{15}$$

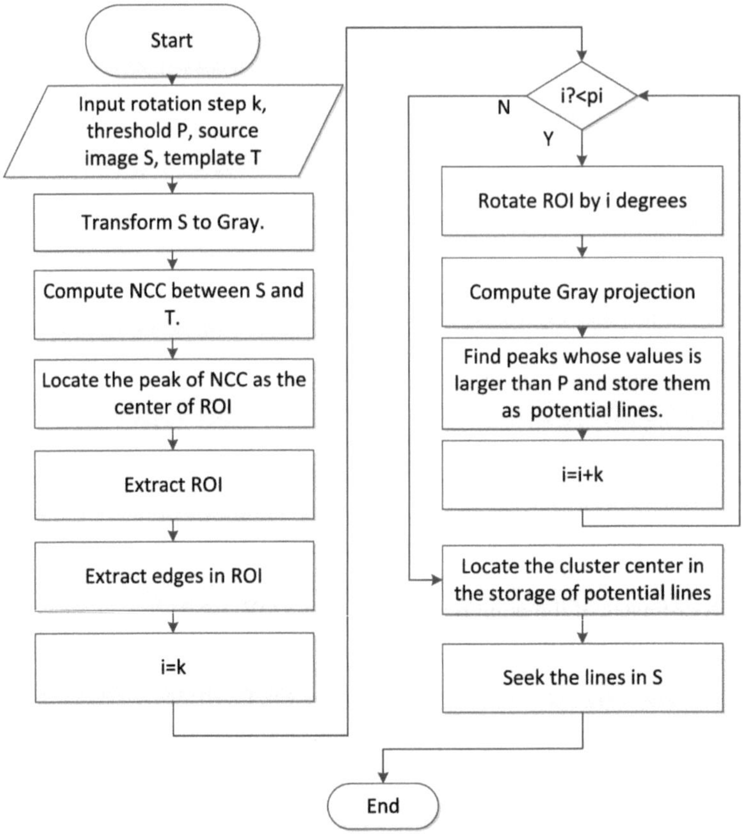

Fig. 3 Flowchart of the runway extraction algorithm for a single frame

where, $x_W = r$, expand (15) we obtain the equation of lines as shown in (16).

$$Y_0 = \cot\theta X_0 - \frac{w}{2}\cot\theta + \frac{h}{2} - \frac{r - W/2}{\sin\theta} \tag{16}$$

The flow chart of runway extraction algorithm for a single frame is shown as Fig. 3.

4 Simulations and Results

The video we used in simulation was generated by OSGVisual, which is a flight visualization tool developed by Institute of Flight System Dynamics (FSD), Technische Universität München. We take some frames of the video as examples to testify algorithms.

4.1 Simulation of Image Preprocessing

The source image we used is the 100th frame of the video streaming, as shown in Fig. 4. We use an artificial image as the template for the matching to get the region of interest (ROI), as shown in Fig. 5.

After the traversal of the source image, the Normalized cross correlation (NCC) between the template and the source image is obtained in Fig. 6, whose peak is treated as the center of ROI. The result of ROI extraction is shown in Fig. 6, in which the red asterisk marks the center of matching template, and the region in the blue rectangle is the ROI.

Fig. 4 Source image

Fig. 5 Template for matching

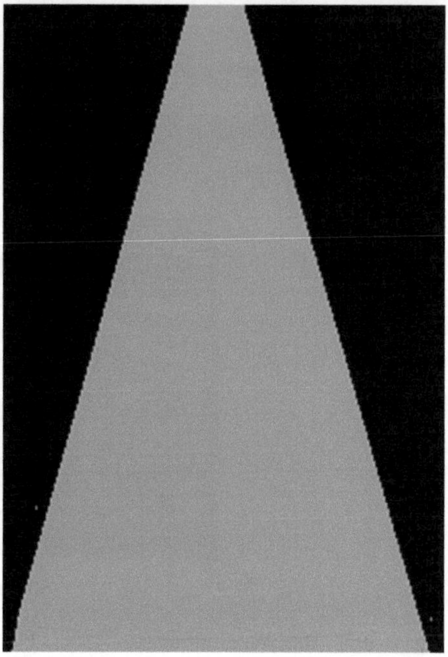

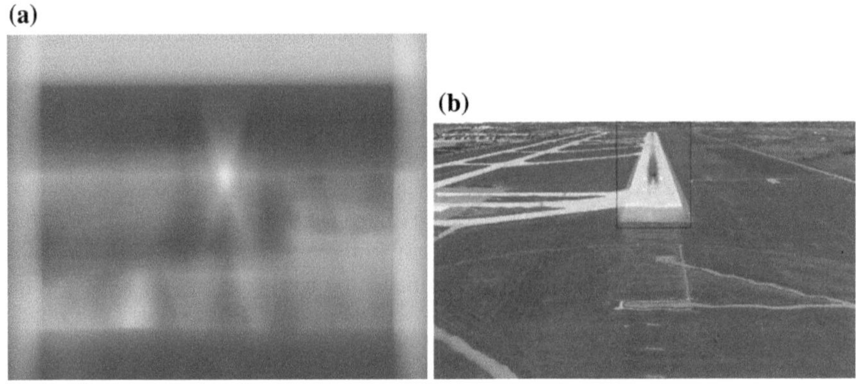

Fig. 6 **a** Normalized cross correlation (NCC). **b** ROI extraction result of 100th frame

Same simulation was done with the images whose runway region is under different scales from 10th frame to 350th frame, the results shows that this method has good matching results at different scales, shown in Fig. 7.

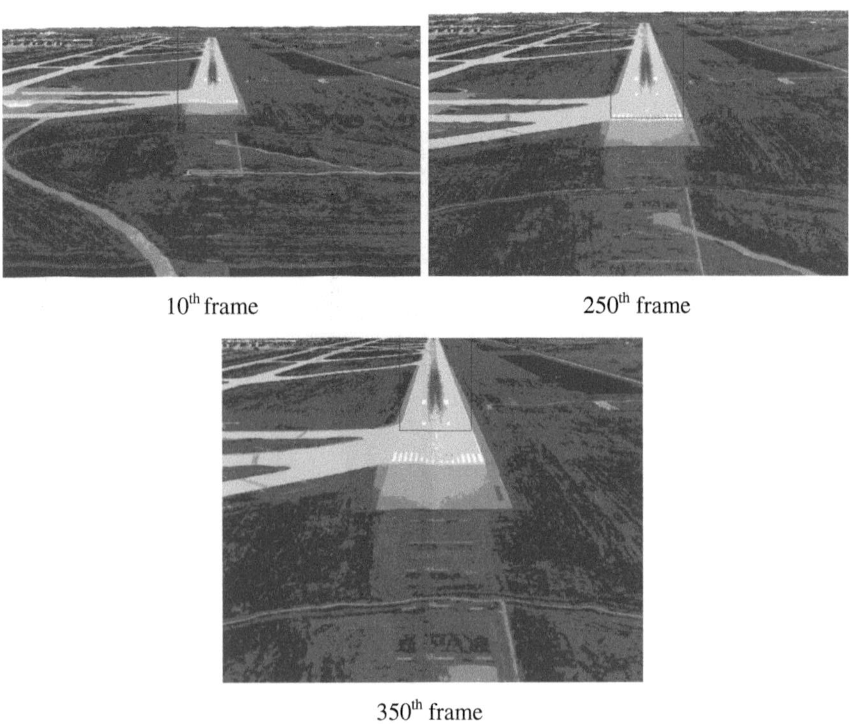

Fig. 7 ROI extraction result of 10th, 250th, 350th frame

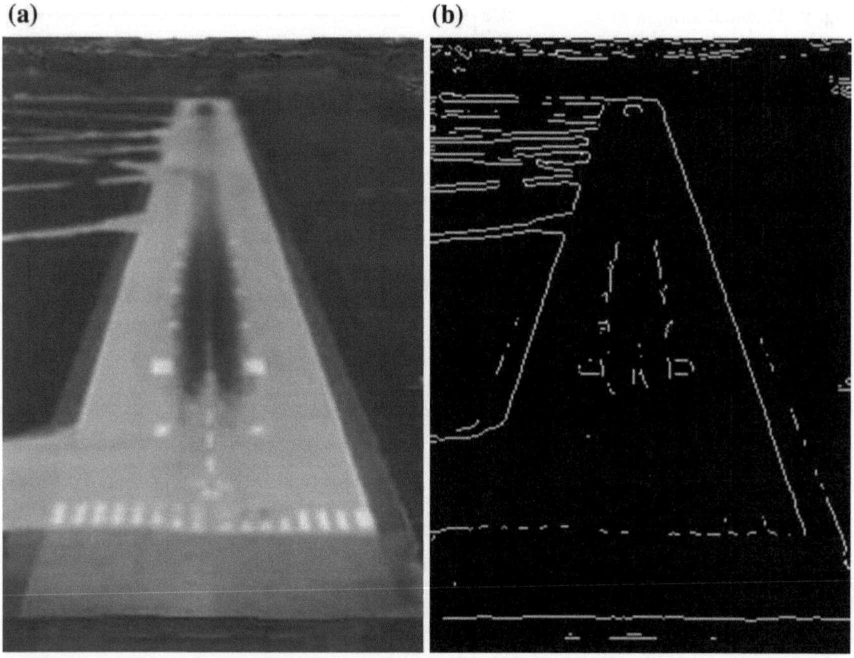

Fig. 8 ROI (**a**) and its edge extraction result (**b**)

And at last, the edge extraction was carried on the image of ROI, whose result is show in Fig. 8.

4.2 Simulation of Rotating Projection Method for Line Detection

The simulation is guided by the flow chart shown in Fig. 3, whose parameters are T = 50 and k = 2°. After the rotating projection algorithm, the potential straight is obtained in (θ, r) space, shown in Fig. 9.

We can observe that the potential straights are distributed into 4 regions as shown in the red rectangular from Fig. 8 obviously. Each region presents a straight in the ROI, and the improved K-means clustering algorithm is employed to locate the cluster centers, shown in Fig. 10.

The matrix of cluster centers is,

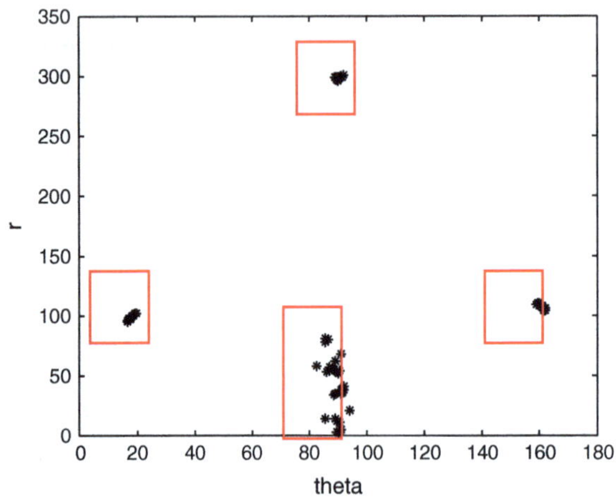

Fig. 9 Potential straight in (θ, r) space

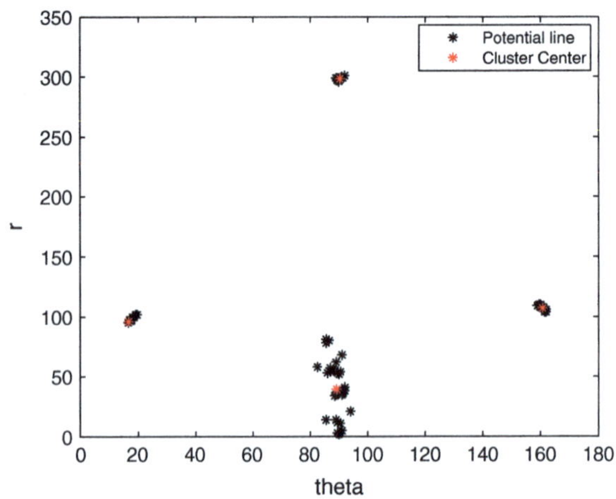

Fig. 10 Locating the cluster centers by improved K-means clustering algorithm

$$Cluster\_center = \begin{bmatrix} 89.1564 & 39.2783 \\ 16.5000 & 96.0000 \\ 90.2719 & 298.3458 \\ 160.7359 & 107.1430 \end{bmatrix} \quad (17)$$

According to (16), the equations of straights is

$$\begin{cases} y = -0.0147x - 276.6078 \\ y = -3.3759x + 24.1507 \\ y = 0.0047x - 17.6789 \\ y = 2.8613x - 622.7757 \end{cases} \quad (18)$$

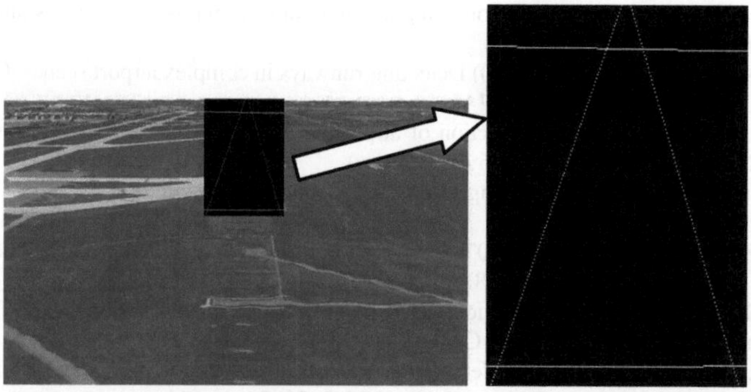

Fig. 11 The runway extraction result

The runway extraction result is shown in Fig. 11.

At last, we made a comparison between Hough transform and method mentioned in this paper. On the Matlab 2012 platform with the Intel i7 core processors, the size of source image is 1024 × 768, the computing time of Hough transform is 0.6734 s, while the computing time of algorithm in this paper is 0.2997, which is half less than the traditional method.

5 Conclusion

In this paper, we bring out a runway extraction method based on rotating projection. Firstly we use an artificial template to locate the ROI which contains the runway area. Then the rotating projection algorithm is employed to seek the potential straights, which will be integrated into the real straights by means of improved K-means clustering method. Simulations are carried out in the end, and results show that the algorithm proposed in this paper can extract the four boundaries of the runway effectively, while it can reduce 50 % of computing time compared with Hough transform.

Acknowledgements The viedo used in the simulation was generated by Dr. Christian Eitner, who works in Institute of Flight System Dynamic (FSD), TUM. So I would like to express my heartfelt gratitude to him here.

References

1. Dryts P, Mees W, Borghys D et al (1997) SAHARA: Semi-automatic help for aerial region analysis. In: Proceeding of the joint workshop of ISPRS working groups I/1, I/3 and IVA: sensors and mapping from Space. Hannover, Germany, pp 267–274

2. Michela A (1998) Airport detection using a simple model, multisource images and altimetric information. SPIE 2315:604–615
3. Huertas A, Cl W, Nevatia R (1990) Detecting runways in complex airport scenes. Comput Vis Graph Image Process 51(2):107–145
4. He Y, Xu X, Sun H (2004) Detection of airport runways in airborne SAR images. J Wuhan Univ (Nat Sci Ed) 50(3):393, 396 (in Chinese)
5. Bao, Fumin, Aiguo Li, and Zheng Qin (2004) Automatic recognition of airfield runway in synthetic aperture radar images. J Xi'an Jiaotong Univ 38(2):1243–1246 (in Chinese)
6. Jia C, Zhou X, Ji K, Kuang G (2007) Extraction of runway in complex synthetic aperture radar image. Sig Process 23(3):374–378 (in Chinese)
7. Chai H, Li H, Png J (2007) Vision-based navigation feature extraction for UAV at night. Comput Eng 33(22):217–219 (in Chinese)
8. Zhu X, Li Y, Yu Q (2009) A Methd to Extract the Runway for Airborne Vision based on Auto-Landing. J Nat Univ Defense Technol 31(2):20–24 (in Chinese)
9. Dong Y, Yuan B, Wang H, Shi Z (2011) A runway recognition algorithm based on heuristic line extraction. Int Conf Image Anal Signal Process 2011:292–296
10. Kaehler A, Bradski GR (2008) Learning OpenCV: computer vision with the OpenCV library. O'Reilly Media, Sebastopol, California

Construction of Corpus and Knowledge Database for Naxi Dongba Manuscripts Based on Internationally Sharing Platform

Haiyan Wang, Hongjun Wang and Xiao Chen

Abstract Naxi Dongba manuscripts were listed as memory of the world by UNESCO and its digital research is urgent. Construction of corpus database for Dongba manuscripts is the first step and basic work of digital research. Knowledge database will be built on the statistical information of corpus and the research experience of experts. The structure of corpus and knowledge base will be discussed and their input and output modes will be studied. Further research will be continued on information retrieval methods including the Naxi words and pictures of manuscripts. All the research aims to motivate the internationally study and protection of Naxi Dongba manuscripts.

Keywords Corpus · Dongba manuscripts · Information retrieval · Knowledge database

1 Introduction

Dongba religion is a unique and original religion of Naxi created about 1000 years ago. The word Dongba is the name for those who preside over the religion ceremony. Dongba manuscripts were written by these people and helped them remember the ritual language on the ceremony. For history reason, the manuscripts are now stored in different organizations of China, the United States, the United Kingdom, France, Japan, Germany, Italy, Canada, Austria and other countries. In 2003 Dongba manuscripts were listed as Memory of the World by UNESCO shown

H. Wang (✉) · H. Wang · X. Chen
School of Electromechanical Engineering, Beijing Information Science and Technology University, Beijing 100192, China
e-mail: asmylady@163.com

H. Wang
Key Laboratory of Modern Measurement and Control Technology (BISTU), Ministry of Education, Beijing 100192, China

E. Qi (ed.), *Proceedings of the 6th International Asia Conference on Industrial Engineering and Management Innovation*,
DOI 10.2991/978-94-6239-145-1_31

Fig. 1 Dongba manuscripts-memory of the World by UNESCO

in Fig. 1 [1]. Their contents cover the fields of philosophy, history, religion, medicine, arts, astronomy, geography, folklore, flora and fauna, military, literature, and so on. Thus they are called the encyclopedia of ancient Naxi [2]. It plays an important role in Chinese history and even the whole human history worldwide.

The manuscripts are written in Dongba character which is a kind of hieroglyph that is still in use and is recognized as the only "living hieroglyph" [3, 4]. As the Dongba manuscripts are written in hieroglyphic characters and most people cannot easily explain them, they have been regarded as "books from heaven". There are about 1400 more or less basic fonts. As the Dongba characters were used to help Dongba remember what to say in the ceremony, they were not written word by word as other hieroglyph such as oracle character. Even if one can read all the characters in the Dongba manuscripts, he cannot explain the meaning except the old Dongbas of Naxi. There are only a few Naxi Dongbas still alive in China and they are getting older and older. So it is an urgent project to build an international platform to gather the Dongba manuscripts that are scattered in different places of the world, explain them by those Naxi Dongbas of China, protect them, study them and share them in world wide range.

There are scholars all over the world studying the Naxi hieroglyphs and the manuscripts such as Fang Guoyu of China and Joseph Charles Francis Rock of America.

2 Digitalization of Naxi Dongba Manuscripts on Sharing Platform

Ancient literature can be processed and digitalized using computer and network technology to form bibliographic databases and full-text database. Thus the rich information in ancient literature can be found and used. Through this way the ancient literature can be protected for the information is digitalized and stored forever [5]. Digitization of ancient literature involves three-dimensional modeling, artificial intelligence, audio and video technology, language processing, optical character recognition and other related technologies. Liu Yang has used the digital technology to solve the problem of inconvenient visit to Dunhuang murals because of far distance [6].

Naxi Dongba manuscripts can also be digitalized through the internationally sharing platform. This sharing platform was completed by teamwork. Group one was responsible for gathering audio and video records of Dongba manuscripts from different places of the world and group two will find the old Dongbas and help them finish the interpretation of these records. Group three will build the hardware and software environment for the sharing of the above work. Group four will build the corpus and knowledge database, input the information into the database and choose the right way for information retrieval.

3 Main Structures of Corpus and Knowledge Database

Corpus database is a computer-based carrier storing the basic linguistic resources. As the research on Dongba manuscripts goes deeper, lots of audio records, video records, pictures and other materials are accumulated. After the materials are cataloged by experts, they can be inputted into the corpus database with different languages for further study.

After the corpus database is based and statistical information is collected from the corpus database along with the experience of Dongba experts, the knowledge database will be built. It consists of the information of word, term, sentence, story and ceremony in Naxi Dongba manuscripts. There are all together about 1400 words used in the manuscripts. These words are the basic component of terms, sentences, stories and ceremonies. As there are variant written patterns for one word, it will be important to collect all the variants and recognize them in the database. The corresponding Chinese word and English word should be concluded in the database for internationally sharing.

The Naxi phonetic and international phonetic should be also marked. Some examples about how to use the word can also be given in the database. There are some fixed collocations of single words and they are called terms. Terms are very often used in Naxi manuscripts and express new meanings. By statistical technology based on corpus database, the terms can be collected by experts and explained by Dongbas. The quantity of terms will grow fast as the corpus database grows and the statistical technology develops. Explanation of terms will make a foundation for interpretation of sentence which is one section divided by "|" in one page of Dongba manuscripts. Continuous sentences connect to each other and produce a story. A story is started and finished in fixed form in Dongba manuscripts. A set of stories will be used in one Naxi ceremony. There are more than 30 ceremonies in Naxi Dongba religion and they can be divided into 3 big groups of praying, exorcism and burial according to ritual objects and targets. Ceremonies are recorded in different manuscripts that are scattered all over the world due to history reason.

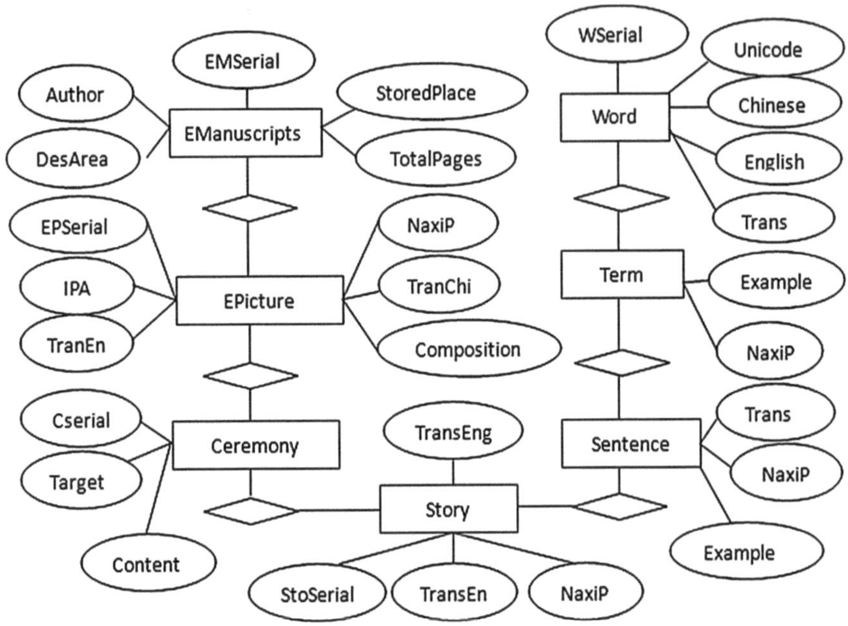

Fig. 2 Skeleton E-R diagram

The manuscripts should also be collected and marked in a uniform standard including the catalog, the storage format, translation model and so on.

The main structure of corpus and knowledge database based on the above analysis is show in Fig. 2.

4 Key Problems and Related Technology

When constructing the corpus database and knowledge database, the key problem and the corresponding technology are as follows.

4.1 The Effective Input Method for Naxi Hieroglyphs

As common people are not familiar with these hieroglyphs and a large scale of materials of Naxi manuscripts are needed to input to the corpus database, effective input methods are needed. According to different level of people familiar with Naxi hieroglyphs, different input methods should be chosen. If one is good at Naxi hieroglyphs such as Dongbas or Naxi experts, they can input the word using IPA, Naxi phonetic or the corresponding Chinese or English word to find the needed

Naxi hieroglyph. If one is not familiar with Naxi hieroglyph, he can input the word using the geometric characters of it such as the connected components count, the hole quantity, the trigeminal-point count, quad-point count etc.

4.2 Dynamic Update of Corpus and Knowledge Database

Various techniques will be used to promote the integration of corpus database and knowledge database. The easy-to-search and easy-to-understand interface for sharing will be designed according to ergonomics rules. Computer-aided interpretation system will help the experts and Dongbas to explain the manuscripts in an easier and quicker way. More interpreted materials will be inputted into the corpus database. The growing corpus database will promote the development of knowledge database. More information of Naxi words, terms, sentences, stories and ceremonies are processed, interpreted, classified, and stored. Thus the construction of corpus database and knowledge database will be in a virtuous cycle to promote the development of each other as shown in Fig. 3.

4.3 Identification of Single Word, Picture and Related Interpretation

Naxi Dongba manuscripts were written by different Dongbas in different period and different area. The identification should consist of distinguishing the different font of one word and its variants. The technology is the image processing combined with knowledge database. When the images are collected from the manuscripts and stored in the database, they can be selected and segmented into regions. One region is chosen and features of words can be extracted. After that the single word selected will be checked with the words in knowledge databases and finally identified. The font, phonetic, translation and samples will be found from the database. The process is shown in Fig. 4.

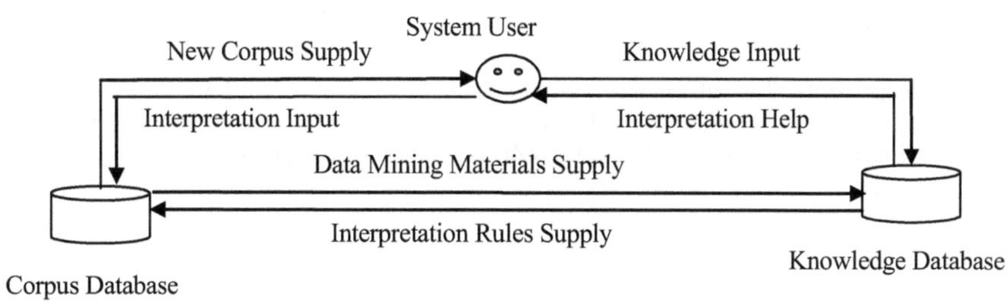

Fig. 3 Development of corpus and knowledge database

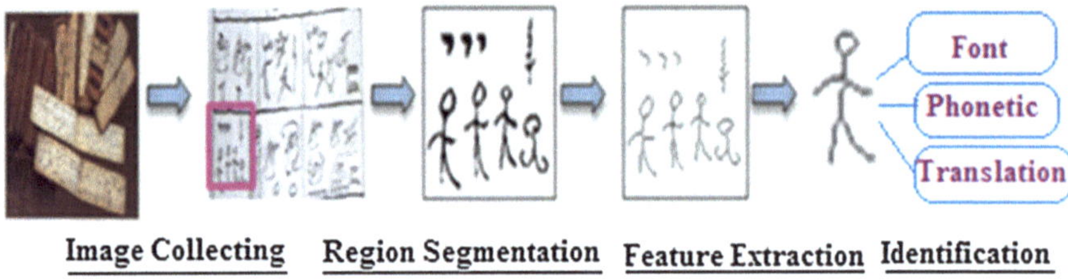

Fig. 4 Identification of single word

As there may be same stories that were recorded in different manuscripts and some of them may be interpreted by different Dongbas, it will be also important to identify the same or analogous pictures and their related interpretations in different manuscripts. The corresponding technology to solve this problem is to use statistical method along with data mining method based on corpus and knowledge database.

5 Conclusion

Construction of corpus and knowledge database is the basic work of digitalization of Naxi Dongba manuscripts. The involved realities and their relationship are discussed and given in the paper. The input method for Naxi hieroglyphs is discussed for further development. The technology of data mining and image processing can be used to resolve the identification of single word, picture and the corresponding interpretation.

With the research going on, more manuscripts will be interpreted and inputted into the corpus. More rules based on corpus will be found and added to the knowledge database which will help the experts and Dongbas explain the manuscripts more easily. Thus the corpus and knowledge database will be promoted by each other and develop in a virtuous way. With the development of corpus and knowledge database, digitalization of Naxi Dongba manuscripts will be finally realized and further study on Dongba culture can be continued.

Further study will be focused on the accomplishment of input methods for Naxi hieroglyphs and the detailed identification methods for the digitalized manuscripts.

Acknowledgements The authors would like to thank the associate editor for his/her constructive comments and valuable suggestions. The research work is supported by Major Programs of National Social Science Funds of China (12&ZD234) and is also supported by Beijing Excellent Young Teacher Program (71A1111143).

References

1. UNESCO (2003) Convention on preserving intangible cultural heritages
2. Gengsheng B (2008) Series of naxi study (30 volumes in total). The Ethnic Publishing House, Beijing
3. Zheng L (2009) The living hieroglyphs: the picture and characters of Naxi Dongba. Art and Design (12):311–313
4. Tatsuo N (1996) The living hieroglyph-naxi nationality's culture. Japan Public Books
5. Wang L (2006) Construction of TRS based ancient books databases. J China Soc Indexers 4(1):31–33
6. Liu Y, Lu D, Diao C, Kuang Y (2004) Dunhuang 285 Cave multimedia integrated virtual exhibit 16(11):1528–1534

The Application of Signed Networks in the Moon-Cake Opinion Network

Wei Li, Ying-hong Ma and Rong-rong Xue

Abstract More and more people are making their opinions available to strangers via the Internet. The relationships between users in the online social networks can be either positive (indicating relations such as friendship, followers, or collaborators) or negative (indicating relations such as opposition or antagonism). In this paper, we first contrast the classical theory of balance with the status balance which proposed in recent years. Then we study the moon-cake opinion dataset from Tianya which is the most popular forum in China to verify these two theories. At last, we conclude that the balance theory is fitter to this opinion network.

Keywords Negative edge · Positive edge · Signed networks · Structure balance · Status theory

1 Introduction

The study of complex networks is a burgeoning research area. Complex networks have related to more and more areas including biology, computer science, economics, mathematics, physics, sociology, and statistics. Social networks are one of the most interesting complex networks. Social interaction on the internet can be both positive and negative, that is to say, people form links to indicate friendship, support, or approval; but they also link to signify disapproval of others, or to express distrust of disagreement of the opinions of others. For a given directed link from user *A* to *B* in a social network, we define its sign to be positive (or negative) if it expresses a positive (or negative) attitude from *A* to *B*. We call such networks

W. Li (✉) · Y. Ma · R. Xue
Department of Management Science and Engineering, Shandong Normal University, Jinan, China
e-mail: xywch1001@126.com

Y. Ma
e-mail: yinghongma71@163.com

E. Qi (ed.), *Proceedings of the 6th International Asia Conference on Industrial Engineering and Management Innovation*,
DOI 10.2991/978-94-6239-145-1_32

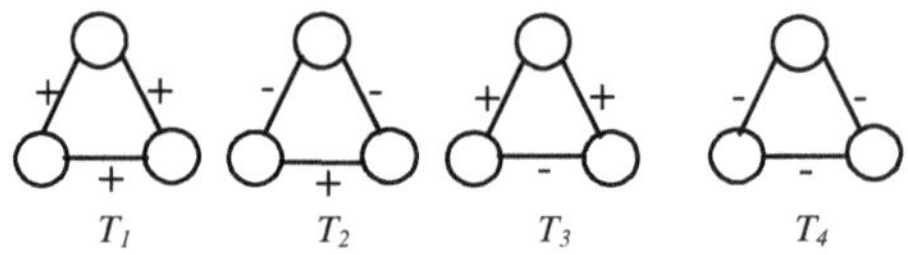

Fig. 1 Undirected signed triads

with both positive and negative links signed social networks. The ties between users give rise to a complex multidimensional web of an aggregated social behavior [1, 2].

There are two theories about signed networks: balance theory and status theory. Structural balance theory was first formulated by Heider [3] in the 1940s. He focused on two types of triads: balanced triads and unbalanced triads. These triads involved three individuals or two individuals and a social object such as a brief. Structural balance theory was modeled in terms of signed graphs by Cartwright and Harary [4]. As we can see in the Fig. 1, based on the number of positive edges, we label triads with odd number of pluses as balanced (T_1, T_2), and triads with even positive edges (T_3, T_4) as unbalanced. The actors in the balance triads are expressed in four aphorisms:" The friend of my friend is my friend"; "The friend of my enemy is my enemy" and "The enemy of my enemy is my friend". Davis [5] proposed the weakly balance theory which further generalized social balance by arguing that in many cases an enemy of one's enemy indeed act as an enemy. In this theory, triads T_1, T_3, T_4 are weakly balanced and triad T_3 is unbalanced in the Fig. 1.

The other theory is status theory, which was implicit in the work of Guha et al. [6] and developed further by Leskovec et al. [7]. Status is a theory of signed link formation based on an implicit ordering of the nodes. In this theory, a positive link from *A* to *B* means that *A* thinks that *B* has higher status than *A*. Similarly, a negative link from *A* to *B* means that *A* thinks *B* has lower status than *B*.

Both theories concern relationships between people. However, in some states, they would behave differently. For example, if user *A* links positively to user *B*, in balance theory we can say user *A* and *B* are friends and *B* should similarly links positively to *A*, but in status theory, it means *B*'s status is higher than *A*'s status, so *B* should links negatively to *A*. In other words, these two theories suggest opposite conclusions in this case.

2 Related Work

By now, there is a large and rapidly growing literature on the analysis of social networks. Signed social networks have attracted more and more attention since R. Guha, R. Kumar, P. Raghavan and A. Tomkins proposed their leading work on trust propagation in signed social networks [6]. With the development of modern communication technology and Internet, especially the advent of Web 2.0, people can collect and observe a large amount of network data which would analysis our behavior patterns, the clustering, emotions and tendency of opinions [8]. Many researches about sentiment analysis [9] have studied online textural data in which individuals can express both positive and negative attitudes towards one another. After the typical signed network datasets of Epinions [10–12], Slashdot [10–13], Wikipedia [10–12, 14] were open to us, a large number of scholars studied these three datasets in different aspects such as prediction [15] and clustering [12].

In recent years, signed networks are applied to many areas. For example, Antal et al. [16] researched the countries in the First World War, they regard the relationship of allied nations as positive edges and the relationship of enemies as negative edges. Szell et al. [17] analyze a vast dataset from a multiplayer online game (Pardus) which encompassed more than 300.000 players. The game allows six types of interactions, of which some have a positive (friendship, communication, trade) and others have a negative association (hostility, aggression, punishment).

3 Dataset Description

Nowadays, more and more people express their opinions in online networks. For example, after the buyers receive their objects from Taobao, they always evaluate the objects if they achieve buyers' satisfaction target. According to their reviews, others would judge whether the objects they like are worth to buy.

This dataset is from Tianya forum which was founded on March 1, 1999. By now, Tianya covers more than 80 million registered users, with tens of millions of user groups with high loyalty and high quality. Users in Tianya can not only express their own opinions but also review other's opinions. Every user could raise topics. There are millions of topics which users can talk about. The topic that we talk about here is moon-cake, landlord pointed out the topic "I think moon-cake is the most unpalatable one of all traditional food." Then others discuss about this topic .People can not only express their own opinions but also review the other people's opinions. We consider users who express their opinions as nodes in our network and consider fields such as agreement, praise and top as positive edges, and fields such as delicious, disagreement and nice as negative edges. We collect this dataset in October, 2013 near the mid-autumn festival from Tianya. In this data, we define the relations of support, agreement, and favor as positive links, and relations of opposite, disagreement and objection as negative links. We use matrix to express this network, it contains $-1, 0, 1$, we made -1 stand for negative link, 0 stand for no link, and 1 stand for positive link.

Table 1 gives statistics for the moon-cake opinion network dataset. This network covered more than five hundred nodes, and less than eight hundred edges. There are more than five hundred users who talked about this topic and more than seven hundred opinions in this network. It is clear that the numbers of positive edges are a little less than the numbers of the negative ones. In these statistics, more than 50 %

Table 1 Dataset statistics

	Opinion network
Nodes	512
Edges	710
+ edges	306
+ probability	43.10 %
− edges	404
− probability	56.90 %
Triads	677

people expressed that they liked to eat moon-cakes. Most people always think that the moon-cake is the standard food of the Mid-Autumn Festival, they taste delicious and we should inherit and carry forward traditional food cultures.

Table 2 summarizes the symbols we would use throughout this whole paper. We use symbols the same as these in [7].

4 Analysis of Opinion Network

4.1 Balance Theory in Opinion Network

We begin to analyze this opinion network as undirected one, where we don't take the directions of links into account.

Table 3 gives the counts of the four possible signed undirected triads; we can find corresponding symbols in Table 2. Here we keep the fraction p of positive edges the same, and shuffle the signs of all edges in the graph, then we let $p_O(T_i)$ denote the expected fraction of triads that are of type T_i after this shuffling. Here we make overrepresentation and underrepresentation as our measure standard. If $p(T_i) > p_O(T_i)$, then triads T_i are overrepresented in the data, or if the sign of $s(T_i)$ is positive, the triads of T_i are overrepresentation.

In the four types of triads, we can clearly find out that triad T_2 is the most heavily overrepresented. Triad T_3 consisting of two enemies with a common friend is the only one which is heavily underrepresented. The fraction of triads T_2 in all triads is more than 50 %, and we can say this structure is the most stable in the moon-cake opinion network. The observations of T_2 and T_3 so far fit well into Heider's original notion of structure balance. Due to the number of people who support landlord is almost as much as the number of ones who opposite landlord, the triads T_i are not representative as we expect, but it still fit well in Heider's notion.

Especially, the number of triads T_4 is 144, which is the second largest number. In the common sense, the case that three individuals who opposite each other is instable, but in this network, it plays an important role. Thus, the Table 3 is more insistent with Davis's weaker notion of balance, where only T_3 is viewed as implausible.

Table 2 Table of symbols

Symbol	Meaning
T_i	Signed triad, also the number of triads of type T_i
Δ	Total number of triads in the network
p	Fraction of positive edges in the network
$p(T_i)$	Fraction of triads T_i, $p(T_i) = T_i/\Delta$
$p_0(T_i)$	A priori prob. of T_i (based on sign distribution)
$E\|T_i\|$	Expected number of triads T_i, $E\|T_i\| = p_0(T_i)\Delta$
$s(T_i)$	Surprise, $s(T_i) = (T_i - E\|T_i\|) / \sqrt{\Delta_1 p_0(T_i)(1 - p_0(T_i))}$
t_i	The number of triads of type t_i
count	Number of contexts t_i
s_g	Surprise of edge initiator giving a positive edge
s_r	Surprise of edge destination receiving a positive edge
B_g	Consistency of balance with generative surprise
B_r	Consistency of balance with receptive surprise
S_g	Consistency of status with generative surprise
S_r	Consistency of status with receptive surprise
$P(Y\|X)$	The probability that the sign X edge is later reciprocated with sign Y edge

Table 3 Numbers of balance and unbalance undirected triads

Triad T_i		$\|T_i\|$	$p(T_i)$	$p_0(T_i)$	$s(T_i)$
T_1	+ + +	122	0.1802	0.0801	4.9716
T_2	+ − −	387	0.5672	0.4186	4.9742
T_3	+ + −	24	0.0355	0.3171	−9.1681
T_4	− − −	144	0.2127	0.1842	1.1946

4.2 Status Theory in Opinion Network

Then we consider the opinion network as directed graphs, incorporating the fact that the links being created go from one user to another, with the sign of a link from A to B being generated by A and received by B.

Here we consider our network as static one. Every edge can be viewed as an object, it contains one generator and one receiver. We make A represent generator and B represent receiver. So we divide this network into sixteen portions. When a user A links to a user B, suppose there is already a user X with the property that X has links to or from A, and also to or from B, and the link from A to X may be positive or negative, so there are $4 \times 4 = 16$ types of links in all, just like we can see in the Fig. 2.

We calculated the count of all types of sixteen triads in this opinion network, and got the probability that A would generate positive link to receiver B. We would refer to the overall fraction of positive signs that a user creates, considering all her links, as her generative baseline, the average of all generative baseline as her probability

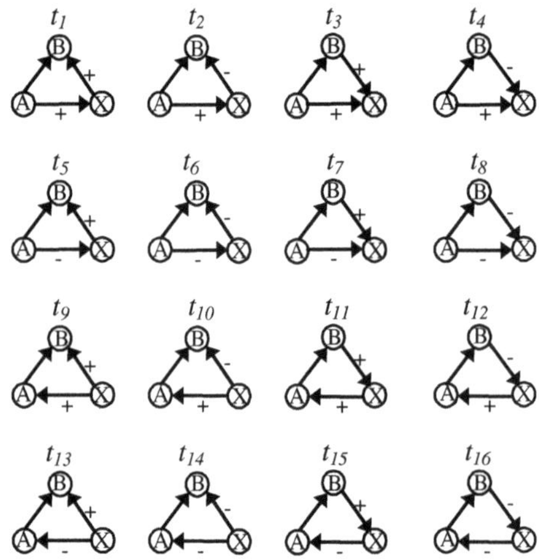

Fig. 2 Directed signed triads

of expected generative number. Similarly, we will refer to the overall fraction of positive signs that a user receives, considering all her links, as her receptive baseline, the average of all receptive baseline as her probability of expected receptive number. In Table 4, we can find out the numbers of each type triad.

We first assign the node *X* a status of 0. Then, if *X* links positively to *A*, or *A* links negatively to *X*, we assign *A* the status of 1; otherwise, we assign *A* the status of −1.We compare the sign of s_g and s_r to the sign of nodes *A* and *B*. We can say that the generative surprise for type t is consistent with status if *B*'s status has the same sign as the generative surprise: in this case, high-status recipient B receive more positive evaluations than would be expected from the generative baseline of the node *A* producing the link. We say that the receptive surprise for type t is consistent with status if *A*'s status has the opposite sign from the receptive surprise, high-status generators of links *A* produce fewer positive evaluations than would be expected from the receptive baseline of the node *B* receiving the link. So we get the Table 5, in this table, we can gain the number of correct predictions the last row in the table shows the number of the correct predictions.

After the analysis of the Table 5, we can clearly draw that the structural balance is much sturdier fit to the data, balance is consistent with generative surprise for all of the 16 types of links, and consistent with receptive surprise for 15 of the 16. Almost more than 95 % types are fit well to the classical structural balance. We can get further insight that the status theory is inconsistent with the data. Status theory is much weaker to the data: there are only 9 or 8 of the 16 types which consistent with generative and receptive surprise. Only half of the predictions are coincident with status theory. Status theory is not fit well to this opinion network.

We just analyze the sum consistency numbers of balance theory and status theory. The details of which situation can not fit well to our two theories are ignored. So we can get the conclusion that balance theory is much sturdier than status theory in this opinion network.

Table 4 Number and related fraction of directed triads

Status	Count	$P(+)$	$p_{Og}(t_i)$	$p_{Or}(t_i)$
t_1	38	0.9737	0.9464	0.7542
t_2	29	0.0345	0.5045	0.4899
t_3	40	0.9250	0.9080	0.7474
t_4	33	0.0303	0.4979	0.2427
t_5	54	0.0556	0.1075	0.5656
t_6	74	0.4324	0.2596	0.3031
t_7	52	0.0192	0.0647	0.3275
t_8	70	0.4000	0.2660	0.4145
t_9	38	0.9737	0.8521	0.7542
t_{10}	29	0.0345	0.2565	0.4899
t_{11}	14	0.7857	0.7458	0.5900
t_{12}	21	0.1429	0.3537	0.3373
t_{13}	35	0.0857	0.2333	0.5478
t_{14}	93	0.5484	0.4176	0.3134
t_{15}	21	0.1429	0.2622	0.4455
t_{16}	36	0.5000	0.2892	0.3471

Table 5 Judgement of balance and status theory

Status	s_g sign	s_r sign	B_g	B_r	S_g	S_r
t_1	+	+	√	√	√	√
t_2	−	−	√	√	√	×
t_3	+	+	√	√	×	√
t_4	−	−	√	√	×	×
t_5	−	−	√	√	×	√
t_6	+	+	√	√	×	×
t_7	−	−	√	√	√	√
t_8	+	−	√	×	√	√
t_9	+	+	√	√	√	×
t_{10}	−	−	√	√	√	√
t_{11}	+	+	√	√	×	×
t_{12}	−	−	√	√	×	√
t_{13}	−	−	√	√	×	×
t_{14}	+	+	√	√	×	√
t_{15}	−	−	√	√	√	×
t_{16}	+	+	√	√	√	√
Number of correct predictions			16	15	8	9

4.3 Reciprocation of Directed Edges

Especially, some links are reciprocal: cases in which there are two nodes *A* and *B* with relations that *A* links to *B* and *B* also links to *A*. In this opinion network data, about 15.5 % edges represent the reciprocation of an existing link, so it's interesting

Table 6 Edge reciprocation

Opinion network	Count	Probability
$P(+\|+)$	37	0.5606
$P(-\|+)$	29	0.4394
$P(+\|-)$	8	0.1818
$P(-\|-)$	36	0.8182

to study these edges. On this occasion, balance theory and status theory shows completely opposite result. If *A* links positive to *B*, balance theory thinks that *B* should links positive to *A* because *A* and *B* are friendly, but status theory thinks that *B* should links negative to *A*, because *A* consider *B*'s status is higher than her and *B* will consider *A*'s status is lower than her.

Table 6 shows the results of the edges reciprocation in this opinion network. Firstly we find that the reciprocation of positive *A*-*B* is a little more consistent with balance than status, but the reciprocation of negative *A*-*B* is closely consistent with balance rather than status. After *A* links positively to *B*, *B*-*A* link is positive over 56 % of the time higher than the roughly 40 % fraction of positive links in the system as whole. After *A* links negatively to *B*, *B*-*A* link is negative well over 80 % of the time much higher than the roughly 57 % fraction of negative links in the system as a whole. So the numbers of same reciprocation are more than those of opposite reciprocation.

Based on the numbers in Table 6, we can get the conclusion that, in this case, balance theory is more suitable in this moon-cake opinion network. That is to say, people always reply the same opinion when they get the review from other people. They often make friends with the people who have the same ideas with them and get hostile to another people who dislike them.

5 Discussion and Conclusion

Social network underlying current social media sites often reflect a mixture of positive and negative links. In this paper, we studied the moon-cake opinion network. Here we have investigated two fundamental theories of signed social network-balance and status. Balance theory is a classical theory from social psychology, which in its stronger form postulates that when considering the relationships between three people in one of the three conditions, in which they have none relation, one relation or all three of the relations should be positive. Status theory is a theory of directed signed networks which postulate that when user *A* makes a positive link to user *B*, it means *A* is asserting that *B* has higher status, when user *A* makes a negative link to user *B*, it means *A* believes *B* has lower status. In this paper, we used different variables to describe the feature value of balance theory and status theory and discussed these variables. By studying this opinion network data, we get the conclusion that balance theory is much fitter to the moon-cake opinion network either as undirected network or as directed network.

Due to the limit of this topic and the time we collected the data, the result of this study can't represented all opinion networks. There are three differences between our study and the early studies of Epinions. Firstly, Epinions is a product view Web site with a very active user community in USA. Users are connected into a network of trust and distrust, which is then combined with review ratings to determine which reviewers are most authoritative. Tianya is the most popular forum in China, there are thousands of users and each one can point out the topic he is interested in. We just study one topic of the website. Secondly, the data spans from the inception of the Epinions site in 1999 until August 12, 2003. Our dataset is from one topic of Tianya forum. It is collected in October 2013. Lastly, the Epinions network contains 119,217 nodes and 841,000 edges. Our moon-cake opinion network just contains 512 nodes and 710 edges, which are much less than these in Epinions. Due to these differences, this result is different from the result of Epinions.

Signed networks are more and more universal in the social networks, such as Youku, Taobao, and Tudou. Users in these networks can express their own opinions not only by words but also by videos. The opinions they expressed would be either positive or negative. For future work, we would like to study more opinion networks and the structure of the networks. The structure of the network is important because on top of direct ties, indirect relations also control behavior and contribute to the establishment of cooperation through reputation mechanisms, such as image scoring.

Acknowledgments This work was supported by NSFC (71071090) and China's Ministry of Education (20133704110003).

References

1. Newman NEJ, Barabási AL, Watt DJ (2006) The structure and dynamics of networks. Princeton University Press, New Jersey
2. Hogg T, Wilkinson D, Szabo G, Brozozwski M (2008) Multiple relationship types in online communities and social networks. In: American association for artificial intelligence-social information processing, pp 30–35
3. Heider F (1946) Attitudes and cognitive organization. J Psychol 21:107–112
4. Cartwright D, Harary F (1956) Structural balance: a generalization of Heider's theory. Psychol Rev 63:277–292
5. Davis JA (1967) Clustering and structural balance in graphs. Hum Relat 20:181–187
6. Guha RV, Kumar R, Raghavan P, Tomkins A (2004) Propagation of trust and distrust. In: Proceedings of the 13th international conference on World Wide Web, pp 403–412
7. Leskovec J, Huttenlocher D, Kleinberg J (2010) Signed networks in social media. In: Proceedings of the SIGCHI conference on human factors in computing systems, pp 1360–1370
8. Li Z, Tang X (2011) Polarization and non-positive social influence: a Hopfield model of emergent structure. Inf Creativity Support Syst Sci 3:15–25
9. Pang B, Lee L (2008) Opinion mining and sentiment analysis. Found Trends Inf Retrieval 2:1–135

10. Facchetti G, Lacono G, Altafini C (2011) Computing global structural balance in large-scale signed social networks. Proc Nat Acad Sci 108:20953–20958
11. Patsonakis C, Roussopoulos M (2013) Asynchronous rumour spreading in social and signed networks. arXiv preprint arXic:1310.6119
12. Chiang KY, Hsieh CJ, Natarajan N, Tewari A, Dhillon IS (2013) Prediction and clustering in signed networks: a local to global perspective. arXiv preprint arXiv:1302.5145
13. Rathore AS, Mutalikdesai MR, Paitil S (2013) Analyzing trust-based mixing patterns in signed networks. In: Digital libraries: social media and community networks. Springer, Berlin, pp 63–72
14. Teng WT, Kao HY (2013) Measuring semantic relatedness using wikipedia signed network. J Inf Sci Eng 29:615–630
15. Leskovec J, Huttenlocher D, Kleinberg J (2010) Predicting positive and negative links in online social networks. In: Procedings of the 19th international conference on World Wide Web, pp 641–650
16. Antal T, Krapivsky P, Redner S (2006) Social balance on networks: the dynamics of friendship and enmity. Physical D 224:130
17. Szell M, Lambiotte R, Thurner S (2010) Multi-relational organization of large-scale social networks in an online World. Proc Natl Acad Sci 107(31):13636–13641

Speckle Filtering Algorithm of PolSAR Imagery Based on Two-Dimensional Polarimetric Subspace ICA

Hao-gui Cui, Gao-ming Huang, Tao Liu and Jun Gao

Abstract To improve the performance of speckle filter on polarimetric synthetic aperture radar (PolSAR) imagery, an ICA algorithm based on two-dimensional polarization subspace is proposed. In this method, the PolSAR data of three channels were divided into three two-dimensional subspaces, and then the speckle noise component and texture component can be separated by the ICA algorithm. The equivalent number of looks (ENL) can be used to evaluate the effect of speckle filter. And an automatic ENL estimation algorithm is introduced to avoid the manual selection of a region with fully developed speckle and no texture. Performance of the novel speckle filter is tested through real data and the results show that the proposed filter is effective and robust.

Keywords Independent component analysis · Polarimetric subspace · Polarimetric SAR · Speckle

1 Introduction

Polarimetric synthetic aperture radar (PolSAR) has become an important source of sensing information because of the ability to identify polarization characteristics of different targets. But the existence of speckle affects the quality of the radar imaging. Strictly speaking, speckle is not a noise, but a phenomenon of electromagnetic scattering. The speckle makes it difficult to interpret the PolSAR image and also causes the target classification performance degraded. So how to effectively suppress speckle has become an important topic of PolSAR image processing. At present, a large number of speckle filtering methods are proposed such as multilook process and space filtering technology. Multilook method is a tradi-

H. Cui (✉) · G. Huang · T. Liu · J. Gao
School of Electronic Engineering, Naval University of Engineering,
717 Jiefang Road, Wuhan 430033, China
e-mail: seachg@163.com

E. Qi (ed.), *Proceedings of the 6th International Asia Conference on Industrial Engineering and Management Innovation*,
DOI 10.2991/978-94-6239-145-1_33

tional and effective speckle filtering technology which is performed by averaging the power from adjacent pixels. But the problem is it will reduce the spatial resolution of the image. The typical spatial filtering technology are Boxcar filter, polarization whitening filter [1] and refined Lee filter [2]. The drawback of Boxcar filter is the retention of details, and there is also a great loss of resolution. The polarization whitening filter has a good performance, but it losts polarization information and provides only a single channel image. The advantage of Lee filter is the good edge preservation, but it is not well enough to maintain the polarization information.

Independent component analysis (ICA) is a new signal processing method which is usually used to solve the problem of blind signal separation. Much reserch have been done on this topic. Chen proposed a speckle filtering algorithm based on ICA algorith [3]. Then, Zhang introduced a kernel function for ICA and done this work using kernel independent component analysis (KICA) [4]. However, both the two algorithms only separated image of one channel which lost the polarization information. To Solve this problem, Chen proposed a sparse coding shrinkage algorithm based on ICA (ICA-SCS) [5]. The ICA-SCS algorithm can retain information from multiple polarization channels. But the disadvantage of ICA-SCS algorithm is that it needs some noiseless image data as training samples which are not easy to obtain. In this paper, a novel ICA algorithm based on two-dimensional polarization subspace is proposed and it can keep polarization information. Real data is used to test the performance and the result shows that the novel filter is effective and robust.

2 Statistical Modeling of Speckle

The speckle in PolSAR imagery is a result of image formation under coherent radiation [6]. Goodman has shown that if the number of scatters per resolution cell is large, a fully developed speckled pattern can be modeled as the magnitude of a complex Gaussian distribution [7]. The multiplicative speckle model is usually used to model the single polarimetric channel SAR image

$$I = x \cdot v \tag{1}$$

where I is the intensity of backscattering signal, x is the true intensity of scattering which is also called texture, and v is the intensity of speckle whose expectation is 1. The two random variables x and v are statistically independent.

After multilook processing, the intensity of speckle can be modeled with Gamma distributing [7]

$$p(I) = \frac{1}{\Gamma(L)} \left(\frac{L}{\sigma}\right)^L I^{L-1} e^{-LI/\sigma}, \quad I \geq 0 \tag{2}$$

When the image intensity is logarithmically transformed, one can readily obtain the PDF of the random variable $D = \log I$

$$p(D) = \frac{L^L e^{DL} e^{-(L/\sigma)e^D}}{\sigma^L \Gamma(L)} \tag{3}$$

Arsenault et al. pointed out that Eq. (3) is approximately a Gaussian distribution with the variance $1/L$ which also makes Eq. (1) become an approximately Gaussian additive model [8].

2.1 *Criterion of the Speckle Reduction Capability*

There are many criterions used to evaluate the speckle reduction algorithm, such as the preservation of the contrast ratio, texture characteristic, edge sharpness and the value of ENL. The ENL describes the degree of averaging during data formation and post processing and its value can be applied to measure the intensity of speckle. The larger the ENL, the smaller the speckle is. For homogeneous regions, the ENL can be defined as

$$\mathrm{ENL} = \frac{E(I)^2}{\sqrt{\mathrm{var}(I)}} \tag{4}$$

Here we use the value of ENL as an indicator of the performance of speckle filtering. Note that the ENL estimation results from Eq. (4) is valid only in homogeneous areas where the intensity of scattering signal is due to the speckle. For this reason, the ENL is usually estimated by identifying homogeneous regions in an image by artificial selection or using computer.

2.2 *Unsupervised Estimation of ENL*

It is not sure that one can find a region with fully developed speckle and no texture in an arbitrary PolSAR image. Although such a subset exists, there may not have enough samples to be calculated to ensure the accuracy. To avoid this disadvantage, Anfinsen designed an unsupervised estimation method which is fully automatic and avoids manual selection of homogeneous regions [9]. The idea is that ENL computed under fully developed will dominate the estimation result. So the ENL of homogeneous regions can be inferred from the peak of the distribution. Here we use a sliding window of size $k \times k$ pixels covering the whole image and then compute the ENL on the pixels of each window.

The distribution of the ENL estimates will depend strongly on both the properties of the image and the window size $k \times k$. On the one hand, the result may be easy to get if the scene is homosexuality and the extent of homogeneous regions is large. In this condition, we can use a large window size to do our estimation which can reduce the estimated variance. On the other hand, a small size of window will be taken if the homogeneous regions are small or wider distributed.

In order to adjust to the unpredictable shape of distribution, a kernel density estimator (KDE) method is proposed. The KDE is a kind of nonparametric estimation and the distribution estimate is defined as

$$\hat{p}(\hat{L}) = \frac{1}{nh}\sum_{i=1}^{n} K_h(\hat{L} - \hat{l}_e(i)) \quad (5)$$

where $\hat{L}$ is the value of ENL, $\hat{l}_e(i)(i = 1, 2, \ldots, n)$ is the estimation of ENL under separate $k \times k$ sliding windows, $K_h(\cdot)$ is the kernel function and h is called bandwidth which determines the degree of smoothing. Here we choose the Epanechnikov kernel function because it is not only simple, but also has a convergence rate of $n^{-4/5}$, which is very closely to the rate of n^{-1}. The Epanechnikov kernel function is defined as [10, 11]

$$K_h(x) = \frac{3}{4}(1 - (x/h))\Phi\{|x/h| < 1\} \quad (6)$$

when the condition $\Omega = |x/h| < 1$ holds, $\Phi\{\Omega\} = 1$, otherwise, $\Phi\{\Omega\} = 1$.

3 Speckle Filtering Algorithm Based on Ica

ICA algorithm is able to separate different source signals from the observed data according to some basic statistical properties of the source signal such as statistically independent and non-Gaussian feature. It also has become a hot topic of signal processing in recent years.

3.1 Basic Principle of ICA

The basic idea of ICA is to find a linear transformation which will make the statistical independence between the signals as large as possible. The model of ICA is very simple

$$\boldsymbol{x} = \boldsymbol{As} = \sum_{i=1}^{n} a_i s_i \quad (7)$$

where $s = [s_1, s_2, ..., s_n]^T$ is the source signal, and $x = [x_1, x_2, ..., x_m]^T$ is the observed vector which is a linear combination of s. The parameters n and m are the dimension of the vectors.

What we would like to do is to find the original signals s from the mixtures x. Here we can get the linear combination of the observed variables

$$y = Wx \tag{8}$$

where $y = [y_1, y_2, ..., y_n]^T$ is the result of the linear combination, W is the transformation matrix, and $x = [x_1, x_2, ..., x_m]^T$ is the observed vector.

In fact, if the signals are not Gaussian, it is enough to determine the coefficients W using the property of statistically independent. From the ICA algorithm we can conclude that if the signals y_i are independent, then $y = [y_1, y_2, ..., y_n]^T$ are equal to the original signals.

3.2 A Fast ICA Algorithm Using Negentropy

Negentropy is a robust measure of non-Gaussianity. It is based on the information-theoretic quantity of differential entropy which is called simply entropy. Negentropy is defined as

$$\begin{aligned} J(y) &= H(y_G) - H(y) \\ &= \int P(y) \lg P(y) dy - \int P_G(y) \lg P_G(y) dy \\ &= \int P(y) \lg[\frac{P(y)}{P_G(y)}] dy - \int [P(y) - P_G(y)] \lg P_G(y) dy \end{aligned} \tag{9}$$

where $H(y_G)$ is the entropy of Gaussian random vector of the same correlation matrix as y. $J(y)$ is always nonnegative and will be zero only if y has a Gaussian component.

In the calculation of negentropy from real data, $P(y)$ is usually unknown and it is generally estimated by the following formula

$$J(y) = [E\{g(y)\} - E\{g(v)\}]^2 \tag{10}$$

FastICA is based on a fixed-point iteration scheme for finding a maximum of the non-Gaussianity. More stringent, it can be derived as an approximate Newton iteration. The non-Gaussianity can be measured by kurtosis, mutual information, negentropy and so on. Here we choose the FastICA algorithm using negentropy that combines the superior algorithmic properties resulting from the fixed-point iteration with the preferable statistical properties due to negentropy.

We must preprocess the signals before using the FastICA algorithm. This step includes center and whiten the data. Firstly we center the data to make its mean zero. Secondly we whiten the data which will weak the correlation between the signals. The whiten process can be defined as

$$\boldsymbol{v} = \boldsymbol{Q}\boldsymbol{x} = \boldsymbol{E}^{-\frac{1}{2}}\boldsymbol{U}^T\boldsymbol{x} \tag{11}$$

where $\boldsymbol{E}$ is a diagonal matrix consisting of the eigenvalues of $\boldsymbol{R}_x = E(\boldsymbol{x}\boldsymbol{x}^T)$, and $\boldsymbol{U}$ is a matrix composed of eigenvectors.

The basic fixed-point iteration in FastICA is defined as [12, 13]

$$\begin{aligned} \boldsymbol{w}^+ &= E\{\boldsymbol{x}g(\boldsymbol{w}^T\boldsymbol{x})\} - E\{\boldsymbol{x}g'(\boldsymbol{w}^T\boldsymbol{x})\}\boldsymbol{w} \\ \boldsymbol{w} &= \boldsymbol{w}^+/\|\boldsymbol{w}^+\| \end{aligned} \tag{12}$$

where $\boldsymbol{w}$ is a m-dimensional matrix and satisfied $E\{(\boldsymbol{w}^T\boldsymbol{x})^2\} = 1$. $g(\cdot)$ is an arbitrary non-quadratic function which is the derivative of $G(\cdot)$. Here we choose $g(u) = \tanh(\alpha u)$.

3.3 Speckle Filtering Based on Polarization Subspace ICA

Equations (1) and (3) show that the multiplicative model of image data is converted to additive model after taking the logarithm of the data. In addition the logarithmic speckle is a Gaussian distribution which is independent with the texture. Thus, the observed PolSAR image can be seen as the linear combination of the Gaussian noise and texture. We can see that the additive model satisfies the applicable conditions of ICA. So the separation results obtained by ICA with the largest ENL is the texture component while the other two are the noise.

The existing FastICA algorithm uses the data of three polarimetric channels as the input and finally obtain an enhanced image with no speckle [3, 4, 14, 15]. In fact, this method has lost the polarization information. In order to remain the polarization information, we firstly divided the PolSAR data of three channels into three two-dimensional subspaces, and then use the ICA algorithm to separate the speckle noise component and texture component. This novel speckle filter method will obtain three images without speckle, and the algorithm process is as follows:

(1) Do the multilook process to the PolSAR data with the nominal look L (usually $L \geq 3$).
(2) Logarithmically transform the multilook data of three polarimetric channels, and divided them into three two-dimensional subspaces Ω_i, $(1 \leq i \leq 3)$, initial $i = 1$.
(3) Center and whiten the data of subspaces Ω_i to make its mean zero.
(4) Separate signals using FastICA algorithm.

(5) Estimate the ENL of each separation result using the unsupervised estimation method. The image with the largest ENL is texture and the others are speckle noise.
(6) If $i = 3$, go to step 7. Otherwise let $i = i + 1$ and go to step 2.
(7) Obtain three texture image without speckle.

4 Discussion

Here we use real data to compare the following algorithms

(1) Refined Lee filter (Spatial filtering method)
(2) FastICA based on three polarimetric channels [3]
(3) FastICA based on two-dimensional polarimetric subspace that is proposed in this paper

We use the PolSAR data of San Francisco Bay area in California which is acquired by the airborne NASA/JPL AIRSAR L-band instrument from 1988. The size of image is 900 × 1024 and the resolution of a pixel is about 10 m × 10 m. The real data are drawn from the software of PolSARPro 4.2. Here, the nominal number of looks is 4 (2 × 2).

Figure 1 shows the independent components which are separated by the ICA algorithm based on two-dimensional polarimetric subspace. In this method, the PolSAR data of three channels was divided into three two-dimensional subspaces. These subspaces are called HH&HV, HH&VV and HV&VV subspace respectively. Using the ICA algorithm we can get two independent components in each subspace. One is the texture image without noise and the other is speckle. From Fig. 1, we can see that Fig. 1b, c, e are the texture images and the others are speckle. Figure 2 is the results of FastICA algorithm based on three polarimetric channels. Figure 2c is the only one texture image and we lost the polarimetric information. Here we use the value of ENL as the criterion to find the texture component based on the assumption that the texture has the largest ENL. Figure 3 shows the ENL of the texture component after filtering with different algorithm using the unsupervised estimation. Each peak of the distributions is the ENL that is calculated from the homogeneous regions of the image.

Here we choose the bandwidth $h = 0.2$ and the size of the sliding windows are $k = \{8, 16, 24, 32, 40\}$. From the panels of Fig. 3, we can study the evolution of the distribution of ENL estimates under different window sizes. On the one hand, if the window size is too small, the variance of the ENL is large because there are too few samples in a window. On the other hand, the variance of the ENL is become smaller but the bias is become larger as the window size increased. In fact, the optimal value of k is closely related to the characteristics of the image itself. We can use a large k when the scene is homosexuality and the extent of homogeneous regions is

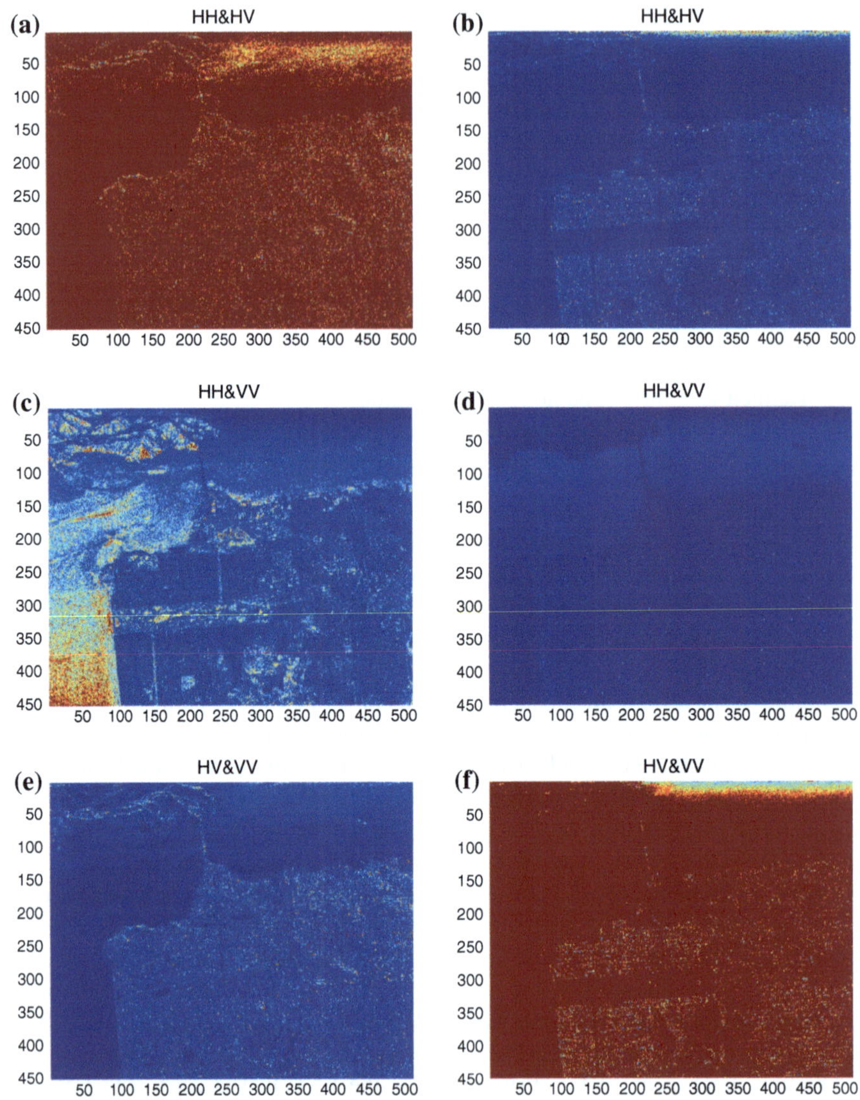

Fig. 1 Independent components separated by the ICA algorithm based on two-dimensional polarization subspace. **a** HH&HV subspace. **b** HH&HV subspace. **c** HH&VV subspace. **d** HH&VV subspace. **e** HV&VV subspace. **f** HV&VV subspace

spacious. From Fig. 3 we also get the conclusion that the ENL from FastICA filter based on the subspace of HV&VV is the largest which also means the best.

Table 1 shows the ENL calculated from the original data or data after Lee filter with the bandwidth $h = 0.2$ and window size $k = 32$. From this figure we can see

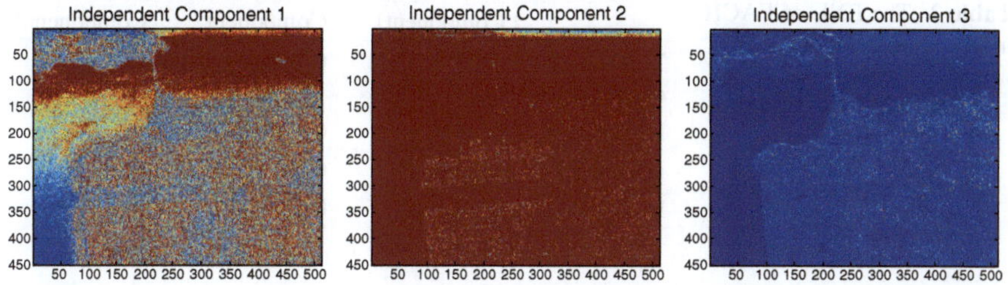

Fig. 2 Independent components separated by the ICA algorithm based on three polarization channels

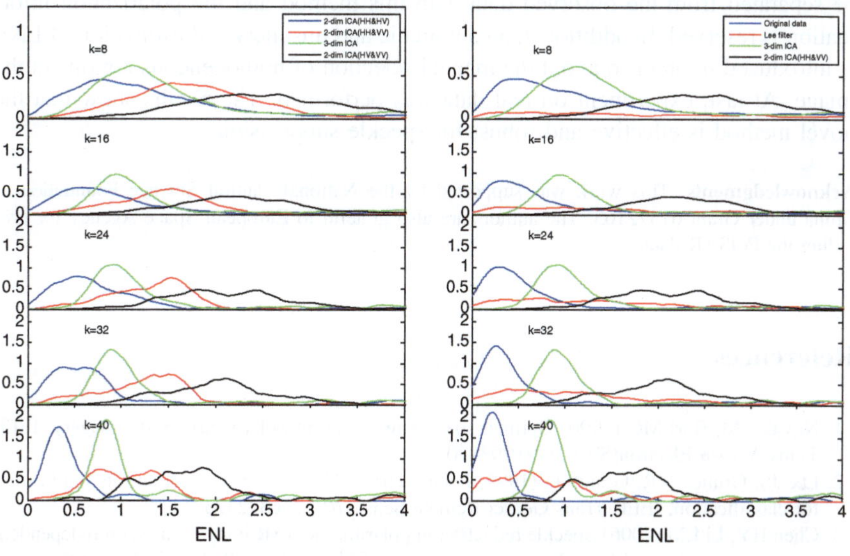

Fig. 3 The equivalent number of looks of the texture component after filtering with different algorithm

Table 1 The ENL of different polarimeteic channels

	HH channel	HV channel	VV channel
Original data	0.2680	0.2200	0.1560
Lee filter	0.3960	0.3640	0.1840

the effect of Lee filter. Table 2 shows the ENL of each independent component under different FastICA algorithm. The largest ENL among the values in a row corresponds to the texture component. So we concluded that FastICA filter based on the subspace of HV&VV is the best one because of its largest value of ENL.

Table 2 The ENL of EACH independent component

	Component1	Component2	Component3
Three channels	0.8980	0.7180	1.3080
HH&HV	0.3900	0.7400	\
HH&VV	1.5140	0.7260	\
HV&VV	2.1100	0.7220	\

5 Conclusion

We have proposed a speckle filter algorithm of PolSAR imagery based on two-dimensional polarimetric subspace ICA in this paper. Three texture images will be separated from the observed data with this method and the polarization information is reserved. In additional, an automatic and unsupervised estimation of ENL is introduced in order to avoid the manual selection of homogeneous regions in the image. At last, experiment on real data was performed. The results show that the novel method is effective and robust for speckle suppression.

Acknowledgments This work was supported by the National Natural Science Foundation of China under Grant 61372165. The authors are also grateful to European Space Agency for providing the PolSAR data.

References

1. Novak LM, Burl MC (1990) Optimal speckle reduction in polarimetric SAR imagery. IEEE Trans Aerosp Electron Syst 26(2):293–305
2. Lee JS, Grunes MR, Grandi G (1999) Polarimetric SAR speckle filtering and its implication for classification. IEEE Trans Geosci Remote Sens 37(5):2363–2373
3. Chen HY, Li LM (2006) Speckle reduction in polarimetric SAR images based on independent component analysis. J Univ Electron Sci Technol China 35(4):480–483 (in Chinese)
4. Zhang ZS, Yu J, Yan Q, Zhao Z (2011) Research on polarimetric SAR image speckle reduction using kernel independent component analysis. Acta Geodaet Cartogr Sin 40(3):289–295 (in Chinese)
5. Chen HY, Fu YS, Pi YM (2007) "Speckle reduction of polarimetric SAR image based on ICA-SCS algorithm. J Electron Inf Technol 29(4):819–821 (in Chinese)
6. Chen PS, Tong QX (1998) Remote sensing information mechanism research. Science press, Beijing, pp 43–50 (in Chinese)
7. Goodman N (1963) Statistical analysis based on a certain multivariate complex Gaussian distribution. Ann Math Stat 34(1):152–177 (an introduction)
8. Arsenault H, April G (1976) Properties of speckle integrated with a finite aperture and logarithmically transformed. JOSA 66(11):1160–1163
9. Anfinsen SN, Doulgeris AP, Eltoft T (2009) Estimation of the equivalent number of looks in polarimetric synthetic aperture radar imagery. IEEE Trans Geosci Remote Sens 47(11):3795–3809
10. Silverman BW (1986) Density estimation for statistics and data analysis. Chapman & Hall/CRC, Boca Raton, pp 103–108

11. Wand MP, Jones MC (1995) Kernel smoothing, Chapman & Hall/CRC, Boca Raton, pp 239–250
12. Hyvarinen A, Oja E (1997) A fast fixed-point algorithm for independent component analysis. Neural Comput 9(7):1483–1492
13. Hyvarinen A (1999) Fast and robust fixed-point algorithms for independent component analysis. IEEE Trans Neural Netw 10(3):626–634
14. Wang H, Pi Y, Liu G, Chen H (2008) Applications of ICA for the enhancement and classification of polarimetric SAR images. Int J Remote Sens 29(6):1649–1663
15. Chitroub S, Amrane H, Boualem S (2000) Compound PCA-ICA neural network model for enhancement and feature extraction of multi-frequency polarimetric SAR imagery. In: Proceedings the 2000 International Conference on Image Processing, Vancouver, British Columbia, Canada, pp 328–331

Mapping the Cortical Activation Changes Induced by Transcranial Direct Current Stimulation: A fNIRS-tDCS Study

Tingju Lei, Ding Ma and Feng Jiang

Abstract Transcranial Direct Current Stimulation (tDCS) is a non-invasive cortical stimulation technology. It uses weak direct currents to modulate the cortical activity below the motor threshold. To better understand the effects of tDCS-induced activation during motor performance, primary motor cortex (M1) activation changes were measures by fNIRS before and after applying tDCS on left M1 with a finger tapping task in eight healthy subjects. This study finds that the effect of tDCS depends on the current polarity. Anodal stimulation excites certain cortical regions and cathodal stimulation produces opposite effect. The finding in this research suggests that the combined use of fNIRS and tDCS can be a useful tool to understand the effect of tDCS. tDCS combined fNIRS can provide insights into changes in activity induced by tDCS, which may find future use guidance on motor rehabilitation.

Keywords Functional near-infrared spectroscopy · Primary motor cortex · Transcranial direct current stimulation

T. Lei (✉)
State Key Laboratory of Cognitive Neuroscience and Learning & IDG/McGovern Institute for Brain Research, Beijing Normal University, 100875 Beijing, China
e-mail: leitingju@126.com

T. Lei
Center for Collaboration and Innovation in Brain and Learning Sciences, Beijing Normal University, 100875 Beijing, China

D. Ma
Institute of Electrical Engineering, Yanshan University, 066004 Qinhuangdao, China

F. Jiang
College of Mechanical Engineering, Yanshan University, 066004 Qinhuangdao, China

E. Qi (ed.), *Proceedings of the 6th International Asia Conference on Industrial Engineering and Management Innovation*,
DOI 10.2991/978-94-6239-145-1_34

1 Introduction

Non-invasive stimulation has become one of the major tools to facilitate plasticity in humans, especially in the patient with brain disorders, such as stroke [1–4]. As far as I know, using fNIRS to detect the effects of tDCS is still a new field. Lately, the HbO changes in resting state under the prefrontal cortex were measured by fNIRS before and after tDCS [5]. And also, the sensorimotor cortex is also be measured by fNIRS before, during, and after tDCS [6, 7]. However, no fNIRS studies have been performed to date to specially explore the effects after tDCS induced by finger tapping task in the primary motor cortex (M1), which is the topic of this study. In this study, the hemodynamics in primary motor cortex regions was measured using fNIRS before and after tDCS. This study presents the combined use of fNIRS and tDCS as a useful tool for studying the cortical plasticity in human subjects. And this method can potentially provide a guide to brain injury rehabilitation of stroke patient.

2 Materials and Methods

2.1 Subjects

We recruited 8 undergraduate volunteersto evaluate the tDCS stimulation effect explored by fNIRS. All subjects were gave their informed consent. The studies were performed was approved by the ethics committee of the State Key Laboratory of Cognitive Neuroscience and Learning, Beijing Normal University. Of the 8 subjects, 4 subjects were accepted anodal stimulation and 4subjects were accepted cathodal stimulation.

2.2 fNIRS-tDCS Setup

A continuous wave optical topography system (ETG-4000, Hitachi Medical Company, Japan) was used to map the hemodynamic changes induced by primary motor cortex (M1) activity before, during, and after bi-hemispheric tDCS. The fNIRS source-detector geometry is shown in Fig. 1a.

TDCS was applied by a battery-driven electrical stimulator (Neuro Conn DC-Stimulator, Germany) connected by electrodes centered over the C3 and FP2 positions according to the EEG International 10/20 system (Fig. 1b). Current was induced by a pair of saline-soaked surface sponge electrodes (7 cm × 5 cm in size, 35 $cm^{2.}$). It is delivered with a output of 1.5 mA. The total stimulation time is 616 s. The effective electrode could be anode or cathode.

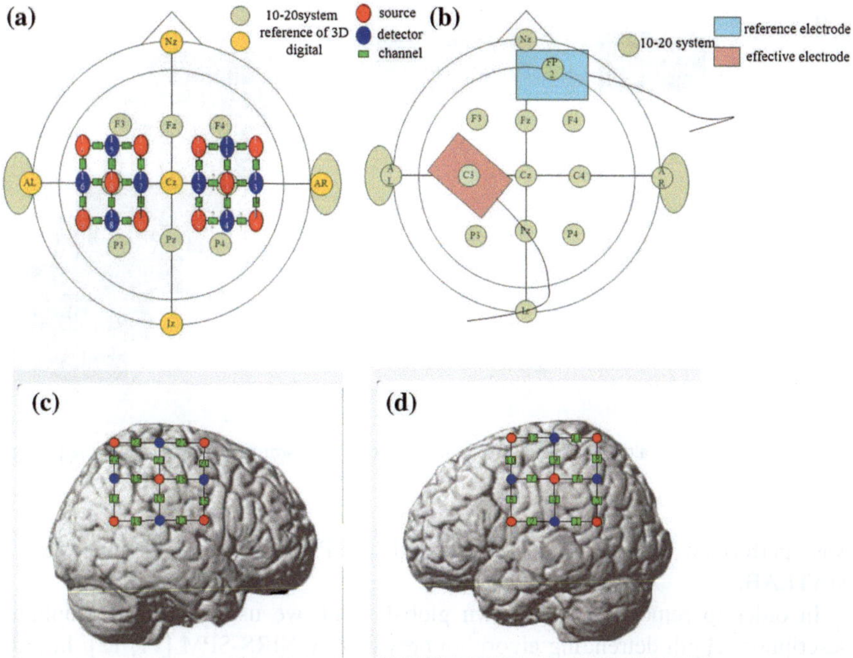

Fig. 1 **a** The geometry of fNIRS sources and detectors; **b** the geometry of tDCS electrodes: the dull-red electrode is the effective electrode; **c** the first optode set place on the right M1; **d** the second optode set placed on the left M1

3 Protocol

For each participant, the experiment was split into two separate blocks: before and after tDCS. In the period of before and after tDCS, we use fNIRS to measure the bilateral motor cortex. The protocol presentation on the computer interface started with 90 s of rest, followed by ten sets of the finger tapping task, and ended with 5 s of rest. Each set of the finger tapping task contained two sections. Every section is 30 s. In total, the protocol presentation lasted for 695 s (11 min and 35 s) (Fig. 2).

4 fNIRS Signal Processing and Image Analysis

Previous studies showed that Hb dynamics are highly correlated with those of HbO [8]. Previous study showed that HbO was the most sensitive indictor of changes in the regional cerebral blood flow in fNIRS measurement [9, 10]. Therefore, in this study we only focused on the changes in HbO concentration. The data processing

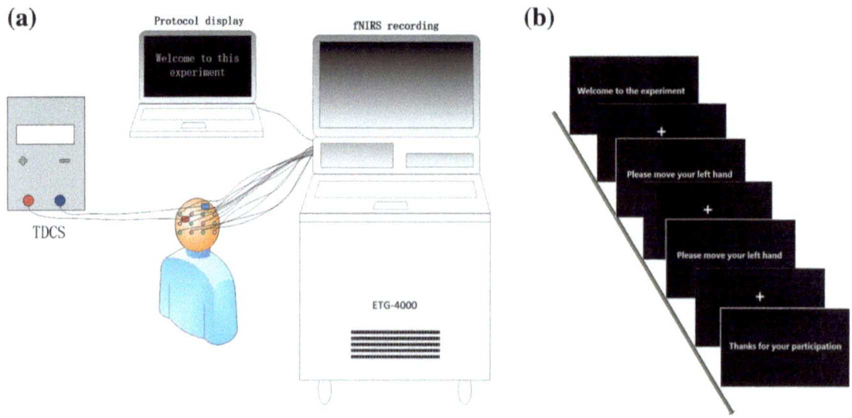

Fig. 2 **a** This figure is a cartoon representation of the overall instrumentation setup; **b** the protocol display

was performed by the open-source NIRS_SPM software implemented in MATLAB.

In order to remove the unknown global trend, we used a wavelet-minimum description length detrending algorithm provided by NIRS-SPM [11, 12]. In order to better illustrate the changes of motor cortex plasticity before and after tDCS, we recorded the spatial registration of NIRS channels using 3D digitizer [13]. Then, the spatial registration of NIRS channels to MNI space with MRI coordinate input is available [14].

5 Results

Performing the finger tapping task before and after tDCS presented cortical activation changes in the M1 (Fig. 3). However, apparent differences were found between anodal and cathodal tDCS stimulation.

The cortical activation after tDCS changed differently depending on the tDCS polarity. This result matched with the previous fNIRS combined tDCS study [7]. When the anode was placed over the M1 area, there was a significant activation increase caused by dominant arm finger tapping. However, when the cathode was placed over the M1 area, the activation changed in the opposite direction. Therefore, these results show that tDCS can affect the cortical regions' sensitivity controlling the finger tapping in health subjects. These representative results matched with previous fMRI studies [15, 16].

In order to better to understand the cortical plastic changes before and after tDCS, a group analysis of the function activation of all eight subjects (four subjects accepted anodal stimulus, and the other cathodal) was performed. The result is in Fig. 4. These results are consistent with reports from fMRI combined tDCS studies

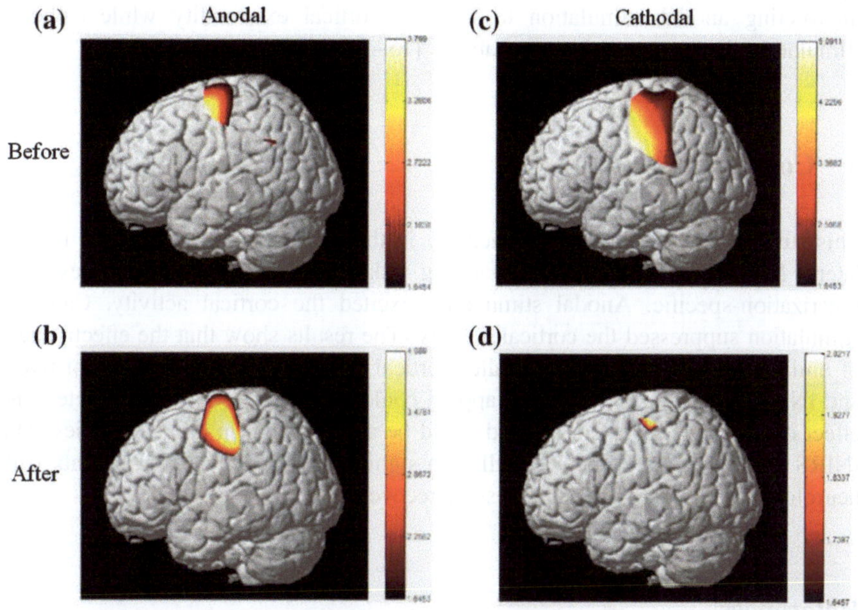

Fig. 3 The individuals activation maps ($p < 0.05$, uncorrected) from dominant arm finger tapping task before and after unilateral tDCS

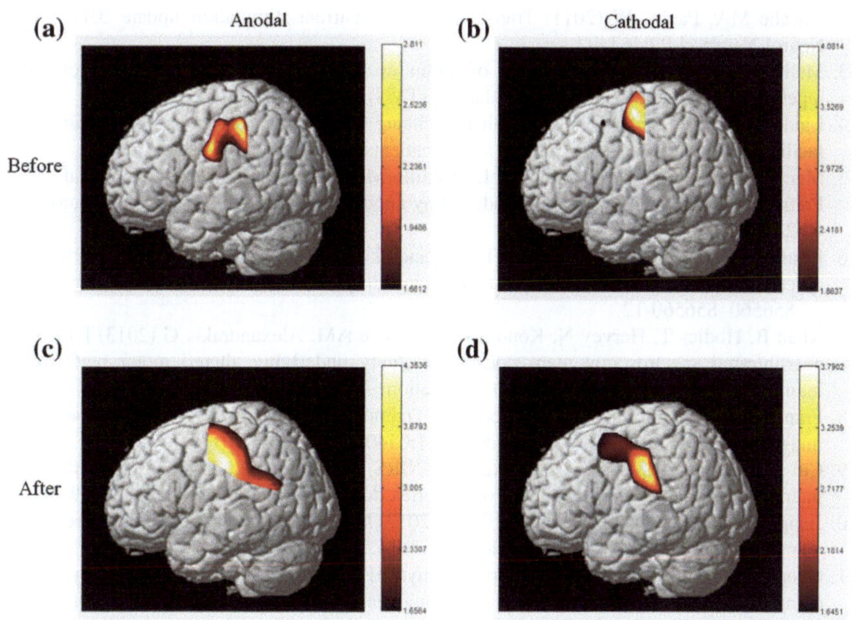

Fig. 4 The group activation maps ($p < 0.05$, uncorrected) from dominant arm finger tapping task before and after unilateral tDCS

interpreting anodal stimulation to increase cortical excitability while cathodal stimulation decreases cortical excitability [15–18].

6 Conclusions

This study compared the cortical activity in the motor cortex in response to unilateral M1 tDCS during a finger tapping task. Cortical activation changes were polarization-specific. Anodal stimulation excited the cortical activity. Cathodal stimulation suppressed the cortical activity. The results show that the effectiveness of unilateral tDCS stimulation to alter cortical activity. The combination of tDCS and fNIRS with the task of finger tapping could become a useful tool to detect the effect of tDCS. The present method could be a useful tool in clinical studies. The fNIRS combined tDCS method will be a significantly beneficial for patients with neurological disorders in the process of recovery.

References

1. Dmochowski JP, Datta A, Huang Y, Richardson JD, Bikson M, Fridriksson J et al (2013) Targeted transcranial direct current stimulation for rehabilitation after stroke. NeuroImage 75:12–19
2. Nitsche MA, Paulus W (2011) Transcranial direct current stimulation–update 2011. Restor Neurol Neurosci 29:463–92
3. Mály J, Dinya E (2008) Recovery of motor disability and spasticity in post-stroke after repetitive transcranial magnetic stimulation (rTMS). Brain Res Bull 76:388–395
4. Lindenberg R, Renga V, Zhu LL, Nair D, Schlaug G (2010) Bihemispheric brain stimulation facilitates motor recovery in chronic stroke patients. Neurology 75:2176–84
5. Merzagora AC, Foffani G, Panyavin I, Mordillo-Mateos L, Aguilar J, Onaral B et al (2010) Prefrontal hemodynamic changes produced by anodal direct current stimulation. Neuroimage 49:2304–10
6. Khan B, Hervey N, Stowe A, Hodics T, Alexandrakis G (2013) Use of functional near-infrared spectroscopy to monitor cortical plasticity induced by transcranial direct current stimulation. pp 856560–856560-12
7. Khan B, Hodics T, Hervey N, Kondraske G, Stowe AM, Alexandrakis G (2013) Functional near-infrared spectroscopy maps cortical plasticity underlying altered motor performance induced by transcranial direct current stimulation. J Biomed Opt 18:116003
8. Franceschini MA, Joseph DK, Huppert TJ, Diamond SG, Boas DA (2006) Diffuse optical imaging of the whole head. J Biomed Opt 11:054007
9. Cui X, Bryant DM, Reiss AL (2012) NIRS-based hyperscanning reveals increased interpersonal coherence in superior frontal cortex during cooperation. Neuroimage 59:2430–7
10. Jiang J, Dai B, Peng D, Zhu C, Liu L, Lu C (2012) Neural synchronization during face-to-face communication. J Neurosci 32:16064–9
11. Cooper RJ, Selb J, Gagnon L, Phillip D, Schytz HW, Iversen HK et al (2012) A systematic comparison of motion artifact correction techniques for functional near-infrared spectroscopy. Front Neurosci 6:147

12. Izzetoglu M, Devaraj A, Bunce S, Onaral B (2005) Motion artifact cancellation in NIR spectroscopy using Wiener filtering. IEEE Trans Biomed Eng 52:934–8
13. Ye JC, Tak S, Jang KE, Jung J, Jang J (2009) NIRS-SPM: statistical parametric mapping for near-infrared spectroscopy. Neuroimage 44:428–47
14. Bai X, Liu Z, Zhang N, Chen W, He B (2009) Three-dimensional source imaging from simultaneously recorded ERP and BOLD-fMRI. IEEE Trans Neural Syst Rehabil Eng 17:101–6
15. Amadi U, Ilie A, Johansen-Berg H, Stagg CJ (2013) Polarity-specific effects of motor transcranial direct current stimulation on fMRI resting state networks. Neuroimage 88C:155–161
16. Sehm B, Kipping J, Schafer A, Villringer A, Ragert P (2013) A Comparison between Uni- and bilateral tDCS effects on functional connectivity of the human motor cortex. Front Hum Neurosci 7:183
17. Kwon YH, Ko MH, Ahn SH, Kim YH, Song JC, Lee CH et al (2008) Primary motor cortex activation by transcranial direct current stimulation in the human brain. Neurosci Lett 435:56–9
18. Antal A, Polania R, Schmidt-Samoa C, Dechent P, Paulus W (2011) Transcranial direct current stimulation over the primary motor cortex during fMRI. Neuroimage 55:590–6

Modeling PolSAR Image with L-Distribution and the Parameter Estimation Method

Hao-gui Cui, Yu-zhong Jiang, Tao Liu and Jun Gao

Abstract The product model, in which the background radar signals are modeled as a product between a texture component and a Gaussian speckle, is an effective method to analyze PolSAR image especially in non-Gaussian case. In this paper, the generalized gamma distribution (GΓD) is chosen as the texture component and the PolSAR image can be modeled with L-distribution provided that the speckle is Gaussian. The parameter estimation method of L-distribution based on matrix log-cumulants (MLCs) is proposed aimed at the computational complexity of the estimation based on moment. The correctness of modeling with L-distribution and the validity of the derivation of the estimator are verified though real data.

Keywords Generalized gamma distribution (GΓD) · L-distribution · Mellin transform · Matrix log-cumulants · Polarimetric SAR

1 Introduction

The knowledge of the exact statistical model of polarimetric synthetic radar data is very important, which is also the basis of many image analysis techniques such as speckle filtering, target detection and segmentation [1]. Gaussian distribution is generally used to model the speckle of low resolution image [2–4]. With the increasing frequency of modern radar, the resolution is improved which also causes the Gaussian distribution fail to model the real images. In this case, researches shown that non-Gaussian model such as Gamma distribution, lognormal distribution and Weibull distributions give better representation of the data [5, 6]. Actually, all these three distributions can be regarded as a special case of the generalized gamma distribution function (GΓD). Meanwhile, a non-Gaussian framework called

H. Cui (✉) · Y. Jiang · T. Liu · J. Gao
School of Electronic Engineering, Naval University of Engineering,
717 Jiefang Road, Wuhan 430033, China
e-mail: seachg@163.com

E. Qi (ed.), *Proceedings of the 6th International Asia Conference on Industrial Engineering and Management Innovation*,
DOI 10.2991/978-94-6239-145-1_35

product model is well-established, in which the background radar signals are modeled as a product between a positive scalar texture component and a complex Gaussian speckle [7, 8]. We can imagine that if we modeling the texture component with GΓD [9], the product model will be more universal and will have a broader application prospects.

Another important aspect of the PolSAR image modeling is the estimation of the parameters, which is also a crucial preprocessing step of the image processing such as speckle filtering, target detection and segmentation. We have studied some traditional parameter estimation methods such as the maximum likelihood estimation and moments based estimation methods. But the problem is that both the derivation and calculation of these two estimation method are usually complicated, which makes it always hard to get the result. Mellin transform (MT) is an effective method to solve the parameter estimation problem, and the transformation based on the statistics of intensity in single polarimetric SAR has already been introduced [10–12]. For full polarimetric SAR, the common method for parameter estimation is obtained by first estimating the parameter of the copolarization channels and then averaging the estimated parameter of different channels [4, 5]. To use the correlation of different polarization channels, Anfinsen et al. introduced MT to PolSAR data and proposed an estimator via the matrix log-cumulants (MLCs) [13–15]. Actually, the nature of log-cumulants is the log-determinant moments via statistics of polarimetric covariance matrix. Doulgeris et al. then introduced this method to the PolSAR image segmentation and also got a good result [16]. Since the MLCs estimation method is effective and reliable, we also use it to estimate the parameters of our proposed L distribution model.

In this paper, we introduce and examine a polarimetric product model so-called L-distribution, which uses the GΓD as the texture component. We derive a closed-form for the pdf of L-distribution under the assumption of reciprocity and symmetry. Taking advantage of the MLCs, we also derive the expression of the MLCs and discuss how the model parameters of L-distribution can be estimated from the PolSAR data. Finally, the validity of the model and the robust of the estimation method are analyzed using both the simulated data and real PolSAR data.

2 Modeling Pol-Sar Image and Parameter Estimation

The product model has been widely used in non-Gaussian modeling. In this model, the background radar signals are modeled as a product between a positive scalar texture component and a complex Gaussian speckle. So we can see that the most important thing is the selection of scalar texture component. Gamma distribution and inverse gamma distribution are usually be used in the modeling of high resolution image [16, 17]. Here we choose the GΓD to be the model of texture component. Because the GΓD covers many known non-Gaussian distributions, we think it will have a wide range of applications.

2.1 Statistical Model of PolSAR Data

The full polarimetric SAR radar separately transmits orthogonal polarized microwaves, and measures orthogonal components of the received signal. For each pixel, the received result can be represented by a matrix of complex scattering coefficients. The scattering matrix can be expressed as

$$\mathbf{S} = \begin{pmatrix} S_{HH} & S_{HV} \\ S_{VH} & S_{VV} \end{pmatrix} \tag{1}$$

where the scattering coefficients S_{ij} are subscripted with the associated receive and transmit polarization. The subscript h denotes horizontal polarization and v denotes vertical polarization. Assuming reciprocity of the target, the cross-polarization terms S_{HV} and S_{VH} can be averaged, then the scattering matrix $\mathbf{S}$ can be reduced to one of the vector with three dimensions

$$\mathbf{s} = \begin{bmatrix} S_{HH} \\ (S_{HV} + S_{VH})/\sqrt{2} \\ S_{VV} \end{bmatrix} \tag{2}$$

the scaling with a factor $\sqrt{2}$ is done to preserve total power of the signal.

It is commonly assumed that the scattering vector elements are jointly circular complex Gaussian. Assuming $\mathbf{s}$ is zero mean and circular complex multivariate Gaussian, the PDF of $\mathbf{s}$ is

$$p_{\mathbf{s}}(\mathbf{s};\boldsymbol{\Sigma}) = \frac{1}{\pi^d|\boldsymbol{\Sigma}|}\exp\left(-\mathbf{s}^H\boldsymbol{\Sigma}^{-1}\mathbf{s}\right) \tag{3}$$

where $|\bullet|$ is the determinant operator, d is the dimension of $\mathbf{s}$, and $\boldsymbol{\Sigma} = \mathrm{E}\{\mathbf{s}\mathbf{s}^H\}$.

Multilook process usually is done in order to reduce the noise and compress the data volume, it is commonly represented by

$$\mathbf{C} = \frac{1}{L}\sum_{i=1}^{L}\mathbf{s}_i\mathbf{s}_i^H \tag{4}$$

where $\mathbf{C}$ is covariance matrix, L is the nominal number of looks and the superscript H represents complex conjugate transpose.

The scaled covariance matrix $\mathbf{Z} = L\mathbf{C}$, follows the nonsingular complex Wishart distribution [11]

$$p_{\mathbf{Z}}(\mathbf{Z};L,\boldsymbol{\Sigma}) = \frac{|\mathbf{Z}|^{L-d}}{|\boldsymbol{\Sigma}|^L\Gamma_d(L)}\exp\left(-\mathrm{tr}\left(\boldsymbol{\Sigma}^{-1}\mathbf{Z}\right)\right) \tag{5}$$

where $\mathbf{\Sigma} = \mathrm{E}\{\mathbf{Z}\}/L = \mathrm{E}\{\mathbf{C}\}$ and $\mathrm{tr}(\bullet)$ is the trace operator. It can be written as $\mathbf{Z} \sim W_d(L, \mathbf{\Sigma})$. The normalization constant $\Gamma_d(L)$ defined as

$$\Gamma_d(L) = \pi^{d(d-1)/2} \prod_{i=0}^{d-1} \Gamma(L-i) \tag{6}$$

where $\Gamma(L)$ is the standard Euler Gamma Function.

The well-known product model decomposes the vector $\mathbf{z}$ as a product of two independent stochastic variables

$$\mathbf{z} = \sqrt{\gamma}\mathbf{y} \tag{7}$$

where γ is the texture random variable and $\mathbf{y}$ represents the speckle. In the multi-look covariance matrix domain the product model becomes

$$\mathbf{Z} = \gamma_L \mathbf{W} \tag{8}$$

where $\mathbf{W} = \sum_{i=1}^{L} \mathbf{w}_i \mathbf{w}_i^H \sim W_d(L, \mathbf{\Sigma_w})$ and $\mathbf{Z} = \sum_{i=1}^{L} \mathbf{z}_i \mathbf{z}_i^H$.

From Eq. (8) we can see the PDF of $\mathbf{Z}$ depends on the multilook texture γ_L, which is related to γ. Here we assume that γ obeys the generalized gamma distribution (GΓD), the PDF is

$$f(\gamma) = \frac{|p|}{a\Gamma(v)} \left(\frac{\gamma}{a}\right)^{pv-1} \exp\left(-\left(\frac{\gamma}{a}\right)^p\right), a > 0, p \neq 0, v > 0, x \geq 0 \tag{9}$$

where $\Gamma(\cdot)$ is the Gamma Function, p is the slope factor, v is the intensity factor and a is the scale factor. There are some special cases for GΓD. When $p = -1$, this distribution is simplified to be the Inverse Gamma distribution. When $p = 1$, this distribution is simplified to be the Gamma distribution. When $v = -1$, this distribution is simplified to be the Weibull distribution.

The texture component is usually assumed to have a unit average, then we arriving at the following equitation

$$a = \Gamma(v)/\Gamma\left(v + \frac{1}{p}\right) \tag{10}$$

Using Eq. (9) in Eq. (8), we can get the PDF of $\mathbf{Z}$ is

$$f_{\mathbf{Z}_C^{(L)}}(\mathbf{Z}) = \frac{2|p|L^{Ld}|\mathbf{Z}|^{L-d}}{a^{Ld}\Gamma_d(L)|\mathbf{C}|^L\Gamma(v)} L\left(p, pv - Ld, \frac{L\mathrm{Tr}(\mathbf{C}^{-1}\mathbf{Z})}{a}\right) \tag{11}$$

where the distribution of texture is GΓD, and $L(p,q,b) = \frac{1}{2}\int_0^\infty t^{-q-1}\exp(-t^{-p} - bt)dt$. This distribution is named L-distribution. When $p = 1$, the PDF of L-distribution is simplified to be the K-distribution.

2.2 Complex Matrix Variable MLCs

The Complex matrix variable MT is defined as [15]

$$\phi_{\mathbf{Z}}(s) = M\{f(\mathbf{Z})\}(s) = \int_{O+} |\mathbf{Z}|^{s-d} f(\mathbf{Z})d\mathbf{Z} \tag{12}$$

where $|\bullet|$ is the determinant operator, and $f(\mathbf{Z})$ is a real-valued scalar function defined on a cone of $d \times d$ Hermitian matrices that are positive define. The vth-order MLCs is defined as

$$k_v(\mathbf{Z}) = \left.\frac{d^v(\log(\phi_{\mathbf{Z}}(s)))}{ds^v}\right|_{s=1} \tag{13}$$

From Eqs. (9) and (8), we can derive the MLCs of GΓD RV is

$$\begin{cases} k_1 = \frac{\psi(0,v)}{p} + \ln a \\ k_j = \frac{\psi(j-1,v)}{p^j} \end{cases} \tag{14}$$

where $\psi(j-1,v) = \frac{d^j \ln\Gamma(v)}{dv^j}$.

For Eq. (8), it has been proven that [15]

$$k_v(\mathbf{Z}) = d^v k_v(\gamma_L) + k_v(\mathbf{W}) \tag{15}$$

The MLCs of a complex Wishart matrix $\mathbf{W}$ is

$$\begin{cases} k_1\{\mathbf{Z}\} = \psi_d(0,L) + \ln|\mathbf{C}| \\ k_j = \psi_d(j-1,L), \quad j > 1 \end{cases} \tag{16}$$

From Eq. (14), (15) and (16), we can get the MLCs of $\mathbf{Z}$ which can be defined as

$$\begin{cases} k_1\{\mathbf{Z}\} = \psi_d(0,L) + \ln|\mathbf{C}| - d(\ln L - \frac{\psi(0,v)}{p} - \ln a) & j = 1 \\ k_j\{\mathbf{Z}\} = \psi_d(j-1,L) + \left(\frac{d}{p}\right)^j \psi(j-1,v) & j > 1 \end{cases} \tag{17}$$

where $\psi_d(j-1,L) = \sum_{i=0}^{d-1} \psi(j-1, L-i)$.

2.3 Parameter Estimation for L-Distribution Based on MLCs

The moments based estimation method for L-distribution has been given by Frery et al., but the calculation was quite complicated [5]. In order to get a fast and accurate estimation method, here we derive the MLCs estimation method. Using the second order and the third order MLCs, Eq. (17) becomes

$$\begin{cases} \tilde{k}_2\{\mathbf{Z}\} = k_2\{\mathbf{Z}\} - \psi_d(1,L) = (\frac{d}{p})^2\psi(1,v) \\ \tilde{k}_3\{\mathbf{Z}\} = k_3\{\mathbf{Z}\} - \psi_d(2,L) = (\frac{d}{p})^3\psi(2,v) \end{cases} \tag{18}$$

Equation (18) leads to the estimators for p and v

$$\tilde{k}_2^3\{\mathbf{Z}\}/\tilde{k}_3^2\{\mathbf{Z}\} = \{\psi(1,v)\}^3/\{\psi(2,v)\}^2 \tag{19}$$

$$p = \frac{-\mathrm{sgn}(\tilde{k}_3^2\{\mathbf{Z}\})d}{\sqrt{\tilde{k}_2\{\mathbf{Z}\}/\psi(1,v)}} \tag{20}$$

where sgn(.) is the sign function. From Eq. (19) we can get the value of v. Then using this value of v in Eq. (19), we can obtain the value of p. At last, the value of $|\mathbf{C}|$ can be obtained from the first order MLCs expression in Eq. (17), as follows

$$|\mathbf{C}| = \frac{\Gamma\left(v + \frac{1}{p}\right)L\exp\left(k_1\{\mathbf{Z}\} - \psi(0,L) - \frac{\psi(0,v)}{p}\right)}{\Gamma(v)} \tag{21}$$

3 Discussion

In the experiment we use the real data to verify the validation of the L-distribution model and the robustness of the parameter estimation method. Here the real data are drawn from the software of PolSARPro 4.2 supplied by NASA/JPL AIRSAR, the nominal number of looks is L = 4(2 × 2). The real data is from a PolSAR image of Flevoland. We firstly estimates the value of v and p using Eq. (19) and (20), then gets the value of from Eq. (21).

KS test is one of the most commonly used goodness of fit tests, which is of great importance in testing whether observed data fit a given probability distribution [6]. Here we use this test to determine whether the observed distribution from three

Table 1 The table of KS distance with the signigicane level α

α	0.25	0.15	0.10	0.05	0.025	0.01	0.005
T_{KS}	1.019	1.138	1.224	1.358	1.480	1.628	1.731

Fig. 1 The three regions of Flevoland for hypothesis testing (Pauli RGB image)

regions of Flevoland can be approximated by the proposed L-distribution. Table 1 shows the threshold value of KS test with the significance level. As shown in Fig. 1, the three regions are samples of bare surface, river and vegetation. The number of samples in each region is 625(25 × 25).

Figure 2 gives the KS test results of HH polarization channel, and it shows that the CDF of both observed distribution are closed to the proposed L-distribution. The estimated parameters and KS distance of three regions of Flevoland is given in Table 2. And it shows that all the calculated values of KS distance is smaller than T_{KS} at $\alpha = 0.01$. So it can be concluded that the samples from three regions of Flevoland is well modeled by L-distribution.

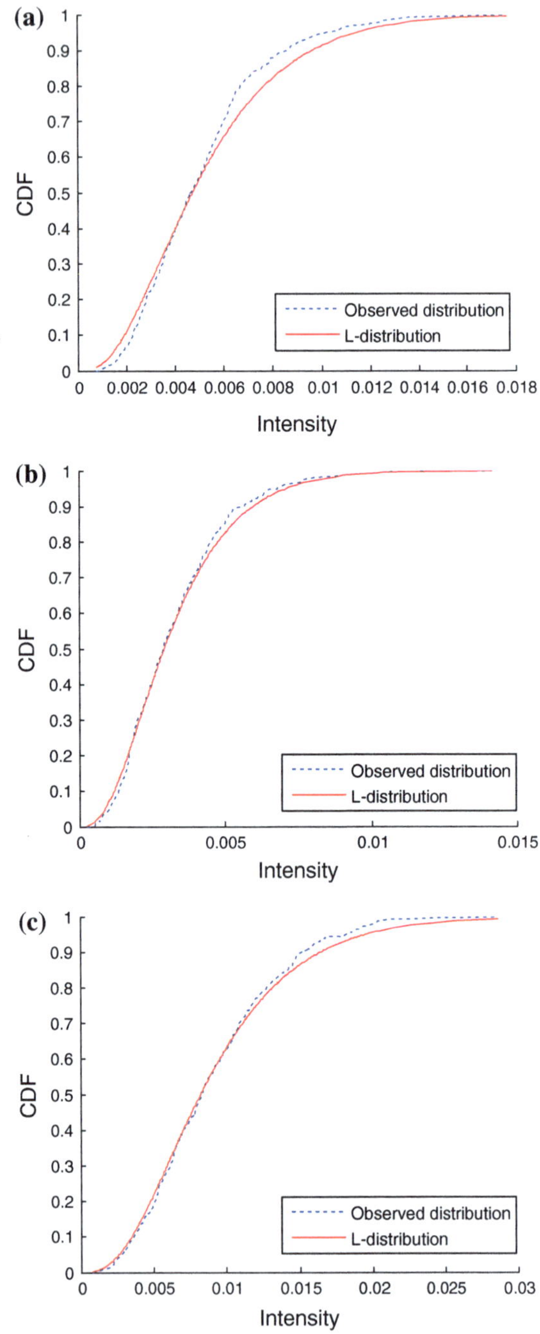

Fig. 2 The CDF of the observed distribution and the proposed L-distribution. **a** Region A of Flevoland. **b** Region B of Flevoland. **c** Region C of Flevoland

Table 2 The estimated parameters and the KS distance of the three regions of Flevoland

	$\lvert\mathbf{C}\rvert$	p	v	KS test distance
Bare surface (A)	HH:0.0016	1413.4	0.001	1.3483
	HV:0.0001			1.5364
	VV:0.0029			0.9876
Vegetation (B)	HH:0.0022	1.0441	82.6550	0.8657
	HV:0.0005			1.2380
	VV:0.0048			0.9470
River (C)	HH:0.0056	2.5759	1.1060	0.8893
	HV:0.0001			1.0235
	VV:0.0093			0.6380

4 Conclusion

In this paper, we have proposed the L-distribution to model the PolSAR image. This distribution is derived based on the well-known product model and using the generalized gamma distribution as the texture component. The parameter estimation method for the L-distribution based on the MLCs is derived, which is faster and more accurate than the moments based estimation method. Experiments with real data from AIRSAR instrument demonstrate the utility of the tests.

Acknowledgments This work was supported by the National Natural Science Foundation of China under Grant 61372165. The authors are also grateful to European Space Agency for providing the PolSAR data.

References

1. Tur M, Chin KC, Goodman JW (1982) When is speckle noise multiplicative? Optics 21:1157–1159
2. Liu T, Wang XS, Xiao SP (2008) Statistical characteristics of the normalized Stokes parameters. Sci China Ser F Inf Sci 51(10):1594–1606
3. Liu T, Huang GM, Wang XS, Shao SP (2010) Statistical analysis of the H/A target decomposition theory. Sci China Ser F Inf Sci 40(1):102–114
4. Liu T, Wang XS, Shao SP (2008) Statistical of the Stokes parameters for multilook SAR image. J Electron Inf Technol 30(5):1027–1031 (in Chinese)
5. Frery AC, Muller HJ, Yanasse CCF (1997) A model for extremely heterogeneous clutter. IEEE Trans Geosci Remote Sens 35(3):648–659
6. Liu T, Huang GM, Wang XS, Shao SP (2009) Statistics of the polarimetric Weibull-distributed electromagnetic wave. IEEE Trans Antennas Propag 57(10):3232–3248
7. Li HC, Hong W, Wu YR, Fan PZ (2010) An efficient and flexible statistical model Based on generalized gamma distribution for amplitude SAR images. IEEE Trans Geosci Remote Sens 48(6):2711–2722
8. Liu T, Huang GM, Wang XS, Shao SP (2009) Modeling polarimetric SAR images with L-distribution. Sci Found China 19(1):89–96 (in Chinese)

9. Song K (2008) Globally convergent algorithms for estimating generalized gamma distributions in fast signal and image processing. IEEE Trans Image Process 17(8):1233–1250
10. Freitas CC, Frery AC, Correia AH (2005) The polarimetric distribution for SAR data analysis. Environmetrics 16(8):13–31
11. Krylov V, Moser G, Serpico SB, Zerubia J (2011) On the method of logarithmic cumulants for parametric probability density function estimation, inria-00605274, version 2, N RR-7666, Aug 2011
12. Nicolas JM (2002) Introduction to second kind statistics: application of log-moments and log-cumulants to analysis of radar images. Traitement du Signal 19(3):139–167 (in French)
13. Anfinsen SN, Doulgeris AP, Eltoft T (2009) Estimation of the equivalent number of looks in polarimetric synthetic aperture radar imagery. IEEE Trans Geosci Remote Sens 47 (11):3795–3809
14. Anfinsen SN, Eltoft T (2009) Application of the matrix-variate Mellin transform to analysis of polarimetric Radar images. IEEE Trans Geosci Remote Sens 49(6):2281–2295
15. Anfinsen SN, Doulgeris AP, Eltoft T (2011) Goodness-of-fit tests for multilook polarimetric radar data based on the Mellin transform. IEEE Trans Geosci Remote Sens 49(7):2764–2781
16. Bateman H (1954) Tables of integral transforms, McGraw-Hill Book Company, New York, pp 298–303
17. Goodman NR (1963) Statistical analysis based on a certain complex Gaussian distribution. Ann Math Stat 34:152–177

Part II
Automation System Theory and Practice

Optimization of Reactive Power for Dual Stator-Winding Induction Generator System

Yong-hua Jiang, Ting Wang and Ling-shun Liu

Abstract The research of induction generator supply system based on static excitation regulator in the fields of aircraft, tank, automobile and renewable energy source has been paid more attention recently. In this paper, the analysis on the reactive power of control winding of the dual stator-winding induction generator based on static excitation regulator under different working conditions is studied. The change laws of reactive power caused by the varying of operating conditions are presented. The reduction to reactive power of exciting winding is the key to the optimization of the whole system. Thus, the optimal strategy to reduce reactive power of exciting winding is proposed which is mainly used to optimize the parameters and operational region of generator. This will provide the guidance and theoretical foundation to the optimal design of generator and system.

Keywords Dual stator-winding induction generator · Optimization of parameters · Operating region · Reactive power

1 Introduction

In recent years, with the increase of electrical equipments and development of automation and advancing of the concept about More Electric Aircraft and All Electric Aircraft, the capacity and quality factor of aircraft power systems have been required to improve more greatly. Furthermore, the power supply system has the trend from low voltage DC to high voltage DC in tank power system. Thus, it is necessary to study 270 V high voltage DC power supply system [1, 2].

Y. Jiang (✉)
Department of Electronic Engineering, Yantai Vocational College, Yantai, China
e-mail: jiangyh89@126.com

T. Wang · L. Liu
Department of Control Engineering, Naval Aeronautical and Astronautical University, Yantai, China

E. Qi (ed.), *Proceedings of the 6th International Asia Conference on Industrial Engineering and Management Innovation*,
DOI 10.2991/978-94-6239-145-1_36

It is well–known that the induction machine has many unique advantages, such as solid structure, high reliability, low cost, high power density, good protection for over loading and short circuit, which people should pay more attention on the fields above [3–6]. The integration of electrical power in stand-alone electrical power system and the rapid development of control technology have made high power density generator system become the trend. The static excitation regulator composed by the power semiconductor replacing the capacitor in the induction generator system may regulate the reactive power continuously [7–9].

The novel dual stator-winding induction generator system based on static excitation inverter which has been put forward in recent years can overcome those drawbacks existing in the traditional three-phase induction generator system above [10–13]. It can provide a new approach for the development and application of aircrafts high DC power supply.

This system is presented schematically in Fig. 1. There are two windings in the stator, one is power winding, which is connected to rectified load. The other one is control winding, or exciting winding, which is connected to static excitation inverter with low volume and weight. This is because it only supplies reactive power without active power. The two windings have the same poles, so that they have the same working frequency. Both of the windings do not have electrical connection but they have magnetic coupling. Therefore, the harmonic from the inverter of control winding has little disturbance on the load, and the electromagnetic compatibility and efficiency have been improved more. Simultaneously, because of some leakage inductance existing in the control winding, it connects to no or low filter inductor. The reliability of the system will be much higher than that of three-phase induction generator system [10–13].

The minimum volume and weight should meet the need of aviation power supply. The compensating capacity demanded by the system is greatly relative to generator parameters, load and the rotating speed of the prime motor. Thereby, the operational range of system should be taken into account when designing the inverter and control winding. How to make the capacity of control winding minimum in the given speed range, accordingly make the inverter's volume smallest to match the feature of aviation, is the key to the optimization of dual stator-winding induction generator system. In this paper, the capacity equation of control winding

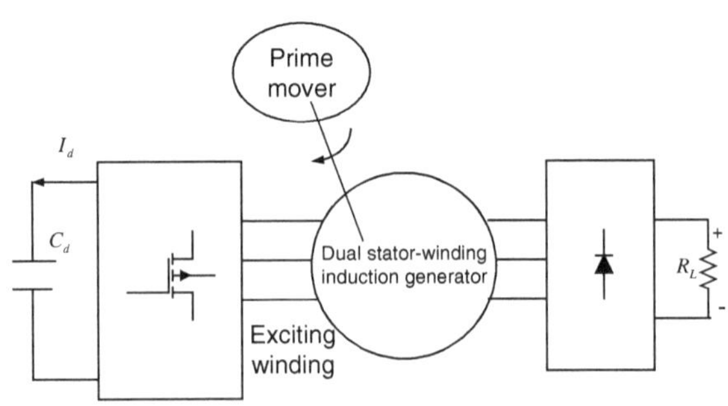

Fig. 1 A novel dual stator-winding induction generator system based on the static excitation regulator

under resistive load is deduced through the equivalent circuit and vector diagram of dual stator-winding induction generator. The relations between the reactive power and the generator's parameters, load and the rotating speed of the prime motor are analyzed. By the means of the optimization of generator's parameters and the working frequency, the determined method of operational ranges to reduce the reactive power of control winding under different rotating speed ratio conditions demanded by system is presented. It will provide the guidance and theoretical foundation to generator optimization in the need of system's performance.

2 Theoretical Analysis of the Reactive Power of Control Winding

2.1 Calculation of the Capacity of Control Winding

In order to analysis the capacity of control winding, such suppositions are given:

(1) The losses of magnetic hysteresis and eddy current are ignored
(2) Control winding only supplies reactive current, which has the same phase as that of exciting current, this can be realized through proper control method
(3) The resistive load is connected to the uncontrolled rectifier bridge while other load can be got by similar method

The equivalent circuit and vector diagram is presented as Fig. 2. Where F is output frequency.

From this equivalent circuit, the exciting current $\dot{I}_m$ can be expressed as bellow

$$\dot{I}_m = \dot{I}_r + \dot{I}_p + \dot{I}_s \tag{1}$$

According to the vector diagram, the I_s can be deduced

$$I_s = I_m + I_r \sin\theta + I_p \sin\gamma \tag{2}$$

where

$$\sin\theta = \frac{FX_{lr}I_r}{E_m},\ \sin\gamma = \frac{FX_{lp}I_p}{E_m} \tag{3}$$

From Eqs. (2) and (3), I_s can be derived as follows

$$I_s = \frac{E_m}{FX_m} + \frac{FX_{lr}I_r^2}{E_m} + \frac{FX_{lp}I_p^2}{E_m} \tag{4}$$

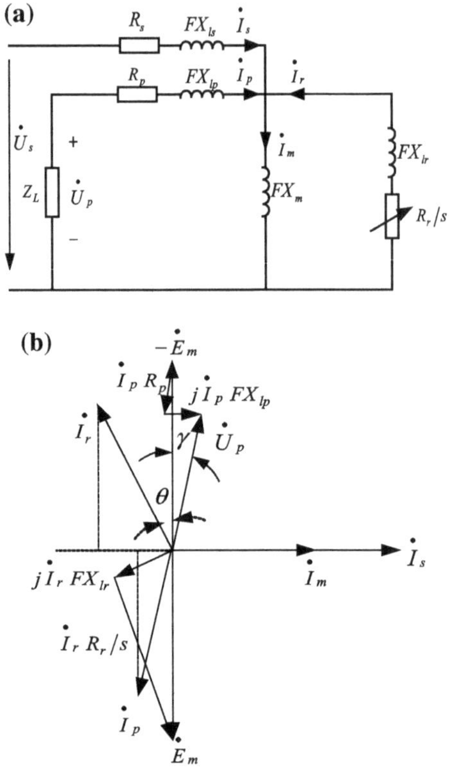

Fig. 2 **a** Equivalent circuit. **b** Vector diagram of dual stator-winding induction generator

Where

$$E_m = \sqrt{(U_p + I_p R_p)^2 + (I_p F X_{lp})^2} \quad (5)$$

$$I_r = \sqrt{\frac{E_m^2}{(\frac{R_r}{s})^2 + F^2 X_{lr}^2}} \quad (6)$$

$$I_m = \frac{E_m}{F X_m} \quad (7)$$

Then the current of control winding can be expressed as

$$I_s = \frac{\sqrt{(U_p + I_p R_p)^2 + (I_p F X_{lp})^2}}{F X_m} + \frac{\sqrt{(U_p + I_p R_p)^2 + (I_p F X_{lp})^2} \cdot F X_{lr}}{(\frac{R_r}{s})^2 + F^2 X_{lr}^2} + \frac{F X_{lp} I_p^2}{\sqrt{(U_p + I_p R_p)^2 + (I_p F X_{lp})^2}} \quad (8)$$

Considering that output voltage U_p of power winding should be constant during its operation, to reduce I_s and improve the power factor of system, the parameters of generator must be optimal designed. Concretely to say, leakage reactance X_{lp}, resistance R_p of power winding and leakage reactance X_{lr} of rotor should be reduced. While exciting reactance X_m should be increased, proper magnetic saturation operational region is chosen. Where the influence of X_m is more prominent in low speed region, that of leakage reactance is greater in high speed region.

The capacity ratio between control winding and power winding in the rated state is expressed as

$$K = \frac{U_s I_s}{U_p I_P} \quad (9)$$

According to the principle of mechanical energy converting to electrical energy, the electromagnetic power transferred is expressed as

$$P = -3 I_r^2 \frac{R_r}{s} \quad (10)$$

Where

$$P = P_N + \Delta P \quad (11)$$

P_N is rated power, slip s is negative, ΔP is the loss of machines. Substituting (6) in (10), it can be achieved

$$P = -3 \cdot \frac{E_m^2}{(\frac{R_r}{s})^2 + X_r^2} \cdot \frac{R_r}{s} \quad (12)$$

Thus $P X_r^2 s^2 - 3 E_m^2 R_r s + P R_r^2 = 0$

Finally, slip s can be deduced as follows

$$s = \frac{-3E_m^2 R_r + \sqrt{9E_m^4 R_r^2 - 4P^2 R_r^2 X_r^2}}{2PX_r^2} \quad (13)$$

Substituting (13) in (6), I_r can be solved, then Substituting it in (3), $\sin\theta$ and $\sin\gamma$ can be got, finally Substituting them in (8), the value of I_s can be calculated under different working conditions.

2.2 Operational Analysis of Different Speed Under Constant Rated Power Conditions

Using the equations stated in the section above, operational analysis of different rotor speed under constant rated power condition is given.

Its parameters are as follows: $P_N = 18\,\text{kW}$, $U_{NP} = 115\,\text{V}$, $R_S = R_P = 0.068\,\Omega$, $R_r = 0.037\,\Omega$, $L_m = 8.73\,\text{mH}$, $L_{ip} = L_{is} = 0.283\,\text{mH}$, $L_{ir} = 0.187\,\text{mH}$, $f = 133.33\,\text{Hz}$, $n_p = 2$ the magnetic saturation is ignored. The results are showed as Fig. 3.

From Fig. 3a, the slip s of the generator is almost not varying with the change of rotating speed under rated load. From Fig. 3b, the phase current of power winding keeps constant. The rotor current is also constant, and the exciting current is inversely reducing with the increase of rotating speed. However, the current of control winding is presenting a concave shape curve with descending firstly and ascending afterward. This is because that the reactive components of both rotor current and power winding current are ascending with the ascent of rotating speed, yet the exciting current is descending with that. From Fig. 3c, the trend of exciting power is descending with the speed ascending. However, the leakage reactance proportionally increases with the rotating speed, thus reactive power of leakage reactance increases accordingly. The total reactive power also shows a concave shape curve.

2.3 Operational Analysis of Variable Load Under Rated Rotating Speed Condition

The calculating results of variable load conditions from no load to rated load at rated rotating speed are presented as Fig. 4.

From Fig. 4a, the slip increases and output frequency changes with the load varying under the rated rotating speed. From Fig. 4b, c, reactive power of exciting reactance increases slowly because the reactive current varies slowly. The current of power winding and rotor increases greatly with that of load, and the reactive power of leakage reactance increases greatly. Therefore, the total reactive power and the current of control winding is presenting ascent.

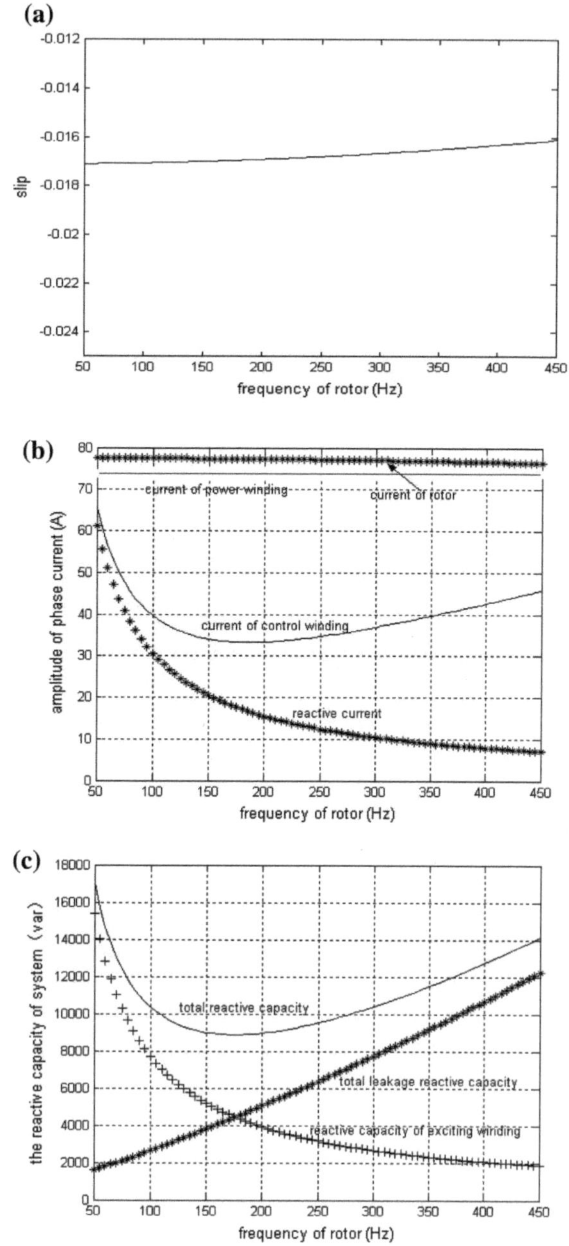

Fig. 3 Operational laws of different speed of rotor under constant rated power conditions. **a** Slip change with the change of frequency of rotor. **b** Amplitude of phase current change with the change of frequency of rotor. **c** The reactive capacity of system with the change of frequency of rotor

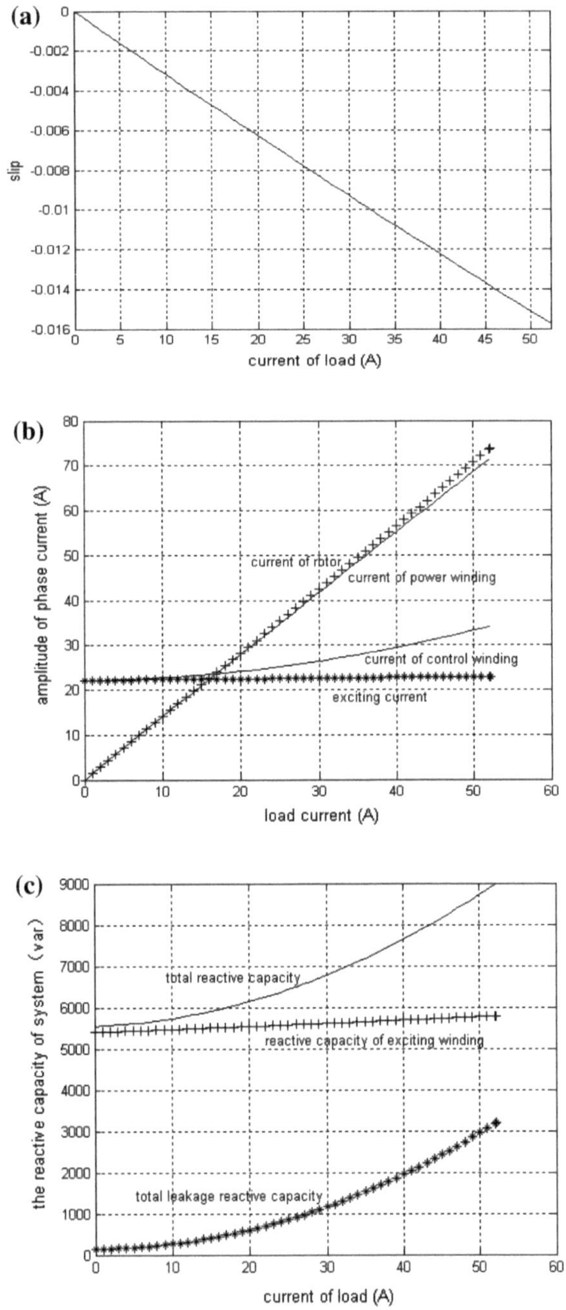

Fig. 4 Calculation results of different load at rated rotating speed. **a** Slip change with the change of load current. **b** Amplitude of phase current change with the change of load current. **c** The reactive capacity of system with the change of load current

2.4 Summarization to Operational Characteristic of System

The conclusions can be drawn from the results of the two working conditions above:

(a) To ensure the normal operation of the system, the choice of reactive power of control winding should be the maximum value throughout the whole operational region, as shown on Fig. 5. If the normal operational region is $f_N \ldots N$ (or A–B), where N is the rotating speed ratio of system. The current of control winding at A is lower, but that at B is larger. For example, the speed ratio is 1:3 in some aviation system, just as the analysis above, the capacity ratio between control winding and power winding is 0.38 at A, whereas it is 0.612 at B. Thus the capacity, volume, weight and current varying range of the inverter will be large because of 0.612 at B. This will not be benefit for the application in the aviation system and the control for inverter, so some measures should be taken to reduce the capacity of control winding.

(b) The current of control winding shows a concave shape curve which is descending firstly and ascending afterward under the constant power operational condition. Thus, how to choose the proper speed region to make the system operate well in the case of minimizing current of control winding is highly worth studying. It can be considered to make the system operate in the region of $f' \cdots N \cdot f'$ (from C to D, Fig. 5).It is easy to find that i_s at C is almost the same as that at D, and it is evidently less than i_s at B. Thereby, the maximum value and varying range of i_s in the whole running region is reduced greatly. The capacity ratio of inverter can be less than 0.612 of above example by optimal calculation. So it is benefit for the control of the system.

(c) The main factors that affect the reactive power of system are exciting reactance and leakage reactance. Larger exciting reactance and lower leakage reactance are more benefit to reduce the reactive power of the system.

(d) According to Eq. (10), it is advisable to reduce the current of switch device by choosing the larger bus rated voltage of inverter properly than that of power winding in engineering. This method is avail to reduce rated current of inverter.

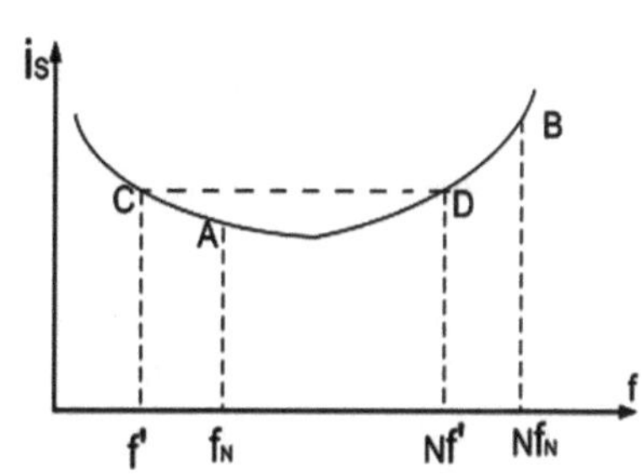

Fig. 5 Regulation of the working region

As a conclusion, the reactive power has much relation with speed, load and parameters of generator. How to make reactive power minimum and the inverter easy to control via choosing proper operational region and optimizing parameters of generator are necessary to research in the future.

3 Optimal Calculation of Reactive Power

Optimal strategy can be adopted based on the formulas and equivalent circuit in part 2.

The basic idea of optimizing reactive power is as follows. Some optimal method is applied to seek for the proper speed region and parameters of machine. As it works in the region of $f' \cdots N \cdot f'$ ($C \cdots D$, presented in Fig. 5), the current of control winding and its variable range are both minimum. Consequently, the minimization of reactive power of inverter can be realized.

Determination of objective function: Making the sum of currents of control winding at f' and $N \cdot f'$ minimum can be acted as objective function. Where N is ratio of rotating speed, it is usually 1:3 or 1:2 in aviation supply system, and larger in automobiles.

Choice of optimal variables: Optimal variables include such seven variables as interior diameter of stator, length of iron core, depth and width of stator and rotor slot, length of gap.

Determination of constraint functions: The constraint functions include performance constraints and variable constraints. Such performances as AJ, J_1, J_2, B_g, B_t, B_j should be considered. Variable constraints are determined to make generator work in proper saturated state [14, 15].

Choice of Optimal algorithm: The objective function is not expressed directly by variables. Thus, the compound form method which needs not derivation to the objective function is chosen.

3.1 Optimal Results of Rotating Speed Ratio of 1:3

After 334 times optimal alternate calculating, the results are as follows:

$$f = 93.42\,\text{Hz}, L_m = 11.946\,\text{mH},\ L_{ls} = 0.249\,\text{mH},$$

$L_{lp} = 0.249\,\text{mH}$, $L_{lr} = 0.1679\,\text{mH}$, the optimal results under rated load are showed as Fig. 6.

The current amplitude of control winding at $f = 93.42\,\text{Hz}$ and $3f = 280\,\text{Hz}$ are $i_s = 34.35\,\text{Hz}$ and $i_s = 35\,\text{A}$ respectively. Now the ratio to the current of power winding is 0.465, yet before optimization the ratio at $3f_N$ is 0.612, which is the

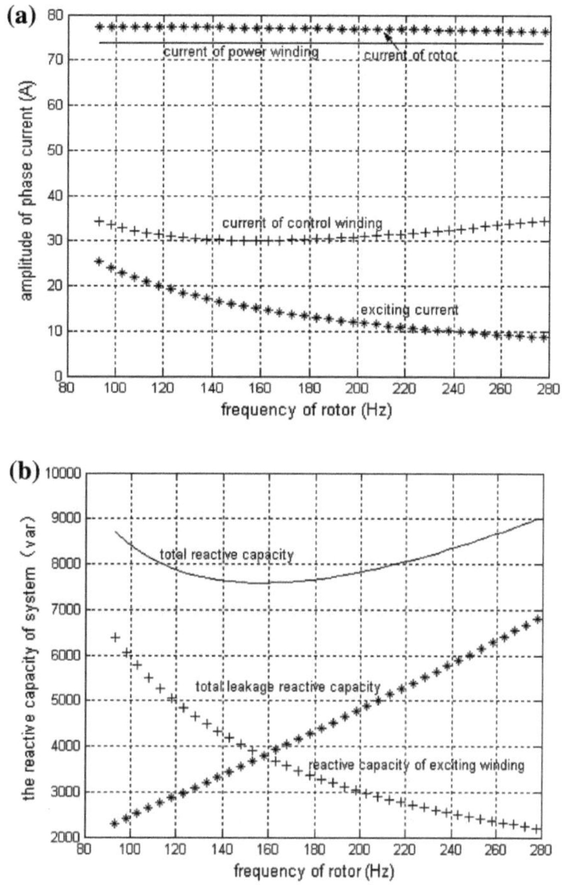

Fig. 6 Optimal results at rotating speed ratio of 1:3. **a** Amplitude of phase current change with the change of frequency of rotor. **b** The reactive capacity of system with the change of frequency of rotor

largest among the whole operational region (presented as Fig. 3). Therefore the capacity, volume, weight and current change of inverter connected with control winding are reduced, which makes inverter easier to control.

3.2 Optimal Results of Rotating Speed Ratio of 1:5

After 371 times optimal alternate calculating, the results are as follows: $f = 75.44$ Hz, $L_m = 11.483$ mH, $L_{ls} = 0.277$ mH, $L_{lp} = 0.277$ mH, $L_{lr} = 0.1653$ mH, optimal results under rated load are showed as Fig. 7. Now the current amplitude of control winding at $f = 75.44$ Hz and $5f = 377.2$ Hz are $i_s = 38.59$ A and $i_s = 39.95$ A respectively. The ratio to the current of power winding is 0.5412, yet before optimization the ratio at the $5f_N$ is 0.824, which is the largest among the whole operational region. Thus the similar effect as that of rotating speed ratio of 1:3 is achieved.

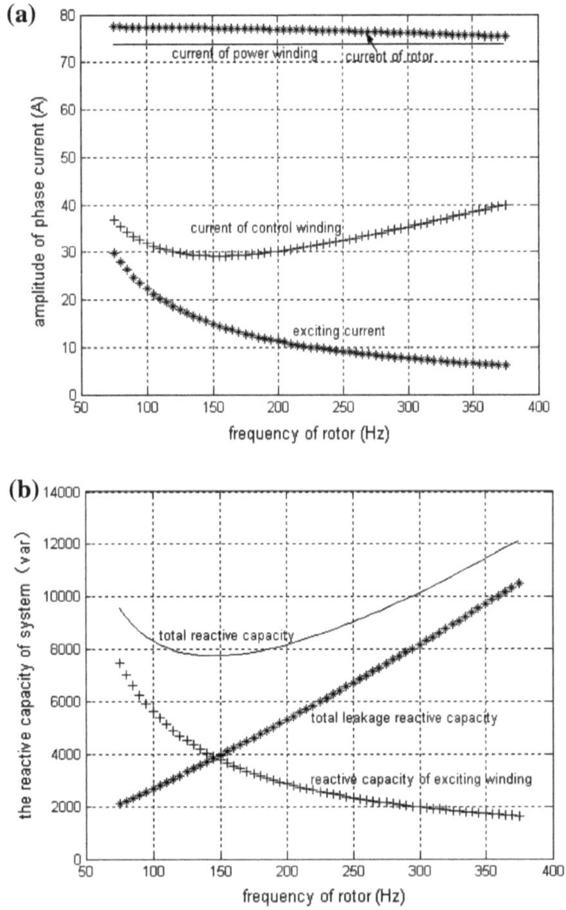

Fig. 7 The optimal results at rotating speed ratio of, **a** Amplitude of phase current change with the change of frequency of rotor. **b** The reactive capacity of system with the change of frequency of rotor

4 Conclusion

In this paper, the capacity calculation of control winding of dual stator-winding induction generator by analytic method under resistive load is analyzed. A general calculation equation is deduced. Moreover, the capacity of system under variable rotor speed and variable load conditions is calculated respectively. The changing laws of reactive requirement of the system are introduced. The effects on the reactive power by parameters and rotor speed of generator are presented. It is much useful for adjusting parameters of generator optimal design.

A novel strategy to reduce reactive power of control winding is presented. That is to optimize the operational region to minimize the reactive power of control winding via adjusting parameters and rotor speed region in the case of matching the rotating speed ratio of system. In this way, the capacity of inverter can be reduced apparently.

The research conclusions and optimal idea can be utilized as the guidance and theoretical foundation for the optimal design of generator and system.

References

1. Huang W (2002) Research on high voltage DC generation system based on squirrel-cage induction generator and power electronic converter. Dissertation for P.h.d., Nanjing University of Aeronautics and Astronautics
2. Naidu M, Walters J (2003) A 4-kW 42-V induction-machine-based automotive power generation system with a diode bridge rectifier and a PWM inverter. IEEE Trans Ind Appl 39 (5):1287–1293
3. Singh GK (2004) Self-excited induction generator research-a surve. Electr Power Syst Res 69:107–114
4. Bansal C, Bhatti TS, Kothari DP (2003) Bibliography on the application of induction generator. Nonconventional Energy Syst IEEE Trans Energy Convers 18(3):433–438
5. Simoes MG, Farret FA (2004) Renewable energy systems design and analysis with induction generators. CRC Press, Boca Raton
6. Mathur HD (2004) Discussion of “Bibliography on the application of induction generators in nonconventional energy systems”. IEEE Trans Energy Convers 19(3): 650–650
7. Bansal RC, Bhatti TS, Kothari DP (2004) Closure on “Bibliography on the application of induction generators in nonconventional energy systems”. IEEE Trans Energy Convers 19 (3):650–651
8. Muljadi E, Lipo TA (1994) Series compensated PWM inverter with battery supply applied to isolated induction generator. IEEE Trans Industry Appl 30(4):1073–1082
9. Singh B, Shilpakar LB (1998) Analysis of a novel solid state voltage regulator for a self-excited induction generator. IEE Proc Gener Trans Distrib 145(6):647–655
10. Wang D, Ma W (2003) Research on dual stator-winding induction generator generator with static excitation regulator. Proc CSEE 23(7):145–150
11. Ojo O, Davidson IE (2000) PWM—VSI inverter-assisted stand-alone dual stator winding induction generator. IEEE Trans Ind Appl 36(6):1604–1611
12. Liu Lingshun H, Yuwen H, Wenxin H (2006) The application of improved genetic algorithm in optimal design of dual stator-winding induction generator with PWM converter. Trans Nanjing Univ Aeronaut Astronaut 23(3):185–194
13. Schiferl RF, Ong CM (1983) Six phase synchronous machine with AC and stator connections. IEEE Trans Power Apparatus Syst 102(8):2685–2701
14. Chen SQ (2002) Electrical machine design. The Publishing House of Machine Industry, Beijing
15. Ye Y, Lin Y (1998) Computer aided electrical machine optimal design and manufacture. Public House of Zhejiang University, Hangzhou

Research on Distributed Networks of the Distribution of Agricultural Products at Minimum Cost

Guang-shu Xu and Hua-lin Sun

Abstract This paper is focused on the design of logistics networks for a variety of agricultural products, including packaging and distribution under the assumption of uncertain of the customers' demands. It discussed how to obtain the optimized distribution volume, and furthermore, with which as the model input factor to seek the best design of the distribution network, by determining the distribution plan with the goal of cost minimization.

Keywords Agricultural products · Distribution cost · Distributed delivery networks

1 Introduction

Technical progress and diversification of consumer's demands makes the products' life cycles shorter and shorter, so it's a crucial factor of to respond to customers' needs rapidly. Kopczak [1] proposed that products' assembly process should be put into the markets where customers are most demand-intensive. And by establishing distributed delivery system, the rapid-respond to demand could be enhanced thus the entire supply chain can become more competitive.

It is distributed delivery system that locates products of assembly in the areas with intensive customers, and when customers' demands are determined, suppliers will send product components to the assembly center, and then the finished goods will be delivered to customers. For example, characteristic agricultural products are

Key project of Philosophy and Social Science Research in Beijing (09AbJG299); Key project of training plan of Beijing young top-notch talent (CIT&TCD201304109).

G. Xu (✉) · H. Sun
School of Logistics, Beijing Wuzi University, Beijing 101149, China
e-mail: guangshuxu@126.com

E. Qi (ed.), *Proceedings of the 6th International Asia Conference on Industrial Engineering and Management Innovation*,
DOI 10.2991/978-94-6239-145-1_37

produced in different places, and then they are delivered to the distribution center for packing and sorting, and finally packed specifically according to the voice of customers.

Currently, packaging is the most common service for distributed delivery system. Based on the necessity of transportation safety, according to the sales strategies manufacturers deliver the bulk products from their major factory to distribution centers, package and label the goods, including some distribution processing.

In a general distribution network of finished products, the volume of product's inflow and outflow in distribution center are balanced. However, owing to the different forms of multiple quantities in variety of products to be assembled, an imbalance between the inflow of components that enter into distribution center and the outflow of finished products certainly exists. So optimizing a distributed distribution network is always more complex than a general distribution network.

For a long time, the design of the distribution network is a very important research content in the field of logistics. Quantitative methods of network design have been further studied at home and abroad. The theory and practice of single-center network layout have been relatively mature, while the study on multi-center network design is more popular.

A lot of literature of current network layout has been reviewed by Hesse and Daskin [2]. Most of the literature gives solutions to network design of minimum cost by using methods of gravity methods, MILP (Mixed Integer Linear Programming) methods, heuristic methods, or simulation methods.

Most studies assume that inflow and outflow of distribution center are balanced, without consideration of in-out proportion of different products or components. Some scholars as Li [3], Nozik [4], Klose [5], do research on the problems of flow equilibrium of distributed distribution centers, and find out the influence on the level of customer service after the orders being merged from different distribution centers. However, most of these studies focus on one-sided, without consideration of different component's assembly proportion to finished product. Furthermore, these studies supposed that the customer demands were determined, so there is no answer of multisource, multi-transport point, multicomponent and multi-user of network design with indeterminate demands.

2 Problem Description

A distribution system may be expressed that $i(i \in m)$ farmers (suppliers) produce $g(g \in w)$ agricultural products, and these products could be packaged as $k(k \in p)$ kinds of finished goods (combined products for sell) in $j(j \in n)$ potential alternative distribution centers and be delivered to $l(l \in u)$ customers. While supply capacity of agricultural products is E_i^g, the demand ratio of finished products to agricultural products is T_g^k, the cost of constructing a distribution center is F_j, and the cost distribution is O_j^k. And there are variations of agricultural products' shipping distance d_{ij}^{1}, transportation speeds v_{ij}^{1}, and unit transport rates c_{ijk}^{1}, with finished products' shipping distance d_{ij}^{2}, transportation speeds v_{ij}^{2}, and unit transport rates c_{ijk}^{2}, which make large variations in the freight of agricultural products and finished goods among different supply chains. In addition, since each distribution center has its own unit working efficiency t_{jk}^{0}, there is a different customer response speed t_{lk}^{2} that could not meet the expectation of expected response time of customers' t_{lk}^{1}. For these reasons, these following contents should be taken into account during the design of distribution network: which distribution center should be selected? How to arrange transportation plan of agricultural products x_{ijg} (A certain amount of agricultural products g that are delivered from supplier i to distribution center j)? How to arrange transportation plans of finished goods y_{ijk} (A certain amount of finished goods k that are delivered from distribution center j to customer l)? At the same time the objective of minimum logistics cost could be attainted on the premise that customer response speed has been satisfied.

3 Analysis on Demand Uncertainty

Distributed distribution system network is based on forecast of market demand. In reality, the supplies S_l^k of finished products will be higher or lower than the market demands D_l^k due to random fluctuations of demands. It may be to cause a certain losses to enterprises in either circumstance. On one hand, excess supplies may lead to excess inventories, and then incurs inventory cost C_{kl}^{+}; On the other hand, short supply can make enterprises lose sales opportunity owing to out of stock, and then cause stock loss C_{kl}^{-}. To this end, when designing distribution networks, overall loss of potential excess inventory and stock outs should be minimized as soon as possible. Meanwhile, it can also transfer the products' supply plan of distribution center into determining optimal products' supplies under certain conditions. It means that optimal supply should be based on minimal expected loss of distribution center. If the density function of customer l's demand for finished product k is $f_l^k(x)$, the expected loss, which incurs when a distribution center delivers S_l^k products k to customer l, is

$$\Phi_l^k(S_l^k) = C_{lk}^{+} \int_0^{S_l^k} (S_l^k - x) f_l^k(x)dx + C_{lk}^{-} \int_{S_l^k}^{\infty} (x - S_l^k) f_l^k(x)dx \tag{1}$$

Take the first and second derivative separately, the result is,

$$\frac{d\Phi_l^k}{d(S_l^k)} = C_{lk}^{+} \int_0^{S_l^k} f_l^k(x)dx - C_{lk}^{-} \int_{S_l^k}^{+\infty} f_l^k(x)dx \tag{2}$$

$$\frac{d^2\Phi_l^k}{d(S_l^k)^2} = (C_{lk}^+ + C_{lk}^-)f_l^k(S_l^k) \tag{3}$$

$C_{lk}^+ > 0, C_{lk}^- < 0, f_l^k(S_l^k) \geq 0$ are known, so $\frac{d^2\Phi_l^k}{d(S_l^k)^2} \geq 0$, as a result, $\Phi_l^k(S_l^k)$ has minimum.

$\frac{d^2\Phi_l^k}{d(S_l^k)^2} = 0$ is known, so the optimal supply of (structured) product is

$$\overset{\wedge}{S_l^k} = F_{lk}^-\left(\frac{C_{lk}^-}{C_{lk}^+ + C_{lk}^-}\right) \tag{4}$$

And F_{lk}^- is an inverse function of the demand function.

Through the above analysis, only when the amount of agricultural product k that the distribution center delivers to customer l is

$$\overset{\wedge}{S_l^k} = F_{lk}^-\left(\frac{C_{lk}^-}{C_{lk}^+ + C_{lk}^-}\right),$$

the overall losses of excess supply and stock outs that might appear will be minimized.

If distribution center plans to supply the customer l in the amount of $\overset{\wedge}{S_l^k}$, the actual amount R_l^k must below the minimum of $\overset{\wedge}{S_l^k}$ and D_l^k, that is

$$R_l^k = \min\left\{\overset{\wedge}{S_l^k}, D_l^k\right\} = \int_0^{\overset{\wedge}{S_l^k}} sf_l^k(x)dx + \int_{\overset{\wedge}{S_l^k}}^{+\infty} \overset{\wedge}{S_l^k} f_l^k(x)dx \tag{5}$$

By formula (5), the number of products that each distribution center delivers to customer can be determined under the circumstance of the lowest expected loss being ensured.

4 Design Network Models

Just as the design of a general logistics network, the objective function for distributed distribution network is the lowest logistics costs under the satisfaction of customer service levels, involved in transportation costs of agricultural products from suppliers to distribution centers, distribution cost of finished-goods from distribution centers to customers, and construction costs of distribution center and distribution variable cost.

$$C = \min\left[\sum_{i=1}^{m}\sum_{j=1}^{n}\sum_{g=1}^{\omega} c_{ijg}^{1}x_{ijg} + \sum_{j=1}^{n}\sum_{l=1}^{u}\sum_{k=1}^{p} c_{jlk}^{2}y_{jlk} + \sum_{j=1}^{n}(F_j + \sum_{l=1}^{u}\sum_{k=1}^{p} o_j^k y_{jlk})z_j\right] \tag{6}$$

$$\left.\begin{aligned} C_{ijg}^{1} &= d_{ij}^{1}c_{ijg}^{1} \\ C_{ijg}^{2} &= d_{jl}^{2}c_{jlk}^{2} \end{aligned}\right\} \tag{7}$$

$$t_{lk}^{2} = \frac{d_{ij}^{1}}{v_{ij}^{1}} + t_{jk}^{0} + \frac{d_{jl}^{2}}{v_{jl}^{2}} \tag{8}$$

$$t_{lk}^{2} \le t_{lk}^{1} \tag{9}$$

$$\sum_{j=1}^{n} x_{ijg} \le E_j^g, i \in m, g \in w \tag{10}$$

$$\sum_{j=1}^{n} y_{jlk} \le s_{jl}R_l^k, l \in u, k \in p, l \in u \tag{11}$$

$$\sum_{k=1}^{p}\sum_{l=1}^{u} y_{jlk}T_g^k = \sum_{i=1}^{m} x_{ijg}, j \in n, g \in w \tag{12}$$

$$\sum_{j=1}^{n} s_{jl} = 1, l \in u \tag{13}$$

$$s_{jl} = \begin{cases} 1 & \text{Candidate j for customer l} \\ 0 & \text{or else} \end{cases} \tag{14}$$

$$z_j = \begin{cases} 1 & \text{Candidate j} \\ 0 & \text{or else} \end{cases} \tag{15}$$

The objective function formula (6)means total cost of distribution network is the lowest. Under the constraint conditions, formula (7) means unit transportation cost is the product of distances and unit transportation rate; Formula (8) means actual response time of customer consists of agricultural product transit time, sorting time and delivery time of finished goods; Formula (9) means the constraints of customer response time; Formula (10) means supplies of agricultural products; Formula (11) means the constraints of customers' demand satisfaction; Formula (12) means the constraints of inflow and outflow in the distribution center; Formula (13) means a distribution center can only serves one customer; The variable chosen by distribution center is 0 and 1 in the formula (14) and (15).

The first two terms of objective function are linear functions which are composed by decision variable x_{ijg} and y_{jlk}, and the first three terms are the combinations of integer programming and linear programming. Therefore, this model belongs to a mixed-integer nonlinear programming model in general.

5 Models Solving

Theoretically, mixed integer programming can be solved by some solutions such as branch-bound method, Lagrange multipliers method and so on. However, it is quite difficult to solve large scale problems because of such many variables and constraints. In addition, the formula (12) contains constraints of distribution coefficient, so it is much more complex to be resolved than the general model of logistics network design and far more difficult to be solved accurately by general integer programming method. Lingo is a package specialized for solving optimization models. As ideal software for solving combination optimal problem, Lingo has advantages of quick run speed, and typing, solving, analyzing easily. This paper is based on Lingo8.0, designing constraint integer programming solver to solve the model.

6 Example Designing

Agricultural products a, b, c, and d are produced by suppliers S1, S2 and S3. Now four distribution centers can be chosen, these are D1, D2, D3 and D4. Finished goods A and B are produced by those agricultural products in distribution centers, and then they are distributed to 12 customers (U1 ~ U12) after being packaged. The unit cost of distribution of agricultural products is 0.02 Yuan/(piece·km), the unit cost of distribution of finished goods is 0.4 Yuan/(piece·km). The average speed of distribution for agricultural products and finished goods are 80 and 50 km/h; Customer response time is 72 h. Excess inventory cost of distribution center is 50 Yuan, and stock out cost is 80 Yuan. Customer demand obeys normal distribution. Under the satisfaction of customer response speed, the distribution networks are designed by minimum cost, included that which distribution centers are chosen, which finished goods should be packaged by each distribution center, which customer would be distributed to.

By Lingo8.0, it have compiled solver for algorithm, finished 832 iterations, taken 4min59 s, and finally got a globally optimal solution. It can get the minimum total cost of 131063.70 Yuan by choosing D1 and D4.

By choosing markets nearby for processing and distribution, the distributed distribution network can not only adapts to the habits of local consumers, but also shortens the supply cycle, at the same time enhances the service standard, and finally becomes an effective competition model of markets. This paper is focused on the design of logistics networks for a variety of agricultural products, including packaging and distribution under the assumption of uncertain of customers' demand. It discussed how to obtain the optimized distribution volume, and furthermore, with which as the model input factor to seek the best design of the distribution network, by determining the distribution plan with the goal of cost minimization. The recursive steps of algorithms and solution speed showed the effectiveness of algorithms when solving nonlinear programming problems.

References

1. Kopczak (1995) Coordinated order delivery under distributed distribution. Stanford University, San Francisco
2. Hesse S, Daskin S (1998) Strategic facility location: a review. Eur J Oper Res 111(3):423–447
3. Li Y, Ma S (2004) An MINLP model for the single-source and p-Median distribution system based on time constraints. Chin J Manage Sci 12(3):86–90
4. Nozik L, Turnguist A (2001) Inventory, transportation, service quality and the location of distribution centers. Eur J Oper Res 129(2):362–371
5. Klose A, Drexl A (2005) Facility location models for distribution system design. Eur J Oper Res 162(1):4–29

Correlation Analysis of Boiler Operation Parameters Based on Mill Ball and Coal Storage

Fang He and Yan Bai

Abstract The working reliability of ball mill directly affects the reliability of coal pulverizing system and even boiler unit. The correlative degree analysis and balanced adjacent degree are utilized to study basic boiler operation parameters with coal storage and kinematics of balls. Based on the physical model for operation parameters of power unit and ball motion, the results of Matlab and Origin show that the trajectory of milling balls varies with coal storage during the coal pulverizing process, and the impact energy of balls can be calculated. The relationship between coal storage and boiler load is better than primary air flow, secondary air flow and main steam pressure. Coal storage is more closely associated with load which reflects boiler efficiency than the other parameters. The approach provides the theoretical and experimental basis for a more convenient and effective control method of coal storage.

Keywords Ball motion · Balanced adjacent degree · Control method · Coal storage · Operation parameters

1 Introduction

Coal fired boiler includes the combustion system, steam-water system, attemperation system, air and flue gas system, pulverizing system and the soot blower system [1]. Ball mill is an important part of the combustion system, and the coal pulverizing and treating process is shown in Fig. 1. The coal falling into coal feeder from coal bunker is grinded up in ball mill coal pulverized system, and the hot air is generally used as a dry atmosphere. The mixture of pulverized coal and air with dry hot air is discharged through the cyclone separator, which is stored in the coal

F. He (✉) · Y. Bai
School of Control and Computer Engineering, North China Electric Power University, Beijng, China
e-mail: hf@ncepu.edu.cn

E. Qi (ed.), *Proceedings of the 6th International Asia Conference on Industrial Engineering and Management Innovation*,
DOI 10.2991/978-94-6239-145-1_38

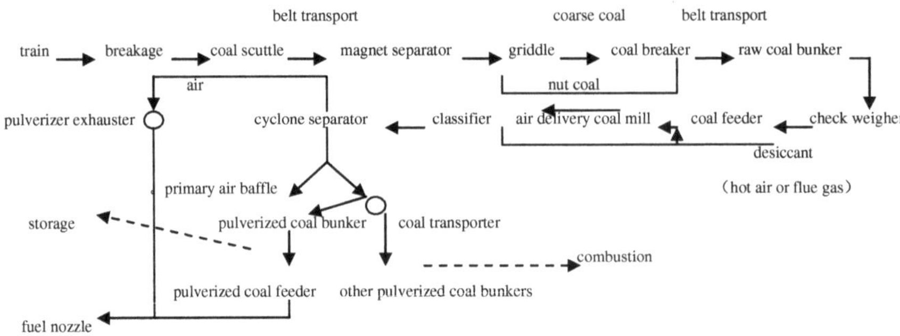

Fig. 1 Coal pulverizing process chart

warehouse for combustion [2–4]. The working reliability of ball mill directly affects the reliability of coal pulverizing system and even boiler unit.

There is not a more accurate and convenient solution pattern for the detection and control of coal storage at present. Variation of operation parameters of the boiler system is analyzed by correlation degree analysis and balance degree method from the perspective of the relationship between energy and basic parameters. The results show that the relevance between load which reflects boiler efficiency and coal storage is better than other parameters. Balls' kinetic energy in different motion states is influenced by coal storage in mill cylinder. Therefore it is established that the closed-loop feedback relationship between the operating parameters, coal storage and real-time kinetic energy of balls.

2 Operating Parameters

Operation parameters of boiler have different characteristics. Some parameters directly affect the operating state of other parameters and even the whole boiler; however, some other parameters that indirectly reflect the state of boiler can only be changed by adjusting the other parameters.

2.1 Parameter Analysis

The operation parameters are divided into two categories as follows.

(1) Controllable parameters are characterized by directly controlling values, which achieve the safe and economic operation in the adjustment process of boiler. Among these parameters, the main steam temperature, pressure, reheat steam temperature and pressure are in relation to the load. However, excess air coefficient and pulverized coal fineness are not associated with the load. From

the point of economy and security, the main steam temperature, pressure and reheat steam temperature can maintain stability at the condition of normal load and constant pressure.

(2) Controlled parameters have characteristics that values are not directly controlled to adjust the combustion performance of boiler, including the coal quality, exhaust gas temperature, carbon content of ash and slag, air leakage, feed-water temperature and attemperating water. Among them, the coal quality is not related to the load.

The boiler efficiency is associated with the variation of boiler load. With the increase of coal quantity, the load and relative excess air coefficient are increased, and the average temperature of furnace rises. Completer pulverized coal combustion is, smaller physical and chemical heat loss is, therefore, higher boiler efficiency is. When the boiler load exceeds the economical load, a large amount of coal makes the furnace temperature increase. The high air velocity and short retention time of pulverized coal in the furnace hearth result in physical and chemical incomplete combustion loss, and even the decreased boiler efficiency [5]. To achieve a more convenient and effective control method of coal storage based on kinetic energy of balls, it is very important to conduct the online monitor of various operating parameters. Consequently, the relation between kinetic energy of balls, coal storage and operation parameters of boiler are fully analyzed in coal pulverizing system, and safety and economy of boiler are supervised.

2.2 *Variation of Parameters*

Figure 2 specifically shows the change trends of the power of mill, import and export differential pressure, output power and kinetic energy of balls with the increase of coal storage [6]. The relationship between coal storage and various parameters is intuitively discussed. The method provides the theoretical basis for optimum control of coal storage.

Fig. 2 Parameters and coal storage diagram

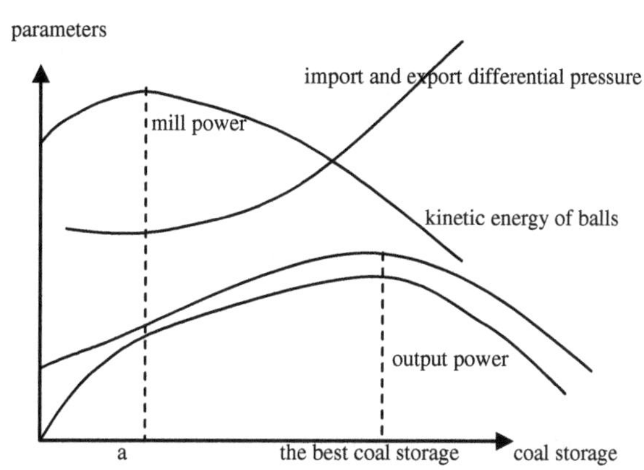

When we input *a* value of coal storage, there exists the maximum value of ball mill power. The power is mainly used for rotating cylinder and lifting balls. Since the gaps between the balls are filled with coal, the grinding mill has been in its most efficient. The output power is the coal storage put into ball mill in unit time. With the increase of coal storage, the energy of pulverizing balls have boosted gradually. Since the ball mill reaches the maximum output, balls have obtained the maximum kinetic energy, and the powder process has achieved the highest efficiency. The corresponding coal storage is known as the best coal storage. The coal pulverizing system usually operates under the negative pressure, which is a result of the suction effect of the mill exhauster. Since the coal storage is increased to a certain extent, the import and export differential pressure has been increasing. The import negative pressure would decrease and even have a positive value, resulting in the rapidly rising import and export differential pressure if the chute blockage happened.

3 Ball Motion Simulation

Grinding parameters have complex effects on ball motion. For the sake of the study of variation law of balls more in line with the coal grinding process, foundational assumptions are made about some parameters [7]. According to the operation characteristics of conventional ball mill and the thermal power operation data of China Shenhua Energy Tianjin Dagang Power Plant, parameters are defined by the following forms: the effective cylinder diameter $D = 4.30$ m, the length $L = 5.95$ m, the working speed $n = 15.60$ rpm and the calculated speed rate $\Phi = 0.76$. Once the condition is successful, the working speed n must be close to the optimal speed n_{zj}. In the above expression, we obtain the optimum ball filling ratio $\Psi_{zj} = 0.19$, so the optimum ball charge is rewritten in the following relationship:

$$G = \rho V \Psi_{zj} = \frac{1}{4} \rho \pi D^2 L \Psi_{zj} = \frac{1}{4} \times 4.9 \times \sqrt{9.81} \times 4.3^2 \times 5.95 \times 0.19 \approx 80.20t \tag{1}$$

The balls identified as 0.04 and 0.03 m in diameter according to the ratio of 3:1 are put into the mill cylinder in Table 1, the ball filling ratio may be calculated as

$$\Psi = \frac{4G}{\pi D^2 L \varepsilon} = \frac{4\rho \Psi_{zj}}{\varepsilon} = \frac{4.9 \times 0.19}{4.76 \times 0.75 + 4.85 \times 0.25} \approx 19.46\% \tag{2}$$

where ε is the ball bulk density (t/m^3),

$$\varepsilon = \sum 0.001 \times \varepsilon_b \times ratio$$

Because the optimum ball filling ratio varies in the range $10\,\% \le \Psi_{zj} \le 30\,\%$, the specific power consumption declines to the minimum value [8–10]. Compared with

Table 1 Ball parameters

Ball diameter D_b(mm)	Ball quality m(kg)	Ball number n_b(/t)	Ball bulk density ε_b(kg/m^3)
30	0.111	9009	4850
40	0.263	3802	4760
50	0.514	1946	4708
60	0.889	1125	4660

kinetic energy of balls, heat consumption and friction loss are ignored in coal grinding process.

It is observed that the ball charge has stratification phenomenon during mill operation. The circular motion and projectile motion equations are derived by Newton's laws. The projectile motion of multilayer balls in the cylinder is simulated using Matlab software, the effect as shown in Fig. 3a. The iterative method of computation and interface design to solve the above equations is proposed in Fig. 3b [11]. Therefore, balls' impact energy can be obtained, and the reasonable position of each layer is verified in real time.

4 Results and Discussion

In order to draw a conclusion that the relationship between kinetic energy of balls, coal storage and load, pulverizing efficiency is better than other operational parameters of boilers, the variation of parameters has been measured every 5 min in the field data collection. Meanwhile, Fig. 4 shows the real-time data of coal storage, primary and secondary air flow, main steam pressure and boiler load from Pingshuo Coal Gangue Power Generation Co., Ltd. The red and grey curves represent the change of coal storage and boiler load separately. It can be clearly seen that boiler load may originally have a time delay resulted from combustion process with the increased coal storage. The change trends may have been almost the same since they are stable. The relevance degree between coal storage, primary and secondary air flow, main steam pressure and boiler load is analyzed by correlative degree analysis and balanced adjacent degree. It further explore the regulation method of coal storage for stable and efficient operation of the coal pulverizing system.

(a)

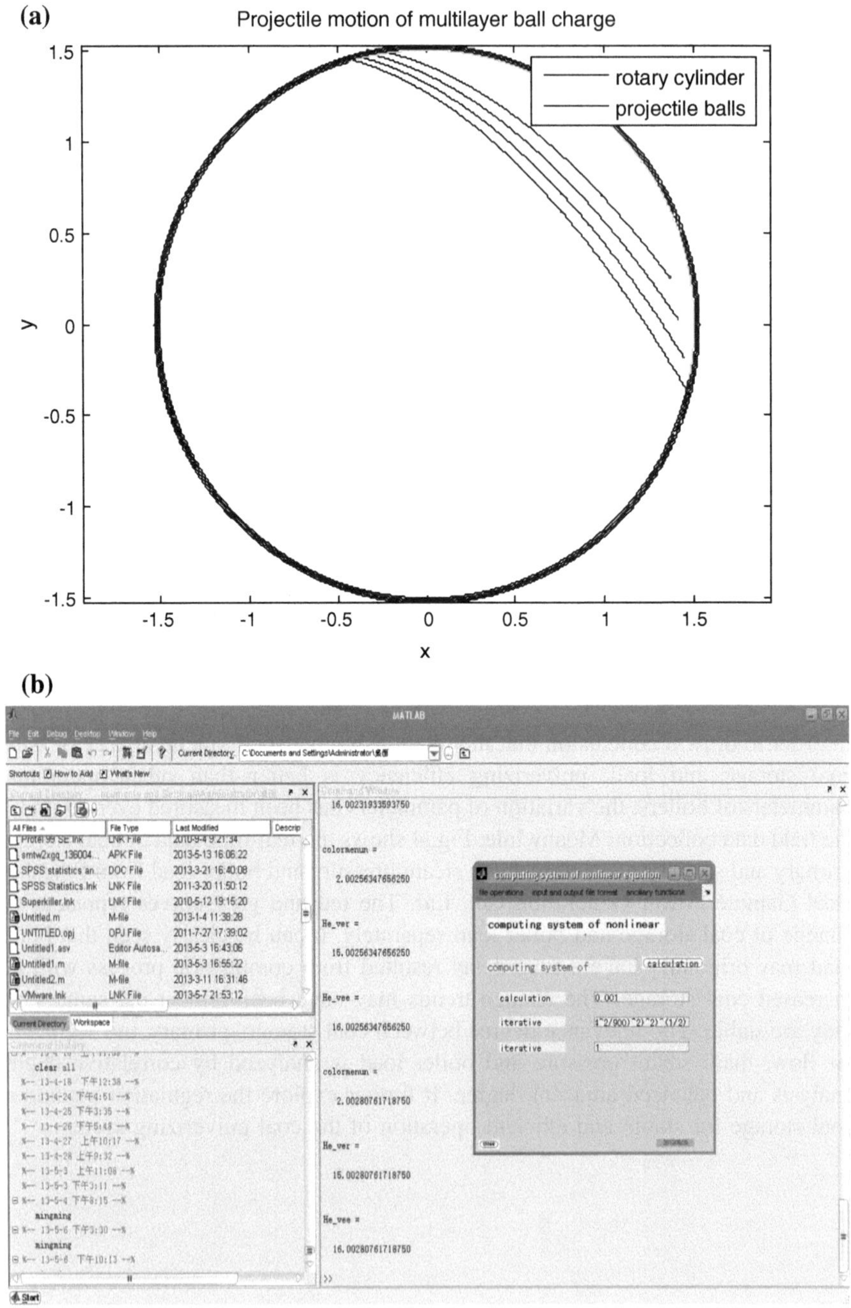

(b)

Fig. 3 Matlab simulation calculation graph. **a** Simulation model of ball motion. **b** Solving surface of nonlinear equation

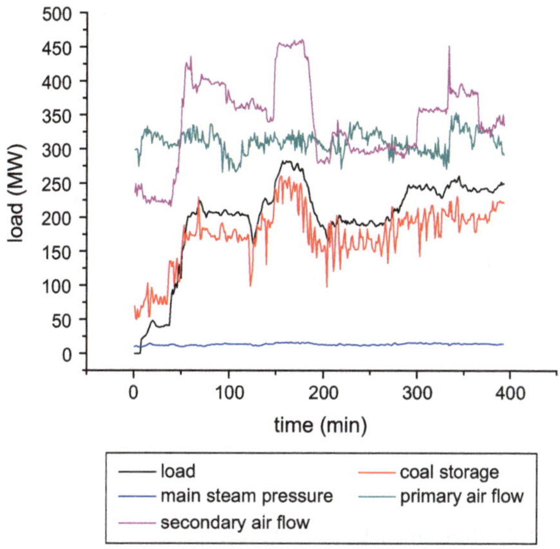

Fig. 4 Scatterplot of boiler load, coal storage, primary and secondary air flow and main steam pressure

4.1 Correlative Degree Analysis and Balance Degree Method

4.1.1 Correlative Degree Analysis

Correlative degree analysis is included in grey system theory, and is appropriate for little data, uncertainty and poor information to seek the law of original sequence. We judge the related degree of reference data sequence and comparison data sequence according to the geometrical similarity of generated data curves. The more approximate the curve trends are, the bigger the correlation degree of the corresponding sequence is, and vice versa.

Correlative degree analysis is to measure the correlation degree of the factor variation trends between two systems with the increase of time or other objects. Reference data sequence reflects the behavioral characteristics of system,

$$x_i(f_k) = \{x_i(f_1), x_i(f_2), \ldots x_i(f_n)\} \tag{3}$$

comparison data sequence affects the factors of system performance,

$$x_j(f_k) = \{x_j(f_1), x_j(f_2), \ldots x_j(f_n)\} \tag{4}$$

the absolute difference between $x_i(f_k)$ and $x_j(f_k)$,

$$\left|x_i(f_k) - x_j(f_k)\right| = \Delta_{ij}(f_k) \tag{5}$$

and the correlation coefficient of $x_i(f_k)$ and $x_j(f_k)$:

$$\xi_{ij}(f_k) = \frac{\Delta\min + \Delta\max}{\Delta_{ij}(f_k) + \Delta\max\eta} \tag{6}$$

Among all sequences, $\Delta_{\max}$ and $\Delta_{\min}$ are the maximum and minimum absolute difference, generally $\Delta_{\min} = 0$. η is distinguishing coefficient, $\eta \in (0, 1)$. The correlation degree of reference data sequence and comparison data sequence is calculated by (7) [12]:

$$\gamma_{ij} = \frac{1}{n}\sum_{k=1}^{n} \xi_{ij}(f_k)$$

$$i = 1, 2, \ldots, m; \quad j = 1, 2, \ldots, n; \quad k = 1, 2, \ldots, n \tag{7}$$

4.1.2 Balanced Degree Method

The correlation degree is to measure the degree of similarity. In order to reduce local points' association tendency, the balance degree is further adopted to measure and compare the correlation degree of data sequence's correlation coefficient series [13, 14]. Supposing the correlation coefficient series of the jth comparison data sequence and the ith reference data sequence is:

$$R_{ij} = \{\gamma(x_i(k), x_j(k))/k = 1, 2, \ldots, n\},$$

then the correlation coefficient distribution map is:

$$Map : R_{ij} \rightarrow P_{ij}$$

$$p_{ij}(k) = \gamma(x_i(k), x_j(k)) / \sum_{k=1}^{n} \gamma(x_i(k), x_j(k)) \tag{8}$$

where $k = 1, 2, \ldots, n, p_{ij} \in P_{ij}$. We can define the following equation based on (8),

$$H_{\otimes}(R_{ij}) = -\sum_{k=1}^{n} p_{ij}(k) \ln p_{ij}(k), \tag{9}$$

The above equation is the correlation coefficient's entropy of the jth comparison data sequence and the ith reference data sequence, and the balance degree is as the following (10):

$$B(R_{ij}) = H_{\otimes}(R_{ij})/H_m(R_{ij})$$

$$X = \{x_i, x_j/i = 1, 2, \ldots, m; j = 1, 2, \ldots, n\}, \tag{10}$$

where $H_m(R_{ij})$ is the largest entropy in the *j*th comparison data sequence, and it's expressed as $L_n^{(\text{sequence number})}$. Then the balanced adjacent degree is:

$$B_a(X_i, X_j) = B(R_{ij}) \times \gamma(X_i, X_j). \tag{11}$$

The correlation degree of each comparison data sequence and reference data sequence is ordered by balance adjacent degree. Finally we can determine the relationship between the comparative parameter and reference parameter, and obtain the effective theory based on parameters' physical significance.

4.2 Method Implementation and Results Analysis

The boiler load is selected as reference data sequence, other characterization parameters such as coal storage, primary and secondary air flow and main steam pressure are chosen as comparison data sequences. The variation of acquisition parameters in real time is as in Fig. 4. The results show that the main steam pressure nearly remains stable, and the change trends of other parameters may be almost the same with the increased time.

Data normalization is utilized to compare parameters because of different units. Normalization methods can be divided into peak normalization and area normalization by applied behavior analysis. The initial values transform method is used to figure out the absolute difference between reference data sequence and comparison data sequence. η reduces the influence of $\Delta_{\max}$ value distortion in (6), and the appropriate value is obtained according to the variation of parameters. More stable the drawn curve is, wider the range of η is. The correlation coefficient curve with programme calculation is shown in Fig. 5 [15].

The correlation degree is the average of correlation coefficients in Table 2. The results of Excel and Matlab show that the relevance of coal storage and boiler load is better than primary and secondary air flow and main steam pressure. It is primarily indicated that coal storage is more closely associated with load which reflects boiler efficiency than the other parameters.

The balanced adjacent degree of reference data sequence and comparison data sequences can be also figured out using (8)–(11), as shown in Table 3. It is the conclusion drawn from Tables 2 and 3 that compared with primary and secondary air flow and main steam pressure, coal storage has a closer correlation with boiler load. The coal storage can better explain the boiler load and reflect the operating efficiency.

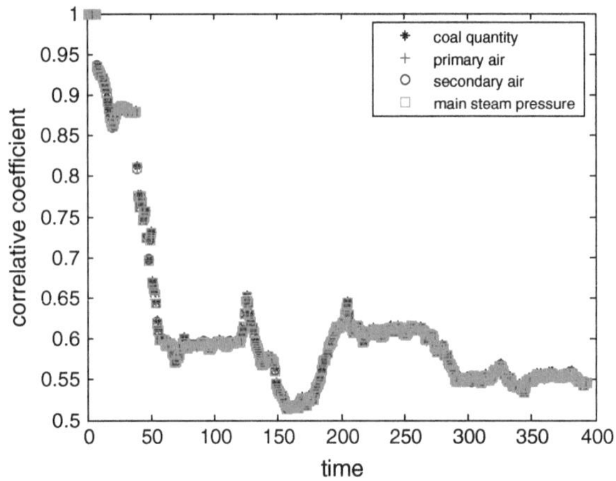

Fig. 5 Correlation coefficient curve of boiler load, coal storage, primary and secondary air flow and main steam pressure

Table 2 Correlative degree of secondary sequences and primary sequence

Parameter	Coal storage	Primary air	Secondary air	Main steam pressure
γ	0.6163	0.6146	0.6150	0.6149

Table 3 Balanced adjacent degree of parameters and boiler load

Parameter	Coal storage	Primary air	Secondary air	Main steam pressure
$B_a\ (X_i, X_j)$	0.6149	0.6132	0.6136	0.6135

5 Conclusions

From the perspective of basic parameters and motion energy, several important conclusions may be drawn as follows.

(1) The ball motion pattern inside a mill cylinder is analyzed and simulated using Matlab software. The track curve of parabolic movement is drawn out, and the space coordinate system is put forward to study impact energy of balls at different moments in the coal grinding process.
(2) Through a detailed analysis of operation parameters of the boiler system, it is showed that the relevance between load which reflects boiler efficiency and coal storage is better than other parameters by mean of correlative degree analysis and balance degree method. This research provides the theoretical and experimental basis for a more convenient and effective control method of coal storage.
(3) The establishment of functional relations between the operating parameters, coal storage and kinetic energy of balls is of great significance for pushing forward the safe and reliable operation of the coal pulverizing system and combustion system.

References

1. Ecevit B, Brian S (2005) Population balance modeling of non-linear effects in milling processes. Powder Technol 153(1):59–71
2. Austin GL, Julianelli K, De Souza AS, Schneider LC (2007) Simulation of wet ball milling of iron ore at Carajas, Brazil. Int J Miner Process 84(1/4):157–171
3. Huang Z, Shi TL (2011) The reliability simulation research of performance parameters in the boiler combustion control system. In: Proceedings of 2011 IEEE 2rd international conference on computing, control and industrial engineering, Wuhan University of Technology, Wuhan, China, pp 78–81
4. Ozkan A, Yekeler M, Calkaya M (2009) Kinetics of fine wet grinding of zeolite in a steel ball mill in comparison to dry grinding. Int J Miner Process 90(1/4):67–73
5. Yan SL, Zhang B, Liu S, Jiao SC (2009) Analysis of the operative intex's character and must-be value in power station boiler under different loads (in Chinese). Energy Eng 6:26–30
6. Xiao B (2008) Multi-sensor information fusion and the application in industry (in Chinese). Ph.M. dissertation, Taiyuan University of Technology, Taiyuan, China
7. Xia EP, Dong WM, Ying LL (2009) Experimental research of the major factors influencing the medium movements in the ball mill(in Chinese). Min Process Equipment 37(19):60–63
8. Zhang CX, Liu WP (2000) Research on the law of the movement in the process of grinding (in Chinese). J South Inst Metall 21(2):99–103
9. Yue JF, Xiao J, Qin P, Su ZM (2011) Experimental study of a dual inlet and outlet coal mill for reducing its milling power consumption (in Chinese). J Eng Therm Energy Power 26 (3):354–358
10. Zheng WD (1996) The calculation and determination of ball filling ratio in the ball mill (in Chinese). East China Electric Power 4:42–44
11. Zhang XM (2010) MATLAB language and application case (in Chinese). Southeast University Press, Nanjing, pp 238–242
12. Wang JG, He F, Di H (2012) Correlation analysis of magnetic field and conductivity, pH value in electromagnetic restraint of scale formation (in Chinese). CIESC J 63(5):1468–1473
13. Zhang QS, Deng JL, Shao Y (1995) A grey correlational analysis by the method of degree of balance and approach (in Chinese). J Huazhong Univ Sci Technol 23(11):94–98
14. Chen Y (2010) A computation method for several new grey proximity incidence degree and its application (in Chinese). J Mianyang Normal Univ 29(8):4–8
15. Lü F (1997) Research on the identification coefficient of relational grade for grey system (in Chinese). Syst Eng Theory Pract 6:49–54

Robust Analysis and Control over Uncertain Networked Control System with Random Time Delays

Yan-peng Wu, Xin-min Wang and Ying Wu

Abstract This paper considered the robust stability problem of networked flight control system, which is a new research field with the widespread applications of avionics networks for the connection of onboard devices of the aircraft. Sufficient conditions for robust stability and stabilization of networked flight control system with random time delays and structured uncertainties are derived based on the Lyapunov-Krasovskii function approach. The memoryless state feedback control law is obtained by the solution of linear matrix inequalities (LMIs) that guarantees the robust stabilization of the uncertain discrete time-delay system. Numerical examples are given to illustrate the effectiveness of design procedure.

Keywords Free weighting matrix · Linear matrix inequalities · Lyapunov-Krasovskii · Networked flight control systems · Robust control · Time delay

1 Introduction

With the rapid development of computer, network and communication technology, widespread applications of avionics networks for the connection of onboard devices of the aircraft have emerged throughout flight control systems. Although network brings lots of benefits to the control system, the proliferation of the networked flight control systems has raised fundamentally new issues in communications and control to deal with the relationship between operations of the network and the performance of the overall system [1–4].

The network-induced delay is the most serious challenge which could be either constant, time varying or random in nature. Although only a few studies research on

Y. Wu (✉) · X. Wang · Y. Wu
School of Information and Control Engineering, Xi'an University of Architecture and Technology, No. 13, Yan Ta Road, Xi'an, China
e-mail: wu.yanpeng123@gmail.com

E. Qi (ed.), *Proceedings of the 6th International Asia Conference on Industrial Engineering and Management Innovation*,
DOI 10.2991/978-94-6239-145-1_39

networked flight control system (NFCS) to solve network-induced issues, many researchers have studied stability criteria for networked control system (NCS) to solve network-induced delay caused by limited communication capability [5–7]. A novel control strategy is proposed for analyzing and compensating network induced unfavorable factors to make system robust based by an observer based control predictor [8]. An augmented state vector method is used for NCS modeling over random time delays [9]. An approach to design of a full-state observer is discussed with network-induced random time delay less than one sample period by introducing the key timing characteristics of control networks [10].

Moreover, in real projects, many uncertainties such as model imperfections or as environmental disturbances come into existence. A certain linear networked flight control system model provides just an approximate description of the real objects [11]. Therefore, the robust analysis of the flight control system with structured uncertainty seems to be emphasized recently. The robust control problem of stochastic systems against bounded disturbances with Application to F-18 aircraft model is studied [12]. A Linear Fraction Transformation (LFT) based model is established an unmanned combat aerial vehicle (UCAV) which suffers from aerodynamic parameter variation [13]. A new robust controller design scheme is designed for good performance of nominal system using nonlinear dynamic inverse control technique [14]. To the best of author's knowledge, almost no research concerning both time-delays and structured uncertainties aspects of the flight control system has been reported.

In this paper, we deal with the problem of robust control for uncertain networked flight control system with random network-induced delays. The parameter uncertainties and delays are assumed to be time-varying but norm-bounded. Sufficient criteria for robust stability and stabilization of the developed uncertain discrete time-delay model are given based on the Lyapunov-Krasovskii function approach. Linear matrix inequalities (LMIs) method is adopted for the memoryless state feedback controller to guarantee the robust stabilization of the uncertain discrete time-delay system.

The remainder of this paper is organized as follows. Section 2 gives problem formulation. Robust stability condition is analyzed and an approximately stabilizing controller is designed in Sect. 3. Simulations and Results are present in Sect. 4 and conclusions are given in Sect. 5.

2 Problem Formulation

To describe a flight dynamics system with norm-bounded uncertainties controlled via a network, a typical abstract mathematical model combined with the characteristics of the networked control system is considered in this paper [15], which is given by the following discrete-time system with a time-varying delay and uncertainties in the state.

$$\begin{cases} x(k+1) = (A+\Delta A)x(k) + (A_d + \Delta A_d)x(k-d(k)) \\ \qquad\qquad + \ (B+\Delta B)u(k) \\ x(k) = 0, \quad d_2 \le k \le 0 \end{cases} \tag{1}$$

where $x(k) \in \Re^n$ is the state vector. $u(k) \in \Re^m$ is the control input vector. $d(k)$ is a time-varying delay and satisfies

$$d_1 \le d(k) \le d_2 \tag{2}$$

where d_1 and d_2 are known constants. A, A_d and B are constant matrix, ΔA, ΔA_d and ΔB denote real-valued matrix functions representing parameter uncertainties with the form of

$$[\Delta A \quad \Delta A_d \quad \Delta B] = DF(k)[E_a \quad E_d \quad E_b] \tag{3}$$

where is an unknown time-varying matrix satisfying

$$F^T(k)F(k) \le I, \ \forall k \tag{4}$$

and E_a, E_d, E_b are constant matrices of appropriate dimensions.

Definition 1 Given integers d_1 and d_2, the uncertain discrete time-delay system (1) is robustly stable, if there exists a linear memoryless state feedback control law $u(k) = Kx(k)$, $K \in \Re^{m\times n}$, such that the resulting closed-loop system is asymptotically stable for all uncertainties which satisfy (3) and (4).

Lemma 1 *(Schur complements) For the given symmetric matrices* $S = S^T = \begin{bmatrix} S_{11} & S_{12} \\ * & S_{22} \end{bmatrix}$, *where* $S_{11} \in R^{r\times r}$

, then the following three conditions are equivalent

$$\begin{aligned} &(1) \quad S<0; \\ &(2) \quad S_{11}<0, \ S_{22} - S_{12}^T S_{11}^{-1} S_{12}<0; \\ &(3) \quad S_{22}<0, \ S_{11} - S_{12} S_{22}^{-1} S_{12}^T<0. \end{aligned}$$

Lemma 2 *For matrices* $Q = Q^T$, H *and* E *of appropriate dimensions, then*

$$Q + HF(k)E + E^T F^T(k) H^T < 0 \tag{5}$$

holds for all time varying matrix function $F(k)$ *satisfying* $F^T(k)F(k) \le I$, *if and only if there exists a positive constant* ε, *such that*

$$Q + \varepsilon^{-1} HH^T + \varepsilon E^T E < 0 \tag{6}$$

3 Main Results

First, we study the delay-dependent stability condition of the uncertain discrete time-delay system (1) with $F(k) \equiv 0$. Applying the state feedback controller $u(k) = Kx(k),\ K \in \Re^{m \times n}$ to the system (1) results in the following closed-loop system:

$$\begin{cases} x(k+1) = A_c x(k) + A_d x(k - d(k)) \\ x(k) = 0, \quad d_2 \le k \le 0 \end{cases} \tag{7}$$

Theorem 1 *Given integers d_1 and d_2, the uncertain discrete time-delay system (7) is asymptotically stable for time-varying delay $d_1 \le d(k) \le d_2$, if there exist symmetric positive-definite matrices $P = P^T > 0$, $Q = Q^T > 0$, $Z = Z^T > 0$, $R = R^T > 0$, $X = \begin{bmatrix} X_{11} & X_{12} \\ * & X_{22} \end{bmatrix}$, and matrices N_1, N_2 such that the following LMIs holds:*

$$\Sigma = \begin{bmatrix} \Phi_{11} & \Phi_{12} & A_c^T P & d_2 A_c^T Z \\ * & \Phi_{22} & A_d^T P & d_2 A_d^T Z \\ * & * & -P & 0 \\ * & * & * & -d_2 Z \end{bmatrix} < 0 \tag{8}$$

$$\Psi = \begin{bmatrix} X_{11} & X_{12} & N_1 \\ * & X_{22} & N_2 \\ * & * & Z \end{bmatrix} \ge 0 \tag{9}$$

where

$$\begin{aligned} \Phi_{11} &= (d_2 - d_1 + 1)R - P + N_1 + N_1^T + d_2 X_{11} \\ &\quad + Q + d_2 Z - d_2 A_c^T Z - d_2 Z A_c \\ \Phi_{12} &= N_2^T - N_1 + d_2 X_{12} - d_2 Z A_d \\ \Phi_{22} &= -R - N_2 - N_2^T + d_2 X_{22} - Q \end{aligned}$$

Proof Define

$$y(l) = x(l+1) - x(l) \tag{10}$$

then we have

$$x(k) - x(k - d(k)) - \sum_{l=k-d(k)}^{k-1} y(l) = 0 \tag{11}$$

Consider the following Lyapunov-Krasovskii functional candidate for the uncertain discrete time-delay system (7): □

$$V(k) = V_1(k) + V_2(k) + V_3(k) + V_4(k) \tag{12}$$

where

$$\begin{aligned} V_1(k) &= x^T(k)Px(k) \\ V_2(k) &= \sum_{\theta=-d_2^k+1}^{0} \sum_{l=k-1+\theta}^{k-1} y^T(l)Zy(l) \\ V_3(k) &= \sum_{l=k-d(k)}^{k-1} x^T(l)Qx(l) \\ V_4(k) &= \sum_{\theta=-d_2+1}^{-d_1+1} \sum_{l=k-1+\theta}^{k-1} x^T(l)Rx(l) \end{aligned} \tag{13}$$

and $P = P^T > 0$, $Q = Q^T > 0$, $Z = Z^T > 0$, $R = R^T > 0$ are to be determined.

Then evaluating the forward deference $\Delta V(k) = V(k+1) - V(k)$, we obtain

$$\Delta V_1(k) = 2x^T(k)Py(k) + y^T(k)Py(k) \tag{14}$$

$$\begin{aligned} \Delta V_2(k) &= \sum_{\theta=-d_2+1}^{0} \sum_{l=k+\theta}^{k} y^T(l)Zy(l) - \sum_{\theta=-d_2+1}^{0} \sum_{l=k-1+\theta}^{k-1} y^T(l)Zy(l) \\ &= d_2 y^T(k)Zy(k) - \sum_{l=k-d_2}^{k-1} y^T(l)Zy(l) \end{aligned} \tag{15}$$

$$\Delta V_3(k) = x^T(k-d(k))Qx(k-d(k)) - x^T(k)Qx(k) \tag{16}$$

$$\begin{aligned} \Delta V_4(k) =& (d_2 - d_1 + 1)x^T(k)Rx(k) - \sum_{l=k-d_2}^{k-d_1} x^T(l)Rx(l) \\ &\leq (d_2 - d_1 + 1)x^T(k)Rx(k) - x^T(k-d(k))Rx(k-d(k)) \end{aligned} \tag{17}$$

Based on the free weighting matrices method, a zero formula is introduced as following form:

$$\begin{aligned} &2[(x^T(k)N_1 + x^T(k-d(k))N_2)]\times \\ &[x(k) - x(k-d(k)) - \sum_{l=k-d(k)}^{k-1} y(l)] = 0 \end{aligned} \tag{18}$$

As to any symmetric semi-positive-definite matrix $X = \begin{bmatrix} X_{11} & X_{12} \\ * & X_{22} \end{bmatrix}$, we have

$$\begin{aligned} &\sum_{l=k-d_2}^{k-1} \eta_1^T(k)X\eta_1(k) - \sum_{l=k-d(k)}^{k-1} \eta_1^T(k)X\eta_1(k) \\ &= d_2\eta_1^T(k)X\eta_1(k) - \sum_{l=k-d(k)}^{k-1} \eta_1^T(k)X\eta_1(k) \geq 0 \end{aligned} \tag{19}$$

where $\eta_1(k) = \begin{bmatrix} x^T(k) & x^T(k-d(k)) \end{bmatrix}^T$

Calculating $\Delta V(k)$,and adding the terms on the left side of Eqs. (18) and (19), then we obtain

$$\begin{aligned} \Delta V(k) =& 2x^T(k)Py(k) + y^T(k)Py(k) + d_2 y^T(k)Zy(k) \\ & - \sum_{l=k-d_2}^{k-1} y^T(l)Zy(l) \\ & + x^T(k-d(k))Qx(k-d(k)) - x^T(k)Qx(k) \\ & + (d_2 - d_1 + 1)x^T(k)Rx(k) - x^T(k-d(k))Rx(k-d(k)) \\ & + 2[(x^T(k)N_1 + x^T(k-d(k))N_2)] \times [x(k) - x(k-d(k)) \\ & - \sum_{l=k-d(k)}^{k-1} y(l)] + d_2\eta_1^T(k)X\eta_1(k) - \sum_{l=k-d(k)}^{k-1} \eta_1^T(k)X\eta_1(k) \\ \leq & \eta_1^T(k)\Sigma\eta_1(k) - \sum_{l=k-d(k)}^{k-1} \eta_2^T(k)\Phi\eta_2(k) \end{aligned} \tag{20}$$

where $\eta_2(k) = \begin{bmatrix} x^T(k) & x^T(k-d(k)) & y^T(l) \end{bmatrix}^T$.

Note that if condition (8) and (9) holds, $\Delta V(k) < 0$. This completes the proof.

Then, we study the delay-dependent stability condition of the uncertain discrete time-delay system(1). Applying the state feedback controller $u(k) = Kx(k)$, $K \in \Re^{m \times n}$ to the system (1) results in the following closed-loop system:

$$\begin{cases} x(k+1) = (A_c + \Delta A_c)x(k) + (A_d + \Delta A_d)x(k-d(k)) \\ x(k) = 0, \quad d_2 \leq k \leq 0 \end{cases} \tag{21}$$

where $A_c = A + BK, \quad \Delta A_c = \Delta A + \Delta BK$.

Theorem 2 *Given integers d_1 and d_2, the uncertain discrete time-delay system (7) is asymptotically stable for time-varying delay $d_1 \leq d(k) \leq d_2$, if there exist symmetric positive-definite matrices $P = P^T > 0$, $Q = Q^T > 0$, $Z = Z^T > 0$,*

$R = R^T > 0$, $X = \begin{bmatrix} X_{11} & X_{12} \\ * & X_{22} \end{bmatrix}$, *and matrices* N_1, N_2 *such that the following matrix inequalities holds:*

$$\Xi = \begin{bmatrix} \Phi_{11} & \Phi_{12} & A_c^T & d_2 A_c^T & -d_2 ZD & E_c^T \\ * & \Phi_{22} & A_d^T & d_2 G A_d^T & 0 & E_d^T \\ * & * & -P^{-1} & 0 & D & 0 \\ * & * & * & -d_2 Z^{-1} & d_2 D & 0 \\ * & * & * & * & -\varepsilon I & 0 \\ * & * & * & * & * & -\varepsilon^{-1} I \end{bmatrix} < 0 \tag{22}$$

$$\begin{bmatrix} X_{11} & X_{12} & N_1 \\ * & X_{22} & N_2 \\ * & * & Z \end{bmatrix} \geq 0 \tag{23}$$

where

$$\begin{aligned} \Phi_{11} &= (d_2 - d_1 + 1)R - P + N_1 + N_1^T + d_2 X_{11} \\ &\quad + Q + d_2 Z - d_2 A_c^T Z - d_2 Z A_c \\ \Phi_{12} &= N_2^T - N_1 + d_2 X_{12} - d_2 Z A_d \\ \Phi_{22} &= -R - N_2 - N_2^T + d_2 X_{22} - Q \end{aligned}$$

Proof Based on Theorem 1, we can easily obtain that the uncertain discrete time-delay system (21) is asymptotically stable when $\Lambda < 0$ and $\Psi \geq 0$ where

$$\begin{aligned} \Lambda &= \begin{bmatrix} \Lambda_{11} & \Lambda_{12} & (A_c + \Delta A_c)^T & d_2 (A_c + \Delta A_c)^T \\ * & \Lambda_{22} & (A_d + \Delta A_d)^T & d_2 (A_d + \Delta A_d)^T \\ ** & -P^{-1} & 0 & \\ ** & * & -d_2 Z^{-1} & \end{bmatrix} \\ \Lambda_{11} &= (d_2 - d_1 + 1)R - P + N_1 + N_1^T + d_2 X_{11} + Q \\ &\quad + d_2 Z - d_2 (A_c + \Delta A_c)^T Z - d_2 Z (A_c + \Delta A_c) \\ \Lambda_{12} &= N_2^T - N_1 + d_2 X_{12} - d_2 Z (A_d + \Delta A_d) \\ \Lambda_{22} &= -R - N_2 - N_2^T + d_2 X_{22} - Q \end{aligned} \tag{24}$$

Replacing ΔA_d with $DF(k)E_d$, and ΔA_c with $DF(k)E_c$, where $E_c = E_a + E_b K$, respectively, then Λ can be written into

$$\begin{aligned}\Lambda &= \overline{\Lambda} + \Delta\Lambda \\ &= \overline{\Lambda} + \begin{bmatrix} -d_2ZD \\ 0 \\ D \\ d_2D \end{bmatrix} F(k)\begin{bmatrix} E_c & E_d & 0 & 0 \end{bmatrix} \\ &\quad + \begin{bmatrix} E_c^T \\ E_d^T \\ 0 \\ 0 \end{bmatrix} F^T(k)\begin{bmatrix} -d_2D^TZ & 0 & D^T & d_2D^T \end{bmatrix} \\ &< 0 \end{aligned} \tag{25}$$

□

By Lemma 1, (25) is guaranteed if there exists a scalar $\varepsilon > 0$,satisfying

$$\begin{aligned}&\overline{\Lambda} + \varepsilon^{-1}\begin{bmatrix} -d_2ZD \\ 0 \\ D \\ d_2D \end{bmatrix}\begin{bmatrix} -d_2D^TZ & 0 & D^T & d_2D^T \end{bmatrix} \\ &+ \varepsilon\begin{bmatrix} E_c^T \\ E_d^T \\ 0 \\ 0 \end{bmatrix}\begin{bmatrix} E_c & E_d & 0 & 0 \end{bmatrix} < 0 \end{aligned} \tag{26}$$

where

$$\overline{\Lambda} = \begin{bmatrix} \overline{\Lambda}_{11} & \overline{\Lambda}_{12} & A_c^TP & d_2A_c^TZ \\ * & \overline{\Lambda}_{22} & A_d^TP & d_2A_d^TZ \\ * & * & -P & 0 \\ * & * & * & -d_2Z \end{bmatrix} \tag{27}$$

$$\begin{aligned} \overline{\Lambda}_{11} &= (d_2 - d_1 + 1)R - P + N_1 + N_1^T \\ &\quad + d_2X_{11} + Q + d_2Z - d_2A_c^TZ - d_2ZA_c \\ \overline{\Lambda}_{12} &= N_2^T - N_1 + d_2X_{12} - d_2ZA_d \\ \overline{\Lambda}_{22} &= -R - N_2 - N_2^T + d_2X_{22} - Q \end{aligned} \tag{28}$$

Applying Schur complement formula, (26) is equivalent to (22), This completes the proof.

Theorem 3 *Given integers d_1 and d_2, the uncertain discrete time-delay system (7) is asymptotically stable for time-varying delay $d_1 \le d(k) \le d_2$, if there exist* *symmetric positive-definite matrices* $U = U^T > 0$, $L = L^T > 0$, $G = G^T > 0$, $H = H^T > 0$, $Y = \begin{bmatrix} Y_{11} & Y_{12} \\ * & Y_{22} \end{bmatrix}$, *and matrices* V_1, V_2, *satisfying*

$$\Omega = \begin{bmatrix} \Omega_{11} & \Omega_{12} & (A\tilde{Z} + BT)^T & d_2(A\tilde{Z} + BT)^T & -d_2 D & (E_a\tilde{Z} + E_b T)^T \\ * & \Omega_{22} & \tilde{Z}A_d^T & d_2\tilde{Z}A_d^T & 0 & \tilde{Z}E_d^T \\ * & * & -\tilde{P} & 0 & D^T & 0 \\ * & * & * & -d_2\tilde{Z} & d_2 D & 0 \\ * & * & * & * & -\varepsilon I & 0 \\ * & * & * & * & * & -\varepsilon^{-1} I \end{bmatrix} < 0 \tag{29}$$

$$\begin{bmatrix} Y_{11} & Y_{12} & V_1 \\ * & Y_{22} & V_2 \\ * & * & G \end{bmatrix} \geq 0 \tag{30}$$

where

$$\begin{aligned} \Omega_{11} &= (d_2 - d_1 + 1)\tilde{R} - H + V_1 + V_1^T + d_2 Y_{11} + \tilde{Q} \\ &\quad + d_2\tilde{Z} - d_2\tilde{Z}A^T - d_2 T^T B^T - d_2 A\tilde{Z} - d_2 BT \\ \Omega_{12} &= V_2^T - V_1 + d_2 Y_{12} - d_2\tilde{Z}A_d \\ \Omega_{22} &= -\tilde{R} - V_2 - V_2^T + d_2 Y_{22} - \tilde{Q} \end{aligned} \tag{31}$$

and the memoryless state feedback controller is given by

$$K = TG^{-1} \tag{32}$$

Proof Replacing A_c in (22) with $A + BK$, pre-multiply and post-multiply $diag\{Z^{-1} \quad Z^{-1} \quad I \quad I \quad I \quad I\}$, then set $Z^{-1}RZ^{-1} = \tilde{R}$, $Z^{-1}XZ^{-1} = Y$, $Z^{-1}QZ^{-1} = \tilde{Q}$, $Z^{-1} = \tilde{Z}$, $Z^{-1}N_1Z^{-1} = V_1$, $Z^{-1}N_2Z^{-1} = V_2$, $T = KZ^{-1}$, $Z^{-1}PZ^{-1} = H$ we can obtain (29) and (30). This completes the proof. □

4 Design Example

Consider the following longitudinal motion equation of the aircraft uncertain discrete-time model:

$$\begin{aligned} x(k+1) &= (A + \Delta A)x(k) + (A_d + \Delta A_d)x(k - d(k)) \\ &\quad + (B + \Delta B)u(k) \end{aligned} \tag{33}$$

$$A = \begin{bmatrix} 0.5 & -0.5 & 0.2 \\ -0.1 & 0.8 & 0 \\ 0.06 & -0.03 & 1.5 \end{bmatrix}, B = \begin{bmatrix} 0.1826 \\ 0.5575 \\ -0.0053 \end{bmatrix},$$
$$A_d = \begin{bmatrix} 0.4193 & -0.3049 & -0.2 \\ -0.0396 & 0.2235 & 0 \\ 0.0349 & -0.0111 & -0.5 \end{bmatrix} \quad (34)$$

$$E_a = diag\{0.01 \quad 0.01 \quad 0.01\},$$
$$E_d = diag\{0.005 \quad 0.005 \quad 0.005\}$$
$$E_b = \begin{bmatrix} 0.01 \\ 0.01 \\ 0 \end{bmatrix}, D = I, F(k) = \sin(k) \quad (35)$$

where $x = [q \quad a \quad \theta]$, q is pitch angular rate, a is attack angle, θ is pitch angle. $u = \xi_e$,ξ_e is elevating rudder deflection angle.

In this example, we choose $\varepsilon = 2.3$, then we can use the Matlab LMI Control Toolbox to solve the LMI feasible problem presented in theorem 3. We can obtain the maximum bound of time delay such that the corresponding system is robustly stabilizable, that is $d_2 = 8$ when $d_1 = 1$, and the memoryless state feedback controller gain is

$$K = [0.5094 - 0.22630.0272] \quad (36)$$

Under this controller, the simulation result of the controlled output is shown in Fig. 1. Figure 2 shows the states trajectories of networked flight control system with $d_2 = 4$.

We can see that both the memoryless state feedback control law obtained with $d_2 = 8$ and $d_2 = 4$ effectively solved the time delay problem of networked flight

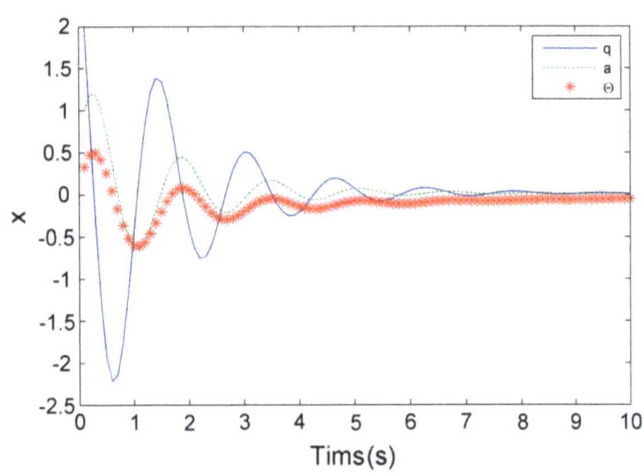

Fig. 1 States trajectories of networked flight control system with $d_2 = 8$

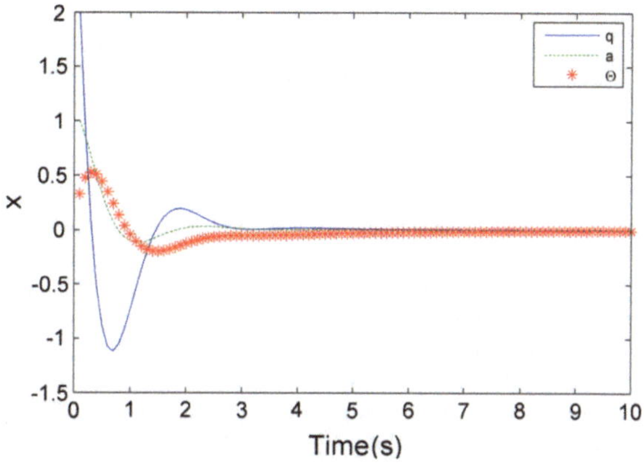

Fig. 2 States trajectories of networked flight control system with $d_2 = 4$

control system with uncertainties, which guarantee the asymptotical stability of the system. Moreover, the factor of time delay impacts the control performance of the flight control system.

5 Conclusion

The problem of robust control for networked flight control system with norm-bounded uncertainties and time-varying delays has been studied in this paper. Sufficient criteria for robust stability and stabilization of the developed uncertain discrete time-delay system are given based on a delay-dependent LMI based approach. A free weighting matrix approach is introduced to obtain the memoryless state feedback controller, which guarantees the asymptotical stability of the resulting closed-loop system. A numerical example has been provided to demonstrate the effectiveness and applicability of the proposed approach.

References

1. Zhi CY (2009) Avionics data bus technology and its applications. National Defense Industry Press, Beijing, pp 1–10
2. Yang TC (2006) Networked control system: a brief survey. IEEE Proc-Control Theory Appl 153(4):403–412
3. Li W, Wang Q, Xu L, Dong C (2010) Exponential stability of a networked flight control system with large-bounded successive packet-dropouts. In: IEEE International conference on intelligent computing and intelligent systems-ICIS, pp 596–605
4. Hespanha JP, Naghshtabrizi P, Xu Y (2007) A survey of recent results in networked control systems. Proc IEEE 95(1):138–162
5. Zeng HB, He Y, Wu M, Xiao SP (2011) Absolute stability and stabilization for Lurie networked control systems. Int J Robust Nonlinear Control 21:1667–1676

6. Sun YG, Gao QZ (2012) Stability and stabilization of networked control system with forward and backward random time delays. In: Mathematical problems in engineering
7. Suzuki T, Kono M, Takahashi N, Sato O (2011) Controllability and stabilizability of a networked control system with periodic communication constraints. Syst Control Lett 60:977–984
8. Wu Y, Yuan ZH, Wu YP (2012) Observer-based control predictor for networked control system with uncertain time delays and packet dropouts. In: 2012 Third International conference on intelligent control and information processing (ICICIP)
9. Ye X, Liu S, Liu PX (2010) Modelling and stabilisation of networked control system with packet loss and time-varying delays. IET Control Theory Appl 4:1094–1100
10. Chen ZP, Liu L, Yin XL (2005) Networked control system with network time-delay compensation. In: Industry applications conference, Fourtieth IAS Annual Meeting
11. Zhang LX, Shi P, Karimi HR (2014) Robust control of stochastic systems against bounded disturbances with application to flight control. In: IEEE Trans Ind Electron 61(3)
12. Li HB, Chow MY, Sun ZQ (2009) Optimal stabilizing gain selection for networked control systems with time delays and packet losses. IEEE Trans Control Syst Technol 17:1154–1162
13. Dai HS, Zhu JH (2010) Uncertainty modeling and robust flight control of unmanned combat aerial vehicle. In: 2010 8th World Congress on intelligent control and automation (WCICA), pp 2583–2587
14. Zhang R, Fan JF, Li JJ (2012) A new robust controller for flight control system of hypersonic flying vehicle. Adv Mater Res 562:1682–1688
15. Yedavalli RK, Belapurkar RK (2009) Design of distributed engine control systems for stability under communication packet dropouts. J Guid Control Dyn 32(5):1544–1549

A Framework for Computationally Expensive Problems with Genetic Algorithm

Ning Zhao, Yong-zhi Zhao and Chen-xi Fu

Abstract A genetic algorithm with fitness estimation mechanisms FEGA is introduced for some computationally expensive problems. A framework based on the concept of fidelity is established to control the timing and frequency of the real fitness evaluation. The fitness of a particular individual can be obtained as averages of other individuals within a certain area. The fitness sharing radius controls the size of the area, and the fidelity thresholds decides whether to use the real fitness functions or not. Besides, mechanisms of the loss of fidelity is used to reduce the estimation errors from spread and accumulation, and mechanisms of removing redundancy individuals in order to reduce the computing consumption is used at the same time. Three benchmark tests have been conducted and the results shows that FEGA achieves satisfactory effects in all the three benchmark functions.

Keywords Fitness estimation · Fidelity · Genetic algorithm · Sharing radius

1 Introduction

Population-based intelligent optimization algorithms, such as genetic algorithms (GA), ant colony algorithm (ACO), particle swarm optimization (PSO), etc., are widely used in the optimal design of complex mechanical systems for its faster speed, better global convergence properties, the insensitivity of the initial value, and the capacity of solving discontinuous, multimodal, high-dimensional optimization problems compared with the traditional optimization algorithms [1]. However, this intelligent optimization algorithm is also facing many new challenges, one of the problems lies in the huge time consumption due to the high complexity of performance analyses for fitness evaluation and the large number of evaluations

N. Zhao (✉) · Y. Zhao · C. Fu
Shaanxi Engineering Laboratory for Transsmissions and Controls, Northwestern Polytechnical University, Xi'an, China
e-mail: zhaon@nwpu.edu.cn

E. Qi (ed.), *Proceedings of the 6th International Asia Conference on Industrial Engineering and Management Innovation*,
DOI 10.2991/978-94-6239-145-1_40

needed in the evolutionary optimization [1–4]. For example, some black-box problem, whose fitness values cannot be computed directly. Usually it requires the use of finite element analysis (FEA) simulation, or computational fluid dynamics (CFD) for fitness computation, where these simulations are very time-consuming. Yaochu Jin described that it takes over 10 h on a high-performance computer for a CFD simulation when it's 3-dimensional in the literature [3]. G. Gary Wang describes that it takes the Ford Motor Company about 36–160 h to perform a crash simulation in the literature [4]. The computational cost of optimizing this problem is not acceptable in the engineering. Therefore, a new method to reduce optimization time and improve the efficiency of optimization is needed.

Currently, the main methods to improve the speed of intelligence algorithm is using estimation strategy, estimating rather than evaluating directly, so as to avoid the time-consuming fitness evaluation. Present investigations on fitness estimation includes [1–5]: polynomials [6], Kriging models [7], Neural Networks [8], Support Vector Machine [9], fitness inherit [10, 11] and population clustering [12]. Polynomials and Kriging models are parameter sample model, the essence of establishing such models is to find the model parameters to minimum the forecasting errors. Neural Networks and Support Vector Machine are non-parameter sample model, the essence of establish- ing such model is acquiring approximation of the original fitness function based on a large amount of sample data. The estimation accuracy of sample models has a great relationship with the samples selected. As the dimension increases, the evolutionary algorithm is misled by the model [1, 8]. The essence of fitness inheritance model is the child generation directly gets the fitness of the father generation in a certain way. The essence of population clustering method is clustering the population into several groups and estimating individual's fitness according to the central ones.

However, no matter what the fitness estimation method is adopted, part of the individuals real fitness evaluation is needed. How to get the optimal solution or approximate optimal solution of the original optimization problem in the minimum times of real fitness evaluation, is the key technology of using fitness estimation strategies. This paper establishes a fitness shared estimation model based on the concept of fidelity, which takes genetic algorithm as an example algorithm.

2 A Fitness Estimation Model

2.1 *Evaluation of fitness*

Each individual i in the population has its fitness $fitness\ (i)$ and the fidelity $R\ (i)$ of the fitness. The fitness can be evaluated using the fitness function, or estimated by the value of other individuals. If $fitness\ (i)$ is evaluated using the original fitness function, the fidelity $R\ (i) = 1$; if $fitness\ (i)$ is estimated, the fidelity $0 \leq R(i) < 1$.

As illustrated in Fig. 1, for each individual i, specify its fitness sharing radius r_{share}. The area in which the dimensionless Euclidean distance between individual i

Fig. 1 Fitness sharing area

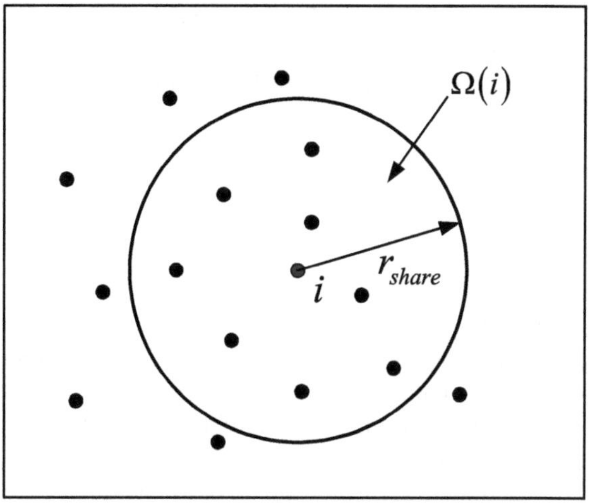

and any other one is no greater than the fitness sharing radius is called fitness sharing area for individual i, expressed as $\Omega(i)$.

Assuming there are m other individuals in the fitness sharing area $\Omega(i)$, which composed a collection $S = \{s_1, s_2, \ldots, s_m\}$. And the evaluation of $fitness(i)$ is as follows:

Firstly, evaluate the fidelity $R(i)$ of individual i according to Eq. (1).

$$R(i) = \frac{1}{m}\sum_{j=1}^{m} R(s_j) \tag{1}$$

where s_j is an individual includes in fitness sharing area $\Omega(i)$, $R(s_j)$ is the fidelity of s_j.

If fidelity $R(i)$ is greater than a given threshold R^*, then the fitness of individual i can be estimated as:

$$fitness(i) = \frac{1}{m}\sum_{j=1}^{m} fitness(s_j) \tag{2}$$

Else if $R(i)$ is less than the threshold R^*, then the fitness of individual i should be evaluated using its original fitness function.

2.2 *Update of database*

To take full advantage of the historical population whose fitness has been evaluated, it is necessary to establish a database of historical populations of coordinates, fitnesses and fidelities. As the population evolves, the database will expand the scale gradually. In order to reduce space complexity and the amount of computation, redundant data need to be eliminated after each generation. The concept of individual redundancy is introduced to determine whether the data is redundant. The redundancy of individual i is defined as:

$$I_r(i) = \sum_{k=1}^{n} \Delta\overline{x}_k(i) \tag{3}$$

where $\Delta\overline{x}_k(i)$ is the coordinate difference (absolute value) between the former individual and the latter one in the kth dimensional. Obviously, the smaller the value of the redundancy of the individual, the greater the degree of redundancy. If the redundancy value of an individual is less than a given threshold, the individual is knocked out.

In addition, since not all of the individuals' fitnesses are evaluated using its original fitness function, the estimated fitnesses are not accurate. Thus, as the population evolves gradually, the fidelities of individuals with estimated fitnesses should gradually decline. Assume that individual i is with the estimated fitness, let the fidelity of individual i in generation t be $R(i,t)$, then the fidelity in generation $t+1$ can be updated as:

$$R(i, t+1) = \beta \cdot R(i,t) \tag{4}$$

where β is fidelity drain factor, and $0 < \beta < 1$.

As the population evolves, the fidelity drops below a given threshold R_0, and the individual should be removed from the database.

2.3 Fitness estimate algorithm flow

S1: Initialize historical population database, set the initial population blank, set fitness and fidelity to 0;
S2: Find the fitness sharing area for each individual i, and find the collection of individuals in the area from the database.
S3: Evaluate the fidelity $R(i)$ of individual i, determine whether $R(i)$ is greater than the threshold R^*. If $R(i) \geq R^*$, estimate fitness of individual i according to function (2); otherwise, evaluate the fitness using its original function and set the fidelity $R(i)$ to 1.
S4: Add individual i to the database;
S5: Update the database as follows: (1) calculate redundancy for all individuals, eliminate all redundant individuals; (2) for all individuals with estimated fitness, update its fidelity according to function (3) and remove all the individuals with low fidelity.

3 Numerical Experiments

There are two purposes of numerical experiments on Fitness Estimation Genetic Algorithm (FEGA): (1) test convergence of the algorithm, which tests whether the algorithm can correctly guide the evolution, so that the optimal solution or approximate optimal solution of the original problem can be obtained; (2) test the effectiveness of the algorithm, namely test what extent the algorithm can reduce the amount of computation.

Three benchmark functions are conducted from literature [1] to test, the three benchmark functions are as follows:

(1) De Jong function 2

$$f(x) = 100\left(x_1^2 - x_2\right)^2 + (1 - x_1)^2 \tag{5}$$

where, $-2.048 \le x_i \le 2.048$.

(2) De Jong function 3

$$f(x) = \sum_{i=1}^{2} \text{int}(x_i) \tag{6}$$

where, $-5.12 \le x_i \le 5.12$.

(3) Rastrigin function

$$f(x) = 6 + \sum_{i=1}^{2} \left[x_i^2 - 3.0 \cdot \cos(2\pi x_i)\right] \tag{7}$$

where, $-5.12 \le x_i \le 5.12$.

The 3D landscape of the three benchmark functions are as illustrated in Fig. 2, where (a), (b) and (c) represent De Jong function 2, De Jong function 3 and Rastrigin function respectively. De Jong function 2 is a unimodal function, while De Jong function 3 and Rastrigin function are multi-modal function with a number of local minima.

The parameters of fitness estimation Genetic Algorithm (FEGA) is setted as follows: the population size is 50, the crossover rate is 0.8, the mutation rate is 0.3, the maximum generation is 200, the fidelity threshold R^* is 0.6, fidelity drain factor β is 0.9. And each benchmark function is tested 20 times independently.

Fitness evolution curves of the three benchmark functions are illustrated as Fig. 3, where (a), (b) and (c) represent De Jong function 2, De Jong function 3 and Rastrigin function respectively. With the evolution of the population, the population quickly converges to a near optimal solution, the average fitness increases steadily.

Percentage of real fitness evaluated in each generation of the three benchmark functions are illustrated as Fig. 4, where (a), (b) and (c) represent Goldstein-Price function, Six-Hump Camel-Back function and Shekel's Foxholes function

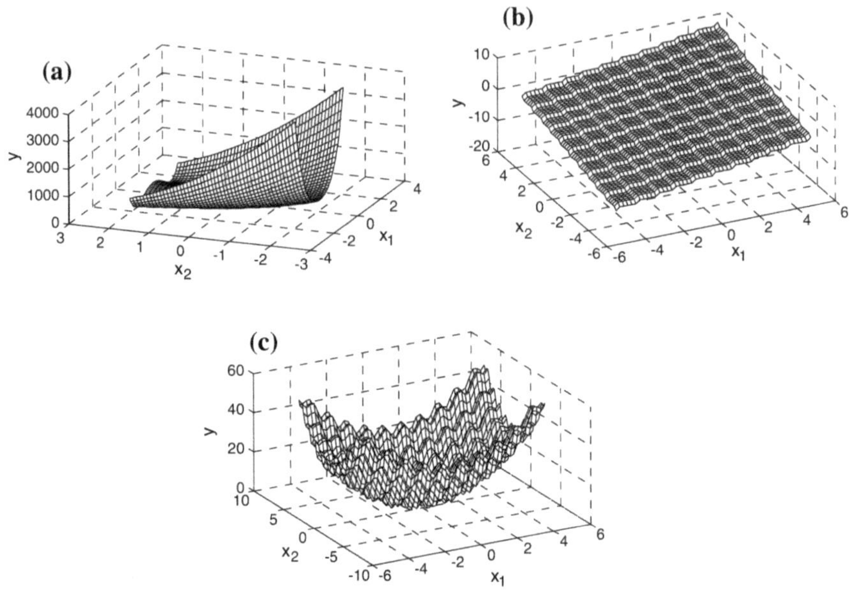

Fig. 2 3D landscape of the three benchmark functions. **a** De Jong function 2. **b** De Jong function 3. **c** Rastrigin function

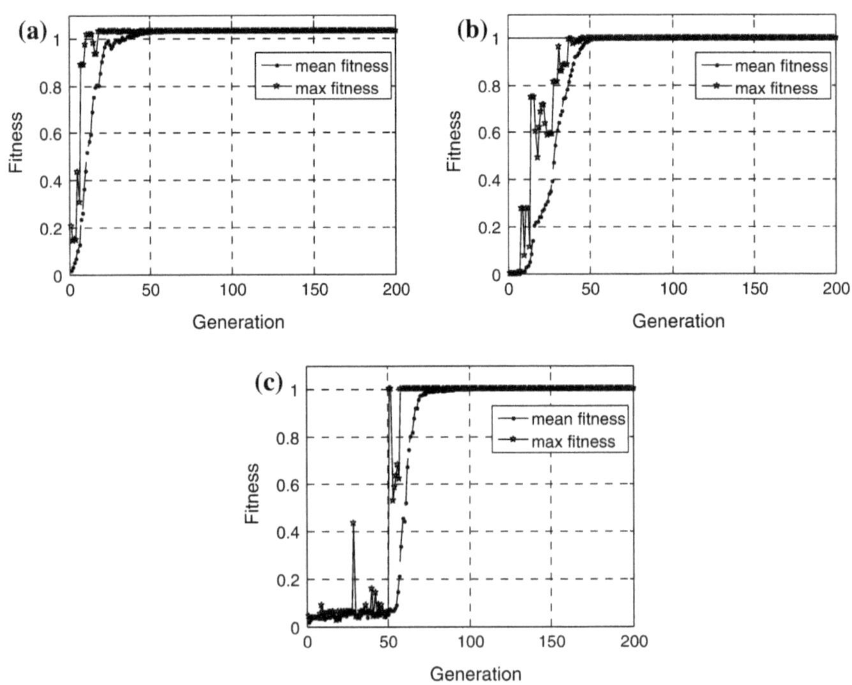

Fig. 3 Fitness evolution curves.3D landscape of the three benchmark functions. **a** De Jong function 2. **b** De Jong function 3. **c** Rastrigin function

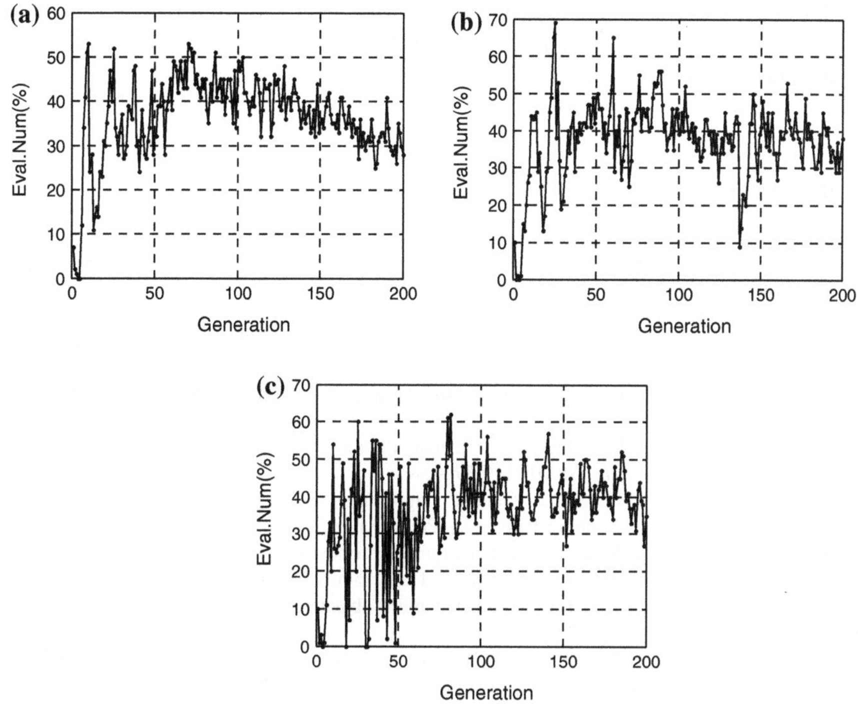

Fig. 4 Percentage of real fitness evaluated.3D landscape of the three benchmark functions. **a** Goldstein-Price function. **b** Six-Hump Camel-Back function. **c** Shekel's Foxholes function

respectively. Percentage of real fitness evaluated in each generation substantially stabilized at 30–50 %. A certain percentage of the individuals' fitness are evaluated using the original fitness function in each generation, so that the evolutionary direction can be guided correctly.

4 Conclusion

The evaluation of the fitness is computationally very expensive for some optimization problems, therefore a genetic algorithm named FEGA with fitness estimation mechanisms is introduced. Some of the individuals' fitnesses are estimated rather than evaluated, in this way the optimize efficiency is improved.

Three benchmark functions are used to test the convergence and effectiveness of FAGA, test results show that FAGA achieves satisfactory optimal solutions in all the three test functions, and more than 60 % of the computation can be reduced at the same time.

Although FEGA shows effectiveness in the given algorithm parameters, the framework still needs further more investigations to reduce the real fitness evaluating times in a greater extent.

References

1. Sun CL (2012) Particle swarm optimization for optimal design of mechanical system. Machinery Industry Press, Beijing
2. Shan S, Wang GG (2010) Survey of modeling and optimization strategies to solve high-dimesional problems with computationally-expensive black-box functions. Struct Multi Optim 41:219–241
3. Jin Y (2005) A comprehensive survey of fitness approximation in evolution computation. Soft Comput 9(1):3–12
4. Wang GG, Shan S (2007) Review of metamodeling techniques in suppot of engineering design optimization. Trans ASME 129:370–380
5. Wang XP, Cao LM (2002) Genetic algorithm—theory, application and software implementation. Xi'an Jiaotong University Press, Xi'an
6. Paenke I, Branke J, Jin Y (2006) Efficient search for robust solutions by means of evolutionary algorithms and fitness approximation. IEEE Trans Evol Comput 10(4):405–420
7. Li M (2011) An improved kriging-assisted multi-objective genetic algorithm. J Mech Des 133:071008(1–11)
8. Jin Y, Olhofer M, Sendhoff B (2002) A framework for evolutionary optimization with approximate fitness functions. IEEE Trans Evol Comput 6(5):481–494
9. Malak RJ, Predis CJJ (2010) Using support vector machines to formalize the valid input domain of predictive models in systems design problems. J Mech Des 132:101001(1–14)
10. Smith R, Dike B (1995) Stegmann S (1995) Fitness inheritance in genetic algorithm. In: Proceedings of the 1995 ACM symposium on applied computing, New York, pp 345–350
11. Sastry K, Goldberg DE, Pelikan M (2001) Don't evaluate, inherit. In: Proceedings of genetic and evolutionary computation conference, pp 551–558, Morgan Kaufmann
12. Kim HS, Cho SB (2001) An efficient genetic algorithms with less fitness evaluation by clustering. In: Proceedings of IEEE congress on evolutionary computation, pp 887–894, IEEE

Research on Improvement of INS Two-Point Calibration in Horizontal Damping State

Wen-ming Shi and Xiao-dong Yang

Abstract In order to satisfy high-accuracy navigation, improve comprehensive calibration accuracy, improved two-point calibration based on Extended State Observer (ESO).Gave a brief analysis on traditional two-point calibration, restrictive navigating state and its shortage, then estimated platform horizontal errors based on EOS, improved the two-point calibration, then gave the algorithm simulation and numerical validate. The results indicate that traditional two-point calibration give a bad estimation result when carrier cannot satisfy navigating state, but new algorithm has the stronger applicability, reduce the restrictive navigating state and improve comprehensive calibration accuracy.

Keywords ESO · Horizontal damping mode · Loosely navigating state algorithm · Platform horizontal errors · Restrictive navigating state · Two-point calibration

1 Introduction

Your goal is to simulate, as closely as possible, the usual appearance of typeset papers in the *Springer*. One difference is that the authors' affiliations should appear immediately following their names—do not include your title there. For items not addressed in these instructions, please refer to a recent issue of a Springer.

Inertial navigation system can provide vector position, velocity, attitude and other information independently, it is not only the protection of the safe navigation, but also the basis of information sources of shipboard weapons systems. Because of the gyro drift and other inertial instruments errors, navigation errors grow with time, comprehensive calibration is the key technology. Comprehensive calibration was usually divided into two-point, three-point and point-point [1]. Document [2] showed that the INS must be in state of horizontal damping and carrier keeps

W. Shi (✉) · X. Yang
Department of Navigation and Communication,
Navy Submarine Academy, Qingdao, China
e-mail: 345967262@qq.com

E. Qi (ed.), *Proceedings of the 6th International Asia Conference on Industrial Engineering and Management Innovation*,
DOI 10.2991/978-94-6239-145-1_41

uniform direct sailing, otherwise system will bring dynamic platform level imbalance angle error; For vehicle motion, without damping comprehensive calibration algorithm was put forward by Yu et al. [3], the algorithm require state switch, guarantee the INS work at undamped state and weaken the influence of maneuvering. Document [4] pointed out that the state switch frequently lead to INS overshoot, which must be compensate by external information.

In recent years, the initial alignment technology of INS has been developed by leaps and bounds: Document [5, 6] introduced the unscented Kalman filter during initial alignment, alignment accuracy and stability are better than the commonly used Kalman filter, but alignment time has not been effectively shorten, then Jin-lai Song proposed ADRC techniques applied to the initial alignment of platform INS, designed suitable feedback control for fast Initial alignment loop [7–11], improved the stabilization loop tracking and anti-jamming capability, greatly shortening the alignment time, the algorithm was simple and had a high accuracy, which is an ideal initial alignment.

The above analysis shows that constraints of the carrier navigation state are so strict that traditional comprehensive calibration cannot guarantee the accuracy and applicability. For underwater platforms INS, in order to reduce the influence of navigation constraints to comprehensive calibration, with traditional two calibration as an example, this paper presents the loosely navigation state algorithm, fast compensate platform misalignment angle in real-time, no state switch, avoid overshoot from state switching, ensure the navigation accuracy, improve system robustness.

2 Shortage of Traditional Two-Point Calibration

The two-point calibration is a certain time interval, twice in a row to obtain accurate position and heading information from the external re-tune, and estimates the gyro constant drift. Typically, the position information was given by the GPS or GLONASS, Heading information obtained from the method of celestial navigation [12, 13].

The carrier movement in space, angle $\boldsymbol{\Phi}$ error is platform coordinate system relative geographic coordinate system (x, y, z), angle $\boldsymbol{\Psi}$ is platform coordinate system to computer coordinate system error, angle $\boldsymbol{\Theta}$ is computer coordinate system relative geographic coordinate system error [1, 2].

$$\begin{cases} \Phi = \Theta + \Psi \\ \Phi = [\alpha \quad \beta \quad \gamma]^{\mathrm{T}} \\ \Theta = [\theta_x \quad \theta_y \quad \theta_z]^{\mathrm{T}} \\ \quad = [-\delta\varphi \quad \delta\lambda\cos(\varphi) \quad \delta\lambda\sin(\varphi)]^{T} \\ \Psi = [\Psi_x \quad \Psi_y \quad \Psi_z]^{\mathrm{T}} \end{cases} \tag{1}$$

In formula: ψ_x, ψ_y, ψ_z are the angles from three-axis of platform coordinate system to computer coordinate system axis; α and β are INS platform level misalignment angles; γ is azimuth offset angle, θ_x, θ_y, θ_z are the angles from computer coordinate system to geographic coordinate system; φ, λ are respectively latitude and longitude; $\delta\varphi$, $\delta\lambda$ are latitude error and longitude error.

The component of geographic coordinate system

$$\begin{cases} \alpha = \theta_x + \Psi_x \\ \beta = \theta_y + \Psi_y \\ \gamma = \theta_z + \Psi_z \end{cases} \tag{2}$$

INS must be in state of horizontal damping when comprehensive calibration, α, β are approximately to 0 when System is stable, so formula (2) turn into

$$\begin{cases} \theta_x = -\Psi_x \\ \theta_y = -\Psi_y \\ \theta_z = \gamma - \Psi_z \end{cases} \tag{3}$$

GPS position and heading signal to determine the system position error $\delta\varphi$, $\delta\lambda$ and heading error C. Consider C = γ, formula (3) rewritten as

$$\begin{cases} \delta\varphi = \Psi_x \\ \delta\lambda = -\Psi_y \sec(\varphi) \\ \delta C = -\Psi_y \tan(\varphi) + \Psi_z \end{cases} \tag{4}$$

Rewrite formula (4) in matrix form

$$\begin{bmatrix} \delta\varphi \\ \delta\lambda \\ \delta C \end{bmatrix} = \begin{bmatrix} 1 & 0 & 0 \\ 0 & -\sec(\varphi) & 0 \\ 0 & -\tan(\varphi) & 1 \end{bmatrix} \begin{bmatrix} \Psi_x \\ \Psi_y \\ \Psi_z \end{bmatrix} \tag{5}$$

Using the Laplace transform to solve the gyro drift see literature [2]. By formula (2), INS work in horizontal damping state, only when carrier uniform direct flights to navigation, the platform can meet Shula oscillation conditions, so (3) established. The downside is in the traditional two-point calibration, uniform direct state is ideal navigation state, which unable to meet the carrier sailed. Submarine must carry out motor cycles to eliminate the blind area when INS re-tunes correction, α, β cannot be ignored, formula (3) does not hold.

3 Estimation for Platform Misalignment Angle

The undamped error equation of INS [14]

$$\begin{cases} \delta\dot{V}_x = \frac{V_y}{R}\tan(\varphi)\delta V_x + (2\Omega\sin(\varphi) + \frac{V_x}{R}\tan(\varphi))\delta V_y + (2\Omega\cos(\varphi) \\ \quad + \frac{V_x}{R}\sec^2(\varphi))V_y\delta\varphi + (\frac{(V_x^2 + V_y^2)}{R} + 2\Omega V_x\cos(\varphi) - g)\beta + \Delta A_x \\ \delta\dot{V}_y = -2(\Omega\sin(\varphi) + \frac{V_x}{R}\tan(\varphi))\delta V_x - (2\Omega\cos(\varphi) + \frac{V_x}{R}\sec^2(\varphi))V_x\delta\varphi \\ \quad + (\frac{(V_x^2 + V_y^2)}{R} - 2\Omega V_x\cos(\varphi) + g)\alpha + \Delta A_y \\ \delta\dot{\varphi} = \frac{\delta V_y}{R} \\ \delta\dot{\lambda} = \frac{\delta V_x}{R\cos(\varphi)} + \frac{V_x}{R}\sec(\varphi)\tan(\varphi)\delta\varphi \\ \dot{\alpha} = -\frac{\delta V_y}{R} - \gamma(\Omega\cos(\varphi) + \frac{V_x}{R}) + \beta(\Omega\sin(\varphi) + \frac{V_x}{R}\tan(\varphi)) + \varepsilon_x \\ \dot{\beta} = \frac{\delta V_x}{R} - \Omega\sin(\varphi)\delta\varphi - \alpha(\Omega\sin(\varphi) + \frac{V_x}{R}\tan(\varphi)) - \gamma\frac{V_y}{R} + \varepsilon_y \\ \dot{\gamma} = \frac{\delta V_x}{R}\tan(\varphi) + (\Omega\cos(\varphi) + \frac{V_x}{R}\sec^2(\varphi))\delta\varphi + \alpha(\Omega\cos(\varphi) + \frac{V_x}{R}) + \beta\frac{V_y}{R} + \varepsilon_z \end{cases} \tag{6}$$

In formula: R is radius of the earth; g is acceleration due to gravity; Ω is the angular velocity of the rotation of the earth; V_x, V_y are east and north velocity; δV_x, δV_y, are the east and north velocity error; ΔA_x, ΔA_y are east and north accelerometer zero bias. Carrier speed is given by Doppler speed log, and has a high precision, carrier position signals are given by the GPS. With speed error as a concept measurements, the measurement equation

$$\begin{cases} \boldsymbol{Y} = \begin{bmatrix} 1 & 0 & 0 & 0 & 0 & 0 & 0 \\ 0 & 1 & 0 & 0 & 0 & 0 & 0 \end{bmatrix} \cdot \boldsymbol{X} + \boldsymbol{v} \\ \boldsymbol{Y} = [Y_1 \quad Y_2]^{\mathrm{T}} = [\delta V_x \quad \delta V_y]^{\mathrm{T}} \\ \boldsymbol{X} = [\delta V_x \quad \delta V_y \quad \delta\varphi \quad \delta\lambda \quad \alpha \quad \beta \quad \gamma]^{\mathrm{T}} \\ \boldsymbol{v} = [v_1 \quad v_2]^{\mathrm{T}} \end{cases} \tag{7}$$

In formula: $\boldsymbol{Y}$ is measurement vector; $\boldsymbol{X}$ is system state vector; $\boldsymbol{v}$ is measurement white Noise vector; calf, Y_1, Y_2 are measured speed errors; v_1, v_2 are measurement noise of speed error.

Horizontal damping state speed error equation is the same with undamping, with east speed error δV_x as the object, in order to get the estimation of β, Set state variables Z_1, Z_2, Z_1 track eastward speed error, Z_2 track differential of eastward speed error, that $Z_1 = \delta V_x, Z_2 = \delta\dot{V}_x$, design following ESO [14–18].

$$\begin{cases} Z_1 = \delta V_x \\ Z_2 = \frac{V_y}{R}\tan(\varphi)\delta V_x + (2\Omega\sin(\varphi) + \frac{V_x}{R}\tan(\varphi))\delta V_y \\ \quad + (2\Omega\cos(\varphi) + \frac{V_x}{R}\sec^2(\varphi))V_y\delta\varphi + (\frac{(V_x^2+V_y^2)}{R} + 2\Omega V_x\cos(\varphi) - g)\beta + \Delta A_x \end{cases} \tag{8}$$

$$\begin{cases} e1 = Z_1 - Y_1 \\ Z_1 = Z_1 + h(Z_2 - B_1 e_1) \\ Z_2 = Z_2 + h(-B_2 f_1) \\ f_1 = \mathrm{fal}(e_1, a, \delta) = \begin{cases} \frac{e_1}{\delta^{a-1}} & |e| \leq \delta \\ |e|^a \mathrm{sign}(e) & |e| > \delta \end{cases} \end{cases} \tag{9}$$

In formula: e is state error; h, a, δ, B_1, B_2 are the parameters of ESO. $\widehat{\beta}$ is the estimation of β, from formula (8)

$$\begin{aligned} \widehat{\beta} =& [Z_2 - \frac{V_y}{R}\tan(\varphi)\delta V_x + (2\Omega\sin(\varphi) + \frac{V_x}{R}\tan(\varphi))\delta V_y \\ &+ (2\Omega\cos(\varphi) + \frac{V_x}{R}\sec^2(\varphi))V_y\delta\varphi]/[\frac{(V_x^2+V_y^2)}{R} + 2\Omega V_x\cos(\varphi) - g] \end{aligned} \tag{10}$$

Steady-state error of $\widehat{\beta}$ is

$$\delta\hat{\beta} = -\frac{\Delta A_x}{\frac{(V_x^2+V_y^2)}{R} + 2\Omega V_x\cos(\varphi) - g} \tag{11}$$

Estimation process of platform level offset angle α is the same with β, with north speed error δV_y as the object, in order to get the estimation of α, take the formula (7), (14) constituting a first order system. Set state variables Z_3, Z_4, Z_3 track northward speed error, Z_4 track differential of northward speed error, that $Z_3 = \delta V_y$, $Z_4 = \delta\dot{V}_y$, $\widehat{\alpha}$ is the estimation of α, only gives the estimated results

$$\begin{aligned} \widehat{\alpha} =& [Z_4 + 2(\Omega\sin(\varphi) + \frac{V_x}{R}\tan(\varphi))\delta V_x + (2\Omega\cos(\varphi) \\ &+ \frac{V_x}{R}\sec^2(\varphi))V_x\delta\varphi]/[\frac{(V_x^2+V_y^2)}{R} - 2\Omega V_x\cos(\varphi) + g] \end{aligned} \tag{12}$$

Steady-state error of $\widehat{\alpha}$ is

$$\delta\widehat{\alpha} = -\frac{\Delta A_y}{\frac{(V_x^2+V_y^2)}{R} - 2\Omega V_x \cos(\varphi) + g} \tag{13}$$

The above analysis shows that, when accurate speed error signal are obtained, it can achieve the accurate estimates of platform level misalignment angle and has nothing to do with the carriers' navigation state.

4 Improvement for Traditional Two-Point Calibration

From the above analysis, the traditional two-point calibration requirements INS must work at the level of damping state, the carrier must make the uniform direct flights to navigation, or the introduction of new dynamic errors, affect the calibration accuracy, even cause the divergence output of INS. When carrier is unable to meet the navigation state demand, if platform offset angle is estimated and complemented in real time, loosely navigation state algorithm for two-point calibration can be realized, which will ensure the accuracy of calibration.

Take ESO estimates $\widehat{\alpha}$ and $\widehat{\beta}$ into Eq. (2) to improve the traditional two- points calibration

$$\begin{cases} \widehat{\alpha} = \theta_x + \Psi_x \\ \widehat{\beta} = \theta_y + \Psi_y \\ \gamma = \theta_z + \Psi_z \end{cases} \tag{14}$$

From formula (14)

$$\begin{cases} \theta_x - \widehat{\alpha} = -\Psi_x \\ \theta_y - \widehat{\beta} = -\Psi_y \\ \theta_z = \gamma - \Psi_z \end{cases} \tag{15}$$

Take formula (15) into formula (5)

$$\begin{bmatrix} \delta\varphi + \widehat{\alpha} \\ \delta\lambda - \frac{\widehat{\beta}}{\cos(\varphi)} \\ \delta C - \widehat{\beta}\tan(\varphi) \end{bmatrix} = \begin{bmatrix} 1 & 0 & 0 \\ 0 & -\sec(\varphi) & 0 \\ 0 & -\tan(\varphi) & 1 \end{bmatrix} \begin{bmatrix} \Psi_x \\ \Psi_y \\ \Psi_z \end{bmatrix} \tag{16}$$

From formula (16), when INS in resetting correction time, estimate and compensate angle in real-time, which has nothing to do with the navigation state, the loosely navigation state algorithm will be realized. The process of Laplace transform to solve gyro drift is the same as traditional two-point calibration.

5 Simulation and Verification

Submarine motion parameter are set to: navigation speed is 10 kn, initial heading is 90°; initial position are latitude 30°, longitude 100°; maneuvering radius is 463 m. INS parameters are set to initial platform error α and β are 3′, γ is 5′; accelerometer zero offset delta ΔA_x and ΔA_y are 10^{-5}g; gyro drift ε_x, ε_y, ε_z are 0.01(°)·h^{-1}.The sampling interval T is 4 s. Parameters h and T of ESO are 4 s, a is 0.5, δ is 5 h, B_1 is 0.25, B_2 is 0.08.

5.1 ESO Estimation of Platform Misalignment Angle

The platform horizontal misalignment angle is estimated by ESO, including maneuvering cycles and direct flights navigation, maneuvering time is 352 s. Figure 1 is the platform horizontal misalignment angle by the maneuvering cycles, Fig. 2 is the ESO estimation. From the platform horizontal misalignment angle's simulation value of Figs. 1 and 2, when submarine maneuvering, the platform horizontal misalignment angle had a greater dynamic error, system will cycle in Shula oscillation movement even direct to the navigation. β angle in Fig. 2 is up to 4, so the platform horizontal offset angle cannot be ignored. Estimation from Fig. 1 and 2 showed that, regardless of the submarine navigation state, ESO can tracking speed signals quickly (estimated steady time is less than 40 s), give the estimates of error angle in real-time, and the estimated results do not diverge. And

Fig. 1 Estimation of α

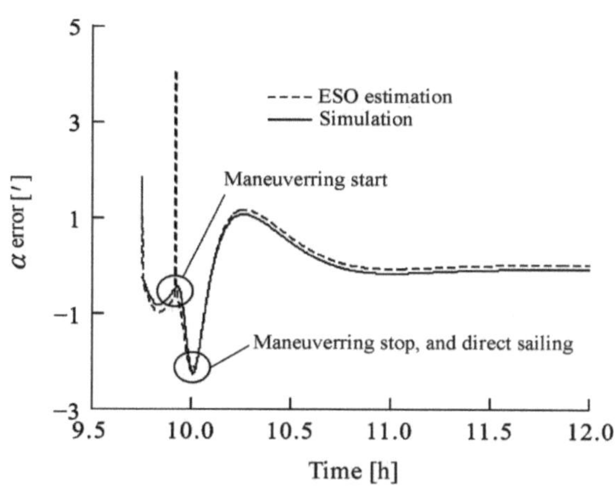

Fig. 2 Estimation of β

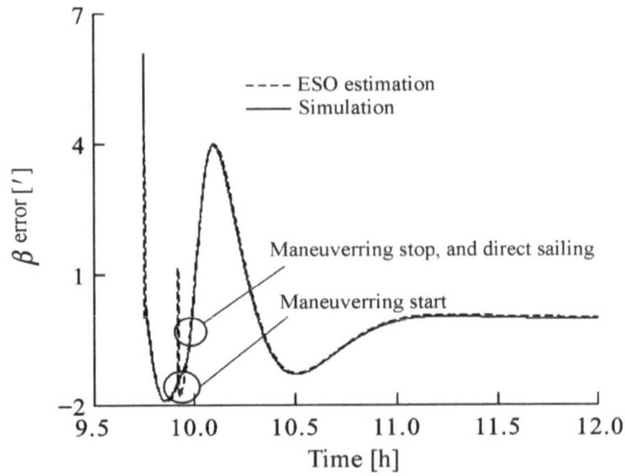

only if the submarine Navigation State changing will make small overshoot, the overshoot can disappear and stable tracking after 40 s.

5.2 *Verification*

Simulation and Verification for Loosely navigation state algorithm, submarine float up to the surface of water for first resetting at a certain moment, and then underwater vehicle, interval 6 h, float up to the surface of water for second readjust and two-point calibration, INS is in the horizontal damping state before and after, and accompanied with Submarine maneuvering. Test 1 simulation conditions: submarine float up to the surface of the water and direct sailing after maneuvering collecting data and two-point calibration. Test 2 simulation conditions: submarine float up to the surface of the water to collecting data and two-point calibration during maneuvering.

Gyro drift estimation results in Tables 1 and 2 showed that, navigation platform does not meet the Shula oscillation condition, and introduct the dynamic error when submarine during maneuvering and direct sailing after maneuvering, which lead to the platform level offset angle cannot be approximated to 0, the traditional two-point calibration cannot be normal used, and has a poor estimation. Gyro drift has a higher estimation accuracy after using ESO to estimate and compensate platform level angle in real-time, impact of platform angle is greatly reduced.

Table 1 Two-point calibration comparison of experiment 1 (°)·h^{-1}

Gyro drift	Setting value	No compensation	After compensation
ε_x	0.01	0.006565	0.010180
ε_y	0.01	0.015116	0.010230
ε_z	0.01	0.011790	0.009730

Table 2 Two-point calibration comparison of experiment 2 (°)·h^{-1}

Gyro drift	Setting value	No compensation	After compensation
ε_x	0.01	0.001740	0.010230
ε_y	0.01	0.005930	0.010030
ε_z	0.01	0.018570	0.010100

5.3 Numerical Verification

Correction errors of uncompensation in test 1 are

$$\Delta\varepsilon_X = \left|\frac{0.006655 - 0.01}{0.01}\right| = 34.3\,\%$$

$$\Delta\varepsilon_Y = \left|\frac{0.015116 - 0.01}{0.01}\right| = 52.2\,\%$$

$$\Delta\varepsilon_Z = \left|\frac{0.01179 - 0.01}{0.01}\right| = 17.9\,\%$$

The correction errors of compensation are

$$\Delta\varepsilon_X = \left|\frac{0.01018 - 0.01}{0.01}\right| = 1.8\,\%$$

$$\Delta\varepsilon_Y = \left|\frac{0.01023 - 0.01}{0.01}\right| = 2.3\,\%$$

$$\Delta\varepsilon_Z = \left|\frac{0.00973 - 0.01}{0.01}\right| = 2.7\,\%$$

Correction errors of uncompensation in test 2 are

$$\Delta\varepsilon_X = \left|\frac{0.00174 - 0.01}{0.01}\right| = 82.6\,\%$$

$$\Delta\varepsilon_Y = \left|\frac{0.00593 - 0.01}{0.01}\right| = 40.7\,\%$$

$$\Delta\varepsilon_Z = \left|\frac{0.01857 - 0.01}{0.01}\right| = 85.7\,\%$$

The correction errors of compensation are

$$\Delta\varepsilon_X = \left|\frac{0.01023 - 0.01}{0.01}\right| = 2.3\,\%$$

$$\Delta\varepsilon_Y = \left|\frac{0.01003 - 0.01}{0.01}\right| = 0.3\,\%$$

$$\Delta\varepsilon_Z = \left|\frac{0.0101 - 0.01}{0.01}\right| = 1.0\,\%$$

Fig. 3 East velocity errors

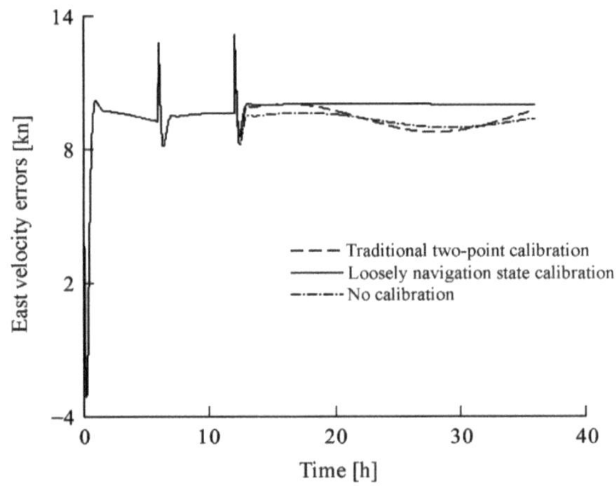

Fig. 4 North velocity errors

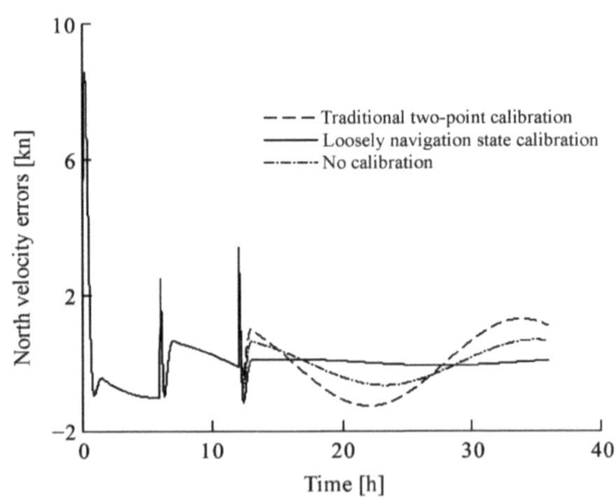

Numerical verification results showed that, estimation error are greatly reduced, Indicated the loosely navigation state algorithm is effective. The algorithms not only ensure the adaptability of calibration, but also reduce navigation state constrict and guarantee the accuracy.

Figures 3, 4, 5, 6, 7, 8, 9 are INS error divergence after loosely navigation state algorithm, navigation accuracy is improved, the respective state error is significantly reduced, simulation conclusion verify the effectiveness of loosely navigation state algorithm.

Fig. 5 Latitude errors

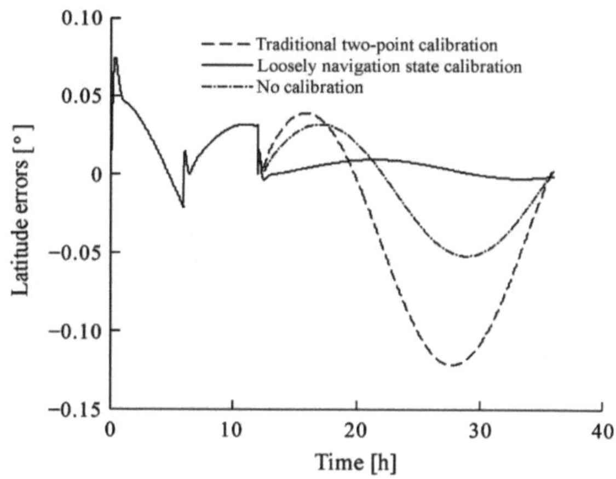

Fig. 6 Longitude errors

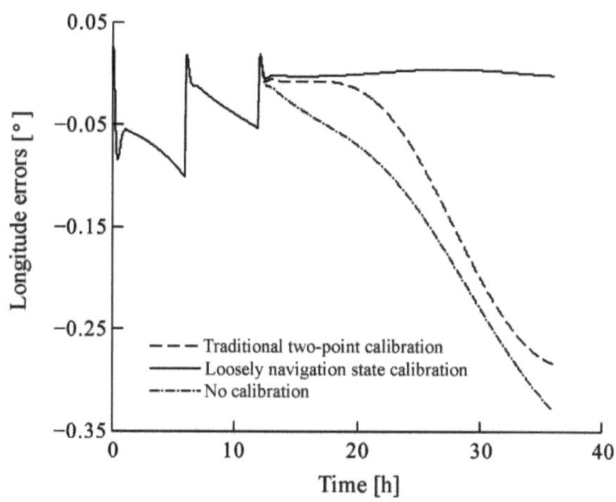

Fig. 7 α errors

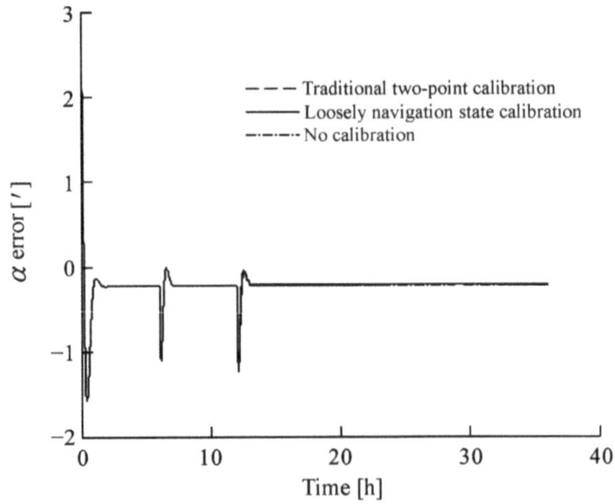

Fig. 8 β errors

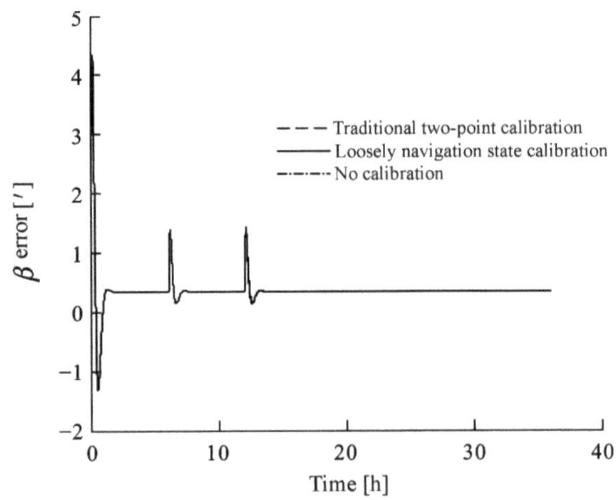

Fig. 9 γ errors

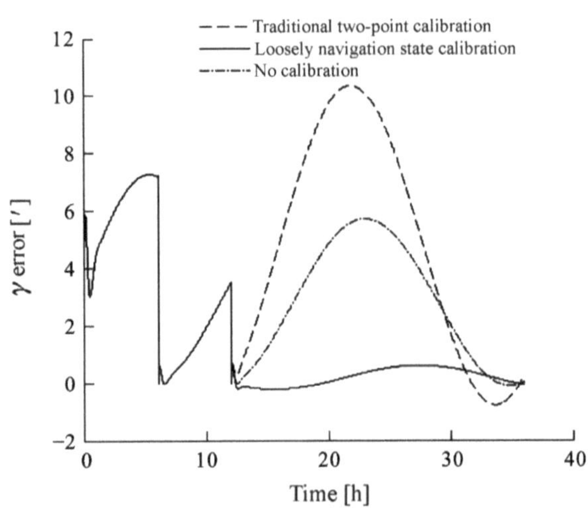

6 Conclusion

Navigation state constraints limit the accuracy of traditional two-point correction algorithm, once the carrier does not meet the navigation state requirements, the traditional two-point calibration accuracy will be greatly reduced and the reliability of the gyro drift estimation results cannot be guaranteed. This paper put forward loosely navigation state algorithm based on estimation and compensation by ESO. Simulation and numerical verification indicated the effectiveness of loosely navigation state algorithm. Gyro drift estimation accuracy is higher than traditional two-point calibration, which has a regardless of navigation state constraints, ensures the accuracy and robustness of the algorithm, and provides an effective way to improve estimation accuracy of inertial device error for the next time.

References

1. Yang X, Deng T (2011) Constant error analysis of two-point comprehensive calibration in gimbaled inertial navigation system. J Traffic Transp Eng 11(3):100–104 (in Chinese)
2. Hartman R, Hawkinson W, Sweeney K (2008) Tactical underwater navigation system. In: IEEE position location and navigation symposium, pp 898–911
3. Yu K, Li L, Liu W (2008) Correction of SINS based on undamped navigation mode. J Chin Inertial Technol 16(6):638–642 (in Chinese)
4. Guan J, Cheng J, Wu L (2008) The state switch technology of gimbaled INS. Ship Build China 49(2):75–80 (in Chinese)
5. Zhang W, Zhang J, Fan J (2007) Research on unscented Kalman filter in initial alignment of inertial navigation systems. Syst Eng Electron 29(4):589–592 (in Chinese)
6. Lu M, Yin Y (2008) Application of unscented particle filter in INS non-linear alignment. Rader Sci Technol 2(1):44–47 (in Chinese)
7. Song J, Han J (2002) Research on a fast initial alignment method for platform inertial navigation system. J Chin Inertial Technol 10(1):25–29 (in Chinese)
8. Wang C, Ma L (2009) Application of the extend state observer on the initial alignment of the inertial navigation system. Ship Sci Technol 31(7):85–88 (in Chinese)
9. Niu L, Li L, Zhao H (2003) Application of active disturbance rejection control technique in initial alignment of SINS on moving base. J Chin Inertial Technol 11(6):27–33 (in Chinese)
10. Zhao W, Zhuang L, Yang G (2007) Application of active disturbance rejection control technique in initial alignment of INS on moving base. Control Decis 22(2):179–183 (in Chinese)
11. Song J, Yang Y, Xu K (2001) Application of active disturbance rejection technique in initial alignment of inertial navigation system. J Chin Inertial Technol 9(4):11–15 (in Chinese)
12. Yang C, Xu X, Wang L (2006) Information fusion technology and its application in INS/GPS/DVL integrated navigation. J Chin Inertial Techno 14(5):39–43 (in Chinese)
13. Li J, Wang H, Liu A (2005) A CNS/INS combined navigation system. Control Theory Appl 24(4):13–15 (in Chinese)
14. Fang JC, Wan DJ (1996) A fast initial alignment method for strap down inertial navigation system on stationary base. IEEE Trans Aerosp Electron Syst 32(4):1501–1505
15. Huang Y, Han JQ (1999) The self-stable region approach for second order systems. Syst Sci Math Sci 12(3):238–245
16. Zhao W, Wu J, Zhuang L (2007) Initial alignments of inertial navigation system on static base with ESO technique. J Data Acquisition Process 22(1):105–109 (in Chinese)

17. Huang Y, Han JQ (2000) Analysis and design for the second order nonlinear continuous extended states observer. Chin Sci Bull 45(21):1938–1944 (in Chinese)
18. Niu L, Li L, Zhuang L (2004) Active disturbance rejection control of stabilizing circuits for inertial platform. J Chin Inertial Technol 12(6):48–51 (in Chinese)

Using Master-Equation Approach to Explore Structural Properties of Scientific Collaboration Network Model

Long Wang

Abstract The master-equation approach is used to describe the dynamic structure property of complex networks. In this paper, we consider the scientific collaboration network model that there are multiple new scientists participant in each time. By master-equation approach, we study the node degree distribution in the scientific collaboration network model with numerical simulations verifying the feasibility of the model. It is proved that the node degree distribution of the model is a right-skewed power-law like distribution. For uncorrelated scientific collaboration network, we get the epidemic threshold of this model by actual data.

Keywords Complex network · Degree distribution · Master-equation approach · Right-skewed power-law · Scientific collaboration network

1 Introduction

In recent years, complex network has gained much attention from many researchers. Many real systems can be described by complex network in which a node represents the system's elementary unit, and an edge represents the interaction or relationship between a pair of nodes. In the late 1990s, Duncan Watts firstly came up with the theory of scale-free networks whose degree distribution is a power-law distribution. These examples include social networks, World Wide Web [1] and biological interacting networks [2]. Complex network research upsurge first derives from Watts-Strogatz small world network model [3] and Barabasi-Albert model (BA model) [4, 5]. The collaboration network, a special complex network, has attracted a great deal of interest in social scientific fields [6–8].

L. Wang (✉)
School of Management Science & Engineering,
Shandong Normal University, Jinan, China
e-mail: w.long163@163.com

E. Qi (ed.), *Proceedings of the 6th International Asia Conference on Industrial Engineering and Management Innovation*,
DOI 10.2991/978-94-6239-145-1_42

A collaboration network is a network consisting of a variety of individuals (e.g. organizations or people) who are largely autonomous, geographically distributed, and heterogeneous in terms of their operating environment, culture, social capital and goals, but that collaborates to better achieve common or compatible goals. The discipline of collaborative networks focuses on the structure, behavior, and evolving dynamics of networks of autonomous entities that collaborate to better achieve common or compatible goals [9]. An example of this kind of networks is scientists related by co-authoring a scientific paper. Scientific collaboration network generally contains "articles" and "authors" denoted by two disjoint node sets [7], and the relationships between articles and authors are usually represented as edges respectively. In scientific collaboration network, scientific research persons are the authors and the published scientific paper. If a paper C is published with two authors, A and B, are cooperated, the edges are (A, C) and (B, C). Obviously, there are two kinds of degree distribution in the collaboration network model, namely paper node degree distribution and author node degree distribution. Paper node degree indicates the size of the article, such as the number of authors in each paper; author node degree indicates the number of authors and articles, such as the number of papers that is published by each author.

Some researchers are usually concerned about the project statistical characteristics in bipartite graphs, bipartite graphs are empirically studied. In this paper, we transform the scientific collation network model into the one-mode projection of the bipartite graph, namely single bipartite graph model. If a paper is published with two authors, A and B, are cooperated, the edges are (A, B). The nodes are the authors and the node degree k is defined as the article number that the author has published. The node degree k is also equal to the number of authors who cooperate with this node. In this model, n authors are created in each time t. The n authors cooperate in a scientific paper. Among the n authors, m authors are 'new' authors who join in the network just now, and the rest of n-m authors are chosen from the 'old' authors who have existed in the network by the preferential attachment mechanism, a probability proportional to the number k of articles.

In this paper, by the master-equation method, we analyze the degree distribution in the scientific collaboration network model. We organize this study as following: In Sect. 2, we give the scientific collaboration network model. By master-equation approach, we find that the network author nodes degree distribution is a right-skewed power-law like distribution, and present some numerical simulations to proof the feasibility of the model. In Sect. 3, for uncorrelated scientific collaboration network, we get the epidemic threshold of this model by actual data.

2 Scientific Collaboration Network Model and Degree Distribution

2.1 Scientific Collaboration Network Model

In this section, we give the Scientific Collaboration Network Model.

Initial Starting with a complete network with $(n - m)$ nodes.

1. At each time step, m new nodes are added into the network. For each new node with $n - m$ edges connect to $n - m$ 'old' nodes, and with $m - 1$ edges connect to $m - 1$ 'new' nodes. We assume that the $n - m$ old author nodes who published the same paper have connected to each other. Thus, each paper has n authors.

2. When the 'old' nodes are chosen to connect the new node, we assume that the probability Π of a new node connecting to an 'old' node i is proportional to the degree k_i of i, that is, $\prod(k_i) = k_i / \sum_j k_j$ called preferential attachment mechanism.

We assume n and m are constants, the average number of participants as well as the average number of new participants per project are $n = \overline{n}$ and $m = \overline{m}$ respectively. At each time step, the total number of articles is t, while the total number of authors increases as $N \rightarrow N + m$.

2.2 Degree Distribution

A basic character of the scientific collaboration network is the author nodes degree distribution $P(k)$. The author nodes degree k is defined as the article number that the author has published. The node degree k is also equal to the number of authors who cooperate with this author. There are $m(n - m)$ edges of m 'new' nodes joining with the $n - m$ 'old' nodes, $m(m - 1)/2$ new edges of m 'new' nodes joining with each other, and there is no new edge among the $n - m$ 'old' nodes at each time step. Therefore, the total degree of the network is $\sum_j k_j = 2(m(n - m) + m(m - 1)/2)$ $t = m(2n - m - 1)t$ at time t.

If $k < n - 1$, then $k/m < (n - 1)/\text{m}$, we have $P(k) = 0$, $P(k - m) = 0$.

If $n - 1 \leq k < n + m - 1$, then $(n - 1)/\text{m} \leq k/m < (n - 1)/\text{m} + 1$, we have $P(k) > 0$, $P(k\text{-}m) = 0$.

If $k \geq n{+}m - 1$, $k/m \geq (n - 1)/\text{m} + 1$, we have $P(k) > 0$, $P(k - m) > 0$.

So we define δ_{km} as a switch variable: $k = n - 1$, then $k/m = (n - 1)/m$, we have $\delta_{km} = 1$. If $k > n - 1$, $k/m > (n - 1)/m$, we have $\delta_{km} = 0$.

Now we set that k_i shows the degree of node i. $P(k,i,t)$ shows the probability that the node i has k edges at time t. At each time, m new nodes and $m(m - 1)/2$ new edges are added in the network. The probability that makes the degree of node i add $n - 1$ is

$$C_{m(n-m)}^{m}\frac{k_i}{\sum_j k_j} = C_{m(n-m)}^{m}\frac{k_i}{m(2n-m-1)t}.$$

Set $\lambda = m(2n-m-1)/C_{m(n-m)}^{m}$, so we can get the master-equation is

$$P(k,i,t+1) = \frac{k-1}{\lambda t}P(k-m,i,t) + (1-\frac{k}{\lambda t})P(k,i,t) \tag{1}$$

The boundary condition is $P(k,t,t) = \delta_{km.}$
For convenience to computation, we can rewrite (1) to

$$\begin{aligned}&\frac{t+1}{t+1}(\sum_{i=1}^{t+1}P(k,i,t+1) - P(k,t+1,t+1))\\&=\frac{k-m}{\lambda t}\sum_{i=1}^{t}P(k-m,i,t) + (1-\frac{k}{\lambda t})\sum_{i=1}^{t}P(k,i,t)\end{aligned} \tag{2}$$

We assume $\lim_{t\to\infty} t(P(k,t+1) - P(k,t)) = 0$. When t $\to \infty$, by (2), we have

$$P(k) - \delta_{km} = \frac{k-m}{\lambda}P(k-m) - \frac{k}{\lambda}P(k) \tag{3}$$

If $k = n - 1$, $\delta_{km} = 1$, we have

$$P(\mathrm{n}-1) = \frac{\lambda}{\lambda + n - 1} \tag{4}$$

For convenience to discussion, we set $\lambda' = \lambda/m$. Thus, the above equation is written by

$$P(n-1) = \lambda'/(\lambda' + \frac{n-1}{m}) \tag{5}$$

If $k > n - 1$, $\delta_{km} = 0$, we have

$$P(k) = \frac{k-m}{\lambda + k}P(k-m) \tag{6}$$

$$\begin{aligned}P(k'm) &= \frac{k'-1}{\lambda' + k'}P((k'-1)m)\\&= \frac{(k'-1)(k'-2)(k'-3)\cdots(\frac{n-1}{m})}{(\lambda'+k')(\lambda'+k'-1)(\lambda'+k'-2)\cdots(\lambda'+\frac{n-1}{m}-1)}P(n-1)\end{aligned} \tag{7}$$

By (5) and (7), the author nodes degree distribution is

$$P(k) = \lambda' \frac{\Gamma(\frac{k}{m})\Gamma(\lambda' + \frac{n-1}{m})}{\Gamma(\frac{n-1}{m})\Gamma(\lambda' + \frac{k}{m} + 1)} \tag{8}$$

To verify the feasibility of author nodes degree distribution in Eq. (8), we simulate the scientific collaboration network model (see Figs. 1 and 2).

The author nodes degree distribution of Eq. (8) is not strictly drift power-law distribution. However, the trend of the distribution in Figs. 1 and 2 is relatively similar to the drift power-law distribution [10]. So we conjecture that the author nodes degree distribution (8) can use the drift power-law distribution to approximate.

Fig. 1 $\lambda' = 2$, m = 1, n = 3, t = 1000, simulation result of (8)

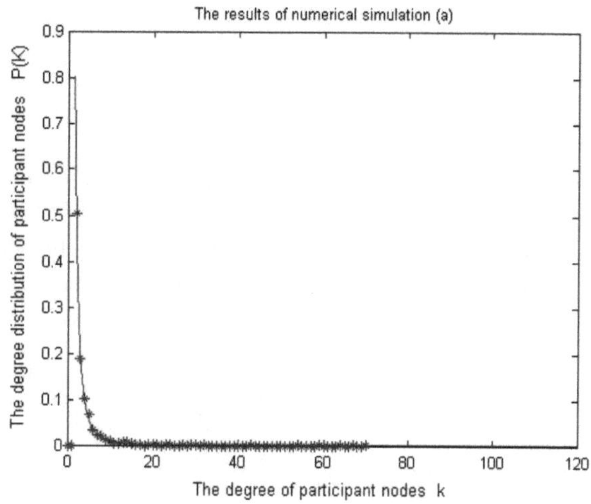

Fig. 2 $\lambda' = 5/6$, m = 1, n = 3, t = 2000, simulation result of (8)

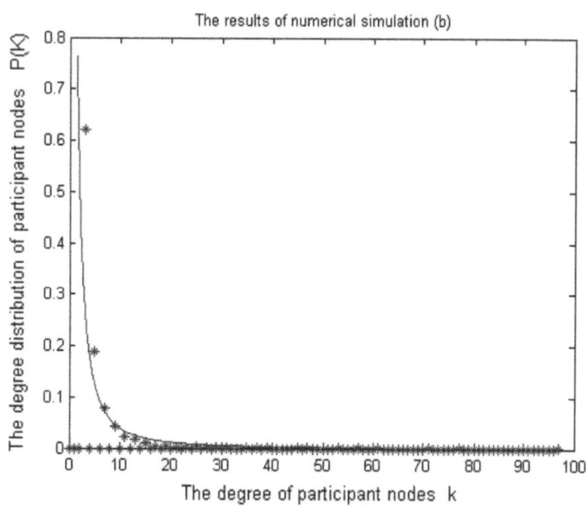

By (8), we have

$$P(k) = \lambda' \frac{\Gamma(\frac{k}{m})\Gamma(\lambda' + \frac{n-1}{m})}{\Gamma(\frac{n-1}{m})\Gamma(\lambda' + \frac{k}{m} + 1)} = \lambda' \frac{\Gamma(k')\Gamma(\lambda' + \frac{n-1}{m})}{\Gamma(\frac{n-1}{m})\Gamma(\lambda' + k' + 1)} \tag{9}$$

The drift power-law distribution is

$$P(k') \propto (k' + a)^{-\eta} \tag{10}$$

By (9) and (10), we assume

$$\frac{A\Gamma(k')}{\Gamma(k' + \lambda' + 1)} = A(k' + a\lambda')^{-(\lambda'+1)} \tag{11}$$

where $A = \lambda'\Gamma(\lambda' + \frac{n-1}{m})/\Gamma(\frac{n-1}{m})$
By (11), we have

$$a = \frac{1}{\lambda'}((\frac{\Gamma(k')}{\Gamma(k' + \lambda' + 1)})^{-\frac{1}{\lambda'+1}} - k') \tag{12}$$

The function image of Eq. (12) is shows in Fig. 3.

In Fig. 3, we found that the two function curves coincide when a is close to 0.5. So when a = 0.5, for different λ' values, (11) was established.

$$P(k) = A(k' + \frac{\lambda'}{2})^{-(\lambda'+1)} = A(\frac{k}{m} + \frac{\lambda'}{2})^{-(\lambda'+1)} \tag{13}$$

where $A = \lambda'\Gamma(\lambda' + \frac{n-1}{m})/\Gamma(\frac{n-1}{m})$ is looked as an approximation of Eq. (8). The node degree distribution of the model is a right-skewed power-law like distribution.

Fig. 3 The function images with two different λ' values of Eq. (12) are shown as the panel. The red curve shows λ' = 2 and the dotted curve shows λ' = 5/6. The X-axe is the values of k', and the Y-axe is the values of a

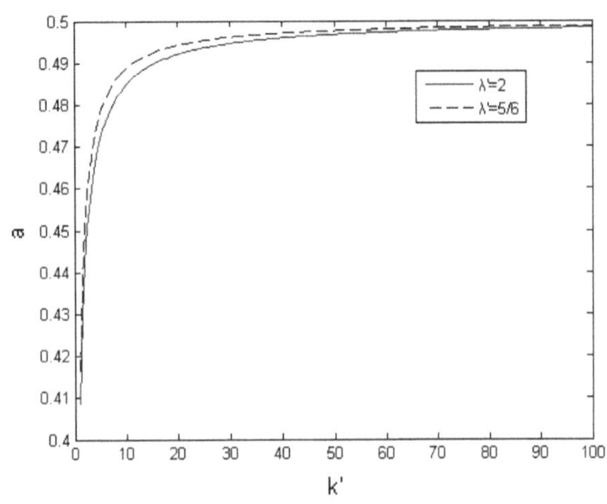

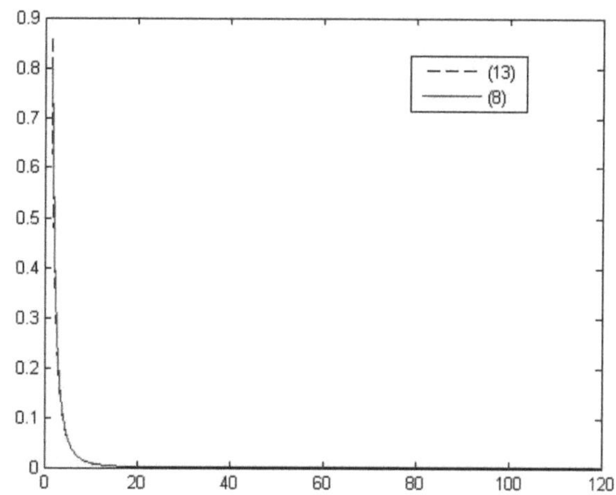

Fig. 4 The comparison of the participant nodes degree distribution (8) and (13), $n = 3$, $m = 1$, $\lambda' = 2$

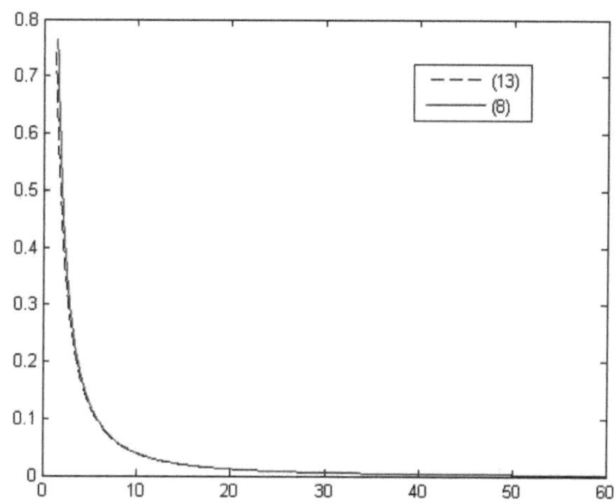

Fig. 5 The comparison of the participant nodes degree distribution (8) and (13), n = 4, m = 2, $\lambda' = 5/6$

In Figs. 4 and 5, we found that the equation images of (8) and (13) are almost entirely fit well to each other.

3 Epidemic Threshold of Uncorrelated Scientific Collaboration Network

Pastor-Satorras and Vespignani consider epidemic threshold of scale-free network [11]. They give the dynamical mean-field equations can thus be written as

$$\frac{d\rho_k(t)}{dt} = -\rho_k(t) + \lambda k[1 - \rho_k(t)]\Theta(\rho_k(t)) \tag{14}$$

where $\rho_k(t)$ express the relative density of infected nodes with given connectivity k; i.e., the probability that a node with k links is infected. The probability $\Theta(\rho(t))$ express that any given link points to an infected node, which has the expression.

$$\Theta(\rho(t)) = <k>^{-1} \sum_k kP(k)\rho_k(t) \tag{15}$$

By solving (14) and (15) in the stationary state $[d\rho_k(t)/dt = 0]$ they obtain the self-consistency equation

$$\Theta = <k>^{-1} \sum_k \frac{\lambda k \Theta}{1 + \lambda k \Theta} \tag{16}$$

where Θ is now a function of λ alone. The self-consistency (16) allows a solution with $\Theta \neq 0$ and $\rho_k \neq 0$ only if the condition $\lambda < k^2 >/<k> \geq 1$ is fulfilled, defining the epidemic threshold

$$\lambda_c = \frac{<k>}{<k^2>}$$

In order to apply the above theory, we consider that if λ' is a positive small value, then $k' + \lambda'/2$ approximately equals to k'. By (13), we obtain

$$P(k) \approx Ak'^{-(\lambda'+1)} = Am^{(\lambda'+1)}k^{-(\lambda'+1)} \tag{17}$$

By Eq. (17), we found that if λ' is relatively small, the author nodes degree distribution is approximately scale-free distribution. In this case, we consider the epidemic threshold of uncorrelated scientific collaboration network.

We consider the actual data of scientific collaboration network (see Table I) [8]. We found that $n = 3.05 \approx 3$, $m = 0.91 \approx 1$ in the scientific collaboration network.

In the scientific collaboration network, $n = 3$, $m = 1$, $n - m = 2$, $\lambda' = 2$. The average degree of this model is

$$<k> = \int_2^{\infty} kP(k)dk = 6, P(k) = 12k^{-3} \tag{18}$$

Table 1 A part of table in [8]

Variable	Coauthors network	Analytic results
t	17828	
N	16258	
n	3.05	3.05
m	0.91	0.91

By (18), we can get

$$\begin{aligned}\Theta(\lambda) &= \frac{1}{6}\lambda\Theta(\lambda)\int_2^\infty k\cdot 12k^{-3}\frac{k}{1+k\lambda\Theta(\lambda)}dk \\ &= 2\lambda\Theta(\lambda)\ln(1+\frac{1}{2\lambda\Theta(\lambda)})\end{aligned} \tag{19}$$

By (19), we have

$$\Theta(\lambda) = \frac{e^{-1/2\lambda}}{2\lambda}(1-e^{-1/2\lambda})^{-1} \tag{20}$$

Next we compute the order parameter

$$\begin{aligned}\rho &= \sum_k P(k)\rho_k = 8\lambda\Theta(\lambda)\int_2^\infty \frac{1}{k^2}\frac{dk}{1+k\Theta(\lambda)} \\ &= 8\lambda\Theta(\lambda)[\frac{1}{2}+\lambda\Theta(\lambda)\ln(1+\frac{1}{2\lambda\Theta(\lambda)})]\end{aligned} \tag{21}$$

By (20) and (21), we have

$$\rho \propto 2e^{-1/2\lambda} \tag{22}$$

Let $2e^{-1/2\lambda} = 0$, we have $\lambda = 0$. So the epidemic threshold of scientific collaboration Network is $\lambda_c = 0$.

4 Conclusion

The main motivation in this study is to apply master-equation approach in scientific collaboration network model. We hope this study would push the application of master-equations approach forward as a useful tool to investigate the structure of dynamic networks.

Acknowledgments The author gives thanks to all the good ideas in the literatures in this study. This work is supported in part by National Science Foundation of China grants 71071090.

References

1. Albert R, Jeong H, Barabási A-L (1999) Internet: diameter of the world-wide web. Nature 401:130–131
2. Hahn MW, Kern AD (2005) Mol Biol Evol: Comparative genomics of centrality and essentiality in three eukaryotic protein-interaction networks. Oxford J 22(4):803–806

3. Watts DJ, Strogatz SH (1998) Collective dynamics of 'small- world' networks. Nature 393:440–442
4. Barabási AL, Albert R (1999) Emergence of scaling in random networks. Science 286:509–512
5. Barabási AL, Albert R, Jeong H (1999) Mean-field theory for scale-free random networks. Phys A 272:173–187
6. Wasserman S, Faust K (1997) Social network analysis: methods and applications. Am Ethnol 24(1):219–220
7. Newman MEJ (2001) The structure of scientific collaboration networks. PNAS 98(2):404–409
8. Ramasco JJ, Dorogovtsev SN, Pastor-Satorras R (2004) Self-organization of collaboration networks. Phys: Rev: E 70(3):036106 (10 pp)
9. Camarinha-Matos LM, Afsarmanesh H (2005) The emerging discipline of collaborative networks. J. Intell Manuf 16(4-5):439–452
10. Chang H, Su BB, Zhou YP, He DR (2007) Assortativity and act degree distribution of some collaboration networks. Phys A 383(2):687–702
11. Pastor-Satorras R, Vespignani A (2001) Epidemic dynamic and endemic states in complex networks. Phys Rev E 63:066117

Distribution Network Service Restoration Interval Number Group Decision-Making Using Grey TOPSIS Method

Tian-lei Zang, Jian-wei Yang, Zheng-you He and Qing-quan Qian

Abstract Considering the participation of multiple dispatchers in decision-making and the influence of load change on power distribution network service restoration, a grey technique for order preference by similarity to an ideal solution (Grey TOPSIS) group decision-making method for distribution network service restoration is proposed in this paper. Firstly, five evaluation indices such as load restoration amount, load capacity margin, switching times, load transfer amount and load balance rate are selected. Interval grey number is used to represent service restoration decision-making matrix. Attribute values of efficiency index and cost index are standardized respectively. On this basis, the weight of each index is determined. Finally, Grey TOPSIS method is used to collect decision-making information of each restoration decision-making expert and obtain service restoration scheme sorting. The test on typical six-feeder power distribution system shows that the proposed method can make scientific service restoration group decision in the case of load change.

Keywords Decision-making · Distribution network · Grey TOPSIS · Load change · Service restoration

1 Introduction

As one of the core key technologies for realizing self-healing control of smart power grids [1], distribution network service restoration [2–11] is an optimal decision-making problem which should consider multiple constraints and objectives such as radial operation, system power losses, load balance, capacity margin and switching times. Due to its numerous constraints, objectives and indices of service restoration candidate schemes [12], it is difficult for power dispatchers to

T. Zang (✉) · J. Yang · Z. He · Q. Qian
School of Electrical Engineering, Southwest Jiaotong University, Chengdu, China
e-mail: zangtianlei@126.com

E. Qi (ed.), *Proceedings of the 6th International Asia Conference on Industrial Engineering and Management Innovation*,
DOI 10.2991/978-94-6239-145-1_43

make optimal restoration decision in a short time. Therefore, the research on practical distribution network service restoration decision-making support system is vital for the safe and economic operation of distribution network.

Grey multiattribute decision-making method [13] is a powerful tool for the comprehensive evaluation of schemes and suitable for distribution network service restoration decision-making. Meanwhile, actual distribution network service restoration decision is often made by multiple power distribution dispatchers. Therefore, it is necessary to study the method for multiple power distribution dispatchers to participate in service restoration decision, i.e. service restoration multiattribute group decision-making method, to meet the development need of smart power distribution in the future. In addition, the load of each node in actual distribution network often changes randomly, causing the optimal operation structure of power distribution network to change [14]. Correspondingly, the loading value of each node in power distribution network might change before and after fault occurrence. Therefore, it is necessary to study distribution network service restoration multiattribute group decision-making method adapting to the load change.

Therefore, five evaluation indices—load restoration amount, load capacity margin, switching times, load transfer amount and load balance rate are selected to evaluate service restoration schemes in this paper. Considering the influence of load change, interval grey numbers of service restoration evaluation indices are given. And considering the participation of multiple power distribution dispatchers in service restoration, a grey technique for order preference by similarity to an ideal solution (Grey TOPSIS) multiattribute group decision-making method is used to evaluate service restoration schemes comprehensively.

2 Distribution Network Service Restoration Evaluation Indices

Considering the main objective of distribution network service restoration, the following evaluation indices are used to evaluate restoration schemes [5]:

(1) Load restoration amount

Load restoration amount index I_1(A) is defined as the sum of load current in the outage areas by transferring the load to other feeders.

(2) Load capacity margin

Feeder load capacity margin refers to the difference value of rated load and actual load of feeder. The minimum value of load capacity margin of each feeder is used as index I_2.

$$I_2 = \min\{L'_{\text{fd1}}, L'_{\text{fd}_2}, \ldots, L'_{\text{fd}_i}, \ldots, L'_{\text{fd}_n}\} \tag{1}$$

where, L'_{fd_i} refers to the load capacity margin of the ith feeder, A; n is the number of feeders.

(3) Switching times

Switching times index I_3 refers to the switching times required for service restoration.

(4) Load transfer amount

Feeder load transfer amount refers to the transfer load amount that can be accepted by a feeder. The maximum value of load current increment of each feeder is used as index I_4.

$$I_4 = \max\{\Delta L_{\mathrm{fd}_1}, \Delta L_{\mathrm{fd}_2}, \ldots, \Delta L_{\mathrm{fd}_i}, \ldots, \Delta L_{\mathrm{fd}_n}\} \tag{2}$$

where, ΔL_{fd_i} is load increment of the ith feeder, A.

(5) Load balance rate

Load balance rate index I_5 refers to the maximum value of load balance rate of all feeders adjacent to the interconnection switch [15].

$$I_5 = \max\{B_1, B_2, \ldots, B_i, \ldots, B_n\} \tag{3}$$

where, B_i is the load balance rate of the ith feeder.

$$B_i = \max[L_m, L_n]/\min[L_m, L_n] \tag{4}$$

where, L_m and L_n are loading values on both sides of the interconnection switch; m and n are the collection of all power supply in feeder couple belonging to interconnection switch T_{S_i}.

3 Interval Grey Number Operations

Definition 1 Assume that there are two interval grey numbers $a(\otimes) \in [\underline{a}, \bar{a}]$ and $b(\otimes) \in [\underline{b}, \bar{b}]$, $\underline{a} > 0$, $\underline{b} > 0$ and β is positive real number, then,

(1) $a(\otimes) + b(\otimes) \in [\underline{a} + \underline{b}, \bar{a} + \bar{b}]$;

(2) $a(\otimes)b(\otimes) \in [\min\{\underline{a}\,\underline{b}, \underline{a}\,\bar{b}, \bar{a}\,\underline{b}, \bar{a}\,\bar{b}\}, \max\{\underline{a}\,\underline{b}, \underline{a}\,\bar{b}, \bar{a}\,\underline{b}, \bar{a}\,\bar{b}\}]$;

(3) $\beta a(\otimes) \in [\beta\,\underline{a}, \beta\,\bar{a}]$;

(4) $\beta + a(\otimes) \in [\beta + \underline{a}, \beta + \bar{a}]$.

Definition 2 Assume that there are two interval grey numbers $a(\otimes) \in [\underline{a}, \bar{a}]$ and $b(\otimes) \in [\underline{b}, \bar{b}]$, the distance between them is:

$$L(a(\otimes), b(\otimes)) = 2^{-1/2}[(\underline{a} - \underline{b})^2 + (\bar{a} - \bar{b})^2]^{1/2} \quad (5)$$

Definition 3 Assume that values of two objects S_1 and S_2 under index set $u = \{u_1, u_2, \ldots, u_m\}$ are interval grey number vectors $a_1(\otimes) = (a_{11}(\otimes), a_{12}(\otimes), \ldots, a_{1m}(\otimes))$, $a_2(\otimes) = (a_{21}(\otimes), a_{22}(\otimes), \ldots, a_{2m}(\otimes))$, the distance between two vectors is:

$$d(a_1(\otimes), a_2(\otimes)) = 2^{-1/2}(\omega_1(a_{-11} - a_{-21})^2 + \omega_1(\bar{a}_{11} - \bar{a}_{21})^2 + \cdots + \omega_m(\bar{a}_{1m} - \bar{a}_{2m})^2) \quad (6)$$

4 Interval Number Grey Decision-Making Model of Service Restoration

4.1 Interval Number Decision-making Matrix and Its Standardization

Assume that there are m restoration schemes $S = \{S_1, S_2, \ldots, S_m\}$ and n evaluation indices $I = \{I_1, I_2, \ldots, I_n\}$, the attribute value of scheme S_i for index I_j is $[\underline{x_{ij}}, \overline{x_{ij}}]$ $(i = 1, 2, \ldots, m; j = 1, 2, \ldots, n)$.

Interval number decision-making matrix $\boldsymbol{X}$ is defined

$$\boldsymbol{X} = \begin{bmatrix} [\underline{x_{11}}, \overline{x_{11}}] & [\underline{x_{12}}, \overline{x_{12}}] & \cdots & [\underline{x_{1n}}, \overline{x_{1n}}] \\ [\underline{x_{21}}, \overline{x_{21}}] & [\underline{x_{22}}, \overline{x_{22}}] & \cdots & [\underline{x_{2n}}, \overline{x_{2n}}] \\ \vdots & \vdots & & \vdots \\ [\underline{x_{m1}}, \overline{x_{m1}}] & [\underline{x_{m2}}, \overline{x_{m2}}] & \cdots & [\underline{x_{mn}}, \overline{x_{mn}}] \end{bmatrix} \quad (7)$$

As index dimension is different, $\boldsymbol{X}$ should be standardized during service restoration decision-making. The specific method is as follows:

If I_j is efficiency type,

$$\underline{r_{ij}} = \underline{x_{ij}} \Big/ \sum_{i=1}^{m} \overline{x_{ij}}, \overline{r_{ij}} = \overline{x_{ij}} \Big/ \sum_{i=1}^{m} \underline{x_{ij}} \quad (8)$$

If I_j is cost type,

$$\underline{r_{ij}} = \frac{1}{\overline{x_{ij}}} \Big/ \sum_{i=1}^{m} \frac{1}{\underline{x_{ij}}}, \overline{r_{ij}} = \frac{1}{\underline{x_{ij}}} \Big/ \sum_{i=1}^{m} \frac{1}{\overline{x_{ij}}} \quad (9)$$

Define standardized decision-making matrix

$$R = \begin{bmatrix} [\underline{r_{11}}, \overline{r_{11}}] & [\underline{r_{12}}, \overline{r_{12}}] & \cdots & [\underline{r_{1n}}, \overline{r_{1n}}] \\ [\underline{r_{21}}, \overline{r_{21}}] & [\underline{r_{22}}, \overline{r_{22}}] & \cdots & [\underline{r_{2n}}, \overline{r_{2n}}] \\ \vdots & \vdots & & \vdots \\ [\underline{r_{m1}}, \overline{r_{m1}}] & [\underline{r_{m2}}, \overline{r_{m2}}] & \cdots & [\underline{r_{mn}}, \overline{r_{mn}}] \end{bmatrix} \tag{10}$$

4.2 Grey TOPSIS Model of Service Restoration

Define decision-making matrix after standardization as $\boldsymbol{R} = (r_{ij}(\otimes))_{n\times m}$. Its attribute vector in each scheme is $r_i(\otimes) = (r_{i1}(\otimes), r_{i2}(\otimes), \ldots, r_{im}(\otimes))$, $i = 1, 2, \ldots n$, where $r_{ij}(\otimes) \in [\underline{r_{ij}}, \overline{r_{ij}}]$ is nonnegative interval grey number between [0,1].

The degree of importance attached to m attributes by each decision maker is different. Assume that the weight of attribute I_i given by decision maker M_k is interval grey number $W_j^{(k)}(\otimes)$, $j = 1, 2, \ldots, m$, i.e. weight information part is known, where $w_j^{(k)}(\otimes) \in [\underline{w_j^{(k)}}, \overline{w_j^{(k)}}](0 \le \underline{w_j^{(k)}} \le \overline{w_j^{(k)}} \le 1, j = 1, 2, \ldots, m)$ and $\sum_{j=1}^{m} \overline{w_j^{(k)}}(\otimes) \ge 1$, $\sum_{j=1}^{m} \underline{w_j^{(k)}} \le 1$, $k = 1, 2, \ldots, l$.

Define $\boldsymbol{y} = (y_{ij}^{(k)}(\otimes))_{n\times m}$ as the weighted standardization decision-making matrix of decision maker M_k, where

$$\begin{gathered} y_{ij}^{(k)}(\otimes) = w_j^{(k)}(\otimes) x_{ij}(\otimes) \in [\underline{y_{ij}}, \overline{y_{ij}}] \\ i = 1, 2, \ldots, n; j = 1, 2, \ldots, m; k = 1, 2, \ldots, l \end{gathered} \tag{11}$$

Assume that attribute vector in each scheme of decision makers' weighted standardization decision-making matrix is $y_i^{(k)}(\otimes) = (y_{i1}^{(k)}(\otimes), y_{i2}^{(k)}(\otimes), \ldots, y_{im}^{(k)}(\otimes))$, $j = 1, 2, \ldots, m$; $k = 1, 2, \ldots, l$.

Define: $^{+}\underline{y_j^{(k)}} = \max\limits_{1\le i\le n}\{\underline{y_{ij}^{(k)}}\}$, $^{+}\overline{y_j^{(k)}} = \max\limits_{1\le i\le n}\{\overline{y_{ij}^{(k)}}\}$, $j = 1, 2 \ldots m$, $k = 1, 2 \cdots l$; $^{-}\underline{y_j^{(k)}} = \min\limits_{1\le i\le n}\{\underline{y_{ij}^{(k)}}\}$, $^{-}\overline{y_j^{(k)}} = \min\limits_{1\le i\le n}\{\overline{y_{ij}^{(k)}}\}$, $j = 1, 2, \ldots, m$; $k = 1, 2, \ldots, l$.

The m-dimensional nonnegative interval grey number vector

$$^{+}y^{(k)}(\otimes) = \{^{+}y_1^{(k)}(\otimes), {}^{+}y_2^{(k)}(\otimes), \ldots, {}^{+}y_m^{(k)}(\otimes)\} \tag{12}$$

is grey positive ideal solution of decision maker M_k, and nonnegative interval grey number vector

$$^{-}y^{(k)}(\otimes) = \{^{-}y_1^{(k)}(\otimes), ^{-}y_2^{(k)}(\otimes), \ldots, ^{-}y_m^{(k)}(\otimes)\} \tag{13}$$

is grey negative ideal solution of decision maker M_k, where $^{+}y_i^{(k)}(\otimes) \in [^{+}\underline{y_j^{(k)}}, ^{+}\overline{y_j^{(k)}}]$, $^{-}y_i^{(k)}(\otimes) \in [^{-}\underline{y_j^{(k)}}, ^{-}\overline{y_j^{(k)}}]$.

Assume that the universe of discourse of grey number $\otimes$ is Ω, $\mu(\otimes)$ is the measure of number field of grey number $\otimes$, $g^{\circ}(\otimes) = \mu(\otimes)\mu(\Omega)$ is the grey level of grey number $\otimes$, g° for short.

Due to all interval grey numbers $r_{ij}(\otimes) \in [0, 1]$ after standardization, it can be considered that the domain of discourse after standardization is Ω=[0, 1].

Assume that there are two interval grey numbers $a(\otimes) \in [\underline{a}, \bar{a}]$ and $b(\otimes) \in [\underline{b}, \bar{b}]$, $\overline{a(\otimes)}$ and $\overline{b(\otimes)}$ are respectively uniform average whitening number of $a(\otimes)$ and $b(\otimes)$, g_a° and g_b° are respectively grey level of $a(\otimes)$ and $b(\otimes)$,

$$d(a(\otimes), b(\otimes)) = \sqrt{|\overline{b(\otimes)} - \overline{a(\otimes)}|^2 + |g_b^{\circ} - g_a^{\circ}|^2}$$

is the deviation degree of interval grey numbers $a(\otimes)$ and $b(\otimes)$. Then,

$$c_k(A_i) = \frac{d_{ki}^{-}}{d_{ki}^{+} + d_{ki}^{-}}, i = 1, 2, \ldots, n; k = 1, 2, \ldots, l \tag{14}$$

is the approach degree of scheme S_i to ideal solution of decision maker M_k, where

$$d_{ki}^{+} = \sum_{j=1}^{m} d(^{+}y_j^{(k)}(\otimes), y_{ij}^{(k)}(\otimes)), i = 1, 2\ldots n;$$

$$d_{ki}^{-} = \sum_{j=1}^{m} d(^{-}y_j^{(k)}(\otimes), y_{ij}^{(k)}(\otimes)), i = 1, 2\ldots n.$$

Matrix

$$\boldsymbol{C} = \begin{vmatrix} c_1(A_1) & c_2(A_2) & \cdots & c_l(A_1) \\ c_1(A_2) & c_2(A_2) & \cdots & c_l(A_2) \\ \vdots & \vdots & \vdots & \vdots \\ c_1(A_n) & c_2(A_n) & \cdots & c_1(A_n) \end{vmatrix}$$

is relative approach degree matrix of group decision-making.

If the status of each decision maker is different, i.e. authority differences exist, but the authority weight of each decision maker is not necessarily clear number and might be interval grey number, i.e. $\alpha_k(\otimes) \in (\underline{\alpha_k}, \overline{\alpha_k}), k = 1, 2, \ldots, l,$ where $0 \leq \underline{\alpha_k} \leq \overline{\alpha_k} \leq 1,\ k = 1, 2, \ldots, l,$ and $\sum_{k=1}^{l} \overline{\alpha_k} \geq 1,\ \sum_{k=1}^{l} \underline{\alpha_k} \leq 1.$ The matrix after weighting is:

$$C_\alpha = \begin{vmatrix} \alpha_1(\otimes)c_1(S_1) & \alpha_2(\otimes)c_2(S_1) & \cdots & \alpha_l(\otimes)c_l(S_1) \\ \alpha_1(\otimes)c_1(S_2) & \alpha_2(\otimes)c_2(S_2) & \cdots & \alpha_l(\otimes)c_l(S_2) \\ \vdots & \vdots & \vdots & \vdots \\ \alpha_1(\otimes)c_1(S_n) & \alpha_2(\otimes)c_2(S_n) & \cdots & \alpha_l(\otimes)c_1(S_n) \end{vmatrix} \tag{15}$$

$I_G^+ = (r_1^+(\otimes), r_2^+(\otimes), \ldots, r_l^+(\otimes))$ is interval grey number group positive ideal solution and $I_G^- = (r_1^-(\otimes), r_2^-(\otimes), \ldots, r_l^-(\otimes))$ is interval grey number group negative ideal solution, where

$$r_j^+(\otimes) \in [\max_{1 \leq k \leq n} \underline{a}_j c_j(S_k), \max_{1 \leq k \leq n} \bar{a}_j c_j(S_k)], j = 1, 2, \ldots, l$$

$$r_j^-(\otimes) \in [\min_{1 \leq k \leq n} \underline{a}_j c_j(S_k), \min_{1 \leq k \leq n} \bar{a}_j c_j(S_k)], j = 1, 2, \ldots, l$$

$$c_G(A_i) = \frac{d_i^-}{d_i^+ + d_i^-} \quad (i = 1, 2 \ldots n) \tag{16}$$

is the approach degree of scheme S_i to group ideal solution, where

$$d_i^+ = \sum_{j=1}^{l} d(r_j^+(\otimes), \alpha_j(\otimes)c_j(S_i)), i = 1, 2, \ldots, n;$$

$$d_i^- = \sum_{j=1}^{l} d(r_j^-(\otimes), \alpha_j(\otimes)c_j(S_i)), i = 1, 2, \ldots, n.$$

Obviously, $0 \leq c_G(S_i) \leq 1$, $i = 1, 2, \ldots, n$, and the higher the $c_G(S_i)$ is, the closer the scheme S_i is to group positive ideal scheme and the farther to group negative ideal scheme.

Combination of the above, grey TOPSIS method of distribution network service restoration is as follows:

Step 1: decision-making matrix $\boldsymbol{X} = (x_{ij}(\otimes))_{n \times m}$ is converted into standardized decision-making matrix $\boldsymbol{R} = (r_{ij}(\otimes))_{n \times m}$ using Eqs. (8) and (9).

Step 2: weighted standardization decision-making matrix $\boldsymbol{y} = (y_{ij}^{(k)}(\otimes))_{n \times m}$ $(k = 1, 2, \ldots, l)$ of each decision maker is established using attribute weight $w_j^{(k)}(\otimes)$ of each decision maker and standardized decision-making matrix $\boldsymbol{R} = (r_{ij}(\otimes))_{n \times m}$ and Eq. (6);

Step 3: positive and negative ideal solutions ${}^{+}y^{(k)}(\otimes)$ and ${}^{-}y^{(k)}(\otimes)(k = 1, 2, \ldots, l)$ of each decision maker are calculated using Eqs. (12) and (13). The approach degree $c_k(S_i)$ of the scheme to ideal solution of decision maker M_k is calculated according to Eq. (14) and then approach degree matrix $\boldsymbol{C}$ is established;

Step 4: weighting matrix $\boldsymbol{C}_\alpha$ of approach degree matrix $\boldsymbol{C}$ is established according to the authority weight of decision makers;

Step 5: group positive and negative ideal solutions are determined according to matrix $\boldsymbol{C}_\alpha$ and the approach degree $c_G(S_i)$ of each scheme to group ideal solution is calculated according to Eq. (16);

Step 6: schemes are sorted in descending order of the value of $c_G(S_i)$.

5 Case Study

The effectiveness of service restoration interval number grey TOPSIS method is verified on six-feeder power distribution network in literature [5, 6]. Suppose that a permanent fault occurs at zone Z_{16} as shown in Fig. 1. Service restoration candidate

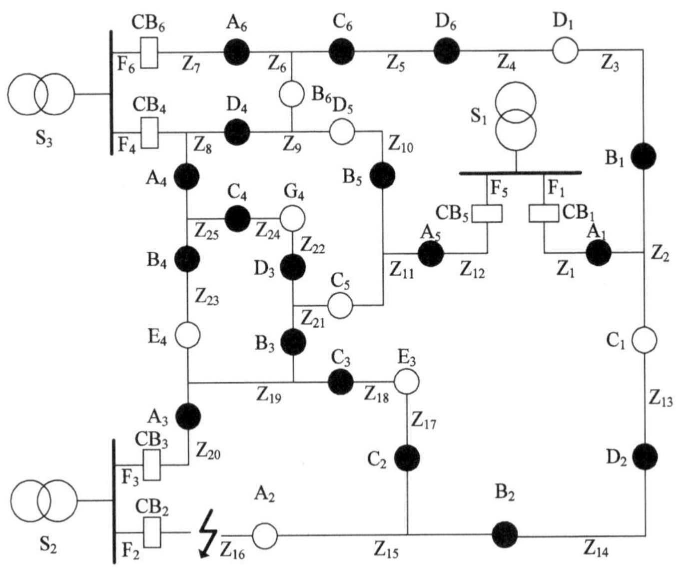

Fig. 1 Six-feeder distribution network

schemes set is formed with the method in literature [5]. Assume that the variation range of load of all feeders is $\delta = \pm 5\%$, i.e. The interval number of load current value I_l is expressed as $[I_l^L, I_l^U] = [0.95I_l, 1.05I_l]$, the interval number of each evaluation index is calculated according to the representing method of interval number, as shown in Table 1.

Among fault restoration evaluation indices, indices I_1 and I_2 are efficiency type and indices $I_3 \sim I_5$ are cost type.

According to Eqs. (6) and (7), standardization decision-making matrix R is obtained:

$$\boldsymbol{R} = \begin{bmatrix} [0.2262, 0.2763] & [0.0665, 0.6379] & [0.3571, 0.3571] & [0.1986, 0.2426] & [0.2418, 0.3606] \\ [0.2262, 0.2763] & [0.0665, 0.6379] & [0.2143, 0.2143] & [0.1986, 0.2426] & [0.2148, 0.3205] \\ [0.2262, 0.2763] & [0.0665, 0.6379] & [0.2143, 0.2143] & [0.2538, 0.3100] & [0.1688, 0.2518] \\ [0.2262, 0.2763] & [0.1595, 0.8724] & [0.2143, 0.2143] & [0.2538, 0.3100] & [0.1934, 0.2885] \end{bmatrix}$$

Assume that the decision-making group is constituted by 5 experts, interval grey number weight vectors of each decision-making expert to five attributes are respectively:

$$w^{(1)}(\otimes) \in ([0.44, 0.46], [0.09, 0.11], [0.09, 0.11], [0.09, 0.11], [0.24, 0.26])$$

$$w^{(2)}(\otimes) \in ([0.29, 0.31], [0.09, 0.11], [0.14, 0.16], [0.14, 0.16], [0.29, 0.31])$$

$$w^{(3)}(\otimes) \in ([0.29, 0.31], [0.09, 0.11], [0.19, 0.21], [0.09, 0.11], [0.29, 0.31])$$

$$w^{(4)}(\otimes) \in ([0.29, 0.31], [0.19, 0.21], [0.09, 0.11], [0.29, 0.31], [0.09, 0.11])$$

$$w^{(5)}(\otimes) \in ([0.59, 0.61], [0.04, 0.06], [0.04, 0.06], [0.09, 0.11], [0.19, 0.21])$$

Assume that authority interval grey number weight vectors of experts are:

$$\alpha(\otimes) \in ([0.20, 0.22], [0.19, 0.21], [0.21, 0.23], [0.18, 0.20], [0.21, 0.23])$$

Positive and negative ideal solutions of each decision-making expert are calculated using Eqs. (12) and (13). And the relative approach degree matrix can be obtained using Eq. (14):

$$\boldsymbol{C} = \begin{vmatrix} 0.5978 & 0.6347 & 0.6901 & 0.3015 & 0.6120 \\ 0.2343 & 0.2273 & 0.2161 & 0.0765 & 0.2788 \\ 0.0996 & 0.1200 & 0.0767 & 0.2338 & 0.1475 \\ 0.5274 & 0.4867 & 0.4254 & 0.7394 & 0.5370 \end{vmatrix}$$

Table 1 Candidate schemes of service restoration and interval number of evaluation indices

Scheme	Status of switch		Evaluation indices				
	Opened	Closed	I_1/A	I_2/A	I_3	I_4/A	I_5
S_1	B_3	C_5 E_3	[218.5,241.5]	[11.25,38.75]	[3]	[218.5,241.5]	[1.448,1.768]
S_2	B_3 D_3	C_5 E_3 G_4	[218.5,241.5]	[11.25,38.75]	[5]	[218.5,241.5]	[1.629,1.990]
S_3	D_2 D_3	C_1 E_3 G_4	[218.5,241.5]	[11.25,38.75]	[5]	[171,189]	[2.074,2.533]
S_4	B_1 B_2	C_1 D_1 E_3	[218.5,241.5]	[27,53]	[5]	[171,189]	[1.810,2.211]

According to the authority weight vector and matrix of decision-making experts:

$$C_\alpha = \begin{vmatrix} [0.1196, 0.1315] & [0.1206, 0.1333] & [0.1449, 0.1587] & [0.0543, 0.0603] & [0.1285, 0.1408] \\ [0.0469, 0.0515] & [0.0432, 0.0477] & [0.0454, 0.0497] & [0.0138, 0.0153] & [0.0585, 0.0641] \\ [0.0199, 0.0219] & [0.0228, 0.0252] & [0.0161, 0.0176] & [0.0421, 0.0468] & [0.0310, 0.0339] \\ [0.1055, 0.1160] & [0.0925, 0.1022] & [0.0893, 0.0978] & [0.1331, 0.1479] & [0.1128, 0.1235] \end{vmatrix}$$

According to matrix $\boldsymbol{C}_\alpha$, group positive and negative ideal solutions can be obtained:

$$I_G^+ \in ([0.1196, 0.1315], [0.1206, 0.1333], [0.1449, 0.1587], [0.1331, 0.1479], [0.1285, 0.1408])$$

$$I_G^- \in ([0.0199, 0.0219], [0.0228, 0.0252], [0.0161, 0.0176], [0.0138, 0.0153], [0.0310, 0.0339])$$

The approach degree of each scheme to group ideal solution can be obtained using Eq. (16):

$c_G(S_1) = 0.8542$, $c_G(S_2) = 0.1915$, $c_G(S_3) = 0.0524$, $c_G(S_4) = 0.7913$.

The corresponding scheme sorting is $A_1 > A_4 > A_2 > A_3$ according to the descending order of $c_G(S_i)$.

Therefore, scheme A_1 is the optimal service restoration scheme.

It can be seen from the case test that interval number grey TOPSIS method can effectively solve the evaluation problem of service restoration schemes under the condition of load change. And the established scheme has a high credibility.

6 Conclusion

An interval number group decision-making method for service restoration based on grey TOPSIS method is studied in this paper considering the participation of multiple power distribution dispatchers in decision-making and load change. The interval grey numbers are used to represent service restoration decision-making matrix. The service restoration group decision-making result is obtained by grey TOPSIS method. The case test on six-feeder system shows that the proposed method can comprehensively consider subjective and objective information, effectively adapt to the load change, make optimal restoration decision and provide scientific reference for power distribution dispatchers. Service restoration evaluation indices can be increased and decreased according to the operation in practical application. The proposed method can also be used to consider power system decision-making support system such as power quality evaluation, reliability evaluation and black start schemes evaluation.

References

1. Alberto V, Mauricio E (2012) Real-time monitoring and economic dispatch of smart distribution grids: High performance algorithms for DMS applications. IEEE Trans Smart Grid 3(2):866–877
2. Huang CM, Hsieh CT, Wang YS (2011) Evolution of radial basic function neural network for fast restoration of distribution systems with load variations. Int J Electr Power Energy Syst 33 (4):961–968
3. Kleinberg MR, Miu KR, Chiang HD (2011) Improving service restoration of power distribution systems through load curtailment of in-service customers. IEEE Trans Power Syst 26(3):1110–1117
4. Srdjan D, Nikola R (2011) An innovative approach for solving the restoration problem in distribution networks. Electr Power Syst Res 81(10):1961–1972
5. Zang TL, Zhong JC, He ZY, Qian QQ (2012) Service restoration of distribution network based on heuristic rules and entropy weight. Power Syst Technol 36(5):251–257 (in Chinese)
6. Zang TL, He ZY, Qian QQ (2013) Distribution network service restoration multiple attribute group decision-making using entropy weight and group eigenvalue. In: 2013 Third International conference on intelligent system design and engineering applications (ISDEA), Hong Kong, pp 602–606
7. Lin CH, Chen CS, Ku TT, Tsai CT, Ho CY (2011) A multiagent-based distribution automation system for service restoration of fault contingencies. Eur Trans Electr Power 21 (1):239–253
8. Zhang HB, Zhang XY, Tao WW (2010) A breadth-first search based service restoration algorithm for distribution network. Power Syst Technol 34(7):103–108 (in Chinese)
9. Singh SP, Raju GS, Rao GK, Afsari M (2009) A heuristic method for feeder reconfiguration and service restoration in distribution networks. Int J Electr Power Energy Syst 31(7–8):309–314
10. Shin DJ, Kim JO, Kim TK (2004) Optimal service restoration and reconfiguration of network using genetic-tabu algorithm. Electr Power Syst Res 71(2):145–152
11. Irving MR, Luan WP, Daniel JS (2002) Supply restoration in distribution networks using a genetic algorithm. Int J Electr Power Energy Syst 24(6):447–457
12. Kumar Y, Das B, Sharma J (2008) Multi objective, multi constraint service restoration of electric power distribution system with priority customers. IEEE Trans Power Deliv 23 (1):261–270
13. Song J, Dang YG, Hua ZM (2010) Study on group decision-making method based on grey cluster model. Control Decis 25(10):1593–1597 (in Chinese)
14. Wang CS, Wang SX (2002) An interval assessment method for distribution network reconfiguration considering load uncertainty. Proc CSEE 22(5):49–53 (in Chinese)
15. Liu J, Cheng HL, Dong HP, Cai JX (2002) Fault section identification and load balancing of distribution network. Autom Electr Power Syst 26(22):34–38 (in Chinese)

The Tracking Filter Models of Kinetic Kill Vehicle's Seeker

Yue Guo, Xin-xue Liu and Rui-dong Li

Abstract This paper seeks to construct a seeker's filter modeling to keep kinetic tracking target stably in the interception progress. The mathematical modeling of tracking in antenna coordinates system has been deduced deliberately. The attitude and centroid filtering algorithm of kinetic kill vehicle has been designed. The results of the simulation have been proved that the online filtering algorithm can make the seeker tracking the target more stably and effectively.

Keywords Filter algorithm · Kinetic kill vehicle · Paper type · Seeker · Technical paper

1 Introduction

In the interception progress, the seeker of kinetic interceptor needs to keep tracking the target stably in the vision of the seeker [1–3]. But there are many errors, such as the positioning error of the interceptor, the target tracking error, the cross coupling of the strapdown system and other nonlinear errors, which have big influence in the stable attitude control, orbit control and interception accuracy [4–6]. In this paper, the strapdown antenna system has been constructed and the error in the interception has been analyzed. Both the mathematical tracking models of the angle of sight and the sight angular velocity and the centroid movement kinetic parameters have been worked out. The simulation of kinetic kill vehicle system and the strapdown antenna platform have been done in matlab simulink. The main works of this paper is to find an effective filter algorithm to constrain or subtract errors in interception progress.

Y. Guo (✉) · X. Liu
Faculty 906, The Second Artillery Engineering University, Xi'an 710025, China
e-mail: guoyue240@163.com

R. Li
Noncommissioned Officers College, The Second Artillery Engineering University, Xi'an 710025, China

E. Qi (ed.), *Proceedings of the 6th International Asia Conference on Industrial Engineering and Management Innovation*,
DOI 10.2991/978-94-6239-145-1_44

2 The Mathematical Modeling of Tracking Target in Antenna Coordinates

The strapdown antenna platform which is connected the interceptor body by gimbal frame has two-degree-of -freedom in lateral and vertical face. The yaw angle outer gimbal contrast to missile body is λ_y and the antenna pitch angle of inner gimbal contrast to outer gimbal λ_z in the Fig. 1 [7–9].

$$M_o^m = \begin{bmatrix} \cos\lambda_y & \sin\lambda_z & -\sin\lambda_y \\ 0 & 1 & 0 \\ \sin\lambda_y & 0 & \cos\lambda_y \end{bmatrix} \tag{1}$$

$$M_i^o = \begin{bmatrix} \cos\lambda_z & \sin\lambda_z & 0 \\ -\sin\lambda_z & \cos\lambda_z & 0 \\ 0 & 0 & 1 \end{bmatrix} \tag{2}$$

The antenna is been fixed in inner gimbal and hasn't rolled, so the antenna has same angular speed with the strapdown platform. The translated matrix can be written as follow.

$$M_a^m = M_i^o M_o^m = \begin{bmatrix} \cos\lambda_y\cos\lambda_z & \sin\lambda_z & -\sin\lambda_y\cos\lambda_z \\ -\cos\lambda_y\sin\lambda_z & \cos\lambda_z & \sin\lambda_y\sin\lambda_z \\ \sin\lambda_y & 0 & \cos\lambda_y \end{bmatrix} \tag{3}$$

$\dot{\lambda}_y$ is he tracking angular rate of outer gimbal. $\dot{\lambda}_z$ is the tracking angular rate of inner gimbal.

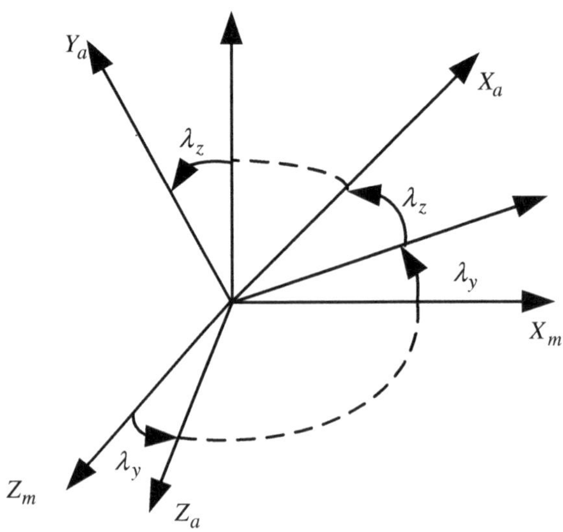

Fig. 1 The relation between antenna coordinates and missile body coordinates

The angular speed in the outer gimbal coordinates which can be expressed as $\omega_o = \left[\omega_{ox}, \omega_{oy}, \omega_{oz}\right]^T$ means the sum of the tracking angular rate of outer gimbal $\dot{\lambda}_y$ and the angular speed of interceptor in the missile body coordinates $\omega_m = \left[\omega_{mx}, \omega_{my}, \omega_{mz}\right]^T$ in outer gimbal coordinates. The equation can be deduced as follow.

$$\omega_o(t) = M_o^m(t)\omega_m(t) + \dot{\lambda}_y(t) = \begin{bmatrix} \omega_{mx}\cos\lambda_y - \omega_{mz}\sin\lambda_y \\ \omega_{my} + \dot{\lambda}_y \\ \omega_{mx}\sin\lambda_y + \omega_{mz}\cos\lambda_y \end{bmatrix} \tag{4}$$

The angular speed in the inner gimbal coordinates which is $\omega_i = \left[\omega_{ix}, \omega_{iy}, \omega_{iz}\right]^T$ means the sum of $\omega_o = \left[\omega_{ox}, \omega_{oy}, \omega_{oz}\right]^T$ in the inner gimbal coordinates and $\dot{\lambda}_z$ which the tracking angular rate of inner gimbal. The equation can be deduced as follow.

$$\omega_i = M_i^o(t)\omega_o(t) + \dot{\lambda}_z(t) = \begin{bmatrix} \omega_{ox}\cos\lambda_z + \omega_{oz}\sin\lambda_z \\ -\omega_{ox}\sin\lambda_z + \omega_{oy}\cos\lambda_z \\ \omega_{oz} + \dot{\lambda}_z \end{bmatrix} \tag{5}$$

From the dynamics theory, the angular speed of antenna in the inner gimbal coordinates is same to the angular speed in inertial coordinates. So the ω_a can been deduced as follows.

$$\begin{aligned} \omega_a = \omega_i &= M_i^o(t)M_o^m(t)\omega_m(t) + M_i^o(t)\dot{\lambda}_y(t) + \dot{\lambda}_z(t) \\ &= \begin{bmatrix} \dot{\lambda}_y\sin\lambda_z \\ \dot{\lambda}_y\cos\lambda_z \\ \dot{\lambda}_z \end{bmatrix} + \begin{bmatrix} \omega_{mx}\cos\lambda_y\cos\lambda_z + \omega_{my}\sin\lambda_z - \omega_{mz}\sin\lambda_y\cos\lambda_z \\ -\omega_{mx}\cos\lambda_y\sin\lambda_z + \omega_{my}\cos\lambda_z + \omega_{mz}\sin\lambda_y\sin\lambda_z \\ \omega_{mx}\sin\lambda_y + \omega_{mz}\cos\lambda_y \end{bmatrix} \end{aligned} \tag{6}$$

Supposed the vector of view angular speed is $\dot{q}(t) = (\dot{q}_x, \dot{q}_y, \dot{q}_z)^T$, while the error of tracking angle is $\varepsilon(t) = (\varepsilon_y, \varepsilon_z)^T$. The equation of tacking angular speed in line of sight coordinates is

$$\dot{q}(t) - D_s^a(t)\omega_a(t) = \dot{\varepsilon}_z(t) + D_{el}^{az}(t)\dot{\varepsilon}_y(t) \tag{7}$$

In the equation, $D_s^a(t)$ is the transition matrix from antenna coordinates to line of sight coordinates and $D_{el}^{az}(t)$ is the transition matrix from the azimuth axes to pitch axes.

$$D_s^a(t) = \begin{pmatrix} \cos\varepsilon_y \cos\varepsilon_z & \sin\varepsilon_z & -\sin\varepsilon_y \cos\varepsilon_z \\ -\cos\varepsilon_y \sin\varepsilon_z & \cos\varepsilon_z & \sin\varepsilon_y \sin\varepsilon_z \\ \sin\varepsilon_y & 0 & \cos\varepsilon_y \end{pmatrix} \tag{8}$$

$$D_{el}^{az}(t) = \begin{pmatrix} \cos\varepsilon_z & \sin\varepsilon_z & 0 \\ -\sin\varepsilon_z & \cos\varepsilon_z & 0 \\ 0 & 0 & 1 \end{pmatrix} \tag{9}$$

Simplified the equation

$$\begin{bmatrix} \dot{\varepsilon}_y \sin\varepsilon_z \\ \dot{\varepsilon}_y \cos\varepsilon_z \\ \dot{\varepsilon}_z \end{bmatrix} = \begin{bmatrix} \dot{q}_x \\ \dot{q}_y \\ \dot{q}_z \end{bmatrix} - \begin{pmatrix} \cos\varepsilon_y \cos\varepsilon_z & \sin\varepsilon_z & -\sin\varepsilon_y \cos\varepsilon_z \\ -\cos\varepsilon_y \sin\varepsilon_z & \cos\varepsilon_z & \sin\varepsilon_y \sin\varepsilon_z \\ \sin\varepsilon_y & 0 & \cos\varepsilon_y \end{pmatrix} \begin{bmatrix} \omega_{ax} \\ \omega_{ay} \\ \omega_{az} \end{bmatrix} \tag{10}$$

So $\dot{\varepsilon}_y$ and $\dot{\varepsilon}_z$ can be deduced as follows.

$$\begin{cases} \dot{\varepsilon}_y = \dfrac{\dot{q}_y}{\cos\varepsilon_z} - \omega_{ay} + \cos\varepsilon_y \tan\varepsilon_z \omega_{ax} - \sin\varepsilon_y \tan\varepsilon_z \cdot \omega_{az} \\ \dot{\varepsilon}_z = \dot{q}_z - \sin\varepsilon_y \omega_{ax} - \cos\varepsilon_y \cdot \omega_{az} \end{cases} \tag{11}$$

By the Coriolis theorem, the formula be deduced as

$$\begin{bmatrix} a_{tx} \\ a_{ty} \\ a_{tz} \end{bmatrix} = \begin{bmatrix} a_{ix} \\ a_{iy} \\ a_{iz} \end{bmatrix} + \begin{bmatrix} \ddot{r} \\ 0 \\ 0 \end{bmatrix} + \begin{bmatrix} 0 & -\ddot{q}_z & \ddot{q}_y \\ \dot{q}_z & 0 & -\ddot{q}_x \\ -\ddot{q}_y & \ddot{q}_x & 0 \end{bmatrix} \begin{bmatrix} r \\ 0 \\ 0 \end{bmatrix} + 2 \begin{bmatrix} 0 & -\dot{q}_z & \dot{q}_y \\ \dot{q}_z & 0 & -\dot{q}_x \\ -\dot{q}_y & \dot{q}_x & 0 \end{bmatrix} \begin{bmatrix} \dot{r} \\ 0 \\ 0 \end{bmatrix} + \begin{bmatrix} 0 & -\dot{q}_z & \dot{q}_y \\ \dot{q}_z & 0 & -\dot{q}_x \\ -\dot{q}_y & \dot{q}_x & 0 \end{bmatrix} \begin{bmatrix} 0 & -\dot{q}_z & \dot{q}_y \\ \dot{q}_z & 0 & -\dot{q}_x \\ -\dot{q}_y & \dot{q}_x & 0 \end{bmatrix} \begin{bmatrix} \dot{r} \\ 0 \\ 0 \end{bmatrix} \tag{12}$$

Simplified the formula [11–15]

$$\begin{cases} \ddot{r} = (\dot{q}_y^2 + \dot{q}_z^2) r + a_{tx} - a_{ix} \\ \ddot{q}_y = -2\dfrac{\dot{r}}{r}\dot{q}_y - \dfrac{a_{tz} - a_{iz}}{r} + \dot{q}_x \dot{q}_z \\ \ddot{q}_z = -2\dfrac{\dot{r}}{r}\dot{q}_z + \dfrac{a_{ty} - a_{iy}}{r} - \dot{q}_x \dot{q}_y \end{cases} \tag{13}$$

3 The Filtering Algorithm of Seeker for Tracking Target

3.1 The Maneuverable Missile Modeling in Free-Flight Trajectory

The most of missile's flight time is in the free-flight trajectory, so accurate targeting model is very important to get the position and velocity of target. So the maneuvering modeling can be expressed as follows [2, 7, 8, 17].

$$\begin{bmatrix} a_{tx} \\ a_{ty} \\ a_{tz} \end{bmatrix} = g \begin{bmatrix} \frac{x_t}{r_t} \\ \frac{y_t + \tilde{R}}{r_t} \\ \frac{z_t}{r_t} \end{bmatrix} + \begin{bmatrix} a_{tjx} \\ a_{tjy} \\ a_{tjz} \end{bmatrix} + \begin{pmatrix} 0 & 2\omega_z & -2\omega_y \\ -2\omega_z & 0 & 2\omega_x \\ 2\omega_y & -2\omega_x & 0 \end{pmatrix} \begin{pmatrix} V_{tx} \\ V_{ty} \\ V_{tz} \end{pmatrix} + \begin{pmatrix} \omega^2 - \omega_x^2 & -\omega_x\omega_y & -\omega_x\omega_z \\ -\omega_x\omega_y & \omega^2 - \omega_y^2 & -\omega_y\omega_z \\ -\omega_x\omega_z & -\omega_y\omega_z & \omega^2 - \omega_z^2 \end{pmatrix} \begin{pmatrix} R_{0x} + x_t \\ R_{0y} + y_t \\ R_{0z} + z_t \end{pmatrix} \quad (14)$$

In the formula

$$r_t = \sqrt{x_t^2 + (y_t + \tilde{R})^2 + z_t^2}$$

fM is gravity constant.$\tilde{R}$ is average radius of the earth. (x_t, y_t, z_t) is the position in the interception coordinates. (V_{tx}, V_{ty}, V_{tz}) is missile's velocity in the interception coordinates. $(\omega_x, \omega_x, \omega_x)$ is rotation speed of the earth in the interception coordinates. $(a_{tjx}, a_{tjy}, a_{tjz})$ is the maneuvering acceleration in the interception coordinates. If the missile's acceleration can be got, the velocity and position can be deduced by the methods of integration.

Stochastic maneuverable acceleration of ballistic reentry vehicle can be valued as a modified Rayleigh-Markov progress. Using the zero mean first-order time-dependent model and considering the cross-coupling of the azimuth and elevation direction, the maneuverable mode can be expressed [10]

$$\begin{cases} \dot{a}_{tjx} = -\alpha a_{tjx} + \omega_{tax} \\ \dot{a}_{tjy} = -\alpha a_{tjy} + \dot{q}_x a_{tjz} + \omega_{tay} \\ \dot{a}_{tjz} = -\alpha a_{tjz} - \dot{q}_x a_{tjy} + \omega_{taz} \end{cases} \quad (15)$$

$\omega_{tax}, \omega_{tay}, \omega_{taz}$ is white noise and the variance can be expressed

$$\sigma_{\omega}^{2}=\frac{2\alpha(4-\pi)}{\pi}\left(a_{tj\max}-E(a_{tj})\right)^{2} \tag{16}$$

In the formula,α is the target maneuvering frequency and $a_{tj\max}$ is the maximum of maneuverable acceleration.

3.2 The Control Acceleration by the Sliding Mode Guidance

By the Sliding Mode Guidance Law, the acceleration of the interceptor can be expressed as [16]:

$$\begin{cases} a_{iy}=(k+1)|\dot{r}|\dot{q}_y+Index1\dfrac{\dot{q}_y}{|\dot{q}_y|+Index2}+a_{ty}+\omega_y \\ a_{iz}=-(k+1)|\dot{r}|\cos q_y\dot{q}_z-Index3\dfrac{\dot{q}_z}{|\dot{q}_z|+Index4}+a_{tz}+\omega_z \end{cases} \tag{17}$$

k is the guiding factor, $IndexI(I=1,2,3,4)$ is parameters of jitter suppression index.

3.3 The Motion Parameters' Error Caused by the Inertial Guidance Instrument

The most of missile's flight time is in the free-flight trajectory, so accurate targeting model is very important to get the position and velocity of target. So the maneuvering modeling can be expressed as follows.

$$\begin{bmatrix} a_{tx} \\ a_{ty} \\ a_{tz} \end{bmatrix}=g\begin{bmatrix} \dfrac{x_t}{r_t} \\ \dfrac{y_t+\tilde{R}}{r_t} \\ \dfrac{z_t}{r_t} \end{bmatrix}+\begin{bmatrix} a_{tjx} \\ a_{tjy} \\ a_{tjz} \end{bmatrix}+\begin{pmatrix} 0 & 2\omega_z & -2\omega_y \\ -2\omega_z & 0 & 2\omega_x \\ 2\omega_y & -2\omega_x & 0 \end{pmatrix}\begin{pmatrix} V_{tx} \\ V_{ty} \\ V_{tz} \end{pmatrix} + \begin{pmatrix} \omega^2-\omega_x^2 & -\omega_x\omega_y & -\omega_x\omega_z \\ -\omega_x\omega_y & \omega^2-\omega_y^2 & -\omega_y\omega_z \\ -\omega_x\omega_z & -\omega_y\omega_z & \omega^2-\omega_z^2 \end{pmatrix}\begin{pmatrix} R_{0x}+x_t \\ R_{0y}+y_t \\ R_{0z}+z_t \end{pmatrix} \tag{18}$$

In the formula

$$r_t = \sqrt{x_t^2 + (y_t + \tilde{R})^2 + z_t^2} \tag{19}$$

fM is gravity constant. $\tilde{R}$ is average radius of the earth.(x_t, y_t, z_t) is the position in the interception coordinates. (V_{tx}, V_{ty}, V_{tz}) is missile's velocity in the interception coordinates. $(\omega_x, \omega_x, \omega_x)$ is rotation speed of the earth in the interception coordinates. $(a_{tjx}, a_{tjy}, a_{tjz})$ is the maneuvering acceleration in the interception coordinates. If the missile's acceleration can be got, the velocity and position can be deduced by the methods of integration.

3.4 The Filtering Algorithm of Interceptor's Attitude Parameters

The discrete state equation of angle tracking system be deduced [16–19]

$$\begin{cases} X_{k+1}^a = \Phi_{k+1,k} X_k^a + \Psi_{k+1,k} U_k^a + \Upsilon_{k+1,k} W_k^a \\ Z_{k+1}^a = H_{k+1,k} X_{k+1}^a + V_{k+1}^a \end{cases} \tag{20}$$

In the formula

$$\begin{cases} X_k^a = (\varepsilon_y, \varepsilon_z, \dot{q}_y, \dot{q}_z, a_{ty}, a_{tz}) \\ U_k^a = (\omega_{ay}, \omega_{az}, a_{iy}, a_{iz})^T \\ Z_{k+1} = (\varepsilon_y, \varepsilon_z)^T \end{cases} \tag{21}$$

$$H = [I_{2\times 2}, O_{2\times 4}] \tag{22}$$

$$\Phi_{k+1,k} = \begin{bmatrix} 1 & \omega_{ax}T & T & 0 & 0 & 0 \\ -\omega_{ax}T & 1 & 0 & T & 0 & 0 \\ 0 & 0 & 1 - 2\frac{\dot{r}}{r}T & \omega_{ax}T & 0 & -\frac{T}{r} \\ 0 & 0 & -\omega_{ax}T & 1 - 2\frac{\dot{r}}{r}T & \frac{T}{r} & 0 \\ 0 & 0 & 0 & 0 & 1 - \alpha T & \omega_{ax}T \\ 0 & 0 & 0 & 0 & -\omega_{ax}T & 1 - \alpha T \end{bmatrix} \tag{23}$$

$$\Psi_{k+1,k} = \begin{bmatrix} -T & -\frac{\omega_{ax}T^2}{2} & 0 & \frac{T^2}{2r} \\ \frac{\omega_{ax}T^2}{2} & -T & -\frac{T^2}{2r} & 0 \\ 0 & 0 & -\frac{\omega_{ax}T^2}{2r} & -\frac{T}{r} - \frac{\dot{r}T^2}{r^2} \\ 0 & 0 & -\frac{T}{r} + \frac{\dot{r}T^2}{r^2} & -\frac{\omega_{ax}T^2}{2r} \\ 0 & 0 & 0 & 0 \\ 0 & 0 & 0 & 0 \end{bmatrix} \tag{24}$$

$$\Upsilon_{k+1,k} = \begin{bmatrix} 0 & 0 \\ 0 & 0 \\ 0 & -\frac{T^2}{2r} \\ \frac{T^2}{2r} & 0 \\ T - \frac{\alpha T^2}{2} & \frac{\omega_{ax}T^2}{2} \\ -\frac{\omega_{ax}T^2}{2} & T - \frac{\alpha T^2}{2} \end{bmatrix} \tag{25}$$

3.5 The Filtering Algorithm of Interceptor's Orbit Parameters

The relative distance is (x_r, y_r, z_r) and the relative velocity (v_{rx}, v_{ry}, v_{rz}) can be expressed as

$$\begin{cases} x_r = x_t - x_i, \ v_{rx} = v_{tx} - v_{ix}, \ a_{rx} = a_{tx} - a_{ix} \\ y_r = y_t - y_i, \ v_{ry} = v_{ty} - v_{iy}, \ a_{ry} = a_{ty} - a_{iy} \\ z_r = z_t - z_i, \ v_{rz} = v_{tz} - v_{iz}, \ a_{rz} = a_{tz} - a_{iz} \end{cases} \tag{26}$$

$$X = (x_r, y_r, z_r, v_{rx}, v_{ry}, v_{rz}, a_{tx}, a_{ty}, a_{tz})^T \tag{27}$$

$$A = \begin{bmatrix} I_{3\times3} & TI_{3\times3} & \frac{T^2}{2} I_{3\times3} \\ O_{3\times3} & I_{3\times3} & TI_{3\times3} \\ O_{3\times3} & O_{3\times3} & I_{3\times3} \end{bmatrix} \tag{28}$$

$$B = \begin{bmatrix} O_{3\times6} & -\frac{T^2}{2} I_{3\times3} \\ O_{3\times6} & -TI_{3\times3} \\ O_{3\times6} & -I_{3\times3} \end{bmatrix} \tag{29}$$

$$U = [O_{1\times6}, a_M]^T \tag{30}$$

$$a_M = [a_{ix}, a_{iy}, a_{iz}] \tag{31}$$

$$W = [\omega_x, \omega_y, \omega_z, \omega_{vx}, \omega_{vy}, \omega_{vz}, \omega_{ax}, \omega_{ay}, \omega_{az}]^T \tag{32}$$

The system equation is

$$X_k = AX_{k-1} + BU_{k-1} + \Gamma W_{k-1} \tag{33}$$

The measurement equation is

$$Z_k = H_k X_k + Y_k + V_k \tag{34}$$

In the equation

$$H_k = \frac{\partial h(x_k, k-1)}{\partial x_k}\Big|_{x_k = x_{k|k-1}} \tag{35}$$

$$Y_k = h(x_{k|k-1}, k) - \frac{\partial h(x_{k|k-1}, k)}{\partial x_k}\Big|_{x_k = x_{k|k-1}} \hat{x}_{k|k-1} \tag{36}$$

$$V = (w_r, w_{q_y}, w_{q_z}, w_{\dot{r}}, w_{\dot{q}_y}, w_{\dot{q}_z})^T \tag{37}$$

In every time update, the angle tracking filter supply $(\dot{q}_y, \dot{q}_z, a_{ty}, a_{tz})$ to the orbit filter. If the stable angular speed of line-of-sight can be got, the precision of the orbit filter will be promoted.

4 The Analysis of the Result

In the simulation, the initial velocity of missile target is $(-1670\,\text{m/s}, -270, -50\,\text{m/s})$ and the initial relative distance is $(138838, 33170, 6320\,\text{m})$, while the initial velocity is $(1317\,\text{m/s}, 354\,\text{m/s}, -8\,\text{m/s})$.

The initial mass of interceptor is 50 kg, while the propellant 20 kg and specific impulse is 2600 N s/kg. The thrust of orbit motor is $F_{gk} = 256\,\text{N}$ and the thrust of attitude motor is $F_{zk} = 50\,\text{N}$.

The moment of inertial of interceptor in missile body coordinates is $(0.76\,\text{kg}\cdot\text{m}^2, 10.25\,\text{kg}\cdot\text{m}^2, 10.25\,\text{kg}\cdot\text{m}^2)$.The initial attitude angle is $(0\,rad, -0.1053\,rad, -01472\,rad)$.

In the Fig. 2, The seeker tracking error angles ε_y, ε_z become stable after filtering and can also provide a stable input signal to the guidance law and avoid noise interference.

From the Figs. 3 and 4, the antenna yaw angle λ_y and the antenna pitch angle λ_z become smaller and tend to the zero. At the same time, angular speeds of λ_y and λ_z also have been stable and tend to zero. So the seeker can track the target more effect and stable by the filtering algorithm. From the simulation, the tracking angle error can be constrained in smaller interval by the online filtering algorithm that means

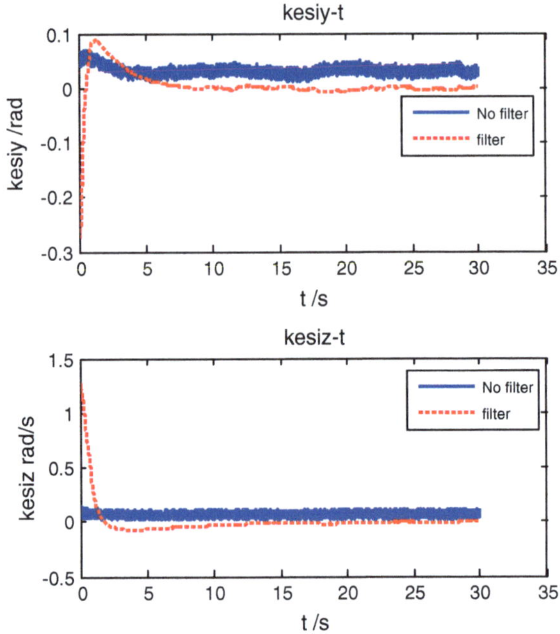

Fig. 2 The tracking angle error in the interception progress

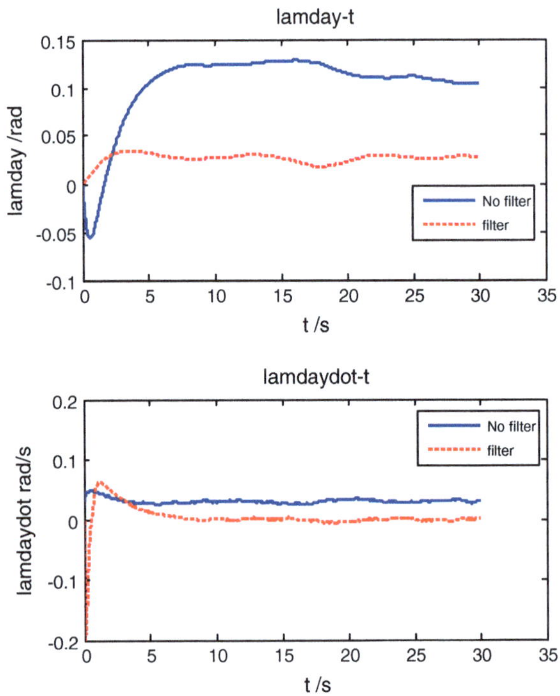

Fig. 3 The antenna yaw angle and the tracking angular speed of outer gimbal in coordinate

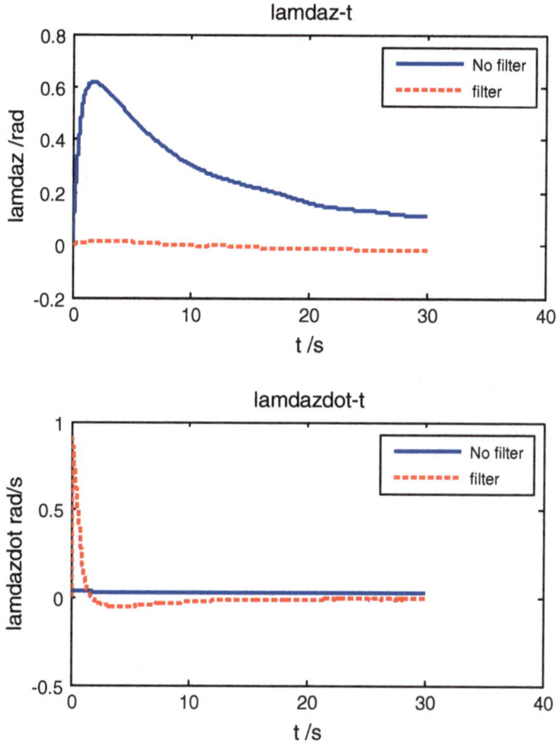

Fig. 4 The antenna pitch angle and the tracking angular speed of inner gimbal in the interception progress

the precision of seeker can be improved more than 30 %. The antenna yaw angle and pitch angle which less than without filtering are more stable and have less fluctuation range. So this seeker's filter modeling can be used for keeping seeker tracking target more stably and precisely.

References

1. Hawkins M, Guo YN, Wie B (2012) Spacecraft guidance algorithms for asteroid intercept and rendezvous missions. Int J Aeronaut Space Science 13(2):154–169
2. Ekstrand B (2001) Tracking filters and models for seeker applications. IEEE Trans Aerosp Electron Syst 37(3):965–977
3. Hough ME (1995) Optimal guidance and nonlinear estimation for interception of accelerating targets. AIAA J Guid Control Dyn 18(5):958–968
4. Sun P, Liu K (2010) Research on launch vehicle attitude control based on direct lateral force. Missile Space Launch Technol 22(2):10–14 (in Chinese)
5. Kong X, Yang M, Wang SY (2009) Attitude control of advanced aero craft under combined control. J Aerosp 30(2):566–571 (in Chinese)
6. Shen MH, Zhou BZ, Chen L, Wu R (2006) Design of the attitude control system of the atmospheric kinetic interceptor with invariable rolling rate. J Natl Univ Def Technol 28(4):133–136 (in Chinese)

7. Rudin RT (1993) Strapdown stabilization for imaging seekers. AIAA-93-2660:1–10
8. Zhou RQ, Lü SW, Liu XH (2003) Design of an angle tracking system with strapdown antenna platform. Syst Eng Electron 25(10):1200–1202 (in Chinese)
9. Zhou RQ, Lu SW, Liu XH (2003) Study of stability for strapdown antenna platform. J Beijing Univ Aeronaut Astronaut 29(6):509–511 (in Chinese)
10. Li XR, Jilkov VP (2000) A survey of maneuvering target tracking: dynamic models. In: Proceedings of 2000 SPIE conference on signal and data processing of small targets, vol 4048, pp 212–235, Orlando, Florida, USA, April 2000
11. Debruin JC, Royalty J, Wand M, Allen E (1996) Feedforward stabilization test bed. SPIE 2739:204–214
12. Mckerly CW (1996) A model for two degree of freedom coupled seeker with mass imbalance. IEEE Proc Control Syst Technol 9(1):84–87
13. Kennedy PJ, Kennedy RL (2003) Direct versus indirect line-of-sight (LOS) stabilization. IEEE Trans Control Syst Technol 11(1):3–15
14. Kennedy PJ, Kennedy RL (2003) Direct versus indirect line of sight (LOS) stabilization. IEEE Trans Control Syst Technol 11(1):3–15
15. Waldmann J (2002) Line-of-sight rate estimation and linearizing control of an imaging seeker in tactical missile guided by proportional navigation. IEEE Trans Control Syst Technol 10(4):556–557
16. Brierley SD, Longchamp R (1990) Applocation of sliding-mode control to air-air interception problem. IEEE Trans Aerosp Electron Syst 26(2):306–325
17. Rong Li X, Jilkov VP (2001) A survey of maneuvering target tracking. Part II: ballistic target models. In: Proceedings of SPIE conference on signal and data processing of small targets, vol 4473, pp 559–581
18. Pearson JB, Stear EB (1974) Kalman Filter applications in airborne radar tracking. IEEE Trans Aerosp Electron Syst 10(5):319–329
19. Yan Z (2007) Improved nonlinear filtering for target tracking. University of Oklahoma Graduate College, Norman, pp 5–14

Research on Route Plan of the Peacetime Military Logistics Distribution Based on FCM&GA

Xue-fu Xiao, Hai-rong Lu and Yang-qi Zhang

Abstract Combined with the research on the route plan of the peacetime military logistics distribution home and abroad, this paper proposes a method to connect genetic algorithm with fuzzy c-means clustering, which firstly clusters the army users with fuzzy math language into several classes according to the distance between the users and the military logistics distribution center, then plans the route with dual objective functions that are time and expense ones by genetic algorithm. Through a case analysis with the method, it proves that the method can decrease the time in a large scale but increase the expense in quite a small scale improving the military effectiveness, which indicates that the method is right and feasible and will provide a new thought for future studies on the problem.

Keywords Dual objective functions · Fuzzy c-means · Genetic algorithm · Military logistics distribution · Route plan

1 Introduction

In order to implement Chairman Xi's instructions about *fight battles, win battles* [1], it is essential and significant for us to improve our army's logistics ability. Military logistics distribution is one of the most important parts of military logistics and shoulders the end of it. Military logistics distribution is to pick out, machine, package, segment, and assemble military supplies and so on, according to the army's demands at some definite areas, and to deliver the supplies to the right position the army users have provided [2]. Military logistics distribution is divided

X. Xiao · Y. Zhang
Technical Support and Service Research Office, Institute of Military Transportation, Tianjin, China

H. Lu (✉)
Postgraduate Training Brigade, Military Transportation University, Tianjin, China
e-mail: 892530462@qq.com

E. Qi (ed.), *Proceedings of the 6th International Asia Conference on Industrial Engineering and Management Innovation*,
DOI 10.2991/978-94-6239-145-1_45

into wartime and peacetime military logistics distribution. The wartime military logistics distribution pays more attention to the security and timelines, but the peacetime focuses on the timelines and economy. Even so, both of the problems depend on each other, and solving either of them will be beneficial to solving another one. So, it will improve the efficiency and capacity of military logistics distribution if the problem about route plans of the peacetime military logistics distribution is solved legitimately, at the same time, it will help solve the other problem. Moreover, to some degree, it will enhance the ability of the military logistics. Consequently, it is of great significance to study route plan of the peacetime military logistics distribution.

2 Requirements Analysis

Because military logistics distribution is playing a more and more important role in supporting the army in the peacetime, it has become a heated focus by many scholars and experts how we can appropriately and efficiently support the army, especially to aim at route plan of the peacetime military logistics distribution. Though there are many theories about route plan, in realty many of them can't be used rationally, which results in much loss in economy, time, manpower, and resources. For the purpose of solving the problem, let's conclude the problem briefly as follows:

Now there are a military logistics distribution center and n army users whose geographic position has been known. According to the principle that vehicles with supplies start from the center, pass by all the users without repetition, and finally return the center, we need to rationally arrange the route to make the expense lowest and the time shortest.

3 Literature Review and Method Study

3.1 *Literature Review*

From the analysis above, route plan of the peacetime military logistics distribution belongs to vehicle routing problems (VRP), which was proposed by the famous scholars Dantzi and Ramser [3]. During the development of more than 50 years, people have designed a lot of algorithms divided into accurate algorithm and approximate algorithm. The former one is mainly made of branch and bound method and dynamic programming. The latter one is made of traditional heuristic algorithm and modern heuristic algorithm. The traditional heuristic algorithm consists of route first-cluster second and cluster first-route second. The modern heuristic algorithm consists of genetic algorithm [4, 5], artificial nerve net [6, 7] and ant colony algorithm [8, 9].

3.2 Method Study

For solving the problem better, there is a necessity of using and designing the algorithm as follows below:

(1) *Fuzzy c-means*

Clustering is to classify things by math methods, and fuzzy clustering as a main method of clustering is to classify things by fuzzy math language [10]. Among the clustering methods based on objective functions, FCM (Fuzzy c-Means) is the most widely used and most perfect in theory [11]. According to established class number for clustering c and stopping threshold ε, firstly initialize prototype pattern $P^{(0)}$, give iteration b a zero, calculate or update dividing matrix $U^{(b)}$, update clustering prototype pattern matrix $P^{(b+1)}$, and then judge whether $|P^{(b)} - P^{(b+1)}|$ is less than stopping threshold ε. If not, return to update dividing matrix, or output dividing matrix $U^{(b)}$ and prototype pattern $P^{(b)}$, and it's over. Moreover, the maximum number of iterations b is usually set 100, and the amount of stopping threshold ε is usually set 1×10^{-5} [12].

(2) *Genetic algorithm*

GA (Genetic Algorithm) was firstly proposed by the famous professor Holland [13], and has been applied in many areas such as economy, logistics and traffic. Combining GA with route plan of the peacetime military logistics distribution is to make the route plan be the process of chromosome adapting for survival, the searching volume map relevant objective function values, under some constraints, and to calculate the adaptability of objective function, then carry out some genetic operations such as crossover, mutation, and parallel selection. The basic thought of the parallel selection is that firstly divide the group into some subgroups pound by pound according to the number of objective function, give every subgroup an objective function, in respective group independently calculate the objective function value, and select some individuals of high adaptability to make a new subgroup, then make all the new subgroups be a new group. So carrying out the operation of *division-selection-combination* continuously helps get the Pareto solutions. Eventually, it will make route plan, including distribution time, distribution expense and distribution route. The maximum number of iterations *Maxgen* is usually set 100–500. The group's number M is usually set 20–100. The precision of variables K is usually set 20–60. Selection operator Φ is usually set 0.8–1.0. Crossover operator Γ is usually set 0.4–0.99. Mutation operator Ψ is usually set 0.0001–0.1 [14].

Though accurate algorithm can gain the accurate solution, with the increase of users' and constraint conditions' number, it needs so much calculation time and storage space that it isn't for the VRP. Therefore, combined with the research situation home and abroad, this paper uses approximate algorithm, which is *cluster first-route second* and *genetic algorithm*, that's to say, firstly uses fuzzy c-means to divide the users into several categories according to the distance between users and the military logistics distribution center, then uses GA to plan the route of every cluster of users, making the total expense lowest and the total time least.

4 Modeling of Route Planning

4.1 Model Assumption

In order to build model conveniently, now make several assumptions as follows:

(1) Time and expense can be the decisive factors of route plan of military logistics distribution, and the other factors can be ignored.
(2) The military logistics distribution can have enough ability of delivering supplies, which supports a lot of the army users.
(3) The distribution expense is in direct proportion to distribution distance.
(4) All vehicles of every fleet should always start and return together, and shouldn't be back independently.

4.2 Variable Definition

In order to describe the model conveniently, now definite the main variables as follows (Table 1):

4.3 Modeling Analysis

Through analyzing VRP above, use the method of cluster-first-route-second, that's to say, firstly divide the users into c categories according to the distance between the military logistics distribution center and the users, which can gain c clusters of users, then make route plan according to every cluster of users.

Firstly, build the cluster model of the fuzzy c-means according to the goal of minimizing the weighted error within class, and the problem can be formulated as follows [15]:

Table 1 Sympol definitions for main variables

Variables	Definition
T	The total time military logistics distribution center costs
C	The total expense military logistics distribution center costs
v	Mean speed of a vehicle
n	Number of the army users
(x_i, y_i)	The geographic position of the ith user
(x_0, y_0)	The geographic position of the military logistics distribution center

$$\begin{cases} \min \quad J_m(U^{(b)}, P^{(b)}) = \sum_{k=1}^{n}\sum_{i=1}^{c} (\mu_{ik}^{(b)})^m (d_{ik})^m, m \in [1, \infty) \\ \qquad s.t. \quad U \in M_{fc} \end{cases} \tag{1}$$

In the formulation above, m is weighting coefficient. b is iteration. d_{ik} is the distance of the samples and can be formulated as follows:

$$d_{ik} = \sqrt{(x_k - p_i)' S^{-1} (x_k - p_i)} \tag{2}$$

S is sample covariance. M_{fc} is dividing interval, and can be formulated as follows:

$$M_{fc} = \{U \in R^{cn} | \mu_{ik} \in [0, 1], \forall i, k; \sum_{i=1}^{c} \mu_{ik}^{(b)} = 1, \forall k; 0 < \sum_{k=1}^{n} \mu_{ik}^{(b)} < n, \forall i\} \tag{3}$$

$\mu_{ik}^{(b)}$ is the subjection degree between the ith user and the kth user in the bth iteration, and can be formulated as follows:

$$\mu_{ik}^{(b)} = \begin{cases} \left\{ \sum_{j=1}^{c} \left[\left(\frac{d_{ik}^{(b)}}{d_{jk}^{(b)}} \right)^{\frac{2}{m-1}} \right] \right\}^{-1} & , d_{ik} = 1 \\ 1 & , d_{ik} = 0 \end{cases} \tag{4}$$

$p_i^{(b)}$ is the clustering prototype vector of the ith class, and can be formulated as follows:

$$p_i^{(b)} = \frac{\sum_{k=1}^{n} \left[x_k (\mu_{ik}^{(b)})^m \right]}{\sum_{k=1}^{n} (\mu_{ik}^{(b)})^m}, \text{where } i = 1, 2, \ldots, c \tag{5}$$

$U^{(b)}$ is the dividing matrix in bth iteration. $P^{(b)}$ is the clustering prototype matrix.

Secondly, according to the clustering, plan the route of every class, and the problem can be formulated as follows:

$$\begin{cases} \min T = \max\{T_1, T_2, \ldots, T_i, \ldots, T_c\} \\ \min C = \lambda \sum_{i=1}^{c} \sum_{j=1}^{c_i+1} d_{ij} \end{cases} \tag{6}$$

In the formulation above, T is the total time, and the maximum value among all the distribution time values.. T_i is the time the ith distribution costs and can be formulated as follows:

$$T_i = \frac{1}{v}\sum_{j=1}^{c_i+1} d_j^{(t_i)} \tag{7}$$

c is the number of the user cluster. c_i is the users' number of the ith user cluster. λ is the price of distribution.

d_{ij} is the distance among distribution points, where $i = 1, 2, \ldots, c$ and $j = 1, 2, \ldots, c_i$. d_{ij} is the distance between the center and the first user of the ith class. d_{ij} (j is from 2 to c_i) is the distance between the users of the ith class. d_{i,c_i+1} is the distance between the center and the last user of the ith class.

4.4 Model Solution

From the analysis above, it is impossible to solve the problem by the way of manual computation because of its complexity and flexibility. The math software Matlab can help solve the problem above by programming the function of Rp that combines genetic algorithm with fuzzy c-means clustering, which can be beneficial to gaining the results and analyzing the results rapidly. The usage of the functions *Rp* is shown as follows:

[T,C,c,Br] = Rp(Gp,P,K,Mg,Ps,Pc,Pm,E)

From the program above, there are eight inputs. Gp is the geographic position of the users and the center. P is the group's number. K is the precision of variables. Mg is the maximum number of iterations. Ps is the probability of selection. Pc is the probability of crossover. Pm is the probability of mutation.

After running the program, there are four outputs, some figures and tables. T is the minimum distribution time. C is the minimum distribution expense. c is the number of the users' class after FCM. Br is the optimal distribution route.

5 Case Analysis

5.1 Case Assumption

As the Table 2 and Fig. 1 show, there is the geographic position of 20 army users, and the military logistics distribution center is situated in (400,400). The mean velocity of fleets is estimated as 80 km/h. The distribution's price λ is estimated as 15 RMB/km.

In Fig. 1, all users and the center are located in the area covering 800 km × 800 km. O represents the users that U is short for, where the digital

Table 2 Geographic position of 20 army users

No.	Geographic position	No.	Geographic position
1	(400, 500)	11	(700, 250)
2	(550, 200)	12	(750, 350)
3	(600, 150)	13	(750, 450)
4	(300, 600)	14	(700, 700)
5	(100, 300)	15	(300, 200)
6	(200, 200)	16	(650, 600)
7	(400, 300)	17	(600, 540)
8	(450, 350)	18	(450, 500)
9	(500, 300)	19	(200, 600)
10	(50, 400)	20	(100, 500)

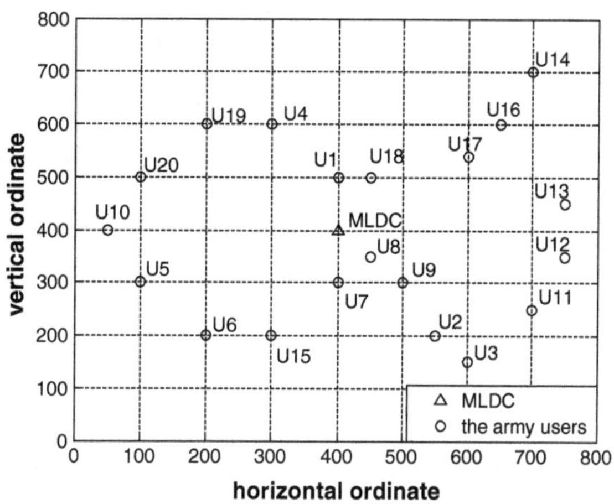

Fig. 1 Geographic position of the MLDC and the army users

number represents every user's sequence. Δ represents MLDC that is short for the military logistics distribution center.

5.2 *Case Solution*

From the analysis and algorithm design above, set the algorithm's operators as the Table 3 shows.

Under the environment where are the system of win7 and the math software of Matlab2007b, it is easy to gain the results such as Fig. 2, 3, 4, 5, 6 and Table 4.

From Fig. 2 and Table 4, it is evident that the users are classified into two classes, and the distribution is U_8-U_9-U_2-U_3-U_{11}-U_{12}-U_{13}-U_{14}-U_{16}-U_{17}-U_{18} and U_8-U_9-U_2-U_3-U_{11}-U_{12}-U_{13}-U_{14}-U_{16}-U_{17}-U_{18}, and the total distribution expense is 3.44×10^4 RMB, the total distribution time is 15.03 h and the total distribution

Table 3 Operators of GA and FCM

No.	Operator	Value	No.	Operator	Value
1	ε	1×10^{-5}	5	Φ	1.0
2	*Maxgen*	100	6	Γ	0.9
3	*M*	40	7	Ψ	0.1
4	K	40			

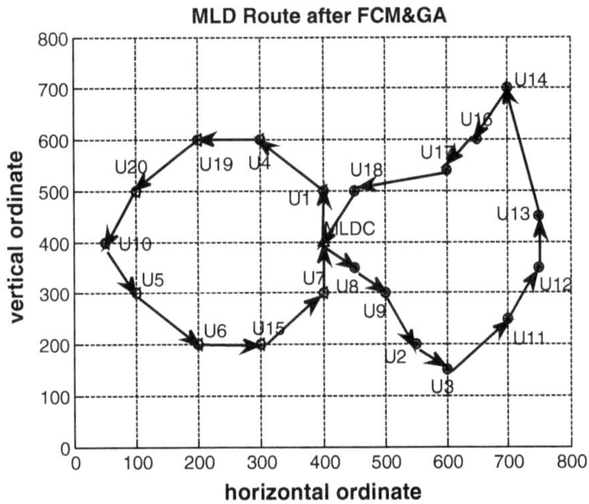

Fig. 2 The best distribution route after FCM&GA

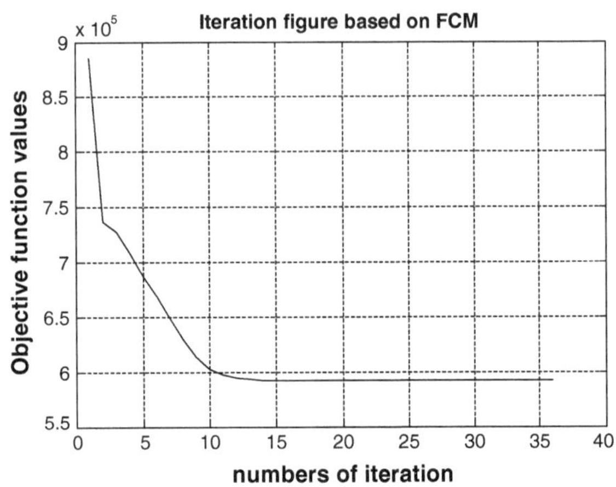

Fig. 3 Iteration curve based on FCM

distance is 2.29×10^3 km. The optimal solution will decrease 47.54 % of the time cost but increase 4.07 % of the expense cost where $c = 2$ than not using the method.

The whole solution is shown above, and the detailed analysis is shown as follows:

In Fig. 3, the decreasing curve shows objective values, based on Formulation 1, converge a constant with the iterations and the objective function is converged.

In Fig. 4, obviously, the users are classified into two classes. The filled dots belong to one class which consists of U_8, U_9, U_2, U_3, U_{11}, U_{12}, U_{13}, U_{14}, U_{16}, U_{17} and U_{18}. The filled dots belong to another class which is U_8, U_9, U_2, U_3, U_{11}, U_{12}, U_{13}, U_{14}, U_{16}, U_{17} and U_{18}.

In Fig. 5, the dotted decline curve represents the change of the mean expense values with iterations; the filled decline curve represents the change of the optimal

Fig. 4 Effect after FCM

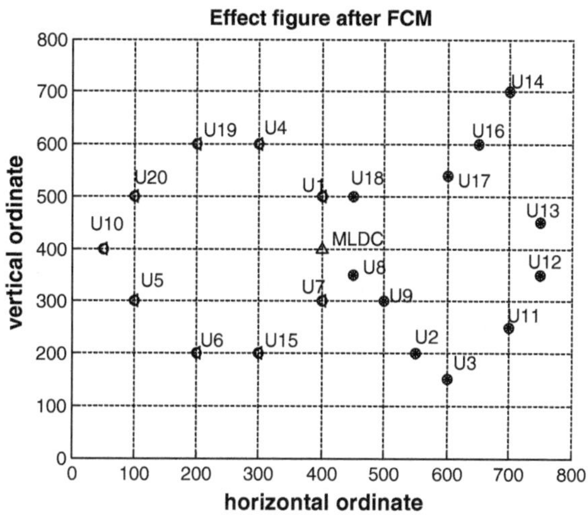

Fig. 5 Iteration effect of the distribution expense

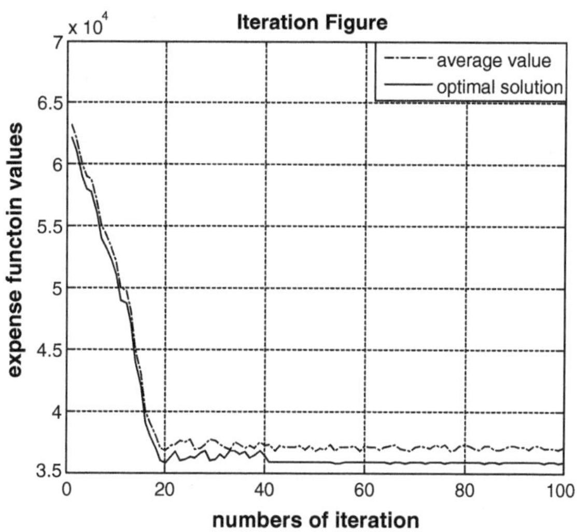

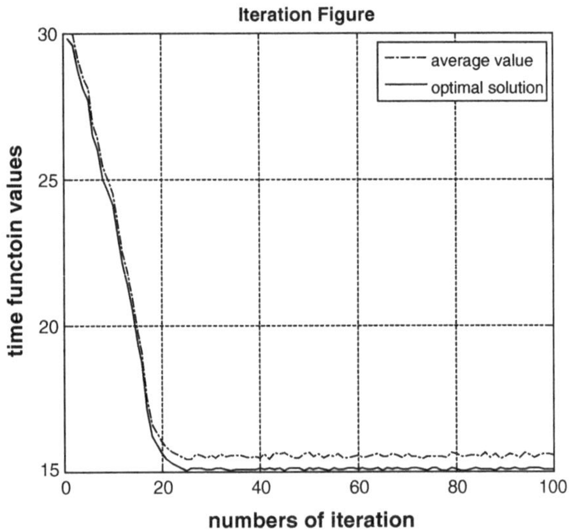

Fig. 6 Iteration effect of the distribution time

Table 4 Solutions output related

c	*D* (km)	*C* (RMB)	*T* (h)	Distribution route
1	2.29×10^3	3.44×10^4	28.65	1-4-19-20-10-5-6-15-7-8-9-2 -3-11-12-13-14-16-17-18
2	2.39×10^3	3.58×10^4	15.03	1-4-19-20-10-5-6-15-7
				8-9-2-3-11-12-13-14-16-17-18

expense values with iterations. Both of the curves converge to a constant. It is obvious that the optimal expense is 3.44×10^4 RMB.

In Fig. 6, the dotted decline curve represents the change of the mean time values with iterations; the filled decline curve represents the change of the optimal time values with iterations. Both of the curves converge to a constant. It is obvious that the optimal expense is 15.03 h.

6 Conclusion

According to studying the route plan of the peacetime military logistics distribution home and abroad, this paper proposes a method that is to combine genetic algorithm with fuzzy c-means clustering, which is to make the number of clustering *c* be a variable as part of the chromosome, then gain the best distribution route through algorithm design such as stopping threshold ε, parallel selection, crossover and mutation and so on. By analyzing a case of the peacetime military logistics distribution, it helps decrease 47.54 % of the time cost but simply increase 4.07 % of the expense cost where $c = 2$ than not using the method mentioned above. This always proves the method right and feasible and will provide a new thought to solve the problem. However, considering the factors inadequately and the precocity of genetic algorithm will become the constraints of gaining the best distribution route, which will become one of the most important parts I need to study.

References

1. The Headquarters of the General Staff (2013) The PLA Military Training Instructions in 2013 (Book style). The Headquarters of the General Staff of PLA, Beijing, pp 1–3 (Chinese)
2. Wang Z, Xu D (2007) Military Logistics (Book style). China supplies house, Beijing, pp 20–50 (Chinese)
3. Hwang HS, Bae S-T, Ha C-S (2007) Three basic algorithms for integrated visual model for multi-depot VRP problem (Presented Conference Paper style). In: Proceedings of the sixth international conference on information and management sciences. Lhasa, Tibet
4. Ayachi I, Kammarti R, Ksouri M, Borne P (2010) A genetic algorithm to solve the container storage space allocation problem (Presented Conference Paper style). In: Intelligent information technology application research association, Hong Kong, China
5. Sataporn N, Suwannik W (2010) Compact genetic algorithm with CUDA (Presented Conference Paper style). In: Intelligent information technology application research association, Hong Kong, China
6. Long TB, Thai LH, Hanh T (2010) Facial expression classification using zernike moment invariant and artificial neural Network (Presented Conference Paper style). In: Proceedings of 2010 international conference on circuit and signal processing and 2010 second IITA international joint conference on artificial intelligence, vol 1. Shanghai, China
7. Punnakitikashem P, Somsuk N, McLean MW, Laosiri T (2010) Linkage between continual improvement and knowledge-based view theory (Presented Conference Paper style). In: Proceedings of 2010 IEEE the 17th international conference on industrial engineering and engineering management, vol 2. Xiamen, Fujian, China
8. Liang Y-C, Lee Z-H, Chan C-L (2009) Ant colony optimization for the permutation flow shop on-line due-date scheduling problem (Presented Conference Paper style). In: Proceedings of the eighth international conference on information and management sciences. Kunming, Yunnan, China
9. Chuang C-P, Chen S-Y, Chen W-H, Huang Y-J (2010) Applying improved ant colony algorithm for multi-robot path planning and task allocation in educational navigation robot (Presented Conference Paper style). In: Proceedings of 2010 third international conference on education technology and training. vol 3. Wuhan, Hubei, China
10. Sivaraman E, Arulselvi S (2009) Gustafson-Kessel (G-K) Clustering approach of T-S fuzzy model for nonlinear processes (Presented Conference Paper style). In: Proceedings of international conference on control and decision, vol 3. Guilin, Guangxi, China
11. Kau S-W, Shiu S-Y, Fang C-H An improvement on observer-based H_∞ control of fuzzy systems (Presented Conference Paper style). In: Proceedings of international conference on control and decision, vol 3. Guilin, Guangxi, China, 2009
12. Xie J, Liu C (2010) Fuzzy math and applications (Book style). Renmin University of China Press, Beijing, pp 20–45 (Chinese)

13. Chang Y-J, Yao M-J (2009) A genetic algorithm for solving the economic lot scheduling problem with reworks (Presented Conference Paper style). In: Proceedings of the eighth international conference on information and management sciences. Guilin, Guangxi, China
14. Yang L, Gao Y, Ling W (2010) Fuzzy math theory and applications (Book style). Renmin University of China Press, Beijing, pp 20–45 (Chinese)
15. Lei Y (2012) Applications of genetic algorithm tool in Matlab (Book style). South China University of Technology press, Guangzhou, pp 27–60

Adaptive Robust Control of Sugar Precipitation for Chinese Medicine Mixed Solution

Qing-wei Li and Hong-jun Duan

Abstract A model dedicated to Chinese medicine sugar precipitation is designed, without consideration of crystal size distribution. Adaptive robust control algorithm is provided for the uncertain nonlinear system based on Lyapunov's stability theory. The mechanism is based on "model decomposition": an adaptive controller is adopted to eliminate the parameter uncertainty; robust controller is adopted to attenuate the unknown dynamics and disturbance; feedback controller is developed to dominate the nominal plant. Three controllers compose the integrated controller for the uncertain nonlinear system. The stability is proved and the output of system tracks the set point well. The algorithm is applied to the sugar precipitation control of Chinese medicine solution. Simulation result sustains the validity of the proposed algorithm.

Keywords Adaptive robust control · Batch process · Chinese medicine · Nonlinearity · Precipitation

1 Introduction

Precipitation from solution is a very important separation and purification process, governed by nucleation and growth rate, which concerns many different industries, for example, pharmaceutical, chemical, and food industries. Due to the highly nonlinearity and complex dynamics of this process, its monitoring and control

Q. Li (✉)
School of Resources and Materials, Northeastern University at Qinhuangdao, Qinhuangdao, China
e-mail: lqwday@126.com

H. Duan
School of Control Engineering, Northeastern University at Qinhuangdao, Qinhuangdao, China
e-mail: dhj@mail.neuq.edu.cn

E. Qi (ed.), *Proceedings of the 6th International Asia Conference on Industrial Engineering and Management Innovation*,
DOI 10.2991/978-94-6239-145-1_46

remain an interesting challenge in terms of quality and global efficiency improvement, low-order, linear models are often inadequate for control purposes. Furthermore, nonlinear black box models, such as neural networks, are also unsatisfactory in many cases because batch crystallizer has such a large operating region. Optimal operation and efficient on-line control strategy of the precipitation process are necessary to enhance their performance in terms of product quality and energy usage [1]. On-line control during batch precipitation offers the possibility of improved crystal product quality, shorter process times, and reduction or elimination of compromised batches [2].

Precipitation of the mixed solution is a batch process which is usually related with the formulation of a control problem in terms of economic or performance objective at the end of the process [3]. Despite the long history and widespread application of batch process, there remain a number of problems related to its control, mainly related the complex nonlinear dynamics with nonideal mixing, and various disturbances characteristic to these systems [4]. In recent years, there has got a growing interest in the optimization control of batch processes, which are suitable for responsive manufacturing of low volume and high value-added products [5, 6]. However, it is usually very difficult to obtain accurate models owing to the complexity of the processes, and it is also difficult to implement them for on-line optimization since optimization based on mechanistic models is usually very time-consuming [7]. The repetitive nature of batch process operations allows that the information of previous batch runs can be utilized to improve the operation of the next batch. The current PID control strategy has proved to be incompetent regarding the strong nonlinearities of the crystallization process. This statement of fact leads to the proposition of alternative controlled variable and a more suitable control algorithm to improve the process control and the process efficiency. Of late, relative algorithms are mainly iterative learning control (ILC) [8, 9] and nonlinear model-based predictive control (NMPC) [10, 11]. The performance and stability of process control depend both on past and present changes, sometimes due to irregular operating conditions: the reduction of available heating steam, the variation of the quality of feeding juices and molasses, etc. If feedback controllers have proved to be suitable in a wide range of operating conditions, advanced strategies may be required to meet steadiness and quality demands, in spite of strong irregularities.

During the recent years, considerable research efforts have been taken to deal with the design of stabilizing controllers for classes of nonlinear plants. However, uncertainty is commonly present in most control systems due to modeling errors, parameter variations, unknown dynamics, and disturbances. To ensure stability and performance, feedback control must be robust. Adaptive control and robust control are effective techniques to treat uncertainty. In simple terms, an adaptive scheme is to estimate unknown but constant parameters in system dynamics via adaptation laws, and robust control is to stabilize an uncertain system by assuming that its uncertainty be bounded in size by a known function. The combination of tools from both robust and adaptive controls may produce better designs than those produced by either method alone [12–15]. In this paper, a novel control strategy is proposed which bases on "model decomposition". Adaptive controller is adopted to eliminate the parameter uncertainty; robust controller is adopted to attenuate the unknown dynamics and disturbances. With a feedback controller to dominate the nominal plant, three controllers compose the integrated controller for the uncertain nonlinear system. The stability is proved by Lyapunov's theorem and the validity of the theoretical result is based on the sugar precipitation control simulation of Chinese medicine solution.

This paper is structured as follows: Sect. 2 focuses on process modeling. Then, adaptive robust controller is developed in Sect. 3. Precipitation control simulation of the mixed solution is dealt within Sect. 4. Finally, conclusion is given in Sect. 5.

2 Process Modeling

A model devoted to sugar precipitation monitoring and control of Chinese medicine solution is developed. The most widely approaches investigated to describe sugar extraction take into account four physical–chemical phenomena: nucleation, agglomeration, growth and dissolution. These investigations lead to the development of models of extreme purity liquors [16]. Little information is provided about the stage where multiple sugars could be co-precipitated. There, low grade Chinese medicine crude solutions (impure solutions) contain a high proportion of impurities, which considerably changes their properties. It is very difficult to adapt the models in all the process. To overcome this difficulty, the most usual approach consists of considering nucleation, agglomeration and dissolution negligible during the growth phase, in supersaturated condition [17]. In other words, crystal growth phenomena are considered to be preponderant.

Sugar precipitation step of Chinese medicine solution is performed through a crystallizer, operating by a continuously agitator. The dynamic model proposed in this paper represents the mass balance for dissolved sucrose and glucose (m_s and m_g), crystals (m_{sc} and m_{gc}), sodium hydroxide solution (m_{sh}) and impurities (m_i). The phase changing from dissolved to crystallize sugar is indicated by a simple and nonreversible conversion. Thus, the evolution of crystal mass is governed by the following expression:

$$\begin{cases} \dot{m}_{sc} = \alpha_1 m_s + \alpha_2 \rho_{sh} F_{sh} \\ \dot{m}_{gc} = \beta_1 m_g + \beta_2 \rho_{sh} F_{sh} \end{cases} \tag{1}$$

where α and β are adjusted crystal parameters of sucrose and glucose respectively, ρ_{sh} and F_{sh} are the density[kg/m^3] and flow rate[m^3/s] of the sodium hydroxide solution respectively.

The evolution of dissolved sucrose and glucose results from the sucrose and glucose are contained in the liquor supply, which depends on the mass fraction of dry substance and on the purity of the solution, and from the disappearance of sucrose and glucose converted into crystals, according to Eq. (1):

$$\begin{cases} \dot{m}_{\mathrm{s}} = \rho_{\mathrm{f}} F_{\mathrm{f}} Bx_{\mathrm{fs}} Pte_{\mathrm{fs}} - \dot{m}_{\mathrm{sc}} \\ \dot{m}_{\mathrm{g}} = \rho_{\mathrm{f}} F_{\mathrm{f}} Bx_{\mathrm{fg}} Pte_{\mathrm{fg}} - \dot{m}_{\mathrm{gc}} \end{cases} \tag{2}$$

where ρ_{f} and F_{f} are the density[kg/m^3] and flow rate[m^3/s] of the feed syrup respectively, Bx_{fs} and Pte_{fs} are the mass fractions of sucrose and its dry substance respectively, Bx_{fg} and Pte_{fg} are the mass fractions of glucose and its dry substance respectively.

The mass balance applied to the sodium hydroxide solution in the crystallizer leads to the following equation:

$$\dot{m}_{\mathrm{sh}} = \rho_{\mathrm{f}} F_{\mathrm{f}} \left(1 - Bx_{\mathrm{fs}} - Bx_{\mathrm{fg}}\right) + \rho_{\mathrm{sh}} F_{\mathrm{sh}} - \dot{m}_{\mathrm{vap}} \tag{3}$$

where m_{vap} is the emitted vapor mass[kg].

The mass of impurities contained in solution is obtained using the following expression:

$$\begin{cases} \dot{m}_{\mathrm{i}} = \rho_{\mathrm{f}} F_{\mathrm{f}} \left(Bx_{\mathrm{fs}} + Bx_{\mathrm{fg}}\right)\left(1 - Pte_{\mathrm{fs}} - Pte_{\mathrm{fg}}\right) \\ m_{\mathrm{i}}(t) = m_{\mathrm{i}}(0) + \rho_{\mathrm{f}} F_{\mathrm{f}} \left(Bx_{\mathrm{fs}} + Bx_{\mathrm{fg}}\right)\left(1 - Pte_{\mathrm{fs}} - Pte_{\mathrm{fg}}\right) t \end{cases} \tag{4}$$

The initial value $m_{\mathrm{i}}(0)$ is achieved using the initial conditions. An energy balance applied to the magma in the crystallizer allows to express the mass of emitted vapor:

$$\dot{H}_{\mathrm{mg}} = \dot{Q} + \rho_{\mathrm{f}} F_{\mathrm{f}} h_{\mathrm{f}} + \rho_{\mathrm{sh}} F_{\mathrm{sh}} h_{\mathrm{sh}} + \lambda_{\mathrm{sc}} \dot{m}_{\mathrm{sc}} + \lambda_{\mathrm{gc}} \dot{m}_{\mathrm{gc}} - h_{\mathrm{vap}} \dot{m}_{\mathrm{vap}} \tag{5}$$

where H_{mg} is the enthalpy of the magma[J], $\dot{Q}$ is the heating power [W], h_{f}, h_{sh}, and h_{vap} are the specific enthalpy of feed syrup, sodium hydroxide solution, and emitted vapor respectively [J/kg], λ_{sc} and λ_{gc} are the specific latent heat of sucrose and glucose respectively [J/kg].

During the precipitation process, the temperature of the magma is available, and its variation trifling compared to the mass variation. The enthalpy variation of the magma can be written as follows:

$$\begin{cases} H_{\mathrm{mg}} = Cp_{\mathrm{mg}} m_{\mathrm{mg}} T_{\mathrm{mg}} \\ m_{\mathrm{mg}} = m_{\mathrm{s}} + m_{\mathrm{sc}} + m_{\mathrm{sh}} + m_{\mathrm{i}} \end{cases} \tag{6}$$

$$\begin{aligned} \dot{H}_{\mathrm{mg}} &= Cp_{\mathrm{mg}} m_{\mathrm{mg}} \dot{T}_{\mathrm{mg}} + Cp_{\mathrm{mg}} T_{\mathrm{mg}} \dot{m}_{\mathrm{mg}} \\ &\approx Cp_{\mathrm{mg}} T_{\mathrm{mg}} \dot{m}_{\mathrm{mg}} \end{aligned} \tag{7}$$

where Cp_{mg}, m_{mg}, T_{mg} are the specific heat capacity [J/(K·kg)], mass [kg], and temperature [°C] of the magma respectively.

From Eq. (5), the mass of vapor can be expressed as follows:

$$h_{\mathrm{vap}}\dot{m}_{\mathrm{vap}} = \dot{Q} + \rho_{\mathrm{f}}F_{\mathrm{f}}h_{\mathrm{f}} + \rho_{\mathrm{sh}}F_{\mathrm{sh}}h_{\mathrm{sh}} + \lambda_{\mathrm{sc}}\dot{m}_{\mathrm{sc}} + \lambda_{\mathrm{gc}}\dot{m}_{\mathrm{gc}} - Cp_{\mathrm{mg}}T_{\mathrm{mg}}\dot{m}_{\mathrm{mg}} \tag{8}$$

Substitute Eqs. (1–4) into Eq. (8), it comes:

$$\begin{aligned}\left(h_{\mathrm{vap}} - Cp_{\mathrm{mg}}T_{\mathrm{mg}}\right)\dot{m}_{\mathrm{vap}} &= \dot{Q} + \rho_{\mathrm{f}}F_{\mathrm{f}}\left(h_{\mathrm{f}} - Cp_{\mathrm{mg}}T_{\mathrm{mg}}\right) \\ &+\rho_{\mathrm{sh}}F_{\mathrm{sh}}\left(h_{\mathrm{sh}} - Cp_{\mathrm{mg}}T_{\mathrm{mg}}\right) + \alpha\lambda_{\mathrm{sc}}m_{\mathrm{s}} + \beta\lambda_{\mathrm{gc}}m_{\mathrm{g}}\end{aligned} \tag{9}$$

Let us denote:

$\boldsymbol{u} = \begin{bmatrix} u_1 & u_2 & u_3 \end{bmatrix}^{\mathrm{T}} = \begin{bmatrix} F_{\mathrm{f}} & F_{\mathrm{sh}} & \dot{Q} \end{bmatrix}^{\mathrm{T}}$ the input vector, $\boldsymbol{x} = \begin{bmatrix} x_1 & x_2 & x_3 \end{bmatrix}^{\mathrm{T}} = \begin{bmatrix} m_{\mathrm{s}} & m_{\mathrm{g}} & m_{\mathrm{sh}} \end{bmatrix}^{\mathrm{T}}$ the state vector. From Eq. (1–3) and (9), the state space representation of the system is nonlinear and can be written as follows:

$$\dot{\boldsymbol{x}} = \boldsymbol{g}(\boldsymbol{x}) + \boldsymbol{b}(\boldsymbol{x})\boldsymbol{u} \tag{10}$$

where

$$\boldsymbol{g}(\boldsymbol{x}) = \begin{bmatrix} -\alpha_1 m_{\mathrm{s}} & -\beta_1 m_{\mathrm{g}} & \frac{\alpha_1\lambda_{\mathrm{sc}}m_{\mathrm{s}} + \beta_1\lambda_{\mathrm{gc}}m_{\mathrm{g}}}{\vartheta} \end{bmatrix}^{\mathrm{T}};$$

$$\boldsymbol{b}(\boldsymbol{x}) = \begin{bmatrix} \rho_{\mathrm{f}}Bx_{\mathrm{fs}}Pte_{\mathrm{fs}} & -\alpha_2\rho_{\mathrm{sh}} & 0 \\ \rho_{\mathrm{f}}Bx_{\mathrm{fg}}Pte_{\mathrm{fg}} & -\beta_2\rho_{\mathrm{sh}} & 0 \\ \rho_{\mathrm{f}}\left(\frac{h_1}{\vartheta} - Bx_{\mathrm{fs}} - Bx_{\mathrm{fg}}\right) & \rho_{\mathrm{sh}}\frac{h_2}{\vartheta} & -\frac{1}{\vartheta} \end{bmatrix};$$

$$\vartheta = h_{\mathrm{vap}} - Cp_{\mathrm{mg}}T_{\mathrm{mg}},\ h_1 = h_{\mathrm{vap}} - h_{\mathrm{f}},\ h_2 = h_{\mathrm{vap}} - h_{\mathrm{sh}}$$

Let $\boldsymbol{m}(\boldsymbol{x}) = \boldsymbol{b}^{-1}(\boldsymbol{x})$, $\boldsymbol{f}(\boldsymbol{x}) = -\boldsymbol{b}^{-1}(\boldsymbol{x})\boldsymbol{g}(\boldsymbol{x})$, the system (10) can be rewritten equivalently by

$$\boldsymbol{m}(\boldsymbol{x})\dot{\boldsymbol{x}} + \boldsymbol{f}(\boldsymbol{x}) = \boldsymbol{u} \tag{11}$$

3 Design of Adaptive Robust Controller

Considering a class of *n*th-order, multi-input multi-output uncertain nonlinear system as the following form:

$$\boldsymbol{m}(\boldsymbol{x})\boldsymbol{x}^{(\mathrm{n})} + \boldsymbol{f}(\boldsymbol{x},\dot{\boldsymbol{x}},\ldots,\boldsymbol{x}^{(\mathrm{n-1})}) + \boldsymbol{h}(\boldsymbol{x},\dot{\boldsymbol{x}},\ldots,\boldsymbol{x}^{(\mathrm{n-1})}) = \boldsymbol{u} \tag{12}$$

where $\boldsymbol{m} \in R^{m\times m}$, $\boldsymbol{x} = [x_1, x_2, \ldots, x_m]^T \in R^m$, $\boldsymbol{f} \in R^m$, $\boldsymbol{u} \in R^m$. The nominal quantity and uncertain quantity of $\boldsymbol{m}(\boldsymbol{x})$ can be expressed as $\boldsymbol{m}_0(\boldsymbol{x})$ and $\tilde{\boldsymbol{m}}(\boldsymbol{x})$, i.e. $\boldsymbol{m}(\boldsymbol{x}) = \boldsymbol{m}_0(\boldsymbol{x}) + \tilde{\boldsymbol{m}}(\boldsymbol{x})$. The unknown dynamics and disturbance are integrated into $\boldsymbol{h}(\boldsymbol{x}, \dot{\boldsymbol{x}}, \ldots, \boldsymbol{x}^{(n-1)})$, there exists a known positive constant η, such that $\|\boldsymbol{h}(\boldsymbol{x}, \dot{\boldsymbol{x}}, \ldots, \boldsymbol{x}^{(n-1)})\| \leq \eta$, $\forall \boldsymbol{x} \in R^m$.

Define error vector $\boldsymbol{e} \in R^{nm}$ and a filtered tracking error $\boldsymbol{s} \in R^m$ as follows

$$\boldsymbol{e} = \left[\boldsymbol{x} - \boldsymbol{x}_d, \dot{\boldsymbol{x}} - \dot{\boldsymbol{x}}_d, \ldots, \boldsymbol{x}^{(n-1)} - \boldsymbol{x}_d^{(n-1)}\right]^T = [\boldsymbol{e}_1, \boldsymbol{e}_2, \ldots, \boldsymbol{e}_n]^T \quad (13)$$

$$\boldsymbol{s} = (\lambda + \frac{d}{dt})^{n-1}\boldsymbol{e}_1 = [\boldsymbol{\Lambda}^T \ 1]\boldsymbol{e} \quad (14)$$

where $\boldsymbol{x}_d$ are the desired trajectories of $\boldsymbol{x}$, $\lambda > 0$, $\boldsymbol{\Lambda} = [\lambda^{n-1}, (n-1)\lambda^{n-2}, \ldots, (n-1)\lambda]^T$.

Again define vector $\boldsymbol{\delta} \in R^m$ as follows

$$\boldsymbol{\delta} = \boldsymbol{x}_d^{(n)} - \boldsymbol{\Gamma}\boldsymbol{e} \quad (15)$$

where $\boldsymbol{\Gamma} = [\lambda^n, n\lambda^{n-1}, (2n-3)\lambda^{n-2}, \ldots, (2n-3)\lambda^2, n\lambda]$.

Then there is the following control law to stabilize the uncertain nonlinear system (12)

$$\boldsymbol{u} = \boldsymbol{m}_0\boldsymbol{\delta} + \boldsymbol{u}_a + \boldsymbol{u}_u + \boldsymbol{f} \quad (16)$$

where

$$\boldsymbol{u}_a = -\boldsymbol{k}\boldsymbol{\delta}\int_{t_0}^{t} \boldsymbol{s}^T\boldsymbol{k}\boldsymbol{\delta}\,d\tau \quad (17)$$

$$\boldsymbol{u}_u = \begin{cases} -\eta\frac{\boldsymbol{s}}{\|\boldsymbol{s}\|}, & \|\boldsymbol{s}\| \geq \varepsilon \\ -\eta\frac{\boldsymbol{s}}{\varepsilon}, & \|\boldsymbol{s}\| < \varepsilon \end{cases} \quad (18)$$

where $\boldsymbol{k} = diag\{k_1, k_2, \ldots, k_m\}$, $k_1, k_2, \ldots, k_m$ are all positive constants;$\|\cdot\|$ denotes the standard Euclidean norm.

Proof The close loop equations of system (12) can be gotten as follows after some manipulations

$$\boldsymbol{m}\dot{\boldsymbol{s}} + \lambda\boldsymbol{m}\boldsymbol{s} = -\tilde{\boldsymbol{m}}\boldsymbol{\delta} + \boldsymbol{u}_a + \boldsymbol{u}_u + \boldsymbol{h} \quad (19)$$

Define a positive definite scalar function by

$$\Omega(\boldsymbol{s}) = \frac{1}{2}\boldsymbol{s}^{\mathrm{T}}\boldsymbol{m}\boldsymbol{s} + \frac{1}{2}\xi^2 \tag{20}$$

where

$$\xi = -(\boldsymbol{s}^{\mathrm{T}}\boldsymbol{k}\boldsymbol{\delta})^{-1}\boldsymbol{s}^{\mathrm{T}}\tilde{\boldsymbol{m}}\boldsymbol{\delta} - \int_{t_0}^{t} \boldsymbol{s}^{\mathrm{T}}\boldsymbol{k}\boldsymbol{\delta}\mathrm{d}\tau \tag{21}$$

The derivative of $\Omega(\boldsymbol{s})$ can be gotten as follows

$$\begin{aligned}
\dot{\Omega} &= \boldsymbol{s}^{\mathrm{T}}\boldsymbol{m}\dot{\boldsymbol{s}} + \xi\dot{\xi} \\
&= -\lambda\boldsymbol{s}^{\mathrm{T}}\boldsymbol{m}\boldsymbol{s} - \boldsymbol{s}^{\mathrm{T}}\tilde{\boldsymbol{m}}\boldsymbol{\delta} - \boldsymbol{s}^{\mathrm{T}}\boldsymbol{k}\boldsymbol{\delta}\int_{t_0}^{t} \boldsymbol{s}^{\mathrm{T}}\boldsymbol{k}\boldsymbol{\delta}\mathrm{d}\tau + \boldsymbol{s}^{\mathrm{T}}(\boldsymbol{u}_u + \boldsymbol{h}) \\
&\quad +(\boldsymbol{s}^{\mathrm{T}}\boldsymbol{k}\boldsymbol{\delta})^{-1}\boldsymbol{s}^{\mathrm{T}}\tilde{\boldsymbol{m}}\boldsymbol{\delta}(\boldsymbol{s}^{\mathrm{T}}\boldsymbol{k}\boldsymbol{\delta}) + (\boldsymbol{s}^{\mathrm{T}}\boldsymbol{k}\boldsymbol{\delta})\int_{t_0}^{t} \boldsymbol{s}^{\mathrm{T}}\boldsymbol{k}\boldsymbol{\delta}\mathrm{d}\tau \\
&= -\lambda\boldsymbol{s}^{\mathrm{T}}\boldsymbol{m}\boldsymbol{s} + \boldsymbol{s}^{\mathrm{T}}(\boldsymbol{u}_u + \boldsymbol{h})
\end{aligned} \tag{22}$$

Substitute Eq. (18) into Eq. (22). If $\|\boldsymbol{s}\| > \varepsilon$, then

$$\begin{aligned}
\dot{\Omega}(\boldsymbol{s}) &= -\lambda\boldsymbol{s}^{\mathrm{T}}\boldsymbol{m}\boldsymbol{s} - \eta\frac{\boldsymbol{s}^{\mathrm{T}}\boldsymbol{s}}{\|\boldsymbol{s}\|} + \boldsymbol{s}^{\mathrm{T}}\boldsymbol{h} \\
&= -\lambda\boldsymbol{s}^{\mathrm{T}}\boldsymbol{m}\boldsymbol{s} - \|\boldsymbol{s}\|(\frac{\boldsymbol{s}^{\mathrm{T}}}{\|\boldsymbol{s}\|}\boldsymbol{h} - \eta) < 0
\end{aligned} \tag{23}$$

If $\|\boldsymbol{s}\| = \varepsilon$ when $t = t_1$, then

$$\Omega(\boldsymbol{s}(t_1)) - \Omega(\boldsymbol{s}(t_0)) \le -\lambda\boldsymbol{s}^{\mathrm{T}}(t_1)\boldsymbol{m}(\boldsymbol{s}(t_1))(t_1 - t_0)$$

s.t.

$$t_1 - t_0 \le \frac{\Omega(\boldsymbol{s}(t_0)) - \Omega(\boldsymbol{s}(t_1))}{\lambda\boldsymbol{s}^{\mathrm{T}}(t_1)\boldsymbol{m}(\boldsymbol{s}(t_1))} \tag{24}$$

From Eq. (24), we can know that the errors get to $\|\boldsymbol{s}\| = \varepsilon$ in finite time.

If $||s|| < \varepsilon$, then

$$\begin{aligned}\dot{\Omega}(s) &= -\lambda s^{\mathrm{T}} m s + s^{\mathrm{T}}(h - \eta \frac{s}{\varepsilon}) \\ &\leq -\lambda s m s^{\mathrm{T}} + s^{\mathrm{T}}(\eta \frac{s}{||s||} - \eta \frac{s}{\varepsilon})\end{aligned} \tag{25}$$

The second item of right side of Eq. (25) achieves a maximum value of $\frac{\varepsilon\eta}{4}$ when $||s|| = \frac{\varepsilon}{2}$. Thus, if $s^{\mathrm{T}} m s > \frac{\varepsilon\eta}{4\lambda}$, there exists the following express

$$\dot{\Omega}(s) \leq -\lambda s m s^{\mathrm{T}} + \frac{\varepsilon\eta}{4} < 0 \tag{26}$$

It indicates that the norm of s is uniformly bounded, i.e. $e_1, e_2, \ldots e_n \rightarrow \mathbf{0}$, as $t \rightarrow \infty$. The controllers proposed in (16)–(18) have preferably robust stability.

4 Precipitation Control Simulation of Mixed Solution

The precipitation performance of the process can be increased by improving the crystal rate. This growth rate can be apprehended by following the crystal mass. In mixed solution, we set sucrose crystal mass and glucose crystal mass in discrete time, respectively, and they are tracked by the proposed control algorithm. Beyond this set point tracking, a subsequent control objective is required, and consists in maintaining the solution in an appropriate supersaturated state, that is a metastable zone, as illustrated in Figs. 1 and 2. Solubility curves are denoted by solid lines, and nucleation curves are denoted by dotted lines.

The controller parameters are as follows: $\varepsilon = 15$, $\boldsymbol{k} = diag\{0.65, 0.5, 0.4\}$. Precipitation tracking curves of sucrose-glucose solution are illustrated in Fig. 3,

Fig. 1 Concentration trajectory of sucrose

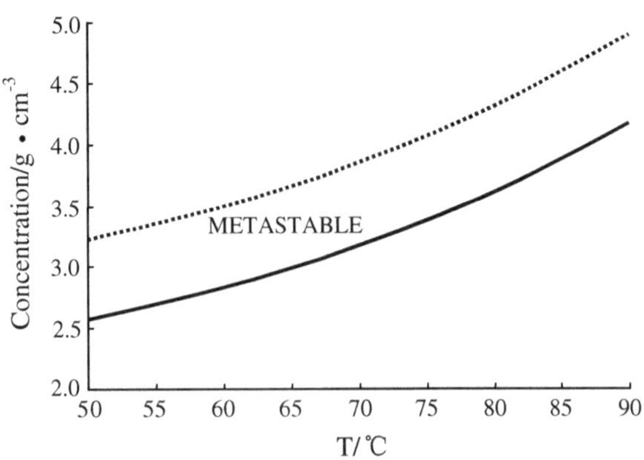

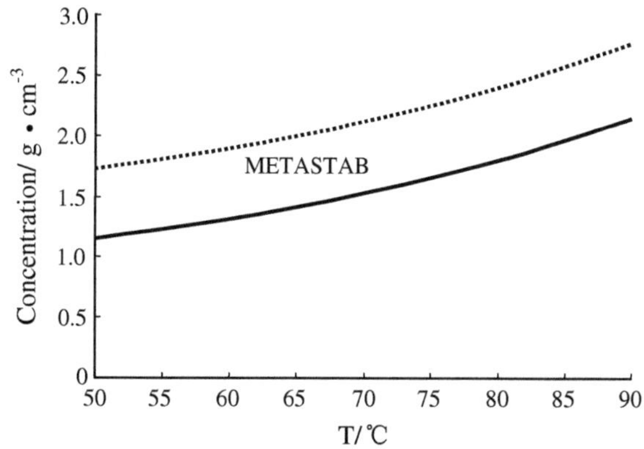

Fig. 2 Concentration trajectory of glucose

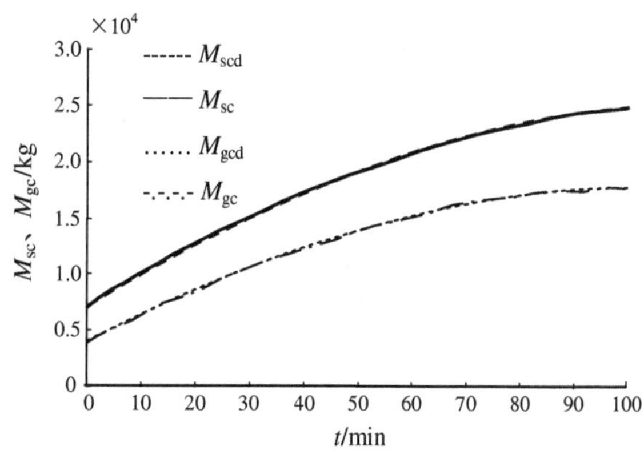

Fig. 3 Precipitation tracking curves

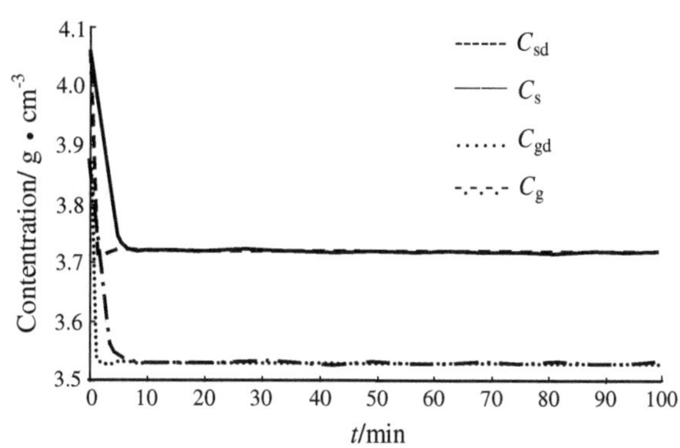

Fig. 4 Concentration tracking curves

where M_{scd}, M_{sc} are the set point and tracking results of sucrose crystal mass; M_{gcd}, M_{gc} are the set point and tracking results of glucose crystal mass. The concentration tracking curves of sucrose-glucose solution are illustrated in Fig. 4, where C_{sd}, C_s are the set point and tracking results of sucrose concentration; C_{gd}, C_g are the set point and tracking results of glucose concentration.

From the simulation results in Fig. 3, we know that the crystal masses of sucrose-glucose track their respective set point commendably. From the simulation results in Fig. 4, we know that the concentrations of sucrose-glucose track their respective set point commendably. As a consequence, the uncertainty is approached in real time, and the control precision and robustness are improved. Furthermore, the system disturbance is restrained by a robust controller.

5 Conclusion

In this paper, the objective is to establish a new control strategy of Chinese medicine sugar precipitation process, based on not only a real-time control of the crystal mass evolution, but also a control of the solution concentration to maintain the magma in supersaturation conditions. In this paper, a novel control strategy is proposed which is built on "model decomposition". Adaptive controller is adopted to eliminate the parameter uncertainty; robust controller is adopted to attenuate the unknown dynamics and disturbances. With a feedback controller to dominate the nominal system, three controllers compose the integrated controller for the uncertain nonlinear system, and the stability is proven. The validity is associated with the precipitation simulation of sucrose-glucose mixed solution. The global performance of the closed-loop precipitation process has been tested with different crystallization set point in terms of energy efficiency.

Acknowledgments This work is supported by the natural science foundation of Hebei province under Grant F2013501041 and F2012501023.

References

1. Paengiuntuek W, Kittisupakon P, Arpornwichanop A (2008) Optimization and nonlinear control of a batch crystallization process. J Chin Inst Chem Eng 39(3):249–256
2. Nagy ZK, Chew JW, Fujiwara M (2008) Comparative performance of concentration and temperature controlled batch crystallizations. J Process Control 18(3–4):399–407
3. Georgieva P, Azevedo SFD (2006) Neural network-based control strategies applied to a fed-batch crystallization process. Int J Inf Math Sci 3(3):224–233
4. Nagy ZK (2009) Model based robust control approach for batch crystallization product design. Comput Chem Eng 33(10):1685–1691
5. Bonvin D (1998) Optimal operation of batch reactors: A personal view. J Process Control 8(5–6):355–368
6. Chen Z, Xu C, Shao H (2003) Batch processes optimization and advanced control-a survey. Control Instrum Chem Ind 30(3):1–6 (in Chinese)
7. Xiong Z, Zhang J (2005) Neural network model-based on-line re-optimisation control of fed-batch processes using a modified iterative dynamic programming algorithm. Chem Eng Process 44(4):477–484

8. Xiong Z, Zhang J, Dong J (2008) Optimal iterative learning control for batch processes based on linear time-varying perturbation model. Chin J Chem Eng 16(2):235–240
9. Jia L, Shi J, Chiu M, YU J (2010) Nonrestraint-iterative learning-based optimal control for batch processes. CIESC J 61(8):1889–1893
10. Damour C, Benne M, Boillereaux L, Grondin-Perez B, Chabriat J-P (2010) NMPC of an industrial crystallization process using model-based observers. J Ind Eng Chem 16(5):708–716
11. Fan L, Wang H, Song Z, Li P (2006) Iterative optimal control for batch process based on generalized predictive control. Control Instrum Chem Ind 33(2):25–28
12. Haddad WM, Hayakawa T, Chellaboina V (2003) Robust adaptive control for nonlinear uncertain systems. Automatica 39(3):551–556
13. Xu H, Ioannou PA (2004) Robust adaptive control of linearizable nonlinear single input systems with guaranteed error bounds. Automatica 40(11):1905–1911
14. Liu Y, Li X-Y (2003) Robust adaptive control of nonlinear systems represented by input-output models. IEEE Trans Autom Control 48(6):1041–1045
15. Qu Z (2003) Adaptive and robust controls of uncertain systems with nonlinear parameterization. IEEE Trans Autom Control 48(10):1817–1823
16. Wright PG, White ET (1974) A mathematical model of vacuum pan crystallization. In: Proceedings of ISSCT 15th congress, pp 1546–1560
17. Barth S (2006) Utilization of FBRM in the control of CSD in a batch cooled crystallizer. Ph.D. thesis, Georgia Institute of Technology, Atlanta

Soft Measurement of Sugar Crystallization Process for Chinese Medicine Mixed Solution

Qing-wei Li and Hong-jun Duan

Abstract This paper deals with the design of model-based soft measurement to improve the process monitoring and control in Chinese medicine sugar crystallization. A model dedicated to Chinese medicine sugar crystallization is designed, without consideration of crystal size distribution. The mass of crystals in the solution, the concentration of dissolved sugar and the purity are important to improve the manufacturing process. As these physicochemical properties are not measurable on-line, a model-based soft measurement is developed. The effectiveness of the soft measurement is demonstrated using real data from a Chinese medicine sugar crystallization process.

Keywords Batch process · Chinese medicine · Sugar crystallization · Soft measurement

1 Introduction

Crystallization is a unit operation for achieving the extraction of a solute from a saturated solution, which is widely promoted in pharmaceutical, chemical and food industries. Due to widespread applications of the crystallization process, optimal operation and efficient on-line control strategy of the crystallization process are necessary in order to improve their performance in terms of product quality and energy usage [1]. On-line control during batch crystallization offers the possibilities

Q. Li (✉)
School of Resources and Materials, Northeastern University at Qinhuangdao, Qinhuangdao, China
e-mail: lqwday@126.com

H. Duan
School of Control Engineering, Northeastern University at Qinhuangdao, Qinhuangdao, China
e-mail: dhj@mail.neuq.edu.cn

E. Qi (ed.), *Proceedings of the 6th International Asia Conference on Industrial Engineering and Management Innovation*,
DOI 10.2991/978-94-6239-145-1_47

for improved crystal product quality, shorter process times, and reduction or elimination of compromised batches [2]. Crystallization of the mixed solution is a batch process, which is usually related with the formulation of a control problem in terms of economic or performance objective at the end of the process [3]. Despite the long history and widespread application of batch process, there remain a number of problems related to its control, mainly related the complex nonlinear dynamics with nonideal mixing, and various disturbances characteristic to these systems [4]. In recent years, there has got a growing interest in the optimization control of batch processes, which are suitable for responsive manufacturing of low volume and high value-added products [5, 6]. However, it is usually very difficult to obtain accurate models owing to the complexity of the processes, and it is also difficult to implement them for on-line optimization, since optimization that based on mechanistic models is usually very time-consuming [7]. Of late, relative algorithms are mainly iterative learning control (ILC) [8, 9] and nonlinear model-based predictive control (NMPC) [10, 11].

In this paper, a soft measurement devoted to Chinese medicine sugar crystallization is designed to improve the process monitoring and control of an industrial process. Since the production objective is limited to a maximization of the sugar exhaustion, a simplified model is available, without consideration for the CSD. This one allows avoiding the solving of the classical population balance. The sugars were precipitated according to their respective crystal rates, and fructose and glucose track their respective concentration set points in real time.

This paper is structured as follows: Section 2 focuses on model design. Then, soft measurement design and its application to a Chinese medicine case study are presented and analysed in Section 3. Finally, conclusion is presented in Section 4.

2 Model Design

A model dedicated to crystallization monitoring and control of Chinese medicine sugars is presented. This investigation leads to the development of models of extraordinary purity liquors [12]. Little information is expected to be available about the stage where multiple sugars could be co-precipitated. There, low grade Chinese medicine crude solutions (impure solutions) contain a high proportion of impurities, which considerably changes their properties.

The most usual approach consists of considering nucleation, agglomeration and dissolution negligible during the growth phase, in supersaturated conditions [13]. In other words, crystal growth phenomena are assumed to be able to preponderant. The crystallization step of Chinese medicine sugar is performed through a water crystallization tank, operating by a continuously agitator, as illustrated in Fig. 1.

The dynamic model proposed in this paper represents the mass balance for dissolved fructose and glucose (m_s, m_g), crystals (m_{sc}, m_{gc}), sodium hydroxide

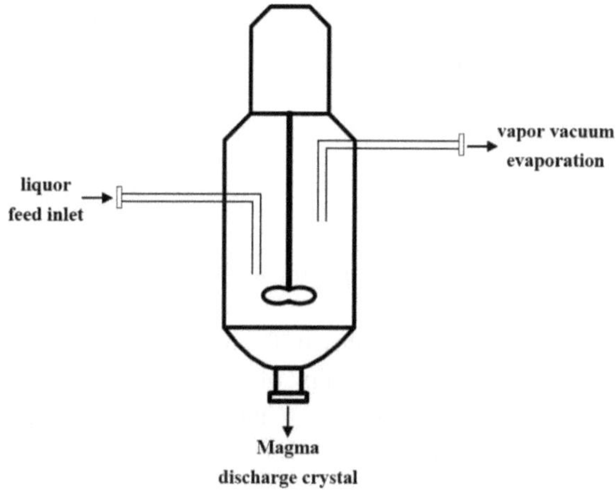

Fig. 1 Water crystallization tank of Chinese medicine

solution (m_{sh}) and impurities (m_{i}). The evolution of the mass of crystals is governed by the following expression:

$$\begin{cases} \dot{m}_{\mathrm{sc}} = \alpha m_{\mathrm{s}} \\ \dot{m}_{\mathrm{gc}} = \beta m_{\mathrm{g}} \end{cases} \tag{1}$$

where α and β are adjusted crystal parameters of fructose and glucose respectively.

The evolution of dissolved fructose and glucose results from the fructose and glucose contained in the liquor supply, which depends on the mass fraction of dry substance and on the purity of the solution, and from the disappearance of fructose and glucose converted into crystals, according to Eq. (1):

$$\begin{cases} \dot{m}_{\mathrm{s}} = \rho_{\mathrm{f}} F_{\mathrm{f}} Bx_{\mathrm{fs}} Pte_{\mathrm{fs}} - \dot{m}_{\mathrm{sc}} \\ \dot{m}_{\mathrm{g}} = \rho_{\mathrm{f}} F_{\mathrm{f}} Bx_{\mathrm{fg}} Pte_{\mathrm{fg}} - \dot{m}_{\mathrm{gc}} \end{cases} \tag{2}$$

where ρ_{f} and F_{f} are the density[kg/m^3] and flow rate[m^3/s] of the feed syrup respectively, Bx_{fs} and Pte_{fs} are the mass fractions of fructose and its dry substance respectively, Bx_{fg} and Pte_{fg} are the mass fractions of glucose and its dry substance respectively. The mass balance applied to the sodium hydroxide solution in the water crystallization tank leads to the following equation:

$$\dot{m}_{\mathrm{sh}} = \rho_{\mathrm{f}} F_{\mathrm{f}} \left(1 - Bx_{\mathrm{fs}} - Bx_{\mathrm{fg}}\right) + \rho_{\mathrm{sh}} F_{\mathrm{sh}} - \dot{m}_{\mathrm{vap}} \tag{3}$$

where ρ_{sh} and F_{sh} are the density [kg/m^3] and flow rate [m^3/s] of the sodium hydroxide solution respectively, m_{vap} is the emitted vapor mass [kg]. The mass of impurities contained in the solution is obtained using the following expression:

$$\begin{cases} \dot{m}_{\mathrm{i}} = \rho_{\mathrm{f}} F_{\mathrm{f}} \left(Bx_{\mathrm{fs}} + Bx_{\mathrm{fg}} \right) \left(1 - Pte_{\mathrm{fs}} - Pte_{\mathrm{fg}} \right) \\ m_{\mathrm{i}}(t) = m_{\mathrm{i}}(0) + \rho_{\mathrm{f}} F_{\mathrm{f}} \left(Bx_{\mathrm{fs}} + Bx_{\mathrm{fg}} \right) \left(1 - Pte_{\mathrm{fs}} - Pte_{\mathrm{fg}} \right) t \end{cases} \tag{4}$$

The initial value $m_{\mathrm{i}}(0)$ is determined by the initial conditions. An energy balance applied to the magma in the water crystallization tank allows expressing the mass of emitted vapor:

$$\dot{H}_{\mathrm{mg}} = \dot{Q} + \rho_{\mathrm{f}} F_{\mathrm{f}} h_{\mathrm{f}} + \rho_{\mathrm{sh}} F_{\mathrm{sh}} h_{\mathrm{sh}} + \lambda_{\mathrm{sc}} \dot{m}_{\mathrm{sc}} + \lambda_{\mathrm{gc}} \dot{m}_{\mathrm{gc}} - h_{\mathrm{vap}} \dot{m}_{\mathrm{vap}} \tag{5}$$

where H_{mg} is the enthalpy of the magma[J], $\dot{Q}$ is the heating power[W], h_{f}, h_{sh}, and h_{vap} are the specific enthalpy of feed syrup, sodium hydroxide solution, and emitted vapor respectively[J/kg], λ_{sc} and λ_{gc} are the specific latent heat of fructose and glucose respectively[J/kg]. The enthalpy variation of the magma can be written as follows:

$$\begin{cases} H_{\mathrm{mg}} = Cp_{\mathrm{mg}} m_{\mathrm{mg}} T_{\mathrm{mg}} \\ m_{\mathrm{mg}} = m_{\mathrm{s}} + m_{\mathrm{g}} + m_{\mathrm{sc}} + m_{\mathrm{gc}} + m_{\mathrm{sh}} + m_{\mathrm{i}} \end{cases} \tag{6}$$

$$\dot{H}_{\mathrm{mg}} \approx Cp_{\mathrm{mg}} T_{\mathrm{mg}} \dot{m}_{\mathrm{mg}} \tag{7}$$

where Cp_{mg}, m_{mg}, T_{mg} are the specific heat capacity [J/(K·kg)], mass[kg], and temperature[°C] of the magma respectively. From Eq. (5), the mass of vapor can be expressed as follows:

$$\begin{aligned} h_{\mathrm{vap}} \dot{m}_{\mathrm{vap}} &= \dot{Q} + \rho_{\mathrm{f}} F_{\mathrm{f}} h_{\mathrm{f}} \\ &+ \rho_{\mathrm{sh}} F_{\mathrm{sh}} h_{\mathrm{sh}} + \lambda_{\mathrm{sc}} \dot{m}_{\mathrm{sc}} + \lambda_{\mathrm{gc}} \dot{m}_{\mathrm{gc}} - Cp_{\mathrm{mg}} T_{\mathrm{mg}} \dot{m}_{\mathrm{mg}} \end{aligned} \tag{8}$$

Substitute the expressions (1–4) into Eq. (8), it comes:

$$\begin{aligned} \left(h_{\mathrm{vap}} - Cp_{\mathrm{mg}} T_{\mathrm{mg}} \right) \dot{m}_{\mathrm{vap}} &= \dot{Q} + \rho_{\mathrm{f}} F_{\mathrm{f}} \left(h_{\mathrm{f}} - Cp_{\mathrm{mg}} T_{\mathrm{mg}} \right) \\ &+ \rho_{\mathrm{sh}} F_{\mathrm{sh}} \left(h_{\mathrm{sh}} - Cp_{\mathrm{mg}} T_{\mathrm{mg}} \right) + \alpha \lambda_{\mathrm{sc}} m_{\mathrm{s}} + \beta \lambda_{\mathrm{gc}} m_{\mathrm{g}} \end{aligned} \tag{9}$$

3 On-Line Monitoring of the Chinese Sugar Crystallization Process

The crystallization performance of the process can be increased by improving the crystal rate. This growth rate can be apprehended by following the crystal mass. In mixed solution, we set fructose crystal mass and glucose crystal mass in different time, respectively, and they are tracked by the proposed control algorithm. Beyond this set point tracking, a second control objective is required, and consists in maintaining the solution in an appropriate supersaturated state, that is a metastable zone, as illustrated in Figs. 2 and 3. Solubility curves are denoted by solid lines, and nucleation curves are denoted by dotted lines.

In the industrial context, when it deals with the gas flow rate measurement, direct measurement may be problematic (large nominal diameter of inlet tubes, turbulent flow). In this case, both the heating steam and the emitted vapor flow rates are calculated by indirect measurements.

The model identification has been performed using a series of five industry data bases, where each data base represents a single experimental run. Three parameters had to be calculated to make the model fit the data. An iterative fitting procedure has been carried out. The algorithm is based on the minimization of a quadratic criterion V expressing the mean squared difference between simulated (sim) and experimental (exp) Bx_{ml}, Bx_{mg} and T_{mg} for the totality of the samples M. All the data are collected at a sampling rate of 30 s. Bx_{mgexp} represents the total dry matter fraction given by a microwave sensor and Bx_{mlexp} represents the dissolved dry

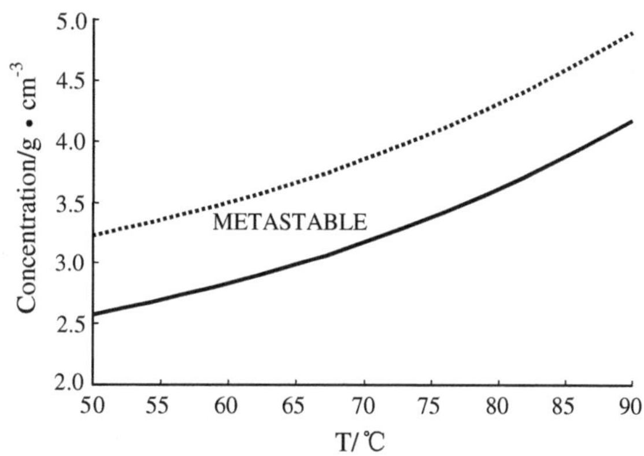

Fig. 2 Concentration trajectory of fructose

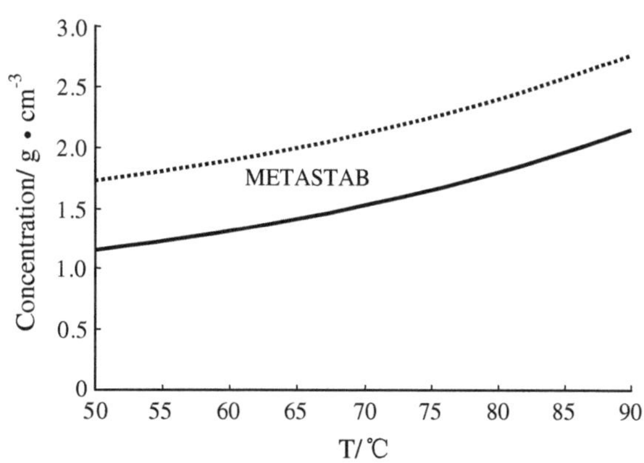

Fig. 3 Concentration trajectory of glucose

matter fraction given by a refract to meter sensor. Bx_{mgsim} and Bx_{mlsim} are obtained by the following linear combination of m_s, m_g, m_{sc}, m_{gc}, m_{sh} and m_i:

$$Bx_{\mathrm{mgsim}} = \frac{m_s + m_g + m_{sc} + m_{gc} + m_i}{m_s + m_g + m_{sc} + m_{gc} + m_{sh} + m_i}$$

$$Bx_{\mathrm{mlsim}} = \frac{m_s + m_g + m_i}{m_s + m_g + m_{sh} + m_i}$$

The validation step is performed using two additional databases. The experimental mass of crystals is obtained by an indirect measurement using Bx_{ml} and Bx_{mg}. Prior to each run, the initial conditions for the state variables are determined from off line measurements for Brix, purity and level. Figures 4 and 5 show simulated versus measured results for m_{sc} and m_{gc}. In comparison with industry data, the soft measurement has proved to be able to calculate m_{sc} and m_{gc} with a relative error less than 5 %.

As previously mentioned, the current control strategy is based on conductivity set point tracking. In other words, m_{sc} and m_{gc} should increase if conductivity κ tracks the desired set point. Experimental investigations confirmed the

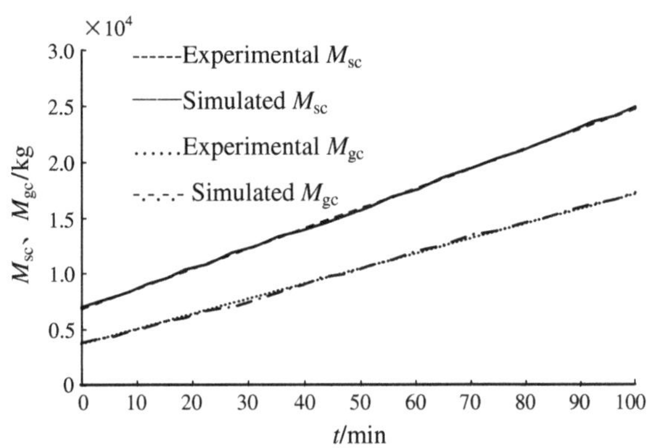

Fig. 4 Crystallization tracking results

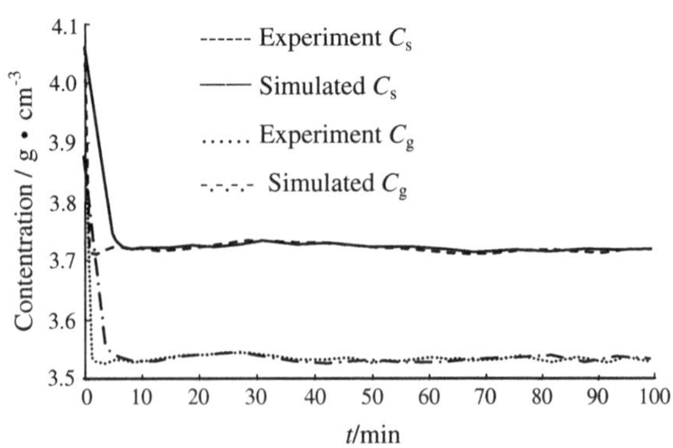

Fig. 5 Concentration tracking results

shortcomings of this current control strategy when it deals with the process. Growth phase operations from solutions containing a high level of impurities reveal m_{sc} and m_{gc} may not increase even if κ tracks the desired set point.

Without mention of the effect of impurities on both measurements and performance, oscillations around the set point are partially due to the use of a linear PID strategy to control a strongly nonlinear process. In these kinds of cases, the soft measurement could improve significantly the process supervision and control and avoid a waste of time and energy for the industrialists.

The crystallization simulated and experimental results of fructose-glucose solution are illustrated in Fig. 4, which are the set point tracking results of fructose and glucose crystal masses; the concentration simulated and experimental results of fructose-glucose solution are illustrated in Fig. 5, which are the set point tracking results of fructose and glucose concentration.

From the Fig. 4, we learn about that the crystal masses of fructose-glucose track their respective set point commendably. From the Fig. 5, we learn about that the concentrations of fructose-glucose track their respective set point commendably.

4 Conclusion

In this paper, the objective is to establish a new monitoring and control strategy of Chinese medicine sugar crystallization process. On the first hand, it bases on a real-time control of the crystal mass evolution, and on the other hand, on a control of the solution concentration to maintain the magma in supersaturation conditions. In this study, a model-based soft measurement has been developed for on-line measurement of the mass of crystals, the concentration of dissolved fructose and the purity of the solution, and it has been used in a Chinese medicine sugar crystallization process. In terms of process control, because of the large number of parameters and the manifold interactions between each other, the traditional strategies have proved to be unsatisfactory. This soft measurement is based on an original model that permits avoiding the solving of the population balance.

The validity was corroborated by the crystallization experiment and simulation of fructose-glucose mixed solution. The global performance of the closed-loop crystallization process has been studied with different crystallization set point in terms of energy efficiency.

Acknowledgments This work is supported by the natural science foundation of Hebei province under Grant F2013501041 and F2012501023.

References

1. Paengjuntuek W, Kittisupakorn P, Arpornwichanop A (2008) Optimization and nonlinear control of a batch crystallization process. J Chin Inst Chem Eng 39(3):249–256
2. Nagya Zoltan K, Chewb Jia W, Fujiwarab Mitsuko, Braatz Richard D (2008) Comparative performance of concentration and temperature controlled batch crystallizations. J Process Control 18(3–4):399–407
3. Georgieva P, Feyo de Azevedo S (2006) Neural network-based control strategies applied to a fed-batch crystallization process. Int J Inf Math Sci 3(3):224–233
4. Nagy Zoltan K (2009) Model based robust control approach for batch crystallization product design. Comput Chem Eng 33(10):1685–1691
5. Bonvin Dominique (1998) Optimal operation of batch reactors: a personal view. J Process Control 8(5–6):355–368
6. Zhi-gang CHEN, Chao XU, Hui-he SHAO (2003) Batch processes optimization and advanced control-a survey. Control Instrum Chem Ind 30(3):1–6 (in Chinese)
7. Xiong Zhi-hua, Zhang Jie (2005) Neural network model-based on-line re-optimisation control of fed-batch processes using a modified iterative dynamic programming algorithm. Chem Eng Process 44(4):477–484
8. Xiong Z, Zhang J, Dong J (2008) Iterative learning control of batch processes based on time varying perturbation models (in Chinese). J Tsinghua Univ (Sci Technol) 48(S2):1771–1774
9. Li JIA, Ji-ping SHI, Min-Sen CHIU, Jin-shou YU (2010) Nonrestraint-iterative learning-based optimal control for batch processes. CIESC J 61(8):1889–1893
10. Damour Cédric, Benne Michel, Boillereaux Lionel, Grondin-Perez Brigitte, Chabriat Jean-Pierre (2010) NMPC of an industrial crystallization process using model-based observers. J Ind Eng Chem 16(5):708–716
11. Lei FAN, Hai-qing WANG, Zhi-huan SONG, Ping LI (2006) Iterative optimal control for batch process based on generalized predictive control. Control Instrum Chem Ind 33(2):25–28
12. Wright PG, White ET (1974) A mathematical model of vacuum pan crystallization. In: Proceedings of ISSCT 15th congress, pp 1546–1560
13. Barth S (2006) Utilization of FBRM in the control of CSD in a batch cooled crystallizer. Ph.D. thesis, Georgia Institute of Technology

Research on Fault Diagnosis of Diesel Engine Based on PSO-SVM

Zhongxing Zhang and Hongyan Guo

Abstract In this study, support vector machine (SVM) theory is introduced into the fault diagnosis of diesel engine, a novel PSO-SVM model has been proposed that hybridized the particle swarm optimization (PSO) and SVM to improve the diagnosis of engine accuracy. Firstly, PSO algorithm is chosen to determine the optimum parameters of SVM, to avoid the blindness of choosing parameters, improve the prediction accuracy of the model. Then, PSO-SVM is applied to predict the fault diagnosis of diesel engine, and compared with the traditional BP neural network (BPNN) and RBFNN network. The result shows that the prediction accuracy of PSO-SVM is greatly improved compared with BPNN and RBFNN, thus PSO-SVM is developed as an efficient tool so that various SVMs can be used conveniently as the core of PSO-SVM for diagnosis of diesel engine.

Keywords Diesel engine · Parameter selection · Particle swarm optimization (PSO) · Support vector machine (SVM)

1 Introduction

With the rapid development of science and technology, structure of equipments is more complex, with increasingly high degree of automation. In the work, the probability of failure is relatively improved, the failure will not only lead to economic losses, but also cause the equipment suffered catastrophic destruction. Because the structure of the diesel engine itself is very complex, additionally the input and output of system is not obvious, it is difficult to describe the mechanism,

Z. Zhang · H. Guo (✉)
Department of Information Engineering, Henan Radio & Television University, Zhengzhou, China
e-mail: ghy@open.ha.cn

Z. Zhang
e-mail: hnddzzx@163.com

E. Qi (ed.), *Proceedings of the 6th International Asia Conference on Industrial Engineering and Management Innovation*,
DOI 10.2991/978-94-6239-145-1_48

function and status light which have great trouble in fault diagnosis. In recent years, with the introduction of pattern recognition and neural network, fault diagnosis technology of the diesel engine has been rapidly developed. The neural network technology provides a new solution for complicated system of diesel engine, especially it has self-organization and learning ability, which make to become a kind of effective method for fault diagnosis.

Recently, Artificial Neural Network (ANN) is one of the common methods of pattern classification. However, there are some problems in ANN which based on traditional statistical theory, such as lack of generalization ability and require a huge number of training samples and so on. To some extent, these problems limit the applications of ANN in small sample size problem such as machinery fault diagnosis. SVM is a type of machine learning method, which adopts the structure risk minimization principle. Based on the structured risk minimization principle, SVM seeks to minimize an upper bound of the generalization error instead of the empirical error as in conventional neural networks [1–3]. Therefore, SVM has been turned into the hot spot on machinery fault research. However, there is no uniform method to determine the selection of the parameters of SVM, which becomes the bottleneck for the improvement of the application of SVM. The particle swarm optimization was found to have the extensive capability of global optimization for its simple concept, easy implementation and fast convergence [4–6]. Therefore, instead of using the above algorithms, a new method, PSO is proposed to optimize the SVM parameters in this study. After the introduction of algorithm, models the simulation experiments were done, the experimental results show that: PSO-SVM can not only improve the correct rate of diesel engine fault detection, but also predict the fault effectively, reduce property loss [7–9].

The remainder of this paper is organized as follows. Section 2 discusses the related work of SVM and Particle swarm optimization. Section 3 describes the proposed method of PSO-SVM in detail. Section 4 shows the experiment results on KDD cup. Section 5 gives some conclusions.

2 Related Work

2.1 SVM Algorithm

A classification technique that has received considerable attention is support vector machine (SVM). This technique has shown promising empirical results in many practical applications, from handwritten digit recognition to text categorization. SVM also works very well with high-dimensional data and avoids the curse of dimensionality.

Consider a binary classification problem consisting of N training examples. Each example is denoted by a tuple (x_i, y_i) $(i = 1, 2, 3, \ldots, N)$, where $x_i = (x_{i1}, x_{i2}, x_{i3}, \ldots, x_{id})T$, corresponds to the attribute set for the ith example. By conversion, let

$y_i \in \{-1,1\}$ denote its class label. The decision boundary of a linear classifier can be written in the following form:

$$y(x) = [w\phi(x)] + b \quad (1)$$

where w is weight vector, b is the threshold. In addition, the coefficients w and b are estimated by the following linear optimization problem:

Constraint conditions:

$$\begin{aligned} y_i - w\phi(x_i) - b \leq \varepsilon + \xi_i (i = 1,2,3,\ldots,n) \\ \phi(x_i)w + b - y_i \leq \varepsilon + \xi_i^* (i = 1,2,3,\ldots,n) \end{aligned} \quad (2)$$

where ξ_i and ξ_i^* is slack variable, $\xi_i > 0,\ \xi_i^* > 0$, since there are no constraints on the number of mistakes the decision boundary can make, the learning algorithm may find a decision boundary with a very wide margin but misclassifies many of the training examples. To avoid this problem, the objective function must be modified to penalize a decision boundary with large values slack variables. The modified objective function is given by the following equation:

$$\min_{w,b,\xi^{(*)}} \frac{1}{2}\|w\|^2 + C\sum_{i=1}^{n} (\xi_i + \xi_i^*) \quad (3)$$

where C is punishment coefficient, C > 0, ξ_i is insensitive loss function, ξ_i^* guarantees the satisfaction of constraint condition; C controls the equilibrium between the complexity of model and training error; ε is a preset constant that for controlling tube size. If ε set too small, then it is easy to lead to over learning, otherwise, it is easy to lead to the owe learning. Formulas is transformed into the corresponding dual problem by duality theory, the results are as follows:

$$\max\left[-\frac{1}{2}\sum_{i=1}^{n}\sum_{j=1}^{n}(a_i - a_i^*)(a_j - a_j^*)K(x_i, x_j) - \varepsilon\sum_{i=1}^{n}(a_i - a_i^*) + \sum_{i=1}^{n} y(a_i - a_i^*)\right]$$
$$s.t \sum_{i=1}^{n}(a_i - a_i^*) = 0 (a_i, a_i^* \in [0, C]) \quad (4)$$

where a_i, a_i^* are both the L dimension support vectors to be solved.

There are many kinds of kernel functions; selecting different kernel functions will construct different regression models. The commonly used kernel functions are: polynomial function, linear kernel function and the Gaussian kernel function etc.

(1) Linear kernel function $K(x, x') = x.x'$
(2) Polynomial kernel function $K(x,x') = ((x.x') + 1)d$
(3) Gaussian kernel function $k(x, x') = \exp(-\|x - x'\|/\sigma^2)$

In this study, we use Gaussian kernel function, where σ is the kernel parameter. The parameter σ precisely defines the structure of high dimensional feature space, thus controlling the complex nature of ultimate solution. In conclusion, it is found that the choice of C, ε and σ has a great influence on the performance of SVM regression estimation. Therefore, how to reasonably and quickly select the above parameters has become a key factor that affects the practical application effects and scope of SVM. In this study, SVM has optimized initialization parameters by using PSO.

2.2 Particle Swarm Optimization

In PSO, each single solution is a "bird" in the search space. We call it "particle". All of particles have fitness values which are evaluated by the fitness function to be optimized, and have velocities which direct the flying of the particles. The particles fly through the problem space by following the current optimum particles.

PSO is initialized with a group of random particles (solutions) and then searches for optima by updating generations. In every iteration, each particle is updated by following two "best" values. The first one is the best solution (fitness) it has achieved so far. (The fitness value is also stored.) This value is called pBest. Another "best" value that is tracked by the particle swarm optimizer is the best value, obtained so far by any particle in the population. This best value is a global best and called gBest. When a particle takes part of the population as its topological neighbors, the best value is a local best and is called lbest.

After finding the two best values, the particle updates its velocity and positions with following Eqs. (5) and (6).

$$\mathrm{v}[] = \mathrm{v}[] + \mathrm{c1} * \mathrm{rand}() * (\mathrm{pbest}[] - \mathrm{present}[]) + \mathrm{c2} * \mathrm{rand}() * (\mathrm{gbest}[] - \mathrm{present}[]) \quad (5)$$

$$\mathrm{present}\ [] = \mathrm{present}\ [] + \mathrm{v}[] \quad (6)$$

v[] is the particle velocity, persent[] is the current particle (solution). pbest[] and gbest[] are defined as stated before. Rand () is a random number between (0,1). c1, c2 are learning factors. usually c1 = c2 = 2.

The pseudo code of the procedure is as follows:

```
For each particle {
Do {Initialize particle
   }
Do{
For each particle {
Calculate fitness value
If the fitness value is better than the best fitness value
(pBest) in history set current value as the new pBest
   }
Choose the particle with the best fitness value of all the
particles as the gBest
For each particle {
Calculate particle velocity according equation (5)
    Update particle position according equation (6)
    }
} While maximum iterations or minimum error criteria is
not attained
```

While maximum iterations or minimum error criteria is not attained

Particles' velocities on each dimension are clamped to a maximum velocity Vmax. If the sum of accelerations would cause the velocity on that dimension to exceed Vmax, which is a parameter specified by the user. Then the velocity on that dimension is limited to Vmax.

3 Fault Diagnosis of Diesel Engine Based on PSO-SVM

3.1 Establish the Model

In this paper, the experimental data to do fault diagnosis are achieved through diesel engine vibration signal in the normal operating condition of diesel engines. Pressure waveform in high pressure pipe contains lots of working information of fuel system in diesel engine, which can be sensitive to distinguish all kinds of fault information. So Pressure waveform in high pressure pipe can represent the fault diagnosis of diesel engine.

Pressure waveform which can reflect the characteristic parameters of diesel engine includes maximum pressure (P_1), least maximum pressure(P_2), wave amplitude(P_3), rising edge width(P_4), wave width(P_5), maximum width of the aftermath (P_6), wave area(P_7), injection pressure(P_8). Therefore, the original data are normalized into [−1,1], so as to obtain input data. In recent years, according to fault sample data of diesel engine, four different types of fault diagnosis are chosen: normal(T_1), Off the oil-cylinder(T_2), Fuel injector stuck(T_3), Fuel injector clogging (T_4), High opening pressure(T_5), Injector spring damage(T_6).

3.2 Data Normalization

Using SVM for fault diagnosis of diesel engine, the training data and testing data need to be normalized. The main purpose of normalization is to avoid attributes in greater numerical ranges dominating those in smaller numerical ranges. Additionally, the normalization could avoid numerical difficulties during the calculation. For a group of real estate data $\{x_{1k}, x_{2k}, \ldots, x_{ik}\}$, it is normalized as $\{X_{1k}, X_{2k}, \ldots, X_{ik}\}$ by the formula:

$$X_{ik} = \frac{X_{ik} - X_k^{\min}}{X_k^{\max} - X_k^{\min}} \tag{7}$$

where X_{ik} is the scaled value, X_{ik} is original value, $X_k^{\max}$ is the maximum value of feature k in the data set, $X_k^{\min}$ is the minimum value of feature k in the data set.

3.3 PSO-SVM Algorithm

The Gaussian kernel function is used to construct SVM regression model in this paper, then the width of the Gaussian kernel, the penalization parameter C, and the ε-insensitive parameter need to be determined. As particle swarm optimization algorithm not only has strong global search ability and but also helps to search for the optimum parameters quickly. Therefore, the PSO method is applied to determine the parameters of SVM, which is shown in Fig. 1. The process of optimizing the parameters of SVM by PSO is represented as follows:

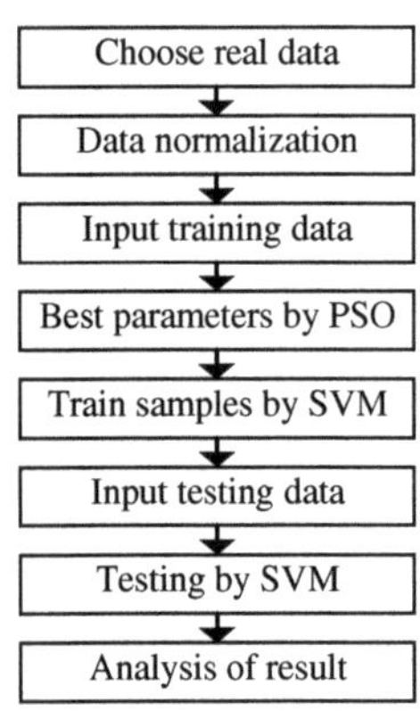

Fig. 1 Model by PSO-SVM

4 Results and Discussion

In this section, we describe the experiments conducted to evaluate the proposed system. The proposed system is tested on an AMD Processor 2.60 GHz with 2G RAM running Windows 7 and coded by Matlab 7.01.

Experiment first needs to extract training set and testing set from the original data, There are 132 training samples and 68 testing samples and then do certain pretreatment (including the initial value of the data set, feature calculation and selection, data normalization preprocessing, dimension reduction pretreatment), second make use of PSO to find optimal parameters c and g, third use the optimal parameters to do SVM training on the training set, last get the model to predict the classification labels of the test set, fault diagnosis samples of diesel engine is as shown in Table 1:

Parameter settings are as follows in PSO-SVM model: parameter of local search capability c1 = 1.6, parameter of global search capability c2 = 1.5, size of population is 20, the number of maximum evolution is 50, the search range of the penalty parameter c is [0.1,1000] in SVM, the search range for the [0.01,1000]. In order to test performance of PSO-SVM algorithm, this paper selected the back-propagation neural network (BPNN) and RBF neural network (RBFNN) to compare the test. In BPNN network, the speed of learning is 0.01, the increasing proportion of learning speed is 1.08, error of target is 0.01; in RBFNN network, mean square error is 0.06, radial basis function distribution coefficient is 72, the number of maximum evolution is 50. The experimental results are shown in Fig. 2 and Table 2.

It could be observed from Fig. 2 that the fault diagnosis algorithm based on PSO-SVM in this paper needs the shortest training time and has the highest operation efficiency among all diesel engine diagnosis models. It indicates that conduct reduction of diesel engine's fault attributes with PSO could eliminate a part of redundant features, decrease the input dimension of SVM, reduce the computed amount and thus improve the instantaneity of fault diagnosis.

It is obvious from Table 2 and Fig. 3 that the fault diagnosis algorithm of diesel engine based on PSO-SVM is most accurate among all fault diagnosis models. It

Table 1 Fault diagnosis samples of diesel engine

Fault type	Number of fault type	Number of training	Number of testing
Normal	Y = T1 = 1	10	7
Off the oil-cylinder	Y = T2 = 2	25	13
Fuel injector stuck	Y = T3 = 3	37	17
Fuel injector clogging	Y = T4 = 4	14	6
High opening pressure	Y = T5 = 5	22	12
Injector spring damage	Y = T6 = 6	24	13

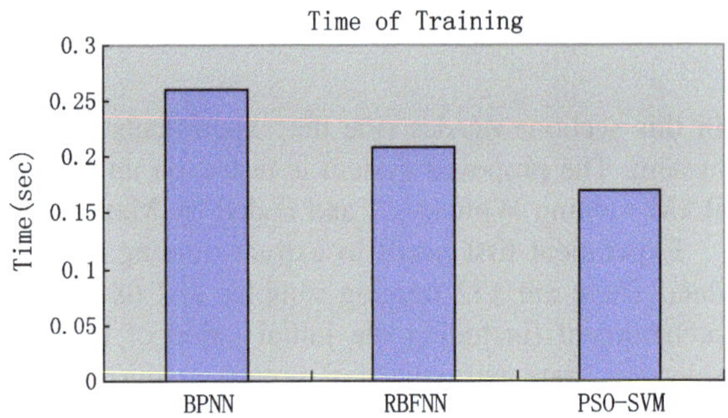

Fig. 2 Training time of three kinds of fault diagnosis models

Table 2 comparison of diagnosis data(testing data)

Test of model	Number of false or error	Rate of correct diagnosis
BPNN network	16	76 % (52/68)
RBFNN network	13	81 % (55/68)
PSO-SVM model	8	88 % (60/68)

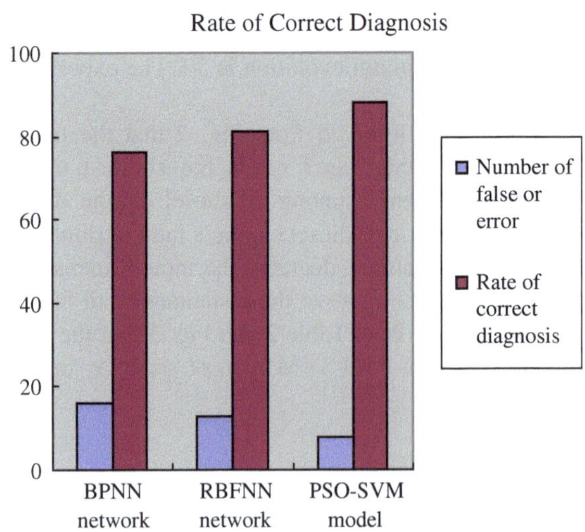

Fig. 3 Rate (Number) of diagnosis

indicates that conduct reduction of the fault attributes with PSO could eliminate some redundant features, reduce the impact of noise information on the diagnosis results, and thus improve the accuracy of fault diagnosis of diesel engine.

5 Conclusion

In this paper, we present a new SVM algorithm framework optimized by PSO algorithm. In previous works the choice of these parameters mainly depends on the experience. In our work PSO algorithm was used to optimize these parameters to form a new SVM framework PSO-SVM. The proposed algorithm was used to forecast fault diagnosis of diesel engine. The experimental results show that the proposed PSO-SVM has higher accuracy the some other NN model such as BPNN network and RBFNN network. So it is an effective means to deal with the similar equipment fault diagnosis problem.

As future work we will apply the proposed method to the larger data set, as well as carry out promotion and practical application. In addition, we will also focus on the research that with the expansion of the data scale, how PSO can maintain its good global search, and make the necessary improvements to improve the local search ability, so as to ensure the accuracy of the forecast. Furthermore, the proposed algorithm will be applied to other datasets.

References

1. Gao C, Bompard E, Napoli R, Cheng H (2007) Price forecast in the competitive electricity market by support vector machine. Physica A 382(1):98–113
2. Hong WC (2009) Hybrid evolutionary algorithms in a SVR-based electric load fore-casting model. Int J Electr Power Energy Syst 31(7–8):409–417
3. Lu CJ, Lee TS, Chiu CC (2009) Financial time series forecasting using independent component analysis and support vector regression. Decis Support Syst 47(2):115–125
4. Kennedy J, Eberhart R (1995) Particle swarm optimization. In: Proceedings of IEEE international conference on neural networks, Perth, Australia, pp 1942–1948
5. Shi YH, Eberhart RC (1999) Empirical study of particle swarm optimization. 1999CEC 99. In: Proceedings of the 1999 congress on evolutionary computation, Indianapolis, USA, pp 1945–1950
6. Boeringer DW, Werner DH (2004) Particle swarm optimization versus genetic algorithms for phased array synthesis. IEEE Trans Antennas Propag 52(3):771–779
7. Subasi A (2012) Classification of EMG signals using PSO optimized SVM for diagnosis of neuromuscular disorders. Comput Biol Med 43(5):576–586
8. Zhang Q, Liu B (2008) SVM parameters rapid optimization method based on PSO and group training. Sci Technol Eng 8(16):4613–4616
9. Kennedy J, Eberhart RC (1995) Particle swarm algorithm. In: Proceedings of the 1995 IEEE international conference on neural networks, New York: IEEE Press, pp 1942–1948

Visual Servo System Design Based on Characteristics of Target's Imaging Center and Area

Dong-bo Zhang, Xi Deng, Dong-jie Zhao and Bin Zhang

Abstract A cartesian robot visual servo system based on image was designed. This system applied target's imaging center and pixel area as characteristics for visual servo. The controlled quantity was gained by comparing differences between real-time images and expected images received after calibration. With the controller of cartesian robot, the robot's space position control could be realized. A personal computer was used as the image processor and a web camera was chose as the image collecting equipment. A Single-Chip Microcomputer (SCM) was used as robot's controller. And serial communication was applied to connect processor and controller. Furthermore, an experiment was conducted indoor using a red circle with its diameter of 60 mm as target. It turned out that the resetting error of the system was ±0.5 mm in both X and Y direction and ±1.0 mm in Z direction. System's locating error was ±1.0 mm in X and Y direction and ±2.0 mm in Z direction.

Keywords Control technology · Computer vision · Robot · Visual servo

1 Introduction

Applying data of computer vision to the feedback element in robot's motion control could form a visual servo control system. In this way, the application potential of visual sensor was able to be developed sufficiently. Robot visual servo control usually aims at robot's position control or trace tracking. Specifically, it uses visual sensors to detect robot's position directly or indirectly. And the position includes both current position and relative position against target. Because of the advantage

D. Zhang · X. Deng · D. Zhao · B. Zhang (✉)
College of Engineering, China Agricultural University, Beijing 100083, China
e-mail: zhangbin64@cau.edu.cn

D. Zhang
e-mail: zdb0204@cau.edu.cn

E. Qi (ed.), *Proceedings of the 6th International Asia Conference on Industrial Engineering and Management Innovation*,
DOI 10.2991/978-94-6239-145-1_49

of machine vision, visual servo system is more flexible and intelligent comparing with traditional robot sensors [1]. According to different definition of error signal and space for calculation, visual servo system could be divided into two kinds: visual servo based on position and visual servo based on image [2]. If the error signal was calculated via geometric model and camera model was defined in three-dimensional (3D) Descartes space, then it's a visual servo based on position [3]. If the error signal was calculated via Jacobian matrix and defined in two-dimensional (2D) image space, then it's a visual servo based on image [4]. Visual servo system could also be divided into open loop and closed loop servo according to the visibility of end actuator. Moreover, the system could be divided into Eye-in-Hand system and fixed camera system [5]. Different composite structures of the servo system decide its controlling structure which directly determines the best method of calculation [6]. System's model and calibration have significant effect to its property and technical parameters. That is to say, the design of visual servo controlling structure should cater to the demand of parameters and application. Thus, the corresponding method of calculation could possibly improve system's control precision and instantaneity, robustness and stability and some other properties [7]. The management of deviation signal or deviation information ran through the whole process of visual servo control. So, it's necessary to manage and control deviation signal and establish the relation between image and actuator's deviation information. And this was the main problem to be solved in the design of servo control system [8]. The visual servo based on image usually needs us to estimate and solve interaction matrix which is known as image Jacobian matrix. In this kind of control method, robot could directly get feedback control from image information [9]. The robot's movement could be controlled by the differences between collected image and expected image [10, 11].

The visual servo system designed for experiment contained a Cartesian robot with three degrees of freedom. And this robot was constructed by an electric translating platform and electric cylinders. Visual servo control based on image was applied to the feedback control process of the robot. Thus, the Cartesian robot's real-time position control could be realized.

2 Visual Servo System Design

2.1 System Construction

The system used electric translating platform driven by stepping motor to realize robot's movement in X-Y plane and electric cylinder to drive Z direction. The stepping motor's stepping angel was 1.8°. Platform's maximum speed was 25 mm/s and limited moving distance was 200 mm. It used a linear bearing slide way with support to translate movement. And its resetting precision was less than 0.005 mm. The electric cylinder in Z direction was driven by a DC motor. Movement in Z

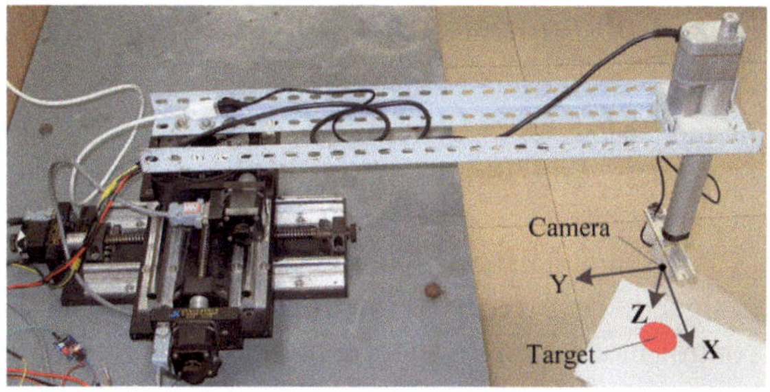

Fig. 1 Cartesian robot system's construction

direction could have a maximum speed of 10 mm/s and moving distance of 200 mm. And the linear bearing slide way with support could make the system stable and realize high precision movement control. Furthermore, the electric cylinder driven by DC motor should cater to the detecting precision of image sensor. System construction was shown in Fig. 1. Camera was set at the end of the robot in order to form an Eye-in-Hand system.

2.2 *Control System Design for the Cartesian Robot*

Due to the application of visual servo, the Cartesian robot's position control did not need complicated sensing and position calculation. So, a 51-series SCM STC89C52RC was used as robot's moving controller. And robot's three degrees of freedom were controlled by stepping motor and PWM speed control DC motor. The SCM communicated with PC via serial port. A key switch was set to change between auto and manual mode. Moreover, there were also movement limit switches to detect limited position in robot's movement and were used for certification of reference point while resetting the system.

In the program design of SCM, timer 0 was used for the output of pulse signal and counting PWM's time for speed control. Timer 1 performed as baud rate generator (BRG) while communicating with PC via serial port. Both timer 0 and 1 were set to operating mode 2. By changing output pulse's circle and duty ratio, stepping motor and DC motor's speed control could be realized. Furthermore, motor's speed could also be controlled by changing parameters' value. While deciding timer's initial value, interrupt response time was also counted. In auto working mode, interrupt response time was three machine cycles. The program adopted transposition and bit manipulation orders to calculate multiplication and division.

A reset function was designed for robot's controller. When the system initialized after power start or limit switch was triggered, reset operation would be started and would locate the system in the middle of its moving range.

2.3 Communication Control Protocol Design

In the visual servo system, PC output motor's speed and motion state to robot controller. And serial communication between SCM and PC was realized by inquiry mode.

A communication scheme with comparatively strong fault tolerance was designed. Data received by robot controller should be divided into two parts: control mode identification and control data bits. Here comes the specific design:

Control mode		Control Data					
CM0	CM1	CD5	CD4	CD3	CD2	CD1	CD0

CM0, CM1 are control mode select bits. Their combinations' meaning in communication is:

CM0CM1 = 00, set X direction's speed as control data;
CM0CM1 = 01, set Y direction's speed as control data;
CM0CM1 = 10, set Z direction's speed as control data;
CM0CM1 = 11, set robot motor's motion state.

In control data, CD5 and CD4 identified motor's motion state in X direction. CD3 and CD2 identified motor's motion state in Y direction. CD1 and CD0 identified motor's motion state in Z direction. When CD(n)CD(n-1) = 00, motor stops. CD(n)CD(n-1) = 01, motor revolves in positive direction. CD(n)CD(n-1) = 10, motor revolves in negative direction.

When the data translated was in the mode of speed control, control data bits CD5CD4CD3CD2CD1CD0 represented motor speed that was set. In this kind of control mode, it was possible to realize motor's speed change control with 63 levels which could satisfy the requirement of system's movement control.

By applying this kind of communication method, system's data translation would be strongly steady, reliable and with good fault tolerance.

3 Visual Servo Control Method Design

3.1 Image Collecting and Processing Facilities

Low price and common used USB2.0 web camera Blue Enchantress M2200 was used as image collecting facility. The camera sensor was 1/5 inch CMOS plate OV7740 with its dynamic resolution ratio of 640 * 480 and maximum frame frequency of 30 frames. It also had functions like auto white balance, automatic exposure and five lens that could be manually focused.

Image processing software used the C/C ++ module in Microsoft Visual Studio 2008 as a development platform. The software was developed combining with OpenCV and Computer Vision Function Library. And the software was operated on a Lenovo Soyang E46 computer with the operating system of Windows7 32 bit. CPU was Intel Core i3 M 380, 2.53 GHz.

3.2 Construction of Visual Servo System

The aim of servo control was to move camera to a certain position right above target's center of gravity under camera view. Target was segmented by image processing technique in computer vision. By controlling Cartesian robot, camera was also moved together which made target's imaging center coincide with camera center. Thus, it would be possible to move camera to the position right above target's center under camera view [12]. Moreover, position control in height direction was decided by calibrating the corresponding relationship between target's pixel area and height [13]. The structure of visual servo was shown in Fig. 2.

This method adopted target's image center and pixel area as characteristics for visual servo. And feedback control applied the difference between current image and expected image of the target which could avoid calculating Jacobian matrix in visual servo based on image. It turned out that the Cartesian robot was able to be controlled reliably and steady.

3.3 Calibration of the Relationship Between Pixel Area and Distance

According to the camera geometric principle, target's pixel area is inversely proportional to the square of distance between camera and target. Actually, due to camera's model error or manufacturing error, the relationship between area and distance would be different from theory. So, in order to improve locating precision in Z direction, a red circle with its diameter of 60 mm was chosen as target.

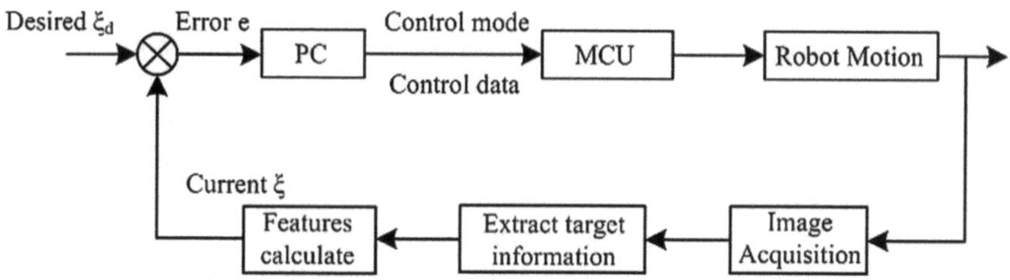

Fig. 2 Visual servo control structure

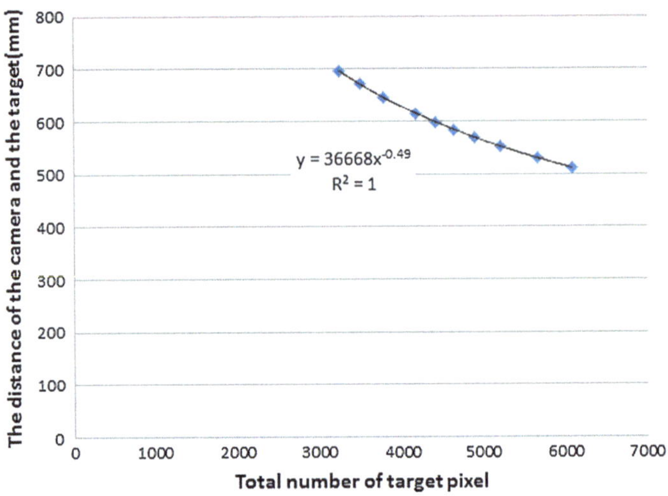

Fig. 3 Relationship between target's pixel area and distance

After collecting images under different distances, target's pixel areas were calculated (that is to count the total pixel number the target occupies). The corresponding relationship between target's pixel area and distance was fitted in Fig. 3.

If h represents the distance between camera and target, S represents the target's pixel area. The calibrated relationship was:

$$h = 36668S^{-0.49} \tag{1}$$

The corresponding relationship between target's pixel area and distance was got after calibration. By transforming depth information that was needed by visual servo control into target's pixel area, we got image information with servo characteristic. And the relationship attained by calibration could improve servo precision in depth direction.

4 Experiment and Analysis of the Result

A red circle with its diameter of 60 mm was chosen as target. The main purpose of this experiment was to move the camera to a certain height right above target's center. System's original point was defined: in X-Y plane, neutral position of X and Y sliding pare's corresponding camera coordinate was set as the world coordinate. The coordinate was defined by right-hand rule (Fig. 1). The circle was decided to be target because of its low interruption while confirming center and its pixel area could be calculated steadily and accurately. The Super Red arithmetic (2R-G-B) was applied to grey the image. Then, the Otsu arithmetic was used to automatically

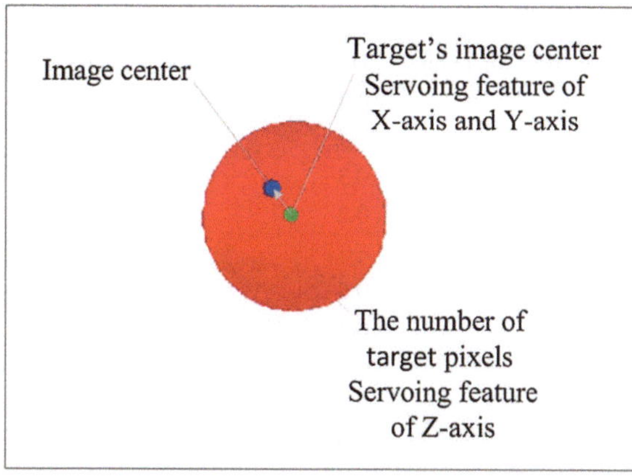

Fig. 4 Target's image and its servo characteristics

choose threshold to binary the grey image [14]. After morphological closed and open operation, target's center and pixel area were calculated for servo control (Fig. 4).

4.1 System Resetting Precision Experiment

In order to test system's resetting precision, the red circle was set at (0, 0, 600) in world coordinate. That is to say, the distance between target and camera was 600 mm and the excursion in X and Y direction was 0. Then, move the Cartesian robot from X, Y, Z's positive direction and negative direction to the target center via visual servo. After the measuring system was stable, locating positions were measured and shown in Table 1.

A steel rule with its minimum scale of 0.5 mm was used as measuring tool. The experiment result turned out that the servo system's resetting error range was: ±0.5 mm in X and Y direction, ±1.0 mm in Z direction.

4.2 System Locating Precision Experiment

Twenty seven points within system's motion space were chosen as locating points in the experiment. Every point was measured by several times of locating. The point cloud of measured data was shown in Fig. 5.

Experiment data are in Table 2. The maximum sum of absolute value of three coordinates' deviation was defined as the point's locating error.

It came out that the system's locating error range was: ±1.0 mm in X and Y direction, ±2.0 mm in Z direction.

Table 1 Servo system's resetting precision (mm)

Movement direction	Deviation in X axis	Deviation in Y axis	Deviation in Z axis
X+	0.3	0.3	0.7
X−	−0.3	−0.3	−0.5
Y+	0.3	0.5	−0.5
Y−	0.3	0.3	−0.5
Z+	−0.3	0.3	−1.0
Z−	0.3	0.5	−0.7
X + Y + Z+	0.3	−0.5	0.5
X + Y + Z−	0.5	0.5	0.7
X + Y − Z+	0.3	0.3	0.0
X + Y − Z−	0.0	0.3	0.0
X − Y + Z+	0.3	0.5	−0.3
X − Y + Z−	−0.3	0.3	−0.7
X − Y − Z+	0.3	0.3	0.0
X − Y − Z−	−0.3	0.5	−0.3

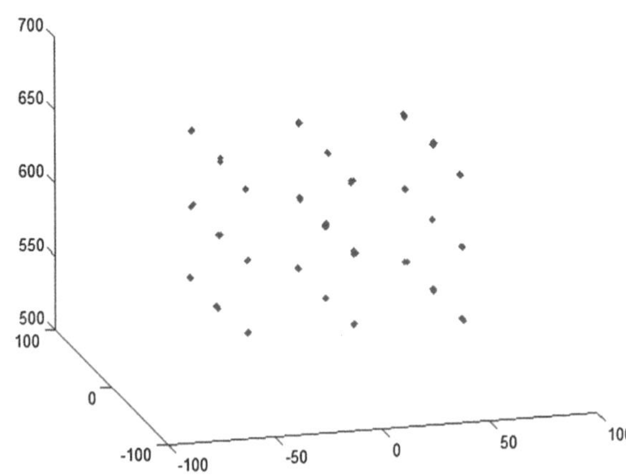

Fig. 5 Locating experiment points and result

4.3 *Analysis of System's Error and Application*

Both system's resetting precision and locating precision were related to executing organization's moving precision and camera's detecting precision. The experiment result was also related to measuring tool's precision. Executing organization's motion error in X and Y direction could be ignored comparing to camera's detecting error. Thus, experiment result in X and Y direction was mainly decided by visual sensor's detecting precision. Executing organization's motion error in Z direction was comparable to camera's detecting error. So, result in Z direction was doubly affected by motion precision and camera's detecting precision. According to

Table 2 Servo system's locating precision (mm)

Locating point coordinate	Deviation in X axis	Deviation in Y axis	Deviation in Z axis
(0.0,0.0,600.0)	−0.3	0.3	−1.0
(−50.0,0.0, 600.0)	0.5	−0.5	−2.0
(−50.0,50.0, 600.0)	0.7	0.5	−1.0
(0.0,50.0, 600.0)	0.5	−1.0	−2.0
(50.0,50.0, 600.0)	0.0	0.5	−1.0
(50.0,0.0, 600.0)	−0.5	−0.5	−2.0
(50.0,-50.0, 600.0)	0.5	0.3	−1.5
(0.0,-50.0, 600.0)	0.0	−0.3	1.5
(−-50.0,−50.0, 600.0)	0.7	0.3	0.5
(0.0,0.0,550.0)	1.0	−1.0	−1.0
(−50.0,0.0, 550.0)	0.7	−1.0	−1.5
(−50.0,50.0, 550.0)	0.5	−0.5	−1.0
(0.0,50.0, 550.0)	1.0	0.5	−0.5
(50.0,50.0, 550.0)	0.5	−1.0	0.5
(50.0,0.0, 550.0)	0.0	−0.5	1.5
(50.0,−50.0, 550.0)	−0.5	−0.5	−2.0
(0.0,−50.0, 550.0)	−1.0	−0.5	−1.5
(−50.0,-50.0, 550.0)	0.5	0.5	−1.5
(0.0,0.0,650.0)	0.0	1.0	−1.0
(−50.0,0.0, 650.0)	0.0	0.5	−2.0
(−50.0,50.0, 650.0)	−0.5	0.5	-1.0
(0.0,50.0, 650.0)	0.0	0.3	1.0
(50.0,50.0, 650.0)	1.0	1.0	−0.5
(50.0,0.0, 650.0)	0.0	0.7	1.0
(50.0 ,−50.0, 650.0)	1.0	−0.3	−1.5
(0.0,−50.0, 650.0)	0.5	0.3	−2.0
(−50.0, −50.0, 650.0)	1.0	0.3	0.7

visual servo characteristics, camera's resolution ratio, camera's intrinsic parameters and distance between target and camera, it was able to find out detecting deviation in three directions. When camera's focal length was about 600 and distance was 600 mm, visual servo's maximum detecting deviation in X and Y direction was 1.0 mm/pixel and 1.0 mm in Z direction. By improving camera's resolution ratio, executing organization's motion precision in Z direction or choosing visual servo with sub pixel precision, it was possible to advance system's precision.

The designed visual servo system could be applied to industrial manufacture. On one hand, it would be able to locate, grab or load and unload products. On the other hand, it could also be used in agricultural production like seeding transplanting, auto drug delivery and other facilities' locating process [15].

5 Conclusion

A Cartesian robot visual servo system based on image was designed. This system applied electric translating platform driven by stepping motor and electric cylinder driven by DC motor to realize executing organization's movement. A 51-series SCM was used to realize executing organization's motion control and communication with upper computer. Target was detected by web camera with its resolution ratio of 640 * 480. And the image processing and servo control arithmetic were dealt with by PC. By choosing standard target's center and pixel area as servo characteristics, experiments were operated to test locating errors. The result turned out that the resetting error was: ±0.5 mm in X and Y direction, ±1.0 mm in Z direction. And system's locating error was: ±1.0 mm in X and Y direction, ±2.0 mm in Z direction. This system could be applied to position control in industrial or agricultural production.

Acknowledgments This work was supported by Research Fund for the Doctoral Program of Higher Education of China under Grant No.20120008110046.

References

1. Hutchinson S, Hager G, Corke P (1996) A tutorial on visual servo control. IEEE Trans. Robot Automat 12:651–670
2. Chaumette F, Hutchinson S (2006) Visual servo control part I: basic approaches. IEEE Robot Autom Mag 13(4):82–90
3. Thuilot B, Martinet P, Cordesses L, Gallice J (2002) Position based visual servoing: keeping the object in the field of vision. In: Proceedings IEEE international conference robot automation, pp 1624–1629
4. Tahri O, Araujo H, Chaumette F, Mezouar Y (2013) Robust image-based visual servo using invariant visual information. Robot Auton Syst 61:1588–1600
5. Lin J, Chen H, Wang Y, Jiang P (2000) Research on robotic visual servoing system. Control Theory Appl 17(4):476–481 (in Chinese)
6. Staniak M, Zieliński C (2010) Structures of visual servos. Robot Auton Syst 58:940–954
7. Chaumette F, Hutchinson S (2007) Visual servo control part II: advanced approaches. IEEE Robot Autom Mag 14(1):109–118
8. Kyrki V, Kragic D, Christensen HI (2006) Measurement errors in visual servo. Robot Auton Syst 54:815–827
9. Kosmopoulos DI (2011) Robust Jacobian matrix estimation for image-based visual servo. Robot Comput Integr Manuf 27:82–87
10. Espiau B, Chaumette F, Rives P (1992) A new approach to visual servoing in robotics. IEEE Trans. Robot Autom 8:313–326
11. Corke P, Hutchinson S (2001) A new partitioned approach to image based visual servo control. IEEE Trans Robot Autom 17(4):507–515
12. Janabi-Sharifi F, Wilson W (1997) Automatic selection of image features for visual servoing. IEEE Trans Robot Autom 13(6):890–903

13. Kondo N, Monta M, Noguchi N (2004) Agri-robot (I)—fundamental and theory. (in Chinese translation from Japanese language) Corona Publishing Co., Ltd, pp 56–58
14. Meyer GE, Neto JC, Jones DD, Hindman TW (2004) Intensified fuzzy clusters for classifying plant, soil, and residue regions of interest from color images. Comput Electron Agric 42:161–180
15. McCarthy CL, Hancock NH, Raine SR (2010) Applied machine vision of plants: a review with implications for field deployment in automated farming operations. Intel Serv Robot 3:209–217

Research on Secondary Arc Current and Recovery Voltage in UHV Lines

Wen-hui WU, Xiang-zheng Xu, Jin-wei Ye and Qiang Hu

Abstract The UHV transmission system failure probability is the largest of the single-phase grounding fault. When a failure occurs in the circuit may exist secondary arc current and recovery voltage of the potential. The ultra high voltage transmission lines of the arc burning a long time, if not to take effective measures to reduce the secondary arc current and recovery voltage, will reduce the single-phase reclosing success rate, affect the safety and stability of UHV system. In this paper, two source model based on 1000 kV, using ATP-EMTP software simulation of various measures to inhibit the effect of secondary arc current and recovery voltage.

Keywords ATP-EMTP · Recovery voltage · Secondary arc current · Ultra high voltage (UHV)

1 Introduction

In UHV transmission system, the single phase earth fault accounted for ninety percent in the short-circuit fault. To improve reliability and stability of the system for power supply, UHV transmission lines in China are mostly use single-phase reclosing system. Duration of the secondary arc current is important for the reclosing time setting [1].

There are many factors that affect the length of the duration of the secondary arc current, Such as the electrical parameters and High impedance values of the transmission line or the measures adopted to suppress the secondary arc current and recovery voltage [2]. Taking different measures will lead to different effect of inhibition. In this paper, doing the research from many angles and using a variety of measures alone or are combined leads to the effect of inhibition.

W. WU (✉) · X. Xu · J. Ye · Q. Hu
School of Electrical and Electronic Engineering,
East China Jiaotong University, Nanchang, China
e-mail: whwu777@163.com

E. Qi (ed.), *Proceedings of the 6th International Asia Conference on Industrial Engineering and Management Innovation*,
DOI 10.2991/978-94-6239-145-1_50

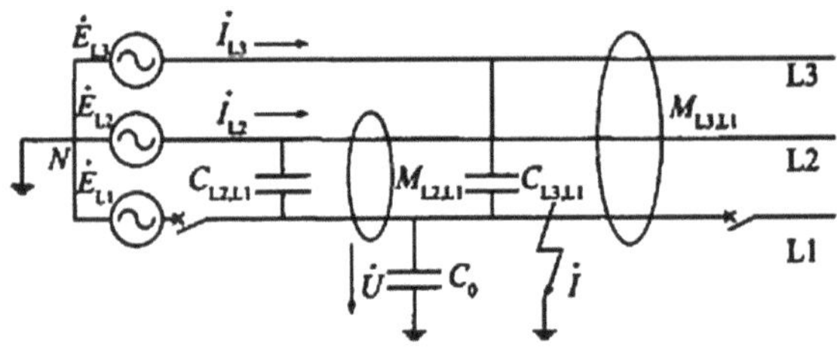

Fig. 1 Secondary arc current diagram

2 The Mechanism of Secondary Arc Current and Recovery Voltage

When the UHV system ground fault occurs, the fault line will be quickly cut by circuit breaker. Due to the coupling effect from mutual inductance and capacitance, a certain value current will flow through Fault phase in the fault point; the current is the form of arc. A secondary arc current diagram will be shown in Fig. 1.

Secondary arc current consists of two parts, capacitance and inductance components. The capacitance component is the current that provided by the normal voltage by capacitance to the point of failure, the load current of normal phase will introduce electromotive force when flows through the mutual phase, The electromotive force provide electric current through relatively capacitance and high formation circuit that is inductive component. Transient fault voltage is called recovery voltage after the secondary arc current is out. Line secondary arc current and recovery voltage related to the parameters of the transmission line, compensation of line, operating voltage at the ends of the line and conveyor trend, the influence of network structure on either side of the line is very small. Secondary arc current and the recovery voltage on the transmission line is the overlay of steady state and transient components, because of the residual charge on the conductor will be released in the fault circuit lead to secondary arc current and transient recovery voltage component. As shown in Figs. 2 and 3.

3 Research on Secondary Arc Current and Recovery Voltage Suppression Measures

When the secondary arc current and the recovery voltage is at a high level in fault line, secondary arc current that to take steps to speed up extinguishing of arc. Now we generally take the following measures.

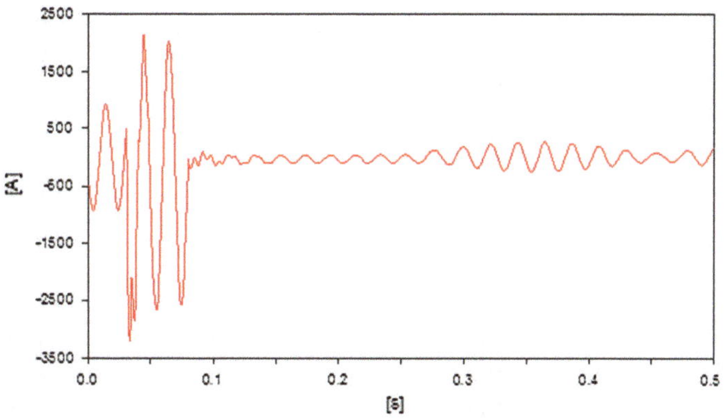

Fig. 2 Secondary arc current waveform

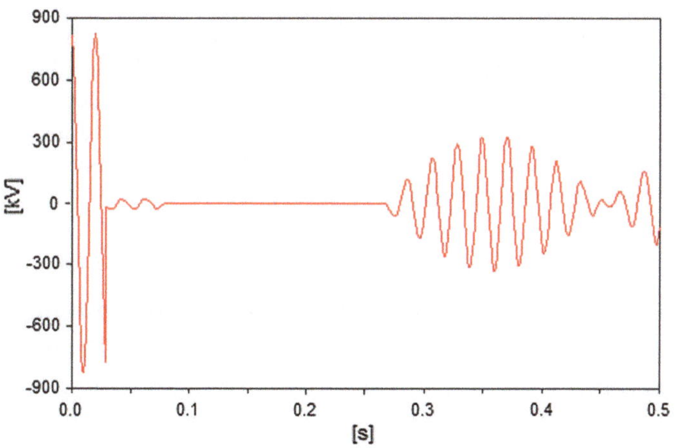

Fig. 3 Recovery voltage waveform

3.1 By Adding Neutral Small Reactance on High Voltage Shunt Reactor

In order to restrict secondary arc current and recovery voltage, we will take a method of taking reactor with high voltage shunt reactor neutral to decrease the secondary arc current and recovery voltage in line with shunt reactor, Small reactance is mainly used for compensating of circuit and capacitance and relatively capacitance to reduce the secondary arc current of the capacitance component. This article research is about the inhibition effect of recovery voltage and secondary arc current when small reactance of the neutral point is different.

Table 1 Small reactance is different the size of the secondary arc current and recovery voltage

Small reactance value neutral/Ω	Secondary arc current/A	Recovery voltage/kV
20	59.4	19.5
50	48.5	20.5
100	37.6	23.2
200	31.1	29.7
300	29.2	34.8
350	28.8	38.4
400	28.1	41.6
450	29.1	44.6
500	29.3	47.3
600	29.4	51.5

From Table 1 we can see the inhibition effect of recovery voltage and secondary arc current when small reactance of the neutral point is different, When the reactance values within the scope of the 20–400 Ω, secondary arc current decreases with the increase of resistance and the resistance value increases, Within the scope of the 400–600 Ω, the secondary arc current will increase as the resistance value increases but increase modestly. The relationship between the voltage and the resistance is generally increases with the increase of resistance. So we consider neutral small reactance value range in 300–450 Ω is reasonable.

3.2 *The Influence of Transmission Lines Transposition for Suppressing Secondary Arc Current and Recovery Voltage*

We know from the knowledge of power system, because of the longer transmission line, we must be sure that it will not lead to the problem of three-phase asymmetry by the transmission and does not affect the secondary arc current and the size of the recovery voltage, The simulation with different length of transmission line are simulated, observing the same length, transposition and transposition of secondary arc current and the recovery voltage size, and in not transposition and transposition in both cases, under different length of the line secondary arc current and the recovery voltage size, the simulation results see the table below.

Can be seen from the simulation data in Table 2: Without transposition the secondary arc current and recovery voltage are increase with the power supply line length. But in the case of transposition, secondary arc current and recovery voltage within 1000 km is lesser than within 1500 km.

Table 2 Line length take different values, the size of the transposition and transposition of secondary arc current and recovery voltage

The length of the line/km	Don't switch		Switch	
	Secondary arc current/A	Recovery voltage/kV	Secondary arc current/A	Recovery voltage/kV
200	107	104	247	25
500	548	452	304	336
700	387	352	214	262
900	323	289	175	132
1000	219	271	184	174
1500	1350	901	588	345

3.3 *High Speed Grounding Switch (HSGS)*

In some cases that the shunt reactor and neutral small reactance can not be used to limit the secondary arc current, for example, static compensation device are used in some lines then high speed grounding switch can be adopted to accelerate the latent power of arc extinguishing. The way of rapid grounding switch be used online to accelerate the latent power of arc extinguishing that is adopted in some countries, circuit breaker in the process and the process of HSGS work with is when single-phase earth fault line, can produce an arc [3, 4]. At this point at the ends of the fault phase circuit breaker tripped, an arc extinguishing, secondary arc. Installed on opposite sides of the fault phase grounding switch fast, secondary arc current is reduced, rapidly since. HSGS, breaker reclosing, transmission line back to normal work.

The simulation is rapid grounding switch arc under different layout parameters. Plan A switch is respectively arranged in the line of the head end full-length 0, 1/9, 2/9, 1/3, 1/2, 2/3, 7/9, 8/9, 1 point; Plan B is switch arrangement in the line of the head end respectively 0,1/10,2/10,3/10,1/2,7/10,8/10,9/10,1 point; Plan C is switch respectively arranged in the line of head end of 0,1/11,2/11,3/11,1/2,8/11,9/11,10/11,1 point. The simulation data are shown in Table 3.

Through the date on the Table 3, it can be seen that the inhibition effect of plan b is the most ideal in a few scheme, the inhibitory effect of plan A is very close to uniform, large differences between the maximum and average show that the amplitude of waveform change big.

Table 3 Secondary arc current and recovery voltage under different configuration scheme

HSGS	Secondary arc current/A		Recovery voltage/kV	
	The maximum	The average	The maximum	The average
Uniform distribution	423.1	107.2	30.5	10.8
Plan A	425.2	123.1	37.2	12.4
Plan B	272.2	98.6	28.7	9.7
Plan C	383.3	102.6	30.8	11.2

Table 4 Different transmission in presence of series compensated line length of arc current and recovery voltage

The length of the line/km	No series compensation		With series compensation	
	Secondary arc current/A	Recovery voltage/kV	Secondary arc current/A	Recovery voltage/kV
200	30.6	82.3	646.7	71.4
300	41.9	93.8	629.3	71.9
500	146.8	100.2	913.6	463.9
700	86.7	78.6	1130.2	496.5
900	102.3	122.8	1397.5	464.1
1000	138.6	130.7	1706.3	267.3

3.4 The Influence of Series Compensation Device

Thyristor controlled series capacitance compensator (TCSC) is the first generation of controllable series compensation in the FACTS family, which has obvious advantages to improve the performance of the electric power system, but it also bring some adverse effects to the system [5, 6] after joining series compensation device. It makes the secondary arc current and the recovery voltage increase, in turn, affects the success rate of the line reclosing. In this article, the influence of the secondary arc current and recovery voltage from series compensation can be observed through the simulation experiment, Table 4 is the different transmission in presence of series compensated line length of secondary arc current and the recovery voltage.

3.5 The Series Capacitance and Rapid Grounding Switch Function at the Same Time [7]

The previous simulation studies on single suppression measure, this simulation experiment show the situation of secondary arc current and recovery of voltage as the series capacitance and rapid grounding switch work together in the line. In Table 5, the secondary arc current and the recovery voltage size can be seen in the lines which have different length.

Table 5 Secondary arc current and recovery voltage under different length of line

The length of the line/km	Secondary arc current/A	Recovery voltage/kV
200	494.7	71.3
500	843.4	463.9
700	1277.2	496.4
900	1567.4	311.8
1000	1607.4	311.7

It can be seen from Table 5 that when the both effect on the lines at the same time, the secondary arc current and recovery voltage are increasing constantly with the increase of the length of transmission line. From their magnitudes of numbers, it can be seen obviously that this combination is very difficult to inhibit.

3.6 High Pick Up Small Reactance and Neutral Point Grounding Switch Quickly

In this simulation experiments by neutral small reactance and rapid grounding switch two measures to work together, the size of secondary arc current and recovery voltage is high. Table 6 is in different length of the line secondary arc current and the recovery voltage size.

3.7 Series Compensation and Neutral Point Grounding and Grounding Switch Quickly

The previous study is about one measure of the single role and two measures of the combined role, the experiment uses three devices of the combined role in the simulation, mainly simulating through the combination role of TCSC, neutral ground switches and HSGS, the size of different circuits length of secondary-arc currents and recovery voltages.

Can be seen from Table 7, the line length increases secondary arc current and the recovery voltage has a trend of increase, but are limited to a reasonable size, the effect of this combination is also obviously better.

3.8 Adaptive Single-Phase Automatic Reclosing

Because of the existence of secondary-arc currents, having effect on fault arc extinction, blocking free arc channel heavily when the short circuit happening, and

Table 6 Secondary arc current and recovery voltage under different length of line

The length of the line/km	Secondary arc current/A	Recovery voltage/kV
200	99.8	72.1
500	153	175.1
700	173.3	182.1
900	163.8	199.8
1000	186.3	205.8

Table 7 Secondary arc current and recovery voltage under different length of line

The length of the line/km	Secondary arc current/A	Recovery voltage/kV
200	96.1	71.3
500	102.2	73.9
700	187.6	264.7
900	212.3	242.5
1000	258.5	269.1

the auto-reclosing could be reclosed successfully only after the fault arc extinction and the recovery of dielectric strength. The reclosing reclosed in failure when the secondary-arc currents are stronger and the time of fault arc extinction is longer. The time of the single-phase auto-reclosing is fixed, and requiring the secondary arc to be out in the required hour. The electric arc will be out on the premise of the stable system when the required hour is longer than the period of the self-be –out secondary arc and the enough recovery hour of medium strength. The truth of the dying of the electric arc is to change the open-ended electric arc into the inside electric arc so that decreasing the secondary-arc to self-be-out easily.

3.9 Adopt Good Conductors Overhead Ground Wire

By using the good conductor to off-contact the ground wire would reduce secondary current and make the arc turn off quick. The analysis of the inductance component in secondary current show that after using the off- contact the ground wire, the zero sequence impedance is reduced and reverse the zero line sequence current, in this way, the total induced electromotive force is smaller, at the same time the inductance component on the arc current is reduced, by this method capacitance and inductance components is smaller, and inductive component is quite different.

Can see clearly from the Table 8, two different types of line capacitance component of current has changed little, and the change of the inductive component is bigger.

Table 8 The secondary arc current using different types of overhead ground wire

Ground wire	The total of secondary arc current/A	Capacitive component/A	Inductive component/A
Aluminum steel wire ground conductors	42.1	26.3	20.7
Steel strand	57.4	26.3	45.8

4 Conclusion

Based on the 1000 kV transmission line model, a variety of methods of extinguishing arc was analyzed through the simulation, such as the methods of shunt reactor neutral small reactance, rapid grounding switch. The advantages and disadvantages of a variety of ways are compared on the basis of the results of simulation. To reduce the secondary arc current and suppress the voltage, it can be seen that the method of Shunt reactor neutral small reactance do relatively well, it also can avoid resonance overvoltage on the line. If the UHV transmission lines are longer and the line charging power is very large, to ensure the success of the single-phase reclosing, we should use the shunt reactor compensation to limit the power frequency overvoltage.

References

1. Guo Z, Sun W, Gao B et al. (2006) A study on restriction of secondary arc current and time coordination of single phase reclosing for 500 kV compact transmission line from Yuncheng to Tai'an. Power Grid Technol 30(17):79–83
2. Lin X, He B, Xu J (2007) Simulation and analysis of secondary arc. High Volt Appar 43(1):8–10
3. Han Y, Fan Y, Shi W (2000) Secondary arc extinction by high speed grounding switches. J Xi'an Jiaotong Univ 34(8):14–17
4. Chen X, Zhang X, Xu Z et al (2012) High speed grounding switch out UHV paralleled double circuit line secondary arc analysis. Power Grid Technol 36(9):25–31
5. Sheng Z (2004) Related to the EHV transmission line with series compensation device system problems and their solutions. Power Grid Technol 28(6):138–143
6. Pan Y (2010) Simulation analysis on secondary arc current and recovery voltage. J North China Electric Power Univ 37(4):11–16
7. Li Z, Xiu M et al (2010) High pressure with double circuit line towers arc suppression measures are. High Volt Technol 36(10):51–56

The Construction and Application of Digital Collaborative Learning System

Dan Tian and He Lin

Abstract Collaborative learning in network education is the group mode of collaborative learning, through the integrated use of computer technology, communication technology and multimedia technology. And as an important means of communication for the practical teaching, the construction and application of digital collaborative learning system broke the operation limit of the traditional group mode of cooperative learning in time, space and personnel scale, expanded the scope of the communication and learning, enhanced the effects of teaching, at the same time, made great innovation in network education practical teaching.

Keywords Collaborative learning · Digital system · Network education · Practical teaching

1 Introduction

Collaborative learning is an important strategy of modern network education practical teaching, and plays an important role in training the comprehensive quality and application skills of in-service employees. Collaborative learning system in the network education environment can make the students and teachers who distributed in different areas to discuss, exchange and learn meanwhile in the same virtual classroom, also by imitate the important features of the true face-to-face teaching, can share an integrated teaching platform based on computer network technology between the students or between teachers and students, which can not only provide some data exchange such as text, graphics, information, more can provide the real-time interaction of language signal and video signal through the support of multimedia communication network [1].

D. Tian (✉) · H. Lin
School of Distance and Continuing Education, Dalian University of Technology, Dalian, China
e-mail: network2@dlut.edu.cn

E. Qi (ed.), *Proceedings of the 6th International Asia Conference on Industrial Engineering and Management Innovation*,
DOI 10.2991/978-94-6239-145-1_51

As the rapid development of the continuous enlargement of the enrollment scale in network education, the continuous improvement of teaching quality and the extensive of the learning source, the construction and application of digital collaborative learning system become the “hot spot” of the practical teaching work. Therefore, the fully understand about the role of the collaborative learning mode in network education has the great significance to the network remote teaching.

2 The Definition, Elements and Mode of Collaborative Learning

2.1 The Definition of Collaborative Learning

Collaborative Learning is a kind of learning strategy in which students study in a group or team. The collaboration of group members is an organic component of achieving the study goals. Group members can share their explorations, discoveries or learning materials with others, even other groups’ members or the whole classmates [2].

In the web education environment, learners can make full use of collaborative learning system based on internet and multimedia technology that can break the limit of time and space, learn from discuss and communicate with others. What’s more important is that they can cultivate their collaborative ability, investigation and knowledge acquiring ability. Therefore, digitalized collaborative learning refers to that learners voluntarily form a group so as to achieve their common learning objectives in the web education environment, to overcome the difficulties and deficiencies brought by facilities, resources, information feedback or learning evaluation, to initiate independent learning, innovation ability and teamwork spirit, in order to build a learning strategy of distance open leaning to improve the web education quality, accomplish the learning objectives.

2.2 The Elements of Collaborative Learning

As a kind of interactive learning mode, collaborative learning has been widely applied in traditional classes and web education learning. Collaborative learning, taking the collaborative learning mode in order to contribute to learners’ comprehension and mastery of knowledge, consists of four elements, including collaborative group, group members, tutors as well as collaborative learning environment in Fig. 1

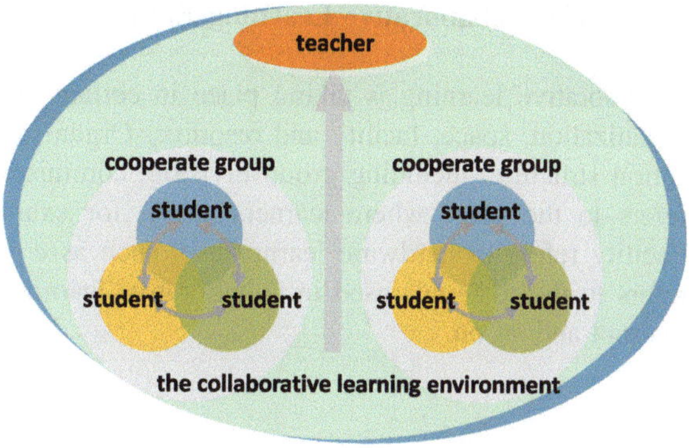

Fig. 1 The basic elements of cooperative learning

2.2.1 Collaborative Group

Collaborative group is the basic component of collaborative learning. The way of group division will directly influence the outcome of collaborative learning. Usually, each group consists of no more than 4–6 people.

2.2.2 Group Members

Group members are those who are sent into a collaborative learning group according to certain strategy. The group division is based on several factors, including learners' score, cognitive structure, cognitive ability, cognitive competence, cognitive style and cognitive approach, etc., which are complementarily used to improve the outcome of collaborative learning. Collocation of high-score students and low-score students is beneficial for the later ones to improve their learning and the former ones to master and digest knowledge; collocation of different cognitive approaches students is helpful for students to give play to their strength in order to mutually reinforce their cognitive styles.

2.2.3 The Tutors

Tutors are essential to collaborative learning whose existence can manage and guarantee the organizing of collaborative learning, efficiency of learners, outcome of collaborative learning, etc. collaborative is demanding to tutors, which requires tutors to own new education concepts which centers on "learning" from the traditional "teaching" and the combination of the two.

2.2.4 The Collaborative Learning Environments

Collaborative learning is taking place in certain learning environment, including organization, space, facility and resource. Organization refers to learner's organization structure, including group division, commission division and so on. Space refers to the place where learners study, for example, classroom, internet, etc. Facility refers to hardware learner use, such as computer, etc. And the resource refers to the materials used in collaborative learning, for example, virtual library, internet and so on.

2.3 *The Mode of Collaborative Learning*

The mode of collaborative learning varies according to the characteristics of learning content or learning commission, for example, the solution mode, mutual mode, debate mode, evaluation mode.

2.3.1 The Solution Mode

In collaborative learning system, learner can ask for help by conveying his problems to others in some way in order to get the correct answers when facing problems. This mode is often applied in collaborative learning, usually in one-to-one situation.

2.3.2 The Mutual Mode

In the teaching process, students are divided into different groups for teamwork spirit cultivation and each group is assigned with corresponding task, which makes every group can finish the task only by collaboration. This mode applied in online learning is called mutual mode. Mutual mode can give full play to their initiative, enthusiasm and creativeness, and secure teachers to carry out colorful group activity via internet to cultivate their teamwork spirit and ability to participate group activities, which help contribute to profound communication among learners, overcome egocentricity and individual socialization development.

2.3.3 The Debate Mode

Debate mode is useful for learners to discuss some hot issues or some tightly prescribed opinions so that they can broaden their horizon and cultivate their debate ability and language organization competence. The topic can be prepared by tutors or learner themselves after discussing. Learners are equal due to their different locations; this allows those shy learners to show their opinions. Anxiety reduction is helpful for learners to try new approach and give play to initiative. At the same time, internet provides learners with affluent materials which can promote them to learn and think.

2.3.4 The Evaluation Mode

Evaluation mode is the judgment and learning among learners. This mode ensures every learner to have access to solution to problems by evaluating the process and outcome of others' tasks, which makes learners have overall comprehension of certain problem and makes learners gain useful knowledge and information to solve his own problem independently.

3 The Construction of Digital Collaborative Learning System

3.1 The Design Ideas

The design of digital collaborative learning system relies mainly on the theory of modern education technology, with the computer technology, network technology and multimedia technology, and works out the teaching strategy around the subject of teaching target, which the emphasis is on integration of all kinds of teaching resources, the development of collaborative learning system based on multimedia resources, also exploring its network teaching pattern and evaluation model [3].

The design philosophy of digital collaborative learning system that emphasizes student-center requires the students from the passive receiver of the past external stimuli and the knowledge object change to the main body of information processing and the meaning of knowledge actively constructing. At this time learning has created by the interaction with students and others, or the external environment, or the outside world. The students in the process of collaborative learning, can not only acquire knowledge, more importantly, in the process with the cooperation of others, can foster their ability to work, the search for knowledge, ability to acquire knowledge.

3.2 The Structure Function

The basic structure of digital collaborative learning system based on the network as the core, using the computer as the media of transmitting information, has build a learning environment which can break through time and space limit, and that the state intervention for collaborative learning, management, supervision, guidance, and other functions to full implementation by tutors in Fig. 2 [4].

Based on the characteristics of collaborative learning, combining the advantages of network education, digital collaborative learning system should be build four modules, front analysis module, prepare module, learning module and evaluation module [5].

3.2.1 The Front Analysis Module

In front-end analysis stage, there have the learning objectives, learning content, and the learning object analysis, and the analysis of the network environment from four aspects. This period is mainly to determine the learning problems and through the solution of the problem requires students to gain the ability to what aspects. In order to obtain these skills, students will already have advance knowledge and to solve the problem of the cooperative learning the basic skills, the existing network environment whether can meet the needs of learning, and then according to the analysis of the learning objectives, establish the corresponding learning content, the structure of the relationship between learning content, analysis of the content is suitable for using collaborative learning.

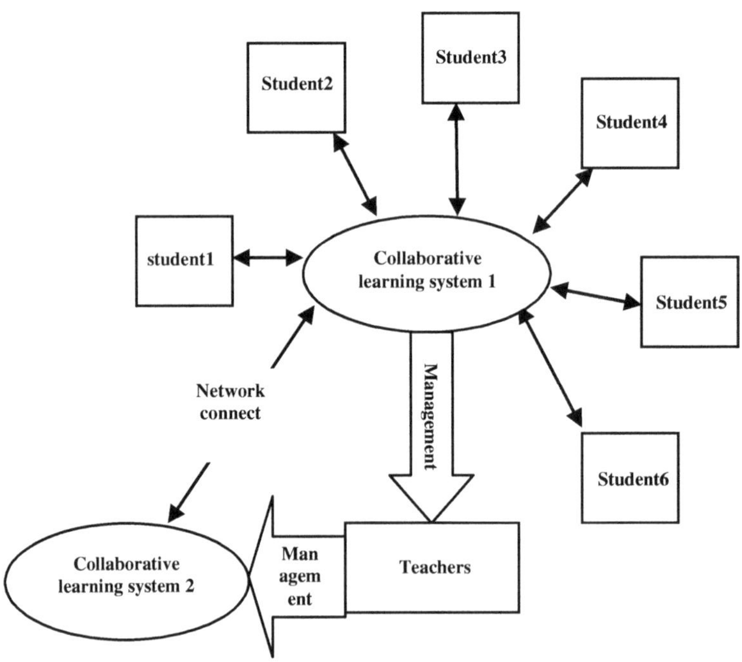

Fig. 2 The structure function of digital collaborative learning system

3.2.2 The Prepare Module

During the preparation phase, there have the establishment of a learning team, the design of learning environment, the design of the information resources from three aspects. The main task of this stage is for the development of collaborative learning to create a good environment.

Study groups to establish reasonable or not will directly affect the quality and effect of learning. Determine the collaboration team structure, want to consider the characteristics of the students, such as cognitive level, thinking characteristics, such as age, interests, try to be heterogeneous and homogeneous between groups, in the group that is conducive to group cooperation and competition between groups, can improve the efficiency of learning and learning the enthusiasm.

Learning environment in addition to including computer and network hardware environment and resource environment, students also contains the synergy group formed between the social cultural environments. By setting up a special chat room system to provide a public learning environment for collaborative learning, students in the virtual community can expand synchronized on a subject of discussion and communication, tutors for students to discuss the problems can be online FAQ, trek phenomena appeared in the process of learning guidance, etc. Email service provides an asynchronous communication between team members, students can be their own problems by E-mail sent to the tutors, tutors and the solution is presented to students by E-mail.

Should ensure that the amount of information to enough information resource, and is closely related to the content to be learning, resources structure more reasonable, it is certain to should have the breadth and depth, at the same time to have layered, in order to meet the needs of different level students. Resources form to diversification, such as text, graphics, animation, video, case, etc., to facilitate the retrieval and search.

3.2.3 The Learning Module

Mainly in the learning process module design and the organization coordinated activities, including guidance teachers and students of various related to learning behavior. Cooperative learning activities centered learning content, and according to the different activities in different mode of learning content. In the learning activity design, cooperative learning emphasizes the collaboration between the student ability and practical ability to solve problems, tutors to make a good organization and process of collaborative activity guidance, causes the student to understand the effectiveness of collaborative learning.

3.2.4 The Evaluation Module

Timely understanding of the problems existing in the learning process, timely feedback on the evaluation results, in order to timely revise stages of collaborative learning and adjustment. For collaborative study effect evaluation of coordinated from the ability of autonomous learning, to the collective contribution, whether to finish the construction of the meaning of knowledge in three aspects such as comprehensive consideration. Throughout the learning process, teachers should not only cannot solve the problems in the study to the student or put forward new problems to give timely guidance and help, and to lead the team work, timely organize the team of formative assessment, timely find the problems existing in the study, the questions of the process in which learners to solve problems and inspiration, help learners knowledge integration, coordination between team and group members, the relationship between the in order to promote the development of collaborative learning. During this period of formative assessment is mainly on the team has achieved and the problems existing in the evaluation, in order to promote better solve the problem for the purpose in Fig. 3.

3.3 The Building Supportive Conditions

The free exchange to support multiple terminals.

The free exchange is the basic requirement of collaborative learning mode. Distributed in different parts of the students can chat freely through the hardware environment, interactive activities such as debates. Digital collaborative learning system more emphasizes the mutual learning between students, should be implemented between any two terminals can be "unlimited".

3.3.1 The Resources That Can Be Shared

Digital collaborative learning system can be understood as based on the common learning resources, so special repository or search resource tools are required. Establish special repository in concrete is feasible to be used in a certain course, but increase the teachers' workload. Find a project can take advantage of the intelligent system in the network information, and the intelligent filter, filter, group, etc.

3.3.2 The Public Platform for Supporting Collaborative Learning

Digital collaborative learning system not only requires students can communicate freely, also can is held some case analysis events coordination analysis, it is needed to have a common platform in the system. In the public platform, students from different terminal with a case study.

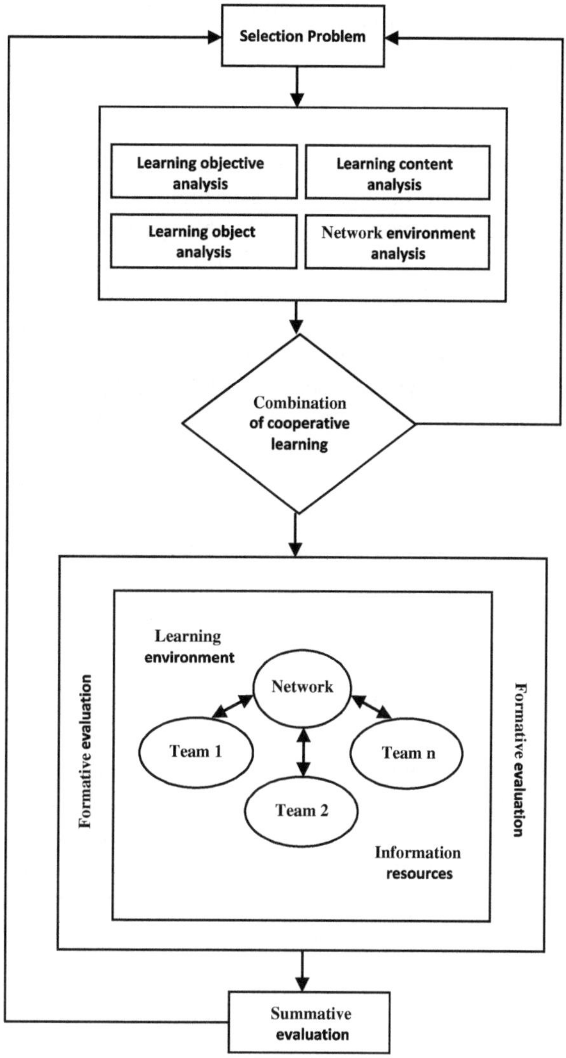

Fig. 3 The building modules of digital collaborative learning system

The function of multimedia information easy to transport.

Digital collaborative learning system should possess perfect multimedia information transmission function; the presence of the multimedia information system construction is an important consideration.

4 The Application Effect of Digital Collaborative Learning System

This system can fully take advantage of the computer network media so that learners can use information effectively. Learners can achieve the collaborative learning with broader scope and higher quality by network

There are more various, flexible group division ways by network. In collaborative learning, group division is a flexible system which varies, including one-people unit, two-people unit, group unit and mass activities and so on, which can improve the learning efficiency most. Group division in collaborative learning is much more convenient, flexible and various by network. Learners can choose group freely and interact with others on line.

This system can cultivate learners' information ability, learning ability and social ability via network, which helpful for their life-long learning.

This system breaks the limits of time and space. Tutors and learners are no longer under the control of time and space; classmates are those who come from different location, even different countries. In this way, students can have different experience.

5 Conclusion

As the 21st century by the era of information technology rapid development, digital collaborative learning system as an important means of network education practice teaching, have good prospects for development. We should pay close attention to modern technology, and make the digital collaborative learning system play a more important role in network education, further promote the intension type development of network education.

References

1. Meng T (1997) Network collaborative teaching system research. MultiMedia World (In Chinese)
2. Zhao J (2000) Under the information technology environment based on the teaching design of collaborative learning. E-education Research (In Chinese)
3. Ma H (2009) Multimedia computer technology principle. Book of Tsinghua University Press (In Chinese)
4. Li H, Wang Y (2003) Based on the problems of collaborative learning mode under the network environment. J Xiangtan Normal Univ (Nat Sci Ed.) (In Chinese)
5. Zhang X (2003) The design of the web-based collaborative learning system model. J China Educ Technol (In Chinese)

An Optimized SVM-Based Hysteresis Current Controller

Wen-xue Jiang and Kai Zhou

Abstract In order to achieve high performance as well as low power loss in high demanding fields, a space vector modulation based hysteresis current controller is optimized in this paper. The paper analyzes the effects of the voltage space vectors on the current error variations, and proposes the principles to optimize the switching pattern. An optimized switching diagram is constructed. Simulation studies are conducted for a permanent magnet synchronous motor in MATLAB/Simulink. The simulation results demonstrate that the proposed method achieves fast dynamic response, improves the tracking performance and reduces the switching frequency significantly under different conditions.

Keywords Hysteresis current controller (HCC) · Optimized switching diagram · Region detector · Space vector modulation (SVM)

1 Introduction

Current control technique plays the most important role in current-controlled PWM inverters which are widely applied in high performance ac drives, active filters, high power factor ac/dc converters, uninterruptible power supply (UPS) systems and ac power supplies [1]. Therefore, various current control techniques have been developed over the past decades, such as synchronous vector controller (PI), state feedback controller, ramp comparison controller, predictive controller and hysteresis current controller (HCC) [1]. Synchronous vector controller, state feedback

Supported by the National Natural Science Foundation of China (No. 51275257) and the Research Fund for the Doctoral Program of Higher Education of China (No. 20110002110079).

W. Jiang · K. Zhou (✉)
Department of Mechanical Engineering, Tsinghua University, Beijing, China
e-mail: jwx08@mails.tsinghua.edu.cn

E. Qi (ed.), *Proceedings of the 6th International Asia Conference on Industrial Engineering and Management Innovation*,
DOI 10.2991/978-94-6239-145-1_52

controller and ramp comparison controller can limit the inverter switching frequency and have a well-defined harmonic spectrum. However, they have inferior dynamic properties and lower bandwidth which hinder their application in high demanding situations [1, 2]. Predictive controller is suitable for a fully digital implementation and performs some optimization on the switching frequency at the expense of extensive hardware and dynamic response [3]. The HCC has been widely used because of its simplicity, absence of tracking errors, robustness to load parameters variations, maximum current limit and fast dynamic response [1]. Nevertheless, due to lack of coordination among three phases of HCC, high switching frequency may happen and the current error may exceed the hysteresis band up to twice the permitted bandwidth [4]. Besides, the conventional HCC has great power loss in high performance demanding fields (e.g., high-demanding servo system) which require very fast dynamic response and high tracking accuracy.

Several methods have been proposed to solve these problems. Hysteresis controller with variable bands and adaptive hysteresis band with constant switching frequency reduce switching frequency at the expense of extra signal processing and control complexity [5–9]. Space vector modulation (SVM) based methods (SVMHCC), which combine advantages of both SVM and HCC, can reduce the switching frequency without jeopardizing the simplicity of conventional HCC. The SVM-based methods having been proposed in literatures [10–16] decrease the switching frequency to some extent. However, these methods do not take full advantage of the voltage space vectors, the problems with respect to the significant contradiction between the high-performance demand and the low power loss requirement have not been solved completely. Thus, further research work needs to be carried out to deal with these key problems.

This paper presents an optimized SVM-based hysteresis current controller. The proposed method makes the most of the voltage space vectors. The effects of the zero voltage space vectors on the current error variations are analyzed elaborately. And the principles to optimize the switching pattern are proposed. Then, simulation studies are conducted to validate the effectiveness of the proposed method.

2 Principle of SVM-Based HCC

A symmetric load is connected to the voltage-source inverter (VSI) as shown in Fig. 1. The states of the inverter legs determine the VSI output voltage vector ($\mathbf{V}_n$) which can be expressed as follows:

$$\mathbf{V}_n = \frac{1}{3} V_{dc}[(2S_1^* - 1) + \exp(j2\pi/3)(2S_3^* - 1) + \exp(-j2\pi/3)(2S_5^* - 1)] \quad (1)$$

where S_x^* indicates the state for the upper switch of each leg. Variable $S_x^* = 1$ denotes the switch $\mathrm{S_x}$ is on, while $S_x^* = 0$ denotes the switch $\mathrm{S_x}$ is off. Two zero

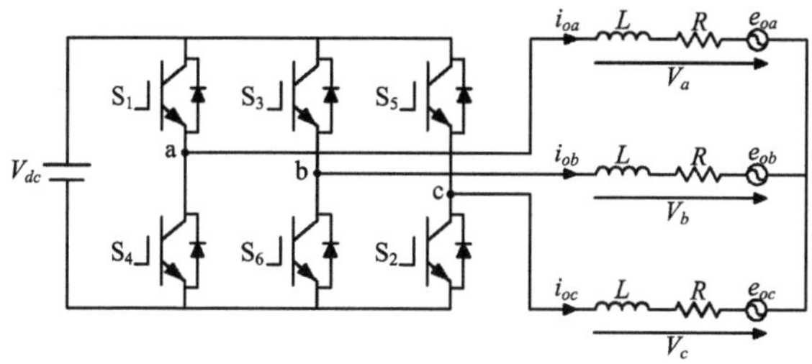

Fig. 1 Circuit diagram of the PWM-VSI connecting a symmetric load

Fig. 2 Current error derivative vectors **a** generated by different output voltage vectors

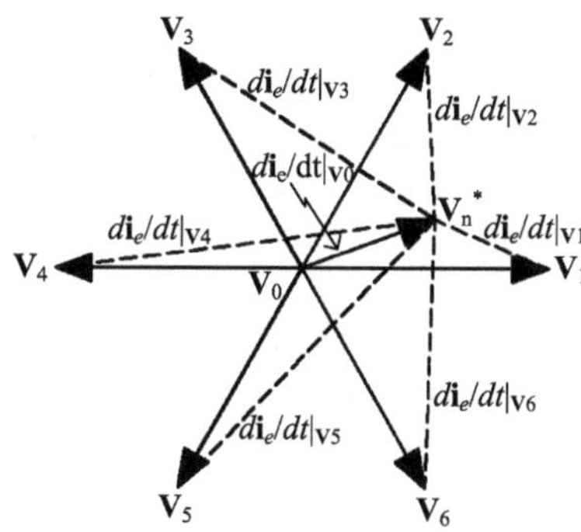

vectors $\left(\mathbf{V}_0^0, \mathbf{V}_0^1\right)$ and six nonzero vectors ($\mathbf{V}_1$–$\mathbf{V}_6$) which separate the space into six regions as shown in Fig. 2.

As shown in Fig. 2, the current error $\mathbf{i}_e$ derivative is determined by the voltage vector $\mathbf{V}_n^*$ and the output voltage vector $\mathbf{V}_n$ [16]:

$$L\frac{d\mathbf{i}_e}{dt} = \mathbf{V}_n^* - \mathbf{V}_n\,, \tag{2}$$

where $\mathbf{i}_e = \mathbf{i}_{ref} - \mathbf{i}_o$, $\mathbf{i}_o$ is the output line current vector, $\mathbf{i}_{ref}$ is the current command vector, $\mathbf{V}_n^* = L(d\mathbf{i}_{ref}/dt) + \mathbf{e}_o$ and $\mathbf{e}_o$ is the counter-emf voltage vector.

3 Switching Pattern Optimizing

The current error derivative ($d\mathbf{i}_e$) can be represented in stationary α-β reference frame and stationary A-B-C reference frame as follows:

Table 1 Current error variations generated by the $\mathbf{V}_0$ vectors in different regions

θ	330°–30°	30°–90°	90°–150°	150°–210°	210°–270°	270°–330°
di_{ea}	+	+	−	−	−	+
di_{eb}	−	+	+	+	−	−
di_{ec}	−	−	−	+	+	+
Sector	①	②	③	④	⑤	⑥

$$\begin{cases} di_{ea} = |d\mathbf{i}_e| \cdot \cos\theta \\ di_{eb} = |d\mathbf{i}_e| \cdot \cos(\theta - 120^\circ) \\ di_{ec} = |d\mathbf{i}_e| \cdot \cos(\theta + 120^\circ) \end{cases} \tag{3}$$

Table 1 gives the current error variations caused by the $\mathbf{V}_0$ vectors which divided the space into six regions labeled ①–⑥ accordingly. Thus, the space has twelve regions separated by the region I–region VI and the region ①–region ⑥ as shown in Fig. 3.

The current error states are defined by a multilevel hysteresis comparator, and the bandwidths of the inner loop and the outer loop are defined as 2bi, 2bo (bo ≤ 2bi) respectively as shown in Fig. 4. The vertical axis n_x in Fig. 4 denotes the state of current error, and the vector [n_a n_b n_c] represents the three phase current error states.

The switching pattern is optimized based on the effects of the $\mathbf{V}_n$ vectors on the current errors and the current error states. To choose the optimal $\mathbf{V}_n$ vector, three rules are proposed where the first two rules are made for the inner current error states and the third one is made for the outer current error states: (1) the $\mathbf{V}_0$ vector is preferred when the $\mathbf{V}_0$ vector decreases the nonzero current error components

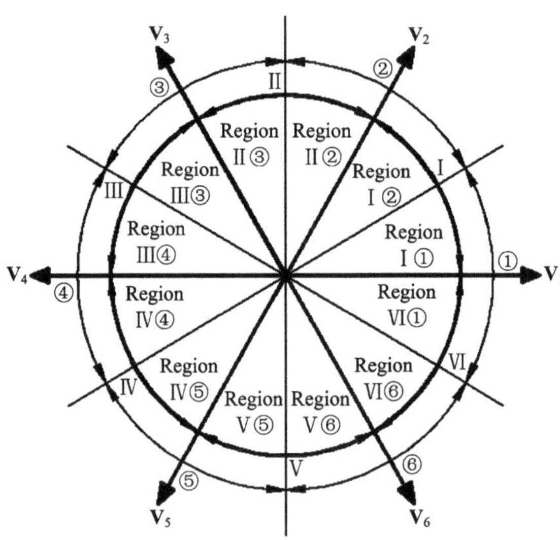

Fig. 3 Twelve regions in space

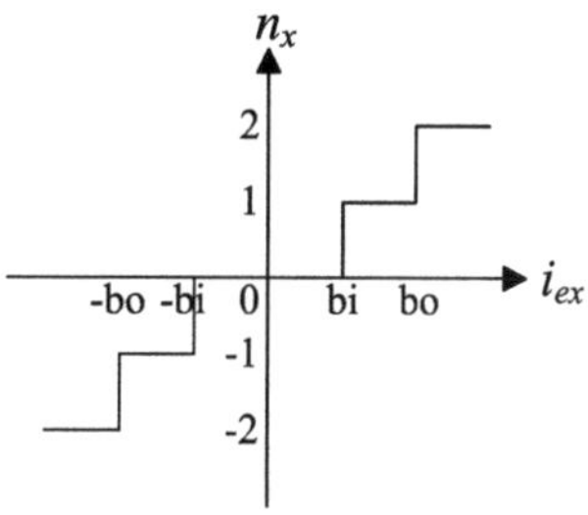

Fig. 4 Multilevel hysteresis comparator

according to Table 1; (2) when the $\mathbf{V}_0$ vector fails the first rule, the nonzero $\mathbf{V}_n$ vector generating the smallest current error variations which decrease nonzero current error components is suboptimal; (3) the nonzero $\mathbf{V}_n$ vector generating the biggest current error variations which decrease nonzero current error components is preferred.

The first rule picks the $\mathbf{V}_0$ vectors as much as possible to decrease the inner current errors effectively as well as hold the switch status for a longer time to reduce the switching numbers to a large extent. The second rule gives a supplement to the first rule, which guarantees the tracking accuracy for all the time. And the nonzero $\mathbf{V}_n$ vector satisfying the second rule reduces the inner current errors regularly, which decreases the switching numbers as well. The third rule selects the nonzero $\mathbf{V}_n$ vector to reduce the outer current errors quickly, which gives fast dynamic response. Based on the proposed rules, the most appropriate $\mathbf{V}_n$ vectors can be selected for all the current error states to form an optimized switching pattern. Besides, keeping the switch status unchanged for the [0 0 0] current error states and applying the vector $\mathbf{V}_0^0$ (0 0 0) or $\mathbf{V}_0^1$ (1 1 1) alternatively reduce switching numbers further.

Based on the proposed rules, the most appropriate $\mathbf{V}_n$ vectors are selected for all the current error states to form an optimized switching pattern. For example, the current error states vector [− 1 1 0] denotes negative current error of phase A and positive current error of phase B, which can be decreased by the $\mathbf{V}_0$ vectors when the $\mathbf{V}_n^*$ vector is located in region I① according to Table 1. While the $\mathbf{V}_n^*$ vector occurs in region I②, the $\mathbf{V}_0$ vectors are not appropriate for current error states [− 1 1 0] anymore. As the $\mathbf{V}_3$ vector and the $\mathbf{V}_4$ vector generate positive di_{ea} and negative di_{eb} when the $\mathbf{V}_n^*$ vector is located in region I②, the $\mathbf{V}_3$ vector with smaller $d\mathbf{i}_e/dt$ should be applied. In the same way, the optimized switching patterns for all the current error states can be achieved.

4 System Stucture of the Proposed SVMHCC

Figure 5 shows the structure of the proposed SVMHCC. The current controller consists of three multilevel hysteresis comparators, a region detector and the optimized switching tables. The comparators give the current error states by

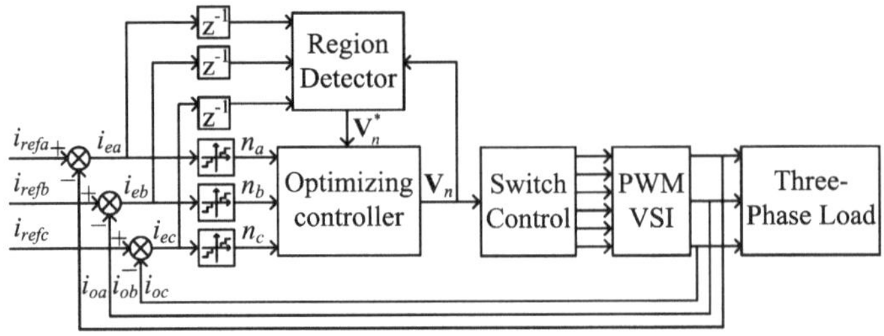

Fig. 5 Block diagram of the proposed SVMHCC

comparing the current errors with the hysteresis bands. The region detector estimates the position of the $\mathbf{V}_n^*$ vector based on the previous $\mathbf{V}_n$ vector and the corresponding current error variations. Then the current controller selects the proper voltage space vector $\mathbf{V}_n$ according to the three optimizing rules on the basis of the current error states and the position of the $\mathbf{V}_n^*$ vector.

5 Simulation Studies

To verify the validity of the proposed current controller, simulation studies are conducted for a permanent magnet synchronous motor in MATLAB/Simulink with the following parameters: stator resistance R = 0.102 Ω, state inductance L = 5 mH, pole pairs p = 2, dc voltage Vdc = 300 V. The current command is set to 30A; the bandwidths of the inner loop and the outer loop for the proposed current controller are 0.8A and 1.2A respectively. And the hysteresis band of the conventional HCC current controller is 0.8A.

Figure 6 compares the output currents of the conventional HCC and the proposed SVMHCC. It is shown that the current waveforms obtained by the proposed method have fewer ripples, and the current errors are kept in the tolerance band all the time.

Tables 2 and 3 give the average switching frequency of the inverter under different conditions for the conventional method and the proposed method. Under the default condition, the average switching frequency generated by the conventional HCC is more than two times the frequency generated by the proposed SVMHCC. When the dc voltage rises as shown in Table 2, the average switching frequency generated by the conventional HCC is increased while the one generated by the proposed method is almost the same as under the default condition. It indicates that the proposed method is insensitive to the dc voltage variation. As shown in Table 3, the average switching frequency generated by conventional HCC increases tremendously when the inductance decreases. In contrast, the average switching frequency generated by the proposed method has much smaller growth, which implies the proposed method works well with various inductance loads.

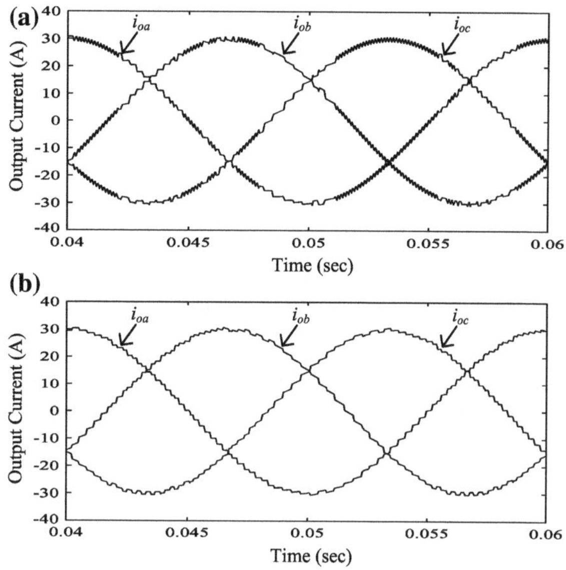

Fig. 6 Output currents of **a** conventional HCC and **b** the proposed SVMHCC

Table 2 Average switching frequency for different dc voltages (in kilohertz)

Method	HCC			Proposed SVMHCC		
Vdc	300 V	450 V	600 V	300 V	450 V	600 V
Switch S_1	9.48	21.19	30.08	4.17	4.27	4.56
Switch S_3	9.71	21.17	30.09	4.76	4.70	4.83
Switch S_5	9.60	21.16	30.13	4.15	4.15	4.20

Table 3 Average switching frequency for different inductances (in kilohertz)

Method	HCC			Proposed SVMHCC		
L	5 mH	1 mH	0.5 mH	5 mH	1 mH	0.5 mH
Switch S_1	9.48	75.41	138.74	4.17	7.56	16.21
Switch S_3	9.71	75.40	138.73	4.76	7.67	15.96
Switch S_5	9.60	75.40	138.74	4.15	7.42	15.54

6 Conclusion

This paper proposes an optimized SVM based hysteresis current controller. The proposed method achieves the same dynamic responding property as the conventional method, and improves the tracking performance. The simulation studies also demonstrate that the proposed method reduces the switching frequency significantly under different conditions. And it is shown that the proposed method is robust to the dc voltage variation and achieves satisfying low average switching frequency under small inductance condition.

References

1. Kazmierkowski MP, Malesani L (1998) Current control techniques for three-phase voltage-source PWM converters: a survey. IEEE Trans Ind Electron 45(5):691–703
2. Mohseni M, Islam SM (2010) A new vector-based hysteresis current control scheme for three-phase PWM voltage-source inverters. IEEE Trans Power Electron 25(9):2299–2309
3. Buso S, Malesani L, Mattavelli P (1998) Comparison of current control techniques for active filter applications. IEEE Trans Ind Electron 45(5):722–729
4. Pan CT, Chang TY (1994) An improved hysteresis current controller for reducing switching frequency. IEEE Trans Power Electron 9(1):97–104
5. Chun TW, Choi MK (1996) Development of adaptive hysteresis band current control strategy of PWM inverter with constant switching frequency. In: Proceedings IEEE APEC'96, San Jose, CA, 1996, PP 194–199
6. Malesani L, Mattavelli P, Tomasin P (1997) Improved constant-frequency hysteresis current control of VSI inverters with simple feedforward bandwidth prediction. IEEE Trans Ind Appl 33(5):1194–1202
7. Rahman KM, Rezwan M, Choudhury MA, Rahman MA (1997) Variable-band hysteresis current controllers for PWM voltage-source inverters. IEEE Trans Power Electron 12(6):964–970
8. Stefanutti W, Mattavelli P (2006) Fully digital hysteresis modulation with switching-time prediction. IEEE Trans Ind Appl 42(3):763–769
9. Tekwani PN, Kanchan RS, Gopakumar K (2007) Novel current error space phasor based hysteresis controller using parabolic bands for control of switching frequency variations. IEEE Trans Ind Electron 54(5):2648–2656
10. Kwon BH, Min BD, Youm JH (1998) An improved space-vector-based hysteresis current controller. IEEE Trans Ind Electron 45(5):752–760
11. Marei MI, El-Saadany EF, Salama MMA (2002) A new contribution into performance of active power filter utilizing SVM based HCC technique. In: Proceedings IEEE power Engineering society summer meeting, vol 2, pp 1022–1026
12. Dianguo X, Jiangjun G, Hankui L (2003) Improved hysteresis current control for active power filter. In: Symposium IEEE ISIE'03, vol 2, pp 836–840
13. Zare F, Ledwich G (2007) A new hysteresis current control for three-phase inverters based on adjacent voltage vectors and time error. In: Conference IEEE PESC, pp 431–436
14. Mohseni M, Islam S (2009) A novel current controller for three-phase voltage-source inverters. In: Conference IEEE IECON'09, pp 76–81
15. Youhua J, Yongwei C (2009) A novel voltage space vector control algorithm and its application in active power filter. In: International conference ICEET'09, pp 47–50
16. Kwon BH, Kin TW, Youm JH (1998) A novel SVM-based hysteresis current controller. IEEE Trans Power Electron 13(2):297–307

Analysis on 3D Detection Method and Software Implementation for Ballastless Track Geometry

Tao Wang, Li-gong Li, Ya Zhang and Li-juan Niu

Abstract Based on the requirement of ballastless track's adjusting and checking, the paper takes high-speed railway engineering survey as an entry point, writers proposed a design scheme about the integration of track internal geometric parameter measurement and external geometric parameter measurement. Then writers gave a mathematical model for track checking and constructed a theoretical system for ballastless track 3D detection. Next the writers produced a design for the ballastless track checking system's architecture, technology, function and system interface in the paper. At last writers developed a ballastless track software system based on Visual Studio.NET platform.

Keywords Ballastless track · High-speed railway · Track detection car · Track 3D checking

1 Introduction

Track geometry state detector (also called track inspection car) is a mainly detection tools that can detect track when no train wheel load static track irregularity [1]. It consists of sensor technology, precision machinery processing technology,

T. Wang (✉) · L. Li · Y. Zhang · L. Niu
Department of Surveying and Mapping, Shaanxi Railway Institute,
Weinan, People's Republic of China
e-mail: survey-tao@sohu.com

L. Li
e-mail: llg.6414@163.com

Y. Zhang
e-mail: ln3719620@163.com

L. Niu
e-mail: 286182381@qq.com

T. Wang
College of Geology Engineering and Geomatics, Chang'an University,
Xi'an, People's Republic of China

E. Qi (ed.), *Proceedings of the 6th International Asia Conference on Industrial Engineering and Management Innovation*,
DOI 10.2991/978-94-6239-145-1_53

computer technology and precision engineering measurement technology and data processing technology, etc. Track inspection car has small weight, easy to carry, simple to operate, high measurement precision, etc. It can be used to real-timely detect the railway track height, horizontal, twist and track alignment, gauge, etc.

The world's most famous static in the field of railway track measurement system mainly include Switzerland Leica company and Amberg measurement technology companies jointly developed Leica GRP multi-function railway track measurement system and German Sinning company GEDO CE track detection system [2]. The above two systems are both successful examples of portable track inspection car the applications of laser measurement technology, but its high price. It makes domestic railway departments prohibitive, only a small number of railway production sectors equipped.

Researching on the track geometry measurement in our country, was originated form solving the disease of ordinary railway tracks. It's the relative measurement methods of track inspection instrument, measuring efficiency is good, but have the problems of measuring precision and reliability. It can't reach the standard of high speed railway track comfortable [3]. We are lagging far behind foreign in terms of rail system for 3D detection, has many problems of technical bottleneck, preventing the construction and development of high-speed railway in our country [4]. In view of this, we take some domestic track inspection car as an example, discuses ballastless track 3D detection system architecture designing, technology, interface and function, and solve the system in the course of the development of automatic observation free stationing, wireless communications and data post-processing and other key technical problems, and in Visual Studio.NET platform software integration and implementation.

2 Ballastless Track of Passenger Dedicated Railway Line Emngineering Survey System

Reviewing the development of the high-speed railway construction, Germany and Japan and other countries set up a suitable for their country's high-speed railway construction engineering survey technology system. German railway departments has specialized to establish a set of independent coordinate system for high-speed railway construction in its territory, its precision is superior to the general national the basic control network, which fully show that they think highly of the measurement system of high-speed railway.

On the basis of absorbing foreign advanced experience, China railway construction sectors has completed Qinhuangdao-Shenyang, Beijing-Tianjin [5], Wuhan-Guangzhou, Zhengzhou-Xi'an, Beijing-Shanghai and other passenger dedicated line successful working [6, 7]. A set of relatively complete, suitable for China's high speed railway engineering measurement system was established by Ministry of Railways of People's Republic of China, drawing on the experience and related technology of foreign, and in 2009 in the foundation of *Passenger transport line temporary provisions of ballastless track high-speed railway engineering survey* [8], to complete the preparation of *Specification for high speed railway surveying* [9], to achieve the establishment of the domestic high speed railway engineering survey system.

3 A Mathematical Model of 3D Detection for Ballastless Track

3.1 The Measuring Principle of Track Internal Geometric Parameters

The internal geometry parameters of railway track including mileage, gauge and superelevation. The main purpose of mileage measurement is determining track inspection trolley forward position by track relative measurement, the principle of which is in track check car measuring wheel is installed on a magnetic rotary encoder, and encoder a week issued 256 pulses, walk through the car wheel radius available measuring wheel circumference. The unit of distance can be determined by one pulse. Gauge measurement results were achieved by displacement sensor track inspection trolley beam. According to the definition of gauge, the displacement sensor readings can be gauge.

Superelevation (level) refers to the elevation difference of two rail surface between the center lines in the vertical plane. The parameters would affect the smoothness of track. The main reason of train vibration roll track is the level irregularity. If it is seriously not horizontal in the curve segment, which would cause the train derailment capsized. Superelevation is measured by the level of chips and A/D conversion chip of Angle sensor, rail detection, obliquity sensor measured about two rail surface and the horizontal plane angle θ, according to gauge sensor tracks of the measured value and track by the width of the two at the top of the rail center distance L, superelevation d calculation by Formula (1) :

$$d = L \cdot \sin\theta \tag{1}$$

In the direction along the route, provisions left curve superelevation is negative, and right curve superelevation is positive.

3.2 Mathematical Model for Measuring the Parameters of Track External Geometry

The outside of the track geometric parameters has two kinds of observations, including the distance and direction, the error equation were as follows [10]:

$$\begin{bmatrix} v_1 \\ v_2 \end{bmatrix} = \begin{bmatrix} A_1 \\ A_2 \end{bmatrix} \hat{x} - \begin{bmatrix} l_1 \\ l_2 \end{bmatrix}, \begin{bmatrix} P_1 & \\ & P_2 \end{bmatrix} \tag{2}$$

In formula (2), v_1, A_1, l_1, P_1 and v_2, A_2, l_2, P_1 were observed value of correct number, design matrix, the observation residual vector and weight matrix respectively; $\hat{x}$ is all the coordinates of the correct number.

$$\text{Let } v = \begin{bmatrix} v_1 \\ v_2 \end{bmatrix}, A = \begin{bmatrix} A_1 \\ A_2 \end{bmatrix}, l = \begin{bmatrix} l_1 \\ l_2 \end{bmatrix}, P = \begin{bmatrix} P_1 & \\ & P_2 \end{bmatrix}$$

Formula (2) Can be simplified as:

$$v = Ax - l \tag{3}$$

External geometric parameters of the track is refers to the actual location of the axis track, In the fact, track laying in lines, before has not yet been determined through precision engineering measurement techniques are needed to determine the position of each point in track, elevation and superelevation, as well as its relationship with the track design value.

3.3 The Track Inspection Car Prism Center Coordinates Acquisition

Track inspection car prism coordinate location plays an important role when the axis was determinate in ballastless track of high-speed railway. Methods of the track inspection car prism coordinate measurement is the function of automatic tracking and locking the car search prism by using total station, real-time 3D coordinate prism point [11, 12], and real-time through wireless communication control program coordinate prism point to track inspection car.

Total station uses the side angle net resection of free station [14, 15], observation of the CPIII branch slope distance, horizontal Angle and vertical Angle, obtained by means of three dimensional adjustment station in construction coordinate system of the 3D coordinates. Observation method is shown in Fig. 1.

After observation, free stationing site to calculate of approximate coordinates. The approximate coordinates calculation is shown in Fig. 2. Assume P_i and P_j is CPIII survey points, and set free site P_k coordinates is shown in Fig. 2, from P_k to P_i and P_j slant distance of observation value is S_{ki} and S_{kj}, and the vertical angle observation value is V_{ki} and V_{kj}, and observation value of horizontal direction is L_{ki} and L_{kj}, P_k point approximate coordinates is (x_k^0, y_k^0, z_k^0), from P_i to P_j of azimuth angle is A_{ij}, from P_i to P_k of azimuth angle is A_{ik}, and P_i,P_j in a horizontal plane though point P_k projection points is P_i' and P_j', form P_i' to P_j' horizontal distance is

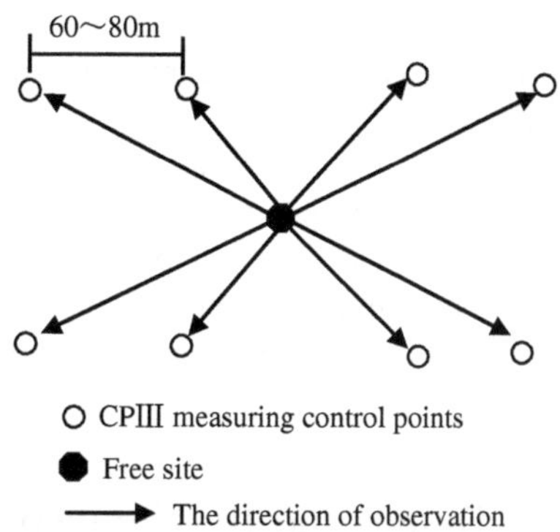

Fig. 1 Free stationing observation of total station [13]

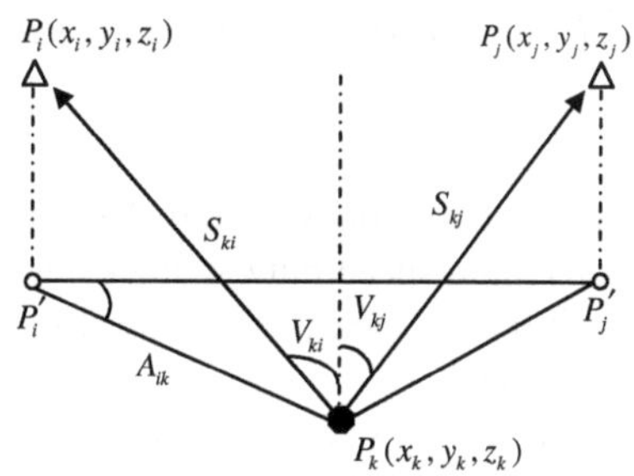

Fig. 2 Station approximate coordinate calculation

$P_i'P_j'$, form P_i' to P_k horizontal distance is $P_i'P_k$, form P_j' to P_k horizontal distance is $P_j'P_k$, so the approximate coordinates calculation formula is:

$$\begin{cases} x_k^0 = x_i + P_i'P_k \cos A_{ik} \\ y_k^0 = y_i + P_i'P_k \sin A_{ik} \\ z_k^0 = z_i - S_{ki} \cos V_{ik} + v \end{cases} \tag{4}$$

among them, $A_{ik} = A_{ij} + \angle P_j'P_i'P_k$, $A_{ij} = \arctan\frac{y_j - y_i}{x_j - x_i}$, $P_i'P_k = S_{ki} \cos V_{ki}$, $P_j'P_k = S_{kj} \cos V_{kj}$, $P_i'P_j' = \sqrt{(x_j - x_i)^2 + (y_j - j_i)^2}$, $\angle P_j'P_i'P_k = \arccos\frac{P_i'P_k^2 + P_i'P_j'^2 - P_j'P_k^2}{2 \times P_i'P_k \times P_i'P_j'}$, v is height for a mandatory prism.

When the approximate coordinates were achieved, next, the station coordinates and azimuths need to compute. Station coordinates and azimuth adjustment contains two aspects: on the one hand is the site coordinate measurement, through the calculation of adjustment of the track center position accuracy of measurement is the basic; On the other hand is the precise coordinates azimuth of station, this is to ensure that the railway center lateral error precision of the main aspects of the position measurement. In order to guarantee the data calculation to achieve high-accuracy, it is necessary to eliminate the influence of gross error on adjustment results. After the adjustment, to station coordinates and azimuth could meet the car prism measurement accuracy and reliability of measurement science and rigorous assessment.

When the station approximate coordinate meets the required precision, the observation on prism coordinates calculation of car, according to polar coordinate method to calculate car prism position. After the track car the prism measuring prism coordinate accuracy of the information in detail, on this basis, the ballastless track geometry will be accurately determinated.

4 Design and Development of Ballastless Track Geometry Inspection System

4.1 Introduce of Ballastless Track Geometry Inspection System

System based on Microsoft Visual Studio.NET platform, developed a set of functional high-speed railway track detection control software. Start interface including engineering, measurement files, measurement, export files statements and tool set five options.

Before track test, assembly track inspection car and install CPIII control points of prism, and test software to open the power supply, placed near the midline of total station in track, the first to track 3D test system software and hardware self-test, and then test track.

Program work interface on the left side of the main function is to display track axis and design axis deviation of two options is responsible for the real-time display on the right side of the car sensors and total station data.

4.2 Work Flow of Ballastless Track Geometry Inspection System

The system of ballastless track geometry inspection should be with functionally such as gauge, superelevation (level), high and low mileage and triangle hole measurement, and so on, under the high accuracy intelligent total station auxiliary, which can quickly get track axis coordinate and elevation about railway track external geometric parameters, real-time computing actual track and track design, the deviation between the results and report generation function, the system working process as shown in Fig. 3.

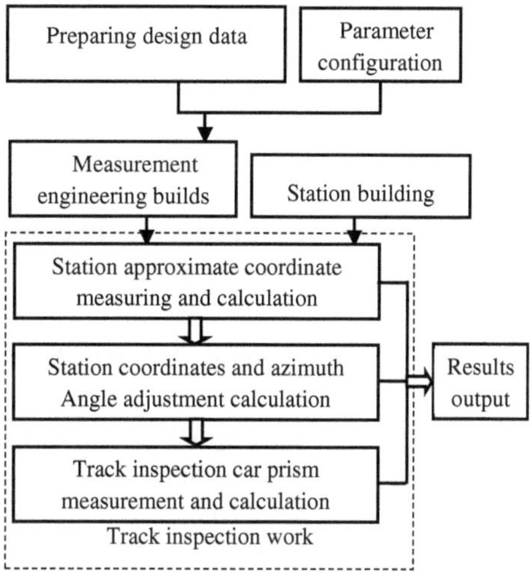

Fig. 3 Work flow of ballastless track geometry 3D detection system

5 Analysis of Ballastless Track Detection System Internal Geometric Accuracy

According to the ministry of railways PR China *Rail tester general technical conditions*, analysis of track inspection car internal geometric parameter measuring accuracy uses the international general the uncertainty analysis. Uncertainty is a measure of measurement uncertainty, track inspection car internal geometrical parameters measurement uncertainty basically has the following sources: ① sensor measurement repeatability is introduced into standard uncertainty components $u_1(x)$; ② the sensor resolution is introduced into standard uncertainty components $u_2(x)$; ③ track inspection car of the introduction of error of the standard uncertainty components $u_3(x)$; ④ uncertainty components introduced by the calibration error. Its including display value error and the error of level by the standard uncertainty, the value is small, can be neglected.

Gauge, superelevation, mileage measurement uncertainty analysis is the track inspection car sensor repeatly test in the experiments. For track inspection car display value, the corresponding single test standard deviation is calculated:

$$s = \sqrt{\frac{\sum_{i=1}^{n} (x_i - \bar{x})^2}{n - 1}} \tag{5}$$

Another a few fixed point, each are done repeatedly test, test results calculated for a single test standard deviation by the formula (5), synthetic standard uncertainty $u_i(x)$ $(i = 1, 2, 3)$ is calculated as follows:

$$u_i(x) = \sqrt{\frac{1}{n}\sum_{i=1}^{n} s_i^2} \tag{6}$$

6 Analysis of Ballastless Track Detection System External Geometric Accuracy

6.1 Free Stationing Accuracy Analysis

Because of the relatively CPIII control point position error m_P is 1 mm, the elevation error of control point $m_H = 0.5$ mm, assuming the ordinate relative error m_x and m_y relative error is equal and independent, its $m_x = m_y = 0.7$ mm. Total station uses Leica TCA2003, its nominal angle measuring precision is $\pm 0.5''$, nominal distance accuracy is $1 + 1$ ppm. In the process of actual measurement, the precision of total station and subject to change with the change of external factors, it is generally believed in a complex environment, one observation set in the direction of observation error is $1''$ for measuring, take its nominal accuracy distance precision is $1 + 1$ ppm.

In car prism measurement process, to improve the efficiency of orbit detection, use only face left observation, then in the direction of observation error is m_{HA} and the distance of observation error is m_L:

$$m_{HA} = \sqrt{2} \times 1 \approx 1.4'' \tag{7}$$

$$m_L = \sqrt{2} \times (1 + 1 \times S_i)mm \tag{8}$$

Among them S_i is slant distance of observations in formula (8).

In free stationing, total station observation three observation sets, according to the law of error propagation, the direction of observation error is m_{HA} and distance error is m_L:

$$m_{HA} = \frac{1}{\sqrt{3}} \approx 0.6'' \tag{9}$$

$$m_L = \frac{1 + 1 \times S_i}{\sqrt{3}} mm \tag{10}$$

6.2 Analysis of the Car Prism Measuring Accuracy

Considering the error of prism coordinate in ballastless track adjust process is mainly affected by angle and distance measuring precision, and the distance accuracy of prism measurement is greatly influenced, laying of ballastless track main transverse direction perpendicular to the center line error. In the accurate adjustment to the linear setting station of total station instrument to perpendicular to the car crossbeam, the distance error mainly affect the height of the point, the longitudinal error is *mm*, the influence value in level, relative to the distance measurement can be ignored, and the influence of value to the nearest point coordinate calculation of horizontal error caused by very small, can be neglected; at this time, the lateral error m_t track accurate adjustment, is mainly affected by the precision of angle measurement, the calculation formula is as follows:

$$m_t \approx \sqrt{m_t'^2 + D_i^2 \cdot (\frac{m_{HA}}{\rho})^2} \tag{11}$$

In formula (11), m_t' is the horizontal precision value of the site, D_i as the station to the car prism from the flat, take $D_i = 60\,\text{m}$,$m_{HA} = 1.4''$, m_t' take its maximum value is 0.35 mm, according to the formula (11), can be adjust in the prism to track the maximum lateral error 0.53 mm.

Prism point height measurement based on the measurement principles and measurement error propagation law. Trigonometric leveling elevation difference at two points in the error calculation formula is:

$$h_{AB} = S_{AB} tg\alpha_A + i_A - V_B + f_{AB} \tag{12}$$

In formula (12), S_{AB} is the distance between two points A and B; α_A is from point A to point B measured vertical Angle; i_A is height point of instrument; V_B is height the goal of point B; f_{AB} is two difference between two points A and B (the earth curvature and atmospheric refraction) correct.

To formula (12) for differential, the error is:

$$m_{h_{AB}} = \pm\sqrt{tg^2\alpha_A m_{S_{AB}}^2 + \frac{S_{AB}^2 \cdot m_{\alpha_A}^2}{\cos^4 \alpha_A \cdot \rho^2} + m_{i_A}^2 + m_{V_B}^2 + m_f^2} \tag{13}$$

Length of be confined to, rail height prism point detection car system measurement and precision analysis are not discussed here, a detailed analysis of the literature [16].

7 Conclusions and Suggestions

(1) Ballastless track as a new kind of track laying form. It is extensively applied in high-speed railway construction in China, its unique characteristics and high precision requirements have brought great challenges to the surveying work.

(2) Single function of the detection system can't meet the requirements of high speed railway track detection, collect a variety of sensors, a variety of functions in one convenient type of track detection system is its development direction, can be considered in ballastless track detection system on the basis of the existing integrated gyro orientation technology, 3D laser scanning technology and photogrammetry technology, etc., to expand the function of the detection system.
(3) High-speed rail track inspection car through the program control measurement of CPIII control points automatically, in the process of reducing the artificial observation error. The type of surveying not only improves the precision of observation but also improves the efficiency of the measurement.
(4) During the study period, writers found that only rely on the measurement in the course of the development of professional knowledge is very difficult to completely solve all problems.

Acknowledgments This work was supported by the National Natural Science Foundation of China under Grant No.41304013, Shaanxi Provincial Education Department under Grant No.14JK1167 and the Shaanxi Railway Institute Natural Science Foundation (2014–28). The authors thanks Dr. Hu-rong DUAN of Xi'an University of Science and Technology for experimental assistance.

References

1. Luo L, Dong Y, Zhu H, Hu Z (2013) Hydrostatic monitoring system and its calibration method used for subgrade settlement monitoring of ballastless track. Railway Stan Des 05:5–9
2. Huang J (2011) Study on 3D checking system for the ballastless track in the high-speed railway and its error analysis. Chang'an University Master Degree Thesis 6
3. Wang G, Gao J, Lu J (2012) Control index and inspection technique of track geometry state in high-speed railway. Railway Invest Surv 1:1–4
4. Zhao Z (2011) Rearch on some technical problems of the track regularity of high-speed railway. Southwest Jiaotong University Master Degree Thesis.6
5. Qing S, Chen S, Hu J (2008) Researches on the key construction equipments and technologies for CRTSIISlab ballastless track for high-speed railway. J Railway Eng Soc 7:22–27
6. Lu J (2010) On precise engineering survey technology system and characteristic for high speed railway in China. High Speed Railway Technol 1(1):31–35
7. An G (2010) Study on technical standard for precise engineering surveying of high-speed railways and its applications. J China Railway Soc 32(2):98–104
8. Ministry of Railways of China (2006) Passenger transport line temporary provisions of ballastless track high-speed railway engineering survey. China Railway Publishing House, Beijing
9. Ministry of Railways of the People's Republic of China (2009) TB10601-2009 specification for high speed railway surveying. China Railway Publishing House, Beijing
10. Cai S, Li B, Shi D, Liu C, Shen Y (2008) Data processing in precision surveying of ballastless track high-speed railway. J Geodesy Geodyn 28(1):114–117
11. Zhu H, Xu R (2009) Research on the 3D precise measurement technology for existing railway line. J Railway Eng Soc 1:49–51

12. Leng D (2009) CPIII control points survey technology for ballastless track of high-speed railway. Tunn Constr 29(2):239–242
13. Xu W (2012) Determination of precision measurement datum for high-speed railway. J Railway Eng Soc 9:8–12
14. Liu C (1995) The principle of three points polar coordinate resection and its precision estimate. J Southwest Jiaotong Univ 30(03):251–256
15. Zhu B, Liu C, Yang T (2004) Resection method and its precision evaluation. J Shandong Jiaotong Univ 12(03):26–29
16. Wang T, Zhang F (2013) Precisely surveying for ballastless track in the passenger dedicated railway line and accuracy analysis. Eng Surv Mapp 22(1):86–89

Monte Carlo Simulation on Automatic Calibration Method of Horizontal Tanks

Jin-guang Zhou and Jian-zhong Zhang

Abstract This paper introduces an automatic calibration method on horizontal tanks in gasoline stations by mining tank gauge data and flow meter data, and evaluates its efficiency by Monte Carlo simulation. Simulation experiments show that this method is effective when the accuracy of tank gauge and flow meters meet the requirement.

Keywords Gasoline station · Horizontal tank · Monte carlo simulation · Tank calibration

1 Introduction

Horizontal tanks are buried underground in gasoline stations and not noticeable, but these volume tables are very import in fuel storage check and market trade. They represent function between a set of height and a set of volume in measurement. Horizontal tank volume tables are traditionally set up by collecting and calculating gasoline in/out data through the Fuel Dispensing Pump and the corresponding height change of gauge level measured by special equipments [1].

The calibration process requires closing business and safety supervision.

Recently a novel but controversial calibration method was proposed through mining historical data in gasoline station daily operation, mainly tank gauge data and corresponding filling flow meter data. This new technique attracts lots of attention for its simplicity, non-interruption of daily operation and low risk of HSE

J. Zhou (✉) · J. Zhang
Research Institute of Safety, SINOPEC, Qingdao, China
e-mail: zhoujg.qday@sinopec.com

J. Zhang
e-mail: zhangjz.qday@sinopec.com

J. Zhou
SINOPEC Sales Corporation, Beijing, China

E. Qi (ed.), *Proceedings of the 6th International Asia Conference on Industrial Engineering and Management Innovation*,
DOI 10.2991/978-94-6239-145-1_54

accidents. However some opponents think this method doesn't work well because the uncertainty of volume table by this may exceed 0.3 %, the threshold of GB/T17605-1998 <*Volumetric calibration of horizontal cylindrical metal tanks* code> [2–4] and error may be even large. So tremendous real verification tests are required in gasoline stations to justify its effectiveness, but it cost so much even oil giant couldn't afford the expense.

This paper avoids trivial deformation of horizontal tanks in real experiments and analyses the effectiveness and uncertainty of the automatic calibration method by Monte Carlo Simulation (MCS) [5, 6]. Various parameters are set up and adjusted in those simulations. Meanwhile 4 evaluation criterions are introduced to show the effectiveness. Thousands of MCS experiments show that uncertainty of the proposed calibration method is less than 0.03 % (96.7 % confidence interval) when the both uncertainty of gauge level and flow meter less than 0.02 %. This result qualified the national standard requirements in appropriate conditions. Finally, some optimal parameters and recommendations are given during compiling volume table.

2 Principle of Auto Calibration

2.1 Volume of Horizontal Tank

Assume the horizontal tank is a standard cylinder, without deformation or HAN materials filled in, and the gauge is vertically installed at the center of tank, so the rotation of the tank wouldn't impair the measurement of liquid height h.

The gasoline underground storage volume V(h) can be shown as in (1)

$$V(h) = \begin{cases} LR^2 \,\mathrm{across}\, \frac{R-h}{R} - L(R-h)\sqrt{2Rh - h^2}, & \le h < 2R \\ \pi LR^2 - LR^2 \,\mathrm{across}\, \frac{(h-R)}{R} + L(h-R)\sqrt{2Rh-h^2}, & R \le h < 2R \end{cases} \tag{1}$$

where L is the length of the tank, R is the radius of the tank. Meanwhile the derivative of V(h) is shown as in (2)

$$\frac{dV(h)}{dh} = 2L\sqrt{2Rh - h^2}, \quad 0 \le h \le 2R \tag{2}$$

2.2 Approximation of Volume Table

For some reasons such as rotation, deformation, etc., a volume table is calibrated regularly instead of using (1) directly. During calibration, we get Eq. (1) by solving Eq. (2), but there is no analytical solution for partial differential Eq. (2).

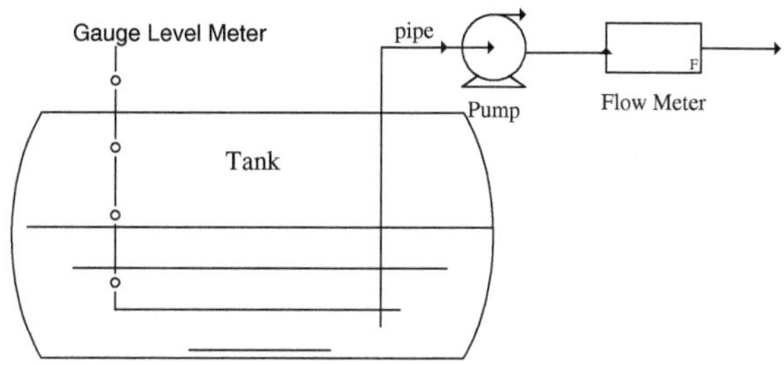

Fig. 1 Filling drops liquid level correspondingly

For simplicity, approximate Eq. (2) by an piecewise polynomial function, as in (3)

$$\frac{dV(h)}{dh} = ah^2 + bh + c \tag{3}$$

where a, b, c are undermined parameters, consecutively the volume table can be represent as in (4)

$$V_s(h) = a_p h^3 + b_p h^2 + c_p h + d_p,\ p = 1, \ldots, P \tag{4}$$

where p is piecewise number, and a_p, b_p, c_p d_p are corresponding parameters.

Batch data of filling flow meter can be taken as dV(h), and the corresponding change in gauge level can be taken as dh, as shown in Fig. 1. Through collecting or generating a lot of batch filling data, get a set of dV(h)/dh and dh, and then determine its polynomial parameters by data regression algorithm, and finally solve Eq. (3), get the volume table.

3 MCS of Auto Carlibration

During simulation, assume Eq. (1) as conventional true value and neglect unimportant factors such as temperate change of fuel, deformation of tank etc., randomly generate a set of filling data and liquid data by corresponding probability density function (PDF), approximate derivative equations by regression algorithm and then solve those derivative equations, finally evaluate the error between conventional true value V(h) and regression value $V_s(i,h)$ in different height and different iteration.

3.1 Monte Calro Simulation in Calibration

Monte Carlo Simulations are a broad class of computational algorithms that rely on repeated random sampling to obtain numerical results by running simulations many times over in order to calculate those same probabilities heuristically just like actually playing and recording your results in a real casino situation.

In this simulations, assume the PDF of filling mounts dV(h) obeying normal distribution and PDF of liquid level h obeying average distribution, randomly generate a set of dV(h) and h, so get a set of between dV(h)/dh and h, then after solving this by Levenberg-Marquardt Regression algorithm [7, 8], get polynomial parameters of $V_s(i,h)$, and finally evaluate the measurement errors. The process can be shown as Fig. 2 and parameters can shown as Table 1.

Initialize Parameters
For i = 1 : Itr
  For h = 1 : 2R
    Generate dV(h) set by normal distribution;
    Generate h set by average distribution;
    Get a group of dV(h)/dh and h;
    Solve Equation (3) by L-M Regression algorithm;
    Get $V_s(i,h)$ by Equation (3);
    Get $Err_s(i,h)$ by $V_s(i,h)$ and Equation (1);
 Evaluate $Err_s(i,h)$

Fig. 2 Procedure of monte carlo simulations of calibration

Table 1 Parameters and its meaning in MCS experiments

Symbol	Meaning	Setting
L	Length of Tank	10 m
R	Radius of the tank	1 m
	Use R as integer from 1 to 1000 for calculation	
A	Gauge level	±0.1 mm
H	Height of liquid level	0.2R–2R
$A_i, i = 1, \ldots, G$	Relative error of flow meter	±0.2 %
dV(h)	Filling mount cause volume change	
P	Piecewise number for polynomial	3–15
N	Number for regression sample	20
Itr	MCS times	10,000
$V(h)$	Conventional true value at h height	
$V_s(i,h)$	Volume at h height in i time	
$Err_s(i,h)$	Error at h height in i time	
U(h)	Uncertainty at height h (95 %)	

3.2 Experiment Parameters

During those experiments, various parameters are set up and adjusted according to the real conditions in gasoline stations.

The Parameters, their meaning and default setting values are shown as Table 1.

3.3 Evalutaion Criterion

How to evaluate the automatic calibration method [9] is the key problem. There is a balance between accuracy and stability in simulation, and this paper proposed 4 criterions to judge it effectiveness as follows:

1. $Err_s(i,h)$ Absolute Error in a simulation. This illustrates accuracy of the automatic calibration method. Define as error between volume table value $V_s(i,h)$ and conventional true value $V(h)$, show as (5).

$$Err_s(i,h) = V_s(i,h) - V(h) \tag{5}$$

2. U(h) Uncertainty of volume table at height h in different simulations. This illustrates the stability of the automatic calibration method. Define uncertainty U (h) at height h as (6).

$$U(h) = \sqrt{\frac{\sum_{i=1}^{Itr} (Err_s(i,h))^2}{Itr - 1}} \tag{6}$$

3. U(i) Uncertainty of volume table in a simulation at different height h as (6). This is average height situation and doesn't care extremely low or high gauge level in calibration.

$$U(i) = \sqrt{\frac{\sum_{h=0}^{2R} (Err_s(i.h))^2}{2R}} \tag{7}$$

4. Maximum Error in Every Simulation. This illustrates the probability of maximums bad situations. Find relative error may exceed 0.3 % as in (8). This is important to show how bad the result when the conditions in gasoline stations are extremely bad.

$$R_{bad} = \frac{\prod_{i=1}^{Itr} \prod_{h=1}^{2R} if(\left| {Err_s(i,h)}/{V(2R)} \right| > 0.3\%)}{\text{Itr*2R}} \tag{8}$$

3.4 Experiment Result

The First experiment to demonstrate accuracy of this automatic method compared with conventional true value as Eq. (5). Assume relative error of flow meter is ±0.2 %, Absolute error of gauge level is ±0.2 mm, piecewise number is 13, Number for regression sample is 20. We can get Error of Volume Table in a simulation as Fig. 3 and that maximum error in this time is 0.062 m^3, which could satisfy real world requirements in this single experiment.

Then second and third experiment are done to demonstrate stability of this calibration method in different height and different iteration as Eqs. (6) and (7), the parameters are identical with the first time. We get Figs. 4 and 5 during the test.

Clearly the uncertainty U(h) couldn't meet the 0.3 % requirement when the liquid level at extremely low or high, but fortunately in real volume table we only care the height form 0.2R to 1.8R, not the extremely situations; Meanwhile no matter what piecewise number n chosen, U(h) at middle height h could satisfy the requirement; as piecewise number p increasing, uncertainty U(h) become lower.

Finally the extreme bad situation are done and get the most bad result $R_{bad} = 0.881\%$.

In some extreme cases it will exceed the standard limit 0.3 %, but this is rather rare.

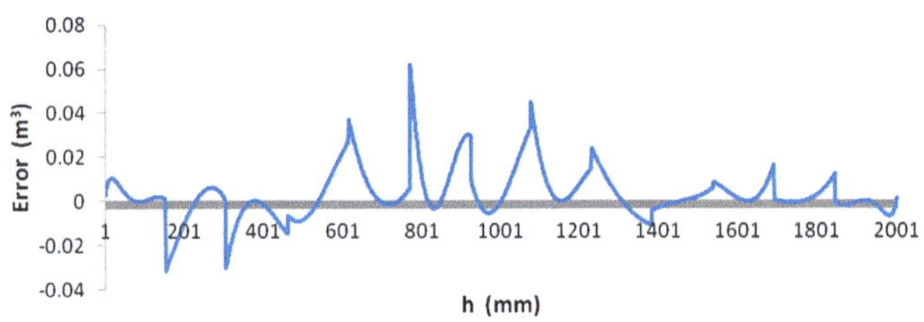

Fig. 3 Error of volume table in a simulation

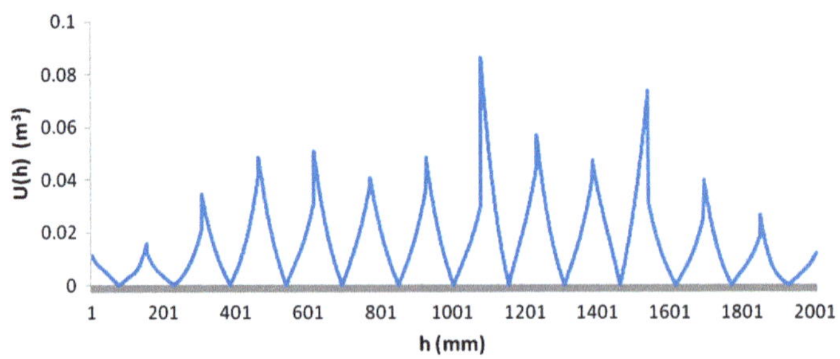

Fig. 4 Uncertainty U(h) in sepcificated height h

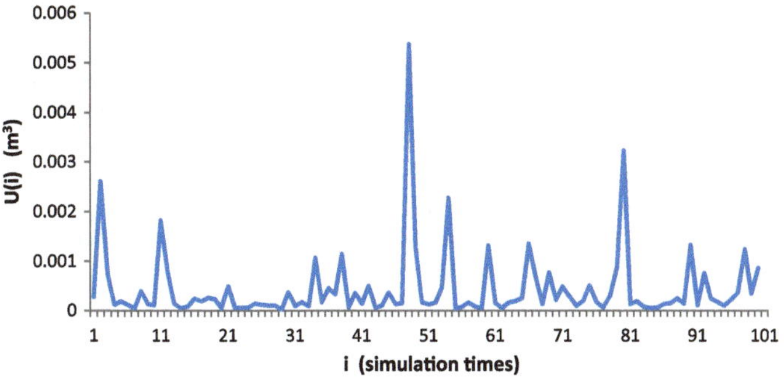

Fig. 5 Uncertainty U(i) in different simulations

4 Conclusion

This paper researches effectiveness of automatic calibration method on horizontal tanks by Monte Carlo Simulation (MCS) through adjusting various experimental parameters. Thousands of MCS experiments show that uncertainty of proposed calibration method is 0.03 % (96.7 % confidence interval) in appropriate conditions, which meets requirement well, and some optimal parameters and recommendation are given during compiling volume table.

Those simulations experiments show that the automatic calibration method could satisfy the code well, from height 0.2R to 1.8R, uncertainty at every height h could be less than 0.3 %, even at most bad conditions, error exceeding 0.3 % less than 0.881 %. This simplicity, non-interruption of daily operation and low risk of HSE accidents would be widely used entire manual operation.

But this method requires accumulative data in daily sales operation and accuracy of tank gauge. The key point is how the gauge could measure the change of liquid level dh, this is more important than measure h.

Acknowledgments Thanks to 44 gasoline stations of SINOPEC for their real test of this calibration algorithms in Tianjin City, and real experiments also show this method is effective.

References

1. Feng D, Dong W, Dong F (2004) R&D on horizontal tank volume table formulation. Oil Depot Gasol Stn 12(2):22–24
2. ISO/IEC Guide 98-3:2008 Uncertainty of measurement—Part 3: guide to the expression of uncertainty in measurement (GUM:1995)
3. GB/T 17605-1998 Petroleum and liquid petroleum products. Volumetric calibration of horizontal cylindrical metal tanks (Manual methods)
4. JJG266-1996 The Volume of horizontal metallic tank

5. Caflisch RE (1998) Monte carlo and quasi-monte carlo methods. Acta Numerica 7. Cambridge University Press. pp 1–49
6. Fishman GS (1995) Monte carlo: concepts, algorithms, and applications. Springer, New York
7. Moré JM (1978) The levenberg-marquardt algorithm: implementation and theory. Lect Notes Math 630:105–116
8. Hall M, Frank E, Holmes G et al (2009) The WEKA data mining software: an update. ACM SIGKDD Explor Newsl 11(1):10–18
9. Leininger M (1994) Evaluation criteria and critique of qualitative research studies. Critical issues in qualitative research methods 95–115

Forecasting and Modeling Engine Air Flow of Transient Condition Based on Neural Network

Huan-chun Gong

Abstract Precise measurement of the air induction flow is the basis of accurate control of air fuel ratio for gasoline engines. During transient conditions, the serious fluctuation of air induction state and the lagging response of the airflow sensor seriously affect the accuracy of air fuel ratio control. The characteristic of air induction flow under transient condition is analyzed in this paper. A method of airflow forecast during transient conditions based on BP neural network is presented in this paper; the topological structure of BP neural network is also established. The model is trained and simulated by using experiment data in the acceleration and deceleration condition of gasoline engine. The results show that this method can accurately forecast the engine induction airflow under transient condition and eliminate the lagging characteristic of the airflow sensor.

Keywords Airflow forcast · Gasoline engine · Neural network · Transient condition

1 Introduction

In the last decade, increasing attention has been devoted to the study of transient phenomena related to the operation and the design of Engine Control Systems, in order to design suitable compensation strategies for mixture strength excursions and to assure good conversion levels for pollutant gases in catalytic converters. In order to accomplish for future exhaust regulations, an accurate control of air fuel ratio has to be guaranteed both in steady state and transient conditions, estimating correctly the air mass flow at the injector location and injecting fuel in the right amount and with the appropriate time dependence. Air fuel ratio is a key index affecting power

H. Gong (✉)
Mechanical and Electrical Engineering College, Yanching Institute of Technology, Yanjiao 065201, Beijing, China
e-mail: gonghuanchun@126.com

E. Qi (ed.), *Proceedings of the 6th International Asia Conference on Industrial Engineering and Management Innovation*,
DOI 10.2991/978-94-6239-145-1_55

performance and fuel economy of the gasoline engine, and the air induction accurate model is the foundation of accuracy air fuel ratio control. This goal can be reached making use of engine control strategies based on both online measurements and dynamic model estimations. Many engine dynamic models have been described in literature, characterized by different structures, goals and complexity as flows. Aquito put forward firstly average value air fuel ratio model of gasoline engine [1], but it is unfit for the condition of variable speed engine; CHO put forward a multipoint eject model targeted to control system developing [2], but the accuracy question of the model was not explained clearly; Hendricks put forward a three-state dynamic model of air fuel radio average value [3], which was an important milestone in average value model developing history and fit for every condition of operating modes, experiment show the stable error of this model is below 3 %, but the accuracy of the model is worse when engine worked during transient condition; Cesare put forward an air fuel ratio model based on feed forward BP neural network [4], where 16.5 h data during engine actual movement conditions is used as the samples, whose quantity is so huge that it influences strongly the study speed and precision of the network.

Automobile gasoline engine works mostly in the transient condition, working parameter of engine such as throttle angle, engine rotate speed and engine load during the practical operating process are variable. The variable of throttle angle and engine speed can lead to the changeable of induction airflow and make gasoline engine in the unstable process. Induction airflow is a key index affecting the accuracy of air fuel ratio control and the performance of exhaust emissions of gasoline engine, whose accurate measurement is the foundation of accuracy air fuel ratio control in the electrical control system. The precision measurement of induction airflow is difficult in the transient condition. Nowadays, the mean value model of induction airflow or airflow sensor is usually applied to measure induction airflow [5]. These two methods have some disadvantages as follows. When engine works during transient condition, the intake system exists in dynamic fill-empty phenomenon, air flow ratio is no longer equal of entering and leaving intake pipe, which results in the great tolerance of the mean value model. If induction airflow senor is used, the sensor also has a considerable time delay, which makes the real time data exist in delay and affects the accuracy of air fuel ratio control during transient condition. Airflow in the intake pipe is unsteady, which is affected by multi-disturbance under engine transient condition. Establishing a physical or mathematical model describing air flow in the intake pipe is hardly. Since, the variable of induction airflow is attracted by the variable of gasoline engine condition, so that the relationship between the variable of induction airflow and gasoline engine condition is absolutely existed. Moreover, this relationship is a kind of multi-variable, time variable nonlinearity, and the physical and mathematical model is also hardly obtained. In this paper, a neural network model between induction airflow and transient condition parameter of gasoline engine is established by using the neural network, whose ability of approximating randomly nonlinear function. This model is applied to predict induction airflow of transient condition after it is learned and trained in some typical transient conditions. The results show that the method will accurately forecast air induction flow during transient condition.

The history of the Mean Value Engine Model (MVEM) for Internal Combustion engine is interesting because it shows directly the importance of simulation in the physical theory of dynamic engine performance and practical engine control. Initially MVEM is constructed in order to understand the dynamics of engines for use in control system analysis and design. As it becomes clearly that all IC engines contain significant nonlinearities which dominate their dynamic behavior, the only way to attain accurate global control of engine emissions is to use the dynamic model in the control system itself. A Mean Value Engine Model (MVEM) consists of three important subsystems which describe the behavior of the fuel mass flow, the intake manifold filling dynamics and the acceleration of the crank shaft. These three subsystems will be described here in their most common forms. A detail derivation of the models and sub-models to be described can be found in the references [6–9].

1.1 Air Intake Subsystem

Exact modeling of the air intake involves fluid mechanics. The model is based on a representation whose distributed parameters cannot be used in real-time for control purposes: local and instantaneous velocity and pressure readings at each point of the manifold consist of a mean value and a pulsed value due to the acoustic stationary wave phenomenon, characterized by pipe geometry and the valve timing. A robust, representative model has been developed on the basis of mean values and leads to a representation using local parameters [10]. In this way, the influence of the acoustic phenomenon will only depend on engine speed (N) represented by a variable parameter in the model. Therefore, a classic air intake system described in average values can be split into three parts [11]:

(1) The throttle and the entire intake system: this section is approximately equivalent to a pressure loss which depends mainly on the throttle position.
(2) The intake manifold: this acts as a buffer volume between the throttle and the engine with uniform average pressure.
(3) The engine air pumping phenomenon.

For the air intake subsystem, input is represented by throttle position ϕ, the state variable is manifold pressure P_c, and output is the air intake ratio r_a defined as a normalized value where M_{CN} the air mass that can be absorbed by each cylinder in normal conditions and M is the actual air mass entering in the cylinder during admission.

$$r_a = \frac{M}{M_{CN}} \tag{1}$$

1.2 Pressure Loss across the Throttle

Pressure loss across the throttle is fairly difficult to characterize. It is caused by restriction due to throttle position and also to the eddy phenomenon resulting from throttle profile. In an initial approach, Bernouilli's equation [12] is applied to calculate frictionless flow of an incompressible fluid through a channel between two areas at different pressures. However, the compressibility of air should be taken into account. The mass air flow through the intake section is determined by

If $\frac{P_c}{P_a} > \gamma_{critical}$ then $D_\phi = \frac{C_D A_T P_a}{\sqrt{RT_0}} \sqrt{\frac{2\gamma}{\gamma-1}\left[\left(\frac{P_c}{P_a}\right)^{\frac{2}{\gamma}} - \left(\frac{P_c}{P_a}\right)^{\frac{\gamma+1}{\gamma}}\right]}$

If $\frac{P_c}{P_a} < \gamma_{critical}$ then $D_\phi = \frac{C_D A_T P_a}{\sqrt{RT_0}} \sqrt{\frac{2\gamma}{\gamma-1}\left[(\gamma_{critical})^{\frac{2}{\gamma}} - (\gamma_{critical})^{\frac{\gamma+1}{\gamma}}\right]}$ with $\gamma_{critical} = \left(\frac{2}{\gamma+1}\right)^{\frac{\gamma}{\gamma-1}}$

where C_D is the discharge coefficient, A_T is the area of the section of the flow restriction, P_a and T_0 are the upstream pressure and temperature (atmospheric conditions), γ is the specific heat ratio, R is the gas constant of the mixture, and P_c is the pressure at the restriction (equal to static pressure immediately downstream of the flow restriction and considered to be mean manifold pressure).

1.3 Engine Air Pumping Phenomenon

A simple relation between manifold pressure, mass air flow and engine speed is sought and several hypotheses are made:

(1) Cylinder pressure is uniform and constant during supply and is equal to manifold pressure.
(2) Cylinder temperature is uniform and constant during supply (T_m).
(3) The inlet valve opens at Top Dead Center and closes at Bottom Dead Center.
(4) A residual volume of burnt gas in the cylinder just before valve opening at pressure P_0 and temperature T_m.

Theoretically, the average air flow entering the engine D_m is given by

$$D_m = \frac{n \cdot \rho_m \cdot N}{120 \cdot R \cdot T_m}(V_{BDC} \cdot P_c - V_{TDC} \cdot P_0) \tag{2}$$

where R is the universal gas constant, n is the number of cylinders, VBDC is the cylinder volume at Bottom Dead Center, VTDC is cylinder volume at Top Dead Center, and ρ_m is air density.

Theoretical manifold pressure is determined by

$$P_c = \frac{A_{th}}{N} \cdot D_m + P_{r_{th}} \tag{3}$$

Experimentally, $R_{m_{th}}$ has been identified by linear fitting as

$$R_m(N) = \frac{A}{N} \tag{4}$$

The relation between P_r and N is not a simple one owing to residual burnt gas and acoustic modes. The mean value model takes the acoustic mode into consideration with the term P_r; a high P_r value means an inefficient supply to the cylinder, whereas a low P_r value means good acoustic phenomena and an efficient supply.

The final relation is determined by

$$P_c = \frac{A}{N} D_m + P_r(N) \tag{5}$$

1.4 Dynamic Phenomenon in the Manifold

This transient phenomenon appears in the buffer volume

$$\frac{dP_c}{dt} = \frac{1}{C}(D_\phi - D_m) \text{ with } C = \frac{M_m \cdot V_a}{R \cdot T_a} \tag{6}$$

C is manifold capacitance, M_m is the molar mass, V_a is the manifold volume, and T_a is the air temperature in the manifold. The fluctuations of C as a function of the temperature have been neglected. Intake depends on throttle opening and engine speed (N). The impact of the other parameters such as inlet air temperature and engine temperature has been neglected. The MVEM has been shown to have a steady state accuracy of about 2 % over the entire operating range of the engine [13]. Other more accurate models have been found, but apply to only a small operating region [14], thus the MVEM is extremely useful for validation of control strategies using simulation. When engine works during transient condition, because intake system exist dynamic fill-empty phenomenon, air flow ratio is no longer equal of entering and leaving intake pipe, which results in the great error of the mean value model. Moreover, some acoustic and non-adiabatic effects are neglected in the mean value model. So, a stochastic model describing the non-linear effects due to random errors in sensors and actuators has not been included in the mean value model.

2 Modeling the Path of Air

The approach used to describe the intake manifold with throttle and inlet valves is the so-called filling and emptying method. Acoustic and non-adiabatic effects are neglected. Thus the interesting state manifold air pressure p_m is determined by the thermodynamic state equation from the states air-mass and air-temperature inside the intake manifold in [15]. Filling and emptying method for modelling the intake manifold is shown in Fig. 1.

$$p_m V_m = m_{am} T_m \tag{7}$$

where V_m is the intake manifold volume, m_{am} is the mass of air in intake manifold, R is the gas constant, T_m is the manifold air temperature. The time derivative of air temperature is negligible in comparison to the time derivative of air-mass; this yields

$$\frac{d}{dt} p_m = \frac{d}{dt} m_{am} \frac{RT_m}{V_m} \tag{8}$$

The time derivative of air-mass $\frac{d}{dt} m_{am}$ can be expressed as the difference between the mass air flow rates to and from the manifold.

$$\frac{d}{dt} m_{am} = \dot{m}_{at} - \dot{m}_{ap} \tag{9}$$

where $\dot{m}_{at}$ is the mass air flow rate at throttle, $\dot{m}_{av}$ is the mass air flow rate into cylinders. By these equations the dynamic storing behaviour of the intake manifold is modelled close to reality.

The mass air flow rate through the throttle into the manifold due to the pressure difference between environment and manifold conditions is expressed as the flow of a compressible medium through a converging nozzle.

Fig. 1 Air intake path model

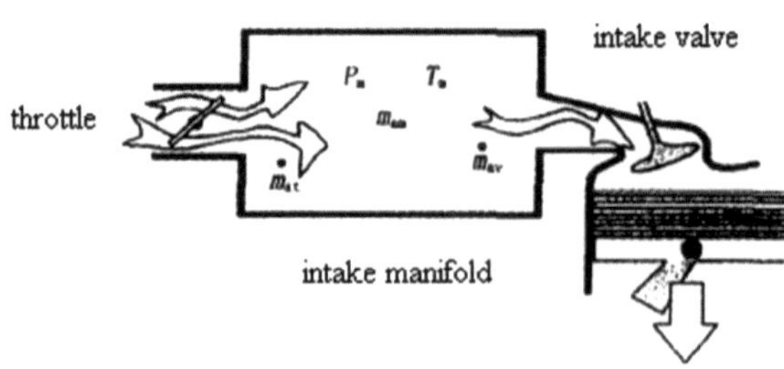

$$\dot{m}_{at} = C_d A_x P_0 \sqrt{\frac{2}{RT_0}} \beta(p_r) \quad (10)$$

where C_d is the throttle discharge coefficient, A_x is the open throttle area, $\beta(p_r)$ is the pressure of induction pipe, P_0 is the ambient pressure, $\beta(p_r)$ is described by equation as fellows.

If $2P_0 \geq P_m > P_0/2$ then $\beta(p_r) = 2\sqrt{P_m P_0 - P_m^2} |P_0 - P_m| / P_0$

If $P_m \leq P_0/2$ or $P_m > 2P_0$ then $\beta(p_r) = |P_0 - P_m|$

The manifold pressure p_m is correlation to engine rotate speed and throttle angle. So, the relationship between the manifold pressure and throttle angle, engine rotate speed is nonlinearity as fellows.

$$P_m = f(\alpha_{th}, n) \quad (11)$$

where f is the nonlinear function, n is the engine rotate speed and α_{th} is the throttle angle.

In Eq. (8), the nonlinear effects of throttle angle α_{th} on nozzle diameter and flow resistance are expressed by a nonlinear function in Eq. (10) including the geometric dependency $A_x = A_x(\alpha_{th})$ of the nozzle area on the throttle angle α_{th}.

$$C_d A_x = f(\alpha_{th}) \quad (12)$$

This nonlinear dependency $C_d(\alpha_{th})$ is described as an isolated nonlinearity. To calculate the mass air flow rate $\dot{m}_{av}$ out of the intake manifold into the cylinders, which are modelled as a nonlinear pump.

$$\dot{m}_{av} = \frac{\eta_{vol} V_D n}{2} \frac{P_m}{RT_m} \quad (13)$$

where η_{vol} describes the nonlinear breathing characteristic due to acoustic and fluid dynamics phenomena. This volumetric efficiency is considered an isolated nonlinearity.

From Eq. (5) to Eq. (13), induction airflow is correlation to engine rotate speed, intake manifold pressure and throttle angle. But the parameter of intake manifold pressure is affected by the parameter of air pressure, specific heat ratio, volumetric efficiency, engine rotate speed and throttle angle and so on. The induction airflow is described in Eq. (14).

$$\dot{m}_{av} = f(\alpha_{th}, P_m, n) \quad (14)$$

Combined with Eq. (9), the induction airflow is decided by the parameter of throttle angle and engine rotate speed. The relationship between these affected parameters and intake manifold pressure is the complex nonlinearity, which is hardly described by the accuracy of mathematical and physical model. During transient condition, the variable of intake manifold pressure is affected by the variable of engine rotate speed and throttle angle. So the main factors of induction airflow variable are the parameter of engine rotate speed and throttle angle.

3 Structure and Algorithm of BP Neural Network

During transient condition, the relationship between throttle angle and rotate speed is complexity and nonlinearity. Neural network is an abstract mathematic model, which is connected with the multi-neuron topology structure. It has high ability of nonlinearity mapping, strong parallel calculating, adaptive learning and gentrification performance in [6]. The multi-layer back propagation neural networks have been successfully employed to model the nonlinear dynamic systems, because it has the ability to represent arbitrary nonlinear input/output relationship. Sigmoid function is usually selected as the neuron transfer function, which can accomplish the nonlinear mapping from input to output. The method of adjusting weight and bias of network is adapted to back propagation algorithm. The network is trained by using input and output sample. Compared with the output data and the target data each train, if the desired error is not obtained, the weight and bias will be adjusted according to the desired algorithm and information to neuron layer. BP neural network has better gentrification performance and nonlinearity mapping after the train is over so that the arbitrary sample input network can generate the ideal output data.

3.1 The Induction Airflow Prediction Topology Structure of BP Neural Network during Transient Condition

The number of input layer of induction airflow prediction BP neural network structure is composed of six nodes. Where is the kth throttle angle $\alpha_{th}(k)$, rotate speed $n(k)$, induction airflow $m_{av}(k)$ and variable rate of throttle angle $\alpha'_{th}(k)$, rotate speed $n'(k)$ and induction airflow $m'_{av}(k)$ by practical test. The main factor of affecting induction airflow is the engine condition parameter of throttle angle and rotate speed, and the variable rate of throttle angle and rotate speed is used as input node in order to embody adequately the dynamic characteristic of gasoline engine transient condition, which can compensate for the drawback of mean value model and increase the precision of network. The number of output node is one, where is the prediction induction airflow $\overline{m}_{av}(k)$ of BP neural network.

It is difficult for determine the hidden neurons. It is usually regarded as the hidden neurons are too fewer, the accuracy and gentrification performance of network is more poor. If the hidden neurons are too much, the time of learning process is too long and the performance of tolerance is not the least. So the experience equation is often selected as deciding the hidden layer n_1 in [11] as fellows.

$$n_1 = \sqrt{n+m} + a$$

where n is input layer, m being the output layer, a is the constant from 1 to 10. The number of hidden layer is ten by trial calculating in this paper.

The training sample of network is not convergence, learning time is too long and network robustness is much poorer when the single hidden layer of BP neural network is applied to trail calculating in this paper. The main reason is that multi-forward network is a general approximation machine, the hidden layer is not enough or the relationship between input and desired output is not obvious. It is enough for choose one hidden layer in theory but its node is too larger to use in practice. So, the hidden layer of network is selected for 2 in this paper. The topology of back propagation network of induction airflow model is 6-10-10-1. The structure of back propagation of induction airflow is shown in Fig. 2.

3.2 *Learning Algorithm of BP Neural Network*

If i $(i = 1, 2, 3, \ldots, n)$ is the input node, x_i is the input vector is transmitted to the hidden layer by control variable.

If j $(j = 1, 2, 3, \ldots p)$ is the hidden node, I_j is the input vector and O_j being the output vector as follows.

$$I_j = \sum_{i=1}^{n} \omega_{ji} x_i + \theta_j$$

$$O_j = f(I_j) = 1/[1 + \exp(-I_j)]$$

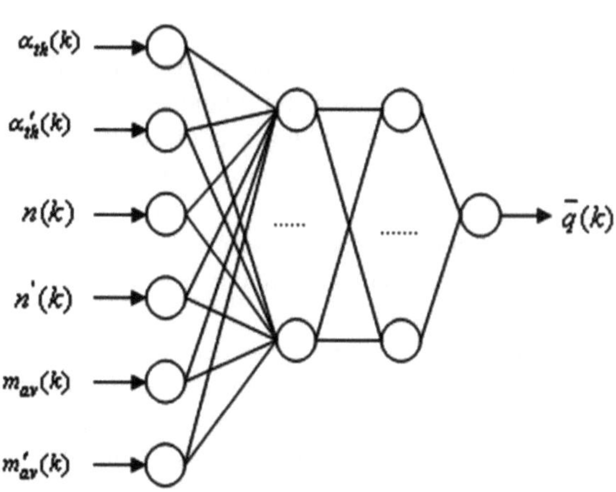

Fig. 2 Induction airflow prediction of BP neural network

In the above formula, ω_{ji} are the hidden node j and the input node i the weight, θ_j is the hidden node j bias, f being the sigmoid function as follows.

$$f(x) = 1/[1 + \exp(-x)]$$

Where k $(k = 1, 2, \ldots m)$ is the output node, I_k is the input vector and y_k is the output vector as follows.

$$I_k = \sum_{j=1}^{p} \omega_{kj} O_j + \theta_k$$
$$y_k = f(I_k) = 1/[1 + \exp(-I_k)]$$

In the above formula, ω_{kj} are the output node k and the hidden node j the weight, θ_k is the output k bias.

Where $(x_{p1}, x_{p2}, \ldots, x_{pn})$ is given to training sample, p is the number of samples $(p = 1, 2, \ldots P)$. The error of network calculating and the target output of sample are demonstrated as follows.

$$E = \frac{1}{p} \sum_{p=1}^{p} E_p$$

In equation $E_p = \frac{1}{2} \sum_{l=1}^{l} (t_{pl} - y_{pl})^2$, p is the sample, t_{pl} is the pth sample and the lth target result of output layer, y_{pl} is the pth sample network calculating result of output layer.

4 Induction Airflow Prediction Model of Information Fusion Based on Transient Condition Parameter

Taking HL495Q gasoline engine as experimental device, whose capacity is 2.84 L, and the calibration power and rotational speed is 75KW/3800r/min, the compression ratio ε is 7.8, the idling rotational speed is 750r/min, the dynamometer is CW260 electricity turbulent flow in [9]. To guarantee real and valid sampling signal, sampling time is 0.01 s.

The respondence time of thermal flow sensor is about 50 ms and has 5 sample time delays when sampling time is 0.01 s. Therefore, neural network produce the fifth and sixth sample data.

Experiment 1 is the engine deceleration experiment, where 5 × 60 groups experimental data were gained by closing the throttle valve at different speed (in 1 or 1.5 or 2 or 2.5 or 3 s) from 60 % throttle degree to idling position when length of sample time is 4 s, which is named as condition 1, 2, 3, 4 and 5. The output signal of throttle position sensor, rotate speed sensor and induction airflow sensor is also gained. Experiment 2 is the engine acceleration experiment, where 5 × 60 groups experiment data were also measured by opening the throttle valve at different speed (in 1 or 1.5 or 2 or 2.5 or 3 s) from initial position to 60 % throttle degree when length of sample time is also 4 s, which is classified as condition 1, 2, 3, 4 and 5. The output signal of throttle position sensor, rotate speed sensor and induction airflow sensor is gained at the same time. The testing data of throttle position sensor

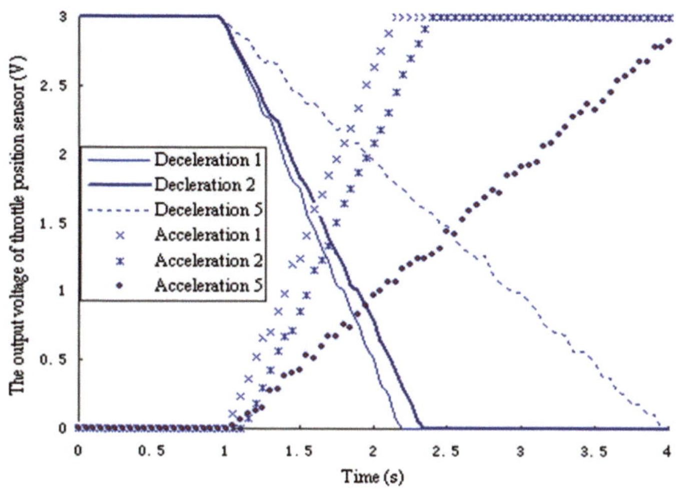

Fig. 3 Experiment data of throttle position sensor in acceleration and deceleration condition

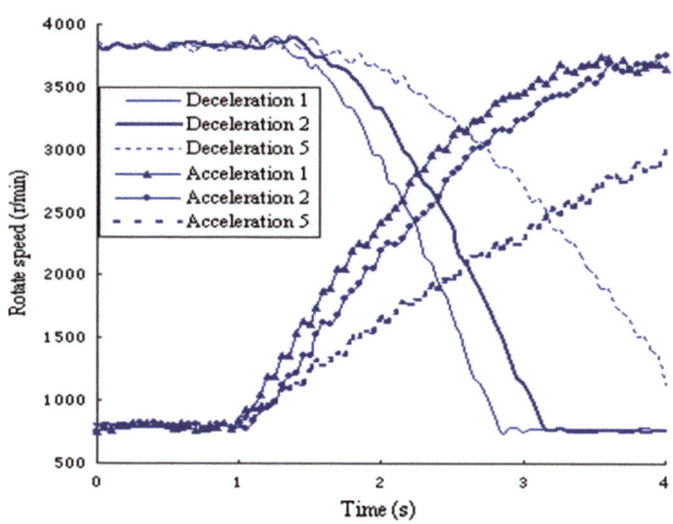

Fig. 4 Experiment data of engine rotate speed

is shown in Fig. 3. The testing data of engine rotate speed is shown in Fig. 4. The voltage experiment value of induction airflow sensor is shown in Fig. 5.

According to the BP neural network section 2.1, the testing data of induction airflow in acceleration and deceleration condition is trained. Making 10 × 60 groups sample data normalized and choosing 6 × 60 groups experiment data of acceleration and deceleration condition in 1, 1.5 and 3 s as training sample, then taking the other 4 × 60 experiment data of acceleration and deceleration condition in 2 and 2.5 s as

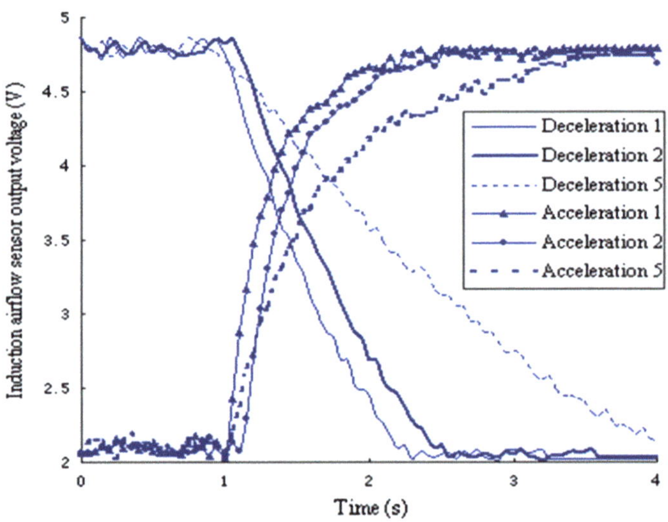

Fig. 5 Experiment output voltage of induction airflow sensor

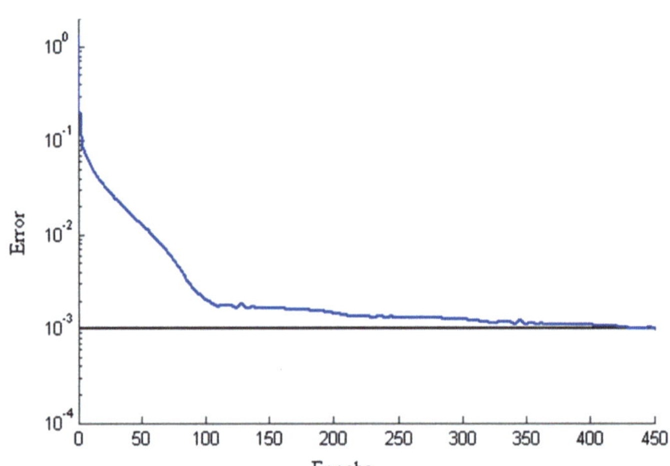

Fig. 6 Variable curve of training error

examine sample, deciding the learning rate of network as 0.05, choosing the training stride length of neural network as 1000 steps, taking the target error as 0.001. Tangent sigmoid function is used as the hidden transfer function and the linear function is selected as the output layer function. The BP neural network is established based on MATLAB software toolbox. The variable curve of training error is shown in Fig. 6 and the training error is 0.000995046. It meets the desired target.

5 Induction Airflow Prediction Result and Analysis During Transient Condition of Gasoline Engine

The induction airflow of acceleration (1, 2, 5) experiment values, the prediction values of back propagation neural network model in deceleration and the calculating value of mean value model in [8] are shown in Fig. 7. The induction airflow of deceleration (1, 2, 5) actual values, the prediction values of back propagation neural network model in acceleration and the calculating value of mean value model in [8] are shown in Fig. 8.

According to the Figs. 7 and 8, gasoline engine works mostly in the transient condition, which is the condition of parameter variable such as throttle angle, engine rotate speed and engine load during the practical operating process. The induction airflow prediction model based on BP neural network of information

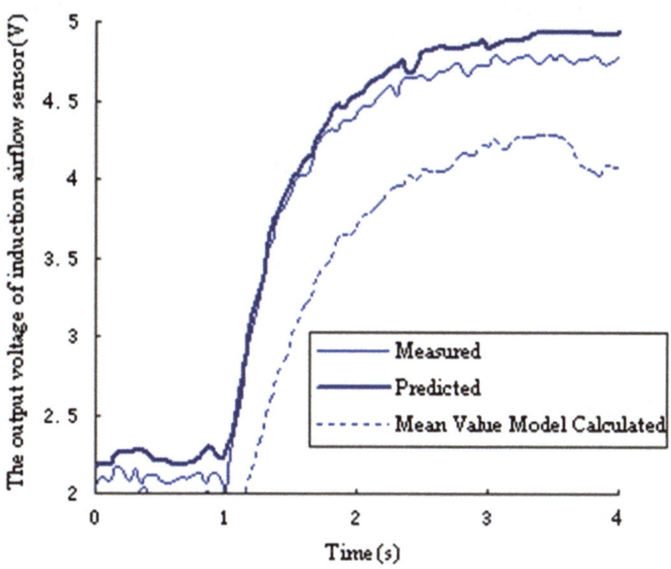

Fig. 7 Comparison with induction airflow predicted, measured and mean value model calculated in acceleration (1, 2, 5)

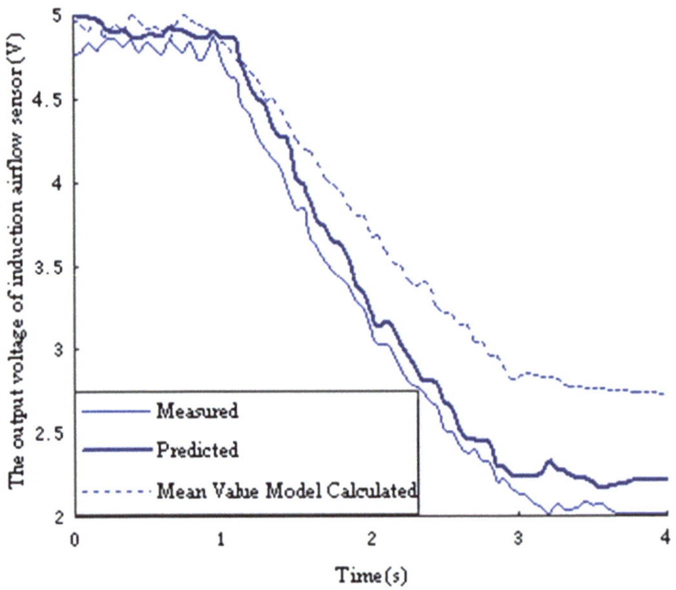

Fig. 8 Comparison with induction airflow predicted, measured and mean value model calculated in deceleration (1, 2, 5)

fusion can accurately forecast the engine induction airflow under transient condition and eliminate the lagging characteristic of the airflow sensor. The respondence time of thermal flow sensor is about 50 ms and has 5 sample time delays when sampling time is 0.01 s. Therefore, neural network produce the fifth and sixth sample data. From Figs. 7 and 8, the prediction value of neural network is lager than the testing data of induction airflow sensor. But the mean value modelling of engine focus on the entire characteristic of the gasoline engine and makes the work situation of engine homogeneous. The dynamic characteristic of engine is hardly described during transient condition. Seen from Figs. 7 and 8 the great error of induction airflow is produced by using the mean value model and the induction airflow prediction model of neural network in the transient condition is superior to the mean value induction airflow model in this paper.

6 Conclusion

(1) The method of induction airflow prediction based on information fusion is presented by analyzing all kinds of affecting the induction airflow of gasoline engine during transient conditions. Some of dynamic parameters are extracted by this method and the induction airflow prediction model of neural network is also established in this paper. The model is trained and simulated by using

experiment data in the acceleration and deceleration condition of gasoline engine. The results show that this method can accurately forecast the engine induction airflow under transient condition.

(2) Because some of dynamic parameters such as engine rotate speed and its variable rate, throttle angle and its variable rate are used in this method, the method can reflect adequately the variable condition of induction airflow during transient condition and eliminate the lagging characteristic of the air-flow sensor.

Acknowledgments The authors gratefully thanks to science and technology bureau of Langfang city science and technology support program funded projects. (Project number: 2011011041).

References

1. Aquino CF (1981) Transient A/F control characteristics of the 5 liter central fuel injection engine. SAE paper 810494
2. Hendricks E (1990) Mean value modelling of spark ignition engines. SAE paper, 960616
3. Cesare A (2003) A neural-network based control solution to air fuel ratio for automotive fuel injection system. In: IEEE transactions on system man and cybernetics-Part C, vol 33(2)
4. Chevalier A, Vigild CW, Hendricks E (2000) Predicting the port air mass flow of SI engines in A/ratio control applications. SAE paper, 2000-01-0260
5. Shayler PJ, Goodman MS (1996) Transient air/fuel ratio control of an S. I. engine using neural networks. SAE paper 960326
6. Hendricks E, Sorenson SC (1991) SI engine controls and mean value engine modelling. SAE paper No.930859
7. Hendricks E, Sorenson SC (1993) Nonlinear transient fuel film compensation. SAE paper No. 930767
8. Gambino M, Pianese C, Rizzo G (1984) Experimental and computation analysis of a dynamic model for control strategy optimization in a spark ignition engine. In: Proceedings of the American control conference Baltimore, Maryland, June 1984, pp 2374–2379
9. Aquino CF (1981) Transient A/F control characteristics of the 5 liter central fuel injection engine. SAE paper 810494
10. Mamie C (1997) Technical report on air fuel ratio control using neural networks, University of Melbourne
11. Rose D, Lodommatos N (1994) In-cylinder mixture excursions in a port injected engine during fast throttle opening. SAE paper No 940382
12. Cho D, Hedrick JK (1988) A nonlinear controller design method for fuel-injected automotive engines. J Eng Gas Turbines Power 110:313–320
13. Pianese C, Rkzo G (1992) Fluid-dynamic and thermal effects on transient performance and emissions of a spark ignition engine with electronic control. In: Proceedings of 25th ISATA symposium, Florence, 1–5 June 1992
14. Rizzo G, Pianese C (1991) A stochastic approach for the optimization of open-loop engine control systems. Ann Oper Res 31:545–568
15. Rizzo G, Pianese C (1989) Optimization of spark ignition engines with stochastic effects in sensors and actuators. In: ASME symposium on engine design, operation and control using computer systems, Dearbom, MI, ICE, vol 9, 15–18 Oct 1989, pp 139–150

Research of Speed Control System for DC Motor Based on Fuzzy PID

Hu-min Yan, Yun-tao Lei, Dang-feng Pang and Xiao Han

Abstract In this paper, based on the application of the fuzzy self-tuning PID controller in the Siemens S7-300 PLC, we design the hardware and software of dc motor speed control system, and studies the promotion contributed by the fuzzy PID controller. Combined with the MATLAB simulation and processing test, we concluded that fuzzy self-tuning PID controller not only remains the good quality of the steady high precision, but also improve quickness and robustness of dc speed regulating system.

Keywords Fuzzy control · PID · Speed control system for DC motor · WinCC

1 Introduction

Dc motor is widely used in automatic control system, but in the closed loop speed regulation system, conventional PID control is mostly applied. If the motor load contains elastic or nonlinear factors, that would make the fixed parameters PID regulator control system poor robustness and stability in the application.

The Fuzzy self-tuning PID control the quality of high steady-state precision and strong robustness of conventional PID. It also can adjust the control parameter with the change of the targeted object parameter, realizing automatic adjustment and meeting the different control requirements. This paper, by using the simulation software, verifies the qualities of the fuzzy self-tuning PID algorithm control of the dc motor speed, and realizes the application of the fuzzy self-tuning PID algorithm

---

H. Yan (✉) · Y. Lei · D. Pang
Department of Engineering Training, Tianjin University of Technology and Education, Tianjin, China
e-mail: tjdangfeng@163.com

X. Han
Department of Psychology, Peking University, Beijing, China

E. Qi (ed.), *Proceedings of the 6th International Asia Conference on Industrial Engineering and Management Innovation*,
DOI 10.2991/978-94-6239-145-1_56

in Simon S7-300 PLC, industrial control, 6RA70 digital dc speed controller and dc motor speed control system.

2 Mathematical Model

2.1 Mathematical Model of Dc Motor

In the controlling process of dc motor, the paper takes the armature voltage as input and the motor speed as output to conduct research. Suppose that the compensation of the dc motor is good. Take no consideration of the dc motor armature reaction and other factors, and set motor excitation mode [1–3], the motor mathematical model is expressed as

$$U_d = L_d \frac{dI_d}{dt} + I_d R_d + Ce\Omega \tag{1}$$

Ud represents the armature voltage; Ld the armature loop inductance; Id as the armature current; Rd the armature loop total resistance; Ce the potential coefficient; Ω the angular velocity; J the output moment of inertia; M the electromagnetic torque; Mf the load torque; Td the armature circuit of electromagnetic time constant; Tm the electromechanical time constant, according to the law of rigid body rotation, the dc motor equation is

$$T_d T_m \frac{d^2 n}{dt^2} + T_m \frac{dn}{dt} + n = \frac{1}{C_e} U_d \tag{2}$$

$$T_m = \frac{JR_d}{C_e C_m} = \frac{GD^2 R_d}{375 C_e C_m} \tag{3}$$

GD2 represents the electric the conversion to the rotational inertia of the motor shaft of the movement part in the power drive system. n represents the motor speed, and the transfer function between voltage and speed can be obtained as

$$WD(S) = \frac{1/Ce}{T_m T_d S^2 + T_m S + 1} \tag{4}$$

The dc motor can be approximately thought to be a two inertial link series composed of two order systems.

2.2 *Mathematical Model of Fuzzy Self-tuning PID Controller*

The key lays in the Fuzzy self-tuning PID controller is setting PID KP, KI, KD three parameters, whose control result have a big effect on the system. Based on the relevant experience, combining with the system control error E and error change rate of EC, by constantly changes in practice, it establishes suitable control rules that could satisfy the requirement of the controller, and follow some basic parameter setting [4–7].

(1) *Calculation of Ke, Kec, Ku*: Suppose that in the actual control system, the system error E continuous span is from eL to eH; the error rate EC from ecL to ecH, control output from uL to uH; sp Ke, Kec, Ku can be expressed as

$$Ke = \frac{2m}{e_H - e_L} \tag{5}$$

$$Kec = \frac{2n}{ec_H - ec_L} \tag{6}$$

$$Ku = \frac{u_H - u_L}{2l} \tag{7}$$

(2) *Calculation of E, EC*: In the practical control system, the control variable is changing. The fuzzy control process is the discrete finite integral domain [8]. Convert generally done by quantitative factors, the two can be obtained by the following formula, the type < > symbol represents the computing results are integer arithmetic.

$$E = <k_e \cdot (e - \frac{e_H + e_L}{2}) > \tag{8}$$

$$E_C = <k_{ec} \cdot (ec - \frac{ec_H + ec_L}{2}) > \tag{9}$$

(3) *Definition the membership function of language variable*: The determination of the membership function in fuzzy control identifies the relation of the fuzzy language variables and the integer variables. the output change with the input more intense, sensitivity high control system; Curve shape is smooth, the input output change caused by the change is not very intense, control characteristic is also gentle, has good system stability, comprehensive membership function curve resolution and system control sensitivity factors, system USES trapezoidal membership function [9], membership function as Fig. 1.

(4) *The general rules of the fuzzy reasoning uses the query expression*: the output voltage U as

Fig. 1 Ladder type membership function

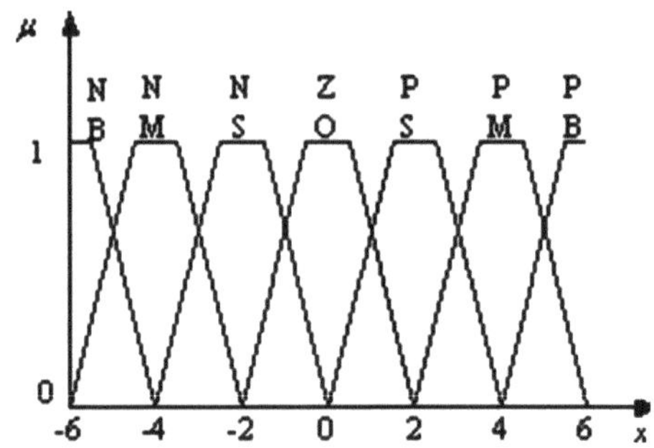

$$u(k) = ku \cdot f[Ke, e(k), Kec, ec(k)] \tag{10}$$

Scale factor is Ku, and the quantitative factors Ke and Kec control rules and membership functions are relevant. Considering the delay of the control of PLC, so the membership function for (3,2,1,0,1,2,3) corresponding to the system deviation rate control deviation E and EC in {NB, NM, NS, ZO, PS, PM, PB} 7 grade fuzzy arithmetic and reasoning, considering the results of the fuzzy arithmetic is between 3 to 3, the actual dc motor speed of 1400 r/min, so scale factor take 1000, 12 quantitative factor in practical applications [10–12].

3 System Simulation

The structure of the fuzzy self-tuning PID controller is shown in Fig. 2. The simulation results of the dc motor speed shows that fuzzy self-tuning PID control the rotating speed of dc motor has high robustness and adaptive ability, when the

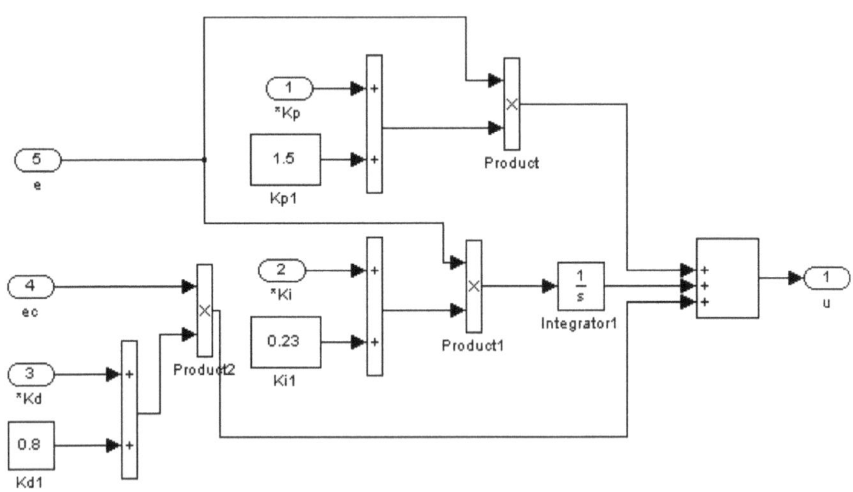

Fig. 2 The controller structure of fuzzy self-tuning PID

Fig. 3 The fuzzy self-tuning PID control simulation results the boot process

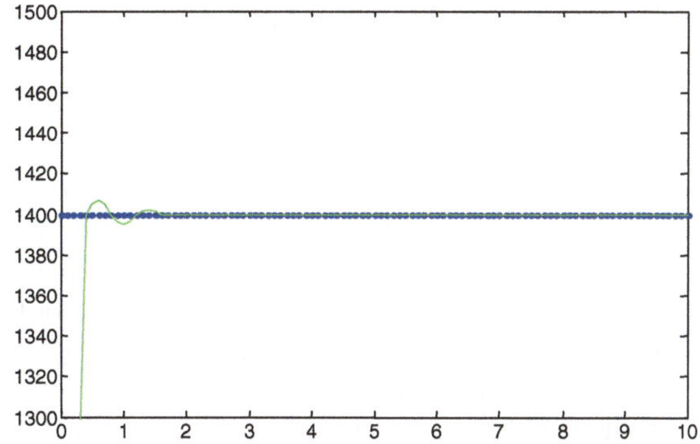

system parameter changes, the PID parameters setting is not change after the optimal PID parameters. Therefore, in this paper, the study on fuzzy self-tuning PID control dc motor speed has a strong feasibility and necessity simulation results as Fig. 3.

4 Hardware Platform Design

4.1 Hardware Wiring Diagram

The hardware connection structure, as is shown in Fig. 4, equipped by: industrial computer, Siemens S7-300 PLC, 6 RA70 dc drive system, dc motor, rotary encoder speed measuring system, PROFIBUS communication bus.

The Siemens S7-300 PLC is the control core of system, mainly completing the speed control. The control of the dc motor double applies the closed loop control, with the loop controlling the speed and the inner ring controlling the armature current. Loop control is realized in PLC system, inner control in 6 ra70 by setting the parameters. Ethernet front-end ports for communication lower place machine and super-ordination machine, DP communication port for the communication machine and 6RA70 dc drive system [13].

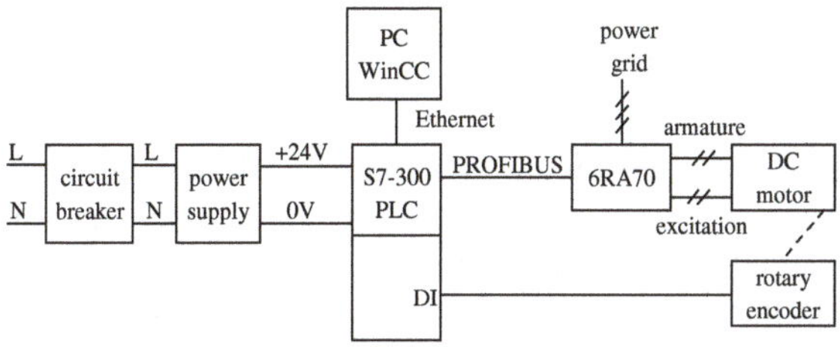

Fig. 4 The system hardware wiring diagram

4.2 Design of Monitor Screen

WinCC main picture on the left side of the set to the button area, the above is set to the title area, through the picture in the picture shown in the main picture process picture, alarm, historical curve, report and picture. Design in the process of the monitoring picture, in order to prevent the picture after there was a flicker and exchange data with the database, in essence, the picture in the main picture, to show and hide the picture by coordinate translation, coordinate translation through script configuration in the event of the buttons in the C.

```
#include "apdefap.h"
void OnLButtonDown(char* lpszPictureName, char* lpszObjectName, char* lpszPropertyName, UINT nFlags, int x, int y)
{
#pragma option(mbcs)
SetLeft("main.pdl","start",190);
SetLeft("main.pdl","alarm",1366);
SetLeft("main.pdl","onlinetrend",1366);
SetLeft("main.pdl","report",1366);
}
```

5 Network Control System

5.1 The System Network

The control system adopts industrial Ethernet and PROFIBUS-DP network, applied to the implementation of the management and monitoring in processing control. Ethernet used in the HMI is to communicate [14, 15]. The topological adopts structure of star. HMI is the man-machine interface, mainly monitoring and controlling, used to monitor the motor parameters, the motor speed of a given curve, alarm, display and historical records and reporting features, dc speed regulating device through the Driver Monitor software parameter settings.

5.2 Control Algorithm

Dc motor speed of fuzzy self-tuning PID controller design, the first thing you need to do is the statistic collection of the actual speed of dc motor, and then calculate the speed deviation and the deviation change rate, and then conduct the fuzzification on the deviation and the deviation rate, fuzzy reasoning, fuzzy and PID arithmetic.

6 Application Results

Compared to the traditional PID control and the fuzzy self-tuning PID, the inanition process has a relatively shorter adjusting time, and the system overshoot volume is smaller. Launching the test system with the rated load [16], by the analysis of simulation results and actual operation result is the same, and significantly improve the quickness and robustness of speed control system, the actual operation results and simulation results are the same.

7 Conclusion

This paper introduce the fuzzy control theory into the Dc speed regulation, combining the PLC and dc speed regulating device, which play their respective advantages of the traditional PID control and fuzzy control, The combining control of the fuzzy control and the PID control is finally realized in the Siemens S7-300 series PLC by the SCL language, making full use of the advantages of PID control and fuzzy control, improving the quickness and robustness of speed control system.

Acknowledgments The work in this paper was supported by "One of Tianjin quality courses, integrated skills training of modern control technology".

References

1. Zhuang M, Atherton DP (1993) Automatic tuning of optimum PID controllers. Proc IEE Pt D 140:216–224
2. Ang KH, Chong G, li Y (2005) PID control system analysis design and technology. IEEE Trans Syst Technol 13(4):559–576
3. Miller RM, Shah SL, Wood RK, Kwok EK (1999) Predictive PID. ISA Trans 38:11–23
4. Marlin TE (1995) Process control. McGraw-Hill International Educations, Singapore, pp 59–66
5. Hersh MA, Johnson MA (1997) A study of advanced control systems in the work place. Control Eng Pract 5(6):771–778
6. Graney Glenn (1994) Batching with SP88-compliant software and PLCs. Instrum Control Syst 34(3):43–47
7. Kyungmoon Nho (1998) Automatic landing system design using fuzzy logic. AIAA-98-4484, Berkeley Univ of California
8. Silva GJ, Datta A, Bhattacharyya SP (2003) On the stability and controller robustness of Some popular PID tuning rules. IEEE Trans Autom Control 48(9):1638–1641
9. Yongchun X, Hongxin W, Zhenduo L (1996) The all-coefficients adaptive control method and its application in spacecraft attitude control. Space Technol 113–205
10. Wang QG, Zou B, Lee TH, Bi Q (1997) Autotuning of multivariable PID controllers from decentralized relay feedback. Automatica 33(3):319–330
11. Astrom KJ, Hagglund T, Wallenborg A (1993) Automatic tuning of digital controllers with applications to HV AC plants. Automatica 29(5):1333–1343

12. Kaya A, Scheib TJ (1998) Tuning of PID controls of different structures. Control Eng 35 (7):62–65
13. Nishikawa Y, Sannomiya N, Ohta T, Tanaka HA (1984) Method for auto-tuning of PID control parameters. Automation 20:321–332
14. Dorbusch R (1993) Policy making in the open economy. Oxford University Press Inc 149
15. Ang KH, Chong G, Li Y (2005) Member PID control system analysis, design and technology. IEEE Trans Control Syst Technol 13(4):559–576
16. Maeda M, Assilian SA (1992) self-tuning fuzzy controller. Fuzzy Sets Syst 51(1):29–40

A Graphical Formalized Deduction for Integrated Production and Logistic Service Flow Monitoring in Social Manufacturing

Kai Ding and Ping-yu Jiang

Abstract As the outsourced manufacturing services flourish in social manufacturing, the production organizing mode has shifted to a horizontal one where manufacturing service providers collaborate to finish product development. So the integrated production and logistic service flow monitoring is urgently needed for product developers to realize closed-loop quality traceability, process efficiency enhancement, real-time inventory feedback and up-to-date logistic tracking. In this paper, a graphical formalized deduction method called RFID driven state block model is built to depict the integrated production and logistic service flow monitoring problem. The minimum unit of state block is a monitoring node and it contains a series of states, which are changing by time. Then two kinds of monitoring nodes are further illustrated. Based on that, the state changing event flow is built to depict the event inducing the state changing. Finally, a simulation example is discussed to verify the feasibility of the proposed model.

Keywords Integrated production and logistics · Monitoring/tracking · RFID · State block · Social manufacturing

1 Introduction

The force of globalization has served to connect companies from all across the globe instantaneously, bringing with it game-changing opportunities to share knowledge and expertise to benefit in a collective manner (sometimes called share-to-gain) [1]. Therefore, many manufacturing service-oriented companies with socialized manufacturing resources (SMRs) and techniques join up to provide all kinds of manufacturing services, as the amount of internal expertise they hold is

K. Ding · P. Jiang (✉)
State Key Laboratory for Manufacturing Systems Engineering,
Xi'an Jiaotong University, Xi'an, China
e-mail: pjiang@mail.xjtu.edu.cn

E. Qi (ed.), *Proceedings of the 6th International Asia Conference on Industrial Engineering and Management Innovation*,
DOI 10.2991/978-94-6239-145-1_57

dwarfed by that held by the global mass of companies that connected through globalization. Furthermore, with the advent of the Internet, the production organizing mode of a product is undergoing a power-shift from once mighty hierarchical business model to a horizontal business model [2], which integrates manufacturing service demanders (SDs), manufacturing service providers (SPs), public warehouse providers (PWs) and logistic providers (LPs) together and supports mass collaboration with them to win agility, creativity, and connectivity that companies require to remain competitive in today's environment. Under that, many product manufacturers are evolving into onefold design companies just like Apple Inc., which outsources most of (even all) the production tasks to the manufacturing service-oriented companies, and focus on the front-end design and the back-end running-and-servicing activities in the value chain. Thus, companies only need to design the product, the production activities are all outsourced to these SPs, PWs and LPs and a product-driven outsourcing service flow network is further formed with them to get closed-loop feedback of the production and logistic processes.

Based on that, a new advanced manufacturing mode called social manufacturing [3, 4] has attracted much attention. Social manufacturing is defined as "an outsourcing driven service-oriented manufacturing model that integrates and virtualizes mass of distributed SMRs, aggregates enterprises into manufacturing communities to provide manufacturing services, promotes intelligent business and all-around transparent order management and monitoring through social networking tools."

To efficiently accomplish the product development under social manufacturing, both the product-centered production and logistic states of all the outsourcing orders from SPs, PWs and LPs are monitored and fed back to the SDs to realize closed-loop quality traceability, process efficiency enhancement, real-time inventory feedback and up-to-date logistic tracking. Note that with the help of radio frequency identification (RFID) technology and data interface technology, the state tracking and monitoring becomes much easier and cheaper. However, in traditional order monitoring methods, only logistic information from the upper inter-company level can be reflected to the SDs and the up-to-date processing information from the lower intra-company level cannot be straightly transferred to the SD, besides, there is no sharing of the upper and lower order state information between SDs and SPs, PWs, LPs.

In this paper, we fill the information gap between LPs, SPs, PWs and SDs, and focus on the modeling of integrated production and logistic monitoring/tracking under social manufacturing, to provide a uniform order monitoring service for SDs. Firstly, the background and motivations of our work is described in Sect. 2. Then, the production organizing and manufacturing service flow model are clarified in Sect. 3. After that, a graphical formalized deduction of the RFID driven state block model is built for order state monitoring. In Sect. 4, a simulation example is studied to verify the feasibility of our model. Finally, discussions and conclusions are reached in Sect. 5.

2 Background and Motivations

Referring to the order monitoring or tracking, much work has been devoted to the logistic tracking in supply chain product visibility [5]. Kelepouris [6] proposed a Bayesian supply chain tracking model using serial-level information to deliver enhanced tracking information to the final user. Ko [7] designed a RFID-based product tracking system that can collaborate with the EPC Network, a suite of network services for RFID data management in the supply chain for product flow management.

On the other hand, when it comes to the production process monitoring/tracking in the shop floor level, Teti [8] reviewed the advanced monitoring methods of machining operations and provided an up-to-date comprehensive survey of sensor technologies, signal processing, and decision making strategies for process monitoring. Das [9] illustrated the process monitoring strategy for a multistage manufacturing facility with the aid of cluster analysis and multiple multi-block partial least squares (MBPLS) models. Jiang [10] discussed an RFID-driven graphical formalized deduction for describing the state changes of WIP material flows in a job-shop floor for order management. However, the tracking information that they focused on is always restricted within the manufacturing company and cannot reflect to the customers.

From the brief review above, it is shown that there is little research on the integrated production and logistic service flow monitoring/tracking in product lifecycle, and it lacks of a uniform modeling approach to formalize the proposed problem.

3 Methodology

3.1 Manufacturing Service Flow Model

Under social manufacturing, the mode of production organizing is different from the traditional one. Few parts or components are produced in the core companies, and there is no inventory in the core companies because all the parts, components and products are stored in the public warehouse. The flow and turning of these physical things are completed by the LPs. According to the bill of materials (BOM) decomposition of a certain product, the SDs make an overall outsourcing planning based on which they establish different outsourcing orders with SPs, PWs and LPs, forming a lifecycle manufacturing service flow network as shown in Fig. 1.

Each node represents a company, each edge stands for a sequential relationship, which means the different manufacturing services are executed in time order. The company node is formalized as:

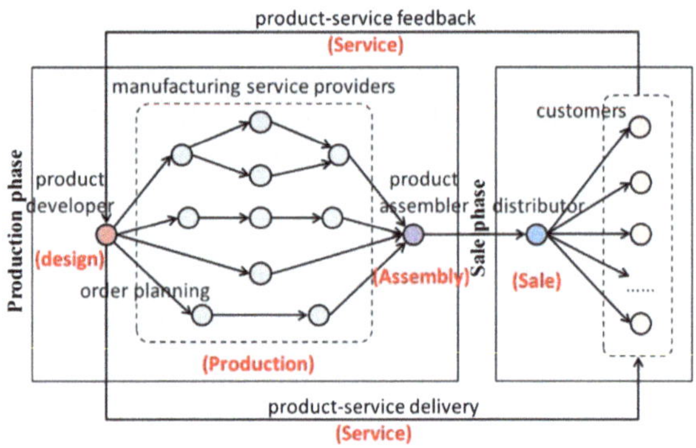

Fig. 1 The production organizing and service flow network

$$Node = \{ID, Type, Ob, T, State, Detail, \mathbf{Next}\} \tag{1}$$

where *Type* represents the company type and $Type \in \{SD, SP, PW, LP\}$, *Ob*, *T* and *State* stand for the service object, total service time and service state of each company node, respectively; **Next** represents the set of next service execution companies. There are 5 kinds of timing sequential relationship in the service flow network as shown in Fig. 2. The meaning of them are listed in Table 1.

In the manufacturing service flow network, we focus on the integrated production and logistic monitoring in the production phase. To implement the monitoring activity, the SPs deploy RFID sensing networks in their workshop to collect real-time production data, the PWs provide provisional inventory information through RFID data acquisition methods, and the LPs upload real-time transportation information gathered from their GIS/GPS systems through data interface.

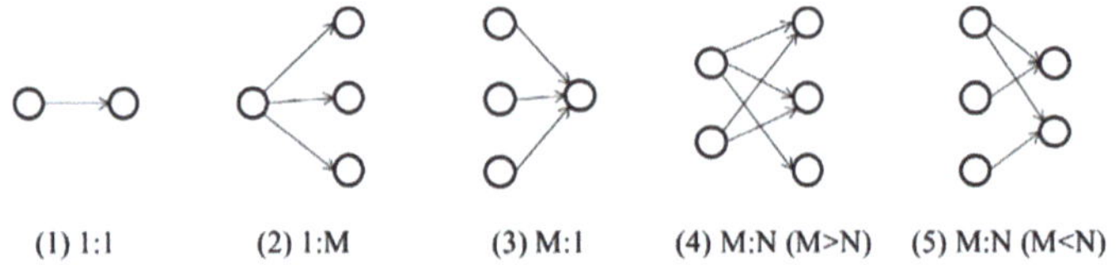

Fig. 2 5 kinds of timing sequential relationships

Table 1 The meanings of all the relationships

Notation	Meaning
1:1	Physical materials from the last company are totally transferred to the next company
1:M	Physical materials are sub-outsourced to the following M companies
M:1	All the physical materials from the last companies are gathered in the following one company
M:N	Cross sequential relationship composed of 1:M and M:1

Accordingly, the SDs could refer to all kinds of the up-to-date order state information for better production decision making and rapid response to the dynamic market.

Based on that, we propose a graphical formalized deduction method called RFID driven state block model to depict the service flow monitoring.

3.2 *RFID Driven State Block Model for Integrated Production and Logistic Service Flow Monitoring*

As RFID technology has taken a big step forward, the RFID applications is carried out increasingly in both academia and industry. According to [10], the nature of RFID technology is to monitor the state of a smart object which is changed by time in manufacturing, so it is in the whole production and logistics phase. The coarse-grained state changing of the production and logistics of a product development scheme under social manufacturing is shown in Fig. 3. There are two kinds of monitoring nodes in the state changing model, i.e., the production nodes and the logistic nodes.

Each order is processed based on the time-orderly process flow in a production node and is transferred along the position flow from one place to another in a logistic node where there can be public warehouses to temporarily store the physical things such as parts or products. As mentioned in the above section, the data gathered through RFID sensing networks in SPs and PWs are analyzed into

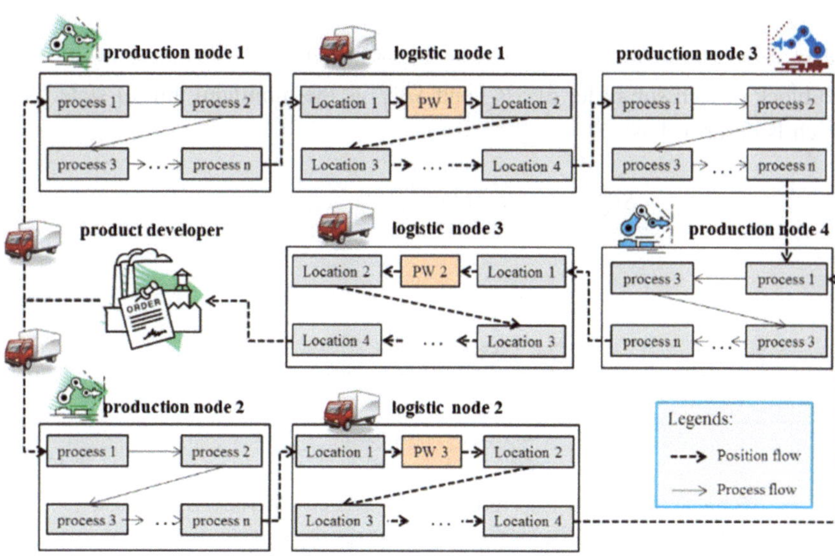

Fig. 3 The state changing model of the production and logistics

information through Auto-ID computing, after that the information is uploaded through data interface to a social manufacturing platform. The LPs provide real-time logistic information from their GIS/GPS systems through the uniform data interface too, all the information are transferred to the SDs through the platform.

3.2.1 Formalization of the State Block Model

Correspond to the state changing model in Fig. 3, a graphical formalized deduction called RFID driven state block model is built in Fig. 4. Each node is mapped into a state block containing multiple state changes. The state blocks are described as follows:

$$SB ::= \{PD, PA, PM, PT\} \tag{2}$$

where *SB* stands for the set of state blocks; *PD*, *PA*, *PM* and *PT* are four types of state blocks as depicted in Fig. 4, *PA*, *PM* belong to the production node, and *PT* belongs to the logistic node.

As for each state block, it could be made up of multiple sub-blocks because of the order decomposition or convergence which is formulated as:

$$\begin{aligned} SB_{i,j} ::= \Big\langle & SB_{i,j}^{1,1}, SB_{i,j}^{1,2}, \ldots, SB_{i,j}^{1,m_1}; SB_{i,j}^{2,1}, SB_{i,j}^{2,2}, \ldots, SB_{i,j}^{2,m_2}; \\ & \ldots; SB_{i,j}^{k,1}, SB_{i,j}^{k,2}, \ldots, SB_{i,j}^{k,m_k} \Big\rangle \end{aligned} \tag{3}$$

where $SB_{i,j}$ represents *j*-th main state block of order *i*; $SB_{i,j}^{k,m}$ stands for *m*-th sub-block of *k*-th sub-order of $SB_{i,j}$ and $SB_{i,j}^{k,m}$ is the minimum unit of state block which is described as:

$$SB_{i,j}^{k,m} ::= \Big\langle S_{i,j}^{k,m}(1), S_{i,j}^{k,m}(2), \ldots, S_{i,j}^{k,m}(p) \Big\rangle \tag{4}$$

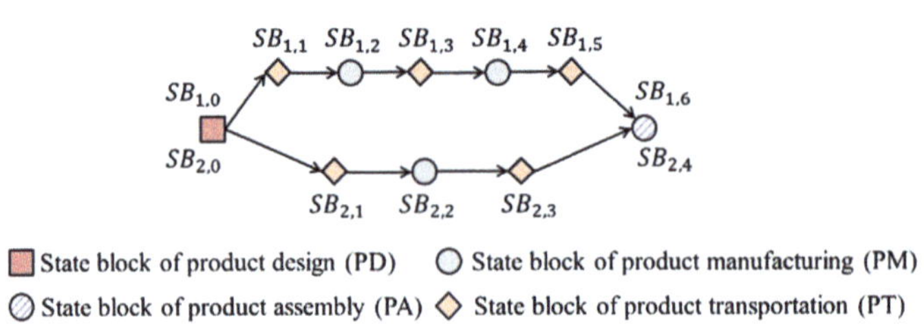

Fig. 4 The state block model

$$S_{i,j}^{k,m}(p) ::= \langle L_p, t_p, t_{p+1}, T_p, N, R_p, Qual, Detail \rangle \quad (5)$$

where $S_{i,j}^{k,m}(p)$ stands for p-th state of $SB_{i,j}^{k,m}$, L_p stands for the location of the tagged physical things related to state $S_{i,j}^{k,m}(p)$; t_p, T_p are the start time and lasting time of the state, $T_p = t_{p+1} - t_p$; N, R_p represent the total number of parts or components and the executant in this state, respectively; *Qual* stands for the corresponding quality information and *Detail* depicts other supplementary information about this state.

Note that the logistic nodes that directly connect the state block of product design (PD) such as $SB_{1,1}$ and $SB_{2,1}$ in Fig. 4 could be abolished in case of the type of processing with supplied drawings.

3.2.2 Formalization of Monitor Nodes

As shown in Fig. 4, all the production nodes and logistic nodes ($SB_{i,j}^{k,m}$) are monitoring modes, which are the representatives of the minimum units of state block, they contain multiple state changes. Further, we formalize these monitor nodes in Fig. 5. For state blocks representing production nodes, starting from the "going-in" state, each process of the physical parts or components is denoted as an "in-process" state, finally it ends with a "going-out" state. Therefore, the whole production activities are arranged in sequence. For state blocks representing logistic nodes, there is a "staying-in or going-through" state in each location and amongst the logistic processes, there could be public warehouses for temporarily storing the physical things (parts, components, products, etc.). In the public warehouse, the "going-in" and "going-out" states are monitored and the inventory information under the "staying-in or going-through" state is reflected. Based on that, the tracking information alongside the integrated production and logistic processes is gathered to SDs for transparent order management and better decision making.

3.2.3 Modeling of Integrated Production and Logistic State Changing Event Flow

Based on the state block model above, a modified event-driven graphical schema according to [11] is proposed to build the state changing event flow model. The schema is shown in Fig. 6. The state changing event can be formalized on the basis of a state-pair $S_{i,j}^{k,m}(p) \rightarrow S_{i,j}^{k,m}(p+1)$ as follows:

$$E_{i,j}^{k,m}(p) = f(S_{i,j}^{k,m}(p), S_{i,j}^{k,m}(p+1)) \quad (6)$$

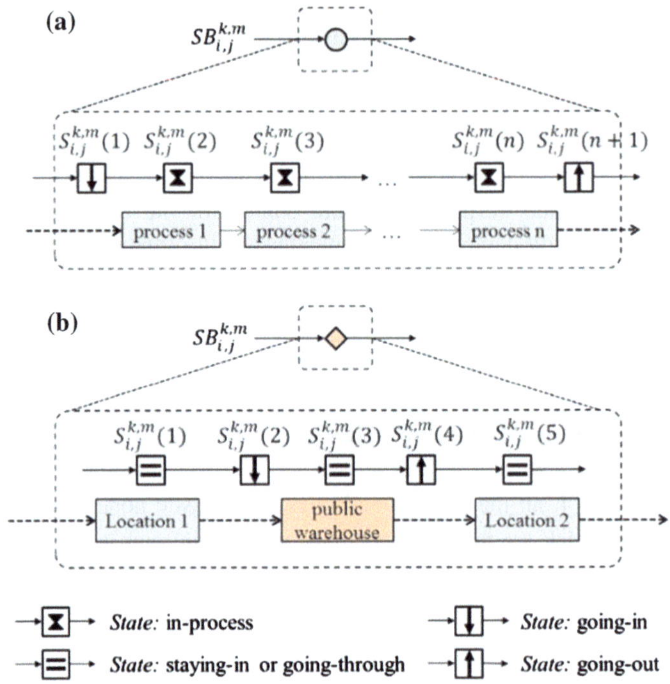

Fig. 5 Formalization of monitor modes. **a** Production node. **b** Logistic node

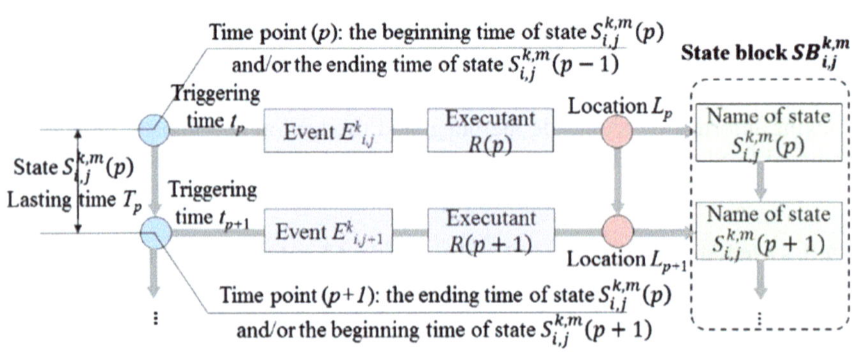

Fig. 6 The modified event-driven graphical schema

where $f(x, y)$ stands for the fictitious mapping function from a state-pair to an event. Therefore, the modified event-driven graphical schema can be defined as:

$$O_{i,j}^{k,m} ::= E_{i,j}^{k,m}(p), S_{i,j}^{k,m}(p), t_p, T_p, R_p,$$

$$S_{i,j}^{k,m}(p+1), t_{p+1}, T_{p+1}, R_{p+1}\Big\rangle \tag{7}$$

4 A Simulation Example

In this section, a simulation example of integrated production and logistic service flow monitoring of a design company is discussed to verify the proposed model. We select 2 outsourcing orders partially from a certain product, forming a 2-branch service flow. The first branch is further divided into 2 branches because of the order decomposition, and all the monitoring nodes in the 2 branches could be denoted as a main state block $SB_{1,3}$. Note that all the outsourcing orders around the product compose a service collaboration network. We formalize the service flow in Fig. 7a.

Based on the state block model, each state block could be monitored further. Here, we take the production node $SB_{2,4}$ as an example, which has 4 processes with 4 states. The state description of it is described in Fig. 7b from which we can vividly see that the state of this order in the fourth state block changes 6 times, and

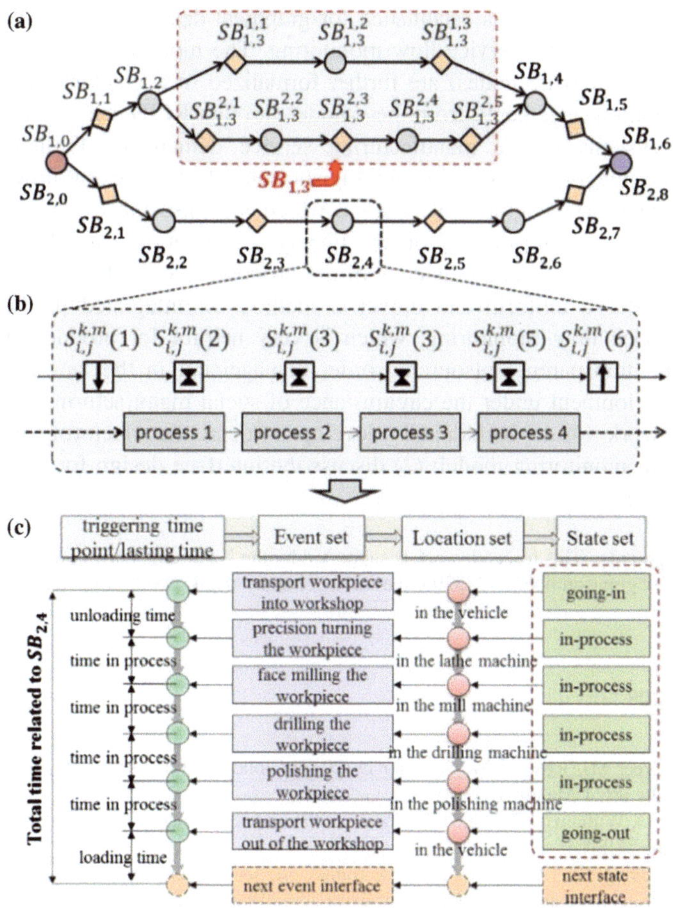

Fig. 7 The formalization of service flow with the modified event-driven graphical schema

the information of quality, scheduling, material flow, location and other order details of each state is informed by focusing on this state. Correspond to the node state description, the modified event-driven graphical schema is illustrated in Fig. 7c.

Through the modeling of the integrated production and logistic service flow and each kind of monitoring modes, together with the detail state changing event of each monitoring nodes, the SDs can clearly monitor all the states of outsourcing orders around a certain product.

5 Conclusion

In this paper, we analyzed the production organizing mode transferring and built a service flow model under social manufacturing circumstance. After that, the RFID driven state block model is established for graphical description of the integrated production and logistic service flow monitoring. The monitor nodes (i.e., production nodes and logistic nodes) are further formalized in a graphical method. The state information is gathered from these nodes through RFID sensing network and GIS/GPS systems to the manufacturing service demanders. Furthermore, we modeled the integrated production and logistic state changing event flow through a modified event-driven graphical schema, which indicates the event inducing the state changes of the outsourcing orders. Finally, a simulation example is described to verify the feasibility and efficiency of the proposed model. It is shown that our model provides a uniform and universal method for integrated production and logistic service flow monitoring, which is very helpful for production decision making and transparent outsourcing order management in the new paradigm of product development under the circumstance of social manufacturing.

Future work would include: (1) develop a social manufacturing platform to visualize our monitoring model; (2) discuss the interface design for GIS/GPS and RFID sensing network.

Acknowledgments This research work is supported by the Natural Science Foundation of China (NSFC) under grant number 51275396, the authors hereby thank NSFC for the financial aid.

References

1. Wu D, Greer MJ, Rosen DW, Schaefer D (2013) Cloud manufacturing: Strategic vision and state-of-the-art. J Manuf Syst
2. Tapscott D, Williams AD (2008) Wikinomics: the art and science of peer production. Penguin Group Inc., USA
3. Ding K, Jiang PY, Zhang X (2013) A framework for implementing social manufacturing system based on customized community space configuration and organization. Adv Mater Res 712:3191–3194

4. Shang X, Liu X, Xiong G, Cheng C, Ma Y, Nyberg TR (2013) Social manufacturing cloud service platform for the mass customization in apparel industry. In: 2013 IEEE International conference on service operations and logistics, and informatics (SOLI), pp 220–224
5. Musa A, Gunasekaran A, Yusuf Y (2014) Supply chain product visibility: methods, systems and impacts. Expert Syst Appl 41:176–194
6. Kelepouris T, Harrison M, McFarlane D (2011) Bayesian supply chain tracking using serial-level information. IEEE Trans Syst Man Cybern Part C Appl Rev 41:733–742
7. Ko JM, Kwak C, Cho Y, Kim CO (2011) Adaptive product tracking in RFID-enabled large-scale supply chain. Expert Syst Appl 38:1583–1590
8. Teti R, Jemielniak K, O'Donnell G, Dornfeld D (2010) Advanced monitoring of machining operations. CIRP Ann Manuf Technol 59:717–739
9. Das A, Maiti J, Banerjee R (2010) A hierarchical process monitoring strategy for a serial multi-stage manufacturing system. Int J Prod Res 48:2459–2479
10. Jiang P, Cao W (2013) An RFID-driven graphical formalized deduction for describing the time-sensitive state and position changes of work-in-progress material flows in a job-shop floor. J Manuf Sci and Eng Trans ASME 135
11. Jiang P, Fu Y, Zhu Q, Zheng M (2012) Event-driven graphical representative schema for job-shop-type material flows and data computing using automatic identification of radio frequency identification tags. Proc Inst Mech Eng Part B J Eng Manuf 226:339–352

Regular Reduction Theory and Application for Rigid Spacecraft with Two Internal Rotors

Hong Wang

Abstract In this paper we introduce briefly the regular point symplectic reduction and Hamilton-Jacobi theorem of regular controlled Hamiltonian (RCH) system, and their applications for the rigid spacecraft with two internal rotors. In the case of coincident centers of buoyancy and gravity, we can give explicitly the motion equation and Type I of Hamilton-Jacobi equation of reduced spacecraft-rotors system on a symplectic leaf by calculation in detail, which show the effect on controls in regular symplectic reduction and Hamilton-Jacobi theory.

Keywords Hamilton-Jacobi theorem · Regular controlled hamiltonian system · Regular point reduction · Rigid spacecraft with two rotors

1 Introduction

The theory of mechanical control systems has formed an important subject in recent years. Its research gathers together some separate areas of research such as mechanics, differential geometry and nonlinear control theory etc., and the emphasis of this research on geometry is motivated by the aim of understanding the structure of motion equations of the system in a way that helps both analysis and design. Thus, it is natural to study mechanical control systems by combining with the analysis of dynamic systems and the geometric reduction theory of Hamiltonian and Lagrangian systems. In particular, we note that in Marsden et al. [1], the authors studied regular reduction theory of controlled Hamiltonian systems with symplectic structure and symmetry, as an extension of regular symplectic reduction theory of Hamiltonian systems under regular controlled Hamiltonian equivalence conditions, and Wang and Zhang in [2] generalized the work to study the optimal reduction

H. Wang (✉)
School of Mathematical Sciences and LPMC, Nankai University,
Tianjin 300071, People's Republic of China
e-mail: hongwang@nankai.edu.cn

E. Qi (ed.), *Proceedings of the 6th International Asia Conference on Industrial Engineering and Management Innovation*,
DOI 10.2991/978-94-6239-145-1_58

theory of controlled Hamiltonian systems with Poisson structure and symmetry by using optimal momentum map and reduced Poisson tensor (or reduced symplectic form), and Ratiu and Wang in [3] generalized the work to study the Poisson reduction theory of controlled Hamiltonian system by controllability distribution when there is no momentum map for our considered system. These research works not only gave a variety of reduction methods for controlled Hamiltonian systems, but also showed a variety of relationships of controlled Hamiltonian equivalence of these systems.

On the other hand, we note that Hamilton-Jacobi theory provides a characterization of the generating functions of certain time-dependent canonical transformations, such that a given Hamiltonian system in such a form that its solutions are extremely easy to find by reduction to the equilibrium, see Abraham and Marsden [4], Arnold [5] and Marsden and Ratiu [6]. In general, we know that it is not easy to find the solutions of Hamilton's equation. But, if we can get a solution of Hamilton-Jacobi equation of the Hamiltonian system, by using the relationship between Hamilton's equation and Hamilton-Jacobi equation, it is easy to give a special solution of Hamilton's equation. Thus, it is very important to give explicitly the Hamilton-Jacobi equations of Hamiltonian system and RCH system. Recently, Wang in [7] studies the Hamilton-Jacobi theory to the regular controlled Hamiltonian system and its regular reduced systems, and describes the relationship between the RCH-equivalence for the RCH systems and the solutions of corresponding Hamilton-Jacobi equations. Now, it is a natural problem if there is a practical controlled Hamiltonian system and how to show the effect on controls in regular symplectic reduction and Hamilton-Jacobi theory of the system. As an application of the regular point symplectic reduction and Hamilton-Jacobi theory of RCH system with symmetry, Wang in [8] considers that the rigid spacecraft with an internal rotor as the regular point reducible RCH system, and gives explicitly the motion equation and Type I of Hamilton-Jacobi equation of reduced spacecraft-rotor system on a symplectic leaf by calculation in detail.

In this paper we introduce briefly the regular point symplectic reduction and Type I of Hamilton-Jacobi theorem of RCH system, and their applications for the rigid space-craft with two internal rotors. It is worthy note that it is different from the model in Wang [8], here we consider all of the rotations and translations of the carrier body. In the case of coincident centers of buoyancy and gravity, we can give explicitly the motion equation and Type I of Hamilton-Jacobi equation of reduced spacecraft-rotors system on a symplectic leaf by calculation in detail similar to the method in [8], which showed the effect on controls in regular symplectic reduction and Hamilton-Jacobi theory. One can show from Marsden et al. [1], Wang [7, 8] for more details.

2 Regular Reduction of RCH System and Hamilton-Jacobi Theorem

Let Q be a smooth manifold and T^*Q its cotangent bundle with the symplectic form ω. Assume that the cotangent lifted left action $\Phi^{T^*}: G \times T^*Q \to T^*Q$ is free, proper and symplectic, and admits Ad^*-equivariant momentum map $\mathbf{J}: T^*Q \to g^*$, where g is a Lie algebra of Lie group G and g^* is the dual of g. Let $\mu \in g^*$ be a regular value of $\mathbf{J}$ and denote by G_μ the isotropy subgroup of the coadjoint G-action at the point $\mu \in g^*$, which is defined by $G_\mu = \{g \in G | Ad_g^* \mu = \mu\}$. Since $G_\mu (\subset G)$ acts freely and properly on Q and on T^*Q, it follows that G_μ acts also freely and properly on $\mathbf{J}^{-1}(\mu)$, so that $(T^*Q)_\mu = \mathbf{J}^{-1}(\mu)/G_\mu$ is a symplectic manifold with the reduced symplectic form ω_μ uniquely characterized by relation $\pi_\mu^* \omega_\mu = i_\mu^* \omega$, where the map $i_\mu : \mathbf{J}^{-1}(\mu) \to T^*Q$ is a inclusion and $\pi_\mu : \mathbf{J}^{-1}(\mu) \to (T^*Q)_\mu$ is the projection. The pair $((T^*Q)_\mu, \omega_\mu)$ is called Marsden-Weinstein reduced space at μ. Let $H : T^*Q \to \mathbb{R}$ be a G-invariant Hamiltonian, the flow F_t of the Hamiltonian vector field X_H leaves the connected components of $\mathbf{J}^{-1}(\mu)$ invariant and commutes with the G-action, so it induces a flow f_t^μ on $(T^*Q)_\mu$, defined by $f_t^\mu \cdot \pi_\mu = \pi_\mu \cdot F_t \cdot i_\mu$, and the vector field X_{h_μ} generated by the flow f_t^μ on $((T^*Q)_\mu, \omega_\mu)$ is Hamiltonian with the associated regular point reduced Hamiltonian function $h_\mu : (T^*Q)_\mu \to \mathbb{R}$ defined by $h_\mu \cdot \pi_\mu = H \cdot i_\mu$, and the Hamiltonian vector fields X_H and X_{h_μ} are π_μ-related. On the other hand, from Marsden et al. [1], we know that the regular point reduced space $((T^*Q)_\mu, \omega_\mu)$ is symplectic diffeomorphic to a symplectic fiber bundle. Thus, we can introduce a regular point reducible RCH system as follows.

Definition 1 (Regular Point Reducible RCH System) A 6-tuple $(T^*Q, G, \omega, H, F, W)$, where the Hamiltonian $H : T^*Q \to \mathbb{R}$, the fiber-preserving map $F : T^*Q \to T^*Q$ and the fiber submanifold W of T^*Q are all G-invariant, is called a regular point reducible RCH system, if there exists a point $\mu \in g^*$, which is a regular value of the momentum map $\mathbf{J}$, such that the regular point reduced system, that is, the 5-tuple $((T^*Q)_\mu, \omega_\mu, h_\mu, f_\mu, W_\mu)$, where $(T^*Q)_\mu = \mathbf{J}^{-1}(\mu)/G_\mu$, $\pi_\mu^* \omega_\mu = i_\mu^* \omega$, $h_\mu \cdot \pi_\mu = H \cdot i_\mu$, $f_\mu \cdot \pi_\mu = \pi_\mu \cdot F \cdot i_\mu$, $W \subset \mathbf{J}^{-1}(\mu)$, $W_\mu = \pi_\mu(W)$, is a RCH system, which is simply written as R_P-reduced RCH system. The function h_μ is called the reduced Hamiltonian, the fiber-preserving map $f_\mu : (T^*Q)_\mu \to (T^*Q)_\mu$ is called the reduced (external) force map, W_μ is a fiber submanifold of $(T^*Q)_\mu$ and is called the reduced control subset. □

In order to describe the dynamics of the regular point reducible RCH system $(T^*Q, G, \omega, H, F, W)$ with a control law $u \in W$, we need to give a good expression of the dynamical vector field of RCH system. By using the notations of vertical lift maps of a vector along a fiber, see Marsden et al. [1], we have the following proposition.

Proposition 1 *The dynamical vector field of a RCH system* $(T^*Q, G, \omega, H, F, W)$ *with a control law* $u \in W$ *is the synthetic of Hamiltonian vector field* X_H *and its changes under the actions of the external force F and control u, that is, for any* $\alpha_x \in T^*_xQ$, *and* $x \in Q$, $X_{(T^*Q,G,\omega,H,F,u)}(\alpha_x) = X_H(\alpha_x) + vlift(F)X_H(\alpha_x) + vlift(u)X_H(\alpha_x)$. *For convenience, it is simply written as*

$$X_{(T^*Q,G,\omega,H,F,u)} = (\mathbf{d}H)^{\#} + vlift(F) + vlift(u). \tag{2.1}$$

If a G-invariant feedback control law $u \in W$ is chosen, and we have that the reduced control $u_\mu : (T^*Q)_\mu \to W_\mu$, then the R_P-reduced system $(T^*Q_\mu, \omega_\mu, h_\mu, f_\mu, u_\mu)$ is a closed-loop regular dynamic system with a control law u_μ. Assume that its dynamical vector field can be expressed as follows $X_{((T^*Q)_\mu,\omega_\mu,h_\mu,f_\mu,u_\mu)} = X_{h_\mu} + vlift(f_\mu) + vlift(u_\mu)$, where $(\mathbf{d}h_\mu)^{\#} = X_{h_\mu}$, and $vlift(f_\mu) = vlift(f_\mu)X_{h_\mu}$, $vlift(u_\mu) = vlift(u_\mu)X_{h_\mu}$, and satisfy the condition $X_{((T^*Q)_\mu,\omega_\mu,h_\mu,f_\mu,u_\mu)} \cdot \pi_\mu = T\pi_\mu \cdot X_{(T^*Q,G,\omega,H,F,u)} \cdot i_\mu$.

In particular, for the regular point reducible RCH system $(T^*Q, G, \omega, H, F, W)$, Wang in [7] can prove the following Type I of Hamilton-Jacobi theorem for R_P-reduced RCH system $(T^*Q_\mu, \omega_\mu, h_\mu, f_\mu, u_\mu)$.

Theorem 1 *(Type I of Hamilton-Jacobi Theorem of* R_P*-reduced RCH System) For a regular point reducible RCH system* $(T^*Q, G, \omega, H, F, W)$, *assume that* $\gamma : Q \to T^*Q$ *is an one-form on Q, and* γ *is closed with respect to the map* $T\pi_Q : TT^*Q \to TQ$, *and* $\tilde{X}^\gamma = T\pi_Q \cdot \tilde{X} \cdot \gamma$, *where* $\tilde{X} = X_{(T^*Q,G,\omega,H,F,u)}$ *is the dynamical vector field of the regular point reducible RCH system* $(T^*Q, G, \omega, H, F, W)$ *with a control law u. Moreover, assume that* $\mu \in g^*$ *is the regular reducible point of the RCH system, and* $\mathrm{Im}(\gamma) \subset \mathbf{J}^{-1}(\mu)$, *and it is* G_μ*-invariant, and* $\bar{\gamma} = \pi_\mu(\gamma) : Q \to (T^*Q)_\mu$. *Then* $\tilde{X}^\gamma$ *and* $\tilde{X}_{h_\mu}$ *are* $\bar{\gamma}$*-related, that is,* $T\bar{\gamma} \cdot \tilde{X}^\gamma = \tilde{X}_{h_\mu} \cdot \bar{\gamma}$, *where* X_{h_μ} *is the corresponding Hamiltonian vector field of* R_P*-reduced RCH system* $((T^*Q)_\mu, \omega_\mu, h_\mu, f_\mu, u_\mu)$, *and the equation* $T\bar{\gamma} \cdot \tilde{X}^\gamma = \tilde{X}_{h_\mu} \cdot \bar{\gamma}$ *is called the Type I of Hamilton-Jacobi equation for* R_P*-reduced RCH system* $(T^*Q_\mu, \omega_\mu, h_\mu, f_\mu, u_\mu)$, *and* $\bar{\gamma}$ *is a solution of the equation.* □

3 The Rigid Spacecraft with Two Internal Rotors

As an application of the theoretical result, in the following we consider a rigid spacecraft (to be called the carrier body) carrying two internal rotors, and assume that the only external forces and torques acting on the space-craft-rotors system are due to buoyancy and gravity. In general, it is possible that the spacecraft's center of buoyancy may not be coincident with its center of gravity. But, in this paper we assume that the spacecraft is symmetric and it has uniformly distributed mass, and the center of buoyancy and the center of gravity are coincident. Denote by O the center of mass of the system in the body frame and at O place a set of (orthogonal) body axes. Assume that the body coordinate axes are aligned with principal axes of the carrier body, and we put two rotors within the spacecraft so that each rotor's rotation axis is parallel to the first and the second principal axes of the body, and the rotor spins under the influence of a torque acting on the rotor, see Bloch and Leonard [9], and Leonard and Marsden [10]. The configuration space is $Q = W \times V$, where $W = \mathrm{SE}(3) = \mathrm{SO}(3) \circledS \mathbb{R}^3$ and $V = S^1 \times S^1$, with the first factor being the attitude and position of the spacecraft and the second factor being the angles of rotors. The corresponding phase space is the cotangent bundle $T^*Q = T^*\mathrm{SE}(3) \times T^*V$, and $T^*V = T^*(S^1 \times S^1) \cong \mathbb{R}^2 \times \mathbb{R}^2$, with the canonical symplectic form.

Let $I = diag(I_1, I_2, I_3)$ be the moment of inertia of the spacecraft in the body-fixed frame. Let $J_i, i = 1, 2$ be the moments of inertia of rotors around their rotation axes. Let $J_{ik},\ i = 1, 2,\ k = 1, 2, 3$, be the moments of inertia of the i-th rotor with $i = 1, 2$ around the k-th principal axis with $k = 1, 2, 3$, respectively, and denote by $\bar{I}_i = I_i + J_{1i} + J_{2i} - J_{ii},\ i = 1, 2$, and $\bar{I}_3 = I_3 + J_{13} + J_{23}$. Let $\Omega = (\Omega_1, \Omega_2, \Omega_3)$ and $v = (v_1, v_2, v_3)$ be the angular and linear velocity vectors of spacecraft computed with respect to the axes fixed in the body, $(\Omega_1, \Omega_2, \Omega_3) \in \mathrm{so}(3)$, $v = (v_1, v_2, v_3) \in \mathbb{R}^3$. Let $\theta_i,\ i = 1, 2$, be the relative angles of rotors and $\dot{\theta} = (\dot{\theta}_1, \dot{\theta}_2)$ the relative angular velocity vector of rotor about the principal axes with respect to the body fixed frame of spacecraft, and $M = diag(m_1, m_2, m_3)$ the total mass of the system. If we introduce the conjugate angular momentum and linear momentum, $\Pi = (\Pi_1, \Pi_2, \Pi_3) \in \mathrm{so}^*(3),\ P = (P_1, P_2, P_3) \in \mathbb{R}^3,\ l = (l_1, l_2) \in \mathbb{R}^2$, then we have Hamiltonian $H(A, c, \Pi, P, \theta, l) : \mathrm{SE}(3) \times se^*(3) \times \mathbb{R}^2 \times \mathbb{R}^2 \to \mathbb{R}$ given by $H(A, c, \Pi, P, \theta, l) = \frac{1}{2}[\frac{(\Pi_1 - l_1)^2}{\bar{I}_1} + \frac{(\Pi_2 - l_2)^2}{\bar{I}_2} + \frac{\Pi_3^2}{\bar{I}_3} + \frac{P_1^2}{m_1} + \frac{P_2^2}{m_2} + \frac{P_3^2}{m_3} + \frac{l_1^2}{J_1} + \frac{l_2^2}{J_2}]$.

4 The Symmetric Reduction and Hamilton-Jacobi Theorem

In order to give the motion equation and Type I of Hamilton-Jacobi equation of the spacecraft-rotors system, in the following we need to consider the symmetry and reduced symplectic structure of the space $Q = \mathrm{SE}(3) \times S^1 \times S^1$. Assume that Lie group $W = \mathrm{SE}(3)$ acts freely and properly on $Q = \mathrm{SE}(3) \times V$ by the left translation on $\mathrm{SE}(3)$, then the action of $\mathrm{SE}(3)$ on the phase space T^*Q is by cotangent lift of left translation on SE(3) at the identity, that is, $\Phi : \mathrm{SE}(3) \times \mathrm{SE}(3) \times \mathrm{se}^*(3) \times \mathbb{R}^2 \times \mathbb{R}^2 \to \mathrm{SE}(3) \times \mathrm{se}^*(3) \times \mathbb{R}^2 \times \mathbb{R}^2$, given by $\Phi((B,s)((A,c),(\Pi,w),\theta,l)) = ((BA,c),(\Pi,w),\theta,l)$, for any $A,B \in \mathrm{SO}(3)$, $\Pi \in \mathrm{so}^*(3)$, $s,c,w \in \mathbb{R}^3$, $\theta,l \in \mathbb{R}^2$, which is also free and proper, and admits an associated Ad^*-equivariant momentum map $\mathbf{J}_Q : \mathrm{SE}(3) \times \mathrm{se}^*(3) \times \mathbb{R}^2 \times \mathbb{R}^2 \to \mathrm{se}^*(3)$ for the left SE(3) action. If $(\Pi,w) \in \mathrm{se}^*(3)$ is a regular value of $\mathbf{J}_Q$, then the regular point reduced space $(T^*Q)_{(\Pi,w)} = \mathbf{J}_Q^{-1}(\Pi,w)/\mathrm{SE}(3)_{(\Pi,w)}$ is symplectically diffeomorphic to the coadjoint orbit $\mathcal{O}_{(\Pi,w)} \times \mathbb{R}^2 \times \mathbb{R}^2 \subset \mathrm{se}^*(3) \times \mathbb{R}^2 \times \mathbb{R}^2$, where $\mathrm{SE}(3)_{(\Pi,w)}$ is the isotropy subgroup of coadjoint SE(3)-action at the point $(\Pi,w) \in \mathrm{se}^*(3)$.

We know that $se^*(3) = \mathrm{so}^*(3) Ⓢ \mathbb{R}^3$ is a Poisson manifold with respect to its Lie-Poisson bracket, and for $(\mu,a) \in \mathrm{se}^*(3)$, the coadjoint orbit $O_{(\mu,a)} \subset \mathrm{se}^*(3)$ has the induced orbit symplectic form $\omega^-_{O_{(\mu,a)}}$, which is coincide with the restriction of the Lie-Poisson bracket on se*(3) to the coadjoint orbit $O_{(\mu,a)}$, and the coadjoint orbits $(O_{(\mu,a)}, \omega^-_{O_{(\mu,a)}})$, $(\mu,a) \in \mathrm{se}^*(3)$, form the symplectic leaves of the Poisson manifold $(\mathrm{se}^*(3), \{\cdot,\cdot\}_{\mathrm{se}^*(3)})$. Let $\omega_{\mathbb{R}^2}$ be the canonical symplectic form on $T^*\mathbb{R}^2 \cong \mathbb{R}^2 \times \mathbb{R}^2$, and it induces a canonical Poisson bracket $\{\cdot,\cdot\}_{\mathbb{R}^2}$ on $T^*\mathbb{R}^2$. Thus, we can induce a symplectic form $\tilde{\omega}^-_{O_{(\mu,a)} \times \mathbb{R}^2 \times \mathbb{R}^2}$ on the smooth manifold $O_{(\mu,a)} \times \mathbb{R}^2 \times \mathbb{R}^2$, and a Poisson bracket $\{\cdot,\cdot\}_-$ on the smooth manifold $\mathrm{se}^*(3) \times \mathbb{R}^2 \times \mathbb{R}^2$, and such that $(O_{(\mu,a)} \times \mathbb{R}^2 \times \mathbb{R}^2, \tilde{\omega}^-_{\mathcal{O}_{(\mu,a)} \times \mathbb{R}^2 \times \mathbb{R}^2})$ is a symplectic leaf of the Poisson manifold $(\mathrm{se}^*(3) \times \mathbb{R}^2 \times \mathbb{R}^2, \{\cdot,\cdot\}_-)$.

From the above expression of the Hamiltonian, we know that $H(A,c,\Pi,P,\theta,l)$ is invariant under the left SE(3)-action. For the case that $(\mu,a) \in \mathrm{se}^*(3)$ is the regular value of $\mathbf{J}_Q$, we have the reduced Hamiltonian $h_{(\mu,a)}(\Pi,P,\theta,l) : O_{(\mu,a)} \times \mathbb{R}^2 \times \mathbb{R}^2 (\subset \mathrm{se}^*(3) \times \mathbb{R}^2 \times \mathbb{R}^2) \to \mathbb{R}$, given by $h_{(\mu,a)}(\Pi,P,\theta,l) = H(A,c,\Pi,P,\theta,l)|_{O_{(\mu,a)} \times \mathbb{R}^2 \times \mathbb{R}^2}$.

From the heavy top Poisson bracket on se*(3) and the Poisson bracket on $T^*\mathbb{R}^2$, we can get the Poisson bracket on $\mathrm{se}^*(3) \times \mathbb{R}^2 \times \mathbb{R}^2$, In particular, for $F_{(\mu,a)}$ and $K_{(\mu,a)} : O_{(\mu,a)} \times \mathbb{R}^2 \times \mathbb{R}^2 \to \mathbb{R}$, we have that $\tilde{\omega}^-_{O_{(\mu,a)} \times \mathbb{R}^2 \times \mathbb{R}^2}(X_{F_{(\mu,a)}}, X_{K_{(\mu,a)}}) = \{F_{(\mu,a)}, K_{(\mu,a)}\}_-|_{O_{(\mu,a)} \times \mathbb{R}^2 \times \mathbb{R}^2}$. Moreover, if we consider the spacecraft-rotors system with a control torque $u : T^*Q \to W$ acting on the rotors, and $u \in W \subset \mathbf{J}_Q^{-1}((\mu,a))$ is invariant under the left SE(3)-action, and its reduced control torque $u_{(\mu,a)} : \mathcal{O}_{(\mu,a)} \times \mathbb{R}^2 \times \mathbb{R}^2 \to W_{(\mu,a)}$ is given by $u_{(\mu,a)}(\Pi,P,\theta,l) = \pi_{(\mu,a)}(u(A,c,\Pi,P,\theta,l)) = u(A,c,\Pi,P,\theta,l)|_{O_{(\mu,a)} \times \mathbb{R}^2 \times \mathbb{R}^2}$, where $\pi_{(\mu,a)} : \mathbf{J}_Q^{-1}((\mu,a)) \to O_{(\mu,a)} \times \mathbb{R}^2 \times \mathbb{R}^2$, $W_{(\mu,a)} = \pi_{(\mu,a)}(W)$. If we assume that the vertical lift of vector field $X_{h_{(\mu,a)}}$ under the action of $u_{(\mu,a)}$ along fibers is given by $\mathrm{vlift}(u_{(\mu,a)})X_{h_{(\mu,a)}}(x) = (U_\Pi, U_P, U_\theta, U_l) \in T_x(O_{(\mu,a)} \times \mathbb{R}^2 \times \mathbb{R}^2)$, where $x = (\Pi,P,\theta,l) \in O_{(\mu,a)} \times \mathbb{R}^2 \times \mathbb{R}^2$, and U_Π, $U_P \in \mathbb{R}^3$, U_θ, $U_l \in \mathbb{R}^2$. Thus, for the above reduced Hamiltonian $h_{(\mu,a)}(\Pi,P) : O_{(\mu,a)} \times \mathbb{R}^2 \times \mathbb{R}^2 \to \mathbb{R}$,in the case of coincident centers of buoyancy and gravity, we can obtain the equations of motion for reduced spacecraft-rotors system with the control torque u acting on the rotors by calculation in detail similar to the method in Wang [8], and the following theorem.

Theorem 2 *In the case of coincident centers of buoyancy and gravity, the spacecraft-rotors system with the control torque u acting on the rotors, that is, the 5-tuple* $(T^*Q, \mathrm{SE}(3), \omega_Q, H, u)$, *where* $Q = \mathrm{SE}(3) \times S^1 \times S^1$, *is a regular point reducible RCH system. For a point* $(\mu, a) \in \mathrm{se}^*(3)$, the regular value of the momentum map $\mathbf{J}_Q : \mathrm{SE}(3)\times \mathrm{se}^*(3) \times \mathbb{R}^2 \times \mathbb{R}^2 \to \mathrm{se}^*(3)$, *the regular point reduced system is the 4-tuple* $(O_{(\mu,a)} \times \mathbb{R}^2 \times \mathbb{R}^2, \tilde{\omega}^-_{O_{(\mu,a)}\times\mathbb{R}^2\times\mathbb{R}^2}, h_{(\mu,a)}, u_{(\mu,a)})$, *where* $O_{(\mu,a)} \subset se^*(3)$ *is the coadjoint orbit,* $\tilde{\omega}^-_{O_{(\mu,a)}\times\mathbb{R}^2\times\mathbb{R}^2}$ *is the orbit symplectic form on* $O_{(\mu,a)} \times \mathbb{R}^2 \times \mathbb{R}^2$, *and* $h_{(\mu,a)}(\Pi, P, \theta, l) = H(A, c, \Pi, P, \theta, l)|_{O_{(\mu,a)}\times\mathbb{R}^2\times\mathbb{R}^2}$, *and* $u_{(\mu,a)}(\Pi, P, \theta, l) = u(A, c, \Pi, P, \theta, l)|_{O_{(\mu,a)}\times\mathbb{R}^2\times\mathbb{R}^2}$, *and the equations of motion of the system are given by*

$$\begin{cases} \dfrac{\mathrm{d}\Pi}{\mathrm{d}t} = \Pi \times \Omega + P \times v + U_\Pi, \\ \dfrac{\mathrm{d}P}{\mathrm{d}t} = P \times \Omega + U_P, \\ \dfrac{\mathrm{d}\theta}{\mathrm{d}t} = (-\dfrac{(\Pi_1 - l_1)}{\bar{I}_1} + \dfrac{l_1}{J_1}, -\dfrac{(\Pi_2 - l_2)}{\bar{I}_2} + \dfrac{l_2}{J_2}) + U_\theta, \\ \dfrac{\mathrm{d}l}{\mathrm{d}t} = U_l. \end{cases}$$

Moreover, in the following we can give the Type I of Hamilton-Jacobi equation for regular point reduced spacecraft-rotors system with coincident centers of buoyancy and gravity. We assume that $\gamma :\ Q = \mathrm{SE}(3) \times S^1 \times S^1 \to T^*Q = T^*(\mathrm{SE}(3) \times S^1 \times S^1)$ is an one-form on Q, and $\gamma^* : T^*T^*Q \to T^*Q$ is symplectic, and γ is closed with respect to the map $T\pi_Q : TT^*Q \to TQ$, and $\mathrm{Im}(\gamma) \subset \mathbf{J}_Q^{-1}((\mu, a))$, and it is $\mathrm{SE}(3)_{(\mu,a)}$-invariant, $\bar{\gamma} = \pi_{(\mu,a)}(\gamma) :\ Q \to \mathcal{O}_{(\mu,a)} \times \mathbb{R}^2 \times \mathbb{R}^2$.

Denote by

$$\begin{aligned} \bar{\gamma}(A, c, \theta) &= (\bar{\gamma}_1, \bar{\gamma}_2, \bar{\gamma}_3, \bar{\gamma}_4, \bar{\gamma}_5, \bar{\gamma}_6, \bar{\gamma}_7, \bar{\gamma}_8, \bar{\gamma}_9, \bar{\gamma}_{10})(A, c, \theta) \\ &\in O_{(\mu,a)} \times \mathbb{R}^2 \times \mathbb{R}^2 (\subset se^*(3) \times \mathbb{R}^2 \times \mathbb{R}^2), \end{aligned}$$

then the function $h_{(\mu,a)} \cdot \bar{\gamma} : \mathrm{SE}(3) \times S^1 \times S^1 \to \mathbb{R}$ is given by

$$\begin{aligned} &h_{(\mu,a)} \cdot \bar{\gamma}(A, c, \theta) = H|_{O_{(\mu,a)}\times\mathbb{R}^2\times\mathbb{R}^2} \cdot \bar{\gamma}(A, c, \theta) \\ &= \frac{1}{2}[\frac{(\bar{\gamma}_1 - \bar{\gamma}_9)^2}{\bar{I}_1} + \frac{(\bar{\gamma}_2 - \bar{\gamma}_{10})^2}{\bar{I}_2} + \frac{\bar{\gamma}_3^2}{\bar{I}_3} + \frac{\bar{\gamma}_4^2}{m_1} + \frac{\bar{\gamma}_5^2}{m_2} + \frac{\bar{\gamma}_6^2}{m_3} + \frac{\bar{\gamma}_9^2}{J_1} + \frac{\bar{\gamma}_{10}^2}{J_2}]. \end{aligned}$$

The dynamical vector field of R_P -reduced spacecraft-rotors system $(O_{(\mu,a)} \times \mathbb{R}^2 \times \mathbb{R}^2, \tilde{\omega}^-_{O_{(\mu,a)} \times \mathbb{R}^2 \times \mathbb{R}^2}, h_{(\mu,a)}, u_{(\mu,a)})$ is given by

$$X_{(\mathcal{O}_{(\mu,a)} \times \mathbb{R}^2 \times \mathbb{R}^2, \tilde{\omega}^-_{\mathcal{O}_{(\mu,a)} \times \mathbb{R}^2 \times \mathbb{R}^2}, h_{(\mu,a)}, u_{(\mu,a)})} = X_{h_{(\mu,a)}} + vlift(u_{(\mu,a)}),$$

where $vlift(u_{(\mu,a)}) = vlift(u_{(\mu,a)})X_{h_{(\mu,a)}} \in T(\mathcal{O}_{(\mu,a)} \times \mathbb{R}^2 \times \mathbb{R}^2)$.

Assume that

$$\begin{aligned} vlift(u_{(\mu,a)} \cdot \bar{\gamma})(A, c, \theta) &= (U_1, U_2, U_3, U_4, U_5, U_6, U_7, U_8, U_9, U_{10}) \\ &\in T_x(O_{(\mu,a)} \times \mathbb{R}^2 \times \mathbb{R}^2), \end{aligned}$$

where $x = (\Pi, P, \theta, l) \in O_{(\mu,a)} \times \mathbb{R}^2 \times \mathbb{R}^2$, and $U_i(A, c, \theta) : \mathrm{SE}(3) \times S^1 \times S^1 \to \mathbb{R}$, $i = 1, 2, \ldots, 10$.. Thus, we can prove Type I of Hamilton-Jacobi equation for spacecraft-rotors system with the control torque u acting on the rotors by calculation in detail similar to the method in Wang [8], and obtain the following theorem.

Theorem 3 *In the case of coincident centers of buoyancy and gravity, for a point* $(\mu, a) \in \mathrm{se}^*(3)$, *the regular value of the momentum map* $\mathbf{J}_Q : \mathrm{SE}(3) \times \mathrm{se}^*(3) \times \mathbb{R}^2 \times \mathbb{R}^2 \to se^*(3)$, *the regular point reduced system of spacecraft-rotors system with the control torque u acting on the rotors* $(T^*Q, \mathrm{SE}(3), \omega_Q, H, u)$, *where* $Q = \mathrm{SE}(3) \times S^1 \times S^1$, *is the 4-tuple* $(\mathcal{O}_{(\mu,a)} \times \mathbb{R}^2 \times \mathbb{R}^2, \tilde{\omega}^-_{\mathcal{O}_{(\mu,a)} \times \mathbb{R}^2 \times \mathbb{R}^2}, h_{(\mu,a)}, u_{(\mu,a)})$. *Assume that* $\gamma : \mathrm{SE}(3) \times S^1 \times S^1 \to T^*(\mathrm{SE}(3) \times S^1 \times S^1)$ *is an one-form on* $\mathrm{SE}(3) \times S^1 \times S^1$, *and* γ *is closed with respect to the map* $T\pi_Q : TT^*Q \to TQ$, *and* $\mathrm{Im}(\gamma) \subset \mathbf{J}_Q^{-1}((\mu, a))$, *and it is* $\mathrm{SE}(3)_{(\mu,a)}$*-invariant, where* $\mathrm{SE}(3)_{(\mu,a)}$ *is the isotropy subgroup of coadjoint* $\mathrm{SE}(3)$*-action at* $(\mu, a) \in se^*(3)$, *and* $\bar{\gamma} = \pi_{(\mu,a)}(\gamma) : \mathrm{SE}(3) \times S^1 \times S^1 \to O_{(\mu,a)} \times \mathbb{R}^2 \times \mathbb{R}^2$. *Then* $\bar{\gamma}$ *is a solution of the Type I of Hamilton-Jacobi equation of reduced space-craft- rotors system, that is,* $T\bar{\gamma} \cdot \tilde{X}^\gamma = X_{h_{(\mu,a)}} \cdot \bar{\gamma}, \tilde{X}^\gamma$ *and* $X_{h_{(\mu,a)}}$ *are* $\bar{\gamma}$*-related, where* $\tilde{X}^\gamma = T\pi_Q \cdot \tilde{X} \cdot \gamma$, $\tilde{X} = X_{(T^*\mathrm{SE}(3), \omega_Q, H, u)}$, *and* $X_{h_{(\mu,a)}}$ *is the corresponding Hamiltonian vector field of* R_P*-reduced RCH system* $(O_{(\mu,a)} \times \mathbb{R}^2 \times \mathbb{R}^2, \tilde{\omega}^-_{\mathcal{O}_{(\mu,a)} \times \mathbb{R}^2 \times \mathbb{R}^2}, h_{(\mu,a)}, u_{(\mu,a)})$.

Remark 1 If we consider that $\lambda = \gamma \cdot \pi_Q : T^*Q \to T^*Q$ is a symplectic map in Theorem 1 and Theorem 3, that is, γ is not closed with respect to $T\pi_Q : TT^*Q \to TQ$, and assume that $\bar{\lambda} = \pi_\mu(\lambda) : T^*Q \to (T^*Q)_\mu$. Then by calculation in detail we can also prove that $\bar{\lambda}$ is a solution of the Type II of Hamilton-Jacobi equation $T\bar{\gamma} \cdot \tilde{X}^\lambda = X_{h_\mu} \cdot \bar{\lambda}$, for R_P-reduced RCH system $(T^*Q_\mu, \omega_\mu, h_\mu, f_\mu, u_\mu)$, if and only if $\bar{\lambda}$ is a solution of the following equation $T\pi_Q \cdot (X_{h_\mu \cdot \bar{\lambda}}) = X_H{}^\lambda$, where $X_H{}^\lambda = T\pi_Q \cdot X_H \cdot \lambda$.

Remark 2 For the case of non-coincident centers of buoyancy and gravity, by using the semidirect product method and by complex calculation, we can also give explicitly the motion equation and Type I and Type II of Hamilton-Jacobi equations of reduced spacecraft-rotors system. □

5 Conclusion

The theory of controlled mechanical systems is a very important subject, following the theoretical and applied development of geometric mechanics, a lot of important problems about this subject are being explored and studied. In this paper, as an application of the symplectic reduction and Hamilton-Jacobi theory of regular control-ed Hamiltonian systems with symmetry on the generalization of a Lie group, in the case of coincident centers of buoyancy and gravity, we give explicitly the motion equation and Type I of Hamilton-Jacobi equation of reduced spacecraft-rotors system on a symplectic leaf by calculation in detail, respectively. Moreover, in the case of non-coincident centers of buoyancy and gravity, we can also give explicitly the motion equation and Type I and Type II of Hamilton-Jacobi equations of reduced spacecraft-rotors system. These research show the effect on controls in regular symplectic reduction and Hamilton- Jacobi theory. But if we define a controlled Hamiltonian system on the cotangent bundle by using a Poisson structure, see Wang and Zhang [2] and Ratiu and Wang [3], and the way given in [1] cannot be used, what and how we could do? This is a problem worthy to be considered in detail. In addition, we also note that there have been a lot of beautiful results of reduction theory of Hamiltonian systems in celestial mechanics, hydro-dynamics and plasma physics. Thus, it is an important topic to study the application of reduction theory of controlled Hamiltonian systems in celestial mechanics, hydrodynamics and plasma physics. These are our goals in future research.

Acknowledgments The author thanks the support of Nankai University, 985 Project and the Key Laboratory of Pure Mathematics and Combinatorics, Ministry of Education, China.

References

1. Marsden JE, Wang H, Zhang ZX (2014) Regular reduction of controlled Hamiltonian system with symplectic structure and symmetry. Diff Geom Appl 33(3):13–45
2. Wang H, Zhang ZX (2012) Optimal reduction of controlled Hamiltonian system with poisson structure and symmetry. Jour Geom Phys 62(5):953–975
3. Ratiu TS, Wang H Poisson reduction of controlled Hamiltonian system by controllability distribution. arXiv:1312.7047

4. Abraham R, Marsden JE (1978) Foundations of mechanics, 2nd edn. Addison-Wesley, Reading, MA
5. Arnold VI (1989) Mathematical methods of classical mechanics, 2nd edn. Graduate texts in mathematics, vol 60, Springer, New York
6. Marsden JE, Ratiu TS (1999) Introduction to mechanics and symmetry, 2nd edn. Texts in applied mathematics, vol 17, Springer, New York
7. Wang H Hamilton-Jacobi theorems for regular controlled Hamiltonian system and its reductions. (arXiv: 1305.3457, to submit to J Geom Mech)
8. Wang H (2013) Symmetric reduction and Hamilton-Jacobi equation of rigid spacecraft with a rotor. J Geom Symm Phys 32:87–111. arXiv: 1307.1606
9. Bloch AM, Leonard NE (2002) Symmetries, conservation laws, and control. In: Newton P, Holmes P, Weinstein A (eds) Geometry, mechanics and dynamic, volume in honor of the 60th birthday of J.E. Marsden. Springer, New York
10. Leonard NE, Marsden JE (1997) Stability and drift of underwater vehicle dynamics: mechanical systems with rigid motion symmetry. Physica D 105:130–162

Discussion About Integrated Operation and Maintenance Framework for Information and Communication Network of Power Grid

Yun Liang, Ji-ming Yao, Li Huang and Jian-bo Liu

Abstract This paper comprehensively analyzed the development of the smart grid, combined with the latest ICT technology development results, demonstrated the important role of information and communication network for new applications in smart grid, put forward the integrated framework of operation and maintenance which is based on resource integration, supported by standardize processes and centered by an integrated evaluation. This paper points out the key technical problems such as the resource model, the monitoring data fusion, intelligent alarm indicator evaluation system, and running state assessment in the building of operation and maintenance system for the integration of smart grid information and communications. The key technologies are pointed out in different aspects: resource model, intelligent warning, evaluation system and visualization.

Keywords Information and communication network · Operation and maintenance · Power grid

1 Introduction

The smart grid is the common choice of the international power industry to actively respond to the challenges in the future. Both China and foreign countries have given a great deal of concern to it. Information and communication is a public platform and an important means of supporting a strong and smart grid construction. It is included in all of the six steps: generation, transmission, substation, distribution, electricity, scheduling. Moreover, information and communication is an important

Y. Liang (✉) · J. Yao · L. Huang
Research Institute of Information Technology & Communication,
China Electric Power Research Institute, Nanjing Jiangsu, China
e-mail: liangyun@epri.sgcc.com.cn

J. Liu
Information System Integration Corporation, NARI Group Corporation,
Nanjing Jiangsu, China

E. Qi (ed.), *Proceedings of the 6th International Asia Conference on Industrial Engineering and Management Innovation*,
DOI 10.2991/978-94-6239-145-1_59

foundation for running strong and smart grid [1, 2]. In order to promote the development and application of smart grid information and communication technology and realize the change from passive to active, from the secondary to the lead and from a single business model to a integration and innovation model, it is needed to keep up with the latest trends, actively design the integration and innovation for the power information and communication network and support the grid operation and business operations in a more "safe, effective and comprehensive" way.

An intelligent power information and communication network is able to control the operational status comprehensively and timely, integrate information and communication business and management in depth, sum the information analysis results of each management system and realize the reasonable allocation of information and communication resources centered on grid operation and business operations. As a result, a power information and communication network which owns advanced technology, reasonable layout, smooth transmission and home extension will be built [3–5]. At the same time, in order to enhance safe production level and customer service capabilities, the power companies have more requirements about operation and maintenance [6]. Moreover, the overall positioning of the power information and communication sector is now changing from previous transformation maintenance to business supporting services.

This paper comprehensively analyzed the development of the smart grid, combined with the latest ICT technology development results and demonstrated the important role assumed by the operation and maintenance system of information and communication in the new application mode of power grid. It is also proposed that the integrated operation and maintenance system of information and communication, which is based on resource integration, supported by standardize processes and centered by an integrated evaluation, is able to adapt to the characteristics and needs of the smart grid in multiple dimensions.

2 Requirement Analysis

Information and communication network is another physical network in addition to the power system. It is the important infrastructure for building a world-class power grid and a world-class enterprise. Power communication network is based on optical fiber, wireless, power line carrier and other technical systems, including data transmission, telephone, video conferencing, power business professional network and etc. By the end of the "Twelfth Five-Year", the communications network of SGCC will include all power networks whose voltage levels change from 1000 kV to 0.4 kV and its scale will be doubled. In addition, more new technology will be used [7]. The rapid development of power grid, the continuous expansion of communications networks, and the increasing complexity of structure, the enlargement of the covering span and the improvement of management difficulty require that the intelligence level of network operation and management for information and communication should be greatly enhanced.

With the evolution of next generation network technology, the division of labor among the existing professional networks will be continuously integrated. The original organizational structure for maintenance which was set by majors will be adjusted with the advancement in technology and development in business [3]. In order to better adapt to the needs of the grid and enterprise development, via using operation and management for information and communication, a standardized, hierarchical and networked operation and management for information and communication should be realized. Moreover, it needs to realize the operation management of business in different levels, pre-control the supervision and safety production vertically, improve the management level of information and communication continuously and further the integration of information and communication. To this end, it is urgent to promote new technologies and business models in the following three aspects.

(1) There is no unified and comprehensive resource management for information and communication network now. In addition, communications and information professionally use relatively independent resource model. This limits the progress of the integration of information and communication. The reliability of the information communication network and the efficiency of the operation and maintenance management can't match the grid, which thereby constrain their ability to support the new smart grid applications of information and communication network.
(2) There are a variety of communication methods for the power of information and communication network [8]. The lack of a unified management tool and the low reasonable usage rate of channels and information will be the bottleneck of the efficient and safe interaction information in smart grid interactive environment.
(3) So far, to a considerable extent, the grid information and communication monitoring system is an isolated part which has not formed a coordinated and efficient integration. As a result, it is hard to conduct a comprehensive analysis and evaluation of the operation of information and communication network based on monitoring data. For example, it is difficult to accurately locate the fault or the alarm occurrence to a specific point. So is the determination of fault channel quickly. All of these can't meet the future power demand in the technical and operational level.

3 Discussion

3.1 Content and Characteristics

Operation and maintenance system for the integration of information and communication is the system which faces to the integrated development trend of information and communication, combines with the current technical means, follows the requirements of the grid safety production and business management. It is a collection of established operation and maintenance of production methods, management methods and management system [9]. In the institutional aspect of operation and maintenance, it expands the monitoring capabilities of communications equipment and business systems, gradually transits from the dispersive and relative concentrated non-modern operation and maintenance system to the improved functional network management and monitoring system. The whole network use highly centralized modern operation and maintenance system. In the aspect of indicating operation and maintenance, in addition to the traditional operation and maintenance assessment indicators, based on the requirements of the modern operation and maintenance, it still need to add two indicators system: the users-oriented perception of network quality management system and network resource management system which can be used to measure the customer-centric service capabilities and network operating ability to create value. In the aspect of network management system, it transits from various professional network management systems to a management system for the entire network and finally becomes a service-oriented management system. In the aspect of customer service, it establishes the unified service assurance system and centered customer service mechanism, standardizes the workflow of customer service and form a linkage system which related to the system. In the aspect of safety protection, it enhances the standardization of information and communication checking process and closed-loop management forming a closed-loop system of safety management.

In summary, operation and maintenance system for the integration of power information and communication owns four characteristics: integrated development, highly centralization, customer perception, operation and maintenance outsourcing. These four main trends not only meet the basis requirement of modern operation and maintenance management system for power information and communication which is standardized and customer centered, but also reflect or mark that the operation and maintenance management level and ability for power information and communication will achieve a relatively high level.

3.2 Framework

The operation and maintenance system is a collection of all elements of which is built to meet the requirements of the operation and maintenance for power information and communication networks. It has operation and maintenance management system and operation and maintenance capacity system. The operation and maintenance management system includes five aspects: operation and maintenance mechanism, organizational development, resource allocation, operation and maintenance processes and management information system. These five aspects are the basic elements to build the core operation and maintenance capabilities of enterprises. Among them, the operation and maintenance mechanism is the core element to determine the operation and maintenance system. The operation and maintenance

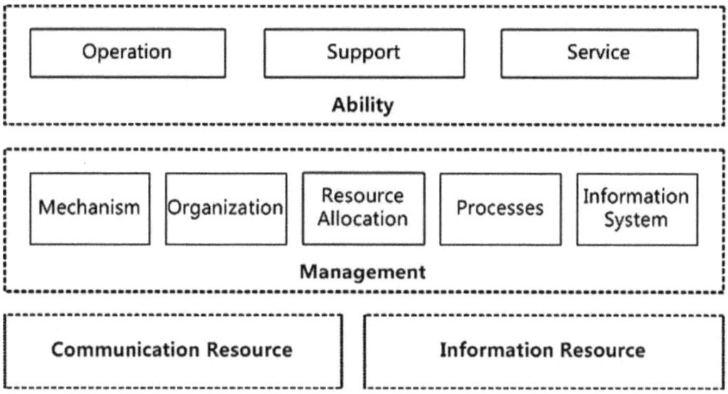

Fig. 1 The integrated framework

capacity system includes three aspects: operation capacity of networks, linkage support capability and service support capability. With the new development of grid, these three capabilities are needed to be enhanced in order to integrate the information and communication and at the same time to improve the level of operation and maintenance system.

In summary, the integrated operation and maintenance framework for information and communication system of power grid is shown in Fig. 1:

3.3 Technology

With the gradually application of large-capacity communications networks, virtualization, cloud computing technology, the power information system changes from the hierarchical deployment to the one–step development and the remote centralized operation and maintenance practices gradually replace most of the on-site maintenance. In addition, operation and maintenance work tends to be centralized and intensified. SGCC has built a comprehensive operation and maintenance of regulatory systems covering the full company information. This breaks the previous construction mode of silos type and gradually forms a horizontal and vertical integrated operation and maintenance management supporting platform. It covers an integrated and basic IT structure including network, host, business applications, safety equipment, desktop terminal infrastructure, and the standardized operation and maintenance processes. In addition, It realizes the goal that "real-time monitoring, careful management, closed-loop control, analysis and planning". The communication management system that covers the entire grid is also in rapid development, this provides the necessary technical means to realize the communication resource optimization, centralized network monitoring and management scheduling.

In technical route, the study of integration should be carried out from two levels [10]. Based on the existing system, which includes network management, security, smart room and information monitoring, and mature centralized information technology, equipment entities, monitoring systems and management applications can be integrated effectively and form an operation and maintenance structure for integrated information and communication which is “based on resource integration, supported by standardized process, centered by comprehensive evaluation”.

At the management level, based on the business system requirements and users' needs, the study of fault management, security management, performance management and resource management should be carried out. At the level of operation and maintenance, the study of service bearing, the state of resources, and equipment positioning should be carried out so as to realize the support of integrated information and communication from three aspects: monitoring system, resource model and equipment entities. In order to improve the integration of the system, the open interfaces will be reserved to meet the integration needs of other information systems (such as geographic information systems GIS) and provide data support for the comprehensive analysis and evaluation of the entire information and communication network. As shown in Fig. 2.

Based on the establishment of an integrated operation and maintenance system and technical support frame, in order to meet the needs of assessment and comprehensive analysis of the operational status of the power information and communication networks, by combining with the actual situation of the power information and communication networks, the performance model and evaluation system will be studied and the power information and communication resources state will be assessed so as to provide key technical support, such as decision, for the efficient operation and maintenance of the power information and communication facilities and realize scientific decision-making and active prevention.

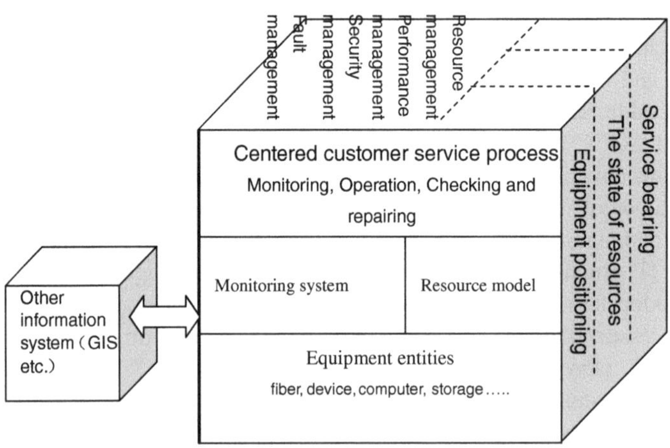

Fig. 2 Technical support for the operation and maintenance of power information and communication

3.4 Research Directions

(1) *Resource model*: Study the categories of information and communication resources and defined rules, and from the aspect of supporting power business, build the integrated information and communication resource model on physical level and business level. Firstly, initialize the classification of the optical fiber, terminal, transmission equipment, switching equipment, main computer, storage device entity and then merge them based on the characteristics of the physical resources, and finally to meet the needs of the business, aggregate and catalog all logic resources related to different business. Via building hierarchical information and communication resources model, it will effectively identify the physical and operational characteristics of the communication equipment and mark it reasonably. As a result, it will provide basis supporting for designing the data interactive interface of the resources operation and maintenance management.
(2) *Comprehensive alarm*: Alarm data is an important basis for the analysis of system failure. They derive from the network management, security management, business systems, terminal management and etc. So far, the alarm formats and types have already been specified within the related monitoring system so that a simple alarm generation, automatic confirmation, processing rules can be customized. However, it still can't meet the requirements of integrated information and communication. Via alarm specification alarm range, alarm processing, alarm rules, alarm analysis, alarm display and alarm restoration, a full range of alarm management can be achieved. Standard specification should clearly describe the level of alarm and urgency, alarm type, alarm content information; optimize alarm generation and notification policy include the configuration of the alarm generation rules, start and stop functions. Alarm notification processing rules include the method arrangement of alarm notification support and the configuration management of alarm notification content; by using the association analysis of repository network, it can realize the function of alarm such as influence analysis, fault location and alarm recombination and compression and so on.
(3) *Evaluation index system*: The index system is the basis of the comprehensive evaluation of operation state of information and communication network and the efficiency of the operation and maintenance work. The indicator system should not be designed too cumbersomely. Once the objective and comprehensive evaluation results can be kept, the indicator system should be as simple as possible via reducing and deleting some indicators which has little effect on the evaluation results. It is easy to collect the required data of evaluation indicators. No matter the evaluation indicators are qualitative or quantitative, their information sources much be reliable and easy to access. Otherwise, the evaluation work will be hard to carry out or too costly. Each evaluation indicator along with the corresponding calculation method and the data must be standardized and specified. In addition, the accuracy and

reliability of the data in the evaluation process need to be controlled. The indicator system focuses on three aspects: the configuration of indicators: matrix indicator system, via making indicator configuration persist, indicator system can be dynamically adjusted; the configuration of indicator evaluation method: via making indicator evaluation method configuration persist, indicator score and weight can be dynamically adjusted; the configuration of indicator display: via the dynamic analysis of indicator system, current indicator system can be dynamically displayed.

(4) *Visualization*: The trend is to refer to the entire network status and trends posed by a variety of information and communication equipment operating conditions as well as business systems and other factors. In the dynamic and complex environment, decision-makers need to use a situational awareness tool to display the continuous change of current environment and then make decisions accurately. The trend visualization displays network monitoring data via graphics. This makes the dispatchers have a comprehensive understanding of the current network conditions so that they can discover the running trend as well as the deeper reasons. In implementation techniques, many problems about visualization technology need to be further solved. For example, the combination of data display methods which are based on different data sources, the determination of unified standard of situation display, the improvement of real-time display and the enlargement of the scale of displayed system as well as the enhancement of the operability of Human-Computer interaction.

4 Conclusion

The fusion of information and communication has become the current development trend. It is widely recognized by the scientific research institutions, telecommunications operators, IT vendors and large enterprises. However, based on the different requirements and basic conditions, the development of integrated information and communication and the content of implementation are different. Operation and maintenance for integrated information and communications is able to build the relationship between power communication network and information system and meet the needs of smart grid and the modern enterprise. In addition, it will give greater value to information and communication resources.

The power information and communication network in one of the two entities as a grid network. Operation and maintenance for integrated information and communications must focus on the development of power production safety and enterprise intensification. So the theoretical researches about it as well as the study and explore the practical tools are the problems need to be solved. New theory supporting and breakage of traditional system structure and supporting tools are needed.

Acknowledgments This work was supported by the Research Program of SGCC (The Research on Integrated ICT Framework of Smart Grid).

References

1. Miao X, Zhang K, Tian S, Li J, Yin S, Zhao Z (2009) Information communication system supporting smart grid. Power Syst Technol 33(17):8–13 (in Chinese)
2. Lee GM, Su DH (2013) Standarization of smart grid in ITU-T. IEEE Commun Mag 90–97
3. Cheng S, Li X, Zhang Z (2011) Entire-grid-area information-sharing and integrated applications in united information system for smart grid. Proc CSEE 31(1):8–14 (in Chinese)
4. Zhang Z, Li X, Cheng S (2010) Structures, functions and implementation of united information system for smart grid. Proc CSEE 30(34):1–7 (in Chinese)
5. Zhong J, Zheng R, Yang W, Felix W (2009) Construction of smart grid at information age. Power Syst Technol 33(13):12–18 (in Chinese)
6. Zhu Kun, Chenine Moustafa, Nordström Lars (2011) ICT architecture impact on wide area monitoring and control systems' reliability. IEEE Trans Power Deliv 26(4):2801–2808
7. Cao H (2011) The summary of '12th five-Year' telecommunication network plan of SGCC. Telecommun Electric Power Syst 32(223):1–6 (in Chinese)
8. Güngör VC, Sahin D, Kocak T, Ergüt S, Buccella C, Cecati C, Hancke GP (2011) Smart grid technologies: communication technologies and standards. IEEE Trans Ind Inf 529–539
9. Chun-cong LU (2010) The tendency of communication operation and maintenance management. ZTE Commun 16(2):34–38 (in Chinese)
10. De-qiang YANG, Bin LI (2008) Discussion on function layered operation and maintenance system in electric power communication network. Telecommun Electric Power Syst 29(186):47–49 (in Chinese)

Vibration Suppression of Flywheel AMB System with Flexible Rotor Using Backstepping Control

Ming Ren, Dong-ji Xuan and Chun-chun Wang

Abstract This paper discusses the model vibration suppression of an Active Magnetic Bearing (AMB) system with flexible rotor. The key problem in the flexible rotor system is how to obtain a precise model considered the gyroscopic effect and design a suitable controller based on the reduced-order model to suppress critical vibration of actual system. Firstly, the vibration modes and mode shapes are analyzed using finite element method (FEM). Secondly, the AMB flexible rotor is stabilized with a PID control. The mode identification is also carried for validating the FEM model. Third, the cholesky decomposition is introduced for the mode separation and a backstepping controller for the flexible model control is designed based on the reduced order model. The effectiveness of the reduced-order model and the designed backstepping controller are verified through the simulation result.

Keywords Active magnetic bearing · Backstepping control · Flexible rotor · Vibration suppression

1 Introduction

Active magnetic bearing with larger flywheel for energy storage has been widely used in many industry domains such as peak load regulation of electric power system, replacement of the lead batteries for vehicles, power and frequency control of wind power generation due to its high energy density, long cyclic life, and wide operating temperature [1–3]. As an energy storage system, in order to obtain the higher energy density possibly, the most effective method is to increase the operating speed of flywheel rotor. However, accompany with the increase of rotation

M. Ren (✉) · D. Xuan
College of Mechanical Engineering, Wenzhou University, Wenzhou, China
e-mail: wzrm@wzu.edu.cn

C. Wang
Library, Wenzhou University, Wenzhou, China

E. Qi (ed.), *Proceedings of the 6th International Asia Conference on Industrial Engineering and Management Innovation*,
DOI 10.2991/978-94-6239-145-1_60

speed, the rotor should be regarded as a flexible rotor. Moreover, the gyroscopic effect and the critical vibration caused by flexible mode will become the main factors that strongly influence the stability of the system. On the other hand, it is difficult to suppress the vibration of the flexible rotor, unless a suitable controller can be designed base on a precise flexible model and the unstable vibration due to the gyroscopic effect at critical speeds can be eliminated.

This paper discusses the vibration suppression of a flexible rotor base on the flexible model which is obtained through the finite element method (FEM) considering the gyroscopic effect. As we know, the FEM model with the gyroscopic matrix is difficult to decouple, here we introduced the cholesky decomposition for the mode separation based on the method of L. Meirovitch. Moreover, In order to ensure the accuracy of FEM model, a closed-loop vibration frequency response base on the rigid model with PID control is carried out and a backstepping controller is also designed base on the flexible.

2 Modeling

2.1 The Research Object

The system used in this work is composed of a larger flywheel assembled at the top of the rotor whose two parts diameter is smaller than others apparently as shown in Fig. 1. Four pairs of radial electromagnets are horizontally assembled at the upper or lower part of the rotor respectively. One pair of axial electromagnet as a thrust bearing which has been controlled by PID is assembled in vertical direction. In general, the influence from radial and axial directions of the AMB is independent each other. Therefore, the vibration suppression of the study only cares about the

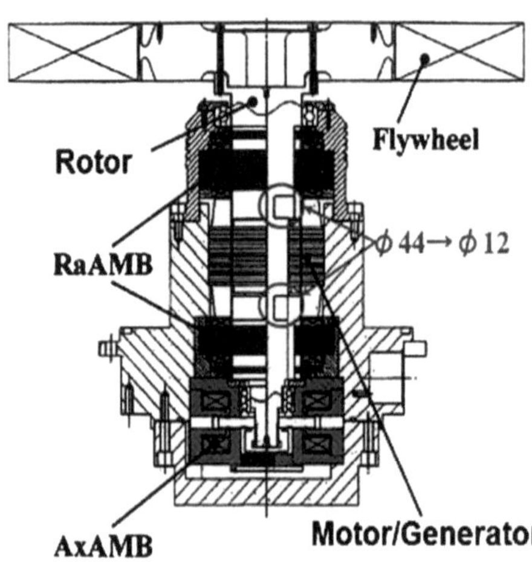

Fig. 1 Flywheel energy storage system

radial direction. A motor/generator is assembled in the middle of the rotor, which provides the possibility of energy conversion between the kinetic energy and the electric energy.

2.2 Analysis Model Based on Flexible Rotor

Since the rotor tends to regard as a flexible rotor due to its high operating speed and thinner size, the gyroscopic effect and the bending mode critical speeds had to be considered for stable control. In general, in order to obtain an accurate model of the rotor, its mode shapes and natural frequencies must be studied in advance. Here, the flywheel rotor is divided into 22 elements based on its shape as shown in Fig. 2. The radial electromagnets lie in the 9th and the 16th nodes respectively. Here, the analysis of mode shape adopts the FEM numerical calculation for controller design, instead of the general ANSYS soft, in other words, the eigenvalues and eigenvectors of the flexible rotor model can be obtained directly through the mass matrix, the stiffness matrix and gyroscopic matrix which are constructed by the lengths, the mass and the moment of inertia based on the size, density, Young's modulus of every elements. Finally, the representation of the flexible rotor system using FEM results in the standard equation of motion as follow [4].

$$\mathbf{M}\ddot{q} + \mathbf{G}\dot{q} + \mathbf{K}q = \mathbf{T} * U \tag{1}$$

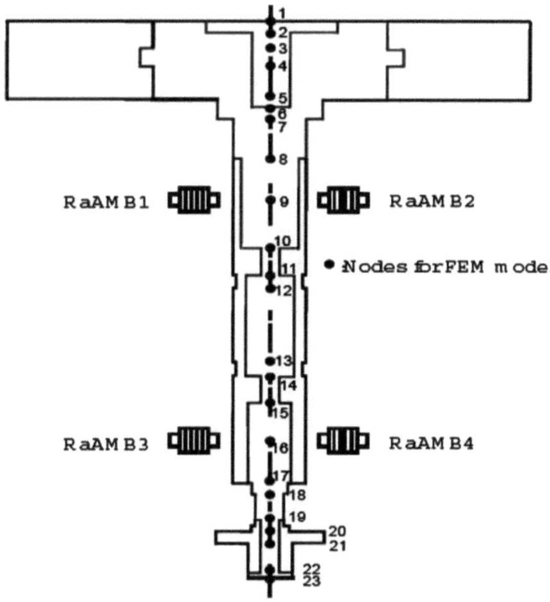

Fig. 2 Schematic of flywheel rotor for FEM

where, M, G and K denote the mass matrix, the stiffness matrix and gyroscopic matrix, T denotes control matrix which indicates the position of electromagnetic in notes, U is external force vector, denotes displacements and the inclination angles of each node. Figure 3 shows the mode shapes of the rotor in free levitation at 0 rpm that the gyroscopic effect is not to be considered and G, U are regarded as zero in (1). From the result the natural frequencies of the 1st, 2nd rigid mode are 21, 73 Hz respectively and the 1st, 2nd bending mode are 162, 565 Hz respectively, where the larger strains occur at the sudden thinner in diameter of the 10th and the 14th elements. Therefore, it is known that the designed controller should be able to suppress the vibration of the first bending mode at least during the 300 Hz range of the rotor operating speed.

2.3 Model Validation

In spite of the natural frequency and mode shapes of the rotor in levitation have been investigated, but the model of the flywheel rotor is variable along with the rotor rotational speed, where the gyroscopic matrix G in (1) is not zero but an anti-symmetric matrix. In order to mach the FEM model with the actual rotor throughout the whole operating frequency range, a mode identification experiment is carried. First, the AMB flexible rotor is stabilized with a PID controller which guarantees to pass through the rigid mode critical speeds. Then, a sweep sinusoidal

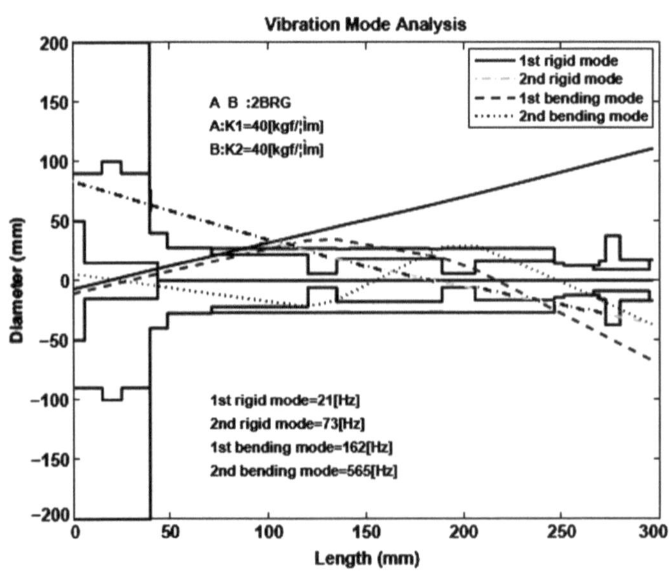

Fig. 3 Mode shapes at levitaion

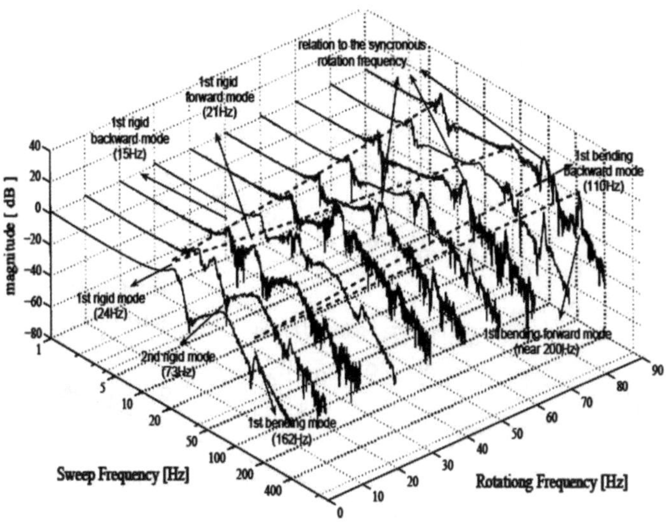

Fig. 4 Vibration analysis of mode identification

signal varying from 0 to 400 Hz is injected into the system as a reference signal. The frequency response obtained in a 3D plot is shown in Fig. 4, where the sinusoidal signal are applied to the system nine times at each rotational frequency of rotor from the levitation to the rotation frequency 90 Hz increased by 10 Hz. It is found that each natural vibration modes at levitation branches to forward mode and backward mode due to the gyroscopic effect in acceleration of rotor.

In order to validate the FEM model, the variable eigenvalues of the flexible model at different rotation frequency of the rotor are also calculated. Here, the spring coefficient in the K matrix and the angular velocity ω in the G matrix are set as same as the experiment value of closed-loop system with PID control. Moreover, the flywheel can be regarded as a concentrated mass, and its polar moment of inertia only effects on the 5th elements of Fig. 2. The relation between natural frequencies and rotation frequencies (also known as the campbell diagram) is shown in Fig. 6 where the theoretical eigenvalues are close to the natural frequencies obtained from the mode identification experiment. Several mismatches are mainly due to the structure simplification of the FEM model and the actual system containing sensors, power amplifier. According to the Fig. 5, the first backward mode tends the low-frequency range rapidly with the acceleration of the rotor and intercross at 110 Hz with the straight line representing the rotation frequency equivalent of the natural frequency. The first bending forward mode is nearly parallel to the straight line. Therefore, it is necessary to design a suitable controller for passing through the critical speed of the bending backward mode at 110 Hz.

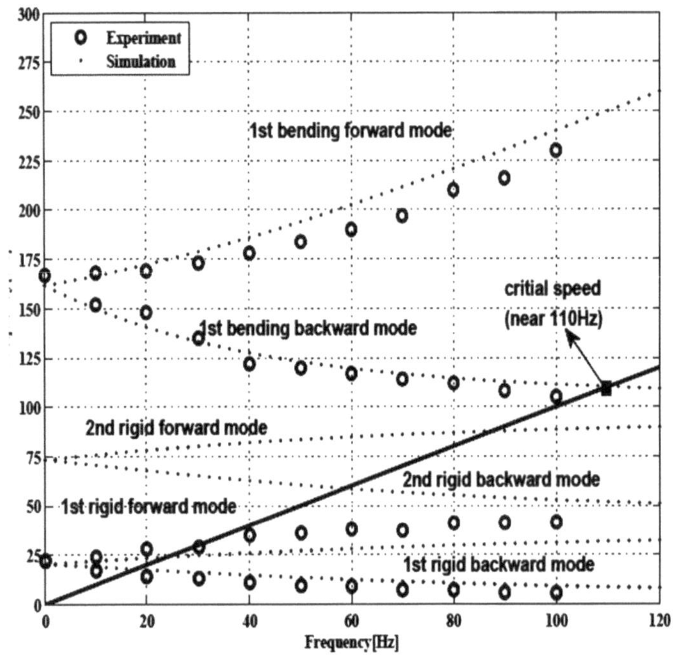

Fig. 5 Relation between resonance frequency and rotational frequency

2.4 *Mode Separation*

In general, the FEM model is unable to be used for controller design due to its higher orders. Here, the order of the designed FEM model considering the gyroscopic effect is 80. Therefore, it is necessary to obtain a reduced order model only including the 1st bending backward and forward mode. On the other hand, since the gyroscopic matrix of the equation of motion in (1) is an anti-symmetric matrix which cannot be decoupled simply by conventional methods. Here, we adopted the method of L. Meirovitch [5, 6] to introduce a new state vector of $z = [q \;\; \dot{q}]^{\mathrm{T}}$. The form of (1) is rewritten as follows.

$$\mathbf{M}^* \dot{z}(t) + \mathbf{G}^* z(t) = \mathbf{T}^* U$$
$$\mathbf{M}^* = \begin{bmatrix} \mathbf{K} & \mathbf{0} \\ \mathbf{0} & \mathbf{M} \end{bmatrix}, \mathbf{G}^* = \begin{bmatrix} \mathbf{0} & -\mathbf{K} \\ \mathbf{K} & \mathbf{G} \end{bmatrix}, \mathbf{T}^* = \begin{bmatrix} \mathbf{0} \\ \mathbf{T} \end{bmatrix} \tag{2}$$

It is well known that the eigenvalue problem of (2) consist of n pairs of pure imaginary complex conjugates, $s_r = \pm i\omega_r$ correspondingly, the eigenvectors also occur in pairs of complex conjugates $z_r = \alpha_r \pm i\beta_r$. Introducing $s_r = i\omega_r$ and $z_r = \alpha_r + i\beta_r$ into $sM^*z(s) + G^*z(s) = 0$ and separating the real and the imaginary parts, therefore, two companion equations in terms of α_r and β_r can be obtained.

Eliminating α_r from the first and β_r from the second, the equation can be obtained as follow.

$$\begin{cases} \mathbf{K}^*\alpha_r = \lambda_r \mathbf{M}^*\alpha_r,\ \mathbf{K}^*\beta_r = \lambda_r \mathbf{M}^*\beta_r \\ \mathbf{K}^* = \mathbf{G}^{*\mathrm{T}}\mathbf{M}^{*-1}\mathbf{G}^*,\ \lambda_r = \omega_r^2 \end{cases} \tag{3}$$

Both the real part α_r and the imaginary part β_r of the eigenvetor z_r satisfy the same eigenvalue $\lambda_r = \omega_r^2$.Though, each ω_r^2 has multiplicity two, the eigenvectors α_r and β_r are linearly independent due to positive matrices in (3). Next, the matrix $\mathbf{M}^*$ is decomposed using cholesky decomposition as $\mathrm{M}^* = RR^T$ and introduced into (3). The following equation is obtained.

$$\begin{cases} \varphi \mathbf{P_r} = \lambda_\mathbf{r} \mathbf{P_r} \\ \varphi = \mathbf{R}^{-1}\mathbf{K}^*\mathbf{R}^{-\mathrm{T}} \end{cases} \tag{4}$$

It has been an eigenvalue problem in standard form and the eigenvector $\mathrm{P}_r = [\alpha_r\ \beta_r]$ corresponds to λ_r. Finally, introduce the linear coordinate transformation $z = \mathrm{R}^{-\mathrm{T}}\mathrm{P}\xi$ into (2), the equation related to the mode vector ξ is obtained as follow.

$$\begin{aligned} &\dot{\xi} = \boldsymbol{\theta}\xi + \boldsymbol{\psi} U \\ &\boldsymbol{\theta} = -\mathbf{P}^{-1}\Lambda\mathbf{P},\ \Lambda = \mathbf{R}^{*-1}\mathbf{G}^*\mathbf{R}^{*-T}, \boldsymbol{\psi} = \mathbf{P}^{-1}\mathbf{R}^{*-1}\mathbf{T}^* \end{aligned} \tag{5}$$

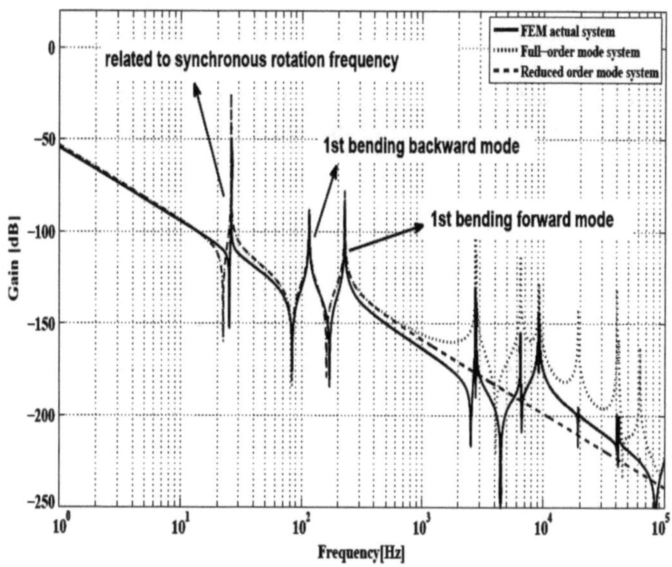

Fig. 6 Mode seperation without rigid modes

Now, all system modes have been separated independently. Next, in order to obtain the reduced order system for controller design, the rigid mode and the 1st bending mode are extracted from full mode system as shown in Fig. 6 in which the full order mode system denoted by the dotted line well matches the FEM model actual system denoted by the solid line and the reduced order mode system denoted by the red dashed line. Here, the rigid modes do not appear because the spring coefficient is set as zero for presenting the free-free rotation rotor.

3 Backstepping Control

3.1 Control Design

The Controller design adopts the reverse recurrence method to obtain a stable feedback control law. It is regarded that the state of each subsystem as a virtual control input from 1 step to n-1 step. The actual control input can be obtained until finishing the n step. The stability of subsystem related to each step is guaranteed by the Lyapunov stable condition, therefore, the stability of the whole system is also guaranteed. Here, In order to design a backstepping controller, the reduced-order model is firstly constructed via mode separation. On the other hand, as we know, it is difficult to guarantee the stability of the actual system using the reduced order model based controller design due to the spillover of high order modes [7]. Therefore, a low pass filter is inserted into the controller output to eliminate the influence of high frequency range. The state equations of reduced order system and low-pass filter are described in (6), whose order numbers are 12 and 8 respectively. Moreover, in order to estimate the unmeasured state variables of the mode system, a full order observer is also added into the system as shown in Fig. 7 which the state structure of closed-loop control system. Here the gain of the observer can be obtained through guaranteeing the estimation error exponentially converges to zero.

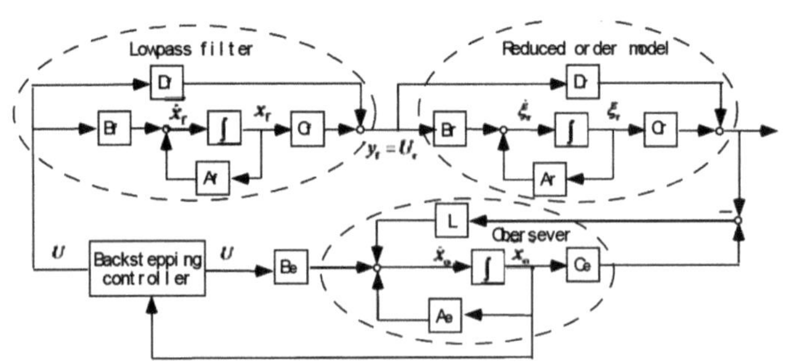

Fig. 7 The state structure diagram of bacstepping control system

$$\begin{cases} \dot{\xi}_r = \mathbf{A}_r\xi_r + \mathbf{B}_r U_r \\ y_r = \mathbf{C}_r\xi_r + \mathbf{D}_r U_r \end{cases}, \quad \begin{cases} x_f = \mathbf{A}_f x_f + \mathbf{B}_f U \\ y_f = \mathbf{C}_f x_f + \mathbf{D}_f U \end{cases} \tag{6}$$

Considering the output of the low-pass filter is the input of the reduced-order model $U_r = y_f$, therefore, the augmented system in terms of the reduced-order system and the low-pass filter can be obtained as follow.

$$\begin{gathered} \begin{cases} \dot{x}_e = \mathbf{A}_e x_e + \mathbf{B}_e U \\ y_e = \mathbf{C}_e x_e + \mathbf{D}_e U \end{cases} \\ x_e = \begin{bmatrix} \xi_r \\ x_f \end{bmatrix}, \quad \mathbf{A}_e = \begin{bmatrix} \mathbf{A}_r & \mathbf{B}_r\mathbf{C}_f \\ 0 & \mathbf{A}_f \end{bmatrix}, \quad \mathbf{B}_e = \begin{bmatrix} \mathbf{B}_r\mathbf{D}_f \\ \mathbf{B}_f \end{bmatrix} \\ \mathbf{C}_e = [\mathbf{C}_r \quad \mathbf{D}_r\mathbf{C}_f], \quad \mathbf{D}_e = \mathbf{D}_r\mathbf{D}_f \end{gathered} \tag{7}$$

In order to design a backstepping controller, the sequence order of the states x_e is rearranged as follow.

$$\dot{x}_e = \begin{bmatrix} \dot{x}_{rf} \\ \ddot{x}_{rf} \end{bmatrix} = \begin{bmatrix} \mathbf{A}_{11} & \mathbf{A}_{12} \\ \mathbf{A}_{21} & \mathbf{A}_{22} \end{bmatrix} \begin{bmatrix} x_{rf} \\ \dot{x}_{rf} \end{bmatrix} + \begin{bmatrix} \mathbf{B}_{rf1} \\ \mathbf{B}_{rf2} \end{bmatrix} U \tag{8}$$

According to the relation of the first row, we can get the equation as follow.

$$x_{rf} = \mathbf{A}_{11}^{-1}(I - \mathbf{A}_{21})\dot{x}_{rf} - \mathbf{A}_{11}^{-1}\mathbf{B}_{rf1}U \tag{9}$$

Take the (9) into the second row of the Eq. (8) and substitute x_{rf}, then, the following equation can be obtained.

$$\begin{aligned} \ddot{x}_{rf} &= \mathbf{M}\dot{x}_{rf} + \mathbf{N}U \\ \mathbf{M} &= \mathbf{A}_{21}\mathbf{A}_{11}^{-1}(\mathbf{I} - \mathbf{A}_{12}) + \mathbf{A}_{22}, \quad \mathbf{N} = \mathbf{B}_{rf2} - \mathbf{A}_{21}\mathbf{A}_{11}^{-1}\mathbf{B}_{rf1} \end{aligned} \tag{10}$$

Assuming the $\dot{x}_1 = x_2 = \dot{x}_{rf}$, then

$$\begin{cases} \dot{x}_1 = x_2 \\ \dot{x}_2 = \mathbf{M}x_2 + \mathbf{N}U \end{cases} \tag{11}$$

Step1. Regarding the x_2 as a virtual control input and assuming $x_2 = -\mathbf{C}_1 x_1$, moreover, the lyapunov function V_1 is defined as $V_1 = \frac{1}{2}x_1^2$, therefore, the derivative of $\dot{V}_1$ is derived as follow.

$$\dot{V}_1 = x_1\dot{x}_1 = x_1 x_2 = -\mathbf{C}_1 x_1^2 < 0 \quad (\mathbf{C}_1 > 0) \tag{12}$$

Here, $\mathbf{C}_1$ is a normal symmetric matrix. A stabilizing function is chosen as $\alpha_1 = -C_1 x_1$ and defining the error variable $z_1 = x_2 - \alpha_1$, then the derivative of the z_1 is obtained as

$$\dot{z}_1 = \dot{x}_2 - \dot{\alpha}_1 = \mathbf{M}x_2 + \mathbf{N}U - \mathbf{C}_1 x_2 \tag{13}$$

$\dot{V}_1 = x_1 x_2 = x_1(z_1 + \alpha_1) = x_1(z_1 - \mathbf{C}_1 x_1) = -\mathbf{C}_1 x_1^2 + x_1 z_1$
Step2. A candidate Lyapunov function for the error variable z_1 is defined as

$$V_2 = V_1 + \frac{1}{2} z_1^2 \tag{14}$$

Then the derivative of the $\dot{V}_2$ is obtained as

$$\dot{V}_2 = \dot{V}_1 + z_1 \dot{z}_1 = -\mathbf{C}_1 x_1^2 + z_1(\mathbf{M}x_2 + \mathbf{N}U - \mathbf{C}_1 x_2) \tag{15}$$

From the Eq. (15), if the control input U can be chosen as

$$\begin{aligned} U &= \mathbf{N}^{-1}(-x_1 - \mathbf{M}x_2 - \mathbf{C}_2 z_1 + \mathbf{C}_1 x_2) \\ &= \mathbf{N}^{-1}\{-(\mathbf{I} + \mathbf{C}_1\mathbf{C}_2)x_1 - (\mathbf{M} + \mathbf{C}_2 + \mathbf{C}_1)x_2\} \end{aligned} \tag{16}$$

Here, $\mathbf{C}_2$ is a normal symmetric matrix. Then, the derivative of the $\dot{V}_2$ is obtained as

$$\dot{V}_2 = \dot{V}_1 + z_1 \dot{z}_1 = -\mathbf{C}_1 x_1^2 - \mathbf{C}_2 z_1^2 \leq 0 \tag{17}$$

It is clear that a global stability is maintained in the final Lyapunov function. On the other hand, the control input U is only the magnetic attractive force. The actual control currents via the electromagnet can be obtained by the zero power current switching method [8].

3.2 Simulation Verification

In order to test the designed backstepping controller whether is able to suppress the critical vibration that is related to the flexible modes, the simulation of impulse disturbance resonance is carried out. Firstly, the reduced-order model based backstepping controller is applied to the actual FEM system of (1). That is, the reduced-order model of Fig. 8 is replaced by the FEM model. Next, an impulse disturbance is exerted into the position corresponding to the electromagnet of the upper of rotor when the simulation time is 0.5 s. Here, the rotation speed of rotor is regarded as variable parameter and set 110 Hz in controller. Figure 8 shows the output displacement of sensor and Fig. 9 shows the control attractive force input. From the results, it shows that the system is still able to keep its stability near the critical speed despite the sudden disturbance.

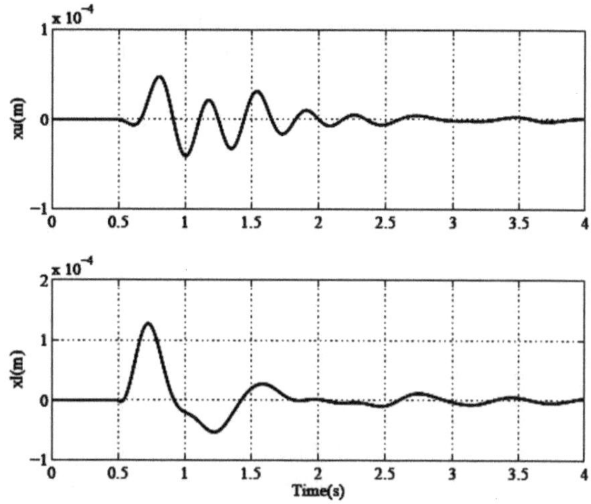

Fig. 8 The displacement of impluse disturbance resonance

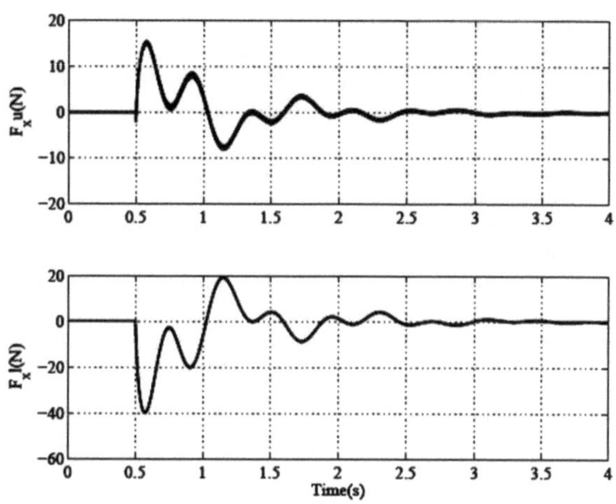

Fig. 9 The control input of impluse disturbance resonance

4 Conclusion

In this study, the stability of a flywheel AMB with flexible rotor considering the gyroscopic effect has been studied. The conclusions of the paper are given as follows

1. In order to obtain a precise mathematic model of flywheel magnetic bearing system with flexible rotor for the controller design, the numerical FEM analysis must be carried out firstly. Meanwhile, the mode indentation via vibration experiment should be done for validating the FEM model.

2. The effectiveness of mode separation using the cholesky decomposition of L. Meirovitch for the actual flexible rotor model with gyroscopic matrix was verified.
3. The effectiveness of backstepping controller based on reduced-order model for the actual flexible rotor was verified by the simulation.

Acknowledgments This project was supported by WenZhou scientific and technological plan projects (No. G20100053).

References

1. Zhang YJ, Nonami KZ, Higasa HM (2001) Sliding mode control of 10 MWh class energy storage flywheel system using superconducting magnetic bearing with gyroscopic effect. Trans Jpn Soc Mech Eng 67(662):3139–3145 (in Japanese)
2. Pichot MA, Kajs JP, Ouroua A (2001) Active magnetic bearings for energy storage system for combat vehicles. IEEE Trans Magn 37(1):318–324
3. Sun CS, Wang YN, Li XR (2008) Synthesized power and frequency control of wind power generation. Proceedings of the CSEE 28(29):111–117
4. Nonami K (1982) Responses of passing through critical speeds for arbitrarily distributed mass flexible rotor systems : 1st report, case without gyroscopic effects. J Jpn Soc Mech Eng 48 (435):1669–1675 (in Japanese)
5. Meirovitch L (1980) Computational methods in structural dynamics. Sijthoff Noordhoff International Publishers B.V., Netherlands
6. Meirovitch L (2012) Modal-space control of distributed gyroscopic systems. J Guidance Control 3(2):140–150
7. Lin SL, Sun T, Xu BH (1997) A survey on spillover problems in the control for flexible structure. Adv Mech 27(4):447
8. Ariga Y, Nonami K, Sakai K (2001) Nonlinear control of zero power magnetic bearing system. Trans Jpn Soc Mech Eng 67(654):654 (in Japanese)

A Study on Stability Control of Vehicle Magnetic Levitation Flywheel Battery Based on Root Locus Analysis

Ming Ren, Tai-hong Cheng and Chun-chun Wang

Abstract The goal of this paper is to research the vibration suppression of a vehicle magnetic flywheel battery system. It is found that the flywheel rotor is always instability near a certain rotation speed controlled on the basis of the rigid model. Accordingly, regarding the rotor as a flexible rotor is investigated based on the FEM model. The original controller used for the rigid model is also applied to the flexible model for analyzing the root locus. The result reveals the poles associated with the first bending backward mode move to the half right complex plane. Finally, a simple cross-feedback control is proposed instead of complex controller design based on the flexible model. The effectiveness is verified through the experiment.

Keywords Bending mode · Cross-feedback control · Flywheel magnetic bearing · FEM · Root locus analysis

1 Introduction

Active magnetic bearing (AMB) attached a larger flywheel as energy storage system equipped in hybrid vehicle has become a research focus instead of conventional lead batteries [1, 2]. On the other hand, In order to promote the continuous marching ability of flywheel battery, the rotation speed of rotor is expected to increase as much as possible. However, the flywheel as the main energy storage device will cause the stronger gyroscopic effect which makes the rotor unstable easily.

M. Ren (✉) · T. Cheng
College of Mechanical Engineering, Wenzhou University, Wenzhou, China
e-mail: wzrm@wzu.edu.cn

C. Wang
Library, Wenzhou University, Wenzhou, China

E. Qi (ed.), *Proceedings of the 6th International Asia Conference on Industrial Engineering and Management Innovation*,
DOI 10.2991/978-94-6239-145-1_61

In this research, a rigid model considering gyroscopic effect is first introduced into a large flywheel AMB system, meanwhile, a backstepping controller is also designed on the basis of the rigid model. Through the experiment, it is found that the rotor is always touchdown near 110 Hz. However, the root locus analysis of closed-loop system shows the system will be stable in whole operating frequency. Aiming at this problem, a FEM model is also derived and the original controller used for the rigid model is applied to the FEM model again. According to the root locus analysis, then the cause of the problem is found. Finally, In order to overcome the critical speed, a simple cross-feedback control is proposed instead of the complex controller designed on the basis of the FEM model. The procession of control can be divided into two steps, one is backsteping control for the vibration suppression of rigid mode, and another is cross feedback backstepping control for the vibration suppression of bending mode. The effectiveness of designed control system is verified through the experiment.

2 Model and Control of Rigid Rotor

2.1 The Rigid Model of Rotor

The schematic diagram of the AMB system in this paper is shown in Fig. 1. The rigid model of such five degree freedom has been researched deeply in several references [3]. Here, the state equation of the rigid model which only considers the motion of the radial direction owing to the independent from axial direction is introduced directly as follows:

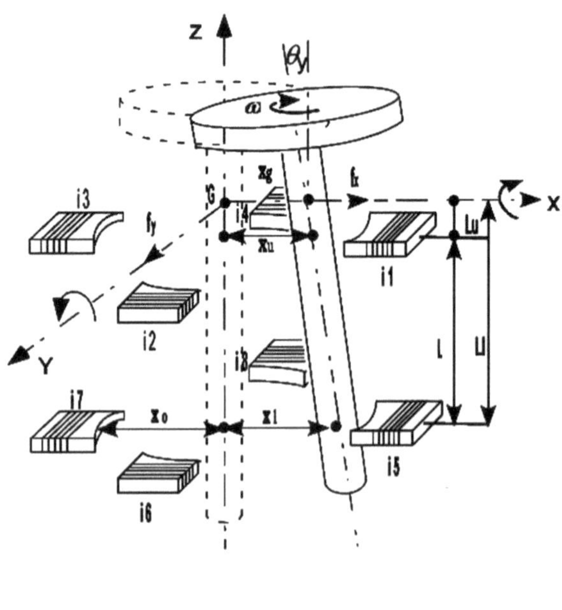

Fig. 1 The schematic diagram of the AMB system

$$\begin{bmatrix} \dot{x}_1 \\ \dot{x}_2 \end{bmatrix} = \begin{bmatrix} 0_{4\times4} & I_{4\times4} \\ 0_{4\times4} & A_{22} \end{bmatrix} \begin{bmatrix} x_1 \\ x_2 \end{bmatrix} + \begin{bmatrix} 0_{4\times4} \\ B_2 \end{bmatrix} U,\ y = Sg * \begin{bmatrix} I_{4\times4} & 0_{4\times4} \end{bmatrix} \begin{bmatrix} x_1 \\ x_2 \end{bmatrix}. \tag{1}$$

$$\mathbf{A_{22}} = \begin{bmatrix} 0 & 0 & \frac{\omega L_u I_a}{L I_r} & -\frac{\omega L_u I_a}{L I_r} \\ 0 & 0 & \frac{\omega L_l I_a}{L I_r} & -\frac{\omega L_l I_a}{L I_r} \\ -\frac{\omega L_u I_a}{L I_r} & \frac{\omega L_u I_a}{L I_r} & 0 & 0 \\ -\frac{\omega L_u I_a}{L I_r} & \frac{\omega L_u I_a}{L I_r} & 0 & 0 \end{bmatrix}$$

$$\mathbf{B_2} = \begin{bmatrix} \frac{1}{M} + \frac{Lu^2}{Ir} & 0 & 0 & 0 \\ 0 & \frac{1}{M} + \frac{Ll^2}{Ir} & 0 & 0 \\ 0 & 0 & \frac{1}{M} + \frac{Lu^2}{Ir} & 0 \\ 0 & 0 & 0 & \frac{1}{M} + \frac{Ll^2}{Ir} \end{bmatrix}$$

Here, $x_1([\text{xu xl yu yl}]^T)$ means the offset of displacement. $U([F\text{xu}\ F\text{xl}\ F\text{yu}\ F\text{yl}]^T)$ is control input associate with the magnetic forces. A_{22} associates with the gyroscopic effect. L_u, L_l denote the distance between the centre of gravity and the upper electromagnet or lower electromagnet, respectively. ω denotes the angle of velocity, M denotes the mass of flywheel rotor. I_a,I_r are the polar and axial moment of inertia.

2.2 Backstepping Control (BC)

As we well know, the advantage of backstepping controller design is that a stable control law can be constructed while finding a Lyapunov function. Here, assuming the reference as r, displacement as ξ_2 velocity as ξ_3, the error of displacement as $e = \xi_2 - r$, and $\dot{\xi}_1 = e$, therefore, the expanded system can be rewritten as follow.

$$\begin{cases} \dot{\xi}_1 = \xi_2 - r \\ \dot{\xi}_2 = \xi_3 \\ \dot{\xi}_3 = A_{22}\xi_3 + B_2 U \end{cases} . \tag{2}$$

According to the step of BC design on the basis of (2), then the final control input U which can guarantee the system as a globally asymptotically stable system is obtained as follows.

$$U = -\frac{1}{B_2}[K_1\xi_1 + K_2\xi_2 + (A_{22}(\omega) + K_3)\xi_3 - r - K_4\dot{r} - \ddot{r}] \tag{3}$$

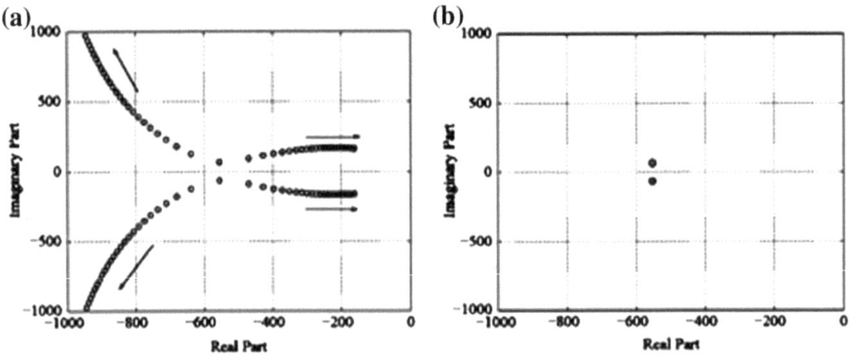

Fig. 2 Root locus analysis of the closed-loop with backstepping controller

Here, $\begin{aligned} K_1 &= C_1 + C_3 + C_1C_2C_3 \\ K_2 &= 2I + C_1C_2 + C_1C_3 + C_2C_3 \\ K_3 &= C_1 + C_2 + C_3 \\ K_4 &= C_1 + C_2 \end{aligned}$, C_1, C_2 and C_3 are the positive definite parameter matrix. The final control currents flowing in the eight electromagnets can be obtained from U by switching method of current [4].

However, in experiment, the rotor always loses its stability at certain rotation frequency although having a good stability in levitation. Therefore, root locus analysis of the close-loop system is carried out to investigate the problem. Here, the rotation speed is increased up to 100 Hz by 5 Hz. Figure 2a shows all poles lie in the left half complex plane and are far from zero point when not considering the gyroscopic effect ingredient in controller. Figure 2b shows all poles are stationary in the complex plane and the gyroscopic effect is eliminated thoroughly when considering the gyroscopic effect ingredient into controller. Therefore, it could be draw a conclusion that the system should keep its stability continuously using designed backstepping controller.

3 Flexible Fem Model

According to the analysis of root locus above, for the rigid model, whether or not considering the gyroscopic effect in a backstepping controller, it does not affect the stability of closed-loop system in all operating frequency range. On the other hand, the rotor always loses its stability near a certain rotation frequency. In order to investigate the conflicting problem, here, the rotor is regarded as a flexible rotor and the flywheel as a concentrated mass. A flexible model is derived by the FEM and the rotor is divided into 8 elements in terms of its shape as shown in Fig. 3. The radial electromagnets lie in the third and the 8th nodes respectively. The standard equation in x-y direction motion of the nth element can be described as follow [5].

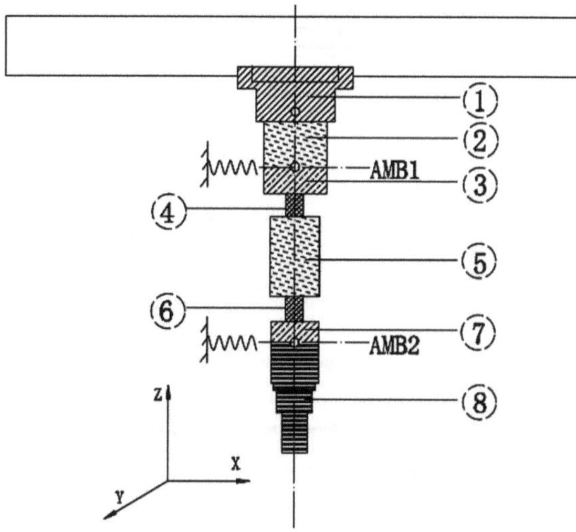

Fig. 3 One dimensional finite element model

$$\begin{bmatrix} \mathbf{M} & 0 \\ 0 & \mathbf{M} \end{bmatrix}\begin{bmatrix} \ddot{X} \\ \ddot{Y} \end{bmatrix} + \begin{bmatrix} 0 & \mathbf{G_{xy}} \\ \mathbf{G_{yx}} & 0 \end{bmatrix}\begin{bmatrix} \dot{X} \\ \dot{Y} \end{bmatrix} + \begin{bmatrix} \mathbf{K_{xx}} & 0 \\ 0 & \mathbf{K_{yy}} \end{bmatrix}\begin{bmatrix} X \\ Y \end{bmatrix} = F_i \tag{4}$$

$$\mathbf{M} = \frac{\mu l}{420}\begin{bmatrix} 156 + M^{L'} & -22l & 54 & 13l \\ -22l & 4l^2 + I_d^{L'} & -13l & -3l^2 \\ 54 & -13l & 156 + M^{R'} & 22l \\ 13l & -3l^2 & 22l & 4l^2 + I_d^{R'} \end{bmatrix}, \quad \begin{aligned} M^{L'} &= \tfrac{420}{\mu l} M^L \\ M^{R'} &= \tfrac{420}{\mu l} M^R \\ I_d^{L'} &= \tfrac{420}{\mu l} I_d^L \\ I_d^{R'} &= \tfrac{420}{\mu l} I_d^R \end{aligned}$$

$$\mathbf{G_{xy}} = \begin{bmatrix} 0 & & & 0 \\ & \omega I_p^L & & \\ & & 0 & \\ 0 & & & 0 \end{bmatrix}, \mathbf{G_{yx}} = \begin{bmatrix} 0 & & & 0 \\ & -\omega I_p^L & & \\ & & 0 & \\ 0 & & & 0 \end{bmatrix}$$

$$\mathbf{K_{xx}} = \mathbf{K_{yy}} = \frac{2EI}{l^3}\begin{bmatrix} 6 + K_{xx}^{L'} & -3l & -6 & -3l \\ -3l & 2l^3 & 3l & l^2 \\ -6 & 3l & 6 + K_{xx}^{R'} & 3l \\ -3l & l^2 & 3l & 2l^2 \end{bmatrix}, \quad \begin{aligned} K_{xx}^{L'} &= \tfrac{l^3}{2EI} K_{xx}^L \\ K_{xx}^{R'} &= \tfrac{l^3}{2EI} K_{xx}^R \end{aligned}$$

Here, **M, G** and **K** denote the mass matrix, the stiffness matrix and the gyroscopic matrix, F_i is the external force on the nth element. μ and l present density and length of each element, $M^{L(R)}$ is concentrated mass of flywheel. $I_p^{L(R)}$ and $I_d^{L(R)}$ are moment of inertia in axis or radial direction respectively, (the superscript L, R means left side or right side of element). $K_{xx(yy)}$ denotes the spring coefficient. Finally, the whole FEM flexible rotor model is as follow.

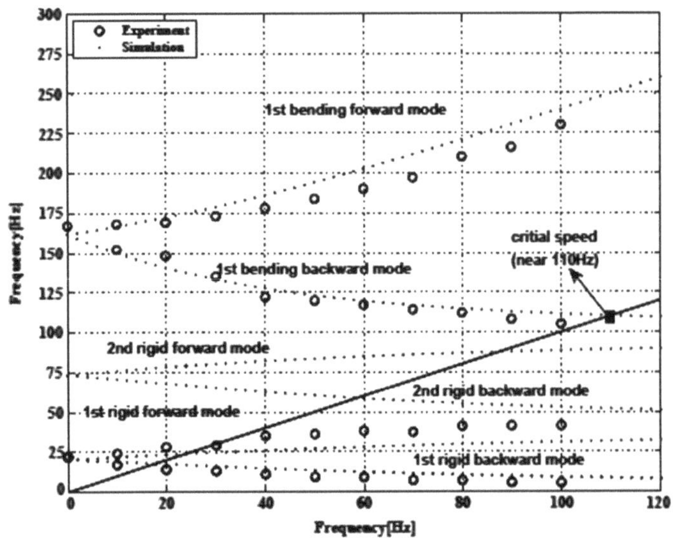

Fig. 4 Relation between the natrual frequency and the rotation frequency

$$\tilde{\mathbf{M}}\ddot{q} + \tilde{\mathbf{G}}\dot{q} + \tilde{\mathbf{K}}q = \tilde{\mathbf{T}} * U \tag{5}$$

Here, $q = [X \quad Y]^T$, matrix $\tilde{T}$ denotes control matrix which indicates position of the electromagnet on the nodes. U is the external force vector. Figure 4 shows the relation between the natural frequency and rotation frequency when the rotor is in free condition. It is clear that the first bending mode branches to the backward mode and the forward mode largely and the eigenvalues associated with the first bending backward mode fast falls to near 110 Hz due to strong gyroscopic effect. From above, it is understood that the rotor gradually close to the critical speed near 110 Hz.

4 Root Locus Analysis on Basis of Flexible Model

In order to investigate the stability of the system deeply, the input of control input $U = -F_p q - F_d \dot{q}$ used for rigid model is applied again to the flexible model for the analysis of root locus. Here, if $A_{22} = 0$ in the (3) means the gyroscopic effect is ignored in the control input U. F_p and F_d are feedback gain matrix associated with position and velocity. Take the control input U into the (5), the follow equation can be obtained.

$$\mathbf{M}^* \ddot{q} + \mathbf{G}^* \dot{q} + \mathbf{K}^* q = 0 \tag{6}$$

$$\mathbf{M}^* = \widetilde{\mathbf{M}},\ \mathbf{G}^* = \widetilde{\mathbf{G}} + \widetilde{\mathbf{T}}\mathbf{F_d}\ \mathbf{K}^* = \widetilde{\mathbf{K}} + \widetilde{\mathbf{T}}\mathbf{F_p}$$

Figure 5a shows the poles associated with the first rigid forward mode have approached to near zero point in complex plant while the rotor rotating at 100 Hz. It can reveal the rotor has a more serious tendency to instability as compared with the result of Fig. 2a. The result considering the gyroscopic effect is shown in Fig. 5b, which reveals the gyroscopic effect is unable to be eliminated compared with Fig. 2b. Figure 5c shows the damping of the poles associated with the first bending backward mode begin to decrease rapidly from 100 Hz.

Through investigating the interior structure of the matrix G^*, it is found that the position of ingredient about the gyroscope effect caused by the flywheel doesn't match with the ingredient of compensating the gyroscope effect in controller. Therefore, it becomes impossible to eliminate the gyroscope effect. The reason could be explained that the rotor is assumed as concentrated mass on gravity in rigid model. On the other hand, the flexible model is assumed as distributed mass in designed controller for the rigid model. From the above, although the backstepping controller considering the gyroscopic effect is designed based on the rigid model,

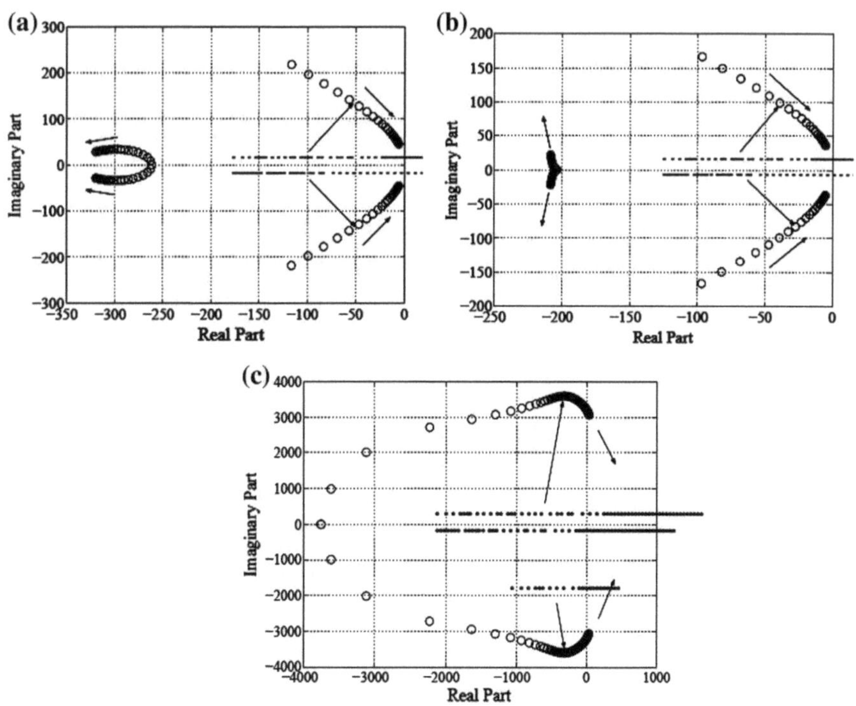

Fig. 5 Root locus of closed-loop of flexible model. **a** Without gyro. **b** With gyro. **c** With gyro

it is clear that the gyroscopic effect is unable to be compensated in actual flexible rotor system due to the critical vibration from the fist bending backward mode.

5 Cross Feedback Backstepping Control (CFBC)

In general, in order to suppress the gyroscopic effect at the critical vibration associated with the bending mode, it is necessary to obtain a relatively precise flexible model. However, the order of the flexible model is usually too high to be used for designing a controller. On the other hand, being existence of the anti-symmetric matrix in the FEM model due to the gyroscopic ingredient,

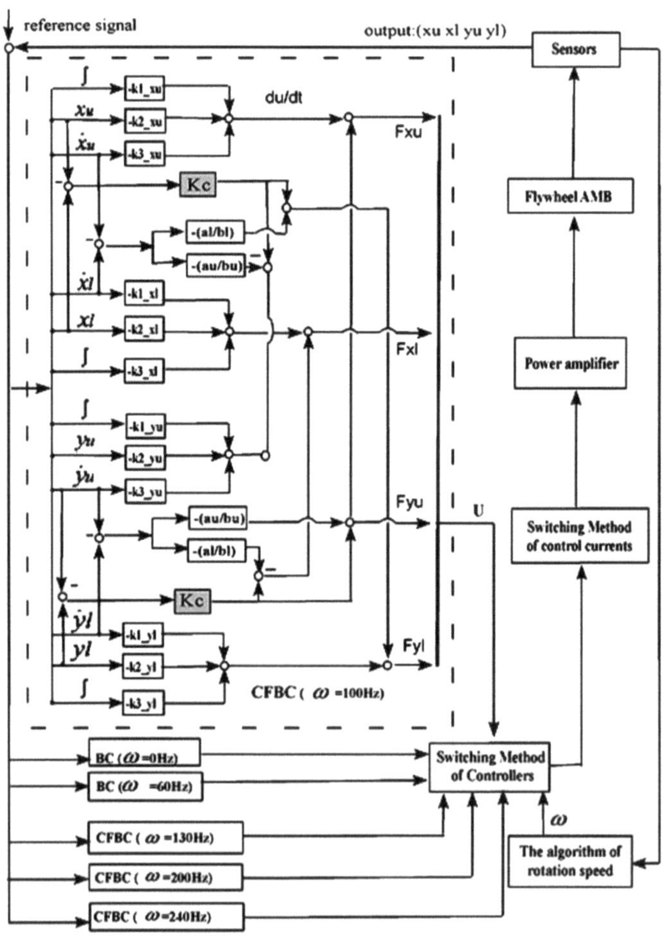

Fig. 6 Diagrams of Cross feedback backstepping control on the basis of switching method

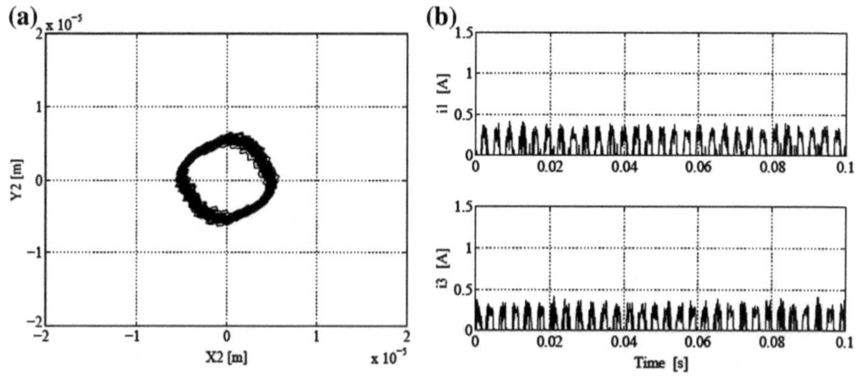

Fig. 7 Locus and currents of rotor at 280 Hz. **a** Rotor locus. **b** Control currents

decoupling of the algorithm for the mode separation is also too complex to implement in actual system. Moreover, it is easy to introduce noise when reducing the order of the model, and the subsequent observer and filter design can also cause serious phase delay.

In this paper, a group of controller added position cross feedback channel is introduced into the original rigid backstepping controller, which is particularly responsible for the suppression of critical vibration associated with the bending mode. The cross-feedback control has been introduced by several references from the theory and the practice [6]. Figure 6 shows the block diagrams of CFBC in which the vibration suppression of the critical frequency is achieved by switching controllers when the rotor speed approaching the each critical frequency of bending modes.

The locus of upper rotor is shown in Fig. 7a and the switching control currents at 280 Hz rotation speed between a pair of electromagnetic shown in Fig. 7b, which can be determined through existing 28 waveforms in 0.1 s. Therefore, the effectiveness of the simple method to control flexible rotor in large rotation speed range could be verified.

6 Conclusion

In this study, the stability of a flywheel AMB considering the gyroscopic effect has been studied. The conclusions of the paper are given as follows

1. The critical vibrations cannot be suppressed only using the gyroscopic effect compensator that is designed on the basis of rigid model.
2. The cause of instability and critical frequency are determined through the analysis of close loop system.
3. The effectiveness of backstepping control with displacement cross feedback channel is verified by the experiment.

Acknowledgments This project was supported by WenZhou scientific and technological plan projects. (No. G20100053)

References

1. Kajs MA, Ouroua A (2001) Active magnetic bearings for energy storage system for combat vehicles. IEEE Trans Magn 37(1):318–323
2. Kuriyama K, Nonami K, Rachmanto B (2009) Modeling and consideration of AMB-flywheel supported by two-axis gimbals. J Syst Des Dyn 3(4):681–693 (in Japanese)
3. Siverioglu S, Nonami K, Saigo M (2004) Low power consumption nonlinear control with H∞ compensator for a zero-bais flywheel AMB system. J Vib Control 10(1077):1151–1166
4. Ariga Y, Nonami K, Sakai K (2001) Nonlinear control of zero power magnetic bearing system. Trans JSME C 67(654):392–406 (in Japanese)
5. Nonami K (1982) Responses of passing through critical speeds for arbitrarily distributed mass flexible rotor systems : 1st report, case without gyroscopic effects. Trans JSME C 48 (435):1669–1677 (in Japanese)
6. Tian XH, Fang JC, Liu G (2006) Gain scheduling cross feedback control approach for magnetic suspending flywheel. J Beijing Univ Aeronaut Astronaut 32(11):1299–1303

Cleaning Process Simulation for Building Façade Maintenance Robot with Built-in Guide Rail

Se-Mi Yoon, S.M. Moon, C.Y. Shin and D. Hong

Abstract Because of frequent construction of high-rise buildings, interest on maintenance and management of the building has been increased recently. With recent development of technology, cleaning outer walls of the high-rise buildings has drawn considerable discussion in terms of human safety and cost. To solve these problems, we've developed an automated cleaning robot and it consists of two parts: horizontal robot and vertical robot. These robots move along the guide rail that is built-in the curtain-wall and clean up the glass with a tool system. The cleaning tool system is comprised of injection nozzle and squeeze-suction module. After the injection nozzle sprays the optimal amount of water, polluted water is sucked by the squeeze-suction. The tool angle and force against the glass surface are important to effectively clean out the contamination. The cleaning process is analyzed in terms of the parameters and a new tool mechanism is introduced.

Keywords Building wall maintenance robot · Cleaning tool system · Squeeze-suction device

1 Introduction

As construction technologies have been developed, the number of high-rise buildings has rapidly grown and most of them have been equipped with glass outer walls. Due to skyscrapers, the risk and costs of the maintenance by human resource have increased. Due to the change of materials of the building-wall, for example glass, the cleaning frequency has also increased. As a result, the maintenance task

S.-M. Yoon · S.M. Moon · C.Y. Shin
Graduate School of Mechanical Engineering, Korea University, Seoul, South Korea

D. Hong (✉)
Department of Mechanical Engineering, Korea University, Seoul, South Korea
e-mail: dhhong@korea.ac.kr

E. Qi (ed.), *Proceedings of the 6th International Asia Conference on Industrial Engineering and Management Innovation*,
DOI 10.2991/978-94-6239-145-1_62

trend has been changed to the introduction of the automation and the robot system to reduce man power [1–13].

In addition, as the number of buildings with the irregular surface increases, it's been needed to prepare a means to enhance the cleaning efficiency about the variable-walls. Thus, both automated cleaning tool mechanism and robot are to be developed for efficient maintenance works.

In this paper, a new cleaning tool mechanism and system for the irregular surface building-wall is introduced, which is installed on the built-in guide rail robot [4–11]. Also, the cleaning process is analyzed in terms of some parameters, such as tool angle, compression force, etc.

2 Building-Wall Maintenance Robot

The Building-wall Maintenance Robot (BMR) is composed with a horizontal robot and a vertical robot which moves along the curtain-wall building (Fig. 1). The vertical robot moves along the mullion rail on the edge of the building and moves up and down using a winch-wire system which is installed on a rooftop [1–3]. The vertical robot system carries the horizontal robot vertically through a docking system. The horizontal robot moves along the transom rail and performs the maintenance task with cleaning tool system.

The cleaning tool system consists of injection nozzle and squeeze-suction module. After the injection nozzle system sprays the optimal amount of water depending on the moving distance of the horizontal robot, polluted water is sucked

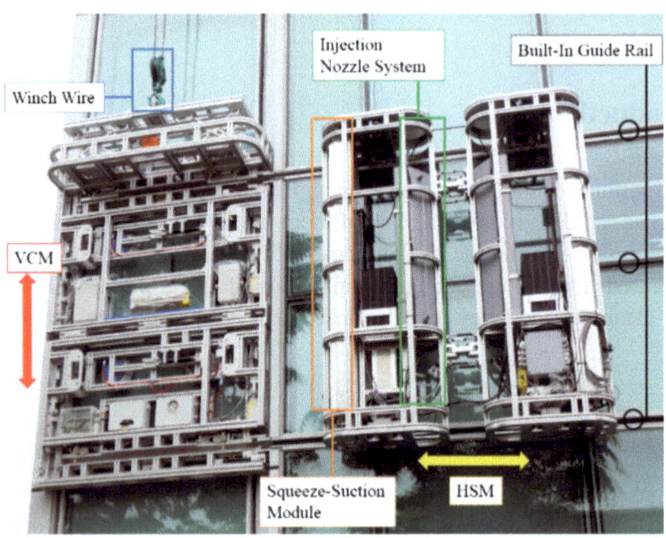

Fig. 1 Configuration of building maintenance robot (BMR)

by the internal structure of the squeeze-suction module. Accordingly, the building-wall becomes cleaned by these tools. Because the rubber wiper which is attached to the squeeze-suction device's endpoint, removes dust and moisture at the same time, we can minimize the movement for cleaning, extra moisture stains and the damage of pedestrian and vehicle which is caused by scattering [14, 15]. By these cleaning system, the cleaning time and expense are reduced and it is possible to simplify the automation system.

3 Mechanism of Squeeze-Suction Device

The squeeze-suction module is composed of four squeeze-suction devices and the modules are attached to both ends of the horizontal robot having bilateral symmetry.

In this module, a pair of the squeeze-suction devices are moved on the x-axis by an actuator. Also, when the horizontal robot moves the curved surface, the squeeze-suction device rotates on the z-axis by a motor (Fig. 2). Each squeeze-suction device can revolve on the y-axis by the damping system. As a result, the squeeze-suction device have 3° of freedom movement about the horizontal robot.

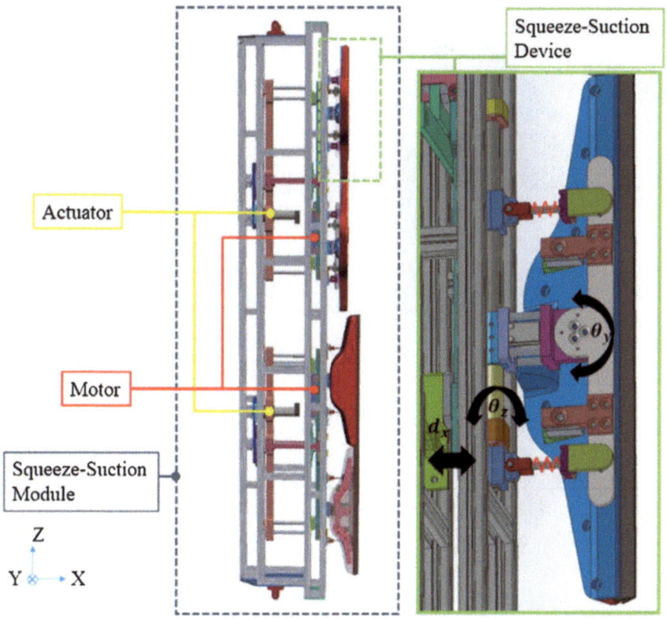

Fig. 2 Configuration of the squeeze-suction module

When the rubber wiper swabs the building-wall, the pressing force and angle between the wiper and glass is important in wiping efficiency. Wrong power and angle applied to the glass, raise wiper-chattering, stripes and noise.

4 Maintaining Proper Angle for Irregular Surface

When the rubber wiper blade contacts the surface, the surface and the wiper blade should form 30° to 60° angle to have high cleaning efficiency. Therefore in this paper, the rubber wiper was designed to get 51° on which the squeeze-suction device sets right angle to the wall. In order to do effective sweeping, the squeeze-suction device's position and angle should change depending on the moving distance of the horizontal robot.

4.1 Inverse Kinematics

It is supposed that α is initial angle of the squeeze-suction device, d_x is moving distance by the actuator and θ_z is the squeeze-suction device's rotation angle by motor (Fig. 3).

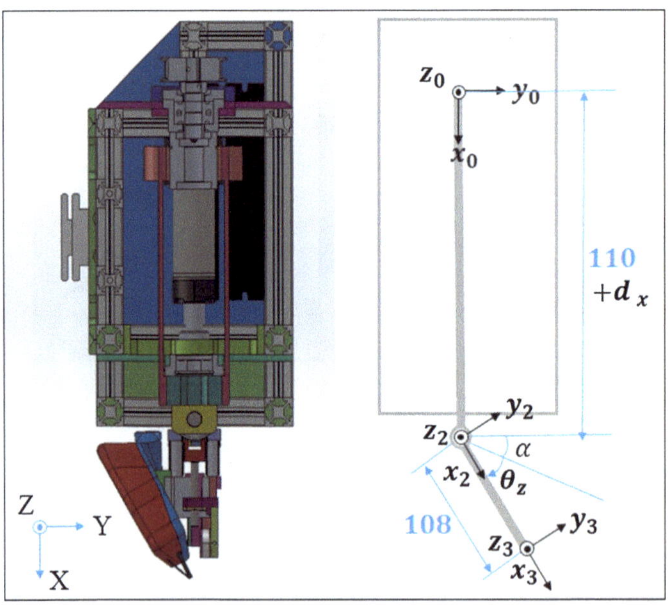

Fig. 3 Schematic of the squeeze-suction module

Table 1 Denavit-Hartenberg's parameters

Link	θ_i	d_i	a_i	α_i
1	0	0	$110 + d_x$	0
2	90°− α	0	0	0
3	θ_z	0	108	0

Forward kinematics of squeeze-suction device's end-point is found using Denavit-Hartenberg notation (Table 1) [13]. By this convention, forward kinematics of the rubber wiper can be determined as

$$\mathrm{X} = d_x + 108\cos\theta_z \sin\alpha - 108\cos\alpha \sin\theta_z + 110 \tag{1}$$

$$\mathrm{Y} = 108\cos\theta_z \cos\alpha + 108\sin\theta_z \sin\alpha \tag{2}$$

Inverse kinematics of the rubber wiper can be determined as

$$\theta_z = d_x - \cos^{-1}\left(\frac{Y}{108}\right) \tag{3}$$

$$d_x = 108\sin\left(-\cos^{-1}\left(\frac{y}{108}\right)\right) + X - 110 \tag{4}$$

4.2 *With Respect to the Irregular Surface*

The façade is segment type at intervals of 1200 mm and 20°. The horizontal robot moves along the transom rail. When the horizontal robot moves–y-axis direction, the squeeze-suction device attached to the left side is used to clean-up. When the robot moves +y direction, the squeeze-suction device attached to the right side is used to wiper (Fig. 4).

Wheels of the horizontal robot is composed with two guide wheels and one support wheel. The guide wheels move along built-in guide rail and determine the moving direction of the horizontal robot. The support wheel is actuator of the robot and supports the weight.

Whenever the robot moves along the rail, the squeeze-suction module's relative position and angle are changed. So it is need to change the squeeze-suction device's moving distance and rotation angle depending on the horizontal robot's position.

For these things, we draw the support wheel moving route about the built-in guide rail and the start point's moving route of squeeze-suction module (Fig. 5). And in each squeeze-suction module's start point, we show traces of the squeeze-suction module and the squeeze-suction device. The relative position of the squeeze-suction module depends on the Support wheel. At the same time, the squeeze-suction device should set right angle to the Building-wall for efficient tasks.

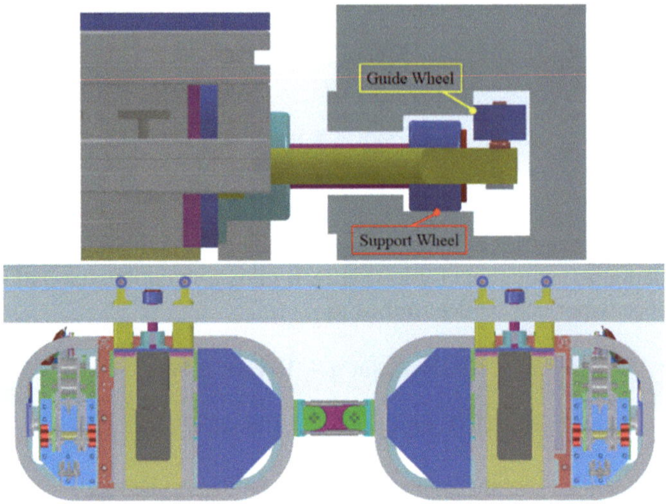

Fig. 4 Cross-section diagram of moving mechanism

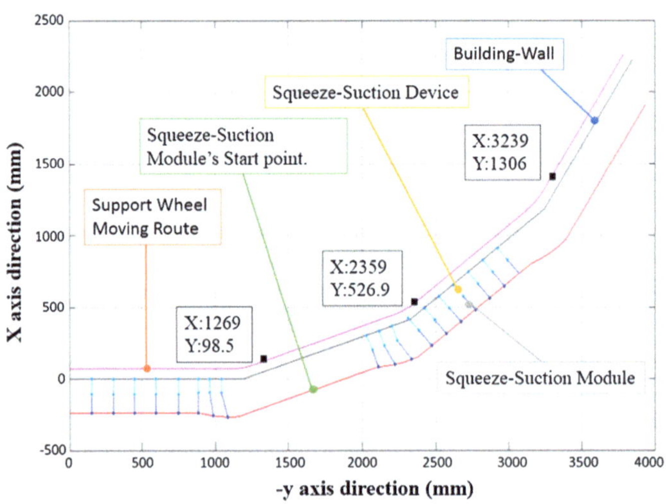

Fig. 5 Simulation of the squeeze-suction device along the rail

From the Fig. 5, we can find d_x which is the moving distance by the actuator and $\theta_z + \alpha$ which is the squeeze-suction device's total rotation angle by motor (Fig. 6). Because the horizontal robot has a rigid body, d_x and $\theta_z + \alpha$ have absolute value according the support wheel moving route.

The start point of the squeeze-suction module has a consistent position pattern about the segment type wall and also the moving distance and the total rotation

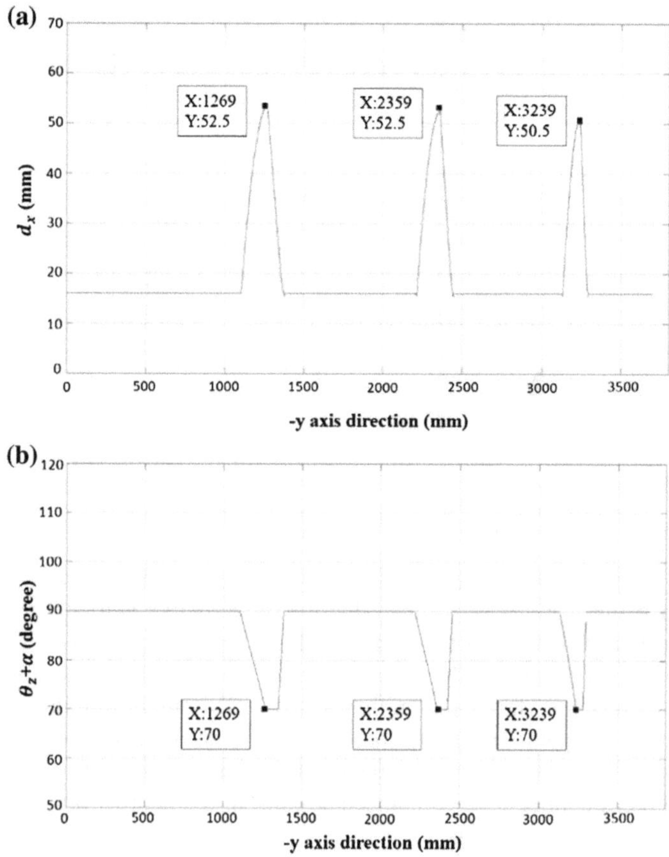

Fig. 6 **a** Simulation result of moving distance, **b** simulation result of the total rotation angle

angle in a regular pattern about the −y-axis. Especially, value of −y of the maximum moving distance and the maximum total angle is the same as the support wheel's −y-axis position which is ending of circular curve.

5 Maintaining the Proper Force for Irregular Surface

5.1 FBD (Free Body Diagram)

After the several types of building has been expanded, some building-walls aren't perpendicular to the ground. To cover this problem, damping system is attached to the squeeze-suction device to allow rotation about the y-axis at point C (Fig. 7). The pressing force between the squeeze-suction device and the glass is determined by actuator and the distribution of power is changed by slope of wall.

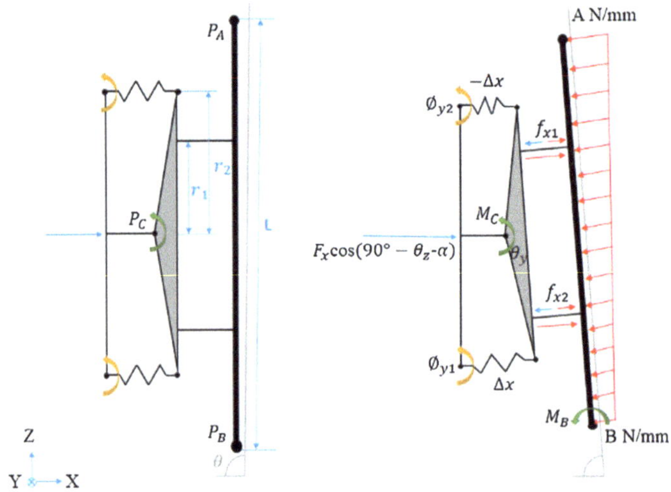

Fig. 7 Free body diagram of the squeeze-suction device

For squeezing work, it is assumed that the rubber wiper is totally touching the building-wall and parallel to the curtain-wall. It is suppose that r_1, r_2 and L are length between each joint of the squeeze-suction device and θ is the slope of wall. AN/mm is maximum force per millimeter and BN/mm is minimum force per millimeter of end-point. F_x is the pressing force by the actuator, Δx is length variation of spring, θ_y is rotation angle about the y-axis at point C and k is spring constant. Forces applied to each contacting point in terms of f_{x1} and f_{x2}.

Force can be determined as

$$F_x = f_{x1} + f_{x2} \tag{5}$$

$$F_x = BL + (\mathrm{A} - \mathrm{B})\frac{L}{2} \tag{6}$$

and moment can be determined as

$$\sum M_C = r_1 f_{x2} - r_2 k(r_2\theta_y) - r_2 k(r_2\theta_y) - r_1 f_{x1} = 0 \tag{7}$$

$$\sum M_B = B\frac{L^2}{2} + (A - B)\frac{L^2}{3} - f_{x1}\left(\frac{L}{2} - r_1\right) - f_{x2}\left(\frac{L}{2} + r_1\right) = 0 \tag{8}$$

using law of force equilibrium and of moment equilibrium. From these equation, we can find the relationship between F_x, θ_y, A, and B.

5.2 Simulation Result

In this squeeze-suction device, the rubber thickness is 1 mm, r_1 is 66 mm, r_2 is 99 mm, L is 410 mm and k is 2.9 N/mm. We assumed that the rubber wiper needs the minimum force per millimeter 0.1 N/mm, which would not cause the wiper chattering. Thus the smallest force of the squeeze-suction device should be more than 0.1 N/mm. The greatest strength of the squeeze-suction device should be below than the yield strength of the NBR rubber wiper and it is assumed 35 % of tensile strength, 3 N/mm in 200 % elongation. For calculations, it is assumed that $\theta_z + \alpha$ is 90°.

In that case, the rubber wiper can cover θ_y from −28.25° to 28.25° (Fig. 8a). However, Δx limited to ±10 mm and θ_y is limited to ±10 degree because of the spring. Thus θ_y is nothing wrong with the rubber wiper. When θ_y is changed, minimum pressing force of F_x is represented in Fig. 8b.

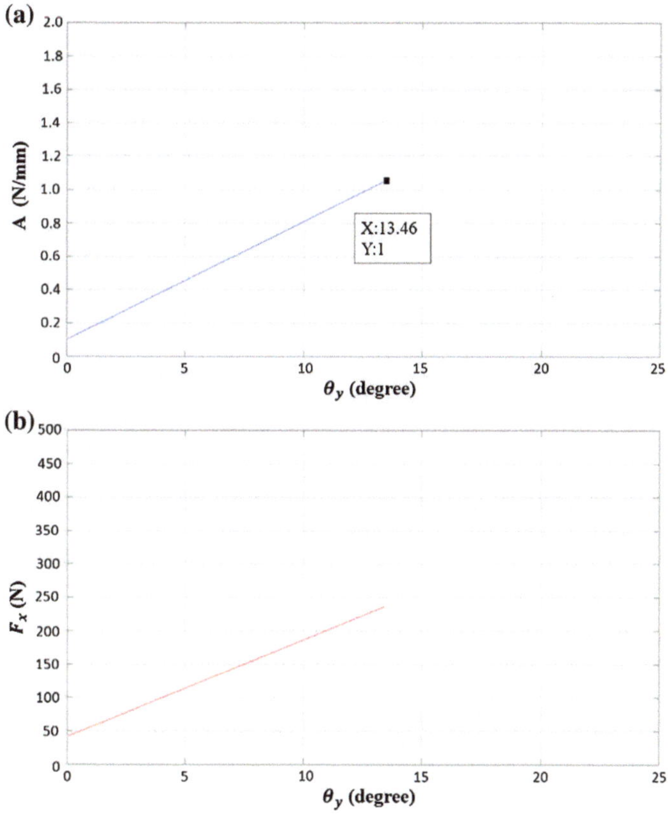

Fig. 8 **a** Simulation result of maximum force per millimeter, **b** simulation result of the minimum pressing force to need

6 Conclusion

In this paper, we developed the squeeze-suction module which is attached to the horizontal robot of BMR system to maintain the curtain-wall with the variable surface. In the irregular surface, the squeeze-suction device should maintain the fixed angle and force for efficient tasks. Through the actuator, motor and damping system, the squeeze-suction devices have excellent adaptability.

In the future, this cleaning tool system is expected to have more efficiency then present tool system in every irregular surface building. This tool system will be a good solution in terms of the risk and costs of maintenance by human resource.

Acknowledgments The work presented in this paper was funded by BMRC (Building-Façade Maintenance Robot Research Center), supported by Korea Agency for Infrastructure Technology Advancement (KAIA) under the Ministry of Land, Infrastructure and Transport (MOLIT).

This work was supported by the Human Resources Program in Energy Technology of the Korea Institute of Energy Technology Evaluation and Planning (KETEP) grant financial resource from the Ministry of Trade, Industry & Energy, Republic of Korea. (No. 20124010203250)

References

1. Yamazaki S (1990) Automatic exterior wall spraying equipment: application and evaluation. In: Proceedings of the 7th international symposium on automation and robotics in construction, pp 610–617
2. Akinfiev T, Armada M, Nabulsi S (2009) Climbing cleaning robot for vertical surfaces. Ind Robot Int J 36(4):352–357
3. Gambao E, Hernando M (2006) Control system for a semi-automatic facade cleaning robot. In: Proceedings of the 2006 international symposium on automation and robotics in construction, pp 406–411
4. Yokota T, Nishiyama Y, Takasaki T (1989) Horizontal moving automatic outer surface cleaning. Nihon Biso Co., Ltd., U.S. Patent, No.4809384
5. Kim YS, Jung MH, Cho YK, Lee J, Jung U (2007) Conceptual design and feasibility analyses of a robotic system for automated exterior wall painting. In: Proceedings of the international journal of advanced robotic system, vol 4, no 4, pp 417–430
6. Monzon JE, Zhang H, Zhang J, Liu R, Zong G (2005) Realization of a service robot for cleaning spherical surfaces. Int J Adv Rob Syst 2(1):53–58
7. Moon SM, Hong D, Kim SW, Park S (2012) Building wall maintenance robot based on built in guide rail. In: Proceedings of the IEEE international conference on industrial technology, pp 509–514
8. Kim SW, Hong D, Moon SM, Huh J (2012) Development of a seismic-wind load effect experimental system for built-in guide rail building maintenance robot. In: Proceedings of the 8th IEEE international conference on automation science and engineering, pp 1016–1021
9. Lee JK, Ryu JH, Lee DJ (2011) An experimental study of automatic cleaning tool and robot for façade in high-rise buildings. In: Proceedings of the 28th international conference climbing and walking robots, pp 1453–1458
10. Kang MS, Lee SH, Chun BI, Shin KS, Traver AE, Han CS (2011) Window contamination detection method for the robotic building maintenance system. In: Proceedings of the 28th international conference climbing and walking robots, pp 1432–1433

11. Elkmann N, Kunst D, Krueger T, Lucke M, Bohme T, Felsch T, Sturze T (2004) SIRIUSc: facade cleaning robot for a high-rise building in Munich, Germany. In: Proceedings of the 7th international conference climbing and walking robots, pp 1033–1040
12. Chu B, Jung K, Han CS, Hong D (2010) A survey of climbing robots: Locomotion and adhesion. Int J Precis Eng Manuf 11:633–647
13. Craig JC (2008) Introduction to robotics: mechanics and control, 3/E. Pearson Education
14. Chevennement CR, Dreher T, Alliot P, Aubry E, Lainé JP, Jézéquel L (2007) Flexible wiper system dynamic instabilities, modelling and experimental validation. Exp Mech 47:201–210
15. Sugita M, Yabuno H, Yanagisawa D (2006) Bifurcation phenomena of the reversal behavior of an automobile wiper blade. Tribol Int 41:27–34

The Research on Hybrid Kernel Function Facing Vehicle Driving Pattern Recognition

Wen Song, Yi Tian, Xin Zhang and Xinn Zhang

Abstract The intelligent energy management strategy of Hybrid Electric Vehicle (HEV) based on driving pattern recognition is created, because there is a close relationship between the driving pattern recognition model and HEV fuel consumption. The Kernel Function (KF) is one of the most important research issues in driving pattern recognition model based on Support Vector Machine (SVM). In this paper, a hybrid KF facing vehicle driving pattern recognition is created by convex combination, using radial basis KF and B-spline KF. And the coefficient of hybrid KF, the boundaries of vehicle driving pattern characteristic parameters are optimized using genetic algorithm at the same time. The hybrid KF facing vehicle driving pattern recognition could improve the recognize accuracy of driving pattern model.

Keywords Driving pattern recognition · Hybrid electric vehicle · Kernel function · Support vector machine

1 Introduction

The intelligent energy management strategy of Hybrid Electric Vehicle (HEV) based on driving pattern recognition was created in many years ago, which could make HEV satisfy different driving situations by justifying the parameters of

W. Song · X. Zhang (✉) · X. Zhang
School of Mechanical Electric and Control Engineering, Beijing Jiaotong University, Beijing, China
e-mail: zhangxin@bjtu.edu.cn

Y. Tian
Department of Mechanical Engineering, Academy of Armored Forces Engineering, Beijing, China

Y. Tian
Railway Management Bureaus, JSLC, Jiuquan, Gansu, China

E. Qi (ed.), *Proceedings of the 6th International Asia Conference on Industrial Engineering and Management Innovation*,
DOI 10.2991/978-94-6239-145-1_63

energy management strategy. So the intelligent energy management strategy based on driving pattern recognition is one of the most important developments of HEV. There are many researches on this. Lin C et al. created a vehicle driving pattern recognition model based on Hamming Neural Network, which included 10 driving pattern characteristic parameters, and recognized American and Korean 6 typical driving patterns [1]. Reza Langari et al. created a vehicle driving pattern recognition model based on Learning Vector Quantization Network using 26 driving pattern characteristic parameters, and recognized American level of service driving patterns [2]. Zhou Nan et al. created a vehicle driving pattern recognition model based on Neural Network, which included 10 driving pattern characteristic parameters, and recognized Beijing, New York, Changchun, Shanghai driving patterns [3]. Zhang Liang et al. created a vehicle driving pattern recognition model based on Support Vector Machine (SVM) using 18 characteristic parameters, and recognized Shanghai and Guangzhou driving patterns [4]. However these papers did not do any researches on the characteristic parameters of driving pattern recognition model, and the performance of model was restricted.

Tian Yi et al. got a best subset including 8 driving pattern characteristic parameters by hybrid search algorithm which was made up of floating search algorithm and genetic algorithm, and created a vehicle driving pattern recognition model based on SVM using these characteristic parameters [5]. But the Kernel Function (KF) in the SVM was not be researched and the KF facing vehicle driving pattern recognition has not be created.

In this paper, a hybrid KF facing vehicle driving pattern recognition is created. Firstly 2 kinds of KF are chosen from normally kinds of KF by comparing the classify results. Secondly a hybrid KF is constituted by these 2 kinds of KF. At the end, the coefficient of hybrid KF, the boundaries of the characteristic parameters are optimized at the same time.

2 Vehicle Driving Pattern Recognition Model Based on Svm

2.1 Optimized Structure

The optimized structure using radial basis KF is shown in Fig. 1 [5].

2.2 Normally KF

The KF of SVM should not be chosen randomly, because the classify accuracy of SVM is disturbed by KF. The normally KF includes linearly KF, polynomial KF, radial basis KF, sigmoid KF, and B-spline KF [6, 7].

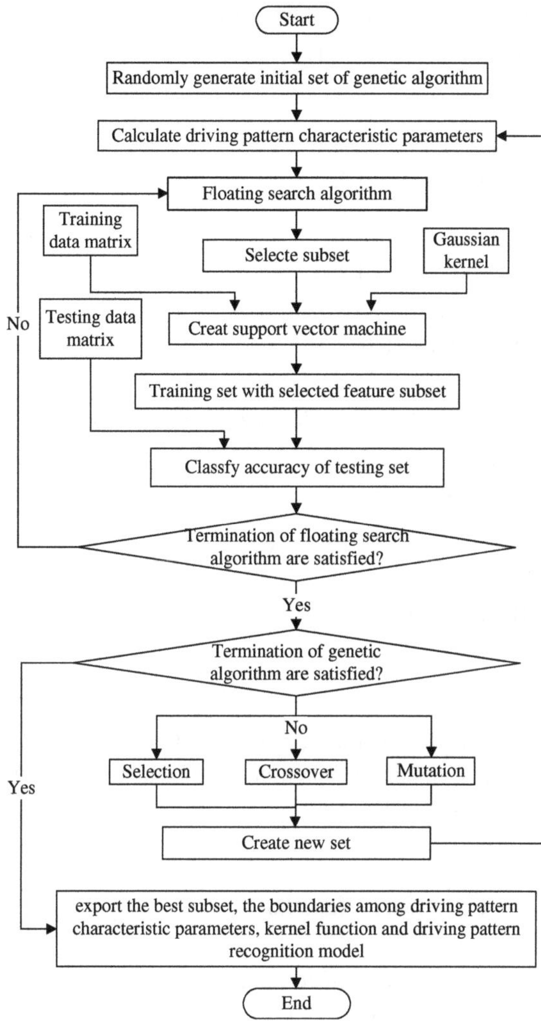

Fig. 1 Optimized structure using Radial Basis KF

Linearly KF

$$K\langle \chi_i, \chi_j \rangle = \chi_i^T \chi_j \tag{1}$$

Polynomial KF

$$K\langle \chi_i, \chi_j \rangle = \left(\chi_i^T \chi_j + 1\right)^d \tag{2}$$

Table 1 Classify results of five kinds of KF

SVM KF	Classify accuracy (%)	Number of characteristic parameters in best subset
Linearly KF	90.6	10
Polynomial KF	90.98	9
Radial basis KF	91.22	8
Sigmoid KF	85.76	10
B-spline KF	91.28	8

Radial basis KF

$$K\langle \chi_i, \chi_j \rangle = \exp\left(-\|\chi_i - \chi_j\|^2 / \sigma^2\right) \tag{3}$$

Sigmoid KF

$$K\langle \chi_i, \chi_j \rangle = \tanh\left(\left(\chi_i^T \chi_j\right) + 1\right) \tag{4}$$

B-spline KF

$$K\langle \chi_i, \chi_j \rangle = B_{2N+1}\left(\chi_i - \chi_j\right) \tag{5}$$

where χ is the sample; i, j are the NO of sample; d is the ladder of polynomial; in this paper, quadratic function is chosen, so $d = 2$; in radial basis KF, σ is the wide parameter, $\sigma = 1$.

2.3 Calculate Results

The classify results of five kinds of KF are shown in Table 1. The input driving pattern includes four Chinese cities' driving pattern and real velocity-time curves that are introduced in preference [5].

The classify accuracies of SVM using different kind of KF are more than 85 %. The SVM using radial basis KF and B-spline KF are better than the others, because the numbers of characteristic parameters in best subset are less and the classify accuracies are greater.

3 Hybrid KF of SVM

3.1 *Create Method*

According to the classify results of normally KF, radial basis KF and B-spline KF are chosen to constitute the hybrid KF. In this paper, the hybrid KF facing vehicle driving pattern recognition is created by convex combination as (1).

$$mixK = \lambda K_1 + (1 - \lambda)K_2 \tag{6}$$

where λ is the constitute coefficient, $0 \leq \lambda \leq 1$; K_1 is the classify accuracy of SVM using radial basis KF, K_2 is the classify accuracy of SVM using B-spline KF. For getting the best hybrid KF, the λ should be optimized.

3.2 *Optimize*

The boundaries among driving pattern characteristic parameters are also should be optimized, so the genetic algorithm is chosen in this paper. The optimized structure using hybrid KF is shown in Fig. 2.

(1) In this paper, the two types of driving situation are urban street and high way; the input data are Beijing, Shanghai, Guangzhou, Wuhan driving patterns and real velocity-time curves [8]. Divide the velocity-time curves by rolling time window and get the segments;

(2) Code the boundaries among driving pattern characteristic parameters v_1, v_2, v_3, r_1, r_2, a_1, a_2 and the constitute coefficient λ of hybrid KF [5]. The restrictions of these parameters are listed in Table 2.
Generate the initial set of genetic algorithm and get 22 driving pattern characteristic parameters according to the boundaries v_1, v_2, v_3, r_1, r_2, a_1, a_2. The driving pattern characteristic parameters include three parts: level measures part, oscillation measures part, distribution measures part, which are introduced in Table 3.
The testing data matrix is calculated by driving pattern characteristic parameters and segments. And the training data matrix is chosen from testing data matrix.

(3) The better subset is gotten by floating search algorithm. For the first time, the starting point is chosen randomly; if not, choose from each individual, and the evaluation function is as (7);

$$F_1 = W_A \times A_{svm} + W_F N_{sw} \tag{7}$$

where A_{svm} is the accuracy of SVM; W_A is the coefficient of A_{svm}; N_{sw} is the number of characteristic parameters in best subset; W_F is the coefficient of N_{sw}.

Fig. 2 Optimized structure using hybrid KF

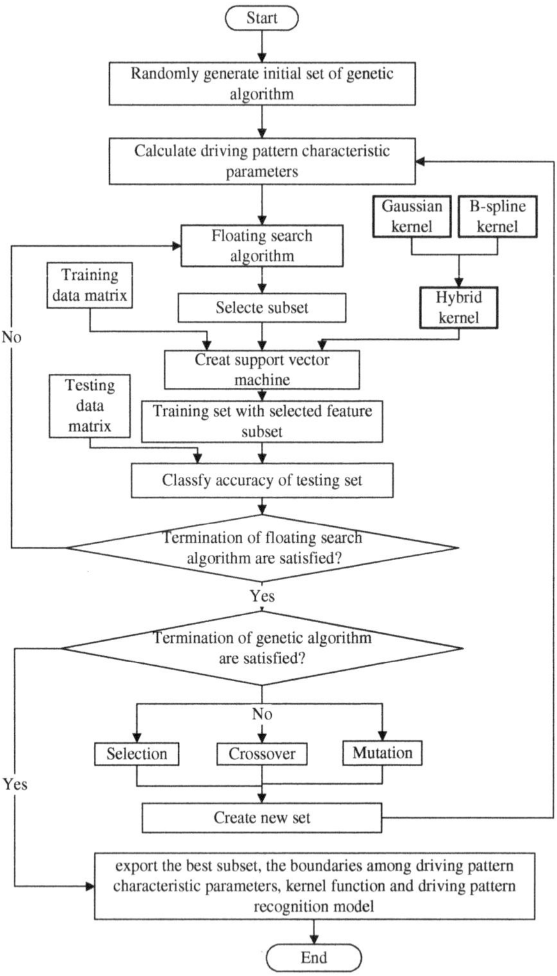

Table 2 Restriction of parameters

Parameters	Restriction
v_1 (km/h)	10–20
v_2 (km/h)	23–35
v_3 (km/h)	43–55
r_1 (m/s^2)	−0.5 to −0.78
r_2 (m/s^2)	−0.8 to −1.18
a_1 (m/s^2)	0.5–0.78
a_2 (m/s^2)	0.8–1.18
λ	0–1

Table 3 Driving pattern characteristic parameters

Type	Parameter	Description	
Level measures	$\bar{v}$	Average of the speed	km/h
	σ_v	Standard deviation of the speed	km/h
	$\bar{a}$	Average deviation of acceleration	**m/s^2**
	σ_a	Standard deviation of acceleration	m/s^2
	$\bar{r}$	Average of deceleration	m/s^2
	σ_r	Standard deviation of deceleration	m/s^2
Oscillation measures	$\overline{a^2}$	Average of a^2	m^2/s^3
	RPA	Relative positive acceleration	m/s^3
	N_{100m}	Frequencies of max and min values > 2 km/h per 100 m	
	LN_{100m}	Frequencies of max and min values > 10 km/h per 100 m	
	N_{100s}	Frequencies of max and min values > 2 km/h per 100 s	
	LN_{100s}	Frequencies of max and min values > 10 km/h per 100 s	
Distribution measures	η_{v1}	% of time in speed intervals $0 \sim v_1$	%
	η_{v1-v2}	% of time in speed intervals $v_1 \sim v_2$	%
	η_{v2-v3}	% of time in speed intervals $v_2 \sim v_3$	%
	η_{v3}	% of time in speed intervals > v_3	%
	η_{r2}	% of time in deceleration intervals < r_2	%
	η_{r1-r2}	% of time in deceleration intervals $r_1 \sim r_2$	%
	η_{r1}	% of time in deceleration intervals > r_1	%
	η_{a1}	% of time in acceleration intervals < a_1	%
	η_{a1-a2}	% of time in acceleration intervals $a_1 \sim a_2$	%
	η_{a2}	% of time in acceleration intervals > a_2	%

The training data matrix is used to train the SVM and the testing data matrix is used to calculate the classify accuracy of trained SVM.

(4) If the termination of floating search algorithm is satisfied, go to step 5; if not, go to step 3;
(5) If the termination of genetic algorithm is satisfied, export the best subset, the boundaries among driving pattern characteristic parameters, hybrid KF and driving pattern recognition model. If not, the better individual should be founded by evaluation function, return to step 6.
(6) Select, crossover, and mutate to get the new set, then return to step 2 [8, 9].

4 Discussion

The CPU of computer is P8400@2.26G, the memory is 3G. The total calculation time is 7.7456e + 005 s. The best subset of SVM using hybrid KF includes 5 characteristic parameters $\bar{v}$, σ_v, $\bar{r}$, N_{100s}, η_{r1-r2} and the classify accuracy is 98.77 %.

Compared with radial basis KF, the best subset of hybrid KF decreases 37.5 % parameters and the accuracy increase 7.55 %. Three characteristic parameters are decreased. They are σ_a in level measures part, LN_{100s} in oscillation measures part, η_{a3} in distribution measures part. Half of parameters in level measures part are kept in best subset. About 60 % of parameters in best subset are in level measures part, which means the level measures part is more important than oscillation measures part and distribution measures part. And the parameters related to velocity $\bar{v}$, σ_v are kept, which means the velocity is most important in driving pattern characteristic parameters.

The optimized results of characteristic parameters are listed in Table 4.

The numbers of characteristic parameters changing curve in best subset using different KF are shown in Fig. 3. Compare with radial basis KF, the hybrid KF could get the best subset more easily and the search steps decrease about 19.4 %, which means the hybrid KF is better than radial basis KF [5].

Table 4 Optimized results of chacteristic parameters

Parameter	Radial basis KF	Hybrid KF
v_1 (km/h)	17.5	19.7
v_2 (km/h)	26.5	26
v_3 (km/h)	48.2	51.5
r_1 (m/s^2)	−0.60	−0.76
r_2 (m/s^2)	−0.96	−1.18
a_1 (m/s^2)	0.70	0.52
a_2 (m/s^2)	1.03	0.88
λ	/	0.44

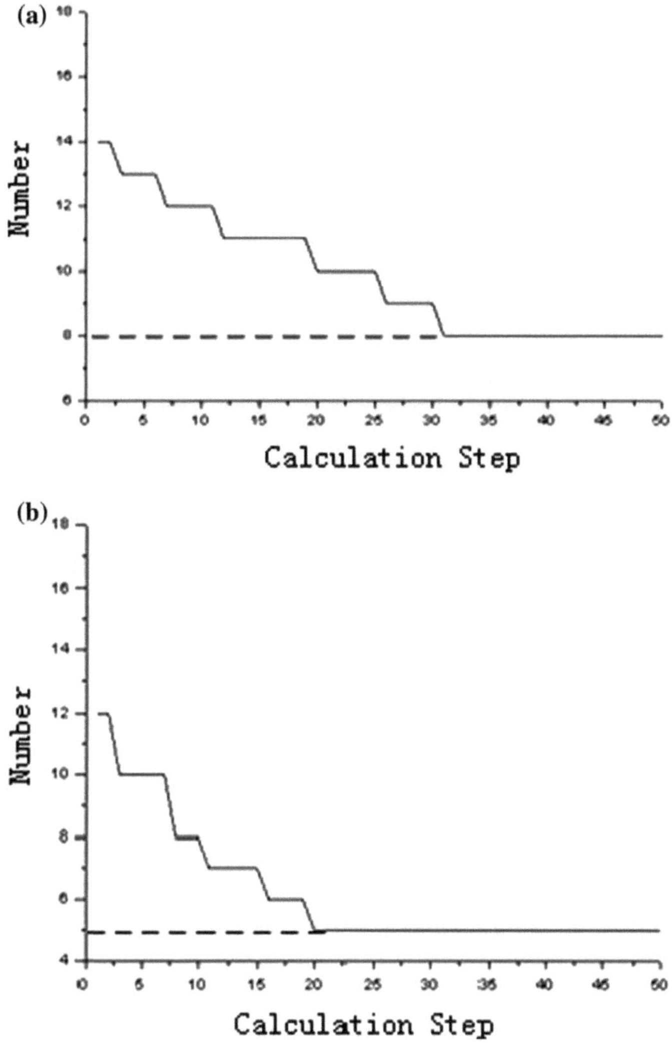

Fig. 3 Numbers of characteristic parameters changing curve using different KF

5 Conclusion

The hybrid KF facing vehicle driving pattern recognition is created by convex combination. Compare with five kinds of KF, the characteristic parameters of hybrid KF are less and the accuracy is greater, which means the hybrid KF is combined by radial basis KF and B-spline KF could reflect the vehicle driving pattern feature really.

Acknowledgments This work is supported by Beijing Natural Science Foundation (4122062).

References

1. Lin C, Jeon S, Peng H et al (2004) Driving Pattern Recognition for Control of Hybrid Electric Trucks[J]. Veh Syst Dyn 42(1–2):41–57
2. Langari R, Won J-S (2005) Intelligent energy management agent for a parallel hybrid vehicle—part I: system architecture and design of the driving situation identification process[C],. IEEE Trans Veh Technol 54(3)
3. Zhou N, Wang Q, Zeng X (2009) Adaptive HEV energy management algorithms based on drive-cycle recognition. J Hum Univ (Nat Sci) 36(9):37–41 (in Chinese)
4. Liang Z, Xin Z, Yi T (2009) Intelligent energy management for parallel HEV based on driving cycle identification using SVM. In: International workshop on IK formation security and application, pp 457–460
5. Yi T, Xin Z et al (2013) Research on vehicle driving situation identification (part I)-driving pattern selection. China Mech Eng 24(9):1258–1263
6. Yi C (2007) The application of hybrid-kernel function support vector machine in credit evaluation. Harbin Institute of Technology (in Chinese)
7. Li W (2008) Study on fuzzy model identification based on kernel method. Shanghai Jiao Tong University (in Chinese)
8. Wei Z-C (2007) Study of robust digital image watermarking based on genetic algorithm. Tianjin University (in Chinese)
9. Zeng Y-J (2007) Satellite constellation design based on genetic algorithm. Huazhong University of Science and Technology (in Chinese)

A Fast Alignment Method for Aircraft Land-Based Takeoff

Hao Liang, Rong-jun Mu and Dan-dan Wang

Abstract Both speed and accuracy are important indexes for initial alignment of inertial navigation system (INS), so rapid and high precision transfer alignment has been widely researched in the past decade. This paper analyzes aircraft's requirements for initial alignment before taking off in modern warfare. A new alignment method is proposed combining optical collimation and transfer alignment to overcome the disadvantages of limited maneuver and poor observable degree in aircraft's initial alignment. A new scheme of transfer alignment is presented, and the optical azimuth transitive relation between SINS (airborne INS) and MINS (master inertial navigation systems) is derived. Both angular rate plus acceleration matching model and relative azimuth obtained by optical collimation are fed into Kalman filter to estimate the installation angle of two INSs. Simulation results show that the new ground alignment method has fast convergence speed and high alignment accuracy to meet the requirements of modern aircraft's initial alignment.

Keywords Aircraft · Fast alignment · Mathematic transfer · Optical collimation · Relative azimuth

1 Introduction

To achieve high precision of initial alignment for aircraft in limited time before they take off are key requirements in modern fast-response warfare [1, 4]. Two major processes are initial alignment and system detection. With the development of electronics technology, system detection requires less time than traditional methods.

H. Liang (✉) · R. Mu
Department of Astronautics, Harbin Institute of Technology, Harbin, China
e-mail: lianghao_hit@163.com

D. Wang
China Academy of Launch Vehicle Technology, Beijing, China

E. Qi (ed.), *Proceedings of the 6th International Asia Conference on Industrial Engineering and Management Innovation*,
DOI 10.2991/978-94-6239-145-1_64

In contrast, because of poor observable degree on static base, traditional initial alignment needs to hold a long time and the azimuth estimation accuracy is low. Therefore, it is a challenging problem for air force rapid response system to shorten time required for initial alignment and to improve precision.

The main alignment approaches of INS include optical orientation, self-alignment and transfer alignment. The main advantage of optical orientation lies in the high precision and executable procedure under stationary state. Nevertheless, because optical orientation requires a mass of matching equipment, it has a relative lower level of automation and costs plenty for alignment time [5, 7]. Self-alignment only depends on inertial device's output information to complete alignment, so it does not need external information. But as a nature of self-alignment its accuracy is limited by the devices' process, pone to environmental disturbance, and thus hard to meet the requirement of aircrafts [8]. Transfer alignment must collocate by using a high precision MINS and it has a special requirement of maneuvering form.

In the paper, a new transfer alignment method with autocollimator is proposed to meet the requirements of speed and accuracy in aircraft's initial alignment. Optical azimuth transitive relation is derived, and mathematical models of transfer alignment are designed by using Kalman filters with the corresponding matching method. Simulation results show the validity and accuracy of new method.

2 Analysis of Fast Transfer Alignment

The most common measurement models of transfer alignment always include calculating parameters matching and measurement parameters matching. The main characteristic of calculating parameters matching such as velocity plus attitude is robust and resistant to interference. However it has a slow alignment convergence speed and demands the carrier to make certain maneuver [9, 12]. Actually, aircrafts are parked in the hangar before taking off that their attitude maneuver is limited. Measurement parameters matching such as angular rate plus acceleration matching has advantages of good real-time performance, rapid and accurate horizontal estimation. Nevertheless, its heading estimation accuracy is low and anti-jam capability is weak, which determine measurement parameters matching is only suitable for slow or smooth motion state [13, 14].

A fast in ground alignment method for aircraft is proposed which combining optical collimation and mathematical transfer alignment. Autocollimator provide relative azimuth relation between SINS and MINS, and the mission of mathematical transfer alignment is accomplishing quick alignment in horizontal direction. In order to determine in ground alignment scheme for aircraft, we choose angular rate plus acceleration matching based on comprehensive consideration of alignment accuracy, convergence speed and maneuverability.

As shown in Fig. 1, the scheme for initial alignment of aircraft mainly includes in ground MINS, optical collimation system, airborne INS, cable, etc. Optical

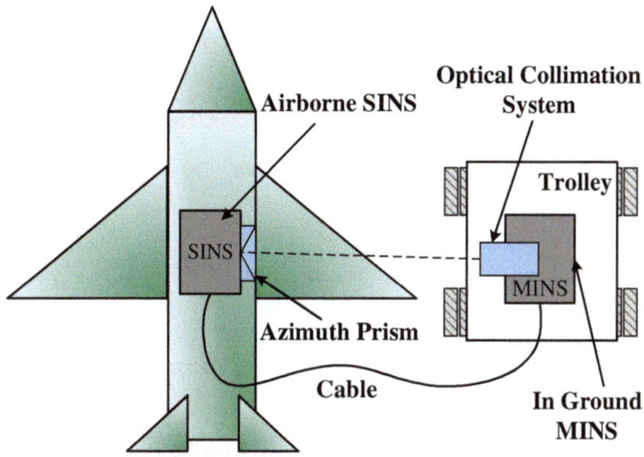

Fig. 1 System composition of initial alignment for aircraft

collimation system consists of collimator and azimuth prism which is installed on the surface of aircraft's underbelly. Airborne SINS and azimuth prism are put on fixed installation as well as MINS and collimator. MINS and collimator are carried by a trolley.

The procedure of fast alignment is devised as follows: In ground, MINS has completed preparatory work such as initial alignment and devices calibration before beginning the warfare business; MINS keeps attitude angle depending on its high precision navigation architecture; Next, MINS-trolley is shifted closed to aircraft which is parked in the hangar stationary, connecting cables and adjusting optical axis of collimation toward azimuth prism; Finally, initial alignment is accomplished through combining transfer alignment and optical azimuth transmission.

3 Transfer Alignment System Model

3.1 Propagation Equation of INS Errors

In almost all kinds of transfer alignment procedures, Kalman filters are used as the primary method of estimation. Therefore, the transfer alignment is designed by using the Kalman filter. There are two set of inertial navigation systems in aircraft's initial alignment which are MINS and SINS, so errors propagation equation of system is different from condition in which only one set of INS exists.

Coordinate systems: a represents master body; b represents slaver body; n represents master inertial navigation coordinate; n′ represents slaver inertial navigation coordinate.

Installation angle between MINS and SINS is define as $\boldsymbol{\varphi}_{\mathrm{a}} = [\,\varphi_{\mathrm{xa}} \quad \varphi_{\mathrm{ya}} \quad \varphi_{\mathrm{za}}\,]^{\mathrm{T}}$. Attitude error vector between calculative navigation coordinate and actual navigation coordinate is define as $\boldsymbol{\varphi} = [\,\varphi_{\mathrm{x}} \quad \varphi_{\mathrm{y}} \quad \varphi_{\mathrm{z}}\,]^{\mathrm{T}}$. Propagation equation of INS errors in geographic coordinate system can be shown as [15, 16]

$$\dot{\boldsymbol{\varphi}}^{\mathrm{n}} = -\boldsymbol{\omega}_{\mathrm{in}}^{\mathrm{n}} \times \boldsymbol{\varphi}^{\mathrm{n}} + \boldsymbol{\Delta\omega}_{\mathrm{in}}^{\mathrm{n}} + \boldsymbol{\Delta\omega}_{\mathrm{c}}^{\mathrm{n}} + \boldsymbol{\Delta\omega}_{\mathrm{r}}^{\mathrm{n}} \tag{1}$$

$$\begin{aligned} \delta\dot{\boldsymbol{V}}^{\mathrm{n}} &= (\mathbf{C}_{\mathrm{n}}^{\mathrm{n}'} - \mathbf{I})\mathbf{f}_{\mathrm{m}}^{\mathrm{n}} - (2\boldsymbol{\omega}_{\mathrm{ie}}^{\mathrm{n}} + \boldsymbol{\omega}_{\mathrm{en}}^{\mathrm{n}}) \times \delta\boldsymbol{V}^{\mathrm{n}} \\ &\quad - \delta\boldsymbol{\omega}_{\mathrm{en}}^{\mathrm{n}} \times \mathbf{V}_{\mathrm{s}}^{\mathrm{n}'} + \mathbf{C}_{\mathrm{b}}^{\mathrm{n}'}\Delta\mathbf{a}_{\mathrm{s}} \end{aligned} \tag{2}$$

$$\delta\dot{\lambda} = \frac{\delta V_x}{R_{\mathrm{N}} + h}\sec L + \frac{V_{mx}}{R_{\mathrm{N}} + h}\sec L \tan L\delta L \tag{3}$$

$$\delta\dot{L} = \frac{\delta V_y}{R_{\mathrm{M}} + h} \tag{4}$$

where $\Delta\omega_{\mathrm{c}}^{\mathrm{n}}$ and $\Delta\omega_{\mathrm{r}}^{\mathrm{n}}$ represent gyro's constant drift and random drift difference between two INSs; $\boldsymbol{\omega}_{\mathrm{in}}^{\mathrm{n}}$ is rotational angular rate of the navigation frame with respect to the inertial frame and $\Delta\boldsymbol{\omega}_{\mathrm{in}}^{\mathrm{n}}$ represents corresponding error; $\mathbf{f}_{\mathrm{m}}^{\mathrm{n}}$ represents master apparent acceleration in geographic coordinate system; $\boldsymbol{\omega}_{\mathrm{en}}^{\mathrm{n}}$ is the turn rotational angular velocity of the navigation frame with respect to the earth; $\mathbf{V}_{\mathrm{s}}^{\mathrm{n}'}$ represents velocity of SINS; $\Delta\mathbf{a}_{\mathrm{s}}$ represents accelerometer error of SINS; R_{M} and R_{N} represent radius of local meridian and local prime vertical respectively; λ, L and h represent geographic longitude, latitude and altitude respectively.

3.2 Propagation System State Equation

State vector is given as

$$\begin{aligned} \mathbf{X} = [&\varphi_x \ \ \varphi_y \ \ \varphi_z \ \ \delta V_x^{\mathrm{n}} \ \ \delta V_y^{\mathrm{n}} \ \ \delta L \ \ \delta\lambda \ \ \Delta\omega_x \ \ \Delta\omega_y \ \ \Delta\omega_z \\ &\Delta a_x \ \ \Delta a_y \ \ \Delta a_z \ \ \varphi_{ax} \ \ \varphi_{ay} \ \ \varphi_{az}]^{\mathrm{T}} \end{aligned} \tag{5}$$

The components in Eq. (5) are, in the following order, misalignment angle, difference of east velocity and north velocity between two INSs, difference of longitude and latitude, slaver gyro's constant drift, zero-bias of accelerometer and installation angle.

System state equation is

$$\dot{\boldsymbol{X}} = \boldsymbol{AX} + \boldsymbol{BW} \tag{6}$$

where $\boldsymbol{A}$ is state transition matrix; $\boldsymbol{B}$ is process noise driving matrix; $\boldsymbol{W}$ represents process noise vector. Each element in $\boldsymbol{A}$ and $\boldsymbol{B}$ is determined by propagation equation of INS errors.

3.3 Angular Rate Plus Acceleration Measurement Model

Angular rate plus acceleration matching main consist of output information of master and slaver INS which are used to estimate installation angles.

The gyro and accelerometer's outputs of MINS can be written as $\boldsymbol{\omega}_{\rm bm}$ and $\boldsymbol{f}_{\rm bm}$, and the gyro and accelerometer's outputs of SINS can be written as $\boldsymbol{\omega}_{\rm bs}$ and $\boldsymbol{f}_{\rm bs}$. MINS is regard as reference datum in transfer alignment, so gyro and accelerometer measurement errors of MINS are not considered. Measurement outputs of MINS and SINS and be wrote as

$$\begin{aligned} \boldsymbol{\omega}_{\rm bs} &= \boldsymbol{C}_{\rm a}^{\rm b}\boldsymbol{\omega}_{\rm bm} + \Delta\boldsymbol{\omega}_{\rm S} \\ \boldsymbol{f}_{\rm bs} &= \boldsymbol{C}_{\rm a}^{\rm b}\boldsymbol{f}_{\rm bm} + \Delta\boldsymbol{a}_{\rm S} \end{aligned} \tag{7}$$

where $\Delta\boldsymbol{\omega}_{\rm S}$ and $\Delta\boldsymbol{a}_{\rm S}$ represent gyro and accelerometer measurement errors of SINS, $\boldsymbol{C}_{\rm a}^{\rm b}$ is installation matrix.

When installation angles are small quantity, installation matrix can be denoted as

$$\boldsymbol{C}_{\rm a}^{\rm b} = \begin{bmatrix} 1 & \varphi_{\rm az} & -\varphi_{\rm ay} \\ -\varphi_{\rm az} & 1 & \varphi_{\rm ax} \\ \varphi_{\rm ay} & -\varphi_{\rm ax} & 1 \end{bmatrix} = \boldsymbol{I} - [\boldsymbol{\phi}_a]^{\times} \tag{8}$$

where $[\boldsymbol{\phi}_a]^{\times}$ is an antisymmetric matrix consist of $\boldsymbol{\phi}_a$. Hence

$$\begin{aligned} \boldsymbol{\omega}_{\rm bs} - \boldsymbol{\omega}_{\rm bm} &= \left(\boldsymbol{C}_{\rm a}^{\rm b} - \boldsymbol{I}\right)\boldsymbol{\omega}_{\rm bm} + \Delta\boldsymbol{\omega}_{\rm S} \\ &= [\boldsymbol{\omega}_{\rm bm}]^{\times}\boldsymbol{\phi}_a + \Delta\boldsymbol{\omega}_{\rm S} \\ \boldsymbol{f}_{\rm bs} - \boldsymbol{f}_{\rm bm} &= \left(\boldsymbol{C}_{\rm a}^{\rm b} - \boldsymbol{I}\right)\boldsymbol{f}_{\rm bm} + \Delta\boldsymbol{a}_{\rm S} \\ &= [\boldsymbol{f}_{\rm bm}]^{\times}\boldsymbol{\phi}_a + \Delta\boldsymbol{a}_{\rm S} \end{aligned} \tag{9}$$

The measurement vector of system is

$$\boldsymbol{Z} = \begin{bmatrix} \omega_{\rm bsx} - \omega_{\rm bmx} \\ \omega_{\rm bsy} - \omega_{\rm bmy} \\ \omega_{\rm bsz} - \omega_{\rm bmz} \\ f_{\rm bsx} - f_{\rm bmx} \\ f_{\rm bsy} - f_{\rm bmy} \\ f_{\rm bsz} - f_{\rm bmz} \end{bmatrix} \tag{10}$$

The measurement equation is

$$\boldsymbol{Z} = \boldsymbol{H}\boldsymbol{X} + \boldsymbol{V} \tag{11}$$

where $\boldsymbol{H}$ represents observation matrix; $\boldsymbol{V}$ represents observation noise.

4 Ins Azimuth Transfer Utilizing Optical Collimation System

As shown in Fig. 2, $O_a - X_aY_aZ_a$ and $O_b - X_bY_bZ_b$ represent MINS and SINS's body coordinates respectively, where Y_a points forwards, X_a points to right, Z_a points to upwards. $O_a - X_SY_SZ_S$ is collimator body coordinate and $\boldsymbol{P}$ represents unit vector in prism ridge direction. Before initiating the optical collimation system, it is need to calibrate prism's installation error angles α, β, and $\boldsymbol{C}_S^a$ which represents installation matrix between collimator and MINS. $\overrightarrow{O_aO_b}$ is a optical axis vector from collimator to prism while conducting optical collimation, and $\overrightarrow{O_aO_b}$ is perpendicular to $\boldsymbol{P}$. A_P and θ_P represent collimator panel's horizontal angle and vertical angle.

Suppose $\boldsymbol{L}$ is the unit vector of $\overrightarrow{O_aO_b}$, then in S coordinate, $\boldsymbol{L}$ can be represented as

$$\boldsymbol{L}^S = [\cos\theta_P \sin A_P,\ \cos\theta_P \cos A_P,\ \sin\theta_P]^T \tag{12}$$

And in a coordinate, $\boldsymbol{L}$ can be represented as

$$\boldsymbol{L}^a = \boldsymbol{C}_S^a \boldsymbol{L}^S \tag{13}$$

Expression of $\boldsymbol{P}$ in b coordinate can be obtain by α and β as follow

$$\boldsymbol{P}^b = [\cos\beta \sin\alpha,\ \cos\beta \cos\alpha,\ \sin\beta]^T \tag{14}$$

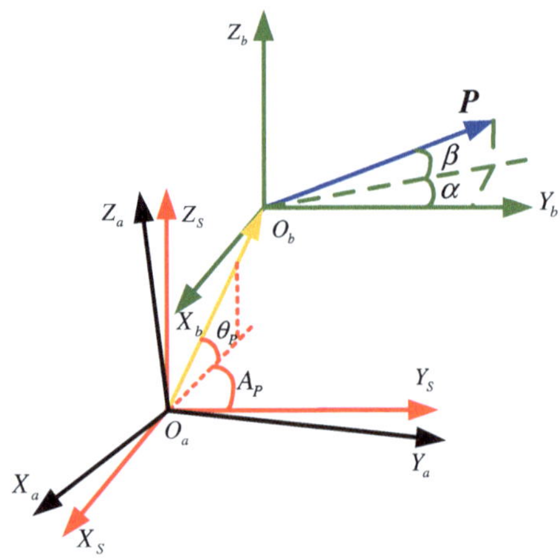

Fig. 2 Coordinate relations of optical azimuth transfer

Then in a coordinate, $\boldsymbol{P}$ is

$$\boldsymbol{P}^{\mathrm{a}} = \boldsymbol{C}_{\mathrm{b}}^{\mathrm{a}} \boldsymbol{P}^{\mathrm{b}} \tag{15}$$

where $\boldsymbol{C}_{\mathrm{b}}^{\mathrm{a}}$ includes three parameters of installation angle. $\varphi_{\mathrm{a}x}$ and $\varphi_{\mathrm{a}y}$ are horizontal installation angles which can be obtain from mathematical transfer alignment. Only φ_{az} is unknown.

$\overrightarrow{O_{\mathrm{a}}O_{\mathrm{b}}}$ is perpendicular to $\boldsymbol{P}$, so

$$\boldsymbol{L}^{\mathrm{a}} \cdot \boldsymbol{P}^{\mathrm{a}} = 0 \tag{16}$$

Combining the Eq. (8), and Eqs. (12–16), we can work out φ_{az}. Installation angle $\boldsymbol{\varphi}_{\mathrm{a}}$ and installation matrix $\boldsymbol{C}_{\mathrm{b}}^{\mathrm{a}}$ can be obtained sequentially.

5 Fast Transfer Alignment Aided with Optical System

After mathematic transfer alignment of angular rate plus acceleration matching has been convergent, relative azimuth φ_{az} obtained by optical collimation system is introduced into measurement equation to assist mathematic transfer alignment.

The improved measurement equation is

$$\boldsymbol{Z} = \begin{cases} \boldsymbol{Z}_1 = \begin{bmatrix} \omega_{\mathrm{bs}} - \omega_{\mathrm{bm}} \\ \boldsymbol{f}_{\mathrm{bs}} - \boldsymbol{f}_{\mathrm{bm}} \end{bmatrix}, & t \leq \mathrm{T}_a \\ \boldsymbol{Z}_2 = \begin{bmatrix} \omega_{\mathrm{bs}} - \omega_{\mathrm{bm}} \\ \boldsymbol{f}_{\mathrm{bs}} - \boldsymbol{f}_{\mathrm{bm}} \\ \varphi_{\mathrm{az}} \end{bmatrix}, & t > \mathrm{T}_a \end{cases} \tag{17}$$

where T_a represents threshold of convergence time which is set beforehand according to prior experiment information.

After installation matrix $\boldsymbol{C}_{\mathrm{a}}^{\mathrm{b}}$ is obtained by mathematic and optical transfer method, we can furthermore calculate SINS's attitude matrix $\boldsymbol{C}_{\mathrm{n}}^{\mathrm{b}}$ utilizing the following equation.

$$\boldsymbol{C}_{\mathrm{n}}^{\mathrm{b}} = \boldsymbol{C}_{\mathrm{a}}^{\mathrm{b}} \boldsymbol{C}_{\mathrm{n}}^{\mathrm{a}} \tag{18}$$

Through extracting attitude angles from $\boldsymbol{C}_{\mathrm{n}}^{\mathrm{b}}$, we can get SINS's attitude information relative to n navigation coordinate and finally achieve correction work of airborne SINS.

6 Results

6.1 Simulation Parameters

(1) INS parameters set:

Gyro constant drift of MINS is 0.003°/h (1σ), acceleration zero-bias is 1×10^{-5}g (1σ). Gyro constant drift of SINS is 0.01°/h (1σ), acceleration zero-bias is 1×10^{-4}g (1σ). Data update cycle of INS is 0.01 s, filtering cycle of transfer alignment is 0.2 s.

(2) Optical device parameters set:

Aiming accuracy of collimator is 10″, calibration error of α and β are 5″, calibration error of installation between collimator and MINS is 5″.

Suppose that installation angle between two INSs are [30′, −50′, 40′], installation angle between collimator and MINS is [20′, 30′, −50′], α is −20′ and β is 30′.

6.2 Result of Mathematic Transfer Alignment

Simulation time of mathematic transfer alignment is 100 s and results are shown as Figs. 3 and 4.

Comparisons of results under different matching models are presented in Table 1, mainly including alignment accuracy and convergence time.

Where $\Delta\varphi_a$ represents estimation error of installation angles.

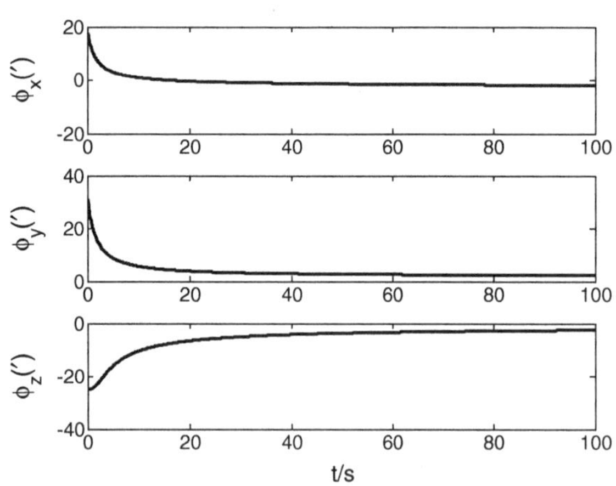

Fig. 3 Estimation error of installation angles utilizing velocity plus attitude matching

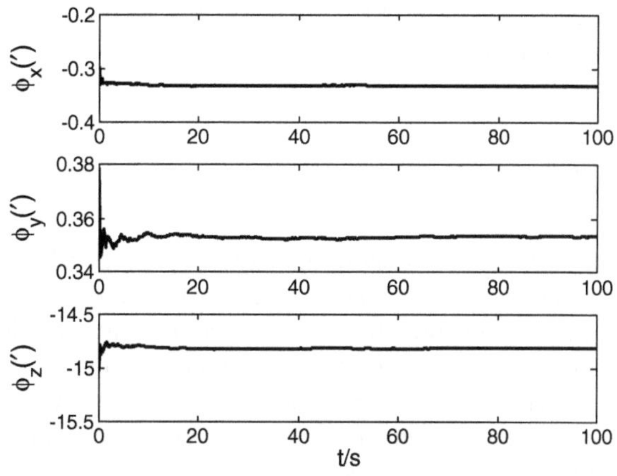

Fig. 4 Estimation error of installation angles utilizing angular rate plus acceleration matching

Table 1 Comparison of simulation results

Matching models	Convergence time /s	$\Delta\varphi_\alpha/(')$
Velocity plus attitude	80	1.9, 2.1, 2.3
Angular rate plus acceleration	25	0.33, 0.35, 14.8

6.3 Result of Transfer Alignment Aided with Optical System

According to Eq. (16), optics assist mathematic transfer alignment is carried out while angular rate plus acceleration alignment is convergent (T_a =25 s). Simulation results are shown in Fig. 5.

Figure 5 illustrates that estimation accuracy of azimuth installation angle is obviously improved benefiting from optical assistance system and estimation error

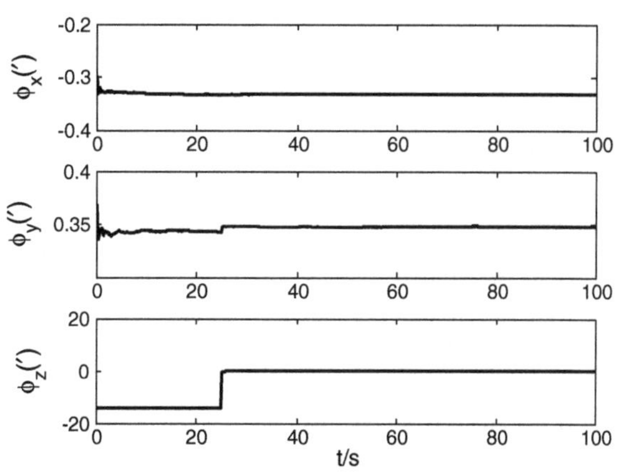

Fig. 5 Alignment error of installation angles aided with optical system

is decreased from 14.8′ of norm method to 12″. Horizontal accuracy does not change much.

6.4 Simulation Analysis

Simulation results show that when utilizing velocity plus attitude matching, the convergence is slow and estimation accuracy is not high owing to lack of attitude maneuver for aircraft in the hangar. In contrast, when utilizing angular rate plus acceleration matching, the convergence is fast (25 s) and horizontal accuracy of installation angle estimation is high (25″). After introducing optical azimuth assistance to system, azimuth accuracy which can reach 12″ is obviously improved and this performance is close to measuring accuracy of optical collimation system meanwhile correction time is much shorter. So improved initial alignment method has a high performance both in time (less than 30 s) and precision (horizontal is 25″, azimuth is 12″).

7 Conclusion

A fast in-ground alignment method combining optical collimation and mathematic transfer alignment for aircraft is proposed. In this research, a transfer alignment system model in geographic coordinate is built. The optical azimuth transitive relation and the calculation method of relative azimuth are derived. The simulation results show that the proposed transfer alignment method implemented with optical system can meet the time and precision requirements of in-ground alignment for aircraft.

Acknowledgments The author wishes to thank the IEEE and Chinese Industrial Engineering Institution for providing this template and all colleagues who previously provided technical support.

References

1. Feng Z, Jiang W, Gao S (2011) Research of rapid transfer alignment and observability analysis. In: The 2nd international conference on intelligent control and information processing, Harbin
2. Zhou N, Wang S, Deng Z (2006) Comparison of three transfer alignment methods in marine missile SINS. In: Proceed of the 25th Chinese control conference, Harbin, Heilongjiang
3. Friedman AL, Gelb A, Sandberg HJ (1963) Alignment of inertial navigators in low-speed vehicles. AIAA J 1(9):2030–2034
4. Hyo S, Won HC (2006) Fast alignment using rotation vector and adaptive Kalman filtering. IEEE Trans Aerosp Electron Syst 42(1):70–83

5. Wang Y, Guo X, Wu K (2002) Development trend of azimuth aiming system of ballistic missiles abroad. Opt Precis Eng 10(1)
6. Kou Y, Huang G, Zhang B (2004) Application of airborne OES in inertial navigation system correction. Electron Opt Control 9(2)
7. Zhang Q (2008) Reflex light-tube in azimuth pointing system of missile. J Appl Opt 29(4)
8. Cheng X, Wan D, Zhong X (1997) Study on initial alignment and calibration of strapdown inertial system on rocking and moving base. J Southeast Univ 27(6):6–10
9. Kai JE, Cloutier JR (1989) Rapid transfer alignment for tactical weapon applications. AIAA-89-3581
10. Hu G, Wang Q, Hu H (1997) Experiment of transfer alignment and analysis of its result. J Astronaut 18(3):71–74
11. Liu X, Wang B, Deng Z (2012) Ship-borne transfer alignment under low maneuver. In: International conference on mechanical engineering and materials, Melbourne, Australia, pp 1155–1158
12. Dai H, Zhou S, Xu Q (2012) Comparison of three rapid transfer alignment algorithms. In: International conference on materials science and information technology, Singapore, pp 3483–3488
13. Wang S, Deng Z (2003) Technique review of transfer alignment of inertial navigation system on moving Base. J Chinese Inertial Technol 11(2):61–67
14. Schultz RL, Keyes CL (1973) Airborne IRP alignment using acceleration and angular rate matching. In: Proceedings of JACC, PP 427–436
15. Han P, Mu R, Cui N (2010) Comparison between transfer alignments of inertial navigation system in two coordinates. J Chinese Inertial Technol 18(3):272–278
16. Liang H, Wang D, Mu R Adaptive filtering algorithm based on observable degree analysis of state parameters in carrier-aircraft transfer alignment. J Chinese Inertial Technol (in press)

Use of Online Temperature Monitoring System for the Surface Accurate Temperature Detection of Open Coal Stockpile

Hong-wei Gu, Ying Li and Wen-hai Xu

Abstract Thermal infrared (TIR) imaging has been successfully applied in the security monitoring and temperature detection of coal stockpile. In this paper, an online temperature monitoring system is designed based on the TIR imaging technique. The temperature correction algorithm is proposed to reduce the affect of weather condition to temperature measurement accuracy. And the temperature correction algorithms composed of infrared imager spatial effect correction, atmospheric radiation transmission correction, and reflect radiation correction. The surface temperature of the coal stockpile in the infrared image can be acquired by calculated these three correction using parameters such as weather condition, geographic coordinates and infrared imaging acquisition time. The experimental results of the temperature correction in different weather conditions indicate that the temperature measurement average error of monitoring system is reduced from 13.88 to 0.97 %. And the temperature correction is an effective method to calculate the accurate surface temperature of coal stockpile.

Keywords Monitoring system · Open coal stockpile · Thermal infrared imaging · Temperature correction

1 Introduction

Coal storage in large opencast coal yard is essential in ensuring continuous supply for large capacity power units. The inner temperature of coal stockpile increased with the prolongation of the storage time and the variation of the weather. Finally, the spontaneous combustion fire in coal stockpile will occur. Spontaneous heating of coal stockpile is a major problem in storage and transport of coal in terms of safety and economy. Detecting of fire source and early predicting the occurrence of

H. Gu · Y. Li (✉) · W. Xu
Information Science and Technology College, Dalian Maritime University, Dalian, China
e-mail: ly86950888@163.com

E. Qi (ed.), *Proceedings of the 6th International Asia Conference on Industrial Engineering and Management Innovation*,
DOI 10.2991/978-94-6239-145-1_65

fire is one of the difficulties and emphases in fire prevention and extinguishing of the coal stockpile storage [1, 2].

Thermal infrared (TIR) imaging has been successfully applied in the thermographic inspection of the coal roadway, the appeared coal column and the coal stockpile [3–5]. TIR imaging detecting technique measures the surface temperature of coal stockpile through the electromagnetic radiation that it emits. Using TIR imaging can monitor the surface temperature variation or anomaly area of coal or mine stockpile [6]. Moreover, the surface temperature variation or anomaly area always reflects inner spontaneous heating of coal stockpile. Hence, TIR imaging detecting technique has been considered as a simple, quick and effective method to detect the hidden fire source of stockpile [7, 8].

However, there still exist some problems in engineering application of TIR imaging detecting technique in the hidden fire source detection of opencast coal stockpile. Firstly, at present, the temperature monitoring location of infrared imager needs to be set artificially. One setting location can only detect limited area, which has related to the focus and field angle of infrared imager. At the same time, the straight-line distance between the central axis of infrared imager and coal stockpile is the influence factor of detection area. However, the coal stockpiles are usually higher than 5 m and longer than 10 m in coal storage yard. Therefore, one coal stockpile needs multiple infrared imagers to monitor a wide range of stockpile [7, 9]. The infrared temperature monitoring system, which can automatically, full-weather, real-time detect multiple location surface temperatures of one or more coal stockpiles, is needed in practical engineering need. And then, characteristics of infrared imager and the processing methods of digital infrared image are mainly concerned problems in infrared imaging technique. There is less research on the impact of weather conditions on the accuracy of infrared temperature measurement. Actually, in normal weather condition except fog and dust storms, when the distance between infrared imager and monitoring target is less than 10 m, the atmospheric radiation transmission has less influence on infrared imaging temperature measurement. The transmission path of infrared radiation continuous increased with increasing distance, consequently, the atmospheric radiation transmission seriously affect the results of infrared temperature measurement [10]. Moreover, the weather condition of coal storage yard is always high temperature, high humidity, floating dust or gale weather. The temperature measurement error is much higher than the nominal error of infrared imager under this adverse weather, and then the detection of hidden fire source often causes deviation. The acquisition of the accurate temperature is the key to analyze the temperature variation of stockpile. At present, the infrared temperature monitoring system, which can automatically reduce the influence of weather condition, has not been reported.

This paper is focused on proposing an effective temperature correction method to calculate the accurate surface temperature of coal stockpile. And, in this paper, an online surface temperature monitoring system of opencast coal stockpile is designed. The designed monitoring system can automatically, real-time detect multiple location temperature of coal stockpile. Moreover, the system is embedded temperature correction algorithm. Therefore, the system can automatically calculate the atmospheric radiation transmission and effectively reduce the affect of weather condition to temperature measurement accuracy.

2 Methodology

2.1 The Online Temperature Monitoring System

The system can be divided into central monitoring station, infrared image acquisition station and meteorological data collection station. Overall plan of system is as shown in Fig. 1.

In the system, wireless communications is designed as all data transmission mode. The wireless remote communication between central monitoring station and infrared image acquisition station is realized by commercial WiFi network, which allows cheaper deployment of local area networks, even in special spaces where cables or fiber cannot be run. And WiFi has the characteristics of real time, rapidity, wide cover range and security, therefore, WiFi is designed as wireless remote communication mode to transmit infrared image and visible video. In the design of meteorological data wireless communications, the GPRS network can be regarded as the communication mode.

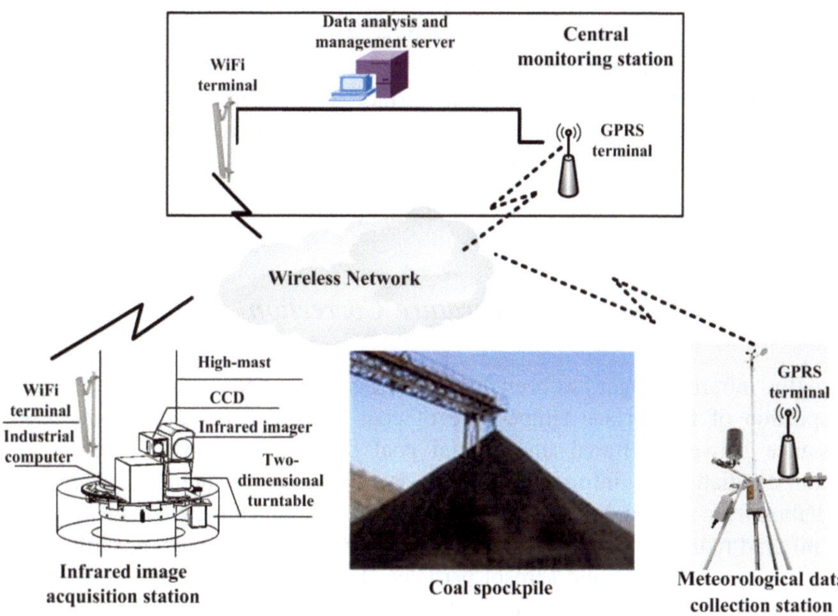

Fig. 1 Online temperature monitoring system structure

The central monitoring station is mainly composed of the data analysis and management server, WiFi terminal and GPRS terminal. The central monitoring station is used to receive the infrared image, visible video and meteorological data, process the image and data, calculate atmospheric radiation transmission in the time of infrared image acquisition, accurate surface temperature of monitoring coal stockpile and send control commend to infrared image acquisition station at the same time. The centre also has a database management system, which can store and query history data.

The infrared image acquisition station consists of five parts: CCD (795pix × 596pix) camera, infrared imager (8–14 μm, 320pix × 256pix), industrial computer, two-dimensional turntable and WiFi terminal. CCD camera is mainly used to monitor and record the opencast coal yard security performance. Infrared imager is the key equipment to detect surface temperature of one or more coal stockpiles. Industrial computer is designed to receive control commands from central monitoring station, and control operation of CCD camera, infrared imager and two-dimensional turntable. Two-dimensional turntable is home-made, and tightly fixed upon a high-mast near the monitored coal stockpiles. The horizontal angle of turntable is widely, which guarantees the infrared imager detection around the high-mast in every horizontal angel (0°–360°) [11]. The detail of two-dimensional turntable has been described in Reference [11].

The meteorological data collection station composes small weather station and GPRS terminal. The small weather station collects weather conditions, such as wind velocity, rainfall, atmospheric humidity, temperature, pressure and visibility. All weather data are important parameter to calculate atmospheric radiation transmission in the time of infrared image acquisition, and transmitted by GPRS network.

The designed online temperature monitoring system is used to automatically and synchronously monitor and record the surface temperature of one or more coal stockpiles, collect the meteorological data in the time of infrared image acquisition, calculate the atmospheric radiation transmission and the accurate surface temperature.

2.2 *The Principle of Temperature Correction*

Usually, infrared imager has been applied in the outdoor noncontact thermographic inspection of the surface temperature of coal stockpile. There are some spatial distance between infrared imager and coal stockpile. Therefore, the received infrared radiations by infrared imager consist of coal self radiation and its reflected radiation from surrounding environment. At the same time, in the transmission path of infrared radiation from coal stockpile to infrared imager, radiation is affected by atmospheric molecular and aerosol particles. The infrared detector receives radiation and changes radiation into an electric signal, then into a gray value, and to scan the field of view of camera in order to form an image. The infrared imager can convert radiation from coal stockpile into thermal image, so that invisibility coal surface temperature can be detected and display in visible image. The surface temperature is easily calculated through the mapping relationship between gray value and temperature [12]. The principle of infrared temperature detection is shown in Fig. 2.

The data of infrared imager detection for the coal stockpile not only include the coal self-radiation, but also include atmospheric information in the path of detection. Moreover, the characteristics of infrared image are the influence factors for the temperature detection of coal.

In this paper, temperature correction consists of three parts: (1) infrared imager spatial effect correction, (2) atmospheric radiation transmission correction, and (3) reflect radiation correction.

In this paper, the gray value transferred by infrared imager can be expressed as Eq. (1).

$$G = K\{(L_{coal} + L_{ref}) \cdot \tau + L_{path}\} \otimes PSF + B + N \quad (1)$$

where G is gray matrix transferred by infrared imager from coal radiation, L_{coal} is self-radiation of coal stockpile, L_{ref} is reflect radiation, τ is atmospheric transmittance, L_{path} is atmospheric path radiation, *MTF* is system transfer function of infrared imager, *PSF* is the Fourier transform of *MTF*, *K* and *B* are calibration coefficients, and *N* is the image noise. In these parameters, only L_{coal} is factor which includes the surface temperature of object.

$$L_{coal} = \frac{1}{\pi}\int_{\lambda_1}^{\lambda_2} \varepsilon \cdot \frac{a}{\lambda^5} \cdot \frac{1}{e^{b/\lambda T_s} - 1} d\lambda \quad (2)$$

where ε is emissivity of the coal stockpile, λ is the wavelength, in this paper, λ_1 is 8 μm, λ_2 is 14 μm, and C_1 and C_2 are the Plank constant.

From the Eq. (1), we can draw a conclusion that the temperature correction is aim to eliminate the interference of other parameters except L_{coal} term by term, and

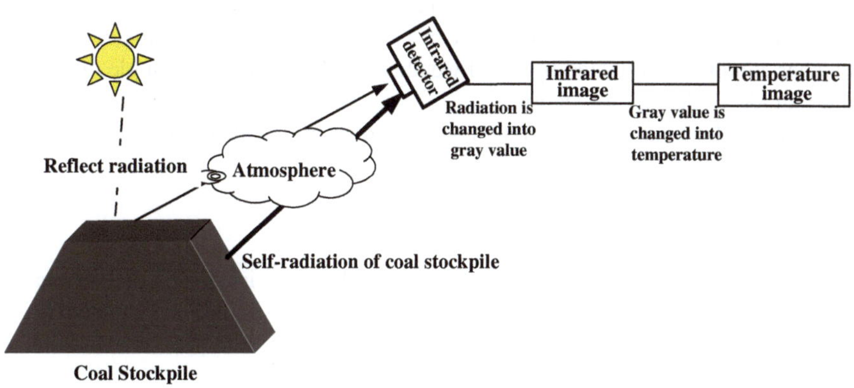

Fig. 2 The principle of infrared temperature detection

calculate the surface temperature which is included in L_{coal}. The aim of infrared imager spatial effect correction is to eliminate the interferences of K, B and PSF. The aim of atmospheric radiation transmission correction is to reduce the affect of τ and to L_{path} to temperature. The reflect radiation correction is used to eliminate the reflect radiation L_{ref}.

2.3 The Temperature Correction Algorithm

In this paper, the temperature correction steps are as follows.

(1) The image noise N is reduced by image averaging.

(2) The K, B and MTF are analyzed from the calibration experiment of infrared imager [13].

(3) The τ and L_{path} are calculated by the models of atmospheric radiation transmission (is described in detail in reference [10]).

The atmospheric transmittance τ can be calculated by Eq. (3).

$$\tau = \tau_1 \cdot \tau_2 \cdot \tau_3 \tag{3}$$

where, τ_1 is the average transmittance caused by atmospheric absorption, τ_2 is the average transmittance caused by atmospheric scattering, and the τ_3 is the average transmittance in adverse weather such as storm, snow and sand storm, etc. These three average transmittances can be calculated using parameters such as detection distance, wind velocity, rainfall, atmospheric humidity, temperature, pressure and visibility [14, 15].

The atmospheric path radiation, we have the formula as [12]:

$$L_{path} = L_{atm} + L_s + L_g \tag{4}$$

where, the L_{atm} is atmospheric radiation, L_s is single scattering of the unscattered direct radiation in the right-hand side, and L_g is multiple scattering of the diffuse intensity. These three radiations also can be calculated using parameters such as weather conditions collected by small weather station, solar zenith angle, longitude and latitude of detection location.

(4) The reflect radiation of coal stockpile equation is calculated as:

$$L_{ref} = k \cdot \frac{E_{env8-14}}{\pi} \tag{5}$$

where k is diffuse reflectance, $E_{\text{env8-14}}$ includes the solar irradiance and sky background irradiance. The parameter k and $E_{\text{env8-14}}$ are similarly calculated using parameters such as weather condition, solar zenith angle, longitude and latitude of detection location.

(5) From steps 1 to 4, all interference of other parameters except L_{coal} can be eliminated term by term. Then, the accurate surface temperature of coal stockpile can be calculated by Eq. (2).

3 Results and Discussion

A verification experiment was carried out in a coal stockpile of Da Lian from 24th July, 2012 to 10th August, 2012. Figure 3 is the thermal image of a detection location of stockpile on a sunny day. Certainly, the chosen detection location should avoid direct sun radiation. Figures 4, 5 and 6 are thermal images of stockpile on haze, rainy and mist day, respectively.

In order to verify the effect of temperature correction, the temperature measured by high precision surface thermometer was taken as the actual surface temperature of detection position. And in each infrared image (from Figs. 3, 4, 5 and 6), we randomly chosen 10 measuring positions to detect the actual surface temperature. The measuring results show that the difference of chosen 10 measuring positions is in 0.04-0.12°C. These differences mean that the actual surface temperature of coal essentially constant in each infrared image. Compared to Fig. 3a, after temperature correction, Fig. 3b is smoother, and the affect of weather is clearly reduced. The same conclusion can be obtained through Figs. 4, 5 and 6. The compared results of

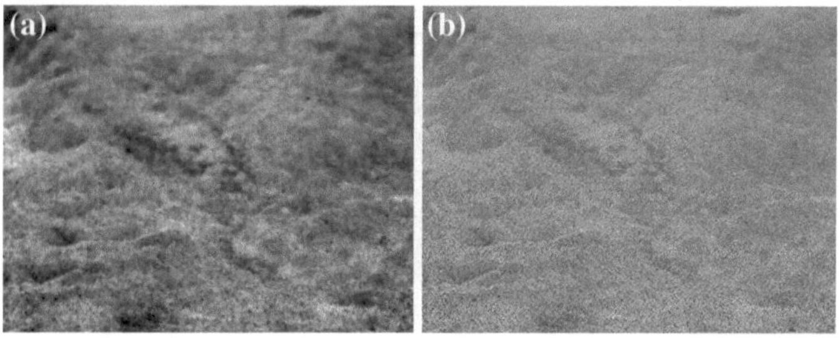

Fig. 3 Infrared image of stockpile on sunny day **a** before correction **b** after correction

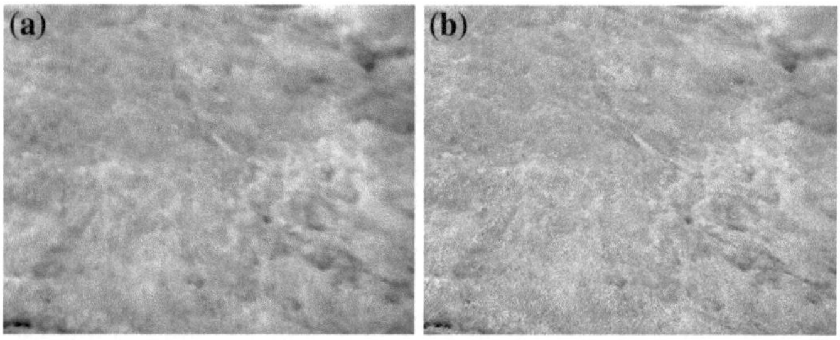

Fig. 4 Infrared image of stockpile on haze day **a** before correction **b** after correction

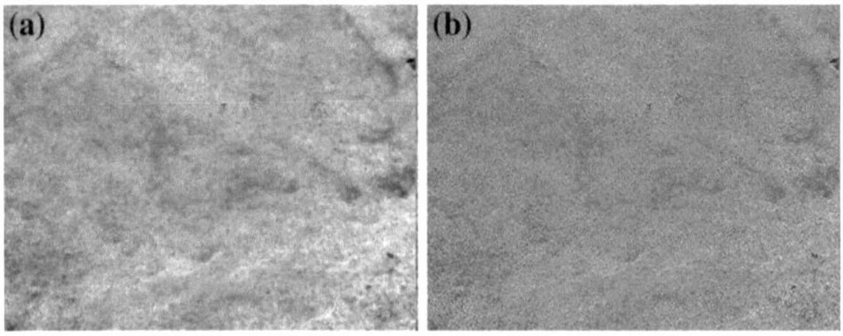

Fig. 5 Infrared image of stockpile on rainy day **a** before correction **b** after correction

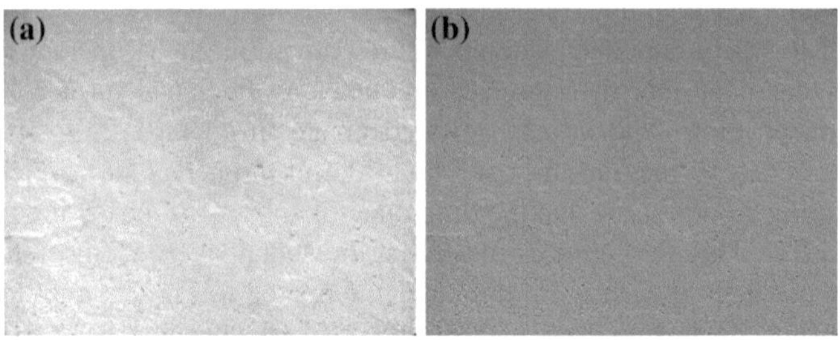

Fig. 6 Infrared image of stockpile on mist day **a** before correction **b** after correction

Table 1 Comparation of temperature correction

Weather	Actual temperature(°C)	Before correction(°C)	After correction(°C)
Sunny	30.65	33.13	30.88
Haze	28.11	25.02	27.89
Rainy	19.41	15.48	19.10
Mist	21.90	18.36	22.06

actual surface temperature, surface temperatures before temperature correction and after correction are listed in Table 1.

The measured results of the temperature correction indicated that the temperature measurement average error of monitoring system is reduced from 13.88 % to 0.97 %.

4 Conclusion

In this paper, a temperature correction method is proposed for calculating the accurate surface temperature of coal stockpile, and an online temperature monitoring system is designed to detect the surface temperature of one or more coal stockpiles at the same time. Compared to the surface temperature before correcting, the temperature after correcting is basically same as the actual temperature, and the average measurement error is reduced to less than 1 %. The experimental results show that the temperature correction method is an effective method to calculate the accurate surface temperature of coal stockpile. And the online temperature monitoring system which is embedded temperature correction algorithm can detect surface temperatures of multiple locations of one or more coal stockpile, reduce the affect of weather condition to the temperature measurement, and calculate the accurate surface temperature.

Acknowledgments This work is supported by the Department of Transportation Construction Projects of China (No. 2013 328 225 080).

References

1. Wang CA, Wang B, Li XQ, Zhao Y, Cui XW (2011) Thermal infrared experimental study of the corresponding relationship between loose coal surface temperature and heat source temperature. J Saf Sci Technol 7(8):38–41 (in Chinese)
2. Carras JN, Young BC (1994) Spontaneous heating of coal and related materials: models, application and test methods. Prog Energy Combust Sci 20(1):1–15
3. Fierro V, Miranda JL, Romera C, Andres JM, Pierrot A, Gomez-Landesa E et al (1999) Use of infrared thermography for the evaluation of heat losses during coal storage. Fuel Process Technol 60:213–229
4. Wang ZM, Cheng WM, Xin S, Song XM, Su ZG (2003) The calculation of close-range coal inflammation position at coal-roads based on infrared detecting and inverse heat conduction technology. J China Coal Soc 28(6):603–607 (in Chinese)
5. Wen H, Ma M, Fei JB (2010) Mine fire disaster control based on infrared imaging technology. Coal Sci Technol 38(1):28–30
6. Ma M (1999) Application research on the technique of IR imaging detecting for hidden fire source of coal seam (Thesis of Dissertation style). M. E. dissertation, Safety Technology and Engineering. Xi&apos. Xi'an, China
7. Zeng FY, Tian YF, Lu GL, Fan SW, Yao HF (2012) Detection and control technology of spontaneous combustion area in coal rejects dump of Xiqu Mine. Coal Sci Technol 40(6):41–44 (in Chinese)
8. Liu CY (2010) Study of IR detection and the treatment technology of coal outcrop and slag fire in Ulan coal mine (Thesis of Dissertation style). M. E. dissertation, Safety Technology and Engineering. Xi&apos. Xi'an, China
9. Wu HJ, Zeng FY, Yao HF, Zheng ZY, Xu CF, Wang YR et al (2013) Danger evaluation and control technology of coal mine gangue spontaneous combustion. Coal Sci Technol 41 (4):119–124 (in Chinese)
10. Gu JL (2012) Research of atmosphere radiative transmission characteristics for the typical weather (Thesis of Dissertation style). Ph.D, dissertation, Comunication and information processing. DMU. Da Lian, China
11. Gu HW, Xu WH, Wang CY, Zou YW, Li Y et al (2013) A kind of monitoring equipment (Patent style). China Patent 201310376546.2 (in Chinese)
12. Li YH (2010) Research on temperature measurement technology and application based on infrared thermal imager (Thesis of Dissertation style). Ph.D, dissertation, Instrument science and technology. HIT, Ha Erbin, China

13. Ji YY, Xu WH, Ma DB, Li Y (2013) Altering integral time method in temperature measurement using mid-wavelength infrared imaging system. Appl Mech Mater 401–403:1519–1522
14. Xu HP (1998) Research into theoretical calculation method on engineering of transmittance of infrared radiation through atmosphere. Opt Precis Eng 6(4):35–43 (in Chinese)
15. Zhou GH, Liu XW, Xu JW (2008) A math model of calculate the atmospheric transmittance of infrared radiation. Infrared Technol 30(6):331–334 (in Chinese)

New Design of the Combined Balance Control System

Peng Guo, Zhen-yong Liu, Cui-jian Zhao and Hong-qiang Sun

Abstract In order to improve the efficiency of the combined balance, this paper proposes a new design of the combined balance control system, providing a new combinatorial calculation algorithm and a hardware system design based on MC9S12XS128 and MEGA16. Experiment results show that compared with traditional design, the new design is more accurate, less costly, and more stable in function.

Keywords Combinatorial calculation · Combined balance · MEGA 16 · MC9S12XS128

1 Introduction

The combined balance, as an electronic weighing apparatus widely used in commodity circulation and food packaging, adopts the multiple-scale combinatorial calculation method, successfully resolving the conflict between accuracy and speed of the weighing system. It works both swiftly and accurately, and is especially handy when it comes to the weighing of irregularly-shaped objects like candy and potato crisps. Its combinatorial calculation technology, with its high speed and accuracy, represents the highest level and future trend of current weighing technology. Combinatorial weighing technology has a relatively short history in China. At first, it was only mastered by several developed countries such as Japan, Germany, the United States and France. Later, owing to the superiority of the combined balance in terms of combinatorial weighing method, it was gradually being researched on and developed by Chinese manufacturers. However, for most of the small and medium-sized food packaging factories in China, the combinatorial weighing technology has not yet been widely used, due to such problems as cost,

P. Guo (✉) · Z. Liu · C. Zhao · H. Sun
College of Physics and Electrical and Information Engineering, Shijiazhuang University, Shijiazhuang, China
e-mail: guopeng993@163.com

E. Qi (ed.), *Proceedings of the 6th International Asia Conference on Industrial Engineering and Management Innovation*,
DOI 10.2991/978-94-6239-145-1_66

speed and accuracy. To improve the situation, this paper proposes a new design that can lower the cost and raise the accuracy of the combined balance.

2 Mechanical Structure

The combined balance consists of two sections: mechanical and electronic. The mechanical section mainly includes storage hopper, linear vibration plate, enter hopper, weighing hopper, discharge hopper, etc., as shown in Fig. 1.

The material is discharged from storage hopper by vibrating of linear vibration plate and is stored in enter hoppers. When last weighing is completed and weighing hoppers are emptied, enter hoppers open and the material enters weighing hoppers. Weighting data are sent to control system. Then, according to combinatorial calculation algorithm, 3-5 hoppers are opened to get the best combinatorial results [1, 2].

3 Control System Hardware Design

The hardware of the combined balance's control system consists of high-accuracy A/D switching circuit, master control circuit, stepping motor driving circuit, electromagnetic vibrating feeding circuit (vibrating machine), photoelectric detection circuit, touch screen, power supply, etc., as shown in Fig. 2.

3.1 High-Accuracy A/D Switching Circuit

A/D switching circuit is a crucial part in the combined balance's control system, whose function determines to a great extent the accuracy and stability of the combined balance. With the switching accuracy, speed and cost taken into consideration, this system decides to utilize CS5532 designed by the US Company

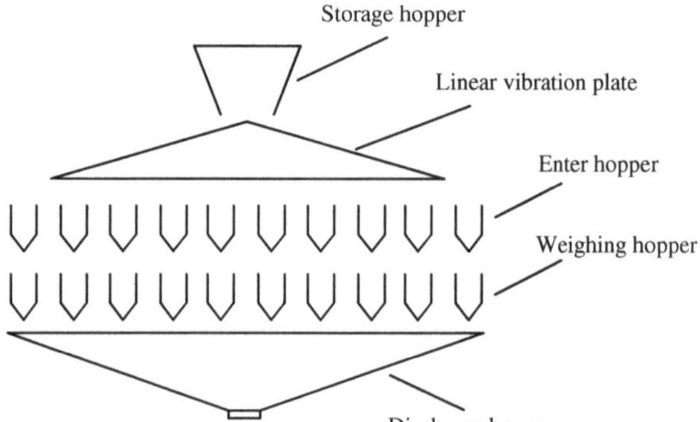

Fig. 1 Mechanical structure of combination scale

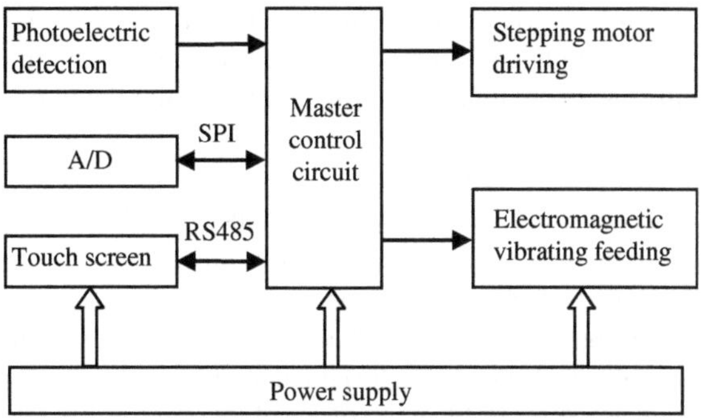

Fig. 2 Diagram of combination scale control system

CirrusLogic as ad chip. CS5532 is a low-noise, multi-channel Δ-∑ analog-to-digital converter, adopting the charge balance technology and the extremely low-noise programmable gain chopping stability measuring amplifier, capable of getting a result of 24-bit resolution. It has advantages such as high accuracy and high dynamic property. As for its application in the combined balance, the internal autocorrect system of CS5532 can calibrate itself and the system, erasing the null-point gain and drift error, and also effectively restraining channel disorders and gain errors generated while the combined balance is working. The differential input end of CS5532 can directly measure the millvolt signals emitted by the weighing sensors, and can support the three-wire serial output mode (compatible with SPI and Microwire), all of which can help decrease the peripheral circuits of the control system and simplify the hardware design, thus lowering the cost and raising the stability of the control system [3, 4].

3.2 Master Control Circuit

The design of the master control circuit is the key element of the combined balance's control system, including the gathering of weighing information, the processing of data, control of hoppers and vibrators, communication, etc. Whether the master control circuit can work accurately and stably directly influences the quality and efficiency of the combined balance. In order to guarantee both the accuracy and the speed of the combined balance, the master control circuit adopts the concept of double-MCU, namely two high-functioning MCU working together to conduct system control. One of the MCU is in charge of data collection and algorithm combination, which is called the principle MCU; the other is in control of hoppers and vibrators, which is called the subordinate MCU. Take the working process of ten-hopper combined balance for example: after the weighing starts, the principle MCU orders the subordinate MCU to fill materials into the weighing hoppers according to the conditions of the previous weighing; and then, the A/D switching circuit corresponding to the 10 weighing hoppers turns the weighing information into digital data, transferring the data to the principle MCU through SPI bus. The principle MCU then conducts combinatorial calculation of the weighing information through combinatorial algorithm, selects 5 qualified weighing hoppers, and informs the subordinate MCU of the numbers of the 5 selected hoppers. The subordinate MCU then drives the 5 hoppers to discharge materials and complete one round of weighing.

Technically speaking, the principle MCU is mainly in charge of data processing and calculation. Take for example the 10-hopper combined balance that works at the speed of 60 packages per minute, after receiving the information about the 10 weighing hoppers, the principle MCU must complete material level detection, data processing and combinatorial calculation in less than one second, select 5 weighing hoppers from 1023 ways of combination, and send the information to the subordinate MCU. Thus it is required that the principle MCU should have extremely high speed and accuracy. Therefore, this system chooses the 16-bit high-functioning MCU MC9S12XS128 of the S12 series developed by the US company Freescale, which has an excellent ability in floating point calculation and low power consumption, with a bus speed of 40 MHz, and with built-in 128 KB ROM Flash and 8 KB DATA Flash, capable of high-capacity storage of programs and data, and equipped with ECC.

The subordinate MCU is relatively simple in function. It is mainly used to control the actuators such as hoppers and vibrators, and it must be able to respond immediately to the orders from the principle MCU and performs reliably according to them. Taking into consideration such factors as performance and cost, this system chooses the 8-bit CMOS MCU MEGA 16 developed by the US Company Atmel, which is based on the enhanced AVR RISC structure. This MCU has an advanced instruction set and single-cycle execution time, a data throughput of 1MIPS/MHz, effectively resolving the conflict between power consumption and working speed. In a 16-KB system, it can program FLASH (capable of both reading and writing, i.e. RWW), 512-bit EEPROM, 1 KB SRAM, 32 general working registers, JTAG port capable of peripheral scanning, and it supports internal debugging and programming.

3.3 Stepping Motor Driving Circuit

When a 10-hopper combined balance is at work, there are altogether 20 valves on the feed hoppers and weighing hoppers. In order to guarantee the smooth opening and closing of these valves, to prevent the materials from being damaged or stuck, the opening and closing of every valve is controlled by a 39BYG two-phase hybrid stepping motor. To guarantee that the motor works steadily, the driver module uses TB6560AHQ stepping motor mini-step drive chip developed by TOSHIBA Company. This chip uses dual-full-bridge MOSFET drive, is capable of withstanding a voltage of 40 V,has a peak current as high as 3.5A,has synchronization, 1/2, 1/8, 1/16 mini-step operating modes to choose from. Compared with its like products, it generates little heat and has a simple peripheral circuit, deserving to be considered a two-phase stepping motor drive chip of low power consumption and high integration.

3.4 Other Circuits

Apart from the above-mentioned major functioning circuits, the hardware circuit also includes electromagnetic vibrating feeding circuit (vibrating machine), photoelectric detection circuit, the touch screen, the power supply, etc. Due to limited space, this paper is not going to discuss these [5–7].

4 Software Design

The system's software design consists of the programs of the principle MCU, subordinate MCU, A/D switching, stepping motor driver, serial communication, touch screen, etc. With product upgrading and maintaining taken into consideration, all programs are written in C language, with easy-to-understand codes and good extendibility.

The core of software design is the program of the principle MCU, used to conduct system condition detection, data collection, filtering algorithm, combinatorial calculation and so on. Due to the vast amount of information involved, the following part will take the combinatorial calculation algorithm for example, as shown in Fig. 3.

After obtaining weighing data, principle MCU will analyze and combine hopper data. In programming, combinatorial calculation is done by a recursive function as follows [8–12].

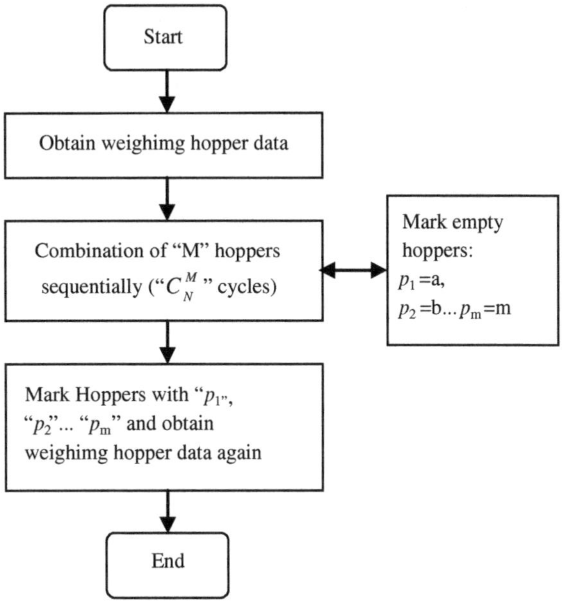

Fig. 3 Combinatorial calculation block diagram

```
cal(int i, int total, int front, int rear, int d[], int n)
{ int j;
    if(front+d[i]<=total)
     {flg[i]=1;
       if(front+d[i]= =total)
       { for(j=0; j<=i; j++)
         if(fig[j]) printf("%d Element\t", j+1, d[j]);
         printf("/n");
         }
    else
      if(i<n-1&&rear+front>=total)
       cal(i+1, total, front+d[i], rear-d[i], d, n);
        flg[i]=0;
        }
  if (i<n-1&&rear-d[i]+front>=total)
    cal(i+1,total,front,rear-d[i],d,n);
}
```

5 Experimental Analysis

In actual application of the combined balance, it is often necessary to increase the total number of weighing hoppers and the number of weighing hoppers involved in the calculation. As a result, the valid number of combinations will increase. Although the principle MCU works very fast, an excessive number of weighing hoppers, both total and that of those involved in calculation, will inevitably lower the working speed and raise the cost [13–15].

In order to test the device performance, we have done a number of experiments in the laboratory. Firstly, we chose 8 weighing hoppers, of which 3 were involved in the combinatorial calculation. Secondly, weighing quality (0 ~ 1000 g) and corresponding measuring voltage were recorded. Finally, using above data, a fitting curve was obtained by Least Squares in MATLAB, as shown in Fig. 4, wherein Y-axis represented the measured voltage signal (unit: V), X axis represented weighing quality (unit: kg).

Fig. 4 Fitting curve of weighing quality and measuring voltage

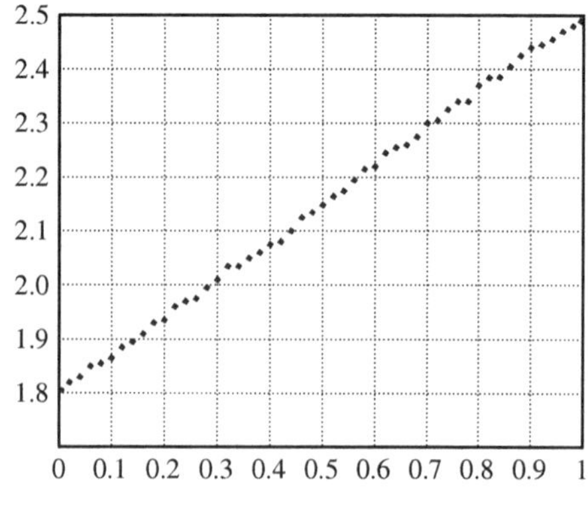

According to the fitting curve, we can know that weighing quality and measuring voltage are linear relationship that proves our circuit design is reasonable.

6 Conclusion

The double-MCU combined balance control system based on MC9S12XS128 and MEGA 16 introduced by this paper, having been tested by experiments, proves to have such advantages as high accuracy, fast speed and low cost. It can replace the traditional auto-packaging balances on a massive scale, and solve its problem of low speed and large error, thus considerably raising the efficiency of the small and medium-sized food packaging factories.

References

1. Quan X (2012) Development analysis of intelligent combination scale market in china. Weighing Apparatus 41(12):1–3
2. Chen Y, Lu H, Hu G (2009) Research and simulation of combinatorial calculation of combined Balance. Packag Eng 30(12):74–76
3. Wu X, Feng W, Wei H (2009) Weighing data acquisition system with 24-bit A/D converter. Electronic Des Eng 17(6):44–48
4. Chen Y, Chen W (2012) Design of highly precise automatic weight measure system based on CS5532. In: Proceedings 2nd conference intelligent system design and engineering application, ISDEA'12, Sanya, CA, pp 1217–1219
5. Gu S, Wang F, Chen Z, Zhang L (2013) The research and design of high-precision strain type electronic balance. In: Proceedings 8th conference industrial electronics and applications, IEA'13, Melbourne, CA, pp 1451–1455
6. Su C, Ye S, Wei C, Xie M (2011) The design and study on the mixture control system of the biogas-gasoline dual-fuel engine. In: Proceedings conference materials for renewable energy & environment, ICMREE'11, Shanghai, CA, pp 517–520
7. Zhang Li-mei, Gao Zhan-bao, Yin Zhi-bing (2013) Dynamic weighing data processing based on equivalent system. Meas Control Technol 32(6):33–34
8. Middendorf L, Bobda C, Haubelt C (2012) Hardware synthesis of recursive functions through partial stream rewriting. In: Proceedings 49th conference design automation conference, DAC'12, San Francisco, CA, pp 1203–1211
9. Ma J, Xiong W, Ding R (2012) Recursive relations of the criterion functions for the recursive least squares algorithms. In: Proceedings 24th conference control and decision conference, CCDC'12, Taiyuan, CA, pp 2069–2074
10. Pont MTS, Garcia Chamizo JM, Mora HM (2013) Resilience modeling by means of a set of recursive functions. In: Proceedings 43rd conference dependable systems and networks workshop, DSN-W'13, Budapest, CA, pp 1–7
11. Chen B, Zhao S, Zhu P, Principe JC (2013) Quantized kernel recursive least squares algorithm. Neural Netw Learn Syst 24(9):1484–1491
12. Arablouei R, Dogancay K (2012) Linearly-constrained recursive total least-squares algorithm. Signal Process Lett 19(12):821–824
13. Sharp I, Yu K (2013) Enhanced least-squares positioning algorithm for indoor positioning. Mob Comput 12(8):1640–1650

14. Zhang Y, Song M (2011) Research and realization of intelligent auto-tracing car for adaptability of environment. In: Proceedings conference electrical and control engineering, ICECE'11, Yichang, CA, pp 1685–1687
15. Chen H, Duan Z, Lian X The design of stepper motor system based on CAN bus. In: Proceedings 2nd conference consumer electronics, communications and networks, CECNet'12, Yichang, CA, pp 3262–3265

A Hybrid Random Bilevel Programming Model and Its PSO Based Algorithm

Hu Huang and Hua Ke

Abstract Considering different management requirements of the decision makers in uncertain decentralized decision making problems, a novel hybrid bilevel programming model is proposed. Furthermore, we apply the model to formulate a decentralized supply chain problem with stochastic coefficients. In order to solve the model, a hybrid intelligent algorithm integrating stochastic simulation and a two-fold particle swarm optimization (PSO) is designed. The numerical experiment demonstrates that the new algorithm is effective.

Keywords Dependent chance model · Particle swarm optimization · Stochastic bilevel programming · Stackelberg equilibrium · Supply chain

1 Introduction

Multilevel programming is a kind of mathematical programming which is developed to solve decentralized planning problems with multiple decision makers in a hierarchical system. The original mathematical formulation for the bilevel programming was proposed in 1970s [1]. Thanks to the popularity of Stackelberg game model which was pioneered by Von [2] in economic field in early 1980s, many scholars were attracted to this field and a great number of significant achievements in theories, algorithms and applications have been made over the next 30 years. By now, the technology has been applied to fields as management, transportation, economic planning, engineering, chemistry and optimal control. Readers can consult [3–6] for more details of multilevel programming.

By now, most work on multilevel programming problems has been done with deterministic assumption, but the real world is full of nondeterministic factors such

H. Huang (✉) · H. Ke
School of Economics and Management, Tongji University, Shanghai 200092, China
e-mail: 13huanghu@tongji.edu.cn

E. Qi (ed.), *Proceedings of the 6th International Asia Conference on Industrial Engineering and Management Innovation*,
DOI 10.2991/978-94-6239-145-1_67

as randomness. In order to deal with the programming problems with stochastic coefficients, there are three commonly used modeling techniques such as expected value (EV) model, chance-constrained (CC) model [7] and dependent chance (DC) model [8] which have been developed to meet different management requirements. The EV model is to optimize the expected objective value subject to some expected constraints. While the CC model, which can be used to handle problems with risk-sensitive decision makers, is to maximize or minimize the objective function under some certain confidence levels subject to chance constraints. And the dependent chance model is to maximize the probabilities of some certain events.

Based on these modeling methods, Wolf and Smeers [9] considered a stochastic stackelberg-nash-cournot model with EV model in the upper level. Patriksson and Wynter [10] first used the EV method in all levels to deal with stochastic decentralized decision-making problem with single follower and discussed the existence of solution and the methods to solving these problem. Gao and Liu [11] presented another two models based on EV model and CC model and raised a hybrid algorithm integrated by stochastic simulation, neural network, and genetic algorithm for finding the Stackelberg-Nash equilibrium. DeMiguel and Xu [12] extended De Wolf and Smeers's model [9] to a multiple-leader one and applied it to formulate supply chain problem in the telecommunication industry. Furthermore, these stochastic bilevel programming models have been applied to fields as transportation network design [13, 14], location problem [15], pricing competition problem [12, 16–18], etc.

However, the traditional models only use single modeling method in all levels ignoring the different management requirements between the upper decision makers and the lower ones. In fact, the difference between these decision makers in various levels may widely exist in practical issues. For instance, in the resource allocation problem, the leader may want to maximize expected profit. While the followers, which may just want to maximize the probability of reaching a certain profit level. If the leaders misjudge the lower level decision makers' behavior, the "optimal solutions" obtained may no longer be optimal in the real world. Therefore, choosing appropriate modeling method is of great significant. Considering the general difference between the decision makers, a hybrid multilevel model integrated with different modeling methods may have wider application prospects.

The contribution of this paper is threefold. First we propose a novel stochastic bilevel programming model notated as EV-DC model which may be suitable for many practical issues such as the principal-agent problem, resource allocation problem, etc. It is known to all that solving multilevel problem is NP-hard [19], and solving nonlinear and uncertain programming are also tough works. Therefore, dealing with the random multilevel problem which nests these three tough works is really a challenging task for us. So we design an intelligent algorithm integrating stochastic simulation and a two-fold PSO to solve the Stackelberg model. Thirdly, we also present an application case study on a supply chain problem in which the novel bilevel model is applied by considering the different management requirements among the members from different levels.

The paper is organized as follows: In Sect. 2, brief introduction of the stochastic modeling method is presented. In Sect. 3, We give our definition of the Stackelberg equilibrium of the EV-DC model. And a hybrid intelligent algorithm integrated by stochastic simulation and particle swarm optimization for finding the Stackelberg equilibrium is introduced in Sect. 4. An application case study on a supply chain pricing problem is provided as a numerical example to demonstrate the effectiveness of the algorithm in Sect. 5. Section 6 concludes this paper.

2 Stochastic Programming Model

The general form of the mathematical programming with random parameters can be denoted like this:

$$\begin{cases} \max\ f(x,\xi) \\ \text{subject to}: \\ \quad g_i(x,\xi) \le 0, i = 1, 2, \ldots, p \end{cases}$$

where x is a decision vector, ξ is a random vector of parameters, $f(x,\xi)$ is the objective function, and $g_i(x,\xi), i = 1, 2, \ldots, p$ is a constraint function set.

However, it is obvious that this random programming is meaningless since ξ is a random vector. We traditionally use the expected value to replace the random parameters. The form is as follows:

$$\begin{cases} \max\ E[f(x,\xi)] \\ \text{subject to}: \\ \quad E[g_i(x,\xi)] \le 0, i = 1, 2, \ldots, p. \end{cases} \tag{1}$$

Here $E[f(x,\xi)]$ means the expected value of the objective function while $E[g_i(x,\xi)], i = 1, 2, \ldots, p$ denote the expected values of the constraints. In the real world, many constraints should be satisfied at a certain confidence level. Hence Liu [20] revised the expected value model with chance constraint which is more in line with the actual situation.

$$\begin{cases} \max\ E[f(x,\xi)] \\ \text{subject to}: \\ \quad \Pr(g_i(x,\xi) \le 0) \ge \beta_i, i = 1, 2, \ldots, p. \end{cases} \tag{2}$$

If the ith constraint is mandatory in any situations, just make $\beta_i = 1$.

Furthermore, the expected value model does not consider the probability and the optimal value may cannot be achieved nearly half the chance. Hence, Liu [8] proposed the dependent chance model to maximize or minimize the probability of satisfying some certain events.

$$\begin{cases} \max \ \Pr(f(x,\xi) \geq 0) \\ \text{subject to:} \\ \quad \Pr(g_i(x,\xi) \leq 0) \geq \beta_i, i = 1,2,\ldots,p. \end{cases} \tag{3}$$

What's more, anther model proposed by Cooper [7] has also been used to deal with the reliability of the stochastic environment:

$$\begin{cases} \max \ \bar{f} \\ \text{subject to :} \\ \quad \Pr(f(x,\xi) \geq \bar{f}) \geq \alpha \\ \quad \Pr(g_i(x,\xi) \leq 0) \geq \beta_i, i = 1,2,\ldots,p. \end{cases} \tag{4}$$

This model is to maximize the objective function under a certain confidence level while the constraints should be satisfied with confidence levels in the same way.

When dealing with bilevel programming problems in uncertain environment, if the upper level decision makers misjudge the follower's behavior, the "optimal solutions" obtained may no longer be optimal in the real world. Therefore, we should choose appropriate stochastic modeling method by considering the risk preferences and management requirements of the decision makers from various levels.

3 EV-DC Model

In this section, a novel hybrid multilevel programming framework denoted as EV-DC model will be discussed. We also present our definition of the Stackelberg equilibrium of the new model in this part. For simplicity, we just consider a two-level decentralized decision system with one leader and one follower. The form of the EV-DC model is as follows:

$$\begin{cases} \max\limits_{x} \ E[F(x,y^*,\xi)] \\ \text{subject to :} \\ \quad \Pr(G(x,\xi) \leq 0) \geq \beta_0 \\ \text{where } y^* \text{ solves problems} \\ \quad \begin{cases} \max\limits_{y} \ \Pr(f(x,,y^*,\xi) \geq \bar{f}) \\ \text{subject to :} \\ \quad \Pr(g_i(x,y^*,\xi) \leq 0) \geq \beta \end{cases} \end{cases} \tag{5}$$

where:

x: the leader's decision variable;
y: the follower's decision variable;

$F(x, y, \xi)$: the leader's objective function;
Pr $(f(x, y, \xi) \geq \bar{f})$: the follower's objective function;
Pr $(G(x, \xi) \leq 0) \geq \beta_0$: the leader's constraint;
Pr $(g_i(x, y, \xi) \leq 0) \geq \beta$: the follower's constraint.

Definition 1 For any x, if the feasible vector (x^*, y^*) satisfies:

$$E[F(x^*, y^*, \xi)] \geq E[F(x, \hat{y}, \xi)] \tag{6}$$

where y^* is the optimal response of the follower with respect to x^* and $\hat{y}$ is the follower's optimal response with respect to x, we call it a Stackelberg Nash equilibrium.

4 Hybrid Intelligent Algorithm

As mentioned above, even the simplest multilevel programming model is NP-hard [19]. The numerical algorithm is an efficient way to solve the problem. In fact, many categories of numerical algorithms have been developed in the past three decades such as the extreme point approaches, branch and bound algorithms and descend methods. But most of them just work with some specific conditions. However, with the development of computer science, using intelligent algorithms to solve this complex problem becomes a hot topic. The intelligent algorithms such as genetic algorithm [21, 22], simulated annealing [23] and ant colony optimization [24] are suitable for solving this complex problem. PSO, which is proposed by [25], is one of the most efficient intelligent algorithms to solve bilevel programming. And many scholars have applied this efficient algorithm to solving the bilevel programming [26–29].

The bilevel model in this paper is even harder to solve while nesting nonlinear and stochastic programming. Without considering any specific conditions, we propose a hybrid algorithm integrated by random simulation and a twofold PSO to compute the Stackelberg equilibrium.

The stochastic functions can be divided into three types as expected value function, α-optimal value function and probability function. Sometimes we are unable to get the explicit value of the complex nonlinear stochastic function. In this case, we can use stochastic simulation to approximate them [20]. The approximation algorithms are as follows:

Algorithm 1. Computing expected value [20]

Step1: Generate $(\omega_1, \omega_2 \ldots, \omega_N)$ according to the distribution of ξ, where N is the number of stochastic simulation;

Step2: Calculate the value of $(f_1, f_2, \ldots, f_N)$ where $f_i = f(x, \omega_i)$;

Step3: Expected Value $= \sum_{1}^{i=N} (f_i)/N$.

Algorithm 2. Computing probability [20]:

Step1: Generate $(\omega_1, \omega_2, \ldots, \omega_N)$ according to the distribution of ξ, where N is the number of stochastic simulation;

Step2: Set counter $i = 1, m = 0$;

Step3: Calculate function value $(f_1, f_2, \ldots, f_N)$ where $f_i = f(x, \omega_i)$;

Step4: $i = i + 1$ and IF $f_i > 0$: $m = m + 1$;

Step5: Repeat steps 3 and 4 until $i > N$;

Step6: Return m/N

Algorithm 3. Computing α-optimal value [20]

Step1: Generate $(\omega_1, \omega_2, \ldots, \omega_N)$ according to the distribution of ξ, where N is the number of stochastic simulation;

Step2: Calculate function value $(f_1, f_2, \ldots, f_N)$ where $f_i = f(x, \omega_i)$;

Step3: Set n as the integer part of αN;

Step4: Return the n th largest element in $(f_1, f_2, \ldots, f_N)$

To solve the problem, we should first derive the follower's optimal response functions given that it has observed the leader's decisions. In most nonlinear cases, we cannot get the analytical solution. Therefore, we can use a PSO to compute follower's optimal response to leader.

Algorithm 4. Computation of the follower's optimal response

Step0: Give a fixed vector x;

Step1: Randomly generate y^j from the feasible region where $j = 1, 2, 3\ldots, popsize$;

Step2: Compute fitness of each particle (x, y^j) through stochastic simulation;

Step3: For each particles, if the fitness value is better than the best fitness value (pBest) in history, set current value as the new pBest and then choose the particle with the best fitness value of all the particles as the gBest;

Step4: Calculate particle velocity and then update particle position;

Step5: Repeat Steps 3 and 4 for a given number of cycles;

Step6: Return the best solution $y^*(x)$;

Algorithm 5. Computation of the Stackelberg equilibrium

Step1: Randomly generate x^j where $j = 1, 2, 3\ldots, popsize$ from the feasible region;

Step2: Compute the followers' optimal response $y^*(x^j)$ for each x^j through the nested PSO;

Step3: Compute fitness of each particle through stochastic simulation;

Step4: For each particles, if the fitness value is better than the best fitness value (pBest) in history, set current value as the new pBest and then choose the particle with the best fitness value of all the particles as the gBest;

Step5: Calculate particle velocity and then update particle position;

Step6: Repeat Steps 3, 4 and 5 for a given number of cycles;

Step7: Return the best solution $(x^*, y^*(x^*))$ and the optimal value

5 Application

In this section, we apply the novel model to a supply chain problem where the manufacturer produces the product and wholesales it to the retailer. The manufacturer is much larger than the retailer and performs as a leader in the supply chain. And the retailer retails the product to customers and performs as the follower. The unit manufacturing costs and the demands are assumed as random numbers. The manufacturer announces wholesale price first and then the retailer decides the retail price by considering the wholesale price respectively.

Most works in the supply chain problem assume that both the leader and follower are risk-neutral, but in this section, we assume that the retailer is risk-averse which just wants to maximize the probability of attaining a certain profit level.

Some notations are given as follows:

p: unit retail price of the product, which is the retailer's decision variable, $p > 0$;
w: unit wholesale price of the product, which is the manufacturer's decision variable, $p > w$;
$\tilde{c}$: unit manufacturing cost of product, which is a random variable;
$\tilde{d}$: the market base of the product which is a random variable;
D: consumer demand of product which is a function of d and p: $D = (\tilde{d} - \beta p)^+$;
Π_m: manufacturer's profit, which is: $\Pi_m = (w - \tilde{c})D$;
Π_r: retailers' profit, which is a function of p and w: $\Pi_r = (p - w)D$.

All the random coefficients are nonnegative and mutually independent.

If the manufacturer wants to maximize the expected profit while the retailer wants to maximize the probability that the profit not less then $\bar{\Pi}_r$, the EV-DC model can be applied:

$$\begin{cases} \max\limits_{w_1,w_2} \ E[(w-\tilde{c})(d-\beta p^*)^+] \\ \text{subject to}: \\ \quad \Pr(w-\tilde{c}\geq 0)\geq 0.9999 \\ \text{where } p_1 \text{ solves problems} \\ \quad \begin{cases} \max\limits_{p} \ \Pr((p-w)(\tilde{d}-\beta p)^+ \geq \bar{\Pi}_r) \\ \text{subject to}: \\ \quad \Pr(\tilde{d}-\beta p\geq 0)\geq 0.9999 \\ \quad p-w>0 \end{cases} \end{cases} \tag{7}$$

We choose the following parameter values to explore a numerical example:

$$\tilde{c} \sim U(5,7),\ \beta = 100,\ \tilde{d} \sim N(3500, 200^2),\ \overline{\Pi}_r = 6400.$$

With these numerical values, we use the intelligent algorithm designed in Sect. 4 to solve this problem. After running the hybrid intelligent algorithm for 100

Table 1 The result of the numerical experiments

Seq	w	c_1	c_2	Pop	Gens	w^*	$p(w^*)^*$	Objective	Err (%)
1	1.6	1.6	0.70	30	100	14.9984	22.9984	10846.82	0.0640
2	1.7	1.7	0.70	30	100	14.9909	22.9909	10840.25	0.1246
3	1.7	1.7	0.80	30	100	14.9591	22.9591	10828.37	0.2340
4	1.6	1.6	0.80	30	100	14.9922	22.9922	10847.39	0.0588
5	1.6	1.6	0.80	20	100	14.9994	22.9994	10853.77	0.0000
6	1.7	1.7	0.70	20	100	14.9989	22.9989	10850.21	0.0328

generations with 30 particles, where the cycle number in the stochastic simulation is 10000 and the lower-level PSO runs for 200 generations with 30 particles, we get a satisfactory solution:

$$w^* = 14.9922, \ p^*(w^*) = 22.9922.$$

And the optimal expected profit of the leader is 10847.39.

In order to illustrate the effectiveness and robustness of the hybrid algorithm, another numerical study is carried out to compare the solutions obtained by the algorithm with various parameter settings, and all the computational results are presented in Table I. For comparison, the quality of solution can be obtained via the deviation from the analytic solutions. The result shows that the deviation denoted by percent error does not exceed 0.5 % when different parameters are selected, which implies the effectiveness and robustness of the algorithm.

6 Conclusion

In this paper, a hybrid bilevel programming model with the revised expected value model in the upper level and the dependent-chance model in the lower level was proposed which may have wide range of applications. We also gave our definition of the Stackelberg equilibrium of the novel hybrid bilevel programming model. And a hybrid intelligent algorithm integrated by stochastic simulation and a two-fold particle swarm optimization for finding the solutions of the model was introduced. Then we applied the models to a supply chain problem with one manufacturer and one retailer. The results of numerical sample demonstrated that the new intelligent algorithm is effect and robust.

Acknowledgments This work was supported by National Natural Science Foundation of China (No.71371141, 71001080).

References

1. Bracken J, McGill JT (1973) Mathematical programs with optimization problems in the constraints. Oper Res 21(1):37–44
2. Von Stackelberg H (1952) The theory of the market economy. William Hodge
3. Vicente LN, Calamai PH (1994) Bilevel and multilevel programming: a bibliography review. J Global Optim 5(3):291–306
4. Dempe S (2002) Foundations of bilevel programming. Springer
5. Colson B, Marcotte P, Savard G (2005) Bilevel programming: a survey. 4OR 3(2):87–107
6. Colson B, Marcotte P, Savard G (2007) An overview of bilevel optimization. Ann Oper Res 153(1):235–256
7. Charnes A, Cooper WW (1959) Chance-constrained programming. Manage Sci 6(1):73–79
8. Liu B (1997) Dependent-chance programming: a class of stochastic optimization. Comput Math Appl 34(12):89–104
9. De Wolf D, Smeers Y (1997) A stochastic version of a stackelberg-nash-cournot equilibrium model. Manage Sci 43(2):190–197
10. Patriksson M, Wynter L (1999) Stochastic mathematical programs with equilibrium constraints. Oper Res Lett 25(4):159–167
11. Gao J, Liu B, Gen M (2004) A hybrid intelligent algorithm for stochastic multilevel programming. IEEJ Trans Electron Inf Syst 124:1991–1998
12. DeMiguel V, Xu H (2009) A stochastic multiple-leader stackelberg model: analysis, computation, and application. Oper Res 57(5):1220–1235
13. Alizadeh S, Marcotte P, Savard G (2013) Two-stage stochastic bilevel programming over a transportation network. Transp Res Part B: Method 58:92–105
14. An K, Lo HK (2013) Ferry service network design with stochastic demand under user equilibrium flows. Transp Res Part B: Method
15. Berglund PG, Kwon C (2013) Solving a location problem of a stackelberg firm competing with cournot-nash firms. Netw Spatial Econ 1–16
16. Bernstein F, Federgruen A (2005) Decentralized supply chains with competing retailers under demand uncertainty. Manage Sci 51(1):18–29
17. Gao Y, Zhang G, Lu J, Wee H-M (2011) Particle swarm optimization for bi-level pricing problems in supply chains. J Global Optim 51(2):245–254
18. Kovacevic RM, Pflug GC (2013) Electricity swing option pricing by stochastic bilevel optimization: a survey and new approaches. Eur J Oper Res
19. Ben-Ayed O, Blair CE (1990) Computational difficulties of bilevel linear programming. Oper Res 38(3):556–560
20. Liu B (2009) Theory and practice of uncertain programming, vol 239. Springer, New York
21. Liu B (1998) Stackelberg nash equilibrium for multilevel programming with multiple followers using genetic algorithms. Comput Math Appl 36(7):79–89
22. Calvete HI, Gale C, Mateo PM (2008) A new approach for solving linear bilevel problems using genetic algorithms. Eur J Oper Res 188(1):14–28
23. Zha W, Sun J (2012) Bi-level programming model for road transport route optimization of dangerous goods based on simulated annealing algorithm. J Highw Transp Res Dev 29 (4):101–106
24. Calvete HI, Galé C, Oliveros M-J (2011) Bilevel model for production–distribution planning solved by using ant colony optimization. Comput Oper Res 38(1):320–327
25. Kennedy J, Eberhart R et al (1995) Particle swarm optimization. In: Proceedings of IEEE international conference on neural networks, vol 4, no 2, Perth, Australia, 1995, pp 1942–1948
26. Li X, Tian P, Min X (2006) A hierarchical particle swarm optimization for solving bilevel programming problems. Artif Intell Soft Comput—ICAISC 2006:1169–1178 Springer
27. Kuo R, Huang C (2009) Application of particle swarm optimization algorithm for solving bi-level linear programming problem. Comput Math Appl 58(4):678–685

28. Jiang Y, Li X, Huang C, Wu X (2013) Application of particle swarm optimization based on chks smoothing function for solving nonlinear bilevel programming problem. Appl Math Comput 219(9):4332–4339
29. Ma W, Wang M, Zhu X (2013) Improved particle swarm optimization based approach for bilevel programming problem-an application on supply chain model. Int J Mach Learn Cybern 1–12

Preemptive Multi-mode Resource-Constrained Discrete Time-Cost-Resource Optimization via Non-dominated Sorting Genetic Algorithm

Wei-min Ma, Yang-yang Che and Hua Ke

Abstract With the current intense competitive circumstance, preemption is taken into consideration in project scheduling gradually, and multi-objective is providing new insights. For the purpose of practical relevance, a preemptive multi-mode resource-constrained discrete time-cost-resource optimization model (P-MRC-DTCRO) is presented in this paper. The integrated model provides interruption extension of the MRC-DTCRO model, which allows activity splitting. Accordingly, a modified non-dominated sorting genetic algorithm (NSGA-II) with preemption validity is proposed. A set of non-dominated schedules involving combinatorial optimization of completion time, cost and resource leveling is determined. This paper based on a valid algorithm provides new perspectives in both preemption and multi-objective problem, which is also propitious to decision-making in project construction.

Keywords Discrete time-cost trade-off · Genetic algorithm · Preemption · Resource leveling · Resource-constrained project scheduling

1 Introduction

The resource-constrained project scheduling problem (RCPSP) has been a widely studied optimization problem since it is initiated. The problem considers the minimization of the makespan, and schedules the project subject to precedence and resource constraints [1]. However, the basic model is, to some extent, restrictive for practical applications, thus it attracts numerous variants and extensions. In multi-mode resource-constrained project scheduling problem (MRCPSP), each

W. Ma · Y. Che (✉) · H. Ke
School of Economics and Management, Tongji University, Shanghai, China
e-mail: 2013emmache@tongji.edu.cn

E. Qi (ed.), *Proceedings of the 6th International Asia Conference on Industrial Engineering and Management Innovation*,
DOI 10.2991/978-94-6239-145-1_68

activity can be performed in one out of several modes, which involves different combinations of resource requirements and durations. A number of methods have been implemented in MRCPSP, including estimation of distribution algorithm [2] and ant-colony-optimization-based methodology [3].

In order to cope with more realistic circumstance, discrete time-cost trade-off problem (DTCTP) was presented. The relationship between time and cost is a trade-off, as completion speed is closely related with cost. Vanhoucke and Debels [4] proposed a meta-heuristic approach for the DTCTP with different realistic settings: time/switch constraints, work continuity constraints, and net present value maximization. Zaman [5] developed an evolutionary search technique, which was effectively tested on large benchmark instances.

Resource leveling, a technique with practical relevance, is to minimize resource fluctuations and results in cost savings. The recent developments made in this area are as follows: El-Rayes and Jun [6] introduced two new metrics to maximize the efficiency of resource allocation by incorporating them in a robust and practical optimization model. Geng [7] presented a directional ant colony optimization algorithm for solving nonlinear resource-leveling problems, which efficiently improved the convergence rate and the quality of solution for real-project scheduling.

Traditionally, a fundamental assumption of project scheduling is that activities in progress are non-preemptive. However, real-world projects are so complex that sometimes interruptions may occur. In most extensions, preemption is discrete and without extra cost. Compared with continuous one, a discrete preemption means activities can be preempted at integer time instants and restart later with no additional cost.

Kaplan [8] first introduced the preemption, established discrete preemption in a single-mode case and adopted dynamic programming for solving the case. Then, branch-and-bound procedures were proposed in both single and multi-mode cases to gain the improvement [9, 10]. Numerous literatures applied heuristic methods towards preemption. Ballestin et al. [11] showed how three basic elements of many heuristics for the RCPSP, codification, serial SGS and double justification, can be adapted to deal with interruption. Alsayegh and Hariga [12] combined particle swarm optimization (PSO) and simulated annealing (SA) search procedures to generate near-optimal project schedules. Recently, Tavana et al. [13] combined interruption with time-cost-quality trade-off problem, and applied multi-objective evolutionary algorithm to solve the model.

Nowadays, along with rapid development of construction project quantity and scale, multi-objective has gradually received significant attention. Xu et al. [14] formulated simultaneous optimization of total project cost, project duration, crashing cost, and environmental impact. Tavana et al. [13] correlated the time, cost, and quality objectives as trade-offs. Zahraie and Tavakolan [15] embedded the time-cost-resource optimization problem in stochastic environment. Ghoddousi et al. [16] considered multi-mode resource-constrained project scheduling problem (MRCPSP), discrete time-cost trade-off problem (DTCTP) and also resource allocation and resource leveling problem (RLP) simultaneously. Both of them implemented non-dominated sorting genetic algorithm to reach the optimization.

Combining the ideas from the aforementioned research, we introduce a preemptive multi-mode resource-constrained discrete time-cost-resource optimization model in this paper, as an extension of MRC-DTCRO [16]. Furthermore, a revised non-dominated sorting genetic algorithm is presented to determine a set of non-dominated optimal solutions. The remainder of this paper is organized as follows: Sect. 2 primarily illustrates the formulation of the proposed P-MRC-DTCRO model. Section 3 is devoted to a detailed description of the revised non-dominated sorting genetic algorithm. Simple model application and computational results are discussed in Sect. 4. In the end, we finalize with some conclusions.

2 Problem Formulation

The P-MRC-DTCRO considered in this paper is described as an activity-on-the-node network $G(N, A)$, where N is the set of activities and A is the set of pairs of activities between which a zero-lag finish-start precedence relationship exists. Dummy activities $i = 0$ and $i = N + 1$ mark the beginning and end of the project. Each activity $i \in N$ performs in a mode m_i, chosen out of a set of M_i different execution modes, $m_i \in \{1, 2, \ldots, M_i\}$. When executed in mode m_i the relevant parameters of activity i are accordingly determined: the duration is denoted as d_{im_i}, the direct cost is appointed as c_{im_i}, and the resource requirement of resource k is indicated by $r_{im_i,k}$. For renewable resource k, the availability R_k is a constant, which is constrained throughout the project. With respect to preemption in this paper, it means an activity can be split into a number of sub-activities with one unit duration or serval at no additional cost. All sub-activities are denoted by $v \in \{1, 2, \ldots, \sum x_{im_i} d_{im_i}\}$ and kept consistent in execution modes selection.

2.1 *Objective Functions*

As a multi-objective problem, the P-MRC-DTCRO is a combined optimization, correspondingly consisting of three objective functions.

The first objective aims at minimizing the total project duration, in accordance with $s_{n+1,1}$, denoted as start time of activity $N + 1$, since it is a dummy end. Therefore the makespan F_t is:

$$F_t = s_{n+1,1}$$

The second objective is to minimize the total project cost, which generally includes two parts: direct and indirect cost. Direct cost is assigned along with the selection of execution modes. While indirect cost is irrelevant to execution modes. Since it involves project management usually, we assume c_{in} as the fixed expenditure per day so that the entire indirect cost varies with completion time. Based on the above-mentioned details, the project cost in P-MRC-DTCRO can be formulated as follows:

$$F_c = \sum_i \sum_{m_i \in M_i} (x_{im_i} c_{im_i}) + s_{n+1,1} c_{in}$$

Decision variable x_{im_i} in this formula is defined as:

$$x_{im_i} = \begin{cases} 1, & \textit{if activity i is excuted in mode } m_i \\ 0, & \textit{otherwise} \end{cases}$$

The third objective is reducing the fluctuation in resource utilization, measured by deviation of x moment M_x^{dev}, $x \in \{1, 2, \ldots, s_{n+1,1}\}$. The metric is based on the average resource usage $\overline{r_k}$ in resource histogram. Taking $\overline{r}_k$ as target value, each resource k can gain squared deviation (SD) with certain resource usage $r_k(t)$ of resource k in period $t \in \{1, 2, \ldots, s_{n+1,1}\}$, collectively evaluating the deviation for resource leveling problem (RLP):

$$F_r = M_x^{dev} = \sum_{k=1}^{k} \sum_{t=1}^{s_{n+1,1}} (r_k(t) - \overline{r_k})^2$$

where $\overline{r_k} = \frac{1}{s_{n+1,1}} \sum_{\tau=1}^{s_{n+1,1}} r_k(t)$.

2.2 Multi-objective Model

The model of P-MRC-DTCRO is the combination of preemption, MRCPSP, DTCTP and RLP. Aiming at multi-objective optimization, non-dominated sorting genetic algorithm introduced by Deb et al. [17] is implemented to reach the optimal solution. The mathematical model of the P-MRC-DTCRO can be established as follows:

$$\begin{cases} \min F_t \\ \min F_c \\ \min F_r \\ \text{subject to:} \\ \begin{cases} \sum\limits_{m_i \in M_i} x_{im_i} = 1, \forall i \in N \\ s_{i,\sum_{m_i \in M_i} x_{im_i} d_{im_i}} + 1 \le s_{j,1}, \forall (i,j) \in A \\ s_{i,v} + 1 \le s_{i,v+1}, \forall i \in N, \forall v \in s_{i,\sum_{m_i \in M_i} x_{im_i} d_{im_i}} \\ \sum\limits_{i=p_t} \sum\limits_{m_i \in M_i} (x_{im_i} r_{im_i k}) \le R_k, k = 1, 2 \ldots k, \\ p_t = \{ i | s_{i,v} < t \le s_{i,v+1} \} \end{cases} \end{cases}$$

In the above model, the objective functions calculated from formulas in the previous section minimize time, cost and resource deviations, respectively. The first constraint ensures that each activity is conducted in exactly one mode m_i. Distinctively, preemption problems contain two constraints, the second and third, to force the generated schedules to satisfy the precedence relationship. In the second constraint, the earliest start time of an activity *j* is forbidden to be smaller than the finish time of its predecessor activity *i*, for the last unit of duration. The third constraint guarantees that for per time instance of activity *i*, the start time has to be at least one time-unit larger than the start time for the previous unit of duration. The last constraint requires, for each time instant *t* and for each resource type *k*, the resource demand to be below the resource limitation.

3 Genetic Algorithm

The P-MRC-DTCRO, is a combinatorial problem of preemption, MRCPSP, DTCTP and RLP. Generally, multi-objectives are in conflict that an improvement in one objective can merely be achieved by sacrificing another. Additionally, it is difficult to predefine preferences since the multi-objectives are incommensurable. Therefore, the model is incapable of reaching a unique optimal solution, but gains Pareto-optimal solutions instead. Pareto-optimal solutions are a set of non-dominated solutions, solved by NSGA-II.

The appropriate representation of preemption is the most critical element, which is an important innovation and improvement in this paper as well. Accordingly, preemption is precisely described by a series of unitary sub-activities, correspondingly related to original activities. We form four information sheets to respectively represent the relationship of activities, sub-activities and modes, information of activities and modes, precedence relationship and preemptive activities. Based on the above, the algorithm can fetch activity information whether preemptive or not. The main steps to obtain Pareto-optimal solutions with modified NSGA-II is described below:

Step 1: Initialization

Randomly generating population with size N as parent population is the first step. Individual creations rely on a double list chromosome representation: an activity list (AL) and a mode assignment list (MAL). In the AL, activities are arranged randomly on the premise that the precedence relationship is satisfied. Modes are selected following activity numbers, sequentially forming the MAL. During the process, the preemptive activities appoint the first sub-activity to make mode selection, to which the remaining coincide.

Step 2: Crossover

Crossover is the main operation to produce the offspring, which doubles the population. We implement one-point crossover in this paper. The randomly chosen mother and father chromosomes via two randomly chosen points respectively delivery informations in the AL and MAL. Taking the first chosen point as divide, the left positions inherit exactly the same activity order from the mother. Subsequently, the remaining right positions are filled according to the father by left-to-right scan, skipping the activities already taken from the mother. Based on the second chosen point, crossover of the MAL is operated in the same way but without considering skipping, benefiting from the activity-numbered sequence of the MAL. Similarly to initialization, mode assignments in preemptive activities still conform to the selection of first sub-activity.

Step 3: Mutation

In consideration of population diversity, a proportion of mutation is significant. For the AL and MAL, two mutation techniques have distinctive features, while both begin with randomly selecting the activity which will be mutated. Mutation applied to the AL is built to keep list precedence feasible. It determines the predecessor and successor of the selected activity first, taking their positions as bound, and then randomly chooses and inserts in a position in the range to generate a feasible child chromosome. As to the mutation in the MAL, the algorithm reselects mode for the selected activity among its selectable modes, then checks the activity if it is preemptive, and updates the remaining sub-activities mode assignment of the preemptive activities.

Step 4: Chromosome Decoding

Based on the above-mentioned, chromosome decoding transforms individuals into desired solutions, containing project schedules and objective function values. We apply serial schedule generation scheme (serial SGS) to decode the two list chromosomes. In this stage, there are N circulations, equaling to number of activities, which means activities will be decoded in sequence of AL. Considering selected mode, schedule generating process schedules activity at the earliest feasible start time in strict accordance with its execution mode, precedence relationship and resource availability. Meanwhile, three objective function values, time, cost and resource utilization deviation, are calculated stepwise.

Step 5: Pareto Front Formation

Pareto front formation contains two missions: rank divide and crowding distance calculation. In the multi-objective optimization, a dominated solution compared to the other one means all its objective function values are superior or equal to the other's; otherwise, it is called non-dominated. Rank bases on dominated relation, and dominated solutions are set at the top of the rank. We implement the fast non-dominated sorting procedure to assign every individual a front number, eventually forming the non-dominated fronts. Afterwards, individuals in each front are sorted by crowding distance. Separately, the algorithm calculates three objectives' deviations, and accumulates three deviations resulting in each individual's crowding distance value.

Step 6: Selection

With the purpose of next iteration of the loop, new parent population requires to be selected from the combined population. Genetic roulette wheel selection based on ranking and crowding distance selects the combined population in two corresponding steps. In rank-based step, individuals in lower front are attached with larger probability to be selected in the roulette wheel. Similarly, in same front, individuals with larger crowding distance are given preference in the crowding-distance-based step. As a consequence, we guarantee the high quality of population, meanwhile avoid the local optimum. Multipliers can be especially confusing.

4 Application and Results

A simple case is discussed in order to verify the validity of the proposed P-MRC-DTCRO model and the revised NSGA-II algorithm. The algorithm is encoded and run in MATLAB. Five activities are included in this simple project instance, whose precedence relationships are indicated in Fig. 1. Each activity possesses two alternative modes with related durations, resource requirement and direct cost, which are presented in Table 1. With the purpose of concision, there is only one renewable resource involved, whose per period availability is 4. The daily indirect cost is supposed to be $500. In consideration that preemption is improbable to occur throughout the project but appears in several activities, we permit activity 3 to be interrupted.

The final Pareto front consisting of five non-dominated solutions is gained, presented in Figs. 2, 3, 4 and 5, which is accomplished as a population size of 10, a mutation proportion of 40, and the generation iteration of 100. Among four solutions of the P-MRC-DTCRO model, the objective values of time, cost and resource moment deviation range respectively from 7 to 9, $16500 to 19500, and 1.7143 to 6. Because of the contradictory relationships between multi- objectives, solution 1 achieves minimum completion time and moment deviation but cost most, while solution 4 is the most time-consuming with most drastic fluctuations, but accompanied by maximum cost saving.

Preemption occurs depending on particular circumstances. In solution 1, optimum is reached without interruption. And in the remaining optimal solutions, activity 3 is split into two parts in different manners to achieve compact and steady resource-consuming schedules. By comparison with non-preemptive case, preemption indeed reveals the superiority and provides new opportunities for project scheduling in the real world.

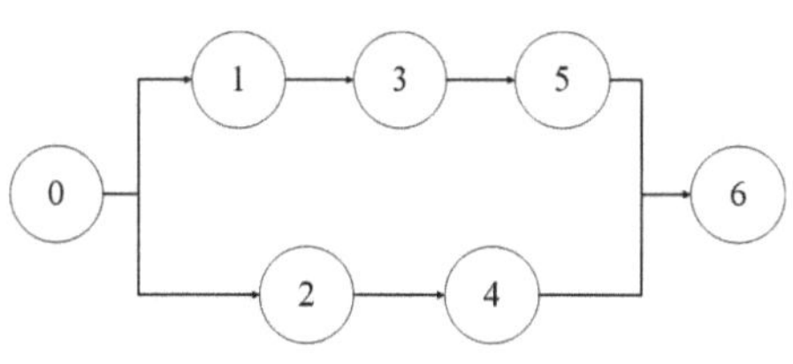

Fig. 1 Network

Table 1 Project information

Act i		Information		
	m_i	d_{im_i}	r_{im_ik}	c_{im_i} ($1000)
1	1	1	4	5
	2	2	3	2
2	1	1	2	3
	2	3	1	3
3	1	3	2	3
	2	4	2	2
4	1	1	3	4
	2	1	4	5
5	1	2	4	1
	2	3	2	3

Fig. 2 Non-dominated Pareto front solution 1

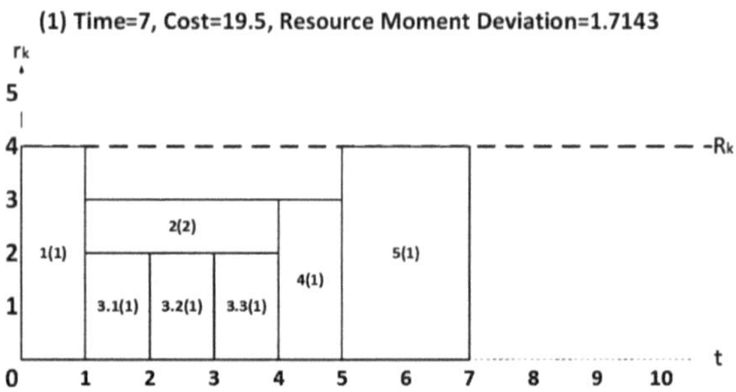

Fig. 3 Non-dominated Pareto front solution 2

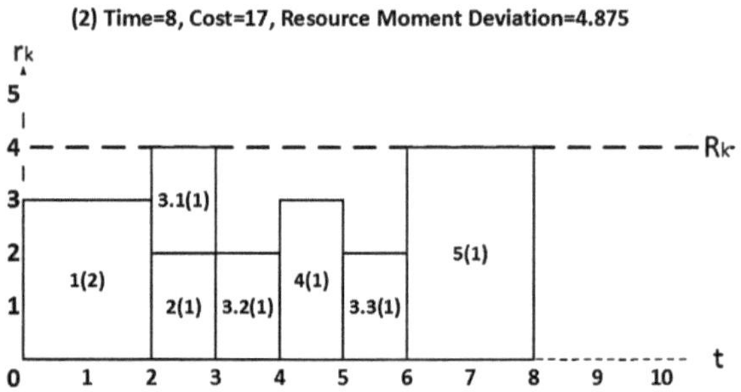

Fig. 4 Non-dominated Pareto front solution 3

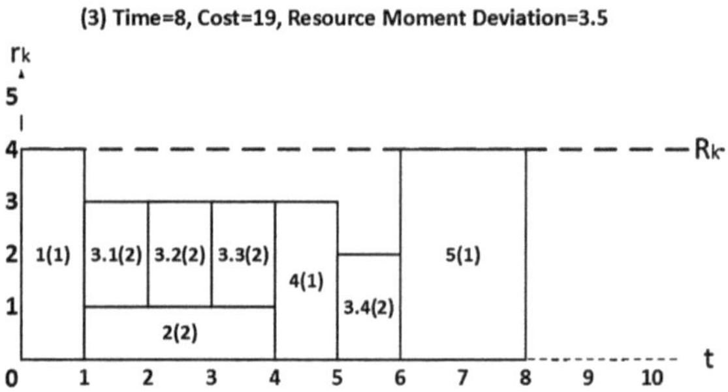

Fig. 5 Non-dominated Pareto front solution 5

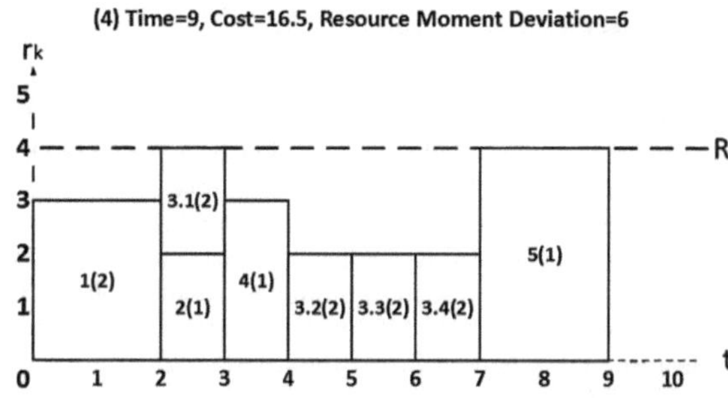

Simultaneously, it is worth mentioning that the Pareto front solutions based on an assumption that no particular preference exists in objectives. In other words, decision makers can choose optimal solution according to his own preference to reach a most suitable trade-off for the reality.

5 Conclusion

The multi-mode resource-constrained project scheduling problem, discrete time-cost trade-off problem, and the resource leveling problem were well-established through the efforts of numerous researchers. While activity splitting is taken into account gradually, and multi-objective is providing new insights. An integrated idea, corresponding to reality, of a preemptive multi-mode resource-constrained discrete time-cost-resource optimization model, based on the aforementioned research was generated, which extended the novel MRC-DTCRO model to preemption.

According to the proposed combinatorial optimization of completion time, cost and resource leveling, a revised non-dominated sorting genetic algorithm was introduced to determine a set of non-dominated optimal solutions. The preemption was appropriately represented and validly executed in the presented algorithm. A simple case was implemented to demonstrate the model capabilities and the algorithm effectiveness.

Towards the P-MRC-DTCRO model, further development can be focused on the adaption of resource leveling metric. Simultaneously, uncertain circumstance can be incorporated into the combinatorial model as well.

Acknowledgments This work was supported by the National Natural Science Foundation of China (No.71371141, 71071113, 71001080), a Ph.D. Programs Foundation of Ministry of Education of China (No.20100072110011), Shanghai Philosophical and Social Science Program (No.2010*BZH*003), and the Fundamental Research Funds for the Central Universities.

References

1. Herroelen W, De Reyck B, Demeulemeester E (1998) Resource-constrained project scheduling: a survey of recent developments. Comput Oper Res 25(4):279–302
2. Wang L, Fang C (2012) An effective estimation of distribution algorithm for the multi-mode resource-constrained project scheduling problem. Comput Oper Res 39(2):449–460
3. Li H, Zhang H (2013) Ant colony optimization-based multi-mode scheduling under renewable and nonrenewable resource constraints. Autom Constr 35:431–438
4. Vanhoucke M, Debels D (2007) The discrete time/cost trade-off problem: extensions and heuristic procedures. J Sched 10(4-5):311–326
5. Zamani R (2013) An evolutionary search procedure for optimizing time-cost performance of projects under multiple renewable resource constraints. Comput Ind Eng 66(2):451–460
6. El-Rayes K, Jun DH (2009) Optimizing resource leveling in construction projects. J Constr Eng Manage 135(11):1172–1180
7. Geng J-Q, Weng L-P, Liu S-H (2011) An improved ant colony optimization algorithm for nonlinear resource-leveling problems. Comput Math Appl 61(8):2300–2305
8. Ann L (1988) Resource-constrained project scheduling with preemption of jobs. Ph.D. dissertation, University of Michigan
9. Demeulemeester EL, Herroelen WS (1996) An efficient optimal solution procedure for the preemptive resource-constrained project scheduling problem. Eur J Oper Res 90(2):334–348
10. Buddhakulsomsiri J, Kim DS (2006) Properties of multi-mode resource-constrained project scheduling problems with resource vacations and activity splitting. Eur J Oper Res 175 (1):279–295
11. Ballestın F, Valls V, Quintanilla S (2008) Pre-emption in resource-constrained project scheduling. Eur J Oper Res 189(3):1136–1152
12. Alsayegh H, Hariga M (2012) Hybrid meta-heuristic methods for the multi-resource leveling problem with activity splitting. Autom Constr 27:89–98
13. Tavana M, Abtahi A-R, Khalili-Damghani K (2014) A new multi-objective multi-mode model for solving preemptive time-cost-quality trade-off project scheduling problems. Expert Syst Appl 41(4):1830–1846
14. Xu J, Zheng H, Zeng Z, Wu S, Shen M (2012) Discrete time-cost-environment trade-off problem for large-scale construction systems with multiple modes under fuzzy uncertainty and its application to jinping-ii hydroelectric project. Int J Project Manage 30(8):950–966
15. Zahraie B, Tavakolan M (2009) Stochastic time-cost-resource utilization optimization using non-dominated sorting genetic algorithm and discrete fuzzy sets. J Constr Eng Manage 135 (11):1162–1171
16. Ghoddousi P, Eshtehardian E, Jooybanpour S, Javanmardi A (2013) Multi-mode resource-constrained discrete time-cost-resource optimization in project scheduling using non-dominated sorting genetic algorithm. Autom Constr 30:216–227
17. Deb K, Pratap A, Agarwal S, Meyarivan T (2002) A fast and elitist multiobjective genetic algorithm: Nsga-ii. IEEE Trans Evol Comput 6(2):182–197

Part III
Management Innovation Method and Practice

Implementing Enterprise Open Innovation Strategy by Effectively Collaborating with External Resources

Hai-jun Wang, J.Z. Feng and H.B. Shi

Abstract Increasing marketing competition forces enterprise to change! Open innovation is especially discussed in this paper focusing on how to manage the external resources rationally, in which the closed innovation is compared with open innovation firstly, then the external resources category is classified by following a flexible mechanism in order to form a self-rotating innovation cycle. Moreover, an open innovation network model is offered in order to coordinate the phase-in as well as phase-out points of external resources involving in enterprise innovation process. Finally, a case concerning Haier Group open innovation practice is presented to further illustrate the adaptability of the methodology given in this paper.

Keywords Open innovation · External resources · Strategic supplier

1 Introduction

Driven by the rapid development of market globalization, enterprises have to make great efforts to fulfil individual demands from market by utilizing global resources. Since the concept of open innovation was present by Chesbrough in 2003, the study of open innovation to facilitate the sustainable development of enterprise is increasingly becoming popular for academics and practitioners alike. Open innovation is defined as the process of strengthen the interaction and collaboration of multiple partners involved in business eco system to jointly deliver innovative product or solutions [1], which means successful open innovation partnerships will

H. Wang (✉) · H.B. Shi
The School of Management, Shenyang University of Technology, Shenyang City
People's Republic of China
e-mail: wanghj77@163.com

J.Z. Feng
School of Management, Hangzhou Dianzi University, Hangzhou City
People's Republic of China

E. Qi (ed.), *Proceedings of the 6th International Asia Conference on Industrial Engineering and Management Innovation*,
DOI 10.2991/978-94-6239-145-1_69

be extremely useful for exploring meaningful and even innovative solutions to satisfy the customers' requirements. Moreover, the sourcing, integration and development of product and business model innovations of enterprise through win-win external partnerships are of great concerns to enterprises in order to capture maximum commercial value for theirs R&D investment [2, 3].

At present, a large number of enterprises from different industries such as consumer goods, electronics and equipment, have greatly benefited from the adoption of open innovation strategy to enrich their internal innovation processes [4]. In this regard, many of them are competing to win through the implementation of open innovation strategy, cooperating with various partners in the value chain—including universities, institute and even other enterprises—by flexibly opening up to external partnerships and increasing reliance on collaborative idea generation [5, 6]. As far as we understand, how to classify innovative resources systematically and integrate the external resources such as university and institute in the right stage of R&D has generally been difficult for enterprise.

Accordingly, this research targets the further understanding of various aspects of open innovation and makes several contributions to improve the innovation capabilities of enterprise combining other stakeholders existing research. Firstly, external resources accessible for the open innovation of enterprise are classified, then key activities during earlier stage of open innovation are analysed in conjunction the involvement of external resources. In the final, a case from Haier Group is given to verify the feasibility of the theory and methodology presented in this paper.

2 Transfer from Closed R&D Innovation to Open Innovation

Nowadays, external business environment variations force enterprise to gradually transfer from an in-house R&D innovation model to open innovation in order to be more competitive and more suitable to new marketing situation [7]. According to Chesbrough's viewpoint, the closed innovation has become faded and been increasingly replaced by the open innovation due to major factors including (1) increased mobility of skilled workers, (2) expansion of venture capital, (3) external options for unused technologies and (4) increased availability of highly-capable outsourcing partners. Based on our observation, there are some additional factors affecting this transition such as (5) individualization of customers' requirements prefers mass customization (customer is core) instead of mass production (enterprise is core), and (6) standardization plays an important role in accelerating industrial cooperation. The differences between closed and Open Innovation principles are further illustrated in Fig. 1.

Meanwhile, it is very essential to assess the risks occurring in the process of implementing open innovation through the following aspects but not limited to [8]:

Closed Innovation Principles	Open Innovation Principles
1. Hire the talents so that the smartest people in the industry work for enterprise	1. Not all the smart people are hired by us – we need to engage with the best inside and OUTSIDE our company
2. To bring new products to the market, enterprise must discover and develop it by itself	2. External R&D can create significant value, in which external resources become indispensable for enterprise innovation
3. We need to discover it ourselves and to get profit from it	3. Products and profit models are designed in both outside and inside of company
4. The company that gets an innovation to market first will usually win	4. Building a better business model is better that getting to market first
5. Industry leaders in R&D investments will discover the best and most ideas, and will come to lead the market as well	5. Companies that make the best use of internal and external ideas will win
6. Control intellectual property to ensure that competitors do not profit from our ideas	6. Company can get profit from others' use of our IP and we should buy others' IP whenever it advances our own business model
7. The enterprise is the core, and mass production is the solution to balance quantity and cost	7. The customer is the core, and mass customization may be adopted to fulfill customers' requirements at a cost -effective way in terms of open innovation
8. No clear initiative to extend the innovation cycle to external resources, and the key concern is to sell out product and win in the market	8. To sustainably develop in the market means more connection and collaboration with external resources, and comply with industrial standards as possible as we can

Fig. 1 Refined comparison between closed innovation and open innovation

1. *Coordination issue*: It is not difficult to seek a partner but not easier to deal with conflicts in an efficient way due to different cultural backgrounds, unbalanced capacities and non-uniform participation purpose.
2. *External resources integration issue*: How to filter right partners through global resources and manage external partnerships in a good manner? Unlimited expansion of partnerships without scientific mechanism will bring up inevitable confusion rather than expected innovation outcomes.
3. *Commercial secret protection issue*: Generally, knowledge is expected to expand to other partners during collaboration, which may leads to leakage of enterprise commercial secrets not intended to be used for the project.
4. *Profit sharing issue*: Also, profits such as intellectual property right are frequently a sensitive issue which is hard to precisely define the contribution degree of each party, and disputes alike may arise and possibly cast negative shade over the sustainability of open innovation.

3 Classification of External Resources and Management Mechanism

Normally, external resources involving in open innovation come from:

1. *External Labs/design Houses*: joint development partners, OEM/ODM, trade partners, contract labs/design houses, certificate institute, start-up, competitors and individual inventors/designers,
2. *Academia/University*: scholarship partners, joint Lab and university student competition,
3. *Research and Management Institutes*: research institute, association and standardization committee,

4. *Innovation Intermediaries*: network platforms, venture capitalist and consumer/employee,
5. *Consultation service providers*: Consulting firm, Law firm.

More importantly, strategic suppliers with proven capacities are also indispensable for earlier involvement in product R&D innovation. The purpose of enterprises' earlier connecting strategic suppliers is to utilize their technology, production and inspection capacities and combine the R&D innovation with supply chain. In our study, commercialization of open innovation findings is exceedingly difficult to realize substantially without the assistance of supply chain.

In order to manage above-mentioned external resources rationally, it is necessary for enterprise to develop relevant mechanism to ensure the quality of participants in the open innovation activity. In principle, there should not be barrier for the introduction of outstanding resources while un-qualified resources have no space to live in the innovation circle. Therefore, a set of guiding principle are given to facilitate the cooperation between enterprise and strategic suppliers (Fig. 2), in addition to this, a funnel model mechanism in conjunction with a fridge compressor case is given herewith which aims at absorbing qualified resources into the innovation process while excluding un-qualified resources based on supply evaluation criteria (Fig. 3). Furthermore, it can be of reference for enterprise to coordinate the phase-in as well as phase-out points of external resources involving in its innovation process. It also indicates that strategic suppliers with right capacities including experiences, solutions, product performances, and cost competitiveness combining volume warranty could be filtered out.

Fig. 2 Strategy supplier appraisal guiding principles

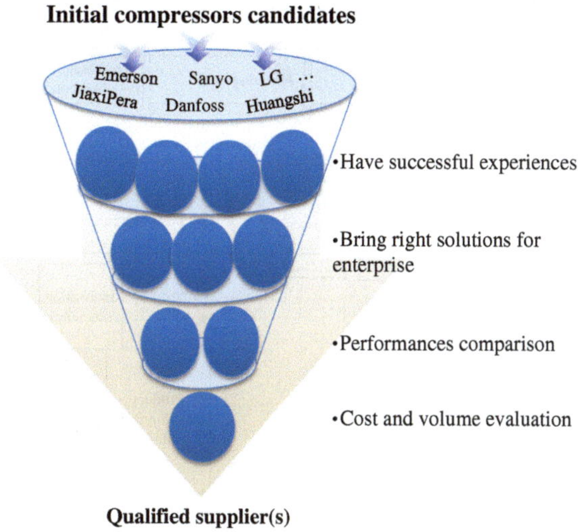

Fig. 3 Funnel model external resources management mechanism: a fridge compressor example

4 Open Innovation Network Model

In this paper, we divide the open innovation into two aspects: basic research and development innovation, and application innovation. The former focuses on prospective and mid-to-long term innovative technology R&D, while application innovation is featuring short-term product R&D to fulfill market requirements in the near future. Normally, basic research and development innovation shall be suitably carried out by enterprise research center in charge of general technology R&D, which is regarded as a corporate and strategic level department. The application innovation is more close to product to come to market and may work better for product development institute. However, the two aspects of innovations shall be rationally coordinated and combined to share external resources and pursue the maximum of outcome of open innovation.

As per above classification, an open innovation network model is presented here (shown in Fig. 4), in which both basic research innovation and application innovation are managed in the integrated R&D process of enterprise and results of collaboration with external partners are positively taken in the open innovation activities. Enterprise research center and product development institute respectively promote external and internal collaborative activities by cross-organizational communities and teams. All of inputs/outputs via external/internal collaboration in the open innovation network model shall be carefully evaluated periodically according to the nature of project and make Go/No Go decisions. For example, Go decision means the corresponding activity to continue as it is or continue but reconsider), while No Go decision would indicate that the input/output of the activity is not in line with evaluation criteria, and shall be suspend or stopped to release resources. To cultivate the knowledge management during open innovation

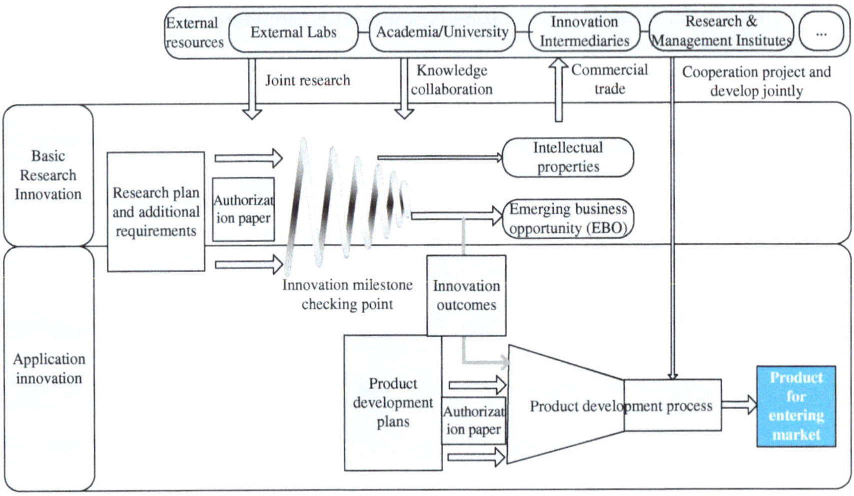

Fig. 4 Open innovation collaboration network model

process, those decisions combining their follow-up actions should be recorded and be reflected to forthcoming strategy and plan.

5 Case Study

The Haier Group open innovation practice is taken as a case to further illustrate the adaptability of the methodology presented in this paper. Being a leader in the home appliance industry, Haier has been seeking potentials for innovation to boost enterprise sustainable development in recent years. Under the severe competition of market, open innovation is applied and promoted by Haier. The following work is done in Haier to implement the open innovation:

1. *Reconstruct R&D system*: Previously, there is no clear obligation definition regarding the research and development work in Haier. The corporate R&D centre of Haier has some overlapping functions with concrete product (fridge, washing machine, A/C, etc.) development institutes, which led to many disputes and quarrels. Upon the open innovation, the corporate R&D centre is in charge of generic technology research and development, while concrete products for entering market are within the scope of product development institutes. Moreover, external resources are commonly utilized by these two departments in open innovation network and the technology innovation transferring mechanism is executed to link them.
2. *External resources integration*: With the help of funnel model, Haier realized the rational management of external resources, which ensures that qualified

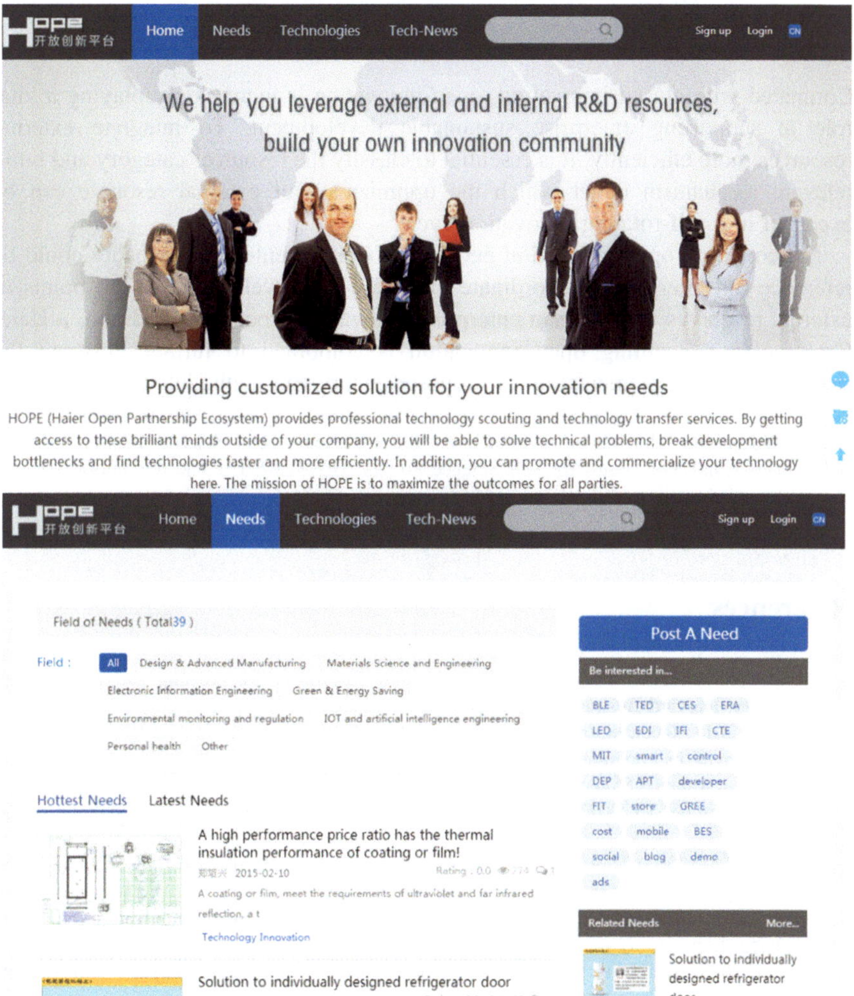

Fig. 5 Haier's open innovation platform

resources could be dynamically absorbed into the innovation process while un-qualified resources is excluded out of innovation cycle.

3. *Open Innovation Platform operation.* To facilitate the interaction and collaboration with external resources, Haier set up an Open Innovation Platform (hope.haier.com, shown in Fig. 5) aiming at providing a community for public, which is proved to be more efficient and accessible for external resources to join in the innovation activities. In our checks, more than 60 percent of innovative ideas and solutions in Haier originate from the open innovation platform since its launch in 2010.

6 Conclusion

Compared with closed innovation, open innovation is increasingly playing a vital role in supporting enterprise sustainable development. To integrate external resources more efficiently, it is essential to classify the resources category and build relevant mechanism under which the optimization of external resource can be executed in a self-rotating innovation cycle.

Moreover, an open innovation network model presented in this paper could be reference for enterprise to coordinate the phase-in as well as phase-out points of external resources involving in enterprise innovation process. In addition, a Haier Group case concerning open innovation is combined to further illustrate the acceptance and implementation of proposed theory and methodology.

Acknowledgments This paper is funded by the Liaoning Social Science Planning Project (L14AJY007) and Shenyang City Soft Science Project (F15-198-5-14), and the author herewith shows appreciation for their financial supports.

References

1. Chesbrough H (2003) Open innovation: the new imperative for creating and profiting for technology. Harvard Business School Press, Cambridge
2. Muhdi L (2011) Open innovation and collaboration for innovations. Eth Zurich
3. The challenges in implementing open innovation in a global innovation-driven corporation. Res-Technol Manage 55(4):32–38
4. Serrono V, Fischer T (2007) Collaborative innovation in ubiquitous system. Int Manuf 18:599–615
5. Leydesdorff L (2003) The mutual information of university-industry-government relations: an indicator of the triple helix dynamics. Scientometrics 58(2):445–467
6. Etzkowita H (2008) The triple helix: university-industry-government innovation in action. Routledge, London
7. Bo Z (2010) Research on the collaboration innovation of mid-small enterprise. Sci Technol Manag Res 2:5–7
8. Fontana R, Geuna A, Matt M (2006) Factors affecting university-industry R&D projects: the importance of searching, screening and signalling. Res Policy 35(2):309–323

Optimal Pricing and Capacity Planning Strategy for Cloud Service

Xiao-hui Gu, Fu-zan Chen and Min-qiang Li

Abstract Cloud computing is a revolutionary paradigm of the next generation IT industry which makes IT resources available to customers in a pay-as-you-go manner. Decision on price and capacity of cloud service (CS) is an essential issue in cloud computing. The optimal strategy on pricing and capacity planning for CS is studied in this paper. From a long-term and dynamic perspective, that issue is formulated as a bi-level programming model, i.e., the upper level is a monopoly CS provider (CSP) deciding price and service capacity to maximize profit and lower level is all the customers making decision on whether to buy the CS according to utility measure. Numerical experiments are investigated to conduct the proposed model. Our findings provide suggestions on optimal policies of price and capacity planning in a monopoly market.

Keywords Cloud service · Capacity planning · Pricing · Bi-level programming

1 Introduction

Cloud computing is an emerging paradigm that evolved from established technologies, such as cluster computing, grid computing, and virtualization [1]. Cloud computing can be defined as a model for enabling on-demand network access to an abstract resource pool [2]. Cloud services typically include the characteristics such as pay-as-you-go and marginal cost non-zero. Furthermore, service-level agreements (SLAs) are negotiation between the providers and consumers, which provide guarantees for service capacity and execution time [3].

Few studies on pricing for cloud services have been conduced. Extant pricing strategies always consider short-term gains from the perspective of providers, and customers just passively accept the price [4]. In addition, CSPs do not take full

X. Gu · F. Chen (✉) · M. Li
College of Management and Economics, Tianjin University, Tianjin, China
e-mail: fzchen@tju.edu.cn

E. Qi (ed.), *Proceedings of the 6th International Asia Conference on Industrial Engineering and Management Innovation*,
DOI 10.2991/978-94-6239-145-1_70

account of SLA impact on service capability planning within a decision-making process [5].

In this paper, the topic is long-run dynamic pricing for monopoly CSP. We contribute a bi-level programming model to represent a global optimization process. The first level is aimed at maximizing the monopoly provider profit, and the second level represents the heterogeneous consumer value perception. We discuss optimal service capacity planning, which can be analyzed in conjunction with SLA constraint. Besides, the variation in optimal decisions with the adjustment of service utilization is explored with a numerical method.

The rest of this article is organized as follows. In Sect. 2, we provide an overview of related work. A bi-level programming model is developed in Sect. 3. Section 4 investigates the optimal pricing and capacity planning strategy with numerical calculations. In Sect. 5, we conclude and discuss directions for future research.

2 Related Work

Pricing cloud services is built on two main streams in academic research: fixed price and dynamic price.

Currently, most cloud providers adopt a fixed-pricing strategy to allocate service capacity for customers. Hsu et al. [6] observed that cloud computing offers innovative pricing strategies. Service capability and perceived value during business operation affect cloud computing adoption. Moreover, a pricing strategy results to a certain degree from the providers' cost-benefit analysis [6]. Zhang and Niu [7] combined fixed price and dynamic capacity in a programming model. They found that exogenously given prices does not influence profit, but opportunistic behavior may decrease customer satisfaction [7]. Basu et al. [8] discussed usage-based plans and fixed-fee plans under different customer utility. This study provided guidance for service pricing from a management perspective. The scholars considered two vectors of customer attributes, which are positive and negative effects in the model of utility [8].

As reviewed above, a fixed-price strategy does not consider SLA constraints [9]. Moreover, a fixed-pricing strategy ignores the provider's long-term benefits. Customers tend to passively accept the price and lack of decision-making initiative in this case [10]. Therefore, dynamic pricing strategies have attracted more and more attention.

Some studies have provided ideas for market-based dynamic pricing strategies that encompass customer feedback. Zaman and Grosu [11] proved that a combinatorial auction-based allocation mechanism is significantly better than the fixed-price mechanism for provider revenue, since the auction mechanism allocates services to the highest valuation user [11]. Tang et al. [12] designed a practical bidding strategy under SLA constraints by using a constrained Markov decision process (CMDP), which aimed to balance service reliability with cost in spot instances [12]. Samimi et al. [13] proposed a novel model called the combinatorial double auction resource allocation (CDARA) to evaluate cloud resource allocation efficiency from an economic perspective; the model considers the benefits for providers and customers. In addition, it encourages participants to reveal their real valuation during the tender process [13]. Baars et al. [14] analyzed the pricing and refund mechanism in shared virtual infrastructures in cloud computing. The authors summarized the influential factors of cloud service chargeback in the cloud, which was pioneering work, and contributed to developing a successful chargeback model. This strategy is combined with SLAs to provide a more effective pricing scheme [14]. Kertész et al. [15] held the view that adaptive and flexible SLA attainment strategies are essential when demand is uncertain. Therefore, they proposed a solution called SLA-based service virtualization for on-demand resource provision; using this method can achieve efficient and reliable service operation in a heterogeneous cloud environment [15]. Xu and Liang [16] have an objective to minimize operational costs and maximize the number of requests submitted by customers in a distributed cloud service environment. The authors incorporated time-varying price into dynamic customer request rates to build an adaptive cost optimization framework and devised an optimal fast approximation algorithm [16].

Dynamic pricing strategies pay more attention to customer-perceived value, but they ignore that individual customers have varying service demands. Additionally, cloud service firms roughly stipulate the penalty policy, since SLA studies on service operation are still in the initial stage. The costs of cloud service are also general and abstract in these studies. Our study is highly related to the prior service pricing studies.

3 Model Setup

This section presents a bi-level programming model to develop cloud service pricing and a capacity planning strategy with SLA constraints in a long term.

We need to clearly demonstrate the stakeholders' interaction process to ensure cloud service achieves its maximum business value. This research builds a simple supply-demand relationship: CSP directly sells its service to customers when company is positioned at the beginning of the cloud computing value chain. Therefore, the stakeholders include monopoly provider and all customers in cloud market. CSP charges a fee to consumers for profit incentives. Consumers consider buying services to reduce in-house hardware cost and gain greater utility. We analyze cloud service from the perspective of provider.

At the first level, the profit is optimized for the monopolist by choosing service capacity with corresponding price. At the second level, all potential customers in cloud computing market make self-selection decisions about buying cloud service or not to gain non-negative utility.

3.1 Provider Perspective

The profit for CSP is equal to the revenue paid by customers minus the costs in service operation process. In this study, cloud service cost from the perspective of provider can be divided into three parts: variable cost (VC), penalty cost (PC), and capacity expansion cost (EC). Actual VC resembles a mixture of power cost, network cost, human cost, etc. PC reflects the SLAs' impact on price and service capability within a decision-making process. Additionally, when the demand for cloud service exceeds supply, expansion cost of additional infrastructure must be paid based on the capacity planning strategy.

As a result, we build the upper model to determine the optimal price and service capacity, to maximize the provider's profit in function (1). This model uses an alternative to extend service capacity or bear the damage for delay when the service is in short supply:

$$\max \pi = p \cdot n - \mathrm{VC} - \mathrm{PC} - \mathrm{m} \cdot \mathrm{EC}. \tag{1}$$

$$(m = 0,1,2,\ldots, M)$$

(1) *Variable cost*:
VC mainly includes the network, power, and labor costs during service operation and maintenance. For the sake of generality, we propose the following Cobb-Douglas function to represent the relation [17]. This function reflects the trends in the change of supply and demand:

$$\mathrm{VC}(n,l) = c \cdot \mu^b n^{1-b} \quad (0<b<1 \quad c>0). \tag{2}$$

In function (2), μ and n are two unrelated variables that represent the change in service capacity and the number of customers in cloud service, respectively. In addition, $0<b<1$ is a real exponent that represents capacity utilization when the service is output, and c is a constant parameter that represents the unit variable cost. To illustrate the limitations with service capacity, consider the sum of the index is 1. This means that the market demand will not have an impact on the provider's inherent computing resources. Figure 1 shows that variable cost is a strictly concave function and increases with rising μ and n.

(2) *Penalty cost*:
We define a simple SLA for each customer i. When the service capacity is not enough to fulfill the agreement, the provider must pay the corresponding compensation to the customer, which is calculated as a percentage of the price. Penalty cost is the sum of compensation for each customer:

$$\mathrm{PC}(n,l) = \begin{cases} p \cdot \sum_{i=1}^{n-\mu} (r \cdot i) & n > \mu \\ 0 & n \leq \mu \end{cases}. \tag{3}$$

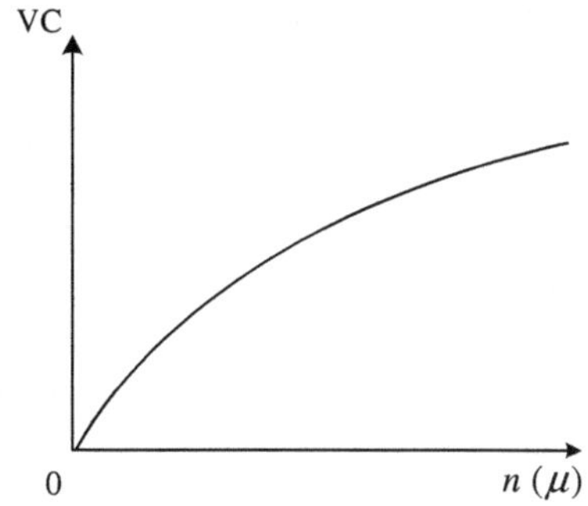

Fig. 1 VC trend with increasing supply and demand

In this function, PC represents the cloud provider penalty for failing to meet SLA. Constant r is default rate used to measure the degree of breach of contract. $r \cdot i$ represents penalty percentage depending on a linear function. CSP negotiates with consumers about the penalty policy, and most of time, default rate does not change after SLA is accepted.

(3) *Capacity expansion cost*:
Cloud computing makes it easier for the enterprise to scale its services capacity according to customer demands. Iyoob et al. (2013) stated that service capacity is discrete, such as the number of virtual machines and software versions [5]. Therefore, expanding data center or upgrading software improves the service capacity step by step.

One key factor in cloud service cost-efficiency is server utilization rate [4]. As cloud service enterprise scale increases, the efficiency of resource allocation decreases. Thus, marginal expansion cost increases along with the rising service capacity, as shown in Fig. 2.

$$EC = a \cdot (m \cdot \Delta\mu) \cdot r_s^{-(\mu + m \cdot \Delta\mu)/\mu_{\max}} \quad (0 < r_s < 1) \tag{4}$$

The capacity expansion cost function is described by function (4), where r_S as server utilization, which is valued between 0 and 1, and m represents the number of capacity expansion blocks. a is unit expansion cost of service. The total service capacity increase is denoted as $m \cdot \Delta\mu$

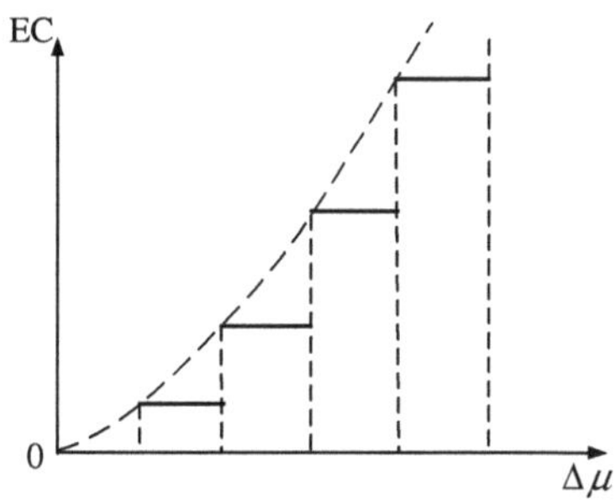

Fig. 2 EC as a step function with discontinuous increases in unit service capacity

3.2 Customer Perspective

The second level represents the heterogeneous customers' value perception. Customers are enterprises or individuals that consume services in cloud market. Because they have limited energy, attention, and time, we assume willingness to pay (WTP) exist upper limit of service capacity. Furthermore, the lower limit is related to whether service has adequate availability. Therefore, this paper represents WTP as a piecewise linear function [18] for heterogeneous service demands in Fig. 3.

In function (5), parameter v indicates customer preference, and all customers are assumed to be uniformly distributed between 0 and 1. μ_R is reservation service capacity, and μ_s means saturation service capacity.

$$W_i(l) = \begin{cases} 0 & \mu < \mu_R \\ v \cdot \mu_s \cdot \frac{\mu - \mu_s}{\mu_R - \mu_s} & \mu_R < \mu < \mu_s \\ \mu_s \cdot v & \mu > \mu_s \end{cases} \tag{5}$$

The customer utility is expressed as function (6). Subscript i denotes different customer preferences for corresponding service. In this model, we assumed that one customer buys only one unit of service at most. Constraints (8) and (9) indicate that customers purchase the service when they have non-negative utility:

$$\mathrm{U}_i(l, \mathrm{p}) = W_i(l) - p, \tag{6}$$

$$n = \sum_{i=1}^{N} k_i \quad i = 1,2, \ldots, N, \tag{7}$$

$$s.t. \quad k = 0 \quad if \quad U_i < 0, \tag{8}$$

$$k = 1 \quad if \quad U_i \geq 0. \tag{9}$$

Fig. 3 Piecewise linear WTP function

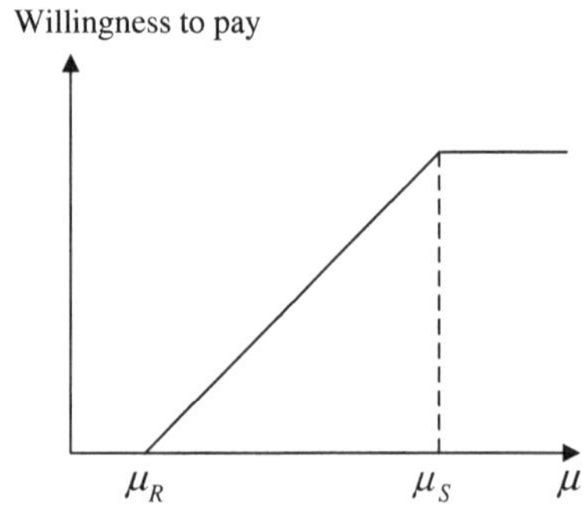

This bi-level model considers monopoly cloud supplier profit and customer's utility, in order to determine what pricing and capacity planning scheme works best. The model can be used to choose and analyze the CSP operation strategy.

4 Experimental Analysis

4.1 Experimental Design

Because the model consists of a complicated profit function and a piecewise linear utility function, getting an analytical solution is impossible. Thus, we suggest using a genetic algorithm to search for the global optimal solution to this bi-level programming model.

During the experiment, the maximum number of iterations is fixed at 1000. For an accurate numerical solution, we repeat 20 times for each benchmark and report the average results. Since we assume that total number of customers is 10000 in potential market, the maximum service capacity is also 10000. The cost parameters in profit function are set as $a = 0.01,\ b = 0.5,\ c = 0.1,\ r_a = 0.1$. Moreover, service utilization rate in the expansion cost which is denoted by r_s changes from 0 to 1. We assume the cloud provider is in a low-end market; thus, the lower model parameters are set as $\mu_R = 0,\ \mu_s = 2000$. This setting helps us obtain more significant results for pricing and capacity planning strategy under SLA constraints.

4.2 Experimental Result and Discussion

The results provide insights into four areas: capacity planning strategy, pricing, optimal profit, and supply-demand relationship.

Figure 4 illustrates the optimal number of capacity expansion blocks and corresponding expansion cost with the increase in service utilization rate μ_s. This figure shows that the capacity planning strategy becomes more useful when service utilization is bigger. The combined effect of both m and r_s causes expansion cost fluctuations. As Fig. 5 shows, the monopolist intends to decrease prices when it has higher service utilization. In addition, the service capacity has an effect on the optimal pricing strategy, which can be compared with the result in Fig. 4. Therefore, Observation 1 is derived.

Observation 1 Numerical results illustrate that cloud provider will increase service blocks to expand its cloud service capacity in order to avoid costly SLA violations, which means that a capacity expansion strategy is effective for provider profit growth. Furthermore, high service utilization makes the service planning strategy more valid. The cloud valuation of cloud diminishes quickly as service capacity increases, as shown in Figs. 4 and 5. Price and service capacity have a complementary relationship to some extent.

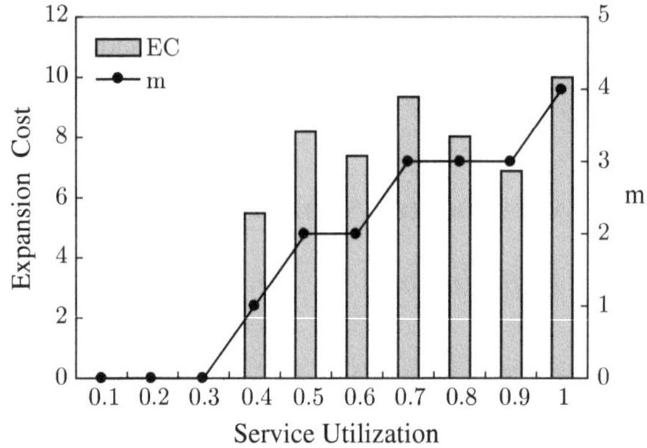

Fig. 4 Optimal number of service blocks and capacity expansion cost

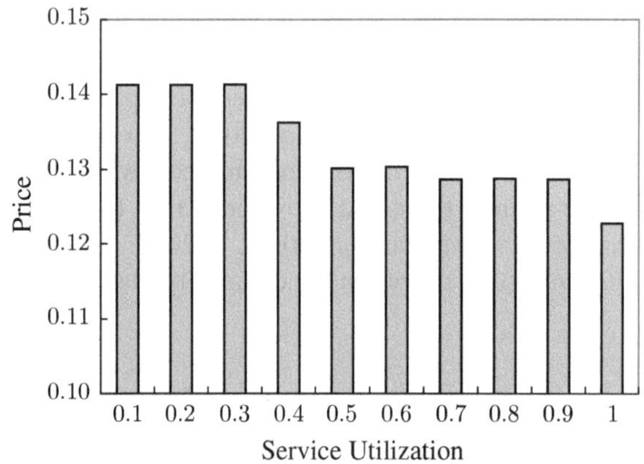

Fig. 5 Cloud service optimal pricing strategy

Then we examined the optimal profit obtained with different coefficients in expansion cost. The experiment results can be clearly seen in Fig. 6. As service utilization rate increases, cloud provider optimal profit keeps increasing, but has a marginal decrease. Meanwhile, the profit remains unchanged when $\mu_s \in \{0.1, 0.2, 0.3\}$. This can be interpreted as the profit optimization strategy does not work under a low service utilization rate. Thus, we note the following observation:

Observation 2 Financial benefit is an important concern for the CSP. Experimental results demonstrate that a monopolist profits from efficient use of some of the core technological components. Moreover, a large organization can learn from best practices to achieve a significantly higher server utilization rate and then improve total ownership profit. This experiment further verifies that capacity planning is useful for CSP profit growth.

Finally, the relationship between service supply and demand is clearly illustrated in Fig. 7. Optimal service capacity grows as the service utilization rate increases. The maximum number of customers is gained when μ_S equals to 1. The difference

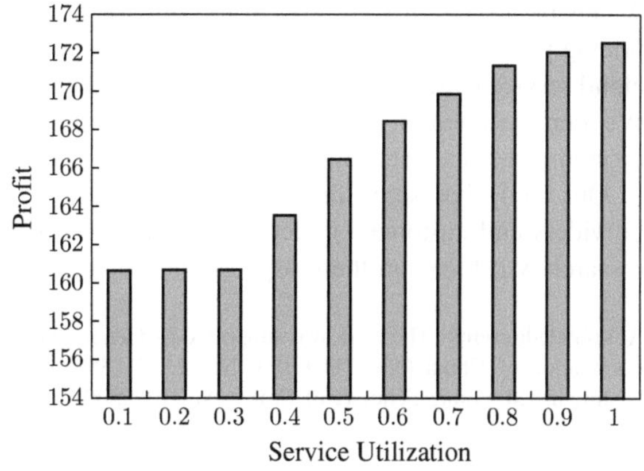

Fig. 6 The maximum profit of the monopoly cloud provider

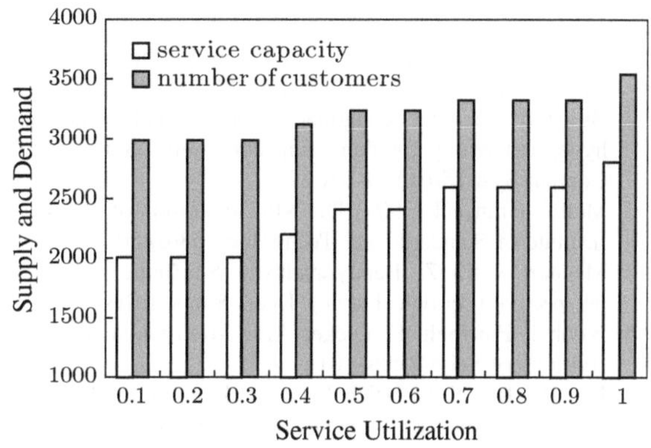

Fig. 7 The optimal decision of supply and demand

between cloud service demand and supply decreases gradually. Thus, Observation 3 is derived.

Observation 3 Numerical results illustrate that cloud service pricing and capacity planning strategy in this paper have an obvious effect on narrowing the trading gap when cloud service are in short supply. In addition, when service utilization rate of the monopolist is high enough, the CSP may have a greater incentive to expand its service capacity to obtain a larger market share.

5 Conclusion

This paper investigated the optimal pricing and capacity planning strategy in cloud computing. We built a bi-level programming model that considers the monopoly provider profit and uniformly distributed customers' utility under SLA constraints.

A genetic algorithm was adopted to compute numerical solutions of this dynamic pricing problem. Our study showed that proper policies have a significant impact on cloud service market in a long term, which also has a certain practicality in industry. We demonstrated in this paper that CSP's decision-making needs to combine economy characteristics and a resource optimization mechanism.

Our study has some limitations. Cloud service stakeholders such as duopoly providers and customers under various distributions were not considered. Further research will focus on these aspects and obtain new managerial insights.

Acknowledgments The work was supported by General Program of the National Natural Science Foundation of China (No. 71101103, No. 71371135) and National Natural Science Fund for Distinguished Young Scholars of China (No. 70925005).

References

1. Iyer B, Henderson JC (2010) Preparing for the future: understanding the seven capabilities of cloud computing (Periodical style). MIS Q Executive 9(2)
2. Buyya R, Yeo CS, Venugopal S (2009) Cloud computing and emerging IT platforms: Vision, hype, and reality for delivering computing as the 5th utility (Periodical style). Futur Gen Comput Syst 25(6):599–616
3. Mell P, Grance T (2009) The NIST definition of cloud computing (Periodical style). National Institute of Standards and Technology 53(6):50
4. Marston S, Li Z, Bandyopadhyay S, Zhang J (2011) Cloud computing—The business perspective (Periodical style). Decis Support Syst 51(1):176–189
5. Iyoob I, Zarifoglu E, Dieker AB (2013) Cloud computing operations research (Periodical style). Serv Sci 5(2):88–101
6. Hsu PF, Ray S, Li-Hsieh Y (2014) Examining cloud computing adoption intention, pricing mechanism, and deployment model (Periodical style). Int J Inf Manage 34(4):474–488
7. Zhang J, Niu B (2014) Dynamic quality decisions of software-as-a-service providers based on customer perception (Periodical style). Electron Commer Res Appl 13(3):151–163
8. Basu S, Chakraborty S, Sharma M (2015) Pricing cloud services—the impact of broadband quality (Periodical style). Omega 50:96–114
9. Huang J, Kauffman R J, Ma D (2014) Pricing strategy for cloud computing: a damaged services perspective (Periodical style). Decis Support Syst
10. Yeo CS, Venugopal S, Chu X (2010) Autonomic metered pricing for a utility computing service (Periodical style). Futur Gen Comput Syst 26(8):1368–1380
11. Zaman S, Grosu D (2013) Combinatorial auction-based allocation of virtual machine instances in clouds (Periodical style). J Parallel Distrib Comput 73(4):495–508
12. Tang S, Yuan J, Wang C (2014) A framework for Amazon EC2 bidding strategy under SLA constraints (Periodical style). IEEE Trans Parallel Distrib Syst 25(1):2–11
13. Samimi P, Teimouri Y, Mukhtar M (2014) A combinatorial double auction resource allocation model in cloud computing (Periodical style). Inf Sci
14. Baars T, Khadka R, Stefanov H (2014) Chargeback for cloud services (Periodical style). Futur Gen Comput Syst 41:91–103
15. Kertész A, Kecskemeti G, Brandic I (2014) An interoperable and self-adaptive approach for SLA-based service virtualization in heterogeneous Cloud environments. Futur Gen Comput Syst 32:54–68

16. Xu Z, Liang W (2015) Operational cost minimization of distributed data centers through the provision of fair request rate allocations while meeting different user SLAs (Periodical style). Comput Netw
17. Pindyck RS, Rubinfeld DL (1998) Econometric models and economic forecasts (Book style)". Irwin/McGraw-Hill, Boston
18. Krishnan V, Zhu W (2006) Designing a family of development-intensive products (Periodical style). Manage Sci 52(6):813–825

Balance Effect of Exploratory and Exploitative Learning Under Incubation Network

Gen-ming Zhang and Qian-ying Wu

Abstract The balance effect of organizational ambidexterity under incubation network is been explored in this paper, and a three dimensional model evaluating "incubator network", "organizational ambidexterity", "entrepreneurial performance", as well as their impact on each other are established. The data and the empirical testing are based on surveys conducted at incubating enterprises, and two conclusions were derived. Firstly, the differential balance between explorative learning and exploitive learning under enclosed space with limited resource have no significant impact on growth and innovative performance, and have negative influence on financial performance. Secondly, the balance induced by the interaction effect of explorative and exploitive learning is beneficial for the performance growth of enterprises.

Keywords Balance effect · Entrepreneurial performance · Entrepreneurial performance · Organizational ambidexterity

1 Introduction

Organizational ambidexterity refers to an organization aiming at achieving two different or even contradictory goals simultaneously. Exploitation in organizations refers to improve with existing resources, and it gains more stable and quicker payback than exploration [1]. But by putting too much focus on exploitation, the organization will experience rigidity, and be less capable of adapting to environmental changes, which leads to capability trap [2]. Exploration refers to experimenting new techniques, discovering new market channels, and improving organizational innovation performance. But by emphasizing on exploration,

G. Zhang · Q. Wu (✉)
Business School, Central South University, Changsha, China
e-mail: thisizjane@foxmail.com

E. Qi (ed.), *Proceedings of the 6th International Asia Conference on Industrial Engineering and Management Innovation*,
DOI 10.2991/978-94-6239-145-1_71

organizations can undergo innovation trap. Thus the balance of exploration and exploitation becomes the focus of problem.

Chinese government has been advocating independent innovation and emphasizing on constructing national innovation system. And incubators are the carriers to transform scientific discoveries into business ideas, and are the intermediary for nurturing small and medium enterprises (SMEs), optimizing industrial structures, leveraging urban competitiveness, stimulating regional economy and providing job opportunities. Therefore the question arises: do incubators enable organizations to gain adequate resources in realizing the balance between exploitation and exploration, and does this result in better performance?

Thus, this paper uses organizational ambidexterity as medium variable, and verifies the balance effect of exploratory learning and exploitative learning under the incubation network using empirical testing, and explores the possibility of organizational ambidexterity.

2 Hypothesis and Conceptual Model

Enterprise incubation networks serve as catalysts for innovation and they provide incubation processes with integrated network platforms. Hansen et al. [3] believe that incubators offer enterprises with networks, and start-up enterprises can use incubation networks to enhance their competitiveness. Moreover, Granovetter [4] puts forward the embedding theory, stating the economic activities are embedded in different social relations and structures. The economic behaviors of network members are embedded into the social network. Gulati et al. [5] proposes that embedding relies on the level of dependencies between enterprises, and embedding has significant impact on enterprise behavior and performance. Therefore, this paper chooses embedding theory to analyze enterprise incubating networks.

Granovetter [4] classifies the network embeddedness into relational and structural categories. Gulati [6] insists that relational embeddedness emphasizes the intimate relationship and level of trust between members, while structural embeddedness emphasizes how the relationship between members interacts with mechanism. McEvily and Zaheer [7] believe structural dispersed embedding will gain heterogeneity resources to help enhance competitiveness. Xu [8] believes relational embeddedness acts through explorative learning and indirectly impacts on innovative performance, meanwhile environment dynamics have positive moderating effect on explorative learning. Liu and Si [9] believes relational embeddedness is positively related to exploitative learning, and negatively related to structural embeddedness. On the contrary, explorative learning negatively affects relational embeddedness and positively affects structural embeddedness. This indicates that exploration and exploitation learning positively impacts on incremental innovation as well as radical innovation. And this paper uses the classic classification of relational and structural embeddedness, combined with the incubating networks characteristics to study the impact of enterprise incubating networks on performance.

March [10] deems that it is difficult for exploration and exploitation to coexist, i.e., exploration and exploitation are contradictory. Menguc and Auh [11] agree with the idea and think that pursuit for ambidextrous balance may have negative impacts on corporate performance since exploration and exploitation correspond to different organizational structures and cultures, and the significant difference would cause high cost in integration. Thus, based on the foregoing theoretical analyses, the following research hypothesis is proposed:

H1 Balance effect of exploitative learning and explorative learning does not have significant impact or holds negative influence on entrepreneurial performance.

Gupta et al. [12] suggests that exploitative and explorative learning may occur at complementary departments where the goals and function of the departments are different, and therefore will not be competing for the same resource all the time. Tushman and O'Reilly [13] also believe that through arrangement of organizational structure and chronologic sequence, exploitative and explorative learning can coexist and supplement each other. Furthermore, He and Wong [14] hold that there is interactive and supplementary effect in ambidexterity and such a kind of balance can bring positive impacts on enterprises.

Therefore, the following research hypotheses are proposed based on the foregoing theoretical analyses:

H2 Exploitative learning has positive impact on explorative learning.

H3a The balance effect of exploitative learning and explorative learning has positive and significant impact on financial performance.

H3b The balance effect of exploitative learning and explorative learning has positive and significant impact on growth performance.

H3c The balance effect of exploitative learning and explorative learning has positive and significant impacts on innovation performance.

Jiang [15] analyzed exploratory factors by using 15 key indexes, carried out confirmatory factor analysis after five indexes had been removed and finally concluded that entrepreneurial performance was composed of three dimensions, i.e., financial performance, growth performance and innovation performance.

In accordance with literature review and research hypotheses, enterprises' incubating network is divided into two dimensions, i.e., relational embeddedness and structural embeddedness; organizational ambidexterity is divided into two dimensions including exploitative learning and explorative learning; and entrepreneurial performance is classified into three dimensions, i.e., financial performance, growth performance and innovation performance. In order to analyze the balance effect of exploitative and explorative learning on entrepreneurial performance under incubating network, a conceptual model is constructed relating to incubating network, organizational ambidexterity and entrepreneurial performance (as shown in Fig. 1).

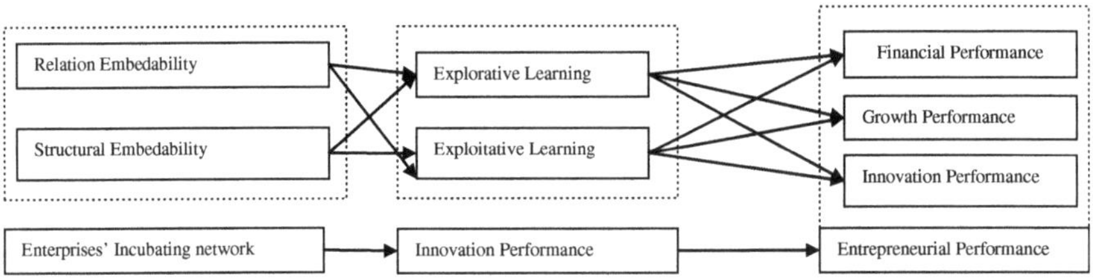

Fig. 1 A conceptual model about enterprises' incubating network and entrepreneurial performance

3 Methodology

The survey was carried out from June 2013 to August 2013. Incubating enterprises in Hunan Province and Guangdong Province were selected, the scope focused on Changsha, Yueyang, Zhuzhou and Pearl River Delta, and respondents mainly involved middle-level and high-level staff in incubating enterprises. Incubating networks were measured using the scale of Nooteboom [16].

Based on sufficient discussions, a preliminary questionnaire was formed and a pre-test was carried out. Finally, a formal investigation was established. In all, 306 questionnaires were delivered during this survey, and 262 of them were withdrawn (with a recovery rate of 85.62 %). After screening, there were 212 valid questionnaires, accounting for 81.92 % of such withdrawn questionnaires and 69.28 % of total delivered ones. Sample enterprises involved many industries, e.g. software and electron and communication equipment manufacturing, new materials and chemical engineering, machinery and electromechanical industry and biological pharmacy. Samples collected in this research included 46 enterprises from software and electron and communication equipment manufacturing industry (22 % of the total enterprises), 40 companies from new materials and chemical engineering (19 % of the total enterprises), 40 enterprises from machinery and electromechanical industry (19 % of the total enterprises), 25 enterprises from biological pharmacy (12 % of the total enterprises), 19 companies from service industry (9 % of the total enterprises), 15 enterprises from light and food industry (7 % of the total enterprises), 15 enterprises from clothing and textile industry (7 % of the total enterprises), and 12 companies from other industries (5 % of the total enterprises). There were 25 enterprises whose registration time was less than 3 years (11.79 % of the total enterprises) and 44 companies with over 10 years' enterprise age (20.76 % of the total number). Enterprises with less than 10 employees accounted for 4.32 % of the total number and companies with over 150 workers held 12.96 % of the total number.

Organizational ambidexterity is measured using the scale of Atuahene-Gima [17]. Entrepreneurial performance was composed of 9 measurement items, and was generated by using the scale developed by Jiang [15] as reference to measure financial, growth and innovation performance. The questionnaire adopted the Lirke

Table 1 Factor analysis and reliability test on variables

		Relational embedability	Structural embedability	Exploitative learning	Explorative learning	Financial performance	Growth performance	Innovation performance
KMO value		0.846	0.828	0.746	0.829	0.745	0.730	0.732
Bartlett sphericity test	Approximate chi-square	357.420	303.327	303.991	456.242	273.526	186.369	299.285
	df	10	10	10	10	3	3	3
	sig	0.000	0.000	0.000	0.000	0.000	0.000	0.000
α coefficient		0.869	0.848	0.833	0.898	0.897	0.847	0.906
Factor loading		0.64–0.81		0.58–0.87		0.80–0.96		
χ^2/df		1.801		2.24		1.46		
GFI		0.91		0.93		0.95		
CFI		0.95		0.93		0.99		
RMSEA		0.062		0.058		0.055		

Five-level Scale. Through reliability analysis, it was found that: Cronbach's α coefficient of all variables is larger than 0.8, which implies that internal consistency is strong and the reliability is good.

Moreover, all KMO values are larger than 0.5 and Bartlett sphericity test is passed, which indicates that factor analysis may be implemented. Then, LISREL8.7 is used to conduct the confirmatory factor analysis on each variable. The result shows that factor loading of all variable items exceeds 0.5, which means the items in the questionnaire can explicitly explain variables since the factor loading represents the relationship between measured items and latent variable. On the other hand, related fit indexes GFI and CFI of all variables are larger than 0.9, RMSEA is less than 0.8 and $\chi2$/df is less than 3. Therefore it can be demonstrated that the fit indexes accord with the specified scope, the model's fitting effect is good and the factor structure passes verification. Results are shown in Table 1.

4 Results

SPSS17.0 software is adopted to conduct data analysis, and structural equation software LISREL8.7 is used to construct SEM, as well as to test the fitting effect of the model by fit indexes, and also to judge whether the research hypotheses are valid or not. The model passes 6 latent variables (GX—relational embeddedness, JG— structural embeddedness, SY—organizational ambidexterity, CW—financial performance, CZ—growth performance and CX—innovation performance), as shown in Figs. 2 and 3.

Model fitting is evaluated according to absolute fit indexes including P value, chi-square value, goodness of fit index, adjusted goodness of fit index and root mean square error of approximation (RMSEA). Besides, the SEM model is

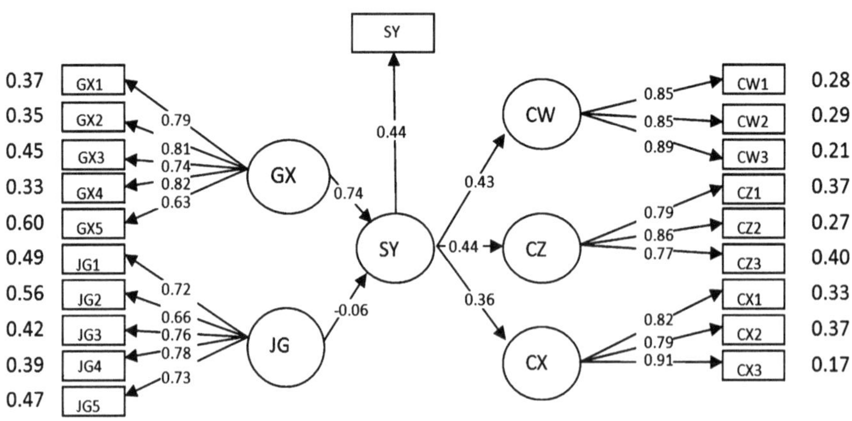

Fig. 2 SEM1 incubating network—organizational ambidexterity—entrepreneurial performance

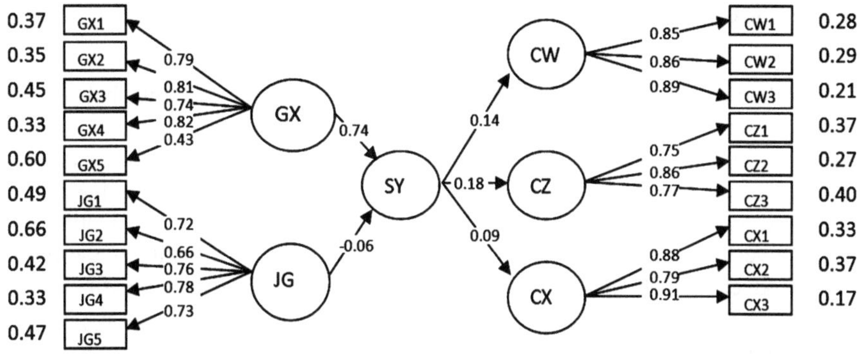

Fig. 3 SEM 2 incubating network—organizational ambidexterity—entrepreneurial performance

compared with standard models in accordance with relative fit indexes covering quasi-fit index, incremental fit index and comparative fit index. Moreover, model simplification is judged based on information standard indexes e.g. parsimonious normed fit index and parsimonious goodness of fit index and combines fit indexes of the three aspects to evaluate the model's goodness of fit. According to the research results in Table 2, fitting results of the model satisfy all standard indication ranges, which mean the model has good fitting results and amendments can be spared.

Table 2 Results of SME's goodness of fit

Index	Acceptable interval	Results of fitting	Whether results of fitting satisfy or not
Absolute fit indexes (model fitting) P	<0.05	0.000	Satisfying
Chi-square value $\chi 2$	/	644.08	/
Degree of freedom df	/	365	/
$\chi 2$/df	≤3	1.765	Satisfying
Goodness of fit index (GFI)	>0.8	0.84	Satisfying
Adjusted goodness of fit index (AGFI)	>0.8	0.83	Satisfying
RMSEA	≤0.05	0.042	Satisfying
Relative fit indexes (model comparison) Normed fit index (NFI)	≥0.9	0.9	Satisfying
Incremental fit index (IFI)	>0.9	0.95	Satisfying
Comparative fit index (CFI)	>0.9	0.95	Satisfying
Information standard indexes (model simplification) parsimonious normed fit index (PNFI)	≥0.5	0.80	Satisfying
Parsimonious goodness of fit index (PGFI)	≥0.5	0.76	Satisfying

5 Discussion

The method developed by He and Wong [14] is adopted in studying the balance and interactive effect, which uses constructs composed of absolute difference value that derived from each measured item (related to exploitative and explorative learning), and eventually they are the measured values of the balancing effect. And as for the measured values of balance effect under interaction (between exploitative and explorative learning), the method utilizes the construct that constitutes the multiplied product term obtained after standardization processing of each measured item.

Under environment of resource closure, there are conflicts between explorative and exploitative learning. Firstly, differential value balance between exploitative and explorative learning is tested to verify authenticity of ambidextrous paradox.

By using SPSS17.0 software, the absolute differential value of exploitative and explorative learning is obtained. Then, the paper implements an overall test on SEM. According to the test in Fig. 2, it is shown that the path coefficients of the three entrepreneurial performance dimensions (i.e. financial performance, growth performance, and innovation performance), are relative to the constructs formed by the absolute difference value between exploitative and explorative learning, and are 0.14, 0.18 and 0.09, respectively. This indicates the difference-value balance between exploitative and explorative learning has no significant impact on growth and innovation performance, and even has negative effect on financial performance. Hence, the research hypothesis3a, b, c is valid. Next, interaction between exploitative learning and explorative learning is tested. Based on the completed SEM 1, paths of exploitative learning and explorative learning are increased. Model 2 is obtained by carrying out an overall SEM test (as shown in Fig. 3).

The test on Model 2 shows that the path coefficient of exploitative learning, which is relative to explorative learning, is 0.44, indicating that exploitative learning has positive effect on explorative learning and therefore the research hypothesis2 is valid. Meanwhile, the path coefficients of the three entrepreneurial performance dimensions are relative to the constructs constituted by the multiplied product term obtained after standardization processing of each measured item, which are 0.43, 0.44 and 0.36, respectively. This implies that the interaction between explorative learning and exploitative learning has significant impacts on entrepreneurial performance and the research hypothesis1 is therefore valid.

Additionally, according to the path coefficient of the incubating networks, which is relative to organizational ambidexterity in the SEM models, it is found that relational embeddedness has positive effect on organizational ambidexterity but structural embeddedness has negative effect on organizational ambidexterity.

In conclusion, on the premise that enterprises face resource closure, the absolute difference-value balance between exploitative and explorative learning has no positive effect on performance, and holds negative influence on financial performance, which reinforces previous scholars' research conclusions about organizational ambidexterity. Besides, in open incubating networks, incubates can obtain external resources freely and exploitative learning has positive effect on explorative learning. At the same time, the product balance between exploitative learning and explorative learning has favorable and positive effect on balance.

Thus, how incubates realize the balance between exploitative and explorative learning is crucial for improvement in incubates' performance.

6 Conclusion

For enterprises to survive and enhance their competiveness, they must efficiently utilize resources and continuously adapt to dynamically changing environment. Based on the premise that the incubating network is open, this thesis studies how enterprises exert the balance effect of organizational ambidexterity on their performance.

According to empirical results, it is shown that exploitative learning has positive accelerating effect on explorative learning, which sufficiently affirms ambidextrous balancing mode. In the ambidextrous balancing mode, calculation of absolute difference value shows that the balance has no obvious positive impacts and even exerts negative influence, which indicates that ambidextrous paradox inhibits incubatee performance. And under the circumstance where explorative and exploitive learning complement each other, product is used to calculate the ambidextrous balance, and the result is satisfying, which shows that ambidextrous balance is beneficial for incubatee performance. It is also found that relational embeddedness has positive impact on organizational ambidexterity, which demonstrates that under open incubating networks, communicating and sharing resources between organizations are beneficial for ambidextrous balance, and such balance can exert interaction effect.

In practice, incubators should strengthen the cultivation of incubating enterprises from the perspective of incubating networks, pay attention to construction of incubating networks and consolidate the construction of incubating external network by establishing long-term cooperation with scientific research institutions, intermediaries and suppliers. At the same time, it is essential for incubatees to focus on establishing relational networks with others social forces, promote resource accessibility and information exchange, and reinforce internal and external network cooperation and joint developments. Also perfecting the credibility mechanism is essential to incubatees, and then mutual benefits can be generated by paying attention to dynamic environment, enhancing competitiveness, innovating technological breakthroughs and collaborating through interactions, and therefore win-win can be realized among subjects of the incubating network.

Acknowledgments Q. Wu thanks the guidance and support received from Central South University, especially from professor G. Zhang of the business school. And also thanks the loving support from family and friends.

References

1. Levinthal DA, March JG (1993) The myopia of learning. Strateg Manag J 14:95–112
2. Levitt B, March J (1988) Organizational learning. Annu Rev Sociol 14:319–340
3. Hansen MT, Chesbrough HW, Noh ria DN (2000) Improving networked incubators performance: some guidelines for success. Am J Small Bus 12(1):11–25
4. Granovetter M (1985) Economic action and social structure: the problem of embedded-ness. Am J Sociol 91(3):481–510
5. Gulati R, Nohria N, Zaheer A (2000) Strategic network. Strateg Manag J 21:203–215
6. Gulati R (1998) Alliances and networks. Strateg Manag J 19(4):293–317
7. McEvily B, Zaheer A (1999) Bridging ties: a source of firm heterogeneity in competitive capabilities. Strateg Manag J 20(12):1133–1156
8. Xu G (2008) Study on the impact of relationship embeddedness on technological innovation performance: based on the mediating role of explorative learning. Zhejiang University, Hangzhou
9. Liu L, Si C (2010) The relationship between embedded, inter-organizational learning and technological innovation. Technol Innov Manag 31(1):9–13
10. March JG (2006) Rationality, foolishness, and adaptive intelligence. Strateg Manag J 27: 201–214
11. Menguc B, Auh S (2008) The asymmetric moderating role of market orientation on the ambidexterity-firm performance relationship for prospectors and defenders. Ind Mark Manage 37(4):455–470
12. Gupta AK, Smith KG, Shalley CE (2006) The interplay between exploration and exploitation. Acad Manag 49(4):693–706
13. Tushman ML, O'Reilly CA (1996) Ambidextrous organizations: managing evolutionary and revolutionary change. Calif Manag Rev 38:8–30
14. He Z, Wong P (2004) Exploration and exploitation: an empirical test of the ambidexterity hypothesis. Organ Sci 15(4):481–494
15. Jiang Q (2007) The study on social capital and innovation performance in incubating enterprises: an analysis based on the intermediary effect of dynamic capability. Shanghai Jiao Tong Univeristy, Shanghai
16. Nooteboom B (2004) Innovation, learning and cluster dynamics. Erasmus Res Inst Manag 12 (3):23–56
17. Atuahene-Gima K (2003) The effects of exploratory and exploitative market learning on new product performance in new technology ventures: a resource-based perspective. City University of Hong Kong Press, Hong Kong

Empiric Study of House Prices Fluctuation and Mortgage on Economic Growth

Liang Xie

Abstract Real estate is an important part of the national economy, financial institutions affect the real estate market by using financial policy and support the rapid development of real estate. Because the real estate bubble effect promoting economic development, studying the relevance of property price fluctuation and mortgage loan from financial institutions impact on economic growth as a starting point, this paper proposed the establishment of a dynamic optimization model of economic growth, through macroeconomic data empirically tested the intrinsic relationship of economic growth between house price and bank mortgage in the in the equilibrium state, thereby provides the theoretical basis for the national macro-control prices.

Keywords House prices · Mortgage loan · Economic growth · Empirical test

1 Introduction

The real estate industry as an important part of our national economy, so there is a supporting each other mutually dependent relationship between the real estate industry and the long-term development of the national economy as well as between the real estate industry and the financial sector. However financial sector affects the real estate market by using the financial leverage, the real estate industry's own development is restricted by the level of economic development, which would ultimately affect the long-term growth of the national economy. Thus the role of the real estate in developing national economy, expanding domestic demand, optimizing the industrial structure is growing gradually, which has become the basis for

L. Xie (✉)
Tibet Vocational Technical College, Lhasa 850000, China
e-mail: xliang516@163.com

E. Qi (ed.), *Proceedings of the 6th International Asia Conference on Industrial Engineering and Management Innovation*,
DOI 10.2991/978-94-6239-145-1_72

a pilot and a pillar industry of the national economy. Through the establishment of optimal dynamic economic growth model, this paper discussed the endogenous relations of house prices, mortgage credit and economic growth, and empirically tested the relations of house prices, mortgage credit and economic growth through the macro data, thereby it has important theoretical and realistic significance for solving the problem of urban residents of the housing, cultivating the effective demand of housing market and promoting the national economy steady and fast growth.

2 Establishing the Dynamic Optimization Model

According to the current literature on research of the model which staying at a single level, Herring and Susan [1] constructed a model of the credit markets to study the internal relations between he prosperity of real estate market and the banking crisis. Hendershott and Shilling [2] put forward a minimum variance of uncertainty model which gave after-tax cost of the assets and liabilities and mortgage demand in the earnings uncertainty. Brueckner [3] did research on consumer demand for the best mortgage by utility maximization model includes limiting condition. Through the empirical research, Rosenthal and Dunsky [4] considered that housing mortgage loan demand and the income level of home buyers were related. Follain et al. [5] believed that non-linear relationship existed between the mortgage and home buyers income, and income elasticity was positively correlated with low earners income level, but was negatively interrelated with high earners income level. From what has been discussed above, this paper embedded the real estate market, the credit market and macroeconomic into the economic growth model, and analyzed the endogenous relationship of the impacts related to house price fluctuations and mortgage on economic growth.

2.1 The Model Assumed Conditions

In order to better analyze the internal mechanism of the impact related to the real estate price and mortgage credit on economic growth, based on the decision-making behavior of consumer utility maximization this paper discussed the microscopic behaviors of consumers in the case of price fluctuations and mortgages, which was economic growth by consumer behavioral responses to the macroeconomic level. Therefore, in order to study the problem more realistic, the assumptions were followed before establishing the model:

(1) Product market split into non-real estate market and real estate market. This paper separately studied the effect of the housing market for the whole macro economy under the condition of invariable in the housing market.

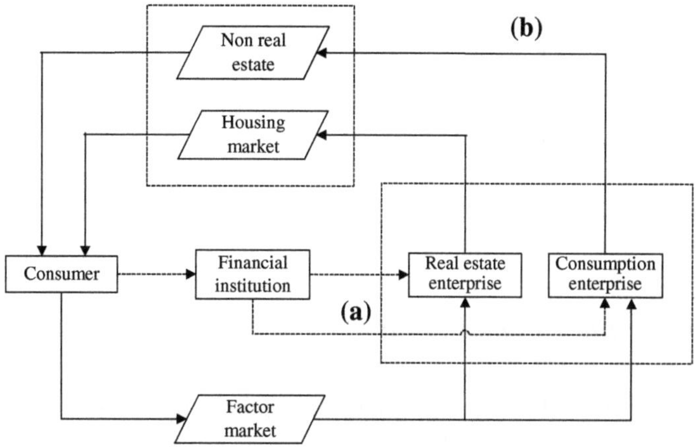

Fig. 1 Economic system diagram. **a** *dotted line*: capital flow direction; **b** *solid line*: elements flow direction

(2) Non-housing consumer goods consumption is constant. In order to facilitate research, the paper set consumption of non-real estate consumer to 0, and studied the housing consumption given the non-real estate consumer consumption.
(3) Production system split into non-real estate consumer goods companies and real estate companies. This paper studies the behavior of the real estate business in the case of a given non-real estate consumer goods manufacturers.
(4) Investment financial system to non-real estate consumer goods production system is constant.

According to the above assumptions, obtain economic system shown in Fig. 1.

2.2 Establishing the Optimal Dynamic Model

According to determine the utility function and constraint equations, this paper embodied the economic variables related to the real estate prices and mortgage loans etc. and established the model by means of dynamic optimization model framework, moreover, solved the model. This paper about the definition of correlation function and explanations can be seen in the reference.

$$U(C_H) = Max \int_0^{\infty} \frac{c_H(t)^{1-\sigma}}{1-\sigma} e^{-\rho t} dt$$

$$\begin{cases} Y = AK^{\alpha}L^{\beta} = C_H + \frac{W+(1-\omega)B+PC_H}{1-r} \\ \dot{L} = nL \\ \dot{K} = I = AK^{\alpha}L^{\beta} - C_H - B \\ \text{Initial Conditions} \\ L(0) = L_0 \\ K(0) = K_0 \end{cases} \quad (1)$$

2.3 *Solving the Optimal Solution*

According to the establishment of dynamic optimization model (1), construct Hamilton function of present value.

$$H = \frac{C_H^{1-\sigma}}{1-\sigma} + \lambda\left[AK^{\alpha}L^{\beta} - C_H - \frac{W + (1-\omega)B + PC_H}{1-r}\right] + \lambda_L nL + \lambda_K\left(AK^{\alpha}L^{\beta} - C_H - B\right) \quad (2)$$

Import the embedding of price and the variables of mortgages into the function; the economic growth rate balanced growth path was obtained according to the constraint conditions of derivation.

$$\frac{\dot{Y}}{Y} = \frac{1}{\sigma}\left[\left(\frac{1-r}{1-\omega}\right)\left(1 + \frac{P}{1-r}\right) + 1\right]\left(A\alpha K^{\alpha-1}L^{\beta}\left(\frac{1-r}{1-\omega}\right) - A\alpha K^{\alpha-1}L^{\beta} + \rho\right) \quad (3)$$

3 The Relationship Between Model Variables and Economic Growth

By establishing a dynamic optimization problem of economic growth, the model was solved by means of optimal control theory, which was an endogenous relational model of long-term economic growth.

3.1 *The Impact of Housing Price Fluctuation on Economic Growth*

Analyzing the relationship between house prices and economic growth, according to (3), (4) was obtained by the partial derivative related to the economic growth rate for real estate price p.

$$\frac{\partial g_Y}{\partial P} = \frac{1}{\sigma}\left(\frac{1-r}{1-\omega}\right)\left(\frac{1}{1-r}\right)\left(A\alpha K^{\alpha-1}L^{\beta}\left(\frac{1-r}{1-\omega}\right) - A\alpha K^{\alpha-1}L^{\beta} + \rho\right) > 0 \quad (4)$$

Because the bank deposit rates are generally less than mortgage rates, we can know that

$$\frac{1-r}{1-\omega} > 0,\ A\alpha K^{\alpha-1}L^{\beta}\left(\frac{1-r}{1-\omega}\right) - A\alpha K^{\alpha-1}L^{\beta} > 0.$$

Because the partial derivative related to real estate prices on the economic growth is a positive value, which indicated that there is a positive correlation between economic growth and real estate prices, it would promote economic growth when real estate prices rise.

3.2 *The Impact of Mortgage Credit Fluctuation on Economic Growth*

There is a $B_t = \beta P_H H_t$ according to mortgages and house prices, housing consumption under optimal state, importing $B_t = \beta P_H H_t$ to the (3), and taking partial derivative of economic growth rate with respect to B_t, (5) was obtained.

$$\frac{\partial g_Y}{\partial B} = \frac{1}{\sigma}\left(\frac{1-r}{1-\omega}\right)\left[\frac{1}{\beta H(1-r)}\right]\left(A\alpha K^{\alpha-1}L^{\beta}\left(\frac{1-r}{1-\omega}\right) - A\alpha K^{\alpha-1}L^{\beta} + \rho\right) > 0 \quad (5)$$

According to (5) we can know that there is a multiplier relationship between the size of mortgage credit H and mortgage percentage β, $\frac{1}{\beta H(1-r)} > 0$. So there is also a positive relationship of the size of bank mortgages, mortgage percentage and economic growth.

4 Empirical Test

The result of the positive relationship between house prices, mortgage and economic growth was obtained only by means of establishing a dynamic optimization model and the qualitative point of view, so we must confirms the correctness of the model through macro data.

4.1 Data Selection and Processing

In order to empirically test the relationship between macroeconomic and real estate price fluctuations, mortgage, this paper chose the gross domestic product (GDP), city commercial housing prices and long-term loans of 3–5 years as a proxy variable model considering the feasibility of the data collected. Therefore we select the National Bureau of Statistics 2005–2013 annual GDP data, and Select GTA database 2005–2013 City House of monthly sales and sales area as price of City House, we obtained the quarterly data by converting the city's commercial housing price per unit area, meanwhile we obtained long-term loans data by converting the People's Bank of China and long-term loans monthly statistics into quarterly data. Finally we get 36 set of 36 set of sample data to empirically test the model.

In order to facilitate quantitative analysis, the paper eliminated the effects of heteroscedasticity by taking all the actual value of all valid data. GDP, LOAN and P each separately represent the gross domestic product, the urban commercial housing prices and long-term loans, while the corresponding values were used for LNGDP, LNLOAN and LNP representation.

4.2 Selecting the Test Method

Unit root test is stationary test of economic and financial data with the time trend carried, the took the unit root test for smoothness of each variable by using ADF method, the regression equation is as follows,

$$\Delta x_t = \alpha_0 + \alpha_1 t + \alpha_2 x_{t-1} + \sum_{i=1}^{k} \alpha_i \Delta x_{t-i} + \mu_t \tag{6}$$

Co-integration reflects a long-term stable equilibrium relationship amongst the variables. Johansen maximum likelihood method is often used to test the co-integration relationship amongst the variables in econometric analysis, there is also two-step test method put forward by Engle and Granger, two-step test procedure is as follows:

First step: Get the regression equation of variable by using the OLS method:

$$y = \alpha + \beta x + \varepsilon.$$

Second step: Stationary test of residuals. If the residuals by stationary test proved to be stable, we can confirm that there is a co-integration relationship between variables.

Table 1 Results of the unit root test

Variables	Test results				Stationary
LNGDP	ADF Test statistic	−1.770313	1 % Critical value*	−3.6228	Not
			5 % Critical value	−2.9446	
			10 % Critical value	−2.6105	
LNLOAN	ADF Test statistic	−1.569988	1 % Critical value*	−3.6228	Not
			5 % Critical value	−2.9446	
			10 % Critical value	−2.6105	
LNP	ADF Test statistic	−0.492468	1 % Critical value*	−3.6228	Not
			5 % Critical value	−2.9446	
			10 % Critical value	−2.6105	

* Significance level is 1%

4.3 Empirical Test

(1) Unit root test

To verify the feasibility of the model, there is approximately the same changing trend among proxy variable selection, it indicated that there may be co-integrated variables. But qualitative judgments are not reliable, we need to determine the relationship amongst them by Metrology and Inspection, we must unit root test before co-integration test among variables, which suggest that it is the same order stationary among variables, there is a co-integration relationship amongst the same order variables. The test results were obtained after taking unit root test on the 36 set sample data which has been processed. The result is as followings (Table 1).

We can know that the values of all the time sequence ADF were higher than the threshold level of 5 % according to the test results, so the original sequence at the 5 % level of confidence to accept the null hypothesis, and we know all the time series are not stationary. Taking the first order difference on the above variables, unit root test results are shown in Table 2.

Table 2 First-order differential unit root test results

Variables	Test results				Stationary
LNGDP(1)	ADF Test statistic	−10.56810	1 % Critical value*	−3.6289	Yes
			5 % Critical value	−2.9472	
			10 % Critical value	−2.6118	
LNLOAN(1)	ADF Test statistic	−7.179106	1 % Critical value*	−3.6289	Yes
			5 % Critical value	−2.9472	
			10 % Critical value	−2.6118	
LNP(−1)	ADF Test statistic	−10.75699	1 % Critical value*	−3.6289	Yes
			5 % Critical value	−2.9472	
			10 % Critical value	−2.6118	

* Significance level is 1%

Table 3 Residuals unit root test results

Variables	Test results				Stationary
Residuals	ADF Test statistic	−6.419258	1 % Critical value*	−3.6353	Yes
			5 % Critical value	−2.9499	
			10 % Critical value	−2.6133	

We can know that ADF statistics for all variables are at the 1 % significance level according to the Table 2, so it suggests that the above sequence after taken on the first order difference is stationary.

(2) Co-integration test

The paper chooses the two-step approach proposed by Engle and Granger for co-integration test. Firstly, we use EViews software for selected Variables regression, the regression equation is as followings:

$$\text{Lngdp} = -0.603729 + 1.330977\text{Lnp} + 0.098392\text{Lnloan} \quad (7)$$

We can conclude that there is a positive correlation of real estate prices, mortgage and economic growth according to the regression equation, but the impact of property prices on economic growth is more significant than the mortgage. According to the result, it suggested that consumers prefer to use their own income for consumption than use loan in consideration of household consumption characteristics.

Secondly, we use EViews software for residuals unit root test to verify the stability of residuals, residuals unit root test results are shown in Table 3.

According to the data in Table 3 test results, we can conclude that residuals passed the unit root test, which suggests that there is a co-integration relationship of house prices, mortgage loans and economic growth.

5 Conclusions

In this paper, a dynamic optimization model of economic growth was introduced to the real estate market and the credit markets, from a theoretical perspective it suggests that economic growth is positively correlated with house prices and mortgage by means of model deduction and solution. Through further empirical test, it proved that economic growth is positively correlated with house prices and mortgage based on the obtained regression equation. The impact of Property price fluctuations on economic growth is more significant than bank loans. In the real economy, the influence of real estate price fluctuations on consumer behavior is obvious, thus it affects the development of the national economy.

References

1. Herring RJ, Susan W (1998) Real estate booms and banking busts: an international perspective. Working paper, the Wharton School of Pennsylvania University
2. Hendershott PH, Shilling JD (1982) Capital allocation and the economic recovery tax act of 1981. Public Finan Q 10:242–273
3. Brueckner JK (1994) The demand for mortgage debt: Some basic results. J Hous Econ 3:1–21
4. Rosenthal SS, Duca JV, Gabriel SA (1991) Credit rationing and the demand for owner-occupied housing. J Urban Econ 30:48–63
5. Follain JR, Dunsky RM (1996) The demand for home mortgage debt and the income tax. Center for Policy Research, metropolitan studies program series occasional paper no. 179, Maxwell School of Citizenship and Public Affairs, Syracuse University

Game Analysis of the New Generation of Migrant Workers Training Risk

Ju-ling Qiao, Qing-qing Cheng and Xue-ting Zhang

Abstract At present, China's new generation of migrant workers have more than 100 million people, who have become an important part of contemporary industrial workers, but also the urgent need to develop realistic and potential human resources. Enhancing the quality of this group through training, which are the common aspiration of enterprise and the new generation of migrant workers. But this training faces many risks. This paper is aimed at game analysis of the new generation of migrant workers training risk, and puts forward some suggestions and corresponding method.

Keywords Generation of migrant workers · Training · Game analysis · Risk management

1 Introduction

Since the reform and opening up, China's migrant workers growing ranks, the total number is more than 500 million people, including 80 after the new generation of migrant workers accounted for one fifth. This group was much welcomed by business practitioners at a time, because relatively speaking, the employer cost is relatively low, requiring relatively small, but also hard. But with the adjustment of industrial structure upgrading, the enterprise demand for the quality of the staff continuously improve and the requirements of migrant workers has also been new changes; Relative to the old generation, the new generation of migrant workers' career expectations are rising, whose pursuit is no longer merely addressing the needs of survival, but also has a higher pursuit; Not only the pursuit of immediate interest, there is a more long-term goals in life.

J. Qiao (✉) · Q. Cheng · X. Zhang
School of Management, University of Jinan, Jinan, People's Republic of China
e-mail: qiao58@126.com

E. Qi (ed.), *Proceedings of the 6th International Asia Conference on Industrial Engineering and Management Innovation*,
DOI 10.2991/978-94-6239-145-1_73

Based on such a background, and strengthen the training of the new generation of migrant workers, has become the consensus of the enterprise is self-evident, but in practice, there are a lot of problems. On the one hand, for the training of cost considerations, some companies do not want to invest too many resources in training the new generation of migrant workers, worrying about investment with no returns, and "do the wedding dress for others "2009 Legislative Affairs Office of the State Council announced the" vocational skills training and firm regulations (draft)", clearing corporate funds for education and training of employees, and the funds shall be withdrawn in accordance with the 1.5–2.5 % of total wages. According to the survey indicated that nearly 80 % of enterprises' training expenditure accounted for the total wages of staff is less than 1.5 %, including below 0.5 % businesses accounted for nearly 40 % [1]. This data is for the purposes of training for all employees, while put into the training for the new generation of migrant workers is much less.

On the other hand most of the companies do not have a comprehensive training system, with single way of training and not paying attention to the training summarization evaluation, which leads to training, just becomes the "face work", the result is not significant. Survey shows that 90 % of corporate training for migrant workers are still remain in the level that should be aware and the training effect is uneven, failed to achieve the desired effect; companies can not obtain the required personnel from training, impacting the enterprise sustainable development, and the new generation of migrant workers also failed to meet scheduled interest demands, which is easy to quit. As a result, for the training of the new generation of migrant workers is, like other investments, with a certain amount of risk, the paper intends to establish a game model from the start, theoretical and practical discussion on resolving this risk.

2 Construction of the New Generation of Migrant Workers Training Game Model

Assuming the pursuit of business is to maximize economic efficiency, the pursuit of individual utility maximization of the new generation of migrant workers, that is, the game two sides enterprises and the new generation of migrant workers are rational economic men. Enterprise training investment and cost is C (cost), after the training gains are R (revenue). Migrant workers staying or quitting depends on whether the company takes incentives or constraints and the reputation cost of migrant workers paid. And if migrant workers choose to quit, many companies can provide jobs, migrant workers are free to choose.

2.1 Construct Game Model of the New Generation of Migrant Workers Training Risk Under no Constraints

Assumed that market information businesses and migrant workers faced are complete, companies have two choices: train or not to train. According to businesses choice, the migrant workers can also make the appropriate choice: stay or quit. Suppose the firm chooses not to train, according to common sense, migrant workers can still receive the original salaries S (salary); if companies choose to train, then will have a certain amount of training expenses C. By this time, due to the migrant worker received training, knowledge and skills have been improved. When he chooses to stay, corporate will therefore obtain benefits R; the migrant worker's salary rose to S' ($S' > S$). When he chooses to quit, corporate will lose C; the migrant worker's salary rose to S'' ($S'' > S$) [2]. Following chart 1 show:

As shown in Fig. 1, in this game model, migrant workers' choice almost entirely depends on the quality of the training of individuals without any constraints. If companies choose to train, while migrant workers choose to stay, then the business and migrant workers are able to obtain gains because the training promotes the improvement of the labor productivity. That is, companies get higher economic benefits, migrant workers get better remuneration packages, to achieve a win-win situation. This is the best results of the game. However, if migrant workers choose to train while staff chooses to quit, then instead of benefiting from the training, the business will suffer huge losses; On the contrary, job-hopping will increase migrant workers' value because of trained migrant workers' knowledge and skills upgraded [3]. How can the migrant workers choose to continue staying after training rather than quit it? As a result, companies began to take into account the increase in training of migrant workers constraints.

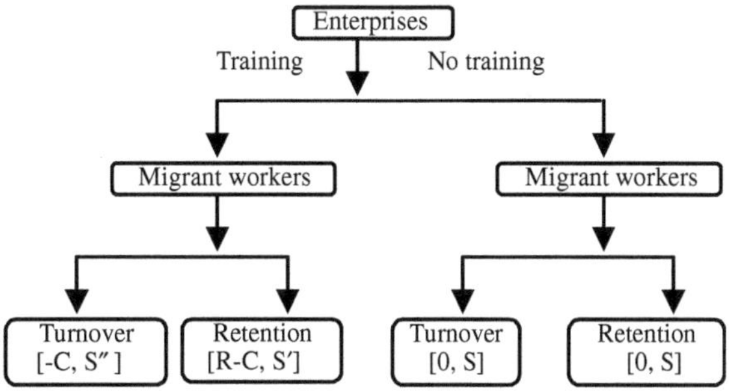

Fig. 1 Dynamic game model of the new generation of migrant workers with no training in risk constraints

2.2 The Perfection of the New Generation of Migrant Workers Training Game Model

(1) *Game analysis under compensation constraint conditions*

By the above analysis shows, When the companies have completed the training of migrant workers in the absence of any constraint forces background, migrant workers choose to quit will make business difficult to recoup their losses. In order to reduce the probability of suffering the loss, we can enable both parties enterprises and migrant workers to sign the training contracts before training, to clear breach of the conditions of trainees to constrain the behavior of migrant workers turnover [4]. In this case the game following Fig. 2.

According to this model, a reasonable strategy of business should be training in the first stage. In the second phase, if migrant workers' choice is to quit, then the third stage companies will choose to ask for compensation. At this point, the company's earnings are (LD-C), and the reason migrant workers have to pay liquidated damages, income becomes (S″-LD). Migrant workers turnover or not depends on enterprise compensation restraint efforts. When S″ >> LD (liquidated damages), the migrant workers will choose to quit; When S″ > LD, migrant workers turnover or not depending on specific circumstances to analyze specifically. So if migrant workers choose to stay in the second stage, it's the best game results. But the migrant workers are still possible to choose to quit in the second stage. It is thus clear that this method can only reduce the migrant workers turnover problem to some extent, but not completely resolved.

(2) *Game analysis under incentive conditions*

Above analysis, we learned companies only constrain migrant workers' behaviors is not enough. So we add on this basis to give incentives to trained migrant workers. Assuming trained migrant workers are able to achieve the training objectives, enterprises in order to retain migrant workers to continue working to reward qualified trained migrant workers, this game is as follows.

In the first stage if companies choose to train, migrant workers can choose to stay or quit. When the migrant worker chooses to quit, he has to pay the cost of LD

Fig. 2 Dynamic Game Model under compensation constraint conditions

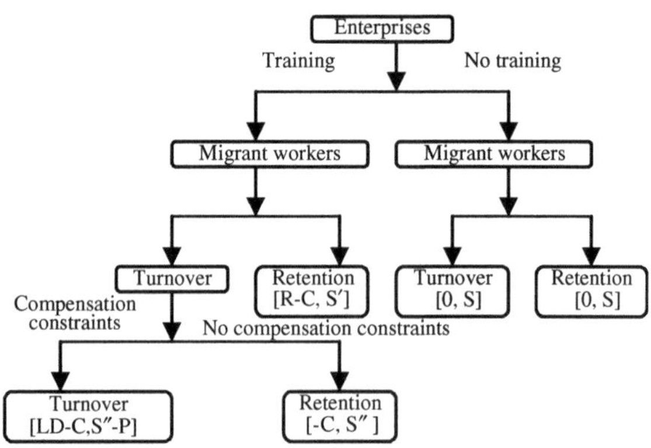

for his job-hopping behavior. And when he chooses to stay, businesses can choose to take the incentive measure or not in the third phase. If companies choose to take the incentive measure, then the migrant worker can obtain after completion of the training bonus, and improved wages; at the same time, companies therefore need to pay a bonus R in addition to the cost of the training costs C. So, if companies choose not to implement the incentive measure to reduce costs and increase revenue. Then in this case the best game for corporate is training, migrant worker is staying. However, there is a precondition, the amount of the incentive payments should be enough to motivate migrant workers to stay in business while labor costs is within the range of controllable. Then companies will choose to implement incentives.

(3) *Infinitely repeated game analysis under reputation mechanism constraints*
Above both Game improvements, we just reduce trained migrant workers' chance of job-hopping. Assuming the first phase of the game companies have chosen to train, while migrant workers not only consider the current interests, but pay more attention to the long-term interests, making infinitely repeated games between enterprises and migrant workers. In this case, the definition of "enterprise" here does not refer to a particular enterprise, but broadly refers to migrant workers now or in the future. Because in the repeated game model, repeating a certain kind of situation more times, so the players will pursue more desirable than a single game results [5]. Obviously, the migrant workers' dominant strategy is to choose to quit after completing the training rather than comply with organizational commitment. As a result, reputation mechanism that enterprises established will play an important role. For non-compliance with commitments and actions need to quit after training, migrant workers have to pay credit costs CC (credit costs). This will reduce the value of migrant workers in the talent market, invisibly also increase the cost of migrant workers turnover. Thus, if in the first game, the migrant worker adopts turnover strategy, then the next time he will be punished because of bad credit; other companies are no longer willing to cooperate with him, which creates a dynamic game situation as follows (Fig. 3).

Fig. 3 Dynamic game under conditions of enterprise implementation of incentive measures

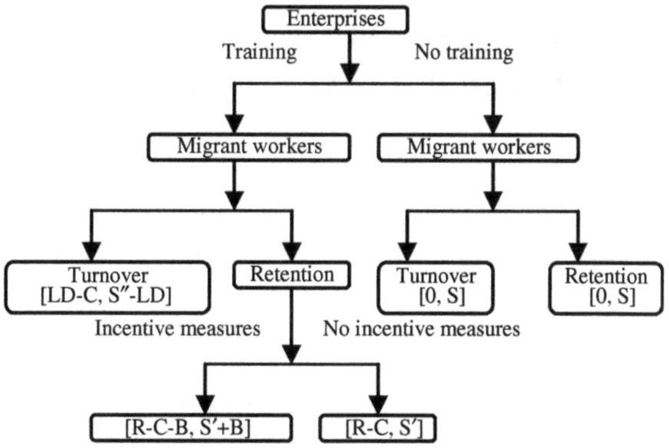

Fig. 4 Dynamic game under the constraint of reputation mechanism

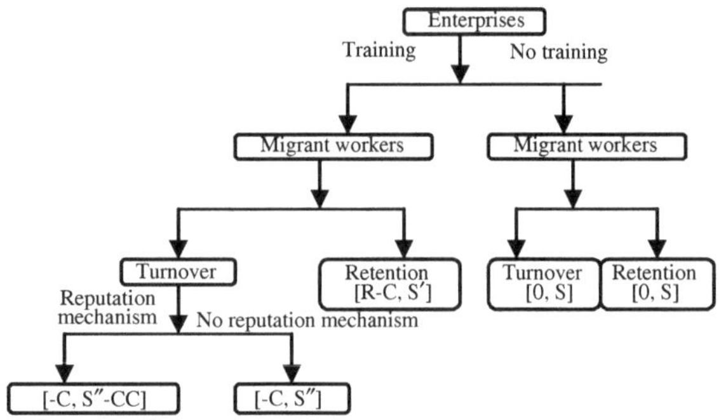

From Fig. 4, we can learn that in this repeated game, the staff will tend to establish a good reputation to choose to stay in order to cooperate businesses with for a long term under the deterrent that there may no more enterprises are willing to corporate with them. Therefore, improvements in this game, it solves the problem that the above two models can not be solved: keeping the enterprise completely away from sources of risk of trained migrant workers' job-hopping.

3 Risk Management Strategy of the New Generation of Migrant Workers Training

At present, with the deepening of reform, companies are facing a new round of restructuring and upgrading, do a good job in human resources management and deal with various aspects of risk is an important issue in the survival and development of enterprises reform. By the above model, this paper has carried out thorough analysis on the new generation of migrant workers training [6]. It is important to build up a set of risk management measures but also on this basis, such as risk protection, risk reduction, risk transfer and so on.

3.1 Risk Protection Strategies of the New Generation of Migrant Workers Training

Companies can establish long-term internal contracts with migrant workers to reduce the likelihood of damage to occur. Credit mechanism will play a role through the establishment of long-term contracts. Enterprises can implement the migrant worker stock ownership plan. After the shareholding migrant workers, the identity change from the migrant worker to the corporate owners of capital, so that the cohesion and loyalty can greatly increase, it will to some extent reduce the probability of the job-hopping conduct of trained migrant workers.

3.2 Risk Reduction Strategy of the New Generation of Migrant Workers Training

Enterprises can follow the "Labor Law" to establish a relatively stable labor relation with migrant workers. When companies have training arrangements, signing training contracts with migrant workers timely, and the two sides businesses and migrant workers reach an agreement on the rights and obligations, which can establish a relatively stable labor relation with migrant workers. If the migrant worker have job-hopping behavior after the completion of training, it is necessary to pay liquidated damages in accordance with the training contract, so as to achieve the goal of reducing the risk of training. In addition, enterprises in the choice of training objects, they should consider migrant workers of the company's loyalty. In general, older migrant workers' loyalty to the enterprise is higher than young migrant workers'. And for the migrant workers of the trial period, not recommending for their training skills, especially the training time is long and more investment training project must be carefully to select.

3.3 Risk Transference Strategy of the New Generation of Migrant Workers Training

According to the idea of risk transfer strategies, in the training process companies can take measures to share the cost of training to constrain training migrant workers job hopping behavior, so that the part of the training risk transfer to the individual migrant worker. Thus, as a business, not only can reduce the likelihood of trained staff turnover, but also reduce the cost spending on migrant workers training; as migrant workers, due to undertaking part of the cost of training, they will naturally increase the training initiative, so as to enhance the quality of training.

We must recognize that the new generation of migrant workers is a huge potential of human resources group, which is also a relatively weak group, needing attention and care from the whole society and government. Training of the new generation of migrant workers, should not only be the responsibility of the enterprise unilaterally, the government should increase the development of appropriate policies to increase funding efforts, so as to ultimately reduce the risk of training, create value for society better.

References

1. Corporate Training large inventory revealed China Corporate Training management practice-China enterprise training task force blue book. Modern Enterp Educ:70–75 (in Chinese)
2. Cui D (2014) Game employee turnover behavior analysis. Knowl Econ:108 (in Chinese)

3. Christopher J (2003) Equipping company managers to attract, retain and employ human capital. Compens Benefits Rev:35
4. Forrier A, Sels L (2003) Flexibility, turnover and training. Int J Manpower
5. Martin JO (2003) Introduction to game theory. Oxford University Press, Oxford
6. Cassidy SM, Dorfman M (1993) Introduction to risk management and insurance, 4th edn. J Risk Insur

Based on the Competency Theory of the New Generation of Migrant Workers Professional Ability Study

Ju-ling Qiao, Xue-ting Zhang and Qing-qing Cheng

Abstract New generation of migrant workers as a new force in social labor, its professional ability promotion is not only related to the economic and social development, but also related to the adjustment of industrial structure and the construction of new urbanization. This paper, supported by competence theory, based on relevant investigation data, analyzes the new generation of migrant workers professional ability influence factor, reveals the explicit and implicit characteristics of their job-hunting ability, and associates these characteristics with the new generation of migrant workers professional ability carry on correlation analysis. Based on this, puts forward to promote the new generation of migrant workers employment ability model, to provide guidance for the new generation of migrant workers to improve professional ability and play its own internal potential.

Keywords New generation of migrant workers · Competency · Professional ability

1 Introduction

Since the 1980s, the new generation of migrant workers has gradually become a new force of migrant workers, and it becomes the indispensable human resources in the construction of our country's economy and society. The concept of the new generation of migrant workers mainly refers to the 80 s migrant workers, among the currently working outside of migrant workers, about one hundred million of the new generation of migrant workers. Such a huge and vibrant workforce, their employment intention, professional ability and personal qualities are very complex and diverse, it is difficult to generalize, so we intend to be in competency theory as a

J. Qiao (✉) · X. Zhang · Q. Cheng
School of Management, University of Jinan, Jinan, People's Republic of China
e-mail: qiao58@126.com

E. Qi (ed.), *Proceedings of the 6th International Asia Conference on Industrial Engineering and Management Innovation*,
DOI 10.2991/978-94-6239-145-1_74

guide and take the reality survey as the basis, on this issue in depth study and discussion.

In order to be more objective understanding of the new generation of migrant workers professional ability, according to the competency model, taking Jinan's new generation of migrant workers to investigate object, designed the questionnaire. A total of 600 questionnaires were distributed, 475 questionnaires conform to the investigation object, and 440 valid questionnaires.

2 Based on the Competency Model of the New Generation of Migrant Workers Professional Ability Analysis

Competence theory is one of the hot spot in the managers in recent years, it is also the direction of strategic human resource management research at present. The concept of competency was proposed firstly by Professor McClelland of the Harvard University in 1973, it is refers to in a job (or organization, culture) to distinguish the outstanding performance and the general performance's personal potential deep characteristics, it can be the motivation, traits, knowledge, cognitive or skills, namely the any individual characteristics which can be reliable measured or counted and can significantly distinguish between performance advantages and disadvantages [1]. Through the above definition, can think of competency is to act as a position and do a good job of this position, should have a several of internal and external quality.

Competency model refers to the sum of the ability elements needed for a certain position, including internal quality and external quality; it is an important tool for human resource management [2]. American scholar Lyle M. Spencer and Signe M. Spencer proposed a "quality iceberg model" which make the competency model more image, they put the individual quality can be divided into two parts: one is the above water of iceberg that is the surface part, referred to as the threshold quality, it is the basic requirement for position, this part can easily be measured by external pathway and changed through training; two is the underwater of iceberg that is the potential part, known as the differentiating quality, it is the deeper requirements for position,this part cannot easily be measured and changed through external pathway, so it is the key factor to distinguish outstanding performance and the general [3]. According to this theory and our investigation study, preliminary established the competency model of the new generation of migrant workers professional ability (see Fig. 1).

New generation of migrant workers the competency model refers to the new generation migrant workers in order to achieve long-term stability of employment, realize their own value, adapt to new working and living environment, and successfully integrated into urban society should have a variety of competency elements [4]. According to the new generation of migrant workers' professional ability quality iceberg model, through the relevant investigation and research, from the

Fig. 1 The new generation of migrant workers' professional ability quality iceberg model

explicit competence characteristic and implicit competence characteristics two aspects to conduct to the new generation of migrant workers professional ability competence analysis.

2.1 *The Explicit Competence Characteristics of the New Generation of Migrant Workers*

The explicit competence characteristics of the new generation migrant workers include professional knowledge, vocational skill and job-hunting ability.

(1) *Professional knowledge*

Professional knowledge refers to the knowledge of employment required for specific areas. Professional knowledge to possess, use and learning ability directly affect the efficiency and effectiveness of work, the new generation of migrant workers professional knowledge can be learned by education level [5]. In the survey, 6.4 % were primary school culture, 38.6 % were junior middle school culture, 30.9 % of high school or technical secondary school, 20 % were junior college, 4.1 % were undergraduate or above (see Fig. 2). Thus, although the new generation of migrant workers' education level generally improved, but the

Fig. 2 New generation of migrant workers education level

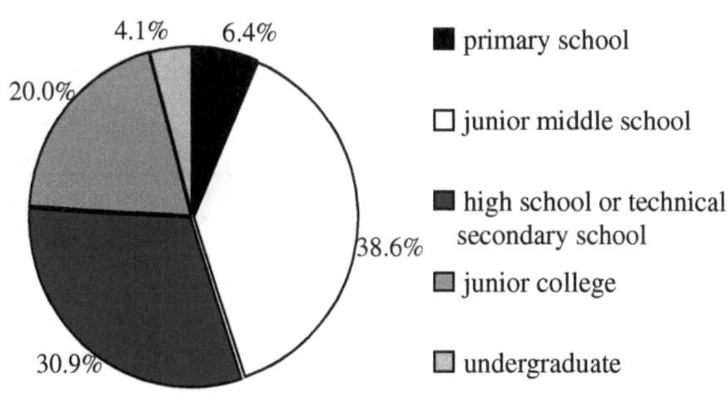

proportion of received professional education was still low, obviously the lack of professional knowledge would undermine their professional ability.

(2) *Vocational skill*

Vocational skill refers to the technology and ability which professional needed, its formation is a dynamic process [6]. With the development of market economy, the economic structure gradually transformed from labor-intensive to technology-intensive, the new generation of migrant workers' vocational skill was still lack of systematic training, many migrant workers can only through the traditional method of apprentice follow master worker to acquire skills. The survey found that 56.8 % had not received technical training, 32.7 % had received a technical training, 10. 5 % had received one or more technical training, 70.5 % of them was the front-line workers or service personnel. Thus, the new generation of migrant workers lack of vocational skill is an important factor to limit their professional ability.

(3) *Job-hunting ability*

New generation of migrant workers from famer into worker, they inevitably be influenced by the original living environment, resulting in their employment awareness and ability is very bad. From the perspective of the job-hunting channels, the survey found that only 13.4 % of new generation of migrant workers had received the job market and career intermediary employment services, 11.4 % obtained employment through the network, 8.6 % by the employer to the local recruitment, 33 % by themselves, the remaining 33.6 % were introduced by friends or relatives work. And from the perspective of the nature of the employment industry, there were 38.6 % in manufacturing, 28.9 % engaged in the catering, accommodation and other services, 19.8 % in construction, 8.3 % engaged in the wholesale and retail industry, 4.4 % chose other (see Fig. 3). Thus, the new generation of migrant workers job-hunting channel is very conservative and lack of diversity, coupled with the lack of job-hunting skills and strategies, resulting in less choice, employment quality is not high and mainly engaged in the physical cost of labor intensive industry [7].

Fig. 3 New generation of migrant workers job-hunting channels

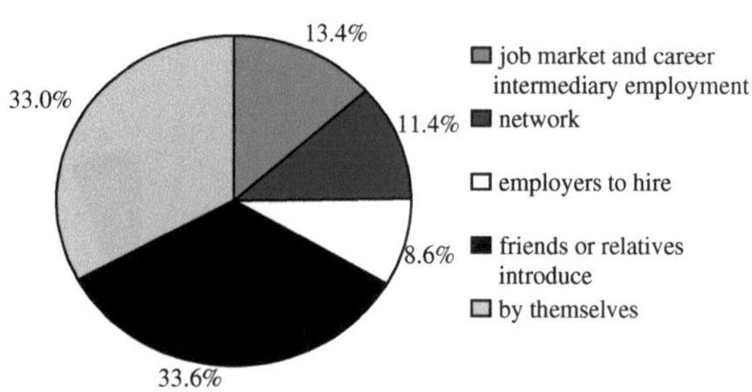

2.2 *The Implicit Competence Characteristics of the New Generation of Migrant Workers*

The implicit competence characteristics of the new generation migrant workers include drive, adaptability, and personal quality.

(1) *Drive*

Drive refers to the intrinsic motivation to push people to engage in certain activities and in one direction [8]. From the point of working drive, the new generation of migrant workers to go out to work were no longer just want to improve their economic capacity, in the survey, 72.7 % of the new generation of migrant workers out of employment are in order to find a suitable job and realize their life values, this shows that the new generation of migrant workers go out motivation gradually transformed from survival to growth-oriented [9]. So that they become more long-term vision and put keen to pursuit better occupation ideal.

(2) *Adaptability*

Adaptability refers to according to the environment to change their behavior, feelings and ways of thinking of the willingness and ability, adaptable workers can position themselves correctly in the uncertain environment and can quickly integrate into the new working environment [10]. For the new generation of migrant workers, moved from the countryside to city, from famer into worker, their living environment and their own career has changed dramatically, if they do not have the necessary adaptability cannot be to obtain employment smoothly. In the survey, 52.5 % of people want to settle down in cities, can be seen most of the new generation migrant workers can well adapt to the city lifestyle and has a certain sense of belonging to city, but there are still 25 % of the new generation of migrant workers has been cannot adapt to the work and life of the city. The main reason is due to the low bear capacity of the new generation of migrant workers, they do not have confidence in the face of setbacks work and life, resulting in employment is not long-lasting. So the adaptability is the key influence factor in the professional ability of the new generation of migrant workers.

(3) *Personal quality*

The new generation of migrant workers although rarely directly engaged in agricultural production, but the grow up in the background of the yellow land also give them a deep mark, relatively grew up in city of the new generation of migrant workers, they have the feature such as enduring hardships and inclusive heart strong, But because of limited growth environment, their vision is not open enough, and their communication skills, legal awareness and self-protection ability is not strong enough, leading to their job satisfaction get lower and lack of confidence in the stable employment [11]. In the survey, 75 % of the new generation migrant workers suffered discrimination or unfair treatment at work, of which only 11 % of new generation migrant workers to use legal or other ways to protect their own interests, 36 % chose from recognize unlucky, 38 % choose to resign to find other work. Thus the personal quality is an important influence factor in the professional ability of the new generation of migrant workers.

3 Based on the Competency Model of the New Generation of Migrant Workers Professional Ability Promotion

According to the above analysis, the explicit competence characteristics compared with implicit competence characteristics, it can be changed by the external behavior, but as for the new generation of migrant workers, only with rich professional knowledge and skills can quickly find a suitable job, so it is the premise and guarantee for new generation of migrant workers to achieve stable employment; the implicit competence characteristics can not be easily change, so it is the key to improve the employment ability of the new generation of migrant workers [12]. At the same time, the two aspects of competence characteristics also influence each other, with high knowledge level or vocational skills of new generation migrant workers can all-round self-understanding, develop their own occupation planning and have the confidence to adapt to city life and work; while high professional identity or adaptable of new generation migrant workers hope to find a better and more suitable for their own development opportunities, thus they have more demand to improve their vocational skills (see Fig. 4).

3.1 To Establish and Improve the New Generation of Migrant Workers Employment Training Mechanism

Firstly, the government should to increase investment in rural education, strengthen the new generation of migrant workers education level from the source; Secondly, rely on social capital to enhance professional ability of new generation of migrant

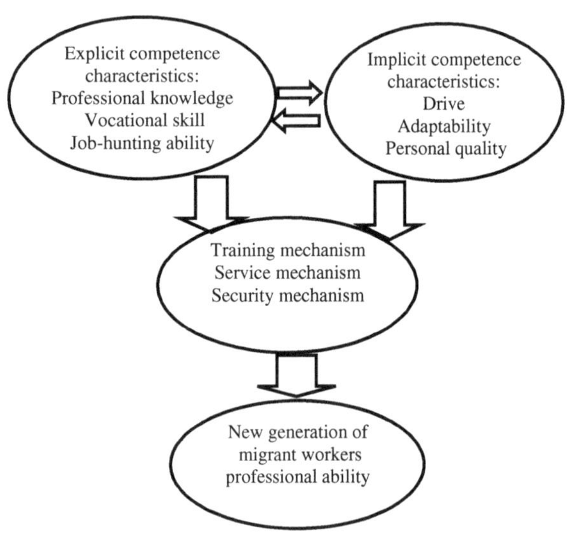

Fig. 4 New generation of migrant workers professional ability promotion model

workers, according to the new generation of migrant workers employment demand set practical training plan and take the market as the guide to implement diversity training, in order to comprehensively improve their professional skills [13]; Finally, we should take the corresponding incentive policy, the new generation of migrant workers should to actively use a variety of learning and training opportunities, take the initiative to improve self professional quality and enhance the competition ability of employment.

3.2 To Establish and Improve the New Generation of Migrant Workers Employment Service Mechanism

To build a suited to the characteristics of the new generation of migrant workers labor information market and establish the new generation of migrant workers employment information service platform; to establish diversified channels of employment information, integrate and use of community, social organizations, private relations and other information dissemination channels, making it possible to obtain a comprehensive job information, broaden the employment channels, expand the scope of career, increase the employment opportunities and upgrade the level of employment [14]. With the aid of social or internal enterprise occupation counseling agency, for the new generation of migrant workers to carry out professional and scientific occupation counseling and improve their psychological endurance capacity, make its can correct positioning themselves and enhance their professional identity; the new generation of migrant workers should to learn to psychological adjustment, with a positive attitude in the face of setbacks at work, to establish a correct concept of employment and strengthen their professional self-confidence [15].

3.3 To Establish and Improve the New Generation of Migrant Workers Employment Security Mechanism

To strengthen employment law enforcement supervision and labor protection, to make the rights to get legal protection and build a good legal environment for the new generation migrant workers employment; the enterprise should establish labor relations in strict accordance with the law and in a fair and impartial attitude towards the new generation of migrant workers, then uses the human-oriented management mode to create a harmonious working environment [16]; the new generation of migrant workers should to strengthen their legal rights awareness and use a variety of opportunities to learn legal knowledge, to improve their rights protection ability in employment.

References

1. McClelland DC (1973) Testing for competence rather than for intelligence. Am Psychol 28:37–38
2. Zheng Y (2013) The building and application of competency quality model. Mechanical Industry Press, Beijing, pp 10–12
3. Spencer LM, Spencer SM (1993) Competence at work: models for superior performance, vol 3. Wiley, New York, pp 48–49
4. Liu F, Yu W (2010) Based on the competency model of the migrant workers flow employment management. J Jiangsu Agric Sci(4):486–470
5. Yao R, Xiong J (2007) Construction on the competency model of the rural labor force transfer. J Anhui Agric Sci (35):1164–1165
6. Heckman JJ (2007) The technology and neuroscience of capacity formation. Proc Natl Acad Sci 104(3)
7. Li X (2009) Study on the improvement of the new generation of migrant worker's employability. J Henan Polytech Univ (Social Sciences) (4):591–592
8. Jung CG (1960) The Psycho genesis of mental disease.International Cultural Press
9. Wang X (2010) Study about the social identity of the new generation of migrant workers. Shandong University, pp 21–22
10. Hall DT, Mirvis PH (1995) The New Career Contract: Developing the Whole Person at Midlife and Beyond [J]. J Vocat Behav 47:269–289
11. Lu G (2014) New generation of migrant workers employment problem analysis. J Shandong Agric Adm College (3):12–13
12. Fu J (2013) Based on the Competency model of the college students employment ability cultivation research. Econ Forum 6:158–159
13. Xu W (2014) Study on the vocational training of new generation migrant workers.Wuhan University, pp 71–72
14. En-li Luo (2010) New generation of migrant workers employment ability research. China Hum Resour Dev 2:4–7
15. Tan M, Fang H (2011) New generation of migrant workers employment ability and improve. J Adult Educ Hebei Univ (2):51–52
16. Zeng L (2011) Research on the employability and its influencing factors of the new generation of the peasant workers in the Pearl River delta region. Guangdong Business School, pp 43–44

Network Governance of Communication Channels Between University-Industry Cooperation

Owusu Ackah, Zheng He and Shuang-shuang Zhou

Abstract The inception of university-industry cooperation over the past decades has been the beacon of technological innovation. This is attributed to the complementary resources that each actor brings to the table during this form of cooperative activity. Despite of the enormous benefits derived from this alliance there have been a number of hindrances affecting the knowledge and information sharing process. The traditional linear way of sharing information and knowledge has proven not to be efficient. Therefore the purpose of this study is to develop a new innovative channel that would improve communication within this alliance by adopting network governance approach. This study proposes a network-communication model that seeks to enable efficient flow of information and knowledge whiles cutting down cost and waste. This model to some extent would improve communication process as compared to the linear method used previously because it promotes interaction between the various actors and sub-actors in university-industry cooperation process.

Keywords Network governance · Network-communication model · University-Industry cooperation · Knowledge and information sharing

1 Introduction

Research in the field of technological innovation and development has revealed the immerse benefits of University-Industry cooperation [5]. This has led to its integration into both regional and national innovation system [1, 13]. The economic significance of this cooperation is attributed to the exchange of information [22] and social interaction among actors [9]. This is possible due to the dyadic form of

O. Ackah · Z. He (✉) · S. Zhou
School of Management and Economics, University of Electronic Science and Technology of China, Chengdu, China
e-mail: hezh@uestc.edu.cn

E. Qi (ed.), *Proceedings of the 6th International Asia Conference on Industrial Engineering and Management Innovation*,
DOI 10.2991/978-94-6239-145-1_75

relationships. Universities and other research facilities serve as the main hub for knowledge innovation, whiles industries are mostly concern with technology commercialization. The interaction nature of this alliance helps improve both the quality of final product and the entire innovation process [20] through knowledge spillover due to geographical proximity [10]. Furthermore, assess to experts in various fields has improve the final output of U-I cooperation [16].

Notwithstanding the increasing number of U-I cooperation due to its performance [15], there have been a number of constraints that affect the performance [29]. The performance of university-industry cooperation has been hindered by a number of hurdles that includes the sharing of information. The lack of effective communication or interaction between parties has resulted in trust distress [6]. The role of communication cannot be underplayed if U-I collaboration can survive. Despite the chain of command and communication channels is not clearly defined in U-I cooperation, this has affected the performance of university-industry cooperation over the years [2, 17, 31].

In order to mitigate the risk of failure that communication poses in U-I cooperation, it is necessary to have a collective effort from parties involved. This collective body as a collective system needs to be governed to enable the achievement of success. In this situation the governance of the communication process needs to be efficient and effective in order to avoid complex interaction channels and promote inter-intra organization and departmental activities. Adopting a network governance approach will aid the efficient governance of this process. Network governance thrives on collective forms of rewards and punishments. Information sharing and exchanges is secured in network governance due to this fact [23].

U-I cooperation can meet it objects when there is an effective sharing of information and communication among various actors involved. That is information should be disseminated timely and accurately. Communication helps build up cordial relationships and trust among various agents [12]. Therefore, there is the need to build robust communication channels to aid this process. Despite the immerse influence this process have on the success or failure of this cooperation, there still remains gaps relating to information and knowledge exchanges. This might arise due to the complex nature of the communication process among actors in this supply chain that creates structural hole or breeds distortion in information sharing before it arrives at its intended recipient. Whiles main actors in this supply chain receive information, there is a gap of information sharing between subgroups. Therefore to achieve success there is a need to bridge this 'communication gap or hole' that cripples the cooperation. The early work conducted by [8] argues that when organizations as a social entity are able to bridge structure holes within its network, they enjoy significant payoffs and benefits since an effective and well-connected network enjoys social capital [3].

Despite of the significances of this process, few studies have be conducted to ascertain how communication such as information and knowledge sharing can be facilitated between both actors and sub-groups in order to reduce transaction cost and improve efficiency.

The focus of this study is to explore how to improve communication channel within actors and other agents in this complex system. From a social network perspective we consider the U-I cooperation as a two-node network with affilate that has different actors and various subgroups. Therefore to have an effective information sharing process there should be an effective communication channel between actors and its subgroups in place that defines the overall objective of the cooperation and specific targets of assigned projects and tasks. A social network approach would help identify the relationship and interactions patterns between the main actors that are the university and industry and various departments within these organizations. Also this study examines how network governance can provide a background for performance evaluation.

The paper is organized into the following sections: Sect. 2 introduces the significances of communication in achieving the objectives of U-I cooperation and current situation of U-I communication. Section 3 focuses on how to adopt network governance to improve the efficiency of U-I communication. Section 4 discusses the model proposed for this study with the conclusion in Section 5.

2 Communication Within U-I Cooperation: Literature Review to Show the Current State

The interaction between University and industries has over the years contributed to improvement and introduction of new innovative products and technologies [24]. Despite of the success of this cooperation there happens to be a high failure rate. This can be attributed to inconsistency and redundancy within the information, knowledge and resource sharing process [25]. Therefore the need to provide effective channels and communication would be beneficial to parties involved. Since scientific research conducted in these research facilities and universities is of immerse value to the industry and vice versa [28], U-I cooperation would enable optimal utilization of complementary resources at the disposure of actors.

In order to achieve the objectives of U-I cooperation either in scientific research and technology commercialization or acquisition of new knowledge effective communication plays a center stage [7]. Therefore having an effective communication mode between all actors and sub actors has a positive impact on the task carried out in this cooperation [30].

Furthermore, U-I as an alliance that is based on significant amount of human interaction is faced with a high level of risk that would hinder it success [29]. As elaborated earlier among the numerous challenges facing the survival of U-I cooperation, information sharing stands out. The ability to break down the complex nature of interaction and a more defined mode of communication will contribute to achieving set targets. The complex nature of communication should be harnessed into a simple continuous process that eliminates any form of redundancy.

For instance when clear roadmaps are defined with efficient interaction modes, it mitigates the risk associated with research diverting from cooperation objective as argued by [27], avoiding ambiguity in targets or culture diversity, and improves trust [17, 31].

Further, there is an inconsistent communication pattern in this form of cooperation which creates a gap in the communication process. Communication is usually between the university and industry and doesn't involve the various faculties and departments. This results in a fractal process and is weak for effective knowledge and technology diffusion. University-industry cooperation being a complex system therefore cannot thrive on a linear communication pattern [18, 19]. This form of communication creates redundancy and limits the amount of relevant information shared among cooperating parties. Also this linear channel of communication creates structural holes within this network, making it difficult to achieve effective outcomes in terms of new product or process development. It further slows down the decision making process and can result in an increase in waste and cost [8, 26].

3 Network Communication Model

Information exchanges comprise of indefinite and logical process between interacting agents. Therefore the need to govern this social interaction is essential. Network governance approach turns to be an effective mode of governing this network as compared to both the market and policy governance respectively [26].

Notwithstanding the essence of communication in U-I cooperation there is the need to have a more efficient communication process in order to aid each party to achieve success in this partnership. Many scholars argue university research usually deviates from the objective of the U-I cooperation expectation in various forms, which affects the cooperation. In order to avoid this situation there is the need to set objectives prior to the forging of the partnership and this should involve both parties and relevant departments. This will aid participants to understand what is expected of them and how it can be achieve effectively. This process will bridge the gap that is formed as a result of inadequate communication to various actors within the cooperation [8]. When there is cordial relationship between actors such as faculty members, industrial experts and others, it aids in trust building and will have a significant impact on the success of the cooperation. The most efficient medium to facilitate information and knowledge transfer is through personal ties that exist in a network [14, 21]. Moreover diffusion of technology is faster when there are strong ties between various actors in a particular network [4].

Despite of the encouragement of personal ties among actors in this network there is the need to consider how information is to be transmitted among actors securely. Information exchanges between two actors with affiliate network turn out to entails indefinite interactions between various parties involved. Therefore there is an urgent need to govern this mechanism [11]. University-industry cooperation is no exemption. Furthermore, university-industry cooperation is a two-node network with overlapping affiliate such as faculty and department members, various schools in the university etc. On the other hand the industry comprises of various sub

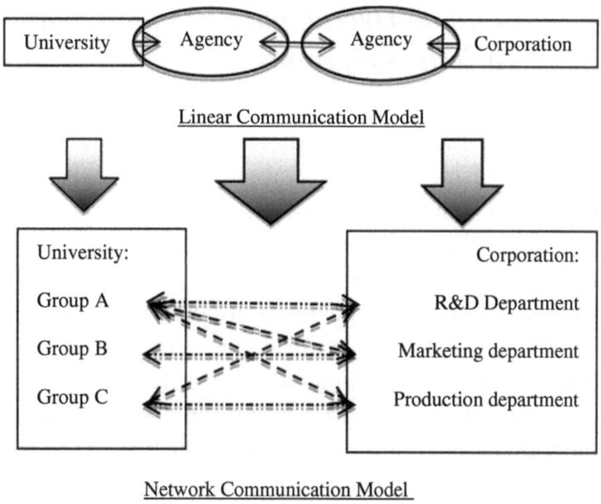

Fig. 1 From linear to network communication

divisions and departments. Therefore to encourage overlap communication between the various subgroups there should be a systematic mode of knowledge and information transfer. In this paper, we adopt a network governance approach as an alternative medium to facility efficient communication.

Figure 1 illustrates how university-industry cooperation can achieve an effective communication process. Two central actors, i.e. the university and industry have several affiliates or members which characterize this model. In other to attain effective communication this should be critical interaction between the various subgroups of both the university and industry and this is what the above model seeks to achieve by promoting overlays in the communication process.

4 Network Governance of U-I Communication

4.1 Promote Cooperation and Interaction Among Members

In order to bridge the structural hole that exists in U-I cooperation there is the essential need for the various subgroups to interact. Not only should these groups interact in every stage of the project but also these groups should be involved in the design and implementation of the objective of such cooperation. When there exist strong ties between various actors and subgroups, it enables each party to derived the best of this cooperation. For instance, faculty members and researchers should be able to interact with industrial experts and other members of the research and development in other to ascertain the industrial expectation of their activities. When this is ensured, researchers have a clear idea of what is its expected of each actor without going through a cumbersome process which mostly leads to distortion of facts therefore deviating from the objectives of the cooperation.

4.2 Promote Overlapping Communication Among Various Subgroups

One major factor that needs to be considered in order to achieve success in university and industry cooperation is whether the objective set out is aligned with individual objectives. When faculty member or industrial community has no interest in the project that leads to the forging of this partnership, there is a high failure rate. Therefore the view and perspectives of various individuals from both parties should be considered. Furthermore, during U-I cooperation faculties should be encouraged to communicate their research findings to industrial experts without going through a cumbersome process as it is elaborated in earlier sections where all communication have to go through the technology transfer office. When these individuals have informal direct contact with each other it enables collective collaboration without restriction of passing a formal process. And this in turn speeds up the process of particular projects and tasks.

4.3 Define Objective and Tasks Clearly

Before the inception of U-I cooperation, there should be a clear roadmap to what the objective is and how it is going to be achieved. Also channels for communication and information sharing should be clearly defined with no ambiguities. In defining these channels the cumbersome process associated with formal communication channels need to be evaluated and revised to encourage various subgroups to cooperate directly without going through complex channels. When different subgroups interact directly it helps eliminate structural hole that exist in organizations with strong team but weak ties as advocated by earlier works of Granovetter [8]. What this communication model seeks to achieve is to have strong teams with strong ties, and this will facilitate an efficient interaction between actors.

4.4 Standard Regulation Should be Set to Govern Exchanges Between Various Subgroups

Insofar as this model promotes, to achieve the direct interaction of various members of subgroups within the central actors, there is the need to set up governing system that will outline the quality and quantity of information. This can be shared outside the technology transfer office that is set for such purposes. When individuals have a clear idea about how much information to share with another individual in regards to specific projects at hand, it will prevent the leak of essential information to unauthorized individuals. This will also serve as a check and balance system for both organization involved in U-I cooperation. Therefore individuals cannot share more than the stipulated amount of information to be shared at a specific period of time.

5 Conclusion

University-industry cooperation evidently has promoted the development and commercializing of most technologies over the years. Despite of these merits associated with university-industry alliances, a number of factors have hindered the performance of this collaboration. One essential factor is the inability of various subgroups to interact with one another. This has hindered the communication process. Therefore the need to enhance these channels is essential if university-industry cooperation is to achieve its targets.

This study focuses on how communication channels can be improved to promote an efficient exchange of information between various actors associated with U-I cooperation. In this paper, we considered U-I cooperation as a two-node social network that has an affiliate network of sub-actors. Furthermore, this paper placed emphasizes on how to enhance communication channel using network governance. A new communication model encouraging the involvement of affiliate network in direct interaction is proposed and its advantages are discussed.

Acknowledgments This research is sponsored by The National Social Science Fund of China (No. 12BGL123), China Scholarship Council and OATF, UESTC.

References

1. Bettis R, Hitt M (1995) The new competitive landscape. Strateg Manag J 16(S1):7–19
2. Black N (2001). Evidence based policy: proceed with care, British Medical
3. Burt RS (1992) Structural holes: the social structure of competition. Harvard University Press, Cambridge
4. Cohen W, Nelson R, Walsh J (2002) Links and impacts: the influence of public research on industrial R&D. Manage Sci 48(1):1–23
5. Eom B, Lee K (2010) Determinants of industry–academy linkages and their impact on firm performance: the case of Korea as a latecomer in knowledge industrialization. Res Policy 39 (5):625–639
6. Florida R, Cohen W (1999) Engine or infrastructure? The university role in economic development. In: Branscomb L, Kodama F, Florida R (eds) Industrializing knowledge: University-industry linkages in Japan and the United States. MIT Press, Massachusetts
7. Giuliani E, Arza V (2009) What drives the formation of valuable university-industry linkages? Insights from the wine industry. Res Policy 38(6):906–921
8. Granovetter M (1973) Strength of weak ties. Am J Sociol 78(6):1360–1383
9. Granovetter M (1985) Economic action and social structure: the problem of embeddedness. Am J Sociol 91:481–510

10. He Z, Rayman-Bacchus L (2010) Cluster network and innovation under transitional economies: an empirical study of the Shaxi garment cluster. Chin Manag Stud 4(4):360–384
11. He Z, Zhou S, Ackah O (2013) Network governance and copycat in industrial cluster: a case study of Zhongshan gas appliance cluster. In: Proceedings of international conference on strategic management
12. Heimeriks K, Duysters G (2007) Alliance capability as a mediator between experience and alliance performance: An empirical investigation into alliance capability development process. J Manage Stud 44(1):25–49
13. Hwang Y, Kim S, Byun B, Lee G, Lee H (2003) Strategies of promoting industry-academia-research institute R&D partnerships to cooperation with new technologies: focusing on industry-research institute inter-firm R&D partnerships, Science & Technology Policy Institute (in Korean)
14. Jaffe A (1989) Real effects of academic research. Am Econ Rev 79(5):957–970 (323, pp. 275–279)
15. Lee Y (2000) The sustainability of university-industry research collaboration: An empirical assessment. J Technol Transf 25(2):11–133
16. Lei D, Slocum J Jr (1992) Global strategic alliances: payoffs and pitfalls. Org Dyn 19(3):44–62
17. Leydesdorff L (2011) Meaning as a sociological concept: a review of the modeling, mapping, and simulation of the communication and knowledge meaning. Soc Sci Inf 50:1–23
18. Luhmann N (1995) Social system. Stanford University Press, Standard
19. Motohashi K (2005) University-industry collaborations in Japan: The role of new technology-based firms in transforming the national innovation system. Res Policy 34 (5):583–594
20. Mowery DC, Nelson R, Sampat B, Ziedonis A (2001) The growth of patenting and licensing by U.S. Universities: an assessment of the effects of the Bayh-Dole Act of 1980. Res Policy 30 (1):99–119
21. Ness H (2009) Governance, negotiations and alliances dynamics: Explaining the evolution of relational practice. J Manage Stud 46(3):452–480
22. OECD (2002) Technology policy and the environment. OECD, Paris
23. Park HW, Leydesdorff L (2010) Longitudinal trends in network of university-industry-government relations in South Korea. Role Program Incent 39:640–649
24. Penrose E (1959) The theory of the growth of the firm. Basil Blackwell, Oxford
25. Perkmann M, Walsh K (2009) The two faces of collaboration: impacts of university-industry relations on public research. Ind Corp Change 18(6):1033–1065
26. Powell WW (1990) Neither market or hierarchy: network form of organisation. Res Org Behav 12:295–336
27. Poyago-Theotoky J, Beath J, Siegel D (2002) University and fundamental research: Reflections on the growth of university-industry partnership. Oxford Rev Econ Policy 18 (1):10–21
28. Rebne D (1989) Faculty consulting and scientific knowledge: A traditional university-industry linkage. Educ Adm Q 25(4):338–357
29. Siegel D, Waldman D, Link A (2003) Assessing the impact of organizational practices on the productivity of university technology transfer offices: an exploratory study. Res Policy 32 (1):27–48
30. Singh K, Mitchell W (1996) Precarious collaboration: business survival after partners shut down or form new partnerships. Strateg Manag J 17(1):99–115
31. Yoon W, Hyun E (2010) Economic, Social and institutional conditions of network governance. Manag Decis 48:1212–1229

Risk Management in Electronic Retail Payment System: An Empirical Study from Ghana

Owusu Ackah and Zheng He

Abstract The introduction of electronic retail payment systems in Ghana has transformed business transactions over the years. The numerous advantages that are associated with this system cannot be doubted, however, there are a number of risks that hinder the success rate of business transactions. The objective of this paper is to investigate into how firms in this industry evaluate these risks and their mitigation strategies. A case study approach was adopted to examine how risks affect the firm's activities. Our findings indicate that a comprehensive risk evaluation of the whole process rather than one or two stages has an important impact on the efficiency of risk management. Furthermore the practice of risk mitigation policies and strategies serve as a critical step to mitigate these risks. At the end of the paper, we suggest a more proactive approach towards risk management whereby risk identification and risk assessment stage of the risk management cycle should be performed in a coherent manner.

Keywords Risk management · Electronic retail payment system · Risk evaluation · Emerging economy

1 Introduction

Over the years, electronic Payment Systems has evolved to be a part of executing business transactions, due to the rapid development of electronic business. The usage of the Internet around the globe is the backbone of electronic commerce and the number of online shopping is forecast to increase in the coming years [1]. This trend of electronic commerce has created a platform for firms to gain competitive

O. Ackah · Z. He (✉)
School of Management and Economics, University of Electronic Science and Technology of China, Chengdu, China
e-mail: hezh@uestc.edu.cn

E. Qi (ed.), *Proceedings of the 6th International Asia Conference on Industrial Engineering and Management Innovation*,
DOI 10.2991/978-94-6239-145-1_76

advantage [2] and gain information about customer's shopping behavior and taste, which serve as a base for managerial decision-making [3].

These forms of electronic payment systems mostly rely on electronic devices and technology and are perceived to be associated with a high probability of risk by the customer [4]. Regardless the numerous advantages in relation to transaction speed, flexibility, and others, there are a number of uncertainties that can make customer lose trust in these systems [5]. The reputation of a firm depends on the trust of its customers therefore firms should be able to gain their trust by educating them on the benefits of its payment system and risk anticipation.

Among the variety of risks reported, there have been a number of them that has been given high priority by either providers or users. A few of these are categorized under security and privacy of client's information, identity theft and card fraud, and legal and operational risks respectively [6]. Many firms work hard to identify the risks they face but do less to evaluate the impact of these risks on their firm's strategies and products. The uncertainties associated with payment systems have forced out most providers and left a few surviving ones. A firm's ability to manage these uncertainties strategically [7] goes a long way to help achieve a firm's long-term goal.

This paper focuses on how retail payment system providers evaluate the extent and impact of these risks by combining risk identification and assessment into a coherent stage and whether mitigation policies outlined are practiced. This study was conducted in an emerging economy, which has seen a rapid development and usage of these electronic payment systems. We conducted our investigation with the providers in context because most of these payment systems in emerging economies are provided through the providers. If providers manage these uncertainties well there would be a high probability of success at all levels involved in the electronic payment system supply chain, which is also one of our sources of motivation.

The paper is organized as follows: Sect. 2 provides the related theoretical background and the methodology used for this study is outlined in Sect. 3; Sect. 4 reports the analysis and findings of our study; Sect. 4.2 discusses the findings of the study and lastly Sect. 5 focuses on the conclusion and further research.

2 Theoretical Background

This paper focuses mainly on how risk is evaluated and mitigated by providers mostly in an emerging economy. To address the problem, there is the need to establish a theoretical background that would serve as the basis for data gathering and analysis. The whole electronic process includes four steps: customer, P.O.S., database and financial institution. The first two steps of the risk management cycle were combined to form a unified coherent stage, which we labeled it as risk evaluation to identify and assess the impact of uncertainties.

One of the important steps in mitigating risk is to be able to identify and assess the type of risk and its degree to which it affects electronic payment activities. Risk evaluation is the cornerstone for effective risk mitigation. When the degree to which the risk affects activities is known, managers can make decisions on what kind of strategies to adapt to control these risks. Risk evaluation is to be approached methodically so as to identify all the risks flowing from all processes. When a risk is being identified it is essential to assess its potency. When the risk is assessed it should also be compared to the estimated risk criteria [8, 9]. In this paper, we emphasize the risk evaluations of the whole process. Thus we come up with the following hypothesis:

H1: *A comprehensive risk evaluation of the whole process has an important impact on the efficiency of risk management.*

The next step after evaluating these risks is to introduce policies and strategies to reduce these risks. Risk mitigation is efforts and measures that are being employed to reduce the probability of uncertainties occurring. This can take the form of sharing risk between parties and includes procedures through which a firm decides to implement to reduce this uncertainty. Despite most firms perform risk evaluation process it turn out that most firms refuse to go a step further in implementing their mitigation plan [10]. Hypothesis was formulated to test whether polices and strategies are implemented;

H2: The practice of a firm's risk mitigation polices during the whole process serves as a positive impact to reduce risk.

3 Methodology

This section provides an overview of the method used for our research and how data for this study were collected and analyzed in order to examine our hypotheses. The main objective of this research is to explore how risk associated with the emerging retail payment system in an emerging economy can be managed. In order to understand and establish a reliable result we adopt both qualitative and quantitative methods. Quantitative approach was used because of the empirical investigation we conduct into this phenomenon. Data for this section was mainly acquired through the administering of questionnaires to be answered by the firm and its customers. Data obtained from the survey was used to test the hypothesis by SPSS software. In addition, in-depth interviews were used to investigate how it happened [11, 12]. This qualitative method can throw up important contributions that enrich the real context. In this paper, a single firm was chosen as case study. The relevant information was acquired through the field survey such as questionnaires of customers and semi-structured interviews of top managers, and secondary archives from customer complaint forms, the company's risk management department as well as other departments.

3.1 Case Selection

The process of selecting a suitable case is an essential step to build theories from case studies. This became important because when unsuitable cases are selected, the result obtained will be misleading and will not help us achieve our research objectives. Appropriate selection of case helps define the limit for generalizing the finding of the study and control waste [13]. Considering the number of cases that can be studied at a particular time choosing a relevant case becomes an essential obligation [14].

The firm sampled as case study for our research was Afric Xpress in Ghana. Afric Xpress is an electronic payment solution provider with the most convenient payment and transfer solutions for the unbanked in Ghana. Ghana is an emerging economy as many developing countries in the world. Financial transaction in these types of economies is predominantly based on cash. With the development of the internet technology and the availability of mobile phones and personal computers, Ghana has embraced the concept of moving traditional retail shops into electronic commerce, leading many firms are finding new ways to run their business activities. These electronic payment systems are innovative ways and solutions of elimination hitches in electronic retailing between parties. Despite the introduction of these payment solutions eliminates of the labor-intensive way of shopping, cheques and fund transfer between parties, it has its own form of uncertainties and risks that disrupts the free flow of information and transaction.

3.2 Data Collection

For the purpose of this study, we combined data from both primary and secondary sources. The designed questionnaires were answered by a cross section of employees of the firm and also its customers. Meanwhile semi-structure in-depth interviews were conducted among top managers and employees of the selected firm. Secondary sources include information and data from company's archives, publication in journals related to this study, firm information available through the Internet and other media forms. In order to measure the accuracy and objectivity of the information and data we had, we compared the answers given by two respondents and this was evaluated during the administrating of our interview questions. The first step was to allow top managers from relevant department including the risk management unit to answer a questionnaire separately, and then an in-depth interview was conducted among these employees. Each interviewee was asked the same set of questions that included their views on risk evaluation and risk mitigation. After gathering and comparing all these answers obtained from the interview section, we evaluated the responses against that of the questionnaire and found out that the responses obtained through the questionnaire were accurate and had a sense of objectivity.

In our survey we targeted a sample of single firm within this industry. We contacted the key informants, which in this study was the director of the Risk Management Unit of the firm. As stated by [15], contacting key informant for a survey enables the researcher to have the right and relevant kind of information and data. Contacts with these key informants were firstly through telephone conversations and then later through electronic mail to explain the objective and purpose of this research so their co-operation can be gained. The survey was conducted within a timeframe of two months in 2012, with 120 questionnaires being sent out to this firm's employees and customers to be completed. After a period, we received 100 effective questionnaires that account for 83 percent of the total questionnaires sent out. We also gathered qualitative data from other units of the firm such as the Technological Unit, which is responsible for the development of the payment platform and Customer Service Representative Department. These qualitative data and other secondary data provide some reliable results and interesting findings.

3.3 Validity and Reliability

The reliability of data used for empirical analysis and hypothesis testing was assessed. The reliability of the data was assured by the use of Cronbach's alpha (numerical value of 0.5 is considered appropriate to show consistency). For our data, the alpha value for risk evaluation is 0.83 and risk mitigation is 0.57 respectively. These alpha results serve as a strong base for the reliability of the data measured. Collinearly diagnosed was tested by the use of Variance Inflation Factor (VIF). A value of more than ten is considered to be a high collinear. In this study the VIF for risk evaluation is 4.8 and risk mitigation is 2.13 respectively and this show that collinear was not a serious problem. The hypothesis formulated for the study was tested by cross-sectional data with the use of statistical software SPSS 13.1. Descriptive statistics and Pearson correlation were generated between variables. Interviews carried out among top managers and employees of the risk management department served as a base to explain the quantitative findings and its validity.

4 Results and Discussion

4.1 Empirical Results

This section reports the statistical analysis of our data. Table 1 reports a summary statistics and Pearson correlation between all variable used. The dependent variable is efficiency of the risk evaluation (ERE). The independent variables include

Table 1 Descriptive statistics and pearson correlation: efficiency of risk evaluation (ERE)

Variables	Mean	SD	2	3	4	5	6	7	8	9
1. ERE	5.20	1.414	0.361**	0.700**	−0.175	−0.033	0.547**	−0.057*	−0.265*	−0.278**
2. ININ	5.22	1.292	1	0.389*	0.302*	−060	0.462**	0.424**	−0.355*	−0.254**
3. CFIT	5.38	1.052		1	0.026	−0.034	0.648**	0.006	−0.488*	−0.382**
4. BDP	3.75	1.175			1	−0.134	−0.290*	0.523**	0.147	0.256
5. DM	3.67	1.223				1	0.226*	0.108	0.335*	0.428*
6. MVS	4.15	1.452					1	−0.155	0.422**	−0.568**
7. STI	4.10	1.059						1	0.0315	0.515
8. BPPE	3.99	1.648							1	0.231
9. VPWE	3.79	0.892								1

Note Sample size = 100, $*p < 0.05$, $**P < 0.001$ (2- tailed)

information insecurity (ININ), card fraud & identity theft (CFIT), breakdown of data processor (BDP), database malfunction (DM), malfunction of verification system (MVS), security of technical infrastructure (STI), breakdown of plants and physical equipment (BPPE) and violation of procedure and working ethics (VPWE).

As shown in Table 1, CFIT has the highest mean (5.38) among eight independent variables, indicating that customers believe card fraud and identity theft is the essential risk during the whole payment processes. Also CFIT has the highest correlation coefficient with the dependent variable at 0.700 at $p < 0.01$ (2- tailed). Information insecurity (ININ) and violation of procedure and working ethics (VPWE) are also significant to efficiency of the risk evaluation (ERE) at 0.360 and 0.265 respectively. Also malfunction of verification system (MVS), security and technical infrastructure (STI) and breakdown of plants and physical equipment (BPPE) are significantly related to risk evaluation with correlation coefficient of 0.547, 0.057 and 0.265 respectively. Only two variables (BDP and DM) are not considered significant by sampled data. There is the probability that information would be leaked into unauthorized hands. As we proposed earlier, when firms are able to evaluate the kind of risk within the whole process of the electronic payment system, it will serves as a base for effective risk management planning, forecasting and mitigation. The empirical results that six variables among eight are significant strongly support H1 states that a comprehensive risk evaluation of the whole process has a positive impact on the efficiency of risk management.

Table 2 reports the descriptive statistics and Pearson correlation of variable used to test the firm's risk mitigation practices. The dependent variable is the firm's commitment to the practice of its risk mitigation policies (PRMP). The independent variables include system security evaluation tools (SSET), risk management education (RME), back-up system availability (BUSA) and monitoring system for transaction (MST). From Table 2, we can see that BUSA has the highest mean (4.76), showing that back-up system has been widely used to mitigate the risk. In contrast, the lowest mean of MST (2.63) indicates that monitoring system for transaction has seldom been used for risk reduction. Meanwhile, system security evaluation tools (SSET), risk management education (RME) and monitoring system for transaction (MST) have a significant correlation with the dependent variable at 0.422, 0.312 and 0.290 respectively. Risk management education (RME) may have

Table 2 Descriptive statistics: risk mitigation

Variables	1	2	3	4	5
1. PRMP	1	0.42**	0.188	0.31**	0.290**
2. SSET		1	0.034	−0.018	0.506**
3. RME			1	0.147	0.116
4. BUSA				1	−0.129
5. MST					1
Mean	4.13	3.14	3.18	4.76	2.63
SD	0.640	0.985	0.744	0.955	1.095

**P< 0.05, ** P< 0.001 (2-tailed)*

positive relationship to dependent variable though it is not significant to PRMP in our sample. The empirical results that three all four independent variables are significant to risk mitigation support our hypothesis 2 (H2) in which the practice of a firm's risk mitigation polices during the whole process serve as a positive impact on risk reduction.

4.2 Discussion

The aim of this paper is on the management of risk in electronic retail payment systems in an emerging economy. We analyze how the sampled firm evaluates and mitigates risks within the whole payment process. Our empirical results show that the risks coming from the whole process rather than one or two process are crucial to the risk evaluation. Among the enormous amount of risk associated with the whole electronic payment system, some are of high priority to both the firm and their clients. For this reason a proper risk mitigation frameworks is put in place to deal with this situation. Within this framework, a proper outline of how risk is evaluated in selected domain of interest with the consideration of client complaints and grievances. Policies are therefore maps to mitigate these risks. We found out that the firm had a positive commitment towards the mitigation of risk within the whole process of the payment system. Furthermore the firm has a specialized risk management department that set out policies and strategies to mitigate and build risk awareness within the organization through the introduction of appropriate risk management education and internal risk policy. Lastly this firm also prepared risk reports for both stakeholders and board members. As the CEO of Afric Xpress said for clients to gain confidences in the patronage and usage of their payment system, risks involved in the whole payment process should to be mitigated to the minimum.

According to Confucius "Virtue is not left to stand alone. He who practices it will have neighbors". Relating this piece of saying to our research objective we went a step further to find out whether these mitigation policies only exist on paper or are being implemented. It was found out that the firm implemented plans outlined as timely as possible through the introduction of additional measures to control fraud and identity theft. In addition, they provided hotline phone call services for clients to send in their complaints. Furthermore, risk management training sessions are provided for employees of different business units. The payment platforms like "tier 111 facilities" and "N + design" are relatively secured, which provide an uninterrupted power supply and internet connection. Also international standard like ISO8583 serves as a feat for the firm in the integration of its platform security and support system. One most important resource in these business activities is information including both personal details of clients and financial transaction. Therefore been an operator or provider of this platform, the firm has to take good care of information at his disposal in other not to result in legal suit and other cost related activities that reduce the revenue of the firm. Through the case study of Afric Xpress, we suggest some areas that can contribute to the effective management of risk in these payment systems.

Firstly, with the promotion of these electronic payment solutions, firms or providers deem it a responsibility to protect client's information. By doing this, the firm's loyalty is at stake here and providers need to have access to their client's information even in the time when most operating servers shut down during an industrial or a natural accident. When this arises there is the need for a back-up system from which clients' information can be retrieved and made available to clients at any point in time. AfriXpress, a provider company in Ghana, has a secured encrypted server communication that is being controlled by authenticating, authorization and accounting, providing virtual private networks and critical access paths within its partners. Tier III facility is relatively a good data center but we suggest a tier IV facility been in place since it performs better than tier III and can meet the new trend of computing technology and data security. The firm to save back up of data accumulated over time should adopt updated technology such as cloud storage to meet the changing of the customer needs.

Secondly, we suggest risk workshops and training sessions be conducted frequently within the organization and its partners to refresh their minds about the rules and regulations governing the discharge of their operational duties. When all partners have an idea about the uncertainties and the measures to deal with risk, they are guided by it in their duties. This training session not only serves as an interaction between partners, but also can serves as a platform on which other forms of risk which affect only the clients can be brought up for the providers to deal with it. When knowledge and skills of the employee's increase, minor mistakes will be avoided drastically, and rules and regulation will be adhered to. Risk education should be given to both employees and customers of these payment solutions. And this will help to reduce risk such as identity theft and card fraud since customers have little knowledge about how to protect their information. On the other side it helps the firm to gain a good brand image and trust in the eyes of their clients.

Finally, we suggest a more proactive approach should be adopted towards risk management. Firms should try their best to anticipate and control risks before they happen. This can be possible when firm consider identification and risk assessment as a coherent body. Our findings show that their interactive processes can help identify risk and the extent to which they affect firms' activities. Some corresponding strategies should be put in place to mitigate these risks. Meanwhile, words should be backed with action in these activities that should be done more frequently since new technology will lead to new form of risks.

5 Conclusion

This study attempts to examine the extent to which electronic payment solution provider evaluates the risk and uncertainties associated with the discharge of their business activities and whether risk mitigation policies are practiced. The research approach was both a qualitative and quantitative methods. Related data and information were acquired from a provider firm in Ghana through questionnaires and interviews. The statistical findings show significantly that the firm is committed to evaluating and mitigating the risks it faces. The hypotheses established for this study were supported by our data. Given the nature of the sampled firm for the study, generalized implications are discussed and some suggestions are given to the sampled firm as well as other firms in retailing payment industry.

Acknowledgments This research is sponsored by The National Social Science Fund of China (No. 12BGL123).

References

1. Forsyle SM, Shi B (2003) Consumer patronage and risk perception in Internet shopping. J Bus Res 56:867–875
2. Amit R, Zott C (2001) Value creation in E-business. Strateg Manag J 22:493–520
3. Doherty NF, Ellis-chadwick FE (2003) The relationship between retailers targeting and E-commerce strategies: an empirical analysis. Interest Res 13:170–182
4. Khai SL, Soo JT E-retailing versus physical retailing. A theoretical model and empirical test of consumer choice. J Bus Res 56:877–885
5. O'Neil O (2002) A question of trust. Cambridge University Press, Sage Publication, Oaks
6. Marilyn G, Miklos V (2002) Electronic commerce: security, risk management and control, 2nd edn. McGraw Hill
7. Andersen TJ (2006) Perspective on strategic risk management. Copenhagen Business School Press, Copenhagen
8. International Strategy For disaster Reduction (ISDR) (2004) Living with risk: a global review of disaster reduction intuitive, vol 1, p 454
9. Internet Banking Handbook (2001) Federal reserve board of Chicago's office of comptroller of the currency
10. Sjoberg LBE, Moen T Rundmo (2004) Explaining risk perception: an evaluation of the psychometric paradigm in risk perception research. Rotunde, Norwegian
11. Royce AS, Bruce CS (1999) Approaches to social research. NY Oxford University Press
12. Yin RK (2009) Case study research: design and methods, 4th edn
13. Eisenhardt KM (1989) Building theories from case study research. Acad Manag Rev 14 (40):532–550
14. Pettigrew A (1998) Longitudinal field research on change: theory and practice. Paper presented at National Science foundation conference on longitudinal research methods in organizations, Austin
15. Campbell DT (1995) The informant in quantitative research. Am J Sociol 60:339–342

Incentive System for High-Level Talents Should Embark on the Institutionalized Road Based on Demand in Enterprise

Cheng-long Yuan and Qiao Hong

Abstract With the advent of the knowledge economy era, high-level talents in companies become the main driving force to promote enterprise development and innovation. As high-level talents has its own particularity, how to design a set of norms, continuous incentives system to keep, encourage them and help the exiting incentives system out of the single, blind, no system situation, which should cause the attention of managers. This article from the demand perspective, using the method of qualitative research, integrating the current incentive system and making innovation, then establishing a incentive system based on the needs of employees.

Keywords Dynamic incentive · Employee needs · High-level talents · Incentive system

1 Introduction

In order to meet the need of the development of today's enterprise, humane management model is widely accepted. As the importance aspect of framework management is directly related to the level of operating efficiency. Enterprises in order to enhance the power of their own operation, which are making a lot of adjustments and innovation.

Mallow's hierarchy of needs theory let us have a clear understand for need level. Now companies make the wage system for their company also pay more attention to this. On the one hand, as employees at different levels led to the excitation level varies, when setting the ratio of excitation level are very different [1]. For junior staff is still based on material incentive, but high-level employees is mainly based on respect and self-motivation [2]. On the other hand, at the same level of incen-

C. Yuan · Q. Hong (✉)
Department of Management Science and Engineer,
Shan Dong Normal University, Jinan, China
e-mail: 863149165@qq.com

E. Qi (ed.), *Proceedings of the 6th International Asia Conference on Industrial Engineering and Management Innovation*,
DOI 10.2991/978-94-6239-145-1_77

tives also vary. For different people, there is great different emphasis for partial incentive, especially for high-level talent.

The system of flexible working hours let the employee have time autonomy, so the assessment system now aim at the quality of work and results, rather than time [3].

Employee benefits and partial pay to choose from. Companies when setting wage system, in addition to a fixed salary, and also according to those who can afford little or more while allowing employees to participate in selecting the pay structure and benefits [4].

High quality of leader-member exchange in the enterprise are attracting more and more attention. A good relationship between leader and staff can create harmonious atmosphere, increase civic behavior of employee and enhance organizational commitment, then can improve organizational performance [5]. Therefore managers now strengthen the art of their management and attaching great importance to establish a good relationship with their subordinates [6].

In addition, regular job rotation, more training opportunities, career planning, broad-band compensation and so on are accepted by more and more enterprises to make adjustment of incentive mechanism according to staff needs [7]. So, mangers take the demand of high level employees into consideration when design system of incentive, which will become a trend. By using these method, enterprise can make incentive system more targeted and reduce the cost of enterprise.

The incentive system construction was referred to the enterprise strategy development schedule. Institutional construction of enterprises gradually matured from daily check to high performance work systems. Specifically, in the daily evaluation, assessment from the indicators select, assessment of personnel selection, implementation of the assessment process, to feedback after the summary from start to finish is a continuous process of institutionalized patterns [8, 9]. In high-performance work systems, in order to achieve the overall system efficiency and orderly, the system from recruitment and hiring, training and development, to compensation and performance and so on. All aspects of the links need to optimize and adjustment [10]. So that the system of enterprises should embark on the road to determine the various aspects of the trend.

By summarizing the research about the high level of talent, this article conclude its particularity as follows:

(1) High personal qualities and accomplishment. Most high-level talents in companies have received better education and mastered professional management knowledge, technical classes, who using high-quality as the basis to create value for the enterprise [11].
(2) The results of their work are difficult to quantify. High-level personnel work tend to have long-term and difficult to measure the results, but most still require community support to complete. Therefore, the lack of improvement in the assessment index system and rear excitation current incentive systems are mostly the result of a lack of flexibility and elasticity, poor incentive effects. Their demand was not met, satisfaction is not high [12].

(3) Get reward pay more attention to their own needs met. The High-level talents in the work, getting paid more attention to their demand. As a result, they demand reward more varied and types. Their incentive to achieve a sense of fairness needs are met by their own needs to achieve.
(4) Outstanding learning and innovation ability. High-level talents unique feature is continuous learning and making use of knowledge [13]. Through access to expertise, they continue to realize innovative applications and create innovation performance for enterprises.
(5) High mobility. High-level talents is unattainable for company, when the balance of their needs cannot be achieved in the enterprise [14], their organizational commitment will be weakened, loyalty will reduce and the inherent relationship between psychological contract will rupture, causing a high turnover of employees rates to businesses and increase the cost of employee turnover [15].

2 Methodology

2.1 The Sample Selection and Data Collection

The study sample sources form four companies 10 executives and 50 MBA students in reading, which accounted for 74.6 % male, 25.4 % female. Involves multiple sectors of real estate, banking, trade, etc. From the view of departments, there are human resources marketing department, finance department, purchasing department, ministry network.

In this paper, we use the method of interviews and collect key word form network. During the interview process, we integrated depth interviews and focus interviews in two ways. Specifically, we use depth interviews for 10 executives using semi-structured questionnaires, limiting time to 20–30 min. Conduction around the following three questions:

(1) How do you like the current enterprise incentive system? Why?
(2) Do you think the current incentive system for high level talents should proceed from demand, and their more options?
(3) Which respect should we set from the high-level talent demand perspective construction of incentive system? What problems should pay attention to?

For the 50 MBA students in reading, we made them into 6 groups using focus group interviews, each group have 8–10 students and discuss how to build incentive system, Each group use 20–30 min and make a record to interviews and focus on group interviews in the process, executives and key information referred to MBA students.

In addition, we use key word search on the web and literature, searching on the library, collecting relevant summary records about the incentive system for high-level talent from the demand aspects.

2.2 *Preliminary Arrangement and Analysis*

First of all, we entry the original records to the computer from the interviews and network, getting the most primitive library projects, including 82 data record form interviews and 57 network record, a total of 139. Then encoded the data, we recorded the general out to refine the phrase form or describe the behavior process with concise language.

Then, merged and classified the code record, which is identical or similar in the project library are to. At present many scholars such as Sun Jinan, Jiao Changquan and Fan Jingli are using data classification method. This paper also use this method to merge 139 original coding data, such as "demand wage" and "enterprises should consider the needs of staff in the setting of wages", "feedback" and "pay attention to feedback" shape and the same record, records are merged and remove irrelevant records, finally get 82 effective record. Then we invited 3 college human resources graduates to make the finial classification. After several rounds of batted they reached an agreement, putting the 82 records into 6 large categories.

2.3 *Reliability Test*

We take the result of classification to consult one associate professor and 2 senior human resources executive in enterprise, testing whether we should put 82 records into 6 categories. According to the principle of at least two agreed can accept. The final result are that there are 49 record 3 people all agree, 2 people pass 29 records, delete 4 records. We keep 78 records and accept the advice of the "evaluation" and "coordination" employee needs merger into "assessment and coordination" employee demand. Finally we divided the 58 records into 5 classes (Table 1).

3 Results

3.1 *The Summary of the Incentive System Based on Employee Demand*

Enterprise dynamic interaction incentive system is based on employee demand that regard demand as the guidance. Through the needs acquisition of employees, evaluating and analyzing the feasibility establishing the needs of staff files, staff demand satisfaction, employee satisfaction survey analysis and other measures to establish incentive system. At the same time, it maintains the dynamic interaction between the enterprises and employees, retaliating mutual communication and close coordination mechanism in the process.

Table 1 Generic analysis result

The serial number	Category	Paraphrase
1	Need acquisition	Understand the needs of employee, through various channels to collect employee needs, let employees to participate in the "demand wage" design, rather than passive acceptance.
2	Assessment and coordination	Evaluate the rationality and feasibility of employee needs, establish a assessment team. Evaluate the results and the staff to do a good job of communication and coordination retain the initial demand of employee and need to quantify.
3	Records build	Mutual combination of employee career development plans, set up the mechanism of employee demand file, keep the real-time performance, do a good job of update and tracking to ensure the effectiveness of the requirements to meet.
4	Meet demand mechanism	According to the need of employees, in the model of satisfy comprehensive material incentive and emotional (psychological) incentives in two ways. Comprehensive for motivation, process and results of incentives to incentive the employees. Ensure that employees can get timely manner to meet demand.
5	Communication and feedback	Satisfaction survey about demand of the staff, understand the employee's sense of fairness, according to the result of communication and feedback to do a good job of improvement.

Through qualitative analysis, we made the result into model, then get the incentive system based on the demand. In the interview process innovative ideas, we put forward to various objects of current researches of scholars as well as excellent enterprise management mode, constructing the dynamic demand of enterprise employees incentives based on the mode of interaction (Fig. 1), and each model has made detailed explanation.

3.2 Need Acquisition

Need acquisition work is the base of the construction of incentive system work. The needs of employees can be divided according to time: short-term needs such as daily demand, weak demand [16]; long-term demand such as mouth demand, quartered mind and occupation development plan. In accordance with the type can be divided into: physical needs, security needs, spiritual needs three dimensional demand.

Liang and Liu (2001) [17] use Mallow's hierarchy of needs theory as the foundation, putting forward three requirements of talents workers: material requirement(MR), (SD)security needs, the knowledge employees, determine its

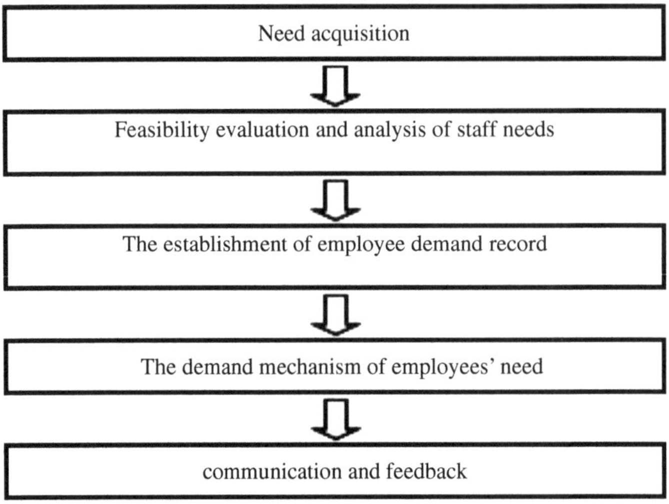

Fig. 1 Incentive mechanism

demand of satisfactory over the same period P point and make adjustment to meet the development needs of staff based on the different stage of constant, and also we follow these good example.

Department of human resources management in the enterprise should make corresponding training and intellectualizing the demand information of employee by the method of the interview questionnaire survey [18]. At last, they can construct system guided by the occupation career planning, combining short-term and long-term needs of the employees.

For the short-term demand of employees like daily or weekly need, employee should fill the demand schedule in time. Employees may have a lot of smaller demand in daily work. But sometimes these demand cannot be achieved because of the enterprise management and considering the factors of self-image. Employees will feel uncomfortable in the psychology, thus affecting working enthusiasm. Real time acquisition staff needs and make the corresponding adjustment can create better working environment for the staff.

The enterprise through continuous acquisition staff needs, which can grasp the direction of employees development and lay a good foundation for the demand motivation.

3.3 Feasibility Evaluation and Analysis of Staff Needs

For the acquisition to the staff demand, short-term demand can be directly deal with by the leader. But, the enterprise must do well feasibility evaluation and analysis work to the medium and long-term demand. This is an important link between the collection of employee demand and establishing employee demand files, directly relating to the staff demand can whether be realized or not. The company can set up a special needs assessment team, assessment of the needs of employees for feasibility, and do the communication and feedback adjustment after work, and ultimately determine the demand category of employee.

3.4 The Establishment of Employee Demand Record

After evaluation and feasibility analysis of the employee needs, the enterprise should classified them in accordance with the categories and find good ways to fufill. Every employee can clearly see their own "3D" demand in each of stage and make communication with, exchange with their enterprise.

Set up employee needs archives need to do the following two aspects of the work:

On the one hand keep the dynamic character, because of the information is timeliness, the value of out-dated information will be weakened or even disappear. The demand information of staffs should update according to the demand of acquisition results, maintaining the timeliness of information.

On the other hand, in order to maintain the interaction, employees can log on demand files to view their demand of archives development and change, which will stimulate the process of feedback to the enterprise, making the enterprise can do adjustment according to employee feedback.

3.5 The Demand Mechanism of Employees' Need

Enterprises should combine the pre-incentive incentive, process incentive and result incentive together to meet the needs of staff in the process, continuing to enhance the organizational commitment.

Pre-incentive is that the enterprise build employee psychological contract from the psychological and physical way, which can let the staff have a negative psychological effect, producing the feeling of debt [19]. The employee will be conscientious to review their own ability and the performance now, judging whether to receive the award [20]. In order to make up the heart of vacancies, employees will actively take measures including training, high work efficiency, learning to achieve phase standards matching the incentive in order to access psychological sense of balance. For the wages, working hours, training and many other quantifiable needs can be addressed by pre-elicitation whole or in part to realize. Zhang Yongmei, Chen Zhen [9] (2005) propinquity of incentive innovation from china traditional culture, analyzing incentive preposition system by using Green Berg's sense of guilt and from the angle of theory, analyzing the debt of reward behavior motivations [21, 22]. Through the analysis and investigation discovery, the staff's psychological feeling of debt can causes stimulation, forming a unique feeling of debt to motivate staff, reflecting the "to return" management philosophy. This is often able to inspire the staff to achieve the goal of greater potential rewards. Therefore the system of pre-incentive for employee is very necessary and critical [23].

We need to determine the time weighting coefficient, which is the employee needs two of the same time period. The emphasis on incentive coefficient, thus letting the staff not only pay attention to the stage of total demand, but also to link the total amount of the upper stage, the formula for calculating is:

$$W = \alpha \times W_1 + (1 - \alpha)W_2$$

W is instead of "total demand wage", α is biased coefficients. W_1 is demand wage of the upper stage, W_2 is demand wage of this stage. α is decided by company as the condition of their own. If the enterprises tend to the former α often set relatively large, on the other hand, set up a relatively small one. This will be the two stages of the demand of wages. Linked to let employees more consider long-term interests. Enterprises can a also be more stage demand wages rise:

$$W = \sum_{i=1}^{n} W_i\alpha_i, \sum_{i=1}^{n} \alpha_i = 1$$

The time to determine the weight coefficient can take the expert method, and then use of AHP to determine the weight. According to the collected employee needs, make corresponding configuration requirements of total wages to employees, so that employees need more targeted and satisfied, thus saving incentive resources.

The work of high-level talents is always difficult to complete, which is a long-term process and result is hard to assess. So process incentive is much more effective. This incentive process we can refer to the process theory of motivation [24]. The process of motivation is a psychological process focuses on people from motivation to action. The main task is to find out some key factors to determine the effects on behavior and make clear the relationship between them. To the behaviors and results for the motivation of knowledge staff, there must be some regulation and intermediate variables, such as internal and external environment, emotional factors [25]. Enterprises can be changed by adjusting the factors such as material conditions. These adjust and change variable will directly affect employee expectation effect and goals to achieve. Which is occurring in the process of employee work, enterprises can from the aspects to motivate employees.

Results incentive is mostly all taken in enterprise, real-time incentives for examination and assessment of employee performance. Li and Wu (2005) [26] referred that the engagement as a way to the assessment of employee performance. In addition to broadband salary, employee stock ownership plan, human capital property right incentive, implementing KPI and 360° performance evaluation indicators, such as improving the result of the traditional incentives.

3.6 Employees Demand Satisfaction Survey

The incentive system of knowledge workers based on demand, enterprises need to do a good job of collection and sorting. Feedback timely access to employees about their demand satisfaction of the way the employees receive from the company to meet their demand and method to improve their satisfaction, such as the opinions and suggestions to adjust the improvement for the next step completes the related work, incentive system to consummate gradually. With these methods, the enterprises can touch the real need of the high-level talents.

4 Conclusion

Aiming at the particularity of the high-level talents, this article use the method of qualitative research and combine with the current domestic and foreign scholars on staff motivation, psychological contract, organizational commitment, organizational performance and so on [27]. we use hierarchy of needs theory and social exchange theory as the foundation, building incentive system based on employee needs. Then, we make high-level talents system of incentive and each link of incentive system has made the summary and the innovation. Do the research, we hope to be able to give some advice to enterprise in developing high-level talents incentive system and let enterprises pay more attention to the design of incentive system, thus improving incentive efficiency and saving the cost of enterprise.

References

1. Sanford S J, Lee C, Bob P (1989) Content, causes, and consequences of job insecurity: a theory-based measure and substantive test. Acad Manag J 32(4):803–830
2. Beau G, Tatum DS, McCoy K, Dorian L et al (2004) Job loss, human capital job feature, and work condition job feature as distinct job insecurity constructs. J Allied Health 33(1)
3. Newman JP (1992) Delay of gratification in Psychopathic and psychopathic offenders. J Abnorm Psychol 101:630–636
4. Dienesch RM, Liden RC (2001) Leader-member exchange model of leadership: a critique and further development. Acad Manage Rev 11:618–634
5. Erdogan B, Bauer TN (2010) Differentiated leader-member exchange: the buffering role of justice climate. J Appl Psychol 95:1104–1120
6. Greenberg J, Ashton-James CE, Ashkanasy NM (2007) Social comparison process in organization. Organ Behav Hum Decis Process 102(1):21–44
7. Hangzhou C, Fang Ll (2012) The psychological contract in the organization. J Manage Sci
8. Julius LX (2007) The employment relationship changes and its influencing factors of empirical examination. Econ Sci 2
9. Zhu X, Wang C (2005) China under the back-ground of research on psychological contract structure of the knowledge-type employees. Prog Sci 23 (chinese)
10. Sun J, Jiao H (2013) The differentiation and integral-on, the trend of study of paternalistic leadership. Psychol Sci 129–2013129 (chinese)

11. Wei J, Zhu G, Liu N (2011) The knowledge employees' psychological contract to research on the effects of organizational citizenship behavior. J Intell 30(2) (chinese)
12. Han Y (2008) Job performance and job satisfaction, organizational commitment and the relationship between goal orientation. J Psychol 40(1):84–91
13. Cai X, Zhao S (2013) loose organization trust influence on job satisfaction, "soft science". J Psychol
14. Being L (2009) Studies of knowledge-based employee com-pen-cation Manage Perspect 5:56–61 (chinese)
15. Liu X, Hao C, Chen J, Cui H (2007) Organizational career management's influence on the professional commitment and job satisfaction. J Psycho 39(4):715–722 (chinese)
16. Long L, Ling W (2002) Studies of organization career management and its effect. J Manage Sci 5(4):61–67 (chinese)
17. Liang Q, Liu X (2001) job performance and job satisfaction, organizational commitment and the relation-ship between goal orientation. J Psychol 40(1):84–91 (chinese)
18. Furien J, Sun JHT (2013) The differentiation and integration, the trend of study of paternalistic leadership. Psychol Sci 1290–1304 (chinese)
19. Epitropaki O, Mar R (1999) The impact of demography on the quality of leader member exchanges an employees' work attitudes and well-being. Formal Occup Organ Psychol 72:237–240
20. Liden RC, Sparrowe RT, Wayne SJ (1997) Leader–member exchange theory: the past and potential for the future. In: Ferris G Research in personnel and human resource management (vol 15, pp 47–119). JAI Press, Greenwich, CT
21. Masterson SS, Lewis K, Goldman BM (2000) Taylor, M. S. Integrating justice and social exchange: the differ of fair procedures and treatment on work relationships. Acad Manage J 43:738–748
22. Stamper CL, Van Dyne L (1999) Diversity at work: do men and women differ in their organizational Citizenship behavior? Perform Improv Q 12:59–76
23. Stamper CL, Van Dyne L (2001) Work status and organizational citizenship behavior: a study of employees. J Organ Behav 22:517–536
24. Alyson B, Julian B, Dupre KE (2014) J Bus Ethics 121:91–106
25. Warr P (2006) Differential activation of judgments in employee well-being. J Occup Organ Psychol 79:225–244
26. Li H, Wu X (2005) From ideal to real: a longitudinal study of the role of implicit leadership theories on LMX and employee outcomes. J Appl Psychol 90(4):659–676 (chinese)
27. Yong J, Wang H (2012) Authorization can assign leadership behavior influence on employee insider identity perception: based on the regulation of self-esteem of the organization. J Psychol 44(10):1371–1382 (chinese)

Prediction of Bauxite Demand Based on Coupled Phase-Space Reconstruction and Neural Network

Xiaojun Yan and Zhiya Chen

Abstract The Chinese over capacity of aluminum causes the obvious elasticity of alumina and it brings great trouble for alumina enterprise in business decision-making. In order to improve the management decisions, the alumina enterprise must carry out the effective predictions of the alumina demands. It will ensure the manufacture and control the inventory. Based on the coupled phase space reconstruction and neural network, we proposed a prediction model of the alumina demand, which first used the raw demand data for the coupled phase-space reconstruction, then trained these reconstructed data with the neural network, finally predicted the alumina demand according to the predicted time. Besides, the alumina quarter demand data at 2006–2014 from a typical aluminum enterprise was used for verifying this prediction model. Results show that this prediction model of the alumina demand is easy to operate and its predicted data is reliable, which can provide theoretical guidance to the scientific and reasonable management decisions.

Keywords Chaos · Coupled phase-space reconstruction · Neural network · Alumina demand prediction

1 Introduction

With the adjustment of construction, the over production of alumina is seriously in China, and the aluminum enterprise has to manufacture flexibly according to the consumers' demand and market change, and it brings huge pressure for these enterprises in decision-making. On the one hand, enterprises should try to satisfy consumers' demand and enhance their satisfaction degree. On the other hand,

X. Yan (✉)
Business School of Central South University, Changsha 410083, Hunan, China
e-mail: xj_yan@chalco.com.cn

Z. Chen
Xidian University, Xi'an 710071, China

E. Qi (ed.), *Proceedings of the 6th International Asia Conference on Industrial Engineering and Management Innovation*,
DOI 10.2991/978-94-6239-145-1_78

enterprises should try to reduce the relative logistics costs because of the demand fluctuation. Therefore, how to effectively forecast consumers' demand, offer scientifically theoretical references for the management decision department, and plan as a whole in the whole enterprise and exert the maximum benefits of various resources in the enterprise will be the important content of the client relation management and the supply chain management.

There are many methods about the demand forecasting at present, such as the regression analysis [1], the Markov prediction [2], the Gray prediction [3], and the artificial neural network (ANN) prediction [4, 5]. The good adaptability, the nonlinearity, and the fault tolerance of ANN have been applied widely, but the train of the network needs sufficient sample space, and the specific forecasting function expression is hard to provide [6]. How to select the forecasting method which needs minimum data and has good fitting nonlinear function is the key problem of the demand forecasting. The phase-space reconstruction technology [7–10] in the chaos theory and the ANN are effective tools to deal with the problems about nonlinear system, and they also could be combined effectively. The phase-space reconstruction technology could display the concealed system state information in the time sequence, so the demand of the original data will be reduced, and the reconstructed data will be fitter for the sequent disposal by ANN.

2 Chaos Theory

The research of the chaos theory by the nonlinear dynamics is most systematic and rigorous up to now. In the nonlinear dynamics, though some definitions of theoretical judgment and standards of actual measurement only define the chaos from the mathematical and physical angle, but they are the base to establish and develop the chaos theory, where the Li-Yorke Theorem is the mathematical definition of chaos with large influences.

In 1975, Li Tianyan and Yorke JA first put forward the word of chaos, and they mathematically define the chaos in the article of "Period 3 means chaos", and it is called the definition of Li-Yorke now.

Li-Yorke Theorem: Suppose $f(x)$ is the continuous self-mapping in $[a, b]$, and if has 3 period points, it has n period points for any positive integer n.

Definition of chaos: Suppose the continuous self-mapping $f : I \rightarrow I$ is contained in R, and I is one subinterval in R, and if the uncountable set S exists and it is contained in I.

(1) S doesn't contain period points.

For any $X_1, X_2 \in S(X_1 \neq X_2)$,

$$\lim_{t\to\infty} \sup|f^t(X_1) - f^t(X_2)| > 0$$

$$\lim_{t\to\infty} \inf|f^t(X_1) - f^t(X_2)| = 0$$

Here, $f^t(\cdot) = f(f(\cdots f(\cdot)))$ denotes t-multiple function relation.
For any $X_1 \in S$ and any period point of f, $P \in I$,

$$\lim_{t\to\infty} \sup|f^t(X_1) - f^t(P)| > 0$$

So, f is chaotic in S.

This definition is based on a set, but it shows the important characters of the chaos. First, countable infinite stable period solutions exist. Second, uncountable infinite stable non-periodic tracks exist. Finally, one unstable non-periodic track exists at least.

3 Modeling Approaches of the NN Forecasting Model of the Phase-Space Reconstruction

3.1 Principle of Phase-Space Reconstruction

The phase-space reconstruction comes from the chaotic dynamics, and its basic principle is to display the system state information containing in the time sequence by the reconstruction technology. The chaotic time sequence signal disposal method is mainly based on the angle of the dynamics system, which is different with the traditional time sequence disposal method which is mainly to transmit and dispose the time sequence, and to acquire the information of the system in the time sequence, Pachard et al. first put forward the phase-space reconstruction technology, and then Taken further developed it and proved it mathematically, and it is called as the phase-space reconstruction theorem (Takens Theorem).

For the dynamic system with n dimensional variables (i.e. a set of n first-order differential equations),

$$\frac{\mathrm{d}x_i}{\mathrm{d}t} = f_i(x_1, \ldots, x_n)(i = 1, 2, \ldots, n)$$

By the method of elimination, it could be transformed as a n-order linear differential equation,

$$x^{(n)} = f(x, x^{(1)}, \ldots x^{(n-1)})$$

After transformation, the track of the system is

$$\overrightarrow{x}(t) = [x(t), x'(t), \ldots, x^{(n-1)}(t)]$$

It describes same dynamic problem, and it will evolve in the phase-space formed by the coordinate $x(t)$ with $(n-1)$-order $x'(t), \ldots, x^{(n-1)}(t)$ and substitute this continuous variable $x(t)$ and its derivative without considering discontinuous time sequence and its $(n-1)$ time-lag displacement.

$$\vec{x}(t) = \{x(t), x(t+\tau), \ldots, x(t+(n-1)\tau)\}$$

Select the time lag τ as the large-scale research of the time sequence, which will ensure the coordinates to be linearly independent. Therefore, if (1) the dynamic system has single one state variable $x(t)$, and (2) its track evolves in the phase-space forming in the time displacement coordinate, so the attractor of the dynamic system could be reconstructed in the new phase-space with the lagged coordinates where its topological character will not be changed.

In the practice, for the appointed time sequence, $x(t_0), x(t_1), \ldots, x(t_i), \ldots x(t_n)$, it will generally be extended to three-dimensional or multiple-dimensional space, which could fully display all information contained in the time sequence, and that is the lag coordinate state space reconstruction method. The above time sequence will be extended as a phase distribution with m-dimensional phase space.

$$\begin{bmatrix} x(t_1), & x(t_2), & \cdots, & x(t_i), & \cdots, & x(t_n), \\ x(t_1+\tau), & x(t_2+\tau), & \cdots, & x(t_i+\tau), & \cdots, & x(t_n+\tau), \\ x(t_1+2\tau), & x(t_2+2\tau), & \cdots, & x(t_i+2\tau), & \cdots, & x(t_n+2\tau), \\ \cdots, & \cdots, & \cdots, & \cdots, & \cdots, & \cdots, \\ x(t_1+(m-1)\tau) & x(t_2+(m-1)\tau), & \cdots, & x(t_i+(m-1)\tau), & \cdots, & x(t_n+(m-1)\tau), \end{bmatrix}$$

Here, τ is the time lag, and each list in the above formula will form one phase point in the m-dimensional phase space, and any one phase point has m-dimensional components,

$$[x(t_i), x(t_i+\tau), \ldots, x(t_i+(m-1)\tau) \quad i = 1, 2, \ldots, n]$$

which denotes the state of certain instant in this system. And the link among phase points would describe the evolvement track of the system in the m-dimensional phase space.

3.2 *Parameter Selection of Phase-Space Reconstruction*

To reconstruct the phase space from the time sequence, the key technology is to correctly select the dimension amount of the embedded space m, and the time lag τ.

The selection of τ is not important principally, but because of the noise and estimation, the selection of proper τ is very important. If τ is too small and the correlation of each coordinate is too strong, tracks in the phase space will extrude to one place, and the information about attractor will not be displayed. If τ is too big, in the chaos and noise, the dynamics state at certain time and the dynamics state at next time will change drastically, so even simple geometric objects seem very complex. In the reconstructed phase space, it is very important to select the time lag τ and the embedded dimension m, and at the same time, this selection is difficult.

3.3 Computing the Delay Time by the Autocorrelation Function Method

The main methods to solve the time lag include the autocorrelation function method and the mutual information method. The autocorrelation function method is selected in this article, and its advantages include that it could ensure the small correlation of various coordinate components in the phase space, and accord with the requirements to reconstruct the phase space, and the computation is simple. The concrete computation method could be described as follows.

For the time sequence of the variable $x(t)$, $x_1, x_2, \ldots, x_n$, the autocorrelation function expression is

$$c(\tau) = \frac{n}{n-\tau} \frac{\sum_{i=1}^{n-\tau} (x_i - \mu)(x_{i+\tau} - \mu)}{\sum_{i=1}^{n} (x_i - \mu)^2}$$

where, τ is the time lag, and μ is the average of the time sequence.

The autocorrelation function denotes the times which subscripts are i, $i + \tau$, and the correlation or the similarity of the movement. The function image of the autocorrelation function about the time τ could be drew, and when the autocorrelation function descends to the first zero value, the value of τ could be the time lag embedded into the phase space. In addition, according to the numerical value experiment result, when the autocorrelation function descends to t $1 - \frac{1}{e}$ of the initial value, the time τ could be the time lag of the reconstructed phase space.

3.4 Computing the Embedded Dimensions by the Cao Method

This method is improved false zero method, for

$$x_i = \left(x_i, x_{i+\tau}, \ldots, x_{i+(m-1)\tau}\right) \in R^m, \quad i = 1, 2, \ldots$$

Define: $a(i,m) = \frac{\left\| x^{(m+1)}(i) - x^{(m+1)}(j) \right\|}{\left\| x^{(m)}(i) - x^{(m)}(j) \right\|}$

where, $x^{(m+1)}(i)$ is the ith phase point in the $(m+1)$-dimensional reconstructed phase space, and $x^{(m+1)}(j)$ is the nearest neighboring point of $x^{(m+1)}(i)$, and $\|\cdot\|$ is the Euclidean distance. The average of all phase points $a(i,m)$ is

$$E(m) = \frac{1}{n - m\tau} \sum_{i=1}^{n-m\tau} a(i,m)$$

$E(m)$ only depends on the embedded dimension m and the time lag interval τ, and to study the change of the embedded dimension from m to $m+1$ in the phase space, define

$$F(m) = \frac{E(m+1)}{E(m)}$$

If $F(m)$ goes to saturation with the increase of m, the value of m is the minimum embedded dimension of the reconstructed phase space.

3.5 *Coupling NN Forecasting*

NN is the adaptive nonlinear system which is associated by concurrent and highly correlative computation disposal unites, i.e. the neural cells, according to certain mode. Though the structure of each neural cell is very simple, but the neural cell system linked by numerous neural cells could implement abundant behaviors. NN has strong nonlinear mapping ability, and because of the concurrent disposal and strong nonlinear mapping ability of NN, the NN could predict the behavior after it is trained by sufficient training samples. Therefore, the phase-space reconstruction technology could be used to confirm the input of the NN model, and the NN forecasting model based on the phase-space reconstruction could be established.

In the practice, according to the embedded dimension m of the chaotic time sequence, $m-1$ could be selected as the node amount of the network input layer, and the node amount of the output layer is 1. The input and the output of the networks respectively are

$$\begin{bmatrix} x(1) & x(2) & \cdots & x(n-(m-1)\tau) \\ x(1+\tau) & x(2+\tau) & \cdots & x(n-(m-2)\tau) \\ \cdots & \cdots & \cdots & \cdots \\ x(1+(m-2)\tau) & x(2+(m-2)\tau) & \cdots & x(n-1) \end{bmatrix}^T$$

$$\begin{bmatrix} x(1+(m-1)\tau) \\ x(2+(m-1)\tau) \\ \cdots \\ x(n) \end{bmatrix}$$

After confirming the input and output structure of the network, the study sample in the known time sequence could be abstracted to compose the training set to train the network until achieving certain error precision.

To further forecast the behavior, the value of $x(n+1)$ needs to be known. Input $[x(n+1-(m-1)\tau), x(n+1-(m-2)\tau), \ldots, x(n+1-\tau)]$ into the trained NN model, the value of $x(n+1)$ could be obtained. In the same way, by changing the input data, the values of $x(n+2), x(n+3), \ldots$ could be obtained.

4 The Application of the NN Forecasting Model Based on the Phase-Space Reconstruction in the Practice

The forecasting example is the alumina demands of one aluminum enterprise from 2006 to 2014, and as seen in Table 1, all data come from the actual survey.

4.1 Reconstructing the Phase-Space

Aiming at the data of demand, the autocorrelation function method and the Cao method are used to compute the time lag and the embedded dimension by the Matlab programming. The time lag computation result of the demand is seen in Fig. 1, and the computation result of the embedded dimension is seen in Fig. 2. From Fig. 1, for the point which first arrives at the zero, because $c(11) = 0.0644$

Table 1 Alumina quarterly demand volume from 2006 to 2011

Year	Demand (10 thousand T)			
	1st quarter	2nd quarter	3rd quarter	4th quarter
2006	108	129	113	120
2007	127	135	137	142
2008	146	150	156	153
2009	157	175	179	171
2010	171	185	179	183
2011	183	192	185	202
2012	204	212	203	198
2013	203	209	212	152
2014	122	159	183	227

Fig. 1 Demand volume autocorrelation function

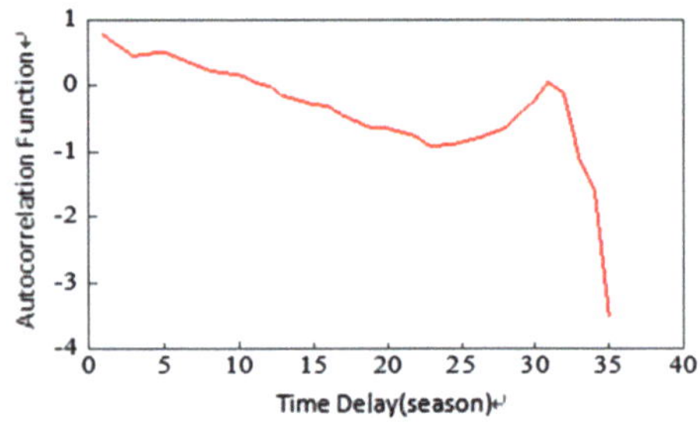

Fig. 2 Demand volume correlation dimension

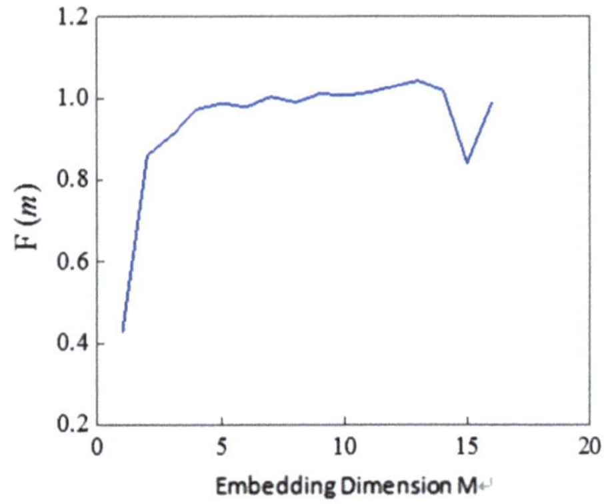

and $c(12) = -0.0221$, if the corresponding time lag τ of the autocorrelation function value which first arrives at zero should be 12, but this value is too big, and after forecasting the time lag of this value in the reconstructed phase space, the value is also too begin the result. Therefore, the descending tendency of the autocorrelation function is adopted to select the time lag τ in the reconstructed phase space. After computing, $c(2) = 0.5863$. After $c(2)$, the autocorrelation function descends slowly with the increase of τ, so in this article, $\tau = 2$. From Fig. 2, when the embedded dimension $m = 4$, the value of $F(m)$ goes to saturation, so $m = 4$.

5 Phase-Space Reconstruction and NN Coupling Forecasting

According to the embedded dimension m (4) and the time lag τ (2), reconstruct the time sequence. Take $m - 1$ as the input node amount of the network, and take the output node amount as 1, and construct one NN model with one input layer, two concealed layers, and one output layer (seen in Fig. 3). Use reconstructed data to train the NN, and the checking error precision is 0.001. Figure 4 is the error change process of the network training and the checking process.

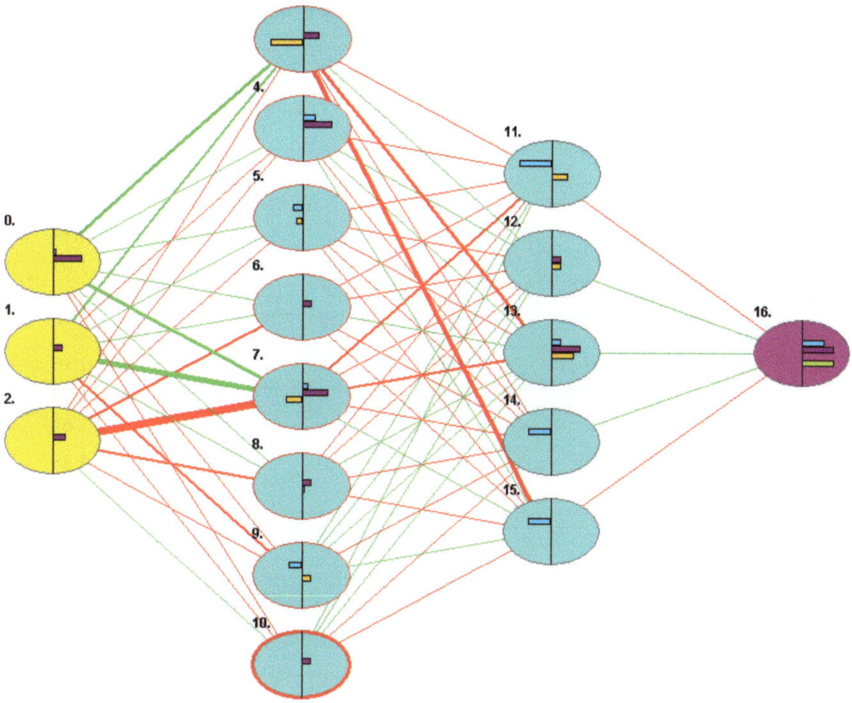

Fig. 3 Neural network structural diagram

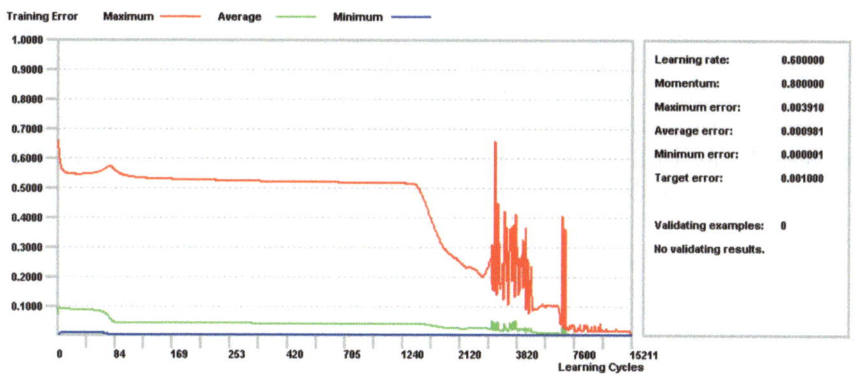

Fig. 4 Network study validation error convergence process curve

When the network is trained successfully, the coupling forecasting model of the phase-space reconstruction and NN is established. Based on this model, forecast the demand, and compute the forecasting values of the demands from 2011 to 2014, and further compute the residuals and relative errors (seen in Table 2).

Table 2 Calculating value of residual and relative error s of original volume and forecasting volume

Year	Quarter	Original volume/ (10 thousand Ton)	Forecasting volume/(10 thousand Ton)	Residual Error/ (10 thousand Ton)	Relative Error/ (%)
2011	1st	183	180.18	2.82	1.54
	2nd	192	188.83	3.17	1.65
	3rd	185	183.63	1.37	0.74
	4th	202	204.10	−2.10	1.04
2012	1st	204	198.93	5.07	2.49
	2nd	212	205.91	6.09	2.87
	3rd	203	200.64	2.36	1.16
	4th	198	194.53	3.47	1.75
2013	1st	203	198.98	4.02	1.98
	2nd	209	206.11	2.89	1.38
	3rd	212	208.66	3.34	1.58
	4th	152	141.84	10.16	6.68
2014	1st	122	126.09	−4.09	3.35
	2nd	159	162.44	−3.44	2.16
	3rd	183	179.28	3.72	2.03
	4th	227	226.88	0.12	0.05

5.1 Analysis of Forecasting Result

The average relative error of the forecasting demand is 1.21 %, and good effect could be achieved. At the same time, the mutation factor could not be display well in the NN forecasting model based on the phase-space reconstruction. For example, the demand descends drastically in the fourth quarter of 2013, and the big gap between the forecasting data with the actual data in the fourth quarter of 2013 obviously exist, and the relative error is 6.68, which far exceeds the average relative error 1.21 %.

6 Conclusion

The demand of alumina is influenced by multiple layers and factors, to exactly forecast the demand, large information needs to be collected, and various influencing factors should be considered. In this article, the coupling ANN model based on the phase-space reconstruction technology in the chaos theory is established, and the forecasting result shows that this model has certain advantages to forecast the demand of alumina. The chaos is a kind of phenomenon and behavior, and NN is a special nonlinear network structure to approach the behavior, and both have their own advantages, and the phase-space reconstruction technology in the chaos theory is introduced in NN to establish the chaotic NN model which could strengthen the ability of NN to study and approach the nonlinear mapping.

References

1. Li SZ, Li DM, Tang WX (2003) Quantitative analysis for single index efficiency of Chinese regional science output. Sci Sci Manage S&T 3:14–18
2. Liu LF (1995) The application of forecasting of Markov chain in enterprise management. Sci Manage 15(2):58–59
3. Peng K (2009) The transaction of railway container based on grey forecasting method (in Chinese). Railway Procurement Logist 6:47–8
4. Zhu QH, Zhang G (2003) Enterprise knowledge management model based on artificial neural network. Sci Sci Manage S&T 8:32–4
5. Yang X, Li JH, Cheng BY (2004) Scientific demand forecasting and research based on artificial neural network. Sci Sci Manage S&T 12:29–32
6. Zhang LM (1993) Artificial neural networks models and its applications. Fudan University Press, Shanghai
7. Li TY, Yorke J (1975) Period 3 implies chaos. Amer Math Monthly 82:985–992
8. Lv JH, Lu JA, Chen SH (2002) Chaotic time series analysis and its application. Wuhan University Press, Wuhan
9. Packard NH, Crutchfield JP, Farmer JD et al (1980) Geometry from a time series. Phy Rev Lett 45(9):712–716
10. Takens F (1981) Detecting strange attractors in turbulence. Lecture notes in mathematics. Springer, Berlin

Enterprise Management Decision-Making Evaluation Model and Its Empirical Study

Xiaojun Yan and Zhiya Chen

Abstract Management decision-making is important to enterprises. To evaluate enterprise decision-making, appropriate evaluation indexes should be established and quantified scientifically. In this paper, the management decision-making evaluation model including the management decision-making quality and speed was built, and results predicted by this model agreed well with sample data from 201 Chinese enterprises.

Keywords Management decision-making · Evaluation model · Empirical study

The decision-making is a key component of enterprise management [1], and main activity of managers [2], which runs through the functions of managers. The incessant changes of internal and external environments make the enterprise decision-making an increasingly complicated task [3]. Enterprises should response quickly and precisely to the changing internal and external environment. Quick and precise decision-making and implement is the precondition of success [4]. Wrong decisions, especially wrong management decisions, make enterprises not only pay the price, but also lose opportunities. Therefore, an effective management decision-making is crucial for enterprises [5].

To evaluate the quality of the enterprise management decision-making and ensure the decisions of managers is important for the enterprise respond quickly and precisely to the dynamic changing internal and external environments, and quire the scientific evaluation of management decision-making. In this paper, the enterprise management decision-making evaluation model was established, and the appropriate evaluation indexes were chosen for the quantitative study with scientific methods.

X. Yan (✉)
Business School of Central South University, Changsha 410083, Hunan, China
e-mail: xj_yan@chalco.com.cn

Z. Chen
Xidian University, Xi'an 710071, China

E. Qi (ed.), *Proceedings of the 6th International Asia Conference on Industrial Engineering and Management Innovation*,
DOI 10.2991/978-94-6239-145-1_79

1 Definition and Connotation of Management Decision-Making

The decision making could be classified by different perspectives. Simon [6] divided the decision-making into the programmed and non-programmed decisions. The programmed decision means the conventional and revolving decision processed by routine program. The non-programmed decision means decision which depends on manual judge, has no fixed pattern, and never happened in the past. Harrison and Pelletier [2] divided enterprise decision-making to strategy decision, management decision and business decision. Strategy decision addresses long-term and overall issue, management decision reallocates resources to implement strategy decision, and business decision means decisions for the enterprise daily production, marketing, and distribution. Teng and Calhoun [7] divided the enterprise decision-making to management decision and operation decision. Management is non-daily decision making, which is generally not structured, ambiguous, hard to process by computer, and depends heavily on human judge.

Based on previous research, this paper believes it would match the research need and enterprise practical situation to divide enterprise decision to operation decision and management decision. The operation decision is enterprise daily and revolving decision and management is non-daily decision making, which is generally not structured, ambiguous, hard to process by computer, and depends heavily on human judge. In practice, management decision is usually made by middle-senior management team and operation decision is made by basic level staff.

2 Establishment of Management Decision-Making Evaluation Model

To evaluate management decision, and to ensure managers' decision respond appropriately to competitive environment, scientific management decision-making evaluation model is required.

Rolf [8] thought the decision evaluation included transparency, consistency, accuracy, improvement, adaption, and speed. Transparency means the clarification degree in the process of decision making. Consistency means to provide same or similar information to achieve same decision. Accuracy means how correct the decisions are on average level. Improvement means the possibility of improve the decision as time goes on. Adaption means the decision can be adjusted to fit the current policy. Speed means how long it takes to made decision. When evaluating decision, the decision maker can give different weight to those respects based on different situation.

Simon [6] paid much attention to two key measurement standards—efficiency and result. Efficiency evaluates the process, while result measures how successful the decision is. Brown et al. [9] thought the quality of decision cannot be assessed only by the final result. Hess and Wells [10] proposed to use effectiveness (decision quality and accuracy) and efficiency (endeavor in process and time cut) to evaluate enterprise decision-making. Desanctis and Gallupe [11] thought the result can be measured by decision quality, decision efficiency, decision satisfaction and process satisfaction. Speier et al. [12] came up with decision accuracy (decision quality) and decision time (decision speed) to measure decision, while Leidner and Elam [13] thought speed and analysis depth should be used. Rajagopalan et al. [14] think the decision output quality can be assessed by speed, acceptability and adaptation to change. Teng and Calhound [7] used several indexes for thorough measurement, like usage of information (information timing, data quantization, and information overload), decision efficiency (forecast accuracy, time horizon, selectivity, and speed), and organizational decision making process (decision communication and participation). Poston and Speler [15] used quality and time to measure decision during individual decision-making laboratory research. Decision time means the time (minutes) to accomplish decision task. Decision quality is measured by deviation between detailed solution and strategic solution, like expected value or effectiveness maximization. Harrison and Pelletier [2] proposed to use external environment acceptance level.

Management decision making performance depends on the quality and how the decision is implemented, which can be measured by two variables, quality and commitment [16]. But, the decision making speed is also critical index in complicated dynamic environment. Vroom [17] believed decision efficiency is generally defined as decision quality and time required, which is decision speed. Trull [18] thought the overall decision success depends on decision-making quality and the implement of the decision. Yan [19] thought decision making has two respects, timeliness and accuracy, and business leaders should be able to make correct decision on operation management.

Based on the above research, this paper believes management decision making can be assessed by two variables, decision making quality and speed. Management decision making quality, which means the contribution to achieve the goal with decision. There are a lot of factors effecting the quality. Those include decision information, decision makers' interests and acceptance level of decision, which mainly divides to environmental factors and process factors. Decision making speed. Eisenhardt [20] defined speed simply as the time required in the decision making process. The decision speed is mainly restricted by the decision maker, decision organization, and corresponding decisional operational environment. Decision making speed is usually related to complexity of the problem, dynamics of external environment and definition of decision process. Appropriate speed means good following performance of the problem, which makes the decision to adjust to the changing internal and external environment. Based on above consideration, this paper builds the model of management decision making with two variables, decision making quality and decision making speed, as shown in Fig. 1.

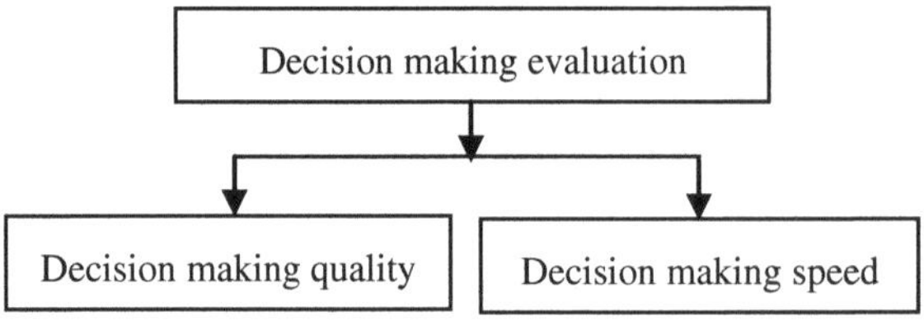

Fig. 1 Decision making evaluation model

3 Variable Measurement and Date Collection

This paper takes Likert5 rating scale for measurement. Decision making is evaluated and measured by two variables, decision making quality and speed.

3.1 Variable Measurement

3.1.1 The Measurement of Decision Making Quality

Decision making quality means the contribution level of management decision to achieve organizational goal. Based on the studies of Schweiger [21], and Dooley and Fryxell [16] used 6 factors to measure the decision quality: whether the decision is based on sufficient information, whether decision is based on valid preconditions, whether decision is in line with the business goal, whether decision is made on current financial positions, and whether the decision contributes to overall strength improvement. Amason [22] used 3 factors: strategic decision overall quality, whether strategic decision achieves the expectation, and the decision's effect to business. The scale tale this paper uses takes reference of Dooley and Fryxell [16], as shown in Table 1.

Table 1 Measurement set of decision making quality

	No	Measurement
Decision making quality	GLZL-1	Mid and high management decision is made on sufficient information
	GLZL-2	Mid and high management decision is in line with development goal
	GLZL-3	Mid and high management decision compares multiple solution
	GLZL-4	Mid and high management decision takes into consideration of pros and cons of the action plan
	GLZL-5	Company uses multiple criteria to make decision

Table 2 Decision making speed measurement set

	No	Measurement
Decision making speed	GLSD-1	New product and service introduce speed
	GLSD-2	Product and service improvement speed
	GLSD-3	New technology adoption speed
	GLSD-4	New market development speed

3.1.2 Decision Making Speed Measurement

Eisenhardt [20] defined speed simply as the time cost in decision making process. Baurn and Wally [23] used 3 factors to measure the speed, procurement decision, new product introduce and new technology adoption decision. The scale table of this paper takes reference of the speed measurement table of Baurn and Wally [23]. The decision making speed in this paper mainly means the speed in decision making process, excluding the implement speed, as shown in Table 2.

3.2 Data Collection

This paper chose Chinese mainland enterprises as the empirical study subject. Subjects are the middle and senior managers with 3 years' service or above and collect a large number of samples after measurement. Questionnaire is issued from August 2012 to August 2014. The release and collection is made by three methods, technology, email and onsite research. Valid questionnaire collected is 201 copies in total.

4 Data Analysis and Result

4.1 Reliability and Validity Analysis

Two hundred and ten samples conform to the Gaussian distribution requirements. Reliability can be estimated by αin Cronbach. The value of α in this paper's variables are above 0.7, which show the scale tale is in good reliability, as shown in Table 3.

The KMO SHOW of 0.871 means strong colinearity in measured subjects. Bartlett sphere test shows x^2 is 692.121 ($P < 0.001$), which means strong colinearity in measured subjects fitting for factor analysis. Meanwhile, exploratory factor analysis test shows scale tale is good structure validity.

Table 3 Variables reliability coefficient

	Measurement	Cronbach α
Decision quality	5	0.826
Decision speed	4	0.779

4.2 Confirmatory Factor Analysis

Validity result of decision making quality and speed measured by AMOS7.0, and confirmatory factor analysis result show the standard loads of tested subjects to potential variable are mostly above 6.0, which means strong consistency and convergent validity. Variance of mean from potential variable indicates that the quality is 52.19 % and speed is 58.59 %, both above 0.5, which means the potential variables has convergence validity.

Confirmatory factor analysis diagram of by AMOS7.0 to analyze management decision making variables is indicated as Fig. 2. The standardization load of tested subjects are above 0.4, and the fitting indexes are all within the acceptance level (X2/df = 1.85, RMSEA = 0.07, NNFI = 0.93, CFI = 0.94, IFI0.94, NFI = 0.89), which means the measurement model of decision making has reasonable construct validity.

From the forgoing, confirmatory result shows decision making can be measured by two variables, decision making quality and speed.

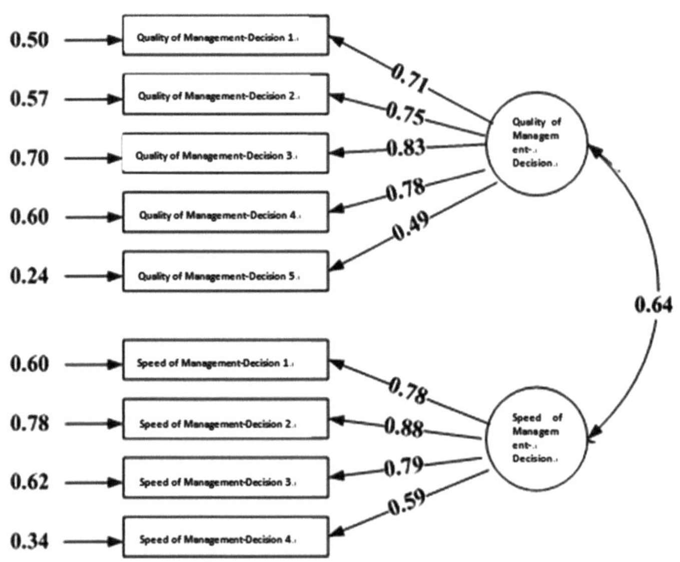

Fig. 2 Confirmatory factor analysis diagram of decision making variables

5 Conclusion

Managers must make decisions to do their duty, and the decision making provides the process to match the enterprise internal power and external business environment. To evaluate the management decision decisions of managers respond suitably to the competitive environment, an effective management decision-making evaluation model was built. The confirmatory study shows the decision making results can be measured by the decision making speed and decision making quality.

References

1. March JG, Olsen JP (1976) Ambiguity and choice in organizations. Universitetsforlaget, Oslo
2. Harrison EF, Pelletier MA (2000) The essence of management decision. Manage Decis 38 (7):462–469
3. Wadhwa S, Saxena A (2007) Decision knowledge sharing flexible supply chains in KM context. Prod Plann Control 18(5):436–452
4. Maclaughlin DI (1995) Strengthening executive decision making. Hum Resour Manage 34 (3):443–446
5. Huber GP, Mcdaniel RR (1986) The decision-making paradigm of organizational design. Manage Sci 32(5):572–589
6. Simon HA (1960) The new science of management decision. Harper and Brothers, New York
7. Teng JTC, Calhoun KJ (1996) Organizational computing as a facilitator of operational and managerial decision making: an exploratory study of managers' perceptions. Decis Sci 27 (4):673–710
8. Rolf CC (2005) Beyond accuracy: how models of decision making compare to human decision making. Master dissertation, Lund University, Lund, Sweden
9. Brown RV, Kahr AS, Peferson C (1974) Decision analysis and overview. Rinehart and Winston, New York
10. Hess TJ, Wells JD (2002) Understanding how metadata and explanations can better support data warehousing and related decision support systems: an exploratory case study. In: Presented at the proceedings of the 35th Hawaii international conference on system sciences, Hawaii
11. Desanctis G, Gallupe RA (1987) Foundation for the study of group decision support system. Manage Sci 33(5):589–609
12. Speier C, Valacich JS, Vessey I (1999) The influence of task interruption on individual decision making: an information overload perspective. Decis Sci 30(2):337–359
13. Leidner DE, Elam JJ (1995) The impact of executive information systems on organizational design, intelligence, and decision making. Organiz Sci 6(6):645–655
14. Rajagopalan N, Rasheed AMA, Datta DK (1993) Strategic decision process: critical review and future directions. J Manage 2:349–384
15. Poston RS, Speler C (2005) Effective use of knowledge management systems: a process model of content ratings and credibility indicators. MIS Q 29(2):221–224
16. Dooley RS, Fryxell GE (1999) Attaining decision quality and commitment from dissent the moderating effects of loyalty and competence in strategic decision-making teams. Acad Manage J 42(4):389–402
17. Vroom V (1973) A new look at managerial decision making. Organiz Dyn 1(4):66–80
18. Trull SG (1996) Some factors involved in determining total decision success. Presented at the managerial decision-making process, Boston

19. Yan JY (1990) Philosophizing of enterprise operation management-decision (in Chinese). Social Sci Jiangxi 2:77–79
20. Eisenhardt KM (1989) Making fast strategic decisions in high-velocity environments. Acad Manage J 32(3):543–576
21. Schweiger DM, Sandberg WR, Ragan JW (1986) Group approaches for improving strategic decision making: a comparative analysis of dialectical inquiry, devil's advocacy, and consensus. Acad Manage J 29(1):51–71
22. Amason AC (1996) Distinguishing the effects of functional and dysfunctional conflict on strategic decision making: resolving a paradox top management teams. Acad Manage J 39 (1):123–148
23. Baurn JR, Wally S (2003) Strategic decision speed and finn performance. Strateg Manage J 24 (11):1107–1129

The Dynamic Analysis of Development: Indigenous Innovation or Technology Importation Based China Provincial 2000–2011 Panel Data

Yuan-chun Yu, Xin Gu and Yi-jun Chen

Abstract Based on the panel data of 30 provinces in 2000–2011, this paper researched the relationship among the independent innovation, the importation of technology and the economic development. The study found that the indigenous innovation played a significant role in promoting China's economic growth, but the role of technology importation is not obvious. And there was no interaction between the indigenous innovation and the importation of technology. But the indigenous innovation and the importation of foreign technology can replace fixed asset to stimulate economic growth. On the basis of the researching conclusion made above, this paper advocated that China should give priority to independent innovation which is supplemented by technology import.

Keywords Economic growth · Indigenous innovation · Panel model · Technology importation

1 Introduction

Chinese national development strategy has been constantly adjusted. From the "march to technology and science" in the 1960s, "science and technology are primary productive forces" in the 1980s, "prospering the nation with science and education" in 1990s, "enhance China's capacity for independent innovation and make China an innovative country" proposed by the 17th CPC National Congress, to the strategy of "innovation-driven development" put forward by the 18th CPC

Y. Yu (✉) · X. Gu
Department of Business School, Sichuan University, Chengdu, People's Republic of China
e-mail: yuanchun1981@163.com

Y. Yu · Y. Chen
Department of Economics and Management, Sichuan University of Science and Engineering, Zigong, People's Republic of China

E. Qi (ed.), *Proceedings of the 6th International Asia Conference on Industrial Engineering and Management Innovation*,
DOI 10.2991/978-94-6239-145-1_80

National Congress, China has been adhering to the road of innovation with Chinese characteristics. The idea of innovation is the fundamental original power for promoting economic and social development has been recognized by the academia and national policy makers. There are two innovative ways, one is the independent innovation, and the other is the technology importation. Some scholars believe that the independent innovation is the original power of the development of a country, and this view has been verified in practice, such as the United States' development strategy of independent innovation which has led the scientific and technological strength of the United States to be the world leader. Depending on the scientific and technological innovation, the U.S. economy grew rapidly and the U.S. economy could recover after the recession quickly. Another view is that technology importation is a shortcut to the rapid economy rising for developing countries. Through technology importation, developing countries will be able to achieve scientific and technological leap-forward development, reduce the risk of innovation, and save the cost of technology development. But which one is the true source of power to promote economic promote economic development, the independent innovation or the technology importation? How do the two innovative ways affect China's economic development? And which kind of power has played a more important role? This article will explore these issues on which the jury may still be out.

2 Literature Review

Innovation is the fundamental driving force of economic development [1]. Robert Solow [2] and Swan [3] introduced the innovation in their own the economic growth model. Their studies suggested that innovation is the decisive factor in economic growth. The economy will stagnate without technological progress. The Arrow [4] constructed a model of learning by doing [4]. Their studies suggest that innovation is an endogenous factor in promoting economic development. Arrow and Scheshinsik's innovation endogenous growth model is better than the exogenous growth model to explain the mechanism of innovation promoting economic growth. The innovation was deemed to be an unconscious behavior in their model, so the model could not explain the continuing role of innovation in driving economy. In view of this, the research and development was introduced into the growth model as a separate innovative production department by Romer [5] to analyses the relationship between innovation and economic growth [5]. Since then, many scholars studied the relationship between innovation and economic growth from different angles, learning from the framework of innovative endogenous growth model, taking developed or developing countries as the research object [6–10]. Many Chinese scholars tried to explain the reasons for China's economic growth from the view of innovation point, and tried to use the innovation to solve the bottleneck problem in China's economic growth, for example the Chinese scholar Wu's [11, 12] studies suggested that technological progress was the key to China's economic growth [11, 12]; Wu [13], Li and Wu [14] analyzed the important role of innovation in economic growth. Lin [15], the representative of some scholars, believes that there is still a lot of space for the technology importation for developing countries like China. So in their opinions, China should take technology importation as a major source of innovation to promote economic growth. At the same time, some scholars had different opinions. Yang [16] believed that from the view of economic costing point, China should give priority to independent innovation in development [16]. Wu [17] found that both independent R&D and technology importation had their significant roles in promoting Chinese production efficiency based on the Chinese industrial panel data from 1996 to 2003 [17]. But Yan-bing Wu also found that the effects of independent R& D and technology importation were different in different regions.

Literature review shows that scholars have reached a consensus on the opinion that innovation plays a positive role in promoting economic growth. Because of the different research perspectives and the one-sidedness of the study objects of the existing research, the researchers had different ideas on the issue that which innovative approaches China should take, the independent innovation, technology transfer, or other innovative approaches. And the empirical researches on interaction of the independent innovation with the technology importation in economic development and the influence of the two innovation approaches working on the other factors promoting economic growth are lack. Also, the theoretical supports for innovative selection are not enough. Above all, this paper will explore the true source of the power promoting China's economy development and research the interaction of independent innovation and the technology importation in economic growth as well as the affects the two innovation approaches working on the other economic promoting factors based on the 2000–2011 panel data of 30 provinces in China to provide a theoretical basis for selecting the innovation path for China's economic development.

3 Variable Definitions, Data Description and Statistical

3.1 Variable Definitions and Data Description

3.1.1 Dependent Variables

The level of economic development: This paper researches whether the independent innovation and technological progress are able to promote regional economy, and therefore the level of economic development is selected as the explanatory variables. GDP represents a total regional wealth, and the academia generally see it as a measure of regional economic development status. Therefore, this study uses GDP as the dependent variable to reflect the economic development status of the various provinces.

3.1.2 Moderator Variables

Independent innovation: currently, domestic and foreign researchers use patents to measure the level of the independent innovation. In this paper, domestic patents are used to measure innovation capability at provincial level.

Technology importation: this paper analyses the role of foreign technology in China's economic development, so the cost importation of foreign technology is used to measure the variable of technology importation.

3.1.3 Control Variables

Investment, consumption and exports are major forces stimulating economic growth in economics. The three variables, investment, consumption and exports, are not the focuses of this study, but they may have effects on the analysis results. Investment, consumption and export which are represented by fixed investments, household consumption and total value of exports respectively are introduced in the study as control variables.

The indicators above are derived from the data of the 30 provinces from year 2000 to 2011 "China Statistical Yearbook", "Science and Technology of China Yearbook," "Foreign Trade and Economic Yearbook" with individual provinces missing data from the Provincial Economic Yearbook and Provincial Technology Yearbook. Tibet is not taken as an analysis object due to a serious lack of data.

3.2 Statistical Processing

3.2.1 Unit Root Test

In this paper, the selected data almost are the time-series ones. The time-series data tend to have the data sequence torque characteristic, that is to say the time series data are often non-stationary. Unit root test is used to test the stationary of time series data. The data having unit root test are non-stationary time series. It will destroy the basic assumptions and result in very significant spurious regression in T, F, R2 test even if two time series do not have any relationships when the non-stationary time series data are used in regression analysis. To avoid the spurious regression, the variables should be tested for unit roots. This paper adopts the ADF test method.

ADF test' the null hypothesis is that the data have unit roots and the null hypothesis should be accepted when the P values are greater. Table 1 shows that the all data have unit roots except technology import fees. To eliminate the variables' instability, each variable will accept first order differential treatment. After the first order differential treatment, GDP' P value is very small, just 0.003. The other variables' P value is 0. So we can reject the null hypothesis. The results of ADF test after the first order differential treatment are shown in Table 2.

Table 1 ADF test

Variable	ADF test t statistical value	P value
GDP	18.9581	1
Fixed investments	7.6812	1
Export	1.1489	1
Household consumption	16.8729	1
Domestic patents	14.3211	1
Cost of foreign technology importation	−5.5994	0

Table 2 Order differential treatment Adf test

Variable	ADF test t statistical value	P value
GDP	18.9581	1
Export	7.6812	1
Household consumption	1.1489	1
Domestic patents	16.8729	1
Cost of foreign technology importation	14.3211	1
Fixed investments	−5.5994	0

3.2.2 Correlation Test

Correlation coefficient matrix reflects the correlation between any two levels. The bigger the correlation coefficient is, the smaller t statistical variable is, and R2 will tend to be 1. In the situation, the statistical result can't reflect the actual situation. The Table 3 shows that these variables' correlation coefficients are small, the highest correlation coefficient between GDP and household consumption is just 0.3616, then the one between GDP and fixed asset investment which is 0.3551, and the smallest one is correlation between exports and investment in fixed assets which is 0.0214. Generally correlation coefficients between independent variables and the

Table 3 Correlation matrix

	GDP	Fixed investments	Export	Household consumption	Technology importation	Patent
GDP	1					
Fixed investments	0.3551	1				
Export	0.3204	0.0214	1			
Household consumption	0.3616	0.2621	0.0361	1		
Technology importation	0.093	0.0234	−0.0451	0.0547	1	
Patent	0.1693	0.1429	0.0084	0.0151	0.1618	1

controlled variables are relatively lower, so the sample data quality is better, multicollinearity is more unlikely to occur, credibility of analysis results is higher.

4 Model Discuss and Empirical Results

4.1 Model Discuss

The multiple regression model is a generally applicable method which is an convenient way to estimate the parameters. But the general multiple regression model can not compare regulatory effect that moderator variables work on the dependent variable regulation. This paper adopts multiple regression analysis method to build comprehensive evaluation model, hierarchical regression analysis to reflect promoting role which innovation and technology transfer playing in economic development and study the interactions that independent innovation and introduction of technology work in the progress of the economic development. There are the specific steps of hierarchical regression as follow: first, place the control variables fixed asset investment, the level of consumption and export into the model; Second, based on the first step, place the independent variable into the model one by one; the third step, place into control variables and all independent variables into the model; the fourth step, based on the third step, place the variables' cross-term into the model, seeing Eq. (1):

$$GDP = \alpha + \beta 1fa + \beta 2con + \beta 3exp + \beta 4ino + \beta 5ti + \beta 6con * exp + \beta 7fa * ino + \beta 8fa * ti + \varepsilon \quad (1)$$

In Eq. (1), GDP is the regional development; fa is fixed assets investment; con is the level of consumption; exp represents exports; ino is innovation, with representation of a number of domestic patent applications; ti is technology importation with the expression of technology importation cost; con * exp is the cross-term of technology importation and innovation which represents the interaction of the independent innovation with technology importation; fa * ino is the cross-term of fixed asset investment and innovation, which means interaction of innovation with fixed asset investment; fa*ti is cross-term of the fixed asset investment and the technology which means interaction of fixed assets investment with technology importation.

4.2 Analysis of Empirical Results

Regression results are shown in Table 4. Model 1 is to make regression analysis of all control variables including fixed asset investment, the level of consumption and export. The results of the regression analysis show that over the past 11 years, the

Table 4 Results of regression analysis

	Model 1	Model 2	Model 3	Model 4	Model 5	Model 6	Model 6
	GDP	GDP	GDP	GDP	GDP	GDP	GDP
fa	0.257***	0.232***	0.255***	0.232***	0.234***	0.219***	0.231***
	(−9.43)	(−8.26)	(−9.35)	(−8.25)	(−8.26)	(7.89)	(8.25)
exp	0.0850***	0.0849***	0.0857***	0.0852***	0.0862***	0.0780***	0.0866***
	(−8)	(−8.1)	(−8.06)	(−8.12)	(−8.12)	(7.47)	(8.28)
con	0.264***	0.266***	0.262***	0.264***	0.264***	0.236***	0.266***
	(−7.18)	(−7.34)	(−7.11)	(−7.28)	(−7.27)	(6.50)	(7.36)
ino		0.0377**		0.0359**	0.0337**	0.0400***	0.0316*
		(−3.16)		(−2.95)	(−2.66)	(3.35)	(2.58)
ti			0.000931	0.000521	0.0697	0.000197	0.00129
			(−1.31)	(−0.73)	(−0.65)	(0.28)	(1.61)
con* exp					0.000205		
					(−0.24)		
fa* ino						−1.320***	
						(−3.82)	
fa* ti							−0.660*
							(−2.08)
R2	0.58	0.60	0.58	0.60	0.60	0.61	0.60

Notes
Standard errors are in parentheses
*, **, *** are significant level at 5, 1 and 0.1 %, respectively

fixed investment, the household consumption and the exports have played a significant role in promoting economic growth. The independent variable innovation is placed in Model 2 to make regression analysis on the basis of Model 1. Model 2 shows that there is a significant correlation between independent innovation and GDP. That means independent innovation has a positive effect on China's economic development. The independent variable is placed in Model 3 to make regression analysis on the basis of Model 1. The results of Model 3 show that the role of technology played in economic development is not obvious. Variables independent innovation and technology importation are placed in Model 4 together. The results show that independent innovation has a significant positive impact on economic development, while the role of technology importation is still not significant. Variables cross-term of independent innovation and technology importation are placed in Model 5 to make regression in order to analyze the interaction of independent innovation with technology importation.

There are two different conclusions on the relationship between independent innovation and technology importation. One view is that independent innovation and technology importation replace each other. The demand of independent innovation will descend when the enterprises are more dependent on technology importation. Another view is that positive interaction exists between independent innovation and technology importation. Some scholars believe that the interaction between independent innovation and technology importation depend on the enterprises characteristics [18] and the industry characteristics [19]. Model 5 analyzed the interaction of independent innovation with technology importation without taking care of the types of business and industry. So the results showing no significant interactions between independent innovation and technology importation are reasonable. The cross-term of fixed asset investment and innovation is placed in the Model 6 based on the Model 4. The regression results show that there is significant substitution effect between fixed asset investment and innovation in promoting economic growth. Although in the past 11 years, China's economic growth model was mainly investment-driven one, the model 6's results show that China economy has the ability to change its development model from investment-driven one to innovation-driven one. The cross-term of fixed asset investment and technology importation is placed in the Model 7 based on the Model 4. The regression results show that there is significant substitution effect between fixed asset investment and technology importation. The regression coefficients display that technology importation has more substitution effect on independent innovation than on fixed assets investment.

5 Conclusion and Implication

This paper research the relationship among the independent innovation, the technology transfer and the economic growth using hierarchical regression method based on the 2000–2012 provincial panel data. The results show that:

(1) In terms of two innovation ways, independent innovation has a significant role in promoting China's economic development, and the role of the technology importation is not obvious.
(2) There is no significant interaction between independent innovation and technology importation. Because this article doesn't compare the differences of the industry, business and region, it is can't be concluded that there is no correlation between innovation and technology importation.
(3) Independent innovation and technology introduction have significant substitution effect on fixed assets investment in stimulating economic growth.

Based on the conclusions above, the following policies are recommended:

(1) The independent innovation is the fundamental driving force of China's economic development. The State should increase scientific and technological innovation funding and staff input to enhance China's capability of independent innovation. The role of technology importation played in promoting economic growth is not obvious. There may be many reasons for the situation, such as the different capacity of technical digestion and absorption in different areas, the different development situation of different industries which import technology.

(2) Change China's economic growth mode from the independent investment-driven one to the innovation-driven one. Seeing that innovation play more important role in promoting economic growth than technology importation, and the independent innovation have more obvious substitution effect on fixed asset investment, China should take the road of independent innovation, supplemented by the introduction of innovative technologies.

Acknowledgments The authors would like to thank Sichuan Key Research Center for Social Sciences Foundation (Grant no. Xq14B09) for financial support for this study.

References

1. Schumpeter J (1942) Capitalism, socialism and democracy. Harper & Brothers Press, New York, pp 10–15
2. Solow RM (1956) A contribution to the theory of economic growth. Q J Econ 70(1):65–94
3. Swan TW (1956) Economic growth and capital accumulation. Econ Rec 32(24):334–361
4. Arrow KJ (1962) The economic implication of learning by doing. Rev Econ Stud 29 (6):155–173
5. Romer PM (1986) Increasing return and long-run growth. J Polit Econ 94(5):1002–1037
6. Fagerberg J, Srholec M (2008) National innovation systems, capabilities and economic development. Res Policy 37(9):1417–1435
7. Howells J (2005) Innovation and regional economic development: a matter of perspective? Res Policy 34(8):1220–1234
8. Youtie J, Shapira P (2008) Building an innovation hub: a case study of the transformation of university roles in regional technological and economic development. Res Policy 37(8): 1188–1204
9. Frenken K, Pyka A, Verspagen B, Windrum P (2010) Innovation, "qualitative change and economic development—special issue in honour of Pier-Paolo Saviotti". Struct Change Econ Dyn 21(1):1–4
10. Hasan I, Tucci CL (2010) The innovation–economic growth nexus: global evidence. Res Policy 39(10):1264–1276
11. Wu J (1995) How can we change economic growth pattern (in Chinese). Econ Res J 11:8–12
12. Wu J (2006) Economic growth and technological progress (in Chinese). Sci Technol Ind China 200(1):27–29
13. Wu J (2007) Innovation and economic growth of China (in Chinese). J Shanxi Financ Econ Univ 29(7):37–41
14. Kang L, Yu-hua W (2005) Research on technological progress model of economic growth (in Chinese). Sci Manag Res 23(2):18–23
15. Lin Y (2005) Well-known economist Lin Yifu said: our ability of independent innovation may become traps (in Chinese). Shanghai Hongkong Econ 12:8–15
16. Yang C (2005) Innovation and technology introduction economic accounting (in Chinese). Sci New Ground 3:16–17
17. Yan-bing W (2008) Indigenous R&D, technology imports and productivity: evidence from industries across regions of China (in Chinese). Econ Res J 8:51–64
18. Cassiman B, Veugelers R (2006) In search of complementarity in innovation strategy: internal R&D and external technology acquisition. Manage Sci 52(1):68–82
19. Sun J, Wu L, Qi J (2009) The relationship between indigenous innovation and technology imports: substitution or complementarity (in Chinese). Stud Sci Sci 27(1):133–138

Study on Ecosystem Model and Decision Making of E-Commerce Based on Multisource Information Fusion

Li Li, Li-fen Zhao and Dong-jun Liu

Abstract With the widespread popularity and the rapid development of e-commerce, e-commerce ecosystem is becoming a hot pot. Structure, the main elements and sources of information of e-commerce ecosystem are analyzed. Sensors from multisource of e-commerce eco-system are defined according to the information fusion technology. The knowledge and model bases of information fusion are determined. The decision support model of e-commerce ecosystem based on information fusion from multisource is developed. The specific fusion algorithms and decision-making process are introduced. This study can provide a new practice idea for dealing with the problems of information fusion from multisource and decision-making.

Keywords E-commerce · Ecosystem model · Information fusion · Decision making

1 Introduction

From the views of symbiosis, cooperation and development of ecological, many participants and partners of E-commerce constitute the basic units of ecological system, and a large, new e-commerce ecosystem formed. In e-commerce ecosystem, technologies of efficient network link, real time sharing and information fusion can provide key support for the effective operation of e-business ecosystem. The information fusion technology has been successfully applied to the many application domains. While a multisource information fusion technology is used in e-commerce ecosystem, information from multisource in e-commerce ecosystem can be treated optimized comprehensively. Useful and accurate information can be

---

L. Li (✉) · L. Zhao · D. Liu
Shenzhen Graduate School Harbin Institute of Technology, Shenzhen, People's Republic of China
e-mail: ximlli@126.com

E. Qi (ed.), *Proceedings of the 6th International Asia Conference on Industrial Engineering and Management Innovation*,
DOI 10.2991/978-94-6239-145-1_81

mined through a certain algorithm, and scientific and reliable estimates can be made according to these information. Information fusion can provide decision support for e-commerce ecological system, and overall performance of e-commerce ecosystem can be optimized. Thus it promotes the evolution and development of e-commerce.

2 Literature Review

With the infiltration of ecology in sociology, economics and other disciplines, the theory and methods of ecology has also been used to discuss the business ecosystem. American management experts Moore [1] presented the concept of "business ecosystem" about new competitive ecology in "Harvard Business Review" for the first time [1]. Based on the theory of enterprise ecological niche, co-evolution and self-organization He suggested that enterprises should saw themselves as a part of business ecosystem like the organisms in biological systems. Moore [2] stated the development process and business ecosystem tasks in each stage in details [2].

The presentation of the concept of commercial ecosystem provided new perspective and ideas for competition, expansion and market operation of enterprises. Pehoniemi and Vuofi [3] studied the basic characteristics of self-organization, emergence, evolution, adaptation in the business ecosystem, and put forward the basic framework of business ecosystem [3]. Iansiti and Levien [4] constructed the theory frame of business ecosystem further, and studied the operation mode of successful business ecosystem according to the investigation of the strategy and innovation process of IBM, HP Microsoft, WAL-MART, AT&T, Dell and other large enterprises [4]. At the same time, with the rapid development of the Internet and network economy, the awareness and interests of e-commerce ecosystem were strengthened. Power and Jerjian [5] argued that business ecosystem consists of integrated electronic business, and emphasized the importance of network because many commercial ecosystems were highly dependent on the network, the network was their virtual life space [5]. The healthiness, structure, and evolution process of business ecosystem were investigated by Moore [6], Kim [7], and Zahra and Nambisan [8].

Sun [9] analyzed the "ecological environment" of electronic commerce, but did not state the definition of ecological environment of electronic commerce [9]. Pan and Huang [10] stated the new business ecosystem, and presented the difference between the natural ecological system and business ecosystem [10]. Peng [11] presented that technology, credit, logistics, humanities, law and economy were the six indispensable measures to establishment the good ecological environment of E-commerce [11]. Yang and Li [12] explained the concepts of E-commerce ecosystem, and pointed out that the competitive strategies of enterprises in e-commerce ecosystem were co-evolution [12]. Liu [13] pointed out that the structure of e-commerce ecosystem [13]. Lv and Wang [14] studied the connotation and characteristics of E-commerce ecosystem [14]. Jiang and Shi [15] analyzed the health of the e-business ecosystem [15].

Information fusion technology is a new subject which rapidly developed from 1970s to now. It was first applied in modern C4ISR systems and various weapons platform [16]. In 1973, America research institutions fused the data from multiple continuous sonars using computer technology In order to realize the automatic detection of enemy submarines position. Information fusion mainly refers to the process of observation and collection of multisource information in space and in time using the computer technology, as well as analysis, synthesis and processing of the information under a certain criteria. Thus the object state can be estimated comprehensively and objectively, and its trend can be predicted as well. The capacities of decision making and control of systems can be improved. Researches on information fusion are very rich, and it is widely used in civil and military. Nevertheless, studies on the electronic commerce ecosystem using information fusion technique are very deficient [17, 18].

Studies on business ecosystem theory were carried out fully now, and the basic concept, characteristics and theoretical framework has been initially formed. Scholars have studied the theory and application of business ecosystem from different aspects. It is known to all that many business ecosystems are highly dependent on the network nowadays. Thus it is of great importance to study the connotation, characteristic and constitution of e-commerce ecosystem. How to acquire, organize and deal with information in e-commerce ecosystem activity is significant. Reasonable and correct decisions can be made according to the information. Harmonious and efficient development of the e-commerce ecosystem can be promoted.

3 Models and Methodology

3.1 Decision Support Model of E-commerce Ecosystem

(1) Definition of sensors: Iansiti and Levien [4] used the biological concepts, and divided the enterprises in e-commerce ecosystem into the backbone enterprises, dominant enterprises, slot type enterprises [4]. Some Chinese scholars divided e-commerce ecosystem into the core layer, support layer, extending layer and environment layer hierarchically, while some scholars divided E-commerce ecosystem into the critical populations, support populations, leadership populations and parasitic population according to their roles [14].

The main components of e-commerce ecosystem were the key participants and assistant participants according to the studies of scholars above. The key participants occupied the central position in the core enterprises in the system, and provided a platform and supervision services for the system. The assistant participants played the role of resource integration and coordination in e-commerce ecosystem. The trading enterprise users who had a key position in the system were the subjects of e-commerce, and they included both the buyers and the sellers in the process of e-commerce. Meanwhile there were also supporting enterprises, such as logistics companies, financial institutions, certification institutions, the network service providers and relevant government agencies, and they provided a variety of support services for e-commerce transactions. Parasitic enterprises also existed, and they were transactions supplier who provided value-added services, including Network marketing service providers, outsourcing companies, electronic business consulting companies, etc.

The key enterprises and the trading subjects were the main factors in e-commerce ecosystem according to the above analysis, while financial institutions, government agencies, logistics companies, certification centers were the supporting parts of the system. In addition there were also enterprises that provided value-added services. These important members were set as the multi-sources of information fusion, and they were defined as the sensors in Table 1.

(2) Knowledge bases and model bases: For e-commerce ecosystem, in order to enable its healthy development, the scales and speeds of the development of the whole system should be kept in a reasonable range. In the process of development of the key enterprises in the system, the roles and influences of the environment and other external factors should be taken into account. The ecological niches in e-commerce ecosystem should be identified appropriately. Ye [19] studies the evolution and the collapse of the e-commerce system, and she found five factors, which were the government, venture capital, the users, the industry competition and capacity of income, played very important roles. These five factors influenced prosperity and decline of each enterprise in the whole system. These five factors were abstracted, and constructed the GVUCI model [19].

The GVUCI model was a model of environmental adaptation, and the strength of the five factors was variable. If one factor of these five factors changed, it would lead to changes of combination. Thus the combination which represented the environment varied accordingly. Different characteristics and evolution trends would exhibit in different circumstances, and this would influence the subjects of the system. The environmental characteristics of various factors in different combinations were shown in Table 2.

Table 1 Definition of sensors

Sensors	Meanings	Category
A	Key enterprises as sources of information	Main factors
B	Transaction subjects as sources of information	
C	Financial institutions as sources of information	Supports
D	Government agencies as sources of information	
E	Logistics companies as sources of information	
F	Certification institutions as sources of information	
G	Enterprises that provided value-added services as sources of information	Others

Table 2 Four typical states of gvuci model

State	Government policy	Users' attitude	Competition intensity	Risk investment	Profitability	Strategy
Best state	Highly supportive	Positive	Moderate	Optimistic	Strong	Shaping the brand and expanding market
Worst state	Indifference	Wait-and-see	Excitation	Indifferent	Weak	Reducing costs, transforming and upgrading
Potential state	Moderate	Wait-and-see	Medium	Optimistic	Strong	Accumulating wealth and expanding brand
Risk state	Moderate	Positive	Excitation	Optimistic	Weak	Saving money and creating products in demand

The GVUCI model could help enterprises to determine the environment state, and guided them to formulate or adjust the strategic in uncertain environment. It could also assist the government to correctly analyze the situation and take positive measures of macro regulation. In this study the GVUCI model was included in the model base of information fusion, and the countermeasures were included in the knowledge base.

(3) Framework of information fusion: Information fusion technology integrated information from multi-sources through a certain fusion algorithm. Reliable decision that the optimal estimation of states was presented based on the observation from information from multisource could be made according to this fusion information. Information was collected from the seven sensors defined in Table 1, and the information of external environment was comprehensive. Then the information collected was collated, and the data might be preprocessed if necessary before the next step of information fusion. Knowledge base and model base were needed to support the information fusion process, and the fusion process included data detection, correlation, estimation and merge.

In order to make the implementation of effective and sustainable use of the models, the participation and feedback of information from enterprises or organizations were needed. While the information fusion results were obtained, the decision results could be judged whether they were feasible according to the specific circumstances of enterprises or organizations. If the information fusion results were reasonable and feasible, the decision could be carried out. Then implementation effect would be evaluated, and the information was feed backed to the knowledge base and model base to improve the model. Otherwise, if the results were not feasible, the viewing angles should be adjusted and the information sources might be changed. The decision support model of E-commerce ecosystem based on information fusion was shown as Fig. 1.

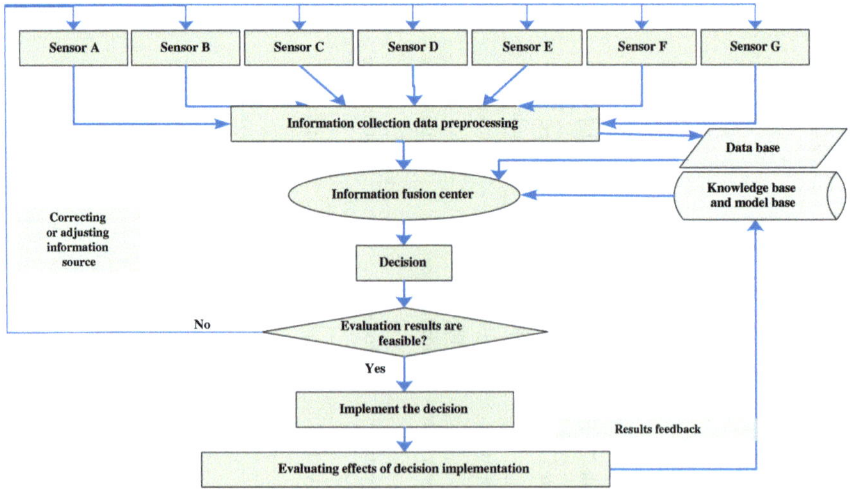

Fig. 1 Decision support model of e-commerce ecosystem based on information fusion

3.2 Method of Information Fusion

(1) Process of information fusion: Information fusion technology is a series of processes that fully use all kinds of environment and objects information which are obtained from a plurality of sensors, analyze and reorganize the information. Thus the complete and effective information can be fused so as to provide support for decision-making. In this study, information fusion process was shown in Fig. 2. The various organizations in e-commerce ecosystem were set as the sensors. Information with different subjects and different types was transmitted to the information processing center. The information processing center filtered and integrated the miscellaneous information, and classified the information in five aspects, that is, government policy, users' attitude, competition intensity, risk investment and profitability, which were required for fusion centers. Various types of information were processed based on the effective fusion algorithm. Information fusion results could be obtained according to the model base. Thus the state type could be determined, which could provide support for the decision making of enterprises.

(2) Information fusion algorithm: In e-commerce ecosystem, enterprises were generally in unknown and changing circumstances. Therefore, the information from multi-sources were fuzzy and uncertainty limited by the environment. Due to the influence of various interference and sensor errors of themselves', all information provided by sensors was generally incomplete, imprecise or vague. Thus the information might hold some uncertainty. Information fusion technology was just an uncertainty reasoning and decision making tool. The core step of information fusion was mainly reflected on the fusion algorithm. For multi-sensor systems, the basic requirements of information fusion algorithm was with capabilities of robust and parallel processing because of the diversity and complexity of information. The D-S evidence theory was suggested to improve the confidence degree of the system.

The basic idea of D-S evidence theory was as follows. Information was obtained from each sensor, and the metric for the proposition was also ready to form the evidences in the theory. The credibility was given to each proposition by constructing basic probability assignment function (mass function) using these evidences. A basic probability assignment function and the corresponding differential

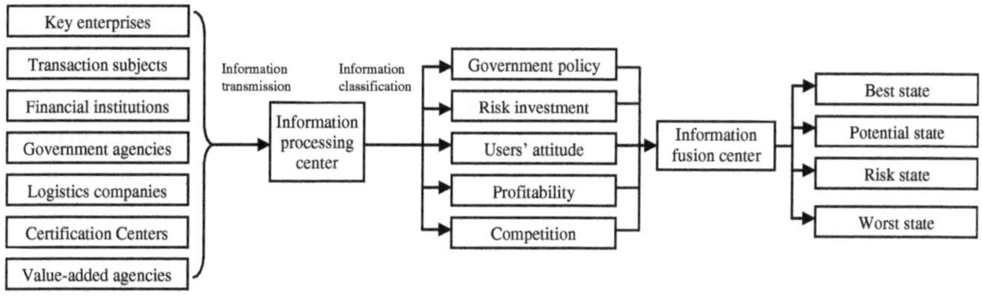

Fig. 2 Information fusion process of e-commerce ecosystem

frame together were combined as the evidence body. Thus the information from each sensor is equivalent to an evidence body [20]. The process of multi-sensor information fusion was to combine all the evidence bodies into a new one using D-S evidence combination method. The new evidence expressed the fused information, and reasonable decision could be made according to the decision rules finally.

D-S evidence theory combined information from multi-sources based on the D-S evidence combination method. The reliable and comprehensive information was generated, and reasonable decisions were made according to the fused information. Definition of D-S evidence combination method was as follow [16].

Bel_1 and Bel_2 were set as two confidence functions in the same recognition framework. m_1 and m_2 were probability assignment values correspondingly. The focal elements were A_1, …, A_k and B_1, …, Br.

We set $K_1 = \sum_{A \cap B = \Phi} m_1(A_i)m_2(B_j) < 1$. Thus we had:

$$m(C) = \begin{cases} \frac{\sum_{A_i \cap B_j = C} m_1(A_i)m_2(B_j)}{1-K_1} & \forall C \subset \Theta \quad C \neq \Phi \\ 0 & C = \Phi \end{cases} \tag{1}$$

where m was identified as a basic probability assignment value, if $K_1 \neq 1$; Otherwise, if $K_1 = 1$, m_1 and m_2 were contradictory, and the two probability assignment values could not be combined.

4 Conclusion and Discussion

E-commerce ecosystem is a complex adaptive system. The collection and utilization of information was the problem. Information fusion technology is just suitable to solve the complex problems of information treatment. The application of information fusion technology in the field of e-commerce could deal with the sources, treatment and integration of the information based on data mining and information fusion. This could provide a new practical idea for the information analysis and decision making problems in e-commerce ecosystem, and it is of great importance for managers to make decision.

Application of information fusion in the field of E-commerce was discussed, by learning from the implementation and methods of the application of information fusion in other fields. Key enterprises, transaction subjects, financial institutions, government agencies, logistics companies, certification institutions and enterprises that provided value-added services were set as the sources of information of the system in this study. The GVUCI model which was an environmental adaptability model was set as the information fusion model base and knowledge base. The D-S evidence theory model was selected as the fusion algorithm, and the decision support model of e-commerce ecosystem was constructed based on information fusion. This model could help enterprises to recognize the situation and forecast the potential changes. It could guide enterprises to make reasonable development strategy, and had a positive theoretical and practical significance for E-commerce enterprises to obtain favorable competitive positions.

E-commerce ecosystem is composed of many enterprises and institutions. Then the system decision support model also required all enterprises and institutions participation and feedback. This model could be carried out effectively and sustainably, while the information in the system was enriched and updated by these organizations continuously. The overall decision and control capability of e-commerce ecosystem could be improved, and the overall performance of the system could be optimized.

The model in this study was the application of basic principle and algorithm of information fusion technology in e-commerce ecosystem. However, the current theory of E-commerce ecosystem is still not mature, and the system itself also has many uncertain factors and characteristics which also need to be carried out further analysis and research. Therefore, the application of information fusion technology in practice is still a complex problem. Thus how to effectively use a variety of information technology and network solutions to deal with information processing and decision-making problem in the e-commerce ecosystem is worth of researching further.

Acknowledgments Research works in this paper are financially supported by Research Planning Foundation in Humanities and Social Sciences of the Ministry of Education of China (Grant No. 13YJAZH044) and National Science Foundation of China (Grant No. 61173052).

References

1. Moore JF (1993) Predator and prey: a new ecology of competition. Harvard Bus Rev 5–6:75–86
2. Moore JF (1996) The death of competition: leadership and strategy in the age of business ecosystems. Harper Collins Publishers, Inc.(USA), p 76
3. Pehoniemi M, Vuofi E (2006) Cluster, value network and business ecosystem. Knowl Innov Approach 8(1):10–18
4. Iansiti M, Levien R (2004) Strategy as ecology. Harvard Bus Rev 82(3):68–78
5. Power T, Jerjian G (2001) Ecosystem: living the 12 principles of networked business. Pearson Education Ltd, London, p 392
6. Moore JF (2005) The rise of a new corporate form. Wash Q 21(1):167–181
7. Kim K, Song CS, Rhee JO (2010) The healthiness of business ecosystem and its effect on SMES performance. International Council for Small Business, pp 1–17
8. Zahra S, Nambisan S (2012) Entrepreneurship and strategic thinking in busines ecosystems. Bus Horiz 55(3):219–229
9. Sun X (2001) Analyzing about ecological environment of e-business. In: 2001 International CATV technology conference, Hangzhou, China, CA, pp 208–214
10. Pan J, Huang X (2004) A new perspective: business ecosystem view. Ecol Econ 8:172–176
11. Peng L, Zhu F (2008) Research of e-commerce ecological environment in China. In: The 13th China information economics association annual conference, Xian, China, CA, pp 271–274

12. Yang Y, Li Q (2008) Competition tactics research among enterprises in e-commerce eco systems. Sci Technol Ind 9:72–74 (Chinese)
13. Liu X (2011) Research on risk evaluation of e-commerce ecosystem. Commercial Times 30:43–44 (Chinese)
14. Lv L, Wang R (2012) Connotation and characteristics of e-commerce ecosystem. Chin Sci Technol Inf 10:175 (Chinese)
15. Jiang Y, Shi W, Sun J (2014) Health evaluation of e—business ecosystem. Ecol Econ 1:205–209
16. Peng L (2010) Information fusion key technology and its application. Metallurgical Industry Press, Beijing, pp 53–64 (Chinese)
17. Hu Q, An S, Yu X, Yu D (2011) Robust fuzzy rough classifiers. Fuzzy Sets Syst 183:26–43
18. Chen K, Zhang Z, Long J (2013) Multisource information fusion: key issues, research progress and new trands. Comput Sci 40(8):6–13
19. Ye X (2010) Research on e-commerce ecosystem. Social Sciences Academic Press, Beijing, pp 32–66; 149–184 (Chinese)
20. Xu Y, Liang X, Jia X (2012) Information fusion based on fuzzy evidence theory and its application in target recognition. J Harbin Inst Technol 44(3):107–111 (Chinese)

Business Model Innovation for Exhibition Industry in China: A Case Study of Zhongshan

Hua Zhu, Zheng He and Xueya Liu

Abstract Convention and Exhibition are essential activities to measure the economic development of a country or regional entity. Taking exhibition industry in Zhongshan as a case, this paper explores its business model innovation. We first discuss the challenges of current government–focused operation mode for exhibition sector. Then using operational ways of Hong Kong and Germany as benchmark, we further analyze how to transfer to market-oriented business model for exhibition industry in China.

Keywords Exhibition industry · Business model · Innovation · Government role

1 Introduction

Convention and exhibition are collective activities organized by firms in a specific geographical location with the aim of promoting interaction among players in a specific industry. These activities take the form of trade fairs, exhibitions, large conferences organized both at home and abroad [1]. Exhibition encourages participants to share information and knowledge aimed at advancing development in

H. Zhu
School of Political Science and Public Administration, UESTC,
Chengdu, People's Republic of China

Z. He (✉) · X. Liu
School of Management and Economics, University of Electronic Science
and Technology of China (UESTC), Chengdu, China
e-mail: hezh@uestc.edu.cn

E. Qi (ed.), *Proceedings of the 6th International Asia Conference on Industrial Engineering and Management Innovation*,
DOI 10.2991/978-94-6239-145-1_82

specific industries. It can be traced 100 year ago with the IWC industry exhibition in London being the primer one in 1851 [2]. Due to recent trends of global economic integration, it has become an essential sector all over the global. Many scholars view exhibition as a critical factor to promote the development of some industries through the display of goods, technology and others [3].

The exhibition industry in China since the 1990s has witnessed a significant growth rate of 20 % [4]. This is attributed to the strategic measures that practitioners have put in place. For instance, in Guangdong province 4603 exhibitions were held in 2009. This sparked the pursuit for exhibition development all over the country. In this paper, we first analyze the operational mode of exhibition industry in Zhongshan and then discuss the challenges faced with current the government–dominated operational model. Our findings indicate that with the change of external environment, present exhibition operation model is not as effective as before and need to innovate in terms of government role and organizing process. Based on the analysis of operational ways of Hong Kong and Germany as benchmark, we further analyzes the role of government and suggest that specific measures needs to be taken in the transformation to market-oriented exhibition business model.

2 Exhibition Operational Model in Zhongshan

In 1999, the first professional trade show was organized by the government of the Zhongshan namely China Guzhen International Lighting Fair, and in 2000 the second professional trade show was also held namely China Casual Wear Expo. After this expo more than a dozen of professional exhibitions such as "Zhongshan International Electronics Exhibition", "Equipment Manufacturing Exhibition" and "Huang Pu food Fair" were organized in 2010 [5]. And these finally promoted the development of exhibition economy in China. In Zhongshan, all these exhibitions were organized by local government. For example, five professional exhibition halls were constructed by government. Until now, local government is still the owner of these halls, whereas the management company of government owns the franchise. There are 27 national specialized productions and manufacturing bases, 14 provincial professional towns and 8 provincial industrial clusters [6]. With many industrial clusters and convenient infrastructure, Zhongshan has hosted a number of exhibitions and gradually becomes one of centers of exhibition industry in China.

Figure 1 shows the exhibition operational model in Zhongshan. From the three processes including pre-shows planning, implementation and post-show activities, it can be seen that the operation model of Zhongshan exhibition is government-dominated one. For example, in the planning stage, it was the local government rather than specific industry that made exhibition plan and then look for partners; during the implementation phase, local government often assigned the task and force the enterprises to take part in the exhibition. After the exhibition, it was again the government that did some investigation and finished a report to summarize the whole process. While this government-dominated model played an

Fig. 1 Operation mode of exhibition in Zhongshan

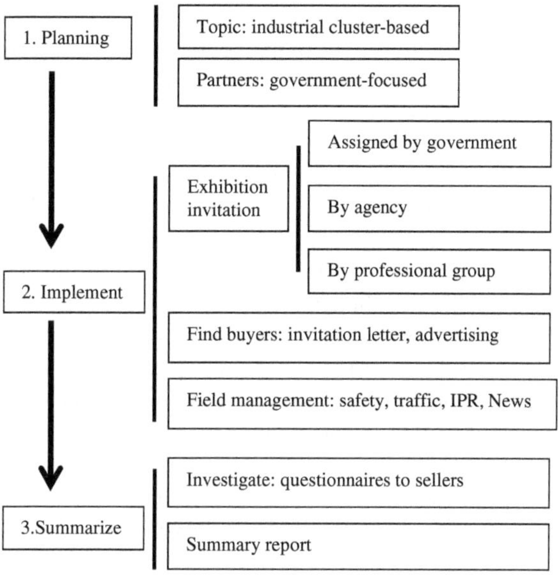

important role on promoting the growth of some clusters, it has gradually shown some disadvantages with the development of industry and radical change of external environment [6].

3 Challenges of Exhibition Industry in Zhongshan

Currently, exhibition activities led by government in Zhongshan is faced with more and more challenges, such as recruiting difficulty, inaccurate positioning, lack of professional management, weak profitability, poor exhibition effect and insufficient awareness among players. In other words, exhibition already cannot adapt to the rapid development of industries context. The challenges affecting this sector are analyzed as follows.

3.1 Overemphasis on Government Administration

Generally speaking, for all Zhongshan exhibitions, the government has been in charge of their operations, even though in recent years, several exhibitions are outsourced to professional companies. For example, Guzhen lights conference and exhibition (Zhongshan) Co., LTD. were set up to operate lamp expo. But this organization was also set up and managed by the government. The chief members are mainly government officers who failed to realize how to commercialize these activities. Even if Electronic Exhibition was cooperated with Chinese Electronic Information Industry Group, the local government still controlled all the organizing process and the operation of the exhibition. As a whole, the behavior of the government still dominates all exhibitions of Zhongshan. The government is not only the managers, regulators, but also the subject of exhibition industry. Under this model, the government is both an athlete and the referee with the role of dislocation.

Government-dominated exhibition tends to easily ignore the inherent purpose of exhibition activities coupled with unsuitable conditions for the development of the exhibition industry. This has led to some insufficient pavilion exhibition projects, serious idle phenomenon, as well as the waste of resources especially associated with overemphasis on the brand image and grand opening ceremony that is luxurious but does not have any realistic impact on the activities. Therefore exhibition has become an image and face activities, not tailoring to fit the requirements of the buyer and seller. Some exhibitions take non-market approaches to recruit exhibitors because of too much administrative participation, let alone the heavy burden of the administrative command to enterprise exhibitors. This phenomenon weakens the effectiveness of exhibition from the early stage of the exhibition. In addition, lack of strict budget as well as cost management made the many exhibitions unprofitable. However government still paid for it and funded it next year regardless of profit or loss. This not only increased the government's financial pressure but also at the same time makes exhibition lack of self-improving motivation mechanism. Because of the intervention of administrative power, the government twist market role in exhibition industry and squeeze the development space for the professional exhibition enterprises. Therefore this not only hinders the growth of professional exhibition organizers to a great degree bur also affects the normal market competition.

In order to promote the local brand, the government organized and paid for all exhibitions. This model is indeed a planning economy, rather than a market operation. Meanwhile, it is likely to make decision according to the leader's idea rather than the market demand. This approach would lead to all the exhibition themes and plans done by government without comprehensive market investigation. Temporary agency or the entrusted institution generally undertakes the government-dominated exhibition. Most of managers lack market-oriented operation experience. As shown in Fig. 1, exhibition invitation mainly depended on administrative orders, and the related government departments divide the booth to subordinate units and then to the enterprises when exhibition effect is not good as expected. At the same time, the investment patterns also give priority to administrative orders. They organized related units to visit the exhibition, sometimes literally pull the audience up for the number of visitors. Although these exhibition may get the support of the local leading enterprises, they are of ostentation and extravagance with actually few effects on industry development. To some degree, the survival of exhibition is greatly linked to the benefit of enterprises. However it's hard to survive independently in the fierce competition of the market lack of government funding.

3.2 Lack of Market Mechanism and Lagged Operating Model

It has been more than ten years since the development of the Zhongshan exhibition industry, but so far there is virtually no exhibition industry association, let alone the exhibition industry norms and self-discipline rules. To be an exception, only Zhongshan torch international exhibition center Co., LTD joins the modern service industry of Zhongshan Association (MSA). Many towns in Zhongshan repeatedly constructed exhibition halls. Even after the commissioning of the expo center of Zhongshan, these halls in the related town still keep exhibition, which limit the size and effect of the exhibition in Zhongshan. The exhibition halls are fragmented with low utilization rate. In 2009, the Zhongshan Torch International Exhibition Center exhibition was actually used for only 21 days, out of which only 6 days was for professional exhibition. As we can see, there exists serious waste of resources. The post-show evaluation is also nonprofessional and usually carried out by the organizers themselves. The assessing ways have already fall into trap, lacking of objective system to evaluate the success or failure of an exhibition. We may find in Fig. 1 that organizer only send questionnaires to the exhibitors, ignoring the feedback of buyers. This method has limited the continuous improvement of exhibition activities in Zhongshan.

4 Business Model Innovation for Exhibition Operation in Zhongshan

4.1 Business Models in Hong Kong and Germany

As one of the largest trading areas in the world, Hong Kong has developed to be one of the global exhibition centers due to its specific geographical location and excellent infrastructure as well as excellent services. The total number of trade exhibitions in Hong Kong reached 87 in 2009, and the area of exhibition was 794,000 square meters [7]. With higher degree of marketization of exhibition industry in Hong Kong, the government makes no restriction on the topic, content, form and so on. So the exhibition industry is a highly competitive market. The Hong Kong SAR government does not directly involve in the organization and management of any exhibition [8].

Unlike the market-oriented exhibition model in Hong Kong, the exhibition model in Germany is government supporting. As a typical sponsor, the German government plays a very important role in the development of local exhibition industry [9]. Even in the financial crisis in 2008, the exhibition industry in German was still the harvest year. Germany has authoritative and national industry associations that play an important role in the exhibition industry. As early as 1907, Germany set up a professional Association of the German Trade Fair Industry

(AUMA), whose headquarter was in Cologne. AUMA was authorized by the German government department and comprises of the exhibitors, buyers, and the exhibition organizers. It is the highest authority organization in the exhibition industry [10]. Under the unified control of AUMA, the goals of exhibition in Germany were very clear with little repetition. AUMA is also a bridge for communication between the government and exhibition organizers. For example, AUMA invite experts to visit all around the world and write report that provides good advice and important reference for the German government to fund the exhibition activities for the domestic enterprises [11, 12].

4.2 Transition From Government-Dominated to Market-Oriented Exhibition Model

4.2.1 Reposition the Role of the Government

From the exhibition operation in Hong Kong and Germany, despite the government's role and position is different in each country, both has a prominent place in the operation of exhibition processes [13]. And this should be considered as reference for operation of exhibition led by government in Zhongshan even in China. From the case of Zhongshan, we may find that in China most of the exhibitions are full of political intervention. More important, a lot of chief leaders in these associations are appointed by the government and nearly all of chairs are government officers, leading the associations under the control of government or even another department of local government. The close relationships with government make these associations difficult to meet the need of all firms in a specific industry. From the enterprise viewpoint, they may consider the association as a part of government administration. However, even though local governments believe the industry association should be organized by enterprises themselves, they still control the daily activities as well as organization structure of association. This independence from government also leads to confused position of industry association in exhibition sector [14].

Thus the government should initiate the transferring process from traditional administration to supporter role for the exhibition associations. Especially in relatively developed provinces like Guangdong, Zhejiang, and Jiangshu, local government should let the private business owners of exhibition industry establish their own alliance voluntarily in order to organize exhibition activities according to market mechanism, rather than the government need [15].

Taking Zhongshan as a case, local government should cultivate a favorable incentive for exhibition industry including soft environment like convenient process for registration and social responsibility. As for the strategic decision of an exhibition like objectives, planning, position and process should leave to the exhibition association. The exhibition industry may take advantage of the various industrial clusters in Zhongshan, which account for the largest contribution to local economy. To provide better service to these clusters, government may decentralize and rely on professional companies to host exhibitions. And the building halls that were invested by the government may sell to the professional enterprise or to rent to exhibition activities [16]. The purpose of this change from authoritative to supportive nature of government is to maintain the independence of the exhibition industry association and make sure the association can actually be on behalf of the enterprise in the industry whiles the government only monitors the activities of exhibition association, providing necessary support to their operation. Since the intangible assets of some successful exhibitions led by government are of considerable value, their names and logos can be used only by the exhibition association of Zhongshan. Government may encourage the exhibition association to apply for patents to protect these intangible assets. To maximize the value of these assets, during the application, the local government also plays important role on providing necessary law support.

4.2.2 Establish Non-Government Exhibition Industry Association

Only if the exhibition association in Zhongshan be a non-government organization that is independent of the local government that seeks to represent the need of most of enterprises in exhibition industry. In general, since industry association has more collective and dynamic information than individual firm, it can be a bridge to connect the need of government and different clusters [17]. More important, it plays an important role in planning and organizing the exhibition. The main purpose of the exhibition association is to organize the exhibition from the beginning to the end, including making exhibition plan, cooperating with the government, financing, organizing sellers and inviting buyers. In addition, exhibition may also set up the evaluation system of exhibition industry and provide necessary training to enhance the capability of people in this industry. For example, exhibition association can organize some visiting and international exchange activities to Hong Kong and Germany to learn their experiences. Exhibition industry should take advantage of their close connections with the various enterprises to provide the support to cluster development and on the other hand gradually build their own brands under the supervision of the government as a whole. Therefore we can see that the exhibition association play a key role on the healthy and orderly development of exhibition industry. It may combine the mandatory function of law and self-discipline function of industry. The establishment of independent exhibition association is an urgent task for the Zhongshan exhibition industry. As we have discussed before, if the director or other positions in the association were still appointed by the local government, then the association was still controlled by government administrative bodies. Thus the organizational structure and human resources are critical factors that influence the functions of the association. From this point of view, we may learn from Hong Kong and Germany to improve a real market-oriented exhibition association.

5 Conclusion

The aim of this study is to explore the business model innovation of the exhibition industry. Along with gradual establishment of socialist market economic system in China and transformation of the functions of the government, the roles that government plays in the exhibition industry must also shift from controller to servicer. The purpose of implementing market-oriented operation model is the need to adapt to new external environment. We highlight the importance of the exhibition association on fostering the sustaibable exhibition economy in Zhongshan. Even though in this paper we only analyze the exhibition industry in Zhongshan, our findings are still valuable for other areas in China, which are still faced with the similar problems. In sum, the dominant position of the government in the exhibition industry should gradually fade out the actual specific groups outreach activities, and concentrate on providing a good development environment for the exhibition market.

Acknowledgments This research is financially supported by The National Social Science Foundation of China (No. 12BGL123).

References

1. Fang Z (2009) The review and prospect of exhibition economy theory research (Periodical style, in Chinese). J Ningbo Univ (humanities and sciences) 22:73–75
2. Guo J (2006) Introduction to exhibition (Book style, in Chinese). Shanghai Jiaotong University Press, Shanghai, pp 10–21
3. Tu C, Chen Z, Yi W (2008) The interactive effect between exhibition industry and urban development (Book Style, in Chinese). Central Compilation Press, Beijing, pp 149–152
4. Milton T, Doffer J, Abby R (2002) The management and service of exhibition (Book style, in Chinese). China Tourism Press, Beijing, pp 33–41
5. Bermejo LR, Cuadrado RJ (1995) Urban hierarchies and territorial competition in europe: exploring the role of fairs and exhibition (Periodical style). Urban Stud 2:379–400
6. Kay AL (2007) International exhibition organizations in china and their performance (Dissertation style). The Hong Kong Polytechnic University
7. Pizey H, Huxham C (1991) 1990 and beyond developing a process for group decision support in large scale event planning (Periodical style). J Mark 8:265–285
8. Wang Y (2010) The analysis of the Chinese government-focused exhibition (Book style, in Chinese), in 2009. In the development of Chinese fairs and exhibitions. Social Science Academic Press, Beijing, pp 99–100
9. Wu X (2007) The development of Zhongshan exhibition economy (Dissertation style, in Chinese). University of Electronic Science and Technology of China
10. Zhang J (2006) Success of Hong Kong exhibition industry. Value Eng 9:11
11. Qiu Z (2009) The characteristics and implication of the protection of intellectual property rights for exhibition industry in Hong Kong. J Guangdong Tech Normal Coll 5:82–85
12. Nocchi L, Zhu S (2010) The overview of German exhibition (Book style). In report of Chinese exhibition economy. Social Science Academic Press, Beijing, pp 116–117
13. www.messefrankfurt.com.cn

14. Shi C (2008) Operation pattern of exhibition economy (Book style, in Chinese). China Social Sciences Press, Beijing, p 119
15. Ma Y, Liang S (2007) Development model and strategic innovation for government-focused exhibition industry (Proceedings style, in Chinese). In: Proceedings of annual conference of academy of Chinese exhibition economy 2007
16. Jiang Y (2008) Government role and strategy in the management of exhibition city (Master's thesis style, in Chinese). East China Normal University, pp 56–64
17. Luo Q (2008) The topic choosing and operation mode for exhibition (Book Style, in Chinese). Nankai University Press, Tianjin, pp 179–185

Emergency Management of Large-Scale Construction Projects Based on Metadata

Shengdeng Xu

Abstract To seek emergency management basis for large-scale construction projects in theory and practice combining metadata modeling and to clarify the emergency management mechanism of large-scale emergencies construction projects and construction projects, based on this, this paper studied the large-scale construction projects Emergency Management system configuration, operation mode, correlation methods. This paper provides new ideas and models to improve modeling accuracy and reliability for large-scale construction projects on project data acquisition, modeling, analysis and summary, which can improve the modeling accuracy and reliability for the future on the research direction of large construction projects emergency management, providing a reference for further study.

Keywords Large-scale construction · Emergency management · Metadata

1 Introduction

Emergency management is a very popular topic with strong allocation, which should not be limited to a single range of disciplines as emergency management is bound to break a single area to a wider filed of development. Theory research on emergency management should go deeper and more comprehensive. So the research of emergency management extends to the large-scale construction project could both expand the research filed and the theory application of emergency management, which also provides a very good direction to researchers to explore.

In the past, emergency management main focuses on the public security, retrieving the relevant literature we can see the very few of them have studied on emergency management in construction projects. The earliest studies of crisis

S. Xu (✉)
Department of Management, Tianjin University, Tianjin, China
e-mail: 604622610@qq.com

E. Qi (ed.), *Proceedings of the 6th International Asia Conference on Industrial Engineering and Management Innovation*,
DOI 10.2991/978-94-6239-145-1_83

management in engineering management is an Australia scholar named Mitroff [1], early in the 20th century, he studied in the field of engineering crisis management, and published a book "Crisis Management in Construction Projects." American scholar Nixonand and Siddons [2] proposed the concept of the project disaster, noting that the project management alone is not enough for risk management, and major mistakes still occur in project risk management and can effectively deal with the idea to have a positive meaning in a variety of project management in his book "Project Disaster and How to Survive Them".

British scholar Bosher [3], Andrew Dainty and others established a unified framework to deal with some aspects of the environment poor conditions to construction agencies caused by floods, climate change, terrorist attacks through the UK construction industry survey. In addition, more attention has been paid to research projects capability evaluation to emergency management, and achieving a lot. Such as: "Richmond evaluation method" proposed by Britain Imperial Chemical Company [4], risk index evaluation method proposed by American Dow Chemical Company Dow and so on. In short, the study program in construction safety management is rich and ripe, but this's little reviewed in the theoretical literature [5].

The large-scale construction project management involved in all aspects of economic life at the same time, to carry out emergency management research project-level of large-scale construction can apply to the economy and people's lives, which has great significance [6]. However, emergency management model large-scale construction is very complex, involving a lot of factors with a huge amount of data modeling, so whether there is a joint point in theory between the emergency management and large-scale construction project management? And whether is there a link between emergency management and large-scale construction project management? There is no answer in the literature study, which is actually also pointed out the direction for seekers to explore, it is necessary to study the theory on project emergency management.

2 Basic Concepts of Construction Project Emergency Management

2.1 Definition of Construction Project Disaster

Construction project refers to one-time task of construction under conditions of limited resources, to achieve the size and quality requirements of the standard within a predetermined period of time. The main purpose of construction project is to form fixed assets, which has a clear construction projects scale, quality standards and the use of the cycle with clear limits on human, financial, and material, usually constituted by the construction, equipment, industrial equipment, technological innovation activities, and related activities. Construction project has some common features include: a huge investment, long construction period, involving a wide surface; with fixity, one-time, multi-constraints and greater risk [7].

Research object that emergency management usually concerns is called emergencies; this study aims at the field of public security, the scope of its coverage beyond the scope of project management. Emergency management of construction projects aimed at unexpected events that occur during construction project implementation, the incident could threaten the successful implementation of the project, including social, natural emergencies, also includes improper management, human mistakes, lack of funds, technical disasters in project management process, so the project-level emergency is defined as "Project disaster", the object of emergency management research projects defined as "construction projects disaster", the main difference lies in the past, unexpected event represents limitation natural disasters, public health emergencies, war and other external disasters [8]. Construction project disasters include not only the scope of the project external environment disasters such as war, floods, etc., but also including other disaster scenarios for the proper implementation of the project disruption or destruction, especially catastrophic consequences of project management itself factors, project internal disasters are the main object of the study on the project of emergency management.

Based on the analysis to the concept above, author intends to give to meaning of "Project disaster", which means some interruption or delay in the in the normal process of construction project implementation due to external and internal reasons, including all serious incidents caused the construction project could not be properly implemented, and these events can be carried out by a certain method to control, prevent and eliminate, or even become a positive impact from the negative effects.

2.2 *Concept, Subject and Target of Construction Project Emergency Management*

Project emergency management refers to a series of elimination and control activities to prevent huge losses in the process of emergency integrated the methods and means of emergency management in the project. Construction project emergency management refers to emergency management in the emergency happened in the construction project, which is the theory application of research projects in the emergency management of construction projects [9].

Construction industry is an important sector of the national economy, because of its production flow, construction diversity, comprehensive coordination and labor-intensive features, which has always been the industry with high risk and high accident, it also has seriously hampered the healthy development of the construction industry, impacting to improve their productivity and quality, and it also a great influence on the overall reputation of the construction industry, so a study of emergency management is essential in the construction project management.

Project emergency management could be divided into emergency management project owner's emergency management and project construction contractor's emergency management, pointing out its essence is to use the lowest cost, using the methods and means of planning, organization, control, reduce the negative impact of unexpected events arising, to maximize the project's objectives.

I believe that the construction project emergency management refers to a series of emergency management activities of the project to ultimately make the project management objectives can be achieved in the construction project decision-making or implementation process. The detailed content of construction project emergency management includes project disaster factor analysis, forecasting, early warning, plans, emergency response, control, and recovery, post-evaluation and so on.

2.3 Object of Construction Projects Emergency Management

Emergency management of construction projects mainly refers to the object of the normal impact of the project implementation, and even external and internal disaster events lead to project failure, the project external disaster means: natural, political, catastrophic events, such as the project itself independent social factors, such as flash floods, typhoon, earthquakes, riots, war, etc. [10]. Project internal disaster refers to project decision-making process, because the unreasonable planning, organization, control, etc., or because of mismanagement, negligence, resulting execution in the project implementation process, which makes the project away from the goal and caused the event in disaster situations. For example: plan disorder, action deviate market demand, technological disasters, capital fracture, quality and safety accidents, group events and more. As the project external disaster has very huge effect, and some emergency management activities have already beyond the scope of the construction project emergency management.

2.4 Target of Project Emergency Management

Construction project is a multi-stakeholder activities, project disaster carriers include owners, suppliers, building contractors, etc. As to different owner, the project disasters are not the same, and the impact on different items subject varying degrees are different. We should select suitable emergency management method according to different subject. At the same time, construction project emergency management should set specific goals for different subject's risks, and ultimately ensure the successful implementation of the construction project.

3 Metadata Management Model on Large-Scale Construction Projects Emergency Information Resource

3.1 Emergency Information Resource Metadata Definitions

Metadata is data that describes data, the use of metadata for emergency information resources management can make vast amounts of information resources related to emergency management organizations rational operation easier. Therefore emergency information resource metadata is used to describe, construct and query other related operations and decision-making required emergency information, emergency information resource metadata overall design has two main aspects: Firstly, the warehouse storage structure defined by the underlying data metadata indicates the field; secondly, it's necessary to build application metadata based on different description of the event and the decision-making.

Emergency information resource metadata consists of the following three parts: attribute concept information, status information and mapping relationship information. Metadata of general application program may include simple data (e.g. program parameters), and the complex data hierarchy structure description data between different subsystems and exchange information. However, due to emergency management in emergencies often do not have prior knowledge of the law and the development of events is not clear, so the metadata used to describe incidents and emergencies metamodeling has flexible structure, used to construct different metadata structures according to different emergencies. Therefore, the development and expansion of field information is essential content in the field of metadata management in emergency management.

3.2 Metadata Organization of Emergency Information Resource

The main task of information resources in the field of emergency management is to effectively describe and rationally organize the collected first-hand information of emergencies theoretical knowledge and accident scene, providing analysis and forecasting of port services. Emergency Information Resource Metadata is used to describe the emergency information resources and provide knowledge support metadata for emergency decision, which is the Meta data of Emergency Management. Therefore, according to the classification of emergency information resources, we can divide the emergency information resource metadata into two categories: application metadata and Core Metadata, the Core Metadata is used to describe the emergency core information resources, mainly for basic public emergencies knowledgeable description of the event, and the integration of application metadata and instantiate metadata, describing specific events in the perspective of the information resources establishment (Fig. 1).

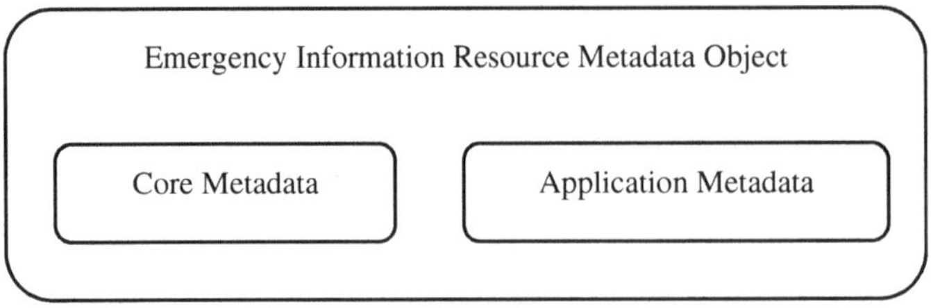

Fig. 1 Metadata object structure of emergency information resource

The emergency metadata could be described as a two-tuples, just like M = <M_{core}, M_{imp}>, here M_{core} stands for core metadata, and M_{imp} stands for application metadata.

M_{core} could be expressed as follows:

$$M_{core} = \{CM_{attr} \cup CM_{stat} \cup CM_{rela}\} \tag{1}$$

CM_{attr} stands for emergency accident attribute of core metadata, CM_{stat} stands for emergency accident state of core metadata and CM_{rela} stands for emergency accident mapping relation of core metadata (Tables 1, 2).

$$M_{imp} = \{IM_{attr} \cup IM_{stat} \cup IM_{rela}\} \tag{2}$$

IM_{attr} stands for emergency accident attribute of application metadata, IM_{stat} stands for emergency accident state of application metadata and IM_{rela} stands for emergency accident mapping relation of application metadata.

Table 1 Element analysis of emergency information resource core metadata

Meta-data type	Meta-data description level	Field content
Core meta-data	Property attribute	Title
		Definition
		Subjection
		Date
	State	Core data type
		Dependent type
		Operation type
	Mapping	Parent ID
		Interface

Table 2 Element analysis of emergency information resource core metadata

Meta-data type	Meta-data description level	Field content
Application meta-data	Property attribute	Accident ID
		Owner
		Parent collection
		Child node ID
		Interface service
	State	Whole
		Part
		Dependent relation
	Mapping	Relation function
		Portfolio template
		Redundancy set
		Redundancy source
		Relate collection

3.3 Metadata Management Model Framework of Emergency Information Resource

Metadata management is a very important part of data management, which provides a hierarchical way for the metadata construction based on the classification to emergency information resource metadata. Metadata can provide a front end support for users to access the storage resources from the bottom of the core metadata management, establishes a connection to the underlying storage resources through the core metadata management. From the view of the top business areas, metadata integrates emergency management process to define application metadata from a practical perspective of emergency accident application metadata, and completes the publishing, sharing and management of application metadata.

4 Large-Scale Construction Projects Metadata Management Model

4.1 Dynamic Management to Metadata of Emergency Management

As to general emergency decision management, metadata management platform has provided a list of available metadata fields for application management activities, including field domains and subdomains, which can solve the basic requirements for transactional work of emergency decision. But as to some unconventional emergencies, the general application metadata may not be able to describe the requirements of the events, thus providing dynamically adding and deleting basic metadata function in this model.

When a user enters the system after establishing an access, metadata management model establishes an access and manipulate metadata strategy in the middle layer. In a single physical domain, new application metadata can request metadata to insert into anywhere in the tree structure domain, which can be divided into emergencies leaf nodes adding or root domain adding. The emergency management system processes this request and add two domains involving different tree restructuring as to emergencies domain node dynamically adjustment. In emergencies domain, adding a node is to add and delete operations in common node leaves of the physical domain metadata; and from the entire application metadata tree domain, the domain is the root of emergency for the entire domain (Fig. 2).

When it comes to fields combination group of multiple applications metadata, only allow a leaf node to join the domain as metadata into the entire application domain, so at this time, only the parent domain increase a leaf node, which has no effect to the already established tree structure (Figs. 3, 4).

4.2 Function and Performance of Emergency Information Resource Metadata Modeling

The study of emergency information resource metadata in Emergency Decision Support System plays an important role in the emergency decision. How to forecast and analyze the existing data faster and more accurate and in accordance with existing data to predict future trends is the main goal of the in-depth study on metadata. Data analysis and processing component could be able to achieve integration data analysis from different dimensions and different levels. It's need to

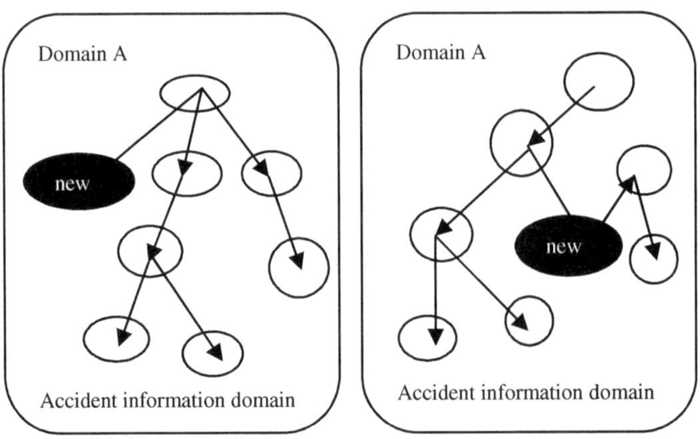

Fig. 2 Change of emergency domain root node

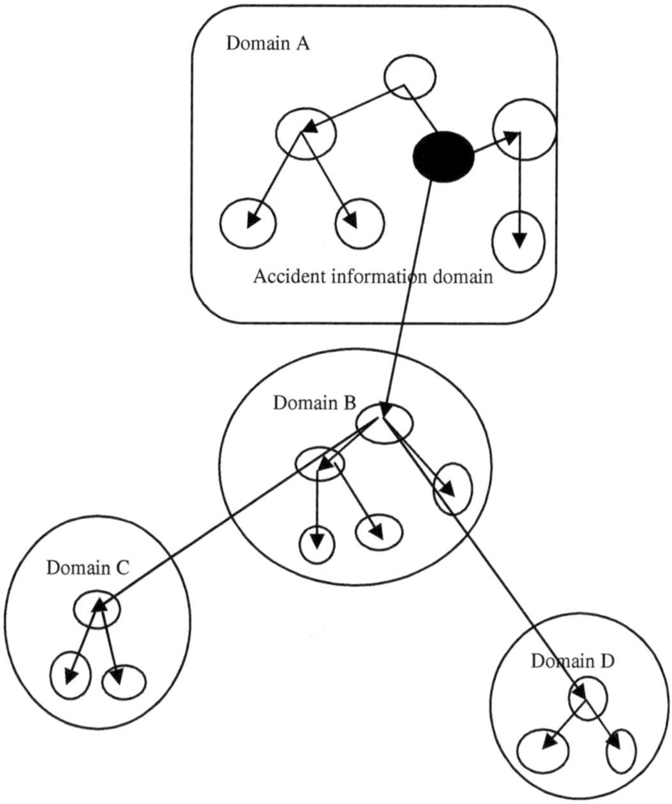

Fig. 3 Add root of application metadata

analyze a large amounts of data for emergency aid decision-making, which involves combination period of cross-table, cross-border, cross-field to abstract field acquisition, template generation and data processing.

Statistical analysis capabilities of emergency information resource metadata management provide the formation of a knowledge-based metadata by analyzing the mathematical model of decision support information returned to the user. When a user needs to make some decision to a contingency issue associated with the knowledge metadata element, the user needs emergency issues raised in the decision-making information, after authentication, the relationship metadata server will navigate on the emergency information, decision support information and emergency support resource information inside database server, to extract character sets emergency information for decision-making, information will be extracted into the statistical analysis, statistical analysis process to be transferred in accordance with the information for decision-making, according to the server to get the information set in predictive decision model.

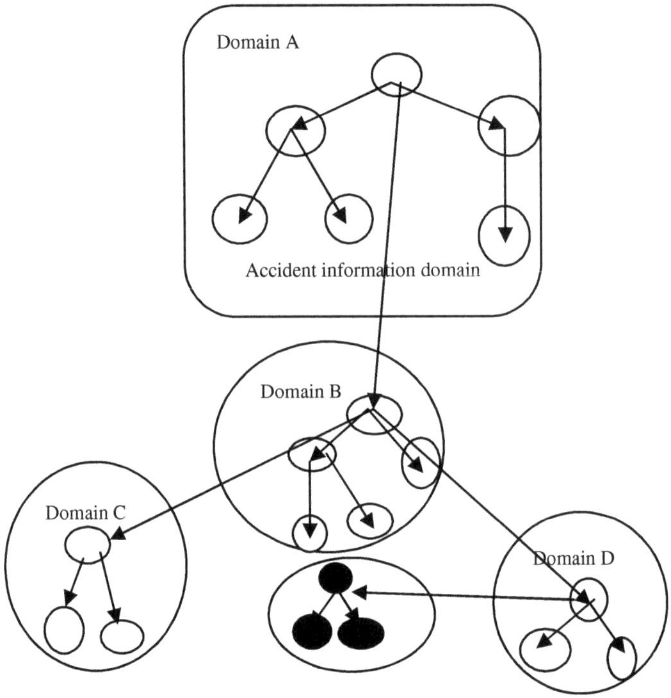

Fig. 4 A single physical domain join in tree structure

In the further development of the emergency decision support system, further research needs to be done on metadata management capabilities with the analysis and prediction model, which could be constructed by refining knowledge element metadata, requiring the development of different application-layer method for users to choose, the core layer needs establishment of core metadata management architecture, optimizing the structure of the core metadata. The tree structure used in Metadata Management is an adaptive restructuring of inquiry and sharing of information, which is a new algorithm with high efficiency.

5 Conclusion

The study subjects are from the project level, re-defined the concept of emergency, and also give the concept of large-scale construction projects Emergency Management. Through theoretical analysis to sort out differences and connections between large-scale construction project management and emergency management, explores the theoretical basis for large-scale construction projects and emergency management combined with the characteristics of the large-scale construction projects. This paper gives the level module of large-scale construction projects emergency management system; establishes a general operation mode of large construction projects Emergency Management. Based on the common characteristics of large-scale construction projects and construction projects disaster emergency management, combining theoretical research stage, this paper designs the construction project of emergency management model, which has universal significance.

This paper takes advantage of metadata modeling and large-scale construction project based on project characteristics emergency management model, combining metadata modeling, which provides new ideas and model to data acquisition, modeling, analysis and summary of large-scale construction engineering projects, improving modeling accuracy and reliability.

References

1. Mitroff II (1988) Crisis management: cutting through the confusion. Sloan management view, pp 212
2. Nickson D, Siddons S (2005) Project disaster and how to survive them. Kogan Page Ltd, London, pp 78–79
3. Bosher L, Dainty A The construction industry and emergency management: towards an integrated strategies framework [EB/OL]. Loughborough University. http://www.grif.umontreal.ca/Pages/BOSHER–%20Lee.Pdf
4. National research council of the national academies facing hazards and disasters understanding human dimensions. The National Academies Press, Washington D.C, USA, 2006
5. Neal DM (1997) Reconsidering the phases of disaster. Int J Mass Emergencies Disasters 15 (2):139–164
6. Boin A (2009) The new world of crises and crisis management: implications for policymaking and research. Rev Policy Res 26(4):367–377
7. van der Peijl J, Klein J, Grass C, Freudenthal A (2012) Design for risk control: the role of usability engineering in the management of use-related risks. J Biomed Inform 4:133–143
8. Lozowicka DH (2012) Problems associated with evacuation from the ship in case of an emergency situation. Int J Autom Comput 2:23–27
9. Kutsch E, Hall M (2009) Deliberate ignorance in project risk management. Int J Project Manage 3:119–128
10. Yanzhang W, Tianwei F (2009) Metadata management model for emergency information resources. In: ACM international conference on web intelligence WI, pp 389–392

Model Construction of Individual Risk Field of Hazards Based on the Theory of Risk Field

Ying Zhang, Yun Luo, Zhu Zeng, Tao Ma and Xiao-xia Yu

Abstract The mature regional risk indexes abroad are individual risk and social risk, whereas few effective methods are to evaluate the influence of hazards on the regional risk in domestic. This paper studies the research of field theory in physics and combines with the definition of individual risk of hazards. By using analogism method of electric field to the risk analysis,the individual risk field of hazards model is established and each parameter in the model is explained. The direct and clear individual risk distribution of hazards is of certain guiding significance for the overall assessment of urban and regional system risk influence.

Keywords Hazards · Individual risk · Risk field model · Risk superposition

1 Introduction

In recent years, with the rapid development of economy, major hazards in China have made rapid development in the area of quantity and scale, and fire, explosion and poison gas leak are the main accidents. Once the accidents happen, the influence area is often beyond the boundaries of industrial park and the construction project [1]. It has not only serious consequences, but also wide destruction, within the scope of the enterprise, and around urban region system. Thus, accurate assessment of risk distribution for hazards of the whole region is very necessary. On this basis, reasonable planning can make to build, rebuild and regulate of major hazards in the area.

The mature regional risk indexes abroad are individual risk and social risk, and the countries using these indexes formulate the corresponding calculation method according to their actual situation, and take them as one of the regional planning or decision-making basis. The roles of individual risk and social risk are different in

Y. Zhang (✉) · Y. Luo · Z. Zeng · T. Ma · X. Yu
School of Engineering and Technology, China University of Geosciences, Beijing, China
e-mail: sunflower8823@126.com

E. Qi (ed.), *Proceedings of the 6th International Asia Conference on Industrial Engineering and Management Innovation*,
DOI 10.2991/978-94-6239-145-1_84

analysis and decision, individual risk pays attention to "point", which is the quantitative risk description on the micro level, aiming at a concrete goal; and social risk focuses on "area", reflecting the public risk, which is the quantitative risk measure on a macro level [2].

At present, there is lack of effective method to evaluate regional hazards impact on the regional risk in China. Therefore, this paper studies the research of field theory in physics and combines with the definition of individual risk of hazards. By using analogism method of electric field to the risk analysis, the individual risk field of hazards model is established and each parameter in the model is explained. The direct and clear individual risk distribution of hazards is of certain guiding significance for the overall assessment of urban and regional system risk influence.

2 Relevant Concepts

2.1 *Field theory*

Field is put forward by theoretical physics, which is a category of physics, and according to the understanding of character, "field" refers to the certain physical quantities in a distribution in the space. Field is used as a material object form, it has a certain relationship with physical exists in the field, for example, the gravitational field exists because of a kind of gravitational mass interface; the existence of electrostatic field because of the physical takes charge; current magnetic field because of the existence of physical objects exist in the current distribution. Physicals with these substances (mass, charge and current) are called field sources. Field and field sources are no primary and secondary, they are two different forms of physicals existing in the space.

Vector Analysis and Field Theory [3] defines the field as, if each point in space corresponds to a certain value of a dangerous amount, so a field of this amount in this space is determined. According to the definition of field, risk field can be defined as risk distribution in the space of hazards. If risk field exists around the hazards, other units and systems within the scope of the field will be threatened by risk of the hazards. Thus the field theory is reasonably introduced to the hazard risk assessment.

2.2 *Individual Risk Field*

Risk may be harm and danger brought by temperature (heat), force (overpressure), electricity, corrosion and toxicity such dangerous variables, when these hazards of different properties in space distribution work, the physical hazard indexes should not be directly added, because of their different risk attributes [4]. As a common attribute indicator of these indicators, individual risk is selected for risk superposition.

Individual risk (IR), as used by the Dutch Ministry of Housing, Spatial Planning and Environment (VROM), is defined as the probability that an average unprotected person, permanently present at a certain location, is killed due to an accident resulting from a hazardous activity [5].

$$IR = P_f P_{d/f} \tag{1}$$

where Pf is the probability of failure and Pd/f is probability of dying of an individual in the case of failure, assuming the permanent unprotected presence of the individual.

A slightly different definition, which considers whether or not the individual is actually present, is used by the Dutch Technical Advisory Committee on Water Defences (TAW) [6] and by Bohnenblust [7]. And another different definition is used by the UK' health and safety executive (HSE). According to this body, the individual risk is the risk that a typical user of a development is exposed to a dangerous dose or worse of toxic substance, heat or blast overpressure [8]. A dangerous dose is likely to cause the person severe distress or injury, but it does not lead to certain death.

Based on the above definitions, the individual risk (IR) in this paper is defined as: the probability that an average unprotected person, permanently presents at a certain location, die or injure due to various potential accidents by all the hazards in evaluation area. This definition reflects the inherent risk level in all parts of the area, considers the multiple hazards existing at the same time, and also the consequences of comprehensiveness.

Individual risk field is the individual risk spatial distribution caused by the hazards, which assesses risk situation from the risk and the risk distance such influence factors, realizing the quantitative risk assessment of hazards, reflecting the real level of risk in all parts of the region. Individual risk field gives a clearer display of individual risk spatial distribution of hazards, showing directly, clearly and completely, and provides a kind of algorithm to define the acceptable risk range in space.

3 Model Construction

3.1 Physics Model

There are many field theories in modern physics, and also a variety of field models. According to the definition and features of individual risk field, electric field model is selected for the analogy analysis.

Table 1 The property of electric field

Name	Force	Field strength	Field strength superposition
Electric field	$F = k\frac{q1q2}{r12^2}$	$E = k\frac{q}{r^2}$	$E' = E_1 + E_2 + \ldots + E_n = \sum_{i=1}^{n} E_i$

Attributes of electric field are shown in Table 1. For electric field, in a vacuum, the interaction force of the two stationary point charges is proportional to product of the absolute value of electricity, and inversely proportional to the distance. The direction is along the two attachment point charge. Field strength E is one of the important characterizations of electric field, and field strength is only related to source charge and spatial location.

3.2 Individual Risk Field Model

In physics research, analogical reasoning is an important exploration method [9]. By comparing two different objects (types), the same or similarities between them are find out, and then based on this, knowledge or conclusion of one object is put to another object (type).

If electric field model is introduced into individual risk model, some of the same or similar nature between them should be identified, and then some attributes of electric field can be introduced to the individual risk field theory. Finding the same or similar attributes for electric field and individual risk field becomes the key point to the construction of individual risk field model. Due to the application of analogical reasoning principle, the similar attributes of the objects are paid attention to, and the differences of objects are not discussed.

According to the concepts of the electric field and individual risk field, contrast of similar properties of two fields is shown in Table 2.

The isomorphism of analogy reasoning method is used [9], when two systems have the one-to-one correspondence elements, so a mathematical model of the describe system can be expressed to the research system, thus a mathematical model is established.

Table 2 Contrast of similar properties of two fields

Electric field	Individual risk field
Point charge	Field source(hazards)
Force value is proportional to the charge size	Individual risk value is proportional to the hazard strength
Force value is inversely proportional to the distance	Individual risk value is inversely proportional to the distance

As well as other physical fields, there is also field strength in individual risk field, representing the effect degree of hazards at every point at individual risk field. The field strength of electric field is a vector, and has direction, but accidents caused by hazards bring damage consequences for the individual, so in this paper the direction is regardless, digital superposition is taken for field strength. The field strength E of a certain point M in individual risk field is:

$$E = k\frac{PL}{r^2} \tag{2}$$

where E is the field strength of individual risk field, k is attenuation coefficient of risk, between 0 and 1, P is possibility of accidents, L is loss of accidents, and r is the distance from hazards(O) to point M(x, y, z).

3.3 *Risk Strength Superposition*

If a point is within the scope of the risk field of two or more hazards at the same time, then it will be subject to the risk of multiple threats. Hazards usually have point-line-surface structure and point structure hazards are relatively more, and the method is simple, the following will be discussed from a point structure hazard superposition, line and surface structure hazards temporarily are not researched. According to the principle of superposition of field strength, individual risk field strength of more hazards to M is:

$$E' = E_1 + E_2 + \ldots + E_n = \sum_{i=1}^{n} k_i \frac{P_i L_i}{r_i^2} \tag{3}$$

where E' is the field strength of individual risk field for many hazards after superposition, n is the number of the hazards, k_i is attenuation coefficient of risk for the ith hazard, between 0 and 1, P_i is possibility of accidents for the ith hazard, L_i is loss of accidents for the ith hazard, and r_i is the distance from the ith hazard to point M(x, y, z). Thus it is available to describe the region's overall quantitative evaluation result of risk condition.

4 The Determination of Parameters

4.1 *Accident Possibility Value*

The accident probability of major hazards is obtained through many years' accumulation of accident data statistics, the model is as follows, see Table 3.

Table 3 Accident possibility value

The frequency of accidents	Score value
Once a day	365
Once a week	52
Once a month	12
Once half a year	2
Once a year	1
Every n year	1/n

$$P = \frac{365}{I} \quad (4)$$

where P is accident possibility value of the hazard and I is accident average interval.

4.2 Accident Loss Value

In the definition of individual risk, accident consequences include two aspects: deaths and injuries, the human body vulnerability model is used to describe the loss. Individual vulnerability model can be established based on the corresponding relationship individual factors exposed to different severity consequences, and it is used to determine the safety and health losses of individual caused by accidents, see Table 4.

According to the model of accident consequences(gas leakage model and vapor cloud explosion model etc.),toxicant concentration values, shock wave overpressure value and thermal radiation flux produced by a certain major accident scene are calculated, and based on individual vulnerability model and probability equation, individual death probability pD can be obtained. According to the corresponding probability variable "Y", the conditional probability of all sorts of consequences (death, all kinds of damage etc.) can be calculated by probability equation [10] due to the exposure dose of D, namely,

$$p_D = \frac{1}{\sqrt{2\pi}} \int_{-\infty}^{Y-5} exp\left(-\frac{u^2}{2}\right) du \quad (5)$$

Table 4 Individual vulnerability model

Factors	Probability variable math model	Dose D
Thermal radiation	$-37.23 + 2.56\ lnD$	$I^{1.33}\ t_{ex}$
Shock wave overpressure value	$5.13 + 1.37\ lnD$	P_s
Poisonous gases	$k1 + k2\ lnD$	$C^a\ t_{ex}$

4.3 Attenuation Coefficient of Risk

After the accident caused by major hazards, firefighting department is responsible for controlling accidents and rescuing people trapped, and timely and effective medical treatment is the key to the wounded man out of danger. Therefore, attenuation coefficient of risk is composed by fire compensation coefficient F and medical compensation coefficient M these two parts [11].

$$k = F * M \tag{6}$$

5 Conclusion

Based on field theory, individual risk concept is established, a new pattern of major hazards' effect to urban public safety is put forward. Learning from the physics model of electric field, individual risk strength model is established, and each parameter in the model is illustrated. Individual risk field research belongs to a new area in the safety engineering, and the related research is rarely. Individual risk strength theory has not related, so this theory has a certain guiding significance for risk research. Individual risk strength mentioned in this article is only a starting point, there is still a lot of work to be further discussed and studied on the model in the application of risk assessment of urban major hazards.

Acknowledgments We gratefully thank the valuable guide of our supervisor and the members of our laboratory.

References

1. Jiang T (1999) Major chemical industry accidents for 50 years in China (in Chinese). Chem Labor Protect 20(11):403–406
2. Ma Y, Li Z, Ni K, Yang L (2012) A quantitative chemical industry area risk assessment model (in Chinese). J Safe Environ 12(5):239–242
3. Shu-yi XIE (1990) Vector analysis and field theory (in Chinese). Higher Education Press, Beijing
4. Huang Y, Li J, Zhang B, Liang D (2008) The research of assessment theory based on risk field (in Chinese). J Safe Sci Technol 4(6):101–105
5. Bottelberghs PH (2000) Risk analysis and safety policy developments in The Netherlands. J Hazard Mater 71:59–84
6. TAW (1985) Technical Advisory Committee on Water Defences, Some considerations of an acceptable level of risk in The Netherlands, TAW
7. Bohnenblust H (1998) Risk-based decision making in the transportation sector. In: Jorissen RE, Stallen PJM (eds) Quantified societal risk and policy making. Kluwer Academic Publishers, Dordrecht

8. Health and Safety Executive (1989) Risk criteria for land use planning in the vicinity of major industrial hazards. HSE Books
9. Yuan X, Gong Y (2003) A brief study on the method of analogy(in Chinese). J Hebei Inst Technol (Soc Sci Ed) 3(11):84–88
10. Papazoglou I, Aneziris O, Bonanos G et al (1996) A computerized toolkit for quantification of the risk from accidental releases of toxic and /or flammable substance. Int J Environ Pollut 6:500–533
11. Chen G, Zhang J, Zhang H, Yan W, Chen Q (2006) Study on regional risk assessment methodology (in Chinese). China Safe Sci J 16(6):112–117

The Study of Impact of FDI on Innovation Capacity in Different Industries

Hong-fei Yin and Qun Yin

Abstract There are more and more FDI to enter Chinese market after China joined the world trade organization in 2001. Now, FDI has become an important driving force for China's economic development, enhancing ability of independent innovation. Based on the related analysis of the difference of innovation capacity influenced by FDI among different industries, this paper discusses the mechanism of FDI impacting on ability of independent innovation, and finally some suggestions using FDI to improve ability of independent innovation of different industries, are given.

Keywords FDI · Ability of independent innovation of industry · Differentiation · Foreign capital policy

1 Introduction

Many of emerging and developing countries cultivate and improve their ability of independent innovation through taking positive policy to attract foreign investment, and eventually catch up with developed country on the technology. Since our country reformed and opened up, government has been putting the introduction of foreign capital as an important policy to promote economic development. In 2013, FDI inflowing our country reached $102.4 billion, second only to USA. As shown in Fig. 1, the amount of actual utilized FDI in our country overall has been growing from 2001 to 2013, and the amount of actual utilized FDI in 2013 is 2.5 times as much as 2001s. There are three main reasons to attract our foreign capital: the rapid development of China's economy, the huge market potential and relatively cheap labor resources. FDI has brought advanced technology, management system and

H. Yin (✉) · Q. Yin
College of Management, Nanjing University of Posts and Telecommunications, Nanjing 210023, China
e-mail: 15951933177@163.com

E. Qi (ed.), *Proceedings of the 6th International Asia Conference on Industrial Engineering and Management Innovation*,
DOI 10.2991/978-94-6239-145-1_85

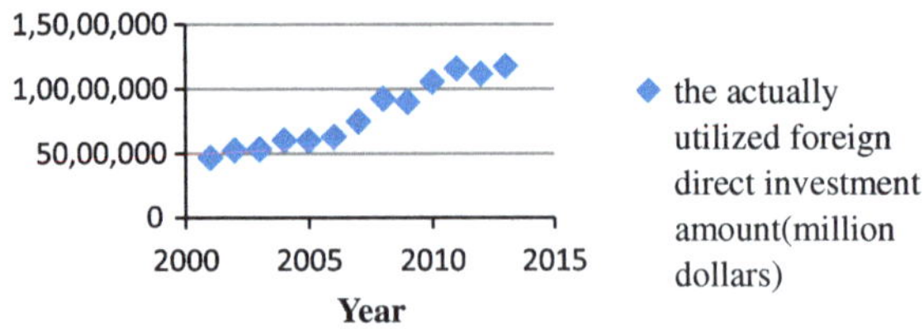

Fig. 1 The map of the changing amount of actual utilized FDI in China from 2001 to 2013

abundant human capital to China, but also created a large number of employment opportunities, promoting China's industry transformation, industry upgrading and the expansion of China's import and export trade scale. In recent years, FDI plays a more and more important role in promoting social economy in China.

Needless to say, the influx of FDI not only inject a large amount of financial capital for China's economic development, but also brought advanced production technology and scientific management methods, and has become a powerful driving force of economic growth and foreign trade development [1]. But the question, can FDI effectively enhance the capability of independent innovation of the host country, has always been the focus of discussions in the academic world.

A few scholars believe that FDI can promote the development of independent innovation ability of the host country. Foreign capital companies which are easy to get domestic bank loans are more innovative than the other companies through studying the relationship between FDI and China innovation activities [2]. The inflow of FDI will cause the foreign capital company get into the high-tech industry and lead the companies with less ownership to do R&D [3]. The entry of FDI has positive spillover effects in the process of innovation [4].The inflow of FDI can significantly improve the technical innovation level in China [5]. FDI has obvious promotion effect on the regional innovation capability of China [6].

There are also a few scholars who believe that the FDI can't obviously improve the independent innovation the capability of host country, and hinder the development of independent innovation ability. Bishop and Wiseman [7] found FDI is negatively related to R&D after doing some research [7]. Fan et al. [8] analyzed the supplement and substitution role which FDI plays on R&D of a developing country theoretically, and the estimation results show that the FDI plays a negative role on domestic R&D input in China [8]. An empirical test of Zhou [9] showed that FDI is negatively related to independent innovation of western regions in China [9]. FDI can inhibit the technological progress of the host country after researching on UK [10].

And others believe that the relationship between the two is not significant. Chen et al. [11] analyzed the effect of FDI on promoting the regional innovation capacity by using the various regions of the country data. The results showed that the inflow of FDI has no significant effect on enhancing the ability of independent innovation and building the ability of science and technology [11]. The influence of FDI on China's innovation output is very limited and FDI did not play a leading role [12].

The influence of FDI has been researched by domestic and foreign scholars in many aspects. These aspects included national perspective, different areas of the economy and the industry point of view. The different research methods, research ideas, model and index have an important reference role on further research. The study on difference of impact of FDI on innovation capacity in different industries is rare not only in China but also in other countries, and this paper mainly starts from this perspective, analyzes reasons for the formation of difference deeply in order for providing reference to enhance the ability of independent innovation of our country industry in the future.

2 Data Collection, Variables Selecting and Model Defining

2.1 Source of Data

This paper uses a time-series data. The period of sample is from 2002 to 2012, and the data is mainly from *China Statistical Yearbook of Science and Technology*, *China Industrial Statistics Yearbook* and China's National Bureau of Statistics website. In this paper, the research object is medium-sized enterprises in 2003–2010, while in 2011 and 2012 this paper uses regulations on Enterprises whose data are processed instead of medium-sized enterprises as the research object due to the statistical abolition of the Statistical Yearbook.

2.2 Variable Definition

What this paper explores is study of difference of FDI on innovation capacity in different industries. One first involving problem is how to measure the ability of independent innovation. The international popular indicators which evaluate scientific output efficiency include: patents, scientific papers, technical trade, high-tech products or technology-intensive products and so on [13]. In China, the academic generally utilize the number of patent applications examined and authorized to measure capability of independent innovation. While, the number of patent applications examined is used more than the number of patent applications authorized, because most scholars believe that later is lack of objectivity since it is easily influenced by other factors. So this paper also selects the number of patent applications examined in the industry as the metrics of the capability of independent innovation.

There are not detailed data of FDI by sector in China, so this paper use paid-in capital of foreign in *China Industrial Statistics Yearbook* to measure FDI by sector, and 2002 is as the research base. Years of foreign paid-up capital of each industry are done comparable correction by CPI in order to eliminate the impact of price changes in the data analysis. As the process of improving industry's capability of independent innovation is not easy, this paper takes a lag period during quantitative analysis. Human capital is the source of innovation [14]. R&D fund is the material base of technology innovation ability into the realistic competitiveness [15]. So there are two main variables of industry innovation input. One is the R&D staff input, which is measured by R&D personnel FTE of each industry. The other is R&D funding, which is measured by the R&D costs of each industry.

2.3 Model Definition

The capability of independent innovation essentially can be considered as the capability of innovation output. So the model references Cobb-Douglas production function which is widely used in the academic circles the empirical models are shown as follows:

$$\text{Patent}_{it} = \theta_{it} \times RDS_{it}^{\alpha_1} \times RDP_{it}^{\alpha_2} \tag{1}$$

From the above analysis we can know, the output efficiency of industrial independent innovation is greatly affected by FDI. So: $\theta_{it} = \beta_1 \times FDI_{it}^{\alpha_3}$. After bringing formula into the Eq. 1 and taking the logarithmic on both sides, the model can the following:

$$\ln(\text{Patent}_{it}) = \alpha_0 + \alpha_1 \ln(RDS_{it}) + \alpha_2 \ln(RDP_{it}) + \alpha_3 \ln(FDI_{it}) + \mu_{it} \tag{2}$$

In order to take interaction into consideration, which is between scientists and the spillover effect of FDI scientists and internal R&D funds. This paper put the product interaction term of FDI_{it} and RDP_{it} into model (2), so the empirical models are shown as follows:

$$\begin{aligned} \ln(\text{Patent}_{it}) = \alpha_0 + \alpha_1 \ln(RDS_{it}) + \alpha_2 \ln(RDP_{it}) + \alpha_3 \ln(FDI_{it}) + \alpha_4 \ln(FDI_{it}) \\ \times \ln(RDP_{it}) + \alpha_5 \ln(RDP_{it}) \times \ln(RDS_{it}) + \mu_{it} \end{aligned} \tag{3}$$

In the model (3), i refers the industry and t refers the time. So RDS_{it} refers the number of the R&D funds of i industry in time period of t. RDP_{it} refers R&D staff input of i industry in time period of t. Similarly, Patent_{it} refers independent industry output capacity. FDI_{it} refers the scale of foreign investment. μ_{it} refers random disturbance.

3 The Results of the Analysis

This paper chooses four industries to analyze. High-Tech industry selects communications equipment, computers and other electronic equipment manufacturing industry. Industry of raw materials and manufacture selects industry of ferrous metal smelting and rolling. General manufacturing industries selects beverage manufacturing industry, while the resource industry selects industry of non-ferrous metal mining.

As shown in Table 1, we can see that a significant degree of each variable coefficient in linear regression model. Obviously, the impact of FDI on the industry of independent innovation is different in different industries. The difference mainly has three aspects. The first aspect is the difference of FDI significant degree. Overall, FDI is positive to the capability of independent innovation of the four industries, but in non-ferrous metal mining industry is not significant. The second aspect is that the degree of influence of FDI on the capability of independent innovation in the other three industries, which are significant. The influence of computer information industry is weakest, followed by beverage manufacturing industry, and the impact of non-ferrous metal smelting and rolling processing industry is greatest. The third aspect is that the influent extent of FDI on the ability of independent innovation is different from the other variables' in the same industry. In the computer information industry, the influence degree of FDI is weaker than scientists' and R&D funds'. In the industry of ferrous metal smelting

Table 1 Analysis results of linear regression of each industry

	Computer information industry	Industry of ferrous metal smelting and rolling	Beverage manufacturing industry	Industry of non-ferrous metal mining
Constant	-59.722^{**} (−5.785)	-190.620^{**} (−6.725)	-68.337^{***} (−6.683)	14.236^{***} (8.244)
$\ln(FDI_{it})$	0.983^{*} (4.071)	15.242^{**} (7.151)	4.810^{***} (6.385)	0.191 (1.932)
$\ln(RDS_{it})$	4.267^{*} (5.134)	−2.429 (−0.127)	−2.879 (−0.539)	22.617 (0.996)
$\ln(RDP_{it})$	3.004^{*} (1.155)	16.558^{**} (6.296)	1.484^{**} (4.795)	-3.184^{***} (−8.536)
$\ln(FDI_{it}) \times \ln(RDP_{it})$	−39.065 (−2.812)	-1.085^{**} (−6.576)	11.007 (0.613)	−8.524 (−2.358)
$\ln(RDP_{it}) \times \ln(RDS_{it})$	-0.264^{**} (−4.528)	-0.240^{**} (−5.631)	-0.104^{**} (−5.333)	0.142^{**} (5.205)
R^2 after the adjustment	More than 0.9	More than 0.9	More than 0.9	More than 0.9

Note The dependent variable is the number of invention patents in corresponding industries, values in brackets are the t statistic for each variable regression coefficient. * represents a significant level of 0.1. ** represents a significant level of 0.05. *** represents a significant level of 0.01

and rolling, the influence degree of FDI is only after the scientists'. In the beverage manufacturing industry, the influence degree of FDI is the most significant. In addition, it can be seen from the table: R&D costs can't effectively promote development of ability of independent innovation. There are some problems in R&D costs of industry and allocation of scientists. FDI and scientists of industry did not effectively combine together to improve the capability of independent innovation.

4 Conclusion and Suggestion

According to the results of empirical analysis, different industry should take the different foreign capital policy. It can be seen from the empirical results: The ability of FDI to promote independent innovation in high-tech industry is limited, unlike other industries which is so remarkable. The paper takes the computer industry for an example because it is the leader of high-tech information industry. With the rapid development of the computer industry in China, China has a number of world-renowned computer companies. Lenovo acquired IBM's notebook computer business, which illustrates our technology has reached the world advanced level. This kind of industry already have the world-class abilities of independent innovation and are able to carry out large-scale independent research. So technology spillover of FDI does not have much effect on improving ability of independent innovation in these industries. Furthermore, the introduction of foreign capital for high-tech industry is not the main means to improve their capability of independent innovation. These companies can cooperate with the developed countries, establish research institutions and joint venture company to further promote the innovation capacity of the domestic industry. In addition, the rapid development of other industries in China cannot do without technology spillover of FDI. In order to further improve the ability of independent innovation of the raw materials industry and general manufacturing industry, the government should further open their markets to attract more FDI into the kind of industry. The introduction of FDI should be based on business and market demand, and the government should play a guiding and safeguarding role. The examination and approval procedure which is used to introduce FDI in China is so cumbersome that the efficiency of utilization of foreign capital is low. Reducing foreign investment approval procedure can attract more foreign capital to enter China's market. Moreover, the government should be fair in the domestic capital. Some industries allow state-owned enterprises and foreign enterprises to enter, but the private enterprises are excluded from the respective industries so that the private enterprises cannot compete with foreign investment. It should allow the private sector to enter the industry when government opens industry. On the one hand, competition in the industry can enhance the vitality, improve R&D initiative. On the other hand, the private enterprises can better absorb foreign technology spillovers because it has more enthusiasm of innovation than state-owned enterprises and ultimately promote ability of independent innovation of whole industry.

Acknowledgments Project Fund: Project supported by the National Natural Science Foundation of China (Grant No.71373133).

Authors: Yin qun (1958-), female, Dean of the Nanjing University of Posts and Telecommunications School of management, professor. Yin hongfei (1991-), male, graduate student of Nanjing University of Posts and Telecommunications School of management.

References

1. Shen Kunrong G (2001) Foreign direct investment, technology spillover and endogenous economic growth–China data econometric test and empirical analysis. Chinese Social Sci 05:82-93206 (Chinese)
2. Girma S (2008) Foreign direct investment, access to finance, and innovation activity in Chinese enterprises. World Bank Econ Rev 22:367–382 (Chinese)
3. Subash S (2011) Foreign direct investment and R&D: substitutes or complements—a case of indian manufacturing after 1991 reforms. World Dev 39(7):1226–1239
4. Vahter P (2011) Does FDI spur productivity, knowledge sourcing and innovation by incumbent firms? Evidence from manufacturing industry in Estonia. World Econ 34(8):1308–1326
5. Wenguang B (2007) Effect of foreign direct investment on technology innovation in China—based on the research of areas. J Finan Econ 06:4–17 (Chinese)
6. Dayong S (2009) Research on the promotion effect of international direct investment on the regional innovation ability—microscopic mechanism and empirical analysis, soft. Science 06:62–65 (Chinese)
7. Bishop P, Wiseman N (1999) External ownership and innovation in the United Kingdom. Appl Econ 31(4):443–450
8. Fan C, Hu Y, Zheng H (2008) Theoretical and Empirical Study of effect of FDI on technological innovation of domestic enterprises. Econ Res 01:89–102 (Chinese)
9. Zhou B (2010) FDI and domestic enterprise innovation in Western China. In: Proceedings of 2010 international conference on regional management science and engineering, Jinan, pp 152–155 (Chinese)
10. Haskel JE (2007) Does inward foreign direct investment boost the productivity of domestic firms? Rev Econ Stat 89(3):482–496
11. Chen J, Chen Y, Yu F (2007) Effect of FDI on promoting regional innovation ability in China. Sci Res Manag 01:7–13 (Chinese)
12. Zeng S, Jiang X, Cai M (2014) The influence of foreign direct investment on multiple dimensions of innovation output. Manag Rev 11:28–38 (Chinese)
13. Chen G (2002) Economic globalization and strategy of technological development in China. Economic Science Press (Chinese)
14. Fu H, Peng C (2014) The impact of human capital on independent innovation ability of the enterprise. Oper Manag 01:145–148 (Chinese)
15. Guo J, Ye W, Wu X, Yuan Y (2014) Research on the evaluation of the comprehensive economic benefits of investment R&D funds based on input output method. Sci Manag 01:9–15 (Chinese)

Study on the Regional Difference of Green Innovation Efficiency in China—An Empirical Analysis Based on the Panel Data

Yue Cheng and Qun Yin

Abstract Nowadays resources shortage and environmental pollution is becoming a serious issue. Green innovation has become an important tool of a country to keep competitive advantage. In this paper green innovation efficiency of different regions are measured by means of DEA (BBC model). According to the research, we find that the green innovation efficiency in our country is growing in the years 2008–2013 on the whole. Moreover, there is a significant difference in the green innovation efficiency of regions showing from strong to weak gradient distribution from the east, central, and west to northeast region. Policy suggestions are put forward on the basis of empirical results.

Keywords DEA · Green innovation · Regional difference

1 Introduction

Since the reform and opening up, Chinese national economy develop rapidly. The emphasis on economy development leads to the serious pollution of natural environment. Haze has dominated recently and environmental pollution became serious. The transformation of economic development mode and industrial re-constructer are the central task of national economy. Therefore, technology innovation became a significant means to speed up the transformation of economic development mode. However, it creates a great pressure to the environment when the traditional technology innovation promoted the development of economy.

Y. Cheng (✉) · Q. Yin
College of Management, Nanjing University of Posts and Telecommunications, Nanjing 210023, China
e-mail: emmacheng0922@163.com

E. Qi (ed.), *Proceedings of the 6th International Asia Conference on Industrial Engineering and Management Innovation*,
DOI 10.2991/978-94-6239-145-1_86

So we should upgrade the traditional theory of technological innovation in order to develop green technology innovation. Green technology innovation is a technological innovation theory, combining ecological theory and technical innovation. Green technology innovation can guide technology innovation to resource conservation and environmental protection for easing resources excessive usage and environment damage.

In recent years, a number of scholars put environment and energy factors into the innovation research. One is the relationship theory of technology innovation and green growth. Hall and Kerr [1] focused on the ecological environment aspect of the green economy and measured green index in 1991 and 1992 [1]. Leipert researched green GDP and welfare indicators [2]. Chen Jin studied the relationship of environmental performance, economic performance and the green technology innovation performance and pointed out that we must persist in the course of ecological innovation to achieve sustainable economic development [3, 4]. Li Can (2007) also explored relationship between technology innovation and green economy [5]. There also has been some research exploring the efficiency of green innovation. Using SBM model, Qian zhengming, Liu Xiaochen have analyzed the regional differences of the green economy efficiency in China that found China's overall economic efficiency of green is at a low level compared with the traditional economic efficiency [6]. Bai Junhong, Jiang Fuxin measured the efficiency of innovation by three stage DEA method, reported that the scale efficiency is the main reason for the our low efficiency of innovation [7]. The discovery of innovation efficiency by Han jing indicated that the entry of foreign capital and adjustment of the structure play a positive role on promoting low green [8]. Liu Mingguang put the amount of sewage and exhaust gas volume as the green output, believed that the innovation efficiency of China's regional innovation system in the area do not exist convergence characteristics [9]. These studies have laid the foundation for this research.

According to the research on the literature, we find that most of research describe the environmental pollution as the undesirable output with the expected output (GDP, innovation output) to the production process [10, 11]. In this way, it only considered the negative effects of the pollution emissions on the growth of economy without thinking that the natural environment, as ecological resources, also has the positive effect on innovative growth. Moreover in view of the selection of environmental pollution indicators, the majority scholars use the original pollution emissions and energy consumption to measure the environmental and energy indicators or by means of principal component analysis method and other static methods [12, 13]. This mode is characterted by lacking dynamic comprehensive evaluation of environmental pollution, which has a great influence on the accuracy of measurement.

Therefore, this paper measures the comprehensive index of environmental pollution by using entropy method and applies it as inputs into the efficiency of green technology innovation model. Then use the method of DEA to measure regional differences of green technology innovation efficiency in the 30 provinces (except Tibet) from 2008 to 2013.

2 Measure of Green Innovation Efficiency

2.1 Research Methods

This paper uses the data envelopment analysis to study Chinese green innovation efficiency. Data envelopment analysis (DEA) is obtained by linear programming method to construct a non-parametric frontier. It aims to measure the decision making unit (DMU), a method of relative efficiency, with multiple inputs and multiple outputs [14]. We made every province as a decision making unit to apply BCC model (a scale variable model) to evaluate green innovation efficiency of each regions. For each DMU, the BBC model is:

$$\text{Min}\text{ß},\ \lambda\text{ß}$$

$$\text{s.t. ß}X_{\text{j}0} - \sum_{j=1}^{j} \lambda_{\text{j}} X_j > 0 \tag{1}$$

$$-\text{Y}_{\text{j}0} + \sum_{j=1}^{j} \lambda_{\text{j}} Y_j > 0 \tag{2}$$

$$\sum_{j=1}^{j} \lambda_{\text{j}} = 1 \quad \lambda_j > 0,\ \ \text{j} = 1,2,\ldots,\text{j},$$

Among them, λ_{j} is the weight coefficient, ß is the target value, the optimal solution can be get according to the linear programming. Where *Xj* represents the input vector elements of each regions, *Yj* represents output vector elements of each regions, *Xj* = *(X1j…Xmj)*, *Yj* = *(Y1j,…Ynj)*. Among them j denotes province, m subscripts input factors, *n* subscripts output factors. *m* = *{1,2…M}*, *n* = *{1…N}*

2.2 Input and Output Index

This paper uses a panel data. The period of sample is from 2008 to 2013, and the data is mainly from *China Statistical Yearbook of Science and Technology, China Environmental Statistics and China's National Bureau of Statistics website.* The research objects are 30 provinces (except Tibet) in China. The division method of china is West and East and northeast of china.

2.2.1 Input Variables

1. The input of production factor: We choose scale industrial enterprises R&D Personnel in each province to represent human resources investment of innovation. In view of capital investment, we select sum of expenditures on R&D and costs of technical renovation.
2. Energy consumption: Energy consumption is measured by energy consumption in GDP unit area.
3. Environmental costs: We choose drainage amount of industrial waste water, the emission of industrial waste gas and industrial solid waste to test environmental pollution Comprehensive index, which was calculated by entropy method. (Due to limited space, omit this part)

2.2.2 Output Variables

This paper selected the number of patents and new product sales income as output indicators. Patents, as the direct output of innovation activities, is an important form of innovation achievements. Sales revenue of new products is innovation achievements on economy. Therefore this paper also selects the sales revenue of new products as the innovation output index [15].

In conclusion, this paper has four items of the input index, two items of output index through the DEA model.

2.2.3 Empirical Results

This paper uses DEA model (BCC) through Deap2.1 software, to calculate green technology innovation efficiency of 30 provinces (except Tibet), shown in Fig. 1 and Table 1.

From the Fig. 1 we can see that from 2008 to 2013 the overall efficiency of innovation of green technology in China is on the upward trend, but there is still great room for improvement. Green innovation efficiency differ from regions, especially between East and West and northeast. The top5 areas of green technology innovation efficiency are Zhejiang, Guangdong, Hainan, to Beijing and

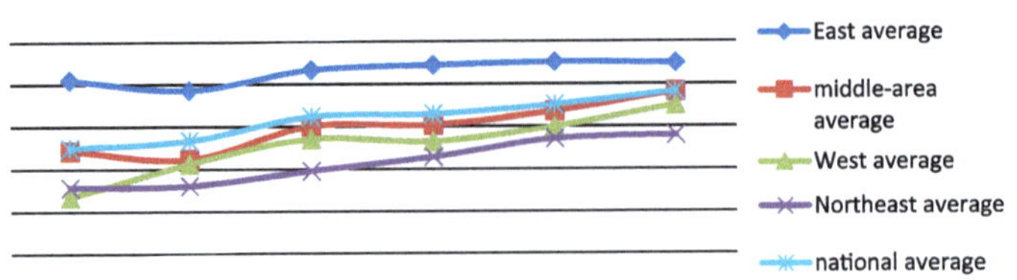

Fig. 1 Regions of the country green technology innovation efficiency ranking

Table 1 Regions of green technology innovation efficiency ranking

Region	2008	2009	2010	2011	2012	2013	Average	Rank
East average	0.818	0.770	0.866	0.888	0.902	0.898	0.857	
Middle-area average	0.484	0.445	0.601	0.606	0.671	0.762	0.858	
West average	0.267	0.423	0.540	0.529	0.598	0.699	0.509	
Northeast average	0.313	0.320	0.390	0.456	0.540	0.557	0.430	
National average	0.499	0.533	0.646	0.657	0.701	0.764	0.687	
Zhejiang	0.962	0.975	0.983	0.990	0.998	0.997	0.984	1
Guandong	0.963	0.975	0.982	0.989	0.997	0.998	0.984	2
Hainan	1.000	1.000	1.000	1.000	1.000	0.868	0.978	3
Beijing	0.876	0.898	1.000	1.000	1.000	1.000	0.962	4
Shanghai	0.852	0.932	0.955	0.987	0.993	1.000	0.953	5
Shanxi	0.315	0.179	0.319	0.310	0.377	0.441	0.857	26
Shanxi	0.538	0.331	0.439	0.435	0.410	0.419	0.429	27
Nengmenggu	0.121	0.277	0.320	0.304	0.336	0.301	0.277	28
Qinhai	0.163	0.239	0.245	0.268	0.282	0.428	0.271	29
Heilongjiang	0.101	0.211	0.272	0.235	0.284	0.342	0.241	30

Shanghai, which concentrated in the eastern region. There are numerous scientific research institutes in this region, making its innovation ability on the cutting edge of national. Regional government advocates innovation activity and industrial upgrading vigorously. Beijing has put "the construction of science and technology innovation city" into their city development strategy and issued a series incentive policies to promote technological innovation. Although environmental pollution index of these areas is higher than other areas, all these regions have introduced a large number of environmental pollution control measures and have increased green innovation technology R&D investment. These measures are directly promoted the 5 provinces the efficiency of green innovation.

Heilongjiang, Qinghai, Inner Mongolia, to Shanxi and Shaanxi are in the ranking of last 5 areas. They are all located in the West and northeast. The average green innovation efficiency of these provinces were less than 0.45, indicating that these regions are facing the double pressure of innovation and upgrading and environmental protection. These regional has simple industrial structure and rely on natural resources to develop economic. Government have proven a number of science and technology project that lack full funding, which has serious negative impact on the growth of Midwest talent. Such conditions further restrict innovation development, giving rise to the economic development rely more on natural resources. It becomes a vicious spiral. There are abundant coal resources in Shanxi, where development rely on coal resources for a long-term, giving rise to serious environmental pollution. Shaanxi and Heilongjiang are both old industrial base. Industrial upgrading is not successful and the consumption of resources is serious, so the green innovation efficiency in these areas in the country the most low-end.

3 Conclusion and Suggestion

Nowadays resources shortage and environmental pollution is becoming a serious issue. Green innovation has become an important tool of a country to keep competitive advantage. This paper uses the method of DEA to test green innovation efficiency in 30 provinces from 2008 to 2013 based on the index of environmental pollution and energy consumption. Conclusions are as follows:

The green innovation efficiency in our country is growing in the years 2008–2013 on the whole. There is a significant difference in the green innovation efficiency of regions showing from strong to weak gradient distribution from the east, central, and west to northeast region. The innovation and the sustainable development ability of the eastern region enhanced continuously. In the West and northeast, the promotion of innovation ability at the cost of more environmental pollution and energy consumption. The gap of green innovation efficiency between central and eastern areas gradually reduced. According to the above conclusions, this paper puts forward policy suggestions as follows:

3.1 Attach Importance to Innovation Investment in the West and Northeast Areas for the Improvement of the Western and North-Eastern Region Innovation Ability

With regard to the limited innovation resources in the West and northeast area, the central and local governments should make great efforts to support the science and technology innovation. For instance, establish a special support fund, set up business loans for the science and technology enterprise specifically. In addition, lower loan standards. Meanwhile, concerning to the shortage of talents in the West and northeast regions, the university and institute must support the enterprise as counterparts, which can strengthen the ability of independent innovation.

3.2 Put More Effort on the Introduction and Development of Green Technology

To improve the efficiency of green innovation, we will increase the introduction of the sophisticated green technology. For instance, we can imitate the Japanese steel industry to strengthen the innovation of production technology and products and utilize more energy-saving equipment at the same time. Compared to 1990, the energy consumption of crude steel unit decreased by 18 % and CO_2 emissions decreased by 20 % resulted from the use of advanced technology and management. Government should strengthen the communication with these countries and support introduction of advanced technology with tax policy.

3.3 *Upgrade the Industry*

The central and western regions are characterted by rich resources, low factor costs, and large market potential, carrying on industry transfer. However, the transferred industry is often with serious pollution and high energy consumption. The government should set up strict industry access system. Applying the resources capacity and ecological environmental capacity as an important basis for the transfer of industry. Eliminate high energy consumption and serious emissions project which do not meet the state's industrial policy. Innovation resources from eastern coastal areas are eagerly welcomed to establish and develop cooperation for improving the innovation ability of central and western regions, such as build industrial parks.

Acknowledgments Project Fund: Project supported by the National Natural Science Foundation of China (Grant No.71373133).

Authors: Cheng Yue (1991-) female, graduate student of Nanjing University of Posts and Telecommunications School of management Yin Qun (1958-), female, Dean of the Nanjing University of Posts and Telecommunications School of management, professor.

References

1. Hall B, Kerr ML (1991) The 1991–1992 green index: a state-by-state guide to the nation's environmental health. Island Press, Washington, DC
2. Leipert C (1987) A critical appraisal of gross national product: the measurement of net national welfare and environmental accounting. J Econ Issues 21(1):357–373
3. Chen J (1991) Construction and analysis of national green technology innovation system. Res Manag Sci 17:38–41 (Chinese)
4. Chen J, Liu J, Yang F (2002) Study on the index measure methods of the green technological innovation auditing. Manag Sci Res 23(2):64–71 (Chinese)
5. Li C (2008) Research on business model innovation based on business process reengineering. Ocean University of China (Chinese)
6. Qian Z, Liu X (2014) The study on regional difference and convergence of China's green economy efficiency. J Xiamen Univ 1:12–48 (Chinese)
7. Bai J, Jiang F (2011) The research of regional innovation efficiency of environmental factors–based on three stage DEA method considering. Finan Econ (10) (Chinese)
8. Han J (2010) Research China high-tech industry innovation efficiency based on SFA. J Beijing Normal Univ 6 (Chinese)

9. Guan J, Chen K (2009) Measure the efficiency of technological innovation of high technology industry in China. Technol Econ Res Quant Econ 10 (Chinese)
10. Bi K, Yang C, Huang P (2011) The influence of FDI on China's green manufacturing technology innovation research, based on an empirical analysis of the industry's panel data. China Soft Sci 9:172–180 (Chinese)
11. Xu X (2014) Evolution of regional green economy efficiency and influencing factors of study. J Bus Res 9:27–33 (Chinese)
12. Qian L, Xiao R-Q, Chen Z-W (2015) Research on the industrial enterprise technology innovation efficiency and regional disparities in china—based on the theory of meta frontier and DEA model. Econ Theory Bus Manag V35(1): 26–43 (Chinese)
13. Wang Y (2012) Regions in our country enterprise green technology innovation performance comparative study. J Technol Econ 31(1):52–59 (Chinese)
14. Yang L, Hu Z (2010) China's green economy based on DEA efficiency of regional differences and convergence analysis. J Econ 2:46–54 (Chinese)
15. Fan H (2012) China's evolution of provincial scientific and technological innovation efficiency and influencing factors of study. J Sci Res Manag 33(1):10–18 (Chinese)

Use of License Duration for Segmenting Software Market

Nan Feng and Yi-yang Miao

Abstract A new service model has emerged which delivers application and services on subscription by several options of license periods. Our paper discusses the optimal license duration for a software vendor to license software: one basic period or two license periods. In the model, we consider the license duration, the discount rate, and the reservation price. We show that two license periods can make software vendors segment their markets more appropriately and how software vendors can maximize their profits through the use of this new licensing policy.

Keywords Discount rate · License duration · Market segmentation · Subscription

1 Introduction

The marketing strategy for selling software is changing rapidly. More and more software vendors deliver their products over the Web on subscription pricing strategy. Subscription is predicted to become increasingly important in most enterprise application software (EAS) markets [1]. To segment market more appropriately, some software vendors such as Adobe, Oracle and McAfee have followed suit into this new business model. They provide more than one license period of the same product for customers. For example, McAfee sells its antivirus products with several license periods (1 Year for $329, 2 Year for $494 and perpetual license for $1102).

This paper addresses the above issue by studying the optimal license duration of a monopolist software vendor. We show that by charging different prices, the monopolist software vendor can segment the customers based on their sensitivity to product license duration and realize second-price discrimination of the customers through their self-selection behavior.

N. Feng (✉) · Y. Miao
Department of Information Management, Tianjin University, Tianjin, China
e-mail: fengnan@tju.edu.cn

E. Qi (ed.), *Proceedings of the 6th International Asia Conference on Industrial Engineering and Management Innovation*,
DOI 10.2991/978-94-6239-145-1_87

Vertical differentiation has become a common practice in information industries. While more and more information goods have little differentiation on qualities, which urges sellers to find another way to differentiate their products and segment customers more appropriately. Then providing more than one license periods is becoming more and more important. Software can be used for a period of time without replacement, but its value always depreciates, which makes customers need a discount for long license period products. In this paper, we assume the discount function is a linear function with license period.

The paper is organized as follows. In Sect. 2, we review the literature most relevant to this topic. Section 3 provides the basic model together with the assumptions and notations. Sections 3.1 and 3.2 study the single license period and two license periods strategy separately. Then we make a comparison between them. Section 4 concludes with findings for firms, points out our limitations and makes directions for our future research.

2 Literature Review

Many researchers [2–4], in such academic fields as economics and marketing have studied versioning strategy. Reference [4] pointed out that information goods had different production economics to physical goods. They found that both the characteristics of an information product and the value that customers placed on them had effects on the number of versions and prices. The versioning strategy brought benefit to firms by serving segmented customers. Reference [5] examined the impact of marginal costs on the optimality of versioning under the linear utility function. They pointed out that multiple versions were suboptimal under their assumptions. In 2008, they derived a set of rules about determining the optimality of versioning. They assumed that customers preferred higher quality and customer utility function increased linearly with the customer type. Reference [3] considered both the initial version and the upgrade version. They put forward a temporal versioning model in these two periods. And finally they demonstrated how demand variability and market uncertainty affected the two-stage design of information products. Customers' preferences on quality were commonly used in these previous researches on the versioning of information products. But there is little study on customers' preferences on products license period.

Another stream of research related to our study is subscription pricing strategy [6–8]. Reference [6] made a comparison between perpetual licensing and subscription with considering publisher's incentive to invest in software quality. He pointed out that the software seller earns more profit and social welfare is higher under subscription. Reference [8] examined the impact of quality and network externalities on a monopoly seller's licensing models: a perpetual license and a subscription contract. And they discussed the optimal strategy between the two pricing schemes. Reference [7] employed a continuous-time utility model and studied service subscription under both perfect and imperfect information dissemination. They explored the underlying consumer heterogeneity in markets for subscription-based IT service with network effects.

Our work also draws on the literature about market segmentation [9–11]. Reference [9] proposed a segmentation method that balances the fit of a segmentation strategy against the corresponding logistics costs. Reference [10] discussed the multi-criterion nature of market segmentation and developed a new mathematical model that addressed this issue. The model complemented existing segmentation methods and was applied and evaluated in two empirical studies. Reference [11] assessed the current status of international market segmentation research and provided a systematic overview of 25 previous empirical studies. And it examined a case study to discuss the various issues and conclude with suggestions for future research.

Building on prior research, in this paper we show that software sellers can use license period to segment markets appropriately. We model a linear discount function with license period. Through the analysis, we provide important insights for software sellers and find some interesting findings.

3 Model

Our analysis focuses on the commonly observed and widely studied pricing scheme: subscription pricing [12, 13]. The strategy, we examine in this paper, can be considered as a special case of two-part pricing. To improve exposition, we will discuss our model in the context of buyers that want to purchase software licenses for different periods.

We model one monopolist seller with no marginal costs [14]. Buyers are perfectly informed of prices, licensing period of each product and there are no transaction costs [15–19]. Thus we model consumers' willingness to pay (WTP) function and the discount function as

$$WTP = \theta T d + \lambda, \tag{1}$$

$$d = -aT + b \tag{2}$$

where θ [0,1] is the preference for product license period (T), d (0,1] is the discount rate of the product licensing time, a (0,1) is the marginal discount rate, b denotes the basic discount rate, and λ (0,1) means the basic willingness to pay or the reservation price of an information product.

And the firm's profit of selling the software product is

$$\pi = \sum PD \tag{3}$$

where P is the product price and D denotes the demand for the product.

We assume that customers' marginal willingness-to-pay as θ is uniformly distributed on the support of $[0, 1]$, b belongs to (0, 1] and a belongs to (0, 1).

3.1 Benchmark Case

We begin by solving one benchmark case where two firms offer only offer one basic license period product. The market segmentation is as Fig. 1 bellow,

To determine the optimal profit, we set U = 0 (it is the boundary point whether consumers buy the product). Then we can get $\theta_{T1} = \frac{p_1 - \lambda}{-aT^2 + bT}$. And in this case, $T_1 = 1$ and $d_1 = 1$.

$$\pi_O = \sum pD = p_1(1 - \theta_{T1}) = p_1(1 - \frac{p_1 - \lambda}{-aT^2 + bT}) \tag{4}$$

Proposition 1 *The optimal price is derived as* $p_1 = \frac{1+\lambda}{2}$ *by solving* $p_1{}^* = \arg\max\{\pi_O, 0 \leq p_1 \leq 1 + \lambda\}$, *and* $\theta^* = \frac{1-\lambda}{2}$, $\pi_O^* = \frac{(1+\lambda)^2}{4}$.

In the following research, we assume that adding one license period has no effect on $p_1{}^*$, θ^* and π_O^*.

3.2 Two License Periods

As shown in Fig. 2, D_{T_1} and D_{T_2} represent customers' demand for the basic licensing time version and the licensing time version separately.

The monopolist profit is

$$\pi_T = \sum pD = p_1(\theta_{12} - \theta_{T1}) + p_2(1 - \theta_{12}) \tag{5}$$

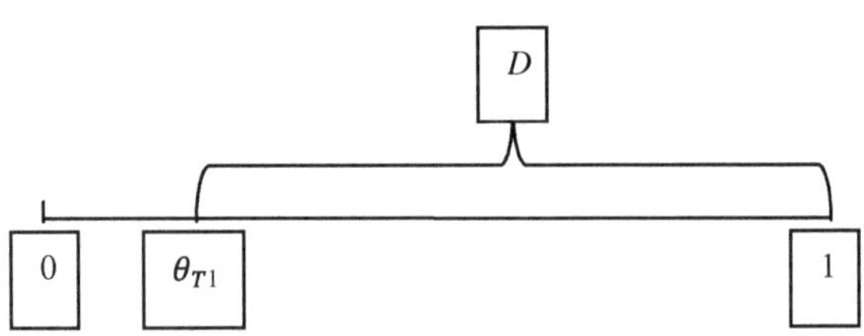

Fig. 1 Market segmentation under one license period

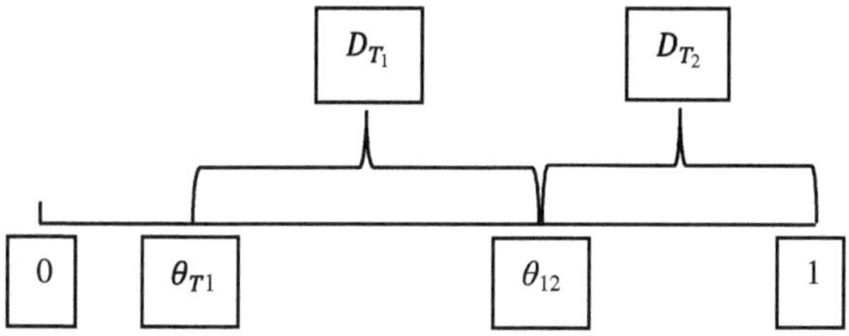

Fig. 2 Market segmentation under two license periods

Meanwhile the monopolist will not always decrease the discount rate with the license duration. This case is shown in Fig. 3 bellow,

As shown in Fig. 3, when the license duration is longer than T_0, the discount rate will remain d_O as a constant.

$$T_0(-aT_0 + b) + \lambda = (T_0 + 1)[-a(T_0 + 1) + b] + \lambda \tag{6}$$

$$\text{s.t.} \begin{cases} 0 < d_0 < 1 \\ 2 \le T_0 \end{cases} \tag{7}$$

Proposition 2 *By solving (6) and (7), we can get* $T_0 = \frac{1}{2a}$ *and* $0 < a \le 1/4$.

The marginal discount rate a is generally close to 1, and customer's utility increases with a. Thus a higher marginal discount rate can make customers get higher utility. But customers are rational, and they know sellers will not give them the biggest discount. What we find $0 < a \le 1/4$ is a rational result. $T_0 = \frac{1}{2a}$ means sellers will not provide a lower discount when the license period still increases.

Fig. 3 Discount rate with license duration

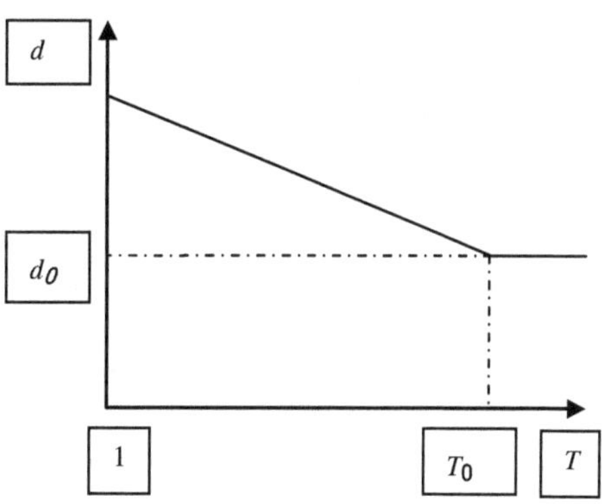

Then we get the inequality constraints of (5).

$$\text{s.t.}\begin{cases} T_2 d_2 > 1 \\ \frac{1+\lambda}{2} < p_2 \\ 1 < T_2 \le \frac{1}{2a} \end{cases} \tag{8}$$

Proposition 3 *The optimal price is derived as* $p_2 = \frac{1}{4} + \frac{1}{8a} + \frac{\lambda}{2}$ *and* $T_2 = \frac{1}{2a}$ *by solving (5) and (8), and* $\pi_T^* = \frac{\lambda^2 + 2\lambda}{4} + \frac{1}{8} + \frac{1}{16a}$.

From Propositions 3 and 2, we can make a comparison between the two strategies. And we find that $\pi_T^* > \pi_O^*$, which means two license periods strategy is better than one basic license period under our research.

4 Conclusion

In our research, we study the effects of information goods license duration on software vendors' profit. We seek to explain the observed diversity in license duration of information goods and whether sellers can make the best decision to earn more profit. The main findings and implications are summarized as follow.

In this paper, we model a linear WTP function considering customers preferences on license duration and reservation price, and a linear discount function discount with license duration. First, we study the benchmark case that the monopolist only offers one license period of his products and point out the optimal price and the maximal profit (π_O^*) in this case. Second, we extend our study to two license periods. We get the maximal profit (π_T^*) in this case. Finally we find that $\pi_T^* > \pi_O^*$.

From Proposition 2, our study shows that the marginal discount rate is not very close to 1, which means that consumers all hope sellers have a discount rate with license duration, but they are rational. The optimal strategies shown in Propositions 1 and 3 reflect that software vendor's profit decreases with marginal discount rate and so is the length of license duration, which indicates that software vendors should know well about their target customer's expected marginal discount rate before they make decisions.

There are some limitations in this paper. In order to study the leasing of software products in a simple and insightful model, we assume the discount function is a linear function. However, this function may not work well to some information goods. In the future, we will examine convex discount function and concave discount function. In this paper, we ignore the quality changing during license duration. And we will take this into consideration in the future research.

Acknowledgments The research was supported by the National Natural Science Foundation of China (no. 71271149) and the Program for New Century Excellent Talents in University (NCET). The authors are very grateful to all anonymous reviewers whose invaluable comments and suggestions substantially helped improve the quality of this paper.

References

1. Alexander B, Thomas H (2011) Opportunities and risks of software-as-a-service: findings from a survey of IT executives. Decis Support Syst 52(1):232–246
2. Bhargava H, Choudhary V (2008) When is versioning optimal for information goods. Manage Sci 54(5):1029–1035
3. Dogan K, Ji Y, Mookerjee VS, Radhakrishnan S (2010) Managing the versions of a software product under variable and endogenous demand. Inf Syst Res 22(1):5–21
4. Shapiro C, Varian HR (1998) Versioning: the smart way to sell information. Harvard Bus Rev 76(6):106–114
5. Png IPL, Wang H (2010) Buyer uncertainty and two-part pricing: theory and applications. Manage Sci 56(2):1090–1112
6. Choudhary V (2007) Comparison of software quality under perpetual licensing and software as a service. J Manag Inf Syst 24(2):141–165
7. Niculescu MF, Shin H, Whang S (2012) Underlying consumer heterogeneity in markets for subscription-based IT services with network effects. Inf Syst Res 23(4):1322–1341
8. Zhang J, Seidmann A (2010) Perpetual versus subscription licensing under quality uncertainty and network externality effects. J Manag Inf Syst 27(1):39–68
9. Turkensteen M, Sierksma G, Wieringa JE (2011) Balancing the fit and logistics costs of market segmentations. Eur Oper Res 213(1):340–348
10. Liu Y, Ram S, Lusch RF, Brusco M (2010) Multicriterion market segmentation: a new model, implementation, and evaluation. Mark Sci 29(5):880–894
11. Steenkamp JBEM, Ter Hofstede F (2002) International market segmentation: issues and perspectives. Int J Res Mark 19(3):185–213
12. Gundepudi P, Rudi N, Seidmann A (2001) Forward versus spot buying of information goods. J Manag Inf Syst 18(2):107–131
13. Jain S, Kannan PK (2002) Pricing of information products on online servers: issues, models, and analysis. Manage Sci 48(9):1123–1142
14. Hitt LM, Chen PY (2005) Bundling with customer self-selection: a simple approach to bundling low-marginal-cost goods. Manage Sci 51(10):1481–1493
15. Huang S, Yang Y, Anderson K (2001) A theory of finitely durable goods monopoly with used-goods market and transaction costs. Manage Sci 47(11):1515–1532
16. Tyagi RK (2004) Technological advances, transaction costs, and consumer welfare. Mark Sci 23(3):335–344
17. Bhargava HK, Choudhary V (2001) Information goods and vertical differentiation. J Manag Inf Syst 18(2):89–106
18. Choudhary V, Tomak K, Chaturvedi A (1998) Economic benefits of renting software. J Organ Comput Electron Commer 8(4):277–305
19. Desai P, Purohit D (1998) Leasing and selling: optimal marketing strategies for a durable goods firm. Manage Sci 44(11):19–34

The Impact of Security Risk on Free Trial Strategy for Software Market

Nan Feng and Zhi-qiang Zhang

Abstract Offering a free trial is a common strategy of software vendors, which has been studied by many researchers focusing on the impact of positive network effects. However, the software is not perfectly security, security vulnerabilities may be exploited and incur negative security externalities among software users. In this paper, we examine the free trial strategy in an environment where both positive network effects and security risk exist. The findings indicate that the negative security externality affects the vendor's decision significantly. Software vendor should offer a free trial when low or moderate negative security externalities and high positive network effects are both satisfied.

Keywords Negative security externality · Network externality · Software free trial

1 Introduction

Information goods (such as software, movies and books) differ from industrial products because of their cost structure: high fixed costs, low (almost zero) variable costs [1, 2]. Because of this special feature, the traditional cost-based pricing strategies cannot be applied to the information goods. Accordingly, the information goods vendors should adopt value-based pricing strategies [1]. A large number of researches on information goods pricing have been studied, including bundling [3, 4], versioning [5, 6], subscribing [7], and free trial strategy [8–10].

The software vendor often provides a free trial to achieve market penetration and take advantage of the network externality [10]. Generally, there are two forms of freeware offered by the vendor: a free trial version with full functionality but with a limited trial time or a freeware without time limit but with limited functionality [9]. Researches on the trade-off between positive network effects and cannibalization

N. Feng (✉) · Z. Zhang
Department of Information System Management, Tianjin University, Tianjin, China
e-mail: fengnan@tju.edu.cn

E. Qi (ed.), *Proceedings of the 6th International Asia Conference on Industrial Engineering and Management Innovation*,
DOI 10.2991/978-94-6239-145-1_88

effects have been studied by [8, 9], and they proved that offering a function-limited software of the commercial version is beneficial under the situation of strong positive network effect.

However, the software users often suffer from the security problem during use because of the flaws existing in the software codes. Specially, for software used in a network environment, the security vulnerabilities of the software can be exploited and lead to negative security externality [11, 12], which means that the unpatched users of the software will face a risk not only from the hackers, but also from other users of the same network software. Similarly, the user may affect others if he/she is attacked by the hackers. Consequently, the users face a security risk positively correlated to the unpatched population. Nevertheless, above researches focusing on the free trial strategy did not consider the issue about the negative security externality, which would influence the vendor's free trial strategy. In this paper, we examine the free trial decision for the software vendor by taking the security factor into consideration.

The remainder of the paper is organized as follows. The next section presents a brief review of the relevant literature. We then present our general model, followed by the presentation of the results. Finally, the paper concludes with a summary of our findings and provides our future research direction.

2 Literature Review

Our research is related to the free trial strategy of software which has been extensively studied. Due to the special cost structure of software, offering a free trial is considered to be feasible from the perspective of economics. Existing researches have studied the free trial strategy by considering the impact of positive network effects. Takeyama develops a model to study the problem of software piracy, and shows that in the presence of network effects, piracy can improve the vendor's profit [13]. Haruvy and Prasad examined the dynamic change of the optimal price-freeware quality group in the presence of positive network effects [10]. However, above researches didn't consider the network effect intensity. Taking the intensity of network effect into consideration, Conner and Rumelt find that manufacturer makes larger profits by tolerating piracy if positive network effects are sufficiently large [14]. Cheng and Tang [9] and Cheng and Liu [8] examined the trade-off between the positive network effect and the cannibalization effect generated by offering a free trial. They found that when the positive network effect is strong, it's profitable for the vendor to offer a function-limited free trial [9], while it's optimal for the vendor to offer a time-limited freeware when the network effect is moderate [8].

Another relevant research work is about the software risk management. Effective risk management in software projects is considered to be a major contributor to project success, because software projects are high risk activities [15]. Bannerman finds that risk management research and risk management as practiced cannot keep up with each other, both researchers and practitioners must continue to learn from each other to reduce project failures and to generate better outcomes [16]. Based on the study of the existing theoretical knowledge and practices and examples of better management, Avdoshin and Pesotskaya provide recommendations on software risk management and give a brief overview of the automated software tools [17]. August and Tunca are among the first to identify negative security externality [11], they investigated the effect of user incentives on software security in a network of individual users under negative network security externalities. They show that users of the network software share not only the benefit of the ability to communicate with a vast number of other users, but also the vulnerabilities that come with it-information security. They construct a model which captures how risk faced by unpatched users is related to the unpatched population in equilibrium.

3 Model

We assume that the commercial software of quality q_h is offered to a continuum of consumers by a monopoly vendor at the price of *P*. If the vendor chooses to provide a free trial version, the quality of the free trial is denoted as q_l, where $q_l < q_h$. We assume that only two products (commercial and freeware) are provided by the software vendor, and this situation is assumed to remain unaltered, i.e., we don't consider the upgrade problem, and the versions are not withdraw.

Following [11], we also assume that the software is used in a network setting. The consumers will get a utility consists of two parts by using the software: intrinsic utility (the utility a user achieves by using the software even if there isn't any other user) and positive network effects (the additional utility that increases with the number of other users). Next, we'll introduce the two parts respectively.

First we consider the situation without positive network effects, following the model setup in [18], let *K* be the size of the total population in market (including all consumers no matter they are interested in the software or not), and *N* be the number of consumers with positive intrinsic valuation for the product, and we can describe the relation between *K* and *N* as an equation $K = a \cdot N$, $a > 1$. Without loss of generality, we also normalize *N* to 1 for the sake of simplicity, like [9] did. We use θ to denote each individual's valuation for the software, and for simplicity, we assume θ is uniformly distributed over the interval [(-*a*-1), 1] (as shown in Fig. 1). Therefore, the corresponding intrinsic utility from using the software with quality q for an individual with valuation θ will be θq, where $q \in \{q_l, q_h\}$.

Now we take the positive network effect into consideration, which increases each person's utility by $\beta Q q$, where Q is the size of the total user base of the software and β is the network effect intensity. As the quality of the freeware is lower than the commercial version, a user will obtain lower positive network effects by using the free trial version than the commercial software. Thus a user can get a total utility of $v = (\theta + \beta Q) \cdot q$, where $q \in \{q_l, q_h\}$.

Fig. 1 Distribution of consumers without positive network effects

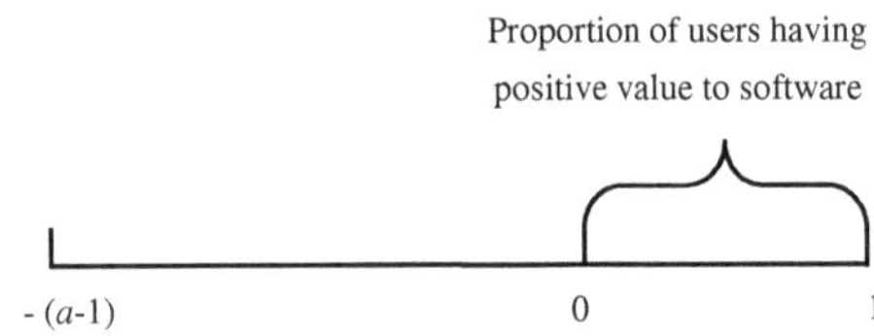

Now we will introduce the net utility based on which the users make their decisions by considering the using cost for a user. Firstly, a price of P must be paid by the users of the commercial version. Secondly, we also consider the aggregate using cost (denoted by c) like [9] did. Besides, if a user chooses to use a free trial version, although he/she doesn't need to pay for the software, he/she may suffer a security risk, depending on the vendor's patching strategy.

3.1 Benchmark

First we consider the benchmark case where the vendor offers the commercial version only, with quality of q_h, pricing at P_0. For simplify, we assume that the security patches for the software are offered to buyers by the vendor without extra fee, and the buyers will install the patches after the patches are released as soon as possible.

In this case, the net utility a consumer gets by using the software is $(\theta + \beta Q)q_h - P_0 - c$, two potential strategies can be adopted by the users, that is, pay for the commercial software of do without. We denote the marginal consumer who obtains zero net utility from the software as θ_0, thus we have

$$(\theta_0 + \beta Q_0)q_h - P_0 - c = 0 \tag{1}$$

The demand for the software is

$$Q_0 = 1 - \theta_0 \tag{2}$$

Since the users' valuation for the quality of the software is uniformly distributed in the interval [-(a-1), 1], the sub-interval $[\theta_0, 1]$ presents the demand of the software, (see Fig. 2).

The vendor's decision objective is to maximize the profit

$$\max_{P_0} \Pi_0 = P_0 Q_0 = P_0 \cdot \frac{q_h - (P_0 + c)}{(1 - \beta)q_h} \tag{3}$$

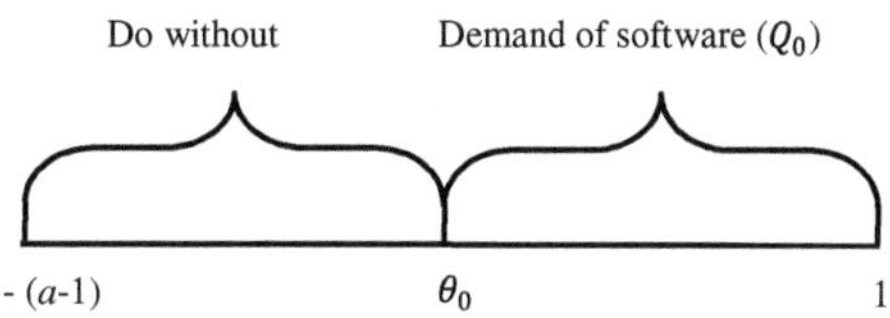

Fig. 2 Demand of software in case of benchmark

Combining Eqs. (1) (2) and (3), we derive the optimal pricing and profit, and the corresponding demand as follows,

$$P_0^* = \frac{q_h - c}{2}, Q_0^* = \frac{q_h - c}{2(1-\beta)q_h}, \Pi_0^* = \frac{(q_h - c)^2}{4(1-\beta)q_h} \tag{4}$$

3.2 *The Vendor's Optimal Decision Under Negative Security Externalities*

We now turn our attention to the case where the software vendor offers both the free trial and commercial software. The vendor will introduce the free trial version due to the profit increasement. Here we assume that the vendor will offer the software patches to the buyers of the commercial version only, so the free trial users will suffer a security risk loss because of the negative security externality.

We assume that the probability that a free trial user is attacked by a security attack is positively correlated with the unpatched population in the network, and the loss that a free software user suffers is positively correlated with his utility from using the software. For simplicity, we assume the correlation is of first order, thus the loss that a user with utility v suffers is $\alpha v Q_u$, where α denotes the negative security externality intensity, and Q_u denotes the unpatched population. So a user of type θ will get $(\theta + \beta Q)q_l \cdot (1 - \alpha Q_u) - c$ net utility from using the free trial.

In this case, there are three potential strategies for the users to adopt: Buy the commercial version, try the free trial, or do without. Thus there are two marginal users, denoted by θ_h and θ_l respectively (see Fig. 3). Similarly, the sub-interval $[\theta_h, 1]$ represents buyers of the commercial software, and the sub-interval $[\theta_l, \theta_h]$

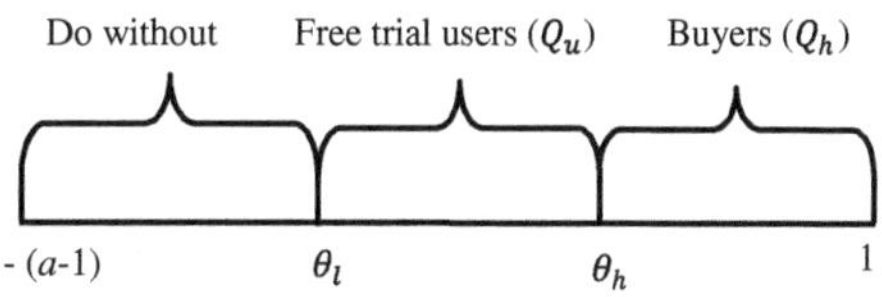

Fig. 3 Demand of free trial and commercial software

corresponds to users of the free trial. Let P_h denote the price of the commercial software under this condition. Then, we have

$$(\theta_l + \beta Q)q_l \cdot (1 - \alpha Q_u) - c = 0 \tag{5}$$

$$(\theta_h + \beta Q)q_h - P_h - c = (\theta_h + \beta Q)q_l \cdot (1 - \alpha Q_u) \tag{6}$$

And the vendor's optimal decision function is

$$\max_{P_h} \Pi_\alpha = P_h Q_h \tag{7}$$

$$Q = 1 - \theta_l, Q_u = \theta_h - \theta_l, Q_h = 1 - \theta_h.$$

Theoretically, by solving Eqs. (5) and (6) for given β, q_l, α and c, we can obtain θ_h as a function of P_h. Since it's difficult to get the analytical form of θ_h, we turn to numerical analyses to find the optimal price and profit, in which we set $q_h = 1, q_l = 0.6, c = 0.2$.

Figure 4 illustrates the optimal profit with respect to the positive network effect under different negative security externalities. From Fig. 4 we can see that, for a software vendor, whether to offer a free trial or not depends on both positive network effects and negative security externalities. Accordingly, two observations are summarized as follows:

Observation 1: Offering a free trial will obtain more profit when the positive network effect is high and the negative security externality is low or moderate are both satisfied.

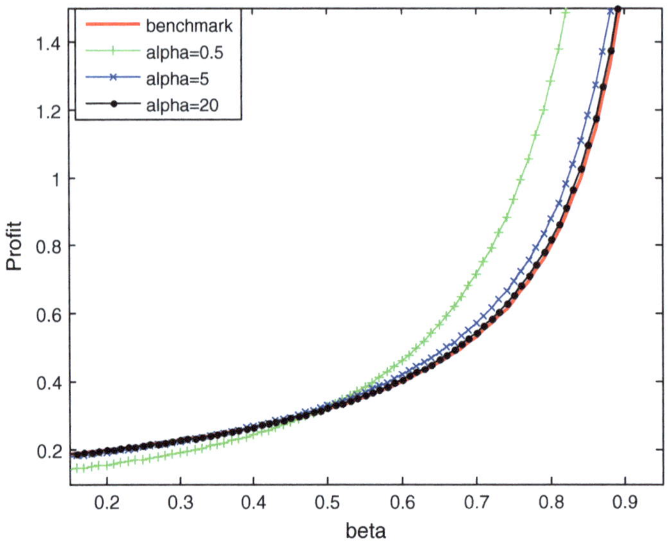

Fig. 4 Optimal price, quantity and profit with respect to β

Observation 2: If the negative security externality is high, the profit by offering a free trial converges to the benchmark, regardless of the intensity of positive network effects.

From Observation 1 and Observation 2, we can conclude that the negative security externality is an important factor to be considered when a software vendor makes his/her free trial decision.

4 Conclusions and Future Research

In this paper, we examine the vendor's free trial decision by considering both positive network effect and negative security externality. We find that negative security externality is an important factor to be considered for the vendor when making a free trial decision, and offering a free trial improves the vendor's profit when low or moderate negative security externalities and high positive network effects are both satisfied.

Patching management is relevant to the security problem, efficient patch management is imperative in an environment with security vulnerabilities. Taking the patching strategies of the software vendor into consideration will make the research more complete, which provides an interesting direction for our future research.

Acknowledgments The research was supported by the National Natural Science Foundation of China (no. 71271149) and the Program for New Century Excellent Talents in University (NCET). The authors are very grateful to all anonymous reviewers whose invaluable comments and suggestions substantially helped improve the quality of this paper.

References

1. Shapiro C, Varian H (1998) Versioning: the smart way to sell information. Harvard Bus Rev:106–114
2. Shapiro C, Varian H (1999) Information rules (Book style). Harvard Business School Press
3. Bakos Y, Brynjolfsson E (1999) Bundling information goods: pricing, profits, and efficiency. Manage Sci 45(12):1613–1630
4. Geng X, Stinchcombe MB, Whinston AB (2005) Bundling information goods of decreasing value. Manage Sci 51(4):662–667
5. Bhargava HK, Choudhary V (2008) When is versioning optimal for information goods? Manage Sci 54(5):1029–1035
6. Wu S y, Chen P y (2008) Versioning and piracy control for digital information goods. Oper Res 56(1):157–172
7. Zhang J, Seidmann A (2010) Perpetual versus subscription licensing under quality uncertainty and network externality effects. J Manage Inf Syst 27(1):39–68
8. Cheng HK, Liu YP (2012) Optimal software free trial strategy: the impact of network externalities and consumer uncertainty. Inf Syst Res 23(2):488–504
9. Cheng HK, Tang QC (2010) Free trial or no free trial: optimal software product design with network effects. Eur J Oper Res 205:437–447

10. Haruvy E, Prasad A (2001) Optimal freeware quality in the presence of network externalities: an evolutionary game theoretical approach. J Evol Econ 11:231–248
11. August T, Tunca T (2006) Network software security and user incentives. Manage Sci 52(11):1703–1720
12. August T, Tunca T (2008) Let the pirates patch? An economic analysis of software security patch restrictions. Inf Syst Res 19(1):48–70
13. Takeyama L (1994) The welfare implications of unauthorized reproduction of intellectual property in the presence of demand network externalities. J Ind Econ 42(2):155–166
14. Conner K, Rumelt R (1991) Software piracy: an analysis of protection of strategies. Manage Sci 37:125–139
15. Charette RN (2005) Why software fails? IEEE Spectr 42(9):42–49
16. Bannerman PL (2008) Risk and risk management in software projects: a reassessment. J Syst Softw 81:2118–2133
17. Avdoshin SM, Pesotskaya EY (2011) Software risk management. In: 2011 7th Central and Eastern European software engineering conference in Russia (CEE-SECR), IEEE, pp 1–6
18. Conner K (1995) Obtaining strategic advantage from being imitated: When can encouraging ‘clones’ pay. Manage Sci 41(2):209–225

The Study on the Effect of Accounting Peculiarities to Financial Conditions of Primary and Middle Schools—on the Background of Carrying Out the New Accounting System for Primary and Secondary Schools

Jian-min Hang, Xia-yuan Li and Ying-xin Wu

Abstract The ministry of finance promulgated the newly revised accounting system for Primary and Secondary Schools on December 27, 2013. By giving out pamphlets and interviewing local school accountants, a better knowledge of the current situations of financial management in primary and middle schools, and a regression analysis model for financial management level in primary and middle schools were obtained during the investigation. Statistics show that the school's financial conditions could be affected by the level of several accounting characteristics of primary and middle schools, like asset-liability ratio, financial informatization and professional quality of accountants. Based on this survey, suggestions are given to primary and middle schools on how to improve the execution of this new system, in order to support the development of primary and secondary schools education and public finance revolution.

Keywords Accounting characteristics · Financial effect · New accounting system for primary and secondary schools · Regression analysis

1 Introduction

The newly revised Accounting system for Primary and Secondary Schools was promulgated by the ministry of finance on December 27, 2013, and executed on January 1, 2014. Educating students is the first mission of primary and middle schools, who bear the burden of cultivating the hopeful youth of our country, consequently, its financial management level could affect the educational level to

J. Hang · X. Li (✉) · Y. Wu
College of Management and Economics, Tianjin University, Tianjin, China
e-mail: lixiayuantju@163.com

E. Qi (ed.), *Proceedings of the 6th International Asia Conference on Industrial Engineering and Management Innovation*,
DOI 10.2991/978-94-6239-145-1_89

some degree. As a department with specific character, the primary and secondary schools have received attentions from relative departments on its financial management issues, while few scholars have done pragmatic researches on these issues. From the survey did in this paper, 200 questionnaires were gave out in primary and middle schools all over T city (158 valid ones were retrieved), and on-site surveys were carried out in 20 representative schools. While investigating new Accounting system for Primary and Secondary Schools executing and problem solved condition, Meanwhile, it also consults related researches on financial management in corporations and higher education institutes and combing the accounting characteristics occurred in the survey, planning to establish a model which shows the factors affecting the financial management in primary and middle schools, in order to offer thoughts and solutions for following study.

2 Literature Review

Currently, few researches had paid attention to factors which affect financial management level in primary and middle schools, many scholars in this field choose to do theoretical analysis research.

Chen [1] (2013) pointed out there are several problems existing on financial management in primary and middle schools, such as lacking management awareness, neglecting department budgeting, and the inadequacy of system Li [2] (2013) striking the problems of accountants' qualities, pointing out some problems still exist in middle school and primary school's accountants, like ambiguous division, low level of expertise, and lack of standardization in bill and certification work. He [3] (2014) emphasized the importance of financial informatization in primary and middle schools, which is an important channel to realize scientific management and reinforce the supervision function of accounting.

There are also many problems in the internal control of financial management in primary and middle schools. Sun [4] (2013) stressed that, the internal control measure or system is not sound currently, and the budgeting is hard to control. Zhang [5] (2013) proposed that one factor that leads to this situation is the lack of motivation, and since internal control is likely to reduce management risk of enterprises, they would certainly strengthen the internal control actively, when it comes to non-profit primary and middle schools, the motivation is less.

Currently the number research on finance evaluation system in primary and middle schools is quite small, Cao [6] (2013) established five evaluation indexes, including debt accrual, net asset accrual, merchandising service payment, the income-to-students percent, the undertaking expenses-to-students percent and liquidity ratio, which is of great guiding significance to this essay.

On this basis, this text establishes financial rating model about primary and middle schools with the reference of enterprises' and universities' financial performance indicators model, starting the empirical analysis by establishing regression analysis.

3 Research Analysis and Hyopthesis

3.1 Budget Issues

Asset-liability ratio is the percent that total liabilities to total assets, which could test the debt condition of a school. Though the definite stipulation emphasized that it is forbidden to borrow money in stage of compulsory education in the new Accounting system for Primary and Secondary Schools, the survey shows that many schools depending on state appropriation as the main income face the problem of lacking expenditure, as a result, there are still many leftover debt problems before the new stipulations were promulgated in some schools. Consequently, Asset-liability ratio can be regarded as a factor to measure the financial level of a school.

So several hypothesis can be built 1: The financial management level in primary and middle schools has positive correlation with Asset-liability ratio.

3.2 Factors of the Accountants

The professional quality of accountants in different schools is unequal, and staffs in schools of less-developed areas generally don't have higher education and professional title, even don't have the basic accounting qualification, particularly in rural schools. Due to the lack of manning quotas of accountants, many schools (especially rural schools) are unable to hire full-time accountant, while acted by teachers, logistical personnel or the retired teachers, who don't have much passion. In the survey, it is noticed that the number of accountants is heavily insufficient, and some schools even have only one or two accountants, who have much working pressure.

In terms of accounting work in primary and middle schools, it is a problem that the number of them is not enough and proficiency is not sound. It is arduous for the tasks like financing, auditing and examining, managing tuition, annual budget and final accounting, consequently, the financial managing level may be affected by the lacking of accountants, related knowledge, and managing ability to some degree. As a result, the ratio of accountants to sum number of teachers and students, and professional qualification of them could affect the financial managing level in school.

On this basis the hypothesis 2 and 3 could be proposed:

Hypothesis 2: The financial management level in primary and middle schools has positive correlation with the ratio of accountants to sum number of teachers and students.

Hypothesis 3: The financial management level in primary and middle schools has positive correlation with the highest accounting professional qualification of accountants.

3.3 The Issue of Financial Informatization

The financial informatization level in different schools of T city is unbalanced. The accountants are lack of the systematic and professional training on the usage of computerization of software, so they have a poor working efficiency; the backups frequency of accounting data are quite low, so the data safety of school finance may be in danger. It is an important trend to achieve cyber-informatization for accounting works, since it can raise the working efficiency, accomplish the real-time monitoring, and unify the working regulations. As a result, the financial informatization level is an indicator to measure the level of schools' financial work.

On this basis, hypothesis 4 could be proposed: the financial management level in primary and middle schools has a positive correlation with the accomplishment level of financial informatization.

4 The Design and Statistics of the Survey

4.1 The Research Model

In cooperation with financial department of T city, 200 questionnaires were given to six schools (public school, excluding voluntary and private school) and retrieved 158 validated ones, a recovery of 85 %. Since the centralized accounting has been adopted by X area and schools would not set up accountants by themselves [7], we just analyzed 102 of questionnaires excluding the 46 ones written by accountants in X area. The informants were mainly the accountants from different schools. Meanwhile, it was selected 20 representative schools to take a visiting survey to have a better knowledge.

4.2 Research Model

A multiple regression model that reflects financial management level has been established which could be affected by 4 accounting characteristics of primary and middle schools: asset-liability-ratio, the ratio of accountants to number of teachers and students, the highest professional qualification of accountants and ;the accomplishment level of financial informatization.

Here is the research model:

$$\mathrm{F}\boldsymbol{R} = \beta_0 + \beta_1\mathrm{L}\boldsymbol{EV} + \boldsymbol{\beta}_2\mathrm{CR} + \beta_3\boldsymbol{H}\mathrm{Q} + \beta_4\boldsymbol{CI} + \varepsilon$$

And the meanings of the variables in this equation are listed on Table 1:

Table 1 Introduction of the variables

Variable	Name	Definition
FR	Finance management level of a school	The management level of a school to finance. By specialist marking methods, related points can be calculated for its finance management. 10 for perfect
LEV	Asset-liability ratio	Year-end total liability is divided by total assets.
CR	The proportion of accountants in the total number of teachers and students	The accountants number is divided by the total number of teachers and students
HQ	The highest professional qualification of the accountants	The professional qualification level of the accountants can be classified for 4 levels: 1. No qualification to be account; 2. Basic certificated qualification of accounting professional; 3. Primary accounting certificated qualification; 4. Middle or senior accounting certificated qualification
CI	Finance informatization level error term	Table 2 shows variables which are designed by Likert Scale method

Table 2 The variable design of CI

Variable	Definition	Form	Likert level
CI (Financial informatization level)	A. Financial accounting by financial accounting software regularly	5 levels of Likert measure	Level 5 is the top, and Level 1 is the lowest
	B. Skillful at financial accounting software for financial accounting		
	C. Participate in the training for finance informatization frequently		
	D. Backup accounting data regular		
	E. Regular back up for finical software		

4.3 Statistical Analysis

A reliability test on the data from the questionnaires was made, which is listed on Table 3, and the reliability value is 0.633. It shows that when the data is tested qualified, it can be started on the following multiple regression analysis.

The result of descriptive statistical analysis is shown on Table 4. The average value of financial management level in school is 4.177, still far from the full mark 10, Standard Deviation number is big, which shows that the financial management level in schools of T city is not ideal and unbalanced in different areas. The average

Table 3 The reliability test

Cronbach's alpha	Number of terms
0.633	5

asset-liability ratio is 0.5438, illustrating that there are many schools in debt; the average value of highest professional accounting qualification is 1.2451, which shows that the professional standards of schools accountants are commonly low, and also uneven among different schools; the average achieving number of financial informatization level is 2.3925, also a big gap to the full mark–5, which reflects that the financial informatization level here needs to be further improved; the ratio of accountants to the sum of teachers and students is 0.0635 %, showing that the accountant is woefully inadequate, since there are only one or two accountants in each school while the number of students and teachers is quite different.

The specific result of the first multiple regression analysis is shown in Table 5. It shows that except the ratio of accountants to students and teachers, the P value of asset-liability ratio, the professional level of accountants, and the accomplishment level of financial informatization are all below 0.05 and pass the test. After rejecting the P value of the ratio of accountants to students and teachers, which not qualifies in the test, the multiple regression analysis is made for the next time. (The result is shown in Table 6). It still shows that the P value of asset-liability ratio, the professional level of accountants, and the achieving level of financial informatization are all below 0.05. Here is the multiple regression equation:

Table 4 The descriptive statistical analysis

	Mean value	Standard deviation	N
FR	4.1777	2.24692	102
LEV	0.5438	0.87104	102
HQ	1.2451	1.93009	102
CI	2.3923	1.19308	102
CR	6.3534	3.62468	102

Table 5 The coefficient form multivariate regression analysis

Model	Unstandardized coefficients		Standardized coefficients	t	Sig.
	B	Standard error			
(Constant)	0.553	0.827		2.871	0.004
LEV	0.153	0.265	0.059	1.983	0.049
HQ	1.105	0.272	0.457	4.263	0.000
CI	0.210	0.188	0.112	3.174	0.001
CR	0.001	0.068	0.001	0.511	0.866

**Dependent Variable* FR

Table 6 The coefficient form multivariate regression analysis

Model	Unstandardized coefficients		Standardized coefficients	t	Sig.
	B	Standard error			
(Constant)	0.553	0.823		2.871	0.004
LEV	0.153	0.262	0.059	1.983	0.050
HQ	1.106	0.252	0.458	4.389	0.000
CI	0.211	0.176	0.112	3.174	0.001

**Dependent Variable* FR

$$\mathrm{FR} = 0.553 + .0153\mathrm{LEV} + 1.106HQ + 0.211CI + \varepsilon \tag{1}$$

5 Conclusion

Through the analysis, 3 indicators that affect the finance evaluation system in primary and middle schools were found, which are asset-liability ratio, professional qualification level of accountants and the achieving level of financial informatization, and their estimated value of regression coefficient are 0.153, 1.106, and 0.211 respectively, which are all below 0.05, showing that they have an evident positive correlation with the financial management level of school. As a result, hypothesis 1, hypothesis 3 and hypothesis 4 passed the test; while hypothesis 2 didn't.

Aiming at the current situation of financial management in T city, improvement suggestions are given to improve financial management level in primary and middle schools.

5.1 The Financial Appropriation Should Be Allocated Properly, and Special Fund Should Be Available in Time

In the appropriation, the competent department should pay attention to the allocation proportion of public funds and personal funds. while when refers to formulating the budget, schools should give a comprehensive thought of situations like policy adjustment and price changes [8], and the department budget should be refined. Special fund should be put into place timely, and relative schools should make overall plans and take all factors into consideration while defining the priority.

5.2 *Improve the Social Benefits and Professional Proficiency of Accountants*

Make a proper arrangement of the manning quota of accountants according to the scale of school, and stipulate the quota with certain rules to raise the benefits of accountants [9]. To separate the organizational affiliation and wage relationship of accountants from school, and the selection, appointment and demission as well as allocation are implemented totally by education bureau. It could help accountants to decrease the disturbance and pressure from the school leadership, and enhance the independence of their financial work [10].To organize excellent staff selection among the accountants in schools regularly. Distribute the quota of people to every district with certain proportion, to select excellent accountant acknowledged by the city, province or even whole nation and give them award to improve the sense of honor [11].

With the solutions above, the wages, benefits, and social standards of accountants could be raised, as well as attract better educated people to work in primary and middle schools on accounting. Therefore, when the middle-aged are retired, they would become the backbone of school.

5.3 *Unify the Computerization Software and Develop "Financial Cloud Sharing" Service*

The accounting systematic software in every district should also be unified and manage the service provided by software developing company, start unified management, installation and training of computerization, update and record it regularly, which can help to achieve the connection of data in different schools of T city; it is better to build "financial cloud sharing" to facilitate competent department to have a better knowledge of the real- time financial data in different schools [12].

5.4 *Pay Attention to the Balanced Development of Accounting in Relative Schools and District, and Increase the Input of Accounting in Rural Schools*

Aiming at the current situation of unbalanced accounting development in different schools, department concerned should increase its support on districts and schools with less developed accounting career [13]. Special funds should also be arranged to help schools in poor areas to raise the informatization standard, as well as achieve the interconnection and interworking of financial data among schools, education authorities and financial sector.

References

1. Chen Q (2013) Research about financial management of primary and secondary schools. Money China 16:89–90 (Chinese)
2. Li H (2013) Countermeasure study to improve internal control construction of primary and secondary schools. Modern Business 15:246–247 (Chinese)
3. He C (2014) Study of problems in financial informatization of primary and secondary schools. Assets Finances Adm Inst 2:117–118 (Chinese)
4. Sun S (2013) Research about internal control system of primary and secondary schools. Assets Finances Adm Inst 12:117 (Chinese)
5. Zhang Y (2013) Contrastive analysis about old and new 'financial system of primary and secondary schools'. Money China 21:48–49 (Chinese)
6. Cao J (2012) The establishment and application of financial evaluation system in compulsory education schools Money China 71:219–220 (Chinese)
7. Chen H (2013) Analysis and exploration of concentrative accounting system in primary and secondary schools. Mod Econ Inf 20:113 (Chinese)
8. Gu Y (2013) Research about internal control construction of primary and secondary schools. Assets Finances Adm Inst 06:115 (Chinese)
9. Wu X (2013) How to strengthen the management of public spending of primary and secondary schools. Assets Finances Adm Inst 05:150–151 (Chinese)
10. Mu Q, Li J (2014) Research of management problems about primary and secondary schools under the new system of finance. Assets Finances Adm Inst 1:56–58 (Chinese)
11. Shen Y (2013) Model of 'school finance controlled by the Bureau' in Shanxi provincial compulsory education. Tangdu J 04:112–118 (Chinese)
12. Liu C (2014) Discussion about the primary and secondary schools financial management under the change of model of 'school finance controlled by the Bureau'. Money China 07:59–60 (Chinese)
13. Jiao X (2014) Financial principal of ministry of education meets the press about 'accounting system of primary and secondary schools'. www.jyb.cn (Chinese)

Carbon Emission Management of Coal Chemical Industry Life Cycle Using System Dynamics Model: Case of Inner Mongolia Region

Pei Zhang, Changqing Li, Yan Zheng and Wei Duan

Abstract This paper analyzes carbon source of coal chemical industry and builds a STELLA model to predict carbon emission of Inner Mongolia in the next 10 years using system dynamics approach. The simulation result shows carbon emission will continue to grow. The result will be good reference for firms and government to manage product life cycle carbon emission and industrial development.

Keywords Carbon emission · Coal chemical industry · System dynamics

1 Introduction

Research shows that there are two approaches that are used to calculate carbon emission of industrial sector. One is the material balance method [1], which uses emission coefficient of different raw materials to calculate carbon emission. This method reflects related data of carbon emission in a static manner. However, it cannot be used to calculate the impact of dynamic factors on the system such as utilization rate of raw materials, energy intensity and economic conditions on carbon emission. The other is the dynamic model approach [2]. Dynamic models are built on linear and non-linear equations to describe the system. The model is built based on the analysis of social-economic system, energy consumption restraint

P. Zhang · C. Li (✉) · W. Duan
Department of Chemical Engineering, Inner Mongolia University of Technology, Hohhot, China
e-mail: imutlcq@126.com

Y. Zheng
Department of Accounting, Inner Mongolia University of Finance and Economics, Hohhot, China

E. Qi (ed.), *Proceedings of the 6th International Asia Conference on Industrial Engineering and Management Innovation*,
DOI 10.2991/978-94-6239-145-1_90

and carbon emission restraint [3]. These models are IO model [4], CGE model [5], COMAP model [6], CO2FIX model [7], F-CARBON model [8]. However, these models focused on the impact of emission reduction policies on the economic system, including GDP, energy mix, and employment rate; other than putting emphasis on the change of raw materials and products when economic system changes, which defect the calculation deviation of carbon emission and hindering the low-carbon development of coal chemical industry.

In respect to these two methods, carbon emission calculation can be divided into energy utilization and industrial process [9], in other words, carbon emitted in energy consumption and manufacturing process. It has been proved that Clean Development Mechanism (CDM) projects in China did not generate ideal results because business unit is in charge. Therefore, this paper adopted system dynamics approach to set up a model for carbon emission calculation and prediction using STELLA. We analyzed carbon emission resource of coal chemical industry and established a STELLA model for calculating carbon emission [10, 11] of the coal chemical industry using dynamic simulation. The model was built after a comprehensive analysis of the system, such as the dynamic change of capital, technologies and policies. By conducting simulation of the change of factors to analyze the changes of carbon emission of coal chemical industry.

2 Methodology and Discussion

This model is built on the basis of system dynamics and system comprehensive dynamic analytics. Then integrated it with the assistance of STELLA simulation software and the carbon emission coefficient published by IPCC in 2012. After comparison and analysis of the status quo of the coal chemical industry in China, we can set the boundary of the system. The main factors affecting carbon emissions includes: science and technology, industry layout, socio-economic, policy and environmental factor. Interactions between these factors affects the behavior of the system. In order to understand these relationships, we need to use causal analysis in SD theory. Carbon emissions of coal chemical industry comes from raw materials and energy consumptions. Raw materials and energy consumption are determined by fixed asset investment and population sector and others. These factors forms the 3 feedback loop of the industry's product yield, as shown in Fig. 1.

There are 3 different types of technologies of coal chemical industry production processes: coal coking, coal gasification and coal liquefaction. In these production processes, coal is used as raw material to produce coke, tar, gasoline, diesel, alcohol and synthesis ammonia. Carbon of these products is discharged through oxidation or reduction reaction in the form of CO_2 and constitutes the main carbon sources in the coal chemical industry, as shown in Table 1. In order to clarify the distribution of carbon sources at various stages of the coal chemical industry, the process flow is indicated in Fig. 2.

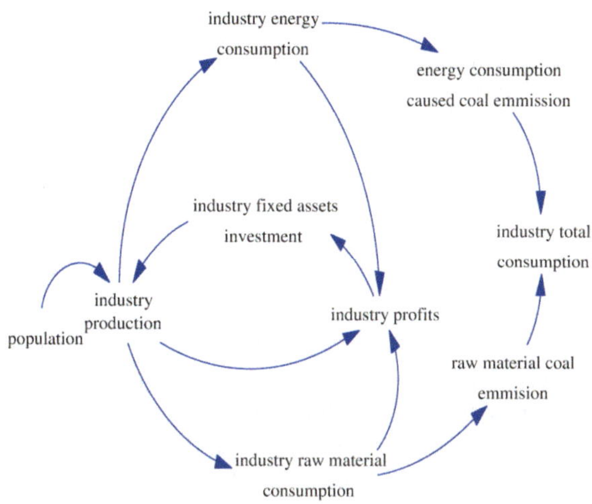

Fig. 1 Main coal emission system factors causality graph

Table 1 Coal sources from coal chemical process

Technic process	Coal coking	Coal gasification	Coal liquefaction
Carbon source	Coke; tar	Methanol; glycol	Gasoline; diesel

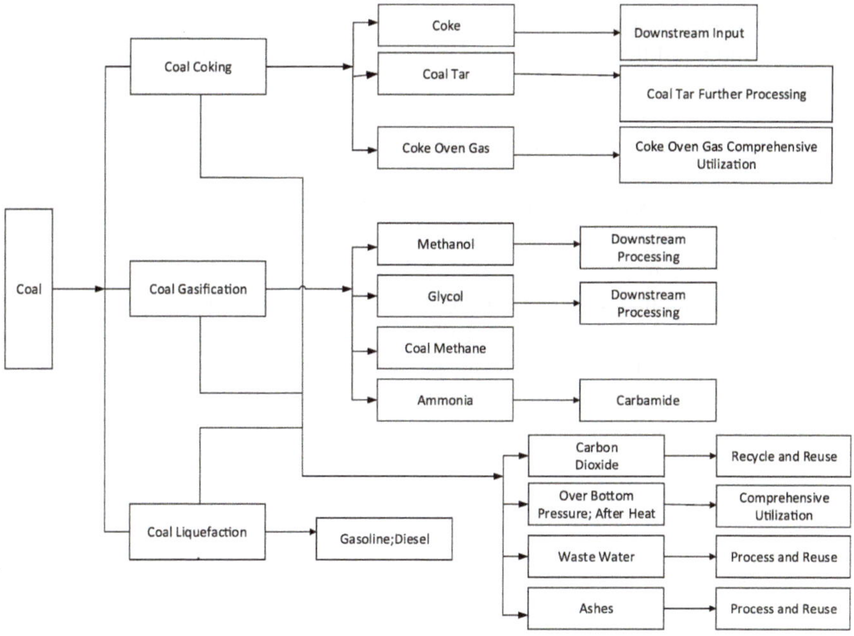

Fig. 2 Coal chemical production process flow

Table 2 Yield ration of different processes of the coal chemical industry in different years

Year Technic	Portion of 1990	Portion of 2000	Portion of 2010	Portion of 2020
Coal coking (%)	99.1	95.4	92.6	89.1
Coal gasification (%)	0.5	3.2	4.5	6.2
Coal liquefaction (%)	0.4	1.4	2.9	4.7

As shown in Fig. 2, each of the three technologies will form carbon sources. And the carbon emission level is high. In these study, carbon emission is estimated through simulation model. By predicting future carbon emission, the firms could control carbon emission during its life cycle and reduce carbon emission. It is needed to emphasize that waste of coal liquefaction and coal coking technology needed to be further processed to reduce carbon emission. However, we assume in this study that the recovery rate of waste gas of this industry is 100 %. In other words, the carbon content contained in energy will be completely emitted in the form of CO_2.

According to causality analysis, the main parameters of the models (as shown in Table 2) are: fixed assets investment, technology investment, regional population and production processes factors (proportion of products by different processes in different years). The first three parameters comes from data of yearbook and industry manuals. Production process factors comes from industry investigation and consultation from experts.

In this model, difference Eq. (1) is used to describe the state of the system at a specific time:

$$Lever(t) = lever(t - dt) + (Inflow(t - dt) - Outflow(t - dt)) * dt \qquad (1)$$

Lever stands for state variable; *Inflow* stands for inflow rate and *Outflow* stands for outflow rate. The main state parameters of the system are: regional population, product price and fixed assets.

We run the simulation model and the results show that the model construction is reasonable. The value of changes in carbon emission of the actual industry and the industrial production are close to the predicted values. Simulation result data is shown in Table 3, we can see the tendency in the next 10 years of coal chemical industry carbon emission in Fig. 3.

Table 3 Results of simulation estimation on coal chemical industry products in Inner Mongolia

Year	Coal chemical products	Actual value (10,000 tons)	Simulation value (10,000 tons)	Error (%)
2000	Coke	173.33	172.08	−0.7237
	Ethanol and tar	46.45	44.53	−4.3135
	Hydrocarbon fuel gas	13.20	12.82	−2.9941
2009	Coke	1770.60	1783.65	0.7316
	Ethanol and tar	233.23	243.26	4.1235
	Hydrocarbon fuel gas	18.81	21.68	13.2273
2010	Coke	1890.00	1856.39	−1.8105
	Ethanol and tar	259.99	255.24	−1.8608
	Hydrocarbon fuel gas	20.34	20.87	2.5581
2011	Coke	2345.00	2451.23	4.3337
	Ethanol and tar	505.99	509.68	0.7237
	Hydrocarbon fuel gas	22.00	21.26	−3.4980
2012	Coke	2569	2653.23	3.1746
	Ethanol and tar	656.25	661.58	0.8056
	Hydrocarbon fuel gas	23.94	22.34	−7.1620

Data is concluded under the analysis of the information listed in Inner Mongolia energy statistical yearbook, industry yearbook, and news data comprehensive analysis. The ethanol and tar product refers to methanol, coal tar and other coal chemical products; the hydrocarbon fuel gas means methane, alkene, natural gas, coke oven gas, and coal chemical gaseous products, the unit is billion cubic meters. (Prior to 2002, Inner Mongolia produced no methanol)

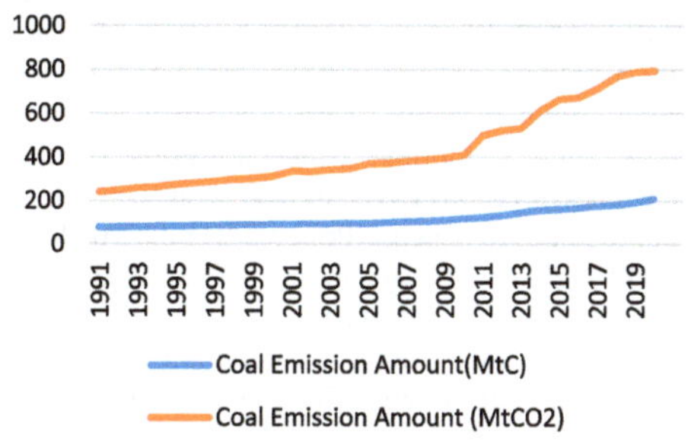

Fig. 3 Trend of carbon emission in coal chemical industry in Inner Mongolia (1990–2020)

3 Conclusion

The coal chemical industry is an important part of Chinese energy industry. This paper discussed the system dynamics approach of carbon emission management of the life cycle of coal chemical industry and predict carbon emission in the next 10 years. Results showed that coal emission will continue to grow in the future. At 2017, the growth rate is expected to be milder. High level of carbon emission comes from the backward of technology. In terms of the nation, it is necessary to adjust industrial structure and eliminate backward technology. In terms of firms, they need to understand that advanced technology is the most important way to product cleaner product and make profits.

Acknowledgments This work was supported by National Natural Science Foundation of China, grant No. 71262022 and the Inner Mongolia autonomous region talents of science and technology project funding.

References

1. Cremona F, Kõiv T, Nõges P, Pall P, Rõõm EI, Feldmann T, Nõges T (2014) Dynamic carbon budget of a large shallow lake assessed by a mass balance approach. Hydrobiologia 731 (1):109–123
2. Bauer N, Bosetti V, Hamdi-Cherif M, Kitous A, McCollum D, Méjean A, van Vuuren D (2015) CO_2 emission mitigation and fossil fuel markets: dynamic and international aspects of climate policies. Technol Forecast Soc Change 90:243–256
3. Llop M, Manresa A (2014) Comparing the aggregation bias in the input–output model and the social accounting matrix model. Appl Econ Lett 21(11):795–800
4. Mandell S (2007) Optimal mix of emissions taxes and cap-and trade. J Environ Econ Manage 12:1–10
5. Bollen J (2015) The value of air pollution co-benefits of climate policies: analysis with a global sector-trade CGE model called WorldScan. Technol Forecast Soc Change 90:178–191
6. Dib L, Silvestro D, Salamin N (2014) Evolutionary footprint of coevolving positions in genes. Bioinformatics 30(9):1241–1249
7. Negash M, Kanninen M (2015) Modeling biomass and soil carbon sequestration of indigenous agroforestry systems using CO2FIX approach. Agric Ecosyst Environ 203:147–155
8. Parry ML (ed) (2007) Climate change 2007: impacts, adaptation and vulnerability: contribution of Working Group II to the fourth assessment report of the intergovernmental panel on climate change, vol 4. Cambridge University Press, Cambridge
9. Grubb M, Butler L, Twomey P (2006) Diversity and secuity in UK electricity generation: the influence of low-carbon objectives. Energy Policy 34(18):4050–4062
10. DeCarolis JF, Keith DW (2006) The economics of large-scale wind power in a carbon constrained world. Energy Policy 34:395–410
11. Lee Y, Tong LI (2011) Forecasting energy consumption using a grey model improved by incorporating genetic programming. Energy Convers Manage

Research of Implementation Mode of Strategic Management Accounting

Yan-fang Gao, Ling Bao and Yu-si Guo

Abstract The more intense market competition, the more we need strategic management accounting. Traditional management accounting focuses on how to make its own management more scientifically sound, rarely pay attention to the changes in the external environment. To obtain long-term competitive advantage for enterprise, collecting internal and external information and making a reasonable strategic decision is crucial. This paper discusses how to collect internal and external information and make useful information for strategic management.

Keywords Strategic management accounting · Implementation mode

1 Introduction

What is strategic management accounting? At present, there is no unified understanding. The main points are summarized as follows:

1. Simmondz' definition and description of strategic management accounting focused on comparison of the firm with its competitions. He advocated collection of information necessary to enable determination of market share and competitors' pricing costs and volume [1].
2. Wilson too criticised traditional management accounting systems as extending to be "reactive and to deal in a programmed way with one-of decisions by contrast a strategic approach needs to be proactive and designed to deal in an unprogrammed way with each phase of strategic decision-making" [2].

Y. Gao (✉)
School of Accounting, Inner Mongolia University of Finance and Economics, Hohhot, China
e-mail: nmgyf512@163.com

L. Bao · Y. Guo
Inner Mongolia University of Finance and Economics, Hohhot, China

E. Qi (ed.), *Proceedings of the 6th International Asia Conference on Industrial Engineering and Management Innovation*,
DOI 10.2991/978-94-6239-145-1_91

3. Bromwich definated strategic management accounting as: The provision and analysis of financial information on the firm's product markets and competitor's cost and those of its competitors in these markets over a number of periods [3].
4. Klarke considers that while most management accounting innovation studies take a demand-side perspective, the supply-side is important because it provides an alternative explanation for the relatively low implementation rate of new management accounting innovations [4].

Reference to the above research results, this paper argues that strategic management accounting is the subject that its goal is to help enterprise obtain long-term competitive advantage, provide the theory and methods for strategic management services. Strategic management accounting starts from understanding the internal and external environment of enterprises, through collecting and processing the internal and external environment information which can make managers clear and definite its environment as well as its own advantages and disadvantages, then determining strategic objectives and implementing rules and phasing planning. Strategic management accounting should also build strategic performance evaluation system that can give a reasonable and comprehensive evaluation on the implementation effects of strategic management accounting.

2 The Basic Implementation of Strategic Management Accounting

The difference between traditional management accounting and Strategic Management Accounting is that Strategic Management Accounting paid special attention to the analysis of supply, quality, market share and competitor, in other words, diversified information is based on export-oriented, in order to determine the presence of competitors and opportunities and threats of the outside world, and then make sound strategic decisions. The external environment is constantly change, strategic management accounting can help the enterprise understand external environment better, to determine the market position and own advantages and disadvantages. To achieve this goal, the application system of strategic management accounting must establish the information warehouse on strategic management, the ways of obtaining information are shown in Fig. 1.

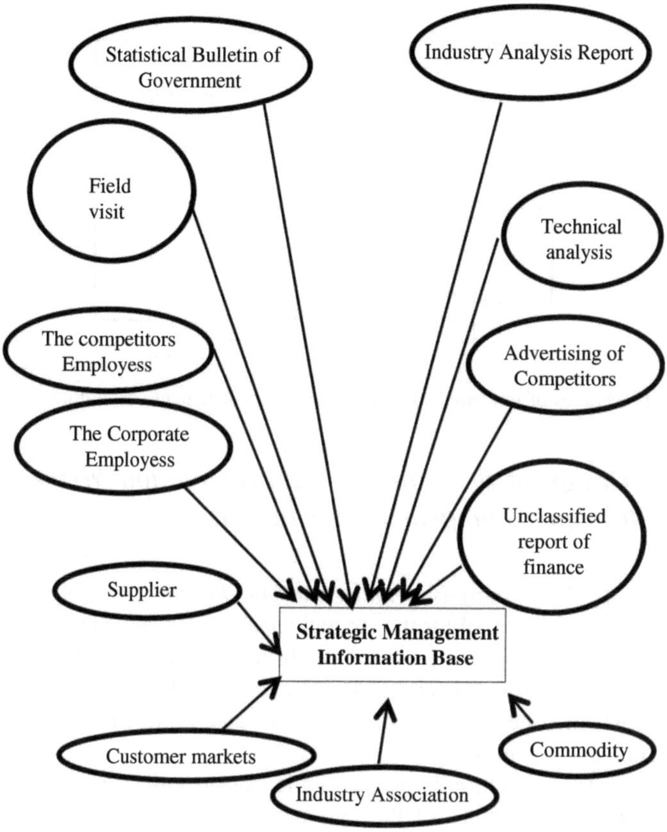

Fig. 1 Access to strategic management information way

3 The Design on Procedures and Method of Strategic Management Accounting

3.1 The Design on Procedures of the Strategic Management Accounting

The implementation procedure is an important part of strategic management accounting system, it makes strategic management accounting can play its functions. Implementation procedure is also a specific step which strategic management accounting can complete its strategic objectives, including collection of internal information and external information, making strategic decisions, operating management and evaluating performance. The implementation procedure of strategic management accounting is shown in Fig. 2.

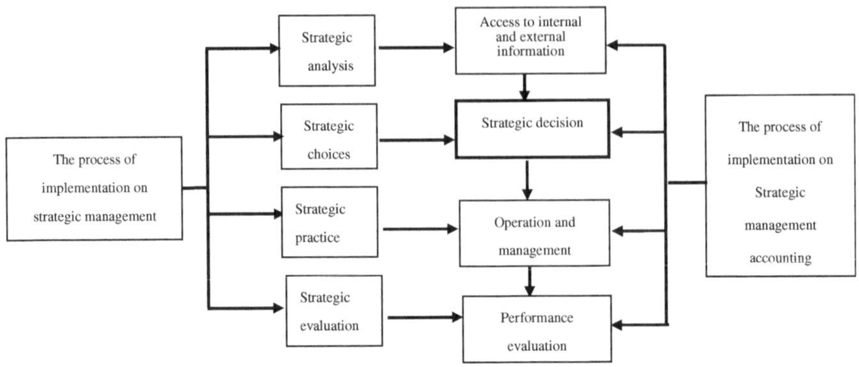

Fig. 2 The procedures of implementation on strategic management accounting

3.2 *The Design of the Concrete Implementation Method of Strategic Management Accounting*

(1) *Through the PEST analysis to understand the business environment*
PEST analysis is a basic tool for the strategic analyses of the external environment, it is advantageous for the enterprise to start from the overall external environment, through the analysis of the macroeconomic environment of the enterprise market, determining whether they take the strategy of the enterprise. PEST analysis consists four elements, as shown in Fig. 3.

P (*Politics*) First of all, enterprises should consider the state system, political system and policy guidelines of the region. If the local authority has a positive attitude to support the development of the industry or enterprise, enterprises should vigorously expand its business, to achieve rapid growth along with government support. If the relevant departments change the policy of industry or enterprise policy changes, enterprises also need to timely adjust business strategy, to adapt to

Fig. 3 PEST analysis diagram

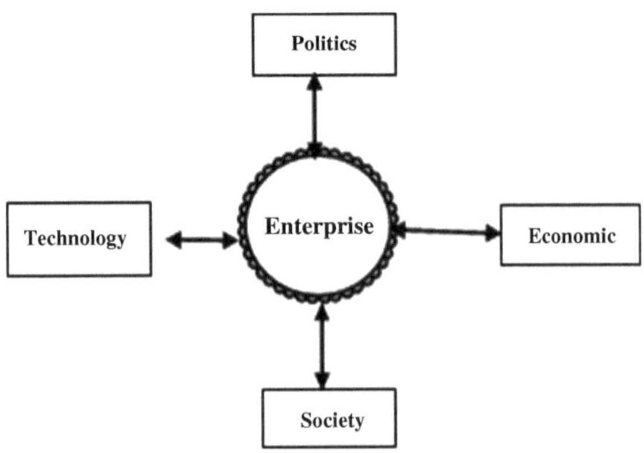

the changes mentioned above, at the same time, enterprises should pay attention to the research on the legal environment of the industry, familiar with relevant laws and regulations, ensuring the lawful operation of enterprises.

E (*Economic*) Enterprises should have a comprehensive understanding of the environmental economic system, economic situation. First of all, enterprises should familiar with the national economic system. At the same time, it should understand the current economic situation. If the economy overheated, the government will tighten monetary policy, enterprises should reduce investment. If the economic situation downturn began to rebound, countries tend to adopt loose monetary policy, enterprise should increase investment.

S (*Society*) Enterprises should understand the customs' culture and customs, income level, risk awareness, investment concepts in the market which the enterprises was in,to divide the different characteristics of customer groups, according to the needs of different groups, design different products. At the same time, enterprises should consider the market characteristics of the social environment, design appropriate products which compliance with local values and cultural traditions, provide not only high quality but also distinctive services.

T (*Technology*) Enterprises also need the support of technology. On the one hand, enterprises need to update and upgrade on electronics system to ensure the improvement of operational efficiency. On the other hand, enterprises should understand new products that can attract customers, moreover its development trend and application prospect, in order to ensure the leading position in the market.

(2) *Using SWOT analysis to determine the company's competitive position in the market*

SWOT analysis is a kind of comprehensive analysis which enable enterprises not only to clear realize their own strengths and weaknesses, but also to define the opportunities and threats in the external environments [5].

S—Strength, it is the organization's strengths, such as adequate capital, increase of market share, skilled workers, strong product development ability.

W—Weakness, it is the organization's own weaknesses, such as shortage of funds, decline of market share, equipment or technical personnel do not adapt to the new materials or new technology, weak development ability and etc.

O—Opportunities, it is a business opportunities offered by external environment, such as good market prospects, the competitor with financial crisis, the supply of raw materials is more and more abundant, new materials or technology appeared.

T—Threats, it is a threats offered by external environment, such as not optimistic about the market prospects, competitors take the lead in using new technology, material supply increasingly tense, higher material prices, more and more high requirements on quality and so on.

The method of SWOT is an important tool that can help enterprises make the decision on strategic aspect. Using SWOT analysis, enterprises obtain objective evaluation of their own comprehensive situation, identify the factors of various advantages, disadvantages, opportunities and threats, in order to make strategic decision to adapt to their internal abilities and external environment. To achieve the goal to maintain the long-term competitive advantage.

(3) *Application ABC to reduce management cost*
ABC (Activity-based Costing) provides information for activity management, it is a method of cost accounting based on activity. Definition activity is a start point of cost accounting and the core of enterprise activity management. It is also the most striking difference between ABC and traditional cost management. Implementation of ABC can be regarded as a major change and breakthrough in the enterprise management. Activity management is to promote the improvement of the value-chain through improving activity-chain, to achieve the purpose of optimizing the structure of activities. Because customers are the most important resource, enterprises can implement ABC, divide the activity into two parts which can increase customer value or not. Those activities that can meet customer needs and enable enterprises to gain more customers are called Value-added activities [6]. However, the characteristics of activities which can't increase customer value will not damage the specific requirements of products, even without them, so are called non-value-added activities. Enterprises should as far as possible to eliminate non-value-added activities. To Value-added activities, enterprises should continue to optimize the operating procedures, reduce consumption of resources, to strive to make every program using the lowest consumption of resources to complete the customer value added.

The production process described by theory of ABC is that products are manufactured by productions, productions lead to the occurrence of activities, activities consume resources, thus activities cost is created. This concept of cost is different from the concept of product cost in traditional costing. In ABC, cost driver is a standard of assigning costs, between resource consumption and product cost, activities are the bridge of separating, induction, eventually forming products cost. Therefore, ABC calculates the cost of various cost objects by two-stage method, using different cost drivers to assign activity cost to the product cost. This method can more accurately calculate product cost. The difference between ABC and traditional cost calculation as shown in Fig. 4.

(4) *Using target costing to enhance the operating profit*
Target costing is a method of strategic cost control which comes from Japan. If Enterprises want to survive in the competitive environment, whether that current activities economic or not are not the things we focus on cost management, but how can we make more economic activities. The lack of a goal in the traditional cost accounting and management systems is many resources should be spent for a new product. This goal is just a key factor, to stimulate designers to design a new product at the lowest cost. Target costing will rule on the future target market and similar products of competitors. This is strategic cost strictly.

Target Costing enables customers' needs at the center of corporate strategy in the process of developing and implementing strategy [7]. The customers' needs including product quality, features and price, these are very important factors in

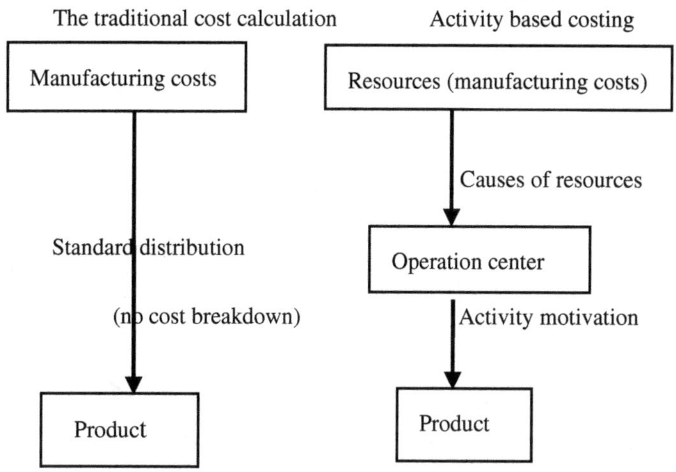

Fig. 4 Comparison chart of ABC and the traditional cost calculation

achieving and maintaining a competitive advantage. In the process of implementation of target costing, first of all, through market survey to collect information, understand customers' willing to pay for the price of this product, as well as to achieve the desired functionality of the product. Moreover, product status of competitors should also be understood. According to the information, after deducting the cost of target profit, enterprises can calculate the maximum cost of the product permitted in the manufacturing, distribution and processing. It is the target cost. Target cost management uses the way of multi-cycle extrusion, through constantly squeezing costs, enterprise can continually reduce costs and attain the target cost.

The basic process is "Setting—decomposition—reach—and then setting—and then decomposition……" Multi-cycle, each cycle is a squeeze on costs, continue to reduce costs in order to reach the target cost before the end of extrusion. Target cost management map as shown in Fig. 5.

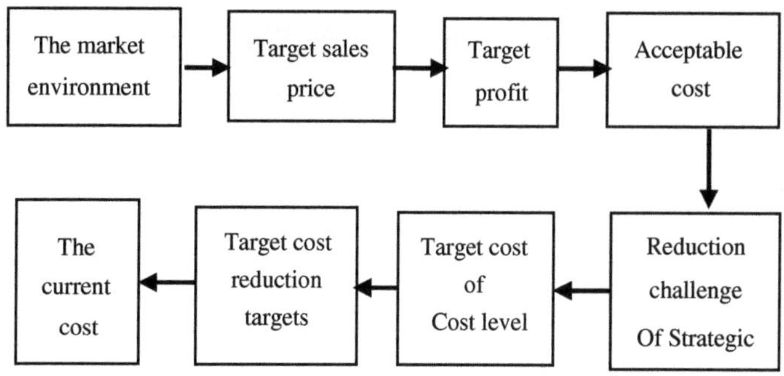

Fig. 5 Target cost management map

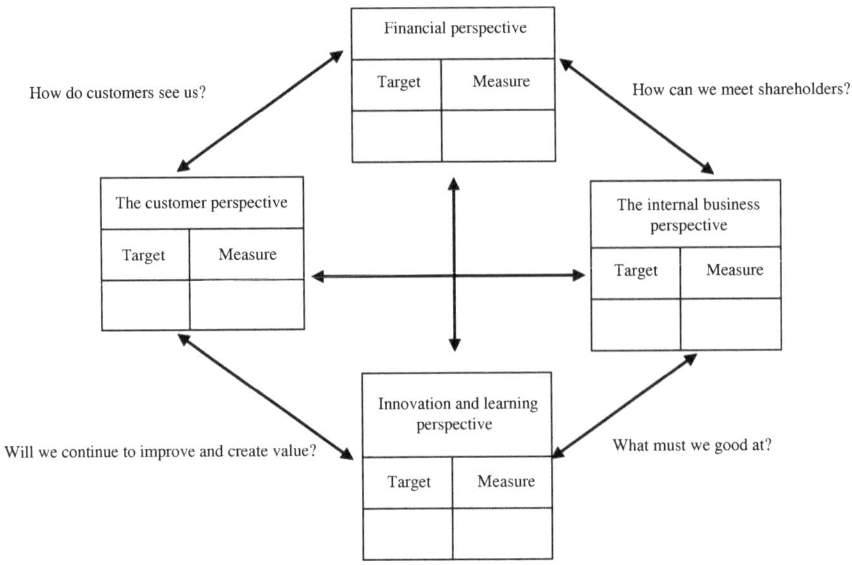

Fig. 6 The balanced scorecard performance measurement schematic diagram

(5) *Using BSC to evaluate strategic performance of enterprise*
BSC (Balanced Score Card) divides the performance evaluation of enterprise into four aspects: finance, customer, internal operation process, innovation and learning [8]. Internal operation process and innovation and learning are driving factors, finance and customer are result factors. These four aspects are not isolated, a mutual promotion relationship exists between them. Learning how to improve operation process makes the process more perfect, the improvement of operation process can improve customers' satisfaction, the improvement of customers' satisfaction will enhance the financial performance of enterprise. Four aspects of Balanced Scorecard establish a kind of balance relationship, it requires that short-term goals and long-term goals should be considered, objective goal and subjective goal should be considered too. Using the Balanced Scorecard evaluates enterprise's performance as shown in Fig. 6.

3.2.1 The Perspective of Finance

Financial indicators reflect the financial results generated by implementation of strategy. These financial indicators used to evaluate whether enterprise's profits can be increased with the help of the strategy. This perspective should answer the question, “How can we meet shareholders?” Financial indicators generally relates to the profit ability, evaluation indicators such as the rate of sales profit, etc.

3.2.2 The Perspective of Customer

This perspective is converting the goal of customer service into specific evaluation indicators, the indicators should be able to reflect correctly various factors associated with the clients. This perspective should answer the question, “How customers see us?” The indicators include the occupation rate of market, the ability of attracting customers, etc.

3.2.3 The Perspective of Internal Operation Process

Internal operation processes affects customers’ satisfaction and financial goals. This perspective should be possible to identify and measure the enterprise’s core capabilities as far as possible, such as the key technology to maintain market leadership. It will answer the question, “How must we good at?”

3.2.4 The Perspective of Innovation and Learning

Whether the enterprise has strong ability of innovation and learning or not, directly related to the value of enterprise. This perspective focus on infrastructure construction enabled enterprises to gain long-term growth and improvement. This perspective should answer the question, “Shall we continue to improve and create value?” These indicators include the satisfaction degree of employees, training time, the potential of the staff, increasing on the capacity of information system, etc.

The four dimensions of BSC constitute a complete cycle, describe the four factors required in the development of the enterprise. Proper management and evaluation in these areas can promote enterprise to achieve sustained and efficient development.

References

1. Simmondz K (1981) Strategy, strategic management accounting and performance: a configurational analysis. Ind Manag Data Syst 112(3):484–501
2. Wilson R (1995) Strategic management accounting. Prentice-Hall, London
3. Bromwich M (1990) The case for strategic management accounting: the role of accounting information for strategy in competitive markets. Organ Soc 15:27–46
4. Klarke Peter J (1995) The old and the new in management accounting management accounting, vol 6, pp 47–49

5. Rauch P (2007) SWOT analyses and SWOT strategy formulation for forest owner cooperations in Austrial. Eur J For Res 126:413–420
6. Cooper R, Slagmulder R (1999) Activity-based cost management system architecture part II. Strateg Finan 12:69–70
7. Cooper R, Slagmulder R (1999) Develop profitable new products with target costing. Sloan Manag Rev 40(4):23–34
8. Kaplan RS, Norton DP (1992) The balanced scorecard: measures that drive performance. Harvard Bus Rev 70:71–79

Research on the Information Management Problems in the Linkage Process of Manufacturing and Logistics Industry

Qiong Li

Abstract With the deepening of manufacturing and logistics industry linkage, information problems gradually become the primary problems hindering linkage development of manufacturing and logistics industry. In this paper, we analyze the information management problems in the linkage process of manufacturing and logistics industry, and give some effectively countermeasures.

Keywords Logistics industry · Manufacturing · Linkage development · Information management

1 Introduction

With the deepening of economic globalization, industrial linkage makes the market competition change from the single enterprise competition to cooperative competition which among in many enterprises in the whole industry chain. Regions competition also changes from industry competition to competition of the industrial cluster and the industrial chain. Therefore industry linkage gets attention from governments and scholars. In the process of economic activities, there are extensive, complex and consanguineous technical and economic relations between every industry. In this technical and economic relation, the most basic link is maintain the reproduction, formed the relationship between supply and demand in the industry.

Fund: Project supported by Jiangsu Province UniversityFoundation of Philosophy and Social Science (Grant No.: 2014SJD454).

Q. Li (✉)
Business School, Hohai University, Nanjing, China
e-mail: qiongber@163.com

Q. Li
Department of Management, Xuzhou Institute of Technology, Xuzhou, China

E. Qi (ed.), *Proceedings of the 6th International Asia Conference on Industrial Engineering and Management Innovation*,
DOI 10.2991/978-94-6239-145-1_92

While the key problem of maintaining this relationship between supply and demand is information management. Therefore, the analysis and management of the information resources of industrial linkage has important theoretical and practical significance. It will improve the information management part in the linkage process of manufacturing and logistics industry; accelerate the organic integration and linkage development between manufacturing and the logistics industry.

At present, the majority of research in our country is still on the analysis and discussion of importance, the present situation and the policy of China's manufacturing and logistics industry linkage development [1]. It is lack of analysis about information management in the linkage process of manufacturing and logistics industry. From the new perspective of information management, this paper analyses the current information management situation of the industrial linkage process, and proposes some strategies response to the information management problem which they will face with.

2 Analysis of Information Management Problems in the Linkage Process of Manufacturing and Logistics Industry

Information management in the linkage process of manufacturing and logistics industry is the management of all information related to various links. It includes collection, analysis and utilization of information. Through the choice, judgment and processing of all information, we can make sure the enterprise to make the optimal decision, optimizing the allocation of resources. It is the essence of information management of manufacturing and logistics industry linkage process.

Information management in the linkage process of Manufacturing and logistics industry Information management have the management function of both external and internal information. On the one hand, it can dispose external information of linkage process, such as change of macroeconomic policy and market environment, etc. This function makes the linkage process of manufacturing and logistics industry always fully understand the external environment of the country, so it is able to respond the policy adjustment timely. On the other hand, information management of manufacturing and logistics have the ability to dispose the internal information of linked process, such as capital, allocation of labor, stock, etc. Internal information management make the manufacturing and logistics industry be linked tightly and exploit the advantages of the linkage to the full, and give play to the ability of manufacturing and logistics industry at the greatest extent.

2.1 *Low Degrees of Information resource Fusion Between Manufacturing and Logistics Industry*

Logistics informatization is the essence of manufacturing and logistics industry linkage development, timeliness of information determines the efficiency of logistics operation. Logistics industry development in our country lags behind, the information system is not perfect, the asynchronies of information and unable share of information resources has become the critical restraints, and restricting the development of manufacturing and logistics industry linkage.

On the one hand, manufacturing logistics demand is scattered, it does not translate into social needs [2]; On the other hand, due to the lack of social demand, and the development of specialized logistics is restricted [3]. Manufacturing is anxious of the deliverability of logistics enterprises; logistic enterprise is blind to the actual demand of manufacturing enterprise, these led to the contradiction between insufficient effective demand and insufficient supply. These cause that manufacturing and logistics industry is difficult to "thanks" to each other and cannot improve the ability and level of manufacturing logistics.

2.2 *Lag of Information Technology Application of Manufacturing and Logistics Industry*

The most of logistics industry follows the traditional operation way and is short of innovation. Its level of modernization is low. Modern logistics is different from traditional logistics. It is based on modern information technology. It is an integrated logistics active mode. It integrates all kinds of functions, such as transportation, packing, loading and unloading, handling, storage, distribution processing, distribution, recycling and processing and logistics information processing and other functions. It aims to maximize the resource utilization and operational efficiency. It emphasizes the systematisms and coordination. Logistics enterprises in China are mostly stay in the traditional economic mode; they do not use logistics information system and lack of material and technological basis to face the new economic model. One of the few logistics enterprises are relatively advanced. They often only have the shape of logistics information system, but they are lack of the essence. Design of the logistics information system few behind, different kinds of information are lack of mutual links and sharing.

2.3 Relatively Slow Construction of Information Platform of Manufacturing and Logistics Industry

The lack of public information platform of manufacturing and logistics industry make their respective information system become an island. Logistic enterprise is blind to the actual demand of manufacturing enterprise, manufacturing is anxious of the deliverability of logistics enterprises; these led to the contradiction between insufficient effective demand and insufficient deliverability. It seriously restrict integration of manufacturing and logistics industry.

2.4 Informationization Standards of Manufacturing and Logistics Industry are not Unified

Information standards of manufacturing and logistics industry are not unified [3]. The overall level of the former is relatively low. The standards of product code, enterprise logistics information interaction standard and logistics information platform is not completely uniform. In the standard selection and implementation process, enterprise is not effectively guided by the government, industry associations and professional bodies. The development of informatization standardization process in the manufacturing and logistics industry is influenced by the lag of enterprise information standardization work.

3 Suggestions for Information Management in Linkage Process of Manufacturing and Logistics Industry

3.1 Focus on Sharing of Information Resources Between Manufacturing and Logistics Industry

At the present, the information-sharing between manufacturing and the logistics industry in our country just achieve the information-sharing between manufacturing enterprises and logistics enterprises. But it doesn't achieve the information-sharing between industries. However, it is important for multi-enterprise to share information in the overall development of manufacturing and logistics industry linkage. According to the problem that the information cannot be shared effectively, some measures should be taken.

(1) Both of the enterprise should pay more attention to strengthen the information management. They should understand that only strengthening the information sharing in the linkage process of the manufacturing and logistics industry, industries linkage can develop faster and better. When linkage of two

industries overall develops, linkage of manufacturing enterprises and logistics enterprises can develop well, which is based on the development of good industry.

(2) Manufacturing and logistics industry should plan as a whole, and establish an information management platform that can used by multiaspect of the linkage. It can realize information-sharing of multi enterprise.

3.2 Strengthen the Application of Information Technology in Manufacturing and Logistics Industry

Information technology is the foundation of the smoothly development of manufacturing and logistics industry linkage. Now, in the linkage process of China's manufacturing and the logistics industry, application of information technology is much lower than that of foreign enterprises. Only a few logistics enterprises apply modern information technology in our country, and only a few kinds of information technology are used in information management. The following kinds of information technology can be used in the process of information management.

3.2.1 Identification and Collection of Information

Logistics information code technique: Logistics information code can be identified information according to certain rules timely; it is convenient and fast to collect information accurately.

Bar codes technique: Bar code technique gives goods a unique code. When we collect information, we should scan code, and data can be automatically transmitted to the computer.

Smart label technique: Smart label technique is a kind of modified bar code technique. It not only has the low cost which is the advantage of the bar code technology, but also it has automatic data collection. It is a kind of low cost wireless identification technology.

3.2.2 Information Transfer

RFID technique: Radio frequency technique is a kind of technology which can send collected data to information management system.

Electronic data interchange (EDI) technology: Electronic data interchange technology standardize information by unified standards, so that information can be transferred in each information system.

3.2.3 Processing and Storage of Information

Information management system: Information management system makes processing and storage of information comes true. Through the information management system, people can get useful information for them and filter the useless information.

3.3 Expanding the Function of Information Management Platform in Linkage Process of Manufacturing and Logistics Industry

Nowadays the majority of the information management platform only has the low-level functions of information collection, information transmission and information sharing in the linkage process, but not have the high-level function like information processing and information analysis. They are at best an information exchange platform, not an information management platform which can realize industries linkage. In order to help the linkage development of manufacturing and logistics industry, it is urging to extend the information management function of information management platform.

3.3.1 The Function of Data Analysis

Recent information management platform only has a simple function of collecting information, but not has the function of data analysis. Through collecting information in the linkage process of manufacturing and logistics industry, data analysis can analysis data and provide management reference.

3.3.2 The Warning Function of Abnormal Data

During the linkage process of manufacturing and logistics industry, there are some things beyond all expectations; information management platform should issue a warning to these abnormal data.

3.3.3 The Functions of the Response to the External Information

The information management in the linkage process of manufacturing and logistics industry is not only collect and process internal information, but also response to the external information. Because the changes of external environment will also impact the linkage of manufacturing and logistics industry.

3.4 *Uniform the Standards of Information Management Platform of Manufacturing and Logistics Industry Linkage*

During the linkage process of manufacturing and logistics industry, various uneven information platforms seriously restricted the linkage development of the manufacturing and logistics industry. Only we unify the standard of information management platform, information can be shared in all aspects and information management of manufacturing and logistics can enter a new stage. When the information management platform standards are drafted, we must take into account of the following aspects:

(1) There are their own local characteristics at every region, most local requirements should be considered.
(2) The formatted standards should be improved as the industry standard level, and standards should be executed to the letter.
(3) When standards formatted, the government should participate in and play a role of organizational leadership.

4 The New Trend of Information Management in Linkage Process of Manufacturing and Logistics Industry

4.1 *The Internet of Things Effectively Solve the Problem of Information Synchronization in the Linkage Process of Manufacturing and Logistics Industry*

On one side, the application of the Internet of things make the collection process of information changed from the traditional semi artificial collection to the process that the information management objects automatically transmitted information to the information management system. It effectively improves the accuracy of the information management system and operating efficiency of information collection. On the other side, it makes the object of information system refined to a specific object by an enterprise; it also makes the scope of information management to be more comprehensive. And all the requirements for information management in the linkage process of the manufacturing and logistics industry are becoming more and more stringent.

With the application of the Internet of things, in the information management of the manufacturing and logistics industry linkage, information can be passed directly from the information management object to the information management platform. It realizes the real-time response of information management platform and solves the problem that information is not synchronized.

4.2 Cloud Computing Vigorously Expanding the Work of the Information Management in the Linkage Process of Manufacturing and Logistics Industry

Nowadays, although the linkage of manufacturing and logistics industry has made certain achievements, but it is still very difficult to take part in the linkage process for a large number of small and medium enterprises in our country. The concept of information management has been accepted by most of the enterprises, but enterprises which real start using information management means are rare. For one thing, Information management is a new management method; it must have to be a long stage to be accepted. For another, the high cost of information management makes the most of our small and medium enterprises step back. The appearance of cloud computing provides better opportunities for China's manufacturing and the logistics industry linkage, and the small and medium enterprises can also participate in the linkage process of manufacturing and the logistics industry. The linkage of the whole manufacturing and the logistics industry is possible to realize.

References

1. Su Q, Zhang Y (2001) Research on status quo and causation of the interactive development of the manufacturing and logistics industries. Soft Sci 25(3):61–64
2. Situation logistics and policy recommendations of Chinese manufacturing enterprise (2007) Task force. Analysis of logistics situation of Chinese manufacturing enterprise. Log Mater Hand 2007(6):31–36
3. Wang M, Liu B (2009) Issue and trend in the development of manufacturing and logistics industry. Modern Manage Sci 3:59–60

Value Chain Optimization of B2C E-Commerce Based on Cloud Computing and ITM

Di Liang, Shuang Wu and Gui-zhi Sun

Abstract B2C e-commerce enterprises have become a mainstream trend of the online retail market. But the talent, technology, mobile terminal and logistics bottleneck problems hinder the development of B2C e-commerce enterprises. This paper, based on value chain theory, optimizes value chain and solves the existing issues of B2C e-commerce business in order to improve the core competitiveness. Cloud computing and ITM interactive transaction model are applied, and the cost of the whole value chain to the B2C e-commerce enterprise is analyzed.

Keywords B2C e-commerce · Cloud computing · Cost analysis · ITM

1 Introduction

B2C trading pattern has become the main trend of online shopping development direction. In the process of B2C e-commerce development, it is affected by a lot of bottleneck problems, mainly reflected in the talent and technology, network security, logistics distribution, after-sale service and other bottlenecks. The links of bottleneck problems make B2C e-commerce enterprises to create value enterprises. It is imperative to solve the bottleneck problem of B2C e-commerce. The value chain optimization of e-commerce has great significance.

In recent years, Chinese and foreign scholars had a certain research results between B2C and electronic commerce value chain. Yang and Yang [1] discussed B2C e-commerce enterprise how to achieve sustainable development and enhance competitiveness in a cloud computing environment on the basis of the value chain

D. Liang · S. Wu (✉) · G. Sun
School of Mechanical Engineering, Shenyang University, Shenyang, Liaoning, China
e-mail: 445391942@qq.com

E. Qi (ed.), *Proceedings of the 6th International Asia Conference on Industrial Engineering and Management Innovation*,
DOI 10.2991/978-94-6239-145-1_93

theory. Pan [2] analyzed the key factors that influenced the B2C e-commerce enterprises core competitiveness according to the Porter value chain model. He [3] discussed mobile e-business three value chain models aimed at the disadvantages of the mobile e-commerce and the future development direction. Slywotzky [4] involved the concept of "value nets" for the first time. Rong [5] analyzed the difference between traditional value chain and value chain under e-commerce. Shank and Govindarajan [6] extended value chain, he considered that value chain covered the whole process of production operation and the scope of the value chain went through the changes from the enterprise to the industry. Rayport [7] and Sviokla [8] extended the concept of virtual value chain related theories based on it.

This paper begins with the theory of B2C e-commerce. First of all, build the B2C electronic commerce value chain and virtual value chain model based on the value chain theory. Then optimize the value chain, reduced the value chain cost, solve the problem of e-commerce bottleneck combined with the theory of cloud computing and ITM interactive transaction mode.

2 The B2C Online Value Chain Optimization in Cloud Computing Environment

Cloud computing has some characters, such as large size, mass storage, on-demand access and the dynamic of extension, it promotes the B2C e-commerce application greatly. Cloud computing system can be divided into three layers according to the different service types. The top is a SaaS (Software as a Service). The middle layer is the PaaS (Platform as a Service). The bottom is IaaS (Infrastructure as a Service) [9].

2.1 B2C e-Commerce Model in Cloud Computing Environment

A B2C e-commerce model, based on cloud computing platform, has to fully meet the demands of customer, platform service, resource acquisition, and payment. Integrate the various elements of the value chain, coordinate the participants of the value chain, and fully embody the strategic thought of "take the customer as the center". B2C e-commerce model based on cloud computing platform is shown in Fig. 1.

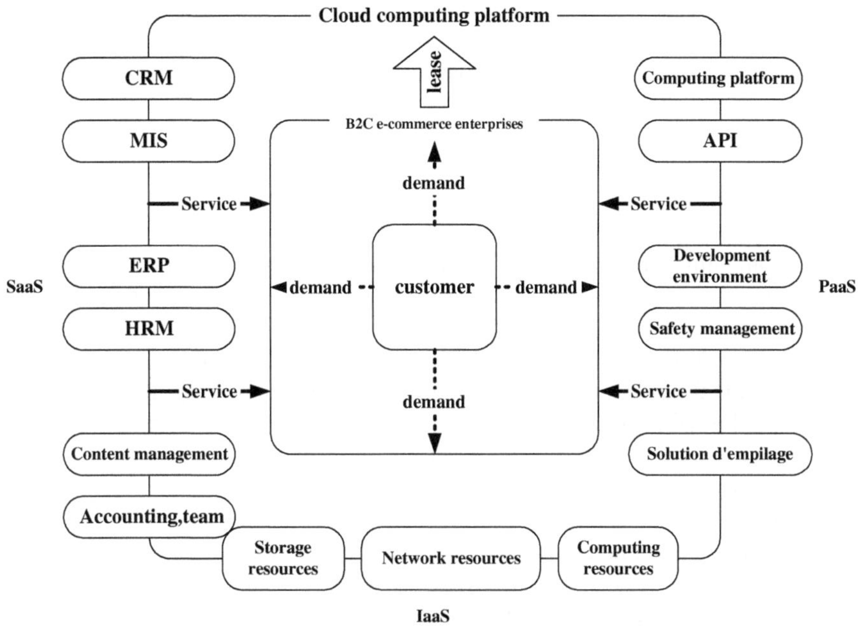

Fig. 1 B2C e-commerce service model based on cloud computing platform

2.2 The Optimization of B2C Electronic Commerce Value Chain in Cloud Computing Environment

2.2.1 Infrastructure Construction and Operation Optimization

Electronic commerce activities can help enterprises to reduce the cost of resources under the cloud computing environment. Cloud computing services can create a virtual data center, it provide services for the user through the network. Cloud hosting technology provides efficient and convenient server resources for enterprises. The applications stored in the cloud reduce the cost of the client software and hardware equipment of the enterprise requirements. In addition, the enterprise must invest enough manpower and financial resources to cope with the increasing speed up the equipment update [10].

2.2.2 Storage Optimization

The e-commerce companies can have real-time communication and transmit information with suppliers under cloud computing environment. Then it can adjust the purchase quantity in order to reduce inventory according to the demand of users timely. In addition, the application of cloud computing makes specialized storage center possible established independent on a single electronic commerce enterprise from economy and technology. The specialized storage center can provide goods storage services for many e-commerce enterprises through the use of advanced information management [11].

2.2.3 Online Trading Mode Optimization

Cloud computing builds efficient advanced defense system to defense malicious attacks based on powerful computing capacity of cloud collection. In the process of cloud payment, personal information and account information are stored in the cloud, so that it reduces the risk of the user information. Users only need a simple mobile terminal based on electronic payment of cloud computing, then can access to the cloud anytime and anywhere. They can finish all kinds of trading activities timely and security. The mobile electronic commerce model based on cloud computing is shown in Fig. 2.

The lack of power that mobile terminal itself handles information and faultiness safety performance hinder the development of mobile e-commerce [12].Cloud computing builds efficient advanced defense system to defense malicious attacks based on powerful computing capacity of cloud collection [13]. In the process of cloud payment, personal information and account information is stored in the cloud, so that it reduces the risk of the user information; Electronic trading has safety management standard and advanced information security technology to ensure convenient and safe of cloud payment [14].

2.2.4 The Optimization of Logistics Distribution Mode

Cloud computing is applied in logistics as "cloud logistics", the platform includes both private and public cloud. Logistics companies, agents, service providers, industry associations, management agencies, and legal institutions are concentrated

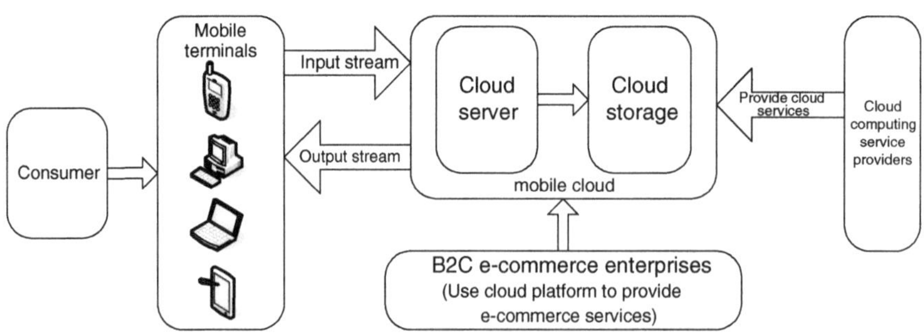

Fig. 2 Mobile e-commerce model based on cloud computing

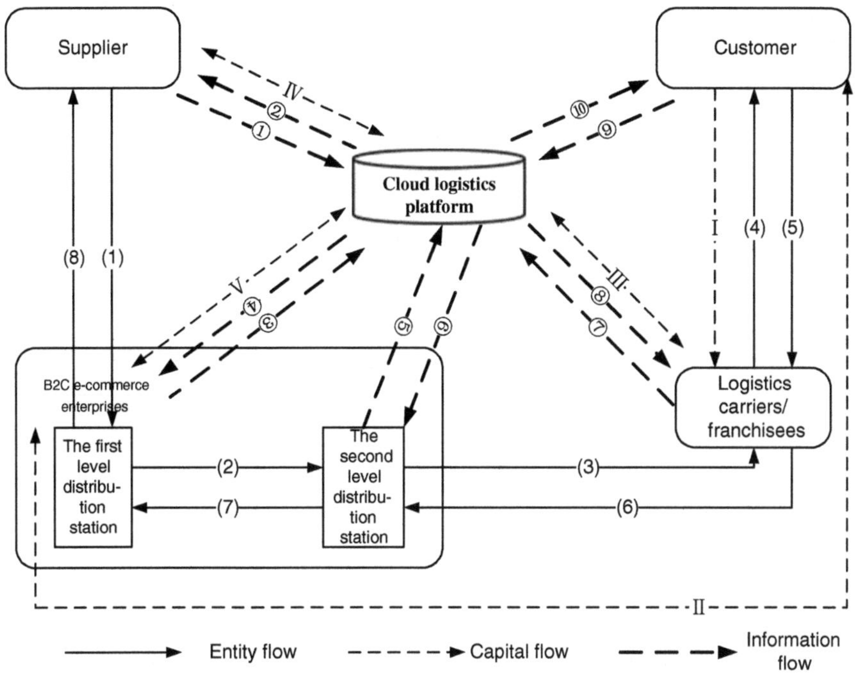

Fig. 3 "Cloud logistics" distribution model

in the cloud to integrate into a resource pool. The various resources communicate on-demand to reduce costs and improve efficiency. The logistics distribution model under cloud computing environment is shown in Fig. 3.

3 Unit the B2C Offline Value Chain Optimization in ITM Mode

ITM entity shop plays a role of bridge as the middle part of the whole service system. It optimizes traditional offline service system from various angles. The current B2Cesses are shown in Figs. 4 and 5.

By comparing two kinds of service process, the ITM mode offline service is optimized from the perspective of value chain analysis:

(1) Distribution service optimization: Many delivery companies take the low level employments; it can reduce the quality of service [15]. In the after-sale system of ITM mode, the customers contact with ITM entity shop, such as the mobile business hall. Logistics enterprises distribute online products to the ITM entity shop, and products are stored in ITM entity shop, ITM store inform customers. After receiving notification, the customers can go to ITM entity shop nearby at any

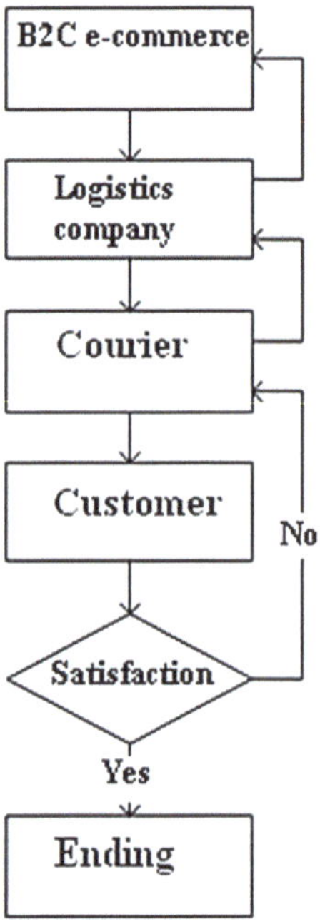

Fig. 4 Current offline service process

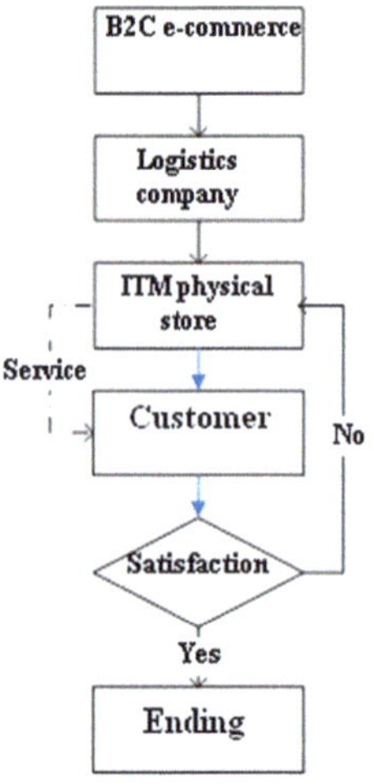

Fig. 5 Offline service process based ITM

time. The introduction of the ITM stores solves the problems existing in the current after-sales service system.

(2) Product service optimization. ITM entity shops meet the demand of more online ITM customer, increase the after-sales service and combine of e-commerce and online shopping together more effective on the basis of retail. ITM is given priority to service. It manages online and offline buyers, sellers of goods quality and

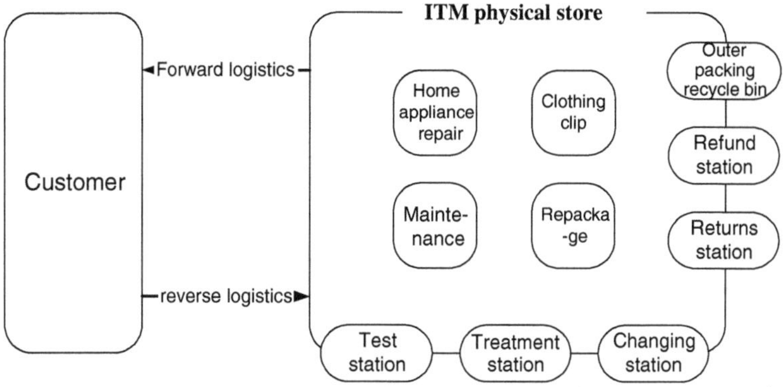

Fig. 6 Reverse logistics model based on ITM

after-sales unified. It has responsibility to undertake the report for consumers and sellers according to facts.

(3) The reverse logistics service optimization. Returned goods problem restricts the development of electronic commerce. B2C electronic commerce reverse logistics system of ITM mode can solve the problem of reverse logistics services. Model is shown in Fig. 6.

The model shows that the ITM shop can act as a third party centralized return centers of B2C e-commerce enterprises. It shorts the commodity cycle of reverse logistics significantly, guarantee reverse logistics in B2C e-commerce system to response rapidly and improve customer satisfaction.

4 The Cost and Benefit Analysis of the Value Chain

4.1 The Cost Analysis of Cloud Computing Data Center and Traditional Data Center

4.1.1 Data Center Cost Factors

The comparative analysis of cloud data center and traditional data center online model shows that they both have the same cost, including hardware, software, maintenance cost, space, power and bandwidth management.

4.1.2 Cost Accounting of Data Center

The various cost factors are showed through the abstract formula expression, it is concluded that the enterprise data center overall cost calculation formula:

$$CTCO = CHW + CPOWER + CMAN \quad (1)$$

CTCO represents the total cost; CHW represents the hardware cost; CPOWER representative electricity costs; CMAN on behalf of the management maintenance costs.

4.1.3 The Cost Analysis of Large B2C e-Commerce Self-built Cloud Computing Center

According to a 2013 IDC a survey analysis is shown in Fig. 7.

According to the data of James Hamilton, the cost comparison table between cloud computing data center and the traditional data center is shown in Table 1.

The cost comparison table of power and cooling is shown in Table 2.

Cloud computing platform can provide flexible services compared with traditional IDC. The cost performance advantage of Cloud computing data center and the traditional data center is shown in Fig. 8:

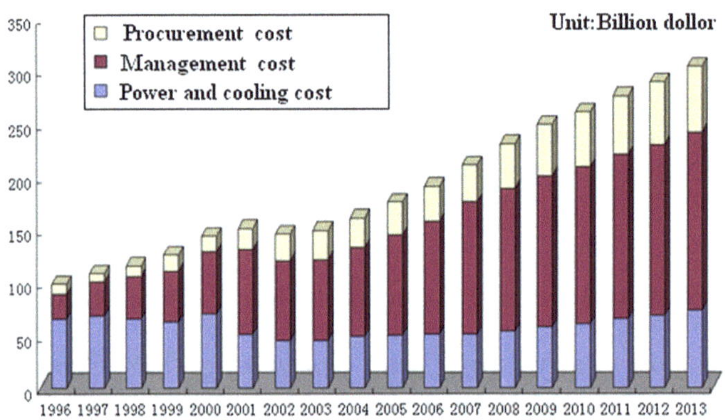

Fig. 7 IT cost development tendency of global enterprise

Table 1 Cost contrast of network, store and management

Technology	Traditional data center	Cloud computing data center	Rate
Internet	$95 per MB/s/month	$13 per MB/s/month	7.1
Storage	$2.2 per GB/month	$0.4 per GB/month	5.7
Management	Each administrator manages 140 servers	Each administrator manages 140 servers	7.1

Table 2 Cost contrast of electric and refrigeration power

	Traditional data center	Cloud computing data center	Rate
Electricity	70 %	38 %	1.9
Power	4 KVA	6 KVA	0.67

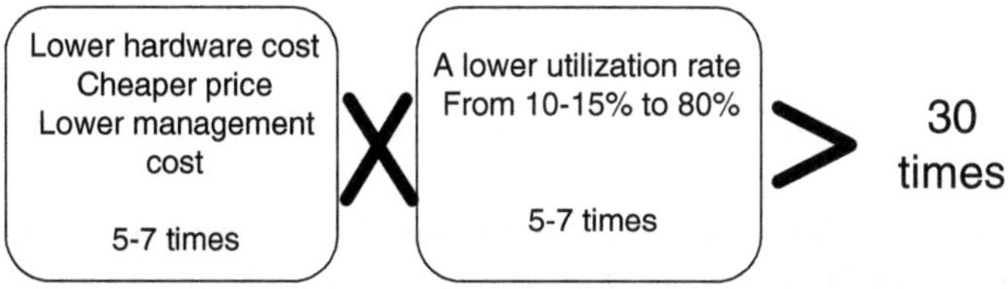

Fig. 8 Contrast performance between cloud computing and tradition pattern

4.2 The Value Chain Benefits Analysis Under the Mode of ITM

This paper make the ITM store project economic efficiency evaluation validate ITM store project with economic rationality from the aspects of investment profit ability.

Value index is used to evaluate the investment project profitability; this paper selects the total net present value of project life period. Calculation formula is shown:

$$NPV = \sum_{t=0}^{n} (CI - CO)_t (1 + i_c)^{-t} = \sum_{t=0}^{n} (CI - CO)_t (P/F, i_c, t) \tag{2}$$

NPV the net present value, reflecting the project life period net income;
IC benchmark yield;
N project life period

Criterion: if NPV $\geqq 0$, the scheme is feasible; If NPV < 0, the plan is not workable.

According to the enterprise income tax law of the People's Republic of China, the electronic equipment depreciation life is minimum for 3 years. Garment industry benchmark yield is 10 %. The sales volume of the second and third year increased to 15 pieces per day. Different market ITM entity shop project net present value calculation is as follows:

The First-line market net present value:

F1 = 60000 − 39000 = 21000; A1 = 90000 – 45000 – 8300 − 700 = 36000; P = 157000

NPV1 = 21000 × 6.814 + 36000 × 8.985 × 0.3186 − 157000 = 8.9148 (Ten thousand yuan) ≧ 0

The second-line market net present value:

F2 = 45000 − 28200 = 16800; A2 = 67500 − 33750 − 5000 − 700 = 28050; P = 107000

NPV1 = 16800 × 6.814 + 28050 × 8.985 × 0.3186 − 107000 = 8.7771 (Ten thousand yuan) ≧ 0

The third-line market net present value:

F3 = 27000 − 16450 = 10550; A3 = 40500 − 20250 − 2500 − 450 = 17300; P = 66300

NPV1 = 10550 × 6.814 + 17300 × 8.985 × 0.3186 − 66300 = 5.5111 (Ten thousand yuan) ≧ 0

Analysis results show that net present value of different market is greater than zero in three years life period, ITM store project is feasible. It has certain ability of investment gains.

5 Conclusion

The B2C e-commerce model based on cloud computing is constructed. Bottleneck problems are solved, such as infrastructure construction and operation, storage, online trading and online, logistics distribution. The enterprise online value chain is optimized.

The ITM entity shop is built. The electronic commerce enterprise offline value chain is optimized from the service of distribution, products and reverse logistics.

The cost of building cloud computing data center is less than traditional e-commerce enterprise data center; ITM entity shop investment project has economic feasibility.

References

1. Yang YH, Yang JQ (2012) B2C e-commerce enterprise value chain optimization based on cloud computing. Internet Econ 5(4):117–120
2. Pan XB (2011) The research of B2C e-commerce business model based on value chain. Modern Econ Inf 12(8):13–14
3. He J (2012) The business model research of mobile e-commerce based on the value chain. Chongqing University, Chongqing
4. Slywotzky A (2000) Find profit area. CITIC press, China, pp 13–16
5. Xi W, Rong F (2000) The analysis of B2B and B2C e-commerce mode from the perspective of value chain. J Tsinghua Univ 4(15):72–76
6. Shank J, Govindarajan V (1992) Strategic cost management and the value chain. J Cost Manage 12(9):5–21
7. Rayport JF, Sviokla JJ (1995) Exploiting the virtual value chain. Harv Bus Rev 13(73):75–99
8. Rayport J, Sviokla J (1996) Exploiting the virtual value chain. McKinney Q 8(1):21–22

9. Hayes B (2009) Cloud computing. Commun ACM 6(7):9–11
10. Edwards J (2008) Cloud computing steals the show at software. Yankee Group 11(5):26–29
11. Hewitt C (2008) ORG for scalable, robust, privacy-friendly client cloud computing. IEEE Internet Comput 2(10):96–99
12. Perkins Earl (2008) Identity services in the cloud. Gartner Res 6(14):17–25
13. Zhang YP (2012) Network operation management system design and implementation based on cloud computing platform. Jilin University, Changchun
14. Bennett K, Layzell P, Budgen D (2000) Service-based software. Future Flexible Softw 21 (6):17–25
15. Zhang Y (2010) Analyze the problems and countermeasures existed in logistics service quality of network shopping. J Modern Bus 6(2):95–99

The Influence of TMT Characteristics on Technological Innovation: Evidence from IT Public Listed Companies in China

Wei-yi Zhu and Qun Yin

Abstract Based on the integration of Upper Echelons Theory and Social Capital Theory, this paper explores the relationship between characteristics of TMT and technological innovation. It collects the data from 117 listed Chinese IT corporations in 2013. The results show that TMT's average age, percentage of female executives have significant negative impact on technological innovation. Technological innovation correlates positively with TMT's education level, output functional experience and social capital.

Keywords Social capital · Top management team · Technological innovation · Upper echelons theory

In the current background environment of vigorously developing "strategic emerging industries", as the main body of innovation enterprises should improve the level of independent innovation. In this case, how to promote enterprises innovation has become the common focus of academic circles and practical circles.

Since Hambrick and Mason [1] have put forward the upper echelons theory to study their effects on the choice of enterprise strategy, the theory of explanatory variables is further extended to the field of technology innovation, many scholars try to reveal the characteristics of TMT background on the impact of technological innovation.

W. Zhu (✉) · Q. Yin
College of Management, Nanjing University of Posts and Telecommunications, Nanjing 210023, China
e-mail: 15905149354@163.com

E. Qi (ed.), *Proceedings of the 6th International Asia Conference on Industrial Engineering and Management Innovation*,
DOI 10.2991/978-94-6239-145-1_94

However, the existing upper echelons theory is based on the demographic characteristics (such as average age, education level, etc.), ignoring the social attribute of TMT behavior [2]. In the view of this, this article considering the social capital of TMT into the range of TMT background characteristics, in order to develop upper echelons theory.[1]

1 Conceptual Background and Hypothesis

1.1 Age

The majority of domestic and international research shows that the average age of TMT is significant with innovation. When they make decisions, older executives are more inclined to the occupation stability and short-term reward. The enterprise R&D activities have longer investment return period and greater uncertainty than other investment activities, and it may probably affect the short-term income, therefore, senior executives may tend to adopt a more conservative strategy when they make decisions, and reduce the innovation [3].

H1: The average age of TMT is negatively correlated with corporate innovation.

1.2 Sex Ratio

On the one hand, female executive may seize the market opportunities, and promote the formation of good team work environment; on the other hand, they also have the tendency to seek stability, risk avoidance [4]. It is worth to study which aspects is the larger influence of innovation.

H2: The percentage of female executives have significant effects on corporate innovation.

1.3 Educational Background

Executives who have higher level of education usually also have stronger absorption ability, cognitive ability, more long-term vision and make long-term

[1]Project Fund: Project supported by the National Natural Science Foundation of China (Grant No.71373133); 2014 Jiangsu province ordinary university graduate students scientific research innovation projects (Grant No. KYZZ_0235).

decision for enterprise to promote a strategic decision for the enterprise innovation [5, 6]. Reviewing previous studies, the level of education is positively related to innovation.

H3: Education level of TMT is positively related to the corporate innovation.

1.4 Functional Experience

Functional experience refers to the different work experience in different industries and companies or in the same companies but in different department. Buyl et al. found that CEO who has the output functional experience such as technology, market etc. is more willing to take risks and has a high level of innovation [7]. However, CEO who has the non-output functional experience such as legal and financial functional experience may turn to focus on process control and choose to be risk averse, and has more conservative attitude towards innovation. With the output functional experience, executives could have a more in-depth understanding on product design and market development, would pay more attention to the enterprise product and market innovation, and also would be more inclined to increase innovation [8].

H4: Members with the output functional experience of TMT is positively related to corporate innovation.

1.5 Social Capital

Hagedoom consider that the social relation network contributed by the enterprise social capital can strengthen the communication between the enterprise and external organizations, reduce transaction costs, increase the resources, knowledge sharing and promote the knowledge transfer and product innovation [9]. Considering Chinese people often put a high value on relationship, Bian and Qiu first put forward the dimensions of social capital, namely, hierarchical, horizontal, and social connections which is suitable for Chinese condition. Later, many scholars make further exploration on the dimensions of social capital [10]. They divide social capital into the competitors, customers, suppliers and other business contacts; scientific research institutes, universities and technology intermediary organization contact, government departments, financial institutions, industry association contact.

According to the classification of the division standard of social capital by Chen et al., this paper divide social capital into market social capital and nonmarket social capital [11].

Through the market social capital, enterprise can gain more market resources, reduce the transaction cost, obtain the valuable market information and related knowledge, and recognize the innovation opportunity early.

H5: Market social capital of TMT is positively related to corporate innovation.

Through nonmarket social capital, enterprises can get more preferential policies, know industry trends in time and grasp more market opportunities and financing opportunities.

H6: nonmarket social capital of TMT is positively related to corporate innovation.

2 Data Collection and Variables Selecting

2.1 Source of Data

IT Public Listed Companies in China form the empirical setting for the study. First, eliminating ST, PT companies. Second, eliminating the company with incomplete innovation data. Finally, firms were eliminated from the sample if their top management team data are incomplete or the TMT existed the appointment and removal of members during 2012–2013. After screening, 117 samples were in accordance with the research request.

2.2 Variable Definition

Dependent variable. Commitment to innovation, the dependent variable in the subsequent regression analysis, is measured by R&D intensity (R&D expenditures/sales). R&D intensity is a common proxy for innovativeness.

2.3 Independent Variables

Based on the current literatures about TMT characteristics, this paper adopt the average age of TMT members to measure TMT age. Sex ratio is measured by the proportion of female executives in TMT. Educational background is typically measured by creating an interval scale covering the possible education levels, expressed by the average value. We coded educational background into one of five levels: (1) senior high school, technical secondary school and below, (2) college degree (3) bachelor degree, (4) master degree, and (5) doctoral degree and above. According to the division standard by Hambrick and Mason [1], TMT functional experience is divided into output functional experience and non-output functional experience. In this paper, functional experience is measured by the proportion of TMT members who have output functional experience in TMT total members. Market social capital of TMT refers to the contacts which TMT members maintain with the competitors, suppliers, customers, which is measured by the proportion of executives who have market social capital in TMT. Nonmarket social capital refers to the enterprise executives who served as the national and local levels CPPCC member or the democratic-parties member; government officials; the person in charge of national and local federation of industry and commerce, industry associations or other social organizations, which is measured by the proportion of executives who have non-market social capital in TMT.

2.4 *Control Variables*

The control variables included in the analysis are the enterprise scale and capital structure function. Enterprise scale is represented by the natural logarithm of total enterprise assets and the capital structure of enterprise is measured by the assets and liabilities rate.

3 Analysis and Results

Table 1 presents the means, standard deviations, and correlations for all variables. As expected, the independent variables and control variables are significantly related with corporate R&D intensity. Table 1 shows that the correlation coefficient between the control variables and the independent variables is relatively low, and by further calculation of VIF values, it was found that VIF value is less than 2, which shows that the model does not exist multi-collinearity problem.

Table 2 reports the regression results of TMT background characteristics and enterprise innovation. All the independent variable's P values are less than 0.05, showed that all the variables in the model have a significant effect on R&D intensity. Among them, the average age of the TMT and the percentage of female executives in TMT both has significant negative correlation with R&D intensity. The standardized regression coefficients of them are −0.202 ($P < 0.01$); −0.169 ($P < 0.05$) respectively. TMT's average education level, output functional experience, market social capital and non-market social capital are positively related to R&D intensity. The standardized regression coefficients of them are namely 0.24 ($P < 0.01$); 0.186 ($P < 0.05$); 0.226 ($P < 0.01$); 0.207 ($P < 0.01$). Hence, hypothesis 1–6 are stated.

Table 1 Correlations and descriptive statics for the sample

	MEAN	SD	1	2	3	4	5	6	7	8	9
R&D	0.11	5.81	1								
Age	44.89	2.83	−0.333**	1							
Female	0.25	0.15	−0.217*	−0.105	1						
Edu	3.54	0.34	0.353**	−0.109	0.066	1					
Back	0.39	0.15	0.495**	−0.62	−0.384**	0.049	1				
MSC	0.41	0.18	0.445**	−0.246**	0.010	0.225*	0.365**	1			
NSC	0.43	0.17	0.349**	0.089	0.019	0.005	0.337**	0.221*	1		
Size	20.81	0.70	−0.403**	0.151	0.112	−0.014	−0.141	−0.192*	−0.185*	1	
CaptStru	0.24	0.14	−0.434**	0.325**	−0.044	−0.091	−0.160	−0.236*	−0.044	0.482**	1

Note *represents a significant level of 0.05. **represents a significant level of 0.01

Table 2 Results of regression analysis. dependent variable: R&D intensity

Regression results	β	t	Sig.
Capital structure	−0.170	−2.571	0.012
Size	−0.176	−2.837	0.005
Age	−0.202	−3.226	0.002
Female	−0.169	−2.597	0.011
Edu	0.240	4.008	0.000
Back	0.186	2.575	0.011
Market SC	0.226	3.141	0.002
Nonmarket SC	0.207	3.212	0.002
R-square	0.646		
Adjusted R-square	0.619		
F-ratio	24.592		
Sig.F-ratio	0.000		

4 Discussion and Conclusion

On the perspective of TMT background characteristics, this paper study the influence factor of enterprise innovation, and expand the upper echelons theory through the integration of social capital theory. This paper draws the following conclusions: (1) The average age of the TMT and the percentage of female executives in TMT both has significant negative correlation with R&D intensity. Older executives may have more conservative ideas and more easily to take action according to their own experience, thus the motivation for them to increase R&D input is small. Female executives in TMT may consider the stable expectations, so they also have little interest in enterprise innovation. (2) TMT's average education level, output functional experience, market social capital and non-market social capital are positively related to R&D intensity. This reflects that TMT who have learning capability, information processing ability, with large knowledge background and broad social contacts would be willing to increase the input of innovation.

This paper enriches the theory of enterprise technology innovation, and has a certain significance of guiding for practice. Enlightenment to promote technological innovation of enterprises is obtained through empirical research: enterprise need to adjust the structure of TMT and optimize the TMT ratio. The specific practices are as follows: (1) Focus on the cultivation of young executives, rationalize the TMT's age structure. Young executives often have a strong spirit to exploration, have high tolerance of risk and also have strong learning ability, therefore they have more sense of innovation. (2) Allocate the ratio of female TMT members properly. Although, the finding shows that female executives may hinder the innovation of enterprise, the participation of more women into TMT is still necessary, as they have different thinking mode and perspective to make decision compared with men, which is also the precondition of enterprise innovation. (3) Improve the education level of TMT members. TMT's education level may have a certain effect on the cognition, judgement and decision making of top managers. It is necessary for enterprise to mining and cultivate knowledge talented person. (4) Increase the proportion of output functional experience of TMT members. With the output functional experience, top managers would have more professional background, know more about the implementation environment for enterprise to innovate than top managers who have other functional experience. (5) Strengthen the social capital of TMT members, encourage TMT members to cultivate rich market and nonmarket social capital. Through the expansion of TMT members' social capital, enterprise may construct the information network, grasp the market information, take the advantage of structure hole and promote the enterprise innovation.

References

1. Hambrick DC, Mason PA (1984) Upper echelons: the organization as a reflection of its top managers. Acad Manag Rev 9(2):193–206
2. Cheng S, Jing R, Liao K (2012) How do social capital, corporate governance effect on private enterprise's R&D investment intensity. Stud Sci Sci 30(6):916–922
3. Li S, Gao L. (2014) Social capital, R&D investment and enterprise performance based on the perspective of instrumental utility heterogeneity. Sci Sci Manage S&T 35(10):106–155
4. Carter DA, D'Souza F, Simkins B et al (2011) The gender and ethnic diversity of US boards and board committees and firm financial performance. Corp Gov 18(5): 396–414
5. Ren T, Wang Z (2010) Female participation in TMT and firm performance: evidence from Chinese private enterprises. Nankai Bus Rev 13(5):81–91
6. Shipilov A, Danis W (2006) TMG social capital strategic choice and firm performance. Eur Manag J 24(1):16–27
7. Buyl T, Boone C, Hendriks W (2011) Top management team functional diversity and firm performance: the moderating role of CEO characteristics. J Manage Stud 48(1):151–177
8. Daellenbach US, McCarthy AM, Schoenecker TS (1999) Commitment to innovation: the impact of top management team characteristics. R&D Manage 29:199–209
9. Hagedoom J (2002) Inter-firm R&D partnerships: an overview of major trends and patterns since 1960. Res Policy 31(4):477–492
10. Bian Y, Qiu H (2000) Social capital of enterprise and its effect. Chinese soc Sci 2:89–99
11. Chen Y, He Y, Chen X (2011) The mediating role of TMT characteristics on the relationship between diversification and firm performance. Forecasting, 30(1):10–17

Personality Trait and Context: An Investigation of Expatriate Adjustment

Ai-jing Ran, Xiao-bing Liu and Xiao-yun Wang

Abstract Expatriates, who are sent by international corporations to work overseas, usually take weeks or months to adjust to an unfamiliar working and living environment. This study aims to explore the relationship between expatriate adjustment and its antecedents (personality, general self-efficacy, and culture distance). In this study, structural equation modeling is used to analyze the data of 196 western expatriates working in mainland China and Hong Kong. The results suggest that expatriate adjustment is positively related to extraversion, but negatively related to general self-efficacy. Culture distance shows no influence on expatriate adjustment. The results provide insights into the research of expatriate adjustment and help international corporations with expatriates selection and training process.

Keywords Culture distance · Expatriate adjustment · Extraversion · Self-efficacy

1 Introduction

Nowadays, the workforce is becoming increasingly diverse with the expansion of globalization. Expatriates, who link domestic and foreign operations, constitute an important part of the workforce, and facilitate the success of corporations in global markets. Expatriates who failed to adjust to the local environment or returned prematurely result in high personal and organizational costs [1, 2]. Normally, the sooner the expatriates adjust to local environment and overseas work, the better they contribute to personal and firm performance. However, the process of

A. Ran (✉) · X. Liu
Faculty of Management and Economics, Dalian University of Technology, Dalian, China
e-mail: ranaijing@sina.com

X. Liu
e-mail: xbliu@dlut.edu.cn

X. Wang
I. H. Asper School of Business, University of Manitoba, Winnipeg, Canada

E. Qi (ed.), *Proceedings of the 6th International Asia Conference on Industrial Engineering and Management Innovation*,
DOI 10.2991/978-94-6239-145-1_95

expatriate adjustment usually goes with problems such as cultural clashes, destructive conflicts, depress, stress, psychological discomfort, low satisfaction and high return rate [3–5]. Therefore, the research on expatriate adjustment has irrevocably implications in both theoretical and practical aspects.

Most research looked into the issue of expatriate adjustment from the expatriate's point of view. Expatriate's personality trait has been studied to explore the way expatriates interact with peers and local people, and further influence their adjustment to local contexts [3, 4]. Tung found the reasons that most expatriates socialize with peer expatriates more than local nationals could be language barrier, culture crash, no occasion to know local nationals, etc. [6–8]. However, there are also some expatriates who actively search for local interactions, which facilitate their adjustment in local environment. A possible explanation accounting for this condition is their personality trait. Moreover, a series of factors from various aspects have been raised to explain expatriate adjustment, including culture factors (e.g., culture novelty), job factors (e.g., role clarity), organizational factors (e.g., selection and training), and personal factors (e.g., personality) [9]. Thus, great attention is needed to better understand the role of these factors in expatriate adjustment.

The aim of the study is to explore the relationships between expatriate adjustment and its important influencing factors. In this study, we choose extraversion (personality trait), general self-efficacy and culture distance as the predictable factors. Theoretically, this study will contribute to expatriate literature by clarifying the relationships among the expatriate adjustment and its antecedents. In practice, the results of study will help multinational corporations select the proper employees to work abroad, design training programs, and better support expatriates' work overseas. Moreover, with the internationalizing process of Chinese corporations, the result is especially helpful for them to choose the committed and competent person for international assignment. This paper begins with literature review and hypotheses. Then, the description of the method and the results are presented. The paper ends with the discussion of results and limitations of the study.

2 Theory and Hypotheses

2.1 *Expatriate Adjustment*

The research on expatriate adjustment has been robust in the last 25 years [1]. How to choose the proper person for international assignment is challenging for both corporations and researchers. The failure of adjustment results in a high cost of corporations on expatriate's training, relocating, and compensation.

Expatriates are defined as "employees of business organizations, who are sent overseas on a temporary basis to complete a time-based task or accomplish an organizational goal [1]." Expatriate adjustment refers to the process through which an expatriate attains psychological comfort in both working and living aspects of oversea life [9–11].

Black and his colleagues suggested that expatriate adjustment involved at least three dimensions: work adjustment, interaction adjustment, and general adjustment [11]. Work adjustment includes the specific work-related responsibilities and performance. Interactional adjustment refers to the extent to which expatriates socialize and interact with local people in a non-work context. And general adjustment includes the surface issues such as living conditions, climate, food, and housing. The multifaceted characteristic of expatriate adjustment indicates that there may be different predictors for different type of adjustment [10]. Thus, in this study, we use three types of adjustment as separate dependent variables to see whether there is a difference among the predictors.

2.2 *Personality and Expatriate Adjustment*

Shaffer and his colleagues examined individual differences as possible predictors of expatriate adjustment [3]. They found that stable personality traits (extraversion, conscientiousness, agreeableness, openness to experience, and neuroticism) and some cross-culture competencies (such as cultural flexibility, task orientation, people orientation, and ethnocentrism) were important predictors for international assignment effectiveness (adjustment process, intentions to quit the assignment, and job performance). Caligiuri examined personality (openness and sociability) as the moderator of the relationship between the interaction with local people and cross-cultural adjustment [4]. She found that when expatriates had a high score on the personality trait of openness, more interactions with local people were associated positively with cross-cultural adjustment. Huang studied three personality traits in a sample of eighty-three US expatriates in Taiwan, and found that the US expatriates' general living adjustment was positively related to their degree of extraversion and openness to experiences, and their interaction adjustment was positively related to extraversion and agreeableness [12].

Given above, this study focuses on the extraversion aspect of personality traits. Merriam-Webster Dictionary defines extraversion as "the act, state, or habit of being predominantly concerned with and obtaining gratification from what is outside the self." A person high on extraversion is assumed to be active and sociable with others, which facilitate the process of expatriate adjustment. Therefore, we predict that,

Hypothesis 1: Extraversion positively relates to expatriates' work adjustment, general adjustment, and interaction adjustment.

2.3 General Self-efficacy and Expatriate Adjustment

Self-efficacy is a key construct in Bandura's social-cognitive theory, which is defined as a belief in one's capabilities to organize and execute courses of action [13, 14]. In her theory, the self-efficacy construct consists of domain-specific self-efficacy and general self-efficacy. This study focuses on general self-efficacy (GSE), which refers to an individual's belief in his or her capacity to use the physical, intellectual and emotional resources needed to succeed [13]. General Self-efficacy is assumed as a stable trait of a person [14]. "If people believe they can deal effectively with potential stressors, they are not perturbed by them [15]."

We assume that expatriates with higher GSE are easier to organize their abroad work and life, and adjust faster to a new environment, since they are more confident and determined in their ability. Specifically, expatriates with higher GSE will be more likely to utilize their communication abilities and the physical, intellectual and emotional resources needed to overcome the difficulties in cross-cultural communications than those with lower GSE. Therefore, we predict that,

Hypothesis 2: General self-efficacy positively relates to expatriates' work adjustment, general adjustment, and interaction adjustment.

2.4 Culture Distance and Expatriate Adjustment

Generally speaking, culture distance refers to the extent to which one country's shared norms and values differ from another country [16]. Ward and Kennedy measured culture distance with a five-dimension scale which includes pace of life, clothing, food, management system, and education. [17]. Culture distance is regarded as a context variable that influences expatriate adjustment [9]. It is argued that dissimilar cultures are more difficult to adjust to than familiar ones [18]. Culture novelty has been found negatively influence interaction and general adjustment [19]. Since cultural barriers may increase the difficulty of expatriate cross-cultural adjustment, we predict that,

Hypothesis 3: cultural distance negatively relates to expatriates' work adjustment, general adjustment, and interaction adjustment. The higher the cultural distance, the harder expatriates adjust themselves to local environment.

3 Methodology

3.1 Samples

The dataset comes from Dr. Xiaoyun Wang's Expatriate Adjustment Project in Canada. The survey was designed to investigate the expatriates' attitudes, work and relationships with other individuals in mainland China and Hong Kong. The 251 expatriates interviewed in the survey come from North America, Europe and Asia. After discarding the missing data, 196 observations were kept. The factors which may influence the expatriate adjustment overseas were measured, such as expatriate adjustment, personality, self-efficacy, cultural distance, psychology well-being and informal social network pattern.

3.2 Measures

Extraversion is measured using the *NEO-Five-Factor Inventory* [20]. The extraversion facet of this scale was 6-point Likert with 1 being "strongly disagree" to 6 being "strongly agree." For example, items are "I really enjoy talking to people," "I like to be where the action is." A higher value represents higher extraversion.

Self-efficacy was measured using the adaptation of *Self-efficacy Scale* [21], which was a 6-point Likert scale with 1 being "strongly disagree" to 6 being "strongly agree." For example, items are "When I make plans, I am certain I can make them work," "When I have something unpleasant to do, I stick to it until I finish it." A higher value represents a higher GSE.

Culture Distance was measured by *Ward and Kennedy's Questionnaire* [17], which was a 5-point Likert scale with 1 being "no different" to 5 being "extremely different." The aspects include pace of life, clothing, food, management system, and education. A higher value represents higher culture distance.

Expatriate adjustment was measured by *Black's Scale* [10], which was a 6-point Likert scale with 1 being "not adjust at all" to 6 being "very well adjusted." For example, the items are "How adjusted are you to your job and responsibilities? (work aspect)," "How adjusted are you to shopping in the host country? (general aspect)," and "How adjusted are you to interacting with locals in general? (Interaction aspect)." A higher value represents a better adjustment.

3.3 Procedures

We use a two-phase process of structural equation modeling to test the causal model. First, exploratory factor analysis is conducted to develop an acceptable measurement model without specifying causal relationships between latent variables. Then the reliability and validity of the measurements are testified. Second, a path analysis with latent variables is performed to test the hypotheses, the significance of predictors, and the model-data goodness of fit assessment. The theoretical model is tested and revised until a theoretically meaningful and statistically acceptable model is found [22, 23]. We use Statistics Analysis System (SAS) to analyze the survey data.

4 Results

4.1 The First Phase: Measurement Model Testing

A measurement model describes the nature of the relationships between the latent variables and the manifest indicator variables [22]. After the exploratory factor analysis, 21 manifest variables were retained to measure 6 latent variables. All of the manifest variables and latent variables as well as their descriptions are listed in Table 1.

The ratio of chi-square and degrees of freedom is 1.23 (<2), providing an acceptable fit of the measurement model. Comparative Fit Index is 0.97. Standardized factor loadings for the indicator variables range from 0.43 to 0.84. The t values obtained from the coefficients are all significant ($p < 0.01$), which indicates the model has a good factor loading.

The result also provides evidence supporting the convergent validity of the indicators. The coefficients of variables were in excess of 0.70, indicating a good level of reliability.

Table 1 Manifest variables and latent variables

Latent variable	Manifest variable
Work adjustment	How adjusted are you to your job and responsibilities?
	How adjusted are you to working with local co-workers?
	How adjusted are you supervising local subordinates?
General adjustment	How adjusted are you to the transportation system in the host country?
	How adjusted are you to the food in the host country?
	How adjusted are you to shopping in the host country?
	How adjusted are you to generally living in the host country?
	How adjusted are you to the entertainment available in China?
Interaction adjustment	How adjusted are you to working with locals outside your company?
	How adjusted are you to interacting with locals in general?
Extraversion	I like to have a lot of people around me
	I really enjoy talking to people
	I like to be where the action is
	I often feel as if I'm bursting with energy
General self-efficacy	When I make plans, I am certain I can make them work
	If I can't do a job the first time, I keep trying until I can
	When I have something unpleasant to do, I stick to it until I finish it
	Failure just makes me try harder
Culture distance	Educational system
	Management system
	Food

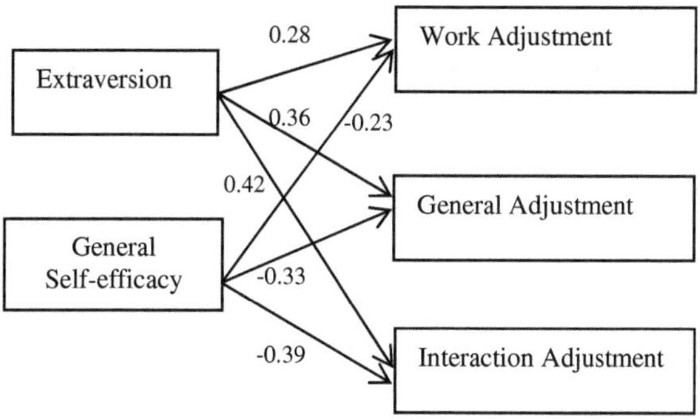

Fig. 1 Causal model with standardized path coefficients

4.2 *The Second Phase: Causal Model Testing*

First of all, a chi-square difference test was conducted by comparing the theoretical model to the measurement model. The chi-square difference is non-significant, indicating that the causal model is successful in accounting for the relationships between the latent variables.

Figure 1 shows the causal model with standardized path coefficients ($p < 0.01$). Chi-square statistic divided by its degrees of freedom is 1.30. CFI is 0.96, and NNFI is 0.95, indicating an acceptable fit of the model to the dataset.

Extraversion is found to have significant path coefficients on all three types of adjustment, which supports *H1*. General self-efficacy is found to significantly relate to the three types of adjustment, but in an opposite direction, which means that general self-efficacy is negatively related to three types of adjustment according to our dataset. Thus, *H2* is not supported.

Culture distance is found no significant influence on either type of adjustment. H3 is not supported. R^2 values show that extraversion and general self-efficacy account for 59.58 % of the variance in work adjustment, 79.26 % of the variance in interaction adjustment. Extraversion, general self-efficacy and culture distance account for 82.46 % of the variance in general adjustment.

5 Discussion

Among the antecedents we studied, extraversion was found to be the most important predicting factor, followed by general self-efficacy whose path coefficients were significant but in an opposite direction. Culture distance was found to have little influence on expatriate adjustment.

Two issues need to be pointed out. The first one is the opposite direction of general self-efficacy. We predict that, expatriates with higher GSE will be more likely to utilize their communication abilities and to use the physical, intellectual and emotional resources needed to overcome difficulties in cross-cultural communications than those with lower GSE. However, the results of this study showed a negative influence of GSE on expatriate adjustment. There is a possible reason that expatriates with higher GSE are lack of flexibility in local environment. When facing difficulties out of their capability (they cannot solve at the moment), they would still stick to it, which bring them more psychological discomfort and depress than expatriates with lower GSE. However, more studies are expected in this area to provide a comprehensive explanation.

Secondly, although we separated three types of adjustment as three dependent variables to see whether there is difference between them and the predictors, similar patterns were found regarding the three predictors examined in this study. That is, for this dataset, we could combine three types of adjustment together, but it does not mean the pattern is always the same. There may be other predictors work differently on each type of expatriate adjustment. Overall, extraversion, GSE, and culture distance were found to have similar causal relationships with work adjustment, general adjustment, and interaction adjustment of expatriates in this study.

6 Limitation and Future Research

First, our dataset only has a moderate sample size (196 observations). Also, we only collect data from the expatriates but no other stakeholders. For example, having their spouse data involved may lead to a better understanding of the variance of expatriate adjustment, because spouses usually have strong influences on the decision-making process of expatriates [11, 24]. Further research should be conducted with a larger sample size and include spouses data to test the causal relationship among those factors.

Second, the residuals of the theoretical model are still large. Most of the large residuals happened between the variables measuring the same constructs. The misspecification is neither in a condition that a variable is assigned to the wrong factor, nor in a condition that the variable is multidimensional. Future research is expected on the residual reduction.

Finally, demographic variables should be considered in the future. Previous research on expatriate adjustment shows that some demographic variables, such as age and gender, also have influences on the way expatriates adjust to local environment [9]. In future research, it is necessary to incorporate demographic data to reveal a more comprehensive understanding of expatriate adjustment overseas.

Acknowledgment This paper is funded under Science and Technology Program of Liaoning, China (Project ID: 2014106007). We also thank to Dr. Xiaoyun Wang's Expatriate Adjustment Project for data source.

References

1. Harrison DA, Shaffer MA, Bhaskar-Shrinivas P (2004) Going places: Roads more and less traveled in research on expatriate experiences. Res Pers Hum Res Manage 23:199–247
2. Harzing AW (2001) Who's in charge? An empirical study of executive staffing practices in foreign subsidiaries. Hum Res Manage 40:139–158
3. Shaffer MA, Gregersen H, Harrison DA, Black JS, Ferzandi LA (2006) You can take it with you: Individual differences and expatriate effectiveness. J Appl Psychol 91:109–125
4. Caligiuri PM (2000) Selecting expatriate for personality characteristics: a moderating effect of personality on the relationship between host national contact and cross-cultural adjustment. Manage Int Rev 40:61–80
5. Wang X (2002) Expatriate adjustment from a social network perspective: theoretical examination and a conceptual model. Int J Cross Cult Manage 2:321–337
6. Tung RL (1981) Selection and training of personnel for overseas assignment. Columbia J World Bus 16:68–78
7. Tung RL (1982) Selecting and training procedures of US European, and Japanese multinational corporations. Calif Manag Rev 25:57–71
8. Tung RL (1998) American expatriates abroad: From neophytes to cosmopolitans. Journal of World Business 33:125–144
9. Black JS, Mendenhall M, Oddou G (1991) Toward a comprehensive mode of international adjustment: an integration of multiple theoretical perspectives. Acad Manag Rev 16:291–317
10. Black JS (1988) Work role transitions: a study of American expatriate managers in Japan. J Int Bus Stud 19:277–294
11. Black JS, Stephens GK (1989) The influence of the spouse on American expatriate adjustment and intent to stay in Pacific Rim overseas assignments. J Manag 15:529–544
12. Huang T, Chi S, Lawler JJ (2005) The relationship between expatriates' personality traits and their adjustment to international assignments. Int J Hum Res Manage 16:1656–1670
13. Bandura A (1986) Social foundations of thought and action: a social cognitive theory. Prentice-Hall, Englewood Cliffs
14. Bandura A (1997) Self-efficacy: the exercise of control. W. H. Freeman and Company, New York
15. Bandura A, Cioffi D, Taylor CB, Brouillard ME (1988) Perceived self-efficacy in coping with cognitive stressors and opioid activation. J Pers Soc Psychol 55:479–488
16. Drogendijk R, Slangen A (2006) Hofstede, Schwartz, or managerial perceptions? The effects of different cultural distance measures on establishment mode choices by multinational enterprises. Int Bus Rev 15:361–380
17. Ward C, Kennedy A (1993) Where's the "culture" in cross-cultural transition?: comparative studies of sojourner adjustment. J Cross Cult Psychol 24:221–249
18. Aycan Z, Kanungo R, Mendonca M, Kaicheng Y, Deller J, Stahl G, Kurshid A (2000) Impact of culture on human resources management practices: a 10-country comparison. Appl Psychol Int Rev 59:192–221
19. Parker B, McEvoy GM (1993) Initial examination of a model of intercultural adjustment. Int J Intercult Relat 17:355–379
20. Costa PT, McCrae RR (1992) NEO-PIR professional manual. Psychological Assessment Resources, Inc., Odessa
21. Sherer M, Maddux JE, Mercandante B, Prentice-Dunn S, Jacobs B, Rogers RW (1982) The self-efficacy scale: construction and validation. Psychol Rep 51:663–671
22. Hatcher L (1994) A step-by-step approach to using SAS for factor analysis and structural equation modeling. SAS Institute Inc., Cary

23. Anderson JC, Gerbing DW (1988) Structural equation modeling in practice: A reviewer and recommended two-step approach. Psychol Bull 103:411–423
24. Gupta R, Banerjee P, Gaur J (2012) Exploring the role of the spouse in expatriate failure: A grounded theory-based investigation of expatriate' spouse adjustment issues from India. Int J Hum Res Manage 23:3559–3577

The Optimal Antidumping Tariff Under Lobbying Competition and Information Searching Activity

Jin-fang Liu

Abstract Based on the model established by Grossman and Helpman, we study how the enterprise lobbying competition and the information searching activity affect the antidumping decision. Through analyzing the participator's utility maximization behavior, we prove the price of imported products and the importance the government attaches to the national welfare is inversely proportional with tariff, while the marginal revenue of domestic enterprise from protection is in proportion with tariff. The final tariff will go down with the number of foreign enterprise lobbying for preferred trade policy, and will go up with the volume of domestic market. After analyzing the affection of the information searching and free riding behavior, we find domestic enterprise would rather directly contribute government officer than collect information, while free riding damage to the enterprise cooperative enthusiasm.

Keywords Lobbying competition · Antidumping tariff · Welfare maximization · Information searching · Free riding

1 Introduction

When asked why free trade is so often preached and so rarely practiced, most international economists blame "politics." Governments will shape the trade policy for their own benefits in response not only to the concerns of the general electorate, but also to the pressures applied by special interests. The decision maker participate in the incentives they face, trading off the financial and other support that comes from heeding the interest group's demands against the alienation of voters that may result from the implementation of socially costly policies.

J. Liu (✉)
Finance and Economics School of Hunan University of Technology, Zhuzhou, China
e-mail: 625526443@qq.com

E. Qi (ed.), *Proceedings of the 6th International Asia Conference on Industrial Engineering and Management Innovation*,
DOI 10.2991/978-94-6239-145-1_96

The outstanding research about the lobbying behavior in trade policy decision is made by Grossman and Helpman. They provide an explanation why antidumping is widely taken instead of free trade and also deduct the equilibrium policy choices taking the governments as an agent and the interest group as a principal. They derivate the equilibrium tariff in a small economy populated by individuals with identical preferences but different factor endowments. Every individual pursue the maximum utility [1]. To successfully simplify the analysis and learn the balancing acts trading contribution against national welfare in policy decision, Grossman and Helpman suppose the national welfare is consists of individuals' utility. Governments' aim is to maximize the sum of political contribution and national welfare with a certain weight. But in actual economic mechanism, it is the enterprise instead of individuals who play a major role in the political contribution, paying tax and providing employment, and then it is the enterprises that get more attention of the government. So I take the sum of enterprise income, consumer surplus and tax income to indicate the national welfare. Follow the model provided by Grossman and Helpman, this paper get what are the major factors influencing the finally antidumping tariff and deduct the equilibrium policy under the lobbying competition between domestic enterprise and foreign enterprise.

The model of protect for sale provide an excellent research framework for explaining the lobbying behavior of interests and the policy choice of governments. Many scholars take their study based on the PFS model.[1] Research on the political and economical elements of trade policy decision based on the PFS can be divided into several approaches. One approach is to test the PFS model with case-level evidence or the empirical investigation. Evans and Sherlund [2] compiled a new, unique database containing information on political donations by the specific firms and labor organizations that have petitioned for antidumping protection from imports, as well as data on the outcomes of their requests. They find that politically active petitioners are more likely to receive protection and that antidumping duty rates tend to be higher for that group. In addition, the relationship between the import penetration ratio and duties imposed depends on whether or not petitioners are politically active—antidumping duties are positively correlated with the import penetration ratio for politically inactive petitioners but negatively correlated for politically active petitioners, consistent with the Grossman-Helpman model's predictions [2]. Eicher and Osang test two models provided by Grossman, Helpman and Gawande, Bandyopadhyay. They find the models performs well against the tariff formation function model where contributions directly influence tariffs, and where non-cooperative solutions do not imply zero net benefits to lobbying as in PFS model [3].

The second approach is to introduce information providing or free riding behavior into the PFS model. Bennedsen and Feldmann [4] identify an information externality that raises the cost of offering contributions and show that this indirect search cost reduces the group's incentive to gather information when contributions

[1]For simplification, here I use PFS model indicate the model of protect for sale.

are allowed. Furthermore they analyze how competition among lobby groups as providers of information and contributions affect the choice and effectiveness of the instruments. They show that the information externality rewards the group that can abstain from information search and focus its influence on contributions [4]. While Gawanda and Magee [5] examines the extent of free riding in lobbying over tariffs in the context of the PFS model in which industry lobbies seek to bend government policy in their favor. In contrast to numerous examples of welfare-reducing policies that have in fact been bought cheaply by special interests, they think the governments are largely welfare-maximizing and care little about campaign contribution because the model assumes away free riding by firms which hinders industry's ability to organize politically [5].

How the lobbying activity influence the trade policy? What will be if there exists lobbying competition? What kind of affection that information providing put on the equilibrium policy formation. Do free riding really reduce the lobbying motive. This paper tries to explain these questions. In the second part I analyze the domestic enterprises' lobbying behavior in shaping the governments' policy choice in a menu auction. It is the basic equilibrium policy model for the coming research. In the third part the lobbying competition between domestic enterprises and foreign enterprises is taken into consideration. I try to figure out the exact influence that lobbying competition carry on the final antidumping tariff. In the fourth part, the information lobbying and free riding affection is taken into consideration and the equilibrium choice is given. In the end the conclusion is given and the future research scheme is drawn out.

2 The Equilibrium Trade Policy Game with Organized Domestic Enterprise

2.1 Formal Framework

Suppose domestic enterprises and foreign enterprises all sell the same good 0 at a domestic price p in the home country. I denote by p^* the exogenous world price of good 0. The whole sales volume is $Q = a - bp$ with quasi-line demand hypothesis in the domestic market. There are i domestic enterprise, while q_i is the output of every domestic enterprise i. So the domestic market's demand is satisfied by

$$Q = \sum_{i \in L} q_i + m \tag{1}$$

Here m is the imports provided by foreign enterprises, L is the subset of domestic enterprises that are willing to lobby for expected trade policy. In the domestic market, the competition pressure forces the domestic enterprise to seek for the protection from the governments through antidumping application. To gain the preferred antidumping tariff, the domestic enterprise will make a contribution. Then the utility of the domestic enterprise is

$$V_i = q_i p - C_i \tag{2}$$

where C_i is the contribution of i. So the whole utility for domestic industry can be summed as $V = \sum_{i \in L} q_i p - \sum_{i \in L} C_i$. Following the common knowledge, the national welfare is defined as the sum of consumer surplus, tariff revenue and domestic manufacturer income [6]. So we can get the welfare

$$w(p) = \sum_{i \in L} q_i p + (p - p^*)m + \frac{Q^2}{2} \tag{3}$$

I restrict the set of policy instruments available to politicians. For now, I allow the government to implement only trade taxes and subsidies. These policies drive a wedge between domestic and world prices. A domestic price in excess of the world price implies an import tariff for a good that is imported and an export subsidy for one that is exported. Domestic prices below world prices correspond to import subsidies and export taxes. The net revenue from all taxes and subsidies is expressed by $(p - p^*)m$.

The incumbent government cares about the total level of political contributions and about the national welfare with a weight estimated by α. The government values contributions, because they can be used to finance campaign spending and improve the officer's living standards. Social welfare will be of concern to the incumbent government if voters are more likely to reelect a government that has delivered a high standard of living. So the government's objective function is

$$G = \sum_{i \in L} C_i + \alpha w(p) \tag{4}$$

Here I suppose the domestic enterprise belong to subset L can be organized to lobby for preferred tariff, we will discuss the free riding problem in the fourth part.

2.2 *The Structure of Protection*

To maximize the G, governments can choose p^0 in domestic price vectors through deciding the antidumping tariff on good 0.

Let ρ denote the set of domestic price vectors from which the government may choose. We bound ρ so that each domestic price p^0 must lie between some minimum $\underline{p}$ and some maximum $\overline{p}$. For the most part, we restrict attention to the equilibrium that lie in the interior of ρ. So the equilibrium to trade policy game can be characterized as follows:

Proposition 1 The sub game perfect Nash equilibrium of the trade policy game if and only if:

(a) C_i^0 is feasible for all $i \in L$;
(b) p^0 maximizes $q_i p - C_i$ on ρ;
(c) p^0 maximizes $\sum_{i \in L} C_i + \alpha w(p)$
(d) p^0 maximize $\alpha w(p) + \sum_{i \in L} q_i p$

Condition (a) restricts each lobby's contribution schedule to be among those that are feasible. Contribution must be nonnegative and no greater than the aggregate income available to the lobby's members. Condition (b) maximizes the benefit of the lobbyist. Condition (c) is the aim of governments to make the best use of policy decision. While condition (d) give an equilibrium solution which can maximize both benefit of the lobbyist and governments.

The fact that p^0 maximize V implies that a first order condition is satisfied at p^0, namely,

$$\nabla V = \sum_{i \in L} q_i - \nabla C = 0 \quad \Rightarrow \quad \sum_{i \in L} q_i = \nabla C \tag{5}$$

The same is for $G + V$, the first order condition satisfied by p^0 can be get from

$$\nabla V + \nabla G = \sum_{i \in L} q_i + \alpha \nabla w(p) = 0 \tag{6}$$

Then substitute (3) and (5) into (6), we get

$$\nabla V + \nabla G = \sum_{i \in L} q_i + \alpha(a - bp)(1 - b) - \alpha b(p^0 - p^*) = 0 \tag{7}$$

Let $\frac{p^0 - p^*}{p^*} = t$, we can transform (7) into

$$t = \frac{1}{p^*}\left[\frac{\nabla C}{\alpha b} - \frac{p^0}{E}(1 - b)\right] \tag{8}$$

Here we lead the price elasticity of demand $E = \frac{\nabla Q}{Q} \div \frac{\nabla p}{p}$ into the Eq. (8). From the equation we can learn the optimal tariff for governments is negative correlation with the world price p^*, or in another words, the foreign enterprise sales price is inverse proportion with the ruling antidumping tariff. To get the intuitive results, I suppose the domestic market demand is in complete linear correlation with the price, that means $b = 1$, then (8) can be rewrite as

$$t = \frac{\nabla C}{\alpha p^*} \tag{9}$$

It is obviously the optimal tariff increase with the marginal contribution ∇C. As Grossman and Helpman have proved that optimal marginal contribution for lobbyist will be equal with his marginal revenue, so the tariff will depend on the lobbyist's consideration how much he can gain from the trade policy contribution. While the tariff also depend on the importance that the government attaches to the national welfare level, When governments care about the reelection and want to win the elector's support from improving the welfare, then α increase and the protective tariff will decrease.

3 Competition Between Domestic and Foreign Enterprises

Now we consider the situation that foreign enterprise also contribute for preferred trade policy which will make the antidumping tariff decrease. I let C_j denote the foreign enterprise's contribution, $j \in H$. So the utility of domestic enterprise is the same as (2), while the government's revenue is

$$G = \sum_{i \in L} C_i + \sum_{j \in H} C_j + \alpha w(p) \tag{10}$$

What will domestic and foreign enterprises contribute under competition? Suppose the enterprise only have a choice of the scalars $\{B_i\}$ in policy bids. They all wish to make B_i as large as possible (and the contribution as small as possible). In Fig. 1, the shaded area represents the set of contribution pairs (C_1, C_2) that the government might collect for all of the various policy choices open to it. Given this shaded opportunity set, a government that cares only about maximizing total

Fig. 1 Excessive contributions

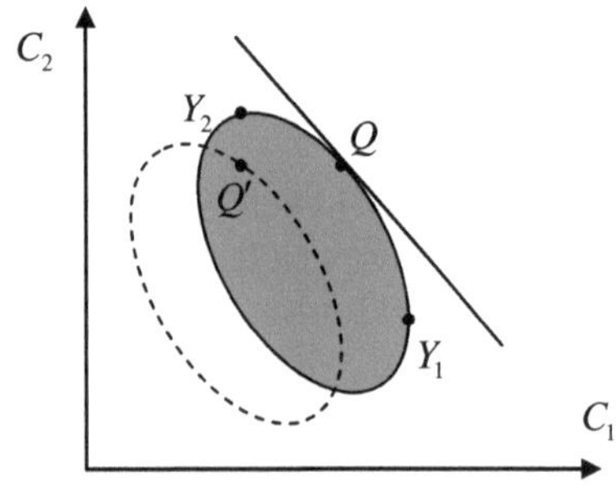

contributions will opt for point Q, where the outer frontier is tangent to a line with a slope of -1.

Now we examine whether lobby 1 might wish to raise B_1 slightly above $\hat{B}_1$. By doing so, it would reduce all of its contributions by the same amount. The shaded area would shift uniformly to the left, to the location indicated by the dotted lines. The government would then be faced with a new set of possibilities and would choose the point Q', a leftward displacement of point Q. But the policy underlying point Q' must be the same as that for point Q, since the relative desirability of different policies has not changed from the government's (political) perspective. Evidently, lobby 1 must benefit from this increase in B_1. Of course, the situation illustrated in the figure affords lobby 2 the same opportunity to improve its net welfare; so Fig. 1 cannot represent an equilibrium situation.

Figure 2 depicts an equilibrium configuration. The government collects the same total donations for choosing any one of Q, Z_1, and Z_2. No other point offers contributions as great as these, so the policies that underlie these three points comprise the set of welfare maximizing choices for the government. The government willingly chooses point Q; and neither lobby wishes to raise its B_i any further, for fear that the government then would select the policy most preferred by its rival.

For the most part, we restrict the attention to equilibrium that lie in the interior of ρ'. Then equilibrium to trade policy game can be characterized as follows:

Proposition 2 The sub game perfect Nash equilibrium of the trade policy game if and only if:

(a) C_i^0 is feasible for all $i \in L$, C_j^0 is feasible for all $j \in H$;

(b) p^0 maximizes both $q_i p - C_i$ and $m_j p - C_j - m_j pt$ on ρ';

(c) p^0 maximizes $\sum_{i \in L} C_i + \sum_{j \in H} C_j + \alpha w(p)$;

(d) p^0 maximizes $\alpha w(p) + \sum_{i \in L} q_i p + \sum_{j \in H} m_j p. - \sum_{j \in H} m_j pt$

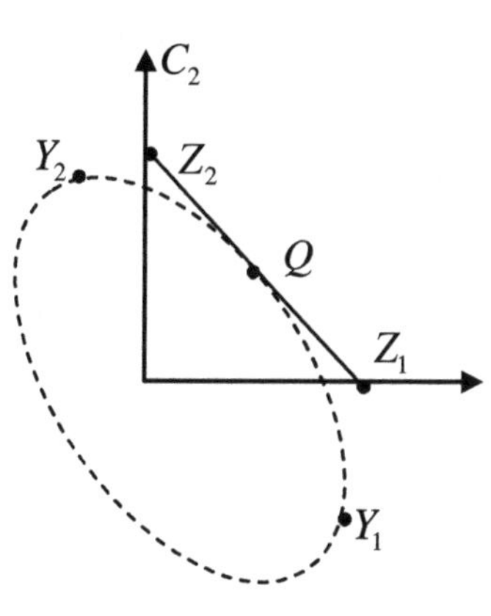

Fig. 2 Equilibrium contributions

Condition (a) is prerequisite for i and j to participate the trade policy lobby game. While condition (b) maximizes the benefits of the domestic and foreign enterprises. Governments' maximization of utility is described in condition (c). Condition (d) give a equilibrium solution with lobbying competition, that is the maximum of $G + \sum_{i \in L} V_i + \sum_{j \in H} V_j$.

Then derivate $\arg\max(G + \sum_{i \in L} V_i + \sum_{j \in H} V_j)$ with p^0, we get

$$\nabla G + \nabla \sum_{i \in L} V_i + \nabla \sum_{j \in H} V_j = 0 \tag{11}$$

From (11), we can derive

$$t = Q \frac{1 + \alpha(1 - b)}{\sum_{j \in H} m_j + \alpha b p^*} \tag{12}$$

To simplify the analysis and get the intuitive results, here I assume the $b = 1$, which means the demand of domestic market is completely liner correlate with price. Then we get

$$t = \frac{Q}{\sum_{j \in H} m_j + \alpha p^*} \tag{13}$$

Comparing with the optimal tariff discussed before, under the lobbying competition between domestic and foreign enterprises, the big difference is domestic market capacity denoted by Q is proportional to the antidumping tariff. In another words, if there is a great domestic market, the governments is more likely to levy high antidumping tariff to protect the home country's industry, which can be proven by the empirical case that the strategic trade protection is widely used today. It is especially frequently carried by great power with huge domestic market. While if the foreign enterprise's sales volume denoted by $\sum_{j \in H} m_j$ grows, in the same time, these companies are well organized, they also can shape the trade policy efficiently by lobbying. Then the tariff will decrease with the foreign lobbying enterprise's sales volume increasing. The degree of attention that government pays to the national welfare is inversely proportion with optimal tariff, while the world price of the products is also inversely proposition with the final tariff. When the products' world price are high or the government may consider the national welfare is important, the final tariff will be low.

4 Consideration of Information Lobbying and Free Riding

In the antidumping decision process, the enterprise whether applying antidumping or appearing all need to provide convincing information. The domestic enterprise need to prove the foreign enterprise's dumping behavior is the major reason causing the damage of domestic industry. To achieve the aim, the enterprise applying for antidumping should collect the sales price, product cost, comparing them to prove there exist dumping. They also need search the profit rate, sales volume, the employment rate etc. to illustrate industry damage and prove it is caused by dumping. While the foreign enterprise need to defend their legal status in export price setting with the information about production cost.[2] The information lobbying play the same function as the contribution discussed before, we can take the information lobbying as a part of contribution.

Johan [7] point out strategically revealing policy-relevant information will make the sender worse off, while the other who do not send information can benefit from the information externality. Gaining preferred information will cost time and money and the more valued information, the more expense it will cost. Will the information investment help the enterprise gain their preferred judgments? We will analyze the problem then.

Taking the information searching cost as S, similar to the previous, the equilibrium of trade policy game can be characterized as follows:

(a) $S_i^0 + C_i^0$ is feasible for all $i \in L$, $S_j^0 + C_j^0$ is feasible for all $j \in H$;

(b) p^0 maximize both $q_i p - C_i - S_i$ and $m_j p - C_j - S_j - m_j pt$ on ρ';

(c) p^0 maximize $\sum\limits_{i \in L} C_i + \sum\limits_{j \in H} C_j + \alpha w(p)$;

(d) p^0 maximize $\alpha w(p) + \sum\limits_{i \in L} (q_i p - S_i) + \sum\limits_{j \in H} (m_j p - S_j - m_j pt)$

The same as discussed before, condition (a) is prerequisite for i and j to participate the trade policy lobby game. Condition (b) is the optimal choice of i and j when they invest in information search. Condition (c) is the Government optimal choice maximizing the sum of interests' contribution and national welfare with information provided by enterprise. The last condition (d) describe the situation the total utility of government and enterprise reach the highest with a given p^0.

[2]Usually, foreign enterprise being accused of dumping will use product differentiation to defend them. They explain as it is not the same products exported by them as Chinese products, it is unreasonable to launch antidumping investigation on them. But this excuse is not agreed by Chinese antidumping ruling authorities. The officers think the export product play the same affections for the consumer although there exists some litter degree difference. And the dumping is the major reason give rise to domestic industry sales volume and profit decline.

To get the optimal choice of the participating enterprise, we can derivate condition (b) with p^0, then get

$$q_i = \nabla C_i + \nabla S_i; \quad m_j(1-t) = \nabla C_j + \nabla S_j \tag{14}$$

Derivate condition (d) with p^0, we can get

$$\sum_{i \in L} (q_i - \nabla S_i) + \sum_{j \in H} (m_j - m_j t - \nabla S_j) + \alpha \nabla w(p) = 0 \tag{15}$$

With (14), we can derive from (15)

$$t = \frac{Q - (\sum_{i \in L} \nabla S_i + \sum_{j \in H} \nabla S_j)}{\sum_{j \in H} m_j + \alpha p^*} \tag{16}$$

The Eq. (16) is the same as (13) except it subtract $\sum_{i \in L} \nabla S_i + \sum_{j \in H} \nabla S_j$ in the numerator. As $\sum_{i \in L} \nabla S_i + \sum_{j \in H} \nabla S_j$ represent the efforts the domestic enterprise and foreigner make for searching information. We can learn from it that the harder the efforts for searching information are, the lower the antidumping duty will be. So the foreign enterprise will try hard to collect benefit information, while this activity is disadvantageous for domestic enterprise as perfect information will more likely lead to the welfare maximum decision. For the aim to max national welfare, the antidumping duty will not reach the expected level of domestic enterprise. So the domestic enterprise will be more crave in contribute directly to judger.

With (14), we can alternate (16) to

$$t = \frac{\sum_{i \in L} \nabla C_i + \sum_{j \in H} \nabla C_j}{\sum_{i \in L} m_i + \alpha p^*} \tag{17}$$

As $\nabla C_i > 0$; $\nabla C_j < 0$, we can learn form (17) that free riding behavior is unfavorable to both domestic and foreign enterprises because it will shorten their desired value. As a consequence they all will lower their contribution when they realize there exists of free riding behavior.

5 Conclusion

The lobbying behavior change the governments optimal antidumping tariff decision, comparing with the former research, we learn the optimal tariff is in proportion with the lobbyists expected revenue. If the revenue is not as much as the lobbyist expectation, interest group usually shorten the contribution or give up, then the final tariff will be lower under this condition. As the low price reveals the strong competitive power and force the domestic enterprise to seek for the protection from trade policy, the foreign enterprise's price is inversely proportional to the tariff. While the weight the governments carry on the national welfare can decrease the protective strength.

Considering the lobbying competition between domestic enterprise and foreigners, we can see the sales volume of foreign enterprise organized form strong influence on government's trade policy decision through lobbying and reduce the final tariff. While for a huge domestic market, the local industry's strength grow strong so the protective tariff will grows with the domestic market. The world price and the weight for national welfare will be inverse proposition with the final tariff.

Information searching motive is different for domestic enterprise and foreign enterprise. The search investment is in proportion to the preferred antidumping decision for foreign enterprise, while it is not the same as domestic enterprise. So domestic enterprise will be more preferred to directly contribution. The free riding problem will do harm to enthusiasm of the enterprise participated in antidumping activity. Especially for domestic enterprise as this behavior will lower the level of effort.

In the future, we will try to figure out whether the equilibrium trade policy be changed by the interests' willingness reports? Is there any effective means to induce the real report? Will the equilibrium trade policy benefit the cultivating of the domestic industry's core competitiveness? All these interesting problems will be studied later.

Acknowledgments The research is supported by Hunan Provincial Natural Science Foundation of China (2015JJ3061), Hunan Provincial social science fund of China (14WTC46), Educational Commission of Hunan Province of China (14c0330), Financial department of Hunan Province of China (Xiangcaijiaozhi [2012]238) and social science program of Hunan University of Technology (2014HSX30).

References

1. Grossman GM, Helpman E (1994) Protect for Sale. Am Econ Rev 84(4):833–850
2. Evans GL, Sherlund SM (2011) Are antidumping duties for sale? Case-level evidence on the grossman-helpman protection for sale model. South Econ J 78(2):330–357
3. Eicher T, Osang T (2012) Protection for sale: an empirical Investigation: comment. Am Econ Rev 92(5):1702–1710
4. Bennedsen M, Feldmann SE (2006) Information lobbying and political contributions J Public Econ 90:631–656
5. Gawanda K, Magee C (2012) Free riding and protection for sale. Int Stud Q 56:735–747
6. Kolev DR, Prusa TJ (1999) Tariff policy for a monopolist in a signaling game. J Int Econ 49:51–76
7. Johan L (1997) Lobbying, information, and private and social welfare. Eur J Polit Econ 13:615–637

The Empirical Research of a Single Commercial Bank Uses Loss Distribution Method to Measure Operational Risk

Zhang-fa Liu

Abstract We use loss distribution method to measure the single commercial bank's (Taking the Industrial and Commercial Bank of China as example) operational risk capital. ICBC's operational risk events loss frequency obey Weibull distribution and loss amount obey lognormal distribution. With the help of Monte Carlo simulate, the analysis of Financial statement data, using loss distribution method to measure operational risk capital, commercial bank's capital adequacy ratio and core capital adequacy ratio are far lower than the level of supervision, so our banks are not suitable for using loss distribution method to measure operational risk at the present stage.

Keywords Loss distribution method · Monte Carlo simulate · Risk measure · The single commercial bank

Fund Project: Yunnan University Shichuan Outstanding Youth Education Fund Key Project "commercial bank operating risk measurement method comparative study" (09KT001); Hunan Province Education Office subject "The research of centralized contiguous destitute areas differentiated sustainable finance poverty alleviation model—taking the Luo Xiao Mountain Area as an example" (15C1298); Hunan province education science plan subject "the Research of centralized contiguous destitute areas differentiated education development poverty alleviation mechanism under a multiple perspective—taking the Luo Xiao Mountain Area as an example" (XJK013CFZ003).

Z. Liu (✉)
Department of Economics, University of Jinan, Guangzhou, Guangdong, China
e-mail: 2001liuzhangfa@163.com

Z. Liu
Department of Economics and Management, University of Xiangnan, Chenzhou, Hunan, China

E. Qi (ed.), *Proceedings of the 6th International Asia Conference on Industrial Engineering and Management Innovation*,
DOI 10.2991/978-94-6239-145-1_97

1 Introduction

December 2010, the Basel Committee released the third edition of the Basel Accord ("Basel Accord III") to further enhance the risk tolerance of the banking system in unfavorable financial and economic environment . In November 2011 the Group of Twenty summit in Cannes, leaders pledged to implement the new regulatory capital standards before January 1, 2013, and full compliance in 2019. Basel III requires that the core capital adequacy ratio① increase from 4 to 6 %, adding 2.5 % of the protective cushion of capital, core capital adequacy ratio should reach 8.5 %, the total capital adequacy ratio② of 8 % maintain unchanged. This agreement gives the long transition period from 2013 to 2019, the core capital adequacy ratio should be 4.5 % in 2013, 5.5 % in 2014, up to 6 % in 2015 during the transition period. Accordingly, the China Banking Regulatory Commission with the domestic banking practice developed a "commercial bank's capital management method (Trial)", which came into effect on January 1, 2013. The "Commercial bank's capital management method (Trial)" with the new international regulatory standards, expanding the scope of risk capital coverage, in addition to credit risk and market risk, but will also incorporate operational risk in capital regulatory framework. Measures apply to commercial banks, rural cooperative banks and rural banks established in the territory of People's Republic of China. Therefore, the measuring of single commercial bank's operational risk capital requirement is very important. There are a lot of operational risk measurement methods in which the loss distribution method by means of historical loss data banks, compared with other methods to better reflect the characteristics of the bank's own operational risk, was the concern of many scholars. At this stage whether the loss distribution method is suitable for our individual banks to measure the bank's operational risk is a question required to answer in a timely manner.

2 Literature Review

Fan and Yang [1], based on 71 loss events (1990–2003) of seven banks in china, with four distributions (normal, lognormal, exponential, Weibull) fitting loss frequency and the loss amount natural logarithm's distribution. Worked out the loss frequency and the loss amount natural logarithm are respectively subject to the Weibull distribution and normal distribution. Zhou et al. [2], based on 96 loss events (2001–2004) from China's banking industry, then assumed directly that the loss frequency and the loss amount respectively obey Poisson distribution and Generalized Pareto distribution, and derived the operational risk capital of the measure object through Monte Carlo simulations. Zhang [3] on the basis of 161 loss events (1990–2005) from China's banking industry, directly assumed that loss frequency obey the Poisson distribution, with three distributions (Pareto, Weibull, lognormal) fit loss amount distribution, draw the loss amount obey the lognormal distribution. Yao [4] according to the bank loss data (2005–2007) of a province in eastern China, directly assumed that loss frequency obey the Poisson distribution, with four distributions (exponential, lognormal, normality, Weibull) fitted loss amount distribution, concluded that the loss amount obey the lognormal distribution. Fei [5] according to 578 loss events (2000–2006)of China's banking industry, with five distributions (normal, Poisson, binomial, lognormal, exponential) fitting loss frequency and the loss amount natural logarithm's distribution, concluded that loss frequency and loss amount natural logarithm respectively obey exponential distribution and the normal distribution. Zheng [6] according to 464 loss events in China's banking industry (1985–2007) with three distributions (normal, Poisson, negative binomial) fitted distribution loss frequency, with two kinds of distribution (positive state, Weibull) fitted the amount of the loss distribution showed by natural logarithm, derived that loss frequency and the natural logarithm of the amount of loss respectively obey negative binomial and Weibull distribution. Xing [7] on the basis of 325 loss datas of Agricultural Bank of China (1994–2009), assuming loss frequency obey Poisson distribution, with four distributions (gamma, lognormal, Pareto, Weibull) fitting the loss amount distribution, draw that the loss amount obey lognormal. Wang [8] using Matlab7.0 fit the 369 operational risk events of China's banking industry from 1994 to 2012 loss frequency and loss amount's distribution, obtained that loss frequency and the loss amount respectively follows negative binomial distribution, lognormal distribution.

Literature review shows, firstly, in most of past domestic literatures scholars used loss distribution method measure some domestic banks' operational risk capital, in Xing Zhiguo's text [7], the measure object is a single commercial banks (ABC), but didn't verify the empirical results; secondly, in previous domestic papers, scholars' fitting results of operational risk loss events' frequency distribution are inconsistent, two-thirds of scholars concluded that the loss amount obey the lognormal distribution (someone directly assumed that loss frequency and the loss amount subject to certain distribution). I tend to think, the possible reason causing scholars hold different analyze results is that the measure objects all are the whole China's bank industry, but the scope of each article's specific banks involved in the bank industries are not the same; thirdly, scholars used a few distribution functions fitting operational risk events loss frequency and loss amount, which will lead to inaccurate fitting results. In order to circumvent problems in the previous literature such as inconsistent sample banks, small distribution methods, and so on, this study only use one single sample (ICBC) as a measure objects, use 16 distributions to fit the operational risk events' loss frequency and the loss amount, use the loss distribution method to measure the operational risk capital of the single sample (ICBC) during a certain period, according to the theory, empirical results and combined with the bank financial statements data to determine the suitability of using the loss distribution method to measure the individual bank's operational risk capital at this stage.

3 Sample Selection and Data Description

3.1 Description of Selecting the Measure Objects

Li [9] counted 171 loss data of China's bank industry from 2000 to 2006, resulted that the four state-owned banks' operational risk events loss frequency is accounting for 59.06 %, the loss amount is accounting for 76.99 % (but he did not make an analysis of the proportion of individual banks' loss frequency and loss amount). This shows that China's large commercial bank is representative, and ICBC's asset is the largest of China commercial bank, it has the largest customer base, its assets is accounting for about one fifth of the total assets of the domestic banking institutions. In addition, in 2004 this bank treated "Operational Risk Management Framework" as part of the "Commercial Bank of China comprehensive risk management framework", marking the ICBC is the first company within our financial system, it formally incorporated the operational risk management into agenda of financial institutions, therefore ICBC's loss event data of operational risk will be more accurate.

To sum up, the ICBC's operations risk profile is the most representative one in China's financial system, so this study choose ICBC as the measure object.

3.2 Data Description

In this study, there are 101 loss events, including 77 loss events data of ICBC from 1995 to 2007 collected by Li [10] and 24 loss events data of ICBC from 2008 to 2011 collected by me from ICBC official website, Baidu and Wind Information. Among them, the maximum loss amount is 7.4 billion and the smallest is 105 yuan. The first 77 loss events data and its specific instructions can be seen in "An Empirical Analysis of China Industrial and Commercial Bank Operational Risk" [10]. The statistics of ICBC loss frequency are illustrated in Table 1 as below.

Table 1 ICBC operational risk event loss frequency Table (1995–2011) (unit: times)

Year	Times	Year	Times	Year	Times
1995	1	2001	6	2007	3
1996	3	2002	7	2008	5
1997	2	2003	10	2009	7
1998	8	2004	13	2010	4
1999	3	2005	6	2011	8
2000	9	2006	6		

Sources Li Xue. An Empirical Analysis of China Industrial and Commercial Bank Operational Risk[D]. Shaxi Province: Northwest Agriculture and Forestry University, 2008. ICBC official website, Baidu, wind information. These data were obtained by statistical processing

4 Empirical Analysis of Commercial Banks Operational Risk Measurement Based on the Loss Distribution Method

In this paper, 16 kinds of distributions (binomial, negative binomial, Poisson distribution, the beta distribution, chi-square distribution, exponential distribution, gamma distribution, half-normal distribution, the Laplace distribution, logistic distribution, lognormal distribution, normal distribution, Pareto distribution, student distribution, Weibull distribution, uniform distribution) were applied to fit loss frequency and loss amount's distribution, test the goodness of fit for each distribution (implement distribution test by SPSS14.0 Eviews5.0) by using J-B test, K-S test and P-P test (Q-Q test).

In the binomial distribution, the events are required to be independent. If the bank just happened operational risk events recently, the management will certainly enhance the management of operational risk, the bank staff will surely be vigilant during their work, the possibility of occurring operational risk events will reduce recently. It can clearly be seen that operational risk events loss frequency are not independent, so the frequency of operational risk loss events do not obey the binomial distribution. Similarly we can see, nor does it obey the negative binomial distribution. So binomial and negative binomial to fit operational risk events' loss frequency and loss amount are not used in the following papers.

By comparing with the results of J-B test, K-S test and P-P test (Q-Q test) under other 14 different distributions, we can draw that: the Weibull distribution is the best to fit loss frequency; lognormal distribution is the best to fit loss amount. Test results of P-P test (Q-Q test) including the loss frequency fitted by Weibull distribution and the loss amount fitted by lognormal are shown in Table 2, Figs. 1, 2 and Table 3 the other test results are omitted.

The probability density function with shape parameter α and scale parameter β of the Weibull distribution is as follows:

$$f(x, \alpha, \beta) = \frac{\alpha}{\beta}\left(\frac{x}{\beta}\right)^{\alpha-1} e^{-\left(\frac{x}{\beta}\right)^{\alpha}}$$

$x > 0$, $\alpha > 0$, $\beta > 0$. From the test results we can know, $\alpha = 1.577$, $\beta = 6.765$. Lognormal distribution density function is:

$$f(x) = \frac{1}{x\sigma\sqrt{2\pi}} e^{-\frac{(\ln x-\mu)^2}{2\sigma^2}}$$

Table 2 Estimated weibull distribution parameters

		VAR00001
Weibull distribution	Scale	6.765
	Shape	1.577

The cases are unweighted

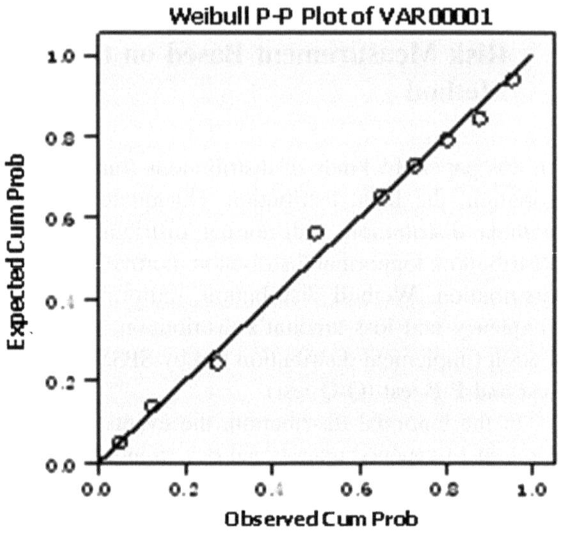

Fig. 1 P-P test results Figure of Weibull distribution fitting Loss frequency (SPSS14.0)

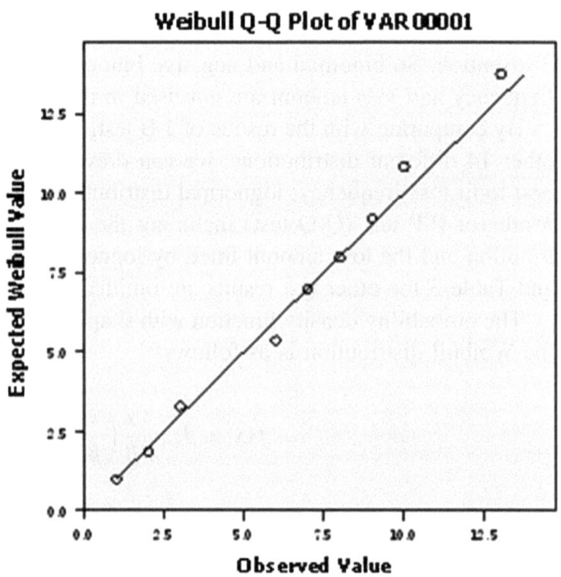

Fig. 2 Q-Q test results Figure of Weibull distribution fitting Loss frequency (SPSS14.0)

Table 3 Estimated lognormal distribution parameters

		VAR00001
Lognormal distribution	Scale	333.193
	Shape	3.720

The cases are unweighted

We can obtain by testing that: $\mu = 5.8087$, $\sigma = 3.71952$.

Monte Carlo simulation principles:

Firstly, from the foregoing, ICBC operational risk events' loss frequency obey Weibull distribution, the loss amount obey lognormal. Do a simulation that n = 10,0000 times (the selection of n is depending on the need for precision), it can generate 100,000 random numbers obeying Weibull distribution through the R software. They are m_1, m_2, …, m_{100000}, and do the next step simulation to every m.

Secondly, let's assume that m taking m_1, the bank loss events may occur m_1 times within a period (one year). Therefore, we do m_1 times simulation for the loss amount. Then we generate m_1 random numbers obeying lognormal, $S_1, S_2, \ldots, S_{m1}$. Those random numbers respectively represent the loss amount of every time of the cycle (a total of m_1)

Thirdly, the possible values of S of operational risk capital can be obtained by adding m_1 loss amount within this period.

Fourthly, repeat the above three steps n =100,000 times, will achieve 100,000 possible values S of operational risk capital.

Fifthly, from the above 100,000 S results based on different confidence level requirements of VaR model (according to "commercial banks operational risk regulatory capital measurement guidelines," released by the CBRC on September 18, 2008, the provisions of Article 22 says using the loss distribution model confidence level should be set to 99.9 %), we can obtain the size of the operational risk capita (That is to range above 100,000 S in ascending order, and take the 99900 number).

With the R software to complete Monte Carlo simulation based on loss distribution method. We can obtain from the output that 99.9 % quantile is 1,972,050,234,000 yuan, get the mean of 100,000 simulations S (ie expected loss) is 18,192,510,000 yuan and the standard deviation is 599,711,930,000 yuan. If the bank has been carried out prevention on the expected loss in their daily operations, the operational risk capital was that 99.9 % quantile minus the expected loss .In 2012, ICBC operational risk capital is 1,953,857,724,000 yuan. According to the meaning of the VaR model, in the provision of 1,953,857,724,000 yuan operational risk capital, ICBC will be able to resist a thousand years' huge operational risk events loss (a thousand years corresponds the confidence level of 99.9 %).

5 Conclusions and Suggestions

5.1 Feasibility Analysis and Conclusions

It is Available from 2012 ICBC's (as of 2012, ICBC has not officially brought operational risk capital in the calculation of capital adequacy ratio) Financial Statements (from ICBC official website) that in 2012, Net capital and core net capital respectively are 1.299014 trillion yuan, 1.010463 trillion yuan of ICBC and the total adjustment capital including credit risk-weighted capital and market risk capital is 9.511205 trillion yuan. According to earlier empirical results, under 99.9 % confidence level, in 2012 ICBC' s operational risk capital is 1.9538577240 trillion yuan, including operational risk capital, the ICBC's capital adequacy ratio and core capital adequacy ratio respectively were 3.83 and 2.98 % [Those are respectively obtained from: 12990.14/(95112.05 + 19538.577240 * 12.5) = 0.0383, 10104.63/(95112.05 + 19538.577240 * 12.5) = 0.0298]. This is much lower than the regulatory standards of 8 % capital adequacy ratio and 4 % core capital adequacy ratio, it is less than the future requirements of 8.5 % core capital adequacy ratio. So the historical loss data collected by individual banks is far less at this stage, it is inappropriate that use the loss distribution method to measure operational risk capital.

5.2 The Proposal

According to Basel Accord III, we can divide the operational risk loss events into seven kinds of event types and eight business lines (There are 7 × 8 = 56 kinds of combinations of events). It is certain that the distribution of 56 kinds of different events combinations's loss frequency and loss amount are different. If there is enough data, we can use Monte Carlo simulation to measure 56 different event portfolio's operational risk capital. Then sum them to obtain the total of the bank's operational risk capital. At this point, during the additive process we can also consider the correlation between 56 different events combinations . Measurement results in this case is more accurate.

The earliest bank in our financial system that bring the operational risk management into the official agenda is only just started from 2004, Historical loss data collected by China's banks is certainly scant at this stage. The premise of the loss distribution method measuring operational risk capital is that the bank must has collected a sufficient number of historical loss data. The number of historical data loss directly impacts on the parameter estimate of the loss frequency distribution function and the loss amount distribution function, thereby affecting the operational risk measurement results.

Our banks should pay full attention to the collection of historical loss data for subsequent loss distribution method can be used to accurately measure the operational risk capital.

Comments

① Core asset adequacy ratio = (core capital − core assets deductions)/(credit risk weighted assets + market risk weighted assets + operational risk weighted assets) = (core capital − core assets deductions)/(credit risk weighted assets + market risk capital requirements × 12.5 + operational risk capital requirements × 12.5)

② Asset adequacy ratio = (Capital − deductions)/(credit risk weighted assets + market risk weighted assets + operational risk weighted assets) = (capital − deductions)/(credit risk weighted assets + market risk capital operations × 12.5 + operational risk capital requirements × 12.5)

References

1. Fan X, Yang X (2005) Monte Carlo simulation estimation of our banking operational risk. Syst Eng Theory Pract 5:12–19 (chinese)
2. Zhou H, Yang X, Nie L (2006) Empirical analysis of bank operational risk measurement. Stat Res 6:47–51 (chinese)
3. Zhang H (2008) Chinese commercial banks operational risk measurement and its application research. Chongqing University, Chongqing (chinese)
4. Chao C (2008) Our banking operational risk capital measurement empirical analysis based on loss distribution method. Reform explore 5:23–60 (chinese)
5. Fei 1 (2008) Chinese commercial banks operating risk theory and empirical research. Wuhan University of Technology, Wuhan (chinese)
6. Zheng H (2008) Empirical study of domestic commercial banks' operational risk measurement. University of Electronic Science and Technology, Chengdu (chinese)
7. Xing Z (2012) The research of Chinese commercial bank operational risk measurement based on loss distribution method. Acc Monthly 5:43–46 (chinese)
8. Wang Y (2014) The research of Chinese commercial bank operational risk Measurement based on loss distribution method. Dongbei University of Finance, Dalian (chinese)
9. LI J (2008) Commercial bank operational risk management. Shanghai Finance and Economics University Press, Shanghai (chinese)
10. Li X (2008) Empirical analysis of ICBC operational risk. Northwest Agriculture and Forestry University, Shanxi (chinese)

The Impact Study of Consumer Demand on the Adoption of Technological Innovations of Enterprise

Yu Zhao and Wei Chen

Abstract Innovation, as the soul of the national progress, is the key to maintain and improve the core competitiveness of enterprises. However, the adoption of technological innovations of enterprises is affected by many factors, this paper is mainly from the perspective of consumer demand to discuss the impact of consumer demand on adoption of technological innovations of enterprises by questionnaires, and makes an empirical analysis on research assumes: (1) consumer demand can promote the adoption of technological innovations of enterprises; (2) the interaction among departments of enterprises can promote the adoption of technological innovations of enterprises; (3) the interaction among departments of enterprises positively regulates the relationship between consumer demand and the adoption of technological innovations of enterprises.

Keywords Consumer demand · The adoption of technological innovations of enterprises · The interaction among departments of enterprises

1 Introduction

Technological innovation, the key factor for modern enterprises to survive in the increasingly fierce market competition, plays an essential part in enterprises. Thanks to technological innovation, enterprises develop and prosper quickly, technologies are propelled forward and the progress of the society is promoted. Let's take Xiao mi as an example. The digital products of this company have gained huge success in the market and in a sense have changed people's life styles as they boast outstanding innovation and at the same time they can cater for the demands of different consumers to manufacture mobile phones at different prices. Against the global, market-oriented, knowledge-based and network background of our times,

Y. Zhao (✉) · W. Chen
School of Economics and Management, Shanghai Institute of Technology, Shanghai, China
e-mail: Zhaoyu491@163.com

E. Qi (ed.), *Proceedings of the 6th International Asia Conference on Industrial Engineering and Management Innovation*,
DOI 10.2991/978-94-6239-145-1_98

enterprises more than ever need to obtain competitive advantage through technical innovation in the cruel and fierce market competition. The government, enterprises and academic circles have been so deeply concerned about how to improve enterprises' technological innovation in a well-targeted manner.

The value and function of technological innovation can't be really played out until companies adopt technological innovation achievements and then gain market acceptance [1]. Enterprises can take a variety of means to adopt the technological innovation. According to a large number of empirical research and analysis about technological innovation since the 1960s, it is the demand that force enterprises to adopt the technical innovation. Then the Demand-Pull comes into being, the adoption of technological innovations of enterprises which comes from Demand-driven refers to the impetus coming from the market and the demand of customers. Therefore, enterprises take new methods, technologies and crafts to meet the demand of customers.

In the process of studying on the domestic and international research, we find that there are few researches about the adoption of the technological innovation from the perspective of consumers while most of the researches are within the enterprises. This paper will make a deep analysis of the relations among the demand of customers, the interaction among departments of enterprises and the adoption of the technological innovation of enterprises, and will prove them with many relevant facts.

2 Literature Review and Hypothesis Development

2.1 Adoption of the Technological Innovation of Enterprises

In the late 1960s and early 1970s, scholars began to do relevant researches on the innovation adoption, Rogers is a famous scholar in researching on the adoption of innovations, in 1962, Rogers, Gross and together with Ryan published Diffusion of Innovations which has pioneering researches on innovation adoption [2]. Rogers believes that the innovation adoption refers to the decision made by individuals or organizations to adopt an innovation. The researchers who are inspired by Rogers have established a lot model of adoption of innovations to describe and explain the relationship between the different variables and the decision-making behavior in the adoption of innovations. According to Mitchell and Singh [3], the enterprise innovation adoption is a complicated process in which the enterprise gains new concepts of technologies and enriches their connotations with its own manufacture, production and management and marketing knowledge and expertise in order to make them become products which can be bought in the market.

The adoption of the technological innovation of enterprises is a complicated process in which the decision behaviors of the enterprise innovation adoption will be influenced by many factors. After analyzing the decision behaviors of the enterprise innovation adoption from the perspective of knowledge, Gopalakrishnan and Bierly think that the enterprise innovation adoption needs to be paid more costs as the knowledge used in it has been increasingly recessive, systematic and complicated [4]. The dynamic factors that influence innovation adoption and contain the features of innovation itself, the characteristics of adopters of innovation and the characteristics of internal and external environment of the enterprise, assuming that they will change with the diffusion of innovation [5]. At the same time, the behavior of innovation adoption is likely to become quite complicated because of the social network relations of adopters [6].

2.2 Demand of Consumers

As the combination of need and the actual purchasing power, the demand of consumers reflects the need and desire of consumers who can afford and are willing to buy some products.

Two centuries ago, Adam Smith proposed the Theory of Demand, in which the most famous economic thought is the rules of supply and demand. Since the "demand" has been widely used as an economic term and scholars from various countries also have made corresponding research on the theory of the demand, as far as Maslow's Hierarchy of Needs is concerned, people's needs are various and hierarchical and the most urgent need of people is the main cause and motivation of human behaviors. There are physiological needs, safety needs, love and belonging needs, esteem needs and self-actualization, according to Maslow's Hierarchy of Needs. Kotler in Principles of Marketing proposes that need, desire and demand is the core concept of the marketing [7]. Qu in Service Advantages and Comprehensive Customer Service System Programme comes up with the Iceberg Theory of the Demand of Consumers. He believes that about a tenth of the iceberg of the demand of consumers is visible. In other words, visible demand accounts for only one tenth of individual demand while the left nine tenths is all invisible demand. Most of the demand do exist only in subconsciousness and it will not appear itself until it is mature. The demand in subconsciousness should be stimulated and guided [8, 9]. Van Trijp proposes in his studies that the mobilization of the demand for sustainable products of consumers is an important prerequisite for sustainable growth [10]. Tian and Huang suggest that the demand of consumers should be integrated to truly satisfy the demand of consumers [11].

Combining with the research results of scholars at home and abroad, we assume that the demand of consumers can promote the adoption of technical innovation of the enterprise. Therefore, Hypothesis 1 is proposed:

Hypothesis 1: the demand of consumers has a positive influence on the adoption of technical innovation of the enterprise.

2.3 Interactive Relationship Among the Departments of Enterprises

Many researches show that the interactive relationship among the departments of enterprises will affect the performance and development of organizations. Souder studies the relationship between research and development departments and marketing departments, which is concerned and quoted by many researchers. It is believed that if marketing departments cannot get on well with research and development departments, then 68 % research and development projects will come to a premature end [12]. According to Souder [12], uncertainty is brought about by the demand of users, technological environment, competition environment and organizational resources can be reduced by the integration of research and development departments and marketing departments. Gupta and some other researchers assume that the demand of the integration of research and development departments and marketing departments comes from the uncertainty of the organizational strategy and external environment and its achievement will be affected by many organizational and personal factors. However, the distance between the integration and the achievement will influence the success of the innovation of organizations.

The researchers also study the specific factors and methods that promote cooperation and overcome the barriers between the sectors of the enterprises. According to Bergen [13], in the process of developing new products, more attention of senior managers, their commitment to research and development, a lot of participation of the production sector in the process of product design and the improvement of the integrity of the information delivered all can greatly enhance the research and development departments and marketing departments, promoting development of new products. If the ideal integration between the design department and the manufacturing department is achieved, then the speed and quality of development of new products will be improved and the enterprises will gain competitive advantages at the same time.

Therefore, we believe that the interactive relationship among the departments of enterprises can accelerate the adoption of technical innovation of the enterprise, Hypothesis 2 is proposed,

Hypothesis 2: the interactive relationship among the departments of enterprises has a positive influence on the adoption of technical innovation of the enterprise.

At the same time, as the change in the demand of consumers and purchasing behaviors and preferences will be likely to influence the model of the adoption of technical innovation of the enterprise and to a large extent, the conveyance of the demand of consumers needs the cooperation of the sectors of the enterprise, we should guarantee the premise that division of labor among sectors of the enterprise is fair, the relationship among sectors of the enterprise will be a top priority in the development of the enterprise especially to tailor the personalized needs of consumers. Therefore, we put forward Hypothesis 3.

Fig. 1 Research model

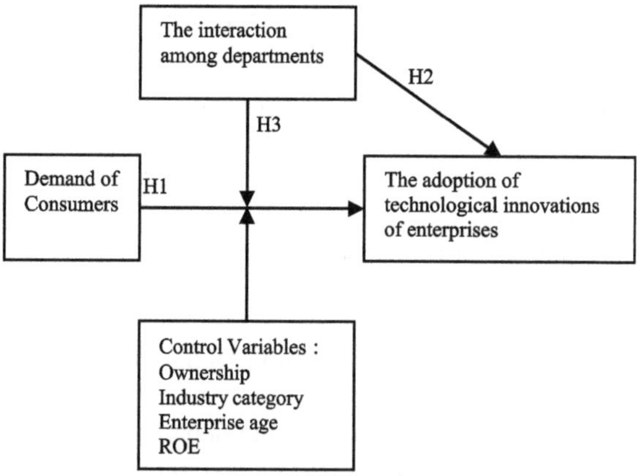

Hypothesis 3: the interactive relationship among the departments of enterprises produces a regulating effect on the demand of consumers and the adoption of technical innovation of the enterprise.

In conclusion, we get hypothesis model of this thesis (Fig. 1).

3 Methodology and Measurement

3.1 Data Collection and the Sample

This research employs an empirical study, which collects data in the way of the questionnaire, the questionnaire participants mainly are the leadership of R & D departments (Technology), product departments and other relevant departments. The number of questionnaires that was distributed is 400, the number of questionnaires that was recovered is 365, and the recovery rate of questionnaires is 91 %, at the same time, this research removed some questionable questionnaires, finally this research gets 314 effective questionnaires, the response rate of questionnaire is 86 %, so the questionnaires meet the need for research.

According to research the related literature, this research selects the frequently-used control variables that affect the adoption of technological innovations of enterprises: (1) the ownership of enterprise, is a dummy variable, state-owned enterprises = 1; non-state-owned enterprise = 2; (2) the age of enterprise, which is 2015 minus established time; (3) industry, is also a dummy variable, Manufacturing industry = 1; non-manufacturing industry = 2; (4) economic indicator, this study mainly uses return on equity (ROE) to research the economic indicator.

3.2 The Selection and Measurement of Variables

The principal method of investigation in this research is questionnaire, the independent variables and dependent variables are interaction among departments, consumer demand, the adoption of technological innovations of enterprises in the questionnaire, the questionnaires adopt 7-point Likert scale, one means strongly disagree; seven means strongly agree.

3.2.1 The Interaction Among Departments

Measuring the interaction among departments mainly uses the scale that Tjosvold Alper, Tjosvold and Law developed, this research slightly improves the scale, and includes one measured item: inter-enterprise sector can often communicate effectively [14].

3.2.2 Demand of Consumers

Measuring consumer demand mainly uses the consumers need for uniqueness (CNFU) scale that Tian developed [15], this research also slightly improves the scale, and includes two measured items: the individual needs of consumers, customized products for consumers.

3.2.3 The Adoption of Technological Innovations of Enterprises

Measuring the adoption of technological innovations of enterprises is diverse, there is no uniform way to measure the adoption of technological innovations, and various measuring methods have some justification on its own. Therefore, this research carries out three measured items based on the related research: (1) the company often introduces new technology; (2) the company often introduces new equipment; (3) the company often introduces new methods.

4 Analysis of the Results

4.1 Data Descriptive Statistics and Correlation Test

Table 1 shows that the descriptive statistics and Pearson correlation test for us, from Table 1, we can see that the consumer demand, the interaction among departments, and the adoption of technological innovations of enterprise exist significant positive

Table 1 Descriptive statistics and pearson correlation test

Variables	Mean	SD	1	2	3	4	5	6	7
1. Ownership	1.89	0.315	1						
2. Industry category	1.24	0.429	−0.320**	1					
3. Enterprise age	17.38	22.029	−0.080	−0.024	1				
4. ROE	4.94	1.474	0.205**	−0.095	−0.014	1			
5. Demand of Consumers	4.61	1.360	0.040	0.003	0.054	0.249**	1		
6. The adoption of technological innovations of enterprises	4.83	1.346	0.093	−0.099	0.099	0.233**	0.351**	1	
7. The interaction among departments	4.98	1.408	0.088	−0.113*	0.063	0.554**	0.351**	0.348**	1

** $p < 0.01$, * $p < 0.05$

correlation ($p < 0.01$). It builds up a base to find further evidence to verify their relationship.

4.2 Hypothesis Test

The results of the data analysis are shown in Table 2, taking the variance inflation factor (VIF) for inspection is to exclude the interference of collinearity, VIF values of variables in the model are greater than 1, but less than 2, in order to determine the models don't exist serious problem of collinearity through correlation analysis. The results of the data analysis are reliable.

Firstly, the first model tests the relationship between the consumer demand and the adoption of technological innovations of enterprises, the result indicates that consumer demand has a positive impact on the adoption of technological innovations of enterprise, the regression coefficient is 0.307 ($p < 0.001$), which proves the first hypothesis. Consumer demand will make enterprise adopt technology innovation.

The second model tests the relationship between the interaction among departments and the adoption of technological innovations of enterprise, the regression coefficient is 0.292 ($p < 0.001$), the results show that the interaction among departments can promote the adoption of technological innovations of enterprise. The second hypothesis is proved.

The third model examines if the interaction among departments plays a leading role in regulating the relationship between consumer demand and the adoption

Table 2 Regression analysis

Variables	Dependent variable: the adoption of technological innovations of enterprises			
	Model 1	Model 2	Model 3	Model 4
Constant	2.689***	2.831***	2.218***	3.390***
Ownership	0.147	0.208	0.190	0.216
Industry category	−0.231	−0.133	−0.175	-0.163
Enterprise age	0.005	0.005	0.005	0.005
ROE	0.130**	0.046	0.027**	-0.237
Demand of consumers	0.307***		0.256***	0.241***
The interaction among departments		0.292***	0.216***	0.209***
Demand of consumers * the interaction among departments				0.054*
Sig.	0.00	0.00	0.00	0.00
df	5	5	6	7
F	11.811	9.623	12.258	10.989
Adjusted R^2	0.147	0.121	0.178	0.183

*, **, and **** represent significance at the 10, 5, and 1 % level, respectively

of technological innovations of enterprise, from the Table 2, the regression coefficients of the model are 0.24 ($p < 0.001$), 0.209 ($p < 0.001$), 0.054 ($p < 0.001$) respectively. The results of the test show that the interaction among departments has a positively moderating effect, in line with expectations. The third hypothesis is verified.

The fourth model is mainly to test that the dual role of the interaction among departments and consumer demand can affect the adoption of technological innovations of enterprise. The regression coefficients of the model are 0.256 ($p < 0.001$), 0.216 ($p < 0.001$) respectively, both of which are significantly positive impact, the results of the test show that the model of moderating effects in this study is stable.

5 Conclusion and Prospect

5.1 Conclusion

The consumer demand can promote enterprise to adopt the technological innovation from the result of this study, which is also consistent with the actual. The starting point that enterprise adopt technological innovation is to gain market competitiveness and meet consumer demand, at the same time, the consumer demand provides the impetus for enterprise to innovate. The interaction among departments will help the dispersal of know-ledge among the departments, improve the efficiency of enterprises, and play a catalytic role in the adoption of technological innovations of enterprise. At the same time, the communication of consumer demand needs to rely on the interaction among departments, the management of the company can accurately adjust the strategy and adopt technological innovation according to consumer demand.

5.2 Insufficiency and Prospect

In this study, the measurement of the interaction between the corporate sectors is relatively simple, this study only takes the indicator that the communication among departments to measure the interaction among departments, the results could be affected by the method of measurement. Meanwhile, the data of the questionnaire may be influenced by subjective factors, the conclusion of this study may exist limitations based on the limited sample data. The further research will obtain more samples, and combine with the case, the next step will expand the model to study systematically.

References

1. Smits R (2002) Innovation studies in the 21st century; questions from a user's perspective. Technol Forecast Soc Change 69(9):861–883
2. Rogers EM (2002) Diffusion of innovations. Central Compilation & Translation Press, Beijing
3. Singh K, Mitchell W (1996) Precarious collaboration: business survival after partners shut down or form new partnerships. Strateg Manage J 17(S1):99–115
4. Gopalakrishnan S, Bierly P (2001) Analyzing innovation adoption using a knowledge-based approach. J Eng Tech Manage 18(2):107–130
5. Warts E, Everdingen YM, Hillegersberg J (2002) The dynamics of factors affecting the adoption of innovations. J Prod Innov Manage 19(6):412–423
6. Hausman A, Stock JR (2003) Adoption and implementation of technological innovations within long-term relationships. J Bus Res 56(8):681–686
7. Kotler P (2001) Marketing Management: The second edition. China Renmin University Press Co Ltd, Beijing
8. Qu Y (1997) Service advantages and comprehensive customer service system programme (in Chinese). Enterprise management Press Co. Ltd, Shanghai
9. Van Trijp, HCM, Arnout RHF (2011) Mobilizing consumer demand for sustainable development. The trans forum model: transforming agro innovation toward sustainable development. Springer, Dordrecht, pp 73–96
10. Oliveira, T, Martins MF (2011) Literature review of information technology adoption models at firm level. Electron J Inf Syst Eval 14.1: 110–121
11. Tian H, Huang Y (2004) On consumers' demand integration (in Chinese). Commer Res (1):24–26
12. Souder, WE, Moenaert RK (1992) Integrating marketing and R&D project personnel within innovation projects: an information uncertainty model*. J Manage Stud 29.4:485–512
13. Bergen SA, McLaughlin CP (1988) The R&D/production interface: a four-country comparison. Int J Oper Prod Manage 8(7):5–13
14. Alper S, Tjosvold D, Law KS (1998) Interdependence and controversy in group decision making: antecedents to effective self-managing teams. Organ Behav Hum Decis Process 74.1: 33–52
15. Tian, KT, Bearden WO, Hunter GL (2001) Consumers' need for uniqueness: scale development and validation. J Consum Res 28.1: 50–66

The Innovative Research on Engineering Master Cultivation Mode Based on University-Enterprise Cooperation

Feng-yuan Wang, Rui-tao Mu, Li-xin Liu, Gang Sun, Guo-qing Ji and Chao-hui Yang

Abstract Engineering master cultivation is an important way for enterprises to train high-level talents. It is an important beginning stage to take advantage of university-enterprise cooperation conditions for talents selection. According to the current culture condition, three typical modes of talent selection were summarized. In the process of personnel training, many companies make full use of university-enterprise cooperation platform. Through the comprehensive guidance of university-enterprise cooperation supervisors, the training links of topic selection, lectures and thesis papers, etc. could cooperatively be completed to improve effectively the talent cultivation quality and management efficiency.

Keywords Engineering master · Talent selection · Talent cultivation · University-enterprise cooperation

1 Introduction

Currently, the basic way to engineering master cultivation basically is that students volunteer to take the test, college enrolls according to admission scores, training units program the training plan for training talents in accordance with the requirements. There is have problem is that the purpose of training is not strong, research topic selection is given priority to with school tutor, students learn research subject difficult to bind to currently engaged in the work, and etc. Many school still exist school and students to distance, guidance in the process of research difficulties, and etc.

F. Wang (✉) · R. Mu · L. Liu · G. Sun
School of Automobile and Transportation, Qingdao Technological University, Qingdao 266520, China
e-mail: fy58wang@126.com

G. Ji · C. Yang
Research Center of Qingte Group Co. Ltd, Qingdao 266106, China

E. Qi (ed.), *Proceedings of the 6th International Asia Conference on Industrial Engineering and Management Innovation*,
DOI 10.2991/978-94-6239-145-1_99

At present, our school establishes a good basis for university-enterprise cooperation in Shandong province, especially in Qingdao region, and establishes good contacts and cooperation with the automotive industry. Therefore, the basis could be used to train talents for the enterprise. Especially in the stage of graduate student, solving the problem of enterprise technology has become a transformative task which has an important practical significance.

2 Domestic Research Status

On this topic, domestic research was mainly concentrated in the school characteristics and professional strengths, some exploration was carried out to combine university-enterprise cooperation with engineering master cultivation. Combining with the actual situation of engineering master education training to improve the quality of graduate student training, Liuhong strengthened the process management, implemented the double tutorial system taking advantage of the university-enterprise cooperation [1]. Heqing analyzed the necessity of local colleges adhering to the cooperation of study, production, research and application in training master of engineering [2]. Wudun thinks that university-enterprise cooperation is an effective way to improve graduates' abilities.

Therefore, it is very important to take advantage of the powerful conditions of university-enterprise cooperation [3]. This paper emphasizes the mode innovative research on enterprise talents selection and cultivation, explores how to utilize the resources and advantages of university-enterprise cooperation, and makes use of the relationship of established university-enterprise cooperation to select graduate students. In order to improve the quality of engineering master we should take advantage of the good cooperation resources in the process of cultivation.

3 The Analysis on Selection Mode of Enterprise Talent Based on University-Enterprise Cooperation

The analysis on selection mode of enterprise talent is based on university-enterprise cooperation which plays an important role in enhancing the cultivation ability and the quality of graduates. From the point of the requirements of enterprise, talent cultivation to select outstanding staff with developing potential to joint the exam, it can improve the purpose of the postgraduate entrance and to ensure the effectiveness of the training. According to the content of different university-enterprise cooperation, there are different selections of recommended companies according to their cultivation. According to our school in recent years, graduate student could sum up three main patterns Selection mode one generally exist in state-owned enterprises, mode two generally presented in outstanding private enterprise and mode three existed in mainly state-owned automobile group company vehicle production.

3.1 Personnel Selection Mode One

Selecting employees to enroll in master of engineering is an annual important work for HR department. Probably about 10 staffs passing the examination have access to enter the universities. The selection process is as following. The individual worker put forward research direction according to the requirements of the recruitment of university. The work department team assesses the application, company working group review it, the company leadership team approve it. After the incorporated exam from the company and the research direction, the approval could be announced by the company. The employees themselves propose master application, where the department recommended; the HR qualification and schools and enterprises both sides hired an expert team on site as a judge in reply, after a comprehensive evaluation and then the results announced, next announced the list of training delivery, organizations to sign up, attend on-the-job personnel of the country's "Graduate Candidate Test" (GCT exam), implement training after admission.

The Personnel Selection Mode One is tight organized and reflects the state-owned enterprises have integrated management concepts and feature.

3.2 Personnel Selection Mode Two

Company selects part of the backbone, generally 5–10 persons in every 2 or 3 years as a cycle, to enter university study for a master of engineering. The selection process begins from the department recommendations based on the performance evaluation of various professionals. The HR organization Group and related departments and external experts recommend evaluation team shall give a comprehensive evaluation based on the evaluation results of high-level approval group. After the leaders agreed with the recommend personnel to take the exam, the staffs take part in the GCT examination. They will enroll the school after pass the examination.

Personnel Selection Mode Two is a relatively simple and practical way, which reflects the flexibility of private enterprise management features and emphasis on talent.

3.3 Personnel Selection Mode Three

Company selection can be carried out by technical center according to the business level. The HR organize the factory, the relevant department heads, external experts to recommend personnel evaluation group to give a comprehensive evaluation. The evaluation results are presented to the plant leader for examination and approval. After factory leadership agreed to assessment, the recommended people could take an examination. Who passed the GCT exams could enter college study.

The Mode Three is characterized by relatively simple and fast. The organization process is relatively tight, long-term perspective, more emphasis on the training and reserve.

4 Program Optimization to Meet the Demand of Enterprise Talent Training Scheme

4.1 Curriculum System Optimization to Meet the Demand of Enterprise Personnel Training

The project team visited enterprises to carry out in-depth discussion related to the enterprises and business-related R&D personnel. After understanding the enterprise needs, the related teacher discussed the previous training scheme whether or not it meets the need of enterprise development to increase the target of the curriculum. The "Engineering Master's Degree Graduate In Vehicle Engineering Field's Training Scheme" is revised to adapt to the training mode based on the needs of university-enterprise cooperation.

According to the basic specification requirements of professional training, professional study should be paid much attention beside the theoretical foundation. Ensure that the fundamental issues and technical course with the current situation closely engaged in professional. Make course to meet the students' understanding of the frontier, and to grasp basic research methods to understand the technical solutions in their field of interest.

In practical teaching process, the communication with the enterprise technology and R&D department shall be strengthened, practical business needs shall be understood, the course shall be set up, and train link of organization shall be reformed. Software class teaching should be reduced, emphasize the frontier of vehicle engineering, vehicle active safety, vehicle vibration and noise control, and modern testing techniques etc. Other practical knowledge, such as intellectual property, document retrieval and thesis writing courses should be set up to improve R&D personnel's awareness of intellectual property. Raise awareness of the intellectual property of R&D personnel, improve the ability of access to information, analysis and summarize and thesis writing.

According to the characteristics of students, universities should take a variety of teaching methods. The course concerning enterprise technology should be taught in enterprise, and that concerning theory should be taught in university. In this way, the hardware resource of the enterprise can be effectively utilized. The graduate student could have plenty of time to study beside the heavy work in enterprise,.

4.2 University-Enterprise Jointly Selected Topics

The university-enterprise tutor group system was implemented to adhere to teamwork. This will give full play to the power of tutor group in the academic tradition and the technology research and development organizations. In the process of guidance and training in engineering, a master of engineering is not only a student of tutor, but also a student of the tutor group. Positive communication between tutors established contact mechanism through the mid-term examination and exchange opportunities for schools and enterprises. Tutor group has been established. The group framework is composed of senior professors, young professors, business leaders, senior executives, chief engineer and other technical staff.

Opening report of research topics were discussed jointly by the school teacher and off-campus tutor. After an analysis on the current enterprise technical and product technical process, the tutor group will help the students to find the scientific problem, and then settle the research topics. Off-campus tutor focuses on the research objectives, implementation plan and technical route of project. School teachers are responsible for thesis research process, research contents, research theory, etc. To ensure the paper research content and the workload is reasonable. Starting from the university-enterprise of graduate education, graduate students have become a link between the university and enterprise cooperation. The relationship between university and enterprise cooperation has been strengthened in the process of graduate education.

4.3 Graduate Thesis Management

The quality of the dissertation is the important part of engineering master's training quality. Guiding students to write high quality thesis, is good to show the level of training for master of engineering. Give full play to the advantages of the training mode based on the university-enterprise cooperation to implement closely with the double tutorial system and the guidance of the tutor group, strengthening the school teacher cooperation, division of labor, guiding students clarify ideas, solving the problem of writing techniques and methods, guiding the engineering master finishing high quality paper.

In the final stage of submitting the thesis paper to take thesis proposal, a variety of methods were taken, such as the repetition rate of detection etc., to encourage students to complete a high-quality paper. In the thesis application presentation sector, the committee chairman was pointed from enterprise with practical experience and comprehensive understanding of the actual situation of the enterprise production. This makes it easier to determine the actual value of the paper, to found paper deficiency, and thus to make a scientific, objective and comprehensive evaluation.

5 Conclusion

This paper analyzes the engineering graduate student selection management mode on the basis of the embodiment of university-enterprise cooperation. Taking advantages of both schools and enterprises, it is beneficial to talents and potential cultivate personnel selection and targeted training. It is useful to explore new ideas of graduate student training, form a new bond, strengthen the university-enterprise cooperation. The cultivation of graduate students made full use of the basic conditions of university-enterprise cooperation, improved the quality of graduate education. Through the university-enterprise cooperation group guidance, graduate thesis writing report, scientific research and innovation ability training have achieved obvious results. Effective use of the practice, processing, the experimental conditions of the enterprise, and effective use of theory, analysis, research, means, methods, ability and human resources of universities will facilitate the graduate students to participate in the enterprise's technical problem research, to achieve the desired effect of the training and research under the university-enterprise cooperation.

References

1. Liu H, Long X, Li W (2009) Assurance of master of engineering education quality. J Chongqing Univ Sci Technol Soc Sci Edn 10:215–216
2. He Q, Hu Z, Li H (2013) Based on "Production use" areas where Mechanical Engineering Master's Universities culture. China Electr Power Educ 35:28–29
3. Wu D, Cai S, Wang B (2014) Based on the university-enterpriser cooperations' on-the-job master of engineering training mode thinking. Stat Manage 2:44–45

The Cause and Solution of Big Enterprise Disease: An Empirical Study from the View of Intellectual Capital and Innovative Culture

Ya-jun Liu and Xiao-mei Wan

Abstract Whether there is a negative effect of Intellectual capital(IC) on technological innovation performance (TIP)? Based on classifying IC to three aspects: human capital (HC), structural capital (SC) and relational capital (RC), the paper introduces innovative culture as a moderator and uses the method of hierarchical regression to analyze the moderating effect of innovative culture on the relationship between IC and TIP with a sample pool consists of 155 high-tech enterprises. Result shows that both HC and RC have positive effects on TIP, and so does proper SC. But too much SC has negative effect on TIP, thus leading to Big Enterprise Disease. Good innovative culture can eliminate or relieve this kind of impact.

Keywords Big enterprise disease · Innovative culture · Intellectual capital · Technological innovation performance

1 Introduction

Big enterprise disease is common in many areas including technological innovation. Despite the abundant intellectual capital they have, many large companies are not good at raising technological innovation efficiency, and keep insensitive to new technologies, leading to difficulty in transformation and upgrading, challenges from competitors as well as fatal blows from new technologies. The failure of Nokia is a negative example while the successful transition to service industry of IBM is a positive one.

Y. Liu (✉) · X. Wan
Business School, Xiangtan University, 411105 Xiangtan, China
e-mail: lyj@xtu.edu.cn; 709441542@qq.com

Y. Liu
Post-Doctoral Station of Theoretical Economics, Xiangtan University, 411105 Xiangtan, China

E. Qi (ed.), *Proceedings of the 6th International Asia Conference on Industrial Engineering and Management Innovation*,
DOI 10.2991/978-94-6239-145-1_100

Knowledge is often considered to be the source of technological innovation and progress of an enterprise. In other words, intellectual capital is of great importance to technological innovation performance. However, consensus has not been reached by researchers about whether or not the enhancement of intellectual capital could necessarily improve technological innovation performance. Many scholars argue that intellectual capital would promote technological innovation to a large extent, and it is also the determinant of innovation performance of enterprises. Others, however, find that the above influence is not significant, or there's no simple positive correlation between the two, through empirical studies. For instance, the result of the empirical research of Shiu [1] doesn't support the hypothesis that intellectual capital can positively affect technological innovation performance. The study of Huang and Liu [2] indicates an "Inverse U Curve" relationship between intellectual capital and technological innovation performance, instead of linear positive correlation.

So, what causes the different effects of intellectual capital on technological innovation? How to adjust that? This paper tries to adopt the intellectual capital classification and introduces moderator to answer those questions and give some suggestions. We take human capital, structural capital, and relational capital as the three basic dimensions of intellectual capital proposed by Hsu and Huang [3] to further examine if there is different influence of the three dimensions on technological innovation performance respectively. Moreover, innovative culture, a subjective initiative factor which can affect technological innovation performance, should be included in the econometric model.

2 Framework and Hypotheses

2.1 The Influence of Intellectual Capital on Technological Innovation Performance

Intellectual capital can be defined as the total knowledge and capabilities that an organization and its members possess to create value or competitive advantages [3]. Intellectual capital can be separated to such three basic aspects as human capital, structural capital, and relational capital.

Being capable of converting knowledge to value, intellectual capital can drive the development of technological innovation. Chen et al. [4] contends that, the more intellectual capital an enterprise has, the stronger its innovation ability, and the better its innovation performance.

High-quality human capital is the prerequisite for technological innovation. With capable staff and experienced administrators, companies would be able to make the most of technologies and improve technological innovation performance. Outstanding managers with knowledge and vision help firms quickly respond to the market in highly turbulent conditions and increase market share and become industry leaders [5].

Edvinsson and Malone [6] argue that, with good structural capital, a learning organization can raise knowledge management efficiency and build up the culture encouraging information sharing and communicating so as to positively affect technological innovation performance.

The positive effects of relational capital on technological innovation performance are manifested in the following three aspects. First, Great structural capital is the source of technological innovation. Networks and alliances with customers, suppliers, competitors and other non-market entities constitute the key source of innovation. Second, Nice relational capital can lower the risks in technological innovation process. Third, good relational capital is capable of providing needed factors and resources for technological innovation.

Based on the discussion above, this study offers the following hypotheses.

H1.1 Human capital positively affects technological innovation performance.
H1.2 Structural capital positively affects technological innovation performance.
H1.3 Relational capital positively affects technological innovation performance.

2.2 *The Moderating Effect of Innovative Culture*

Innovative culture, or innovation-based culture, is a culture type including such humanistic connotations and concepts as values, beliefs, attitudes related to innovation, and a culture rewarding innovation and encouraging adventuring, thus providing relaxing and harmonious work environment for employees. Innovative culture is partly characterized by the pursuit of challenging tasks by employees of the enterprises. A large number of researches find that demanding mission arrangement would transfer human capital to technological innovation, thus forming a totally different advantage. Generally speaking, people would get tired of simple and repetitive work, and challenging work assigned to them can motivate their creativity, thus promoting technological innovation of enterprises. In operation process, with company scale expanding and structural capital increasing, the procedures and routines in a company would become detailed and complicated. On one hand, as a result, technological innovation efficiency can be improved, cost may be reduced and economies of scale can be achieved. On the other hand, however, rigid process and redundant organization can be new problems. Therefore, in technological innovation process, an organization should adjust its strategies, structures, routines and procedures, all of which requires the support from innovative culture [7].

Innovative culture is also related to pressure resistance. With innovation-based cultural atmosphere, enterprises are willing to suffer from huge environmental stress, and will seek to innovate and make breakthroughs in technologies. A stressful environment would help relational capital contribute to technological innovation through two ways, namely, first, obtain information from customers, suppliers and competitors, as well as know about the trend of related market and technological development by scanning the environment; second, cooperate with external organizations to get the resources that innovation requires while a company lacks.

This study offers the following hypotheses based on the above discussion.

H2.1 The greater the innovative culture in enterprises, the stronger the positive influence of human capital on technological innovation performance.
H2.2 The greater the innovative culture in enterprises, the stronger the positive influence of structural capital on technological innovation performance.
H2.3 The greater the innovative culture in enterprises, the stronger the positive influence of relational capital on technological innovation performance.

3 Methodology

3.1 Variable Definitions and Measures

The intellectual capital scale this study adopted referred to the research of Subramaniam and Youndt [8] and Dzinkowski [9].The measurement for technological innovation performance adopted the following three indicators: the yearly number of patents, the annual percentage of new product in total enterprise product, and the proportion of new product sales, adapted from Guan [10], who contributed a lot to the evaluation system of technological innovation performance. Our measurement for the innovation propensity adopted the following six indicators: management features, organizational leadership, staff management, organizational cohesion, strategic goals and success criteria, combining the research on corporate culture assessment of L-PCAI, the evaluation questionnaire of corporate culture tendency invented by Redetac Management Consulting Co. Ltd.

The study set the industry and company size as control variables to see whether they would influence technological innovation performance or not. Industries were classified to four types based on whether they're capital-intensive or knowledge-intensive: computer and communication device; software; bio-engineering and new medicine, new material, new energy, resource and environment technology; other industries.

To sum up, through the initial questionnaire design and the pilot test, our formal questionnaire covered such five latent variables as technological innovation performance, innovative culture, human capital, structural capital and relational capital, scales of which included 22 indicators altogether.

3.2 Sample and Data Collection

In order to guarantee the validity of samples, we concentrated on privately-run high-tech manufacturing enterprises and required all of the respondents were senior administrators or technicians who had a good knowledge of technological innovation and research and development of their own companies. The sample pool consisted of 180 privately-run high-tech manufacturing enterprises located in 9 provinces in China. By convenient sampling, all informants, including engineers and major executives received questionnaires via the relationship network of MBA students from four colleges and universities.

The formal investigation process distributed about 280 questionnaires and 177 informants responded to this wave of data collection. Removing 22 invalid responses, these efforts elicited 155 usable ones, creating a final usable response rate of 87 %.

3.3 *Quality Assessment of Samples*

The study first examined the validity and reliability of questionnaire items using SPSS 15.0. Through factor analysis we found that the cumulative explained variation greater than 0.5, showing content construct validity for the measures.

In this study. The Cronbach's α value for all constructs exceeded the 0.7 criteria, indicating good reliability of measurement. Therefore, we can further analyze the measures utilized in this study.

4 Results and Discussion

Using SPSS 15.0, this work continued the analyses of correlations among variables and moderating effect of innovative culture on the basis of the test of measurement reliability and validity. When analyzing moderating effect of innovative culture, we divided the sample enterprises into two groups according to the amount of their structural capital so as to show the different effects of structural capital on technological innovation performance respectively.

4.1 *Correlations*

Correlations analysis shows that neither industry nor company size showed correlations with other latent variables, while structural capital exhibited weak, negative relationship with technological innovation performance.

4.2 Moderating Effect of Innovative Culture

The paper used the method of hierarchical regression to analyze the influence of the three dimensions of intellectual capital on technological innovation performance, and the moderating effect of innovative culture in the relationship between intellectual capital and technological innovation performance. We designed model I to model V, among which model I only examined the effects of the two control variables, industry and company size, on technological innovation performance. Then four latent variables were included in model II to test the effects of human capital, structural capital and relational capital as well as innovative culture on technological innovation performance. Furthermore, model III to model V studied the moderating effect of innovative culture in the relationships between the three aspects of intellectual capital-i.e. human capital, structural capital and relational capital, -and technological innovation performance respectively. The results are shown in Table I.

The result of model I indicated that neither industry nor company size would affect technological innovation performance. The regression analysis of the four latent variables including human capital, structural capital, relational capital and innovative culture (Model II) showed that structural capital exhibited a weak, negative effect on technological innovation performance, with which human capital and structural capital showed significantly positive relationships, thereby providing support for H1.1 and H1.3, while the result is opposite to what we predicted in H 1.2.

Table 1 Hierarchical regression analysis

Dependent variable: technological innovation performance (N = 155)					
Independent variables	Model I	Model II	Model III	Model IV	Model V
Industry	0.03	0.01	0.01	0.01	0.01
Company size	0.11	0.10	0.06	0.06	0.06
Human capital		0.49 **	0.49 ***	0.49 ***	0.45 ***
Structural capital		-0.12 *	-.12 *	-0.13 *	-0.09
Relational capital		0.40 **	0.39 ***	0.39 ***	0.37 ***
Innovative culture		0.05	0.01	-0.01	0.02
Human capital* Innovative culture			-0.06		
Structural capital* Innovative culture				0.14 **	
Relational capital* Innovative culture					0.15
ΔR2		0.58	0.01	0.03	0.03
R2	0.01	0.59	0.60	0.62	0.62
F	2.01	44.07	31.62	34.33	34.17

Note (1)*p < 0.05, **p < 0.01; (2) The intercepts are not listed here

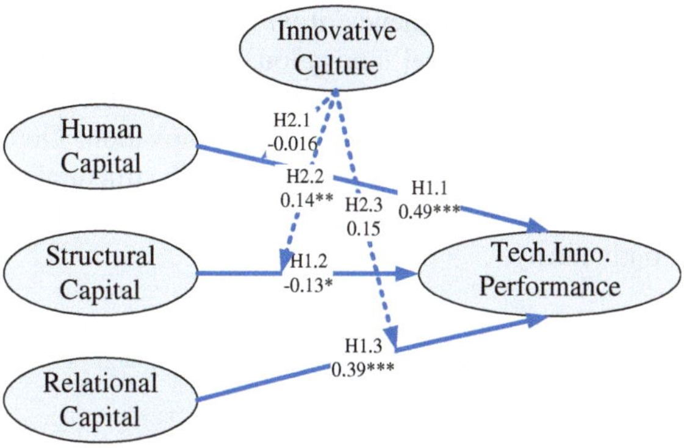

Fig. 1 The model and the result

The crossing effects of model III, model IV and model V suggested that innovative culture, as a moderator, had a significantly positive effect on the structural capital-technological innovation performance relationship, while it didn't have the same effect in the relationships between the other two aspects of intellectual capital and technological innovation performance, thus providing support for H2.2 and no support for H2.1 and H2.3.

Based on the above analysis, intellectual capital of an enterprise would not always positively affect technological innovation performance, probably for the existence of structural capital (Fig. 1).

5 Conclusions and Implications

Based on the above empirical study, our hypothesis H1.1, H1.3, H2.2 are supported.

Based on the above analyses and conclusions, the paper contributes to theoretical innovation and provides important implications for management practice and they're shown as follows.

Firstly, overabundant structural capital is the knowledge source of Big Enterprise Disease in the process of technological innovation, to which intellectual capital is of great importance. Given that intellectual capital consists of three dimensions, however, it won't always affect technological innovation performance positively. There's no doubt that human capital and relational capital have positive effects on technological innovation performance, but excessive structural capital has the opposite influence. This may explain the common Big Enterprise Disease, which causes the poor performance of innovation. During the operation and management process, high-tech enterprises keep accumulating structural capital, and organization procedures, routines and regulations would be continuously mounted in quantity and tend to be stabilized and solidified in quality. Although the increasing structure capital may contribute to the automatic operation of organization and the enhancement of efficiency, it will also bring organizational rigidity and inertia and hamper technological innovation, thus causing Big Enterprise Disease.

Secondly, good innovative culture is able to remove the adverse impact of structural capital on technological innovation. Organizational rigidity and inertia often exist in enterprises possessing high structural capital, and innovative culture is needed to solve the problem so as to promote technological innovation performance continuously.

Acknowledgments This work is supported by the National Natural Science Foundation of China (No. 71272224), National Social Science Foundation of China (No. 14BGL154).

References

1. Shiu H (2006) The application of the value added intellectual coefficient to measure corporate performance: evidence from technological firms. Int J Manage 23(2):356–365
2. Huang CJ, Liu CJ (2005) Exploration for the relationship between innovation, IT and performance. J Intell Capital 6(2):237–252
3. Hsu YH, Fang W (2009) Intellectual capital and new product development performance: the mediating role of organizational learning capability. Technol Forecast Soc Change 76(5):664–677
4. Chen YS, Lin MJ, Chang CH (2006) The influence of intellectual capital on new product development performance—the manufacturing companies of Taiwan as an example. Total Quality Manage Bus Excell 17(10):1323–1339
5. Zhang H, Lv S (2014) The study of relationships among intellectual capital, innovation type and product innovation performance. Sci Sci Manage S. & T(monthly) 02:162–168
6. Edvinsson L, Malone MS (1997) Intellectual capital: realizing your company's true value by finding it's hidden roots. Harper Business, New York
7. Liao S, Fei WC, Liu CT (2008) Relationships between knowledge inertia, organizational learning and organization innovation. Technovation 28(4):183–195
8. Subramaniam M, Youndt MA (2005) The influence of intellectual capital on the types of innovative capabilities. Acad Manag J 48(3):450–463
9. Dzinknowski R (2000) The measurement and management of intellectual capital: An introduction. Manage Acc 78(2):32–36
10. Guan J, Shi X (2004) Study on the relationship between technological innovation capability and innovation performance. China Mech Eng 15(11):1000–1004

Research on the Integration of Lean Construction and BIM and a Case Study in Shanghai Tower Project

Xing Bi and Xi-mu Jia

Abstract Lean Construction and building information modeling (BIM) technology have been produced since the last century, and have been spread rapidly in foreign countries in the 21st Century. In China, disadvantages in building industry are constantly displayed: non-guaranteed project time, low quality, and huge energy consumption. Lean Construction and BIM are advanced concepts and tools for construction industry, both of which are designed to reduce waste, improve building efficiency and create the greatest value. This thesis explores the integration of Lean Construction and BIM, proposes a basic integration framework and analyzes specific technology. Then the thesis gives a case study of Shanghai Tower, applying fuzzy-grey comprehensive evaluation method to calculate quantitative effect of the integration of Lean Construction and BIM.

Keywords BIM · Fuzzy-grey comprehensive evaluation method · Lean construction · Shanghai tower

1 Introduction

Construction industry has become the main power to drive the Chinese economic growth. Data from the *National Statistics Yearbook* shows that the gross output value of the construction industry has doubled in the past ten years.

Although the construction industry is in the era of rapid development, there still exist many problems: few construction personnel are experienced, technology innovation is weak, serious waste phenomenon and rework caused by the limitation of management level are frequent. New technology like Lean Construction and BIM are regarded as effective solution to the low efficiency, high energy consumption and the serious environmental impact.

X. Bi · X. Jia (✉)
College of Management and Economics, Tianjin University, Tianjin, China
e-mail: jiayipei_257.com@163.com

E. Qi (ed.), *Proceedings of the 6th International Asia Conference on Industrial Engineering and Management Innovation*,
DOI 10.2991/978-94-6239-145-1_101

Successful projects abroad have proved that Lean Construction and BIM brought a revolutionary change in the construction field. Research of foreign scholar on the relationship between Lean Construction and BIM says that the two items are independent to each other. It means that each doesn't rely on the implementation of another system or platform [1]. But if both can be effectively integrated, it will bring benefits for building industry.

2 Methodology

2.1 Lean Construction Theory System

Lean production mode originated in manufacturing industry. But there exists significant differences between construction industry and manufacturing industry, shown in Table 1. So, full consideration to the construction industry characteristics should be given, and suitable changes should be made to make the principle and tools applicable for the construction industry.

In the construction industry, the most important lean tool is the tool derived by construction enterprises in order to improve the schedule and the reliability of the construction process planning—the last planner system, LPS. In LPS the construction enterprises encourage devolution of power on schedule, the project manager invite all parties including professional subcontracting, the builder, and even shift supervisors to participate deciding the weekly plan [3].

Table 1 The differences between construction industry and manufacturing industry

Comparison	Construction industry	Manufacturing industry
Product	Single, design changes a lot	Repeatable production, the design is basically remain unchanged
Production features	One-off and not repeatable	Large-scale, mass production on the assembly line
Production cycle	Long	Fixed, relatively short
Production site	More projects	Fixed workshop
Degree of automation	Low	High
Outside influence	Affected by the weather	Almost unaffected by weather
The work environment	Dirty, disorderly and poor	Relatively clean workshop
The owner	Confirmed before production	Using marketing tool to find consumers
Workers	I High liquidity II Often appear shortage, employment difficult III Relative low quality	I Liquidity is relatively low II Permanent staff III Relatively high quality
Relationships with suppliers	Through bidding, temporary	Establish a long-term partnership through negotiations

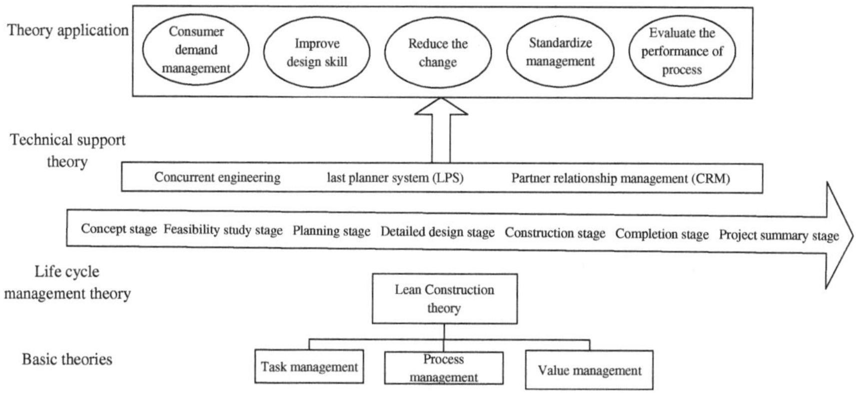

Fig. 1 The construction of Lean Construction theory

Lean Construction is based on the production management theory, guided by the lean thinking principle. Shown as the Fig. 1, it redesigns the process of engineering project management, and is the new project management pattern with the target of to build and to transfer project with high quality, shortest construction period and least amount of resources consumption [2].

2.2 *The Foundation and Development of BIM*

Building Information Modeling (BIM) is a cutting-edge technology in recent years. BIM is a process, gathering geometry and non-geometry information through the 3D numerical modeling, to utilize information resources and management of the construction project in the whole-life-cycle: from planning, design, construction and operations to dismantle.

The earliest prototype of BIM is architecture description system, proposed in 1975 by the Charlie Eastman, who taught at Carnegie Mellon University. He put forward the ideas related to BIM, such as: from the relevant description of the same item, get the floor plan, profile plan and perspective plan. Any layout change will lead to all figure's accordingly update. So, all drawings from the same item automatically stay consistent [4].

BIM spread quickly in the construction industry at this century. J. Lucas and W. Thabet discussed about using BIM platform and tools for recording, operating and generating information, to support the whole project life cycle process [5]. And Ham, Nam-Hyuk presented the conception and function of BIM planning by analyzing the planning process and whole lifecycle [6].

Participants in the traditional construction industry interact little in every stage. The information separation occurs between owners, designers, builders, suppliers and users. BIM provides a platform, on which participants in different stages can cooperate well, use three-dimensional digital model input relating views, simulating the construction process and operation process.

3 Integration Analysis

3.1 Integration Framework of Lean Construction Key Technology and BIM

Lean Construction minimizes waste and maximizes value by reducing the non-value waste during the whole project cycle. The implementation of Lean Construction must be attached to all participations' comprehensive and in time understanding of the project-related information. And BIM can solve problem of traditional data storage method. Figure 2 shows the integrated frame:

The BIM information platform establishes several databases, and each database is used to manage the whole project, resources, technological process and 4D models, etc. The ideas of Lean Construction can provide decision support for the entire project. Their same aim is to reduce waste and achieve maximum value. Through effective integration, it realizes the visual control management, multiple-proposal generating and argumentation, and the projects online communication platform, etc. [7].

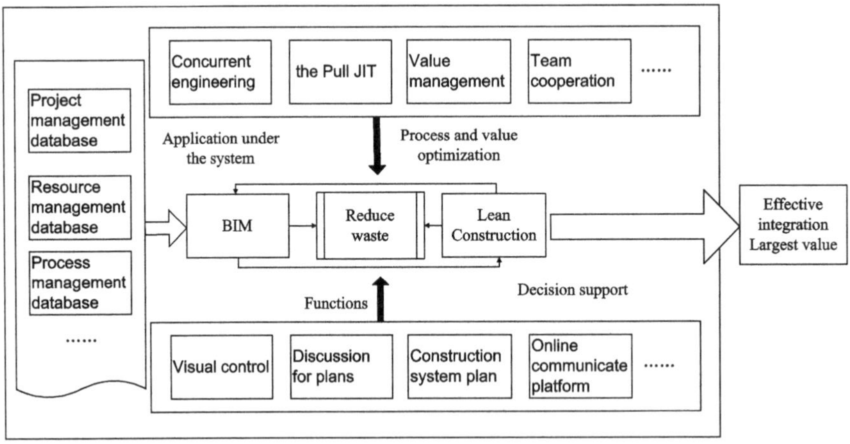

Fig. 2 The integrated frame of Lean Construction key technology and BIM

3.2 *Integration Analysis of Lean Construction Key Technology and BIM*

3.2.1 Concurrent Engineering and BIM

Concurrent engineering is a method of engineering to parallel and integrated design product; it is a process model of a systematic work. It designs architectural product for the whole lifecycle. The developers consider various factors of whole project life cycle, such as quality, cost, schedule and a market demand from the very start [8]. It requires the establishment of the information platform for in-time information feedback. So the work can be paralleled.

BIM is a platform for huge amounts of decision-making information within the whole life cycle of a project. This makes it an information platform for concurrent engineering, with all information timely summarized and stored.

Under the concurrent engineering, the advantage of BIM is reflected as the design-construction integration. Design-construction integration emphasizes the cooperation of designer and contractor. They can exchange and share information on an interactive platform, complete the concurrent workflow, and achieve the perfect combination of design and construction.

3.2.2 Pull Just-in-Time Production and BIM

Pull just-in-time production is the core of Lean Construction. In the Lean Construction, the last planner system (LPS) is implied to pull the next working procedure. Namely, the value stream is from downstream customers to upstream builder. LPS is a kind of new engineering management operating system, proposed by Glenn Ballard. The professional engineer set the pull schedule at various stages, in which the construction plan of the specific milestone back off. Then, experienced management personnel determine the final sequence, operation time and buffer measures to guarantee the stability of the milestone plan. The specific milestone tasks are decided by line workers and managers.

3.2.3 Kanban Management and BIM

Unlike the traditional ERP system, lean production mode gives the specific information of when need, need what and how much. Based on lean thought, a new tool —Kanban management appears. "Kanban" is actually a kind of tool to transmit information, a kind of "plate" to record information. It can be a white board or a card, with the most basic function to record the production information, pass work instructions, and avoid waste.

Kanban system explains an important principle of lean production—visual principle, and this point is the same with the pursuit for visualization of BIM.

While, BIM is a platform for information in the total project life cycle from all parties to get together. So, it is an indispensable tool.

Similar to the Kanban management in manufacturing, KANBIM is a system process of applying the BIM in the Lean Construction system. It is a process management model under the LPS system supported by the powerful dynamic database of BIM.

4 Case Study

Shanghai Tower has completed in December 2014 relied on the BIM and the thought of Lean Construction. This part is quantitative evaluation for the integrated effect in the project based on the fuzzy-grey comprehensive evaluation method [9].

4.1 Integrated Performance Evaluation Research

4.1.1 Determine the Evaluation Index Set

Establish the evaluation index system of integrated effect by recognizing integrated effect factors. For R = {lean degree R_1, information circulation R_2, partners intimacy degree R_3, project process evaluation R_4, owner satisfaction R_5}.

The integrated effect hierarchy system is shown as Fig. 3 [10].

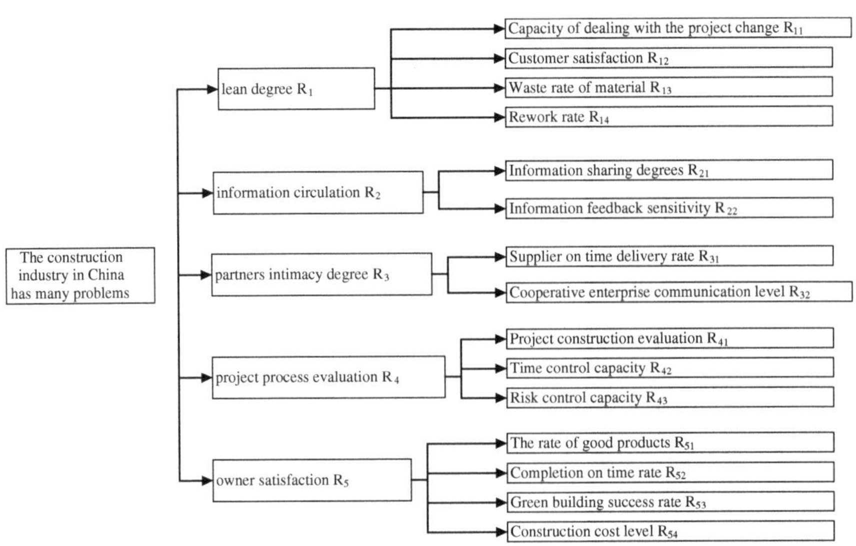

Fig. 3 Integrated effect hierarchy system of Lean Construction and BIM

4.1.2 Determine the Weight Set

W = $\{w_1, w_2, w_3, w_4, w_5\}$ is the first level weight set, set w = $\{w_{i1}, w_{i2}, w_{i3}, \ldots, w_{is}\}$ as the second level wet set, i = 1,2,3,4,5. Referring to relevant data, the numerical values are as follows:

$$\begin{aligned} w &= \{0.29,\ 0.15,\ 0.12, 0.26,\ 0.18\} \\ w_1 &= \{0.13,\ 0.26,\ 0.37, 0.24\}; w_2 = \{0.58,\ 0.42\}; \\ w_3 &= \{0.36,\ 0.64\};\ \ w_4 = \{0.35,\ 0.38,\ 0.27\}; w_5 = \{0.28,\ 0.26, 0.34, 0.12\} \end{aligned}$$

4.1.3 Evaluate the Index of the Second Level

1. Determine the rating of evaluation
   Evaluation sets V = $\{V_1, V_2, V_3, V_4, V_5\}$ = {excellent, good, moderate, qualified, poor}, the corresponding score is V = {9, 7, 5, 3, 1}.
2. Constitute a matrix of sample evaluation
   Using the scoring method, separately score for each single index from 0 to 100. Invite five people to evaluate the related index level. Table 2 shows the results.
3. Calculate the gray statistics

$$b_{11} = \sum_{i=1}^{5} f_1(d(i,1)) = f_1(8) + f_1(9) + f_1(8) + f_1(9) + f_1(6) = 4.556$$

So, $b_{12} = 3.000$, $b_{13} = 1.800$, $b_{14} = 1.429$, $b_{15} = 1.000$, $b_1 = b_{11} + b_{12} + b_{13} + b_{14} + b_{15} = 11.784$

4. Calculate grey evaluation weight, get the matrix

$$r_{11} = \frac{b_{11}}{b_1} = 0.387, r_{12} = \frac{b_{12}}{b_1} = 0.255, r_{13} = \frac{b_{13}}{b_1} = 0.153,$$
$$r_{14} = \frac{b_{14}}{b_1} = 0.121, r_{15} = \frac{b_{15}}{b_1} = 0.085$$

Table 2 Score of each evaluation index

Expert	Evaluation index														
	R_{11}	R_{12}	R_{13}	R_{14}	R_{21}	R_{22}	R_{31}	R_{32}	R_{41}	R_{42}	R_{43}	R_{51}	R_{52}	R_{53}	R_{54}
E_1	8	9	8	9	9	8	8	7	9	6	9	9	8	8	7
E_2	9	7	7	9	8	9	5	5	8	8	7	8	9	9	6
E_3	8	9	9	9	7	9	9	9	9	8	9	6	9	8	8
E_4	9	8	6	8	8	7	8	7	9	9	8	8	9	9	8
E_5	7	7	8	9	8	8	9	8	9	9	7	8	8	9	7

In the same way, evaluation weights of R_{12}, R_{13}, R_{14} can be determined, weight matrix of R_1 are:

$$R_1 = \begin{bmatrix} 0.387 & 0.255 & 0.153 & 0.121 & 0.085 \\ 0.361 & 0.271 & 0.162 & 0.116 & 0.090 \\ 0.315 & 0.299 & 0.179 & 0.107 & 0.100 \\ 0.480 & 0.196 & 0.118 & 0.140 & 0.065 \end{bmatrix}$$

Then, determine the other single evaluation weight matrix:

$$R_2 = \begin{bmatrix} 0.361 & 0.271 & 0.162 & 0.116 & 0.090 \\ 0.387 & 0.255 & 0.153 & 0.121 & 0.085 \end{bmatrix}$$

$$R_3 = \begin{bmatrix} 0.361 & 0.271 & 0.162 & 0.116 & 0.090 \\ 0.295 & 0.311 & 0.187 & 0.103 & 0.104 \end{bmatrix}$$

$$R_4 = \begin{bmatrix} 0.480 & 0.196 & 0.118 & 0.140 & 0.065 \\ 0.361 & 0.271 & 0.162 & 0.116 & 0.090 \\ 0.361 & 0.271 & 0.162 & 0.116 & 0.090 \end{bmatrix}$$

$$R_5 = \begin{bmatrix} 0.337 & 0.285 & 0.171 & 0.111 & 0.095 \\ 0.446 & 0.218 & 0.131 & 0.133 & 0.073 \\ 0.446 & 0.218 & 0.131 & 0.133 & 0.073 \\ 0.277 & 0.323 & 0.194 & 0.099 & 0.108 \end{bmatrix}$$

5. Calculate the fuzzy comprehensive evaluation matrix, $A_i = W_i * R_i$:

$$A_1 = W_1 * R_1 = (0.3759, 0.2613, 0.1566, 0.1191, 0.0871)$$

$$A_2 = W_2 * R_2 = (0.3719, 0.2643, 0.1582, 0.1181, 0.0879)$$

$$A_3 = W_3 * R_3 = (0.3188, 0.2966, 0.1780, 0.1077, 0.0990)$$

$$A_4 = W_4 * R_4 = (0.4027, 0.2448, 0.1466, 0.1244, 0.0813)$$

$$A_5 = W_5 * R_5 = (0.3952, 0.2494, 0.1498, 0.1228, 0.0834)$$

Calculate the evaluation results of the second level from weight matrix $R = (A_1, A_2, A_3, A_4, A_5)^T$

$$C = R \times V = \begin{bmatrix} 0.3759 & 0.2613 & 0.1566 & 0.1191 & 0.0817 \\ 0.3719 & 0.2643 & 0.1582 & 0.1181 & 0.0879 \\ 0.3188 & 0.2966 & 0.1780 & 0.1077 & 0.0990 \\ 0.4027 & 0.2448 & 0.1466 & 0.1244 & 0.0813 \\ 0.3952 & 0.2494 & 0.1498 & 0.1228 & 0.0834 \end{bmatrix} \times \begin{bmatrix} 9 \\ 7 \\ 5 \\ 3 \\ 1 \end{bmatrix}$$

$$= \begin{bmatrix} 6.4342 \\ 6.4304 \\ 6.2575 \\ 6.5254 \\ 6.5034 \end{bmatrix}$$

4.1.4 Evaluate the Index of the First Level

Do the secondary fuzzy comprehensive evaluation based on the first level weight set and the matrix C:

$$E = [0.29 \quad 0.15 \quad 0.12 \quad 0.26 \quad 0.18] \times \begin{bmatrix} 6.4342 \\ 6.4304 \\ 6.2575 \\ 6.5254 \\ 6.5034 \end{bmatrix} = 6.4486$$

The evaluation score is 6.4486, the integrated effect is good, showing that the integration is useful in the project.

4.2 Instance Summary

From the quantitative analysis of Shanghai center Tower project, we can see the integration of Lean Construction and BIM technology has a promoting effect to the project.

5 Conclusion

The technologies in Lean Construction perform well under BIM platform, guaranteeing the project schedule, cost and quality. At the same time, the guidance of Lean Construction system makes the application environment for BIM better. The integration is win-win.

For project, the integration can solve problems, which only can be found in construction stage in traditional ways, in an early stage. So it reduces the waste caused by rework and slowdown. At the same time, it guarantees the construction quality by visual design and management of whole life cycle. For enterprises, the integration can change the enterprise internal management mode and organization structure, improving the operation efficiency and competitiveness. For the whole industry, development of BIM is an indispensable content of Lean Construction. BIM plays an important role in the communication between design units and construction units, making the construction higher quality and lower cost. The integration of Lean Construction and BIM provides a new way for sustainable development.

Integration of Lean Construction and BIM solved problem of controlling information flow and quantitating value flow in the process of building. Lean Construction provides an advanced frame work system for BIM. But in the real application, due to the short time of the integration's promotion and the imperfection of the relevant system, further exploration is needed.

References

1. Sacks R, Dave B, Koskela LJ (2009) Analysis framework for the interaction between Lean Construction and building information modelling. In: Proceedings of the 17th annual conference of the international group for Lean Construction, pp 221–233
2. Jiang S, Su Z (2004) Lean Construction: an advanced construction system. Constr Optim 2004, 25(3):11–13 (Chinese)
3. Liu Y, Lu H (2010) Research on the sustainable construction project management under Lean construction system. J Eng Manage 24(4) (Chinese)
4. Liserly J, Wang X (2011) History of BIM. J Archit Creation 6:146–150
5. Lucas J, Thabet W, Bowman D (2009) Analyzing capacity of BIM tools to support data use across project lifecycle. In: Proceedings managing IT in construction/managing construction for tomorrow
6. Ham N-H (2008) A study on application of BIM(building information modeling) to pre-design in construction project. In: Third international conference on convergence and hybrid information technology, ICCIT'08, IEEE, vol 1, 2008
7. Mihindu S, Arayici Y (2008) Outlook of building information modeling within UK construction industry. School of Built Environment University of Salford, Greater Manchester, UK
8. Zhang P, Su Z (2005) Real estate development mode based on concurrent engineering. Constr Optim 26(3):16–19 (Chinese)
9. Zhang K (2009) Application of grey fuzzy comprehensive evaluation method in the research of the port construction project post-evaluation. Port Waterw Eng 4:69–72 (Chinese)
10. Ding G, Chen J (2010) Research on risk assessment for green construction based on the grey system theory. J Eng Manage 2:182–185

Research on Influence Model of High-End Equipment Manufacturing Industry Innovative Performance

De-feng Song, Wei-ming Wang and Hao Chen

Abstract Impact model of innovative performance can found key factors of the innovation performance of high-end equipment manufacturing industry. Firstly, based on previous research of open innovation enterprise, combined with the characteristic of our country manufacturing industry, an initial structural equation model including 21 exogenous manifest variables can be set up to calculate 6 exogenous latent variables. Then, according theory relation between path correction index MI and its variables to correct initial model, and to further narrow the gap between the calculation results and Ideal fit index values, correction index are used to revise the model. The final correction model reveals the structural relations between latent variables. The Research of this paper provides direction of improving innovative performance of High-end equipment manufacturing industry in China.

Keywords High-end equipment manufacturing industry · Innovative performance · Structural equation model

1 Introduction

Under the background of diverse sources of technology, high-end equipment manufacturing enterprise can't also don't need to master all the required technical innovation, using the external technology to implement open innovation is a realistic and effective choice. Based on the innovation efficiency of high-end equipment manufacturing industry shows diverse characteristics in different periods, so we need to explore the causes of this efficiency results in a deeper level, to find the path

D. Song (✉) · H. Chen
Office of Science and Technology Research, NUST, Nanjing 210094, China
e-mail: sdefe@163.com

W. Wang
School of Economics and Management, NUST, Nanjing 210094, China

E. Qi (ed.), *Proceedings of the 6th International Asia Conference on Industrial Engineering and Management Innovation*,
DOI 10.2991/978-94-6239-145-1_102

of how to promote the innovation efficiency of high-end equipment manufacturing industry.

In recent decades, scholars examines the factors that affect enterprise performance and innovation performance from various perspectives, the results show that influencing factors can be divided into three systems: Internal organization affect subsystem, inter-organizational network subsystem and regional innovation subsystem. Meanwhile, Each system includes a number of aspects, but only has considerable limitations about analysis of the influence of different factors, in order to be able to comprehensively receive each factor and the function of these factors of the high-end equipment manufacturing industry innovation efficiency, we need to establish a corresponding model to analyze [1, 2].

This paper is based on the analysis of the high-end equipment manufacturing industry open innovation activities, put forward the initial structural equation model is used to describe the innovation performance, and revised the model according to the results of the calculation, the revised model can reveal the key factors which influencing the high-end equipment manufacturing industry innovation performance, and the quantitative analysis of the degree of their influence on innovation performance.

2 Methodology

2.1 *The Building of Initial Model*

Based on the analysis and research of open innovation in enterprises at home and abroad [3–5], combined with the present situation of the high-end equipment manufacturing industry in our country, this study sets up the initial structural equation model, which is shown in Fig. 1. This model measures six exogenous latent variables (absorption ability, technical skills, openness, external knowledge properties, external cooperation network relations, regional environment) by 21 exogenous manifest variable, Set up four endogenous show variable (new product sales income, patent applications, the condition of product quality improvement and shorten the production cycle, etc.) to measure an endogenous latent variable (innovation performance). In addition, this study also set up a control variable (scale).

2.2 *The Fitting Evaluation of Initial Model*

To estimate whether theoretical model fits the data, it is needed to calculate sample variance covariance matrix and the theory variance covariance matrix S, if the value of the difference between the two covariance matrixes Σ is close to zero, it can be thought the data is reasonable, it means that the model is fitting the data. On the

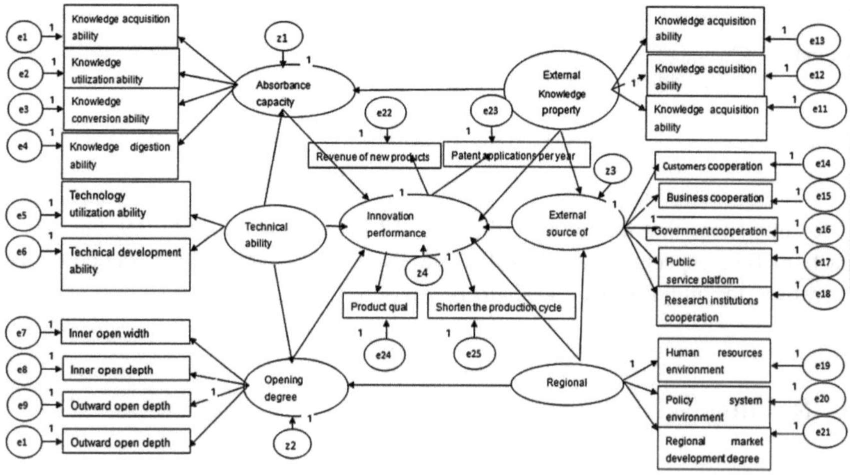

Fig. 1 Initial model

other hand, the fitting degree of the date can be measured by the simulation fit index.

In view of the complexity of the model is different, and the size of the sample is different, it is needed to adopt the corresponding type of fit index to describe the fitting degree of data. Table 1 shows various model fit index which is putting forward by Amos.

In this paper, the calculation results about initial model of the innovation performance influence factors in the high-end equipment manufacturing industry are shown in Table 2.

Table 1 The common fitting index [6]

Index name		Evaluation criterion[a]
Absolute fit index	χ^2	The smaller the better
	GFI	Greater than 0.9
	RMR	Less-than 0.05, the smaller the better
	SRMR	Less than 0.05, the smaller the better
	RMSEA	Less than 0.05, the smaller the better
Relative fit index	NFI	Great than 0.9, the closer to 1, the better
	TLI	Great than 0.9, the closer to 1, the better
	CFI	Great than 0.9, the closer to 1, the better
Information index	AIC	The smaller the better
	CAIC	The smaller the better

[a]The fit index in the form is the optimal criteria, such as for RMSEA, its value is less than 0.05 means model fitting is good, between 0.05–0.08 (Browne and Cudeck 1993) means the model fitting is fair. Therefore, in the actual study, according to the specific situation analysis

Table 2 The common fitting index

Fit index	Chi-square value (degree of freedom)	CFI	NFI	IFI	RMSEA	AIC	BCC	EVCI
Results	1342.4 (180)	0.853	0.839	0.853	0.116	2133.441	2139.378	2.834

Results show that the model CFI, NFI, IFI indexes are close to 0.9, but do not reach to 0.9, especially the RMSEA shows 0.108, beyond the intended scope. Therefore, the initial model needs to be revised.

3 Results

Initial model fitting results not ideal is a common phenomenon in structural equation model, can be modified to improve the model. AMOS software can calculate the Modification Indices (MI), it can provide some useful information about the decrease of the chi-square value fit index. This study will be according to the path of the modified index MI and the theory relationship between variables to modify the model, to reach a better fitting effect.

In the initial structure model, the greatest path of modified index is the influence of external knowledge attributes to absorptive capacity, this may be due to the external knowledge attribute is the existence of an objective fact, and absorptive capacity of enterprise is also an objective phenomenon for the enterprise itself, thus to make the directly affect of external knowledge attributes about the absorptive capacity shows indifference. Therefore, the first step in the model correction, we need first consider to remove the path of external knowledge attributes and absorptive capacity in the structural equation model in this paper, meanwhile, retain the other paths. Table 3 is for the revised results.

The revised model of CFI, NFI, IFI index has certain growth, other parameters including chi-square value has a certain decline, according to the standard of Table 1, it shows that the fit index is improved, but the CFI equivalent does not reach to 0.9, it illustrates that we need a further revision.

In the following paper, it considers to modify the model by modified index, the MI value of e22 and e22 is the maximum, the data is 151.102 (although the e3 and e6 MI value equal to 26.746, but they do not belong to the same latent variable factor, so we can't consider increasing the correlation path, same below), in order to

Table 3 Revised common fit index

Fit index	Chi-square value (degree of freedom)	CFI	NFI	IFI	RMSEA	AIC	BCC	EVCI
Results	929.5 (145)	0.878	0.862	0.884	0.109	1109.541	1014.278	2.274

Table 4 Final model common fit index

Fit index	Chi-square value (degree of freedom)	CFI	NFI	IFI	RMSEA	AIC	BCC	EVCI
Results	690.8 (145)	0.981	0.952	0.977	0.053	345.909	348.402	0.865

further reduce the chi-square values, we can consider the related path about residuals between e22 and e33. In reality, the annual sales revenue of the new products really associated with annual patent applications quantity as a matter of fact, imagine if the new product sales income increased, in order to protect the intellectual property rights or the monopoly position of new products, it is bound to apply for a patent.

After amendment, by calculating we can find that the e18 and e17 MI value is larger, the value is 104.644, same as above, it can increase the relater path about residual between them, it also explains that local policy institutional environment would be correlated with the local combination involving production.

Once again, after amendment, by calculation, we found e7 and e8 MI value is 87.326. Actually, the external resources type which the enterprise have used in the process of innovative is correlated with the frequency of external cooperation network, therefore consider increasing a dependency path between the e7 and e8. The revised final model uses part of the results of Maximum Likelihood Estimation (MLE) in Amos, as shown in Table 4.

Can be seen from Table 4, each final model fit index is improved greatly, it can consider that the data of the model are reasonable.

4 Discussion

By analyzing the model path coefficient, we can receive the relationship between all the variables (latent variables and the measurable variables), for this model, the relationship is as shown in Table 5.

4.1 Direct Effect

It means, because of the external or internal factors, it causes the internal variables, it uses the path coefficient to represent the condition in the model. According to Table 5, it shows that the direct effect of technical ability to absorptive capacity is 0.591, the direct effect of technical ability of the openness to the enterprise is 0.396, the direct effect of firms' innovation performance is 0.311; the direct effect of regional environment to external cooperation network is 0.186, the direct effect of openness to the enterprise is 0.387, the direct effect of firms' innovation

Table 5 The gross effect, direct effect, indirect effect of all latent variable (the standardized consequence)

Effect		Technical capacity	Regional environment	External knowledge property	External cooperation network tie	Opening degree	Absorbing capacity
External cooperation network tie	Direct effect		0.186	0.147			
	Indirect effect						
	Total effect		0.186	0.147			
Opening degree	Direct effect	0.396	0.387				
	Indirect effect						
	Total effect	0.396	0.387				
Absorbing capacity	Indirect effect	0.591					
	Indirect effect						
	Total effect	0.591					
Innovation performance	Direct effect	0.311	0.209	-.064	0.556	0.400	0.882
	Indirect effect	0.521	0.222	-.052			
	Total effect	0.832	0.431	-.116	0.556	0.400	0.882

performance to 0.209; the devoted direct effect of external knowledge property to external cooperation network is 0.147, the direct effect of innovation performance is 0.064; the devoted direct effect of external cooperation network to innovation performance is 0.556; the direct effect of enterprises opening degree to innovation performance is 0.400; the direct effect of absorptive capacity to innovation performance is 0.882.

From the direct effect of innovation performance, the forward direction influence of absorptive capacity to innovation performance is up to 0.882, it means that the absorption capacity increases 1 units, it can promote the innovation performance to increase 0.882 units. Secondly, it is openness, external cooperation network investment, enterprise technical ability, the regional environment respectively.

4.2 Indirect Effect

It means that, by the change of causal variable (external or internal factors), caused a change of the intermediate variable, finally it has influenced the outcome variables (internal variables). For the indirect effect which only has an intermediate variable, its value is the product of two path coefficient. As shown in Table 5, the influence of technical ability to absorptive capacity is 0.591, the influence of absorptive capacity to innovation performance is 0.882, so the influence of technical ability to impact on innovation performance is $0.591 \times 0.882 = 0.521$. Similarly, when other condition is constant, the latent variables of "external knowledge attribute" each increasing 1 units, it will promote the "external cooperation network tie inputs" enhancing 0.147 units, " innovation performance" will improve 0.052 units; Increase 1 unit of "regional environment", will prompt "external cooperation network tie inputs" increases 0.186 units, at the same time, it will lead to "opening degree" increases 0.387 units, which will indirectly from two aspects to promote the "innovation performance" 0.067 units and 0.1548 units respectively, so the total indirect effect of "regional environment" to the "innovation performance" is 0.222.

From the indirect effect of innovation performance, first of all, the indirect effect of technical capacity affecting the absorptive capacity and then influencing the innovation performance is the greatest, secondly, regional environment through both external cooperation network tie and opening degree influences the innovation performance indirectly, finally, external knowledge attributes through the external cooperation network tie inputs to affect the innovation performance indirectly.

4.3 Total Effect

Synthesize the above two kinds of effects, we can get the ultimate influence about the causal variables to the outcome variables. According to Table 5, the direct effect of technical capability to innovation performance is 0.311, indirect effect is 0.521, the total impact is 0.832. By the same token, the total effect of regional environment to innovation performance is 0.431, the general impact of external knowledge attribute, external cooperation network tie investment, opening degree, and enterprise absorptive capacity to innovation performance is 0.116, 0.556, 0.400 and 0.882 respectively.

5 Conclusion

Seen from the total effect of the innovation performance, the sequence of influence degree is absorption ability, technical skills, external cooperation network tie, regional environment, opening degree and external knowledge attributes which from big to small.

References

1. Chen A (2008) The restriction and breakthroughs of self-dependent innovation in China's equipment manufacturing industry—based on analysis of the competitive perspective in global value chain. Nanjing Coll J (Philos Human Stud Soc Sci) 45(1):10
2. Chen Q (2009) Research summary of open innovation about foreign enterprise. Sci Technol Progr Countermeas 26(23):5
3. Xu B, Li X, Zhang X (2011) The mechanism research of the regional environment impact to the enterprise innovation performance. Sci Res Manage 9:147–156
4. Chen Y, Chen Q (2010) The mechanism research about the growth of the small and medium-sized scientific and technical enterprises which is based on the constitutive equations. Sci Sci Technol Manage 04:156–161
5. Wang C (2009) Knowledge property, network characteristics and enterprise innovation performance. Shandong University, Shandong
6. Hou J, Wen Z, Cheng Z (2004) Structural equation model and its applications. Educational Science Press, Beijing

Using Multi-interest Model to Enhance Collaborative Filtering Performance

Yang Zhou, Jin Tian and Minqiang Li

Abstract Personalized recommender systems have become an essential part of online businesses since hundreds of thousands of personalized services and products are provided to customers over the Internet. Capturing customer interests is a critical task for designing recommendation methods. Usually, customers have multiple interests at one period of time. In this paper, a novel collaborative filtering (CF) method is proposed by considering user's multiple interests. We firstly build a novel user multi-interest model by using clustering technique. Then, we combine the multi-interest model with CF method. Finally, the top-N strategy is employed to generate the recommendation list. Experiment results show that the proposed method can achieve higher precision, recall and F1 compared with some conventional CF methods.

Keywords Collaborative filtering · Multiple interests · Personalized recommendation · E-commerce

1 Introduction

The bloom of the internet has changed the way we touch the world. Obtaining information seems to be much easier than before. However, we are also suffering from serious information overload problem [1]. When we search information online, we have to face thousands of choices simultaneously. To get the information we want, we need spend huge amount of time filtering out irrelevant information. The recommender systems are designed to help users find what they like quickly [2]. By analyzing the historical behaviors of users, the recommender systems can identify the current interests of users and recommend products that can match users'

Y. Zhou · J. Tian (✉) · M. Li
College of Management and Economics, Tianjin University,
Tianjin, People's Republic of China
e-mail: jtian@tju.edu.cn

E. Qi (ed.), *Proceedings of the 6th International Asia Conference on Industrial Engineering and Management Innovation*,
DOI 10.2991/978-94-6239-145-1_103

interests [3]. Many websites like Amazon or Netflix have been integrated with recommender systems and can provide personalized services and products. Among various recommendation methods, the collaborative filtering (CF) is one of the most popular methods [4, 5]. The basic idea behind CF is that the suggestions recommended from similar users will match the active user's taste [6].

However, users tend to own multiple interests simultaneously in some situations. For instance, a user may like comedy movies, horrible movies and science fiction movies. In traditional CF methods, the similarity between users is computed globally which may ignore local but strong and valuable similarity [7]. If this user rates many comedy movies and few horrible and science fiction movies, the ratings on comedy movies will dominate the similar neighbor selection process in traditional CF methods. In another word, users with similar ratings on comedy movies are more likely to be chosen as neighbors for this user compared with users with similar rating patterns on the other two movie types. As a consequence, the recommendations provided by these neighbors may not cover the horrible movies and science fiction movies. To mitigate this problem, the user's multiple interests should be involved in the design of recommendation algorithm.

There have been several studies that considered user's multiple interests. Yu et al. refined the traditional user-based CF [8] and involved the concept of multiple interests in their recommendation model. They thought users had multiple interests and items contained multiple contents. They believed that the computation of similarity between active user and other users should be only based on items similar to the target item. Li et al. proposed a system called Farseer that could figure out the unique and diverse interests of users and deliver content that fits for them [9]. They tried to solve the limitation of traditional CF methods that had a high bias on popular content and neglected the low popular ones. This method constructed a user interest model but it relied partly on the content data which may not be obtained easily. The matrix factorization method derives user latent factors and item latent factors from the rating data. Each latent factor represents a character of user or item [10]. This is another way to present user's multiple interests, but the latent factors are often hard to explain.

In this article, we present a novel CF model by considering user's multiple interests (MCF) to enhance recommendation performance. Firstly, we apply SOM neural network to cluster items into several categories. The rating on items in a category can represent an interest of the user. Then, we design a novel method to measure user's preference in multiple categories by using the information of rating score and rating amount. When selecting neighbors, we modify the traditional method by considering user's preference instead of rating similarity. Finally, the top-N strategy is employed to generate the recommendation list.

The remainder of this paper is organized as follows: Sect. 2 describes the details of the MCF algorithm. The experiments and results are discussed in Sect. 3. Section 4 is about the conclusion and future work.

2 Methodology

The proposed method contains three stages. At stage 1, we build user's multi-interest model. It includes the item clustering process and the preference computation process. At stage 2, based on user's each interest, we select the corresponding neighbors. At stage 3, the recommendations are generated by these neighbors.

2.1 Item Clustering

First, we cluster items in the system into groups. We use the self-organizing map (SOM) neural network [11] to cluster items. SOM has an input layer and an output layer. The weight of the neurons on the output layer can be tuned according to the input. Finally, the neurons on the output layer can reflect the distribution of the input data.

Let U represent the set of users, I denote the set of items, and C stands for the set of clusters derived from the SOM. Each input of SOM represents all users' ratings for an item i. When the SOM converges, each item belongs to a cluster c. Let I_c be the set of items in cluster c. The items in the same cluster usually present similar characteristics because of all users' similar rating patterns for them. So, we regard each cluster c as a category that users may prefer. Based on it, we can build user's multi-interest model.

2.2 Preference Computation

After clustering the items into categories, we design a method to compute user's preference for each category. In our method, the user's preference is classified into local preference and global preference. The former aims to evaluate a user's preference for different categories, while the latter tends to compare the preference between different users for a certain category.

The computation of preference depends on the value of ratings and the number of ratings since both of them can reflect the preferences of users to a certain category. Thus we combine them to design the final local preference.

Before combination, we normalize these two parts and make them range from 0 to 1. The number of ratings is normalized as follows and is dominated as local coverage.

$$local\_cover_{u,c} = \frac{M_{u,c}}{M_u} \tag{1}$$

where $M_{u,c}$ is the number of items in category c that user u has given ratings and M_u is the total number of items that user u has rated.

Moreover, the average value of ratings is normalized as follows:

$$n\_\bar{R}_{u,i\in I_c} = \frac{\bar{R}_{u,i\in I_c}}{R_{MAX}} \tag{2}$$

where $\bar{R}_{u,i\in I_c}$ denote the average rating score of user u for items in category c, and R_{MAX} is the max value of ratings in the recommender system. In this article, R_{MAX} is 5.

Thus, the local preference can be computed as the average of the two parts:

$$local\_pref_{u,c} = \frac{local\_cover_{u,c} + n\_\bar{R}_{u,i\in I_c}}{2} \tag{3}$$

A similar concept of global coverage can be defined as follows, which aims to derive the preference strength of a user among all users that have rated in category c.

$$global\_cover_{u,c} = \frac{M_{u,c}}{\max\limits_{u\in U} M_{u,c}} \tag{4}$$

where $\max\limits_{u\in U} M_{u,c}$ is the max number of ratings that users rate in category c.

Thus the global preference is designed as follows:

$$global\_pref_{u,c} = \frac{global\_cover_{u,c} + n\_\overline{R_{u,i\in I_c}}}{2} \tag{5}$$

2.3 Neighbor Selection

The neighbor selection process is different from traditional CF methods, which usually identify neighbors by computing the similarity with some similarity measure, such as Pearson correlation [12], on common ratings. However, the proposed MCF method doesn't select neighbors based on the common ratings, but based on user's multiple preferences directly. Here we use global preference to determine neighbors because we need to compare the preference between different users. We assume that if there is a big difference between two users' preference, they won't be regard as similar to each other. So the neighbors will be the users whose preferences are near to the active user's preference. Let $global\_pref_{u,c}$ denote the active user's global preference for category c. A constant β is used to tune the similar preference range and $\beta \in [0, 1]$. A user is regarded as a candidate of neighbors only if his global preference in category c is in the range of $[global\_pref_{u,c} - \beta, global\_pref_{u,c} + \beta]$.

The candidates whose preferences are low mean that they tend to rate less items, or give low rating scores. The recommendations from them may not match the active user's interests. In comparison, the candidates with high preferences have more chances to recommend what the active user like.

The neighbor candidates are ranked according to their global preferences and the top-k neighbors are finally selected.

2.4 *Top-N Recommendation*

For each item i in category c we calculate a score.

$$score_{u,i,c} = \frac{\sum_{v \in N_{u,c}} global\_pref_{v,c} \times R_{v,i}}{\sum_{v \in N_{u,c}} global\_pref_{v,c}} \tag{6}$$

where $N_{u,c}$ is the neighbor set of user u in category c. and $R_{v,i}$ is the rating to item i rated by a neighbor, user v. $global\_pref_{v,c}$ is the global preference of user v.

In order to compare the items in different categories, we multiply these score with the active user's local preference to get the final score.

$$fins_{u,\mathrm{i}} = local\_pref_{u,c} \times score_{u,\mathrm{i},c} \tag{7}$$

Finally, we choose the top N items according to their final scores and recommend to the active user.

3 Experiments and Results

3.1 *Dataset*

Experiments are conducted on Baidu movie recommendation dataset to testify the recommendation performance of the proposed method. Baidu dataset contains users' ratings for different movies. The items in the dataset are about 8,000 and we randomly select 8,000 users to conduct experiments. The average rating of each user is about 140.

We randomly choose 20 % of the ratings as test set. The rest of the ratings are set as training set. The experiments are conducted 10 times and the average results are recorded.

3.2 Evaluation Metrics

The precision, recall and F1 are often used to evaluate the recommendation accuracy [13]. In the experiments, if an item is in the test set of user u and the assigned rating is no less than 4, this item is regarded as relevant to user u. Let L be the length of the recommendation list. The precision for user u is calculated as follows:

$$Precision_u(L) = \frac{|hit_u(L)|}{L} \tag{8}$$

where $hit_u(L)$ is the recommendation hit list of user u and refers to the items relevant to user u in the recommendation list.

The precision for all users is computed as follows:

$$Precision(L) = \frac{1}{|U|}\sum_{u \in U} Precision_u(L) \tag{9}$$

The recall for user u is defined as:

$$Recall_u(L) = \frac{|hit_u(L)|}{|Rel_u|} \tag{10}$$

where Rel_u is the items in the test set that are relevant to user u.

The recall for all users is computed as follows:

$$Recall(L) = \frac{1}{|U|}\sum_{u \in U} Recall_u(L) \tag{11}$$

F1 score is the tradeoff between precision and recall.

$$F_1(L) = \frac{2 \times Precision(L) \times Recall(L)}{Precision(L) + Recall(L)} \tag{12}$$

Moreover, there are other important metrics to evaluate the recommendation performance. The metric of coverage stands for the percentage of recommended items that a recommendation model offers [14]. The coverage in this paper is computed as:

$$Coverage(L) = \frac{\left| \bigcup_{u \in U} R_u(L) \right|}{|I|} \tag{13}$$

where $R_u(L)$ is the items in the recommendation list for user u and $|I|$ is the total number of items in the dataset.

To measure the recommendation diversity, we use the method presented in [15]. Let $s(i,j)$ be the Pearson correlation between item i and j. Then the dissimilarity between them is defined as $d(i,j) = 1 - s(i,j)$. The diversity of recommendation for user u is defined as Eq. (14) and the total diversity is computed in Eq. (15).

$$D_u = \frac{1}{L(L-1)} \sum_{i \in R_u(L)} \sum_{j \in R_u(L), j \neq i} d(i,j) \tag{14}$$

$$D = \frac{1}{|U|} \sum_{u \in U} D_u \tag{15}$$

To measure the "non-obviousness" of the recommendation, we use the novelty metric defined in [16]. The popularity of item i is defined as follows:

$$p_i = \frac{pop_i}{|U|} \tag{16}$$

where pop_i is the number of users who rate item i, and $|U|$ is the total number of users in the dataset.

The novelty for user u can be computed in Eq. (17) and the total novelty is defined in Eq. (18).

$$pnov_u = -\sum_{i \in R_u(L)} p_i \times log(p_i) \tag{17}$$

$$nov = \frac{1}{|U|} \sum_{u \in U} pnov_u \tag{18}$$

3.3 Experiments Setup

In the experiments, we compare the MCF with the traditional user-based CF and item-based CF.

The neighbor number of user-based CF and item-based CF is set as 20. In comparison, the neighbor number in the MCF is also set as 20. The number of clusters is set as 40 and β is set as 0.1.

3.4 Recommendation Accuracy

Figure 1 presents the recommendation performances of the three methods. As the figure shows, the proposed MCF outperforms both the user-based CF and

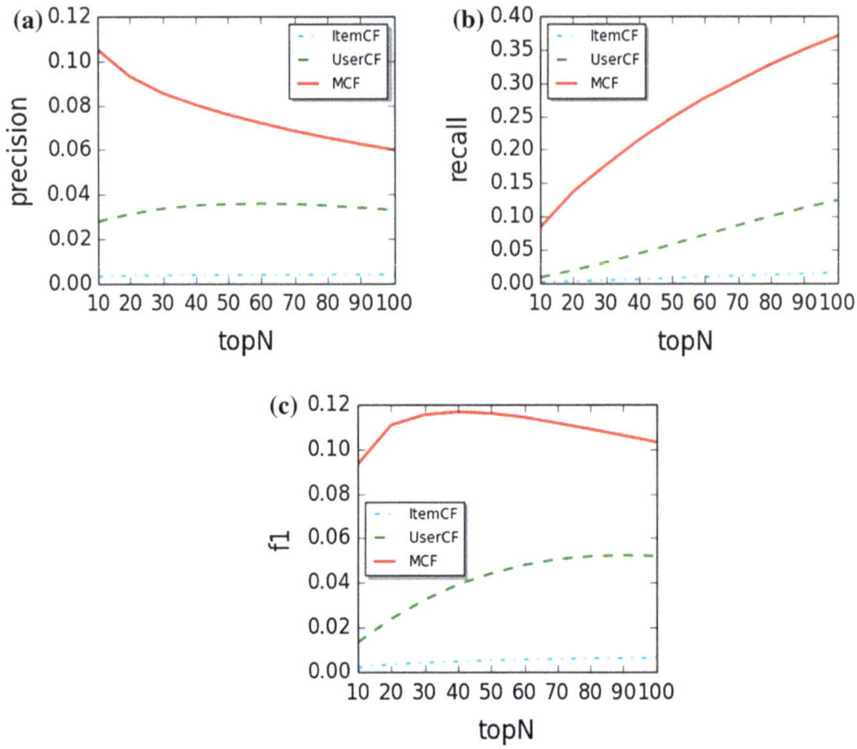

Fig. 1 Comparison with ItemCF and UserCF on precision, recall and F1. **a** Precision. **b** Recall. **c** F1

item-based CF. For the precision, although the precision of MCF declines slightly as the value of N increases, there is still a huge gap between MCF and the other two methods. For the recall, we can notice that as the recommendation list becomes longer, the recall turns to be higher. And also MCF performs much better than the other two methods. Finally, we focus on the results of F1 score. The F1 score is the combination of precision and recall. So it can reflect the comprehensive accuracy of each method. As the result shows, the MCF performs best in the three comparison methods. Thus we can conclude that the MCF model can indeed improve the recommendation performance.

3.5 Coverage, Diversity and Novelty

We further conducted experiments to evaluate the performance of the three methods on coverage, diversity and novelty. For simplicity, we only list the results on top 100 in Table 1.

Table 1 Performance on coverage, diversity and novelty

Algorithms	Coverage	Diversity	Novelty
ItemCF	0.9906	0.4926	2.539
UserCF	0.7566	0.4309	15.99
MCF	0.6130	0.4285	26.21

As the result shows, the three methods present three different tendencies. The item-based CF gets the highest coverage and diversity, but obtains the lowest novelty. Compared to item-based CF, user-based CF gets lower coverage and diversity while the novelty of user-based CF is higher than item-based CF. The proposed method achieves the highest novelty. Basing on the performance on accuracy stated before, we may deduce that novel items can match user's multiple interests and lead to more accurate recommendations.

4 Conclusion

In this paper, we design a novel CF model by considering user's multiple interests. We firstly use SOM neural network to cluster items into different groups and regard each group as an item category. Then we combine the information of user's rating score and rating amount to evaluate user's preference for each category. This is a way to represent user's multiple interests. For user's each interest, we select neighbors by user preference and use them to make predictions for items. Finally, we get all the predictions together and generate a top N recommendation list. We conducted experiment on real-world dataset to testify the performance of the proposed method. The results show that the proposed method has better recommendation accuracy than traditional user-based CF and item-based CF. This implies that the involvement of multi-interest model really enhance the recommendation performance. Besides, the novelty of the MCF is much higher than the two comparison methods, which means the MCF can generate more "non-obvious" recommendations to the active user.

Many possible extensions can be pursued relative to our work in this paper. In some situations, user's interests tend to change over time. One possible future work is to extent the proposed multi-interest model into a changeable interest scenario by involving the temporal information. Several studies have stated that user may own similar interests with social friends. Thus, how to using social interactions to improve the proposed model is also one of the possible future work.

Acknowledgment The work was supported by the General Program of the National Science Foundation of China (Grant No. 71471127, 71371135, 71001076, 71101103, and 71271149). Authors are very grateful to the editor and all anonymous reviewers whose invaluable comments and suggestions substantially helped improve the quality of the manuscript.

References

1. Li Y, Wu C, Lai Y (2013) A social recommender mechanism for e-commerce: Combining similarity, trust, and relationship. Decis Support Syst 55(3):740–752
2. Zhao X, Niu Z, Chen W (2013) Interest before liking: two-step recommendation approaches. Knowl Based Syst 48:46–56
3. Kim HN, Ha I, Lee KS, Jo GS, El-Saddik A (2011) Collaborative user modeling for enhanced content filtering in recommender systems. Decis Support Syst 51(4):772–781
4. Carrer-Neto W, Hernández-Alcaraz ML, Valencia-García R, García-Sánchez F (2012) Social knowledge-based recommender system: application to the movies domain. Expert Syst Appl 39(12):10990–11000
5. Martín-Vicente MI, Gil-Solla A, Ramos-Cabrer M, Blanco-Fernández Y, López-Nores M (2012) Semantic inference of user's reputation and expertise to improve collaborative recommendations. Expert Syst Appl 39(9):8248–8258
6. Huete JF, Fernández-Luna JM, Campos LM, Rueda-Morales MA (2012) Using past-prediction accuracy in recommender systems. Inf Sci 199:78–92
7. Cantador I, Castells P (2011) Extracting multilayered communities of interest from semantic user profiles: application to group modeling and hybrid recommendations. Comput Hum Behav 27(4):1321–1336
8. Yu L, Liu L, Li X (2005) A hybrid collaborative filtering method for multiple-interests and multiple-content recommendation in e-commerce. Expert Syst Appl 28(1):67–77
9. Li D, Lv Q, Xie X, Shang L, Xia H, Lu T, Gu N (2012) Interest-based real-time content recommendation in online social communities. Knowl Based Syst 28:1–12
10. Koren Y, Bell R, Volinsky C (2009) Matrix factorization techniques for recommender systems. Computer 42(8):30–37
11. Kohonen T (1990) The self-organizing map. Proc IEEE 78(9):1464–1480
12. Lü L, Medo M, Yeung CH, Zhang YC, Zhang ZK, Zhou T (2012) Recommender systems. Phys Rep 519(1):1–49
13. Bobadilla J, Ortega F, Hernando A, Gutiérrez A (2013) Recommender systems survey. Knowl Based Syst 46:109–132
14. Herlocker JL, Konstan JA, Terveen LG, Riedl JT (2004) Evaluating collaborative filtering recommender systems. ACM Trans Inf Syst 22(1):5–53
15. Hurley N, Zhang M (2011) Novelty and diversity in top-n recommendation–analysis and evaluation. ACM Trans Internet Technol 10(4):1–14
16. Bellogín A, Cantador I, Castells P (2013) A comparative study of heterogeneous item recommendations in social systems. Inf Sci 221:142–169

Entrepreneurial Passion Definition, Antecedent, Outcome, and Model Building

Xi-feng Lu, Biao-an Shan and Biao Chen

Abstract The study analyzes the definition and finds that entrepreneurial passion' dimensions include passion for developing, passion for founding, passion for inventing. Entrepreneur's individual characteristics (gender and previous experience etc.), identity and self-perception are key affecting factors of entrepreneurial passion. This study also finds that entrepreneurial passion has important influence on opportunity development, resources acquisition and employee commitment. Based on existing research, we build a model of the impact of entrepreneurial passion on entrepreneurial activity and propose future research direction. Particularly, we emphasis on the important role of entrepreneurial learning and entrepreneurial skills.

Keywords Entrepreneurial passion · Identity · Entrepreneurial emotion

1 Introduction

Most of literatures on entrepreneurship are based on Timmons Model (Timmons 1999) [1], which relied on entrepreneur/team, opportunity and resource to explain the process of new venture creation, growth and failure. However, whether the research on entrepreneur/team or on opportunity and resource, that both neglected the entrepreneurial emotion. Most of current researches suggest that entrepreneurship is uncertain (Cai and Shan 2013) [2], such as Cardon et al. (2005) remarked [3], entrepreneurship is mysterious and unpredictable, which is influenced by many irrational emotion factors. The uncertainty of entrepreneurial activity due to the

X. Lu · B. Shan (✉) · B. Chen
School of Management, Jilin University, Changchun, Peoples Republic of China
e-mail: shanbiaoan@163.com

B. Shan
Postdoctoral Research Station of Management Science and Engineering,
Jilin University, Changchun, Peoples Republic of China

E. Qi (ed.), *Proceedings of the 6th International Asia Conference on Industrial Engineering and Management Innovation*,
DOI 10.2991/978-94-6239-145-1_104

rapid change of external environment, the asymmetry of information, the unpredictable nature of entrepreneurship difficulties, and so on, may lead to the high failure rate of entrepreneurial activities (Hu et al. 2010) [4]. So, uncertainty, failure and frustration are along with entrepreneurial activities (Zhu et al. 2011) [5], which requires entrepreneurs have strong and continuous motivation to conduct entrepreneurship practice (Shane et al. 2003) [6]. The motivation is entrepreneurial passion, which is a kind of strong and proactive intention and identity, and it is manifested by entrepreneur during the process of creating new venture, and developing new venture (Cardon et al. 2013) [7].

Entrepreneurial passion is considered to be a key element to push forward entrepreneurial activity, especially when the environment is uncertain and the resource is limited (Chen et al. 2009) [8]. We can easily find many entrepreneurs are passionate. Passion is important for entrepreneurs to take risk, recognize opportunity, stimulate creativity, attract key employees and investment (Foo 2011) [9]. Furthermore, although passion is important for the creation and growth of new venture, there are some gaps for current researches, such as the concept, affecting factors, and the affecting mechanism of entrepreneurial activity. Therefore, based on previous studies, this study attempts to systematically summary the concept, dimensions, antecedents and consequences of entrepreneurial passion and tries to build a model to reveal the role of entrepreneurial passion on entrepreneurial activity.

2 Concept of Entrepreneurial Passion

Entrepreneurial passion derives from the conception of "passion" in psychology field. Baron and Hannan (2002) defined entrepreneurial passion as belonging and self-identity of individuals to new ventures [10]. Breugst et al. (2012) suggested that entrepreneurial passion is intense feeling that entrepreneurs consciously participate in entrepreneurial activities [11], which closely associates with self-identity; Like "a fire of desire", entrepreneurial passion of successful entrepreneurs is particularly obvious, which promotes them to engage in daily work with passion. Chen et al. (2009) argued that passionate entrepreneurs have intense positive feelings for desired goals, will not stop thinking and discussing their ideas, and are busy with motivating resources to convert their ideas into reality [8]. Compared to entrepreneurs who lack passion, passionate entrepreneurs usually have "cartoon-like" expressions, energetic body movement, and abundant body language (Klaukien et al. 2010) [12].

Based on reviewing the existing literatures, Cardon et al. (2005) found that scholars' understanding of the conception of entrepreneurial passion can be summarized as follows: passion is the whole or part of intensive feelings with various differentiated and complicated feelings, which aims at or around a specific goal, and has motivational effects [3]. The viewpoint of Cardon clearly put forward that passion is not only a intensive postive feeling, but also the positive feeling which is expressed by something that is meaningful for self-identity of individuals.

Subsequently, based on exisiting researches, Cardon et al. (2005) analysed the conception of entrepreneurial passion in depth, and defined it as the intensive positive feeling and self-identity of entrepreneurs when they explore new approach, found and develop new ventures [3]. And they suggested three types of entrepreneurial passion: the passion for inventing new things; the passion for creating new ventures and the passion for developing new ventures (Cardon et al. 2013) [7]. These three types of etrepreneurial passion comprehensively summarize the connotation of entrepreneurial passion, and subsequent empirical studies also show that they have significant discriminant validity.

Firstly, passion for inventing. Entrepreneurs who are passionate about inventing new things are good at observing external market environment to search for new opportunities, and are enthusiasm about developing new products and pursuing new ideas persistently (Laaksonen et al. 2011) [13], thereby discovering new opportunities that can't be recognized by others (Grichnik et al. 2010) [14]. These passionate entrepreneurs often have innovative thinking, and consider more comprehensively. Secondly, passion for founding. Passion for founding is the passion that entrepreneurs experience when they take advantage of opportunities to commercialize (Cardon et al. 2013) [7]. Entrepreneurs who are passionate about founding new ventures often utilize a variety of ways to acquire and intergrate critical resources to participate in activities associated with founding new ventures, such as capital and human resources (Baum and Locke 2004) [15]. Thirdly, passion for developing. Passion for developing refers to the passion entrepreneurs experience when they explore new market positively and create value for customers to continuously expand new ventures once the venture has been created (Cardon et al. 2013) [7].

3 Antecedents of Entrepreneurial Passion

The most important distinction between entrepreneurial passion and emotion is that the former is continuous, not a temporary state (Cardon 2008) [16]. Based on current researches, factors that impact entrepreneurial passion is composed of individual characteristics (such as gender and previous experience), identity and self-perception of entrepreneur. So, this part we intend to introduce the current researches on the affecting factors of entrepreneurial passion, which consists of entrepreneur's individual characteristics, identity and self-perception of entrepreneur.

Firstly, individual characteristics of entrepreneur. Gender and previous experience are typical individual characteristics that affect entrepreneurial passion. On one hand, Cardon et al. (2013) pointed out [7], compared to female entrepreneur, male entrepreneur is more aggressive, and the stronger need for achievement makes male entrepreneur more passionate. On the other hand, some scholars suggest that people who have entrepreneurial experience may have more entrepreneurial passion than those don't have, because they know more about the creation procedure and resource requirement of new venture. So, people who have entrepreneurial experience may express more entrepreneurial passion.

Secondly, identity. Many scholars agreed that there is a close relationship between entrepreneurial passion and identity (Murnieks et al. 2012) [17]. The identity reflects a kind of self-understanding of his role as a "entrepreneur", and is presented as a action norms in entrepreneurial activity (Murnieks et al. 2012) [17]. Therefore while one realize that the role as an entrepreneur is more important than other role, he will focus on this role and try to achieve it (Farmer et al. 2011) [18]. So this identity make entrepreneur show more entrepreneurial passion in the entrepreneurial process.

Thirdly, self-perception. There is some distinction between entrepreneur's identity and self-perception, such as the former reflects that one's understanding and attention level to the role as a entrepreneur, and the latter reflects one's assessment that whether he can become an entrepreneur (Farmer et al. 2011) [18]. McMullen and Shepherd (2006) pointed out [19], when one finds a chance, he will conduct a self-perception, or he may stay out and miss the chance. Thus, once one's self-perception fits with its role norms as an entrepreneur in a situation, it will stimulate his passion to become an entrepreneur (Farmer et al. 2011) [18].

4 Consequences of Entrepreneurial Passion

In entrepreneurial process, entrepreneur may face high uncertain and need to overcome various difficulties, such as the lack of financial resources and intense market competition. So why some can persist, but others give up? We regard entrepreneur's entrepreneurial passion as the key factor to explain the phenomenon (Zhu et al. 2011) [5]. Entrepreneurial passion may increase the entrepreneurial interest, even make entrepreneur take new venture as part of him. This helps entrepreneur persist on his behavior and keep high enthusiasm (Cardon et al. 2005) [3]. So entrepreneurial passion is important in the entrepreneurial process, especially for opportunity development, resources acquisition and employee commitment.

The role of entrepreneurial passion on the opportunity development process. Entrepreneurial passion is the key affecting factor in the opportunity recognition, because it makes entrepreneurs have stronger capability that can help them solve problems creatively. The capability helps entrepreneur explore new things actively and pursue potential market opportunity (Wilson and Stokes 2005) [20]. Meanwhile, passion can make entrepreneur more sensitive to external environment and gather information unconsciously to find if the market opportunity exist. This will help entrepreneur recognize opportunity. So, entrepreneur's entrepreneurial passion will encourage him to recognize and develop opportunity (Farmer et al. 2011) [18].

The role of entrepreneurial passion on resource acquisition. Resource is the key factor to put entrepreneurship forward in opportunity development process, and how to acquire resource is the obstacle that entrepreneur need to overcome, especially the shortage of resources, funding strand breaks which decide whether entrepreneurship can success. Entrepreneur with entrepreneurial passion will do everything he can to solve these problems in creative ways, such as acquire financial resource, human resource and social resource. In their research, Baum and Locke (2004) pointed [15], entrepreneurial passion can stimulate entrepreneur's continuous learning to promote entrepreneurial skills, and acquire resource to conduct entrepreneurial practice.

The role of entrepreneurial passion on employee commitment. Employee commitment is one of the most important determinants to decide company's success, especially for new venture, which needs employee's support (Baron and Hannan 2002) [10]. The entrepreneurial passion of entrepreneur is the key factor that influences employee commitment. Gross (1998) noted that entrepreneur's passion may affect others' behavior in the same environment [21]. Cardon (2008) found that entrepreneur's passion can pass to employee through emotional mimicry and social comparison [16]. Entrepreneurs with entrepreneurial passion can express attention and enthusiasm to new venture, and pass it to employee, which makes employee perceive entrepreneur's determination that he will success. And then, employees will work harder (Klaukien et al. 2010) [12].

The role of entrepreneurial passion on other entrepreneurial activities. Apart from the affect on opportunity development, resources acquisition and employee commitment, entrepreneurial passion also may affect perceived difficulty, the growth of new venture, the goal-setting and the communication with others. Such as Grichnika et al. (2010) suggested that strong and positive emotions may decrease entrepreneur's perceived difficulty [14], and make he be optimistic. Baum and Locke (2004) found that entrepreneurial passion can prompt entrepreneur to communicate with others and affect new venture growth.

5 Construct the Model of the Impacts of Entrepreneurial Passion

Current researches focus more on whether entrepreneurial passion has impacts on entrepreneurial activities, ignored the underlying mechanism. This is the important issuethat research on entrepreneurial passion can not be avoided, and it also reveals the role of irrational factors. Entrepreneurial passion is belonging to interdisciplinary field, involveing the field of management science and psychology, which makes it difficult for scholars to research on it. For practice, successful entrepreneurs often show great passion in entrepreneurial activity. It is the passion that leads them to engage in daily work with great enthusiasm, and infect people around them (including employees). In founding stage, the role of entrepreneurial passion is more salient. Resource scarcity, competition, high uncertainty of enviroment and lack of entrepreneurial experience make entrepreneurs face difficulties which were never been considered before. If entrepreneurs have no passion, they lack the belief of persistence, and show helpless in face of difficulty. Thus, it is difficult to obtain the support of resource owner. On the contrary, passion makes entrepreneurs highly motivated, reduces the difficulty, and then make entrepreneurs resolve problems in creative ways. Meanwhile, passionate entrepreneurs will spend more time and energy on entrepreneurial activities, show stronger entrepreneurial capability, and can attract the support of resource owners.

Although the study of entrepreneurial passion has important value whether from theoretical level or practice level, many key problems still need further researches. This research follows the logic: affecting factor-entrepreneurial passion-function route-outcome, built the model of the impacts of entrepreneurial passion (see Fig. 1), and classify entrepreneurial passion into three types: passion for inventing, passion for founding and passion for developing, and then reveal the mechanism of entrepreneurial passion on entrepreneurial output. In terms of affecting factors, this research considers characteristics of entrepreneur, self-identity and self-perception. In terms of function routes, we consider resource/information acquisition, employee commitment, entrepreneurial capability and entrepreneurial learning. In terms of entrepreneurial output, we consider opportunity recognition, founding new venture, developing new venture and performance of new ventures.

The research focuses on the impact role of entrepreneurial learning and entrepreneurial capability on the process, which involves how entrepreneurial passion affect entrepreneurial output. New venture needs to face the high uncertain, liabilities of newness and failure risk. So if they only rely on entrepreneur's previous experience and knowledge and his own entrepreneurial passion is not enough to overcome these problems. Entrepreneurial learning emphasizes the accumulation of knowledge is through ongoing observation, imitation and practice, so it can help entrepreneur adapt to external environment. On the other hand, entrepreneurial capability reflects entrepreneur's capability to recognize opportunity and integrate resource (Man and Lau 2000) [22]. Therefore, entrepreneur needs to equip with appropriate capability to develop opportunity and resource during the creation process of new venture. In sum, entrepreneurial learning and entrepreneurial

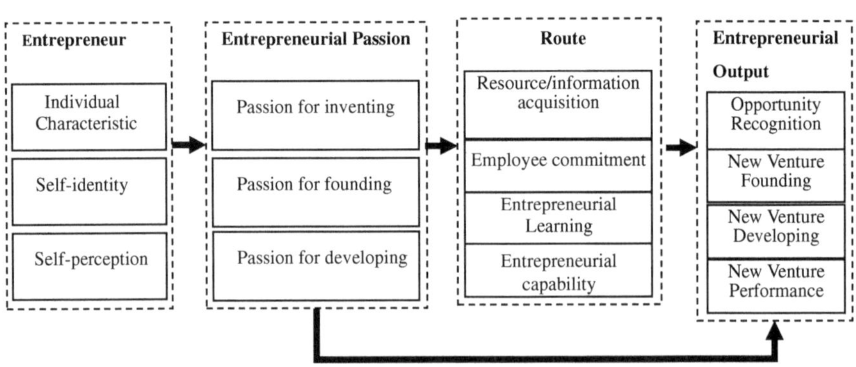

Fig. 1 The model of the impact of entrepreneurial passion on entrepreneurial activity

capability play an extremely important role in the entrepreneurial process, and is the key way to transform entrepreneurial passion into performance, create new venture successfully and promote the growth of new venture.

6 Conclution

Based on above analysis we can note that as important irrational affecting factor in entrepreneurial process, entrepreneurial passion is attracting more and more scholars. As cutting-edge research question in entrepreneurship research field, current researches on entrepreneurial passion have some shortcoming, such as, there is no one studying the in-depth affecting mechanism. So, this study builds a model to reveal the role of entrepreneurial passion on entrepreneurial activity, and points out the direction for future research. First, we should discuss the intrinsic relationships among different kinds of entrepreneurial passion and their dynamic effects on entrepreneurial activity. Second, we should discuss other affecting factors, and research their effect on entrepreneurial passion. Third, future research should explore how entrepreneurial passion affects entrepreneurial learning and entrepreneurial capability. Forth, we should reveal the path that entrepreneurial passion affects the entrepreneurial output. Last, the research method and measurement system needs further expansion. As one important trend in entrepreneurship research field, entrepreneurial passion belongs to cross-disciplinary research, and is important emotion factor that effects on entrepreneurial activity. To explain related phenomenon, we exploit emotion theory from psychology. Therefore, if we solely rely on general conceptual analysis and empirical studies when we research entrepreneurial passion, there may be some shortcoming. And it is the main obstacle of how to study entrepreneurial passion. Future research should exploit experimental research method and consider more situational factors in the research on entrepreneurial passion. Meanwhile we can analyze how it affects information acquisition and opportunity recognition to obtain actual and reliable conclusions.

Acknowledgments This work was supported in part by the National Natural Science Foundation of China (NNSFC) under Grant 71402064 and China Postdoctoral Science Foundation Funded Project (No. 2014M550176).

References

1. Timmons JA (1999) New venture creation: entrepreneurship for 21st century, 5th edn. McGraw-Hill, Homewood, Illinois
2. Cai L, Shan B (2013) Entrepreneurship research under Chinese context. Manag World 12:160–169
3. Cardon MS, Zietsma C, Saparito P (2005) A tale of passion: new insights into entrepreneurship from a parenthood metaphor. J Bus Ventur 20(1):23–45

4. Hu W, Zhang, Yang YJ (2010) Explorering the relationships between entrepreneurial orientation and performance in new venture based on capability. Foreign Econ Manag 2:1–8
5. Zhu F, Chen R, Li M (2011) The divergent effects of autonomy and passion on persistence in new venture creation. Acad Manag Proc 1:1–6
6. Shane S, Locke EA, Collins CJ (2003) Entrepreneurial motivation. Hum Resour Manag Rev 13:257–279
7. Cardon MS, Gregoire DA, Stevens CE, Patel P (2013) Measuring entrepreneurial passion: conceptual foundations and scale validation. J Bus Ventur 28(3):373–396
8. Chen XP, Yao X, Kotha S (2009) Entrepreneur passion and preparedness in business plan presentations: a persuasion analysis of venture capitalists' funding decisions. Acad Manag J 52 (1):199–214
9. Foo MD (2011) Emotions and entrepreneurial opportunity evaluation. Entrepreneurship Theor Pract 35(2):375–393
10. Baron JN, Hannan MT (2002) Organizational blue prints for success in high-tech start-ups: lessons from the Stanford project on emerging companies. Calif Manag Rev 44(3):8–36
11. Breugst N, Domurath A, Patzelt H, Klaukien A (2012) Perceptions of entrepreneurial passion and employees' commitment to entrepreneurial venture. Entrepreneurship Theor Pract 36 (1):171–192
12. Klaukien A, Breugst N, Patzelt H (2010) Entrepreneurs' displays of passion and employees' commitment to new ventures. Acad Manag Proc 1:1–6
13. Laaksonen L, Ainamo A, Karjalainen TM (2011) Entrepreneurial passion: An explorative case study of four metal music ventures. J Res Market Entrepreneurship 13(1):18–36
14. Grichnik D, Smeja A, Welpe I (2010) The importance of being emotional: how do emotions affect entrepreneurial opportunity evaluation and exploitation. J Econ Behav Organ 76(1):15–29
15. Baum J, Locke E (2004) The relationship of entrepreneurial traits, skill, and motivation to subsequent venture growth. J Appl Psychol 89(4):587–598
16. Cardon MS (2008) Is passion contagious? The transference of entrepreneurial passion to employees. Hum Resour Manag Rev 18(2):77–86
17. Murnieks CY, Mosakowski E, Cardon MS (2014) Pathways of passion: identity centrality, passion, and behavior among entrepreneurs. J Manag 40(6):1583–1606 (in press)
18. Farmer SM, Yao X, Kung-Mcintyre K (2011) The behavioral impact of entrepreneur identity aspiration and prior entrepreneurial experience. Entrepreneurship Theor Pract 35(2):245–273
19. McMullen JS, Shepherd DA (2006) Entrepreneurial action and the role of uncertainty in the theory of the entrepreneur. Acad Manag Rev 31(1):132–152
20. Wilson NC, Stokes D (2005) Managing creativity: the challenge for cultural entrepreneurs. J Small Bus Enterp Dev 12(3):366–378
21. Gross JJ (1998) The emerging field of emotion regulation: an integrative review. Rev Gen Psychol 2(3):271
22. Man TWY, Lau T (2000) Entrepreneurial competencies of SME owner/managers in the Hong Kong services sector: a qualitative analysis. J Enterp Cult 8(3):235–254

Pricing Cloud Service Considering Heterogeneous Customers

Shixian Lei, Fuzan Chen and Minqiang Li

Abstract The increasingly popularity of cloud computing paradigm brings out the challenging problem of optimal strategy on pricing cloud services. Most of nowadays studies fail to take the interaction between cloud service providers (CSPs) and customers into consideration. In this paper, the optimal strategy on pricing cloud service in a monopoly market from a long-term and dynamic perspective. We develop a bilevel programming model which aims at maximizing CSP's profit while customer's utility is relatively maximized. In this model, the upper level is a monopoly CSP making decision on price and service rate of tiered services to maximize profit and lower level is all the customers deciding which level of service to buy to maximize their utility. With analytical solution for linear customer utility function, we demonstrate that the optimal strategy for the monopoly CSP is to provide nontier service and the optimal price can be calculated through derived formula under certain conditions.

Keywords Cloud computing · Cloud service · Pricing · Bilevel programming

1 Introduction

With cloud computing paradigm, IT resources such as hardware, software or operating system is provided as commodity through Internet [1]. The emergence of cloud computing paradigm has largely changed both the manner of customers acquiring IT resources and the manner of providers delivering their IT resources. For example, online content provider Netflix is able to quickly deploy thousands of

S. Lei · F. Chen (✉) · M. Li
College of Management and Economics, Tianjin University, Tianjin, China
e-mail: fzchen@tju.edu.cn

E. Qi (ed.), *Proceedings of the 6th International Asia Conference on Industrial Engineering and Management Innovation*,
DOI 10.2991/978-94-6239-145-1_105

servers and thus support seamless global service by partnering with Amazon Web Service (AWS) for services and delivery of content.[1]

With the increasing popularity of cloud computing, pricing cloud service has been becoming crucial in the operation of cloud computing [2]. A number of researches concentrate attention on the optimal strategy of pricing cloud service [3 −7]. On the aspect of fixed pricing strategy, Chun and Choi (2013) present optimal pricing strategies for monopolistic CSPs based on subscription and usage-based schemes considering total cost of services and customer utilities [3]. They indicate that when the total cost is relatively high, it is more profitable if the CSP adopt subscription pricing strategy. He and Sun (2011) have explored the optimal pricing strategy for customized SaaS services with the constraints of fixed cost and presented that the optimal prices are related to the scale of customization [4]. Katzmarzik (2011) illustrates a decision model which aims at maximizing SaaS provider's profits and deduces the optimal granularity through changing granularity of its service based on the idea of versioning [5]. Basu et al. (2015) compare fixed fee pricing plan and usage based pricing plan with the assumption that each customer is affected by two sets of parameter—one is positive while the other is negative [6]. They find that fixed fee plan is better for both CSP and customers who have strong demands. On the aspect of dynamic pricing strategy, Huang et al. (2014) compare the one-service-only strategy (such as fixed-price reserved service only, spot-price on-demand only) and a hybrid pricing strategy which combines the fixed price and spot price from the perspective of CSP profit, customer surplus and social welfare [7]. They find that in most cases, a hybrid pricing strategy outperforms one-service-only strategy. Even though this stream of researches considers a short-term dynamic pricing, they fail to consider the long-term interaction between provider and customers.

Moreover, customers' willingness to pay (WTP) has great impact on the selection of the appropriate pricing plan and thus indirectly affects providers' profit [6]. Zhang and Niu (2014) develop a dynamic quality decision model considering heterogeneous customer perception [8]. When modeling customers` perception, they incorporate two effects of different customers' behaviors which consist of the experience of service quality and negative time preference. In our context, we follow their steps and characterize heterogeneous customers by different perception of service quality.

Furthermore, tiered strategy [9, 10], characterized as a set of tiers offering gradually higher levels of service, has become an important scheme used by providers or investor to satisfy customers' diverse requirements and eventually expand their market share or maximize their profits. Under the application of tiered strategy, customers are encouraged to subscribe the best appropriate level of service among a set of tiers with gradually growing capacities and prices [9, 11]. Either in respect to cost or from the perspective of efficiency, it is essential issue in the business of cloud service to optimize the capacity partitioning and price the tiered services.

[1]http://aws.amazon.com/cn/solutions/case-studies/netflix/.

As stated, pricing and capacity planning of cloud service are significant for the improvement and expansion of cloud computing. This study aims to investigate the optimal strategy on pricing cloud service in a monopoly market from a long-term and dynamic perspective, i.e., the equilibrium price of the cloud service should be negotiated by the CSP and customers. First of all, the pricing interaction between the monopoly CSP and all the customers is formulated as a bilevel programming model where price of each tier as well as its service rate is designated as a decision variable to maximize the CSP's profit while each user maximizes his/her utility. Either a non-linear cost function of CSP or a utility function for customer is designed. Finally, optimal pricing strategy is reached through analytical solution.

The rest of the paper is organized as follows. Section 2 proposes a bilevel programming model for pricing cloud services. And then we investigate the optimal pricing strategy in a monopoly market where customers' utility function is linear in Sect. 3. Section 4 concludes the paper and indicates some future work.

2 Economic Model of Pricing Cloud Services

2.1 *Bilevel Programming Model of Pricing Cloud Services*

Consider a real scenario of a monopoly cloud service market. In such a market, the monopolist announces their services and the corresponding prices online at first. We present some assumptions for the services, including:

(1) There are $K(K \geq 1)$ tiers of services. When $K = 1$, the CSP provides nontier service. When $K \geq 2$, the CSP provides tiered services.
(2) The tiered services offer the identical functionality but vary with service rate μ_i, representing the quality of service (QoS) of the ith level ($i = 1,2,\ldots,K$) and the corresponding price of each tier is labeled as p_i.
(3) The price p_i increases with the service rate μ_i.
(4) The unit price of service rate p_i/μ_i is decreasing with the increase of service rate μ_i.

After the services announced, the customer $j(j = 1,2,\ldots,M)$ who desires to obtain the computing resources chooses an appropriate service plan to maximize his utility. The hypothesis of consumers' choices is presented as follows.

(1) The utility of customer j for service tier i is $U_{ij}(\mu_i, p_i)$ and the customer will choose the tier with maximal $U_{ij}(\mu_i, p_i)$. If $U_{ij}(\mu_i, p_i) < 0$, the customer wouldn't subscribe to service tier i.
(2) Each one can only subscribe to less than one class of services and the choice is measured by x_{ij} where i represented for the service. When $x_{ij} = 0$, the customer wouldn't subscribe to service tier i. When $x_{ij} = 1$, the customer would accept service tier i.

(3) Each customer's demand is same as others. For example, each one who would like to use the software provided by SaaS CSPs pay for the same services.
(4) Once the choice is made, there's no changes. In this way, we can estimate the access rate of each service as n_i.

In all, the specific model is listed as follows.
(1) First level: Monopolist's decision

$$\max \pi(\mu, p) = \sum_{i=1}^{K} (p_i n_i - VC_i) \tag{1}$$

$$\text{s.t.}\, p_i < p_j \text{ } if \text{ } \mu_i < \mu_j, i \neq j, K \geq 2 \tag{2}$$

$$\frac{p_i}{\mu_i} > \frac{p_j}{\mu_j} \text{ } if \text{ } \mu_i < \mu_j, i \neq j, K \geq 2 \tag{3}$$

(2) Second level: Customer's decision

$$\max u_j(x) = \sum_{i=1}^{M} U_{ij}(\mu_i, p_i) x_{ij} \tag{4}$$

$$\text{s.t}\, x_{i_1 j} x_{i_2 j} = 0 \text{ } if \text{ } i_1 \neq i_2 \, i_1, i_2 = 1, 2, \ldots, M \tag{5}$$

$$x_{ij} = 0 \text{ } if \text{ } U_{ij}(\mu_i, p_i) < 0 \tag{6}$$

$$x_{ij} = 0 \, or \, 1 \tag{7}$$

We consider the behaviors of both customers and the CSP through the bilevel programming model. In general, the trade process of online services consist of two parts—the release of service quality and corresponding price, the selection of the appropriate service on the basis of user behavior. Since pricing strategy directly affects users' choices which have great impact on CSP's profits, the provider may need to change the pricing strategy if the prime prices are inappropriate. In this model, (1) presents for the CSP's profit while (4) presents for customer's utility. The constraints of (2) (3) ensures the prices are relatively appropriate and the constraints of (5) (6) (7) represents the choice of customer will be less than one tier.

2.2 *Profit Function of CSP*

As it is known that profits are generated by revenues under the constraints of cost, the profit function of CSP comes from the whole revenues of services substracting all the cost. Since the access rate of each tier is n_i, the revenue of each tier is $p_i \cdot n_i$.

From the perspective of trade process, the cost of cloud service can be classified into two parts based on the specific time when they occur [12]. In this model, the first part consists of coordination cost, maintenance cost, agency cost and IT service cost, which occur prior to the conclusion of a contract. Meanwhile, the second part involves negotiation cost, adoption cost and allocation cost, which happen after the contract.

Another aspect to develop cost model is the factors which influence operation process of the services [13]. These factors can be classified as capital cost (fixed cost), operation cost and allocation cost. In addition, the cost of a cloud data center consist of servers cost (compute, storage, software), infrastructure cost (power distribution and cooling), power draw cost (electrical utility costs) and network cost (links, transit, equipment) [2].

In our model, we consider three parts of cost based on the combination of recent studies—maintenance cost, coordination cost and allocation cost. Since IT service cost happens when the company sets up, the fixed cost can be treated as sunk cost. In addition, negotiation cost is nearly zero because the majority of computing resources are traded online. The reason why we don't consider agency cost is that we assume there is no agency between the CSP and customers. Table 1 shows the impact factors of cost for cloud service.

When modeling operation cost, we take Cobb-Douglas function into consideration. The cost of digital goods is considered discontinuous by Huang and Sundararajan (2011) [14], while cost function presented by Lv and Rouskas (2011) is linear [9]. Note that operation cost consists of maintenance cost, coordination cost and allocation cost. It is observed from Table 1 that maintenance cost is highly related to service capacity while allocation cost and coordination cost are affected by the number of users. To combine three forms of cost as a whole, it is assumed that operation cost is positively correlated with either service rate or arrival rate, which is

$$\frac{\partial VC_i}{\partial \mu_i} > 0, \frac{\partial VC_i}{\partial n_i} > 0 \tag{8}$$

Table 1 The specific cost of our model

Specific costs	Impact factors
Allocation cost	Usage duration, peak and off peak usage, usage of memory, storage, processing power etc., bandwidth consumption, overall allocation capacity
Negotiation cost	Negotiation cost is nearly zero since the majority of computing resources are traded online
Agency cost	Agency cost is nearly zero since we assume there is no agency between the CSP and customers.
Coordination cost	Service delivery (human resource, software, etc.), network bandwidth requirements, miscellaneous (insurance, legal, training etc.)
Maintenance cost	Facilities maintenance (power, cooling costs etc.), hardware maintenance, software maintenance

Considering economy of scale, we also point out that the operation cost function is concave, which is

$$\frac{\partial^2 VC_i}{\partial \mu_i^2} < 0, \ \frac{\partial^2 VC_i}{\partial n_i^2} < 0 \tag{9}$$

The function that satisfy (8)(9) is presented in (10), in which $\alpha \in (0,1)$ represents the impact factor of service rate and c means the coefficient of cost.

$$VC(n_i, \mu_i) = c\mu_i^{\alpha} n_i^{1-\alpha} \ (0 < \alpha < 1) \tag{10}$$

2.3 Utility Function of Customer

As mentioned in Sect. 1, from the perspective of a cloud service provider, customers are heterogeneous in the experiences of service quality. We characterize customers' sensitivities of QoS by $\vec{v} = \{v_1, v_2, \ldots, v_M\}$, which is randomly distributed from 0 to 1. Note that the higher quality of service is provided, the more willingness customers would like to pay for the service, which is

$$\frac{\partial WTP_{ij}}{\partial \mu_i} > 0 \tag{11}$$

For example, Microsoft provides three versions of operating system Windows XP which differs in quality. Customers who would like to obtain the best experience need to pay more. Assuming the WTP function is linear, which is

$$WTP_{ij} = v\mu_i \tag{12}$$

the utility function can be denoted by (13), which is similar to the function presented by Lee, Jeong and Seo (2013) [10].

$$U_{ij}(\mu_i, p_i) = WTP_{ij} - p_i = v\mu_i - p_i \tag{13}$$

3 Optimal Pricing Strategy

In this section, we analytically investigate the monopolist's optimal tiered pricing policy when customer's utility function is linear. To derive the optimal policy, we analyze the selection process at first and obtain the following propositions.

Proposition 1 *Customers would choose either none of all the services or the one that gets the minimal unit price of service rate under the constraint of* (3).

Proof Consider any two classes of all the services which are denoted by (μm, pm), (μn, pn). Then the difference in utility between those two services for any customer is,

$$\begin{aligned} U_{mj}(\mu_m, p_m) - U_{nj}(\mu_n, p_n) &= v(\mu_m - \mu_n) - (p_m - p_n) \\ &= (\mu_m - \mu_n)(v - \frac{p_m - p_n}{\mu_m - \mu_n}) \end{aligned} \tag{14}$$

Let us suppose $\mu_m > \mu_n$. Given constraints presented in (2) and (3), we can prove

$$\frac{p_m - p_n}{\mu_m - \mu_n} < \frac{p_m}{\mu_m} < \frac{p_n}{\mu_n} \tag{15}$$

As a result, when $v \geq p_m/\mu_m$, there are $U_{mj}(\mu_m, p_m) - U_{nj}(\mu_n, p_n) \geq 0$ and $U_{mj}(\mu_m, p_m) > 0$. In this situation, customer would choose service m. Otherwise, when $v < p_m/\mu_m$, $U_{nj}(\mu_n, p_n) < 0$ and $U_{mj}(\mu_m, p_m) < 0$ are reached and customer won't require any service. □

Proposition 2 *The market share is negatively correlated with the minimal unit price.*

Proof According to Proposition 1, when user's sensitivity exceeds the minimal unit price, he will select one service. If the lower bound increases while other conditions remain the same, the occupancy of this market will decrease □

Proposition 1 suggests that the optimal policy for the monopolist is to provide only one class of service and we assume the corresponding service rate is 1, which means the service could meet the needs of the whole market. Note that arrival rate can be measured by market share if there is only one service. Considering Proposition 2, we suppose the market share is $q = r(1\text{-}\theta)$, where θ denotes the unit price of service rate and r represents the decreasing rate of the market share with respect to θ. For our purposes, we add the constraints that is $0 < r < 1$ and $0 \leq \theta \leq 1$. As a result, the variable cost is $c[r(1\text{-}\theta)]^{1-\alpha}$ and (1) changes into the (16). In addition, the first and second derivative of profit function $\pi(\theta)$ are presented in (17) and (18) respectively.

$$\pi(\theta) = \theta \cdot \mathrm{r}(1-\theta) - c[\mathrm{r}(1-\theta)]^{1-\alpha} \; (0 < r < 1, 0 \leq \theta \leq 1) \tag{16}$$

$$\pi'(\theta) = r\{1 - 2\theta + c(1-\alpha)[r(1-\theta)]^{-\alpha}\} \tag{17}$$

$$\pi''(\theta) = -2r + c\alpha(1-\alpha)r^{1-\alpha}(1-\theta)^{-\alpha-1} \tag{18}$$

Observation 1 Based on our assumptions, we derive the following results:

(1) When $c \geq (\frac{\alpha}{2(\alpha+1)})^{\alpha+1} * \frac{2r^{\alpha}}{\alpha(1-\alpha)}$, there is no profit for the provider
(2) When $0 < c < (\frac{\alpha}{2(\alpha+1)})^{\alpha+1} \cdot \frac{2r^{\alpha}}{\alpha(1-\alpha)}$, we can get an extreme point θ* through $\pi'(\theta^*) = 0$ and $\theta^* \approx \frac{\pi'(0.5)\theta_1 - 0.5\pi'(\theta_1)}{\pi'(0.5) - \pi'(\theta_1)}$ where $\pi''(\theta_1) = 0$
   (a) If $\frac{1}{2} < \theta^* < \frac{1}{2 - r(1-\alpha)}$, the provider can make the maximal profit at the extreme point θ*,
   (b) Otherwise if $\frac{1}{2-r(1-\alpha)} \leq \theta^* < 1$, the provider will get nothing.

Proof To reach the optimal prices, we have to meet

$$\pi''(\theta) = 0 \tag{19}$$

and the satisfied value is

$$\theta_1 = 1 - (\frac{2r^{\alpha}}{c\alpha(1-\alpha)})^{-\frac{1}{\alpha+1}} \tag{20}$$

It is clear that $\theta_1 < 1$, and

$$\pi''(\theta) > 0 \, if \, \theta > \theta_1 \tag{21}$$

$$\pi''(\theta) \leq 0 \, if \, \theta \leq \theta_1 \tag{22}$$

□

In this way, when $\theta_1 \leq 0$ which is $c \geq \frac{2r^{\alpha}}{\alpha(1-\alpha)}$, $\pi'(\theta)$ increases monotonically with θ since $\theta > 0 > \theta_1$. As a result,

$$\pi'(\theta) > \pi'(0) = r(1 + c(1-\alpha)r^{-\alpha}) > 0 \tag{23}$$

which means $\pi(\theta)$ increase monotonically when $\theta \in [0,1)$. Thus,

$$\pi(\theta) < \pi(1) = 0 \tag{24}$$

Consequently, when $c \geq \frac{2r^{\alpha}}{\alpha(1-\alpha)}$, there is no profits for the provider.

Otherwise, when $\theta_1 > 0$ which is $c < \frac{2r^{\alpha}}{\alpha(1-\alpha)}$, it is identified that $\pi'(\theta)$ increases monotonically with θ when $\theta \in [\theta_1, 1)$ and decreases monotonically with θ when $\theta \in [0, \theta_1)$. The minimum value of $\pi'(\theta)$ is

$$\begin{aligned} \pi'(\theta_1) &= r[1 + \frac{2}{\alpha} - (2 + \frac{2}{\alpha})\theta_1] \\ &= r[(2 + \frac{2}{\alpha})(1 - \theta_1) - 1] \end{aligned} \tag{25}$$

If $\pi'(\theta_1) \geq 0$ which is $c \geq \left(\frac{\alpha}{2(\alpha+1)}\right)^{\alpha+1} * \frac{2r^{\alpha}}{\alpha(1-\alpha)}$, the result is as same as $\theta_I \leq 0$. If $\pi'(\theta_1) < 0$ which is $0 < c < \left(\frac{\alpha}{2(\alpha+1)}\right)^{\alpha+1} \cdot \frac{2r^{\alpha}}{\alpha(1-\alpha)}$,

$$\pi'(0) = r(1 + c(1-\alpha)r^{-\alpha}) > 0 \tag{26}$$

$$\lim_{k\to 1^-} \pi'(\theta) = r \lim_{k\to 1^-} (1 - 2\theta + c(1-\alpha)(r(1-\theta))^{-\alpha}) \to +\infty \tag{27}$$

The roughly appearance of $\pi'(\theta)$ is illustrated in Fig. 1a, which means the relative maximum is $\pi(\theta^*)$. Since $\pi'(0.5) = c(1-\alpha)r^{1-\alpha}2^{\alpha} > 0$ and $\pi'(\theta_1) < 0$, $\theta^* \in (0.5, \theta_1)$. Assuming the curves of AB in Fig. 1a is linear, we can estimate the value of θ^* through

$$\frac{\theta_1 - \theta^*}{|\pi'(\theta_1)|} \approx \frac{\theta^* - 0.5}{\pi'(0.5)} \text{ and } \theta^* \approx \frac{\pi'(0.5)\theta_1 - 0.5\pi'(\theta_1)}{\pi'(0.5) - \pi'(\theta_1)} \tag{28}$$

If $\pi(\theta^*) \leq 0$, that is $\frac{1}{2-r(1-\alpha)} \leq \theta^* < 1$, providing services can't gain any profits as can be seen in Fig. 1b. Otherwise, as is shown in Fig. 1c when $\frac{1}{2} < \theta^* < \frac{1}{2-r(1-\alpha)}$, $\pi(\theta^*) > 0$ which means it's profitable for CSPs and the maximal profit is $\pi(\theta^*)$.

Fig. 1 The roughly appearance of $\pi'(\theta)$ and $\pi(\theta)$

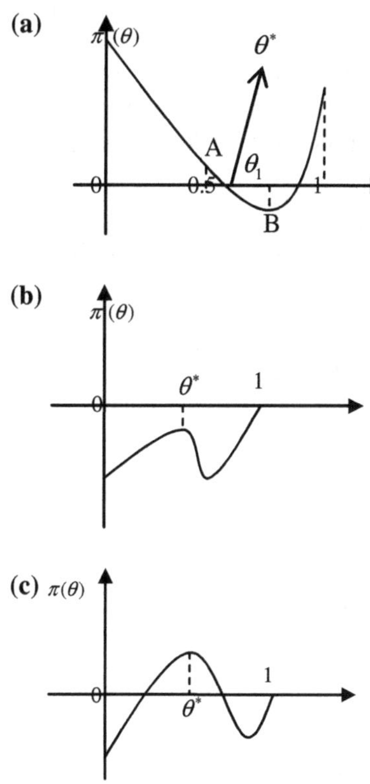

In all, we identify that if $c \geq (\frac{\alpha}{2(\alpha+1)})^{\alpha+1} \cdot \frac{2r^{\alpha}}{\alpha(1-\alpha)}$, the profit $\pi(\theta)$ is positively correlated with the parameter θ. Thus, the maximal profit is $\pi(1) = 0$, which shows that the service is not profitable if its variable cost is great.

If $0 < c < (\frac{\alpha}{2(\alpha+1)})^{\alpha+1} \cdot \frac{2r^{\alpha}}{\alpha(1-\alpha)}$, the relative maximum is $\pi(\theta^*)$. If $\pi(\theta^*) \leq 0$, there is no profits and otherwise it's profitable.

An interesting finding indicated by Observation 1 is that the optimal strategy for the CSP is to provide nontier service which is different from the result illustrated by Li, Feng and Chen (2012) [15]. They concentrate on optimal versioning strategy for information goods and illustrate that two-version scheme is superior to one-version scheme under linear utility function. However, we suggest that the CSP should adopt nontier service and the price is calculated through certain formulas.

4 Conclusion

In this research, we study a monopolistic cloud provider's pricing problem in the presence of heterogeneous customers. A key feature of our model is the dynamic interaction between the monopolist and users. In addition, the operation cost is also taken into consideration in our model. With analytical solution for linear customer utility function, we demonstrate that the optimal policy for the CSP is to provide nontier service which is valuable for a monopolistic CSP.

An economic model considering the impact of service level agreements (SLAs) may be an interesting future research. Moreover, taking network externality into consideration could be another direction of our future study.

Acknowledgments The work was supported by General Program of the National Natural Science Foundation of China (No.71101103, No.71371135) and National Natural Science Fund for Distinguished Young Scholars of China (No. 70925005).

References

1. Mell P, Grance T (2011) The NIST definition of cloud computing (draft). NIST Spec Publ 800 (145):7
2. Iyoob I, Zarifoglu E, Dieker AB (2013) Cloud computing operations research. Serv Sci 5 (2):88–101
3. Chun SH, Choi BS (2013) Service models and pricing schemes for cloud computing. Cluster Comput 5:1–7
4. He Y, Sun Y (2011) Customized pricing strategy for software-as-a-service. In: 2011 4th IEEE international conference on broadband network and multimedia technology (IC-BNMT), pp 375–378
5. Katzmarzik AD (2011) Product differentiation for software-as-a-service providers. Bus Inf Syst Eng 3(1):19–31
6. Basu S, Chakraborty S, Sharma M (2015) Pricing cloud services—the impact of broadband quality. Omega 50:96–114

7. Huang J, Kauffman RJ, Ma D (2014) Pricing strategy for cloud computing: a damaged services perspective. Decis Support Syst. http://dx.doi.org/10.1016/j.dss.2014.11.001
8. Zhang J, Niu B (2014) Dynamic quality decisions of software-as-a-service providers based on customer perception. Electron Commer Res Appl 13(3):151–163
9. Lv Q, Rouskas GN (2011) On optimal tiered structures for network service bundles. In: 2011 IEEE global telecommunications conference (GLOBECOM 2011), pp 1–5
10. Lee SH, Jeong HY, Seo SW (2013) Optimal pricing and capacity partitioning for tiered access service in virtual networks. Comput Netw 57(18):3941–3956
11. Lv Q, Rouskas GN (2010) An economic model for pricing tiered network services. Ann Telecommun 65(3–4):147–161
12. Martens B, Teuteberg F (2012) Decision-making in cloud computing environments: a cost and risk based approach. Inf Syst Front 14(4):871–893
13. Kozhipurath J (2012) Cloud service costing challenges. In: 2012 IEEE international conference on cloud computing in emerging markets (CCEM), IEEE, 2012
14. Huang KW, Sundararajan A (2011) Pricing digital goods: discontinuous costs and shared infrastructure. Inf Syst Res 22(4):721–738
15. Li M, Feng H, Chen F (2012) Optimal versioning and pricing of information products with considering or not common valuation of customers. Comput Ind Eng 63(1):173–183

Explicit Incentive Mechanism of the Agent Construction System for Government Investment Projects: A Model Study Based on Asymmetric Information

Da-shuang Dai, Shuang Zhang, Zi-xin Liu and Chao Feng

Abstract The incentive mechanism of the agent construction system for government investment projects is not well developed in China, which seriously affects the working enthusiasm of the construction agent. Aimed at improving the explicit incentive mechanism of the agent construction system and exploring how the construction agents' comprehensive strength and the government investors' centralization cost affect the expected utility of the government investors and the best incentive intensity to reach the relationship between expected utility of government investors and the best incentive intensity, respectively. This paper constructs a principal agent model based on asymmetric information. The research results show that the construction agents' comprehensive strength plays an important role to the project output. Government investors can increase incentive intensity appropriately based on the centralization cost to encourage construction agents to work hard and to increase expected utility.

Keywords Agent construction system · Explicit incentives · Information asymmetry · Principal agent model

1 Introduction

Government investment projects refer to the government fixed assets investment projects mainly. The project funds come from government financial infrastructure funds or loans whose guarantee is the government's financial department and whose

---

Foundation item: National Natural Science Foundation (71272091) "Study on the Governance Structure Design and Application of the Agency Construction System for Government-funded Projects".

D. Dai (✉) · S. Zhang · Z. Liu · C. Feng
Faculty of Management and Economics, Dalian University of Technology, Dalian, China
e-mail: daids@dlut.edu.cn

E. Qi (ed.), *Proceedings of the 6th International Asia Conference on Industrial Engineering and Management Innovation*,
DOI 10.2991/978-94-6239-145-1_106

debtors are the state-controlled enterprises [1]. The implementation of the agent construction management for these projects is a new breakthrough of the reform of investment management system in China. In recent years, the management modes of the agent construction system can be summarized as three kinds: government agent construction mode, marketization agent construction mode and middle mode [2]. Whereas the marketization agent construction mode is widely adopted in local governments of provinces and cities. Compared with the traditional mode of government project management, the marketization agent construction mode has more stakeholders and more complicated principal-agent relationships. Therefore, this paper focuses on the research on the marketization agent construction mode.

The incentive of the agent construction system for government investment projects refers to the government investors giving construction agents material and nonmaterial satisfaction, which can encourage the construction agents to select action plans consistent with the objectives of the government investors, based on their own utility maximization [3]. However, the agent construction system is still in the exploratory stage in China and its incentive mechanism is not perfect, which is present as a low standard for agent construction management fees, a little incentive intensity, a high degree of government intervention, a heavy liability for breach of contract and so on [4–6]. These are very prone to lead to moral hazards, which make the government investors suffer losses.

Currently, scholars' researches on the incentive mechanism of the agent construction system focus on constructing an incentive model. Lin and Yin improved the traditional model starting from two perspectives: contract price quantification and punishment coefficient, which makes the incentive effect more obvious [7]. Hua and Sha thought that the government investors' and construction agents' attitudes towards risk have three combinations: "neutrality-aversion", "preference-aversion" and "neutrality-preference" [8]. However, the "neutrality-aversion" risk combination is used widely now. Wei [9], Tang [10], and Li [11] constructed principal agent models for the incentive mechanism based on the "neutrality-aversion" risk attitude. Cao et al. constructed an incentives-supervision model based on the assumption of complete rationality and fairness preference [12]. Scholars have also considered the impact of the projects' three goals on motivation strategies and constructed a "cost–schedule–quality" model [13, 14].

The authors believe that current incentive models include incomplete influence factors. Therefore, this paper adds new elements to construct an incentive model based on the principal agent theory to study the principal agent relationship between the government investors and the construction agents. This model will be solved and analyzed under asymmetric information and the results will provide a basis for improving explicit incentives.

2 Methodology

Construction and management of the government investment projects will be entrusted to the civil construction unit through the contract negotiation. Thus, the government investor is a principal and the civil construction unit is an agent. In the process of constructing a project, the government investor cannot supervise the whole process of agent construction all the time. Therefore, the government investor is unable to grasp the real effort level of the construction agent. And the construction agent tends to have more personal information than the government investor, so the information asymmetry must occur. If the incentive mechanism, which the government investor develops, is reasonable, it will motivate the construction agent to work hard and increase the utility of the government investor. But if the incentive mechanism is unreasonable, it will lead the construction agent to being slack and directly affect the interests of the government investor.

2.1 *Hypothesis*

Based on the principal agent theory and the characteristics of the government investment projects of the agent construction system, the authors suggest the following hypothesis:

(1) *Hypothesis of information asymmetry*: The construction agent tends to have more personal information, which the government investor needs to pay for in order to obtain it. It is likely to produce an adverse selection problem before the contract comes into effect, because it is difficult to know the real situation of the construction agent for the government investor, he can only pay the agent construction fees in accordance with the market's average. As a result, it is possible that the selected construction agent does not meet the project's requirements. It is likely to produce moral hazards during the execution of the contract. The government investors cannot grasp the real effort level of the construction agent, which they can only judge according to the results. For his own interests, the construction agent will choose actions that go against the government investors' interests.
(2) *Hypothesis of the rational economic man*: The construction agent and the government investor are rational economic men, and they both seek to maximize their own interests. For the construction agent, he will make a choice to make his own input minimum and the expected return maximum because he will pay for his hard work.
(3) *Uncertainty hypothesis*: The results of the construction agent's work are not only determined by his level of effort, but also affected by many uncertain factors that cannot be controlled by the construction agent and the government investor.
(4) *Hypothesis of risk preference*: The government investor is risk-neutral and the construction agent is risk-averse.

The construction agent's utility function has the characteristics of constant absolute risk aversion. It is expressed as $\mu = -e^{-\rho w}$, where μ refers to the construction agent's utility, ρ refers to the absolute risk aversion variable, that indicates the level of risk aversion, and w refers to the real income [15].

2.2 Model Construction

Based on the hypotheses, this paper selects the following variables:

(1) I refers to the initial estimated investment amount of the government.
(2) $\alpha(\alpha > 0)$ refers to the adjustment coefficient for adjusting the expenditure of profits.
(3) x refers to the effort level of the construction agent. It is a continuous variable of one dimension. The effort level affected by the external factors is not static.
(4) b refers to the coefficient of effort cost. The effort cost function is expressed as $C(x) = bx^2/2$. It needs to meet that the effort cost of the construction agent increases as the effort level increases and the marginal cost of effort is incremental which can be expressed as $\partial C/\partial x > 0$ and $\partial^2 C/\partial^2 x > 0$.
(5) A refers to the project management ability of the construction agent.
B refers to the organizational ability of the construction agent.
C refers to the performance level of the construction agent.
M refers to the comprehensive ability of the construction agent. It can be expressed as $M = A + B + C$.
(6) a refers to the agent construction management fees. As the fixed income of the construction agent, it has nothing to do with the output. It is calculated based on the initial estimated investment amount of the government and the specific proportion is determined according to the local Agent Construction System Management Methods and the negotiation results between the government investor and the construction agent.
(7) $y(y \in (0, 1))$ refers to the excitation intensity coefficient, that is the output proportion shared by the construction agent, and c refers to the centralization cost coefficient. When the construction agent shares the output, he should also assume a proportionate risk and the risk premium formula is $\rho y^2\sigma^2/2$. When $y = 0$, the government investor assumes all of the risk. Because the government investor is risk-neutral, he has no risk premium. When $y = 1$, the construction agent assumes all of the risk. Whereas $1 - y$ also reflects the centralization level of the government investor. The higher the centralization level of government is, the less output proportion shared by the construction agent is. The government investor needs to pay for the centralization. The centralization cost function can be expressed as $C(y) = c(1 - y)^2/2$.
(8) θ refers to the exogenous uncertainties. It is a normal distribution with a random variable, which meets the mean in 0 and the variance is σ^2.

According to the set variables, the paper constructs the following objective function. To facilitate the study, these formulas do not consider the excess of the actual expenses and the most unfavorable outcome is that real spending is equal to the expected investment.

The output of the construction agent (λ') is determined by the effort level of the construction agent, the project management ability, the organizational ability, the performance level, the exogenous uncertainties and the adjustment coefficient for adjusting the expenditure of profits. So the function can be expressed as:

$$\lambda' = \alpha I(A + B + C)x + \theta \tag{1}$$

The expectation of the output can be expressed as:

$$\lambda = \alpha I(A + B + C)x \tag{2}$$

Agent construction payment consists of agent construction management fees and reward or punishment. The expectation of the agent construction payment can be expressed as:

$$S(\lambda) = a + y\lambda \tag{3}$$

The expected utility of the government investor can be expressed as:

$$Ev(\lambda - S(\lambda)) = (1-y)\alpha I(A + B + C)x - a - c(1-y)^2/2 \tag{4}$$

The real income of the construction agent can be expressed as:

$$w = S(\lambda') - C(x) = a + y(\alpha I(A + B + C)x + \theta) - bx^2/2 \tag{5}$$

The certainty equivalence income of the construction agent refers to the real income subtracting the risk cost. Thus, it can be expressed as:

$$W = a + y\alpha I(A + B + C)x - bx^2/2 - \rho y^2\sigma^2/2 \tag{6}$$

The goal of the government investor is to maximize the expected utility and the goal of the construction agent is to maximize the certainty equivalence of income. Meanwhile, the government investor faces two constraints from the construction agent, namely, participation constraint (*IR*) and incentive compatibility constraint (*IC*).

The participation constraint refers to the construction agent choosing a scheme that will make him obtain more expected utility than other discarded schemes in the face of multiple alternatives, namely, the retained earnings *R* or the opportunity cost. It can be expressed as:

$$a + y\alpha I(A + B + C)x - bx^2/2 - \rho y^2\sigma^2/2 \geq R \tag{7}$$

The incentive compatibility constraint refers to the construction agent always choosing the effort level maximizing his own utility. It can be expressed as:

$$\begin{aligned} & a + y\alpha I(A + B + C)x - bx^2/2 - \rho y^2\sigma^2/2 \\ & \quad \geq a + y\alpha I(A + B + C)x' - bx'^2/2 - \rho y^2\sigma^2/2 \\ & \quad \forall x' \in X \end{aligned} \tag{8}$$

In summary, the principal agent model between the government investor and the construction agent can be expressed as:

$$Max\,Ev = Max(\lambda - S(\lambda)) = Max\left[(1-y)\alpha I(A + B + C)x - a - c(1-y)^2/2\right]$$

s. t. (*IR*)

$$a + y\alpha I(A + B + C)x - bx^2/2 - \rho y^2\sigma^2/2 \geq R$$

(*IC*)

$$\begin{aligned} & a + y\alpha I(A + B + C)x - bx^2/2 - \rho y^2\sigma^2/2 \\ & \quad \geq a + y\alpha I(A + B + C)x' - bx'^2/2 - \rho y^2\sigma^2/2 \\ & \quad \forall x' \in X \end{aligned} \tag{9}$$

3 Results

The key of the model is to select an appropriate a and y to solve the optimization problem. Under the condition of optimization, the equality of participation constraint holds:

$$a = R - y\alpha I(A + B + C)\,x + bx^2/2 + \rho y^2\sigma^2/2 \tag{10}$$

To obtain the effort level x that maximizes the expected utility of the construction agent, this paper needs to take a derivative with respect to x of the incentive compatibility constraint equation and make the derivative zero. The derivation result is as follows:

$$x = y\alpha I(A + B + C)/b \tag{11}$$

This paper plugs (10) and (11) into the objective function of (9). It can be expressed as:

$$Max\,Ev(\lambda - S(\lambda)) = Max(y\alpha^2 I^2 M^2/b - y^2\alpha^2 I^2 M^2/2b - \rho y^2\sigma^2/2 - R - c(1-y)^2/2 \quad (12)$$

For deriving one-step condition of (9), this paper takes a derivative which is 0 with respect to y. The derivation result is as follows:

$$dEv/dy = \alpha^2 I^2 M^2/b - y\alpha^2 I^2 M^2/b\rho y\sigma^2 + (1-y)c = 0 \quad (13)$$

$$y^* = 1/\left(1 + b\rho\sigma^2/(\alpha^2 I^2 M^2 + bc)\right) \quad (14)$$

This paper plugs (14) into (11) and (12). It can be expressed as:

$$x^* = \alpha IM/b\left(1 + b\rho\sigma^2/(\alpha^2 I^2 M^2 + bc)\right) \quad (15)$$

$$Ev^* = (\alpha^2 I^2 M^2 + bc)/\left(2b\left(1 + b\rho\sigma^2/(\alpha^2 I^2 M^2 + bc)\right)\right) - R - c/2 \quad (16)$$

4 Discussion

Equation (4) shows that the harder the construction agent works, the higher the expected utility of the government investor is under the same conditions.

For analyzing how the comprehensive strength of the construction agent affects the best incentive intensity and the expected utility of the government investor, this paper takes the partial derivatives with respect to A, B and C of (15) and (16). Derivation results are as follows:

$$\partial y^*/\partial A = \partial y^*/\partial B = \partial y^*/\partial C = 2b\rho\sigma^2\alpha^2 I^2 M^2 \Big/ \left(\alpha^2 I^2 M^2 + bc + b\rho\sigma^2\right)^2 > 0 \quad (17)$$

$$\partial Ev^*/\partial A = \partial Ev^*/\partial B = \partial Ev^*/\partial C = \alpha^2 I^2 M^2\left(\alpha^2 I^2 M^2 + bc + 2b\rho\sigma^2\right) \times \left(\alpha^2 I^2 M^2 + bc\right)\Big/\left(b\left(\alpha^2 I^2 M^2 + bc + b\rho\sigma^2\right)^2\right) > 0 \quad (18)$$

The results suggest that the comprehensive strength of the construction agent, namely, the project management ability, the organizational ability and the performance level, shows a positive correlation with the best excitation intensity and the expected utility of the government investor.

This paper takes the partial derivatives with respect to ρ of (15) and (16) to analyze the effect of the degree of risk aversion of the construction agent on the optimal incentive intensity and expected utility of the government investor. The derivation results are as follows:

$$\partial y^{*}/\partial\rho = -b\sigma^{2}\left(\alpha^{2}I^{2}M^{2}+bc\right)\Big/\left(\alpha^{2}I^{2}M^{2}+bc+b\rho\sigma^{2}\right)^{2}<0 \tag{19}$$

$$\partial Ev^{*}/\partial\rho = -\sigma^{2}\left(\alpha^{2}I^{2}M^{2}+bc\right)^{2}\Big/\left(2\left(\alpha^{2}I^{2}M^{2}+bc+b\rho\sigma^{2}\right)^{2}\right) \tag{20}$$

The results suggest that the degree of risk aversion of the construction agent shows a negative correlation with the best excitation intensity and expected utility of the government investor.

This paper takes the partial derivatives with respect to b of (15) and (16) to analyze the effect of the effort cost coefficient on the optimal incentive intensity and the expected utility of the government investor. The derivation results are as follows:

$$\partial y^{*}/\partial b = -b\sigma^{2}\alpha^{2}I^{2}M^{2}\Big/\left(\alpha^{2}I^{2}M^{2}+bc+b\rho\sigma^{2}\right)^{2}<0 \tag{21}$$

$$\partial Ev^{*}/\partial b = -\alpha^{2}I^{2}M^{2}\left(\alpha^{2}I^{2}M^{2}+bc\right)\left(\alpha^{2}I^{2}M^{2}+bc+2b\rho\sigma^{2}\right)\Big/\left(2b^{2}\left(\alpha^{2}I^{2}M^{2}+bc+b\rho\sigma^{2}\right)^{2}\right)<0 \tag{22}$$

The results suggest that the effort cost coefficient of the construction agent shows a negative correlation with the best excitation intensity and the expected utility of the government investor.

This paper takes the partial derivatives with respect to σ^2 of (15) and (16). And the derivation results are as follows:

$$\partial y^{*}/\partial\sigma^{2} = -b\rho\left(\alpha^{2}I^{2}M^{2}+bc\right)\Big/\left(\alpha^{2}I^{2}M^{2}+bc+b\rho\sigma^{2}\right)^{2}<0 \tag{23}$$

$$\partial Ev^{*}/\partial\sigma^{2} = -\rho\left(\alpha^{2}I^{2}M^{2}+bc\right)^{2}\Big/\left(2\left(\alpha^{2}I^{2}M^{2}+bc+b\rho\sigma^{2}\right)^{2}\right)<0 \tag{24}$$

The results suggest that exogenous uncertainties negatively affect optimal excitation intensity and expected utility of the government investor.

This paper takes the partial derivatives with respect to the centralization cost coefficient c of (15) and (16). And the derivation results are as follows:

$$\partial y^{*}/\partial c = b^{2}\rho\sigma^{2}\Big/\left(\alpha^{2}I^{2}M^{2}+bc+b\rho\sigma^{2}\right)^{2}>0 \tag{25}$$

$$\partial Ev^{*}/\partial c = -b^{4}\rho\,\sigma^{4}\Big/\left(2\left(\alpha^{2}I^{2}M^{2}+bc+b\rho\sigma^{2}\right)^{2}\right)<0 \tag{26}$$

The results suggest that the centralization cost coefficient positively affects optimal excitation intensity but negatively affects the expected utility of the government investor.

At last, this paper takes a partial derivative with respect to R of (16). And the derivation result is as follows:

$$\partial Ev^{*}/\partial R = -1 < 0 \tag{27}$$

So the more retained earnings the construction agent wants to get, the less expected utility the government investor will get.

The above results are summarized in Table 1. It shows that when the centralization cost coefficient c remains unchanged, changes in the variables associated with the construction agent make the expected utility of the government investor and the optimum excitation intensity change with the same tendency. Whereas when the centralization cost coefficient changes, the expected utility of the government investor varies inversely with the optimal excitation intensity. The centralization cost coefficient increases as the management intensity increases. The optimal excitation intensity increases when c increases, because the government investor wants to take advantage of the decentralization to reduce the increasing range of the centralization cost, and the construction agent also expects a higher incentive intensity to reduce the loss, which behavioral restrictions, where the bad behavior is easily found, lead to with the increase in management intensity. But this approach will reduce the proportion of investment under-expenditure shared by the government investor. Increase in the optimum excitation intensity will motivate the construction agent to work hard to increase the output, but the expected utility of the government investor does not increase. It is indicated that the increase in output caused by hard work cannot offset the loss of the government investor. The reason may be that the comprehensive strength of the construction agent restricts the increase in output caused by effort.

Table 1 Effect of variables on expected utility of government investor and excitation intensity

Variables	Trends	Expected utility of government investor	Excitation intensity
A	↑	↑	↑
B	↑	↑	↑
C	↑	↑	↑
ρ	↑	↑	↓
b	↑	↓	↓
σ2	↑	↓	↓
c	↑	↓	↑

5 Conclusion

According to the above discussion results, they show that the evaluation of bidders should pay more attention to comprehensive ability when the government investor selects the construction agent by call for bids. And the government investor should determine the management strength and breadth according to his own management ability and centralization cost. The incentive intensity should increase appropriately as the management intensity increases.

According to the current problems of the incentive mechanism of the agent construction system for the government investment projects and the analysis results of the model study, this paper makes the following recommendations for improvement.

(1) *Designing a reasonable selection mechanism for the construction agent*:

A basic success condition of an agent construction project is to select a construction agent with strong comprehensive ability. The government investor should establish a dedicated information base, including the basic situation, the qualification status and the previous performance levels and so on, for the excellent construction agents in the market. This information should be open to the community for community supervision and restraints on the behavior of the construction agent. The government investor should select an appropriate construction agent from the information base. In advance, the government investor should set the evaluation index and proportion. And then, the experts will use the method of integrated scoring for bid assessment. Finally, the bidder with the highest comprehensive score is chosen as construction agent. This solution can make the government investor know the construction agent fully and increase the probability of successful selection. Meanwhile, it can avoid some of the drawbacks of the open tender, for example: the high bidding costs, the government investor cannot know the bidders fully and so on.

(2) *Establishing a sound system of agent construction fees*:

According to the problem of low agent construction management fees, this paper draws lessons from the PMC remuneration design. The agent construction management fees include a fixed part and a risk-sharing part reflecting the responsibility and the risk of the construction agent [16].

A fixed part: It is paid according to milestone schedule requirements. Only when the construction agent completes these milestone tasks can he receive remuneration for the corresponding parts.

A risk-sharing part: The government investor sets some project objectives, a clear assessment method, criteria and a certain percentage (Please see Table 2) in the contract. Only reaching the objectives can he get the remuneration for the corresponding parts. The aim of setting the risk-sharing part is to encourage the construction agent to save costs rationally, but to pursue the investment savings at the expense of quality blindly.

Table 2 Project goal percentage distribution table

No	Project objectives	Percentage (%)
1	Quality	25
2	Progress	20
3	Equipment performance and applicability	20
4	Health safety and environment management	20
5	Sustainability	15

Acknowledgments All the authors of this paper thank the sponsors, Chinese Industrial Engineering Institution, CMES and Taiwan Industrial Engineering Institute and the organizer, Tianjin University, for providing a high-level international platform for experts, scholars and entrepreneurs. We thank all the staff of this conference. Meanwhile, we are grateful for the help of the other members of our research group. At last, we sincerely hope that the experts put forward their valuable advices.

References

1. Fan Y (2013) Research on agent-construction management mode and incentive mechanism of government-invested project. J Lanzhou Jiaotong Univ 32(3):101–104 (in Chinese)
2. Song J, Fu Y (2010) A comparative study of agent construction mode for government investment projects. Constr Econ 8:27–29 (in Chinese)
3. Holmstrom B (1979) Moral hazard and observability. Bell J Econ 10(1):74–91
4. Yan L, Zhou G (2009) The state and problems of agent construction system of the government investment projects in China. J Beijing Inst Technol (Soc Sci Ed) 11(5):137–142 (in Chinese)
5. Ke H (2007) Research on the improvement on public project management performance based on market agent construction system. Ph.D. dissertation, Management Science and Engineering, University of Tianjin, Tianjin, China, 2007 (in Chinese)
6. Zhou X, Wang P, Chen Y, Shi J (2012) Operating condition investigation of agent construction system for the government investment projects. Urban Probl 205(08):36–40 (in Chinese)
7. Lin G, Yin Y (2010) Improvement study on the cost control model of agent construction of governmental investment project. J Beijing Inst Technol (Soc Sci Ed) 12(4):53–56 (in Chinese)
8. Hua D, Sha K (2011) Risk attitudes of parties and incentive strategies of construction projects. Proj Manage Technol 9(10):21–25 (in Chinese)
9. Wei X (2009) The study on agent construction contract's incentive terms of government investment projects. M.A. dissertation, Business and Management, Huazhong University of Science and Technology, Wuhan, China, 2009 (in Chinese)
10. Tang G (2012) Research on incentive and restraint mechanism of railway project management. M.A. dissertation, Civil Engineering, University Central South, Hunnan, China, 2012 (in Chinese)
11. Li C (2013) Study on the agent-construction system for government investment project based on principal agent theory. M.A. dissertation, Management Science and Engineering, University of Chongqing, Chongqing, China, 2013 (in Chinese)
12. Cao Q, Sheng Z, Zhou J, Liu H, Qian Li (2014) Motivation and supervision mechanism of agent construction system for government investment projects based on fairness theory. China Soft Sci Mag 10:144–153 (in Chinese)

13. Yin Y, Zhao J, Zhong W (2012) Cost-progress-quality incentive model based on utility theory. J WUT (Inf Manage Eng) 34(3):390–393 (in Chinese)
14. Wu Z (2013) Study on supervision system of agent-construction project funded by government. Ph.D. dissertation, Management Science and Engineering, Beijing Jiaotong University, Beijing, China, 2013 (in Chinese)
15. Sappington D (1991) Incentives in principal agent relationships. J Econ Perspect 5(2)45–66
16. Liu J, Chen Y, Qi G (2005) Project management contract—theory and practice of PMC. Post & Telecom Press, Beijing, pp 243–256 (in Chinese)

The Evaluation of Undergraduate Accounting Experimental Teaching

Xia Li and Sheng-dao Gan

Abstract The experimental teaching, as an essential part of undergraduate accounting teaching, is very important for cultivating qualified accounting personnel in the society. Scientific evaluation of undergraduate accounting experiment teaching is conducive to the improvement of teaching quality. This paper discusses the whole evaluation in the process of undergraduate accounting experimental teaching, and points out that the evaluation must be based on the talent training objectives and other teaching elements of all kinds of universities.

Keywords Accounting · Experimental teaching · Classroom teaching · Teaching evaluation

1 Introduction

According to modern accounting teaching concept, accounting knowledge system comes from practice, and is used in practice, combining theory with practice is the basic rule of accounting teaching [1]. Therefore, the experimental teaching should be the necessary link in accounting teaching, it play a pivotal role in training students' practical ability, innovation ability and developing the quality education, so attentions must be paid.

Narrow sense of accounting experimental teaching (excluding off-campus practice activities) includes two aspects, computerization experiment teaching and manual experimental teaching. Among them, the manual accounting experimental teaching is one kind of teaching activities or practice teaching methods, which takes the economic business, and related internal and external economic ties, accounting

X. Li (✉) · S. Gan
Business School, Sichuan University, Chengdu 610064, China
e-mail: 15955203990@163.com

X. Li
Business Institute, Anhui University of Finance and Economics, Bengbu 233041, China

E. Qi (ed.), *Proceedings of the 6th International Asia Conference on Industrial Engineering and Management Innovation*,
DOI 10.2991/978-94-6239-145-1_107

procedures in a certain accounting occurring of a business unit as simulating object, in accordance with the accounting standards and the accounting system requirements, using authentic accounting vouchers, accounting books, accounting statements to undertake accounting practice drills, causes the student to obtain intuitive and systematic understanding for accounting work; Computerization experiment teaching makes computerized accounting process in practice simulated so as to introduce it into classroom. Accounting computerization is an abbreviation that applies contemporary electron technology and information technology to practical accounting, the core of which is electronic compute, it is an accounting information system using the electronic computer to implement. It realizes data processing automation, makes the traditional manual accounting information system develop and evolve into electronic accounting information system.

Undergraduate accounting experimental teaching evaluation of colleges is the process to judge and evaluate the teachers' teaching and students' learning in the process of undergraduate accounting computerization and manual experimental teaching of universities, generally including the evaluation of many factors such as teachers, students, teaching contents, teaching methods, teaching environment and teaching management and etc. in teaching process [2], but mainly aiming at the evaluation of students' learning effects and teachers' teaching process. Its core is to evaluate teachers' teaching work and students' learning effects. This is a whole behavior, including preparation before class, classroom teaching and after-school evaluations, taking classroom teaching assessment for the center.

2 Classroom Teaching Evaluation

The objects of class teaching evaluation are the teaching activities of teachers, and the learning activities of students. The effect evaluation of teachers' teaching is based on classroom teaching elements, which can be divided into several evaluation index, such as teaching aim, teaching content, teaching process, teaching methods and teaching effects and so on; the evaluation index of students' learning effect can be divided into several parts, for example, the transferring degree of students' enthusiasm and initiative, mastering degree of the classroom knowledge, the improvement of students' innovation consciousness, mastering degree of the practice ability and so on.

For each class teaching element, we need develop detailed evaluation index and give accurate weights, but it seems lack of feasibility, and facing many difficulties now. For example, in the process of the assessment of teaching contents in every chapter, the content selected and emphases determined are decided by students' knowledge structure, the entire course system, and other aspects, and it is difficult to score and weight with concrete measures. Therefore, the evaluation for teachers should adopt combining method with quantitative evaluation and qualitative evaluation, which can be implemented by several subjects such as teacher himself, student, and third-party and so on. First, we should formulate objective and scientific evaluation criteria, list the evaluation standard. Considering the characteristics of accounting experimental teaching, we can take the following items as evaluation points. (1) teaching contents are close to the practice, language is accurate and conforms to specifications; (2) the teaching goal is clear, the requirements are specific, which accord with general outline and reflect social needs for the skills of accounting personnel; (3) teaching language is vivid and attractive, with strong strain capacity and regulation ability in classroom; (4) teaching skill is proficient, teaching method is agile and diversity, using teaching media is freely; (5) trying best to foster students' creative thinking, stimulate students' learning interest, cultivate students' sentiment, and develop learning ability [3]; (6) processing textbooks has integrity, highlight, explaining difficulty pellucid [4]; (7) choosing teaching method properly, paying attention to combining theory with practice, so as to combine teaching and practice together, speaking refined and practicing earnestly [5]; (8) teaching atmosphere is relaxed and harmonious, etc. Then, we can organize activities of teachers' self-assessment, students' classroom teaching assessment, and the third party's assessment. We also can organize evaluation panel, the leader should has high theoretical level, having certain research on class teaching evaluation, serious and responsible in work, and strong organization ability. Members should also have certain accounting knowledge and professional knowledge. Team members give a mark on the basis of earnestly studying evaluation standard, recording the teaching contents earnestly and filling in evaluation table. Finally, according to the evaluation results from tripartite, we can get a comprehensive evaluation. For some qualitative projects which are not easily quantified, we write objective and impartial comments.

For the students' evaluation and suggestions in the classroom teaching, we can primarily take the way of students' self-assessment., we can design evaluation table, let the students fill by themselves, then collect and analyze. The design of evaluation table should reflect the overall principle, objectivity principle and dynamic principle or developmental principle, show some different levels, and the standard and index of the evaluation also can be readjusted at any time.

3 After-Before Class Evaluation

Accounting experimental classroom teaching is a process with elaborate design and organization by teachers, no matter how much dynamic generative class teaching has, it cannot deny that it is the product of a prior design on purpose by teachers, so in order to evaluate classroom teaching, we should first evaluate the design in teaching aim, teaching content, teaching method, teaching resources before the class. For instance, the situation of keeping abreast of teaching syllabus, whether document writing is normative and with practicability, whether experiment project design is close to reality, whether media is prepared adequately, etc. On the other hand, understanding degree for correlated knowledge of the student before class, the student's professional knowledge structure, the computer level, will have influence on the effect of accounting experiment classroom teaching, so we need carry on the baseline evaluation. The evaluation before the class is mainly a diagnostic evaluation. Diagnostic evaluation means before teaching activities evaluating the preparation degree of evaluation objects (including teachers and students), in order to take corresponding measures to make teaching plan smooth, and effective, it is such a determination assessment. Before-class evaluation, is not only used in daily evaluation, but also used in all terms or the beginning of the semesters.

After classes, according to before-class and class teaching evaluation and their own experience of the teaching effect, teachers should summarize the experience, find out the deficiency, in order to improve the teaching quality. Especially for the results from others, teachers need learn modestly, analyze the problems earnestly, accept others' good suggestions and opinions to improve their teaching level.

For after-class accounting experiment evaluation of students, we need take diversified methods to assess. After each laboratory course, we can require students to write experimental analysis report, summarize the harvest and deficiencies, and encourage students to develop their individual thinking. In the midterm and after the semester, various universities can refer the implementation of the credit system, use level evaluation method for evaluation results, namely outstanding, good, passed, failed, the four levels, to distinguish different ability level and at the same time, not cause students to pay much attention to score, to let the students have more space for self-development and self-innovation; we can evaluate with the practice training results priority, supplemented by test scores, allow to use post test result substitute for midterm and final exams, or based on the simulation of the accounting vouchers, use real accounting business process of enterprises to replace the traditional accounting evaluation form, such as calculation topic, the entry form, in this way, students can use the evaluation forms such as passing accountant post test, job performance, post homework assignments grades, social practice evaluation to comprehensively evaluate their experiment results.

4 We Should Care for Two Accounting Experimental Teaching Evaluation Problems

4.1 Accounting Experimental Teaching Evaluation Should Combine Talents Training Target

The accounting experimental teaching evaluation should be accordance with accounting personnel training goals of domestic and international society and the colleges. Current society requires the relevant accounting personnel not only to make simple accounts but also have certain abilities of analyzing and processing various reality problems, including communication ability, interpersonal skills, management ability, etc. Therefore, the selected evaluation index should be able to support the realization of applied, composite and innovative talent training goal. Evaluation methods should be diversified, overall, and can really reflect all the necessary quality and ability of qualified accounting personnel. On the other hand, accounting experimental teaching evaluation should also be combined with talents cultivation target of each undergraduate colleges. At present, our country accounting degree education system is divided into technical secondary school, college, undergraduate, postgraduate (including master and doctor) four basic levels, training targets of all levels is put forward from the early 1980s until now, and there is nothing substantial changes [6]. The goal of undergraduate teaching is to cultivate integrity and senior specialized talents who can be engaged in accounting practical work and the professional teaching and research work in enterprises and institutions, certified public accountant offices, economic management departments, schools and research institutions. Specifically, the various universities have different position. Therefore, before the accounting experimental teaching evaluation, we should make it clear that the college is teaching-centered or research-centered or integrated for teaching and scientific research, and then according to the talent localization of universities, we evaluate experimental course, targeted to cultivate accounting talents. Orientation is different; the weight of each indicator will come to a great disparity.

4.2 Accounting Experimental Teaching Evaluation Should be Integrated with Other Teaching Elements Evaluation

Classroom teaching is constantly interactive process with teachers' teaching and students' learning occurring simultaneously, teaching situation, teaching resources, teaching organization such variable and teaching elements can also affect the teaching result, they are teaching process variables, such as the behaviors of teachers [7], the behaviors of students, and interacting between teachers and students, thus, for teaching process evaluation, we will also focus on the behavior of teachers, the behavior of students, interacting between teachers and students and teaching situation, teaching resources, and other factors.

The teaching management standardization evaluation and accounting experimental teaching evaluation are closely related. Standard teaching management process, comprehensive and reasonable laboratory system and regulations are the basis of high efficiency experimental teaching, so, in the process of accounting experimental teaching evaluation, we should consider various teaching resources such as hardware and software.

In the process of accounting experimental teaching evaluation, all the assessment for teachers and students may be affected by teaching assessment system of universities strongly [8]. Some universities have established single and unified assessment methods, facing all courses, this will limit the evaluation mode of experimental course, and reduce the quality of assessment.

So, accounting experimental teaching evaluation, as an activity with the whole teaching process, is not independent, and it is under the influence of various factors [9]. We should distinguish the particular situation and problems, treat and analyze particularly, so as to find out the assessment system suitable for this academy.

References

1. Li XC, Wu QY (2008) The research on the operation mechanism of accounting experimental teaching. J Qiongzhou Univ (2):62–64 (in Chinese)
2. Du J (2010) Accounting experimental teaching: situation, problems and countermeasures—research review of accounting experiment teaching. J Account 4:112–114 (in Chinese)
3. Gan M (2014) discussion on teaching model of exploratory accounting experimental teaching. Finance Account Mon 3:126–128 (in Chinese)
4. Lou JC (2010) Concerning the construct of accounting experimental teaching system. J Ningbo Eng College (1):110–114 (in Chinese)
5. Yu YL (2011) The new process of accounting experimental teaching. Friends Account 2: 117–120 (in Chinese)
6. Cheng XC (2011) Study on the problems of accounting experimental teaching in application-oriented universities. J Haerbin School Econ Finance 5:88–89 (in Chinese)
7. Xue J (2008) Constructing normalization accounting experimental teaching system—the thinking of our accounting experimental teaching. Econ Financ Circles (Acad Ed) 2:10–11 (in Chinese)
8. Tan Y, Xu X, Cheng H (2005) Evaluation system of experiment teaching quality. J Guangdong Univ Technol 9:308–310 (in Chinese)
9. Qie M (2014) Establishing system of accounting experimental teaching based on ERP. Adm Asset Finance 4:225–226 (in Chinese)

Teaching Model Reform and Practice of Distribution Center Design and Optimization

Yuepeng Yao, Ying-de Li and Chao-qun He

Abstract In order to solve the teaching embarrassment of Distribution center design and operation, which include the shortage and out date of teaching resource, the scarcity in practice projects and the simplicity of the assessment mode, we design the teaching reform solution to enhance the understanding and application capability of students whose mayor is logistics engineering. Some foreign advanced textbook and latest research achievement are brought into the course contents, some practices projects which come from the real research projects are brought into the project-based teaching mode, and a new comprehensive assessment mode are explored to replace the simply test mode. The experiments show that the reform achieves promising achievements.

Keywords Comprehensive assessment mode · Distribution center design and optimization · Project-based teaching model · Teaching reform

1 Introduction

The quality of talent training is the first embodiment of the higher education, is the basis existence and development for a university, and the professional courses is a fundamental part of it. So to make the content of education thought become the content

Y. Yao
IT Department, Zhejiang Red Dragonfly Shoes Corporation Limited, Wenzhou 325100, Zhejiang, People's Republic of China

Y. Li (✉)
Institute of Industrial Engineering, Zhejiang University of Technology, Hangzhou 310014, Zhejiang, People's Republic of China
e-mail: liyingde2008@sina.com

C. He
Faculty of Mechanical Engineering & Automation, Zhejiang Sci-Tech University, Hangzhou 310018, Zhejiang, People's Republic of China

E. Qi (ed.), *Proceedings of the 6th International Asia Conference on Industrial Engineering and Management Innovation*,
DOI 10.2991/978-94-6239-145-1_108

of teaching and learning, We must take the curriculum as the link, through planning the course content, design it and achieve the goals of it, make the education thought to be materialized [1]. In 2002, many universities of our country opened the major of logistics engineering, but due to different school have different orientate, so it was divided into different subject categories, like logistics, management, mechanical, engineering and other categories, it's reason why basic teaching curriculum and planning is inconsistent [2]. Many scholars have carried out the different reform practice in logistics engineering professional course teaching.[1]

Huang et al. according to the teaching commands and its characteristics of Logistics Engineering in research universities, the teaching mode of Logistics Engineering to develop innovation ability and research ability was explored [3]. Pan and Xue carried out the reform of logistics course from the content of the curriculum and practice teaching system, researched on teaching method innovation through the teaching-methods on multimedia teaching, video-teaching cases-teaching and Web-Based classroom [4]. Jiang et al. took the course «Logistics Information Technology and Application» as the research object, in order to train the student's practical ability of logistics information, they put the case which is closely integrated with the enterprise into the teaching process [5]. Mai took the course «Logistics Information Technologies» as the research object, based on the basic theories and methods, she puts forward the teaching objective, planned the task contents and technical objective, and gave an designed example of special teaching task content [6]. Liu et al. took the teaching reform of the course «Logistics System Planning and Design» as the research object, combined with the characteristics of this course and the present learning situation of students, they probed the reformation of teaching methods and assessment methods for the improvement of student's learning efficiency and the best learning effect [7]. In the study of introducing advanced foreign teaching mode, some scholars discuss and summarize the advanced experience of professional course teaching. Guo and Han according to a short-time international study on freshman seminar course and undergraduate specialized course in Lehigh University, and summarized and discussed the advanced experience of foreign engineering education [8]. Li took the undergraduate education mode of the Worcester Polytechnic Institute as the research object, and got inspiration from it [9]. Liu and Lu summarized the evolution of 1861–2010 America higher engineering education curriculum policy, and provided reference of China Higher Engineering Education reform [10].

"distribution center design and optimization" is a core course of logistics engineering, it main contents include data acquisition and analysis of modern logistics center, layout and storage mode design of store, order processing and order picking design, box type chosen and dismantle the chosen layout and design

[1]This research was supported by Zhejiang University of technology teaching reform Foundation of under Grant No. JG1211 and by Zhejiang university graduate student teaching reform Foundation under No. JXY201203 and by Zhejiang provincial National Science Foundation of China under Grant No. LY13G020019 and by Zhejiang University National Science Foundation under No. 2012XZ001.

operations, order picking path optimization, capacity calculation and equipment selection, KPI performance evaluation and other professional knowledge, Its content is closely related with all kinds of logistics center design and daily operation, and have many characteristics like multiple of concept, strong theory and multiple of model, multiple quantitative calculation, multiple of formula, practicability and maneuverability, so it not only need classroom explanation also requires students to complete the course design. The level of knowledge is directly affect their employment and career prospect who major in logistics engineering.

Given that fact, this paper will take the "distribution center design and optimization" as reform target, put forward the reform scheme of teaching mode innovation on he course, introduction of advanced foreign teaching contents and methods, enhance the reform and perfection of this knowledge system, and become a scientific knowledge system. Then students can overall, system and standardize the core knowledge of the distribution center and achieve the purpose of teaching.

A. Analysis of current teaching situation as follows.

(a) Domestic teaching content is difficult to meet the teaching goal and practice demand of modern logistics planning and design. At present, the content of textbooks and teaching domestic emphasis on the macro strategy and management concept, lack of quantitative analysis of specific problems and the content of optimizing design with advanced methodology. At the same time, there is distance between students' theoretical knowledge and the actual operation, and the textbook knowledge is so dull and abstract that students lost an interest in course, and students can't application of what they have learned at last.
(b) The teaching method is unitary and lack of the training of professional project with systematic and comprehensive. On the one hand, the students passively accept the teaching content in the traditional classroom which based on teaching, and participate in discussions of case in the classroom which mainly by the descriptive analysis, so, it's difficult for them to improve their ability of knowledge application and innovation. On the other hand, lack of opportunities for students to participate in real projects and nuclear how it will work in practice, lead to a problem that teaching theory and the teaching practice come apart and students lack of the ability to apply the knowledge to solve practical problems.
(c) The single curriculum assessment methods lead to students lack the motivation to study. We usually have one test at the end of the term about the curriculum, and students take the examination as the final purpose of learning, so they have less pressure to learn and the poor learning motivate, finally, it's difficult to achieve the purpose of teaching and even influence teaching effect.

Therefore, in order to solve the above problems, this paper analysis and learn the successful teaching experience, teaching resources and frontier development results from foreign high university courses, then according to the actual needs of our school syllabus and enterprises, reform the teaching model including the content of teaching material, classroom experiment, practice project and examining mode, and put forward the curriculum reform and practice model of the "distribution center design and optimization" which suit for logistics engineering students in our college students.

2 Reform Ideas and Thoughts

For existing problems in the professional instruction and the basic conditions of our students, we made the following reform ideas.

2.1 Introduction of Advanced Materials and Technology Achievements and Reform the Teaching Content

The teaching content is critical elements of knowledge and ability structure, so it need change with the times that mean we not only ensure students master the basic theory and basic principle but also need introduce of the new knowledge, new technology and new process to classroom, then bring students to the forefront of technological development.

Hence, the reform take the teaching quality and cultivating the ability of students as the goal, so we introduction of foreign advanced teaching materials, scientific research and industry's latest technology into the course contents, combine with the actual characteristics of students and the teaching of our school to adjusting the teaching content, and combine with the teaching content to increase the comprehensive and innovative experiment and design of the teaching-aided software etc. (as shown in Fig. 1).

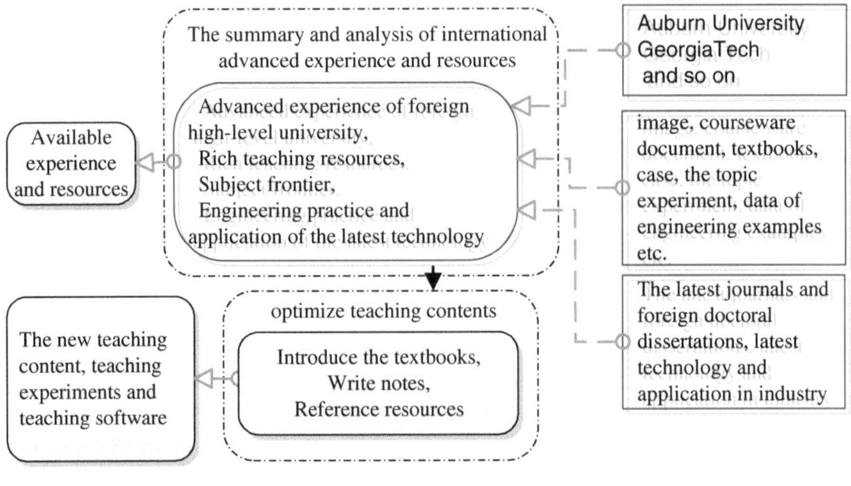

Fig. 1 The reform ideas about content of courses

2.2 The Practice of Scientific Research Project for Teaching, Design Project-Based Teaching Mode

In the present case teaching, teachers have no conscious that to train the ability of students to identify problems from examples and then to analyze and solve problems, it lead to they just appear examples, case and the operation of knowledge, without careful design the way and process of examples show, then the teaching goal of logistics engineering cannot achieve. The curriculum of the "distribution center design and optimization" involves many content which theory combined with practice like data analysis, process analysis, layout design, equipment capacity design, picking method design, picking the organization, personnel performance evaluation etc., it's need to seek suitable project module and integrated project from practice to teaching. Therefore, the requirements of the curriculum reform based on student mastery of the basics, we import some strong practical and participation modular practical projects or integrated projects timely and appropriate, then to guide students to learn with problems and professional knowledge, For a variety of teaching methods, we need design classroom activities with the right rhythm and process of project activities so that the theory and practice teaching to achieve the perfect combination (as shown in Fig. 2).

2.3 Improve Assessment Methods and Implementation of the Comprehensive Evaluation Mechanism

The teaching of traditional evaluation methods are based on the final exam scores to judge the quality of students and the degree of mastery of knowledge, it's so narrow that students can pass the exam by forcing memory, then they easy to develop simple learning habits and lead to loose and cope with daily study, so traditional assessment and evaluation methods cannot evaluate the extent of students' understanding and application of knowledge.

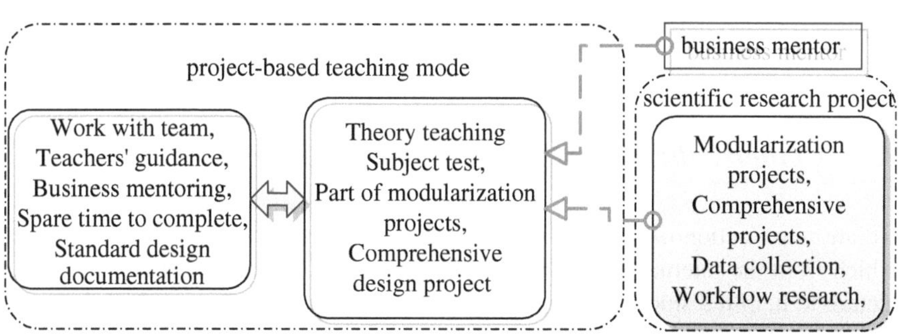

Fig. 2 Project-based teaching mode

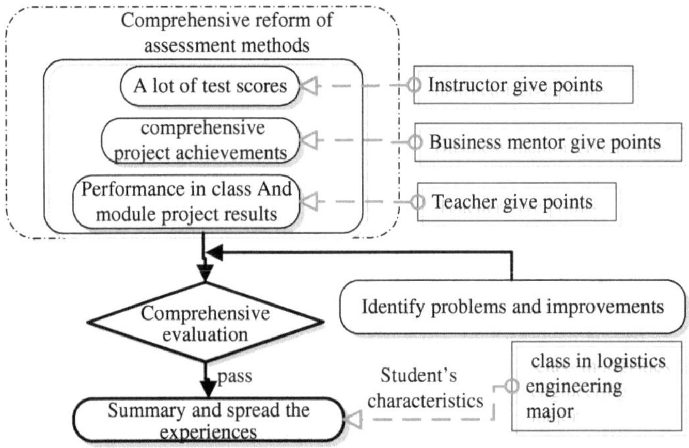

Fig. 3 Comprehensive reform of assessment methods

Therefore, the reform concept requires innovative existing assessment methods, change the method of one-time final exam determine, using the way of examination "multiple examinations and projects", assessment needs integrating theories with practice and bring on the heat, modest increase in students' daily learning motivation and pressure, and formation of a stable evaluation mechanism for the promotion of the professional class in logistics engineering (as shown in Fig. 3).

3 The Content of Reform Teaching

Form September 2011 to January 2014, we have reformed and practiced the course of the "distribution center design and optimization" by means of teaching content, project practice, assessment methods, etc. under the guidance of the reform theory and though, supported by the teaching project in our school, and borrow successful teaching experience from a high-caliber university and our 3 years of curriculum reform in logistics engineering.

3.1 Optimize the Teaching Content and Methods

we are introduction of the English textbook "*Warehouse & Distribution Science*" which from the international top university Georgia tech and preparation of a new lecture to help the students obtain the basic theoretical knowledge which based on the latest research achievements of famous international journals and the latest doctoral thesis material at home and abroad.

Adjustment and optimization of teaching as follows:

(a) The module of layout and channel design in unit load type warehouse: The introduction of the latest Flying-V, Fishbone, V, Leaf and other advanced layout mode. Under the mode of dual-task operation, introduction of a job matching design and collaborative optimization design. We also use the position optimization model and application in lateral channel.
(b) The module of FCL picking space design and optimization: introduce the design method of picking space and SKU decision optimization method, and design the teaching software to assist it.
(c) The module of unbundled picking space design and optimization: introduce the design methodology of unbundled area and all kinds of design methods of order picking operation.
(d) The module of design and optimize the picking method: organization design method is used to relay chosen mode; introduce the assignment and order picking path optimization.
(e) Cross-docking design module: introduce the layout and design methods of the library system.
(f) E-commerce logistics module: introduce the design method and case study of electric business logistics distribution center.

3.2 *Classroom Experimental Design*

Combined with lectures, the experiments including “warehouse channel direction and position” “correlation slotting and packing optimization” “books-picking path optimization” “picking method Bucket Bridge” “ants picking system” is designed with comprehensive and innovative, but the maximum time allowed for each experiment is 20 min, and then students can enhance the understanding of knowledge by participate in classroom experiments with a team.

3.3 *Teaching Software Design*

This course needs carry on data calculation and analysis, like the judgment of SKUs’ correlation, orders characterization, SKUs selection rule in logistics center, throughput and turnover rate calculation, EIQ analysis, SKUs selection of picking space, etc. So, we design of a number of small teaching software used to aid the classroom teaching and project design. At the same time, it can improve the students’ awareness of the application of modern information technology in logistics.

3.4 Project-Driven Reform

The characteristics of logistics professional requires us to focus on training students' practical skills in logistics education, therefore, on the basis of the integration of teaching content and customized research content, we combine with research project, cooperation with the enterprises, create the enterprise professor institution, and from the start of the course we import of modular projects and integrated projects. In the project, students do excises with team in the modularization projects and comprehensive project, then the enterprise mentors and instructors will encourage students to discover the spot problems and give them some advises, so, in this way the teachers and students can analysis and discussing even debating around the specific issues, and it also can help students to develop their creative thinking. We introduce the following items on the basis of research projects.

Modularization projects: distribution center design of a enterprises that their shoes and clothes is delivered to the store, business logistics center location planning and design of a enterprises; Comprehensive project: a cross-docking design of a courier company, elevated storage space layout design of a medical distribution center, unbundled picking space picking methods and work flow design of a medical distribution center, functional requirements of the WMS design of a electricity business distribution center.

3.5 The Reform of Comprehensive Assessment

One-time final exam system is ended, and change it to the comprehensive examination system that three exam (about once a month, 20 % of the total score every test) and project training. Assessment content should pay attention to the combination of theory and practical application, appropriate to increase the students' learning pressure, and improve the learning enthusiasm and initiative of students. The final total score consists of the following components.

There times exam and each test account for 20 % of the total score, the test is main to assess student's level of textbook theoretical knowledge.

Twice comprehensive project performance and each project, he score determined by their accomplishment and innovation. This part pay attention on the participation, research spirit and innovation ability of students.

The mark of homework and it account for 10 % of the total score, and students must complete five written assignments on time.

Modular project performance and it account for 10 % of the total score, it's main to evaluate the performance of students in daily study, and according to the requirements of teaching arrangement and learning assessment, and the students must complete five modular design projects.

It is worth noting that comprehensive project and homework and modular project is required to complete out of class time.

4 Practice and Achievements

The key to cultivate the logistics engineering applied talents is to reform the teaching content and teaching methods of their professional courses. Review the curriculum reform implemental process, we found that the reform improved the studying interests of students and enhanced their independent thinking, problem solving skill and the enthusiasm of class through the research which includes the discussion with students, classroom behavior and the effect of the comprehensive practice course. Student evaluation of curriculum reform from teaching evaluation system show that it's achieved good effects, and students learned useful knowledge, and then the teaching evaluation index also gained high evaluation in teaching resources, teaching content, teaching method, teaching effect and teaching services etc.

References

1. Zhang X, Zeng H, Wang D (2013) Strengthen the construction of top-quality course and promote the cultivation of innovative talents. China High Educ Z3:41–43 (in Chinese)
2. Lv C, Zhang L, Liu S (2013) Research of modern logistics equipment course reform based on professional programming. Logist Eng Manage 07:200–201 (in Chinese)
3. Huang Q, Chen L, Li N (2012) Study on teaching mode of logistics engineering in research universities. Logist Eng Manage 34(2):112–113+132 (in Chinese)
4. Pan L, Xue D (2011) Practice and investigation on the revolution of logistics course of industrial engineering specialty. Logist Eng Manage 04:181–182 (in Chinese)
5. Jiang X, Deng Z, Mi Z (2010) Curriculum reform of logistics information technology and application based on working procedure of logistics. Vocat Tech Educ 35:30–32 (in Chinese)
6. Mai W (2009) Research on curriculum reform of logistics information technologies teaching based on tasks. Logist Eng Manage 12:147–149 (in Chinese)
7. Liu X, Lan P, Liu W (2009) Research on the reform of the course logistics system planning and design. Logist Sci-Tech 11:130–132 (in Chinese)
8. Guo L, Han L (2013) Case analysis on teaching methodologies of freshman seminar course in USA. Res High Educ Eng 01:170–176 (in Chinese)
9. Li J (2013) Student-centered, Project-driven and result-oriented in WPI. Res High Educ Eng 05:115–119 (in Chinese)
10. Liu H, Lu Y (2013) The evolution characteristics of American higher engineering education from 1861 to 2010. Res High Educ Eng 01:147–152+180 (in Chinese)

A Research on the Construction of Distinctive Teachers in Teaching and Research University

Kang Shao

Abstract It is crucial to construct a high-level profession-specific faculty for the development of teaching and researching university. Through the analysis of the research status on the construction of faculty both at home and abroad and in light with the growth theory and motivation theories of teachers in universities and colleges, this thesis intends to analyze the features and components of the distinctive faculty in teaching and research university and put forward such construction thoughts as "Creating a distinctive faculty by elevating the level of introduction, enhancing the training process and optimizing the overall structure", which has achieved preliminary effects in terms of training young teachers and advancing their capacity for teaching and scientific research through the exploratory practice in the teaching process of management and engineering disciplines. This has enriched the experience for building up the faculty.

Keywords Characteristics of teachers · Construction of distinctive teachers · Talent development · Teaching Research University

1 Introduction

Talent has to be one of the most precious treasures in the 21st century and university has to be the major place where talent is born. With the constant advancing of the reform and opening-up policy, large quantities of integrated

This research was supported by the Anhui University of Science and Technology great teaching reform project in 2013 (2013zdxm44).

K. Shao (✉)
School of Economic and Management, Anhui University of Science and Technology, Huainan, China
e-mail: seeke@163.com

E. Qi (ed.), *Proceedings of the 6th International Asia Conference on Industrial Engineering and Management Innovation*,
DOI 10.2991/978-94-6239-145-1_109

talents with advanced degrees are wanted in the construction of socialist cause. Therefore, a group of teaching and research universities whose level should be high enough to compete with those world first-class universities is highly in demanded to bear such kind of talents. Except the qualified hardware facilities, an excellent faculty is the main consideration to judge the standard of a Teaching and research university. Hence how to construct a high-level profession-specific faculty has become the main issue we are facing with at present [1].

According to the investigation and survey we've made, during the popularization of higher education, problems like unreasonable level of teacher's education background, similar learning disciplines, "big-pot" distribution system, iron bowl, brain drain, non-academic factors considered in the promotion of professional titles haven't been fundamentally solved all along. All these explain the phenomenon like irregularity in teacher quality, inability in knowledge sharing, imperfection in the construction of teaching system, ambiguous motivation in the development of teacher profession, lack of the awareness in autonomous development and weak consciousness in cooperation among teachers, which have affected a great deal the campus cultural accumulation and the course to construct teaching and research universities [2].

2 The Features of the Faculty in Teaching and Research Universities

Through effective integration in disciplines, a university of teaching and research type, normally with a certain period of teaching history is the one that attaches great importance to discipline construction and scientific research. It intends to foster such advanced and specialized talents with certain broad and compound knowledge, of high strain capacity, with excellent comprehensive quality, with pioneering spirit and with research potential who are majored in technology development and application. As for the construction of faculty, special attention must be paid not only to training and introducing those academic leaders who has significant impact on respective disciplines, but also to structuring the education background, the professional titles, age, and similar discipline reasonably. Meanwhile, an excellent quality as well as vitality is also required [3].

According to the National Guideline on Medium and Long-Term Program for Science and Technology Development (2006–2020) in China, "It is truly essential to accelerate the development of a group of high-level, especially world famous high-level teaching and research universities for China to speed up the innovation of science and technology and build an innovation-originated nation" [4]. There are various scientific measurement indexes to build up a Teaching and research university, among which the most important one is the issue of constructing the faculty. Since a strong team of teachers represents the core competence in a Teaching and research university, only can those top teachers gain the top scientific research projects through competition and achieve great research output, carry out the research-oriented teaching activity and thus foster those innovative talents with high quality.

3 Thought on the Construction of Distinctive Faculty in a Teaching and Research University

In light with the result of the investigation and survey we made and with the experience on building up faculty in universities both at home and abroad, [5] this thesis intends to put forward such thought as "Creating a distinctive faculty by elevating the level of introduction, enhancing the training process and optimizing the overall structure", and as shown in Fig. 1, improve the overall level of the teaching team on the principle of harmonious development among hierarchy, structure and sustainability.

Principle of hierarchy. Aiming at building up a top-level staff, a university should firstly formulate the related police and measures to introduce high-level talents. The talents introduced shall be divided into four parts: leading figures, foregoers and backbones in the area of teaching and scientific research as well as members in the academic groups. Secondly, talents shall be brought in both in full-time and flexible ways. Preferential treatment must be given to those who have been introduced in so that a talent team can take shape.

Principle of structure. The construction of the faculty shall meet such structural conditions as balanced age, compatible disciplines, relatively high education background, and multiple academic and reasonable professional titles.

Principle of sustainability. The construction of the faculty is a systematic project which is also long-term, constant, integral, coordinating and sustainable. Therefore a long-term planning must be prepared.

Fig. 1 Thought on constructing a distinctive faculty

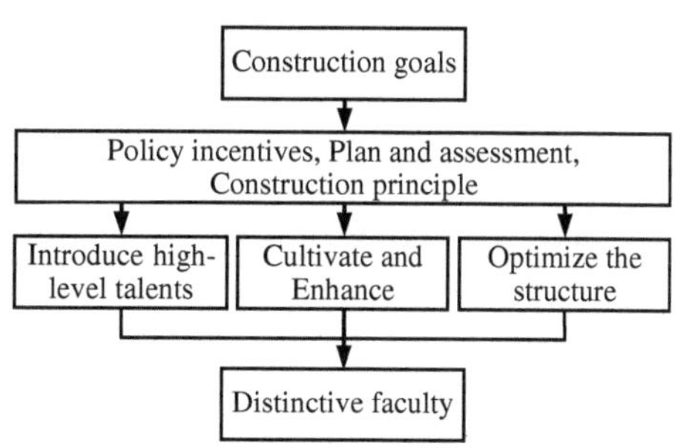

4 Theoretical Basis of Constructing the Faculty

Fessler once divided the growing process of a teacher into three stages according to various demands she or he needs in different phases and different problems about education and teaching she or he pays attention to in different periods [6]. There are the concern about the survival, about the situation and about the student. In the stage concerning about survive, since a teacher just enters a university and his or her career as a teacher just begins, the feature is that he or she must face and adapt to the new profession and focus on how to acclimatize to it from the bottom part. After this period, with the gradually mastering of the daily teaching and basic knowledge and technology about scientific research, confidence will grow. So the teacher will shift his or her concerns greatly to the teaching situation from the survival stage before. As for the phase concerning about the student, a teacher will consider more about the individual difference and be aware that since students of different levels have different needs, some teaching materials and methods may not suit all.

By studying the growth theory of teachers in universities and colleges and analyzing the developmental features in different growing phases, a corresponding incentive principle and method of teachers' growth can be probed, which is of positive guiding significance to the construction of distinctive faculty in universities [7].

Phonologically enthusiasm is defined as a kind of behavioral motivation performance in a dynamic mental state. The main influencing factors are desire, ambition and the motivation of an individual. Incentive theory is the core theory applied to dealing with the relationship among desire, motivation, ambition and behavior. According to the behavioral science, an individual's motivation comes from desire which determines his or her behavioral objective while incentive acts on one's inner world by stimulating, driving and strengthening his or her behavior [8]. As indicated by the experience, the theory of incentive has a fine implement environment and sound stimulating effect in the area of educational administration. Thus a serious study and an effective practice of this theory can leads to a great improvement in teachers' enthusiasm [9].

According to the theory of incentive, we should firstly learn all that teachers' need and guide them to build up a career interactive between teaching and scientific research by carrying out the motivating system of "material reward" and "spiritual reward" and founding a fair and reasonable system of rewards and penalties. The principle of people first shall be insisted so that teachers' enthusiasm can be aroused [10].

5 The Practice of Constructing Distinctive Faculty

Financially aided by the School of Economics and Management in Anhui University of Science and Technology, an "E-commerce Innovation Team", constructed as a group of characteristic faculty was founded in 2009 and carried into valid practice in the discipline of Management Science and Engineering which is a key discipline in this university. Achievements have been made in the aspects of training young teachers and improving teachers' ability in teaching and scientific research, which enriches the experience of constructing faculty up to a point.

This teaching group of about 20 members has witnessed great progress in aspects of teaching, scientific research and talent training since its foundation several years ago. Among them, 16 are Master Tutors, 12 are with the doctoral degrees, 13 of which have been promoted to the senior professional and technical titles and still there are five who are pursuing the doctorates right now. Since the year 2010, members of the team have conducted over 40 scientific research projects supported by enterprises, universities, research institutions, and even by the province and the nation, among which, 10 national-level ones are supported by the National Natural Sciences Foundation and National Social Sciences Fund, 7 by the Science Foundation of Ministry of Education of China and another 10 by funds like natural science, soft science, humanities and social science, philosophy and social science in Anhui Province. Also, more than 30 high-quality academic papers and 5 monographs have been published during this period.

As for the training of talents, the team actively explored the personal training mode of innovational and practical type. In the aspects of educational reforming and discipline building such as curricula construction, practice and training base construction, the team has made great progress in the training of fully-developed internationalized innovative talent through the cooperation with those excellent IT enterprises at home and the efforts of both sides. Furthermore, the talent training mode and practical teaching system of "originality, innovation, entrepreneurship and creation" has been proposed for the purpose of cultivating students' ability in professional practice, reforming and advancing the practice and teaching activities in the department. The college students of the e-commerce major where this team belongs to, with their outstanding performance, have achieved highlighted success from the year 2012 to the year 2014 by undertaking more than 30 College Students' Innovation and Entrepreneurship Training Projects of National Level.

6 Conclusion

Through the analysis of research status both at home and abroad and in light with the relative theories, this thesis intends to study what kind of impact a faculty might have upon the development of a Teaching and research university. Regular research conclusions have been achieved through optimizing the structure of faculty, mastering the essential factors of constructing a characteristic teaching group in a Teaching and research university, exploratory practice in the process of the professional teaching activities and the estimation of the effect a characteristic faculty might have upon the construction of colleges and universities, especially universities of teaching and research type. The conclusions are:

(1) Through the comparative analysis between the experience of constructing faculty in high-level universities at home and abroad, faculty is proved to be the key factor for a university to transform successfully from a teaching one to a teaching and research one.
(2) Characteristic, as the key competitiveness, is the main feature of a university to distinguish itself from the others. Those provincial focused universities must construct their own distinctive faculty according to their own special fields and lay the teaching foundation to foster a group of innovative talents of high quality through improving the comprehensive quality of the whole teaching group.
(3) The structure of the staff, in terms of student-teacher ratio, education background, professional tiles, related education, age and master teachers shall be as scientific and reasonable as possible.
(4) The effect of building up a characteristic faculty is closely related to the corresponding appraisal and examining mechanism. Therefore, a matched and relevant administration system shall be studied and constructed so as to perfect the management of teaching motivation.
(5) Those engineering dominant universities shall construct the faculty in line with their character of engineering practice, which means adopting the construction model of double-track pricing system so as to enhance the staff's ability in engineering practice.

References

1. Tong Y (2014) Some thoughts on university teachers' team construction in the new period. J Liaoning Adm Coll 16(3):58–60 (in Chinese)
2. Fan J (2013) The study on lab faculty team construction and development in Beijing Polytechnic Universities. Mod Educ Technol 23(9):62–66 (in Chinese)
3. Zou C, Gan C, Chen Q (2013) Study on construction of faculty in the transition from a teaching-oriented university to a teaching and research oriented university. J Chongqing Jiaotong Univ (Soc Sci Ed) 13(5) (in Chinese)
4. http://www.gov.cn/zwgk/2006-02/26/content_211553.htm
5. You Y (2013) On the screening mechanisms in building faculty teams: an analysis of five US world-class universities. Peking Univ Educ Rev 11(4):87–97+187 (in Chinese)
6. Zhang J (2009) Higher education psychology, Chapter 4. Higher Education Press, Beijing, pp 117–131
7. Wang P, Chen S (2007) A review on theoretical constructing of teachers' teaching competencies abroad. Stud Foreign Educ 34(10):28–32 (in Chinese)
8. Zhang W (2012) Improving teachers' enthusiasm by using the theories of incentive. J Liaocheng Univ (Soc Sci Ed) 2:89–91 (in Chinese)
9. Xu G, Zhang X, Ji X, Xing H (2010) On the characteristics and cultivation of practical ability of teachers. J Anhui Univ Sci Technol (Soc Sci) 12(1) (in Chinese)
10. Xu Y, Rui B (2009) Discuss simply the motivation theories application in the work of college teachers incentive. Heilongjiang Res High Educ 1:100–102 (in Chinese)

A New Government Procurement Pricing Method Based on Copula Model

Pin-yue Wang, Jiang Shen, Yu-Bin Zhang and Dan Lei

Abstract This article is relied on the market price monitoring system in the cloud platform of the Government Procurement in Tianjin and using the Copula model to establish the dependency relationship between the market reference price, the actual bid price and maximum likelihood method to estimate the unknown parameters. Through comparing fitting effect between the Empirical function and five kinds of Copula defined by different distributions containing Gaussian, t, Clayton, Gumbel and Frank, we can draw a conclusion that the two price series have a high level of dependency, and the Gaussian-Copula fit to the data most. However, any change in one price could not cause significant fluctuations to the other price by examining the tail of the correlation coefficient. From the model research applied in the dynamic changes of the procurement price, the government is able to monitor the reasonableness and accuracy of bids, and a certain preventive mechanism is established for inflated prices.

Keywords Government procurement · Copula model · Procurement management · Market price

1 Introduction

In recent years, in most Europe and America countries, the monitor in the price of government procurement has been developed into a set of engineering theory combined by theory, practice and management all together. Due to reduce the

P. Wang · J. Shen · Y.-B. Zhang · D. Lei
Department of Management and Economics, Tianjin University, Tianjin, China

P. Wang (✉)
Department of Basic Courses, Tianjin Sino-German Vocational Technical College, Tianjin, China
e-mail: pinyuewang@yahoo.com

E. Qi (ed.), *Proceedings of the 6th International Asia Conference on Industrial Engineering and Management Innovation*,
DOI 10.2991/978-94-6239-145-1_110

risk of anomalous high price, we introduced an analysis system of dual Copula to establish price correlation model where the two series of data specified as market preference prices and actual bidding prices, and the comparison of the best fitting of different Copula functions is allowed to predict the effect of market preference price fluctuations. There would be introduced maximum likelihood to the estimation of the dependence parameter, on the condition of the consistent first stage estimates of the marginal distributions. Firstly, the applicability of the normal distribution characterized in terms of the data of normality test and the Q-Q plot is discussed. And then, spline interpolation method describing the cumulative distribution of the two series would be used to build connection with the five different copula functions through the Euclidean distance. Finally, the tail correlation coefficient in financial time series is illustrated, and it is adopted to express the degree of influence between the sequences changes.

2 Literature Review

Since Sklar linked the multivariate distributions function with Copula function by theorem and Nelson (2006) defined n-Copula function by theorem, many Domestic and foreign scholars continued to complete a variety of application types of Copula function, and did a lot of basic empirical researches [1]. Patton (2006) introduced the concept of condition Copula function and establish timing Copula function, to research the problem of asymmetric rate [2]. Freesa et al. (2005) applied the theory of Copula to the measures of credit, established linkages between reliability and profit and loss using general linear model [3]. Based on the traditional model, Hafner et al. (2010) proposed a new semi-parametric Copula model, whose marginal distribution was specified as the GARCH distribution and can be changed by non-parametric methods on a period of time [4]. After constructing vine Copula [5], Guégan et al. (2010) proposed a new method of constructing vine Copula, which can construct more n-vine Copula and was quantifiable and flexible in multiple functions [6].

The researches about the application of Copula are mainly concentrated in the financial futures market in the past, de Melo Mendes et al. (2010) used different kinds of Copula to research Portfolio Risk Management [7]. Empirical studies show that different kinds of Copula have different flexibility on the construction of multivariate dependence structure. Portraying symmetric or asymmetric dependence structure and whether having the characteristics of tail distribution or not can affect the accuracy of the calculation of VaR. Chollete L et al. (2011) used a static Copula to constructed dependence on the international market assets, finding that the dependence of assets enhanced over time, and the East Asian market yields have little downside risk [8]. Dias et al. (2010) used timing Copula function to research exchange rate volatility between three different currencies and used Fisher's transformation to establish dependency structure, and found that changes in exchange rates depend on the period of past earnings [9]; Wang et al. (2011) construct Copula between China's stock market returns and other six international markets, finding that China, Japan and the Pacific's financial markets have a high degree of dependence, and attract investors by investing in diverse international markets [10]. Embrechts (1999) introduced Copula into risk management in the financial markets for the first time [11]; Reboredo et al. (2013) used the Copula function to establish the relationship between the dollar trend and gold price fluctuation, and found gold to improve the effectiveness of monetary investment portfolio risk through the tail dependence coefficient [12]. Kresta (2012) Establish elliptic Copula model through the international stock portfolio risk, and the data were estimated using the Student Copula function, finding that the condition coefficient of tail dependence and dependence can better reflect the actual investment. On the other hand [13], Badanidiyuru et al. (2012) used function to establish the purchase price of the budget system by the symmetric and non-symmetric submodular utility, and put it in different price mechanisms [14]. Neupane et al. (2012) researched the government procurement automation and audit tracking function, finding that it can increase the transparency of government procurement processes and ensure effective accountability and can greatly improve the ability of the government procurement's anti-corruption [15].

3 Basic Theory and Model

3.1 Dealing with the Original Sample Data

Select a particular brand of portable computers totaling 425 sets of data, including the market research price P_1 and the government procurement bidding price P_2;

Class the selected data for the nature of itself, and each transaction price would be multiplied by the corresponding number: $WP_{ij} = P_{ij} * n_j$, weighted average number WP_{ij}; procurement price $P_{ij}(i = 1, 2; j = 1, 2, \ldots, 425)$; the number of each procurement $n_j(j = 1, 2, \ldots, 425)$;

Calculate the average monthly price based on the unit of month: $AP = \sum (WP_{ij}/n_{j)})$, the average price of each month AP; weighted by number of each month $\sum WP_{ij}$; the total number purchasing of each month $\sum n_j$;

Calculate the rate of change of the price of each transaction: $r_i = (P_i - AP)/AP$, the rate of change r; procurement price P; the average price AP.

3.2 Copula Function Model

Reference [1] proposed the properties of the function of N-Copula as follows:

Suppose 1: The function $C(u_1, u_1, \ldots, u_N)$ increases monotonically for each variable;

Suppose 2: The function $C(u_1, u_2, \ldots, 0, \ldots, u_N) = 0, C(1, \ldots, 1, u_i, 1, \ldots, 1) = u_i$;

Suppose 3: For any $ui, vi \in [0, 1](i = 1, 2, \ldots, N)$, $C^{-}(u_1, u_2, \ldots, u_N) = \max\left(\sum_{i=1}^{N} u_i - N + 1, 0\right)$;

Suppose 4: $C^{-}(u_1, u_2, \ldots, u_N) = \max\left(\sum_{i=1}^{N} u_i - N + 1, 0\right)$, $C^{+}(u_1, u_2, \ldots, u_N) = \min(u_1, u_2, \ldots, u_N)$, then, for any

$u_i \in [0, 1](i = 1, 2, \ldots, N)$,

$C^{-}(u_1, u_2, \ldots, u_N) \le (u_1, u_2, \ldots, u_N) \le C^{+}(u_1, u_2, \ldots, u_N)$, it could be recorded that $C^{-} \prec C \prec C^{+}$, and defined the C^{-} is the Frechet lower bound, the C^{+} is the Frechet higher bound. And more, when $N \ge 2$, C^{+} is a function of N-Copula. However, when $N \ge 2$, C^{-} is not a function of N-Copula.

Suppose 1: If $U_i \sim U(0, 1)(i = 1, 2, \ldots, N)$ is independent for each other, then $C(u_1, u_2, \ldots, u_N) = \prod_{i=1}^{N} u_i$.

For the sample data of this paper, the function dual Copula can be carried out effectively to build analysis model, then the expression of five different Copula function would introduced as follows:

For the function of dual Gaussian-Copula, let ρ be the linear correlation coefficient of dual variables, then the expression of the function is that:

$$C^{Gaussian}(u, v; \rho) = \int_{-\infty}^{\phi^{-1}(u)} \int_{-\infty}^{\phi^{-1}(v)} \frac{1}{2\pi\sqrt{1-\rho^2}} \exp\left[-\frac{s^2 - 2\rho st + t^2}{2}\right] dsdt$$

For the function of dual Gaussian-Copula, let ρ be the linear correlation coefficient of dual variables and k be the degrees of freedom, then the expression of the function is that:

$$C^{t}(u, v; \rho, k) = \int_{-\infty}^{t_k^{-1}(u)} \int_{-\infty}^{t_k^{-1}(v)} \frac{1}{2\pi\sqrt{1-\rho^2}} \exp\left[1 + \frac{s^2 - 2\rho st + t^2}{k(1-\rho^2)}\right]^{-(k+2)/2} dsdt$$

For the function of dual Gumbel-Copula, let ∂ be the relevant parameter, when $\varphi(t) = (-\log t)^{\partial}$ becomes the Meta-function, then the expression is that:

$$C^{Gumbel}(u, v, \partial) = \exp(-((-\log u)^{\partial} + (-\log v)^{\partial})^{1/\partial})$$
$$\partial \in [1, +\infty]);$$

For the function of dual Cayton-Copula, let ∂ be the relevant parameter, when $\varphi(t) = (t^{\partial} - 1)/\partial$ becomes the Meta-function, then the expression is that:

$$C^{Clayton}(u, v, \partial) = \max((u^{-\partial} + v^{-\partial} - 1)^{-1/\partial}, 0)$$
$$\partial \in ((-1, 0) \bigcup (0, +\infty))$$

For the function of dual Frank-Copula, let ∂ be the relevant parameter, when $\varphi(t) = -\log\frac{e^{\partial t}-1}{e^{\partial}-1}$ becomes the Meta-function, then the expression is that:

$$C^{Frank}(u, v, \partial) = -\frac{1}{\partial}\log\left(1 + \frac{(e^{-\partial t} - 1)(e^{-\partial v} - 1)}{e^{-\partial} - 1}\right)$$
$$\partial \in (-\infty, 0) \bigcup (0, +\infty)$$

In order to facilitate the selection of the higher fitting of the Copula function, the empirical Copula should be introduced and compared with different type of Copula function, and let $(x_i, y_i)(i = 1, 2, \ldots, n)$ becomes the representatives of sample data of the dual totality. When the empirical distribution function of X, Y are $F_n(x)$ and $G_n(y)$, then the empirical function of Copula describing (x_i, y_i) is that:

$\overset{\Lambda}{C}_n(u, v) = \frac{1}{n}\sum_{i=1}^{n} I_{[F_n(x_i) \le u]} I_{[G_n(y_i) \le v]}$ $u, v \in [0, 1]$, and $I_{[\bullet]}$ in this function is the indicator function, when $F_n(x_i) \le u, I_{[F_n(x_i) \le u]} = 1$, otherwise $I_{[F_n(x_i) \le u]} = 0$.

Introduce the concept of Euclidean distance, and select the type of Copula function could meet the dual function of sample data distribution best, then the Euclidean distance can be defined that:

$$d_r^2 = \sum_{i=1}^{n} \left| \overset{\Lambda}{C}_n(u_i, v_i) - \overset{\Lambda}{C^q}(u_i, v_i) \right|^2, \text{ when } \quad r = 1, 2, 3, 4, 5,$$

$\overset{\Lambda}{C^q}$ is the Copula function of Gaussian, t, Gumbel, Clayton and Frank, when any Copula function can be adopted, then its Euclidean distance is the minimum of all compared with any other Copula function.

3.3 Estimation of Unknown Parameters

Use the maximum likelihood method to estimate the unknown parameters of the function. If the general case would be considered, the marginal distributions of the random variables are $F(x; \theta_1)$ and $G(y; \theta_2)$ for the two-variable model of X and Y, and the marginal density function are $f(x; \theta_1)$ and $g(y; \theta_2)$ (θ_1 and θ_2 are the unknown parameters of the marginal distribution). The Copula density function is

$c(u,v;\alpha) = \frac{\partial^2 C(u,v;\alpha)}{\partial u \partial v}$ (α is the unknown parameter of the Copula function), then the joint distribution function of both sides is that:

$$H(x,y;\theta_1,\theta_2,\alpha) = C[F(x;\theta_1), G(y;\theta_2);\alpha]$$

The joint density function of (X,Y) is that:

$$h(x,y;\theta_1,\theta_2,\alpha) = \frac{\partial^2 H}{\partial x \partial y} = c[F(x;\theta_1), G(y;\theta_2);\alpha]f(x;\theta_1)g(y;\theta_2)$$

Then the likelihood function of the sample $(X_i, Y_i)(i = 1, 2, \cdots, n)$ can be calculated:

$$L(\theta_1,\theta_2,\alpha) = \prod_{i=1}^{n} h(x_i, y_i;\theta_1,\theta_2,\alpha) = \prod_{i=1}^{n} c[F(x;\theta_1), G(y;\theta_2);\alpha]f(x;\theta_1)g(y;\theta_2)$$

Make the description of the logarithmic scale of the above functions, the result as follows:

$$\ln L(\theta_1,\theta_2,\alpha) = \sum_{i=1}^{n} c[F(x;\theta_1), G(y;\theta_2);\alpha] + \sum_{i=1}^{n} f(y;\theta_1) + \sum_{i=1}^{n} g(y;\theta_2)$$

Solve the maximum point of the function values, then the marginal distribution and the unknown parameters can be found out, and the following formula is that:

$$\hat{\theta}_1, \hat{\theta}_2, \hat{\alpha} = \arg\max \ln L(\theta_1,\theta_2,\alpha)$$

4 Analysis of Data

The data in this paper are quoted from the database of Tianjin Municipal Government Procurement Center of a certain brand of portable computers from March 2014 to August 2014, divide into the market reference price time series (X) and the actual bid price time series (Y) totaling 425 sets of data, where the market reference price is the data raised by sourcing center of electronic research platform based on the actual market situation, and the actual bid price was generated by the bidding price. In order to determine the actual form of the distribution of the data, establish the price fluctuations histogram based on the two sets of data in the way of 1.1, as the Figs. 1 and 2.

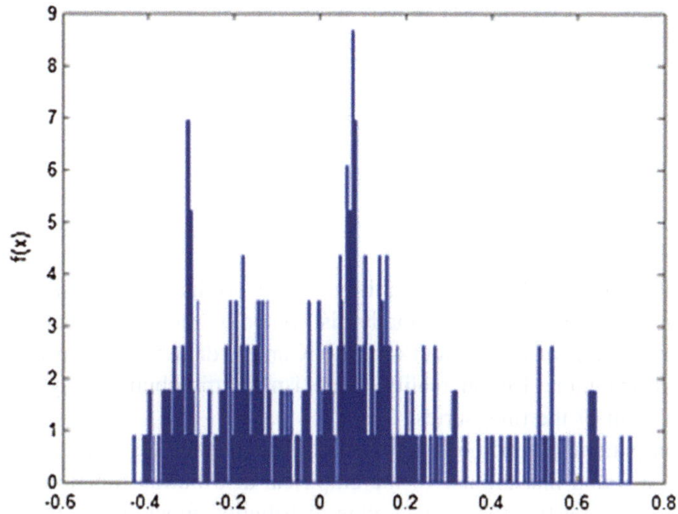

Fig. 1 The change rate of the market price histogram

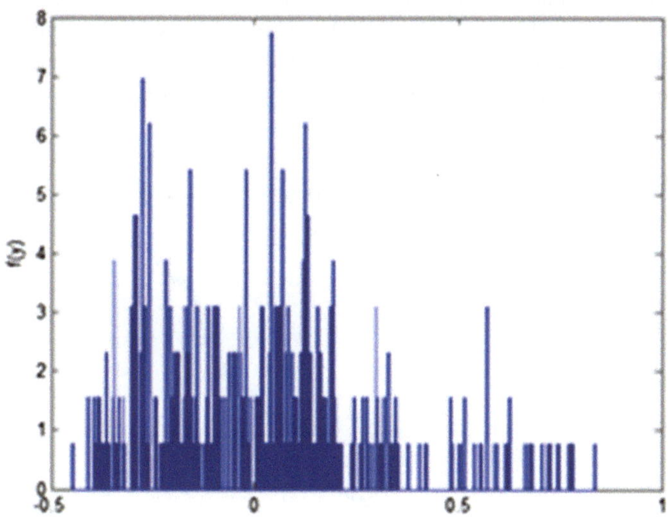

Fig. 2 The change rate of the auction histogram

By the preliminary observations of the drawings, it cannot draw the conclusion that the two sets of data are in line with the normal distribution. In order to verify the truth of the assumption one, other two types of price data calculated parameters are shown in Table 1, and the normality test of the three types as follows.

Through the above test procedure, we found that the P values of two types of data in normality test are less than 0.05 in probability level. And it showed lower

Table 1 Preliminary description and analysis of data

Type	Skewness	Kurtosis	J-B test	K-S test	Lilliefors test
Market reference price (X)	0.5472	2.9983	0.0013 (h = 1)	0.012 (h = 1)	0.001 (h = 1)
Actual bid price (Y)	0.7892	3.4257	0.001 (h = 1)	0.0235 (h = 1)	0.001 (h = 1)

levels of the situation by the probability of the original hypothesis, and it also describes that X and Y are not normally distributed by the rate of change as shown by QQ plot analysis of Figs. 3 and 4 of the X and Y data, which showed peak and thick tail distribution. But in reality, there isn't a distribution way can fit in a distributed form of the time series.

Use non-parametric methods to estimate the overall distribution of the two sequences. It found that the use of the empirical distribution function method and kernel distribution estimator derived edge distribution function is almost the same by studying, and the function values are quite close which are shown as the Figs. 5 and 6. Then select one of them to establish two variables frequency histograms and two variables frequency histograms are shown in Figs. 7 and 8 as follows:

The joint density function of (X, Y) can be observed from Fig. 8 (Copula function chart) have a symmetrical tails, initially considered the Gaussian-Copula function and t-Copula function able to fit better the effect of the data model, but for

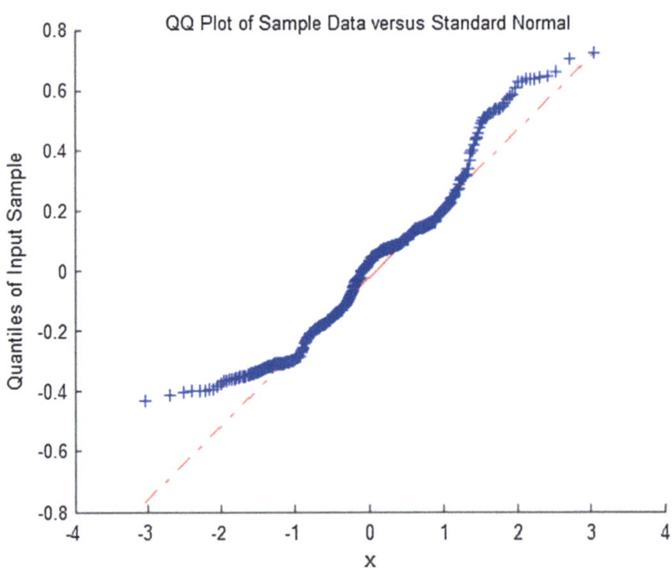

Fig. 3 QQ plot of the market reference price

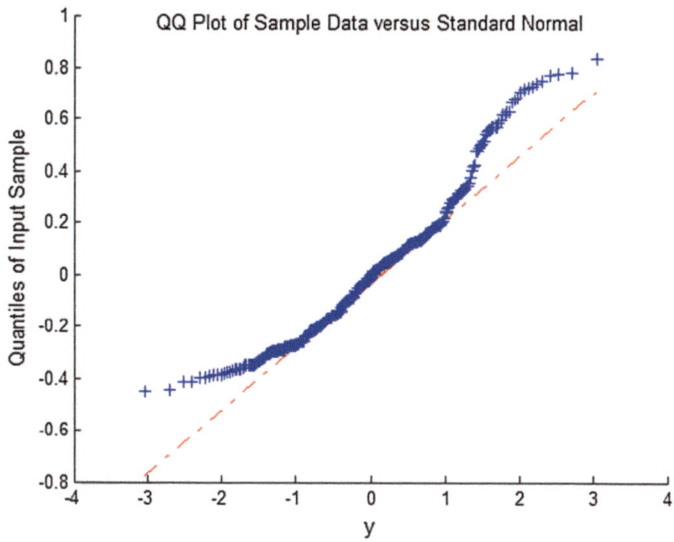

Fig. 4 QQ plot of the bid price

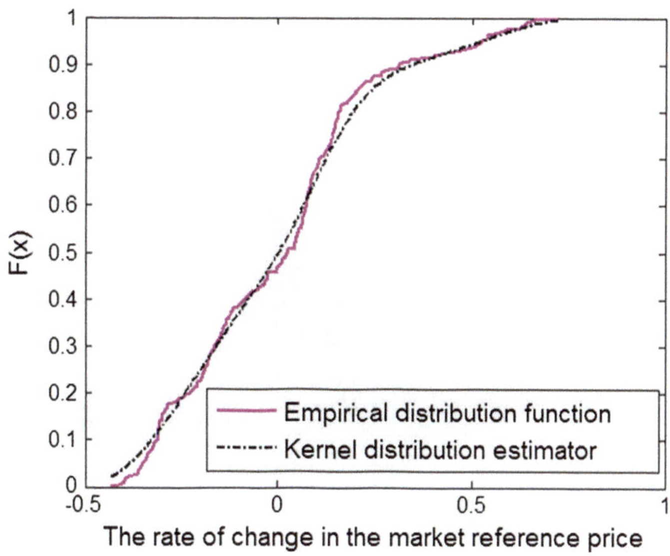

Fig. 5 The market reference price marginal distribution function

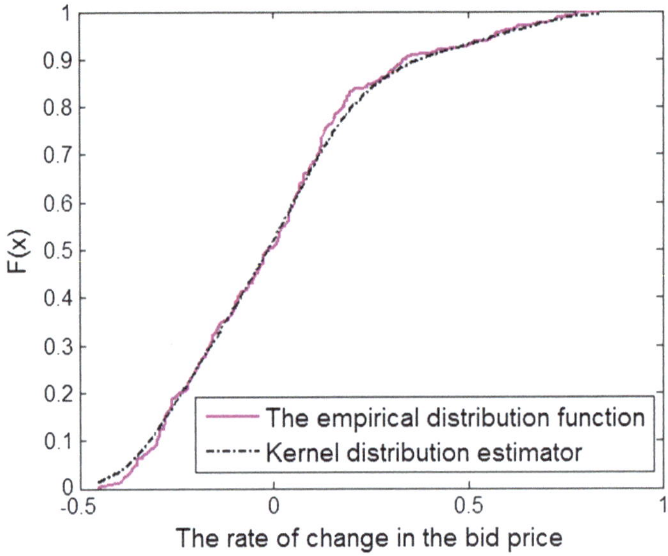

Fig. 6 The actual bid price marginal distribution function

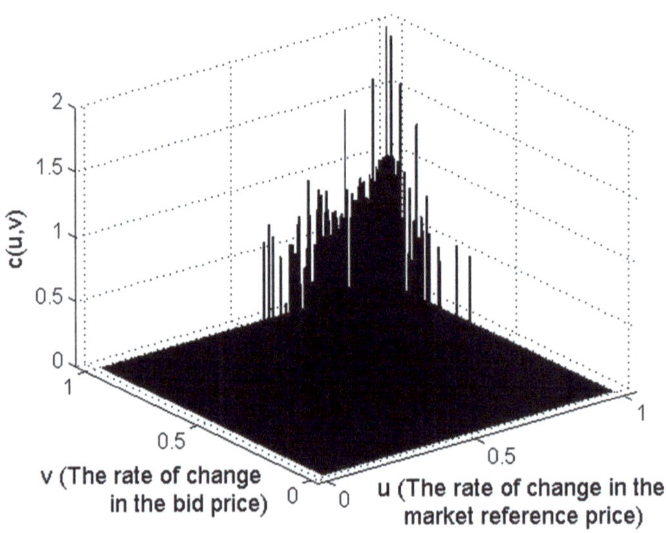

Fig. 7 Two variables frequency histograms

other Archimedes Copula function fitting the merits, the below is the Archimedes Copula Function Modeling for this model, find five kinds Copula function parameters are shown in Table 2.

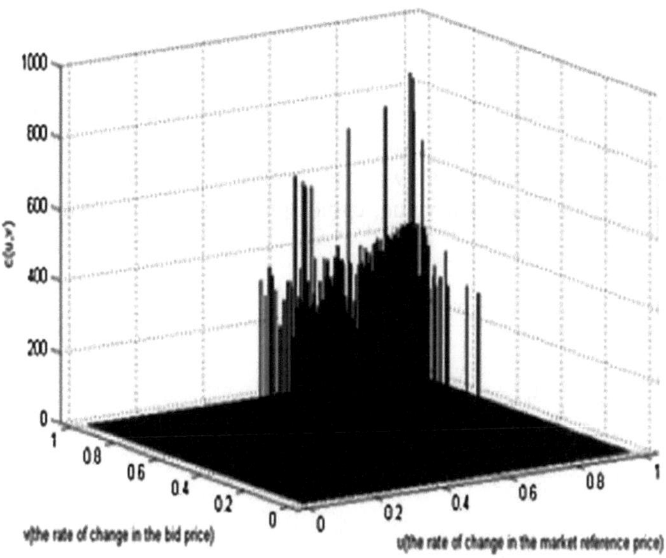

Fig. 8 Copula distribution histograms

Table 2 The parameters of five types fuctions

Copula type	Experience	Gaussian	t	Gumbel	Clayton	Frank
Parameters		0.9342	0.9439 (K = 19.8419)	4.2703	4,5525	15.3976
Kendall rank correlation coefficient	0.7581	0.7678	0.7857	0.7658	0.6948	0.7680
Spearman rank correlation coefficient	0.9174	0.9282	0.9387	0.9224	0.8699	0.9325
The tail correlation coefficient			0.4517	−2.2703		
the end of the correlation coefficient			0.4517		0.2064	
Euclidean distance		0.0372	0.0418	0.0516	1.1356	0.1747

For this model variable feature, Nelson [1], Durante et al. [16] defined the tail correlation coefficient was as follows:

$$\lambda_{UU} = \lim_{u\to 1} P\left[Y \geq F_Y^{-1}(u) \middle| X \geq F_X^{-1}(u)\right] = \lim_{u\to 1} \frac{1-2u+C(u,u)}{1-u}$$

$$\lambda_{LL} = \lim_{u\to 1} P\left[Y \leq F_Y^{-1}(u) \middle| X \leq F_X^{-1}(u)\right] = \lim_{u\to 1} \frac{C(u,u)}{u}$$

To test the function goodness of fitting in the obtained Copula, calculate cumulative distribution function of the two yield time series separately. Using spline interpolation method on two data points to obtain the original sample empirical distribution function value, cubic spline interpolation method is defined as follows:

$$S_i(x) = \frac{Z_{i+1}(x-x_i)^3 + Z_i(x_{i+1}-x^3)}{6h_i} + \left(\frac{y_{i+1}}{h_i} - \frac{h_i}{6}Z_{i+1}\right)(x - x_i) + \left(\frac{y_{i+1}}{h_i} - \frac{h_i}{6}Z_i\right)(x_{i+1}-x)$$

And $h_i = x_{i+1} - x_i$

$$\begin{cases} Z_0 = 0 \\ h_i Z_{i-1} + 2(h_{i-1}+h_i)Z_i + h_i Z_{i+1} = 6\left(\frac{y_{i+1}-y_i}{h_i} - \frac{y_i - y_{i-1}}{h_{i-1}}\right) \\ Z_n = 0 \end{cases}$$

It is defined in the space of Sobolev $H^2([a;b])$, which is the function of minimize $J(f) = \int_a^b |f''(x)|^2 dx$ functional, and the curvature of $f(x)$ is: $k = \left|\frac{f''(x)}{(1+f''(x)^2)^{\frac{3}{2}}}\right|$

Through the analysis and comparison in the distance of European square in Table 2, Gaussian-Copula function for the two data sets of this model has the best fitted distribution: Market reference price fluctuations would not have the same impact on the bid price, but the comparison of the two rank correlation coefficient shows that the correlation coefficient of the two prices presented on a high level, with a high degree of dependency.

5 Conclusion

Through the establishment of Copula model of two kinds of data in this paper, we find that the reference price and the bidding in the marketing price system have a high degree of correlation. And through the comparison of 5 kinds of Copula function for fitting goodness, we find that a kind of price's rise or fall cannot cause another price's rise or fall. Due to the fluctuations of the market reference price does not yield the same degree of fluctuations of procurement bidding price, bidding price has high independence. Therefore, when the government forecasts the purchasing price referenced the market reference price system, they should take the market price as the purchase price of the benchmark, but should not trend the fluctuation of market price as the benchmark to determine the amplitude of fluctuation and the quantitative analysis of purchasing price.

This paper is based on the market forecast Platform of Tianjin Municipal Government Procurement Center in China, establishing Copula model and making dependency analysis about the two kinds of price, with introducing 5 kinds of Copula model into the actual purchase price mechanism of deterministic construction, and draws the corresponding conclusion. However, the final conclusion shows that the fluctuation of market price cannot produce the same effect on the actual feedback of price fluctuations. Due to the limitations of the model, we put forward two kinds of impact on the volatility of the dependent factors, to provide ideas for later scholars:

(1) The instant sensitivity of the actual bidding price to market price changes is not high, with existing a certain period of time lag. Then, we may establish ARMA or GARCH model on the actual prices, to find the optimal lag period and the variation characteristics of error term.
(2) By the influence of the game behavior between the purchasers (the governments) and the bidders in commodity procurement bidding process cannot be ruled out. We can establish multiple return matrix bid among the purchaser, the bid winner and other bidders, based on the actual market price, seeking the optimal Nash equilibrium solution, and then compare it to the market price, observing the effect of the two kinds of price volatility.

References

1. Nelson RB (2006) An introduction to Copulas. Springer Press, New York, pp 158–166
2. Patton AJ (2006) Modelling asymmetric exchance rate dependence. Int Econ Rev 2:527–555
3. Freesa EW, Wang P Ph.D (2005) Credibility using Copulas, North Am Actuarial J 9(2):32–48 (in UK)
4. Hafner CM, Reznikova O (2010) Efficient estimation of a semiparametric dynamic Copula model. Comput Stat Data Anal 54(11):2609–2627 (in Louvain-la-Neuve, Belgium)
5. Bedford T, Cooke RM (2001) Probability density decomposition for conditionally dependent random variables modeled by vines. Ann Math Artif Intell 32:245–268 (in Scotland)
6. Guégan D, Maugis PA (2010) Prospects on vines. Forthcoming in Insurance Mark Co: Anal Actuarial Comput 1(1):4–1 (in France)
7. de Melo Mendes BV, Semeraro MM, Leal RPC (2010) Pair-Copulas modeling in finance. Financ Mark Portfolio Manage 24(2):193–213 (in Rio de Janeiro, Brazil)
8. Chollete L, De la Pena V, Lu CC (2011) International diversification: a Copula approach. J Bank Finance 35(2):403–417 (in Norway)
9. Dias A, Embrechts P (2010) "Modeling exchange rate dependence dynamics at different time horizons. J Int Money Finance 29(28):1687–1705 (in UK/Switzerland)

10. Wang K, Chen YH, Huang SW (2011) The dynamic dependence between the Chinese market and other international stock markets: a time-varying Copula approach. Int Rev Econ Finance 20(4):654–664 (in Taiwan)
11. Embrechts P, McNeil A, Straumann D (1999) Correlation: pitfalls and alternatives. Risk-London-Risk Mag Ltd 12:69–71
12. Reboredo JC (2013) Is gold a safe haven or a hedge for the US dollar? Implications for risk management. J Bank Finance 37(8):2665–2676 (in Spain)
13. Kresta A, Tichy T (2012) International equity portfolio risk modeling: the case of the NIG model and ordinary Copula functions. Czech J Econ Finance 62(2):141–161 (in Ostrava)
14. Badanidiyuru A, Kleinberg R, Singer Y (2012) Learning on a budget: posted price mechanisms for online procurement. In: Proceedings of 13th conference on ACM conference on electronic commerce. University of Pennsylvania, Philadelphia, PA, pp 128–145
15. Neupane A, Soar J, Vaidya K et al (2013) Role of public e-procurement technology to reduce corruption in government procurement. In: Proceedings of 5th conference on international public procurement conference. Public Procurement Research Center, Seatle, WS, pp 304–334
16. Durantea F, Kolesárová A, Mesiarc R, Sempi C (2008) Semilinear Copulas. Fuzzy Sets Syst 159(1):63–76 (in Austria)

Causal Effect Analysis of Visual Management on Customer Satisfaction and Repurchase Intention Using McDonald's as an Example

Fong-Fan Wang and Tsan-Ming Chang

Abstract In this paper we explore the causal effect of fast-food industry using tentative constructs via some McDonald's stores located around central Taiwan as the studied objects. The empirical study shows that the two dimensions: "visual management inside the store" and "visual management outside the store" have significant correlation with each other and both management methods are equally important. Further, both management methods have equally significant effect on "satisfaction level". Empirical results show that regardless of internal or external visual management, customer satisfaction is higher whenever the visual management is clearer. The result reveals positive relationship between "customer satisfaction" and "repurchase intention". That is, higher customer satisfaction results into higher repurchase intention. Thereby our study confirms the role of the customer satisfaction variable as a mediation, which affects customer repurchase intention.

Keywords Visual management · Customer satisfaction · Repurchase intention · SEM

1 Introduction

McDonald's is a well-known fast-food industry leader. They make customized ordering so efficiently that their customers can eat fresh and delicious food at McDonald's within as short as a couple of seconds. In fact, only change of system processes together with truly revolutionary measures make today's McDonald's a successful practice. The changes include process planning, kitchen equipment, IT systems, visual management processes and recycling procedure. Since McDonald's

F.-F. Wang (✉) · T.-M. Chang
Department of Industrial Engineering and Management, Hsiuping University of Science and Technology, Taichung, Taiwan
e-mail: frank@hust.edu.tw

E. Qi (ed.), *Proceedings of the 6th International Asia Conference on Industrial Engineering and Management Innovation*,
DOI 10.2991/978-94-6239-145-1_111

uses original process, skilled staff and advanced kitchen equipment to make fresh fast-food, the revolutionary process has taken a long time to become what it is today. Traditionally, quality and efficiency are on the opposite sides, i.e. when the quality is achieved, speed is bound to lose. However, the customer can't stand waiting for a long time. Therefore the whole procedure must be re-planned as a whole, which reflects on a variety of measures. In order to achieve efficiency, McDonald's think how to use systems planning, work in progress management in kitchen process to quickly produce finished products.

McDonald's innovates in enhancing its competitiveness through several measures which include effective visual management. For example, McDonald's starts the reform process in order to help its consumers when they travel around. When consumers travel to other countries and they do not speak the local languages, the consumers may feel helpless because they do not know how to order meals. Different ordering habits and table manners would cause further confusion. However, the implementation of internal and external visual management using standardization and color management within McDonald's have effectively reduce consumers' confusion and unnecessary search. Furthermore the thinking of lean service as well as effective visual management on the shop floor has reduced consumers' shopping times. The improvement through the lean supply and internal visual management within McDonald's store operation results in more efficient performance. Since currently there is rare related literature about the complex relationship between the visual management, customer satisfaction and re-purchase intention within a fast-food supply system, we are therefore motivated to conduct this research using some stores of McDonald's located somewhere around central Taiwan as an example.

In this paper we define two visual management methods: visual management inside the store and visual management outside the store. Visual management inside the store include using scoreboards, production control charts, or any other visual management displays to keep vital information flowing between management and employees such that quality control and production indication can be implemented on the shop floor [7]. On the other hand, visual management outside the store is that enterprises try to use marketing strategies such as advertisement, sales promotion, brand identity etc. to communicate with its potential customers in hoping that customers' long-term loyalty relationship with the enterprises can be established [4, 12] studied Taiwan's fast-food industry leader McDonald's for nearly 4 years. The author found that for the fast-food industry who wishes to establish long-term and stable relationships with its consumers the most important job is to implement the customers' overall satisfaction in order to create outstanding customer value and loyalty, in addition to improving corporate image. Lin [13] conducted a comparative analysis through ways of defining the market using demographic variables in the market segment, he finds there are not significant preference difference between consumers of McDonald's and MOS Burger. However, market segmentation with higher brand loyalty has significant differences. The author showed McDonald's brand consciousness, through advertising and the image of quality of their products has a significant impact on consumer purchase intention. Through strengthening advertising effect and maintaining product quality, customers' brand awareness is invoked, resulting in higher purchase intention and willingness to buy again. The customers even recommend others to buy. Other researches pertaining to service satisfaction to achieve customer loyalty are referred to [11, 12]. However, the above studies only investigate the relationship between customers' satisfaction or brand loyalty and purchase intension, lacking a comprehensive study from the service generation to customer evaluation. The objective of this study is thus to establish a causal relationship between service provisioning and customer perception via visual management inside the store (abbreviated as in-store) and outside the store (abbreviated as out-store) of some stores of McDonald's located around central Taiwan as research objects. This study explores whether the two dimensions, which form effective management as well as eye-catching advertisement from the service providers, will affect the customer satisfaction and customer repurchase intention. We research through consumers' questionnaire redesign first, followed by investigation of consumers' perception on McDonald's visual management, customer satisfaction level, and re-purchase intention. We look forward to exploring causal model using the statistical structure equation modeling (SEM) to reveal consumers' perceptions about the efforts of McDonald's and hope that the results provide fast-food industry as a useful reference model. SEM is a useful causal relationship research tool. For more research on the causal relationship research line using SEM we refer to [6, 10].

2 Literature Review

It is known from the definition of visual management that enterprises can use form, image, and adequate color scheme to manage their production activities such that the productivity can be enhanced [5]. Visual management is characterized by a management skill integrating public information and visual presentation. It is also called "a visible management". There are some principles to follow when one uses visual management. For example, according to [15] it should arouse the employee's participating willingness. It must be easy to see and easy to understand. Signs and Kanban must be put on significant places. The visual tools have to be cleaned and repainted periodically. Internal on job training has to be carried out for the related responsible personnel continuously such that they can understand the importance of visual management and know how to utilize the visual management skill. To create more profits, enterprises have to maintain their customers in hoping that they repurchase again and again. The more rate of return of their loyal customers means the service quality is acknowledged by these customers. In order to maintain a sustainable growth and increase the profit continuously, the companies have to keep their current customers in addition to creating more new markets. According to [11] the cost of searching for a new customer is five times that of maintaining an old customer. According to [3], customer satisfaction is the cause of customer loyalty. Therefore, nowadays every corporate is very concern about the customer relationship hoping that through maintaining a good relationship with their customers and providing good service the customers can repurchase again [14]. Researched the relation between customer loyalties and repurchase intention and found there is a positive relation between them [16]. Found that customer satisfaction creates customer loyalty and such feeling of loyalty cause the customer willing to repurchase again [12]. Also agrees that more customer satisfaction enhance customer repurchase intension.

3 Research Method

In this section we first propose research framework followed by research hypotheses.

3.1 Hypothesized Relation Chart

We conjecture that the two dimensions: "in-store visual management" and "out-store visual management" have positive causal effect on "customer satisfaction level". And "customer satisfaction level" has positive causal effect on "repurchase intention". Figure 1 shows the hypothesized relation chart of this study.

3.2 Research Hypotheses

Our research hypotheses are as follows:

H1: The correlation between "In-store visual management" and "out-store visual management," is significant.

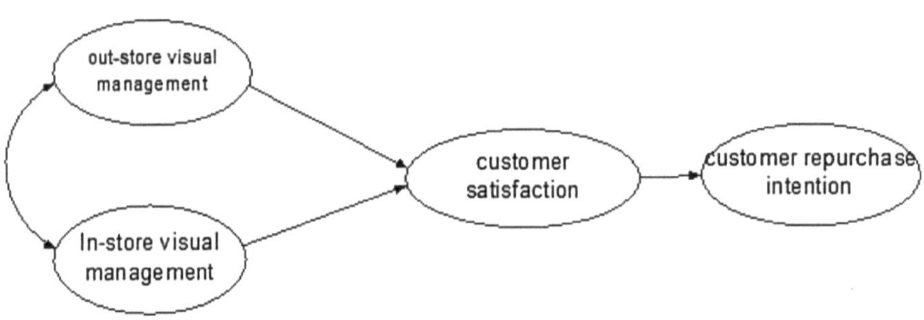

Fig. 1 The study chart of this research

H2: "The In-store visual management" has significant influence on "customer satisfaction".
H3: "The ut-store visual management" has significant influence on "customer satisfaction".
H4: "customer satisfaction" has a positive influence on "customer repurchase intention".

3.3 Research Design

This research is initiated with the questionnaire redesign used as a measuring tool for visual management. Questionnaire is designed as: the visual management of shop inside and outside the store, customer satisfaction level, customer-repurchase intention. Due to the insufficiency and ambiguity of domestic and abroad questionnaires related to fast-food industry under study we resort to design the research questionnaire ourselves to enhance the quality of the questionnaire. The questionnaire we designed used Likert's scale of five points: "strongly agree", 5, "agree, 4, "ordinary", 3, "disagree", 2 and "strongly disagree", 1 to represent the extent of the consent level faced by the interviewees.

The formal questionnaire investigation is conducted several times in 2011. The official questionnaires are dispatched randomly to 400 correspondents who finished their meals at some McDonald's stores located around central Taiwan. The number of valid questionnaires is 299 copies. The measuring parameters of the four dimensions of the formal questionnaire are shown from Tables 1, 2, 3 and 4 respectively.

Table 1 McDonald's in-store visual managemen

1. I agree that "sign of business hours inside store" is clearly and easily identified
2. You agree that "sign of escape safety equipment inside store" is clearly and easily identified
3. You agree that "recycling signs inside store" is clearly and easily identified
4. You agree that "meal pricelist" is clearly and easily identified
5. You agree that "pricelist of meal" is clearly and easily identified
6. You agree that "McDonald's meal boxes differentiated by color signs" is clearly and easily identified
7. You agree that "sign of restroom" is clearly and easily identified
8. You agree that "sign of children's play-ground" is clearly and easily identified
9. You agree that "safety signs for children's play-ground" is clearly and easily identified
10. You agree that "ads of new promotional specials" is clearly and easily identified

Table 2 McDonald's out-store visual management

1. You agree that "McDonald's business logo" is clearly and easily identified
2. You agree that "sign of drive-through order" is clearly and easily identified
3. You agree that "drive-through checkout line" is clearly and easily identified
4. You agree that "pick up your order at drive-through course" is clearly and easily identified
5. You agree that "parking lot sign" is clearly and easily identified

Table 3 Customer satisfaction with McDonald's visual management

1. I think the "the clear identification of the inside store billboard" is very satisfactory
2. I think "the clear identification of the outside store billboard" is very satisfactory
3. I think the "instructions set of signs inside store" is very satisfactory
4. I think "instructions set of signs outside store" is very satisfactory
5. As a whole, "Clear and easy identification of McDonald's overall visual management" is very satisfactory

Table 4 Repurchase intention at McDonald's

1. Even McDonald's meals price is raised up, I will still continue to shop at McDonald's for dinner
2. Compared with fellow fast-food restaurant, I will still choose to shop at McDonald's for dinner
3. To organize similar meetings such as family gatherings, I'll still choose to shop at McDonald's for dinner
4. Even though I experience service problems at McDonald's, I would like to shop at McDonald's once again for meal

4 Causal Analysis

This section begins with the path analysis of the hypothetic model, followed by hypothesis testing results, and analysis of goodness of fit (GOF), including inner and overall GOF.

4.1 Path Analysis

SEM is a very good tool to understand the causal model, which is aimed at testing whether the evidence supports the hypothesis of causality or not. SEM measures the

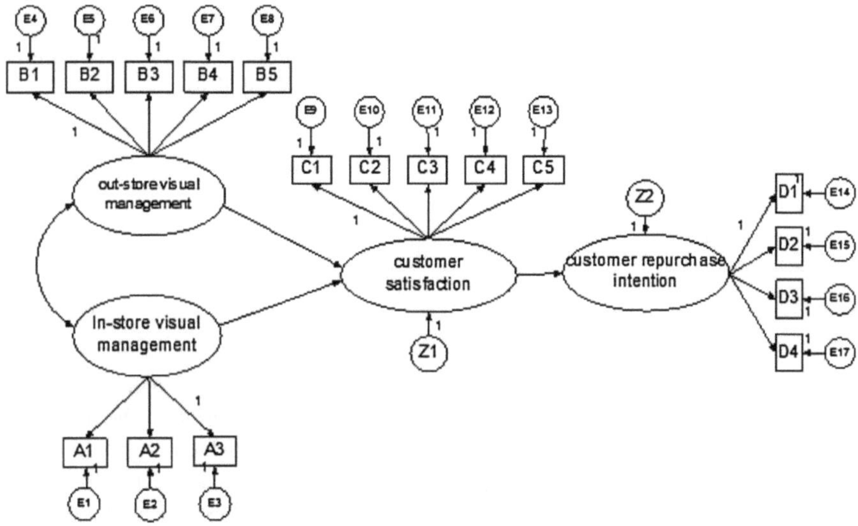

Fig. 2 Relational schema

detailed relationships between variables and errors. The original relational model of the whole causal map is drawn in Fig. 2, consisting of a diagram to understand the entire causal pattern: the study's four-constructs, i.e. internal and external visual management, level of customer satisfaction, and customer repurchase intention. B1 to B5 represent measurement variables for out-store visual management. A1 to A3 denote measurement variables for in-store visual management. Measurement variables of customer satisfaction are coded from C1 to C5. Measurement variables of customer repurchase intention are denoted from D1 to D4. E1 to E17 denote errors of each measurement variable respectively. Z1 and Z2 account for the model error.

4.2 Analysis of the Inner Goodness of Fit

Chang [4] proposed an index named construction reliability (CR). The minimum threshold value for CR should be 0.7. Reliability refers to measurements of the degree of stability. It means with higher reliability, the more reliable the results are. Cronbach's α is by far the most commonly used reliability index. According to Cronbach definition, if the α coefficient is less than 0.35, it indicates that the reliability is low; if $0.35 < \alpha < 0.7$ the reliability is acceptable; If the α coefficient is greater than 0.7, it is a high reliability. Chang [4] suggested that average extraction (AVE) must be at least 0.5, to ensure the convergent validity of the dimensions. Bentler [2] believed that to attain fit requirement the standardized factor loading of individual observed variables must be greater than 0.45 and the total amount of explanation (R2) must be above 0.2. The inner goodness of fit is shown in Table 5.

Table 5 Validity and reliability result

	Average changing an extraction (AVE)	Construct reliability (CR)	Cronbach's α
Out-store visual management	0.5276	0.8422	0.822
In-store visual management	0.4729	0.7285	0.828
Customer satisfaction	0.5578	0.8629	0.862
Customer repurchase intention	0.408	0.731	0.725

From Table 5, the construct validity and α reliability is greater than 0.7 in each dimension. The above results are in line with the recommendation of the threshold. On average, an AVE of visual management and customer satisfaction should be 0.5. Our results show that in-store visual management and customer satisfaction have AVE of 0.4729 and 0.408 respectively, which are near 0.5, the threshold suggested by the literature. Further, the factor loadings of customer satisfaction and customer repurchase intention are both greater than 0.45, which is in line with [2]. Based on the above results we are confident in this study of cause and effect with a certain reliability and validity.

4.3 Analysis of the Overall Goodness of Fit

The overall goodness of fit is shown in Table 6. From Table 6, GFI Value is 0.913, CFI Value is 0.929, AGFI Value is 0.884, NFI Value is 0.877. All are not far from the ideal value, 0.9. RMR Value is 0.035, which is very close to ideal value, 0. Due to the overall fit of the above statistics, it justifies the causal model of this study.

4.4 Hypothesis Testing Results

The SEM results are presented in Fig. 3. Table 7 lists the original hypotheses with their results. Below the description of the results corresponding to the hypotheses is further stated.

H1: There is a significant correlation between "In-store visual management" and "out-store visual management". From Fig. 3 and Table 7, the correlation coefficient between McDonald's in-store and out-store visual management is as high as 0.48. The hypothesis H1 is justified.

H2: "In-store visual management" has significant influence on "customer satisfaction". Figure 3 and Table 7 indicate that the path from "in-store visual management" to "customer satisfaction" is significant with path coefficient 0.27 (t value

Table 6 The overall Goodness of Fit of this study

Statistical measures	Standard of fit	Suggestion in the literature	The results	Conform to the index?
Index of the absolute Goodness of Fit				
Chi-square degree of freedom	It is recommended CHI/DF < 3 to indicate good overall GOF	[8]	2.183	Yes
GFI	The closer to 1, the better as a whole, it is recommended that GFI > 0.9	[8]	0.913	Yes
AGFI	The closer to 1, the better as a whole, it is recommended that AGFI > 0.8	[1]	0.884	Yes
RMR	The smaller the better as a whole, it is recommended that RMR < 0.05	[8]	0.035	Yes
RMSEA	The smaller the better as a whole, it is recommended that RMSEA < 0.1	[8]	0.063	Yes
Index to Goodness of Fit				
IFI	The closer to 1 the better as a whole, it is recommended that IFI > 0.9	[10]	0.930	Yes
NFI	The closer to 1 the better as a whole, it is recommended that NFI > 0.9	[7]	0.877	No
RFI	The closer to 1 the better as a whole, it is recommended that RFI > 0.9	[7]	0.855	No
CFI	The closer to 1 the better as a whole, it is recommended that CFI > 0.9	[9]	0.929	Yes
NNFI	The closer to 1 the better as a whole, it is recommended that NNFI > 0.9	[9]	0.916	Yes
Simple index of Goodness of Fit				
PGFI	The closer to 1 the better as a whole, it is recommended that PGFI > 0.5	[8]	0.686	Yes
PNFI	The closer to 1 the better as a whole, it is recommended that PNFI > 0.5	[8]	0.742	Yes

is 4.55). T value is greater than 1.96, indicating highly positive correlation. The hypothesis H2 is justified.

H3: "Out-store visual management" has significant impact on "customer satisfaction". Figure 3 and Table 7 indicate that the path from "out-store visual management" to "customer satisfaction" has significant influence with path coefficient, 0.46

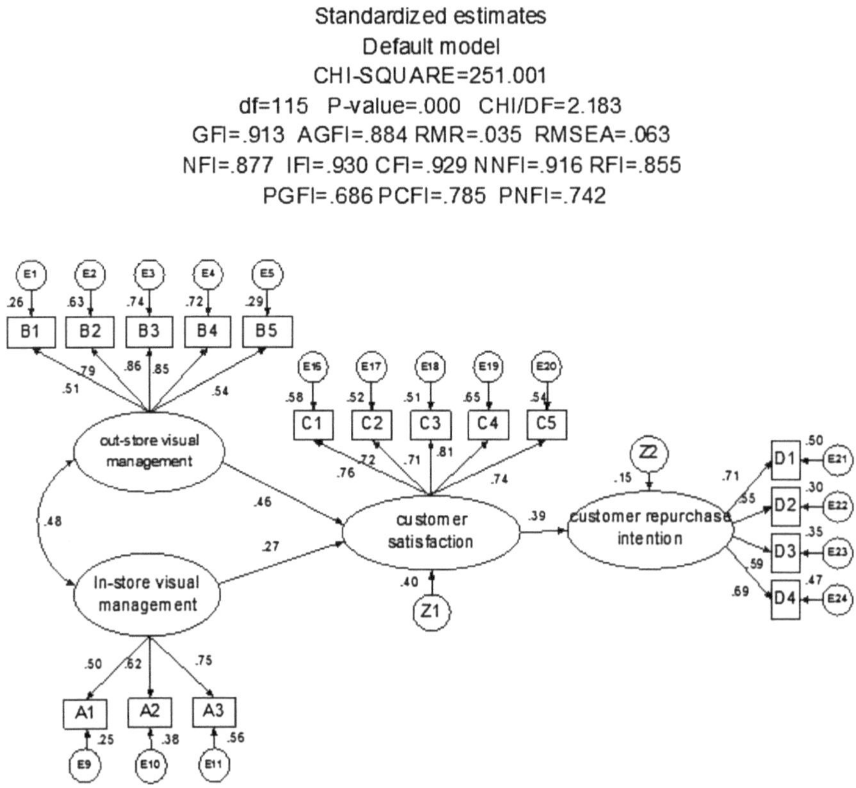

Fig. 3 Overall relational path coefficient map

Table 7 Empirical results of research hypothesis

Research hypotheses		t	P	Empirical results
H1	The correlation between "In-store visual management" and "out-store visual management" is significant	5.222	***	Establishment of significance
H2	The management of "in-store visual management" has significant influence on "customer satisfaction"	4.55	***	Establishment of significance
H3	"Out-store visual management" has significant influence on "customer satisfaction"	4.789	***	Establishment of significance
H4	"Customer satisfaction" has a positive relationship with "customer repurchase intention"	5.199	***	Establishment of significance

(t value of 4.789). T value is greater than 1.96, indicating highly positive correlation. The hypothesis H3 is justified.

H4: "Customer satisfaction" has a positive relationship with "customer repurchase intention". Figure 3 and Table 7 indicate that the path from "customer satisfaction" to "customer repurchase intention" has a positive relationship with path coefficient 0.39 (t value of 5.199). T value is greater than 1.96, indicating highly positive correlation. The hypothesis H4 is justified.

Combing the above results, visual management inside and outside the McDonald's shops has significant impacts on the customer satisfaction, which acts as an intermediary variable to impact customer repurchase intention.

5 Conclusion

Whether to go to a fast-food restaurant for consumption again is dependent on the pleasure and convenient consumption experience and less time in waiting. This may be obtained through proper visual management as well as other management measures. This research verifies visual management inside and outside some stores of McDonald's located around central Taiwan significantly affect customer satisfaction, thereby affecting the repurchase intension. The results obtained from this research may be provided to other fast-food industries as a reference model hoping that the managers may utilize their visual management more effectively such that more profits can be created. However, it is also to be noted that this study is based only on a case study on McDonald's using some stores located around central Taiwan as research samples, to apply the results found here to all of the fast-food industry may not guarantee a unanimous success.

References

1. Bagozzi RP, Yi Y (1988) On the evaluation of structure equation models. J Acad Mark Sci 16 (1):74–94
2. Bentler PM, Wu EJC (1993) EQS/windows user's guide. BMDP Statistical Software, Los Angeles
3. Binter MJ (1990) Evaluating service encounters: the effects of physical surroundings and employee responses. J Mark 54:69–82
4. Chang CK (2008) The research of relationship among corporate image, consumer behavior, customer satisfaction and customer royalty-an empirical study on McDonald in Taiwan, National Cheng Kung University, Master's thesis (in Chinese)
5. Chou K (2000) 72 methods for visual management, corporate synergy development center, Taipei (in Chinese)
6. Fornell C, Larcker D (1981) Evaluating structural equation models with unobservable variables and measurement error. J Mark 18(1):39–50
7. Greif M (1991) The visual factory: building participation through shared information. Productivity Press, Portland

8. Hair JF, Anderson RE, Tatham RL, Black WC (1998) Multivariate data analysis. Macmillan, New York
9. Hair JF, Black WC, Babin BJ, Anderson RE, Tatham RL (2006) Multivariate data analysis. Pearson Education, New Jersey
10. Hu L, Bentler PM (1999) Coutoff criteria for fit indexes in covariance structure analysis: Conventional criteria versus new alternatives. Struct Equ Model 6(1):1–55
11. Kotler P (2000) Marketing management the millennium edition. Prentice Hall, New Jersey
12. Kotler P (2003) Marketing management: analysis, planning, and control. Prentice-Hall, New Jersey
13. Lin YC (2011) The study of the relation between lifestyle segmentation and brand royalty of franchised fast food restaurant-a comparative analysis of McDonald's and MOS Burger, National Kaohsiung University of Hospitality and Tourism, Master's thesis (in Chinese)
14. Oliver RL (1980) A cognitive model of the antecedents and consequences of satisfaction decision. J Mark Res 17(11):460–469
15. Practice Management Research Association of Japan (Provisional translation) (2000) Seven tools for shop floor management, corporate synergy development center, Taipei (in Chinese)
16. Richheld FF, Sasser WE (1990) Zero defections: quality come to service. Harvard Bus Rev 68:105–111

MIX
Papier aus verantwortungsvollen Quellen
Paper from responsible sources
FSC® C105338

If you have any concerns about our products,
you can contact us on
ProductSafety@springernature.com

In case Publisher is established outside the EU,
the EU authorized representative is:
Springer Nature Customer Service Center GmbH
Europaplatz 3, 69115 Heidelberg, Germany

Printed by Libri Plureos GmbH
in Hamburg, Germany